Handbooks of Crop Diversity: Conservation and Use of Plant Genetic Resources

Series Editor

Anurudh Kumar Singh, Department of Genetics, M.D. University,
Scholar-in-Residence, Rohtak, Haryana, India

Agrobiodiversity refers to the variation found among plants, animals, fish, insects, microbes, avian etc., species used directly or indirectly for food and agriculture. In case of crop and cultivated plant species, it mainly refers to the genetic diversity of plant genetic resources (PGR), found in the form of landraces and varieties of crops species used for food, fodder, fibre, fuel, pharmaceuticals, etc., and breeds of livestock. As per the FAO definition, it may also include the diversity of non-harvested species that support production system (soil micro-organisms, predators, pollinators) of these species and those in the wider environment support that agroecosystems (agricultural, pastoral, forest and aquatic) of their cultivation/production. However, the controlled genetic improvement of the cultivated species to facilitate continued breeding of cultigens with greater resilience to stresses and productivity is mainly dependent on overall genetic variation found in individuals belonging to the cultivated species and/or the ancestral species related to cultivated species. As these are the plant sources from where transfer of gene (genetic introgression) is possible through controlled breeding process of genetic improvement, incorporating desirable features, the accessibility to the information about these plant genetic resources is key to the success of the breeding efforts. Therefore, comprehensive information is required about these resources to facilitate their conservation and long-term sustainable use in research and crop improvement. The present effort of bringing out a series of volumes dealing with different crop groups containing comprehensive filtered information regarding these genetic resources and the availability of genetic diversity in each crop species of the group is aimed at facilitating their priority conservation in gene banks, and research and use in crop improvement.

Plant genetic resources are basic building blocks in the crop improvement and the quality of information and access to PGR is key to scientific efforts of genetic engineering of crops species, developing better cultigens, mitigating various challenges posed by increasing population and climatic change. The proposed series of volumes compiles information regarding the available genetic diversity among various cultivated species belonging to diverse crop groups and within a crop species of food and agriculture, from all possible perspectives of conservation and use (in genetic improvement of cultivated species). It would be able to enlist most genetic resources available globally, offering opportunities to be used in genetic engineering of crop species, and that would facilitate more predictive and productive genetic engineering programmes to breed new cultivars. The content of these volumes on agro-biodiversity would be very useful for technology-poor developing countries as this is not going to cover only major food crops but the whole range of crop/economic.

S. K. Datta • Youdh Chand Gupta
Editors

Floriculture and Ornamental Plants

With 199 Figures and 72 Tables

Springer

Editors
S. K. Datta
Floriculture Division
CSIR-National Botanical Research Institute
Lucknow, India

Bose Institute
Kolkata, India

Department of Plant Science and Biotechnology
Assam University
Silchar, Assam, India

Youdh Chand Gupta
Department of Floriculture and Landscape Architecture
Y.S. Parmer University of Horticulture and Forestry
Solan, HP, India

ISBN 978-981-15-3517-8 ISBN 978-981-15-3518-5 (eBook)
ISBN 978-981-15-3519-2 (print and electronic bundle)
https://doi.org/10.1007/978-981-15-3518-5

This Springer imprint is published by the registered company Springer Nature Singapore Pte Ltd.
The registered company address is: 152 Beach Road, #21-01/04 Gateway East, Singapore 189721, Singapore

Foreword

Crop diversity, the source of genetic variation, provides the building block for genetic enhancement/improvement of crop species to overcome productivity constraints and increase productivity per se of economic components.

Availability of useful genetic variation commonly referred to as plant genetic resources (PGR) is key to genetic improvement crops. Management of these genetic resources therefore becomes very important for research and assessment of their potential value, which facilitate their effective use in crop improvement for sustainable agriculture development of any specific nation and the world at large. Therefore, the review progress in relation PGR management activities, such as collection, characterization, evaluation, conservation, and documentation of generated information and mechanism evolved for sustainable handling of these resources for their effective use in cultivar development to achieve food and nutritional security at national and global level at regular intervals, becomes very essential to keep pace with the development in this regard. Such efforts also help to develop awareness regarding the potential value of crop diversity and identification of gaps.

Thus, more than the availability of these resources, generation of information about the potential value of these genetic resources through systematic scientific research, involving their characterization and evaluation using appropriate technologies, is more important. Access to this generated information is a must for researchers and plant breeders globally to facilitate their effective use in crop improvement, that is, to raise the productivity of target crop species.

Since the time of Vavilov (1935), realizing the importance crop diversity in genetic improvement of crop species, a significant progress has been made since the early twentieth century towards collection, characterization, evaluation, conservation, and information documentation regarding PGR at the national and international level in search of new gene(s)/sources and their use in crop improvement. These efforts got a further boost with the establishment of the Consultative Group on International Agricultural Research (CGIAR) and International Board of Plant Genetic Resources (IBPGR) in 1974 under its auspices. The establishment commodity International Agricultural Research Centers (IARCs), with a Plant Genetic Resources Unit, which functioned as centralized repositories, led to effective conservation of PGR. These initiatives/efforts resulted in assembly of a large number of collections/accessions in most mandated crops, particularly food and forage crops through the participation of the international community in collection missions and donation of national collections. However, the enforcement of CBD (1993), which made the nations sovereign owners of PGR found within their political boundaries restricted access to PGR, which are now governed either by the bilateral systems between the nations or the provisions of multilateral systems, such as International Treaty on Plant Genetic Resources for Food and Agriculture (ITPGRFA, 2001) and Nagoya Protocol (2011).

During this period, supporting systems, including databases with information on global PGR for food and agriculture, were also developed for information dissemination about PGR to promote use in crop improvement. For example, system-wide information network on genetic resources (SINGER) was developed about the conserved at the IARCs and the Global Information System, based on a central registry of cooperators (contracting parties) developed under auspices of ITPGRFA. However, these are mechanical systems without much specific analysis or tailor-made filtered information on PGR to support greater use. Consequently, the estimated use of PGR in crop improvement and research at the national and global level has almost remained static. Access to this information has been a major stumbling block in the underdeveloped and developing world, limiting their use in genetic improvement of crop species, particularly in case of crops of marginal importance or under-utilized crops that are the staple food or economic backbone of many of these regions, and adversely affecting food and nutritional security and economy of these nations.

For these reasons, it was felt there is a need to bring out a comprehensive publication with additional accessory and supplementary information with analyses and specifically filtered information about genetic resources of breeding value that can help promote greater research, accelerate the search for new gene(s)/allele(s), revealing the opportunities available for exploitation of PGR in genetic engineering of new cultivars to meet upcoming challenges of crop improvement and diversification. Further, such analyses will help identify gaps in relation to geographical regions and genetic diversity, for future search of genes/collections.

The series Handbook of Crop Diversity: Conservation and Use of Plant Genetic Resources is an attempt in this direction to fill existing gaps in access to information/dissemination for underdeveloped and developing countries, which do not have access or expertise to handle recent development/revolution in information technologies and Internet facilities carrying vast information. The volume on floriculture and ornamental crops, which are relatively neglected crops, because of being considered accessory crops for industry, trade, and economy, becomes a very important milestone in this regard. The effort may present new opportunities to researchers and breeders to develop improved cultivars. Whereas improved and diverse cultivar (key attraction) and yield can contribute greatly to providing additional income to farmers, thereby helping in poverty alleviation and improving national economy. I congratulate the volume editors and chapter authors for their effort and hope it will serve as a compendium of information for concerned researchers, breeders, teachers, and students.

Series Editor

Preface

Floriculture is a fast-emerging and highly competitive industry. It is one of the important high-value agricultural industries in many countries of the world. Commercial floriculture has become a very popular and important sector which provides livelihood security to small and marginal farmers besides providing ample opportunities for domestic and export markets. Floriculture business covers all aspects related to germplasm collection/exploitation and enhancement, cut flowers and foliage, dry flowers and plant parts, potted flowering and foliage plant, flower seeds, bulbs, tubers, corms, tissue culture plants, essential oils, and flower perfumes. The book comprises chapters having an authoritative review on various components plant genetic resources activities of most important floriculture/ornamental crop species of national, regional, and global value. The chapters account for most aspects of current interest and progress in the field of floriculture. The volume contains chapters on respective important floriculture crops, written by specialists and subject experts in the floriculture discipline. The chapters also attempt to provide an illustrated account of the milestone achieved, gaps, and the future direction in which the management of PGR should move for more sustainable conservation and use to engineer new cultigens to achieve the national target of productivity in an overall environment friendly manner that would contribute to livelihood support and poverty alleviation of rural poor, besides the national income. The editors acknowledge the contributions of the respective authors, who have tried to bring out the information in a uniform structure for a focused presentation.

The book covers all important ornamental crops under various chapters as per their economic importance and available information. Chapters provide in-depth information under the headings – introduction, botany and distribution, origin/domestication, and spread of various genetic resources belonging to various gene pools, including wild relatives, status of their collections, characterization, evaluation, and conservation. Also, they refer to major constraints in crop production. List of genetic resources with genes conferring resistance to various biotic and abiotic stresses and desired commercial traits, and extent of their use in breeding programs for development of new genetic stocks with desired traits, and cultigens, etc.

It is hoped that this handbook will stand by its definition and serve as a handy reference book with filtered information that would be useful to genetic resources managers, researchers, policy makers, teachers, and research scholars. The information, particularly on availability of PGR globally, along with specific source, wherefrom they can be procured for use in breeding programs, particularly to overcome various crop production constraints and to improve productivity and quality, should be of great help in developing resilient and diverse cultivars of ornamental plant species. The book will function not only as reference book on collection, characterization, evaluation, conservation, and use of plant genetic resources in present and future breeding programs, concentrating on increased resilience against various stresses and productivity, but also for value addition and diversification of the concerned crop species and domestication or direct use of potential wild and weedy relatives. To make the effort dynamic, Springer is going to have provisions for regular updates of the chapters with new information by the concerned author(s) to ensure its present and future use for many years to come.

Editors

Contents

About the Editors

Dr. S. K. Datta, Ph.D., D.Sc., is an internationally acclaimed expert in floriculture and mutation breeding. Dr. Datta was engaged in both basic and applied research for the improvement of floriculture for more than 30 years at CSIR-NBRI, Lucknow. His main field of research was induced mutations and improvement of ornamental plants for development of new and novel varieties and promotion of floriculture. The main field of work was on ornamental crops (Amaryllis, Asiatic hybrid Lily, Bougainvillea, Carnation, Chrysanthemum, Dahlia, Gladiolus, Gerbera, Hibiscus, Marigold, Narcissus, Orchid, Rose, Tuberose, *Lantana depressa*). His main research field was floriculture on different aspects related to cytomorphology, in vivo and in vitro mutagenesis, conventional breeding, tissue culture, management of chimera, conservation, molecular characterization, dehydration of flowers, floral craft, etc. which are internationally competitive. As an outstanding breeder in floriculture, his contributions in developing varieties and enriching knowledge-based floriculture have been laudable. Dr. Datta has published series of research papers (approx. 338) on different aspects related to floriculture and induced mutagenesis. He has published three books, edited four books and eight bulletins. Dr. Datta retired from NBRI in July 2007 as Scientist "G", Coordinator and Head, Botanic Garden and Floriculture Division. Further, Dr. Datta continued his research on dehydration of flowers and foliage and floral craft as Council of Scientific and Industrial Research (CSIR), Emeritus Scientist at Bose Institute, Kolkata, from January 2008 to January 2013. Dr. Datta did extensive rural development program during this period.

Dr. Datta visited Berlin for two months (under CSIR-DAAD Scientists Exchange Program) and Korea for one month (INSA-KOSEF Scientists Exchange Program). Considering the quantum of work done and published literature on induced mutagenesis, International Atomic Energy Agency (IAEA), Vienna, selected Dr. Data as "Expert on Mission" for evaluation of mutation breeding projects sponsored by IAEA to Philippines, Jakarta, Indonesia, for project evaluation mission. Dr. Datta organized international training program as supervisor on induced mutagenesis sponsored by IAEA, Vienna, at Central Research Institute for Horticulture, Cipanas, Indonesia, for students, teachers, and researchers. Dr. Datta was invited to present papers in different international symposium – at Sixth International Congress of SABRAO held at Tsukuba, Japan; International Nuclear Conference held at Putra World Trade Centre, Kuala Lumpur, Malaysia; International Symposium on "Underutilized Plant Species for food, nutrition, income and sustainable development" held at Arusha, Tanzania; FAO/IAEA International Symposium on the contribution of Plant Mutation Breeding to Crop Improvement, held at International Atomic Energy Agency, Vienna, Austria – four times. He was deputed to Germany (Humboldt Universitat Zu Berlin, Berlin) under a CSIR-DAAD Scientists Exchange Program for two months. He visited Korea (South) under INSA-KOSEF scientist exchange program for a period of 21 days to gather knowledge on bioreactor technology and recent advancement in floriculture. He also visited Bangkok to explore the orchid and dry flower market.

Youdh Chand Gupta is Professor and Head, Department of Floriculture and Landscape Architecture, Dr. Y. S. Parmar University of Horticulture and Forestry, Solan, India. He has guided many M.Sc. and Ph.D. students as major advisor in the discipline of floriculture and landscaping, published more than 60 research papers in various journals of repute, and received best paper/poster awards in different national and international seminars. Besides this, he has also authored books/manuals, technical bulletins, and more than 50 extension articles and handled more than 15 externally funded research projects as principal investigator. Prof.

Gupta developed various hybrids including hybrids of gladiolus, marigold, antirrhinum, and chrysanthemum and also developed cheaper technology for growing rose carnation and gerbera in Himachal Pradesh. He also established a hi-tech floriculture farm for production of rose, gerbera, alstroemeria, and carnation—a unique project in the country with cool chain facility including refrigerated van service.

Prof. Gupta received the Dr. Manmohan Attawar Gold Medal Award in Floriculture (2016) and also the ISOH Fellowship and Lotus Puraskar from the Indian Society of Ornamental Horticulture (ISOH) for outstanding contribution and commitment to research and development of ornamental horticulture.

Contributors

Fernando B. Aurigue Department of Science and Technology-Philippine Nuclear Research Institute, Quezon City, Metro Manila, Philippines

P. Kiran Babu ICAR– National Research Center for Orchids, Pakyong, India

Namita Banyal IARI, New Delhi, India

Bhavya Bhargava CSIR, IHBT, Palampur, Himachal Pradesh, India

Sukanta Biswas Department of Horticulture, Institute of Agricultural Science, University of Calcutta, Kolkata, West Bengal, India

Suhrita Chakrabarty Das AICRP on Floriculture, Department of Post Harvest Technology, F/Horticulture, Bidhan Chandra Krishi Viswavidyalaya (BCKV), Kalyani, West Bengal, India

Stephen F. Chandler School of Applied Sciences, RMIT University, Bundoora, VIC, Australia

S. L. Chawla Navsari Agricultural University, Navsari, Gujarat, India

T. K. Chowdhuri Department of Floriculture and Land Scape Architecture, Bidhan Chandra Krishi Viswavidyalaya, Kalyani, West Bengal, India

S. K. Datta Floriculture Division, CSIR-National Botanical Research Institute, Lucknow, India

Bose Institute, Kolkata, India

Department of Plant Science and Biotechnology, Assam University, Silchar, Assam, India

M. Dayamma ICAR– National Research Center for Orchids, Darjeeling Campus, Darjeeling, India

M. R. Dhiman ICAR-INARI, Regional Station, Katrain, Kullu-Valley, Himachal Pradesh, India

M. Ganga TNAU, Coimbatore, Tamil Nadu, India

Upashana Ghissing Natural Product Biotechnology Group, Agricultural and Food Engineering Department, Indian Institute of Technology Kharagpur, Kharagpur, West Bengal, India

Youdh Chand Gupta Department of Floriculture and Landscape Architecture, Y.S. Parmer University of Horticulture and Forestry, Solan, HP, India

T. Janakiram Horticultural Science Division, Indian Council of Agricultural Research, New Delhi, India

M. Jawaharlal TNAU, Coimbatore, Tamil Nadu, India

R. Jayanthi Department of Horticulture UAS, GKVK, Bangalore, India

Bharati Kashyap UHF, Nauni-Solan, HP, India

Rajiv Kumar Division of Flower and Medicinal Crops, ICAR-Indian Institute of Horticultural Research, Bengaluru, Karnataka, India

Kiran Kumari College of Horticulture, SDAU, Jagudan, Gujarat, India

Poonam Kumari IHBT, CSIR, Palampur, India

Pablito M. Magdalita Institute of Crop Science, College of Agriculture and Food Science, University of the Philippines Los Baῃos, College, Los Banos, Philippines

Institute of Plant Breeding, College of Agriculture and Food Science, University of the Philippines Los Baῃos, College, Los Banos, Philippines

Moumita Malakar Department of Horticulture & Floriculture, Central University of Tamil Nadu, Thiruvarur, Tamil Nadu, India

Uzma Mehraj IARI, New Delhi, India

Adinpunya Mitra Natural Product Biotechnology Group, Agricultural and Food Engineering Department, Indian Institute of Technology Kharagpur, Kharagpur, West Bengal, India

Kalkame Ch. Momin College of Horticulture & Forestry, Pasighat, Arunachal Pradesh, India

Ram Pal ICAR– National Research Center for Orchids, Darjeeling Campus, Darjeeling, India

Sapna Panwar IARI, New Delhi, India

Sudha Patil Navsari Agricultural University, Navsari, Gujarat, India

R. D. Pawar RCSM College of Agriculture, MPKV, Kolhapur, Maharashtra, India

Parul Punetha Dept of Agriculture Sciences, Doon Group of Institutions, Rishikesh, India

R. Sadhukhan Department of Genetics and Plant Breeding and In-Charge, All India Coordinated Research Project on Floriculture, Directorate of Research, Bidhan Chandra Krishi Viswavidyalaya, Kalyani, West Bengal, India

Tarak Nath Saha ICAR-Directorate of Floricultural Research, College of Agriculture Campus, Pune, Maharashtra, India

Swati Saha ICAR-Indian Agricultural Research Institute, Pune, India

Alangelico O. San Pascual Institute of Plant Breeding, College of Agriculture and Food Science, University of the Philippines Los Baῆos, College, Los Banos, Philippines

Panchal Sangmesh Dr. YSPUHF, Nauni-Solan, Himachal Pradesh, India

Neha Sharma Department of Floriculture and Landscape Architecture, Y.S. Parmer University of Horticulture and Forestry, Solan, HP, India

Puja Sharma Dr. YSPUHF, Nauni-Solan, Himachal Pradesh, India

S. S. Sindhu Division of Floriculture and Landscaping (School of Horticulture), ICAR- Indian Agricultural Research Institute, New Delhi, India

Ranjan Srivastava Department of Horticulture, G.B. Pant University of Agriculture & Technology, Pantnagar, India

Neelam Thakur Department of Floriculture and Landscape Architecture, Y.S. Parmer University of Horticulture and Forestry, Solan, HP, India

S. P. Thamaraiselvi TNAU, Coimbatore, Tamil Nadu, India

David Tribe School of Biosciences, University of Melbourne, Parkville, VIC, Australia

Himanshu Trivedi School of Agricultural Sciences and Engineering, IFTM University, Moradabad, India

Ujala Academy of Scientific and Innovative Research, New Delhi, India

Marigold

1

Youdh Chand Gupta, Sapna Panwar, Namita Banyal,
Neelam Thakur, and M. R. Dhiman

Contents

Y. C. Gupta (✉) · N. Thakur
Department of Floriculture and Landscape Architecture, Y.S. Parmer University of Horticulture and Forestry, Solan, HP, India

S. Panwar · N. Banyal
IARI, New Delhi, India

M. R. Dhiman
ICAR-INARI, Regional Station, Katrain, Kullu-Valley, Himachal Pradesh, India

S. K. Datta, Y. C. Gupta (eds.), *Floriculture and Ornamental Plants*, Handbooks of Crop Diversity: Conservation and Use of Plant Genetic Resources,
https://doi.org/10.1007/978-981-15-3518-5_1

Abstract

Marigold (*Tagetes* sp.) belonging to Asteraceae family is an economically important flower crop grown worldwide for cut and loose flowers, bedding and pot plants, essential oils, carotenoid pigments for nutraceutical and pharmaceutical purpose, etc. Marigold seed is the economic part that is being used for collection and conservation at low temperature storage and cryopreservation. Most of the accessions are conserved by OPGC (Ornamental Plant Germplasm Center) repository, USDA internationally, and ICAR-NBPGR, India, at national level. *Tagetes* species were also characterized for various traits such as essential oils, carotenoid pigments, thiophenes, flower yield related traits, biotic and abiotic stress related traits, etc. for further use in specific breeding purpose. Besides, marigold is also characterized using molecular markers RAPD, ISSR, UPR, SSR markers, and corelated with other economically important traits. Transcriptome assembly and digital gene expression profiling were also done by researchers in *T. erecta* to generate expression profiles of MS (Male sterile) and MF (Male fertile) plants. AGAMOUS subfamily proteins, encoded by MADS box family genes, play key roles in the determination of reproductive floral organs such as stamens, carpels, and ovules of marigold. The first chloroplast genome sequence of *T. erecta* was given in 2020 having 152,065 bp chloroplast genome size with 37.4% GC content. Transgenic technology was also successfully employed in marigold for modifying the various traits. Selection of appropriate cultivars for different seasons as each genotype performs differentially in different agroclimatic conditions, susceptibility to water logging, very low and high temperature, etc. are some of the major constraints in crop production.

Keywords

Tagetes spp. · Carotenoids · Conservation · Molecular markers · Transcriptome · Genome · Transgenic

1.1 Introduction

Marigold, member of Asteraceae family, is one of the economically important ornamental crops grown worldwide. Genus *Tagetes* comprises of about 55 species (Hernandez and Ham 2007), out of which, African marigold (*Tagetes erecta* L.) and French marigold (*Tagetes patula* L.) are of commercial importance while *Tagetes minuta* (wild marigold) is rich in essential oils. It is the choicest crops among growers because of its wider adaptability to varied agro climatic conditions, short duration habit, longer blooming period, diversity in flower color and form, more shelf life of flowers, etc. Being ornamental in nature, it also holds a prominent position in landscaping and mostly utilized as a bedding and pot plant. It is also the best choice for herbaceous borders. French marigold is mostly suitable for hanging baskets, window boxes, edging, and rockery whereas African marigold varieties especially those that bear long straight stalk and globular flowers are nowadays used as cut flowers for interior decoration. Marigold is majorly grown as loose flower crop in India, which is being mostly utilized in marriages and religious functions. It has also emerged as a major natural source of carotenoid pigments, which are widely used as dietary supplements in poultry industry to enhance chicken skin color and egg yolk pigmentation. Lutein is the major xanthophyll (70–88%) found in the marigold petals (Quackenbush and Miller 1972; Hadden et al. 1999). Moreover, lutein fatty acid esters from marigold are known for its readily soluble nature in vegetable oils as compared to other FDA-approved synthetic carotenoids (Timberlake and Henry 1986), thereby finding its applications in various industries especially food colors. Dietary carotenoids are also used in the treatment of cancer and other photosensitivity diseases (Gau et al. 1983; Park et al. 1998). Marigold produces aromatic sulfur containing secondary compounds mainly in their roots known as thiophenes which is known to be effective in keeping the nematode population under control, and it is also grown as an antagonistic crop in vegetable crops to control nematodes. Essential oil of marigold is known for its antimicrobial and insecticidal properties. *Tagetes minuta* is one of the rich sources of essential oils. Major components of essential oil in *Tagetes minuta* are (Z)-ocimenone (15.9%), (E)-ocimenone (34.8%), (Z)-beta-ocimene (8.3%), limonene (2.3%), (Z)-tagetone (1.8%), dihydrotagetone (1.4%), and an unidentified dimethylvinylketone derivative (20.6%) (Ali et al. 2014). Moreover, *Tagetes lucida* and *Tagetes filifolia* have been traditionally used in the preparation of native teas of Mexico (Laferriere 1991). The decoctions of the leaves of *Tagetes erecta* and *Tagetes patula* have been traditionally used against malaria (Rasoanaivo et al. 1992).

1.2 Botany and Distribution

Marigold is an herbaceous plant which is mostly annual in nature but few species are also perennial. It belongs to Asteraceae family. The genus *Tagetes* contains about 55 species of which two species, that is, *Tagetes erecta* (African marigold) and *T. patula* (French marigold) are known for their ornamental uses worldwide. *Tagetes* species are endemic to the Americas, primarily Mexico and Central America (Kaplan 1960; Neher 1965).

Botanical Classification

Kingdom	Plantae – Plants
Subkingdom	Tracheobionta – Vascular plants
Super-division	Spermatophyta – Seed plants

(continued)

Division	Magnoliophyta – Flowering plants
Class	Magnoliopsida – Dicotyledons
Subclass	Asteridae
Order	Asterales
Family	Asteraceae – Aster family
Genus	*Tagetes* L. – Marigold

Source: ITIS, the Integrated Taxonomic Information System (https://www.itis.gov)

1.3 Origin, Domestication, and Spread

Tagetes species are endemic to the Americas, primarily Mexico and Central America (Kaplan 1960; Neher 1965). It has been cultivated in Mexico for centuries and probably spread to the rest of Central and South America where now it can be found growing in naturalized conditions (Standley and Steyermark 1949; Davidse et al. 2018; Heuzé et al. 2017). It is believed that the Aztec people were the first to use the marigold and attributed magical, religious, and medicinal properties to the plant. The first recorded use of marigolds is in the De La Crus-Badiano Aztec Herbal of 1552. In 1500s, Spanish explorers were believed to have taken the marigold seeds from the Aztecs to Spain, and the crop was brought into cultivation and also found growing in monastery gardens. Thereafter marigold seeds were further transported from Spain to France and northern Africa. The taller marigolds, known as African American, became naturalized in North Africa. In Mexico and Latin America, marigold flowers were used in celebration of special occasions like All Saints Day and All Souls Day. In India, it was believed to be introduced from Portugal in the late sixteenth century and now it is one of the most commonly cultivated ornamental flower crops in urban and rural India (Shukla and Thakur 2018).

1.4 Plant Genetic Resources

1.4.1 Geographic Distribution

The genus *Tagetes* is native to America but some of its species, viz. *Tagetes erecta* and *Tagetes patula*, were naturalized in India, North Africa, and Europe as early as in sixteenth century. Some researchers suggest that both species reached India anciently through pre-Columbian transoceanic voyages (Sorenson and Johannessen 2004). Marigold was introduced to Georgia from India, and its ground dried petals became one of the most popular local spices (Akhalkatsi et al. 2012). Both *Tagetes erecta* and *Tagetes patula* are grown in Georgia as spice and dye plants (Beridze et al. 1990) and is recognized for their health beneficial properties (Esaiashvili et al. 2011).

Mexico and Guatemala are considered native places of *Tagetes erecta.* It is said to be naturalized in rest of the Central America and the western Andes of South America. It is found in countries of Central America and the Caribbean Belize,

Bolivia, Colombia, Costa Rica, Cuba, Ecuador, El Salvador, Guatemala, Guyana, Honduras, Jamaica, Nicaragua, Panama, Puerto Rico, and Venezuela (Villaseñor and Espinosa-Garcia 1998). Presently *Tagetes erecta* is widely cultivated in the tropical zone of Africa, Asia, and Oceania, from sea level up to 2000 m altitude as a popular garden ornamental. *Tagetes erecta* is commercially used for dye making in Latin America; in Africa on a small scale in Zambia and South Africa. It was also introduced into Europe for ornamental purposes. The species are found wild in the Balsas basin and western Mexico and live in diverse types of ecosystems, such as tropical deciduous forests, thorny forests, cloud forests, and pine-oak forests. In the wild, it is found absconding in heavily disturbed places at altitudes of 800–2300 m (Heyden 1987). The species are also found as introduced ones (cultivated) in China, India, Zambia, South Africa, and Australia. *Tagetes minuta* is native to temperate forests and mountain regions of most of the countries in the world. It originated in South America and has spread throughout the world as a weed (Singh et al. 2003).

1.4.2 Primary Gene Pool

Tagetes erecta and *T. patula* are the basis of almost all of the modern garden cultivars.

Descriptions of some of the important species are as follows:

***Tagetes erecta* (African marigold)**: Diploid ($2n = 2x = 24$), plants are mostly tall (about 90 cm), erect, and branched. Leaves are pinnately divided and leaflets are lanceolate in shape and serrated. It bears flowers ranging from single, semidouble, and double in nature. Semidouble and double flowers are generally large in size and globular in shape. The flower color varies from lemon yellow to yellow, golden yellow or orange.

***Tagetes patula* (French marigold)**: Tetraploid ($2n = 4x = 48$), plants are generally dwarf statured (about 30 cm) and bushy in appearance. The leaves are pinnately divided and leaflets are linear lanceolate in shape and serrated. It bears small flowers ranging from single, semidouble, and double in nature. The flower color varies from yellow to mahogany red.

***Tagetes tenuifolia/T. signata* (Signet marigold)**: Diploid ($2n = 2x = 24$), plants are annual and generally have a branching habit. Glossy green foliage, nearly pinnatifid, and divided into 12 oblong, linear, sharply serrated segments. The plant produces small, terminal, yellow or orange single daisy flowers and round to obovate in shape.

***Tagetes lucida* (Sweet scented marigold)**: Diploid ($2n = 2x = 22$), plants are perennial in nature. The plant produces small leaves which are linear to oblong, sessile, and lanceolate in shape. Small flowers are yellow with 2–3 rays and are produced in dense, terminal corymbs. *Tagetes lucida* flowers possess much more pleasant odor than other species.

***Tagetes lemmonii* (Mexican marigold/Mountain marigold)**: Diploid ($2n = 2x = 24$), an evergreen shrub, 60–70 cm tall. Leaves are slender, opposite, with 2–3 cm long leaflets. Leaves are strongly fragrant when rubbed or bruised and smell like a blend of marigold, mint, and lemon. The plant produces many small flower heads, golden orange in color, in a flat-topped array, each head with 3–8 ray florets and 20–30 disc florets.

1.4.3 Wild Genetic Resources and Others

Wild/Weedy Relatives of Crop Plants: *Tagetes minuta* L./*T. glandulifera* Schrank: It is an annual herb, aromatic, branched, and mostly tall in nature (1–2 m). Leaves are pinnately dissected with lanceolate shaped leaflets and are serrated. Undersurface of the leaves bears glands, which emit aroma when ruptured. Flower heads are yellow and small surrounded by green involucre bracts. It grows wild in western Himalaya and is one of the important sources of essential oils and has domesticated by Natural Plant Products Division, Institute of Himalayan Bioresource Technology, Palampur, HP, India, not only in subtemperate but also in subtropical zones (https://www.ihbt.res.in).

1.5 Collections and Conservation

1.5.1 Methods

In marigold, seed is the economic part that is being used for collection and conservation purposes. Seeds are easy to store and transport so forms the basis of germplasm collections. Seed storage does not require sophisticated facilities and moreover marigold seeds have good viability and germination percentage.

1.5.2 Status of Collections (National, Regional, and Global with Appropriate Listing)

1.5.2.1 International

The Germplasm Resources Information Network (GRIN) documents the various collections which include the animal, microbial, and plant collections through informational pages, searchable databases, and links to USDA-ARS projects that curate the collections. GRIN is operated by the National Germplasm Resources Laboratory in Beltsville, MD. Most of the accessions are maintained by OPGC (Ornamental Plant Germplasm Center) repository. The information with respect to *Tagetes* species is as follows:

S. N.	Accession	Name	Taxonomy	Origin
1	PI 667535	Mission Giant Yellow Stone	*Tagetes* spp.	United States, California
2	PI 667536	All-Double Mixed Colors	*Tagetes* spp.	United States, California

(continued)

S. N.	Accession	Name	Taxonomy	Origin
3	NSL 15430	Superjack	*Tagetes* spp.	United States, California
4	PI 667537	Eldorado	*Tagetes* spp.	United States, California
5	PI 667538	Spry Hybrids	*Tagetes* spp.	United States, California
6	PI 667541	Orange Peony	*Tagetes* spp.	United States, Pennsylvania
7	PI 667543	Mediterranean Moon	*Tagetes* spp.	United States, Minnesota
8	PI 667545	Big Almost White	*Tagetes erecta*	United States
9	PI 667546	Aztec Gold	*Tagetes erecta*	United States, California
10	PI 667547	Aztec Orange	*Tagetes erecta*	United States, California
11	PI 667548	Aztec Yellow	*Tagetes erecta*	United States, California
12	NSL 91410	Redcoat	*Tagetes patula*	United States
13	NSL 108688	Firelight	*Tagetes* spp.	United States, California
14	PI 600737	Happy Red	*Tagetes patula*	United States
15	PI 600753	Ice Crush	*Tagetes erecta*	United States
16	PI 600813	Scarlet Sophia	*Tagetes patula*	United States
17	PI 600864	Xantho-Orange	*Tagetes erecta*	United States
18	Grif 12616	Aurora Fire	*Tagetes patula*	United States, California
19	Grif 12617	Aurora Gold	*Tagetes patula*	United States, California
20	Grif 12618	Janie Harmony	*Tagetes patula*	United States, California
21	PI 316078		*Tagetes erecta*	United Kingdom, England
22	PI 316079	18341	*Tagetes subulata*	Mexico
23	PI 326200	22516	*Tagetes* spp.	Mexico
24	PI 387868	Cacalahua	*Tagetes* spp.	Bolivia
25	PI 387869	Suico	*Tagetes* spp.	Bolivia
26	PI 438928	Flor de muerto	*Tagetes* spp.	Mexico
27	PI 438930	Musa	*Tagetes* spp.	Mexico
28	PI 442397	Zempasuchitl	*Tagetes* spp.	Mexico
29	PI 442399	Zempasuchitl	*Tagetes* spp.	Mexico
30	PI 442400	Zempasuchitl	*Tagetes* spp.	Mexico
31	PI 442401	Zempasuchitl	*Tagetes* spp.	Mexico
32	PI 442402	Zempasuchitl	*Tagetes* spp.	Mexico
33	PI 442403	Zempasuchitl	*Tagetes* spp.	Mexico
34	PI 442405	Zempasuchitl	*Tagetes* spp.	Mexico
35	PI 442406	Zempasuchitl	*Tagetes* spp.	Mexico
36	PI 442407	Zempasuchitl	*Tagetes* spp.	Mexico
37	PI 442409	Zempasuchitl	*Tagetes* spp.	Mexico
38	PI 442410	Zempasuchitl	*Tagetes* spp.	Mexico
39	PI 442411	Coronilla	*Tagetes* spp.	Mexico
40	PI 442412	Zempasuchitlenano	*Tagetes* spp.	Mexico
41	PI 586960	1306	*Tagetes patula*	Mexico, Puebla

(continued)

S. N.	Accession	Name	Taxonomy	Origin
42	PI 667302	Ff01-003	*Tagetes lucida*	Mexico, Chihuahua
43	PI 667327	Cempoalxochitl Marigold	*Tagetes erecta*	United States, New Mexico
44	W6 56932	Az930-409	*Tagetes lemmonii*	United States, Arizona

Source: https://npgsweb.ars-grin.gov/gringlobal/search

1.5.2.2 National

In India, ICAR-National Bureau of Plant Genetic Resources is maintaining 364 accessions of marigold (www.nbpgr.ernet.in).

1.5.3 Conservation

1.5.3.1 Low Temperature Storage

Seed conservation at low temperature is one of the methods to conserve the seeds. Here, seeds can be dried up to 1.5% moisture content without losing viability. However, seed moisture content between 3% and 6% is safe for long-term storage at low temperature (−20 °C). In a study of seeds of two released cultivars of African marigold (*Tagetes erecta* L.), Pusa Narangi Gainda and Pusa Basanti Gainda did not behave as true orthodox when ultra-desiccated seeds (<6% moisture content) and stored at ultra-low (−196 °C) temperature (Singh et al. 2004). It was suggested that the seed size is an important attribute to estimate the health and quality of seed lots in marigold. Though seed deteriorated with increase in storage period and moisture content, seed leachate conductivity is not a good indicator of loss of viability if marigold seeds have been stored at low temperature.

1.5.3.2 Cryopreservation

Seed material of marigold has a limited shelf life, loses viability in a short time. Therefore, modern method of long storage of genetic resources is the cryopreservation in liquid nitrogen. Cryopreservation allows freezing plant seed at the critical low temperatures with maintaining viability for a long period (Babu et al. 2012; González-Benito et al. 2009). High seed quality can be maintained for about 8 years in marigold by following cryopreservation (https://www.iihr.res.in/division-plant-genetic-resources).

Status of Plant Genetic Resources (general germplasm, base collection, active collection, breeder's collection, genetic stocks, pre-breeding material (including interspecific derivatives, etc.))

African Marigold **(Types)**

S. No.	Type	Characteristics
1.	**Carnation flowered**	75 cm tall Flowers: 10 cm across Orange, deep orange, golden yellow, lemon yellow

(continued)

S. No.	Type	Characteristics
2.	**Chrysanthemum flowered**	Tall Double Various shades of orange and yellow Luxor series (improved cupid type) Rexor series (improved spun type)
3.	**Tall F_1 hybrids**	Tall (3 m) Flowers: large, fully double (12 cm across) F_1 Gold coin series F_1 climax series
4.	**Semi-tall F_1 hybrids**	Hedge type – uniform and compact growth Double (10 cm across) Lemon, golden yellow, light orange
5.	**Dwarf F_1 hybrids**	15–40 cm high Compact growth Inca series (large, fully double, and uniform) Space age series (uniform and early flowering) Galore series (uniform and long flowering duration)
6.	**F_1 triploids**	Early Very free flowering Golden yellow

French Marigold **(Types)**

S. No.	Type	Characteristics
1.	**Dwarf double**	20–30 cm high Orange, yellow, and mahogany red
2.	**Dwarf double- Scabious flowered**	Flowers with crested center Wide range of colors with spots and markings
3.	**Dwarf double petite**	Very dwarf (15–20 cm) Numerous attractive flowers
4.	**French dwarf single**	20–35 cm high Single compact flowers
5.	**Dwarf triploid F_1 hybrids**	25–40 cm high Profuse flowering Extremely early
6.	**Dwarf double**	Dwarf all saints (20 cm) Dwarf chrysanthemum flowered (20 cm)
7.	**Tetraploid hybrids**	Fully double carnation type flowers Early flowering

Source: Commercial Floriculture, T. K. Bose

Varieties

Tagetes erecta: Golden Jubilee, Giant Double African Orange, Giant Double African Yellow, Crackerjack, Climax, Doubloon, Golden Age, Chrysanthemum Charm, Crown of Gold, Spun Gold.

Tagetes patula: Valencia Yellow, Dainty Marietta, Bolero Red, Gulzafri Orange, Gulzafri Yellow, Boy O Boy, Red Brocade, Spanish Brocade, Rusty Red, Butter Scotch, Succana, etc.

Tagetes tenuifolia: Golden Gem, Lulu, Pumila, Ursula.

Indian varieties developed by public organizations

S. No	Variety	Flower color	Developing institute/university
Tagetes erecta			
1.	Pusa Narangi Gainda	Orange	ICAR-Indian Agricultural Research Institute, New Delhi
2.	Pusa Basanti Gainda	Sulfur yellow	ICAR-Indian Agricultural Research Institute, New Delhi
3.	Pusa Bahar	Yellow	ICAR-Indian Agricultural Research Institute, New Delhi
4.	Arka Bangara-2	Yellow	ICAR-Indian Institute of Horticultural Research, Bengaluru
5.	Arka Agni	Orange	ICAR-Indian Institute of Horticultural Research, Bengaluru
6.	Bidhan Marigold-1	Yellow	Bidhan Chandra Krishi Visvavidhalaya, Kalyani
7.	Bidhan Marigold-2	Orange	Bidhan Chandra Krishi Visvavidhalaya, Kalyani
Tagetes erecta			
8.	Pusa Arpita	Orange	ICAR-Indian Agricultural Research Institute, New Delhi
9.	Pusa Deep	Dark Red	ICAR-Indian Agricultural Research Institute, New Delhi
Tagetes minuta			
10.	Him Gold' Essential oil rich	Pale yellow	CSIR-Institute of Himalayan Bioresource Technology, Palampur

1.6 Characterization and Evaluation

1.6.1 Characterization for Essential Features and Classification

1.6.1.1 Essential Oils

Tagetes spp. have been evaluated for the essential oil's times and again. The oils are isolated by steam distillation from the flowers and leaves, and analyzed by GC and GC/MS. The flowers and the leaves of each species often show very similar qualitative oil compositions but the leaves, on average, are richer in oil contents. *Tagetes erecta, T. minuta, T. patula*, and *T. tenuifolia* comprise the same pool of components (dihydrotagetone, tagetones, ocimenones, and piperitone) which showed different and typical ratios in each species and piperitone, (E)-tagetone, terpinolene, and (E)-ocimenone were more abundant respectively. In *Tagetes filifolia* and *T. lucida* methyl chavicol was found as the main compound (Mauro Marotti et al. 2004).

GC and GC/MS analysis were done in *Tagetes erecta* (var. Pusa Narangi Gainda) essential oil for its antioxidant and antimicrobial activity. In the aerial part extract

43 constituents were identified, representing more than 83% of the total detected. The major components were identified as *cis*-ocimene (18.46%), (E)-oscimene (8.65%), l-limonene (11.16%), (E)-tagetone (10.56%), b-caryophyllene (6.9%), and dl-limonene (4.16%) (Tripathi et al. 2012).

Characterization and evaluation of bioactive compounds of *Tagetes erecta* L. by GC-MS, was done by Hosea Jaya Edy et al. (2017). Characterization of extracts was done for the determination of moisture content, total ash value, and the total value of acid insoluble ash. Evaluation of the content of bioactive compounds is done using gas chromatography-mass spectrometry (GC-MS). Characteristics of the extract obtained results of moisture content $8.28 \pm 0.540\%$ v/w, the value of total ash content of $2.54 \pm 0.038\%$ w/w, and the value of acid insoluble ash content of $0.98 \pm 0.064\%$ w/w. GC-MS identification produced 17 types of compounds with a retention time of 31.284 to 50.614. Three compounds with the greatest abundance are Neophytadine 43.88%, 9, 12, 15-Octadecatrienoic acid-methyl ester 13.45%, and hexadecanoic acid-methyl ester 13.24%.

1.6.1.2 Carotenoids

Commercially prepared marigold flower (*Tagetes erecta*) extract was saponified and analyzed for carotenoid composition by W. Leigh Hadden et al. (1999). HPLC analyses were performed on two normal-phase columns (β-Cyclobond and silica) and on a C_{30} reversed-phase column. The extract contained 93% utilizable pigments (detected at 450 nm), consisting of all-*trans* and *cis* isomers of zeaxanthin (5%), all-*trans* and *cis* isomers of lutein, and lutein esters (88%). Also, insignificant levels (<0.3%) of lutein oxidation products were detected in the saponified extract. This study presented the evidence that commercial marigold flower extracts contain *trans*-lutein as the main carotenoid component with several *cis*-lutein isomers as minor components.

A novel quantitative procedure for the analysis of lutein esters in marigold flowers (*Tagetes erecta*) was described by Gregory et al. (1986) using high-performance liquid chromatography. They developed a new solvent system consisting of methanol and ethyl acetate suitable for the separation of carotenoid hydrocarbons and esters on reversed phase adsorbent, for the analysis. Lutein ester concentrations in fresh marigold flowers varied from 4 pg/g in greenish yellow flowers to 800 pg/g in orange brown flowers. The method was found to be suitable as a general procedure for carotenoid analysis in fruits and flowers where the hydroxylated carotenoids are acylated.

Pigments of *Tagetes erecta* petals have been separated by Alam et al. (2011) using multiple fractional extractions and thin-layer chromatography and analyzed by spectral methods. The petals showed a predominance of xanthophylls (98.7%) and only a small amount of carotenes. On the basis of absorption studies, lutein (64.1%), antheraxanthin (31.1%), a-cryptoxanthin (3.1%), 8-carotene (0.6%), phytofluene (0.4%), and a-carotene (0.15%) have been characterized.

For the analysis of carotenoid composition commercially prepared marigold flower (*Tagetes erecta*) extract was saponified. HPLC analyses were performed on two normal-phase columns (β-Cyclobond and silica) and on a C_{30} reversed-phase column. The extract contained 93% utilizable pigments (detected at 450 nm), consisting of all-*trans* and *cis* isomers of zeaxanthin (5%), all-*trans* and *cis* isomers of lutein, and lutein esters (88%). All were identified by chromatographic retention, UV – visible spectra, and positive ion electrospray mass spectrometry in comparison to authentic standards (W. Leigh Hadden et al. 1999). Quackenbush (1972) found little variation in pigment distribution in the different varieties and in commercial marigold petal meals when marigolds of different types and colors were grown locally from seed and the carotenoids of their petals were extracted, saponified, and chromatographed. Out of the 17 different pigments which were separated, lutein was preponderant; lutein plus zeaxanthin constituted 88–92% of the total. Epoxy pigments were less than 3% of the total.

1.6.1.3 Thiophenes

The species of genus *Tagetes* produce thiophenes, polyacetylenic compounds having strong biocidal activity, thus making *Tagetes* plants very useful for suppressing nematode populations in the soil and as sources of natural pesticides. Jacobs et al. (1995) screened more than 300 plants from a mutagenized M, population using high-performance liquid chromatography analysis of root extracts and two mutants of *Tagetes erecta* displaying aberrant thiophene composition were identified. Both mutants, which may have originated from the same mutational event, contained high amounts of the C_{13}monothiophene 2-(but-3-en-I -ynyl)-5-(penta1,3-diynyl)-thiophene that was previously not found in *T. erecta* and also high amounts of two C_{13} bithienyls that were absent or present at low concentrations in the wild type.

In a study by Marotti et al. (2010), *Tagetes* species (*T. erecta, T. filifolia, T. lucida, T. minuta, T. patula,* and *T. tenuifolia*) grown in northern Italy were evaluated for their morphophenological parameters and thiophene pattern in different plant parts (roots, shoots, and flowers). It was found that roots had the highest diversity and contents of thiophenes, with 5-(3-buten-1-ynyl)-2,2 -bithienyl (BBT) as the main component. *Tagetes lucida* and *T. tenuifolia* were found to possess the highest amounts of total thiophenes (6717.3 and 6452.5 mg kg^{-1} dry weight, respectively), while *T. minuta* contained highest total thiophene yield (518.8 mg m^{-2}). Hence concluding that considering both thiophene concentrations and biomass yields, *T. minuta* and *T. lucida* appeared to be the most promising *Tagetes* species, with high potential for use as biocidal crops for the implementation of pest control practices that are less harmful to human health and natural resources.

Extracts of flowers of some *Tagetes* species grown in Belgorod (RF) *T. erecta* L. (cv. Gavaii), *T. patula* L. (cv. Petit orange), *T. tenuifolia* Cav. (unknown cv.), and *T. lucida* Cav. were investigated by Victor et al. (2014) using thin-layer chromatography (TLC) and high-performance chromatography (HPLC) with diode-array (DAD) and mass spectrometric detector (MS). It was found that C13 bithienyl (instead of C12-derivatives being common for hairy roots) was a major thiophene derivative of all extracts under investigation, and that thiophenes are accumulated mainly in the parts of flowers hidden from sunlight by a flower's sepals.

1.7 Evaluation of Genetic Diversity for Desired Traits

An investigation was conducted by Lydia et al. (2019) to assess the existence of genetic variability in divergent genotypes of marigold (*Tagetes erecta*). The pattern of distribution of the genotypes studied in various clusters showed considerable amount of genetic divergence among these genotypes for all the vegetative and flowering characters studied. They suggested that as these genotypes have better means for yield and yield contributing characters and are placed in different clusters showing great genetic diversity, these can be used us parents in hybridization program for the higher yield.

Santhosh et al. (2018) evaluated 33 genotypes of African marigold (*Tagetes erecta*) for 27 characters contributing for yield, carotenoid and lutein. Phenotypic coefficient of variability (PCV) and genotypic coefficient of variation (GCV) were found to be maximum for fresh petal meal per flower and dry petal meal per flower. High broad sense heritability coupled with high genetic gain was observed for number of secondary branches, fresh petal meal per flower, dry petal meal per flower, dry weight per flower, flower weight per plant, flower number per plant, zea-xanthin content, lutein content, total carotenoid content, seed number per flower, 100 seed weight, number of seeds per gram, and shelf life. The heritability was found to be likely due to additive gene effects.

Latha and Dharmatti (2018) evaluated 26 genotypes of marigold for 12 growth, flowering, and yield attributes to study their genetic parameters such as variability, heritability, genotypic coefficient of variation (GCV), and phenotypic coefficient of variation (PCV). All traits showed significant difference among the genotypes. They recorded maximum value of GCV and PCV for the number of petals per flower and plant height. The highest broad sense heritability was recorded for flower yield (t/ha), flower yield (g/plant), number of flowers per plant, flower diameter, and number of petals per flower. Highest genetic advance over mean was recorded for plant height, flower yield (t/ha), flower yield (g/plant), flower diameter, and internodal length.

Panwar et al. (2016) carried out genetic diversity of marigold (*Tagetes erecta* L.) on the basis of morphological traits and reported significant amount of variability in the germplasm that will be needed for breeding marigold for various qualitative and quantitative traits.

1.8 Available Sources of Breeding Value

1.8.1 Salt Tolerance

Salinity is one of most significant environmental stresses. Plants exposed to salt stress undergo changes in their environment. Marigold is moderately tolerant to salinity stress. Salinity stress has a significant impact in terms of micronutrient and macronutrient uptake in marigold. Under saline conditions, the Ca^{2+}, Mg^{2+}, K^{+}, and Na^{+} uptakes of marigold are important parameters in terms of revealing the effects of stress (Koksal et al. 2016). Marigold "French Vanilla," "Flagstaff," and "Yellow

Climax" may be used as bedding plants and "Yellow Climax" as specialty cut flower production as well as in landscape sites when ECw (electrical conductivity) is lower than 8 dSm^{-1} with minimal effects on plant quality. Although quality of marigold flowering stems is reduced as a result of their sensitivity to treatment irrigation waters, all three cultivars will produce acceptable plants in landscape sites where high salinity and high pH co-occur as dual stress factors (Valdez-Aguilar et al. 2009). *Tagetes erecta* is moderately tolerant to salt stress. After 2 weeks treatment fresh biomass of plant was found to increase significantly ($P < 0.001$) in 50 and 100 mM NaCl. Plant height, root length, no of leaves, fresh and dry biomass, chlorophyll a, b, and carotenoids reduced under higher concentration of NaCl (Sayyed, Aqib et al., 2013). Saline water irrigation causes significant yield losses in crop plants. However, short-term saline application might cause fewer negative effects on yield yet at the same time improve quality aspects of edible products. Short-term saline exposure of *Tagetes* plants activated metabolic processes, and as a result there was an accumulation of minerals such as N, P, Na, and Zn on edible flowers (Chrysargyris et al. 2018).

Sun et al. (2018) carried out relative salt tolerance of eight marigolds "Discovery Orange," "Discovery Yellow," "Taishan Gold," "Taishan Orange," "Taishan Yellow" of African marigold (*Tagetes erecta*) and "Hot Pak Gold," "Hot Pak Orange," "Hot Pak Yellow" of French marigold (*Tagetes patula*). All marigold cultivars showed moderate sensitivity to salinity. In EC 6, all "Hot Pak Gold" and "Taishan Orange" marigolds died, whereas only one of nine "Hot Pak Orange" and "Hot Pak Yellow" plants survived and all other survived marigold cultivars experienced severe foliar salt damage. Even in EC 3, "Taishan Gold" and "Taishan Orange" marigolds had moderate foliar damage. Among all tested marigold cultivars, Discovery Orange, Taishan Yellow, Discovery Yellow, and Taishan Gold were relatively more tolerant than the remaining four cultivars to the salinity of irrigation water in this study.

1.8.2 Drought Tolerance

Drought is an important abiotic stress that limits the plant growth and productivity. A pot experiment was conducted by Atif Riaz et al. (2013) who evaluated morphological and physiological attributes that can be used for characterization of drought tolerance in two varieties of Marigold (Super Giant & Inca F1). Four drought levels at 100% (control), 80%, 70%, and 60% field capacity were maintained throughout the experiment. Morphological characteristics including plant height, number of leaves/plants, leaf firing percentage, leaf area, plant quality, root length, shoot fresh and dry weight, root fresh and dry weight, and root-shoot ratio for fresh and dry weights were studied. Several physiological parameters like net CO_2 assimilation rate, transpiration rate, stomatal conductance, sub-stomatal conductance, leaf water potential, water use efficiency, and chlorophyll content were also studied. Results showed that overall plant quality of varieties decreased with the progression of drought stress where 70% FC (field capacity) can be considered appropriate for acceptable plant quality. Also, Inca F1 was shown to perform better compared to Super Giant for all attributes studied.

A study was conducted by Mohamed et al. (2000) aimed to use somaclonal variation to select drought tolerant plants of *Tagetes*. After growth in greenhouse conditions for 2 months, one clone of *T. minuta* selected under drought appeared to show drought tolerance and had higher proline content and soluble sugars than the non-stress-selected clone even after growth in non-stress conditions for 6 months. This clone was developed from shoot clumps selected on medium with mannitol and exhibited a significant tolerance in vitro in medium containing 90 mM mannitol; this medium completely inhibited growth of control plants. This clone had significantly higher proline content and soluble sugars than the non-stress-selected clone when cultured on medium containing 0 or 30 mM mannitol. When this clone was tested for drought tolerance (growth at 40% soil field capacity), it showed a significant tolerance compared with other regenerated and control plants and revealed lower water potential, greater accumulated biomass and a higher relative growth rate.

Physiological (growth inhibition) and biochemical (proline levels) responses to water stress of five cultivars of *Tagetes patula* and five cultivars of *T. tenuifolia* was studied by Cicevan et al. (2014). They found the most drought-tolerant cultivars to be "Bolero" in *T. patula* and "Luna gold" in *T. tenuifolia*, while "Orion" and "Luna orange" were found to be the most sensitive for each species. With the aim of determining the best performing cultivar under water/drought stress. The response of two marigold cultivars (Inca and Bonanza) was examined under different regimes of drought stress by Younis et al. (2018). The results revealed that increasing water stress adversely affected plant height, in both cultivars. Both cultivars showed a decreasing trend to the number of flowers under water stress. Total chlorophyll contents including a and b also showed reduction under prolonged drought treatment in both cultivars from (2.7 mg g^{-1} FW) to (1 mg g^{-1} FW). However, the performance of cultivar "Inca" was found satisfactory under overall water stress regimes.

1.9 Molecular Characterization

1.9.1 Molecular Markers

Shahzadi et al. (2016) characterized *Tagetes* species by using RAPD and STS markers. Two primer systems including 25 RAPD and 3 STS (limonene gene) were used to ascertain genetic diversity of 15 *Tagetes* genotypes belonging to different species. The findings concluded that PCR-based assay such as RAPD and STS could be used successfully for estimation of genetic diversity of different genotypes of *Tagetes* that can be used for selection of parents for improvement of the species.

Namita et al. (2013) assessed the genetic diversity of 15 genotypes of *Tagetes erecta* L. and *Tagetes patula* L. using PCR-based RAPD (Random Amplified Polymorphic DNA) and ISSR (Inter Simple Sequence Repeat) markers. Molecular characterization in marigold germplasm was done by using RAPD, ISSR, and UPR markers Panwar et al. (2016) and Panwar et al. (2017). All these marker systems were found to be effective in distinguishing genotypes at molecular level.

Mor et al. (2008) characterized nine genotypes of marigold, selected from two species, *Tagetes erecta* L. (3 genotypes) and *Tagetes patula* L. (6 genotypes), through electrophoresis of protein and using RAPD markers. It was concluded that the SDS-PAGE of total seed soluble proteins and RAPD-based DNA fingerprinting or polymorphism can differentiate the marigold genotypes. Whankaew et al. (2014) developed simple sequence repeats (SSRs) specific to marigold using the inter-SSR technique and an SSR-enriched genomic DNA library. In total, 38 polymorphic markers with 112 observed alleles were identified in 20 African marigolds (*Tagetes erecta* L.) consisting of 14 commercial varieties and six Thai landraces, and six French marigolds (*Tagetes patula* L.). Qi et al. (2007) evaluated the genetic relationships of 12 *Tagetes patula* inbred lines, 4 *T. patula* F_1, and 2 *T. erecta* with ISSR markers and morphological traits. In this study, simple sequence repeats (SSRs) specific to marigold were developed using the inter-SSR technique and an SSR-enriched genomic DNA library. The results of the study indicate that the SSRs developed are effective for genetic diversity analysis, species classification, and individual identification.

AFLP- and SCAR-based linkage map in marigold was constructed by He et al. (2010) and reported that SCAR 4 marker has been linked to male sterility which is placed at a distance of 0.3 cM distance from the sterility locus on the linkage map showing that it is tightly linked to sterility gene. He et al. (2008) in their studies on marigold used the inter-simple sequence repeat (ISSR) and sequence-related amplified polymorphism (SRAP) techniques combined with bulked segregant analysis to develop markers linked to the male sterility trait. From a survey of the 38 ISSR primers and 170 SRAP primer combinations they found that only one SRAP marker that was closely linked to the target trait was identified and successfully converted into sequence characterized amplified region (SCAR) marker that was located within 2.4 cM from Tems locus. Hence, they found that SCAR marker permits the efficient marker-assisted selection of male sterile individuals in breeding programs of marigold and will greatly facilitate the breeding of F1 cultivars.

1.9.2 Genomics

APETALA2 is a gene involved in flower development, which has been characterized in *Arabidopsis thaliana*. Apetala2 is a transcription factor associated with meristem determination, flower identity, and seed development. Alegría-Mundo et al. (2012) conducted a molecular analysis of the APETALA2 ortholog in marigold during flower development. The ortholog of *APETALA 2* was identified in *Tagetes erecta* floral organs, *APETALA 2* showed a 70% homology with other *APETALA2* genes. The *teAPETALA2* mRNA showed a differential expression, mainly in floral meristems, although expression was detected also in roots and leaves.

Ai et al. (2016) in their studies, performed transcriptome assembly and digital gene expression profiling to generate expression profiles of MS (Male sterile) and

MF (Male fertile) plants in *T. erecta*. Taking the transcriptome (from the generated cDNA library) as reference, they detected 125 differentially expressed genes in both developmental stages of MS and MF flower buds. MADS-box genes were presumed to be highly related to male sterility in *T. erecta* based on histological and cytological observations. Twelve MADS-box genes showed significantly different expression levels in flower buds (4 mm in diameter), whereas only one gene expressed significantly different in flower buds (1 mm in diameter) between MS and MF plants. This work is the first attempt to sequence and assemble a reference transcriptome in *T. erecta* using Illumina sequencing technology.

AGAMOUS subfamily proteins are encoded by MADS-box family genes. They have been shown to play key roles in the determination of reproductive floral organs such as stamens, carpels and ovules. However, they also play key roles in ensuring a fixed number of floral organs by controlling floral meristem determinacy (Dreni and Kater 2013). A research conducted by Zhang et al. (2020) isolated four AG subfamily genes of marigold and phylogenetically grouped into C class (TeAG1 and TeAG2) and D class (TeAGL11-1 and TeAGL11-2) genes according to the ABCDE flower organ development model. And these four genes were highly expressed in reproductive organs of marigold. Moehs et al. (2001), in their studies, examined the expression of carotenoid genes in marigold petals, for which they cloned the majority of the genes in this pathway and used these to assess their steady-state mRNA levels in four marigold cultivars with extreme differences in carotenoid content. They also cloned genes encoding early steps in the biosynthesis of isopentenyl pyrophosphate (IPP), the precursor of all isoprenoids, including carotenoids, as well as two genes required for plastid division. Differences among the marigold varieties in the expression of these genes suggested that differences in mRNA transcription or stability underlie the vast differences in carotenoid synthesis and accumulation in the different marigold varieties.

Zhao Song et al. (2020) reported the first chloroplast genome sequence of *T. erecta*. Their studies showed the chloroplast genome size to be 152,065 bp with GC content of 37.4%, including a large single-copy (LSC) of 83,895 bp, a small single-copy (SSC) of 18,065 bp, and a pair of 25,048 bp IR (inverted repeat) regions. A total of 132 genes were annotated including 87 protein-coding genes, 37 tRNA genes, and 8 rRNA genes. The phylogenetic analysis revealed that *T. erecta* belongs to the subfamily Asteroideae.

1.9.3 Transgenics

Transgenic technology based on genetic engineering is the main breeding method for many crops. Vanegas et al. (2006) have succeeded in genetically modifying leaf explants of "Alcosa" marigold using a gene gun. However, the tissue culture technology for marigold is not very mature. Also, since marigold is rich in lutein hence its explants are prone to browning, leading to a low plant regeneration rate. The floral dip transformation method (wherein the exogenous gene is directly introduced into the reproductive organs of the target plant to obtain transgenic plants.) avoids tissue culture and regeneration processes (Feldmann and Marks 1987; Bechtold et al. 1993; Chang et al. 1994; Katavic et al. 1994; Clough and Bent 1998). *Agrobacterium tumefaciens* EHA105-mediated transformation method of transformation was used in marigold to transfer the binary vector PCB260 containing the screening gene basta, the reporter gene GFP, and a 4 × 35S enhancer into marigold (*Tagetes erecta*) using the floral dip method. After herbicide resistance screening and genomic PCR testing, four transgenic lines were obtained in T0 generation. In the T1 generation, 15 transgenic plants showed fluorescence and were GFP-positive with phenotypic changes (Xi Cheng et al. 2019).

1.10 Major Constraints in the Crop Production

1. The major constraint in crop production is the selection of appropriate cultivars for different seasons as each genotype performs differentially in different agro climatic conditions.
2. The crop is very sensitive to water-logged conditions hence proper care is required to maintain proper moisture regimes in soils.
3. The occurrence of frost in winters also affects the growth and development of the plant.
4. The crop is highly cross-pollinated in nature, and hence proper pollination management techniques are required for maintenance of germplasm.

References

Acevedo-Rodríguez P, Strong MT (2012) Catalogue of the seed plants of the West Indies. Smithsonian Institution, Washington, DC. 1192 pp

Ai Y, He YH, Hu Y, Zhang Q, Pan C, Bao MZ (2014) Characterization of a novel male sterile mutant of *Tagetes patula* induced by heat shock. Euphytica 200(2):159–173

Ai Y, Zhang Q, Wang W, Zhang C, Cao Z et al (2016) Transcriptomic analysis of differentially expressed genes during flower organ development in genetic male sterile and male fertile *Tagetes erecta* by digital gene-expression profiling. PLoS One 11(3):e0150892

Ai Y, Zhang CL, Sun YL, Wang WN, He YH, Bao MZ (2017) Characterization and functional analysis of five MADS-box B class genes related to floral organ identification in *Tagetes erecta*. PLoS One 12(1):e0169777

Akhalkatsi M, Ekhvaia J, Asanidze Z (2012) Diversity and genetic erosion of ancient crops and wild relatives of agricultural cultivars for food: implications for nature conservation in Georgia (Caucasus), chapter 3. In: Tiefenbacher J (ed) Perspectives on nature conservation – patterns, pressures and prospects. InTech, Rijeka, pp 51–92

Alam A, Couch J, Creger C (2011) The carotenoids of the marigold, *Tagetes erecta*. Can J Bot 46: 1539–1541. https://doi.org/10.1139/b68-213

Alegría-Mundo H, Yong L, Cruz-Ramírez A, Herrera-Estrella L, Cruz-Hernández A (2012) Molecular analysis of marigold (*Tagetes erecta*) *apetala2* in flower development. Acta Hortic 929:293–298

Ali N, Ab A, Sharopov F, Al-kaf A, Hill G, Arnold N, Al-Sokari S, Setzer W, Ludger D (2014) Composition of essential oil from *Tagetes minuta* and its cytotoxic, antioxidant and antimicrobial activities. Nat Prod Commun 9:4

Asaf S et al (2016) Complete chloroplast genome of *Nicotiana otophora* and its comparison with related species. Front Plant Sci:7. https://doi.org/10.3389/fpls.2016.00843
Babu KN, Yamuna G, Praveen K, Minoo D, Ravindran PN, Peter KV (2012) Cryopreservation of spices genetic resources. In: Current frontiers in cryobiology. InTech, pp 16–484. https://doi.org/10.5772/35401
Baciu A, Pamfil D, Sestras A, Mihalte L, Sestras R (2012) Heterosis studies for response to *Aphis fabae* attack in Calendula. Bull Univ Agric Sci Vet Med Cluj-Napoca. Horticulture 69(1):40–47
Banyal N, Panwar S, Sonah H, Singh K, Sharma T (2013) Genetic diversity analysis of marigold (*Tagetes* sp) genotypes using RAPD and ISSR markers. Indian J Agric Sci 83:484–490
Base Collection In NGB (2020). http://www.nbpgr.ernet.in/Research_Projects/Base_Collection_in_NGB.aspx
Bechtold N, Ellis J, Pelletier G (1993) In planta *Agrobacterium* mediated gene transfer by infiltration of adult *Arabidopsis thaliana* plants. C R Acad Sci Paris Life Sci 316:1194–1199
Beridze RK, Hanelt P, Kandelaki VN, Mandzgaladze D, Schultze-Motel J (1990) Collecting plant-genetic resources in the Georgian SSR (Kartli, Meskheti) 1989. Die Kulturpflanze 38(3):157–171
Birky CW (1995) Uniparental inheritance of mitochondrial and chloroplast genes: mechanisms and evolution. Proc Natl Acad Sci U S A 92:11331–11338. https://doi.org/10.1073/pnas.92.25.11331
Botelho FBS, Rodrigues CS, Bruzi AT (2015) Ornamental plant breeding. Ornament Hortic 21(1): 9–16
Burpees Seeds and Plants. Marigold history. https://www.burpee.com/gardenadvicecenter/annuals/marigolds/marigoldhistory/article10006.html
Chaney L, Mangelson R, Ramaraj T, Jellen EN, Maughan PJ (2016) The complete chloroplast genome sequences for four *Amaranthus* species (Amaranthaceae). Appl Plant Sci 4:1600063. https://doi.org/10.3732/apps.1600063
Chang SS, Park SK, Kim BC (1994) Stable genetic transformation of *Arabidopsis thaliana* by *Agrobacterium* inoculation in planta. Plant J 5:551–558
Chatterjee S, Prakash T, Kotrsha D, Rao NR, Goli D (2011) Comparative efficacy of *Tagetes erecta* and *Centella asiatica* extracts on wound healing in albino rats. ChinMed 2:138–142. 3
Cheng X, Huang C, Zhang X, Lyu Y (2019) Establishment of transgenic marigold using the floral dip method. Acta Physiol Plant 41:147. https://doi.org/10.1007/s11738-019-2937-3
Cho KS et al (2015) Complete chloroplast genome sequence of tartary buckwheat (*Fagopyrum tataricum*) and comparative analysis with common buckwheat (*F. esculentum*). PLos One 10: e0125332. https://doi.org/10.1371/journal.pone.0125332
Choi KS, Chung MG, Park S (2016) The complete chloroplast genome sequences of three Veroniceae species (Plantaginaceae): comparative analysis and highly divergent regions. Front Plant Sci 7:355. https://doi.org/10.3389/fpls.2016.00355
Chrysargyris A, Tzionis A, Xylia P, Tzortzakis N (2018) Effects of salinity on *Tagetes* growth, physiology, and shelf life of edible flowers stored in passive modified atmosphere packaging or treated with ethanol. Front Plant Sci 9:1765. https://doi.org/10.3389/fpls.2018.01765
Cicevan R, Hassan M, Vicente O, Boscaiu M, Sestras A, Sestras R (2014) Drought tolerance in several *Tagetes* L. cultivars. Bull Univ Agric Sci Vet Med Cluj-Napoca. Horticulture 71. https://doi.org/10.15835/buasvmcn-hort:10553
Clough SJ, Bent AF (1998) Floral dip: a simplified method for *Agrobacterium*-mediated transformation of *Arabidopsis thaliana*. Plant J 16:735–743
Collecting Plant Diversity: Technical Guidelines, Luigi Guarino, V. Ramanatha Rao and Robert Reid (Editors) (1995) CAB International, International Plant Genetic Resources Institute (IPGRI), the Food and Agriculture Organization of the United Nations (FAO), the World Conservation Union (IUCN) and the United Nations Environment Programme (UNEP). Chapter 6: Strategies for the collecting of wild species, Mauricio Parra-Quijano, José María Iriondo and Elena Torres Lamas (Editors) (2011) Article in: Crop gene bank knowledge base
Darwin C (1877) The different forms of flowers on plants of the same species. Murray, London
Davidse G, Sousa-Sánchez M, Knapp S, Chiang F, UUoa Ulloa C, Pruski JF (2018) Flora Mesoamericana, Volumen 5, Parte 2: Asteraceae [ed. by Davidse, G., Sousa-Sánchez, M., Knapp, S., Chiang, F., UUoa Ulloa, C., Pruski, J. F.]. Missouri Botanical Garden Press, St. Louis. xix + 608

Deineka VI, Tret'yakov MY, Lapshova MS, Deineka LA (2014) Thiophenes of *Tagetes* flowers and partial purification of xanthophyll esters. Univ J Agric Res 2(3):101–106
Dong W, Liu J, Yu J, Wang L, Zhou S (2012) Highly variable chloroplast markers for evaluating plant phylogeny at low taxonomic levels and for DNA barcoding. PLoS One 7:e35071. https://doi.org/10.1371/journal.pone.0035071
Dong W, Xu C, Cheng T, Zhou S (2013) Complete chloroplast genome of *Sedum sarmentosum* and chloroplast genome evolution in saxifragales. PLoS One 8:e77965. https://doi.org/10.1371/journal.pone.0077965
Ludovico D, Kater MM (2013) *MADS* reloaded: evolution of the *AGAMOUS* subfamily genes. New Phytol 201: 717–732
Edy HJ, Marchaban SW, Nugroho AE (2017) Characterization and evaluation of bioactive compounds of extract ethanol *Tagetes erecta* L. by GC-MS. Int J ChemTech Res 10(2):172–175. CODEN (USA): IJCRGG. ISSN: 0974-4290. ISSN(Online):2455-9555
Esaiashvili M, Gogia N, Meladze T, Chkhikvishvili I, Sephashvili A, Gongadze M (2011) The effect of marigold extract on experimental diabetes mellitus. Exp Clin Med 1:44–47
Feldmann KA, Marks MD (1987) *Agrobacterium*-mediated transformation of germinating seeds of *Arabidopsis thaliana*: a non-tissue culture approach. Mol Gen Genet 208:1–9
González-Benito ME, Aguilar N, Avila T (2009) Germination and embryo rescue from *Passiflora* species seeds postcryopreservation. CryoLetters 30(2):142–147
Gregory GK, Chen T-S, Philip T (1986) Quantitative analysis of lutein esters in marigold flowers (*Tagetes erecta*) by high performance liquid chromatography. J Food Sci 51:1093–1094
GRIIS (2018) Global register of introduced and invasive species. http://www.griis.org/
Hadden WL, Watkins RH, Levy LW et al (1999) Carotenoid composition of marigold (*Tagetes erecta*) flower extract used as nutritional supplement. J Agric Food Chem 47(10):4189–4194
Hamayun M, Sohn EY, Khan SA, Shinwari ZK, Khan AL, Lee IJ (2010) Silicon alleviates the adverse effects of salinity and drought stress on growth and endogenous plant growth hormones of soybean (*Glycine max* L.). Pak J Bot 42(3):1713–1722
He Y, Ning G, Sun Y, Qi Y, Bảo M (2008) Identification of a SCAR marker linked to a recessive male sterile gene (Tems) and its application in breeding of marigold (*Tagetes erecta*). Plant Breed 128:92–96
He YH, Ning GG, Sun YL, Qi YC, Bao MZ (2009) Identification of a SCAR marker linked to a recessive male sterile gene (*Tems*) and its application in breeding of marigold (*Tagetes erecta*). Plant Breed 128(1):92–96
He YH, Ning GG, Sun YL, Hu Y, Zhao XY, Bao MZ (2010) Cytological and mapping analysis of a novel male sterile type resulting from spontaneous floral organ homeotic conversion in marigold (*Tagetes erecta* L.). Mol Breed 26(1):19–29
He YH, Sun YL, Zheng RR, Ai Y, Cao Z, Bao MZ (2016) Induction of tetraploid male sterile *Tagetes erecta* by colchicine treatment and its application for interspecific hybridization. Hortic Plant J 2(5):284–292
Hernandez G, Ham ML (2007) Marigold biotechnology: tissue culture and genetic transformation. Transgenic Plant J 1(1):169–174
Heuzé V, Tran G, Hassoun P, Lebas F (2017) Mexican marigold (*Tagetes erecta*). In: Feedipedia, INRA, CIRAD, AFZ and FAO. https://www.feedipedia.org/node/90
Heyden D (1987) Symbolism of ceramics from the Templo Mayor. The Aztec Templo Mayor: a symposium at Dumbarton Oaks, 8th and 9th October 1983 (Dumbarton Oaks). ISBN 9780884021490
Hong SY et al (2017) Complete chloroplast genome sequences and comparative analysis of *Chenopodium quinoa* and *C album*. Front Plant Sci 8. https://doi.org/10.3389/fpls.2017.01696
Howe CJ et al (2003) Evolution of the chloroplast genome. Philos Trans R Soc Lond Ser B Biol Sci 358:99–107
Hu H et al (2016) Species delimitation and interspecific relationships of the genus *Orychophragmus* (Brassicaceae) inferred from whole chloroplast genomes. Front Plant Sci 7. https://doi.org/10.3389/fpls.2016.01826

Huang H, Shi C, Liu Y, Mao S-Y, Gao L-Z (2014) Thirteen *Camellia* chloroplast genome sequences determined by high-throughput sequencing: genome structure and phylogenetic relationships. BMC Evol Biol 14:151. https://doi.org/10.1186/1471-2148-14-151

Idrees M, Irshad M (2014) Molecular markers in plants for analysis of genetic diversity: a review. II

Ishmuratova MY et al (2020) Cryopreservation of seeds of wild plants of Premorsky Krai. EurAsian J BioSci 14:501–505

Jacobs JJ, Arroo RR, De Koning EA, Klunder AJ, Croes AF, Wullems GJ (1995) Isolation and characterization of mutants of thiophene synthesis in *Tagetes erecta*. Plant Physiol 107(3):807–814

Jain R, Katare N, Kumar V, Samanta AK, Goswami S, Shrotri CK (2012) In vitro anti-bacterial potential of different extracts of *Tagetes erecta* and *Tagetes patula*. J Nat Sci Res 2(5):84–90. 4

Kassahun B (2017) Conservation and cultivation of medicinal plants. https://doi.org/10.13140/RG.2.2.28153.16488

Katavic V, Haughn GW, Reed D (1994) In planta transformation of *Arabidopsis thaliana*. Mol Gen Genet 245:363–370

Koksal N, Alkan-Torun A, Kulahlioglu I et al (2016) Ion uptake of marigold under saline growth conditions. Springerplus 5:139. https://doi.org/10.1186/s40064-016-1815-3

Kolodner R, Tewari KK (1979) Inverted repeats in chloroplast DNA from higher plants. Proc Natl Acad Sci U S A 76:41–45. https://doi.org/10.1073/pnas.76.1.41

Krak K et al (2016) Allopolyploid origin of *Chenopodium album* s. str. (Chenopodiaceae): a molecular and cytogenetic insight. PLoS One 11:e0161063. https://doi.org/10.1371/journal.pone.0161063

Latha S, Dharmatti PR (2018) Genetic variability studies in marigold. Int J Pure Appl Biosci 6(3): 525–528

Leigh Hadden W, Watkins RH, Levy LW, Regalado E, Rivadeneira DM, van Breemen RB, Schwartz SJ (1999) Carotenoid composition of marigold (*Tagetes erecta*) flower extract used as nutritional supplement. J Agric Food Chem 47(10):4189–4194

Li FR, Zhang JC, Xu JR, Zhou JH (2005) Studies on the cross-breeding of *Tagetes erecta* L. × *Tagetes patula* L. and the sterility of hybrid. Inner Mongolia Agric Univ 26(2):51–54

Li QQ, Zhou SD, Huang DQ, He XJ, Wei XQ (2016) Molecular phylogeny, divergence time estimates and historical biogeography within one of the world's largest monocot genera. AoB Plants 8:plw04

Lydia J, Ponnuswami V, Thangaselvabai T (2019) Genetic diversity analysis in *Tagetes erecta* L. genotypes. Int J Chem Stud 7(3):2766–2768

Marotti M, Piccaglia R, Biavati B, Marotti I (2004) Characterization and yield evaluation of essential oils from different *Tagetes* species. J Essent Oil Res 16(5):440–444

Marotti I, Marotti M, Piccaglia R, Nastri A, Grandi S, Dinelli G (2010) Thiophene occurrence in different *Tagetes* species: agricultural biomasses as sources of biocidal substances. J Sci Food Agric 90:1210–1217. https://doi.org/10.1002/jsfa.3950

Martin W, Deusch O, Stawski N, Grunheit N, Goremykin V (2005) Chloroplast genome phylogenetics: why we need independent approaches to plant molecular evolution. Trends Plant Sci 10: 203–209. https://doi.org/10.1016/j.tplants.2005.03.007

Moehs CP, Tian L, Osteryoung KW et al (2001) Analysis of carotenoid biosynthetic gene expression during marigold petal development. Plant Mol Biol 45:281–293

Mohamed MA, Harris PJ, Henderson J (2000) *In vitro* selection and characterization of a drought tolerant clone of *Tagetes minuta*. Plant Sci: Int J Exp Plant Biol 159(2):213–222. https://doi.org/10.1016/s0168-9452(00)00339-3

Mor VS, Deswal DP, Mann A, Dahiya BS, Beniwal BS (2008) Characterization of marigold (*Tagetes* spp.) genotypes using SDS-PAGE and RAPD markers. Seed Sci Technol 36:757–766. https://doi.org/10.15258/sst.2008.36.3.25

Namita N, Singh KP, Bharadwaj CP, Prasad KV, Raju DVS (2009) Studies on character association and path analysis of quantitative traits among parental lines of marigold (*Tagetes erecta* and *T. patula*) and their interspecific F1 hybrids. Indian J Hortic 66(3):348–352

Namita N, Singh KP, Bharadwaj C, Sharma TR, Sonah H, Raju DVS, Deshmukh RK (2011) Gene action and combining ability analysis for flower yield and its component traits in interspecific hybrids of marigold (*Tagetes* spp). Indian J Agric Sci 81(9):807–811

Nesterova SV (2004) Thesis. Vladivostok. 150 p
Nikishina TV, Popova E, Vakhrameeva MG, Popov AS (2007) Cryopreservation of seeds and protocorms of rare temperate orchids. Russ J Plant Physiol 54(1):121–127. https://doi.org/10.1134/S1021443707010189
Nishanthi J, Anuradha R (2013) Efficacy of *Tagetes erecta* on lead acetate induced oxidative injury in rat kidney. Int J PharmTech Res 4(4):1377–1382. 2012
Panwar S, Singh KP, Janakiram T, Namita (2016) Assessment of genetic diversity of marigold (*Tagetes erecta* L.) genotypes based on morphological traits. J Ornament Hortic 17(3&4):77–81
Panwar S, Singh KP, Janakiram T, Bharadwaj C, Namita HS, Deshmukh RK, Sharma TR (2017) Molecular characterization of African marigold (*Tagetes erecta*) genotypes using RAPD markers. Indian J Agric Sci 87(5):663–668
Panwar S, Singh KP, Namita T, Janakiram HS, Sharma TR (2018) DNA fingerprinting in African marigold (*Tagetes erecta* L.) genotypes using ISSR and URP markers. Indian J Hortic 75(1): 105–110
Parida AK, Das AB (2005) Salt tolerance and salinity effects on plants: a review. Ecotoxicol Environ Saf 60(3):324–349. https://doi.org/10.1016/j.ecoenv.2004.06.010
Pedapati A, Yadav SK, Vandana T, Singh SP, Ranga SS, Binda PC, Brahmi P (2018) Ornamental germplasm: potential new resources for floriculture industry. Int J Curr Microbiol Appl Sci 7(12):1731–1742
PIER (2018) Pacific Islands ecosystems at risk. HEAR, University of Hawaii, Honolulu. http://www.hear.org/pier/index.htmi
PROTA (2018) PROTA4U web database. Plant Resources of Tropical Africa, Wageningen/Nairobi. https://www.prota4u.org/database/
Qi YC, Ning GG, Bao MZ (2007) Evaluation of genetic relationships of *Tagetes patula* inbred lines using ISSR markers and morphological traits. Sci Agric Sin 40(6):1236–1241
Quackenbush FW, Miller SL (1972) Composition and analysis of the carotenoids in marigold petals. J Assoc Off Anal Chem 55(3):617–621
Rasoanaivo P, Petitjean AM, Ratsimamanga-Urverg S, Rakoto-Ratsimamanga A (1992) Medicinal plants used to treat malaria in Madagascar. J Ethnopharmacol 37:117–127
Raven JA, Allen JF (2003) Genomics and chloroplast evolution: what did cyanobacteria do for plants? Genome Biol 4:209–209
Reed BM (2002) The basics of in vitro storage and cryopreservation. National Clonal Germplasm Repository, Corvallis, pp 34–46
Riaz A, Younis A, Hameed M, Kiran S (2010) Morphological and biochemical responses of turf grasses to water deficit condition. Pak J Bot 42(5):3441–3448
Riaz A, Younis A, Taj AR, Karim A, Tariq U, Munir S, Riaz S (2013) Effect of drought stress on growth and flowering of marigold (*Tagetes erecta* L.). Pak J Bot 45(S1):123–131
Santhosh N, Tejaswini SKS, Seetharamu GK, Gadre A (2018) Genetic diversity for morphological characters and biochemical components in African marigold. Int J Chem Stud 6(6):624–627
Sayyed A, Gul H, Ullah Z, Hamayun M (2013) Effect of salt stress on growth of *Tagetes erecta* L. J Life Sci 02(03/04):96–106, 2014
Shahzadi I, Ahmad R, Waheed U, Shah MM (2016) Genetic diversity analysis of *Tagetes* species using PCR based molecular markers. Pak J Bot 48:1593–1599
Shukla A, Thakur R (2018) First report of septoria leaf spot on marigold (*Tagetes erecta* L.) from Himachal Pradesh, India. Int J Curr Microbiol App Sci 7(1):1744–1748
Singh V, Singh B, Kaul VK (2003) Domestication of wild marigold (*Tagetes Minuta* L.) as a potential economic crop in Western Himalaya and North Indian Plains. Econ Bot 57(4):535–544
Singh RK, Jain SK, Singh AK (2004) Seed storage of African marigold (*Tagetes erecta* L.) for ex-situ conservation. Seed Sci Technol 32:503–509
Song Y et al (2017) Development of chloroplast genomic resources for *Oryza* species discrimination. Front Plant Sci 8. https://doi.org/10.3389/fpls.2017.01854
Song Z, Li T-Q, Liu Z-H, Jiang K, Xiao-Bin O (2020) The complete chloroplast genome of *Tagetes erecta* (Asteroideae), a wildly cultivated ornamental plant. Mitochondrial DNA Part B 5(1):79–80
Sorenson JL, Johannessen CL (2004) Scientific evidence for pre-Columbian transoceanic voyages to and from the Americas. Sino-platonic papers, vol 133. University of Pennsylvania, Philadelphia

Sreekala C, Raghava SPS (2003) Exploitation of heterosis for carotenoid content in African marigold (*Tagetes erecta* L.) and its correlation with esterase polymorphism. Theor Appl Genet 106(4):771–776

Standley PC, Steyermark JA (1949) Flora of Guatemala. VI. Fieldiana: Bot 24(6):440

Sun Y, Niu G, Perez C, Pemberton H, Altland J (2018) Responses of marigold cultivars to saline water irrigation. HortTechnologyhortte 28(2):166–171. Retrieved Sept 3, 2020, from https://journals.ashs.org/horttech/view/journals/horttech/28/2/article-p166.xml

Thormann I, Dulloo M, Engels J (2006) Techniques for ex situ plant conservation. https://doi.org/10.1300/5546_02

Tripathi B, Walia S, Kumar B (2012) Chemical composition and evaluation of *Tagetes erecta* (var. Pusanarangigenda) essential oil for its antioxidant and antimicrobial activity. Biopesticides Int 8:138–146

USDA-ARS (2018) Germplasm Resources Information Network (GRIN). Online database. National Germplasm Resources Laboratory, Beltsville. https://npgsweb.ars-grin.gov/gringlobal/taxon/taxonomysimple.aspx

USDA-NRCS (2018) The PLANTS database. National Plant Data Team, Greensboro. https://plants.sc.egov.usda.gov

Valdez-Aguilar L, Grieve C, Poss J (2009) Salinity and alkaline pH in irrigation water affect marigold plants: I. Growth and shoot dry weight partitioning. HortScience 44. https://doi.org/10.21273/HORTSCI.44.6.1719

Vijay K, Bhingare CL, Balasaheb SR, Yuvraj NR, Janardhan PM (2013) Pharmacognostic, physicochemical and phytochemical investigation of *Tagetes erecta* flowers (asteraceae). J Biol Sci Opin 1(1):21–24

Villar-Martínez AAD, Vanegas-Espinoza PE, Paredes-López O (2010) Marigold regeneration and molecular analysis of carotenogenic genes. In: Jain S, Ochatt S (eds) Protocols for In vitro propagation of ornamental plants. Methods in molecular biology (methods and protocols), vol 589. Humana Press

Villaseñor JL, Espinosa-Garcia FJ (1998) Catalog of weeds of Mexico. National Autonomous University of Mexico. National Consultative Phytosanitary Council. Fund of Economic Culture. Mexico City

Wambugu PW, Brozynska M, Furtado A, Waters DL, Henry RJ (2015) Relationships of wild and domesticated rices (*Oryza* AA genome species) based upon whole chloroplast genome sequences. Sci Rep 5:13957. https://doi.org/10.1038/srep13957

Whankaew S, Hasthanasombut S, Thaikert R, Tappiban P, Smith D, Triwitayakorn K (2014) Molecular characterization and genetic relationship of marigolds (*Tagetes* spp.) based on simple sequence repeat markers. Plant Genet Resour 12(3):317–322

Wicke S, Schneeweiss GM, dePamphilis CW, Müller KF, Quandt D (2011) The evolution of the plastid chromosome in land plants: gene content, gene order, gene function. Plant Mol Biol 76:273–297. https://doi.org/10.1007/s11103-011-9762-4

Younis A, Riaz A, Qasim M, Mansoor F, Zulfiqar F, Tariq U, Ahsan M, Naseem M, Bhatti Z (2018) Screening of marigold (*Tagetes erecta* L.) cultivars for drought stress based on vegetative and physiological characteristics. Int J Food Allied Sci 3(2):56–63

Yuyan AN, Zongsuo L, Ruilian H, Guobin L (2007) Effects of soil drought on seedling growth and water metabolism of three common shrubs in Loess Plateau, Northwest China. Front For China 2:410–416

Zhang HL, Dong AX, Wang T, Zhao LJ, Xin HB (2014) A new cultivar of marigold 'Jingyue 1'. Acta Hortic Sin 41(7):1521–1522

Zhang H, Xin H, Cong R, Li Z, Song L, Wei Z (2019) Cross compatibility analysis to identify suitable parents of *Tagetes erecta* and *T. patula* for heterotic hybrid breeding. Not Bot Hortic Agrobot 47(3):676–682. https://doi.org/10.15835/nbha47311388

Zhang C, Wei L, Wang W, Qi W, Cao Z, Li H, Bao M, He Y (2020) Identification, characterization and functional analysis of AGAMOUS subfamily genes associated with floral organs and seed development in marigold (*Tagetes erecta*). https://doi.org/10.21203/rs.3.rs-35253/v1. PPR: PPR182039

Carnation

2

Sapna Panwar, Youdh Chand Gupta, Poonam Kumari, Neelam Thakur, and Uzma Mehraj

Contents

S. Panwar · U. Mehraj
IARI, New Delhi, India

Y. C. Gupta (✉) · N. Thakur
Department of Floriculture and Landscape Architecture, Y.S. Parmer University of Horticulture and Forestry, Solan, HP, India

P. Kumari
IHBT, CSIR, Palampur, India

S. K. Datta, Y. C. Gupta (eds.), *Floriculture and Ornamental Plants*, Handbooks of Crop Diversity: Conservation and Use of Plant Genetic Resources,
https://doi.org/10.1007/978-981-15-3518-5_3

Abstract

Carnation (*Dianthus caryophyllus* L.) belonging to family Caryophyllaceae is an important commercial cut flower worldwide. Besides cut flowers, it is used for growing in beds, pots, borders, edging, rock gardens, etc. Wild carnation is known to be originated from the Mediterranean regions of Greece and Italy, but the long time in cultivation makes it difficult to confirm its precise origin. Mostly carnations are generally diploid ($2n = 30$) plants but tetraploid forms ($4n = 60$) have also been identified in the genus. The genetic diversity within carnation is relatively poor, and hence the breeding potential for new flower colors and patterns as well as resistance to biotic and abiotic stresses is also very limited. The carnation germplasm collections originating from 189 countries and 35 biodiversity hotspots of which 74% of taxa represents endemic, endangered or species with economic, ecological or scientific value was conserved in the Millennium Seed Bank and at United States Department of Agriculture (Agriculture Research Serviced) GRIN. Different ornamental *Dianthus* species and cultivars and one endemic taxa (*Dianthus giganteus* ssp. banaticus) were conserved using different techniques of cryopreservation including two-step cooling, encapsulation-dehydration and encapsulation-vitrification, PVS2-droplet freezing, DMSO-droplet freezing, and aluminum cryo-plate vitrification. Carnations do not naturally produce delphinidin-based anthocyanin pigments responsible for blue- and purple-colored petals, whereas different shades of purple have been generated using technique of co-expression of the flavonoid 3′ 5′-hydroxylase (F3′5′H) gene from petunia alongside a petunia cytochrome b_5 gene and by downregulation of endogenous dihydroflavonol-4-reductase (DFR) in carnations using RNA interference (RNAi). The carnation (*Dianthus caryophyllus* L.) transcriptome and genome sequence was published in 2012 and 2014, respectively, providing novel opportunities to explore similarities in genetic structure and understanding of genetics with respect to polyploidy events and floral developmental genes in carnations.

Keywords

Carnation · Germplasm · Traits · Conservation · Gene pool · Genome · Sequencing · Transcriptome

2.1 Introduction

Carnation (*Dianthus caryophyllus* L.) is one among the commercially important cut flowers traded worldwide. The genus *Dianthus* has about 300 species of annual, biennial, and perennial herbs of which *Dianthus caryophyllus* is the most important species commercially. The genus *Dianthus* in Greek means "Flower of Zeus" or "Divine

Flower." Linnaeus chose the species name, *caryophyllus*, after the genus of Clove, as the fragrance from Carnation is reminiscent of clove. It is genetically a quantitative long day plant based on its photoperiod requirement (Blake 1955). The *Dianthus* species are adapted to the cooler Alpine regions of Europe and Asia and are also found in Mediterranean coastal regions. Carnations were the first flowers to be cultivated on commercial scale in high altitude areas of tropics for export to Europe and North America.

It is cultivated throughout the world. The largest growing areas of carnation are in Bogota and Colombia. Carnation is also grown in Italy, Spain, Kenya, Sri Lanka, Canary Islands, France, Holland, Germany, Israel, and the USA. In India, main production areas are located around Pune and Bangalore. It is also being grown in Solan, Shimla, Mandi, Kullu, Chamba, Sirmour, Bilaspur, and Hamirpur districts of Himachal Pradesh. As per the State Department of Horticulture – Himachal Pradesh, carnation occupied 44.07 hectares out of 705.77 hectare total area under floriculture in the state during 2018–2019. The acreage of carnation is still expected to increase because of the favorable climatic conditions for its cultivation in different parts of Himachal Pradesh and other hilly states of the country.

In India, depending upon the regions, there is a wide difference in temperature, light intensity, and humidity which not only affect the yield and quality of the flowers but also limit their availability for a particular period of the year. To produce quality flowers, carnation need to be grown under cover, that is, in greenhouse which provides the plants with the optimum condition of light, temperature, humidity, and carbon dioxide for proper growth and to achieve maximum yield of best quality flowers. Though, there are different types of the greenhouses, naturally ventilated polyhouses are preferred in mild climate in which temperature is reduced by ventilation (Ryagi et al. 2007).

2.2 Origin and Description

Carnation belongs to the family of Caryophyllaceae and is believed to be the native of Mediterranean region. The center of origin extends from Eurasia to South Africa. It is a half hardy perennial with branching stems and tumid joints. Leaf blades are simple, entire, and usually narrow. Leaves are linear, glaucous, in opposite or decussate pairs. Stipules are scarious, entire, or lacerate. Each stem forms a terminal flower, and hence inflorescence is generally a terminal cyme, sometimes racemiform, bracteates. Flowers are bisexual, occasionally unisexual. Calyx free or united below, cylindrical with scaly bracts at base. Most carnations are diploid (2n = 30). Tetraploid carnations are also available which produce larger flowers, but less productive. The hybrids continue flowering for longer periods which produce blooms continuously in mild weather.

2.3 Importance and Uses

Carnation is excellent for cut flowers, due to its excellent keeping quality, wide range of colors, ability to withstand long distance transportation, and remarkable ability to rehydrate after continuous shipping. It can also be used for bedding purposes, pots, borders, edging, and rock gardens. Though cut carnations are traded in the world market year round, they are in particular demand for the Valentine's Day, Easter, Mother's Day, and Christmas. Miniature carnations are now gaining popularity for their potential use in floral arrangement. In India, places having cool climate like Kalimpong, Kodaikanal, Bangalore, Pune, Nasik, Himachal Pradesh, etc. are most suitable areas for the production of carnation cut flowers, which may also be exported to Europe.

2.3.1 Classification

Based on the availability of large number of varieties and diversified cultural requirements, carnations are classified as:

(a) **Chabaud or Marguerite Carnations**: These are annual carnations developed by crossing of *D. chinensis* with *D. caryophyllus*. Flowers are single or double, propagated by seeds. Flowers are large with fringed petals and have shorter post-harvest life. The various kinds of Chabaud are Giant Chabaud, Compact Dwarf Chabaud, Entant de Nice, Fleur de Camelia, and Margarita.
(b) **Border and Picotee Carnations**: The flowers of border-type carnations are symmetrical and are the easiest to grow. The flower color varies from single to blended color with irregular markings. The petals are broad and smooth edged. The flowers are generally frilled with open centers. In Picotee type, ground color is without spot or bars. The edges are regular and of bright color.
(c) Malmaison Carnations: These are strong, sturdy, and stiff plants with broad leaves. Flowers are large, double with well-filled centers and are mainly pink colored with good fragrance.
(d) Perpetual Carnations: These are hybrids of different *Dianthus* species flower round the year. The plants are not hardy and are generally treated as cool greenhouse plants. They bear flowers round the year with long stalks which make them suitable for cut flowers. Perpetual flowering and greenhouse carnations are of modern origin.

The commercially important perpetual carnations can be grouped into four major groups: standard, spray, dianthini, and mignon or micro types. Out of these, the large flowering standard types are the most commercially important types.

(i) **Standard types**

The standard type was the first group of carnation used for large-scale production. In this type, flower buds formed on short lateral shoots arising from the axis of the upper leaves are removed to leave one large, terminal flower on a long leafy stem; therefore, it has one large flower on an individual stem. Flowers of standard types are available in various color combinations.

(ii) **Spray types**

In spray or miniature type, terminal flower bud is removed at an early stage to encourage more even development of the lateral flowers, which then produce a multiple flowered stem. The spray types produce many flowers of smaller size and are better adapted to higher temperatures than the standard types.

(iii) **Dianthini types**

These varieties more or less closely resemble the spray types in appearance. However, these plants have longer stems, smaller flowers with higher production. Market of these types of carnation is small.

(iv) **Mignon or Micro types**

Mignon or micro types also have much similarity with the normal spray varieties. The difference is that these plants have shorter stems and yield is higher than spray varieties. However, demand for these types in the market is less.

2.4 Plant Genetic Resources

2.4.1 Geographic Distribution

After *Silene* L., *Dianthus* L. is the second largest genus of Caryophyllaceae. This genus, containing approximately 300 species, is mainly distributed in the Mediterranean region of Europe and Asia (Reeve 1967; Bittrich 1993) including Japan, and other minor species are found in North America and in upland regions in Africa (Galbally and Galbally 1997). The *Dianthus* species are adapted to the cooler alpine regions of Europe and Asia and are also found in the Mediterranean coastal regions. Carnation is not seen in the wild except in some Mediterranean countries. This is consistent with records on floras indicating that the natural distribution of carnation is restricted to the Mediterranean regions of Greece, Italy, Sicily, and Sardinia. It has been cultivated for over 2000 years, and today commercial cultivation is the result of 200 years of improvement and breeding. It is believed that carnations were cultivated by the Muslims of Africa and introduced to Europe from Tunis in the thirteenth century. During sixteenth century, its improvement work was started in various countries. The gardens of Italy, 4 France, Germany, the Netherlands, and England have contributed much to the development of cultivars. The major breakthrough in the carnation flower industry came with the evolution of the cultivar William Sim in 1938 by William Sim of the USA. From this red variety there have been mutations to white, pink, orange, and different variegated forms. The first genetically modified carnation variety "Moon Dust" was developed in Australia by a plant biotechnology company called "Florigene."

a. Primary gene pool: Carnation is one of the most valuable commercial cut flowers around the globe. Commercial carnation cultivars are distributed in two main groups: standard SIM group and spray carnation (Hughes 1993). Standard or single flower SIM is characterized by a big single flower with a high and solid flower stalk. In spray carnation, the axillary flower bud is removed and adventitious flower buds are preserved, which leads to the spray flower architecture.

The genetic variability within carnation is relatively poor therefore; the breeding potential for new flower colors and patterns as well as resistance to biotic and abiotic stresses is also very limited. Carnation is a vegetatively propagated plant, which further reduces its genetic pool availability.

2.4.2 Wild Genetic Resources and Others

The great diversity of flowering plants is generally viewed as a result of key innovations and (co-)evolutionary processes with pollinators. There are over 20 species of wild Dianthus used for breeding carnations (*D. caryophyllus* L.) and pinks (*D. chinensis* L. and others) (Galbally and Galbally 1997). *Dianthus superbus* L. *D. superbus* var. *superbus* f. *leucanthus* T. Shimizu (96), *D. superbus* var. *speciosus* Rchb (95), *D. hungaricus* Pers (59), *D. pyrenaicus* Pourret (86), *D. banaticus* (Heuffel) Borbás (11), and *D. knappii.*

Dianthus carthusianorum L. (Caryophyllaceae) is a gynodioecious perennial herb. The protandrous flowers consist of a calyx tube, protruding unfused petals, stamens, and stigma lobes (Fig. 1). The flowers are pollinated mainly by butterflies (Knuth 1898, Hegi 1979, Bloch et al. 2006). Protandry allows convenient separation of male and female function under experimental conditions. Natural populations of *D. carthusianorum* show continuous variation in flower depth that is distinct between individuals, populations, and sympatric congeners.

Dianthus superbus: The fringed pink or large pink, *Dianthus superbus* L. (Caryophyllaceae), is a perennial herb with a scattered Euro-Siberian distribution. This species was initially described from low-altitude grasslands, with an unresolved infraspecific taxonomy adding an alpine, a forest, and an East-Asian subspecies (Hardion et al. 2019).

2.5 Collections

Wild *Dianthus caryophyllus* is likely to have originated from the Mediterranean regions of Greece and Italy, but the long time in cultivation makes it difficult to confirm its precise origin. The genus *Dianthus* contains several species that have been cultivated for hundreds of years for ornamental purposes (Ingwerson 1949). The Caryophyllaceae comprise over 80 genera and 3000 species, in a mostly holarctic (i.e., temperate to arctic portions of Eurasia and North America) distribution (Harbaugh et al. 2010). Carnations are generally diploid ($2n = 30$) plants (Carolin 1957). Tetraploid forms ($4n = 60$) have also been identified. Triploid

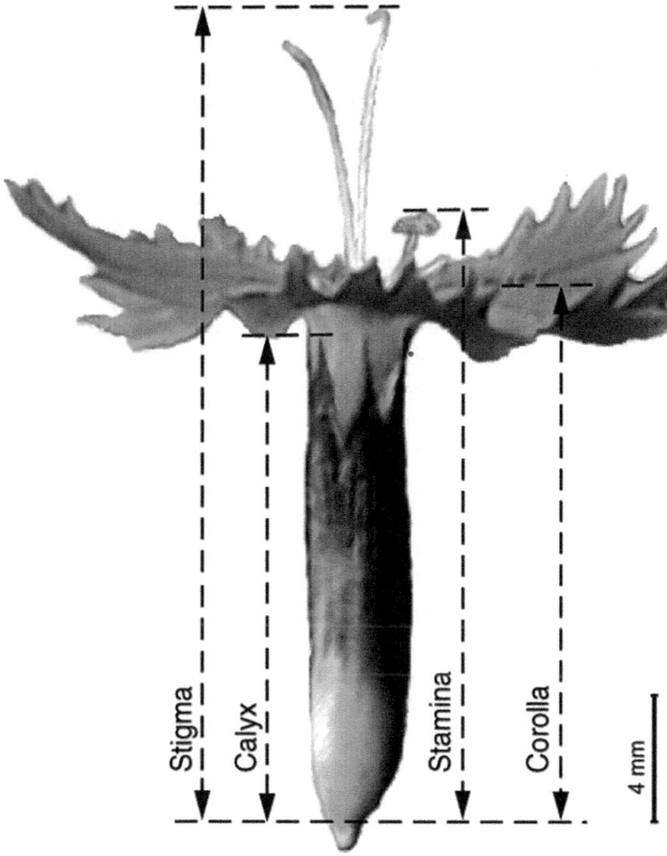

Fig. 1 Floral traits of protandrous and tubular flower of *Dianthus carthusianorum.* (Source: Bloch and Erhardt 2008)

carnations were produced for commercial purposes, but the resulting plants were mostly aneuploid (Brooks 1960). The majority of available cultivars in Australia and Europe are diploid (Galbally and Galbally 1997). The carnations grown and sold in floriculture today are very different from the wild *D. caryophyllus* growing in Mediterranean regions. Flowers of wild-type *D. caryophyllus* are single and five-petaled.

2.5.1 Status of Collections

Carnation (*Dianthus caryophyllus* L.) is one of the commercially important cut flowers of the world and ranks second in the cut flower trade after rose. It is preferred by several exporting countries, on account of its excellent keeping quality, wide range of shapes, sizes, and colors and ability to withstand long distance transportation (Salunkhe et al. 1990). In 2014, it was ranked 14th among the top 25 cut flowers sold in the Dutch auction with a turnover of €25 million (FloraHolland 2015). It was fourth among the top ten imported cut flower to the Netherlands after roses, St John's wort, and gypsophila (FloraHolland 2015). *Dianthus caryophyllus* is one of the major floricultural crops in Japan and worldwide. Many *Dianthus* species are distributed throughout Europe and Asia, and the distribution of this genus extends to arctic North America and to mountainous sites in Africa. Several species, including *D. caryophyllus*, *D. barbatus*, *D. chinensis*, *D. plumarius*, *D. superbus*, and their hybrids, are widely used as horticultural cultivars.

In India, sim carnation is being cultivated in many cities like Nasik, Pune, Bengaluru, Ludhiana, Solan, and Shimla. It is commercially grown in many countries like Columbia, Kenya, Canary Islands, Italy, Spain, Holland, the USA, etc. Production of carnations is confronted with many problems that affect quality including calyx splitting and shorter stem length for some varieties. Conventionally, growers obtain the quality parameters such as stem length and girth, flowers size and number by heavy application of inorganic fertilizers and synthetic plant growth regulators. Although this results in increased production, it adversely affects soil productivity and the environment (Niyokuri et al. 2017).

Carnations have been extremely popular cut flowers since the eighteenth century, with large-scale production of flowers beginning in the mid-1800s (Galbally and Galbally 1997). Colombia is the largest exporter of carnations. Carnation seeds are commercially available in Australia for cultivation in gardens and have been grown commercially as a flower crop since 1954. Currently, increased enforcement of the European codes of practice on good agricultural practices and environmental standards are affecting cut flower trade in the European market (Rikken 2011).

2.5.2 Gaps in Collections, Both Geographical and Genetic

Over half of its species occur in small, geographically restricted ranges. Previous studies have highlighted that climatic preferences have played prominent roles in the evolution and diversification of *Dianthus*. In addition, most *Dianthus* species in the Irano-Turanian region live in rocky mountain habitats that may be affected severely by changing environmental conditions, land-use changes, and human activities. Therefore, many such montane *Dianthus* species, given their restricted ranges, may be highly sensitive to climate change (Behroozian et al. 2020).

Populations of the species are impacted severely by disturbance, so identifying the potential geographic distribution of this species and predicting how climate change will affect its geographic range can be useful for its conversation and management. Previous studies have focused on phylogeny and taxonomic status, yet little is known as regards the details of its geographic distribution or the environmental factors associated with that distribution. The carnation genome was published in 2014, which provides novel opportunities to explore similarities in genetic structure of carnations and other ornamental plants (Yagi et al. 2014). Improved sequencing methods and selection of markers to assist breeding can be used to increase understanding of carnation genetics with respect to polyploidy events and floral developmental genes (Zhang et al. 2018). Early experiments with carnation established plant tissue culture regeneration systems – a necessary precursor to successful transformation.

In 2012, the carnation transcriptome was published. Over 300,000 unique sequences were obtained, including genes involved in flower chlorophyll and carotenoid metabolism, anthocyanin biosynthesis, and ethylene biosynthesis (Tanase et al. 2012). Interspecific hybrids between carnations and *D. capitus* have resulted in plants that are highly resistant to bacterial wilt caused by *Pseudomonas caryophylli*. Crosses between carnations and other *Dianthus* species to generate progeny with desirable floral characteristics such as color patterns, bud number, flower arrangement, and improving year-round flowering (for northern Europe) have also been successful (Umiel et al. 1987).

2.6 Conservation

Originated from subtropics and clonally propagated in commercial, carnation was referred to a limited life span plant in intact condition (Nugent et al. 1991). In the tropic, the conservation of such plant in active growth and in vivo conditions would be very laborious, expensive, and accompanying risks associated field such as pathogens, pests, climatic perturbation, and human error.

Currently, there are two common complementary strategies for the conservation of plant genetic resources, in situ conservation where plant species are maintained in their natural habitat, and ex situ conservation where plant species are conserved outside their natural habitats, in botanical gardens and gene banks, ranging from seed genebanks to in vitro conservation facilities, including cryopreservation genebanks and DNA libraries (Engelmann 2012; GSPC 2012; Peres 2016; O'Donnell and Sharrock 2018). In the selection of the most appropriate conservation approach for a specific plant gene pool, various criteria should be considered, including both the storage characteristics of the species and the reliability of the chosen methods (Cruz-Cruz et al. 2013). The Millennium Seed Bank conserves seed collections originating from 189 countries and 35 biodiversity hotspots of which 74% of taxa represents endemic, endangered, or species with economic, ecological, or scientific value (Liu et al. 2018). In vitro propagation techniques have been successfully applied for the short- to medium-term conservation of certain plant species that might be at risk or are grown in threatened habitats (Cristea 2010; Laslo et al. 2011; Hammond et al. 2019; Kulus 2019). Likewise, cryopreservation (−196 °C) is a feasible alternative for both orthodox seeded species, to complement other conservation strategies and for plant species that do not produce viable seeds or the seeds are "recalcitrant" being desiccation sensitive (Izgu¨ and Mendi 2017). Currently, cryopreservation is the only available technique for the safe long-term storage of clonally propagated plant species

(Kulus and Zalewska 2014; Umesha 2019). The South-Eastern Carpathians have a high distribution of endemic taxa including numerous *Dianthus* species (Mra'z and Ronikier 2016). *Dianthus* genus (Family: Caryophyllaceae) has importance due to the large number of endemic and/or threatened species, some of them with economic value due to its potential ornamental use (Jarda et al. 2011). The Romanian flora comprises approximately 21% endemic taxa of this genus (Cioca^rlan 2009) of which 30% are on the Red List (Dihoru and Negrean 2009). Studies related to micropropagation and medium-term preservation of some endangered Dianthus 1,233,446 Biodiversity and Conservation (2020) 29:3445–3460 species have been reported (Cristea 2010; Cristea et al. 2006, 2010, 2013, 2014; Holobiuc et al. 2009a, b; Jarda et al. 2011; Markovic´ et al. 2013). Cryopreservation studies have been reported mainly for different ornamental *Dianthus* species and cultivars and only for one endemic taxa (*Dianthus giganteus* ssp. *banaticus*) using different techniques, including two-step cooling (Fukai et al. 1991), encapsulation-dehydration and encapsulation-vitrification (Tannoury et al. 1995; Tannoury and Vintejoux 1997; Halmagyi and Lambardi 2006; Halmagyi and Deliu 2007), PVS2-droplet freezing, DMSO-droplet freezing (Halmagyi and Lambardi 2006), and aluminum cryo-plate vitrification (Sekizawa et al. 2011), and vitrification-based methods (Jarda et al. 2011). However, there are numerous cryopreservation protocols developed for different endemic species worldwide (Turner et al. 2001; Coste et al. 2012; Kaczmarczyk et al. 2013).

Considering biodiversity decrease, the high extinction threats, and the potential loss of these endemic species, conservation measure must be undertaken to support their long-term storage. Based on this significant prerequisite, the major aim of our study was to optimize genotype-specific cryopreservation protocols with high regrowth frequencies following cryostorage for seven endemic and endangered *Dianthus* taxa (*D. callizonus*; *D. glacialis* ssp. *gelidus*; *D. henteri*; *D. nardiformis*; *D. spiculifolius*; *D. tenuifolius*; *D. trifasciculatus* ssp. *parviflorus*) from a 2 years active in vitro collection.

2.6.1 In Vitro Conservation

In vitro conservation was considered to be one of the promising methods in preserving the collection and reducing the limitation of in vivo conservation. This method was usually conducted using cell growth inhibitor and protectant. The growth inhibitor suppressed the cells growth and controlled the plant size in culture flask, while cell protectant kept the cells from all the factors which would affect the viability of the cell or plantlet in low temperature during storage (Panis and Lambardi 2005). Among several available methods, in vitro conservation is considered to be one of the promising methods in preserving the collection and reducing the limitation of in vivo conservation. This method is usually conducted using cell growth inhibitor and protectants. The growth inhibitor suppresses the cells growth and controls the plant size in culture flask, while cell protectant keeps the cells from all the factors which would affect the viability of the cell or plantlet in low temperature during storage (Panis and Lambardi 2005).

In vitro conservation of carnation accessions in low temperature was successfully conducted by Kurniawan Budiarto and Budi Marwoto (2009), using solidified media containing ½MS + 3% DMSO and ½ MS + 3% DMSO + 3% sucrose without significant variation on all the accessions tested up to 10 and 12 months respectively. The increase of death plantlets in the media of ½MS + 3% DMSO was detected after 6 months storage with significant decrease in viability, while those in ½MS + 3% DMSO + 3% sucrose were lengthened. Supplemental carbon source in DMSO media induced root formation and plantlet resistance to low temperature during storage.

2.6.2 Ex Situ Conservation

Carnation does not like extreme temperature and has been classified as cool crop. The ideal range of temperature for carnation is considered to be 10–20 °C and exposure of plant to 7 °C favors development of branches. Sunlight for 12 h and humidity in the range of 50–60% is considered very favorable for growth of plants as well as flower development. Carnations require sufficient amount of light and proper ventilation to produce high-quality flowers, and therefore design and orientation of greenhouse are of greater importance (Jose et al. 2017). The performance of carnation varieties varies with region, season, genotypes, and growing environment. In India, there is a wide fluctuation in temperature, light intensity, and humidity which not only affect the yield and quality of flowers but also limit their availability for a particular period of a year (Gharge et al. 2012). Cristea et al. (2013) performed ex situ conservation of three endangered species, that is, *Dianthus callizonus*, *D. glacialis* ssp. *Gelidus*, and *D. spiculifolius* are the species that we seek to preserve in “Alexandru Borza” Botanical Garden of Cluj-Napoca (Romania).

Carnation (*Dianthus caryophyllus*) is one of famous ornamental cut flowers in the world. It belongs to Caryophyllaceae family. The family consists of 80 genera and 2000 species which are either annual or perennial. The *Dianthus* species are adapted to the cooler Alpine regions of Europe and Asia and are also found in Mediterranean coastal regions, and up to this moment over 300 *Dianthus* species have been identified (Jurgens et al. 2003). The desirable characters of carnation are improved, while some unpreferred characters were systematically swept out. These dynamic trends have also accelerated the genetic erosion on such plants (Halmagyi and Deliu 2007). Therefore, preserving genetic diversity is considered of great importance for the future breeding to important characters (Poulos 1993). In the tropic, the conservation of such plant in active growth and in vivo conditions would be very laborious, expensive, and accompanying risks associated field such as pathogens, pests, climatic perturbation, and human error (Fukai 1992).

2.6.3 Conservation Via Cryopreservation

Among different approaches to the conservation of germplasm, only storage at ultralow temperature (in liquid nitrogen) can ascertain long-term preservation of material while ensuring good genetic stability at the same time. Cryopreservation has been successfully applied to meristems of several species (Kartha 1985), including carnation (Seibert 1976; Seibert and Wetherbee 1977; Uemura and Saka 1980). Methods developed for greenhouse plants, for example, are generally not suitable for plants of the same species grown in vitro (Dereuddre et al. 1987). Axillary shoot tips have been less used in cryopreservation (as compared to apical shoot tips) except by Towill (1981, 1983) and Bajaj (1978), in the case of potato.

In one of the studies J. Dereuddre et al. (1988) compared the resistance to liquid nitrogen of apical and axillary shoot tips of carnation plantlets produced by in vitro culture. The ability of shoot tips from carnation cultured in vitro to develop resistance to freezing in liquid nitrogen depends on the physiological state of the cell material and the pretreatment conditions. Regrowth rates close to 100% have been obtained with apical shoot tips isolated from 2-month-old stems, precultured on medium supplemented with sucrose (0.75 M) and treated with dimethylsulfoxide (5% or more). Resistance of axillary shoot tips decreased progressively as a function of their distance from the apical shoot tip. During the development of the stem from axillary buds (obtained by cutting), progressive increases in the regrowth rate of frozen apices were noted, from 30% before cutting (axillary buds) to 98% after 3 weeks of culture.

Axillary shoot tips were found more sensitive to deep freezing in LN than apical shoot tips sampled from the same 2-month-old plantlets. This sensitivity of axillary shoot tips was found to increase with distance from the apical shoot tip which may be closely related with physiological differences caused by correlative inhibitions induced by apical dominance. With carnation, reactivation of axillary buds on cuttings was rapid (less than 2 days), but was not sufficient to increase survival of cryopreserved shoot.

Cryopreservation using an aluminum cryo-plate was successfully applied to in vitro-grown carnation (*Dianthus caryophyllus* L.) shoot tips by Shin-ichi Yamamoto et al. (2012). The shoot tips (1–1.5 mm × 1 mm) were dissected from the shoot and precultured at 25 °C for 2 days on MS medium containing 0.3 M sucrose. The precultured shoot tips were placed on the aluminum cryo-plate containing ten wells embedded with alginate gel. Osmoprotection was performed by immersing the cryo-plates in loading solution (2 M glycerol and 1.4 M sucrose) for 90 min at 25 °C. Then, dehydration was performed by immersing the cryo-plates in PVS2 for 25 min at 25 °C. After storage in liquid nitrogen, shoot tips attached to cryo-plate, were directly immersed into 2 ml 1 M sucrose solution for regeneration. Using this procedure, the average regrowth level of vitrified shoot tips of four carnation cultivars reached 95%. Adela Halmagyi et al. (2007) successfully cryopreserved carnation germplasm by shoot tips excision from in vitro cultured plants of *Dianthus caryophyllus* L. (cv. Pallas, cv. Pink Candy and cv. Wanessa) using an encapsulation-vitrification method.

2.7 Status of Plant Genetic Resources

Carnation cultivars are highly heterozygous to avoid inbreeding depression (Tanase et al. 2012). Consequently, vegetative propagation is preferred to produce plants with the same phenotype as their parents, with plants only grown from seed for selection of new varieties. Over 500 years of cultivation has resulted in highly modified morphology of the border and perpetual-flowering cultivars. Border carnation cultivars may have double flowers with as many as 40 petals (Bird 1994). Breeding in the perpetual-flowering floriculture carnations has similarly resulted in large flowers with many petals. When grown in gardens, flowers grow to between 6 and 8.5 cm in diameter. Some greenhouse-grown plants, disbudded for exhibition, have flowers up to 10 cm in diameter (Galbally and Galbally 1997).

Carnations do not naturally produce delphinidin-based anthocyanin pigments responsible for blue and purple colored petals. By inserting genes involved with the biochemical pathway for production of the pigment delphinidin into white carnations, purple carnations can be produced (Tanaka et al. 2009). Carnations in different shades of purple have been generated using different techniques, including co-expression of the flavonoid 3′ 5′-hydroxylase (F3′5′H) gene from petunia alongside a petunia cytochrome b_5 gene and by down-regulation of endogenous dihydroflavonol-4-reductase (DFR) in carnations using RNA interference (RNAi) (Tanaka and Brugliera 2014).

The intraspecific breeding procedure typically comprises hybridization, self-pollination, and selection (Holley and Baker 1992). If the desired trait is recessive, it may not be expressed in the F_1 progeny. By self-pollinating the F_1 generation and growing a large population of F_2, selection of one or more individuals with desirable traits is possible. In the absence of self-pollination, continuous hybridization has inadvertently resulted in highly heterozygous carnation varieties. Mutation breeding is used to create new color mutants. The development of double haploidy techniques has permitted breeders to accelerate breeding and selection. Dwarf carnations, which have been commercialized as alternatives to potted chrysanthemums, have also been generated by breeding programs (Holley and Baker 1992).

The mutation breeding performance in *Dianthus* species can significantly accelerate many breeding endeavors, which have proven difficult with classical *Dianthus* breeding techniques (Roychowdhury and Tah 2011). Genetic linkage maps of the carnation genome have been constructed and used to identify quantitative trait loci responsible for resistance to carnation bacterial wilt. With the aid of next-generation sequencing technology, large-scale transcriptome analysis has been conducted and revealing 300,740 uni-genes consisting of 37,844 contigs and 262,896 single-tons.

2.8 Gaps in Available Diversity

Two large gaps (1997–2006 and 2016–2020) in which no carnation cryopreservation studies were published, and it requires future studies to cover new knowledge to fill gaps in information. Carnation cryopreservation research would benefit from testing a wide range of in vitro explants, new techniques such as the cryo-mesh, improved regeneration protocols for post-cryopreserved material, and the use of low-temperature storage as a mid- to long-term complementary germplasm storage strategy (Teixeira da Silva et al. 2020).

While not a food, carnation can be used as a garnish. Wild-type *D. caryophyllus* (and other members of the genus) may have a clove scent and can be crystallized or used as a garnish in salads or for flavoring many foods including fruit, fruit salads, butter, lemonade, vinegars, conserves, and syrups (Facciola 1990; Hughes 1993). Modern floriculture carnations have little or no scent, and scent loss is often correlated with increased vase-life in cut flowers (e.g., roses) (Chandler and Brugliera 2011). Carnation petals can be used as an ingredient for a tonic to perfume the skin (Pieroni et al. 2004) and can be crushed for oil used in perfumery (Lim 2014).

Carnations have been used in European traditional herbal medicine for coronary and nervous disorders and previously used to treat fevers (Bown 1995; Lim 2014). Compounds from carnation buds have exhibited in vitro activity against several bacteria, including *Bacillus cereus*, *Listeria monocytogenes*, *Staphylococcus aureus*, and *Escherichia coli*. Furthermore, antiviral compounds have been isolated from the leaves and seeds of carnation.

2.8.1 Germplasm Availability

The following accessions are conserved at the United States Department of Agriculture (Agriculture research serviced) GRIN

Accession	Name	Taxonomy	Origin	Repository
NSL 15526	Chabauds giant	*Dianthus caryophyllus* L.	California, United States	OPGC
PI 618700	No. 915	Dianthus hybr.	Pest, Hungary	OPGC
PI 618701	No. 919	Dianthus hybr.	Pest, Hungary	OPGC
PI 618702	No. 923	Dianthus hybr.	Pest, Hungary	OPGC
PI 618703	No. 926	Dianthus hybr.	Pest, Hungary	OPGC
PI 618704	No. 932	Dianthus hybr.	Pest, Hungary	OPGC
PI 667491	Ames 19027	*Dianthus caryophyllus* L.	Kazakhstan	OPGC
PI 667492	Ames 19028	*Dianthus caryophyllus* L.	Kazakhstan	OPGC

2.9 Characterization and Evaluation

2.9.1 Characterization for Essential Features and Classification

2.9.1.1 Essential Oils

Essential oil production is highly integrated with the physiology of the whole plant and so depends on the metabolic state and present development differentiation program of the synthesizing. The chemical composition and the essential oil and its antioxidant activity of the carnation flowers (*Dianthus caryophyllus*) were studied by A.H. El-Ghorab et al. (2006). A total of 12 volatiles were identified by gas chromatography-mass spectrometry (GC-MS) as the main components of carnation flower fragrance signature. Phenyl ethyl alcohol, eugenol, hexyl benzoate, hexenyl benzoate (z), benzyl benzoate, benzoin, nootkatone, benzyl salyicylate, m-cresyl phenyl acetate, hexadecanoic acid, and eicosane were found to be the major components of the volatiles found. Vitaliy Kirillov et al. (2016) studied the chemical composition of the essential oil of *D. acicularis* Fisch. ex Ledeb from Northern Kazakhstan. The major components of the oil were found to be 2-pentadecanone (26.9–32.2%) and 2-tridecanone (4.7–17.7%). It was inferred that the presence of methyl ketones of 2-pentadecanone and 2-tridecanone in the essential oil of carnation coniferous (*D. acicularis* Fisch. ex Ledeb) provides its resistance to different insects and pathogens, including the resistance to the bacterial wilt, a serious disease caused by *B. caryophylli* (*P. caryophylli*).

2.9.1.2 Evaluation of Genetic Diversity for Desired Traits

Abbas Saidi et al. (2018) evaluated genetic diversity in carnation, *Dianthus caryophyllus* (family: Caryophyllaceae), using molecular markers and morphological traits. Fifteen cultivars were examined using seven conserved DNA-derived polymorphism (CDDP) and seven directed amplification of minisatellite DNA (DAMD) markers. Morphological traits such as flower longevity, flower diameter, and number of branches were statistically significant at 1 percent probability level. CDDP and DAMD markers produced an average polymorphic index content of 0.38 and 0.29 for CDDP and DAMD markers, respectively. Genetic distance matrix ranged from 0.02 to 0.84 in CDDP and from 0.09 to 0.78 in DAMD marker analysis. Cluster analysis for the two markers revealed that cultivars were grouped in three clusters. Results of molecular evaluation showed that CDDP and DAMD markers have high ability to differentiate carnation cultivars through assessment of the genetic diversity. In a vast study conducted by Tanase et al. (2012), an EST database was developed to enable broad characterization of the carnation transcriptome. They detected 17,362 potential simple sequence repeats (SSRs) in 14,291 unigenes and identified transcripts corresponding to genes associated with carotenoid biosynthesis, chlorophyll biosynthesis and degradation, anthocyanin (flavonoid) biosynthesis, and ethylene biosynthesis and signaling. This collection of transcripts from carnation will be useful for the annotation of the forthcoming carnation genome sequence and provide a remarkable resource for genomics studies in Caryophyllaceae.

Morphological variation and genetic diversity were recently studied in carnation by Ram Lakhan Maurya et al. in 2020. The objective of the study was to assess the morphological variation and genetic divergence among carnation accessions based on agro-morphic traits. The wide spectrum of variation was observed in carnation germplasm for growth and flowering traits. Divergence analysis of 25 genotypes of carnation was grouped in to three different clusters. Cluster II contained highest number of (11 genotypes) followed by, cluster I with 10 genotypes, and cluster III had minimum 4 genotypes. The maximum inter-cluster distance was observed between cluster III and I (4.682), followed by cluster III and II (3.391), and minimum intercluster distance was recorded between cluster II and I (2.459). Cluster mean values of different cluster showed differences between the characters. Therefore, crosses between members of clusters having high cluster means for important characters coupled with high inter-cluster distance between them are likely to be more rewarding. Clustering pattern in this situation exhibited considerable differences which indicated that it would be logical to examine genetic divergence in this environment individually for genetically reliable information.

2.9.1.3 Genome Sequencing

The whole-genome sequence of carnation (*Dianthus caryophyllus* L.) cv. 'Francesco' was determined using a combination of different new-generation multiplex sequencing platforms by Masafumi Yagi et al. (2014). The total length of the non-redundant sequences was 568,887,315 bp, consisting of 45,088 scaffolds, which covered 91% of the 622 Mb carnation genome estimated by k-mer analysis. The N50 values of contigs and scaffolds were 16,644 bp and 60,737 bp, respectively, and the longest scaffold was 1,287,144 bp. The average GC content of the contig sequences was 36%. A total of 1050, 13, 92, and 143 genes for tRNAs, rRNAs, snoRNA, and miRNA, respectively, were identified in the assembled genomic sequences. For protein-encoding genes, 43,266 complete and partial gene structures excluding those in transposable elements were deduced. Gene coverage was ~98%, as deduced from the coverage of the core eukaryotic genes. Intensive characterization of the assigned carnation genes and comparison with those of other plant species revealed characteristic features of the carnation genome. The results of these studies will serve as a valuable resource for fundamental and applied research of carnation, especially for breeding new carnation varieties.

2.9.1.4 Characterization and Analysis for Stress Tolerance

Drought, salinity, and low temperature are common adverse environmental factors encountered by land plants (Inzé and Van Montague 1995; Ishitani et al. 1997). Water deficit caused by drought and high salinity have a major selective force in plant evolution. The most important environmental constraint for plant distribution on land is most likely low temperature. In order to survive in these environmental stresses, plants respond and adapt to stress by means of physiological, developmental, and biochemical changes, including the synthesis of a number of proteins and the induction in gene expression of many genes.

The excess production of active oxygen species (AOS), such as superoxide ($O_2^{\cdot -}$), hydrogen peroxide (H_2O_2), and hydroxyl radicals ($OH^{\cdot}$), is one of the important mechanisms by which plants are damaged during adverse environmental. Such oxidative stress has been shown to occur in plants exposed to high or low temperature, particularly in combination with high light intensity (Inzé and Van Montague 1995). H_2O_2 level could be a signal that induces antioxidant enzymes, such as catalase, that subsequently protect the plant from excess H_2O_2 production. The

H_2O_2 produced from the oxidative response functions as a local trigger of programmed cell death of challenged cells and causes a rapid cross-linking of cell wall proteins (Kurepa et al. 1998). Tolerance to a wide variety of environmental stress conditions has been correlated with increased activity of antioxidant enzymes and levels of antioxidative substances. Convincing evidence now suggests that the manipulation of antioxidant capacities is a valuable route to obtain stress-tolerant plants.

Wi et al. (2006) used transgenic technology in an attempt to evaluate their potential for mitigating the adverse effects of several abiotic stresses in plants. Sense construct of full-length cDNA for *S*-adenosylmethionine decarboxylase (SAMDC), a key enzyme in PA biosynthesis, from carnation (*Dianthus caryophyllus* L.) flower was introduced into tobacco (*Nicotiana tabacum* L.) by *Agrobacterium tumefaciens*-mediated transformation. Several transgenic lines overexpressing *SAMDC* gene under the control of cauliflower mosaic virus 35S promoter accumulated soluble total PAs by 2.2 (*S16-S-4*) to 3.1 (*S16-S-1*) times than wild-type plants. H_2O_2-induced damage was attenuated by spermidine treatment. Transcripts for antioxidant enzymes (ascorbate peroxidase, manganese superoxide dismutase, and glutathione *S*-transferase) in transgenic plants and GUS activity transformed with *SAMDC* promoter::*GUS* fusion were induced more significantly by stress treatment, compared to control. These results that the transgenic plants with sense *SAMDC* cDNA are more tolerant to abiotic stresses than wild-type plants suggest that PAs may play an important role in contributing stress tolerance in plants. Li et al. (2019) characterized the *Hsf* genes (Heat shock transcription factors (Hsfs) are a class of important transcription factors (TFs) which play crucial roles in the protection of plants from damages caused by various abiotic stresses) in carnation (*Dianthus caryophyllus*). They identified a total of 17 non-redundant *Hsf* genes from the *D. caryophyllus* genome. Specifically, the gene structure and motifs of each *DcaHsf* were comprehensively analyzed.

2.10 Molecular Studies.

2.10.1 Marker Studies for Vase Life

ut flowers are marketed based upon their ornamental characteristics, which include color, shape, plant architecture, and post-harvest life. In carnations, vase life is one of the first and most important traits considered by breeders. In a study conducted by Yu and Bao (2004), the cultivar "Mabel" of carnation was transformed with direct repeat gene of ACC oxidase (the key enzyme in ethylene synthesis), driven by the CaMV35S promoter mediated by *Agrobacterium tumefaciens*. Hygromycin phosphotransferase (HPT) gene was used as selection marker. The transformants were obtained by transferring explants to selection medium supplemented with 5 mg/L hygromycin (Hyg) and 400 mg/L cefotaxime (Cef). After being transplanted to soil, transgenic plants were reported to grow normally in greenhouse. Ethylene production of cut flower of transgenic T257 line was 95% lower than that of the control, and that of T299 line was reduced by 90% than that of the control, while that of transgenic T273 line has no of significantly different from control. Vase life of transgenic T257 line was reported to be 5 days longer than that of the control line at 25 °C.

RAPD analysis was used for the identification of molecular markers that could be associated with flower vase life by De Benedetti et al. (2003). Two cultivars (Roland and Milady) with different flower longevity and their F1 and backcross offspring were analyzed. Sixty random primers were initially tested on a few genotypes. DNAs from 70 F1 and 36 backcross offspring were then analyzed with 10 selected primers. For each RAPD band tested, the progenies were divided into two groups according to the similarity of the band pattern with Roland or Milady. The statistical significance of the differences in vase life between the two groups was evaluated. The analysis revealed several bands that allow to discriminate significantly the individuals with extended longevity. Statistical analysis showed a positive correlation between the score of each progeny (number of RAPD bands similar to Roland) and flower longevity.

With the aim to verify the general use of these markers for assisted selection, 12 commercial varieties of carnation were collected and analyzed with the RAPD technique by De Benedetti et al. (2005). The 23 fragments produced with 10 decamers were not able to discriminate the genotypes with greater vase life. In order to identify more effective markers, preliminary analyses were also conducted on four genotypes, using 30 primer sets designed to amplify internal sequences from ethylene biosynthesis and response pathway gene PCR products were obtained with 22 primer pairs, and some polymorphic fragments were observed even in the agarose gels.

Various studies have reported that copper (Cu) salts can cause either positive, negative, or neutral effects on the vase life of different cut flower species (Halevy and Mayak 1981). In one such study, Van Meeteren et al. (1999) observed that one of the positive effects of tap water on the vase life of flowers may be due to the presence of the low concentration of Cu^{+2} ions. Celikel et al. (2011) also found that Cu sulfate ($CuSO_4$) at 2 mM can assist in reducing the blockage of the xylem, and hence result in the improvement in the relative fresh weight (RFW), vase solution uptake (VSU), and vase life of *Acacia holosericea* and *Chamelaucium uncinatum* flowers.

A research was carried out by Nahid Rashidiani et al. (2020), to evaluate the effects of copper nanoparticles (CuNPs) on the post-harvest physiology of carnation and chrysanthemum cut flowers. Synthesized CuNPs were applied at 10 and 20 mg L^{-1} to both cut flowers by the pulsing method. In this study, the pulse treatment of CuNPs had positive effects such as relative fresh weight, vase solution uptake, membrane stability index, total soluble carbohydrate, and hydrogen peroxide. These effects were also observed in activities of POD (peroxidase) and SOD (superoxide dismutase) which, in turn, can ultimately prolong the vase life of the two cut flowers significantly, as compared with the control. Using the pulsing treatment by CuNPs at 20 mg L^{-1} was, therefore, recommended when aiming at prolonging the vase life of chrysanthemum and carnation.

2.10.2 Marker Studies for Disease Resistance

A simple sequence repeat (SSR)-based genetic linkage map of carnation was constructed using an F2 population of 90 plants derived from a cross between a highly resistant line (85–11) and a susceptible cultivar (Pretty Favvare). To develop a large number of SSR markers, four new SSR-enriched genomic libraries were constructed and expressed and sequence tag analysis was conducted. Mapping of 178 SSR loci into 16 linkage groups was done. The map covered 843.6 cM, with an average distance of 6.5 cM between two loci. This is the first report of a genetic linkage map based mainly on SSR markers in the genus *Dianthus*. Quantitative trait locus (QTL) analysis identified one locus for resistance to bacterial wilt in linkage group (LG) B4. The locus explained 63.0% of the phenotypic variance for resistance to bacterial wilt. The SSR markers CES1161 and CES2643 that were closest to the QTL were efficient markers for selecting lines with resistance derived from line 85-11. A positional comparison using SSR markers as anchor loci revealed that LG B4 corresponded to LG A6 in a previously constructed map. It was reported that the position of the resistance locus derived from line 85-11 was similar to that of the major resistance locus observed for a highly resistant wild species, *Dianthus capitatus* ssp. andrzejowskianus (Yagi et al. 2012).

Homozygous dihaploid lines (DHLs) from standard carnation (*Dianthus caryophyllus* L.) clones, resistant to *Fusarium oxysporum* f. sp. *dianthi* could be used to accelerate breeding for new resistant varieties. In the studies conducted by Dolcet-Sanjuan et al. (2001), fertile DHLs ('D220', 'D504', and 'D524'), resistant to *F. oxysporum*, proved to be homozygous for the three genes involved in this resistance after transferring this character to all the progeny of a susceptible clone.

Traditionally, resistance to diseases has been attained by crossing the appropriate parents and selecting resistant progeny following inoculation with the fungus. Regardless of whether inoculation is artificial or natural, these approaches are cumbersome and/or time-consuming. DNA markers, linked to resistance loci, offer an easy and quick solution to scoring for resistance at the genotype level. Therefore, the adoption of a MAS strategy is an efficient solution to modern-day breeding for resistance. In one of such studies conducted by Scovel et al. (2001), linkage was established by examining the entire F2 family (consisting of 107 segregants) using a standard t-test. F3 populations derived from selected F2 segregants were used to verify linkage and to identify F2 segregants homozygous for the marker. The RAPD fragment was sequenced and used to synthesize sequence-specific primers in an attempt to establish a PCR-based, SCAR marker analysis for resistance. The same fragment was also used as a probe for a Southern blot analysis, and an RFLP marker was established that co-segregated with the resistant progeny. By exploiting the existing F2 family, containing the tagged locus, it should now be possible to develop an additional marker. This work thus paves the way toward tagging the loci involved in resistance, thereby enabling efficient selection for resistance at the genotype level in carnation.

References

Bajaj YPS (1978) Crop Improv 5:137–141

Behroozian M, Ejtehadi H, Peterson AT, Memariani F, Mesdaghi M (2020) Climate change influences on the potential distribution of *Dianthus polylepis* Bien.ex Boiss. (Caryophyllaceae), an endemic species in the Irano- Turanian region. PLoS One 15(8):1–18

Bird R (1994) Border pinks, 1st edn. Timber Press, Portland

Bloch D, Erhardt A (2008) Selection toward shorter flowers by butterflies whose probosces are shorter than floral tubes. Ecology 89(9):2453–2460

Bown D (1995) Encyclopaedia of herbs and their uses. Dorling Kindersley, London

Brooks T (1960) Cytological and genetical studies of the carnation, *Dianthus caryophyllus*, with special reference to the production of triploids. PhD thesis, University of Connecticut

Budiarto K, Marwoto B (2009) Medium term conservation of several carnation accessions *via in vitro* culture. J Nat Indones 13(2):174–177

Carolin RC (1957) Cytological and hybridization studies in the genus Dianthus. New Phytol 56:81–97

Celikel FG, Joyce DC, Faragher JD (2011) Inhibitors of oxidative enzymes affect water uptake and vase life of cut *Acacia holosericea* and *Chamelaucium uncinatum* stems. Postharvest Biol Technol 60:149–157

Chandler SF, Brugliera F (2011) Genetic modification in floriculture. Biotechnol Lett 33:207–214

Cristea V, Palada M, Jarda L, Butiuc-Keul A (2013) Ex situ in vitro conservation of *Dianthus spiculifolius*, endangered and endemic plant species. Stud Univ Babeş-Bolyai Biol 58(1):57–69

De Benedetti L, Burchi G, Bruna S, Mercuri A, Schiva T (2003) Use of molecular markers to improve cut flowers longevity in carnation. Acta Hortic 624:343–348. https://doi.org/10.17660/actahortic.2003.624.47

De Benedetti L, Braglia L, Bruna S, Burchi G, Mercuri A, Schiva T (2005) PCR-based markers and cut flower longevity in carnation. Acta Hortic 683:437–444. https://doi.org/10.17660/actahortic.2005.683.58

Dereuddre J, Galerne M, Gazeau C (1987)

Dereuddre J, Fabre J, Bassaglia C (1988) Resistance to freezing in liquid nitrogen of carnation *(Dianthus caryophyllus* L. var Eolo) apical and axillary shoot tips excised from different aged in vitro plantlets. Plant Cell Rep 7:170–173

Dolcet-Sanjuan R, Clavería E, Llauradó, Ortigosa A, Arús P (2001) Carnation (*Dianthus caryophyllus* L.) dihaploid lines resistant to Fusarium oxysporum f.sp. dianthi. Acta Hortic 560:141–144. https://doi.org/10.17660/actahortic.2001.560.22

El-Ghorab AH, Mahgoub MH, Bekheta M (2006) Effect of some bioregulators on the chemical composition of essential oil and its antioxidant activity of Egyptian carnation (*Dianthus caryophyllus* L.). J Essent Oil Bear Plants 9(3):214–222. https://doi.org/10.1080/0972060X.2006.10643494

Facciola S (1990) Cornucopia: a source book of edible plants. Kampong Publications, Vista

FloraHolland (2015) Facts and figures, 2014

Fukai S (1992) Studies on the cryopreservation of shoot tips of Dianthus and Chrysanthemum. Mem Fac Agric Kagawa Univ 56:1–7

Galbally J, Galbally E (1997) Carnations and pinks for garden and greenhouse. Timber Press, Portland

Gharge CP, Angadi SG, Basavaraj N, Patil A (2012) Performance of standard carnation (*Dianthus caryophyllus* L.) varieties under naturally ventilated poly house. Karnataka J Agric Sci 24(4): 487–489

Halevy AH, Mayak S (1981) Senescence and postharvest physiology of cut flowers-part 2. Hortic Rev 3:59–143

Halmagyi A, Deliu C (2007) Cryopreservation of carnation (*Dianthus caryophyllus* L.) shoot tips by encapsulation vitrification. Sci Hortic Amsterdam 113(3):300–306. https://doi.org/10.1016/j.scienta.2007.04.002

Harbaugh DT, Nepokroeff M, Rabeler RK, McNeill J, Zimmer EA, Wagner WL (2010) A new lineage-based tribal classification of the family Caryophyllaceae. Int J Plant Sci 171:185–198

Hardion L, Pouvreau M, Schwoertzig E, Hoff M, Nguefack J, Combroux I (2019) Performance, genetic and ecological insights for the conservation of the endangered large pink, *Dianthus superbus* L.(Caryophyllaceae) in semi-natural grassland. Bot Lett 166(1):104–112

Holley WD, Baker R (1992) Breeding for better varieties. In: Holley WD, Baker R (eds) Carnation production II. Colorado State University, pp 21–30

Hughes S (1993) Carnations and pinks. The complete guide. The Crowood Press, Ramsbury, p 221

Ingwerson W (1949) The Dianthus, a flower monograph. Collins, London

Inzé D, Van Montague M (1995) Oxidative stress in plants. Curr Opin Biotechnol 6:153–158

Ishitani M, Xiong L, Stevenson B, Zhu J (1997) Genetic analysis of osmotic and cold stress signal transduction in *Arabidopsis*: interactions and convergence of abscisic acid-dependent and abscisic acid-independent pathways. Plant Cell 9:1935–1949

Jose DA, Fatmi MU, Singh D, Benny JC (2017) Valuation of carnation (*Dianthus caryophyllus* L.) varieties under naturally ventilated polyhouse. Plant Arch 17(2):1262–1266

Jurgens A, Witt T, Gottsberger G (2003) Flower scent composition in *Dianthus* and *Saponaria* species. Biochem System Ecol 31:345–357

Kartha KK (1985) IAPTC Newsl 45:2–22

Kirillov V et al (2016) Chemical composition of the essential oil from carnation coniferous (Dianthus acicularis Fisch. ex Ledeb) growing wild in Northern Kazakhstan. Natural Product Research

Kurepa J, Smalle J, Van Montagu M, Inzé D (1998) Oxidative stress tolerance and longevity in *Arabidopsis*: the late-flowering mutant *gigantea* is tolerant to paraquat. Plant J 14:759–764

Li W, Wan X-L, Yu J-Y, Wang K-L, Zhang J (2019) Genome-wide identification, classification, and expression analysis of the *Hsf* gene family in carnation (*Dianthus caryophyllus*). Int J Mol Sci 20:5233. https://doi.org/10.3390/ijms20205233

Lim TK (2014) *Dianthus caryophyllus*. In: Lim TK (ed) Edible medicinal and non-medicinal plants: volume 7, flowers. Springer Science, Dordrecht, pp 684–693

Maurya RL, Kumar M, Malik S, Yadav MK, Sengar RS (2020) Morphological variation and genetic diversity in carnation (*Dianthus caryophyllus* L.) using agro-morphic traits. Int J Agric Environ Biotechnol 13(3):299–304

Niyokuri AN, Nyalala S, Mwangi M (2017) Effects of bioslurry and plant biostimulant Hicure on yield, flower quality and vase life of carnation (Dianthus *caryophyllus* L.). J Appl Hortic 19(1):29–34

Nugent G, Wardley-Richardon T, Lu CY (1991) Plant regeneration from stem and petal of carnation (*Dianthus caryophyllus* L.). Plant Cell Rep 10:477–480

Panis B, Lambardi M (2005) Status of cryopreservation in plants (crops and forest trees). In: Symposium on the role of biotech. Villa Guallino, Italy, 5–7 March 6–9p

Pieroni A, Quave CL, Villanelli ML, Mangino P, Sabbatini G, Santini L, Boccetti T et al (2004) Ethnopharmacognostic survey on the natural ingredients used in folk cosmetics, cosmeceuticals and remedies for healing skin diseases in the inland Marches, Central-Eastern Italy. J Ethnopharmacol 91:331–344

Poulos JM (1993) Germplasm evaluation and utilization. In: Germplasm collection, evaluation, documentation and conservation. AVRDC, Shanhua, Taiwan, ROC 69–74p

Rashidiani N, Nazari F, Javadi T, Samadi S (2020) Copper nanoparticles (CuNPs) increase the vase life of cut carnation and chrysanthemum flowers: antimicrobial ability and morphophysiological improvements, Ornam. Horticulture 26(2):225

Rikken M (2011) The global competitiveness of the Kenyan Flower Industry. Technical paper Prepared for the Fifth Video Conference on the Global Competitiveness of the Flower Industry in Eastern Africa

Roychowdhury R, Tah J (2011) Mutation breeding in *Dianthus caryophyllus* for economic traits. Electr J Plant Breed 2(2):282–286

Saidi A, Jabalameli Z, Ghalamboran M (2018) Evaluation of genetic diversity of carnation cultivars using CDDP and DAMD markers and morphological traits. Nucleus 61:129–135. https://doi.org/10.1007/s13237-018-0238-7

Salunkhe DK, Bhat NR, Desai BB (1990) Postharvest biotechnology of flowers and ornamental plants. Springer, New York, pp 28–46
Scovel G, Ovadis M, Vainstein A, Reuven M, Ben-Yephet Y (2001) Marker assisted selection for resistance to *Fusarium oxysporum* in the greenhouse carnation. Acta Hortic 552. https://doi.org/10.17660/ActaHortic.2001.552.16
Seibert M (1976) Science 191:1178–1179
Seibert M, Wetherbee PJ (1977) Plant Physiol 59:1043–1046
Tanaka Y, Brugliera F (2014) Metabolic engineering of flower colour pathways using cytochromes P450. In: Yamazaki H (ed) Fifty years of cytochrome P450 research. Springer, Tokyo, pp 207–229
Tanaka Y, Brugliera F, Chandler S (2009) Recent progress of flower colour modification by biotechnology. Int J Mol Sci 10:5350–5369
Tanase K, Nishitani C, Hirakawa H, Isobe S, Tabata S, Ohmiya A, Onozaki T (2012) Transcriptome analysis of carnation (*Dianthus caryophyllus* L.) based on next-generation sequencing technology. BMC Genomics 13:292
Teixeira da Silva JA, Wicaksono A, Engelmann F (2020) Cryopreservation of carnation (*Dianthus caryophyllus* L.) and other *Dianthus* species. Planta 252:105
Towill LE (1981) Plant Sci Lett 20:315–324
Towill LE (1983) Cryobiology 20:567–573
Uemura M, Saka A (1980) Plant Cell Physiol 21:85–94
Umiel N, Dehan K, Kagan S (1987) Genetic variations in carnation: colour patterns of petals, number of buds and the arrangement of flower buds on the stems. Acta Hortic 216:355–358
Van Meeteren U, Van Gelder H, Van Ieperen W (1999) Reconsideration of the use of DIW as vase water in postharvest experiments on cut flowers. Postharvest Biol Technol 17:175–187
Wi SJ, Kim WT, Park KY (2006) Overexpression of carnation *S*-adenosylmethionine decarboxylase gene generates a broad-spectrum tolerance to abiotic stresses in transgenic tobacco plants. Plant Cell Rep 25:1111–1121. https://doi.org/10.1007/s00299-006-0160-3
Yagi M, Kimura T, Yamamoto T, Isobe S, Tabata S, Onozaki T (2012) QTL analysis for resistance to bacterial wilt (*Burkholderia caryophylli*) in carnation (*Dianthus caryophyllus*) using an SSR-based genetic linkage map. Mol Breed 30:495–509. https://doi.org/10.1007/s11032-011-9639-x
Yagi M, Kosugi S, Hirakawa H, Ohmiya A, Tanase K, Harada T, Kishimoto K, Nakayama M, Ichimura K, Onozaki T, Yamaguchi H, Sasaki N, Miyahara T, Nishizaki Y, Ozeki Y, Nakamura N, Suzuki T, Tanaka Y, Sato S, Shirasawa K, Isobe S, Miyamura Y, Watanabe A, Nakayama S, Kishida Y, Kohara M, Tabata S (2014) Sequence analysis of the genome of carnation (*Dianthus caryophyllus* L.). DNA Res 21(3):231–241
Yu YX, Bao MZ (2004) [Prolonging the vase life of carnation "Mabel" through integrating repeated ACC oxidase genes into its genome]. Zhi Wu Sheng Li Yu Fen Zi Sheng Wu Xue Xue Bao 30(5):541–545. Chinese
Zhang X, Wang Q, Yang S, Lin S, Bao M, Bendahmane M, Wu Q (2018) Identification and characterization of the MADS-Box genes and their contribution to flower organ in carnation (*Dianthus caryophyllus* L.). Genes 9:193

Gladiolus

3

M. R. Dhiman, Neelam Thakur, Youdh Chand Gupta, and Neha Sharma

Contents

M. R. Dhiman
ICAR-INARI, Regional Station, Katrain, Kullu-Valley, Himachal Pradesh, India

N. Thakur · Y. C. Gupta (✉) · N. Sharma
Department of Floriculture and Landscape Architecture, Y.S. Parmer University of Horticulture and Forestry, Solan, HP, India

S. K. Datta, Y. C. Gupta (eds.), *Floriculture and Ornamental Plants*, Handbooks of Crop Diversity: Conservation and Use of Plant Genetic Resources,
https://doi.org/10.1007/978-981-15-3518-5_5

Abstract

Gladiolus is a genus of perennial herbaceous bulbous flowering plants of high economic importance valued both as an ornamental garden plant and as a cut flower crop. There are about 255 species and most of the modern day cultivars came from diverse genetic parentages. Most members of this genus are tetraploids having very small chromosomes ranging from $2n = 30$–120. Breeding of novel gladiolus cultivars by either traditional cross or genetic engineering is possible only when valuable genetic resources are available. In recent years, due to globalization and urbanization many wild species of ornamental geophytes have become endangered. So, germplasm conservation is very important as a source of genetic variation for breeding, research, and to prevent rare species from becoming extinct. Many in vivo and in vitro techniques have been employed to conserve gladiolus germplasm, out of which cryopreservation is a way to maintain the plant lines in a minimal amount of space and to eliminate the labor needed to routinely transfer plants growing in culture or in the greenhouse. Biotechnological tools like tissue culture, molecular markers, and recombinant DNA technology have played vital role for the development of gladiolus cultivars with horticultural qualities.

Keywords

Gladiolus · Germplasm · Genetic diversity · Tetraploid · Cryopreservation · Biotechnology

3.1 Introduction

Gladiolus (*Gladiolus grandiflorus* L.) popularly known as "Queen of bulbous flowers," member of family Iridaceae, is one of the most important bulbous ornamental for cut flower, which has both domestic and international market. It is a winter crop in the plains of India and at far off places in hills during summer for flowering, corm and cormel production. There are about 226 recorded species, majority of which are found in South Africa, but some are spread all over Africa and Mediterranean region. Species of African origin, popularly known as the "Cape Species," are normally diploids (2n = 30), whereas those belonging to Europe are polyploids (2n = 60 to 130). Recently the genus *Acidanthera* was merged with *Gladiolus* (Mukhopadhyay 2002).

3.2 Botany and Distribution

Gladiolus is an herbaceous plant bearing terminal spike. Flowering stem is aerial, terete, simple or branched, erect or flexed downward or often above the sheath of the uppermost leaf. Leaves are contemporary, usually born on the same shoot as the flower. Sometimes leaves are borne earlier or later than the flowers, on separate shoots. The number of such shoots varies, born either aboveground or at the base, having sheathing base and having isobilateral, unifacial blade. The blades are linear or lanceolate, either plane with margins with midrib or with vains lightly thickened and hyaline or sometimes even winged. In some cases mid ribs and margins are much thickened and blade is terete with four narrow longitudinal grooves. In some plants blade is reduced making leaf party to entirely sheathing. Inflorescence of gladiolus is called a spike which is reduced to a single flower. Flowers are borne in distichous manner. Each floret has two bracts, usually green. The inner one is smaller than the outer, usually notched apically for 1–2 mm or entirely (Cantor and Tolety 2011).

The genus *Gladiolus* is a petaloid monocot belonging to the family Iridaceae Juss., which is among the largest families of the order Asparagales (Goldblatt et al. 2001).

The taxonomical hierarchy of *G. hybridus* is as follows:
Subkingdom: Viridaeplantae
Phylum: Tracheophyta
Class: Liliopsida
Order: Iridales
Family: Iridaceae
Subfamily: Ixioideae
Botanical name: *Gladiolus* × *hybridus* Hort.

3.3 Origin, Domestication, and Spread

The majority of the ~260 *Gladiolus* species originate from South Africa and the Cape of Good Hope is considered to be the center of diversity of this genus (Goldblatt 1996; Valente et al. 2011). The second center of this species richness is in the Mediterranean region with only about seven species originating from this region (Valente et al. 2011).

History reveals that gladioli is known since 1578, as evidenced by a record in Lyte's *Nievve Herball;* they were first introduced into France. Afterwards, they spread to England, Germany, Holland, and North America. In 1596, John Gerard, gardener to Lord Buleigh, who also owned his own "Physic Garden" in Holborn, had two European species, *G. communis* from Mediterranean region and *G. segetum* from South Europe, the canaries and Mediterranean region; as is evidenced by the catalog of some 1030 plantlets he grew during that period. In 1597, Gerard published his famous *Herball, Geeral Histrie of Planets* describing these two species.

Prior to 1730, the major garden species in England were *G. communis, G. segetum*, and *G. byzantius*, the latter being introduced in 1629 from Constantinople. With the establishment of trade routes from England to India via the Cape of Good Hope, several South African species were sent to England starting in 1737. The expedition of the Siomon van der Stel, Governor of the Dutch colony at the Cape in 1685 (De Wet and Pheiffer 1979), conducted a botanical exploration and an artist Claudius, who accompanied the expedition, painted two species of *Gladiolus* (*G. esculentus* syn. *G. speciosus* and *G. aquilegia* syn. *G. caryophyllaceaus*).

In 1753, the Swedish scientist Carl Linnaeus published six species of *Gladiolus* in *Species Plantarum,* the modern system of naming plants: two Eurasian and four from the Cape of Good Hope. Just one of these later four, *Gladiolus angustus*, still belongs to the genus as it is understood today. The other three have long since been shifted to different other genera.

By 1753 however, several more species of *Gladiolus* from southern Africa were already known in Europe, although their generic affiliation was not always clear. These species included *G. alatus, G. blandus, G. recurvus*, and *G. tristis.* These species were formally named according to the binomial system in the later years. Phillip Miller had the honor to find the first flower, a South African *Gladiolus, G. tristis,* in England in 1745.

During 1772–1775, Carl Peter Thunberg in South Africa collected hundreds of new species and sent thousands of specimens to Europe. Linnaeus's son and many more botanists described some and the results were remarkable. Where handful of species of Cape *Gladiolus* were known in 1770, the *Flora Capensis*, published in various editions from 1807 to 1823 (Thunberg 1823), included some 50 species, almost one-third of the total recognized today. *Gladiolus aethiopian*, described first by Cornutus, is from Cape region bearing white red-scarlet flowers. It was used in Cape region about the middle of the eighteenth century until the first South African species were introduced. *Gladiolus tristis*, a sweet-scented species closely related to *Gladiolus grandis*, was first introduced in Chelsea Physic Garden, US, in 1745 from Cape region and Natal. In 1772, Kew Gardens sent overseas the first of its numerous plant collectors, the Aoerdonian, Francis Masson who traveled principally in Cape region and collected many wild growing species. *Gladiolus citricus* is native to Mediterranean region, which later on was found growing in England as the area of diversity. *G. grandis*, in 1749, was naturalized in Spain from South Africa. In 1784, C.P. Thunberg published *Dissertatio de Gladiolo*, which may be taken as the starting point of post- Linnaean era. In 1789 W. Aiton published his *Hortus Kewensis* describing two *Gladiolus* species brought by Masson during 1773–1775.

Gladiolus cultivation in India dates back to the nineteenth century as "Firmingers Manual of Gardening in India" published in 1863 mentions that Charles Gray of

Coonoor grew some gladioli from corms and seeds in his garden. The major states growing gladiolus are Uttar Pradesh, Himachal Pradesh, Haryana, Delhi, Karnataka, Maharashtra, Punjab, West Bengal, Odisha, Chhattisgarh, Assam, Sikkim, Meghalaya, and Uttarakhand (Naresh et al. 2015; Verty et al. 2017; Pattanaik et al. 2015). Thus gladiolus is considered as winter crop in plain areas and summer crop in hilly areas. Gladiolus was introduced in Bangladesh around 1992 from India (Mollah et al. 2002; Hossain et al. 2011).

3.4 Plant Genetic Resources

3.4.1 Geographic Distribution

On the basis of the total number of species, it has been observed that England, France, Albania, Yugoslavia, Italy and Sardinia, Corsica, Sicily, Cyprus, former USSR, Iraq, Israel, Lebanon, Ghana, Ivory Coast, Mali, Sierra Leone, Reunion Island, Uganda, each contains one species of *Gladiolus*. However, *G. communis* is common in Italy, Spain, Greece, Corsica, Sicily, Sardinia (Italian island in the west in the Mediterranean), Portugal, Mediterranean area, and North Africa; while *G. illyricus* is common in Lebanon, Israel, Turkey, and England; *G. palustris* in Central Europe, Albania, and Yugoslavia; and *G. atroviolaceus* in Greece, Iran, and Turkey. *G. natalensis* (Eckl.) Reinw. is common in western Arabia, Ethiopia, Botswana, Namibia, Nigeria, Zimbabwe, and South Africa. Countries like Spain, Greece, Portugal, Mediterranean areas, tropical Africa, Guinea, Botswana, Namibia, and Somali contain two species each. Romania, Turkey, East Africa, Cameroon, and Madagascar contain three species each; Iran, Ethiopia, Kenya, and Zambia contain four species each. Nigeria and Zimbabwe contain five species each; whereas Mozambique contains 10 species; Zaire 11 species; Tanzania 13 species; Malawi 14 species; and Angola 16 species. About 20 species have been described with uncertain origin. South Africa is the only country from where more than 114 species have been recorded, out of which 62 are said to be winter flowering and 35 as scented ones: *G. acuminatus, G. alatus, G. arcuatus, G. brevifolius, G. brevitubus, G. carinatus, G. caryophyllaceus, G. ceresianus, G. engysiphon, G. equitans, G. esxilis, G. floribundus, G. gracilis Jacq., G. guthriei, G. jonquilliodorus, G. lewisiae, G. liliaceus* (night scented), *G. longicollis* (night scented), *G. maculates, G. marlothii, G. mutabilis, G. odoratus, G. orchidiflorus, G. permeabilis, G. pillansii, G. pritzelii var. pritzelii, G. recurvus, G. rebertsoniae, G. tenellus, G. tristis* (night scented), *G. uysiae, G. vaginatus, G. viridifflorus*, and *G. watermeyerl*; however, *G. callianthus* Marais is a scented species from Ethiopia (Misra 1995).

3.4.2 Primary Gene Pool

3.4.2.1 Gladiolus x Hybridus

The present-day *Gladiolus* cultivars are complex hybrids and following species are mostly involved in breeding: *G. cardinalis* Curtis., *G. dalenii* van Geel.,

G. oppositiflorus Herb., *G. papilio* Hook. f., *G. carneus* Delaroche, *G. cruentus* Moore., *G. tristis* L., and *G. saunddersii* Hook. f.

3.4.3 Wild Genetic Resources and Others

Genera Gladiolus is divided into four groups by Baker in 1892. The species belonging to different groups along with their center of origin and flower colors are listed below:

3.4.3.1 Plurifoliati

Twenty-two species having 5–8 or more well-developed leaves, distichously arranged to form fan shape.

Gladiolus cardinalis (Cape), *G. sempervirens* (Cape), *G. cruentus* (Natal and Lesotho), *G. saundersii* (Natal and Lesotho) – all of red or scarlet color; *G. oppsibiflorus* (Cape) – flowers white or pink; *G. eeliotii* (Natalk Botswaa and Rhodesia) – white, speckled with purple; *G. sericeo-villosus* (Cape, Natal, Swaziland, and Transvaal) – pink and/or yellow flowers, arranged in opposite directions; *G. virus* (Transvaal) – pink: *G. ochroleucas* (Cape) – cream to yellow; *G. ecklonii* (Cape) – white red and mauve; *G. buckerweldii* syn. *Antholyza buckerveldii, Petamenses buckerveldii* (Cape) – yellow; *G. matelensis* (Cape and Natal) – red, orange, yellow, and or greenish; *G. papilio* syn. *G. purpuireoauarntus* (Transkei to Transvaal) – yellow; all have one side facing spike; *G. hollandii* (Transvaal) – white, speckled with pink and mauve; *G. densiflorus* (Transvaal) – white, with redspots; *G. invenustus* (Swaziland and Natal) – white to mauve, with blotches; *G. calcaratus* (Transvaal) – white; *G. appendiculatus* (Transvaal, Natal, and Swaziland) – white, with mauve marking and *G. pole- evansii* (Transvaal) – pinkish lilac, hairy on all vegetable parts.

3.4.3.2 Paucifoliati

Thirty-four species, having 2–5 well-developed leaves, distichously or spirally arranged.

Gladiolus stellantus (Cape) – fragrant, whitish to pale mauve; *G. gueinzii* (Crape and Natal) – pink, marked with purple; *G. acuminatus* (Cape) – fragrant dull greenish yellow; *G. lapeirousiodes* (Cape) – white, with the lower lobes each bearing two red blotches; *G. leptosiphon* (Crape) – cream to cream- brown; *G. vigilans* (Cape) – pale rose-pink; *G. carneus* (Cape) – white, cream, pink, or mauve; *G. macneilii* (Transvaal) – salmon-pink; *G. microcarpus* (Natal) – whitish, pink, or mauve; *G. buckerveldii* (also descried under Plurifoliati); *G. angustus* (Cape) – white to stream-yellow; *G. floribundus* (Cape) – fragrant, with white, cream pink, or mauve, closing at night; *G. udulantus* (Cape) – greenish to cream-white or pink; *G. carlyophyllacleus* (Cape) strongly scented, flowers pale to mauve; *G. lewisiae* (Cape)- fragrant, creamy white, flushed with pink; *G. involutus* (Cape) – pale pink or white; *G. sclullyi (Cape)* fragrant, cream, yellow, lime-green, pink, mauve, or maroon; *G. arcuatus* (Cape) – fragrant, mauve, or purple; *G. salteri*

(Cape) – pink; *G. kamiesbergensi* (Cape) – fragrant, mauve, or purple; *G. permeabillis* (Cape) – fragrant, white to cream, mauve pink, or brown; *G. vernus* (Transvaal) – pale magenta, pink; *G. pretoriensis* (Transvaal) – dark purple to pink; *G. mostertiae* (Cape) – pale pink. *G. rufoarginatus* (Transvaal) – white or cream-yellow, densely flecked with red or purple; *G. orchidiflorus* (Cape) – fragrant, gray- green, densely flecked with red or purple; *G. watermeyeri* (Cape) – fragrant, cream or greenish cream, flushed with dull purple; *G. irescenss* (Cape) – fragrant, cream or greenish cream, flushed with dull purple; *G. virescens* (Cape) – fragrant, cream or greenish cream, flushed with dull purple; *G. seresianus* (Cape) – fragrant dull or greenish yellow; *G. uysiae* (Cape) – fragrant, cream or greenish; *G. alatus* (Cape) scented, red, orange, or salmon but lower leaves greenish; and *G. equitans* (Cape) – fragrant, orange, red, or vermillion, with yellow-green patches on laterals and falls.

3.4.3.3 Unifoliati

Twenty-nine species, having one well-developed basal leaf, rest of the leaves, bract-like with short blades spirally arranged on the scapes.

Gladiolus brevitubus (Cape) – pale pink or red; *G. quadrangulus* (Cape) – fragrant, white pale, blue, pinkish, mauve; *G. citrinus* (Cape) – bright yellow; *G. tenellus* (Cape) – fragrant, yellow, cream or white tinged purple: *G. oreocharis* (Cape) – white to mauve; *G. inflatus* (Cape) – pale, pink or mauve; *G. roberso- niae* (Transvaal) – fragrant, mauve or white tinged pale mauve; *G. cylindraceus* (Cape) – pale pink or creamy pink, flushed with salmon-pink; *G. debilis* (Cape) – white to pale pink, with red dots in throat; *G. longicollis* (Cape, Natal, Lesotho, Orange Free State, Swaziland and Transvaal) – evening-fragrant, white-cream or yellow; *G. tristis* (Cape) – night fragrant, white to buff-cream or pale yellow tinged green; *G. liliaceus* (Cape) – night-fragrant, yellowish to dull yellow, flecked with brown, pink, red or purple; *G. hyalinus* (Cape) – slightly fragrant toward evening, yellow-brown, yellowish cream or greenish brown, spotted with brown and purple; *G. recurvgus* (Cape) – fragrant, pale greenish grey or greenish mauve to pale yellow or cream; *G. symonsii* (Natal) – pink; *G. punctulatus* (Cape) – pink to mauve; *G. carnatus* (Cape) – fragrant, pale grayish blue to pale pink; *G. viridiflorus* (Cape) fragrant, pale green; *G. blommes- teinii* (Cape) – pale pink or mauve: *G. ornatus* (Cape) – pale to deep pink: *G. gracilis* (Cape) – fragrant, pale mauve or pink, sometimes white or yellow; *G. exilis* (Cape) – slightly fragrant, pale blue, mauve, or white tinged pale mauve; *G. mutabilis* (Cape) – fragrant, grayish mauve to cream, brown, or yellow; *G. violaceo- lineatus* (Cape) – mauve; *G. comptonii* (Cape) – bright yellow with reddish streaks; *G. rogersii* (Cape) – purple; *G. bullatus* (Cape) – deep mauve; and *G. pritzelli* (Cape) – fragrant, yellow sometimes tinged red or gray outside.

3.4.3.4 Exfoliati

Twenty-four species, having well-developed leaves (or leaf) absent at the time of flowering, the scape bearing 1–3 (or 4) sheathing bracts without or with small free blades.

Gladiolus stefaniae (Cape) – red; *G. stokoei* (Cape) – scarlet: *G. nerineoides* (Cape) – pale salmon-pink to deep golden red; *G. guthriei* (Cape) – fragrant, deep pink or reddish; *G. carmineus* (Cape) – deep rose-pink or carmine, with a white or cream median stripe; *G. monticola* (Cape) – cream color tinged pink; *G. maculatus* (Cape) – fragrant, yellow, pink or white, streaked and spotted with brown, purple or red; *G. engysiphon* (Cape) – strongly and sweetly scented, white with a crimson median line on three lower and upper lateral lobes: *G. bilineatus* (Cape) – white or cream, faintly flushed with pink; *G. emiliae* (Cape) – fragrant, yellow, densely speckled with red, brown, or purple; *G. odoratus* (Cape) – fragrant, dull or brownish yellow, speckled and striped with purple, red or maroon; *G. vaginatus* (Cape) – fragrant, mauve to grey-mauve, rarely white; *G. brevifolius* (Cape) – scented, pale to pink, rarely white, mauve, brownish green, or brown; *G. pillansii* (Cape) – fragrant, mauve or pink; *G. mar- tleyi* (Cape) – white, suffused with rose-pink; *G. subcaeruleus* (Cape) – mauve or pinkish; *G. jonquil- liodorus* (Cape) – fragrant, cream-colored or pale yellow, flushed with pink or mauve; *G. aurantiacus* (Natal, Swaziland, and Transvaal) – golden yellow or orange and yellow, streaked with red; *G. brachyphyllus* (Transvaal, Swaziland, and Mozambique) – pink to mauve, with red spots; *G. unguiculantus* (Sierra Leone and Transvaal) – mauve, with purplish blotches on lower lobes; *G. woodii* (Natal, Transvaal, and Swaziland) – dark brown; and *G. parvulus* (Cape, Natal, and Transvaal) – mauve, pink, or white (Cantor and Tolety 2011).

3.5 Collections and Conservation

3.5.1 Methods

In gladiolus, corm is used for collection and conservation purpose. Other than the corm, seeds can also be used for collection and conservation purpose.

3.5.2 Status of Collections (National, Regional, and Global with Appropriate Listing)

National

Gladiolus spp.	IC296733	Green Wood peeker x Oscar	Capsicum red flowers, lip petals chartreuse yellow petals ruffled
	IC296774	Vink's Glory x Eurovision	Brick red petals having dresden yellow on lip petals nicely ruffled
	IC296775	Snow princess x Her Majesty	Hybrid, with white petals having deep purple violet edges
	IC296776	Snow princess x Her Majesty	Reddish purple petals and dark reddish purple blotches on the edges of both sides of petal, lip with white stripes magenta petal

(continued)

G. grandiflorus Andrews	IC0584125	Snow princess x Her Majesty	Purple-violet having red-purple margin with yellow-green blotch
	IC0584126	Green bay x Gold Medal-412	Yellow lower lip with Red blotch
	IC0584127	Water melon Pink x Aarti	Red floret having red margin and white line on tepals with yellow blotch and resistant to *Fusarium* wilt
G. hybridus Hor	IC0611879	A cross between P-16-1 and Eurovision	Early blooming (76–80 DAP) and red florets with dark red stripes on inner two tepals, and also red spots on outer throat, each mother corm produces more than 2.00 corm
G. hybridus Hor	IC0611878	A selection among the open pollinated seedlings of the variety "Melody"	Very early blooming (74 DAP), red florets with two red spots on two central tepals. Corm multiplications rate is higher and produces more than 2.00 corms from each mother corm

www.nbpgr.ernet.in:8080/registration/InventoryofGermplasm.aspx

International

There are 556 accessions of *Gladiolus* registered in GENESYS as follows:

Gladiolus hybr.	219
Gladiolus x hybridus hort.	104
Gladiolus italicus	32
Gladiolus sp.	21
Gladiolus X hybridus	15
Gladiolus x gandavensis	11
Gladiolus atroviolaceus	7
Gladiolus crassifolius	6
Gladiolus dalenii	6
Gladiolus illyricus	6

Accession browser (genesys-pgr.org)

Regional

Total number of gladiolus varieties reported from each center for passport data is as follows – Maharana Pratap University of Agriculture and Technology, Udaipur = 63; PAU, Ludhiana = 36; G.B. Pant University of Agriculture and Technology, Pantnagar = 42; Sher-e-Kashmir University of Agricultural Sciences and Technology, Srinagar = 60; Mahatma Phule Krishi Vidyapeeth, Pune = 50; Acharya N.G. Ranga Agricultural University, Hyderabad = 27; Birsa Agricultural University, Ranchi = 15; IARI, New Delhi = 47; IIHR, Hessaraghatta = 74; Indian Agricultural Research Institute Regional Station, Katrain = 29. CSIR-NBRI Lucknow reported total germplasm 105. But, there are duplications. Name of varieties maintained at different centers, varieties are grown extensively in many parts of the world, varieties for cut flowers in world trade, Indian bred varieties, etc. have been mentioned in different publications (Datta 2001; Arora et al. 2002; Sharga and Roy 2002; Bhattacharjee et al. 2002).

3.6 Gaps in Collections, Both Geographical and Genetic

3.6.1 Planning of Explorations and Collection

Well decisive plan of exploration for germplasm collection is must. The time of visit and the number of visits are to be predecided keeping in mind the life cycle and growth behavior of the plant. The time should correlate with the expression of important traits of the species like spike emergence, floret opening, etc. Inappropriate planning will lead to missing out on important observations and thus loss of considerable part of the diversity.

3.6.2 Problem in Identification and Characterization

Identification of a species and its proper characterization requires adequate knowledge of biology and systematics of the genera. Adequate characterization leads to effective utilization of that germplasm in the pre-breeding programs.

3.6.3 Extensive Evaluation, Documentation, and Information Management

Extensive exploration programs are usually being carried out at central and state level. Most of these programs are non-coordinated at a single platform and, therefore, multiple submissions of the same genotype are generally made. Therefore proper evaluation of the collected germplasm is done to prevent duplication. Such duplication leads to wastage of resources and time. The evaluated genotype if found new needs proper documentation at the conservation agency level and uploaded to a web platform so that the interested candidates can utilize this material as and when required.

3.6.4 Geographic Considerations

The center of diversities expresses maximum diversity of a crop as a rule; however, when a crop evolves at different geographical platforms, different mutations are sustained in it. Therefore limiting the exploration programs only to the areas of center of diversity is a wrong strategy and should include all the areas suitable for its growth and cultivation. This would allow the collection of maximum amount of alleles of the important genes especially of abiotic and biotic stresses and quality.

3.6.5 Phytosanitary Regulations

Pathogen and pest infected and contaminated accessions may cause several problems in germplasm management. This affects the seed longevity and poses problems in evaluation of the material. If conserved as such, the progeny may be distributed to other accessions and may lead to the failure of the gene bank and destroy susceptible accessions, and pathogens may be distributed to new sites along with the germplasm. If not properly dealt with, infection will pose quarantine problems that negatively impact on germplasm flow. Therefore, adequate management measures should be taken by the genebank to eliminate infection and contamination or at least reduce it to acceptable levels.

3.6.6 Effective Utilization in Pre-breeding Programs

To narrow the gap between raw germplasm and commercial crop genotypes and to combine various alleles into a single population, the accessions from the genebank can be used. However, plant breeders and other users are often hesitant about using material directly from a genebank as it is often too unadaptable. Therefore, along with conservation a pre-breeding program can be devised for proper utilization of these lines for which some physiological traits also have to be understood including seed dormancy, germination, and storability; or in the case of vegetative materials, growth behavior in tissue culture under in vitro conditions. Some of this information exists and some needs to be generated. It is generally advisable to conduct a comprehensive literature search before determining details of the management procedures for species for which information on optimal conservation conditions is unavailable.

3.6.7 Loss of Genetic Integrity

In cross-pollinated crops like Gladiolus, to maintain the genetic makeup of a line, large isolation distance is required which is not only tedious but also expensive. Often in such species gene flow is very high therefore, during regeneration, maintaining genetic integrity is hard if vegetative propagation is not followed.

3.7 Conservation

3.7.1 Methods

Conservation in gladiolus can be done by reproduction through corms and seeds. Gladiolus is cross-pollinated crop due to self-incompatibility. Long-term storage of pollen in gladiolus is necessary to expedite artificial pollination for crossing between spring-flowering species and summer-flowering species and varieties that bloom at different times. Pollen fertility was tested for all germplasm collections

(CSIR-NBRI, Lucknow). It was interesting to note that 88% of cultivars had pollen grain fertility above 50%. Maximum cultivars had pollen fertility between 80% and 90% (Datta 2001). Pollen of the gladiolus can be stored in plastic vials for longer duration under freezer condition (–4 °C) up to 11 weeks of storage without any alarming loss of viability (Johnson et al. 2009; Kaur and Dhatt 2019) and also in liquid nitrogen (Balaram et al. 2000).

3.8 Cryopreservation of *Gladiolus*

Prolonging pollen viability through pollen storage is reported in many crop species as a means for overcoming asynchrony in flowering, scheduling hybrid seed production and haploid gene pool conservation. In *Gladiolus*, pollen storage for duration of 730 days at −40 °C has been reported by Koopwitz et al. (1984) for five species. This demonstrates the feasibility of cryogenic storage of gladiolus pollen and establishment of cryo banks for conservation of species.

Many plant lines that are genetically different are generated while genetically engineering plants. These plants must be screened to determine which lines have the same phenotype as the non-transformed plant and are most resistant to the pathogen of interest. Cryopreservation of *Gladiolus* is a way to maintain these plant lines in a minimal amount of space and to eliminate the labor needed to routinely transfer plants growing in culture or in the greenhouse. Only four bulb crops have been cryopreserved by vitrification including lilies (Bouman et al. 2003), taro (Takagi et al. 1997), garlic (Baek et al. 2003), and gladiolus (Joung et al. 2007).

b. Status of Plant Genetic Resources (general germplasm, base collection, active collection, breeder's collection, genetic stocks, pre-breeding material (including interspecific derivatives, etc.))

3.9 Horticultural Classification

For horticultural purposes, gladioli are grouped as follows:

3.9.1 Large Flowered (Grandavensis) Hybrids

The flower spike grows to a height of 120–150 cm, is strong and erect, with florets of 10–15 cm across, closely arranged, triangular and symmetrical flowers, and produces flowers late in the season.

3.9.2 Primulinus

They bear hooded flowers which are smaller in size; the spike grows to a height of 100 cm and bears florets of 5–10 cm across. It is mid-season flowering type.

3.9.3 Miniatures

The spike grows to a height of 60–100 cm, with florets 5–7.5 cm across, good for cutting and early flowering type. They include Butterfly hybrids, characterized by waved and fluffy petals with rich markings in contrasting shades.

3.9.4 Peacock Hybrids

These are good for cutting, dwarf in height, multicolored sorts with reflexed petals.

3.9.5 Star Flowered

These types bear flat star-like flowers, a race raised by Unwins of Cambridge, England.

3.9.6 Face Ups

These are quaint, spike grows 60–90 cm, with florets 5–6 cm and face upwards.

3.9.7 Ochideola

This is a new group developed in Israel; spikes are light in weight, the smaller florets on shorter stems.

(Source: Commercial Floriculture, T.K. Bose)

Indian varieties developed by public organization

(i) **Varieties evolved at IARI, New Delhi**

Varieties	Parentage	Remarks
Agnirekha	Sylvia Seedlings (1980)	Fire red with saffron yellow blotch and scarlet florets, mid-season variety
Mayur	Sylvia Seedlings (1980)	Florets lilac-purple with dark purple throat, mid-season variety
Suchitra	Sylvia Seedlings X Jo Wagenenaar	Florets camellia rose with vermillion and purple blotch, mid-season variety
Neelam	Sylvia X Patricia (1987)	Deep mauve florets, mid-season variety
Pusa Suhagin	Sylvia Seedlings (1987)	Florets ruby-red with barium yellow streaks on the lower tepals, late season variety
Anjali	Sancerre X Rose Spire (1997)	Florets are scarlet pink with yellow dusting on falls, mid-season variety
Archana		Florets are scarlet pink with yellow dusting on falls, mid-season variety

(continued)

Varieties	Parentage	Remarks
	Creamy Green X American Beauty (1997)	
Bindiya	Ratna Butterfly Seedling (1997)	Florets are yellowish cream with fan-shaped red coloration on 2 side falls, it is also a mid-season variety
Chirag	Cygnet X Little Fawn (1997)	Florets are orange in color with deeper throat and mid-season variety
Sarang	White Oak Seedlings (1997)	Florets purple red and a mid-season variety
Shweta	Wind Song X Pink Frost Seedlings (1997)	Florets frilled white with green-yellow throat, mid-season variety
Sunayna	George Mazure X Eurovision (1997)	Florets pink with red throat, early variety
Vandana	George Mazure X Eurovision (1997)	Orange colored variety, early mid-season variety
Chandni	Green Woodpecker X White Butterfly (1997)	Florets are greenish white and early season variety
Dhanvantri	Jr.Prom X Lucky Star (1995)	Florets are light yellow and mid-season variety
Kamini	Ava X Christian Jane (2000)	Floret color orange-red with fan-shaped purple red lip on light yellow base on 2 side falls, early mid-season variety
Lohit	Creamy Green X American Beauty (2000)	Floret color is red with white mid-ribs on 2 side falls, early mid-season variety
Mohini	Ave X Christian Jane (2000)	Floret color red-purple with fan-shaped deep purple color on yellow on 2 side fall, early mid-season variety
Rangmahal	Red Bantam X Flaura Belli (2000)	Florets ruffled, red-purple and compactly arranged, mid-season variety
Sukanya	Salmon Queen Seedlings (2000)	Floret color white with Scarlet ring in the lip, early mid-season variety
Swapnil	Viola Seedlings (2000)	Florets violet with creamy throat, early blooming
Urmil	Tinker Belle X Break O'Dawn (2000)	Florets violet with creamy throat, early blooming
Swarnima	Dhanvantari spontaneous mutant (2000)	Florets coopery yellow, mid-season variety

(ii) **Varieties evolved at IIHR, Hessaraghatta, Bangalore**

Varieties	Parentage	Remarks
Aarti	Shirley X Melody (1980)	Florets poppy-red with purple-red and canary-yellow blotch, it is a mid-season variety
Apsara	Black Hack X Friendship (1980)	Florets ruby-red with barium yellow flecks in throat

(continued)

Varieties	Parentage	Remarks
Darshan	Watermelon Pink X Shirley	Floret color ruby-purple with red-purple margins having white blotch
Kum Kum	Watermelon Pink X Lady John (1993)	Florets are red with yellow blotch
Meera	G.P.I.X Friendship (1979)	Florets are white
Nazrana	Black Jack X Friendship (1979)	Florets are cardinal-red with barium yellow flash in throat
Poonam	Geliber Herald X R. N.121 (1979)	Florets are yellow, spikes 98 cm long with 17 florets
Sagar	Melody X Wild rose (1994)	Florets are yellow, spikes 98 cm long with 17 florets
Sapna	Green Woodpecker X Friendship (1979)	Florets are greenish yellow
Shakti	Wild rose mutant (1981)	Florets are pink yellow throat
Sindur	(1994)	Florets are red with darker blotches and yellow splashes

(iii) Varieties evolved at NBRI, Lucknow

Varieties	Parentage	Remarks
Archana	Sylvia X Friendship (1984)	Floret color purple with yellow blotch and white mid-rib, mid-season variety
Arun	Sylvia X Fancy (1984)	Florets vermillion and it is a late blooming variety
Basant Bahar	Unias Challenge Seedling	Florets are yellow with magenta specks in throat
Dhiraj	Beauty Spot X Psittacinus Hybrid (1993)	Florets are purple with deeper and yellow blotch
Gazel	White Friendship Seedlings	Floret are pink with darker lips and linear shading, having yellow throat
Hans	Friendship X G.tristis (1985)	Florets white with falls having mid-rib
Indrani	Friendship X G.tristis (1985)	Florets crimson with white mid-rib
Jwala	Psittacinus Hybrid Seedling	Floret vermillion with blotched yellow
Kalima	Sylvia Seedlings	Florets red with 2 side falls blotched yellow
Kohra	Sylvia X King Lear	Florets pink with smoky-violet streaks and yellow throat
Manhar	Friendship X G.tristis (1983)	Florets with rosy tips and yellow throat
Manisha	Friendship X G. tirstis (1983)	Florets yellow with purple splashes at tips and this is a late blooming variety
Manmohan	Friendship X G.tristis (1982)	Florets yellow with purple and splashes at tips and it is a late blooming variety

(continued)

Varieties	Parentage	Remarks
Manohar	Friendship X G.tristis (1982)	Floret are purple and throat yellow and it is a late blooming variety
Mohini	Friendship X G.tristis (1982)	Florets white splashed with rose and throat yellow and late bloomer
Mridula	Friendship X G.tristis (1985)	Florets purpled specked at edges with yellow throat and white mid-ribs, mid-season bloomer
Mukta	Friendship X G.tristis (1981)	Florets sulfur-yellow splashed with purple, late bloomer
Pitamber	Friendship X G.tristis (1985)	Florets light green with purple steaks in the throat, a mid-season bloomer
Priyadarshini	Lavanesque seedling	Florets mauve and throat white
Sada Bahar	Sylvia seedling	Florets specked purple with sulfur yellow petal mid-ribs
Sanyukta	Friendship X G.tristis (1984)	Florets rose with primrose-yellow throat and mid-season bloomer
Smita	Lavanesque seedling	Floret rose with dark margins
Triloki	Friendship X G.tristis (1984)	Florets rose with yellow throat and it is a mid-season blooming variety

(iv) **Varieties evolved at HETC, Chaubattia**

Horticulture Experiment and Training Centre, Chaubattia (Almora), developed four varieties:

Chaubattia Ankur: (Oscar X Friendship)
Chaubattia Arunima: (Oscar X Motherfisher)
Chaubattia Shobhit: (Meria Goretti X Tropic Sea)
Chaubattia Tripti:(Sunny Boy X Oscar)

(V) **Varieties evolved at PAU, Ludhiana**: Punjab Dawn (Suchitra X Melody), Punjab Morning (Sancerre X White Prosperity) and Sher-e-Punjab (Suchitra X Melody)

(VI) **Varieties developed by GB Pant University of Agriculture and Technology, Pantnagar**

Shubangini: A mutant of 'Fidelio' developed through gamma radiation. Spikes are 95–100 cm long each with 15–18 florets. Florets are white, slightly ruffled and 12 cm across. Very good multiplier.

(VII) **Variety developed by MPKV, Pune**

Shree Ganesh: This variety possesses yellow white floret. Spike length is 115 to 120 cm with nearly 19 florets on each spike. The diameter of floret is 10–11 cm. Each corm produces 2 corms and 70–80 cormels.

Two more varieties of gladiolus at pre-release stage from this center are Preeana (GK-GL-94–42) and Neelrekha (GK-GLK-94-55).

(VIII) **Varieties developed by B.B.S. Bhadri in Himachal Pradesh**

Bhadri (1963) developed some varieties suitable for cultivation in H.P. Some of these varieties are as follows:

Bhadri May Blossom	Bhadri Dwarf
Bhadri Blue	Bhadri Bicoloured
Bhadri Purple Striped	Bhadri Salmon Glow
Bhadri Deep Purple Splashed	Cherry Glow
Bright Red Primulinus	Rose of Heaven
Morning Kiss	Bhadri's Red Giant
Bhadri Early Peace	Bhadri Yellow Crest
Zakir Hussain (6-petalled)	Border Gem
Zakir Hussain (8-petalled)	Bhadri's Simla Sunset
Light Mauve (deep mauve strips)	Bhadri's Milky Way
Light Salmon Pink (throat)	Yellow Beauty
Bhadri Violet Beauty	Bhadri Jupiter
Bhadri Elite	Bhadri Velvet
Bhadri Orange Glow	Bhadri Dazzier
Bhadri Queen of Pink	Bhadri Scarlet
Raj Niwas Pride	Bhadri's Baby Doll (lighter purple)
Bhadri Red & White	Bhadri's Red Prince
Bhadri Lemon Queen	Bhadri' Liliput
Bhadri Rose (deep rose, throat veined red)	Bhadri's Love Song
Bhadri Blazing Star	Bhadri Pearl
Bhadri Indian Chief	Bhadri Tricolour
Rare Colour	Bhadri Royalty
Glory of Raj Niwas	Bhadri Bouquet
Bhandri Enchantment	Bhadri Oriental Charm
Bhadri Giant Flowered	Bhadri's Fire Dream
Bhadi Celestial	Bhadri Morning Glory
Bhadi Peach Glow	Bhadri's Souvenir

(ecourses)

3.10 Characterization and Evaluation

3.10.1 Characterization for Essential Features and Classification

3.10.1.1 Pigments

Flower color and its commercial value are determined by flower pigments, among others. Pigment composition is very important as it provides adequate message for practical breeding program. The anthocyanin compositions in the flowers of different gladiolus genotypes were assessed and wide variation was observed about their total and individual anthocyanin constituents. The results may be useful for various chemical breeding program improving desired organoleptic and nutritional quality characteristics and helpful for colorant industries (Takemura et al. 2005; Islam 2016). Twenty-three anthocyanins have been detected from the flowers of *Gladiolus* cultivars (Yatomi and Arisumi 1968; Akaviaa et al. 1981). 3-galactosides of kaempferol, quercetin, myricetin, laricitrin, and syringetin as well as other flavonoids were found from the flowers of a wild species, *Gladiolus tristis*. The major anthocyanins of the red flowered cultivar 'Shukuho' and the pink 'Naples' were different from the purple flower cultivars 'Ariake' and 'Blue Lagoon'. 'Naples' contained many other anthocyanins. White cv. 'Saiun' and yellow 'Jacksonville Gold' did not contain anthocyanin. The major anthocyanin of the red 'Shukuho' was suggested to be a cyanidin derivative. The major flavonoid was kaempferol. The purple flower cultivars 'Blue Lagoon' and 'Ariake' contained many flavonoids compared with other flower colors suggesting that those flavonoids represent the more purplish color as co-pigments (Scheffeldt and Hrazdina 1978). The six anthocyanidins in four types of glycosylation were found in *Gladiolus* petals. Six monoglucosides appear in minute quantities, whereas any of the other 18 anthocyanins can serve as the major contributor to the coloration of *Gladiolus* petals (Akaviaa et al. 1981). Number of anthocyanins reported by different authors in *G. x gandavensis* varies – six (Robinson and Robinson 1931; Macek et al. 1946), two on two hybrids (Robinson and Robinson 1934; Cohen et al. 1986), seven (Shibata and Nozaka 1963), 13 (Yatomi and Arisumi 1968), 16 (Arisumi and Kobayashi 1971). Seilleur (1977) studied quantitative relationship of the pigment between induced mutants of the cultivar 'Hawaii' and its progenitor. Four types of glycosylation of the six common anthocyanidins – pelargonidin, cyanidin, peonidin, delphinidin, petunidin, and malvidin – with the sugars glucose and rhamnose, were reported in commercial tetraploid *Gladiolus* cultivars. Pigments of each group may appear individually or in conjunction with pigments from either or both groups, all pigments contributing their share to the visible color in the petals. Flowers having a single pigment in their petals were also found. The presence and absence and the relative amounts of specific pigments of each group affect the results of crosses between selected parents. Specific parental clones containing a single pigment produce offspring lines having no pigmental variation and little or no difference in the visible color. This opens the possibility for F1 hybrid production in gladiolus. Understanding the inheritance of pigments determining colors is very important for application of chemical color analysis data for breeding. Gladiolus breeders should consider all these information for identifying parents when selectively breeding for color. There is need to develop sound knowledge on different groups of pigments in existing varieties, their biosynthetic pathways, etc.

3.10.1.2 Fragrant Gladiolus

Many original species that produced modern gladiolus were fragrant. These fragment species were crossed among themselves and with standard gladiolus. Crosses produced varieties carrying qualities of modern gladiolus and also the fragrance of the species. The maximum used species was *G. tristis*. The main drawback to this natural cross is that the fragrant quality had been neglected and lost during selection for flower size, color, and form. The size of floret and spike are small in the fragrant species and also spring flowering which is the main bottle neck of breeding and producing normal summer-flowering gladiolus. However, many common varieties still carry recessive strains of fragrance from their original ancestors. An early white with a red spot in the throat variety 'Mibloom' was probably identified as a fragrant gladiolus. 'Incense' a small pink with yellow throat variety was forerunner for its fragrance. The next cultivar 'New Era' was identified as slightly fragrant. 'Frilled Fragrance', a bud sport of 'New Era' was introduced for sale. Another small ruffled rose pink variety 'Summer Fragrance' ('Queen of Bremen' seedling) had light fragrance. 'Mora' (a 'Shaylor' seedling) is light pink with a large contrasting red throat that has an odour rather than a fragrance. All these fragrant varieties were of unknown parentage or developed as chance crosses. Under chance crosses category 'Diadem' ('New Era' and 'Vista Bonita' parentage) was one of the largest in the fragrant class. Another light fragrance variety was 'Gwen' (a seedling of 'Rose Marie Pfitzer' and 'Meanly'). A ruffled cream white with a rose dart in the throat variety 'Perfume' is still one of the most fragrant gladiolus in commerce, by out-crossing with 'Mibloom' and then recrossing those seedlings. Spencer selected more than 150 lines with varying amount of fragrance. One particular trait found in wild species is floral scent. Previously studied wild species and their major scent compounds are *G. liliaceus* (eugenol); *G. alatus*, *G. maculates*, *G. recurvus,* and *G. tristis* (linalool); *G. jonquilliodorus*, *G. orchidiflorus*, *G. patersoniae,* and *G. scullyi* (geraniol or geraniol acetate, nerol and citronellol); and *G. virescens* (B-ionone; Goldblatt et al. 1998; Goldblatt and Manning 2002). Suzuki et al. (2008) analyzed floral scent from 9 wild species and detected 20 scent compounds of which 9 were benzenoids, 7 monoterpenes, and 4 sesquiterpenes. The species were divided into four groups based on their major scent compounds like – Lonalool/Benzenoid group (*G. recurvus* and *G. tristis*), Ione group (*G. trichonemifolius*, G. uysiae and G. watermeyeri), and Ocimene/caryophyllene group (*G. gracilis*; Suzuki et al. 2008).

3.10.1.3 Mini Cultivar

Miniatures are extremely versatile because of their small plant size which can be grown easily in beds and most favorable in landscape plantings, lightweight, suitable for density planting, reduced transport costs. A new group of miniature gladiolus was developed to meet the market demand which was named 'Orchidiola' (orchid glads) by the Israeli flower industry. Mini forms were developed from seedling selections from intercrossed of various ecotypes of the diploid species *G. tristis* and several tetraploid commercial varieties (*G. x grandiflorus*). Separate group of miniature varieties ('Pixiola') was developed in America (Barnard 1972; Halevy et al. 1984, Cohen and Barzilay 1991; Hort et al. 2012). The dwarf gladioli, 'Nanus' hybrids' (=*G. nanus*) is an interspecific hybrids from *G. tristis*, *G. carneus*, *G. primulinus*, and *G. cardinalis* (Goldblatt 1996; Goldblatt and Manning 1998).

3.10.2 Development/Identification of Gene Pools and Core Collections

The present-day *Gladiolus* cultivars are complex hybrids and following species are mostly involved in breeding: *G. cardinalis* Curtis., *G. dalenii* van Geel., *G. oppositiflorus* Herb., *G. papilio* Hook. f., *G. carneus* Delaroche, *G. cruentus* Moore., *G. tristis* L., and *G. saunddersii* Hook. f. Hybridization between wild species and cultivars of *Gladiolus* is an effective means of producing individuals (genotypes) with desirable features (Ohri and Khoshoo 1983b; Cantor and Tolety 2011). Moreover, *G. communis* subsp. *byzantinus* appears to hybridize with *G. illyricus* within its core range in southern Spain producing an evenly graded range of morphological intermediates (Cantor and Tolety 2011). *Gladiolus palustris* Gaudin and *G. imbricatus* L. indigenous to Europe were recorded for *Gladiolus* species (Hamilton 1980). Natural, interspecific hybridization between these species was possible due to their close genetic relatedness. Putative hybrids, *G. palustris X G. imbricatus* has been reported in Poland (Kaminski 2012; Cieslak et al. 2014). Interspecific hybridization between a modern cultivar of *Gladiolus x grandiflora* hort. (2n = 60) and the wild species *G. tristis* L. (2n = 30) was made to introduce characteristics of the wild species into the cultivated one.

3.10.3 Evaluation of Genetic Diversity for Desired Traits

The main objectives of gladiolus breeding are to develop varieties with agronomic characteristics of high productivity, high rate of germination of the cormlets, vigorous corms which grow under several conditions of soil and climate, early flowering, attractive/novel floret colors and floret arrangement on the spikes, large sized florets, long spike, increased number of florets, good stem strength, prolonged flowering period, resistance to major pest and diseases, long vase life, flowers with high ornamental and commercial qualities, resistance to the great variation of temperature and luminosity, flowers easy to manipulate, pack, store, and capacity to flower after 2–3 days of dry storage, good corm multiplication rate, winter flowering capability, flower fragrance, etc.

Genotypic correlation was of higher magnitude than phenotypic correlation in most of the cases. The results of correlation coefficient reveal that for yield improvement through selection, much emphasis should be given on the characters like spike yield, corm yield, plant height, spike length, days to spike initiation, days to floret initiation, number of florets, corm weight, spike weight, corm diameter, etc. (Pattanaik et al. 2015; Sundaram and Nambisan 1991; Maitra and Satya 2004). Corms weight per plant was significantly and positively correlated with spike weight, number of cormels per plant, number of corms per plant, spike initiation, number of floret per spike, and number of leaves per plant (Rashmi and Kumar 2014).

Highest PCV and GCV coefficients of variation were observed for the spike length followed by number of floret per spike, days taken for the spike emergence was positively and highly significant with the days taken for bud initiation, days taken for first floret to show color, days taken for first floret to open, number of floret open at a time, diameter of the floret and spike length. Days taken for spike emergence was negatively and significantly correlated with number of floret per spike, number of shoot per plant, and vase life. High heritability and high genetic advance in spike length indicate the presence of additive gene effects in these traits and their amicability for direct selection (Kispotta et al. 2017a, b). Genotypic correlation coefficients were higher than the phenotypic correlation (Balaram and Janakiram 2005; Kumar et al. 2012). Significant positive association of number of florets per spike was observed with spike length (Monika et al. 2008; Choudhary et al. 2011; Lahijie 2012; Pal and Singh 2012; Bhatia and Grewal 2009; Jhon et al. 2002; De and Misra 1994; Patra and Mohanty 2015). High PCV and GCV for rachis length and low for number of spike, floret diameter, floret length and number of leaves, leaf width, plant height, spike length, spike weight, weight of daughter corm and number of cormels per corm, corm weight per plant were observed (Anuradha and Gowda 1990; Balaram et al. 2000; Geeta et al. 2014; Rashmi and Kumar 2014). More variability in corm weight and spike weight is indicated by high value of PCV along with GCV. High GCV noticed for number of side shoots, number of cormels per plant, plant height, number of florets per spike, and days to flowering (Balamurugan and Arumugam 2002; Bichoo et al. 2002; Pratap and Manohar 2006). Low GCV observed for longevity of individual florets and duration of first floret (Balamurugan and Arumugam 2002). Significant positive correlation between plant height with spike length has been reported (Gowda 1989; Neeraj and Jha 2001). Misra and Saini (1990) reported positive significant correlation of number of florets per spike with height of the plant. Deshraj and Mishra (1998) and Neeraj and Jha (2001) reported significant positive association of number of florets per spike with plant height. Maitra and Satya (2004) reported that days to flower bud initiation exhibited high negative correlation with plant height, spike length, and number of florets spike-1. Naresh et al. (2015) studied eight hybrids for 20 characters and the genetic variability, heritability, genetic advance, and correlation for identifying suitable strains for coastal conditions of Andhra Pradesh, India. The magnitude of phenotypic coefficient of variation (PCV) was higher than genotypic coefficient of variation (GCV). Heritability used in conjunction with genetic advance provides better information for selecting the best individuals than the heritability alone. Grafius (1965) mentioned that the optimal genetic level for each component would differ depending on the type of the environment. The results of maximum correlation coefficient investigations indicate that spike yield and corm yield are the two most important characters which should be considered for yield improvement of gladiolus population. Hence, these characters may be considered as selection indices in gladiolus breeding program.

3.10.4 Available Sources of Breeding (Listing of Genetic Resources Available for Various Biotic and Abiotic Stresses and Nutritional Traits, Other Desirable Traits, Quantitative Traits, Particularly Related with Yield, etc.)

3.10.4.1 Abiotic Stress

a. There are several environmental factors that affect the winter hardiness trait, including low temperatures, variable snow/ice cover, low light periods, and secondary invasion by pathogens (Takatsu et al. 2002; Ali et al. 2016). Winter hardiness is a desirable trait for gladiolus as well as consumers for breeding. *Gladiolus* is a genus that has not been studied to any great extent in the area of winter hardiness. Bettaieb et al. (2007) found that low temperature stress of 8 °C caused increased catalase (CAT) activity and lower hydrogen peroxide (H_2O_2) levels in gladiolus, but such information has not resulted in breeding for winter-hardy gladioli. Literature clearly indicates that majority of commercial gladioli have been developed for providing cut flowers in the summer. These varieties produce low-quality flowers when grown in subtropical and Mediterranean climates during winter (Magic and Cowperthwaite 1954; McKay et al. 1981; Shillo and Halevy 1976a, b). Different environmental factors like low light intensities, short photoperiods, and the combined effect of low irradiance with low night temperatures have been identified as the reasons for the failure to grow high-quality gladiolus during winter months (Halevy et al. 1984; Shillo and Halevy 1963, 1975, 1976a, b, 1981). This motivated scientists to develop cultivars to produce high quality flowers under winter conditions (Cohen and Barzilay 1991).

3.10.4.2 Biotic Stress

Quite a large number of diseases cause serious damage to gladiolus and floriculture trade suffers massive losses. Extensive work has been carried out to ward off/minimize the losses. About 20 pathogenic fungal species belonging to 15 genera have been reported from all over the world on different varieties of gladiolus causing leaf spot to corm rot. The details about the symptoms, main causal agents, extent of losses, and the control measures of some of the major diseases (wilt, neck rot, corm rot, leaf and flower blight, viral diseases, Aster yellows; pest–thrips, cut worm, leaf caterpillar, mite, mealy bug, etc.) under open field and greenhouse cultivation have been worked out and the major approaches in controlling these diseases using chemical, cultural methods, and breeding for disease resistance discussed (Datta 2001; Sharma and Bhattacharjee 2002; Shanmugam et al. 2011; Nasir et al. 2012). From gladiolus breeding stock few valuable traits like resistance or immune to *Botrytis gladiolorum, Fusarium oxysporum f. gladioli, Curvularia trifolii,* and *Stemphylium* sp.; resistance to cold weather, flowering adaptation to short winter days, etc. were identified. Varieties (‘Florida pink’, ‘Picardy’, ‘Spic’ and ‘Span’) released during 1958 were very promising (Magie 1960).

Different diagnostic methods (based on RT-PCR, immunodiagnostics, molecular biology, DAS-ELISA, DTBIA, ISEM, use of DNA/RNA probes, the polymerase chain reaction, etc.) for detecting various viral diseases (Bean yellow mosaic virus (BYMV), Cucumber mosaic virus (CMV), Arabis mosaic (ArMV), Broad bean wilt (BBWV), Tobacco mosaic (TMV), Tobacco necrosis (ToNV), Tobacco black ring (TobBRV), Tobacco ringspot (TobRSV), Tomato ring spot (TRSV), Tomato spotted wilt (TSWV), Soybean mosaic (SMV), Strawberry latent ringspot, etc.) have been standardized and their management has been recommended (Raj et al. 2010; Katoch et al. 2004; Singh et al. 2015a, b).

3.10.5 Breeding Options

3.10.5.1 Hybridization

The present-day *Gladiolus* cultivars are complex hybrids and following species are mostly involved in breeding: *G. cardinalis* Curtis., *G. dalenii* van Geel.,

G. oppositiflorus Herb., *G. papilio* Hook. f., *G. carneus* Delaroche, *G. cruentus* Moore., *G. tristis* L., and *G. saunddersii* Hook. f. Hybridization between wild species and cultivars of *Gladiolus* is an effective means of producing individuals (genotypes) with desirable features (Ohri and Khoshoo 1983b; Cantor and Tolety 2011). However, Eurasian species have not been used in developing modern cultivars of gladiolus even though they can be valuable because of their relative hardiness and low sensitivity to fungal diseases (Ohri and Khoshoo 1983b; Rakosy-Tican et al. 2012). Hybridization between closely related species (*G. imbricatus*, *G. italicus,* and G. illyricus) has occurred as a result of migrations due to climatic changes (Van Raamsdonk and de Vries 1989). Moreover, *G. communis* subsp. *byzantinus* appears to hybridize with *G. illyricus* within its core range in southern Spain producing an evenly graded range of morphological intermediates (Cantor and Tolety 2011). Low seed set in a few intervarietal crosses and failure of seed set in some interspecific crosses are apparently realized as major problems in breeding of gladiolus (Van Tuyl 1997). The first *Gladiolus* hybrids were raised by Dean Herbert in the 1820s, producing hybrids between various Cape species, including hardy ones derived from *G. angustus*, *G. cardinalis*, *G. carneus*, and *G. tristis*, and tender wintergrowing ones. Others also started breeding Gladiolus, including Colville's of Chelsea, who crossed *G. cardinalis* and *G. tristis* to make G. x colvillei, described in 1826 and still cultivated (Chis et al. 2010). In 1874, Max Leichtlin obtained the first *Gladiolus* hybrid from the cross of *G. gandavensis* Van Houtte with *G. saundersii* Hook. The modern hybrids are designated as G. X grandiflorus, which is a complex of about 11 species like *G. carneus*, *G. dalenii*, *G. oppositiflorus*, *G. gandavensis*, *G. primulinus*, *G. papilio*, *G. saundersii*, etc. In varietal assessment experiments a number of varieties have been identified which are less susceptible to the day length, more resistant to the low temperature, produce long spikes, more adapted to the winter production, well adapted to summer climate, etc. (Anderson and Park 1989; Ohri and Khoshoo 1985b; Tombolato et al. 2002). New varieties have been developed with extended vase life, floral novelty, extended flowering periods, etc. (Takatsu et al. 2002). Winter-hardy species from Russia include *G. imbricatus* and *G. palustris*. In addition, several species are adaptable to cultivation, including *G. alatus*, *G. angustus*, *G. cardinalis*, *G. carmineus*, *G. carneus*, *G. dalenii*, *G. ochroleucus*, *G. pritzelii*, *G. saundersii*, and *G. sempervirens* (Duncan 1982). Modern gladioli are primarily grown as summer-growing cut flowers and tender annuals. They are derived from summer-growing species, including *G. dalenii*, *G. oppositiflorus*, *G. papilio*, and *G. saundersii*. *Gladiolus cardinalis*, a winter growing species (winter rainfall region), has also been used in hybridization. In South Korea breeding work started in 1986 and the first reddish cultivar ('Hongkwang') and the bright orange-based ('Hongeun') were released in 1995. More than 49 new cultivars were released from 1995 to 2008 comprising new colors, very early and very late flowering, resistance to thrips ('Pink Smile'), etc. (c.f. DH Goo, HY Joung, YI Kang, YJ Choi, HK Shin, Gladiolus Breeding in Korea). Desirable qualities (flower form and shape, color diversity or scent, etc.) of individuals might have produced through hybridization between wild species and cultivars of *Gladiolus* (Ohri and Khoshoo 1983b; Cantor and Tolety 2011). *Gladiolus palustris* Gaudin and *G. imbricatus* L. indigenous to Europe were recorded for *Gladiolus* species (Hamilton 1980). Natural, interspecific hybridization between these species was possible due to their close genetic relatedness. Putative hybrids, *G. palustris* X *G. imbricatus*, have been reported in Poland (Kaminski 2012; Cieslak et al. 2014). Both species are cold-tolerant and their seeds and corms require a cooling period to germinate (Jogar and Moora 2008). Natural hybridization between *G. palustris* and *G. imbricatus* was confirmed by chloroplast (psbA-trnH and rpl32-trnL) DNA and nuclear ribosomal DNA (ITS1) sequences, AFLP markers and macro-, micromorphological, and reproductive characters (Szczepaniak et al. 2016). CSIR-NBRI, Lucknow, developed varieties ('Manmohan', 'Manohar', 'Mukta', 'Manhar', 'Manisha', 'Mohini, 'Urvashi', 'Neelima', etc.) suitable for cultivation in Indo-Gangetic Plains involving *Gladiolus* cultivar 'Friendship' ($2n = 60$) and *Gladiolus tristis* ($2n = 30$). Breeding resulted in development of large number of triploid ($2n = 45$) hybrids (Sharga and Sharma 1984; Sharma et al. 1988). In hybridization program, seed setting is very important for intervarietal crosses. Different aspects like seed setting, early capsule maturity, highest seed germination, higher number of seeds per capsule and per cross were assessed in a number of cross combinations (Kumari et al. 2016). Interspecific hybridization between a modern cultivar of *Gladiolus x grandiflora* hort. ($2n = 60$) and the wild species *G. tristis* L. ($2n = 30$) was made to introduce characteristics of the wild species into the cultivated one. *Gladiolus x grandiflora* is a summer-flowering species, and *G. tristis* flowers in winter. The effect of storage temperature on pollen viability was tested, as long-term storage of pollen was necessary to facilitate crossing these two species. Pollen of *G. tristis* could be stored at -20 °C for $\approx$ year, and was more practical than storage at -80 °C. Air temperature affected pollen tube growth, fertility, and fruit set in the cross between *G. x grandiflora* and *G. tristis*, and low temperatures (15 °C–20 °C) were best. The morphological data and flow cytometric analysis showed that the F1 plants were hybrids between *G. x grandiflora* and *G. tristis* (Takatsu et al. 2002). Modern cultivars of *G. x grandiflora* were bred originally from only six species (Barnard 1972), and considerable genetic potential exists for developing new kinds of gladiolus using wild species. As many wild species flower during winter and modern cultivars of *G. x grandiflora* flower in summer, storing the pollen is necessary to facilitate hybridization. In addition, flower abortion of *G. x grandiflora* occurs in the winter because of low light intensity, and the species is not cold resistant (Imanishi 1989). Winter-flowering wild species are tolerant of low light intensity and cold temperatures. Such interspecific hybridization between *G. x grandiflora* and *G. tristis* was initiated to introduce some characters of the latter species, such as winter flowering, cold tolerance, and fragrance into the former hybrid. This cross combination has been reported in Israel with some miniature cultivars (Cohen and Barzilay 1991). This study demonstrated that long-term storage of pollen of wild gladiolus is possible -20 °C and that low temperature has a positive effect on interspecific crossing. These results suggest that interspecific hybridization between a modern cultivar and a wild species can be made more efficiently with these techniques and can be a useful tool to create new products for the cut flower market (Takatsu et al. 2002). Excellent parents can give poor progenies, since the phonotypical value of parents does not permit a prediction of the value offspring. Experiment was conducted to compare the morphological characters of five genotypes and results indicated significant variation among the genotypes with respect to morphological characteristics as well as with yield, yield attributes, and plant height and white genotype has been mentioned as the best planting materials (Hossain et al. 2011). Cantor and Chis (2009) developed two promising varieties 'Candida Ali' (early flowering cyclamen 'Early Riser' x early flowering pink 'Priscilla') and 'Excelsa' (pink 'Priscilla' x move-cyclam 'Speranta'). 'Candida Ali' showed vigorous growth, early flowering, multiple middle sized bright pink cyclamen florets with good simultaneous flowering and 'Excelsa' was very early flowering, red coral floret. *Gladiolus palustris* is believed to be one of the rarest elements in the Polish flora and at the greatest threat of extinction. AFLP studies of the genetic diversity of *G. palustris* from ex situ conservation at the Botanical Garden of Wroclaw University, Poland, revealed putative hybrids between *G. palustris* and *G. imbricatus* (Kamiñski 2012; Cieœlak et al. 2014). Anthers of gladiolus open and release their pollen a day or two before the stigma becomes receptive which is a constraint for inbreeding. Some admirable varieties have been created by inbreeding. Ervin H. Doerr developed 'Bridesmaid' (multi-colored with intense ruffling) inbreeding 'Boise Belle' through 17 successive generations (c.f. Paul H. Franklin, Fine Gardening). *G. communis* subsp. *byzantinus* appears to hybridize with *G. illyricus* in southern Spain (Cantor and Tolety 2011). Spontaneous hybridization between *G. communis* and *G. italicus* has occurred in Malta (Mifsud and Hamilton 2013).

Interspecific hybridization
Inter-varietal hybridization

3.10.5.2 Mutation Breeding

Appreciable amount of induced mutagenesis work has been done on gladiolus using both physical and chemical mutagens. Mutation experiments generated huge amount of basic and applied knowledge on different parameters of mutation and resulted in development of mutants with new flower color, improved flower form, resistance to lodging or disease, etc. Mutations in flower color mostly arise as chimera in gladiolus. The size of the mutant chimera varies from a narrow streak on a tepal to the entire floret. Such chimeric mutant tissues cannot be isolated using the available propagation technique. Mutation in gladiolus can be established in pure form when mutated cells participate in corm/cormel formation in subsequent generations. Therefore, many new flower color/shape mutants are lost. Management of such chimera and in vitro mutagenesis for induction of solid mutants have been well standardized in chrysanthemum. Direct shoot regeneration will help to retrieve mutants from chimeric tepals. So induced mutation combined with in vitro technique is must in gladiolus to change one or a few characters of an otherwise outstanding cultivar without altering the remaining and often unique of the genotypes (Datta 2014, 2017b).

3.10.5.3 Polyploidy Breeding

Colchicine polyploidy (mixoploids + octoploids) and polyploidy: plants were produced after treatment with different concentrations of colchicines (0.1%, 0.2%, and 0.3%). These putative polyploids may be helpful for further improvement in ornamental and horticultural value of gladiolus (Manzoor et al. 2018). *Gladiolus* is highly heterozygous which makes it an ideal material for polyploidy induction that results in the production of new forms in which one or few traits are improved without changing the whole genome.

Chromosome Studies: *Gladiolus* is a representative ornamental where various stages of transformation from wild to cultivated condition largely affected by hybridization and polyploidy are well preserved. A meaningful picture of origin and evolution of garden gladiolus has been worked out on the basis of cytogenetic mechanisms which is very helpful for further improvement. Interspecific hybridization is possible regardless of ploidy or species level due to its flexible genome. Crosses between 2x and a 6x species produced a highly heterozygous tetraploid hybrid. Many cycles of further hybridization involving some other species resulted in amazing variability in present-day cultivars (Ohri 2013). Chromosomal studies on species and cultivars illustrate a range of ploidy from diploid ($2n = 30$) to dodecaploid ($2n = 180$) but all the garden cultivars are tetraploid ($2n = 60$). Meiotic behavior and pollen stainability of diploid and tetraploid taxa exhibited predominant bivalent formation along with pentavalent, quadrivalent, trivalent, and univalent. Studies revealed that the species native to Cape Winter-Rainfall region (*G. tristis*, *G. carneus*, *G. lapeirousioides*, etc.) are diploid and those occurring in Summer-Rainfall region and Tropical Africa (*G. oppositiflorus*, *G. primulinus*, *G. natalensis*, etc.) range from diploid ($2n = 30$) to hexaploid ($2n = 90$); while Mediterranean, European, and West Asian species (*G. italicus*, *G. byzantinus*, *G. illyricus*, *G. segetum*, *G. communis*, etc.) are all polyploidy, ranging from tetraploid to dodecaploid (Ohri and Khoshoo 1985a, b; Cantor and Tolety 2011; Valente et al. 2011). The majority of the African species of *Gladiolus* are diploids ($2n = 2x = 30$) whereas the European species are polyploids ($2n = 60–130$; Cantor and Tolety 2011; Valente et al. 2011; Delpierre and Du Plessis 1973; Lewis et al. 1972; Ohri and Khoshoo 1985a; Tombolato et al. 2002).

3.10.6 Present Status of Use or Incorporation of Desired Traits

3.10.6.1 Biotechnological Approach

Molecular Characterization: PCR-based techniques (RAPD, RFLP, SSR, STS, AFLP, DAMD, ISSR, etc.) have been applied to study the genetic variability, germplasm characterization, and relationships among *Gladiolus* cultivars. The information generated will facilitate choosing the appropriate breeding program to incorporate beneficial genes from genetically divergent cultivars in desirable genotypes lacking the particular trait (Singh et al. 2015a, b; Gonai et al. 2005; Pragya et al. 2010a, b; Dallavalle et al. 2002; Jingang et al. 2008; Ranjan et al. 2010; Raycheva et al. 2011; Nasir et al. 2012).

Tissue Culture: Ziv et al. (1970) reported first micropropagation of *Gladiolus* and subsequently the technique has been enriched for different species and varieties by a number of workers. Tissue culture procedure applied effectively in breeding program to develop virus-free plants and for fast and reliable micropropagation of elite clones. In vitro selection at cellular level to assess genetic variation reduces time to breed and improve variety over conventional breeding method. Liquid culture and simple bioreactors systems were standardized to reduce production cost and to enhance multiplication rate. Efficient systems for the shoot regeneration and rapid mass shoot propagation, shoot and corm regeneration, 'bioreactor' system for production of uniform and vigorous corm-bearing shoots, high frequency in vitro bud multiplication, cell suspension cultures, selection of the gelling agent, standardization of conditions for in vitro corm formation, etc. have been optimized (Ruffoni et al. 2012; Choudhary et al. 2010; Menon et al. 2016; Nhut et al. 2004; Sinha and Roy 2002; Roy et al. 2006). Biotechnology New variety of bulbous ornamentals takes 5–10 years to be introduced into the market through natural vegetative propagation. Micropropagation is essential for development of large scale quality planting material and introduction of new variety within short period. To overcome the restrictions of classical breeding application of biotechnology is a necessity for future ornamental breeding. Molecular breeding is anticipated to have enormous potential for developing desired varieties. Cormel derived cells were useful for inducing an embryogenic cell suspension and in vitro selection of somaclonal variation (Remotti and Loffler 1995). Regeneration of gladiolus has been standardized using gelling agent and kanamycin for successful plant transformation (Chauvin et al. 1999). Nhut et al. (2004) reported an economic in vitro 'bioreactor' system for production of long-term production of uniform and vigorous corm-bearing shoots. Transgenic gladiolus plants have been developed via microprojectile bombardment of various regenerative tissues (sliced cormels, cell suspension, and callus; Kamo et al. 2005). Kamo (2003) reported highest levels of GUS expression in callus, shoots, and roots of plants carrying the bar-uid. A fusion gene under control of the CaMV 35S promoter and in shoots and roots of greenhouses grown plants that contained the rolD promoter. Transformation protocol has been developed using embryogenic cells and particle bombardment with gold particles coated with a construct harboring the gus reporter gene and the pat (resistance to the herbicide Basta) selection gene. An *Agrobacterium*-mediated transformation protocol using pre-wounded in vitro derived shoot tips bombarded with 1.6 m naked gold particles by the biolistic delivery system has been developed (Babu and Chawla 2000). Molecular breeding is a good preference to develop new flower color and form using sense and anti-sense strategy or incorporation of new colored genes in ornamentals.

3.10.6.2 Looking Forward or Future Perspective

1. Establishment of a "Pollen Cryobank" through which pollen of desired species could be consolidated and obtained for breeding without any seasonal or geographic barrier.

2. Molecular genetics–related research to provide a sound basis for future advances in the species.
3. International efforts to sequence the *Gladiolus* genome.
4. Future genetic mapping in *Gladiolus* will rely on integration of sequence from other bulbous plants.
5. Development of an interspecific map will facilitate the introgression of genes from wild germplasm into cultivated *Gladiolus* species, which will help in studying of gene flow from genetically engineered plants into the wild, and allow the processes of domestication to be elucidated if the progenitor species is included.
6. Development of genomic resources will be important for future breeding work. Improvement of *Gladiolus* cultivars will be directed towards ornamental characteristics and disease resistance through traditional and molecular-aided breeding methods. Some of these traits may be improved by genes from wild *Gladiolus* species that can be used as a reservoir for identifying and screening closely related species for resistance factors.
7. Genetic transformation is potentially a valuable tool for improving gladiolus. Genetic transformation of selected *Gladiolus* cultivars, with genes conferring specific traits of interest may create new phenotypes useful for different purposes of the grower and the consumer.

References

Accession browser (genesys-pgr.org)

Akaviaa N, Strackb D, Cohen A (1981) The coloration of Gladiolus I. Survey of anthocyanins in petals of Gladiolus. Z Naturforsch 36(1):378–382

Ali Z, Shabbir M, Qadeer A, Ahmad HM, Qasim M, Aziz O (2016) Performance evaluation of gladiolus varieties under diverse climatic conditions. Plant Gene Trait 7(4):56–48

Anderson W, Park R (1989) Growing gladioli. Timber Press, Portland

Anuradha S, Gowda JVN (1990) Genetic variability in gladiolus. Progress Hortic 22(1):155–159

Arisumi K, Kobayashi Y (1971) Studies on the flower colours in gladiolus II. Bull Fac Agric Yamaguti Univ 22(4):157–170

Arora JS, Misra RL, Singh K, Singh P, Bhattacharjee SK (2002) Gladiolus. AICRP on floriculture, technical bulletin no. 14

Babu P, Chawla HS (2000) In vitro regeneration and agrobacterium mediated transformation in gladiolus. J Hortic Sci Biotechnol 75(4):400–404. https://doi.org/10.1080/14620316.2000.11511258

Baek et al. (2003) Garlic

Balamurugan RP, Arumugam T (2002) Variability studies in gladiolus. J Ornament Hortic 5(1):38–39

Balaram MV, Janakiram T (2005) Correlation and path coefficient analysis in Gladiolus. J Res Birsa Agric Univ 17(2):215–222

Balaram MV, Janakiram T, Vasantha K, Choudhary E, Ramachandran N, Ganeshan S (2000) Genetic variability among gladiolus genotypes exploring the gladiolus in India. In: Proceedings of national conference on Gladiolus, pp 17–33

Barnard TT (1972) On hybrids and hybridization. In: Lewis GJ, Obermeyer AA, Barnard TT (eds) Gladiolus, a revision of the South African species. J S Afr Bot 10(10):304–310

Bettaieb T, Mahmoud M, De Galarreta JIR, Du Jardin P (2007) Relation between the low temperature stress and catalase activity in gladiolus somaclones (*Gladiolus grandiflorus* Hort.). Sci Hortic 113(1):49–51

Bhatia R, Grewal HPS (2009) Genetic variability and heredity studies in gladiolus. Indian Agric 53(1/2):1–6

Bhattacharjee SK, Pushkar S, Kumar PN (2002) Passport data of flower crop germplasm. AICRP on floriculture, technical bulletin no 23

Bichoo GA, Jhon AO, Wani SA (2002) Genetic variability in some quantitative characters of gladiolus. J Ornament Hortic 5(1):22–24

Bose TK, Yadav LP (1989) Gladiolus. In: Commercial flowers. Naya Prakash, Calcutta

Bouman H, Tiekstra A, Petuschning E, Homan M, Schreurs R (2003) Cryopreservation of *Lilium* species and cultivars. Acta Hortic 612:147–154

Cantor M, Chis LM (2009) New gladiolus cultivars homologated at USAMV Cluj-Napoca. Bull UASVM Hortic 66(1):1856

Cantor M, Tolety J (2011) Wild crop relatives: genomic and breeding resources. In: Chittaranjan K (ed) Plantation and ornamental crops. Springer, Heidelberg

Chauvin EJ, Marhadour S, Cohat J, Nard LM (1999) Effects of gelling agents on in vitro regeneration and kanamycin efficiency as a selective agent in plant transformation procedures. Plant Cell Tissue Organ Cult 58(1):213–217

Chis ML, Cantor M, Eugenia HE (2010) Realizations and new trends in breeding of Gladiolus hybridus at fruit research station Cluj. Bull UASVM Hortic 67(1):1843–5394

Choudhary D, Agarwal G, Singh VP, Arora A (2010) In vitro micropropagation of Gladiolus grandiflora (var. Snow Princess) flower from cormel explant. Indian J Plant Physiol 15(1):90–93

Choudhary M, Moond SK, Kumari A (2011) Correlation studies in gladiolus. Res Plant Biol 1(4): 68–72

Cieslak E, Szczepaniak M, Kaminski R, Heise W (2014) The condition of Gladiolus paluster (Iridaceae), critically endangered species in Poland: analysis of genetic variation of individuals in ex situ culture in the Botanical Garden of Wroc[3]aw University in the context of ongoing conservation efforts. Fragm Florist Geobot Polonica 21(4):49–66

Cohen A, Barzilay A (1991) Miniature gladiolus cultivars bred for winter flowering. HortScience 26(2):216–218

Cohen A, Akavia N, Umiel N (1986) The identification of anthocyanin pigments in the petals as an aid to the breeding of Gladiolus. Acta Hortic 177(53):375–384

Course: Breeding and Seed Production of Ornamental Crops (iasri.res.in)

Dallavalle ED, Aulireo AZ, Verardi E, Bertaccini A (2002) Detection of RAPD polymorphisms in gladiolus cultivars with differing sensitivities to *Fusarium oxysporum* f. sp. Gladioli. Plant Mol Biol Rep 20(1):305a–305f

Datta SK (2001) Exploring the gladiolus in India. In: Proceedings of the national conference on Gladiolus, January, 2000. CSIR-NBRI, Lucknow

Datta SK (2014) Induced mutagenesis: basic knowledge for technological success, Chapter 5. In: Tomlekova NB, Kozgar MI, Wani MR (eds) Mutagenesis, exploring genetic diversity of crops. Wageningen Academic Publishers, pp 97–139

Datta SK (2017a) Breeding of ornamentals: tuberose (*Polianthes tuberosa* L.). Curr Sci 113(7): 1255–1263

Datta SK (2017b) Improvement through induced mutagenesis: ornamental crops, Chapter 4. In: Malik CP, Wani SH, Kushwaha HB, Kaur R (eds) Advanced technologies for crop improvement and agricultural productivity. AGROBIOS (International), Jodhpur, pp 49–86

De LC, Misra RL (1994) Studies on correlation and path coefficient analysis in gladiolus. J Ornament Hortic 2(1):14–19

De Wet GC, Pheiffer RH (1979) Simon van der Stel's journey to Namaqualand in 1685. Human & Rousseau, Cape Town

Delpierre GR, Du Plessis NM (1973) The winter-growing Gladiolus of South Africa. Tafelberg Publishers, Cape Town/Johannesburg

Deshraj RL, Mishra HC (1998) Stability analysis in gladiolus II floral characters. J Ornament Hortic 1(1):61–65

Duncan G (1982) Gladiolus ochroleucus – a desirable species for pot culture. Veld Flora 68(4):112–113

Geeta SV, Shirol AM, Nishani S, Shiragur M, Varuna KJ (2014) Assessing genetic diversity of gladiolus varieties using SRAP markers. Res J Agric Sci 5(4):658–661

Goldblatt P (1996) Gladiolus in tropical Africa: systematics, biology & evolution. Timber Press, Portland

Goldblatt P, Manning J (1998) Gladiolus in Southern Africa. Timber Press, Portland

Goldblatt P, Manning J (2002) Evidence for moth and butterfly pollination in *Gladiolus* species (Iridaceae: Crocoideae). Ann Mo Bot Gard 89:110–124

Goldblatt P, Manning JC, Bernhardt P (2001) Radiation of pollination systems in Gladiolus (Iridacae: Crocoideae) in southern Africa. Ann Mo Bot Gard 88(4):713–734

Gonai T, Nogi M, Kasumi M (2005) Interspecific hybridization of wild Gladiolus species using a dendrogram based on RAPD analysis. Acta Hortic 673(1):475–480

Gowda JVN (1989) Genotypic and phenotypic variability and correlation in quantitative and qualitative characters in gladiolus. Crop Resour 2(1):235–237

Halevy AH, Kofranek AM, Besemer ST (1984) Photoperiodic response of miniature gladiolus cultivars. HortScience 19(1):858–860

Hamilton AP (1980) Gladiolus L. In: Tutin TG, Heywood VH, Burges NA, Moore DM, Valentine DH, Walters SM, Webb DA (eds) Flora Europaea, Vol. 5, Alismataceae to Orchidaceae (Monocotyledones). Cambridge University Press, Cambridge, pp 101–102

Hort D, Cantor M, Buta E, Andriescu I (2012) Researches regarding intraspecific hybridization of Gladiolus L. species in order to obtain novel ornamental varieties. Bull UASVM Hortic 69(1): 125–135

Hossain MD, Talukder KH, Asaduzzaman M, Mahmud F, Amin N, Sayed MA (2011) Study on morphological characteristics of different genotypes of Gladiolus flower. J Sci Found 9(1&2):01–08

Imanishi H (1989) Collected data of plant genetic resources. In: Matsuo T (ed) Gladiolus. Kodansya Scientific, Tokyo

Islam S (2016) Anthocyanin compositions in different colored Gladiolus Species: a source of natural food colorants. Am J Food Sci Technol 4(4):109–114. https://doi.org/10.12691/ajfst-4-4-4

Jhon AQ, Bichoo GA, Wani SA (2002) Correlation studies in gladiolus. J Ornament Hortic 5(1):25–29

Jingang W, Ying G, Daidi C, Shenkui L, Chuanpin Y (2008) ISSR analysis of 26 general species of Gladiolus hybridus Hort. J Northeast Agric Univ 15(1):6–10

Jogar Ü, Moora M (2008) Reintroduction of a rare plant (*Gladiolus imbricatus*) population to a river floodplain – how important is meadow management? Restor Ecol 16(1):382–385

Johnson SD, Harris F, Proches S (2009) Pollination and breeding systems of selected wildflowers in a southern African grassland community. S Afr J Bot 75(1):630–645

Joung YH, Cantor M, Ellis D, Kamo K (2007) Vitrification of *Gladiolus* shoot tips from cormels. Hortic Environ Biotechnol 48(4):251–255

Kaminski R (2012) Mieczyk blotny Gladiolus paluster Gaudin. In: Perzanowska J (ed) Monitoring gatunków roslin. Przewodnik metodyczny. Czesc II [Monitoring of plant species. Methodological guide. Part II]. G3ówny Inspektorat Ochrony Srodowiska, Warszawa, pp 169–183

Kamo KK (2003) Long-term expression of the uidA gene in Gladiolus plants under control of either the ubiquitin, rolD, mannopine synthase, or cauliflower mosaic Virus promoters following three seasons of dormancy. Biomed Life Sci 21(1):797–803

Kamo K, Gera A, Cohen J, Hammond J, Blowers A, Smith A, Eck JV (2005) Transgenic Gladiolus plants transformed with the bean yellow mosaic Virus coat-protein gene in either sense or antisense orientation. Plant Cell Rep 23(1):654–663

Katoch M, Abdin MZ, Zaidi AA (2004) First report of tobacco rattle virus occurring in Gladiolus in India. Plant Pathol 53(1):236

Kaur S, Dhatt KK (2019) Response of pollen viability in Gladiolus (*Gladiolus grandiflorus* L.) to storage method and period. Int J Curr Microbiol App Sci 8(2):1625–1631
Kispotta LM, Jha KK, Horo P, Tirkey SK, Misra S, Sengupta S (2017a) Studies on combining ability and heterosis in Gladiolus (Gladiolus hybridus). Int J Sci Environ Technol 6(1):420–442
Kispotta LM, Jha KK, Horo P, Tirkey SK, Misra S, Sengupta S (2017b) Genetic variability and heritability in Gladiolus hybridus. Int J Sci Environ Technol 6(1):519–528
Koopwitz H, Voss R, Neil O (1984) Long term storage of gladiolus pollen. HortScience 19:513–514
Kumar R, Kumar S, Yadav YC (2012) Genetic variability, heritability, genetic advance, correlation coefficient and path analysis in gladiolus. Indian J Hortic 69(3):369–373
Lahijie MF (2012) Pheotypic variation of Gladious species of Iran. Int J Agric Crop Sci 4(11):721–725
Lewis GJ, Obermeyer AA, Barnard TT (1972) Gladiolus. A revision of South African species. J S Afr Bot 10(1):316
Macek J, Robinson BM, Robinson R (1946) Anthocyanins of gladiolus. Nature 158(4):342
Magic RO, Cowperthwaite WG (1954) Commercial gladiolus production in Florida. Fla Agric Exp Station Bull 535(1):1–67
Magie RO (1960) Breeding Gladiolus for Florida. Fla Agric Exp Station J Ser Fla State Hortic Soc 1133(1):375–378
Maitra S, Satya P (2004) Studies on genetic parameters of some off-season planted gladiolus genotypes in humid subHimalayan region. J Ornament Hortic 3–4:57–61
Manzoor A, Ahmad T, Bashir MA, Baig MMQ, Quresh AA, Shah MKN, Hafiz IA (2018) Induction and identification of colchicine induced polyploidy in Gladiolus grandiflorus 'white prosperity'. Folia Hortic 30(2):307–319
McKay ME, Byth DE, Tommerup JA (1981) Environmental responses of gladioli in south-east Queensland. Sci Hortic 14(2):77–92
Menon N-u-N, Wahocho NA, Miano TF, Leghari MH (2016) Propagation of Gladiolus corms and cormels: a review. Afr J Biotechnol 15(32):1699–1710
Mifsud S, Hamilton AP (2013) Preliminary observations from long-term studies of Gladiolus L. (Iridaceae) for the Maltese Islands. Webbia 68(1):51–56
Misra RL (1995) Advances in horticulture. In: Chadha KL, Bhattacharjee SK (eds) Advances in horticulture, vol 12, part 1: Ornamental plants. Malhotra, New Delhi, pp 93–202
Misra RL, Saini HC (1990) Correlation and path coefficient studies in gladiolus. Indian J Hortic 47(1):127–132
Mollah MS, Khan FN, Amin MM (2002) Gladiolus. In: Landscape, ornamental and floriculture division. HRC, BARI, Gazipur
Monika S, Sehrawat SK, Gupta AK, Suresh K, Dahiya DS (2008) Studies on phenotypic correlation coefficient in gladiolus cultivars. Haryana J Hortic Sci 3(1/2):82–84
Mukhopadhyay A (2002) Gladiolus, 7th edn. M/s Vinayak Press, New Delhi, p 4
Naresh S, Dorajee Rao AVD, Vijaya Baskhar V, Paratpara Rao M, Uma Krishna K (2015) Genetic variability, heritability and genetic advance in Gladiolus hybrids. Plant Arch 15(1):377–381
Nasir IA, Jamil A, Rahman Z, Husnain T (2012) Molecular analyses of Gladiolus lines with improved resistance against Fusarium wilt. Pak J Bot 44(1):73–79
Neeraj M, Jha HP (2001) Correlation and path coefficient analysis in gladiolus. J Ornament Hortic 4(1):74–78
Nhut DT, Jaime ATS, Phan XH, Paek KY (2004) The importance of explant source on regeneration and micropropagation of Gladiolus by liquid shake culture. Sci Hortic 102(1):407–414
Ohri D (2013) Cytogenetics of domestication and improvement of garden Gladiolus and Bougainvillea. Nucleus (India) 56(3):149–153
Ohri D, Khoshoo TN (1983a) Cytogenetics of garden Gladiolus. III. Hybridization. Pflanzenzüchtg 91(1):46–60
Ohri D, Khoshoo TN (1983b) Cytogenetics of garden Gladiolus. IV. Origin and evolution of ornamental taxa. Proc Indian Natl Sci Acad 49(1):279–294

Ornduff R (1969) Ecology, morphology, and systematics of Jepsonia (Saxifragaceae). Brittonia 21(1):286–298
Pal A, Singh R (2012) Correlation and path coefficient analysis in gladiolus. Ann Hortic 5(1):103–107
Parthasarathy VA, Nagaraju V (1999) Gladiolus. In: Bose TK, Maiti RG, Dhua RS, Das P (eds) Floriculture and landscaping. Naya Prakash, Calcutta
Patra SK, Mohanty CR (2015) Genetic divergence in Gladiolus. Int J Agric Sci Res 5(2):1–6
Pattanaik S, Paul A, Lenka PC (2015) Genotypic and phenotypic variability and correlation studies in gladiolus. J Crop Weed 11(1):113–119
Pragya R, Bhat KV, Misra RL, Ranjan JK (2010a) Analysis of diversity and relationships among gladiolus cultivars using morphological and RAPD markers. Indian J Agric Sci 80(9):766–772
Pragya R, Bhat KV, Misra RL, Singh SK, Ranjan JK (2010b) Genetic relationships of gladiolus cultivars inferred from fluorescence based AFLP markers. Sci Hortic 123(1):562–567
Pratap M, Manohar A (2006) Assessment and variability studies in Gladiolus. J Ornament Hortic 9(1):145–147
Raj SK, Snehi SK, Khan MS, Asif MH, Bag SK, Roy RK, Goel AK (2010) Molecular characterization of two phytoplasma isolates associated with malformation and twisting of floral spike of Gladiolus in India. J Plant Pathol 76(1):389–394
Ranjan P, Bhat KV, Misra RL, Singh SK, Ranjan JK (2010) Genetic relationships of gladiolus cultivars inferred from fluorescence based AFLP markers. Sci Hortic 123(1):562–567
Rashmi, Kumar S (2014) Estimation of genetic variability, correlation and path analysis in gladiolus (Gladiolus species L.). Int J Plant Sci 9(1):186–189
Raycheva T, Stoyanov K, Denev I (2011) Genetic diversity and molecular taxonomy study of three genera from Iridaceae family in the Bulgarian flora based on ISSR markers. Biotechnol Biotechnol Equip 25(1):2484–2488
Remotti PC, Loffler JM (1995) Callus induction and plant regeneration from gladiolus. Plant Cell Tissue Organ Cult 42(1):171–178
Robinson GM, Robinson R (1931) A survey of anthocyanins. I. Biochem J 25(5):1687–1705
Robinson GM, Robinson R (1934) A survey of anthocyanins. IV. Biochem J 28(5):1712–1720
Roy S, Gangopadhyay G, Bandopadhyay T, Modak BK, Dattas, Mukherjee K (2006) Enhancement of in vitro micro corm production in Gladiolus using alternative matrix. Afr J Biotechnol 5(12): 1204–1209
Ruffoni B, Marco M, Barberini S (2012) Biotechnological support for the development of new Gladiolus hybrids. Floric Ornament Biotechnol 6(1):45–52
Seilleur P (1977) Bulletin research agronomy. Gembloux 12(1–2):97
Shanmugam V, Kanoujia N, Singh MK, Singh S, Prasad R (2011) Biocontrol of vascular wilt and corm rot of gladiolus caused by *Fusarium oxysporum* f. sp. gladioli using plant growth promoting rhizobacterial mixture. Crop Prot 30(1):807–813
Sharga AN, Roy RK (2002) Gladiolus. NBRI Bull 8
Sharma P, Bhattacharjee SK (2002) Plant protection in ornamental crops. AICRP on Floriculture, ICAR, New Delhi
Shibata M, Nozaka M (1963) Paper chromatographic survey of anthocyanins in Gladiolus-flowers I. Bot Mag Tokyo 76(1):317–323
Shillo R, Halevy AH (1976a) The effect of various environmental factors on flowering of gladiolus. I. Light intensity. Sci Hortic 4(1):131–137
Shillo R, Halevy AH (1976b) The effect of various environmental factors on flowering of gladiolus. II. Length of the day. Sci Hortic 4(1):139–146
Shillo R, Halevy AH (1981) Flower and corm development in gladiolus as affected by photoperiod. Sci Hortic 15(1):187–196
Singh N, Mohar KS, Verma S, Meena B, Roy RK, Tewari SK, Goel AK, Rana TS (2015a) Molecular analysis of genetic variability and relationship among Gladiolus cultivars. Indian J Biotechnol 17(1):118–127

Singh PJ, Kakade DS, Majumder N, Sridhar V, Girish KS, Prabha K (2015b) Disease and pest management in flower crops under polyhouse. ICAR-Directorate of Floriculture Research, DFR Bulletin No. 12

Sinha P, Roy SK (2002) Plant regeneration through in vitro cormel formation from callus culture of *Gladiolus primulinus* Baker. Plant Tissue Cult 12(2):139–145

Soorianath Sundaram K, Nambisan KMP (1991) Studies on variability and certain genetic parameters in gladiolus. South Indian Hortic 39(1):207–209

Suzuki K, Oyama-Okubo N, Nakayama M, Takatsu Y, Kasumi M (2008) Floral scent of wild *Gladiolus* species and the selection of breeding material for this character. Breed Sci 58:89–92

Takagi H, Tien Thinh N, Islam OM, Senboku T, Sakai A (1997) Cryopreservation of in vitro-grow shoot tips of taro (*Colo- casia esculenta* L. Schott) by vitrification. 1. Investigation of basic condition of the vitrification procedure. Plant Cell Rep 16:594–599

Takatsu Y, Kasumi M, Manabe T, Mikio HM (2002) Temperature effects on interspecific hybridization between *Gladiolus grandiflora* and *G. tristis*. HortScience 36(2):341–343

Takemura T, Takatsu Y, Kasumi M, Marubashi W, Iwashina T (2005) Flavonoids and their distribution patterns in the flowers of Gladiolus cultivars. Proceeding of IXth international symposium on flower bulbs, eds

Thunberg CP (1823) Flora capensis. JA Schultes, Stuttgart

Tombolato AFC, de Castro JL, Matthes LAF (2002) Brazilian breeding program on Gladiolus spp. – history and first results. Proceedings of 8th international symposium on flower bulbs, eds. Littlejohn G et al. Acta Hortic 570(2):219–224

Valente LM, Savolainen V, Manning JC, Goldblatt P, Vargas P (2011) Explaining disparities in species richness between Mediterranean floristic regions: a case study in Gladiolus (Iridaceae). Glob Ecol Biogeogr 20(2):881–892

Van Raamsdonk LWD, de Vries T (1989) Biosystematic studies in European species of Gladiolus (Iridaceae). Syst Evol 165(2):189–198

Van Tuyl JM (1997) Interspecific hybridization of flower bulbs: a review. Acta Hortic 430(4): 465–475

Verty P, Prasad VM, Collis JP, Nazir M (2017) Correlation analysis in Gladiolus (*Gladiolus grandiflorus* L.). Agric Res Technol 10(4):555792. https://doi.org/10.19080/ARTOAJ.2017.10.555794

Yatomi T, Arisumi K (1968) Studies on the flower colours in Gladiolus I. Lab. Hortic Fac Agric Yamaguchi Univ 19(2):1183–1196

Ziv M, Halevy AH, Shilo R (1970) Organ and plantlet regeneration of Gladiolus through tissue culture. Ann Bot 34(4):671–676

Lilium: Conservation, Characterization, and Evaluation

4

M. R. Dhiman, Puja Sharma, and Bhavya Bhargava

Contents

M. R. Dhiman (✉)
ICAR-INARI, Regional Station, Katrain, Kullu-Valley, Himachal Pradesh, India

P. Sharma
Dr. YSPUHF, Nauni-Solan, Himachal Pradesh, India

B. Bhargava
CSIR, IHBT, Palampur, Himachal Pradesh, India
e-mail: bhavya@ihbt.res.in

S. K. Datta, Y. C. Gupta (eds.), *Floriculture and Ornamental Plants*, Handbooks of Crop Diversity: Conservation and Use of Plant Genetic Resources,
https://doi.org/10.1007/978-981-15-3518-5_6

Abstract

Lilies are spectacular bulbous plants grown worldwide for cut flower and pot plant production. There are more than 110 species of *Lilium* which are divided into seven sections, and since species belonging to the same section have relatively high interspecific crossing abilities, interspecific hybridization is the prime method of lily breeding. The interspecific hybrids within the sections especially those within the sections Leucolirion, Archelirion, and Sinomartagon represent the most important breeding groups which are Longiflorum hybrids (L genome), Asiatic hybrids (A genome), and Oriental hybrids (O genome). Most of the natural lily species are diploid ($2n = 24$), but few species are triploid ($2n = 3x = 36$) that are sterile. Breeding of novel lily cultivars by either traditional cross or genetic engineering is possible only when valuable genetic resources are available. In recent years, lily habitats have been suffering serious destruction, and many wild species have become endangered. So, germplasm conservation is very important as a source of genetic variation for breeding and research and to prevent rare species from becoming extinct. Many in vivo and in vitro techniques have been employed to conserve *Lilium* germplasm, out of which cryopreservation, i.e., storage of living cells, tissues, or organs at extra low temperatures, usually that of liquid nitrogen (−196 °C), has been recognized as an ideal means for long-term storage of *Lilium* germplasm. The lily (*Lilium longiflorum*, Easter lily) genome size is 36 GB which is one of the largest among all plants, i.e., ~ 550x of *Arabidopsis thaliana* (135 Mb). Biotechnological tools like tissue culture, molecular markers, and recombinant DNA technology have played vital role for the development of *Lilium* cultivars with improved traits.

Keywords

Interspecific hybridization · Germplasm · Triploids · Cryopreservation · Genome size · Molecular markers

4.1 Introduction

Lilium is one of the most important ornamental bulbous crops worldwide grown as cut flower and pot plant and for park, garden, and landscape plant because of its great diversity in plant architecture, flower color, shape and size, fragrance, long shelf-life, bulb morphology, and resistance to biotic and abiotic stresses (Bakhshaie et al. 2016). The bulbs are mainly produced in the Netherlands with acreage of 6400 hectares followed by the United States, Japan, China, and Southern Hemisphere. The assortment of lilies consists of thousands (>10,000) of cultivars which can be classified into about nine groups (https://www.rhs.org.uk/plants/plantsmanship/plant-registration/Lily-cultivar-registration/Lily). According to FAO agro-biodiversity is the part of biodiversity that nurtures people and is nurtured by the people. The importance of biodiversity for humankind has been well recognized in the recent decades, and many would argue that diversity is essential for allowing sustainable development of various human activities. Lilies are one of the economically important plants that holds wider diversity in genetic resources which provided valuable genes for breeding of novel cultivars. There is a great demand for novel lily cultivars from the production industry (Bakhshaie et al . 2016). The availability of and easy access to diverse plant genetic resources are essential to provide the valuable genes necessary to develop novel cultivars. The bulbs of some *Lilium* species like *L. davidii* var. unicolor, *L. brownii* var. viridulum, and *L. lancifolium* which contain high levels of protein, amino acids, minerals, and dietary fiber are widely consumed as vegetables (Wu 2017). Other *Lilium* species including *L. regale*, *L. concolor*, *L. pumilum*, *L. davidii* var. *unicolor*, and *L. lancifolium* rich in biochemical compounds such as alkaloids, polysaccharides, saponin, and colchicine that have antioxidant activities are traditionally used in Chinese medicines. Extensive explorations and collections, heavy grazing, forest planting, and urban expansions have threatened wild *Lilium* species. *L. tsingtauense*, *L. polyphyllum*, *L. pomponium*, and *L. maculatum* var. bukosanense are listed as endangered species in China, Italy, India (Rana and Samant 2011), and Japan, respectively. As a result, it is necessary to preserve and protect *Lilium* genetic resources. In this chapter, we will look through the origin, distribution, domestication, utilization, and conservation of available *Lilium* genetic resources to develop novel cultivars and briefly describe the morphological features of lily species to help breeders and readers understand how the genus *Lilium* has come out and what characters have attracted gardeners for a long time.

4.2 Botany and Distribution

Lilium belong to order Liliales of the family Liliaceae and subclass Monocotyledonae. It is herbaceous perennial. All organs of a lily originate from the basal plate of a bulb, which makes lily morphogenesis particularly interesting. The inflorescence may be a raceme, an umbel, or a single terminal flower. Flowers are perfect and contain six petaloid tepals, six stamens, and a superior three-celled ovary with a three-lobed stigma. Flowers are characterized by showy color and fragrance and may be upright or upfacing, outfacing, or pendent. Lily pollen varies greatly in color from species to species and among hybrids, ranging from soft yellow to dark brown. The center of the flower contains the pistil composed of ovary at the base, a long style, and a three-lobed stigma at the tip. At the base of each tepal, there is a narrow groove, the nectary furrow.

The form of the lily seed capsule differs among species and hybrids, ranging from the relatively short capsule of *Lilium candidum* to the long, slender capsule of *L. formosanum*. All capsules are divided into three 2-part sections with papery dividing walls, inside which the numerous flat seeds are stacked like corns in a wrapper. All lily seeds are flat except *Lilium auratum*, which has large "wings," and *L. polyphyllum* has a little wing tissue.

Lily bulbs consist of imbricating scales, which are morphologically specialized leaves that contain nutrients and water reservoirs. The bulb is always without any protective tunic or coat, and the growth type is sympodial. The apical meristem produces the inflorescence, whereas the axillary meristems produce the side bulbs that continue the growth after the main shoot perishes. The lily bulbs may be concentric (*L. medeoloides*), rhizomatous (*L. pardalinum*), sub-rhizomatous (*L. washingtonianum*), stoloniferous (*L. canadense*), or stoloniform (*L. wardii*). In many species, bulblets are produced on the underground part of stem, above the bulb and leaf axils. The stem is erect and unbranched. A mature lilies' flowering stem may be as short as a few inches in *Lilium nanum* or other high alpine species or as tall as 250 cm, as in *Lilium leucanthum* var. centifolium or *L. superbum*. Some stems rise straight from the bulb, as in *L. martagon* and *L. regale*; others travel horizontally underground before emerging, as in *L. lankongense* and *L. nepalense*. The leaves range from the narrow grass-like foliage of *L. pumilum* to the broad, lanceolate leaves of *L. auratum* var. platyphyllum. Some species, such as *L. taliense*, produce a naked, asparagus-like stem that rises 30 cm or more before the leaves expand. *Lilium martagon*, *L. hansonii*, and their hybrids, as well as several North American lilies, bear their leaves in regular whorls around the stem, with gaps between the leaves. Roots are adventitious fibrous.

The genus *Lilium* L. comprises approximately 110 species around the world that are distributed between the 10^0 and 60^0N in Asia (50 to 60 species), North America (approximately 24 species), and Europe (approximately 12 species) (McRae 1998). Most species of the Sinomartagon sections, such as *L. dauricum*, *L. maculatum*, *L. concolor*, *L. leichtlinii*, *L. davidii*, and *L. cernuum*, are distributed in East Asia. China, the central distribution center of *Lilium* species in the world, holds extremely abundant germplasm resources, including 55 wild species and 30 cultivars (Wu et al. 2006). The distribution of the lily is mainly based on environmental conditions like altitude, temperature, humidity, light intensity, etc. The habitat of Lilium is the mountainous area located in the European part of Russia, in the Carpathians, in Transcarpathia, and in the south of Western and Eastern Siberia. It grows on the slopes of mountains and on edges, glades, and meadows. It occurs singly in mixed, deciduous, and small-leafed forests on rich medium-moistened soils in the mountains – from the lower to the upper mountain belt.

4.3 Origin, Domestication, and Spread

All species of *Lilium* are endemic to mountainous region of the Northern Hemisphere up to South Canada and Siberia, and their Southern limit is Florida and the Nilgiri hills of India. *Lilium*, which is taxonomically and phylogenetically regarded as an important clade of the core Liliales, appears to have evolved in the Himalayas approximately 12 million years ago, despite the lack of fossil records (Gao et al. 2015).

China is considered to be the main biodiversity center of this genus, especially Sichuan, Yunnan, and Tibet (Wu et al. 2006), with a total of 55 species and 32 varieties (Wu et al. 2014). The east coast of Asia, the West Coast of North America, and the Mediterranean region are the three most richly garnished places. Far East is the home of almost half of the world's lilies and majority of our finest garden lilies originated there. Three interesting species originate from tropical or subtropical countries, i.e., *L. neilgherrense* from south of India, *L. formosanum* from Taiwan, and *L. longiflorum* from south of Japan and Taiwan. A few species have been found in Northeast India (*Lilium wallichianum*, *L. nepalense*, and *L. polyphyllum*) also. The northeastern hilly regions of India, comprising Nagaland, Manipur, Mizoram, Tripura, and parts of the Cachar district of Assam, possess diverse topographies with hilly sloping terrain and are rich in species and crop genetic diversity. *Lilium mackliniae* is known as Manipur lily. A species may have a limited distribution, for example, *Lilium regale* is found only in one steep-sided Chinese valley; others can be wanderers like the champion globe-trotter like *Lilium martagon* known from Siberia across to Poland and down to the Balkans. In Taiwan, the trumpet lily (*Lilium formosanum*) called after the island's former name grows from sea level to some 3600 m. Some *Lilium* species restricted to tiny areas such as *L. hansonii* are found only on an island off the South Korean coast or *L. pekinese* which occurs in one area of marsh in California. *Lilium bosniacum* is a rare species endemic to the Balkan peninsula that grows on serpentine soil at an altitude from 1200 to 1300 m. Molecular investigation reveals that lilies arose in Eurasia, nearly 68 million years ago, and then spread throughout the world, and until now more than 10,000 cultivars have been developed. *Lilium* spread in nine floristic regions including the circumboreal floristic region, the eastern Asiatic regions, and Mediterranean floristic region (north-western China). *L. fargesii* migrate to Bashan from the north edge of the Sichuan Basin, Mt. Wushan in Chongqing, to Shennongjia forest region of Hubei (Hao et al. 2017).

4.4 Plant Genetic Resources

4.4.1 Geographical Distribution

Geographical distribution of *Lilium* species were investigated by many researchers. Kim et al. (2006) studied the habitats and geographic distribution of 367 populations of diploid and triploid *L. lancifolium* grown in islands and mainland areas covering a whole country of South Korea. Among all populations investigated, 185 (50.4%) and 182 (49.6%) were diploid and triploid populations, respectively. The diploid cytotype has a very narrow distribution, occurring in western and southern coastal areas of the Korean Peninsula and in the Japanese Islands of Tsushima and Iki, although it has also been once cited in Russia, near Vladivostok, in Primorsky Region. Compared to the relatively limited distribution of diploid *L. lancifolium*, triploid forms would have a much broader distribution occurring in the inland areas of the Korean Peninsula, in coastal areas of eastern Korea, and in Jeju and Tsushima

Table 1 Origin and native distribution of *Lilium* hybrids and species

Section	Species	Native distribution
Martagon	*L. distichum*	Amur and Vladivostok regions in Siberia and to Manchuria and Korea
	L. hansonii	Ulleungdo and Takeshima Islands off the coast of Korea and the Diamond and Negita mountains in mainland Korea
	L. martagon	Eurasia
	L. medeoloides	Honshu and Hokkaido in Japan, north to Sakhalin, the Kurile Island off Kamchatka, and Cheju Island off southern Korea
	L. tsingtauense	China and Korea
American		
a:	*L. bolanderi*	Southern Oregon and northern California
	L. columbianum	Western North America
	L. humboldtii	United States
	L. kelloggii	United States
	L. rubescens	Pacific Coast of the United States
	L. washingtonianum	Western American mountains
b:	*L. maritimum*	Coastal California in the United States
	L. occidentale	Pacific Coast of the United States
	L. pardalinum	Pacific Coast of the United States
	L. parryi	Southwestern United States
	L. parvum	Western United States
c:	*L. canadense*	Eastern North America
	L. grayi	North Carolina, Tennessee, and Virginia in the United States
	L. iridollae	Southern Alabama and north-western Florida in the Southeastern United States
	L. michauxii	Southeastern United States
	L. michiganense	United States
	L. superbum	United States
d:	*L. catesbaei*	Southeastern United States
	L. philadelphicum	North America
Candidum	*L. bulbiferum*	Europe
	L. candidum	Eastern Mediterranean
	L. carniolicum	Balkans
	L. chalcedonicum	Albania and Greece
	L. monadelphum	Northern Caucasus
	L. polyphyllum	Himalayas (India)
	L. pomponium	Alps-Maritimes of France
	L. pyrenaicum	Europe, Turkey, and the Caucasus
	L. alexandrae	Japan

(continued)

Table 1 (continued)

Section	Species	Native distribution
Oriental	*L. auratum*	Japan
	L. brownie	
	L. japonicum	Southern China
	L. nobilissimum	Japan
	L. rubellum	Japan
	L. speciosum	Japan
	L. davidii	Japan, Taiwan, China
Asiatic		
a:	*L. duchartrei*	China
	L. henryi	China
	L. lancifolium	China
	L. lankongense	Japan, Korea, China
	L. leichtlinii	China
	L. papilliferum	Japan
	L. amabile	Korea
b:	*L. callosum*	Eastern Asia
	L. cernuum	Korea, Russia, China
	L. concolor	China, Japan, Russia, Korea
	L. pumilum	South and North Korea, Russia, Mongolia, northern China
	L. bakerianum	Burma, Nepal, China
c:	*L. mackliniae*	Burma
	L. nepalense	Himalayas of Nepal, Bhutan, and Kumaon
	L. ochraceum	Nepal
	L. sempervivoideum	China
	L. taliense	China
	L. wardii	South-eastern Tibet
Trumpet section		
a:	*L. leucanthum*	China
	L. regale	China
	L. sargentiae	China
	L. sulphureum	China
b:	*L. formosanum*	Taiwan
	L. longiflorum	Japan
	L. neilgherrense	India
	L. philippinense	Philippines
	L. wallichianum	Himalayas
Dauricum section	*L. dauricum*	Asia
	L. maculatum	Japan

islands. *L. martagon* is widespread with its distribution area from East Russia to West Spain. *L. hansonii* is very peculiar to have its only habitat in Ulleungdo Island, Korea. Dhyani et al. (2018) reported that *L. polyphyllum* has 80 percent geographical distribution in India, thus it represents global population and conservation status.

4.4.2 Primary Gene Pool

Irrespective of all the attempts that have been made before to classify *Lilium* species, the taxonomic work of Comber (1949) is by far found the most comprehensive. He classified the genus *Lilium* into seven taxonomic sections, viz., *Martagon*, *Pseudolirium*, *Liriotypus*, *Archelirion*, *Sinomartagon*, *Leucolirion*, and *Daurolirion*. This classification was based on 15 phenotypic and physiological characteristics, and then Jan de Graaff classified only hybrid lilies into different sections, but most recent classification is based on the genetic makeup. Besides this, information that is relevant to phylogenetic considerations has also emerged from the crossability information of species within and among different taxonomic sections (Lim et al. 2007) as well as meiotic studies. Molecular phylogeny of the genus *Lilium* has also been carried out in addition to all the other approaches adopted for classification.

4.4.2.1 Section *Lilium*

This section includes all European, Turkish, and Caucasian species with the exception of *L. martagon* which belongs to section *Martagon* (İkinci et al. 2006). They have scattered leaves, entire bulb scale, turk's cap-shaped flower (except *L. candidum*), and epigeal (except *L. polyphyllum* and *L. monadelphum*) and delayed (*L. candidum*) germination with numerous entire bulb scales, which bloom during June to July. Involved species in this section are *L. candidum*, *L. carniolicum*, *L. chalcedonicum*, *L. monadelphum*, *L. polyphyllum*, *L. pomponium*, *L. pyrenaicum*, *L. albanium*, *L. ledebourii*, *L. kesselringianum*, *L. jankae*, *L. ponticum*, *L. ciliatum*, *L. akkusianum*, *L. bosniacum*, and *L. szovitisanum*. The species are important from the hybridizer's point of view, as *L. candidum* crosses readily with *L. chalcedonicum* and *L. monadelphum*. *L. polyphyllum* in most of its characteristics are very close to *L. monadelphum*. Interspecific crosses were made between *L. henryi* and *L. longiflorum* with *L. candidum* (Fig. 1).

4.4.2.2 Section *Martagon*

The involved species are *L. tsingtauense*, *L. miquelianum*, *L. distichum*, *L. hansonii*, *L. martagon*, and *L. medeloides*. Most of them are distributed in Korea, Japan, China, Manchuria, and Russia and just have a restricted distribution in the countries and islands along the East Sea of Korea and Japan. China and Korea host *L. distichum* and *L. tsingtauense*. The Russian area around the Vladivostok and up along the Amur River is also the home of *L. distichum*. In terms of geographical distribution and phenotype, *L. hansonii* harbours only in Ulleungdo, Island of Korea, and that is the only known area of native growth of this species (Pelkonen and Pirttila

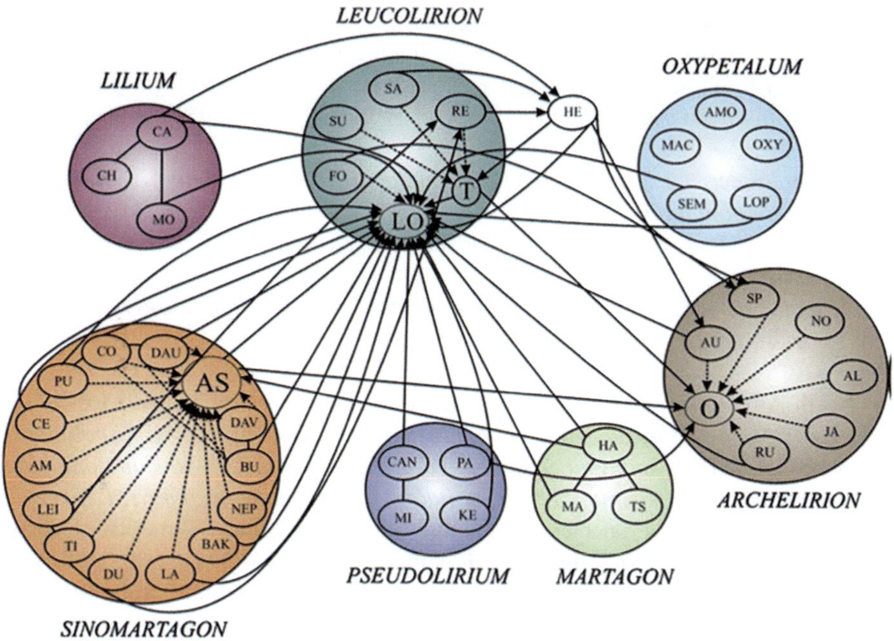

Fig. 1 Crossing polygon of the genus *Lilium* including all the successful crosses of species between different sections of the genus *Lilium*. In this figure, the connections between the Asiatic, Trumpet, and Oriental hybrid groups (large ellipses) are shown by dotted lines. In successful crosses between species (small circles) of different sections (large circles), the arrows point toward the female parent. Abbreviations: AL, *L. alexandrae*; AM, *L. amabile*; AMO, *L. amoenum*; AS, Asiatic hybrids; AU, *L. auratum*; BAK, *L. bakerianum*; BU, *L. bulbiferum*; CA, *L. candidum*; CAN, *L. canadense*; CE, *L. cernuum*; CH, *L. chalcedonicum*; CO, *L. concolor*; DAU, *L. dauricum*; DAV, *L. davidii*; DU, *L. duchartrei*; FO, *L. formosanum*; HA, *L. hansonii*; HE, *L. henryi*; JA, *L. japonicum*; KE, *L. kelloggii*; LA, *L. lankongense*; LEI, *L. leichtlinii*; LO, *L. longiflorum*; LOP, *L. lophophorum*; MA, *L. martagon*; MAC, *L. mackliniae*; MI, *L. michiganense*; MO, *L. monadelphum*; NEP, *L. nepalense*; NO, *L. nobilissimum*; OXY, *L. oxypetalum*; O, Oriental hybrids; PA, *L. pardalinum*; PU, *L. pumilum*; RE, *L. regale*; RU, *L. rubellum*; SA, *L. sargentiae*; SEM, *L. sempervivoideum*; SP, *L. speciosum*; SU, *L. sulphureum*; T, trumpet hybrids; TI, *L. tigrinum*; TS, *L. tsingtauense*

2012). Flowers of the section Martagon are nodding to upright, horizontal, and pendant with a range of 6–12 cm in diameter. They possess a range of perfume-like fragrances as in *L. distichum*, sweet smell as in *L. hansonii*, and no smell as in *L. tsingtauense*. There is no clear criterion for discriminating the four highly related species: *L. distichum* and *L. medeoloides* and *L. tsingtauense* and *L. miquelianum*. All flowers showed pendant thick tepals with slightly unbalanced tepals like *L. distichum* found in other Korean region. Plants are vigorous and tall to reach about 1 m. They possess strong odour, like sweet mixed with other unpleasant smells. *L. distichum*, called Choson lily (Choson is one of Korea's dynasties), is native in China, Russia, and Korea.

4.4.2.3 Section *Pseudolirium*

This section consists of about 21 species having rhizomatous bulbs and whorled leaves. Most of the American species originate from southwestern part of the continent. Involved species are:

2a. *L. bolanderi*, *L. columbianum*, *L. kelloggii*, *L. humboldtii*, *L. rubescens*, *L. washingtonianum*
2b. *L. maritimum*, *L. nevadense*, *L. occidentale*, *L. pardalinum*, *L. parryi*, *L. parvum*, *L. roezlii*
2c. *L. canadense*, *L. grayi*, *L. iridollae*, *L. michauxii*, *L. michiganense*, *L. superbum*
2d. *L. catesbaei*, *L. philadelphicum*

North America is one of the centers of worldwide diversity of lily. Most of the species are distributed along the American West Coast (*L. bolanderi*, *L. columbianum*, *L. kelloggii*, *L. humboldtii*, *L. rubescens*, *L. washingtonianum*, *L. maritimum*, *L. kalleyanum*, *L. occidentale*, *L. pardalinum*, *L. parryi*, *L. parvum*, *L. wigginsii*). A few species are native to eastern North America such as *L. canadense*, *L. grayi*, *L. superbum*, *L. catesbaei*, *L. michiganense*, *L. michauxii*, and *L. iridollae*. In natural condition, the species can be found from sea level (*L. columbianum*) to areas at high elevation of around 3000 m (*L. parvum* and *L. parryi*). Species of North American lily have a wide variation in their flower tepals' color (from white to peach, orange, yellow, pink, red, purple, scarlet), flower shape (Turk's cap-, trumpet-, bowl-, bell-shaped) with or without spotting. Uttermost species are characterized by nodding flowers with the exclusion of two species with erected flowers (*L. philadelphicum* and *L. catesbaei*). Plants' height can range from 30 cm (in *L. catesbaei*) to 3 m (in *L. superbum*). The number of the successful hybridizations that have been done among species of *Pseudolirium* section created the group of American species hybrids. Successful crossing has been made between *L. pardalinum* and *L. humboldtii*, *L. kelloggii* and *L. parryi*, *L. pardalinum* and *L. bolanderi*, as well as *L. canadense* and *L. michiganense*.

4.4.2.4 Section *Archelirion*

Seven species (*L. alexandrae*, *L. auratum*, *L. brownii*, *L. japonicum*, *L. nobilissimum*, *L. rubellum*, and *L. speciosum*) were included in this section. Species of this section are native to Japan with the repudiation of *L. speciosum*, which is also found in Taiwan and South-eastern China (McRae 1998). Most species of this section are resistant to *Botrytis elliptica* and viral diseases. They have generally scattered leaves, distinctly petiolate, entire scales, and trumpet-shaped flowers with hypogeal and delayed germination. In natural condition the species can be found on the margins of hillside woods and well-drained slopes (*L. auratum*), on the steep cliffs of the coast (*L. nobilissimum*) and in shaded and moist places in forest, grassy slopes (*L. speciosum*) in high mountain meadows (*L. rubellum*), and moist places (*L. japonicum*). *L. rubellum* and *L. nobilissimum* are threatened species and are provided with special protection measures in Japan. *L. auratum* is an endemic species in Japan and is an important genetic resource for Oriental hybrids.

4.4.2.5 Section *Sinomartagon*

This section consists near about 30 Chinese species and is divided into three sub-sections. Involved species are:

5a. *L. dauricum*, *L. maculatum*, *L. wilsonii*, *L. bulbiferum*, *L. davidii*, *L. duchartrei*, *L. henryi*, *L. tigrinum*, *L. lankongense*, *L. leichtlinii*, *L. papilliferum*, *L. rosthornii*
5b. *L. amabile*, *L. callosum*, *L. cernuum*, *L. concolor*, *L. pumilum*, *L. fargesii*
5c. *L. arboricola*, *L. bakerianum*, *L. euxanthum*, *L. majoense*, *L. nepalense*, *L. ochraceum*, *L. paradoxum*, *L. poilanei*, *L. primulinum*, *L. sherifiae*, *L. souliei*, *L. stewartianum*, *L. taliense*, *L. wardii*

Interspecific hybridizations among species (*L. bulbiferum*, *L. dauricum*, *L. tigrinum*, *L. amabile*, *L. cernuum*, *L. concolor*) and their hybrids have eased out the release of commercial hybrid cultivars by virtue of being ancestors of the modern Asiatic lily hybrids. The large-scale diversity in phenotypical and physiological aspects in this section is, for example, shown in the flowers with array of colors such as orange, pink, red, purple, yellow, white, and green, upfacing to downward-facing. The flower shape is bowl, flat, and recurved. *L. tigrinum* shows a wide range of habitats in most areas of East Asia, distributed at elevation of 400–2500 amsl. This species shows diploid and triploid plants in the natural habitat. It is very tolerant to the different environmental conditions in China, Korea, and Japan.

4.4.2.6 Section *Leucolirion*

The species in this section is marked by the funnel-shaped, large, white or whitish flowers. The leaves are scattered and with epigeal germination. The six species in this group can be divided into two subgroups by their natural occurrence and morphology: *L regale*, *L. sargentiae*, *L. sulphureum*, *L. leucantum*, and *L. leucanthum* originated from Central and Southern China and *L. longiflorum*, *L. formosanum*, *L. wallichianum*, *L. neilgherrense*, and *L. philippinense* originated from Southern Himalayan regions, Northern India, and Taiwan.

4.4.2.7 Section Daurolirian (*L. bulbiferum* and Dauricum Group)

Lilium bulbiferum is a familiar lily in European gardens. The origin of the cultivars is plausibly Central Europe as the wild forms. The wild forms of *L. bulbiferum* reproduce via seeds, but the domesticated strains do not produce any seeds and are generally propagated vegetatively. The inability for sexual reproduction may be due to strong self-incompatibility that can result from extensive breeding.

In general, wild species within each section are relatively easy to cross, and the hybrids are fertile (Van Tuyl et al. 2002a, b). Fig. 2 describes the crossing polygon of the genus *Lilium* including all successful crosses of species between different sections of the genus *Lilium* developed at Plant Research International, the Netherlands. The interspecific hybrids within the sections especially those within the sections *Leucolirion*, *Archelirion*, and *Sinomartagon* represent the most important breeding groups which are:

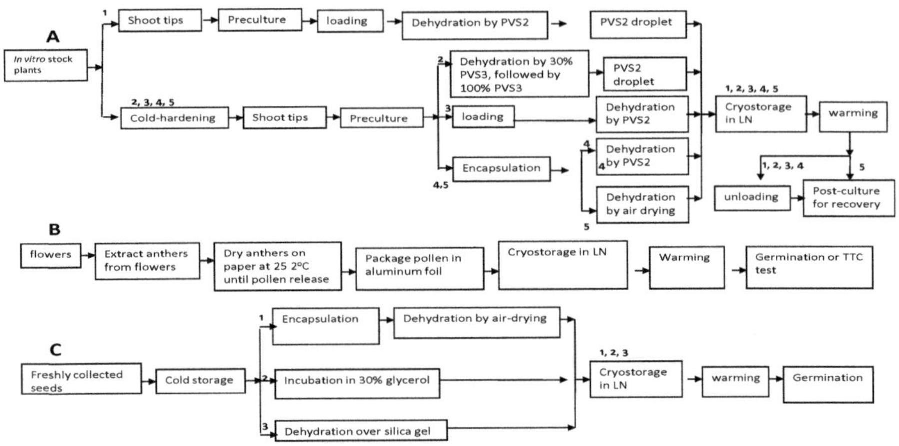

Fig. 2 Major steps of *Lilium* cryopreservation for shoot tips (A), pollen (B), and seeds (C) by droplet vitrification

1. **Longiflorum hybrids (L genome)**: They originate from intra- or inter-specific hybridization with *L. longiflorum* in the Leucolirion section, have trumpet-shaped, pure white flowers, a distinctive fragrance, year-round forcing ability and, mostly outward-facing flowers.
2. **Asiatic hybrids (A genome):** They are derived from interspecific crosses among at least 12 species of the Sinomartagon section. Their cultivation can be traced to the early 1800s in Japan. Cultivars of Asiatic hybrid lily have a wide color variation in their tepals (orange, white, yellow, pink, red, purple, and salmon) and early to late flowering. Some species in this section show resistance to *Fusarium* and viruses.
3. **Oriental hybrids (O genome):** They result from hybridization among five species of the Archelirion section. Generally, Oriental hybrids are late-flowering, with big and showy flowers with a pleasant fragrance (McRae 1998). Some species are resistant to *Botrytis elliptica* that affects most of the lilies from the other sections.

4.4.3 Wild Genetic Resources

There are several wild species of *Lilium* found in and around the place of origin. These wild species are of great value and importance, not only as basic and virtual material for plant taxonomy but also as a starting point for horticulture and hybridization as a seemingly inexhaustible source of hybrid vigor. Fifty species, 1 subspecies, 34 varieties, and 1 forma of *Lilium* are native to China and distributed in 27 provinces and autonomous regions. Yunnan, Sichuan, Tibet, and Guizhou in southwest are the center of wild lilies. Some wild species including *L. lancifolium* and *L. brownii* are distributed in 17 provinces and *L. brownii* var. viridulum in 16 provinces. In Yunnan provinces, about 25 species and 9 varieties are found in wild habitat and grow in thickets and edge of open woodland, at an altitude ranging from 400 to 4800 m above sea level and latitude between 23^0 and 27^0 N (Wu et al. 2012). Species of *Lilium* and their varieties including *L. longiflorum*, *L. japonicum*, *L. rubellum*, *L. alexandrae*, *L. nobilissimum*, *L. auratum*, *L. platyphyllum*, *L. dauricum*, *L. maculatum*, *L. callosum*, *L. lancifolium*, *L. maximowiczii*, *L. medeoloides*, and *L. speciosum* are some wild genetic resources distributed in Japan. *L. pyrophilum* is endemic to a small area in the Florida Panhandle and adjacent Alabama. *Lilium grayi* is a regional endemic of the southern Appalachian Mountains with majority of the global population occurring in North Carolina.

The changing climate is leading to loss of genetic diversity in the wild relatives. The importance of conservation of wild species and their genetic potential have been recognized by many workers. Many species of *Lilium* are threatened by loss of their habitat (Table 2). Recent reports have indicated that six species of *Lilium*, viz., *L. polyphyllum*, *L. pomponium*, *L. chalcedonicum*, *L. jankae*, *L. ciliatum*, and *L. rhodopeum*, have been listed as extinct even in its wild habitat (Saha et al. 2015; Gargano 2015; Lansdown 2018; Bilz 2011; Ikinci 2014 and Petrova and Bazos 2013). Species endemic to China, *L. paradoxum* Stearn, *L. medogense* S. Y. Liang, *L. pinifolium* L. J. Peng, *L. saccatum* S. Y. Liang, *L. huidongense* J. M. Xu, *L. matangense* J. M. Xu, *L. stewartianum* I. B. Balfour et W. W. Smith, *L. habaense* F. T. Wang et Tang, *L. jinfushanense* L. J. Peng et B. N. Wang, *L. xanthellum* F. T. Wang and Tang, and *L. fargesii* Franch., have been put on the China Species Red List (Wang and Xie 2004).

4.5 Collection and Conservation

Germplasm collections serve as an important source for the crop improvement, more so in ornamental crop species like *Lilium* with long duration of juvenile period. The crop genetic resources play a key role in crop development and are considered as the basic materials for germplasm innovations and crop breeding. Genetic materials for ornamental plants are not centrally collected and maintained anywhere in the world. Different institutes and universities at international and national levels collected and conserve the genetic resources of *Lilium*. In China, about 32 species, 1 subspecies, and 7 varieties, such as *L. amabile* Palibin, *L. amoenum* E. H. Wilson ex Sealy, *L. bakerianum* Collett & Hemsley, *L. concolor* var. sinicum Lindley & Paxton, *L. dauricum* Ker Gawler, are conserved at the Institute of Vegetables and Flowers, Chinese Academy of Agricultural Sciences. There are 13 species and 5 varieties at Beijing University of Agriculture, including *L. concolor* var. megalanthum F. T. Wang & Tang, *L. dauricum*, *L. lancifolium*, and *L. leichtlinii* J. D. Hooker var. maximowiczii (Regel) Baker. Various institutes in Shenyang, Kunming, Nanjing, and Yangling (Shaanxi province) maintain a germplasm center for conservation. There are 12 species and

Table 2 Threatened *Lilium* spp. in IUCN Red List

S. no.	Scientific name	Species authority	Red List category	Year published	Population	Geographic range	Country of occurrence	Habitat and ecology	Conservation action	References
1.	*Lilium polyphyllum*	D.Don	Critically Endangered	2015	Decreasing	India, Afghanistan, Pakistan	Afghanistan, India (H.P., J&K, Uttaranchal), Pakistan	North-east aspect with communities of *Cedrus deodara*, terrestrial	In situ and ex situ conservation	Saha et al. (2015)
2.	*Lilium jankae*	A. Kern	Data deficient	2011	Unknown, fragmented	Bulgaria, Romania, and Serbia	Bulgaria; Romania; Serbia	Subalpine belt in communities of *Pinus mugo* and *Juniperus sibirica* and montane and alpine grasslands, terrestrial	European Wildlife and Natural Habitats (Bern Convention)	Bilz (2011)
3.	*Lilium pomponium*	Lis de Pompone, Lis Turban	Least Concern	2015	Decreasing	Southwestern Alps	France and Italy mainlands	Sunny calcareous rocky pastures, terrestrial	Seed conserved in seed banks	Gargano (2015)

4.	*Lilium chalcedonicum*	Linnaeus	Least Concern	2018	Stable	Albania, Greece and the Greek region of Macedonia	Albania; Greece (Greece (mainland)); Macedonia,	Terrestrial	Species occurs in Natura 2000 sites	Lansdown (2018)
5.	*Lilium ciliatum*	P.H.Davis	Endangered B1ab(ii,iii, v) + 2ab(ii, iii,v); C2a(i)	2014	Rare, decreasing	NE Anatolia, Turkey, Zigana Bulancak and areas in Giresun, and Gumuşhane	Turkey	Shrubland and diverse meadows above the treeline, on acid soils, terrestrial	Collection is prohibited in the wild at national level by bulbous plant regulation	İkinci (2014)
6.	*Lilium rhodopeum*	Delip.	Vulnerable B1ab(ii, iii) + 2ab(ii, iii)	2013	Stable	Bulgaria and Greece, native to Rhodope mountains	Bulgaria; Greece (Greece (mainland))	Mesophilous mountain meadows, grassy mountain slopes, terrestrial	Ex situ conservation of seeds and plants	Petrova and Bazos (2013)

5 varieties such as *L. distichum* Nakai ex Kamibayashi, *L. pumilum* Redouté, at the Shenyang Agricultural University. The Flower Research Institute, Yunnan Academy of Agricultural Sciences, has nine species and one variety, including *L. lancifolium*, *L. regale* E. H. Wilson, and *L. duchartrei* Franchet. There are 11 species and 1 variety, including *L. dauricum*, *L. lancifolium*, and *L. pumilum* at the Nanjing Forestry University. There are five species and two varieties, including *L. pumilum* and *L. leucanthum* var. leucanthum (Baker) Baker, at the Northwest Agricultural and Forestry Science and Technology University.

The diverse germplasm collections of Lilium are also available at different global portals and gene banks. Two hundred fifteen accessions are accessible on GENESYS, 102 on EURISCO, and 35 on GRIN portals (Pedapati et al. 2018). Species of economic importance such as *Lilium pumilum*, *L. bulbiferum*, *L. philadelphicum*, *L. pensylvanicum*, *L. distichum*, *L. michiganense*, *L. pyrenaicum*, *L. pardalinum*, *L. iridollae*, and *L. sargentiae* are conserved at PGRC, Germany; Centre for Genetic Resources, the Netherlands Plant Research International; and Research Institute of Pomology and Floriculture, Poland. Ornamental Plant Genetic Conservation (OPGC) (Ohio State University) and Herbaceous Ornamental Germplasm Conservation within the National Plant Genetic System (NPGS) (United States) collected and maintain 11 accessions of *Lilium* at NPGS maintenance site, Ames, Iowa. Columbia conserves the genetic diversity of herbaceous ornamental crop plants including *Lilium*. Fifteen species of *Lilium* that grow naturally are documented in Korea.

Field gene bank of *Lilium* is conserved at Royal Botanic Garden, Kew, and National Botanical garden, Cape Town, South Africa. In India, field gene bank is maintained at Dr. Yashwant Singh Parmar University of Horticulture and Forestry, Nauni, Solan, Himachal Pradesh, and ICAR-Indian Agricultural Research Institute, Regional Station, Katrain, Kullu, Himachal Pradesh. Gaps for wild species are Indian Himalayan Region (IHR), Bosnia and Herzegovina, southwestern China, South Korea, North Carolina, and Florida.

4.6 Conservation

In recent years, the genus *Lilium* is facing a serious threat of genetic erosion like other endangered plant species, due to habitat fragmentation, inbreeding, weaker immune system against pathogens and pests, and climate change. Also, the lily bulbs have to face different storage disease like *Fusarium*, viruses, and gray mold, and different physiological disorders are real threat to *Lilium* germplasm. This may lead to the loss of the bulbs, poor-quality flowers, and virus infection. Many wild species have become endangered species. So, germplasm conservation is very important as a source of genetic variation for breeding and research and to prevent rare species from becoming extinct which required careful attention for its long-term conservation and preservation. The conservation of lily germplasm is divided into two main strategies: in situ and ex situ.

4.6.1 Ex Situ Conservation

4.6.1.1 Cold Storage of Bulbs

Collections of ornamental geophytes are usually preserved in the field or greenhouse by annually planting, harvesting, and storing of the bulbs, with high investments of labour, space, and risk of losses caused by diseases. Due to unique and heterozygous nature of the crop, using seeds for preservation would break up unique genetic combinations, and the characters would show segregation. Increasing the storage duration of the bulbs would make the maintenance of a field collection more efficient. Temperature is the most important factor in lily bulb storage. Normally, the lily bulbs are stored at 4 °C. If a longer storage is required, the temperature must be decreased to -2 °C, and bulbs should be put in moist peat. Lower temperature (less than -2 °C) would further minimize growth conditions, and therefore bulbs could be stored for a maximum time period. Freezing tolerance can be increased by cold acclimatization, abscisic acid treatment, partial dehydration, or low atmospheric pressure. Controlled atmospheric storage (CAS) and modified atmosphere packaging in closed bags (MA package) were also used to decrease the metabolism of clonal material in bulbous flowers. Ion leakage from lily bulbs measured by the electrical conductivity of external solution increased with damage caused by frost, heat, or dehydration and with viability loss during storage. Therefore, ion leakage could be used for measuring the viability and to estimate maximum storage duration of lily bulbs.

4.6.1.2 Conservation in Gene Banks

Gene banks are a type of bio-repository that safeguards genetic material. In case of plants, it could be by freezing cuttings from the plants or stocking the seeds (e.g., in a seed bank). In gene banks, as a means of preserving genetic material, the storage of seed is the preferred method, but clonal germplasm and recalcitrant seed species can also be kept in field plantings as field gene banks. The lily gene bank at Plant Research International has maintained several thousands of lily genotypes for more than 35 years.

4.6.2 In Situ Conservation

4.6.2.1 Tissue Culture as In Vitro Gene Banks

An alternative method to preserve lily collections is conservation in vitro. In vitro stored collections need relatively small amounts of space, medium components can be used that minimize growth, plants can be multiplied quickly, and there is often a possibility to eliminate viral diseases. Lily material could be stored more than 28 months at 25 °C before transfer to new medium is required; however, at each transfer event, there is a risk of contamination with microbial organisms. Furthermore, the establishment of an in vitro collection is labor-intensive, and genotypes may react differently under identical conditions. A combination of high sucrose concentration and a low salt concentration proved to minimize metabolic activity, the best for in vitro storage of lily.

4.6.2.2 Cryopreservation in Liquid Nitrogen

Cryopreservation is a long-term low temperature storage technique to preserve any biological material in liquid nitrogen (LN, – 196 °C) or liquid nitrogen vapour (LNV, approx. – 165 to –190 °C) without deterioration for at least several thousands of years. Cryopreservation of biological tissues can be successful only if intracellular ice crystal formation is avoided. Crystal formation can be disallowed through vitrification. Two requirements must be met for a cell to vitrify: rapid freezing and a concentrated cellular solution. Sugars play a very important role in the acquisition of resistance to desiccation and to freezing in liquid nitrogen. Kaviani et al. (2009) reported 75% survival of *Lilium Iedebourii* (Baker) Bioss seeds treated with sucrose and dehydration. Shoot tips of adventitious shoots derived from bulb-scale segments of five *Lilium* genotypes *(L. longiflorum*, *L. longiflorum* × Oriental 'Triumphator', L. Oriental hybrid 'Siberia', *Lilium* Asiatic hybrid 'Elite', and *L. davidii* var. unicolor) were cryopreserved by droplet vitrification (Yi et al. 2014). Major steps of *Lilium* germplasm cryopreservation by pollen, seeds, and shoot tip by droplet vitrification reported by Li et al. (2019) are shown in Fig. 2, and a list of successful cryopreservation of *Lilium* spp. shoot tips, pollens, and seeds are presented in Tables 3 and 4.

4.7 Characterization and Evaluation

4.7.1 Characterization for Essential Features and Classification

Germplasm characterization is the description of each accession of a collection with a set of stable (non-plastic) plant characters that are unique to the crop. Morphological characterization is the easiest and simplest characterization process that allows for the study of plant variation using visual attributes, variability for growth, and development parameters and helps in identifying elite varieties as well as superior donor parents for different floricultural traits. Plants in genus *Lilium* have evolved with a fair amount of phenotypic diversity despite the genetic similarities, and this diversity of characteristics is of considerable evolutionary significance. Comber (1949) exploited 15 morphological characteristics (e.g., seed germination form, leaf order, leaf shape, species size, bulb form, flower nectary, flower form, bulb color, stem property, stigma size, appearance of stem root, etc.) to classify naturally grown lilies into seven sections. Experts familiar with the crop develop a crop descriptor. They select plant characters, which are stable and show variations among accessions in a collection or genus. Stable characters are those diagnostic features of the plant, e.g., leaf shape, that remain the same when grown in different environments. However, other characters such as plant height and number of days to flowering, which are affected by the environment and physiology, are also used. IPGRI (1995) and the International Union for the Protection of New Varieties of Plants (UPOV) have developed standardized descriptors for most

Table 3 A list of successful cryopreservation of shoot tips of *Lilium* spp.

Species or hybrids	No. of genotypes tested	Cryo-methods	Shoot re-growth (%)	References
L. japonicum	1	Encapsulation-dehydration	88	Matsumoto and Sakai (1995)
L. japonicum	1	Encapsulation-vitrification	94	Matsumoto and Sakai (1995)
L. japonicum	1	Vitrification	92	Matsumoto and Sakai (1995)
L. longiflorum, *L. formosanum*, *L. henryi*, *L. auratum*, Oriental hybrids and Asiatic hybrids	2 cultivars for *L. longiflorum*, one for each of *L. formosanum*, *L. henryi*, *L. auratum*, 3 Oriental hybrids and 2 Asiatic hybrids	Vitrification	53–89 for *L. longiflorum*, 49–71 for *L. formosanum*, 5–19 for *L. henryi*, 51–66 for *L. auratum*, 53–93 for three Oriental hybrids and 53–89 for two Asiatic hybrids	Bouman et al. (2003)
Oriental hybrids 'Siberia'	1	Vitrification	51	Chen et al. (2007)
L. lancifolium, *L. x longiflorum* and Oriental hybrid 'Siberia'	3	Droplet vitrification	67 for *L. lancifolium* and 35 for *L. x longiflorum* and 60 for Oriental hybrid 'Siberia'	Chen et al. (2011)
Lilium spp., Asiatic hybrid and Oriental hybrid	20 *Lilium* species, 1 Asiatic hybrid, and 1 Oriental hybrids	Droplet vitrification	52 for *L. amabile*, 67 for *L. auratum*, 43 for *L. bulbiferum*, 67 for *L. callosum*, 62 for *L. candidum*, 42 for *L. concolor*, 85 for *L. davidii*, 41 for *L. distichum*, 48 for *L. formolongi*, 86 for *L. hansonii*, 38 for *L. henryi*, 41 for *L. lancifolium*, 40 for *L. leichtlinii*, 69 for *L. leucanthum*, 43 for *L. longiflorum*, 41 for *L. pyrenaicum*, 83 for	Yi et al. (2014)

(continued)

Table 3 (continued)

Species or hybrids	No. of genotypes tested	Cryo-methods	Shoot re-growth (%)	References
			L. regale, 43 for *L. tsingtauense*, 41 *L. washingtonianum,* 43 for Asiatic hybrid, 73 for Oriental hybrid	
L. davidii var. *unicolor*, *L.* × *formolongi, L. longiflorum* × Oriental 'Triumphator', Oriental hybrids and Asiatic hybrids	3 species, 2 Oriental hybrids, and 2 Asiatic hybrids	Droplet vitrification	80 for *L. davidii* var. *unicolor*, 53 for *L.* × *formolongi,* 43 for *L. longiflorum* × Oriental Triumphator', 73 for Asiatic hybrids 'Elite', 68 for Asiatic hybrids 'Pollyanna' and 88 for Oriental hybrid 'Siberia'	Yin et al. (2014)
L. martagon	1	Droplet vitrification	81–89	Urbaniec-Kicpura and Bach (2017)

Table 4 A list of successful cryopreservation of *Lilium* pollens and seeds

Species or hybrids	Cryo-methods	Survival or germination (%)	References
Pollens			
Oriental hybrid 'Siberia'	Rapid cooling	59	Xu et al. (2014)
Oriental hybrid 'Siberia'	Vitrification	70	Zhao et al. (2014)
L. concolor var. pulchellum	Desiccation + rapid cooling	21–92 (according to the conservation duration)	Zhao et al. (2014)
Seeds			
L. ledebourii	Pregrowth dehydration	75	Kaviani et al. (2009)
L. ledebourii	Encapsulation-vitrification	10	Urbaniec-Kicpura and Bach (2014)
L. ledebourii	Vitrification	97	Mohajeri et al. (2014)
	Desiccation	95	
	Encapsulation-dehydration	69	
	Glycerol pretreatment	98	
L. martagon	Dehydration	93–100	Gudeva and Tarj Kova (2015)

food and industrial crops. UPOV standards also cover many ornamental crops, including *Lilium* (UPOV 2005).

The biochemical markers using isozyme profile and cluster analysis based on allele frequencies were used to reveal genetic variation and show relationships within and between species and cultivars. The combined morphological and molecular techniques were used for delimiting native and exotic varieties, clustering varieties/putative hybrids based on geographical origin within country and from outside country, and confirming the phylogenetic relationship and geographic distribution of different *Lilium* spp. Principal component analysis for the morphological characters between inland and island reveals partly related to their geographical origins. Quantitative traits were also found important in determining the groupings and working out the relationships among the various cultivars and hybrids. Ornamental characteristics and potential utilization of nine *Lilium* species evaluated using analytic hierarchy process reveal that species *L. lancifolium*, *L. cernuum*, *L. pumilum* and *L. concolor* var. buschianum had better ornamental value and utilization potential (Rong et al. 2011). The species *L. lophophorum* possesses a yellowish-green flower for good breeding potential. *L. martagon* var. pilosiusculum has whorled leaves; *L. cernuum*, *L. amoenum*, and *L. wardii* have obvious purple flowers; and *L. fargesii* has greenish-white flowers, which make these species good parents for breeding plants with colored flowers. *L. leichtlinii* var. maximowiczii, *L. pumilum*, *L. davidii*, and *L. davidii* var. willmottiae have strong propagation abilities, and *L. davidii* and *L. davidii* var. willmottiae have been widely used as food (Du et al. 2014).

4.7.2 Development/Identification of Gene Pools and Core Collections

The core collection concept was proposed in the 1980s with the goal to minimize the cost of germplasm conservation while ensuring the preservation of maximum genetic diversity due to the rapid increase in the number of accessions in collections of major food crops like wheat, rice, corn, potatoes, etc. A set of criteria must be established to select a "core" collection of species that captures the maximum amount of genepool of the genus. Some of the selection criteria proposed are:

1. Specific unique genotypes
   - Pest and disease resistance
   - Stress tolerance, e.g., drought, frost, and heat tolerance
   - Adaptation characteristics, e.g., cultural input efficient genes
   - Physiological characteristics, e.g., photoperiod response
   - Specific marketing/commercial traits, e.g., aesthetic genes and shelf-life genes
2. Cultivar groups and genetic diversity
   - Morphological and DNA data
3. Geographical and ethnological distribution
4. Balance representation of related species (secondary and tertiary genepool as described in the above section on "genepool concept")

Van Hintum (1996) modified the concept and defined it as "a germplasm collection optimally representing specific genetic diversity" to allow flexibility in the assembly of core collections and to justify the formation of multiple core collections of a target species in space and time. This definition is equivalent to breeder collections where individual breeders assemble and manage their own distinct collections. In recent years, many approaches including random sampling, stratified sampling, phenotypic analysis, genetic markers, and coefficient of parentage have been proposed to establish core collections.

Harlan's primary (1°), secondary (2°), and tertiary (3°) crop genepool concepts was used to define the targeted germplasm for collection and conservation. The total genepool of a crop is complex especially when the species concerned is of hybrid origin. Germplasm of direct progenitors and all the contributing parents and related species have to be collected and conserved. In a gene bank, germplasm is stored as seeds in a seedbank, as living plants in a field gene bank, as propagules such as bulbs in humidified coolers, as meristem culture in in vitro collections, and seed, dormant buds, and tissue culture in liquid nitrogen cryopreservation banks. Table 5 provides a summary of the species of all levels of Harlan's genepool that need to be conserved in *Lilium*.

4.7.3 Evaluation of Genetic Diversity for Desired Traits

Evaluation of genetic diversity is the screening of a collection for specific genes, such as resistance to important pests and diseases, abiotic stresses, adaptation, and

Table 5 The primary (1°), secondary (2°), and tertiary (3°) genepools and origins of some common herbaceous ornamental plants (Herbaceous Ornamental Crop Germplasm Committee Report, Sept. 1, 1995, http://www.ars-grin.gov/npgs/cgc_reports/herbscgc1995.htm)

Crop	Origin	1° Genepool	2°/3 ° Genepool
Lilium (70)	N. America, Asia, Europe		
American	N. America	*L. Parryi*, *L. pardalinum*	
Asiatic	East Asia	Complex hybrids	
Candidum (Madonna)	Europe	*L. candidum*	
Longiflorum (Easter Lily)	East Asia	*L. longiflorum*	
Martagon	Europe and East Asia	*L. martagon*, *L. Hansonii*	
Oriental	East Asia	*L. auratum*, *L. speciosum*	
Tiger	East Asia	*L. lancifolium*, *L. tigrinum*	
Trumpet	East Asia	As Asiatic	

other commercial traits. These traits are included in all well-formulated standard crop descriptors, e.g., IPGRI (1995). The prioritization of the traits for evaluation depends on researcher needs and, thus, is location-specific. *Lilium* possess great genetic diversity in its growth habit, flower color, form, shape, size, as well as persistence. This diversity in species of agronomic traits offers a substantial germplasm and opportunities for the development of hardy and healthy varieties for variable climatic zones (Anderson et al. 2010).

Cluster analysis in 12 populations of *Lilium longiflorum* employing morphological and Grower's similarity coefficient indicated three phenomes, and the observed variation was the result of genetic and environmental variations. Significant genotypic x environmental variation throughout the entire life cycle exists within *L. formolongi* hybrids which would complicate flowering of this crop for specific weeks or target holidays. Combining ability and gene action study of ten quantitative traits in *Lilium* × *formolongi* reveals that the diameter of stem, number of leaves, width of leaf, number of flowers, diameter of flower, and length of bud predominantly showed dominance variance, whereas the height of plant, length of leaf, days to flowering, and attitude of floral axis demonstrated additive gene action (Rai et al. 2018). Plant Research International conducted research for the evaluation of disease resistance against *Fusarium* and *Botrytis*, and useful screening techniques have been developed. A high degree of *Fusarium* resistance and strong propagation ability were found in some Asiatic hybrid cultivars and species like *L. dauricum*.

The morphological characters of some wild lily species showed morphological diversities in bulb, stem, leaves, and flower organs. Some subtle features were observed, such as color variation of bulbs, nectar, and colors of filament and style. Two ecotypes of yellow flower *L. leucanthum* were found for the first time. The color variation of tepals and trichomes of *L. pumilum* was also recorded. The genetic diversity of Zunyi and Mouding population of *L. sargentiae* generally came from individual plants instead of populations which were also observed in *L. nepalense*. Considerable variation exists in different populations of *Lilium ledebourii* for different agro-morphological characteristics. Agro-morphological and multivariate analysis of major quantitative traits (six vegetative and nine reproductive traits) of the Oriental lily seedling population and *L. ledebourii* (Sayadalian et al. 2014) reveals considerable variations among the studied characteristics which can be taken into account for the selection of plants of desirable traits.

4.7.4 Available Sources of Breeding Value

A wide diversity of Lilium genetic resources has provided valuable genes for breeding of novel cultivars. Many important horticultural characters are present in the different *Lilium* species. Great importance had been continuously devoted to *L. lancifolium*, *L. cernuum*, *L. pumilum*, and *L. concolor* var. buschianum. *L. leichtlinii* var. maximowiczii had no special traits in ornamental characteristic but has strong propagation ability and high medicinal and edible value and can be used in *Lilium* breeding program. Some species like *L. dauricum* has outstanding *Fusarium* and leaf blight resistance. *L. distichum* has a certain shade tolerance. *L. concolor* var. megalanthum has some wet endurance. Liang et al. (2018) examined the phenolic compounds and antioxidant capacities in different organs of wild *Lilium pumilum* and found that epicatechin was the most abundant phenolic compound, and salicylic acid was the most abundant phenolic acid. Some well-known examples of valuable characters among species are listed in Table 6.

4.7.5 Molecular Characterization of Genetic Resources

Genetic diversity is a prerequisite for successful plant breeding. Characterization and assessment of genetic relationships among diverse accessions allow breeders to select desirable genes from different sources and to accumulate those genes in one cultivar. A lot of approaches have been applied in plant genetic diversity analysis which includes the use of morphological, physiological, biochemical, and agronomical DNA-based markers. But the selection of approaches is decided by objectives, required information, and resources. Several attempts have been undertaken to utilize molecular markers in lily, for example, using restriction fragment length polymorphisms (RFLPs) to trace the parents of hybrids, RAPD (randomly amplified polymorphic DNA) for genetic fidelity tests and diversity analysis and to study the variation pattern from different latitudes and different geographical locations, inter simple sequence repeats (ISSRs) to detect mutants (Xi et al. 2012), simple sequence repeats (SSRs) to assess gene flow among populations (Wang et al. 2011), and SNPs to identify suitable loci to construct a mapping population (Shahin et al. 2012). The genetic relationships among 13 *Lilium* species/cultivars and genetic

Table 6 Characteristics and some useful traits for the commercialization of *Lilium* species

Species	Characteristics for breeding	
	Desirable	Undesirable
L. longiflorum	Low temperature tolerance, flower shape, white	Susceptible to *Fusarium*, virus
L. formosanum	Year-round forcing, upright, growth vigor, fragrance	Weak stem, virus susceptible
Trumpet hybrid	Upright, yellow color, fragrance, flower type	Susceptible to *Fusarium*, virus, weak stem
L. nepalense	Pea-green flower color	Susceptible to virus, late flowering
L. henryi	Growth vigor, virus, and *Fusarium* resistance	Flower shape, weak stem
L. concolor	Upright flower, flower shape and size	Weak stem, leaf and growth vigor
L. tigrinum	Growth vigor, resistance to virus, large flower, bulbil formation, resistance to *Fusarium*	Hair, spots
L. callosum	Small, many flowers per stem, flower color	Late flowering, weak growth vigor
L. davidii	Resistance to *Fusarium* and virus	Short stem
L. dauricum	Fusarium resistance	Short plant height
L. auratum	Large flower, fragrance, growth vigor, disease resistance, early flowering	*Fusarium* susceptible
L. speciosum	Pink color, fragrance	Spots, late flowering
L. nobilissimum	Pure white flower, fragrance, sturdy stem, upright	Late flowering
L. rubellum	Very early flowering, pink flower color, fragrance	Short stem, susceptible to *Fusarium*
L. candidum	Low-temperature and low-light intensity tolerance, pure white, fragrance	Susceptible to virus, weak growth vigor
L. candidum	Low-temperature and low-light intensity tolerance, pure white, fragrance	Susceptible to virus, weak growth vigor
L. hansonii	Many flowers, long vase life	Flower fragrance, short stem, weak growth vigor, susceptible to virus
L. martagon	Purple, small flower, long vase life, many flowers	Fragrance, susceptible to virus
L. tsingtauense	Resistance to *Botrytis*	Short stem, weak growth, fragrance

diversity among 2 populations of *L. regale* E. H. Wilson, cultivar of 'Regale' and 'Regale Album', and genetic diversity among 2 populations of *L. henryi* Baker and cultivar of 'Henryi' are assessed through target region amplification polymorphism (TRAP) polymorphic marker technique (Hu et al. 2020). Simple Sequence Repeat (SSR)-based studies have focused on characterizing the genetic diversity and structure of *L. philadelphicum* in the mid-western United States and of *L. japonicum*

Thunb. var. abeanum (Honda) Kitam, one of the rarest plants in Japan (Kawase et al. 2010). Yong et al. (2019a, b) characterized a MYB-related homolog (LlMYB3) nuclear protein from tiger lily (*Lilium lancifolium* L) as a positive regulator in plant stress resistance and might be involved in anthocyanin biosynthesis pathway in response to cold stress. Jeknic et al. (2014) cloned the gene for capsanthin-capsorubin synthase (Llccs) from flower tepals of *L. lancifolium* cv. 'Splendens' by the rapid amplification of cDNA ends (RACE) with a heterologous non-degenerate primer that was based on the sequence of a gene for lycopene b-cyclase (lcyB).

Abe et al. (2002) used RAPD and ISSR markers to construct parental linkage maps with 95 and 119 markers, respectively, in a cross between the two Asiatic cultivars Montreux and Connecticut King to elucidate the genetics of floral anthocyanin pigmentation. AFLP markers are used in an Asiatic backcross population to map disease resistance against two important diseases: *Fusarium oxysporum* and LMoV (lily mottle virus). Four quantitative trait loci (QTL) for *Fusarium* resistance were identified, whereas LMoV resistance was controlled by a single locus. Shahin et al. (2011) used three different molecular marker systems (AFLP, DArT, and NBS profiling) for generating linkage maps for Connecticut King. Additionally, resistance to lily mottle virus (LMoV) was mapped as a locus on LG AA10. For *Fusarium* resistance, the Kruskal-Wallis test identified six putative quantitative trait loci (QTL) in the AA population of which one QTL (explaining 25% of the variation in resistance) could be confirmed by interval mapping. Complete chloroplast genomes were used to explore the phylogeny of *Lilium* genus (Du et al. 2017). They have sequenced nine *Lilium* chloroplast genomes and retrieved seven published chloroplast genomes for comparative and phylogenetic analyses. Biswas et al. (2018) developed and characterized a novel set of SSR markers from candidate TF-gene transcriptome sequences. They confirmed that these TFSSR markers provide valuable information on the level of polymorphism and diversity in lily. They also identified a set of TFSSR markers that are useful for hybrid identification of lily. A total of 30 EST-based simple sequence repeat (EST-SSR) markers derived from trumpet lily (*Lilium longiflorum*) were used across 11 native lily species for their genetic relationship (Kumari et al. 2019). Among these 30 markers, 24 SSR markers that showed polymorphism were used for evaluation of diversity spectrum.

In lily, transformation efficiency is quite low making it very laborious to generate a population of transgenic plants. Transgenic pollen-less lily plants were developed by introducing *rol* genes in *L. longiflorum*. Azadi et al. (2010) developed two transgenic plantlets of *Lilium* x *formolongi* engineered by *Agrobacterium*-mediated transformation with the plasmid pCrtZW-N8idi-crtEBIY, which contains seven bacterial enzyme genes of carotenoid biosynthesis pathway, as shown by their orange leaves. *Lilium* cv. 'Acapulco' were transformed with a defective cucumber mosaic virus replicase gene, and four plant lines were found to show increased levels of virus resistance. Wang et al. (2012) compared the *Agrobacterium*-mediated transformation efficiencies of eight ornamental lily cultivars, and the highest was 1.4% for callus regenerated from filaments of the cultivar 'Santander'. Lily transformation was achieved by gene gun co-bombardment using both a pBluescriptbased vector containing the cystatin gene and pDM307 that contains a bar gene for phosphinothricin selection. Both soybean hairy roots and lilies overexpressing the OcIDD86 transgene exhibited enhanced resistance to *Pratylenchus penetrans*. Two stable and efficient genetic transformation systems based on somatic embryogenesis and adventitious bud regeneration were established in two *Lilium* species (*Lilium pumilum* DC. Fisch. and the *Lilium longiflorum* 'White Heaven'). Transgenic plants and T-DNA insertion lines were confirmed by β-glucuronidase (GUS) assay, polymerase chain reaction (PCR), and Southern blot. The LpPDS gene in the two *Lilium* species was knocked out by CRISPR/Cas9 (clustered regularly interspaced short palindromic repeats (CRISPR)-associated protein 9) technology (Yan et al. 2019). Vieira et al. (2015) reported that expression of a cystatin transgene in *Lilium longiflorum* conferred resistance to the root lesion nematode *Pratylenchus penetrans*. Resistance to herbicide was obtained in transgenic *Lilium longiflorum* expressing the bialaphos-resistance gene PAT under the constitutive CaMV35S promoter.

4.8 Use of Plant Genetic Resources

4.8.1 Major Constraints in the Crop Production

A wide diversity of *Lilium* genetic resources has provided valuable genes for breeding of novel cultivars. There are more than 300 new registrations of *Lilium* cultivars each year and approximately 10,000 cultivars in total globally. There is a demand for novel lily cultivars from the production industry. Lilies have a wide variety of valuable characters such as flower size, color, flowering time, and resistance to different pathogens. Combining these vital horticultural traits into one cultivar by crossing is almost the only way to obtain introgression of traits, since genetic transformation approaches are not well developed for lily yet. Possibilities for cross-combinations in *Lilium* between the species of the seven sections are limited by incompatibility and incongruity which are due to pre-fertilization and post-fertilization barriers. To overcome these barriers, integrated methods such as grafted style, in vitro pollination, embryo rescue, and ovule culture techniques are required (Van Tuyl et al. 1991). Using these methods, many lily interspecific hybrids have successfully been made. However, most of these interspecific hybrids tend to be sterile (Van Tuyl et al. 2002a, b). Chromosome doubling and 2n gametes (gametes with somatic chromosome numbers) have been used to restore the fertility of interspecific hybrids in lily.

4.8.2 Common Sources Used to Overcome Production Constraints

Until relatively recently, most of the cultivars in all the three major hybrid groups of lilies were diploid ($2n = 2x = 24$), and most importantly, they were intrasectional species hybrids such as Longiflorum (L), Asiatic (A), Oriental (O), Trumpet (T), etc., in which closely related species were involved as parents. However, in recent years intersectional species hybrids such as LA, OT, and LO involving distantly related cultivars/species are most successfully cultivated. These are not only hybrids between the cultivars of different sections, but also they are polyploids possessing distinctly differentiated genomes (i.e., allopolyploids). Obviously, these allopolyploid hybrids are most ideal for combining desirable horticultural traits available in the distantly related cultivars/species into new cultivars.

A recent study of the LA group of the Dutch lily cultivars has shown that they are predominantly triploid ($2n = 3x = 36$) with one genome of L and two genomes of A (i.e., LAA) constitution. The predominance of triploid cultivars also indicates that this particular ploidy level is the most ideal threshold for a successful cultivar. An important drawback of triploid cultivars is that they are not suitable for use in further breeding. Therefore, starting from diploid intersectional cultivar/species hybrids (e.g., LA, OA), different breeding strategies can be envisaged including (1) breeding at the diploid level through the production of diploid BC progenies; (2) use of unreduced (2n) gametes to produce triploid cultivars (unilateral sexual polyploidization); (3) use of bilateral sexual polyploidization; and (4) use of somatically doubled allotetraploids.

4.8.3 Breeding Options

4.8.3.1 Breeding at the Diploid Level

Hybrids between two diploid cultivars/species from two different taxonomic sections are, in almost all cases, completely sterile, and they cannot be used as parents. Rarely however, such F_1 hybrids do produce normal haploid gametes, and they can be utilized to generate diploid BC progenies as has been demonstrated in the case of LA hybrids (Khan et al. 2009). This has opened up the prospect of breeding and selection at the diploid level after which the selected genotypes can be used to produce triploid cultivars. For this synthesis, either unilateral sexual polyploidization or diploid (synthetic) or tetraploid crossing can be used.

4.8.3.2 Use of Unilateral Sexual Polyploidization

Although F_1 hybrids between cultivars of different sections are mostly sterile, a small percentage of them can produce 2n gametes in reasonable frequencies. It is easy to detect such genotypes because the presence of larger pollen grains is a reliable indication. Detection of genotypes that produce 2n eggs is more difficult because it requires crossing with normal pollen and testing whether it leads to fruit and seed set.

One of the important advantages of using unilateral sexual polyploidization is that the resulting progenies possess intergenomic recombination. As a consequence of the presence of recombinant segments in the BC_1 progenies, there is scope for the expression of recessive loci that might become nulliplex in the BC_1 triploids (Barba-Gonzalez et al. 2005). This means that selection can be effective in the BC_1 generation itself. A cardinal feature of 2n gametes in distant hybrids such as LA/OA genotypes is that they originate predominantly through first division restitution (FDR), and because of this the heterozygosity of the parental hybrids is largely preserved in the 2n gametes and transferred to the progenies. This is in contrast to the use of 2x gametes derived from the somatically doubled tetraploids, which can lead to "inbreeding depression" giving rise to weakly performing polyploid progenies.

The important advantages of using 2n gametes are that they help to overcome the F_1 hybrid sterility, transfer hybrid vigor, facilitate intergenomic recombination, and directly give rise to triploid progenies that are preferred for cultivar selection.

4.8.3.3 Bilateral Sexual Polyploidization

As was pointed out earlier, only very few genotypes of F_1 Lilium hybrids produce either 2n pollen or 2n eggs in reasonable frequencies, but none of the genotypes that have been examined so far produce both types of 2n gametes in appreciable frequencies. Therefore, it has never been possible to obtain a tetraploid through the functioning of 2n pollen and 2n egg from the same diploid hybrid parent. However, by using two separate LA hybrids as parents, one donating 2n egg and the other 2n pollen, it has been possible to produce tetraploid progenies (Khan et al. 2010). Such bilateral sexual (tetraploid) progenies have certain advantages for using them as parents in breeding. In the first place, they are expected to be reasonably fertile because of their allotetraploid constitution (LLAA). Secondly, they will not have the drawback of being "permanent hybrids" in which no recombination can occur. On the contrary, because they have originated through 2n gametes, these allotetraploids do possess recombinant segments in some pairs of chromosomes. This means such genotypes have the potential for segregation of genetic traits that might be present on the distal parts of the crossover segments. Thus, there is scope for selection of sexual tetraploids, which may be repeatedly used as parents in order to produce triploid progenies.

4.8.3.4 Somatic Chromosome Doubling

Doubling of chromosome number and production of allopolyploids that are mostly fertile is one of the extensively used methods for overcoming the F_1 sterility of distant hybrids. One of the shortcomings of allopolyploids produced through somatic doubling is that there will not be any intergenomic recombination between the parental genomes due to autosyndetic pairing of chromosomes (Lim et al. 2000). Nevertheless, this method has been successfully used for producing polyploid cultivars of lily.

Apart from the use of colchicine or oryzalin, the use of nitrous oxide (N_2O) for chromosome doubling in the germinal cells has proven to be effective in producing 2n gametes with certain amount of intergenomic recombination (Barba-Gonzalez et al. 2006). This method can be a substitute in order to induce 2n gametes in those genotypes that normally never produce such gametes (or only in very low quantities). The potential of this method must be further evaluated for large-scale application. So far, lily breeding has been carried out through traditional methods. These approaches are time-consuming, especially in this crop, because its generation time is about 2–3 years from seed germination to maturity of fruits and seeds. In such situation, it is attractive to apply molecular methods of tagging desirable traits and practice the so-called marker-assisted selection, which might reduce time. In this context, linkage maps have been constructed using amplified fragment length polymorphism (AFLP) and diversity array technology (DArT) markers (Shahin et al. 2010). Like in other crops, these molecular methods might be potentially useful in lily as well.

4.8.4 Present Status of Use or Incorporation of Desired Traits

Aside from morphological or phenotypic resources, few molecular aspects have also been investigated in *Lilium*. Zhang et al. (2008) characterized one of the protease inhibitors, a trypsin inhibitor (17 kDa), in the bulb of *Lilium brownii*. The amino acid sequence of this protease showed similarity to a short fragment of a known trypsin inhibitor from *Populus tremula* and a presumed trypsin inhibitor from *Arabidopsis thaliana* and sporamin B from sweet potato. Trypsin (protease) inhibitors are quite important compounds due to their role as defense proteins against pests.

Another compound "free mannose" has been indicated in *L. longiflorum* bulbs. In addition to the starch, glucomannan is known to be the main storage of carbohydrate in *Lilium* that has been recorded to encompass about 15% of the carbohydrate in the bulb. "Free mannose" is a water-soluble polysaccharide that is considered as a dietary fiber and used in food as an emulsifier and thickener.

Many studies were carried out in order to understand the complex process of transportation of sperm to egg cell keeping in view the importance of sexual reproduction mechanism in plants. Kim et al. (2003) identified a chemotropic molecule "chemocyanins" in lily stigma, which is a small basic protein that shows sequence similarity to plantacyanins, cell wall proteins of unknown function, and that belongs to the ancient phytocyanin family of blue copper proteins. Genes that encode histone proteins have been reported in male gametic cells within the pollen grain of *L. longiflorum*. Histones are highly conserved throughout the evolution and are encoded by multigene families. H_3 and H_2B have been identified as potential tissue-specific histones in the generative cell of lily. These two genes (gcH_2A and gcH_3) are expressed specifically in the generative cell but not in microspores undergoing pollen mitosis-I or in other dividing cells of *Lilium* somatic tissue (Xu et al. 1999).

Many physiological processes in plant cells are highly correlated with actin cytoskeleton, such as the elongation of pollen tubes tips. The dynamics of actin cytoskeleton Rop1Ps and its importance for pollen tube elongation characteristics in *L. davidii* were investigated. Many compounds were detected in the pollen including ABP29, LlLIM1, and LLA23. *Lilium* ACTIN BINDING PROTEIN29 (ABP29, 29 kDa) is the smallest identified member of the villin/gelsolin/fragmin superfamily, and it is a splicing variant of *Lilium villin* that plays important roles in remodeling of the actin cytoskeleton (Xiang et al. 2007). LlLIM1, also an actin-binding protein (ABPs), was identified in *L. longiflorum* pollen. It plays an important role in regulating the actin microfilaments, which is essential for polar cell tip growth

(Wang et al. 2008). LLA23 is an abscisic acid-, stress-, and ripening-induced protein, which was isolated from *L. longiflorum* pollen.

Two other MADS box genes (LMADS2 and EgMADS1) characterized in *L. longiflorum* showed expression in the carpel, mainly in ovules and partly in style tissues, whereas they were absent from other flower organs or vegetative leaves (Tzeng et al. 2002). LMADS3 and LMADS4 are two AGL2-like MADS box genes also characterized in *L. longiflorum*, and their expression was detected in the inflorescence meristem and floral buds at different developmental stages and in all four whorls of the flower organ. LMADS5, LMADS6, and LMADS7 have been isolated and characterized from *L. longiflorum* (Chen et al. 2008). The expression of these three genes was similar, and their RNAs were detected in vegetative stem and inflorescence meristem.

Very little sequence information is available for *Lilium*. Some regions of *Lilium*'s chloroplast DNA (trnT–trnL, trnL–trnF, and atpB–rbcL) have been sequenced. These sequences were used to study the phylogenic relationships among different cultivars of the *Lilium* and *Archelirion* sections (Nishikawa et al. 2002). In the near future, a considerable amount of DNA sequences will be available as they are currently regenerated for genetic mapping purposes.

4.9 Future Perspective

The understanding of functioning of flowering genes at the cellular level and in the plant as a whole is the present-day challenge in the research of geophytes' floral biology. It is crucial in order to understand the molecular regulation of flowering diversity under various environmental conditions and during flower senescence. One of the reasons that only limited molecular information is available on lily is the huge size of the genus's genome. Investigations in the coming decade are expected to improve our understanding of this process, thus contributing to the development of breeding techniques and the production of new ornamental hybrids.

In addition to all other horticultural traits, introduction of disease resistance into new cultivars of lily has to be taken into thoughtful consideration by breeders, if it has to survive as a crop. Fortunately, the genus *Lilium* includes about 100 species that are distributed widely in the northern hemisphere reaching up to some tropical areas as well. This indicates that there might be scope to discover more useful genetic variation than hitherto has been done. Keeping this in mind, the given lines of work may be contemplated for the future:

(1) More in-depth screening of wild germplasm for (partial) disease resistance
(2) Interspecific hybridization
(3) Screening for haploid and 2n gametes in distant hybrids
(4) Perfecting methods for polyploidization
(5) Development of molecular linkage maps and tagging of useful traits

1. There is a wide diversity within and between wild species and the current cultivated assortment. For many traits (disease), it has not been yet fully explored leaving significant perspective for crop improvement under current and new culture conditions.
2. The importance of creating new breeding material through interspecific hybridization cannot be overemphasized. Fortunately, methods for producing hybrids between distantly related species have been well developed and fairly successful in lily (Van Tuyl et al. 2000). Aside from the three important hybrids of groups of lily, viz., Longiflorum, Asiatic, and Oriental, it might be desirable to extend to other groups as well and pay attention to the use of wild species of other sections that might possess disease resistances and other desirable traits.
3. Based on the experience gained from LA hybrids, it is apparent that distant hybrids can produce normal haploid gametes in spite of their genome differentiation. This certainly depends on the genotypes of the parents. Once such genotypes are identified, the benefits can be highly rewarding because it can open up the way for breeding at the diploid level and can be used for producing polyploids of desired level, i.e., triploids. Besides haploid gametes, it might be worthwhile to screen for genotypes that produce 2n gametes. This is because, as in numerous other ornamental plants (Van Tuyl et al. 2002a, b), polyploid cultivars in lily will be the most successful ones.
4. Not all distant F_1 hybrids might be able to produce genotypes with 2n gametes. In such cases, it might be desirable to use traditional methods or preferably, through the use of nitrous oxide treatment, to restore fertility. But this method may have to be refined in order to fetch more predictable results.
5. It is imperative that molecular genetic maps might be potentially useful in breeding. At present, however, none of the molecular markers are assigned to chromosomes as yet, and the number of linkage groups exceeds the basic chromosome number (i.e., x ¼ 12). In this context, the cytological maps of the three genomes of lily cultivars (Khan et al. 2009) might be useful for assigning linkage groups to the respective chromosomes. Regardless of the availability of general high-quality linkage maps, however, it might be useful to develop markers to tag useful quantitative traits and increase the efficiency of selection in breeding.

References

Abe H, Nakano M, Nakatsuka A, Nakayama M, Koshioka M, Yamagishi M (2002) Genetic analysis of floral anthocyanin pigmentation traits in Asiatic hybrid lily using molecular linkage maps. Theor Appl Genet 105:1175–1182

Anderson NO, Younis A, Sun Y (2010) Intersimple sequence repeats distinguish genetic differences in Easter lily 'Nellie White' clonal ramets within and among bulb growers over years. J Am Soc Hortic Sci 135:445–455

Azadi P, Otang NV, Chin DP, Nakamura I, Fujisawa M, Harada H, Misawa N, Mii M (2010) Metabolic engineering of *Lilium* x *formolongi* using multiple genes of the carotenoid biosynthesis pathway. Plant Biotechnol Rep 4:269–280

Bakhshaie M, Khosravi S, Azadi P, Bagheri H, van Tuyl MJ (2016) Biotechnological advances in *Lilium*. Plant Cell Rep 35:1799–1826

Barba-Gonzalez R, Lim KB, Ramanna MS, Visser RGF, Van Tuyl JM (2005) Occurrence of 2n gametes in the F1 hybrids of Oriental x Asiatic lilies (Lilium): relevance to intergenomic recombination and backcrossing. Euphytica 143:67–73

Barba-Gonzalez R, Miller CT, Ramanna MS, Van Tuyl JM (2006) Nitrous oxide (N_2O) induces 2n gametes in sterile F1 hybrids between Oriental x Asiatic lily (Lilium) hybrids and leads to intergenomic recombination. Euphytica 148:303–309

Bilz M (2011) Lilium jankae. The IUCN red list of threatened species 2011: e.T165213A5990909. https://doi.org/10.2305/IUCN.UK.2011-2.RLTS.T165213A5990909.en

Biswas M, Nath U, Howlader J, Bagchi M, Natarajan S, Kayum MA, Kim HT, Park JI, Kang JG, Nou IS (2018) Exploration and exploitation of novel SSR markers for candidate transcription factor genes in Lilium species. Genes 9(2):97

Bouman H, Tiekstra A, Petutschnig E, Homan M, Schreurs R (2003) Cryo preser- vation of Lilium species and cultivars. Acta Hortic 612:147–154

Chen H, Chen XL, Chen LQ, Lu XX (2007) Cryopreservation of shoot tips from in vitro plants of our flower of lily (Lilium L.) by vitrification method. J Plant Gene Res 2:170–173

Chen MK, Lin IC, Yang CH (2008) Functional analysis of three lily (*Lilium longiflorum*) APETALA1-like MADS box genes in regulating floral transition and formation. Plant Cell Physiol 49:704–717

Chen XL, Li JH, Xin X, Zhang ZE, Xin PP, Lu XX (2011) Cryopreservation of in vitro-grown apical meristems of Lilium by droplet-vitrification. S Afr J Bot 77:397–403

Comber HF (1949) A new classification of genus *Lilium*. Lily Yearb N Am Lily Soc 13:86–105

Dhyani A, Nautiyal BP, Nautiyal MC (2018) Distribution, status and conservation of Lilium polyphyllum (Liliaceae), a critically endangered medicinal plant from India. Plant Biosyst 152(4):608–611

Du Y, Heng-bin H, Zhong-xuan W, Chi W, Shuang L, Gui-xia J (2014) Investigation and evaluation of the genus Lilium resources native to China. Genet Resour Crop Evol 61:395–412

Du YP, Bi Y, Yang FP, Zhang MF, Chen XQ, Xue J, Zhang XH (2017) Complete chloroplast genome sequences of Lilium: insights into evolutionary dynamics and phylogenetic analyses. Sci Rep 7(1):5751

Gao YD, Harris AJ, He XJ (2015) Morphological and ecological divergence of *Lilium* and *Nomocharis* within the Hengduan Mountains and Qinghai-Tibetan Plateau may result from habitat specialization and hybridization. BMC Evol Biol 15:147. https://doi.org/10.11826/s12862-015-0405-2

Gargano D (2015) Lilium pomponium. The IUCN red list of threatened species 2015: e. T190872A70290522. https://doi.org/10.2305/IUCN.UK.2015-1.RLTS.T190872A70290522.en

Gudeva KL, Trajkova F (2015) The effect of plant growth regulators on morphogenesis in tissue culture of some agricultural species. In: The 6th international scientific agricultural symposium, Agrosym, pp 238–244

Hao H, Wu ZH, Chen G, Lin QL (2017) Geographical distribution and morphological diversification of Lilium fargesii. Acta Hortic 1171:349–356

Hu Y, Ren J, Liu Y, Zhang M, Moe TS, Khan MS, Du Y, Zhang X (2020) Evaluating the genetic relationship of *Lilium* species/cultivars based on target region amplification polymorphism (TRAP). Genet Resour Crop Evol 67:503–513

İkinci N (2014) Lilium ciliatum. The IUCN red list of threatened species 2014: e. T200273A2646311. https://doi.org/10.2305/IUCN.UK.2014-1.RLTS.T200273A2646311.en

İkinci N, Oberprieler C, Güner A (2006) On the origin of European lilies: phylogenetic analysis of *Lilium* section *Liriotypus* (*Liliaceae*) using sequences of the nuclear ribosomal transcribed spacers. Willdenowia 36:647–656

IPGRI (1995) Descriptors for capsicum (Capsicum spp.), IPGRI, ISBN 92-9043-216-0, Via delle Sette Chiese, 142, 00145 Rome, Italy

Jeknic Z, Jeffrey T, Morre S, Sladana J, Subotic A, Chen THH (2014) Cloning and Functional Characterization of a Gene for Capsanthin-Capsorubin Synthase from Tiger Lily (*Lilium lancifolium* Thunb. 'Splendens'). Plant Cell Physiol 53(11):1899–1912

Kaviani B, Abadi DH, Torkashvand AM, Hoor SS (2009) Cryopreservation of seeds of lily (*Lilium ledebourii* (baker) bioss): use of sucrose and dehydration. Afr J Biotechnol 8:3809–3810

Kawase D, Hayashi K, Takeuchi Y, Yumoto T (2010) Population genetic structure of *Lilium japonicum* and serpentine plant *L. japonicum* var. *abeanum* by using developed microsatellite markers. Plant BioSyst 144:29–37

Khan N, Zhou S, Ramanna MS, Arens P, Herrera J, Visser RGF, Van Tuyl JM (2009) Potential for analytic breeding in allopolyploids: an illustration from Longiflorum x Asiatic hybrid lilies (Lilium). Euphytica 166:399–409

Khan N, Barba-Gonzalez R, Ramanna MS, Arens P, Visser RGF, Van Tuyl JM (2010) Relevance of unilateral and bilateral sexual polyploidization in relation to intergenomic recombination and introgression in Lilium species hybrids. Euphytica 171:157–173

Kim S, Mollet JC, Dong J, Zhang K, Park SY, Lord EM (2003) Chemocyanin, a small basic protein from the lily stigma, induces pollen tube chemotropism. Proc Natl Acad Sci U S A 100:16125–16130

Kim JH, Kyung HY, Choi YS, Lee JK, Hiramatsu M, Okubo H (2006) Geographic distribution and habitat differentiation in diploid and triploid Lilium lancifolium of South Korea. J Fac Agric Kyushu Univ 51(2):239–243

Kumari S, Kim YS, Kanth BK, Jang JY, Lee GJ (2019) Genetic diversity assessment of lily genotypes native to Korea based on simple sequence repeat markers. J Plant Biotechnol 46(3):158–164

Lansdown RV (2018) Lilium chalcedonicum. The IUCN red list of threatened species 2018: e. T13142534A18614450. https://doi.org/10.2305/IUCN.UK.2018-1.RLTS.T13142534A18614450.en

Li JW, Zhang XC, Wang MR, Bi WL, Faisal M, Jaime A, da Silva T, Volk GM, Wang QC (2019) Development, progress and future prospects in cryobiotechnology of *Lilium* spp. Plant Methods 15(125):1–12

Liang ZX, Zhang JZ, Sun MY, Zhang YL, Zhang XH, Li H, Shi L (2018) Variation of Phenolic Compounds and Antioxidant Capacities in Different Organs of Lilium pumilum. Nat Prod Commun 13(6):717–722

Lim KB, Chung JD, Van Kronenburg BCE, Ramanna MS, De Jong JH, Van Tuyl JM (2000) Introgression of *Lilium rubellum* Baker chromosomes into *L. longiflorum* Thunb.: a genome painting study of the F1 hybrid, BC1 and BC2 progenies. Chromosome Res 8:119–125

Lim KB, Barba-Gonzalez R, Zhou S, Ramanna MS, Van Tuyl JM (2007) Interspecific hybridization in lily (*Lilium*): taxonomic and commercial aspects of using species hybrids in breeding. In: Teixeira da Silva JA (ed) Floriculture, ornamental and plant biotechnology: advances and topical issues. Global Science Books, Ltd., Isleworth, pp 138–145

Matsumoto T, Sakai A (1995) An approach to enhance dehydration tolerance of alginate-coated dried meristems cooled to − 196 °C. Cryo Letters 16:299–306

McRae EA (1998) Oriental lily hybrids in lilies, a guide for growers and collectors. Timber Press, Portland, pp 239–257

Mohajeri MR, Zare AG, Shahab M, Jari SK (2014) Seed germination of Lilium ledebourii (Baker) Boiss. after cryopreservation. J Rangeland Sci 4:279–285

Nishikawa TK, Okazaki K, Nagamine T (2002) Phylogenetic relationships among *Lilium auratum* Lindley *L. auratum* var. *platyphyllum* Baker and *L. rubellum* Baker based on three spacer regions in chloroplast DNA. Breed Sci 52:207–213

Pedapati A, Yadav SK, Tyagi V, Singh SP, Ranga SS, Binda PC, Brahmi P (2018) Ornamental Germplasm: Potential New Resources for Floriculture Industry. Int J Curr Microbiol App Sci 7(12):1731–1742

Pelkonen V, Pirttila A (2012) Taxonomy and phylogeny of the genus *Lilium*. Floricult Ornam Biotechnol 6:1–8

Petrova A, Bazos I (2013) Lilium rhodopeum. The IUCN Red List of Threatened Species 2013: e. T165256A5996953. https://doi.org/10.2305/IUCN.UK.2011-1.RLTS.T165256A5996953.en

Rai R, Shrestha J, Kim JH (2018) Combining ability and gene action analysis of quantitative traits in *Lilium* × *formolongi*. J Agric Life Environ Sci 30:131–143

Rana MS, Samant SS (2011) Population biology of *Lilium polyphyllum* D. Don ex Royle—a critically endangered medicinal plant in a protected area of Northwestern Himalaya. J Nat Conserv 19:137–142

Saha D, Ved D, Ravikumar K, Haridasan K, Dhyani A (2015) Lilium polyphyllum. The IUCN red list of threatened species 2015: e.T50126623A79918170. https://doi.org/10.2305/IUCN.UK.2015-4.RLTS.T50126623A79918170.en

Sayadalian M, Naderi R, Fattahi M, Mohammad R, Padasht D, Mohammad N (2014) An evaluation of some different populations of *Lilium ledebourii* (BAKER) BOISS, employing agro-morphological characteristics and multivariate analyses. Iran J Hortic Sci (Iran J Agric Sci) 4: 379–387

Shahin A, Arens P, Van Heusden AW, Van Der Linden G, Van Kaauwen M, Khan N, Schouten H, Van De Weg E, Visser R, Van Tuyl JM (2010) Genetic mapping in Lilium: mapping of major genes and QTL for several ornamental traits and disease resistances. Plant Breed 130:372–382

Shahin A, Arens P, Van Heusden AW, Van Der Linden G, Van Kaauwen M, Khan N, Schouten HJ, Van De Weg WE, Visser RG, Van Tuyl JM (2011) Genetic mapping in Lilium: mapping of major genes and quantitative trait loci for several ornamental traits and disease resistances. Plant Breed 130(3):372–382

Shahin A, Martijn VK, Danny E, Joachim W, Van Tuy JM, Richard GV, Paul A (2012) Generation and analysis of expressed sequence tags in the extreme large genomes *Lilium* and *Tulipa*. BMC Genomics 13:640

Tzeng TY, Chen HY, Yang CH (2002) Ectopic expression of carpel-specific MADS Box genes from lily and lisianthus causes similar homeotic conversion of sepal and petal in Arabidopsis. Plant Physiol 130:1827–1836

UPOV. International Union for the Protection of New Varieties of Plants (2005) Test guidelines – English Version. http://www.upov.int/en/publications/tg-rom/tg_index.htm

Urbaniec-Kiepura M, Bach A (2014) Effects of pre-storage on *Lilium martagon* L. seed longevity following cryopreservation. Cryo Letters 35:462–472

Urbaniec-Kiepura M, Bach A (2017) Cryopreservation of Lilium martagon L. meristems by droplet-vitrification and evaluation of their physiological stability. Cryo Letters 38(2):78–89

van Hintum TJL (1996) Core collections in germplasm conservation, evaluation and use. In: Scoles G, Rossnagel B (eds) V International oat conference & VII International barley genetics symposium: proceedings. University Extension Press/University of Saskatchewan, Saskatoon, pp 113–119

Van Tuyl JM, Van Dien MP, Van Creij MGM, Van Kleinwee TCM, Franken J, Bino RJ (1991) Application of *in vitro* pollination, ovary culture, ovule culture and embryo rescue for overcoming incongruity barriers in interspecific *Lilium* crosses. Plant Sci 74:115–126

Van Tuyl JM, van Dijken HS, Chi HS, Lim KB (2000) Breakthroughs in interspecific hybridization of lily. Acta Hortic 508:83–88

Van Tuyl JM, Chung MY, Chung JD, Lim KB (2002a) Introgression with Lilium hybrids: introgression studies with the GISH method on *L. Longiflorum* x Asiatic, *L. longiflorum* x *L. rubellum* and *L. auratum* x *L. henryi*. Lily Yearb N Am Lily Soc 55(17–22):70–72

Van Tuyl JM, Lim KB, Ramanna MS (2002b) Interspecific hybridization and introgression. In: Vainstein A (ed) Breeding for ornamentals: classical and molecular approaches. Kluwer Academic Publishers, pp 85–103

Vieira P, Wantoch S, Lilley CJ, Chitwood DJ, Atkinson HJ, Kamo K (2015) Expression of a cystatin transgene can confer resistance to root lesion nematodes in Lilium longiflorum cv. 'Nellie White'. Transgenic Res 24(3):421–432

Wang S, Xie Y (2004) China species red list, vol 1. Higher Education Press, Beijing

Wang HJ, Wan AR, Jauh GY (2008) An actin-binding protein, LlLIM1, mediates calcium and hydrogen regulation of actin dynamics in pollen tubes. Plant Physiol 147:1619–1636

Wang WK, Liu CC, Chiang TY, Chen MT, Chou CH, Yeh CH (2011) Characterization of expressed sequence tags from flower buds of alpine *Lilium formosanum* using a subtractive cDNA library. Plant Mol Biol Rep 29:88–97

Wang Y, van Kronenburg B, Menzel T, Maliepaard C, Shen X, Krens F (2012) Regeneration and Agrobacterium-mediated transformation of multiple lily cultivars. Plant Cell Tissue Organ Cult 111:113–122

Wu XM (2017) Studies on nutrient composition of *Lilium davidii* var. *unicolor* derived products. Food Nutr China 23:77–80

Wu XW, Li SF, Xiong L, Qu YH, Zhang YP, Fan MT (2006) Distribution situation and suggestion on protecting wild lilies in Yunnan Province. J Plant Genet Resour 7:327–330

Wu XW, Wang LH, Wu LF, Qu SP, Suh JK, Wang JH (2012) Native species of the genus *Lilium* and the closely related *Nomocharis* in Yunnan, China. In: Floriculture and biotechnology. Global Science Books, pp 28–38

Wu XW, Tian M, Wang LH, Cui GF, Yu RP, Lu ZH, Jia WJ, Qu SP, Gui M, Wang JH (2014) Native species of the genus *Lilium* in China. Acta Hortic 1027:27–40

Xi M, Sun L, Qiu S, Liu J, Xu J, Shi J (2012) In vitro mutagenesis and identification of mutants via ISSR in lily (Lilium longiflorum). Plant Cell Rep 31:1043–1051

Xiang Y, Huang X, Wang T, Zhang Y, Liu Q, Hussey PJ, Ren H (2007) ACTIN BINDING PROTEIN29 from Lilium pollen plays an important role in dynamic actin remodeling. Plant Cell 19:1930–1946

Xu H, Swoboda I, Bhalla PL, Singh MB (1999) Male gametic cell-specific expression of H2A and H3 histone genes. Plant Mol Biol 39:607–614

Xu J, Liu Q, Jia M, Liu Y, Li B, Shi Y (2014) Generation of reactive oxygen species during cryopreservation may improve Lilium ×Siberia pollen viability. Vitro Cell Dev Biol Plant 50: 369–375

Yan R, Wang Z, Ren Y, Li H, Liu N, Sun H (2019) Establishment of Efficient Genetic Transformation Systems and Application of CRISPR/Cas9 Genome Editing Technology in Lilium pumilum DC. Fisch. and Lilium longiflorum White Heaven. Int J Mol Sci 20(12):2920

Yi JY, Lee GA, Lee SY, Chung JW (2014) Development of cryo-banking system of Lilium species. Flower Res J 22:185–189

Yin ZF, Bi WL, Chen L, Zhao B, Volk GM, Wang QC (2014) An efficient, widely applicable cryopreservation of Lilium shoots tips by droplet vitrification. Acta Physiol Plant 36:1683–1692

Yong Y, Zhang Y, Lyu Y (2019a) A MYB-related transcription factor from Lilium lancifolium L. (LlMYB3) is involved in anthocyanin biosynthesis pathway and enhances multiple abiotic stress tolerance in Arabidopsis thaliana. Int J Mol Sci 20:3195. https://doi.org/10.3390/ijms20133195

Yong Y, Zhang Y, Lyu Y (2019b) A stress-responsive NAC transcription factor from tiger lily (LlNAC2) interacts with LlDREB1 and LlZHFD4 and enhances various abiotic stress tolerance in Arabidopsis. Int J Mol Sci 20:3225. https://doi.org/10.3390/ijms20133225

Zhang X, Wang H, Bun NT (2008) Isolation and characterization of a novel Trypsin inhibitor from fresh lily bulbs. Planta Med 74:546–550

Zhao P, Zhang QW, Shi BS, Liu Y (2014) Studies on cryopreservation of Lilium concolor var. pulchelium. J Agric Univ Hebei 37:54–58

Alstroemeria: Conservation, Characterization, and Evaluation

5

M. R. Dhiman and Bharati Kashyap

Contents

M. R. Dhiman (✉)
ICAR-INARI, Regional Station, Katrain, Kullu-Valley, Himachal Pradesh, India

B. Kashyap
UHF, Nauni-Solan, HP, India

S. K. Datta, Y. C. Gupta (eds.), *Floriculture and Ornamental Plants*, Handbooks of Crop Diversity: Conservation and Use of Plant Genetic Resources,
https://doi.org/10.1007/978-981-15-3518-5_7

Abstract

Alstroemeria (Alstroemeriaceae), the Inca lily or Lily of the Incas, is endemic to South America, mainly in Chile and Brazil, represented by around 90 species that occupy a diverse range of habitats. It is potential cut flower, pot plant, and bedding plant. The slow process of breeding delays the introduction of new cultivars to the commercial market.

The species classification in *Alstroemeria* is based on an evaluation of morphological traits of the flower, stem, leaf, fruit, and rhizome. The available biosystematic information on *Alstroemeria* species is restricted to the Chilean species. Little is known about the classification of the Brazilian species.

Objectives for breeders of this plant include vigorous growth forms, winter hardiness, continuous year-round flowering, fragrant flowers, novel flower colors, and removal of allergens. Winter-hardy hybrids were developed by using the Chilean species *Alstroemeria aurea*; fragrant hybrids were developed by using the Brazilian species, *A. caryophyllaea* with the assistance of in vitro techniques such as ovule embryo rescue, micropropagation, and somatic embryogenesis. *Alstroemeria* mutagenesis has been focused in changing few traits on outstanding cultivars. The allergen content can be reduced by mutagenesis with γ rays. Possibilities for cross combinations in *Alstroemeria* between the species are limited due to embryo abortion. In vitro embryo rescue has been demonstrated to be necessary in order for interspecific hybridization of *Alstroemeria* to be successful. However, in some of the species open pollination is possible.

Various methods of breeding, characterization and conservation of germplasm have been discussed.

Keywords

Alstroemeria · Breeding methods · Conservation · Characterization

5.1 Introduction

Alstroemeria, also known as the lily of the Incas, Peruvian lily, or Inca lily, is a member of Alstroemeriaceae and is known for its showy, long-lasting cut flowers that are available in a variety of vibrant colors of purple, lavender, red, pink, white, orange, peach, and yellow and bicolors. The cut flowers have a very long postharvest vase life that lasts for more than 2 weeks. *Alstroemeria* is named after Swedish botanist Baron Klas Van Alstomer who collected seeds of this flower from Spain in

1753, which in fact came from Chile and Brazil in South America. Later, he bought this plant and introduced it in Europe.

In Europe and the United States, *Alstroemeria* flowers began to become popular in the 1970s, primarily as a cut flower crop due to its ability to grow under low greenhouse temperatures. Recently, this plant became popular as a potted plant (Olate et al. 2000) and as a landscape garden flower. Columbia is number one in the world as far as growing area of *Alstroemeria* is concerned. *Alstroemeria* flowers are third in terms of popularity and foreign exchange earnings in Kenya after roses and statice.

Native species of *Alstroemeria* exist mostly in Chile and Brazil, but other South American countries as well (Bridgen et al. 2000; Sanso et al. 2005). The species are indigenous to a variety of habitats including the snowline of the Andes plateau, the river valleys, highland forests, and the western coastal desserts of Chile (Healy and Wilkins 1985; Verboom 1979). Although very diverse, the species are of little commercial value because the plants go dormant for part of the year and the flowers are often small and insignificant, and the postharvest life is not always optimal. The variation among and within species, the opportunity to collect native species in the wild (Bridgen 2001), and the possibility for species hybridization allow a wide range of new cultivars to be bred as cut flowers, potted plants, and garden plants (Bridgen 2006). New *Alstroemeria* cultivars are the results of years of inter- and intraspecific breeding and irradiation to induce mutations (Bridgen et al. 2009; Broertje and Verboom 1974). Objectives for breeders of this plant include vigorous growth forms, continuous year-round flowering, fragrant flowers, and novel flower colors. Many other wild species, that are barely known so far, are also of potential use in breeding programs. Emphasis has been taken to take species of Brazilian origin with potential ornamental value and plants of low height for use as new pot plants. In this chapter, we will look through the origin, distribution, domestication, utilization, and conservation of available *Alstroemeria* genetic resources to develop novel cultivars and briefly describe the morphological features of *Alstroemeria* species to help breeders and readers understand how the genus *Alstroemeria* has come out and what characters have attracted gardeners for a long time.

5.2 Botany and Distribution

Alstroemeria belongs to the order Liliales of the family Alstroemeriaceae and subclass Liliopsida. It is herbaceous perennial. *Alstroemeria* plants are multiplied by splitting of fleshy rhizomes. The roots vary from thick and tuberous to thin and fibrous and produce thickened cylindrical storage roots that mainly contain starch and are edible (Bridgen et al. 1989). Aerial stems are erect or decumbent. The leaves are often resupinated, that is, twisted from the petiole or the leaf blade so the lower surface becomes functionally the upper surface; sometimes the leaves form basal rosettes; leaf blades thin or thick, sometimes with papillae; the blade varies in shape from linear to elliptic or ovate; fertile stems usually have reduced leaves but sterile stems have well developed leaves.

The inflorescences are terminal and umbeliform. In the flower structure, the perianth consists of two whorls of three petals. The petal of the outer whorl has a different size and shape compared to the petal of the inner whorl. Moreover, spots and streaks are associated with the signaling of pollen vectors (De Jeu et al. 1992). The ovary is pseduo-epigyn with three carpels forming a tripartite ovary in which an axial placenta is present.

An inflorescence of *Alstroemeria* (Cv. Tiara) and various floral parts

In each cavity of the ovary, two rows of ovules are located next to each other along the central placenta. The total number of ovules varies from 24 to 36 depending on the genotype (De Jeu et al. 1992). The *Alstroemeria* has a protandrous flowering, which means that the anthers dehisce before the stigma is receptive. Therefore, self-pollination within a flower is difficult. In total, six anthers are situated in two whorls. Two days after anthesis, the first anther dehisces; at that moment, the style is still short and undeveloped. Four days after anther dehiscence, the anthers become dried and the filaments curl towards the lowest petal, at a distance of the developing style. Two days after all six anthers have wilted, the stigma becomes receptive, producing droplets of exudate on the papillae. This is the exact moment for pollination of the stigma. The pollen grains are only able to germinate a wet stigma. The pollen tubes grow between the papillae and within 24 h grow through the cavity into the ovary (Chevalier 1994). In general, after compatible fertilization, it takes about 2 months before the round seeds are scattered with force out of the ripe fruits.

The genus *Alstroemeria* L comprises of 80 species endemic to South America. Its northern limits are southern Venezuela and north-west Brazil (3–4°N) while it extends as far south as Tierra del Fuego in Argentina (52–53°S), with two main centers of distribution, one in Chile (extending into Peru, Bolivia, and Argentina) and the second throughout Brazil, Paraguay, and Argentina. Native species occupy a diverse range of habits from high Andes to marshy lands. In Chile, *Alstroemeria* is geographically distributed from a latitude of 22°S, near Tocopilla (Region II) in the north (*A. paupercula* Phil.), where the vegetation is sparse, with various species of cacti and other xerophytic plants, to 51°S in the south (*A. patagonica* Phil.). The central region of Chile has the greatest number of species. South of the 36° latitude the number of species is limited to *A. aurea*, the most widespread species in Chile, and *A. presliana* Herb (Oogaard and Kristiansen 1998). The Brazilian species occur at different habitats: forest, savanna, high fields, marsh, "campos rupestres" and "caatinga," from a height of 300 m in the Amazon up to 2300 m at Serra do Itatiaia. With exception of *A. isabellana* Herb., *A. gardneri* Baker, *A. apertiflora* Baker, *A. cunha* Vell., *A. stenopetala* Schenk, *A. psittacina*, and *A. longistyla* Schenk, most species have a restricted distribution (de Assis 2001).

5.3 Origin, Domestication, and Spread

The centers of diversity are the Mediterranean zone of Central Chile and the mountains of south eastern Brazil. The genus can be divided into two groups: the Brazilian and the Chilean group. The flowers of the Chilean group are more open than the Brazilian species. In Chile, *Alstroemeria* represents one of the most diverse genera of vascular monocotyledons, comprising more than 50 recognized or accepted taxa (36 species, 11 subspecies, and 10 varieties) from which ca. 82% are endemic to the Mediterranean zone of central Chile, one of the world's diversity hotspots. In Peru, five species have been mentioned, one of which is also present in Chile: *A. violacea*. In Bolivia, three species are found, one of which is grown in Chile: *A. aurea* (Jorgensen et al. 2014), under *A. aurantiaca*. In Argentina, there are 10 species, five shared with Chile: *A. andina* var. *venustula*, *A. aurea*, *A. patagonica*, *A. presliana* subsp. *presliana*, and *A. pseudospathulata*. Thus, more than 88% of the genus is represented by taxa endemic to Chile. In Chile, *Alstroemeria* spreads from 20°S (Tarapacá Region) to 53°S (Magallanes Region) (Muñoz-Schick and Moreira 2003). Most taxa have a very restricted distribution in Chile. The vast majority of the species are distributed in north (Tarapacá-Coquimbo) and central (Valparaíso-Biobío) Chile; only six species are grown in southern Chile (Araucanía-Magallanes). The most boreal taxa are *A. lutea* and *A. violacea* that reach the Region of Tarapacá (20°S) in northern Chile. *Alstroemeria lutea* is restricted to the coast of the Tarapacá Region (Iquique) whereas *A. violacea* extends southern to 28°S in the Atacama Region; this species if known also from Peru (Arequipa) (Muñoz-Schick and Moreira 2003). The regions with the largest number of taxa are Atacama (14 taxa), Coquimbo (26), Valparaíso (19), and Metropolitan. The number of taxa decreases abruptly in southern the Maule Region where 12 taxa are found; in Los Ríos and Los Lagos, only one species has been collected (*A. aurea*) and *A. patagonica* is found in Aysén and Magallanes being the most austral species of the genus *Alstroemeria* in the world. The latter species grow from 46°30′S to 52°45′S and also in Argentina (Neuquén to Tierra del Fuego). *Alstroemeria aurea* is the species with the widest distribution in Chile (this species spreads over 10 regions, from the O'Higgins

Region, 34°12′S to Torres del Paine National Park, Magallanes Region, 51°21′S). *Alstroemeria revoluta*, the second widely distributed species, spreads from Valparaíso (La Campana National Park, 32°57′S) to Araucanía Region (Traiguén-Galvarino, 38°16′S) and *A. versicolor* ranges from the Metropolitan Region (Rio Clarillo National Reserve, 33°40′S) to Araucanía (Malleco, Renaico, 37°48′S). With these exceptions, most species show very narrow distribution, some of them being confined to a single region, such as *A. lutea* (Tarapacá Region), *A. kingii*, *A. philippi* var. *albicans* and *A. polyphylla* (Atacama Region), *A. andina* var. *venustula*, *A. hookeri* subsp. *maculata*, *A. magnifica*, *A. schizanthoides* var. *alba* and *A. mollensis*, *A. traudliae* (Coquimbo Region), *A. marticorenae* (Valparaíso Region), *A. achirae* (Maule Region), *A. hookeri* subsp. *sansebastiana* (Biobío Region), *A. presliana* subsp. *Australis* (Araucanía Region, Nahuelbuta National Park). Altitudinally, the genus *Alstroemeria* spreads from the sea level to nearly 4000 m. a.s.l., although most species are found below 2000 m.a.s.l. *Alstroemeria andina*, *A. crispata*, *A. exerens*, *A. pallida*, *A. parvula*, *A. spathulata*, and *A. umbellata* can be found above 3000 m of elevation.

5.3.1 Growth Habit of *Alstroemeria*

In general, soil temperature seems to be a crucial factor for the growth of *Alstroemeria* species. In many *Alstroemeria* species flowering is highly dependent on a period of cool soil temperature (Healy and Wilkins 1982, 1986). Cool temperatures are also important for seed germination of many species of *Alstroemeria* (Hannibal 1942), especially the ones that grow high up in the mountains or in coastal areas. *A. campaniflora* is adapted to tropical marshy areas, whereas *A. parvula* is found in an alpine area. Surprisingly, *A. polyphilla* and *A. graminea* are found in the desert (Aker and Healy 1990).

5.3.2 Chromosome Studies of *Alstroemeria*

The species of *Alstroemeria, Bomarea*, and *Leontochir* are mainly diploid with a basic chromosome number of $n = 8$ for *Alstroemeria* species and $n = 9$ for *Bomarea* (Whyte 1929). This is critical for discriminating between the genera *Alstroemeria* and *Bomarea* that have similar morphological features. *Bomarea* genus has smaller chromosomes than that of *Alstroemeria. Alstroemeria* may have derived from *Bomarea* by translocation of essential genetic material to other chromosomes, loss of centromere, and addition of repetitive DNA in all chromosomes (Hunziker and Xifreda 1990). However, the commercial cultivars are not only diploid, but also triploid ($2n = 3x = 24$), tetraploid ($2n = 4x = 32$), and even aneuploid (Hang and Tsuchiya 1988; Tsuchiya et al. 1987). The most attractive cultivars are triploid and tetraploid with big-sized flowers and a variety of colors. *Alstroemeria* has special interest because of its large chromosomes and its asymmetric karyotypes. Chromosome analysis of *Alstroemeria ligtu* hybrids revealed that they contained two pairs of

Table 1 Chromosome numbers in 25 cultivars of *Alstroemeria*

Chromosome number	Cultivars
2n = 2x = 16	Orchid, Canaria, Zebra, Eureka
2n = 3x = 24	Pink Perfection, Regina, Carmen, Marina, Campfire, Red Surprise, King Cardinal, Apple Blossom, Pink Triumph, Mona Lisa, Yellow King, Rosita
2n = 3x + 1 = 25	Orange Beauty
2n = 4x − 1 = 31	Luciana
2n = 4x = 32	Jubilee, Arizo, Orego, Alnba, Texas, Neva
2n = 4x + 1 = 33	Rosario

satellite metacentric chromosomes that were not found in any *Alstroemeria* cultivars (Rustanius et al. 1991).

Chromosome numbers in 25 cultivars of *Alstroemeria* are given in Table 1.

5.3.3 Genetics

The CYCLOIDEA (CYC) gene controls the development of zygomorphic flowers in *Alstroemeria aurea, Alstroemeria magenta*, and *Alstroemeria pelegrina var. rosea* (Hoshino et al. 2014). *Alstroemeria* has two layers of petaloid tepals, in which the often spotted narrow inner tepals can be distinguished easily from the wider outer tepals. AlsDEFb and AlsGLO are expressed in whorls 1, 2, and 3 (outer tepals, inner tepals, and stamens, respectively), whereas AlsDEFa expression is detected only in whorls 2 and 3 (Hirai et al. 2007). Results from interspecific crosses of Brazilian species demonstrated that fragrance is genetically dominant and can be expressed in the F1 hybrids. Genes controlling the fragrance are probably nuclear and can be transmitted through pollen grains; hybrids that were produced when *A. caryophyllaea* was the pollen parent were all fragrant (Bridgen et al. 2009). Two 6-hydroxypelargonidin glycosides were isolated from the orange-red flowers of *Alstroemeria* cultivars, and determined to be 6-hydroxypelargonidin 3-*O*-(b-d-glucopyranoside) and 3-*O*-[6-*O*-(a-l-rhamnopyranosyl)-b-d-glucopyranoside], respectively, by chemical and spectroscopic methods. In addition, five known anthocyanidin glycosides, 6-hydroxycyanidin 3-malonylglucoside, 6-hydroxycyanidin 3-rutinoside, cyanidin 3-malonylglucoside, cyanidin 3-rutinoside, and pelargonidin 3-rutinoside were identified in the flowers (Tatsuzawa et al. 2003).

5.3.4 Breeding History of *Alstroemeria*

In the early 1950s, three *Alstroemeria* species were released into Europe – *A. pelegrina, A. ligtu*, and *A. aurea.* Since then, the interest in *Alstroemeria* as an ornamental has increased. The commercial quality of this first *Alstroemeria* was poor due to the short flowering period and bad quality of stem and leaf. Mutation techniques have been used for *Alstroemeria* breeding since 1970 to increase variation in flower color, stripes of the inner petal, flower size, and height of plants. After irradiation of actively grown rhizomes with X-rays, a variety of mutants were obtained. Some of these mutants were selected and vegetatively propagated and then developed into a new cultivar (Broertje and Verboom 1974). Up till now, more than 60 species/genotypes have been released onto commercial markets by applying conventional breeding techniques. One problem found in conventional breeding is the lack of useful genes in *Alstroemeria* germplasm for use in further breeding. The majority of the *Alstroemeria* cultivars are polyploid, which makes breeding time consuming (Chevalier 1994). However, new cultivars have been produced by using interspecific hybridization in the last decades (De Jeu and Jacobsen 1995). Furthermore, cross-hybridization does not always lead to seed set, although some hybrids were produced by using embryo rescue techniques. The slow process of breeding delays the introduction of new cultivars to the commercial market.

5.3.5 Importance and Uses

Used as a cut flower in bouquets and arrangements, and as a pot plant. Hardy garden varieties are used in landscaping. Roots of many members of this genus are edible and a source of starch that is very nutritious (Bridgen et al. 1989).

5.3.6 Varieties of *Alstroemeria*

Variegated *Alstroemeria*: Rock n Roll, *Alstroemeria psittacina variegata*: a Japanese Selection, Ivory Halo, Phoenix and Spitfire: Garden & Patio Variegated Varieties (Parigo Horticultural Co. Ltd., UK) and Glory of the Andes: variegated sport of Sweet Laura.

Fragrant cultivars: Sweet Laura (Mark Bridgen, Cornell University, USA).

Commercial varieties:

Pot varieties: HilverdaKooij started the marketing and sales of young plants of pot alstroemeria in 2006. The brand name of these pot alstroemeria is "Inticancha." These are a genetically compact and free-flowering pot alstroemeria. Growth control is not required. Some of these varieties are Maya, Paola, Cabana, Dark Purple, Bandit, Mystic, White Pink Blush, and Indigo.

Konst Alstroemeria developed the Inca Collection which are compact pot varieties. Some of these varieties are Inca Fire, Inca Exotica, Inca Replay, and Inca Sweety.

The Garden *Alstroemeria* varieties are hardy down to −15 °C, and this trait makes them unique. These varieties are Indian Summer, Summer Breeze, and Summer Break.

Spray varieties: Pradiso, Charmelia, Charmelia White, and Charmelia Bridesmaid. Charmelia Bridesmaid is the most wanted *Alstroemeria* in the world. Won

Dutch Flower Award 2014 and won FloraHolland Glass Tulip Award "Oscar of the Flower Business" in 2015.

Alstroemeria "Las Olas" is a semi-dwarf, tetraploid hybrid selection for hot climates.

5.4 Plant Genetic Resources

5.4.1 Geographical Distribution

Alstroemeria is one of the most diverse genera of the Chilean flora, and it is represented by 35 species, most of them distributed between 28° S and 39° S (Muñoz-Schick and Moreira 2003) in the Mediterranean zone. Eleven species of this genus contain complexes of two to four intraspecific taxonomic entities (subspecies and varieties). Representatives in each of these complexes grow in the Chilean Mediterranean zone and some of them are endemic to this region, with a very restricted distribution (Bayer 1987; Muñoz-Schick and Moreira 2003). *Alstroemeria presliana* Herb. includes two subspecies: subsp. *presliana* and subsp. *Australis* Bayer. They grow in a restricted fashion in Chile from the cordillera of Curicó (35°27′S) to the cordillera of Antuco (37°25′S), and from 1500 to 2000 m elevation; they also occur in Neuquén Province of Argentina. *Alstroemeria presliana* subsp. *australis* is endemic to Chile, occupying a narrow geographic distribution, from Curanilahue (37°23′S) south to the river Cautin (38°29′S), and from 200 to 1500 m elevation (Muñoz-Schick and Moreira 2003). Baeza et al. (2008) completed a comparative karyotype study of one population of *A. presliana* subsp. *Presliana* and one population of subsp. *australis*, which revealed two different karyotype formulae. It would, therefore, be very interesting to analyze more populations of both subspecies to determine the stability of karyotype structure within each. Both subspecies have attractive pink flowers that can be differentiated primarily by the size and color of their tepals, in addition to geographic distribution. This species represents, therefore, a high potential for development as an ornamental plant that has so far not been successfully developed.

5.4.2 Primary Gene Pool

The species classification in *Alstroemeria* is based on an evaluation of morphological traits of the flower, stem, leaf, fruit, and rhizome. The available biosystematic information on *Alstroemeria* species is restricted to the Chilean species. Little is known about the classification of the Brazilian species. Furthermore, morphology-based identification is rather difficult because morphological characteristics can vary considerably in different environmental conditions. The immense genetic variation present in the genus *Alstroemeria* offers many opportunities for the improvement and renewal of cultivars. Therefore, identification of genetic relationships at the species level could be very useful for breeding in supporting the selection of crossing combinations from large sets of parental genotypes, thus broadening the genetic basis of breeding programs. Molecular phylogeny of the genus *Alstroemeria* has also been carried out in addition to all the other approaches adopted for classification and identification.

5.4.3 Classification of *Alstroemeria* Based on Origin

Brazilian species are vigorous and rustic plants, mostly evergreen with large leaves. They flower all the year around and are usually diploids (2n = 16). Chilean species have the largest and most spectacular flowers with small, short-lived leaves. Plants present interesting flower shapes and in almost all colors. These are also usually diploids (2n = 16) (Meerow and Tombolato 1996).

5.4.4 Classification of *Alstroemeria* Based on Floral Morphology

Genus *Alstroemeria* is divided into three types based on floral morphology (Park et al. 2010). These are **Orchid Types**, which are diploid (2n = 2x = 16), almost sterile, easily propagated in vitro, have open flowers, 3–5 months of major flower production, have tall growth habit, therefore, unsuitable for pot culture. **Butterfly Types** resulted due to crossing between Chilean and Brazilian species. These are allotetraploid (2n = 4x), produce viable 2x gametes, 9–12 months of major flower production, have shorter growth habits and larger, more open flowers. Used for potted plant production and can be used as cut flowers. The **Hybrids Types** are created by several crossings between various species and cultivars.

5.4.5 Wild Genetic Resources

The distribution of the genus *Alstroemeria* was studied by Bayer (1987) and Aker and Healy (1990). Hofreiter and Rodríguez (2006) summarized the countries in which these species occurred and the number of wild species of *Alstroemeria* based on previous reports by Sanso (1996), and Muñoz and Moreira (2003). At present, there are 39 species in Brazil; 33 species in Chile; 10 species in Argentina; 2 species in Peru; and 1 species in Bolivia, the Guianas, Paraguay, Uruguay, and Venezuela (Hofreiter and Rodríguez 2006). Among the wild species of *Alstroemeria*, *A. ligtu*, *A. pelegrina*, *A. aurea*, *A. magenta*, *A. psittacina*, *A. inodora*, *A. pulchra*, *A. violacea*, *A. versicolor*, and *A. caryophyllaea* are presumed to be the parents of extant cultivars or candidates for further breeding. More than 88% of the genus is represented by taxa endemic to Chile. The regions with the largest number of taxa are Atacama (14 taxa), Coquimbo (26), Valparaíso (19), and Metropolitan. The number of taxa decreases abruptly southern the Maule Region where 12 taxa are found; in Los Ríos and Los Lagos, only one species has been collected (*A. aurea*) and *A. patagonica* is found in Aysén and Magallanes being the most austral species of the genus *Alstroemeria* in the world.

5.5 Collection and Conservation

Germplasm collection and conservation is the protection of the genetic diversity of a targeted crop and related species as seeds or living plants for future use. Germplasm collections serve as an important source for the crop improvement, more so in ornamental crop species like *Alstroemeria*. It is realized by the higher authority of the nations, planners, researchers, and users at various levels all over the world that ornamental plant conservation for future consumption is important. The market for ornamental plants is constantly increasing with each passing day but at the same time is subjected to periodic trend-driven changes. Indeed, every year, hundreds of new cultivars, replacing the current assortment, are produced. However, changes in consumer preferences mean that cultivars unfashionable today may in the future once again be attractive for potential buyers. Genetic materials for ornamental plants are not centrally collected and maintained anywhere in the world. Nevertheless, it is difficult for breeders and horticulturalists to provide enough space and funds for traditional cultivation of such numerous cultivars, which is laborious and threatened with biotic and abiotic stresses (Sekizawa et al. 2011). Traditional genetic conservation in the field or greenhouse requires intensive care of pot cultures or carefully separated field plots (Reed 2006). Haploids (important in breeding) and transgenic cultivars, which are gaining popular among ornamental plants (*Rosa* L., *Dianthus* L., *Gladiolus* L.), require isolation to protect them from cross-breeding (Joung et al. 2006).

During the past decade, conservation of plants at international and national levels received excellent momentum, which is reflected in the establishment of different plant genetic resource centers; for example, in India NBPGR established its 10 Regional Stations and 59 National Active Germplasm Sites (NAGS), comprising ICAR Institutes, Project Directorates, NRCs, AICRPs, SAUs, KVKs, etc. Gene banks/germplasm banks refer to a place or organization where germplasm is conserved in the living state and the germplasm can be stored in the form of seeds, pollens, in vitro cultures, or as plants growing in the field. Different institutes and universities at international and national levels collected and conserve the genetic resources of *Alstroemeria*. The germplasm collections of *Alstroemeria* are also available at different global portals and gene banks. Two accessions of *Alstroemeria* are accessible on OPGC herbaceous ornamental plant accessions in the GRIN database. In India, field gene bank is maintained at Dr. Yaswant Singh Parmar University of Horticulture and Forestry, Nauni, Solan, Himachal Pradesh, and ICAR-Indian Agricultural Research Institute, Regional Station, Katrain, Kullu, Himachal Pradesh.

5.6 Conservation

As in most zones with Mediterranean climate of the world, in central Chile there is a large number of species with restricted distribution and with high probability of destruction. Moreover, this is also the zone of the country that concentrates the highest human population, accompanied by intense anthropic intervention due to agriculture and forest activities, as well as urbanization. The genus *Alstroemeria* is facing a serious threat of genetic erosion like other endangered plant species, due to habitat fragmentation, inbreeding, weaker immune system against pathogens and pests, and climate change. Many wild species have become endangered species. So, germplasm conservation is very important as a source of genetic variation for breeding, research, and to prevent rare species from becoming extinct and required careful attention for its long-term conservation and preservation.

5.6.1 Ex Situ Conservation

Ex situ conservation is an effective tool for maintaining and safeguarding plant diversity. It is possible to preserve samples representative of the genetic diversity of populations of threatened species, as a safeguard against extinction in situ. It is also possible to provide germplasm to establish and reintroduce populations, or to augment existing populations. These ex situ methods thus help to ensure medium and long-term survival, and prevent the extinction of wild populations and species. Due to unique and heterozygous nature of the crop using seeds for preservation would break up unique genetic combinations, and the characters would show segregation.

5.6.2 Field Gene Banks

Field gene banks, also called plant gene banks, are areas of land in which germplasm collections of growing plants are assembled. This approach is also ex situ conservation of germplasm. Those plant species that have recalcitrant seeds or do not produce seeds readily are conserved in a field gene bank. In field gene banks, germplasm is maintained in the form of plants as a permanent living collection. The soil temperature seems to be a crucial factor for the growth of *Alstroemeria* species. In many *Alstroemeria* species flowering is highly dependent on a period of cool soil temperature (Healy and Wilkins 1982, 1986). Cool temperatures are also important for seed germination of many species of *Alstroemeria*, especially the ones that grow high up in the mountains or in coastal areas. *A. campaniflora* is adapted to tropical marshy areas, whereas *A. parvula* is found in an alpine area. Surprisingly, *A. polyphilla* and *A. graminea* are found in the desert (Aker and Healy 1990). *Alstroemeria* germplasm could be conserved effectively in Field gene banks.

5.6.3 In Situ Conservation

5.6.3.1 Tissue Culture as In Vitro Gene Banks

An alternative method to preserve *Alstroemeria* collections is conservation in vitro. In vitro, stored collections need relatively small amounts of space, medium components can be used that minimize growth, plants can be multiplied quickly, and there is often a possibility to eliminate viral diseases. Furthermore, the establishment of an in vitro collection is labor-intensive and genotypes may react differently under identical conditions. A combination of high sucrose concentration and a low salt concentration proved to minimize metabolic activity the best for in vitro storage of *Alstroemeria* germplasm.

5.6.3.2 Cryopreservation in Liquid Nitrogen

Cryopreservation is a long-term low temperature storage technique to preserve any biological material in liquid nitrogen (LN, − 196 °C) or liquid nitrogen vapor (LNV, approx. −165 °C to −190 °C) without deterioration for at least several thousands of years. Cryopreservation of biological tissues can be successful only if intracellular ice crystal formation is avoided. Crystal formation can be disallowed through vitrification. Two requirements must be met for a cell to vitrify: rapid freezing and a concentrated cellular solution. Sugars play a very important role in the acquisition of resistance to desiccation and to freezing in liquid nitrogen.

5.7 Characterization and Evaluation

5.7.1 Characterization for Essential Features and Classification

A priority goal in conservation is to evaluate levels of apportionment of genetic diversity in targeted species, given the association between population genetic diversity and their potential for local adaptation and evolutionary resilience. Genetic variability is the result of the dynamics of gene flow, for which a homogeneous distribution of allelic frequencies is expected under high levels of gene flow among populations (Grant 1991). Interestingly, this situation is rarely found in nature, since the strong effect that geographic isolation and selection represents for local populations of plants. As a result, it is not surprising that peripheral populations tend to increase gene differentiation and population structure levels; hence, contributing to the local isolation that eventually could result in different isolated species. Plants in genus *Alstroemeria* have evolved with a fair amount of phenotypic diversity despite the genetic similarities, and this diversity of characteristics is of considerable evolutionary significance. Experts familiar with the crop develop a crop descriptor. They select plant characters, which are stable and show variations among accessions in a collection or genus. Stable characters are those diagnostic features of the plant, for example, leaf shape, that remain the same when grown in different environments. However, other characters such as plant height and number of days to flowering, which are affected by the environment and physiology, are also used. IPGRI (1995) and the International Union for the Protection of New Varieties of Plants (UPOV) have developed standardized descriptors for most food and industrial crops. UPOV standards also cover many ornamental crops, including Alstroemeria (UPOV 2005).

The biochemical markers using isozymes profile and cluster analysis based on allele frequencies were used to reveal genetic variation and show relationships within and between species and cultivars. The combined morphological and molecular techniques were used for delimiting native and exotic varieties, clustering varieties/putative hybrids based on geographical origin within country and from outside country, and confirming the phylogenetic relationship and geographic distribution of different *Alstroemeria* spp. Aros et al. (2006) revealed that commercial varieties were clustered closer together than wild species, suggesting they share a relatively narrow and common genetic background. To explore the feasibility of identifying parental genotypes of *Alstroemeria* cultivars, and to access the genetic diversity of various species and accessions (12 *Alstroemeria* species; two F1 hybrids, one cv. "Jubilee"; an anther-cultured plant from cv. "Jubilee," and two related outgroup species, *Bomarea salsilla* and *Leontochir ovallei*) AFLP marker technique (Lee and Han 2006) was used. The results revealed that the AFLP marker technique appears to be a satisfactory tool for identifying the parental genotypes of interspecific hybrids in *Alstroemeria*.

5.7.2 Development/Identification of Gene Pools and Core Collections

The core collection concept was proposed in the 1980s with the goal to minimize the cost of germplasm conservation while ensuring the preservation of maximum genetic diversity due to the rapid increase in the number of accessions in collections of major food crops like wheat, rice, corn, potatoes, etc. A set of criteria must be established to select a "core" collection of species that captures the maximum amount of genepool of the genus. Some of the selection criteria proposed is:

1. Specific unique genotypes
   - Pest and disease resistance
   - Stress tolerance, for example, drought, frost, heat tolerance
   - Adaptation characteristics, for example, cultural input efficient genes
   - Physiological characteristics, for example, photoperiod response
   - Specific marketing/commercial traits, for example, aesthetic genes and shelf-life genes
2. Cultivar groups and genetic diversity
   - Morphological and DNA data
3. Geographical and ethnological distribution
4. Balance representation of related species (secondary and tertiary genepool as described in the above section on "genepool concept")

van Hintum (1996) modified the concept and defined it as "a germplasm collection optimally representing specific genetic diversity" to allow flexibility in the assembly of core collections, and to justify the formation of multiple core collections of a target species in space and time. This definition is equivalent to breeder collections where individual breeders assemble and manage their own distinct collections. In recent years, many approaches including random sampling, stratified sampling, phenotypic analysis, genetic markers, and coefficient of parentage have been proposed to establish core collections.

Harlan's primary (1°), secondary (2°), and tertiary (3°) crop genepool concepts was used to define the targeted germplasm for collection and conservation. The total genepool of a crop is complex especially when the species concerned is of hybrid origin. Germplasm of direct progenitors and all the contributing parents and related species have to be collected and conserved. In a genebank, germplasm is stored as seeds in a seedbank, as living plants in a field genebank, as propagules such as rhizomes/bulbs in humidified coolers, as meristem culture in in vitro collections, and seed, dormant buds, and tissue culture in liquid nitrogen cryopreservation banks.

5.7.3 Evaluation of Genetic Diversity for Desired Traits

Evaluation of genetic diversity is the screening of a collection for specific genes, such as resistance to important pests and diseases, abiotic stresses, adaptation, and other commercial traits. These traits are included in all well-formulated standard crop descriptors, for example, IPGRI (1995). The prioritization of the traits for evaluation depends on researcher needs and, thus, is location-specific. *Alstroemeria* possess great genetic diversity in its growth habit, flower color, form, size, and as well as in persistence. This diversity in species of agronomic traits offers a substantial germplasm and opportunities for the development of hardy and healthy varieties for variable climatic zones. Han et al. (2002) used an F1 population derived from an interspecific cross between two *Alstroemeria aurea* accessions, to map quantitative trait loci (QTL) involved in ornamental and morphological characteristics. Five species of *Alstroemeria* and two species of *Bomarea* were characterized by Kashihara et al. (2011) and compare the flower and tepal sizes, and tepal colors. They reported that the flower shape in *Alstroemeria psittacina* Lehm. was found to be similar to that in *Bomarea coccinea* (Ruiz & Pav.) Baker. The length/width ratio of *B. salsilla* (L.) Mirb. was intermediate compared with that of the other species. These preliminary data will be useful in selecting wild species in order to examine interspecific or intergeneric hybridizations in the breeding of Alstroemeriaceae plants.

Donoso et al. (2021) evaluated flower color in the native Chilean geophyte *Alstroemeria pallida,* by using three different approaches. Visual evaluation of *A. pallida* flower color identified five accessions, ranging from white to pink. Moreover, this visual evaluation of the accessions correlated highly with the CIELab parameters obtained by colourimetry. An anthocyanidin corresponding to a putative 6-hydroxycyanidin was identified, which was least abundant in the white accession. Although *CHS* was not expressed differentially between the accessions, the expression of *anthocyanidin synthase* (*ANS*) was significantly higher in the accession with pink flowers.

The ornamental value of new hybrids of *Alstroemeria* has been usually assessed using morphological descriptors considering flower size and color, stem length (Aros et al. 2015), and vase life (Leverentz et al. 2002), which are the main characters chosen by ornamental plant breeders. Although an important character, well appreciated by consumers, floral scent has been poorly studied in this species. Aros et al. (2019) morphologically characterized 18 flowering plants of new interspecific hybrids of *Alstroemeria* originated from *A. caryophylleae* scented lines through embryo rescue. Flower stem length ranged from 25 to 83 cm and most of the flowers showed pink/white colors with stripes over the inner tepals. Only three hybrids were perceived as scented and one of them was evaluated through GC-MS analysis, detecting nine VOCs, all of them monoterpenes.

5.7.4 Available Sources of Breeding Value

A wide diversity of *Alstroemeria* genetic resources has provided valuable genes for breeding of novel cultivars. Many important horticultural characters are present in the different *Alstroemeria* species. Modern varieties of *Alstroemeria* have been obtained by the hybridization of several Chilean and Brazilian species, and the genetic background of these cultivars still seems to be narrow using both molecular and morphological markers (Aros et al. 2006). *Alstroemeria psittacina, A. caryophyllaea, A. aurea, A. ligtu, A. versicolor, A. revoluta, A. pelegrina, A. violacea* Phil., and *A. hookeri* Schultes have been used, among other, in crossings from which several commercial cultivars have been obtained. Several of the wild species carries genetic traits that wide-open possibilities for improvement and development of more attractive new varieties. For example, some of the Brazilian taxa have rigid leaves, instead of the very tender foliage that possess most of the species; some Patagonian ones have rather short plant height, which is important for potted plants. Among the wild species of *Alstroemeria*, A. *ligtu, A. pelegrina, A. aurea, A. magenta, A. psittacina, A. inodora, A. pulchra, A. violacea, A. versicolor*, and *A. caryophyllaea* are presumed to be the parents of extant cultivars or candidates for further breeding. The variation among and within species opens up the possibility for production of a wide range of new cultivars suited for various purposes, such as cut flowers, pot plants, and garden and landscape uses (Sanso et al. 2005).

Some selected Brazilian *Alstroemeria* species that have more potential use in breeding programs include *A. isabellana, A. longistaminea, A. sellowiana* (an odoriferous species, just as *A. caryophyllaea*), *A. foliosa* (one of the showiest species from Brazil), *A. speciosa* (showy species), *A. variegata* (species with crowded leaves), *A. plantaginea, A. gardneri, A. malmeana*, and *A. stenophylla* (charming species because of its vegetative characteristics). Plant height also varies considerably in the genus but the species with lower height are found among the Chilean or Argentine-Chilean taxa. Some of these species with potential use as pots plants are: *A. pygmaea, A. patagonica, A. andina* subsp. *venustula, A. pseudospathulata, A. spathulata, A. garaventae, A. umbellata, A. hookeri, A. graminea, A werdermannii* and *A. polyphylla*.

5.7.4.1 Species of *Alstroemeria*

Alstroemeria aurea **(syn.** ***Alstroemeria aurantiaca*****)** is one of the most common parental species for commercial cut flower cultivars because of its large and attractive orange or yellow flowers.

Alstroemeria caryophyllaea is one of the Brazilian species so little known. It is a forest plant, evergreen and makes a very fine dwarf ground cover. It is aromatic (smelling like carnations). It is photoperiodic and flowers under short days.

Alstroemeria diluta **spp.** ***diluta*** produces compact winter foliage, low and mounded; flowers appear when the foliage is mostly dried.

Alstroemeria exserens is a high-elevation species, and stays small in cultivation and tolerates temperatures at least down to 20 °F (–6.6 degree centigrade).

Alstroemeria violacea is a species of delicate appearance with purple flowers that are not heavily marked.

Alstroemeria leporina is characterized by the lack of dark markings on the flowers.

Alstroemeria ligtu hybrids (LH) are from the Andes in Chile, subspecies are subsp. *ligtu* 50 cm tall with pink flowers streaked with dark red, subsp. *incarnata* 1 m tall with large pink flowers, subsp. *simsii* 1.6 m orange red streaked red on a gold background and subsp. *splendens*. They are believed to have originated from a natural crossing between *A. ligtu* and *A. ligtu* ssp. *simsii.* They are widely utilized for cut flowers in Japan, because of the flowers with favorable light colors.

Alstroemeria pelegrina *var.* ***rosea*** (PR) is a species from Peru and Chile that grows to 60 cm. The plants grow in a rocky area close to the shore, and are usually less than 12.5 cm (5 in.) tall. It is dwarf and has large flowers; it is thus a valuable gene source for improving commercial *Alstroemeria* as a cut flower or pot plant.

Alstroemeria philippii is a species from Chile that grows in coastal areas.

Alstroemeria pulchella (syn. *Alstroemeria psittacina*) from Brazil has flowers that are dark red marked with green and blotched dark purple. It has variegated leaves.

Alstroemeria werdermannii has flowers rather small and narrow, and the leaves are distinctively wavy and densely placed on the stems. Yellow-flowered forms have been distinguished as var. flavicans. This *Alstroemeria* grows in deep sand on stabilized dunes.

Alstroemeria isabellana is a lovely species with a distribution from eastern/southern Brazil to northeastern Argentina.

Alstroemeria presliana **subsp.** ***australis*** (Baeza et al. 2008): Chilean species

A. calliantha (de Assis 2009): Brazilian species

A. marticorenae (Negritto Maria et al. 2015): Chilean species. The plants are small, reaching up to 35 cm tall, the flowers are pink, the upper inner tepals are rhomboidal and narrow, with a yellow stripe and a pattern of purple lines in the two upper tepals, varying from yellow to white in the basal portion; the anthers are yellow. It grows in hard reddish soils.

A. albescens (de Assis 2009)

5.8 Molecular Characterization of Genetic Resources

Recently, an increasing accessibility to molecular data and the development of a huge range of bioinformatics analysis has favored the successful implementation of genetic tools in the identification and conservation of biological diversity. Dominant molecular markers based on random fragment alleles, for example, Inter Simple Sequence Repeat (ISSR), Random Amplified Polymorphic DNA (RAPD), and Amplified Fragment Length Polymorphism (AFLP), have been used for characterizing genetic diversity in *Alstroemeria* (Han et al. 2000), becoming the marker of choice for the identification of cultivar/varieties with ornamental value (De Jeu et al. 1995; Dubouzet et al. 1997), and conducting population genetic analyses. The use of DNA sequences has been more related to the construction of phylogenetic hypotheses and establishment of bio-geographic patterns. Interestingly, nearly 30% of the Chilean species of *Alstroemeria* form species complexes, comprising from two to four infraspecific taxa each. This pattern is likely explained by adaptation to a wide range of environmental heterogeneity present in Chile, which is possibly driving processes of microevolutionary divergence. Given the complexity of interpreting the integrity within and among these species complexes, several initiatives started for applied genetic studies with the purpose of disentangling the discernibility of intraspecific patterns of divergence, especially in groups highly regarded for their ornamental and conservation value.

Alstroemeria hookeri represents a species complex that comprises four subspecies, two of them (subspecies *recumbens* and *maculata*) distributed in North-Central of Chile and two (subspecies *hookeri* and *sansebastiana*) in southern Chile. Based on ISSR (Inter Simple Sequence Repeat) markers, high levels of population structure were found among southern subspecies (Fig. 1, Table 2) (Ruiz et al. 2010). Similarly, high levels of within population diversity was found in *A. presliana* complex using AFLP markers (Table 2), exhibiting significant levels of among population variability and moderate levels of genetic population structure (Table 2). This complex comprises of two varieties (var. *presliana* and var. *australis*), both separately distributed across Coastal and Andean mountain ranges in Chile. The results from both complexes showed the existence of two heterogeneous genetic groups with no evident spatial congruence suggesting genetic differentiation among varieties or subspecies.

The most recent and comprehensive phylogenetic studies in genus *Alstroemeria*, based on DNA sequences and cytological data conducted by Chacón et al. (2012). Quantitative traits were also found important in determining the groupings and working out the relationships among the various cultivars and hybrids. Brazilian species *A. sellowiana* and *A. variegata* had better ornamental value and utilization potential. Some Chilean or Argentine-Chilean species like *A. hookeri*, *A. polyphylla*, *A. umbellata*, and *A. patagonica* had dwarf plant height and potential use for pot culture. Molecular analysis using RAPD markers clustered two main groups: (A) hybrids with *A. pelegrina* as parental line and (B) hybrids coming from the crossing UC05 × C3 and its reciprocal and this analysis was possible to subcluster hybrids coming from different seasons.

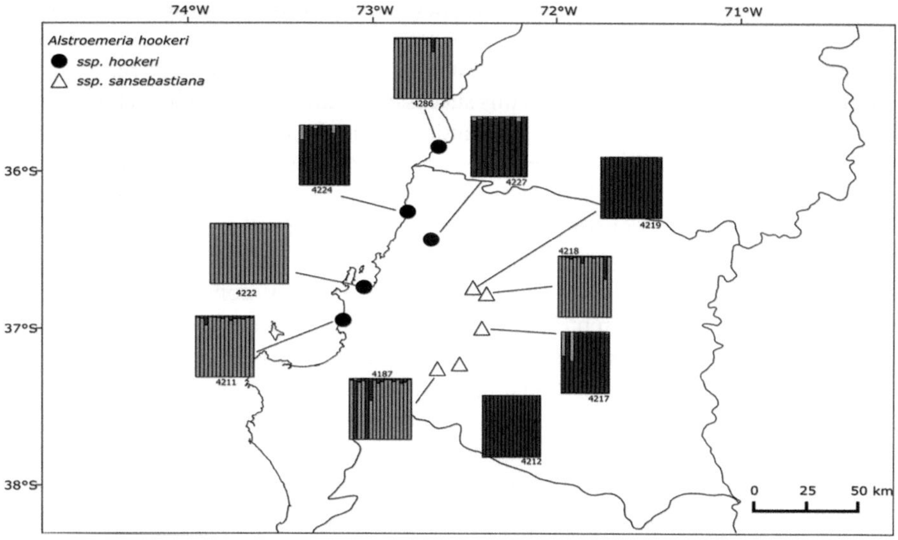

Fig. 1 Population structure inferred with ISSR in the *A. hookeri* complex. Bar plots colors represent levels of genetic membership (k = 2) in each individual per sampled population, as inferred under Bayesian admixture inference criterion with the program structure. (Source: Finot et al. 2018)

Table 2 Genetic diversity values obtained with allozymes and DNA markers (fragments analyses), for *A. hookeri* and *A. presliana* complexes. (Source: Finot et al. 2018)

Species	Subspecies	He	Fst	AMOVA (%) Within pop.	AMOVA (%) Among pop.	Marker/source
A. hookeri	*hookeri*	0.052	0.582	41.71	58.28	Allozymes/Ruiz et al. (Lin et al. 2000a)
	hookeri	0.248	0.415	58.47	41.53	ISSR/unpublished
	sansebastiana	0.246	0.36	63.99	36.01	ISSR/unpublished
A. presliana	*presliana*	0.200	0.171	82.91	17.0	AFLP/unpublished
	australis	0.198	0.179	82.10	17.9	AFLP/unpublished

Karyotypes and variation of RAPD markers have been investigated in 15 geographically isolated populations of five species of *Alstroemeria* (*A. aurea*, *A. hookeri*, *A. ligtu*, *A. pelegrina*, and *A. presliana*) collected in Chile. Tandemly repeated DNA sequences – 5S and 18/25S rDNA genes and the sequence A001-1 were used to characterize karyotypes by fluorescence in situ hybridization (FISH). Ten somatic metaphases per population were used for measurement of chromosome length. Differences in RAPD marker bands were used for characterization of populations, creating a similarity index. FISH with all three DNA probes shows a high degree of polymorphism among and sometimes also within accessions of

A. aurea, *A. hookeri*, and *A. ligtu*. The number of chromosome pairs showing 5S rDNA signals is more different for the investigated species *A. aurea*, *A. hookeri*, *A. ligtu*, *A. pelegrina*, and *A. presliana* with 5, 7, 5, 3, and 7, respectively, than the number of 18/25S rDNA signals in this succession with 7, 7, 6, 5, and 7 chromosome pairs, showing a high evolutionary dynamics within the genus. Furthermore, among the four populations of *A. hookeri*, accession 4181 was different in arm length of chromosome 3. RAPD markers (index of similarity) also showed a greater genetic distance of accession 4181 from the other three accessions of *A. hookeri* (Baeza et al. 2007).

The delimitation of species is a fundamental step for conducting natural and applied sciences, as they represent the main study unit for most areas of research (global evaluations of biodiversity, assessment and initiatives for biological conservation, etc.). In this sense, molecular approaches in taxonomy have been used under the assumption that observed divergent patterns of genetic variation are the direct result of breaks in gene flow, leading to phenotypic and genotypic differences that sustain isolated and differentiated species and populations. The simultaneous use of multiple molecular markers and criteria of delimitation has improved the taxonomic work, particularly helping to contextualize the role of microevolutionary processes in the species generation. *Alstroemeria* species share attributes that could heavily influence patterns of microevolutionary isolation and divergence, such as restricted seed and pollen dispersal and vegetative reproduction by rhizomes. Therefore, the wide distribution of taxonomic complexes in areas with contrasting topography and climatic conditions implies the existence of restrictions for gene flow, where substantial effects of ecogeographic isolation and divergence are expected in local diversification patterns (Ruiz et al. 2010).

Recent molecular and phenotypic integrated studies conducted in *Alstroemeria* complexes resulted beneficial when multiple sources of evidence are placed to solve questions about the integrity or the validity of previous taxonomic treatments. For example, when morphometric, cytogenetic, and molecular data were employed in *A. hookeri* complex taxa (Baeza et al. 2010), all analyzed characters were partially consistent with the recognition of the new subspecies *Alstroemeria hookeri* subsp. *sansebastiana*, and supported the hypothesis of Muñoz & Moreira of elevating *Alstroemeria hookeri* ssp. *cummingiana* to species level. Subsequent investigations were also conducted in other three complexes (*A. ligtu, A. magnifica*, and *A. presliana*), eliciting similar evidence with significant taxonomic impact. In the *A. magnifica* complex, evidence from morphology, colorimetry, and cytology support the change of the taxonomic status of *A. magnifica* var. *magenta*. Preliminary analyses based on chloroplast sequences (*rpl*32-*trn*L) also supported this observation, validating the separation of var. *magenta* from the other *A. magnifica* varieties. In the *A. ligtu* complex, a new entity was discovered based on cytogenetic data, and its taxonomic status was redefined (Baeza et al. 2016). The molecular data, based on chloroplast DNA (*rpl*32-*trn*L region), support the separation of Coastal populations of *A. ligtu* subsp. *ligtu* from populations of the inland distribution range.

5.9 Use of Plant Genetic Resources.

5.9.1 Major Constraints in the Crop Production

A wide diversity of *Alstroemeria* genetic resources has provided valuable genes for breeding of novel cultivars. There is a demand for new, novel *Alstroemeria* cultivars from the production industry. *Alstroemeria* has a wide variety of valuable characters such as flower size, color, flowering time, and resistance to different pathogens. Combining these vital horticultural traits into one cultivar by crossing is almost the only way to obtain introgression of traits. However, genetic transformation approaches are also not developed in *Alstroemeria*. Possibilities for cross combinations in *Alstroemeria* between the species are limited due to embryo abortion. In vitro embryo rescue has been demonstrated to be necessary in order for interspecific hybridization of *Alstroemeria* to be successful (Bridgen 1994; Lu and Bridgen 1996, 1997). Using this technique, many interspecific hybrids have successfully been made.

Interspecific hybrids that have been rescued by embryo culture are often sterile because of the genotypic differences of their parents. To improve and restore the fertility of these hybrids, chromosome numbers can be doubled by using chemical treatments to obtain tetraploids or other polyploids (Lu and Bridgen 1997).

5.9.2 Common Sources Used to Overcome Production Constraints

In the early 1950s, three *Alstroemeria* species were released into Europe – *A. pelegrina, A. ligtu*, and *A. aurea*. Since then, the interest in *Alstroemeria* as an ornamental has increased. Up till now, more than 60 species/genotypes have been released onto commercial markets by applying conventional breeding techniques. One problem found in conventional breeding is the lack of useful genes in *Alstroemeria* germplasm for use in further breeding. The majority of the *Alstroemeria* cultivars are polyploid, which makes breeding time consuming (Chevalier 1994). However, new cultivars have been produced by using interspecific hybridization in the last decades (De Jeu and Jacobsen 1995). Furthermore, cross-hybridization does not always lead to seed set, although some hybrids were produced by using embryo rescue techniques. The slow process of breeding delays the introduction of new cultivars to the commercial market. In *Alstroemeria* different breeding strategies can be envisaged, including (1) Introduction, (2) Hybridization: Techniques for overcoming post-zygotic incompatibility barriers, Bridging species techniques, Embryo rescue technique, In vitro/in vivo embryo culture, Ovary culture, Ovule culture, In vitro pollination, fertilization, and ovule culture, (3) Mutation Breeding, (4) Polyploidy Breeding, (5) Biotechnological tools: Tissue culture, Genetic Engineering: Agrobacterium-mediated transformation system and Particle bombardment, and (6) Advanced Selection Methods: Marker-assisted breeding – RAPD, AFLP, Cell Sorting (Individual cells are selected by leading them in a liquid flow through an apparatus that can detect certain desired traits by measuring fluorescence or light scattering, the flow cytometer (FCM)).

5.9.3 Breeding Options

Breeding Objectives in *Alstroemeria* may consist of development of varieties with:

- Novel color.
- Winter hardy cultivars.
- Dwarfness for pots, suitable for landscaping and gardening. So far, some interesting hybrids have been developed by conventional breeding and embryo rescue but the main challenge ahead is the flower shape and its size.
- Variegated leaves.
- Longer duration of flowering in field.
- Increase in the heat tolerance of *Alstroemeria* clones by using heat tolerant Brazilian species in breeding programs.
- Varieties with special flower aroma, a new trait for this crop, using hybrids with *A. cryophyllaea.* There have been developments of breeding specimens with the aroma, but lacking the desired flower shape that could make this flower competitive.
- New scented pot plants.
- With a distinctive character, to commercialize them not only as part of bouquets but also as single flowers.
- Longer vase life.
- Removal of allergen content.
- Resistance to virus and diseases. To develop a transformation system to produce resistant *Alstroemeria* plants against *Alstroemeria Mosaic Virus* (AlMV), which is one of the most dangerous and endemic viruses in *Alstroemeria.*
- Resistance to insects and pests.

5.9.3.1 Introduction

Alstroemeria and *Bomarea* (Alstroemeriaceae) were evaluated for different horticultural traits in Japan. The flower shape in *Alstroemeria psittacina* Lehm. was found to be similar to that in *Bomarea coccinea.* The length/width ratio of *B. salsilla* (L.) Mirb. was intermediate compared with that of the other species. These preliminary data will be useful in selecting wild species in order to examine interspecific or intergeneric hybridizations in the breeding of Alstroemeriaceae (Kashihara et al. 2010). Aros and Rogers (2013) evaluated new fragrant lines of *A. cryophyllaea* in Chile. The study revealed that some of the characters observed were below the standards suggested by the market for *Alstroemeria* cut flowers (i.e., stem length). However, these lines showed a distinctive, and appreciated floral scent, and these might therefore be suitable as a new commercial product and considered as a promising starting point for further breeding through other methodologies.

5.9.3.2 Hybridization

Mainly two types of interspecific hybrids of *Alstroemeria* have been used for the development of cultivars. These include inter-Chilean species hybrids and hybrids between the Chilean and Brazilian species. The variation among and within species opens up the possibility for production of a wide range of new cultivars suited for various purposes, such as cut flowers, pot plants, and garden and landscape uses (Hoshino 2008). By using fluorescence microscopy, de Jeu and Jacobsen (1995) observed the entry of pollen tubes into ovules and the subsequent fertilization and embryo development in interspecific diallel crossing among *A. aurea, A. hookeri, A. pulchra, A. ligtu, A. pelegrina, A. inodora*, and *A. brasiliensis*. They described the post-fertilization barrier-induced incompatibility in many interspecific crosses of *Alstroemeria*. This result indicated that embryo rescue before abortion is effective in obtaining successful interspecific hybridizations. To ensure success of embryo rescue, many culture conditions were examined, such as using various sucrose concentrations, culture temperatures, and basal media. The most reliable embryo rescue techniques for the production of interspecific hybridizations in *Alstroemeria* are summarized in Table 3.

Bridgen et al. (2009) developed breeding procedures to breed novel, commercially valuable cold-hardy and fragrant flowered cultivars of *Alstroemeria* with the assistance of in vitro techniques such as ovule embryo rescue, micropropagation, and somatic embryogenesis. Winter-hardy hybrids were developed by using the Chilean species *Alstroemeria aurea*; fragrant hybrids were developed by using the Brazilian species, *Alstroemeria caryophyllaea*. At Dr YS Parmar University of Horticulture and Forestry, Solan, HP; new genotypes with attractive and unique colors are being developed through open pollinated seedling selection in selected cultivars.

5.9.3.3 Mutation Breeding

At the moment *Alstroemeria* is one of the major products in the cut flower industry and the breeding of new cultivars of this species have been based on the induction of mutagenesis through irradiation of rhizomes with x-rays and gamma rays. *Alstroemeria* mutagenesis has been focused in changing few traits in outstanding cultivars. Even though most of mutations are recessive and negative, variation in interesting traits are still relatively easy to obtain by mutation induction in *Alstroemeria*. Mutation in case of *Alstroemeria* can only be applied at actively growing rhizomes of young plants. Even though the biochemistry base of mutation is still not completely clear, it is thought that mutagenic treatments can generate chromosome rearrangements or produce changes in some genes to other allelic forms, explaining the phenotypic variation. In vitro culture techniques can be considered as an appropriate tool for inducing mutations because it can supply a large amount of homogeneous and virus free individuals to be irradiated. In addition, in vitro culture also provides an efficient propagation system for irradiated material allowing assessment of mutant stability and easy later selection.

Mutations breeding in *Alstroemeria* are used for the induction of different traits like color and size of flowers, shape, and stripes on the tepals, plant vigor,

Table 3 List of Interspecific Hybridizations in *Alstroemeria*

Cross Combination	(Harvest of Ovule) Days after pollination	Culture Condition	Reference
Diallel cross *A. aurea, A. pelegrina, A. magnifica, A. inodora, A. psittacina*	14	1/4× MS macronutrients, MS micronutrients and vitamins 400 mg l^{-1} casein hydrolyzate 60 g l^{-1} sucrose, 21 °C dark, liquid culture on rotary shaker	Buitendijk et al. 1995
Diallel cross *A. aurea, A. hookeri, A. pulchra, A. ligtu, A. pelegrina, A. inodora, A. brasiliensis*	2	MS medium, 9% sucrose 21 °C, 12 h photoperiod	De Jeu and Jacobsen 1995
A. violacea × A. pelegrina var. Alba A. violacea × A. pelegrina var. rosea A. violacea × A. ligtu ssp. Incarnata A. pelegrina var. alba × A. Violacea A. violacea × A. 'Rosy Wing' A. 'UC12' × A. violacea A. gayana × A. violacea A. 'ER292' × A. pelegrina var. alba A. 'UC12' × A. pelegrina var. alba	7	MS medium, 146 mg l^{-1} glutamine, 30 g l^{-1} sucrose, 6.5 g l^{-1} agar, 18 °C ± 1 °C dark	Lu and Bridgen 1996
A. ligtu hybrid × A. pelegrina var. rosea	7, 14, 21, 28, 35	MS medium, 3% sucrose, 2 g l^{-1} gellan gum, 20 °C, 16 h photoperiod	Ishikawa et al. 1997
A. pelegrina var. rosea × A. magenta	7–14	MS medium 3% sucrose, 2 g l^{-1} gellan gum 20 °C, 16 h photoperiod	Ishikawa et al. 2001
Diallel cross *A. angustifolia, A. aurea, A. diluta, A. garaventae, A. hookeri, A ligtu, A. magnifica, A. magenta, A. pelegrina, A. presliana, A. pulchra, A. versicolor, A. zoellneri, A. inodora, A. psittacina*	14	1/2 MS medium 3% sucrose, 2 g l^{-1} gellan gum 20 °C, 16 h photoperiod	Shinoda and Murata 2003

productivity and blooming time, early flowering, extended duration of flowering, and tolerance to low light conditions. A study was conducted to find out appropriate dosage of gamma irradiation on *Alstroemeria aurea* G. in vitro rhizomes for breeding purposes (Aros et al. 2012). In vitro cultured rhizomes of *Alstroemeria aurea* were irradiated with a gamma source using different dosages to evaluate the direct effect produced. The results revealed that both mortality and weight increased depending on dosage. All irradiated rhizomes showed early sprouting in comparison with control (0 Gy) and no significant difference in final number of shoots after 61 days among irradiated treatments was observed. Bleaching and necrosis was observed in all irradiated rhizomes and was more evident at higher doses. LD_{50} was

Table 4 Polish commercial mutant cultivars of *Alstroemeria*

Name of cultivars	Conditions Parentage	Main improved attributes of cv (in comparison to parent)
Catalina	Gamma rays, 3 Gy Sterile mutant	Bright violet flowers (brown flowers)
Ines	Gamma rays, 6 Gy Sterile mutant	Salmon pink flowers, all year round bloom (red flowers, seasonal bloom)
Juanita	Gamma rays, 3Gy Sterile hybrid	Orange red flowers (dark red flowers)
Carlota	Fast neutrons, 3Gy Sterile mutant	Orange brown flowers with violet blush on the outer whorl petals (brown flowers)
Isabel	Gamma rays, 5 Gy Sterile mutant	Yellow flowers (brown flowers)
Matilde	Gamma rays, 6 Gy Sterile mutant	Purple red flowers (dark red flowers)
Azucena	Gamma rays, 10 Gy Sterile hybrid	White-pink flowers with reduced stripping (pink flowers)
Erendira	Gamma rays, 5 Gy Sterile hybrid	Red-violet flowers (brown-red flowers)

established at about 40 Gy and the optimum dosage to induce mutation was suggested between 2.5 and 5 Gy, when the growth was reduced in 50%, and probably this dosage could be used for breeding purposes. Actively growing rhizomes of *Alstroemeria* sterile hybrids and sterile mutants were irradiated with gamma rays (3 to 7 Gy) and fast neutrons (3 to 6 Gy) for creation of new *Alstroemeria* genotypes (Przybyla 2000) (Table 4). There are 35 released varieties in *Alstroemeria* developed through induced mutations as reported by Schum (2003).

Some of the commercial mutant varieties of *Alstroemeria* evolved by breeders/companies by X-ray are given in Table 5.

Mutation breeding has also been utilized for reducing allergin content in *Alstroemeria. Alstroemeria* may cause severe allergic contact dermatitis for a large part of the personnel working in the cut flower industry. For sensitized (allergic) persons, *Alstroemeria* contact dermatitis normally develops on the hands and finger tips ("tulip fingers"), although severe eczematous reactions at the forearms, face, and neck have also been observed. The allergen content in *Alstroemeria* was determined by high performance liquid chromatography (HPLC) in order to investigate breeding possibilities for reducing the risk of contact dermatitis caused by *Alstroemeria* (Kristiansen and Christensen 1998).

Table 5 Commercial mutant varieties of *Alstroemeria* evolved by breeders/companies

Variety	Parent	Type of mutagen
Canaria	Orchid Flower	X-ray
Capitol	Carmen	X-ray
Fanfare	Carmen	X-ray
Harlequin	Paringo's Charm	X-ray
Harmony Stabrons	Regina	X-ray
Result	Carmen	X-ray
Rosali Staliro	Starosa	X-ray
Rosita Stareza	Regina	X-ray
Trident	Carmen	X-ray
Valient	Carmen	X-ray
White Wings	Orchid Flower	X-ray
Yellow Tiger	Orchid Flower	X-ray
Zebra Stazeb	Orchid Flower	X-ray
Zenith	Carmen	X-ray

A study was conducted by Kristiansen and Christensen (1998) to reduce allergin content in *Alstroemeria* by use of mutation breeding. The investigated plants consisted of 28 Chilean and 1 Brazilian species, a large number of interspecific hybrids presumably at the diploid level, "Butterfly" hybrids (*A. pelegrina* x *A. pulchella*) such as *A.* "Rosy Wings" at the tetraploid level, and hybrids with *A.* "Rosy Wings" presumably at the triploid level. The variation in allergin content within populations of some Chilean species was analyzed. One "Butterfly" hybrid clone was transplanted from in vitro conditions monthly to estimate environmental influences (season and growth conditions). Further, other clones were analyzed several times. In vitro plants from three unnamed tetraploid Butterfly hybrids were irradiated with 0, 5, or 10 Gy ϒ rays. Plants consisted of a rhizome piece with approximately two aerial shoots and after radiated they were subcultured in vitro twice prior to root formation and flower induction. Plants were then transplanted to greenhouse conditions and grown until flowering. Flowering stems were collected for chemical analysis. The results revealed that allergin concentrations were highest in flowers and lowest in leaves. 6-tuliposide A, D, and E as well as tulipalin A were found, whereas tuliposide B was not detected. Amount of 6-tuliposide A exceeded that of other allergens in leaves and stems. Tulipalin A was found in the flowers. Environmental effects (season and growth conditions) on the allergen content were significant. Positive correlations for 6-tuliposide A between plant parts and between 6-tuliposide A and tulipalin A within leaves and stems were found. Tuliposide D was not correlated with the other allergens. Significant narrow sense heritabilities were found for total allergen content as well as for 6-tuliposide A and tuliposide D, but not for tulipalin A. The heritability was higher for total allergen content and tuliposide D as compared to 6-tuliposide A. After ϒ-radiation of three tetraploid clones, plants with a reduced content of 6-tuliposide A in leaves and/or flowers were found in one of the clones. Tulipalin A content was not affected after ϒ-radiation in neither of the clones. Four months after the first determination, plants with low content of allergens were still found to have less than the control plants.

5.9.3.4 Polyploidy Breeding

Interspecific hybrids between *A. ligtu* hybrids (LH) and *A. pelegrina* var. *rosea* (PR) were produced by ovule culture, the hybrids (LHxPR) exhibited an extremely low pollen fertility and failed to produce normal seeds by self-pollination. The sterility associated with meiotic abnormalities in the F1 hybrid may be overcome through amphidiploidization by chromosome doubling effected by colchicine treatment. Ishikawa et al. (1999) described two methods of colchicine treatment of the interspecific hybrid LH X PR: by in vitro treatment of ovules and by the treatment of the rhizomes of the mature plants. Lu and Bridgen (1997) studied the Chromosome doubling and fertility study of *Alstroemeria aurea* × *A. caryophyllaea*. Self-pollinations of diploid (2n = 2x = 16) interspecific hybrid *Alstroemeria aurea* × *A. caryophyllaea* resulted in no seeds. Backcross of the hybrid with parent *A. aurea* did not produce any seeds. In an attempt to restore the hybrid fertility, an efficient in vitro procedure has been developed and applied effectively in the chromosome doubling of the diploid hybrid. Results revealed that 41% of the treated plants were proven to be truly tetraploid by chromosome counts and stomatal measurements after applying 0.2 to 0.6% colchicine for 6 to 24 h. Over 87.5% of these colchicine-induced tetraploids were stable and retained their tetraploidy after one year of growth. Cytological studies on the pollen mother cells (PMCs) of the sterile diploid hybrids revealed abnormal meiotic behaviors. In addition, aneuploid chromosome numbers, ranging from 2n = 1 to 2n = 18, were observed in over 45% of the PMCs examined. PMCs of the colchicine-induced tetraploids showed that meiotic chromosome pairings were normal in most cases. These results indicate that the sterility of this hybrid is not only caused by parental chromosome differences, but other complex fertility/sterility-regulating mechanisms are involved too.

5.9.3.5 Biotechnological Approach

Somatic Embryogenesis

Conventional breeding techniques have been used in *Alstroemeria* to create new and attractive cultivars with new colors, longer vase life, high yield of flowers, and resistance to diseases. However, the genes for many agriculturally useful traits are not present in the *Alstroemeria* gene pool. In those cases, genetic modification can be used. However, genetic modification requires regeneration protocols with high efficiency and reproducibility. In the past decades, multiplication of *Alstroemeria* via tissue culture was carried out mainly by rhizome splitting (Lin et al. 1998). Breeding companies and farmers use rhizome splitting in commercial propagation. Some *Alstroemeria* species, especially the "Butterfly type," have shown significantly low propagation efficiency (Buitendijk et al. 1992). Therefore, a "Butterfly type" was used in this study. A large number of new cultivars have been developed through embryo rescue. However, neither embryo rescue techniques nor rhizome division is suitable for genetic modification due to the low efficiency and the non-adventitious character of the regeneration system. Adventitious regeneration is a precondition for genetic engineering.

Lin et al. (2000a) developed a plant regeneration system from leaf and stem explants. Organogenesis from leaf and stem has not been considered possible in monocotyledons. The development of culture system from leaf and stem in *Alstroemeria* is valuable, although successful organogenesis from leaf and stem is strongly genotype-dependent at the present. The approach described by Lin et al. (2000a) had several disadvantages including the low frequency of FEC initiation, lack of repeatability, and the long period of time required to initiate FEC. The leaves in the first and second positions with axil in the shoots produced the highest frequency of embryogenic callus. By utilizing the embryogenic callus culture in *Alstroemeria*, isolation of protoplasts, subsequent protoplas culture, and regeneration of plants from protoplasts were achieved (Kim et al. 2005). Such techniques will promote research on somatic hybridization between cross-incompatible species as well as on direct gene transfer with electroporation.

Protoplast Fusion

In some cases cross-incompatibility was overcome by ovule culture, breeders are still looking for new techniques to transfer genes of interest to current *Alstroemeria* cultivars. One such technique, used successfully in other crops to overcome sexual incompatibility, is protoplast fusion. Protoplast fusion requires the development of a regeneration system from protoplasts. Furthermore, an efficient and reliable regeneration system from protoplasts opens up new opportunities for genetic transformation via direct DNA uptake or electroporation, as was shown in many other crops. Friable embryogenic callus (FEC) proved to be the best source for protoplast isolation and culture when compared with leaf tissue and compact embryogenic callus. Protoplast-derived plants showed more somaclonal variation than vegetatively propagated control plants. Somatic hybridization and genetic modification may enable us to enhance the genetic variation of ornamental plants and improve them by employing these techniques in combination with conventional breeding (Kim et al. 2005).

In Vitro Fertilization

Novel regeneration systems established in *Alstroemeria* from isolated egg cells, in vitro-fertilized egg cells, and zygotes can be expected to be used for further breeding of the species of this genus through direct gene transfer and in vitro fertilization between sexually incompatible species. In vitro fertilization is considered to be a useful approach for the generation of interspecific hybrids within this genus. To facilitate future study of in vitro fertilization, procedures were developed for the isolation of egg cells and zygotes produced in vivo in *Alstroemeria aurea,* using enzymatic treatments and microdissection with glass needles. These procedures may be utilized for in vitro fertilization, enabling artificial fusion between a single isolated egg cell and sperm cell. This concept can be applied for obtaining crosses of sexually incompatible *Alstroemeria* species through in vitro fertilization.

Genetic Transformation

Particle bombardment and *Agrobacterium*-mediated procedures were applied for genetic transformation, resulting in the generation of transformed plants containing the marker genes. Commercially important traits such as virus resistance and delayed post-harvest leaf yellowing, the essential genes have not yet been detected in the *Alstroemeria* gene pool. In these cases, gene transformation mediated by *Agrobacterium* or particle bombardment is considered to be more promising because target genes from other unrelated species can be used. Foreign gene expression examined with a β-glucuronidase (GUS) assay indicated the reducible transformation conditions in which the GUS gene was driven by the maize ubiquitin promoter; suitable parameters for the assay equipment have been reported. Although stable transformed calluses were obtained after visual selection based on luciferase activity, no plant regeneration was observed during the 2 years of maintenance. Subsequently, Lin et al. (2000b) succeeded in producing transgenic plants by particle bombardment. They used a luciferase gene from a firefly as the reporter gene under the control of the maize ubiquitin promoter (Ubi1) and attempted transferring this gene into friable embryogenic calluses and proembryos. Following visual selection, transgenic plantlets were regenerated through embryogenesis. Moreover, another selection by phosphinothricin (PPT) helped in obtaining bar and GUS gene-incorporated plants after particle bombardment. The successful transformation obtained by Lin et al. (2000b) appears to be due to the selection of elite embryogenic callus lines with a high regeneration potential and maintaining these cells till abiolistic procedure was performed. Hoshino et al. (2000) reported the genetic transformation of *Alstroemeria* via co-cultivation of embryogenic suspension cells with *Agrobacterium tumefaciens* using ovule culture of an interspecific cross between *A. pelegrina* and *A. magenta*. The suspension cells were co-cultivated with *A. tumefaciens* strain EHA101/pIG121Hm or LBA4404/pTOK233. Transgenic plants were obtained from co-cultivated cells after selection on a hygromycin-containing medium. Examples of the production of transgenic plants in *Alstroemeria* are given in Table 6.

Meristem Culture

Alstroemeria plants are primarily vegetatively propagated, and therefore, viral infections are a serious concern. Although several reports have appeared describing the

Table 6 Examples of the production of transgenic plants in *Alstroemeria*

Methods[a] (used explant[b])	Reporter gene[c]	Selectable marker[c]	Reference
P.B. (FEC)	Luc, GUS	PPT	Lin et al. 2000b
A.T. (FEC)	GUS	Hyg and Kan	Akutsu et al. 2004a
A.R. (FEC)	GUS	Hyg and Kan	Akutsu et al. 2004b

[a]P.B, Particle bombardment; A.T, *Agrobacterium tumefaciens*-mediated transformation; A.R, *Agrobacterium rhizonenous*-mediated transformation
[b]FEC (Friable embryogenic callus)
[c]Hyg (Hygromycin), Kan (Kanamycin), Gus (β-glucuronidase), Luc (luciferase), PPT (phosphinotricin)

attempts to produce virus-free *Alstroemeria* plants using a meristem culture (Chiari and Bridgen 2002), the presence of viruses continues to be an important problem in *Alstroemeria* species, as meristem-derived *Alstroemeria* plants can be re-infected during harvest or transport. A number of viruses are currently considered critical in *Alstroemeria*. Examples include Alstroemeria carla virus (AlCV), Alstroemeria mosaic virus (AlMV), Cucumber mosaic virus (CMV), Tomato spotted wilt virus (TSWV), Alstroemeria streak virus (AlSV), and Impatiens necrotic spot virus (INSV). Of these viruses, AlMV is considered to be the most endemic and is difficult to detect and prevent in *Alstroemeria* culture (Bellardi et al. 1994).

Alstroemeria plants were transformed with viral sequences [coat protein (CP) gene and 3′-nontranslated region (3′-NTR)] of Alstroemeria mosaic virus (AlMV) using an improved particle gun-mediated transformation system (Kim 2020). Friable embryogenic callus (FEC) induced from the leaves with axil tissues of *Alstroemeria* plant was used as target tissue. As a result, two bombardments with a pre- and post-bombardment culture (8 h before and 16 h after transformation) resulted in the greatest efficiency of transformation and recovery of transgenic lines. The phenomenon of virus resistance was observed at different levels when transgenic *Alstroemeria* plants containing the viral sequences were evaluated for infection after challenging with AlMV. Establishment of an efficient transformation system in *Alstroemeria* will allow the insertion of transgenes to acquire resistance to viral and fungal pathogens.

5.10 Future Perspective

The understanding of functioning of flowering genes at the cellular level and in the plant as a whole is the present-day challenge in the research of geophytes' floral biology. It is crucial in order to understand the molecular regulation of flowering diversity under various environmental conditions and during flower senescence. One of the reasons that only limited molecular information is available on *Alstroemeria* is the huge size of the genus's genome. Investigations in the coming decade are expected to improve our understanding of this process, thus contributing to the development of breeding techniques and the production of new ornamental hybrids.

In addition to all other horticultural traits, introduction of disease and virus resistance into new cultivars have to be taken into thoughtful consideration by breeders, if it has to survive as a crop. Fortunately, the variation among and within species and the many possibilities for species hybridization opens up the possibility for production of a wide range of new cultivars suited for various purposes, such as cut flowers, pot plants, and garden and landscape uses. Besides, several of the wild populations are nowadays threatened and with risk of extinction. Most of the species occupies reduced geographical areas. Species that inhabit plain regions are vulnerable because those fields are burned and cultivations are introduced. Patagonian species like *A. pseudospathulata* or *A. patagonica* need to be protected especially, from the grazing of the animals, which causes their declination. The cultivation of the native plants would help to the ex situ conservation. Reintroduction in some natural areas might also play an important role in order to maintain the landscape value. This indicates that there might be scope to discover more useful genetic variation than hitherto has been done. Keeping this in mind, the given lines of work may be contemplated for the future:

- Wild *Alstroemeria* species like, *A. ligtu*, *A. pelegrina*, *A. aurea*, *A. magenta*, *A. psittacina*, *A. inodora*, *A. pulchra*, *A. violacea*, *A. versicolori,* and *A. caryophyllaea* are presumed to be candidates for further breeding.
- Development of hybrids by cross-combinations between distantly related genera such as *Alstroemeria* and *Bomarea*.
- Production of *Alstroemeria* cultivars totally without allergens must be considered very difficult or even impossible.
- An efficient in vitro chromosome doubling technique and cytological studies on the mechanisms of hybrid sterility are valuable to the genetic and breeding studies on *Alstroemeria*.
- There is a need to develop precise and appropriate embryo rescue or ovule culture systems.
- Novel regeneration systems established in *Alstroemeria* from isolated egg cells, in vitro-fertilized egg cells, and zygotes can be expected to be used for further breeding of the species of this genus through direct gene transfer and in vitro fertilization between sexually incompatible species.
- The first generation of virus-resistant *Alstroemeria* plants containing viral sequences was produced by particle bombardment holds clear promises for the future but more research is needed to obtain virus-resistant plants, which also might be of practical value.
- Further research is needed before useful genes can be successfully incorporated into *Alstroemeria* and the safety of genetically modified plants can be evaluated.

References

Aker S, Healy E (1990) The phytogeography of the genus *Alstroemeria*. Herbertia 46:76–87

Akutsu M, Ishizaki T, Sato H (2004a) Transformation of the monocotyledonous *Alstroemeria* by *Agrobacterium tumefaciens*. Plant Cell Rep 22:561–568

Akutsu M, Ishizaki T, Sato H (2004b) Transformation of the monocot *Alstroemeria* by *Agrobacterium rhizogenes*. Mol Breed 13:69–78

Aros D, Rogers H (2013) Floral scent as a new attribute for *Alstroemeria*: evaluation of new fragrant lines of *A. caryophyllaea*. Acta Hortic 1002:207–214

Aros D, Meneses C, Infante R (2006) Genetic diversity of wild species and cultivated varieties of alstroemeria estimated through morphological descriptors and RAPD markers. Sci Hortic 108(1):86–90

Aros D, Olate E, Valdes S, Infante R (2012) Gamma irradiation on *Alstroemeria aurea* G. *in vitro* rhizomes: an approach to the appropriate dosage for breeding purposes. Rev FCA UNCUYO 44(1):191–197

Aros D, Spadafora N, Ventur IM, Nunez-Lillo G, Meneses C, Methven L, Rogers H (2015) Floral scent evaluation of segregating lines of *Alstroemeria caryophyllaea*. Sci Hortic 185:183–192

Aros D, Suazo M, Rivas C, Zapata P, Ubeda C, Bridgen M (2019) Molecular and morphological characterization of new interspecific hybrids of alstroemeria originated from *A. caryophylleae* scented lines. Euphytica 215:93. https://doi.org/10.1007/s10681-019-2415-4

Baeza C, Schrader O, Budahn H (2007) Characterization of geographically isolated accessions in five *Alstroemeria* L. species (Chile) using FISH of tandemly repeated DNA sequences and RAPD analysis. Plant Syst Evol 269:1–14

Baeza C, Schrader O, Ruiz E, Negritto M (2008) *Alstroemeria presliana* HERB. (Alstroemeriaceae) in Chile from a cytogenetic perspective. Chilean J Agric Res 68(4):328–333

Baeza C, Ruiz E, Negritto M (2010) Comparative karyotypic analysis in the *Alstroemeria hookeri* Lodd. (Alstroemeriaceae) complex sensu Bayer (1987). Genet Mol Biol 33:119–124

Baeza CM, Finot VL, Ruiz E, Carrasco P, Novoa P, Stuessy T, González A (2016) Comparative karyotypic analysis and cytotaxonomy in the *Alstroemeria ligtu* L. (Alstroemeriaceae) complex of Chile. Rev Bras Bot 39:305–313

Bayer E (1987) Die Gattung *Alstroemeria* in Chile. Mitt Bot Staatssamml (München) 24:1–362

Bellardi MG, Bertaccini A, Betti L (1994) Survey of viruses infecting alstroemeria in Italy. Acta Hortic 377:73–80

Bridgen MP (1994) A review of plant embryo culture. HortScience 29:1243–1246

Bridgen M (2001) A nondestructive harvesting technique for the collection of native geophyte plant species. Herbertia 56:51–60

Bridgen MP (2006) Using traditional and biotechnological breeding for new plant development. Comb Proc Int Plant Propag Soc 56:307–310

Bridgen MP, Langhans R, Graig R (1989) Biotechnological breeding techniques for Alstroemeria. Herbertia 45:93–96

Bridgen MP, Olate E, Schiappacasse F (2000) Flowering geophytes from Chile. Acta Hortic 570: 75–80

Bridgen MP, Kollman EE, Lu C (2009) Interspecific hybridization of Alstroemeria for the development of new, ornamental plants. Acta Hortic 836:73–78

Broertje C, Verboom H (1974) Mutation breeding of Alstroemeria. Euphytica 23:39–44

Buitendijk JH, Ramanna MS, Jacobsen E (1992) Micropropagation ability: towards a selection criterion in *Alstroemeria* breeding. Acta Hortic 325:493–498

Buitendijk JH, Pinsonneaux N, van Donk AC, Ramanna MS, van Lammeren AAM (1995) Embryo rescue by half-ovule culture for the production of interspecific hybrids in *Alstroemeria*. Sci Hortic 64:65–75

Chacón J, de Assis MC, Meerow AW, Renner S (2012) From East Gondwana to Central America: historical biogeography of the Alstroemeriaceae. J Biogeogr 39:1806–1818

Chevalier F (1994) Hybrid detection in *Alstroemeria* by the use of species specific repetitive probes. MSc thesis, Wageningen University, Wageningen, p 58

Chiari A, Bridgen MP (2002) Meristem culture and virus elimination in Alstroemeria. Plant Cell Tissue Organ Cult 68:49–55

de Assis MC (2001) *Alstroemeria* L. (Alstroemeriaceae) do Brasil. PhD thesis, Universidade de Sao Paulo, Sao Paulo

de Assis MC (2009) New species of *Alstroemeria* (Alstroemeriaceae) from Brazil. Novon: J Bot Nomencl 19(2):145–149

De Jeu MJ, Jacobsen E (1995) Early postfertilization ovule culture in *Alstroemeria* L. and barriers to interspecific hybridization. Euphytica 86:15–23

De Jeu MJ, Sasbrink H, Garriga Calderé F, Piket J (1992) Sexual reproduction biology of Alstroemeria. Acta Hortic 325:571–575

De Jeu M, Lasschuit J, Chevalie RF, Visser RGF (1995) Hybrid detection in *Alstroemeria* by use of species specific repetitive probes. Acta Hortic 420:62–64

Donoso A, Rivas C, Zamorano A, Peña Á, Handford M, Aros D (2021) Understanding Alstroemeria pallid flower colour: links between phenotype, anthocyanins and gene expression. Plan Theory 10:55. https://doi.org/10.3390/plants10010055

Dubouzet JG, Murata N, Shinoda K (1997) RAPD analysis of genetic relationships among *Alstroemeria* L. cultivars. Sci Hortic 68:181–189

Finot VL, Baeza CM, Ruiz E, Toro O, Carrasco P (2018) Towards an integrative taxonomy of the genus Alstroemeria (Alstroemeriaceae) in Chile: a comprehensive review, selected studies in biodiversity, Bülent Şen and Oscar Grillo. IntechOpen. https://doi.org/10.5772/intechopen.71823

Grant V (1991) The evolutionary process. A critical study of evolutionary theory, 2nd edn. Columbia University Press, New York, p 487

Han TH, De Jeu M, Van Eck H, Jacobsen E (2000) Genetic diversity of Chilean and Brazilian *Alstroemeria* species assessed by AFLP analysis. Heredity 84:564–569

Han T-H, van Eck HJ, De Jeu MJ, Jacobsen E (2002) Mapping of quantitative trait loci involved in ornamental traits in *Alstroemeria*. HortScience 37(2):585–592

Hang A, Tsuchiya T (1988) Chromosome studies in the genus Alstroemeria II. Chromosome constitutions of eleven additional cultivars. Plant Breed 100:273–279

Hannibal LS (1942) *Alstroemeria* and *Bomarea* from seeds. Herbertia 9:183–187

Healy WE, Wilkins HF (1982) Reponse of *Alstroemeria* 'Regina' to temperature treatments prior to flower-inducing temperatures. Sci Hortic 17:383–390

Healy WE, Wilkins HF (1985) Alstroemeria. In: Haley AH (ed) Handbook of flowering 1. CRC Press, Boca Raton, pp 419–424

Healy WE, Wilkins HF (1986) Relationship between rhizome temperatures and shoot temperatures for flower initiation and cut flower production of *Alstroemeria* 'Regina'. J Am Soc Hortic Sci 111:94–97

Hirai M, Kamimura T, Kanno A (2007) The expression patterns of three class B genes in two distinctive whorls of Petaloid Tepals in Alstroemeria ligtu. Plant Cell Physiol 48(2):310–321

Hofreiter A, Rodríguez EF (2006) The Alstroemeriaceae of Peru and neighbouring areas. Rev Peruana Biol 13:5–69

Hoshino Y (2008) Advances in *Alstroemeria* biotechnology. Flori Ornamentl and Plant Biotech, vol 51. Global Science Books

Hoshino Y, Sato H, Murata N, Shinoda K (2000) Genetic transformation of the interspecific hybrid of *Alstroemeria* via co-cultivation of embryogenic suspension cells with *Agrobacterium tumefaciens*. The Asia Pacific Conference on Plant Tissue Culture and Agribiotechnology, 93 pp

Hoshino Y, Igarashi T, Ohshima M, Shinoda K, Murata N, Kanno A, Nakano K (2014) Characterization of CYCLOIDEA-like genesincontrollingfloral zygomorphy. In the monocotyledon Alstroemeria. Sci Hortic 169:6–13

Hunziker JH, Xifreda CC (1990) Chromosome studies in *Bomarea* and *Alstroemeria* (Alstroemeriaceae). Darwin 30(1/4):179–183

IPGRI (1995) Descriptors for capsicum (*Capsicum* spp.), IPGRI, ISBN 92-9043-216-0, Via delle Sette Chiese, 142, 00145 Rome, Italy

Ishikawa T, Takayama T, Ishizaka H, Ishikawa K, Mii M (1997) Production of interspecific hybrids between *Alstroemeria ligtu* L. hybrid and *A. pelegrina* L. var. *rosea* by ovule culture. Breed Sci 47:15–20

Ishikawa T, Takayama T, Ishizaka H (1999) Amphidiploids between *Alstroemeria ligtu* L. hybrid and *A. pelegrina* L. var. *rosea* induced through colchicine treatment and their reproductive characteristics. Sci Hortic 80:235–246

Ishikawa T, Takayama T, Ishizaka H, Ishikawa K, Mii M (2001) Production of interspecific hybrids between *Alstroemeria pelegrina* L. var. *rosea* and *A. magenta* Bayer by ovule culture. Euphytica 118:19–27

Jorgensen PM, Nee MH, Beck SG (2014) Catálogo de las plantas vasculares de Bolivia. Monographs in systematic botany from the Missouri Botanical Garden, vol 127. Missouri Botanical Garden Press, pp 1–1744

Joung HY, Cantor M, Kamo K (2006) Cryopreservation of *Gladiolus* cultivars. Acta Hortic 760: 225–231

Kashihara Y, Hirano T, Murata N, Shinoda K, Araki H, Hoshino Y (2010) Evaluation of pre-fertilization barriers by observation of pollen tube growth and attempts for overcoming post-fertilization barriers in intergeneric hybridization between *Alstroemeria* and *Bomarea* by ovule culture. Acta Hortic 855:159–164

Kashihara Y, Shinoda K, Murata N, Araki H, Hoshino Y (2011) Evaluation of the horticultural traits of genus *Alstroemeria* and genus *Bomarea* (Alstroemeriaceae). Turk J Bot 35:239–245

Kim JB (2020) Production of transgenic Alstroemeria plants containing virus resistance genes via particle bombardment. J Plant Biotechnol 47:164–171

Kim JB, Bergervoet JEM, Raemakers CJJM, Jacobsen E, Visser RGF (2005) Isolation of protoplasts, and culture and regeneration into plants in *Alstroemeria*. In Vitro Cell Dev Biol Plant 41:505–510

Kristiansen K, Christensen LP (1998) Allergen contents in *Alstroemeria* can be reduced by breeding. Euphytica 101:367–375

Lee JH, Han TH (2006) Identification of parental species of the Alstroemeria cv. 'Jubilee' using AFLP marker technique. Sci Hortic 111:63–67

Leverentz MK, Wagstaff C, Rogers HJ, Stead AD, Chanasut U, Silkowski H, Thomas B, Weichert H, Feussner I, Griffiths G (2002) Characterization of a novel lipoxygenase-independent senescence mechanism in *Alstroemeria peruviana*floral tissue. Plant Physiol 130(1):273–283

Lin HS, de Jeu MJ, Jacobsen E (1998) Formation of shoots from leaf axils of *Alstroemeria*: the effect of the position on the stem. Plant Cell Tissue Organ Cult 52:165–169

Lin HS, van der Toorn C, Raemakers KJJM, Visser RGF, de Jeu MJ, Jacobsen E (2000a) Development of a plant regeneration system based on friable embryogenic callus in the ornamental *Alstroemeria*. Plant Cell Rep 19:529–534

Lin HS, van der Toorn C, Raemakers KJJM, Visser RGF, de Jeu MJ, Jacobsen E (2000b) Genetic transformation of *Alstroemeria* using particle bombardment. Mol Breed 6:369–377

Lu C, Bridgen MP (1996) Effects of genotype, culture medium, and embryo developmental stage of the *in vitro* responses from ovule cultures of interspecific hybrids of *Alstroemeria*. Plant Sci 116: 205–212

Lu C, Bridgen MP (1997) Chromosome doubling and fertility study of *Alstroemeria aurea* X *A. caryophyllaea*. Euphytica 94:75–81

Meerow AW, Tombolato AFC (1996) The *Alstroemeria* of Itatiaia. Herbertia 51:14–21

Muñoz-Schick M, Moreira A (2003) Alstroemerias de Chile. Diversidad, Distribución y Conservación. Taller La Era, 140 p

Negritto Maria A, Baeza Carlos M, Eduardo R, Patricio N (2015) *Alstroemeria marticorenae* (Alstroemeriaceae), a New Species from Central Chile. Syst Bot 40(1):69–74

Olate E, Ly D, Elliott G, Bridgen MP (2000) Influence of the timing of propagation and cold storage on the growth and development of Alstroemeria pot plants. Proc Int Plant Propag Soc 50:379–391

Oogaard M, Kristiansen K (1998) *Alstroemeria* in Chile. Herbertia 53:29–39

Park TH, Han I, Kim J (2010) Review on the development of virus resistant plants in *Alstroemeria*. J Plant Biotechnol 37:370–378

Przybyla A (2000) Mutagenesis in creation of new *Alstroemeria* genotypes. Acta Hortic 508:351–356

Reed BM (2006) Cryopreservation of Bermuda grass germplasm by encapsulation dehydration. Crop Sci 46(1):6–11

Ruiz E, Balboa K, Negritto MA, Baeza CM, Briceño V (2010) Variabilidad genética y morfológica yestructuración poblacional en *Alstroemeria hookeri* subsp. *hookeri* (Alstroemeriaceae), endémica de Chile. Rev Chil Hist Nat 83:605–616

Rustanius P, Hang A, Hughes HG, Tsuchiya T (1991) Chromosome analysis of *Alstroemeria ligtu* hybrids. HortScience 26(7):902–904

Sanso AM (1996) El genero *Alstroemeria* (Alstroemeriaceae) en Argentina. Darwin 34:349–382

Sanso AM, de Assis MC, Xifreda CC (2005) Alstroemeria: a charming genus. Acta Hortic 683:63–77

Schum A (2003) Mutation breeding in ornamentals: an efficient breeding method. Acta Hortic 612: 47–60

Sekizawa K, Yamamoto S, Rafique T, Fukui K, Niino T (2011) Cryopreservation of in vitro-grown shoot tips of carnation (*Dianthus caryophyllus* L.) by vitryfication method using aluminium cryo-plates. Plant Biotechnol 28:401–405

Shinoda K, Murata N (2003) Cross-compatibility in interspecific hybridization of fifteen Alstroemeria species. J Jpn Soc Hortic Sci 72:557–561

Tatsuzawa F, Saito N, Murata N, Shinoda K, Shigihara A, Honda T (2003) 6- Hydroxypelargonidin glycosides in the orange-red flowers of Alstroemeria. Phytochemistry 62:1239–1242

Tsuchiya T, Hang A, Healy WE Jr, Hughes H (1987) Chromosome studies in the genus Alstroemeria. I. Chromosome numbers in 10 cultivars. Bot Gaz 148(3):519–524

UPOV. International Union for the Protection of New Varieties of Plants (2005) Test guidelines – English Version. http://www.upov.int/en/publications/tg-rom/tg_index.htm

van Hintum TJL (1996) Core collections in germplasm conservation, evaluation and use. In: Scoles G, Rossnagel B (eds) V International oat conference & VII international barley genetics symposium: proceedings. University Extension Press, University of Saskatchewan, Saskatoon, pp 113–119

Verboom H (1979) Alstroemerias and some other flower crops for the future. Sci Hortic 31:33–42

Whyte RO (1929) Chromosome studies. I. Relationship of the genera Alstroemeria and Bomarea. New Phytol 28:3119–3344

Rose

6

S. K. Datta

Contents

Abstract

Rose is an important cut flower crop throughout the world in the floriculture industry. The rose has been grown for millions of years. It is a versatile plant adapted to various climatic conditions. Within one genus, we have every stage from the purely wild species, the early cultivated forms to the most highly evolved garden forms of today. The basic chromosome number is seven

S. K. Datta has retired.

S. K. Datta (✉)
Floriculture Division, CSIR-National Botanical Research Institute, Lucknow, India

Bose Institute, Kolkata, India

Department of Plant Science and Biotechnology, Assam University, Silchar, Assam, India

S. K. Datta, Y. C. Gupta (eds.), *Floriculture and Ornamental Plants*, Handbooks of Crop Diversity: Conservation and Use of Plant Genetic Resources,
https://doi.org/10.1007/978-981-15-3518-5_8

(haploid), and in the ordinary diploid condition 14 chromosomes are present, but there are also tetraploids, hexaploids, and octaploids form. Most of the modern roses are the result of hybridization, selection and bud sports, and induced mutations. The chapter will provide maximum information on different aspects with special reference to hybridization. Voluminous literature is now available on rose breeding artificially created by the concentrated efforts of many great rosarians using different technology.

Keywords

Rose · Germplasm · Genetic diversity · Hybridization · New variety · Floriculture

6.1 Introduction

The rose has been grown for millions of years. 30 million years old fossils of rose have been found in Oregon and Colorado. The Greek and the Roman mythology illustrate roses in their ancient civilization. Rose was the symbol of Venus, a deity of love and peace. The world's most popular and appreciated flower is the rose due to its long history, symbolism, color, fragrance, and sheer elegance of form. Among all the flowers, the rose has hypnotized mankind the most and has, therefore, attained a unique status in human hearts. Within one genus, one can find every stage from the purely wild species, the early cultivated forms to the most highly evolved garden forms of today, artificially created by the concentrated efforts of many great rosarians. The genetic basis of the modern rose cultivars was developed due to extensive hybridization among the Chinese, European, and Middle-Eastern roses. The basic chromosome number is seven (haploid), and in the ordinary diploid condition 14 chromosomes are present, but there are also tetraploids, hexaploids, and octaploids form. Genus *Rosa* comprises about 200 species and thousands of cultivars and more than 150 species have been tabulated (Wylie 1955). Nowadays, roses are available in numerous attractive colors and shades. All the present-day colorful roses are the result of extensive hybridization, spontaneous and induced mutations, and selections.

6.2 Distribution

As per available literature, the rose originated in Central Asia, dating back to between 60 and 70 million years – the period known as the Eocene epoch and gradually spread all over the northern hemisphere. Roses were cultivated extensively by the Egyptians, Chinese, Greeks, Romans, and Phoenicians as early as 5000 years back. Chinese roses were introduced to Europe by Missionaries in the fourteenth century. It is difficult to theorize when rose cultivation started in India.

The rose plays a very important role as part of the social, medical, cultural, and religious fabric from time immemorial as confirmed in the medical monographs of Charaka and Susruta (Matthews 2004).

6.3 Uses

Rose is one of the most important commercial flowers in the floriculture industry. It is mostly used as cut flowers for decoration. Rose is also most popular due to its fragrance. Damask rose (*R. damascene*) and Edouard rose (*R. bourboniana*) flowers are used for the preparation of rose attar and a wide range of perfumery products. ***Rose water*** – Rose water is prepared by water distillation of flowers. Flowers should be collected in the early hours of the morning and distilled as quickly as possible after harvesting. The extract is stored for 2–3 months for maturation of odor and then diluted with distilled water in different proportions and bottled. ***Rose attar*** – Rose flowers are water distilled and the distillate is collected over sandal wood oil. ***Gul-roghan*** – This is a type of hair oil produced by the maceration of rose flowers with warm sesame seed oil. ***Rose oil or otto of Roses or Ruh-e-Gulab*** – It is produced by the water distillation of rose flowers, by redistilling the distillate two or three times till it gets saturated with the oil dissolved in it. Then it is chilled and the oil drops floating on the surface of water are removed. ***Gulkand*** – It is prepared by mixing the rose petals with white sugar in different proportions (1:1 or 1:2) and keeping the mixture in the sun. It is a very good tonic and has laxative properties. ***Pankhuri*** – It is prepared from dried rose petals. Dried petals are used for the preparation of cool summer drinks. Pankhuri is used for making potpourri, conserve, rose vinegar, rose petal wine, jams, jellies, syrups, etc. Rose fruits (hips or berry) are very rich in vitamin C, A, B, and F. Crystallized petals are eaten in sandwiches. Rose liquors and rose honey are made from roses. Rose water is used in making sherbets, perfumes, for flavoring syrups, soft drinks, and confectionary. Damask roses are used in cakes, sauces, etc. Perfume of rose petals cannot be completely exhausted by water distillation which can be used in the preparation of agarbattis and incense after drying and grinding. Pruned rose branches and shoots are used as fuel.

6.4 Botany

Rose belongs to the family Rosaceae and all species of this flower with minor exceptions belong to the genus *Rosa*. The genus contains about 200 species, and there are more than 30,000 cultivars differing in form, shape, size, color, fragrance, and flowering habit in cultivation. Roses are perennial, erect or climbing herbs, shrubs, and trees. Roots are fibrous, tap, and adventitious. Stems are herbaceous or woody covered with various shapes and sizes of thorns/prickles. Leaves are petiolate, alternate, oval, simple or pinnately compound, stipulate. Inflorescence mostly cymose or racemose, and some are solitary. Flowers are pedicillate, bractate, hermaphrodite, perigynous, or epigenous. Sepals 4 or 5; wild roses have five free petals and multiple of five petals in cultivated roses; numerous filamentous versatile stamens; fruits are fleshy, berrylike red to orange in color; hip and seeds are nonendospermic. Majority of wild roses are polyploidy and chromosome numbers vary from $2n = 2x = 14$ to $2n = 8x = 56$ (Vamosi and Dickinson 2006). *R. canina* is pentaploids showing unusual asymmetric meiosis. *R. prealucens* represent the highest naturally occurring ploidy (decaploidy) (Jian et al. 2010). Cytological information about the parental lines is very important in planning the rose breeding program. The genus is known to exhibit well-marked inter- and intraspecific chromosomal diversity, both at diploid and polyploid levels. The basic chromosome number n = 7 and many of the spp. Like *R. moschata*, *R. gigantean*, *R. multiflora*, *R. wichurarina*, and *R. chinensis* are diploid 2n = 14 while spp. like *R. galica*, *R. foetida*, *R. damascene*, and *R. centifolia* are tetraploids with 2n = 28 and species crosses have triploid with 2n = 21 and with spontaneous origin of 4n forms developed with 2n = 28. Many of the species are diploid (2n = 14), tetraploid (2n = 28), hexaploid (2n = 42), and octaploid with 2n = 56 chromosomes. *R. spinosissima* is octaploid found wild in Siberia. The Indian diploid spp. are *R. moschata*, *R. brunonii*, *R. clienophylla*, *R. eglanteria, R. fortida*, *R. gigantean*, *R. lescchnaueliana*, *R. longicuspis*, *R. macrophylla*, *R. sericea*, *R. webbiana*, while *R. laevigata* and *R. multiflora* and *R. ecae* have since been naturalized by extensive use as hedges. There has been crossing between tetraploid spp. Like *R. gallica*, *R. foetida* of European origin with those of Asian origin which are mostly diploid like *R. gigantean* resulting into triploids or aneuploids (c.f. Kaicker 1995).

6.5 Classification

Roses are known in various stature, form, and use as shrub or bush, standard, climber and rambler, hedge and edge, miniature, pot plants, rockery cut flowers, etc. The horticultural classification of roses involves the grouping of cultivars into reasonably well-defined type with similar characteristics and uses. Such a classification is a very handy tool for both the amateur and professional rosarians, because a rose belonging to a given class will be similar in habit of growth, method of flowering, and use to all others in the same class (Datta 2002). At the initial stage rose derived from different wild species which were natives of various countries in the Northern Hemisphere. Different rose species which are reported to grow wild in India are *R. eglanteria*, *R. foetida*, *R. involucrata*, *R. leschenaultiana*, *R. longicuspis*, *R. macrophylla*, *R. moschata*, *R. odorata*, *R. sericea*, *R. webbiana*, *R. brunonii*, *R. ecae*, *R. gigantean*, *R. beggariana*, *R. laevigata*, *R. banksii*, *R. bracteata*, etc. According to available literature, 15–18 species have played an important and central role in the evolution of garden roses of the world, of which eight are very important – *Rosa chinensis*, *R. damascena*, *R. foetida*, *R. gallica*, *R. gigantea*, *R. moschata*, *R. multiflora*, and *R. wichuraiana*. Today's modern roses are mainly classified under the following classes:

(i) **Hybrid Perpetuals**: The China roses played a very important role in the advancement of rose culture. A number of hybrid china roses of perpetual flowering habit were developed through repeated crossing between *R. chinensis* (Crimson china) and *R. gigantea*. These hybrids were further crossed in England and France with the autumn Damask (*R. damascena bifera*). The main object of this cross was to develop roses which flowered more than once in the season. Prior to about 1800 A.D., roses used to flower only once in a season except *R. damascena bifera*. The Bourbon roses were also included in the crosses. These crosses led to the production of hybrid perpetuals with complicated parentages, but the principal ancestors are the hybrid chinas, Bourbon roses, and products of cross between *R. damascena* and *R. gallica*. Hybrid perpetuals do not bloom perpetually despite its name. Examples of this class are 'Frace Karl Druschki', 'General Jacqueminot', 'Hagh Dickens', etc.

(ii) **The Tea Roses**: The name of this group of roses is supposed to have their similarities in aroma to that of the china tea chests. The aroma is certainly very delicate and the Musk rose (*R. moschata*) has a good influence on it. The Tea Roses are not as hardy and resistant to cold as the Hybrid Perpetuals. Examples of this class are "Anna Olivier', 'Lady Hillingdon', 'Molly Sharman Grawford', 'Mme Falcot'.

(iii) **Hybrid Teas**: Hybridization work continued for development of better roses. Attempts were made to combine the desirable characters of hybrid perpetuals and the tea roses. The first hybrid Tea rose 'La France' (large-flowered silvery-pink with fragrance) was developed in 1867 by a French breeder named Guillot. The second achievement in rose breeding was development of yellow hybrid tea 'Soleil d'or' in 1885. This was the success of a cross between bright yellow Austrian briar (*R. foetida*) and hybrid perpetuals by Pernet-Ducher (French rose breeder).

A new class 'Pernetianas' was developed due to further crossing with 'Soleil d'or'. In spite of a good color range (yellow, orange, bi-color, etc) they possessed a few undesirable characters of *R. foetida* like thorny nature, lack of perfume, susceptibility to disease, etc. Subsequently this group was merged and all present-day modern Hybrid tea varieties developed due to repeated back crossing. This is one of the most popular rose types of present days which include the large-flowered bedding and exhibition roses. Bloom mostly single on long stems with tall and erect growth habit. Some roses of this group contain strong fragrance.

(iv) **Polyanthes**: This group of roses was developed parallel to the hybrid teas. These roses developed through crossing between *R. polyantha* (now called *R. multiflora*) and dwarf type of *R. chinensis* (also known as *R. lawranceana* or the Fairy Rose). Varieties developed from these crosses were further crossed with tea roses for their further improvement. Examples are 'Cameo', 'Chatillion Rose', 'Baby Faurax', 'Echo', etc.

(v) **Floribundas**: A completely new class of roses was developed by a Danish rose breeder Poulsen through a hybridization program between Dwarf Polyantha and Hybrid Tea. 'Rodhatte', 'Else poulsen', 'Kirsten poulsen', etc. are first introduced roses of this class. Roses of this group are characterized by their bushy habit and blooms are in clusters. In recent years, there has been remarkable improvement in this group of roses.

(vi) **Grandiflora**: This group was developed by crossing the Floribundas with the Hybrid Teas. But this terminology has not been well accepted as it is difficult to differentiate between Grandiflora and Hybrid Teas. Flowers are large and vigorous in small clusters, for example, 'Queen Elizabeth', 'Elite', 'Una Wallace'.

(vii) **Damask Roses**: It is considered that these roses have been introduced from Damascus. Some highly scented roses belong to this group and are used for the manufacture of otto of roses.

(viii) **Miniatures**: At present, Minis are the fastest-growing segment of the rose market. This has become a very popular group of rose. The flowers and leaves are small and beautiful. Miniature roses are hardy, intriguing, and delightful. Miniatures are more uniform in their bloom size and bush habit. Flowers are solitary or in bunches. Mini roses are a fascinating group of roses with all the characteristics of a large rose reduced to mini proportions. Miniature roses have gained much popularity mainly because of the diverse and interesting flower forms and color (except true blue or black) and growth habits (Datta 2011).

There are a number of roses which have been classified under Bourbon roses, Cabbage roses, Moss roses, French roses, Albas, Musk roses, Noisette roses, Rugosas, Austrian Briars, Ramblers, etc.

6.6 Cultivars

The International Registration Authority of Roses is located in the USA which registers the newly evolved roses in the world and lists them in the book "Modern Roses" which is updated periodically. In India, the National Registration Authority of Roses under the International Registration Authority is at the Indian Agricultural Research Institute, New Delhi, which undertakes registration of newly evolved Indian rose cultivars. It is difficult to predict the exact number of rose cultivars. The cultivars under different classes differ in form, shape, size, color, fragrance, etc. (Datta 2002; Duha 1999).

(i) **Hybrid Tea**: ***Red*** – 'Atida', 'American Dream', 'Candella', 'Cara Mia', 'Dallas', 'Gladiator', 'Crimson Glory', 'Papa Meilland', 'Mister Lincoln', 'Avon', 'Happiness', 'Amercian Pride', 'Christian Dior', 'Gladiator', 'Matt God', 'Norma', 'Oklahoma', 'Olympiad', 'Red Planet', 'Royal Canadian', 'Torowyant', 'Ace of Hearts', 'Pace Maker', 'Achanta', 'Alec's Red', 'Black Lady', 'Black Pearl', 'Dilruba', 'Flying Tata', 'Forgotten Dreams', 'Grand Masterpiece', 'Jennifer Heart', 'Kardinal', 'Sugandha', 'Summer Fragrance', 'Toro', 'Victor Hugo', 'Deep Secret', 'Matt Gold', 'Ruby Wedding', 'Hidalgo', 'Mattgod', 'Bambey', 'Precious Plantinum', 'Vivekanand', 'Monica', 'Mother and Baby', 'Smooth Velvet', 'Taboo', 'Tadoroki'. ***White*** – 'Anastasia', 'Cabaret', 'Virgo', 'John F. Kennedy', 'Garden party', 'Evening Star', 'Honor', 'Matterhorn', 'Pascali', 'Pristine', 'Chablis', 'Ivory Tower', 'Jawahar', 'Jotiba', 'Louisiana', 'Rajhans', 'Sinjun', 'White

Comet', 'White Lightning', 'White Masterpiece', 'Matterhorn', 'Polar Star', 'Honor', 'Francis Phoebe', 'Grand Moghul', 'Joy of Life', 'White Delight'. ***Pink*** – 'Bridges Dream', 'Countess Sonja', 'Kiss', 'Koppies', 'Queen Elizabeth', 'Pink Parfait', 'Appreciation', 'Mrinalini', 'Eiffel Tower', 'Country Two', 'Colour Magic', 'Cynthia', 'Esmeralda', 'First Prize', 'Folklore', 'George Sand', 'Koppies', 'Panorama Holiday', 'Peter Frankenfeld', 'Rajni', 'Sonia Meilland', 'Sweet Heart', 'Tribute', 'Dr. B.P. Pal', 'Eterna', 'Harmony', 'Jadis', 'Maharani', 'Margaret Thatcher', 'Pristine', 'Priyadarshini', 'Rebecca Claire', 'Sheer Bliss', 'Sweet Surender', 'Cynthia', 'Dr. Darley', 'ICO Beauty', 'Koppies', 'Miss All American Beauty', 'Velvet Lustre', 'Wind Sounds', 'Esmerelda', 'Fancy Princess', 'Friendship', 'Ellora Caves', 'Eterna', 'Mullard Jubilee', 'Peach Beauty', 'Sylvia', 'Pink Charm', 'Drima Donna', 'Signature'. ***Yellow*** – 'Bellissima', 'Chalis Gold', 'Excitement', 'Gina Lollobrigida', 'Aalsmeer Gold', 'Apollo', 'Gold Medal', 'Golden Star', 'Gold Dot', 'Helmut Schimd't', 'Golden Jubilee', 'Austra Gold', 'Freedom', 'Landora', 'Mabella', 'Midas', 'Vasant', 'Beroliana', 'Dutch Gold', 'Elina', 'Golden Giant', 'Hokotu', 'King's Ransom', 'Sunblest', 'Pusa Sonia', 'Sunbright', 'Fragrant Gold', 'Shensei', 'Pot O Gold', 'Grandpa Dickson', 'Selfridges', 'Golden Moments', 'Goldie', 'Morning Sun', 'Texas'. ***Mauve*** – 'Big Purple', 'Blue Moon', 'Blue Diamond', 'Heirloom', 'Lady X', 'Lilac Time', 'Paradise', 'Ajanta', 'Anurag', 'Blue Nile', 'Blue Ocean', 'Blue Perfume', 'Indraneel', 'Jacaranda', 'Kasturi Rangan', 'Mystic', 'Blue river', 'Blue Sky', 'Jadis', 'Madame Violet', 'Mauve Melody', 'Mystic'. ***Orange and Vermilion*** – 'Abbaye de cluney', 'Baceardi', 'Harvest Sun', 'Hiogi', 'Apricot', 'Arizona', 'Basildon Bond', 'Brondy Butter', 'Angelique', 'Raja Rammohan Roy', 'October', 'Roklea', 'Valencia', 'Super Star', 'Montezuma', 'Summer Holiday', 'Dalject Futura', 'Flrorada', 'Atoll', 'Angelus', 'Coalite Flame', 'New Year', 'Troika', 'Command Performance', 'Dolly parton', 'Fortuna', 'Harmonie', 'Jogan', 'Interflora', 'Lovers Meeting', 'Orient Express', 'Shreveport', 'Sun Fire', 'Ambassador', 'Arizona', 'Brandy', 'Doris Tysterman', 'Jost Joey', 'Medallion', 'Julia Tysterman', 'Remember Me', 'Sunset Song', 'Adora', 'Bing Crosby', 'Cancan', 'Cannes Festival', 'Dolly Parton', 'Fragrant Cloud', 'Lokmanya', 'Orient Experss', 'Raja of Nalagarh', 'Banaras Dawn', 'Golden Afternoon', 'Julia', 'Maid of Honour', 'Medallion', 'Queen Beatrix', 'Sun Fire', 'Bing Crosby', 'Lady Rose', 'Our Love', 'Lambada', 'New Year'. ***Bicolored*** – 'Kiss of Fire', 'Double Delight', 'Paradise', 'Love', 'Granada', 'American Heritage', 'Abhisarika', 'Anvil Sparks', 'Bajazzo', 'Careless Love', 'Montreal', 'Supriya', 'Alleluia', 'Disco', 'Fortunata', 'Las Vegas', 'Mascotte', 'Osiria', 'Summer Dream', 'Surkhab', 'Trade Winds', 'Siddartha', 'Sahashradhara', 'Gallivarda', 'Abhaya', 'Souviens Toi'.

(ii) **Floribunda**: ***Red*** – 'America's Choice', 'Imperator', 'Santa Maria', 'Europeana', 'Nordea', 'Marienda', 'Showbiz', 'Jantar Mantar', 'Brown Velvet', 'Tequila', 'After Glo', 'Wee Jock', 'Love Potion', 'Merecedes'. ***White*** – 'Dimples', 'Iceberg', 'Akito', 'Saratoga', 'Margaret', 'Merril', 'Bharat Ram',

'French Lace', 'White Koster'. ***Pink*** – 'Angelica', 'Frolic', 'Prema', 'Arunima', 'Sada Bahar', 'Rose Sherbet', 'Letchworth Garden City', 'Pink Parfait', 'Pink Puff', 'Sea Pearl', 'Tiki', 'Young Mistress', 'Annesley Dickson', 'Bonica', 'Gene Boerner', 'Queen Elizabeth', 'Baby Talk', 'Cherish', 'Kerryman', 'Mahak', 'Junior Miss', 'Pink Iceberg', 'Play Girl', 'Sexy Rexy', 'Simplicity'. ***Yellow*** – 'Arthur Bell', 'Bright Smile', 'Bellona', 'Golden Times', 'Anne Harkness', 'Princess Michael', 'Sun Flare', 'Kanak', 'Friesia', 'Gold Bunny', 'Mitzi', 'Bianco', 'Candeur', 'Chandrama', 'French Lace', 'Judy Garland', 'Yellow Belinda'. ***Mauve*** – 'Africa Star', 'Azure', 'Angel Face', 'Neelambari', 'Deep Purple', 'ICO', 'Shocking Blue', 'Akash Nartaki', 'Intrigue', 'Pillow Talk'. ***Orange and Vermilion*** – 'Contempo', 'Carbet', 'Doris Norman', 'Zambra', 'Zorina', 'Anisley', 'Dickson', 'City of Belfast', 'Courtoisie', 'First Edition', 'Flamenco', 'Prominent', 'Amber Queen', 'Anne Harkness', 'Brown Velvet', 'Mahak', 'Mohini', 'St. Boniface', 'Evening Centinel', 'Impatient', 'Orange Sensation'. **Bicolored** – 'Charisma', 'Daily Sketch', 'Masquerade', 'Paint Box', 'Red Gold', 'Banjaran', 'Golden Rays', 'Green Sleeves', 'Mahak', 'Rare Edition', 'Strawberry', 'Rangoli', 'Saroja', 'Sea Pearl', 'Judy Garland', 'Laminuette Vital Spark', 'Double Talk', 'Fantasia', 'Confetti', 'Courtosic', 'Sheilas Perfume'.

(iii) **Miniature**: ***Red*** – 'Beauty Secret', 'Red Elf', 'Cupcake', 'Hombre', 'Janna', 'Lady Eve', 'Red Cascade', 'Red Det', 'Red Flush', 'Tiffic', 'Valerie Jeane', 'Black Jade', 'Holiday Cheers', 'Starina', 'Galaxy', 'Wee Jock', 'Swedish Doll', 'Acey Deucy', 'Deep Velvet', 'Don Don', 'Don Marshall', 'Jingle Bells', 'Little Crimson', 'Sorcerer', 'Teddy Bear', 'Perla de Alcanada', 'Robin'. ***Pink*** – 'Cri-Cri', 'Windy City', 'Cuddles', 'Cupeake', 'Mini Pearl', 'Peaches n cream', 'Wind Jammer', 'Fairy Tale', 'Heartbreaker', 'Perla de montserrat', 'Simple Sinon', 'Sweet Fairy'. ***Yellow*** – 'Baby Gold Star', 'Calgold', 'Corn Silk', 'Rise'N Shine', 'Center Gold', 'Free Gold', 'Golden Century', 'Softee', 'Bit O'Gold', 'Arizona Sunset', 'Swinger', 'Yellow Doll'. ***White*** – 'Green Ice', 'Jet Trail', 'Pushkala', 'Cotton Tail', 'Minnie Pearl', 'Snow Bride', 'Chattem Centennial', 'Cinderella', 'White Maddona', 'Pour Toi'. ***Orange*** – 'Angela Rippons', 'Petite Folice', 'Autumn Fire', 'Mother's Day', 'Swedish Doll', 'Pandenong', 'Hot Shot', 'Heartland', 'Chatten Centennial', 'Mark One', 'Sun Blaze', 'Orange Honey', 'Pride N'Joy', 'Colibri'. ***Mauve*** – 'Mr. Blue Bird', 'Silver Tips', 'Lavender Jewel', 'Valerie Jeane', 'Winsome', 'Bharani', 'Lavender Lace'. ***Bicolored*** – 'Jeanie Williams', 'Over the Rainbow', 'Flying Colours', 'Glori Glo', 'Little Jackie', 'Magic Carrousel', 'Puppy Love', 'Star Trail', 'Stars'N Stripes', 'Strawberry Swirl', 'Red Tag', 'Party Girl', 'Rainbows End', 'Sierra Sunrise', 'Earth Quake', 'Jewel Box', 'Cooper Sunset', 'Hoot N'Holler', 'Jitterbug', 'New Beginning', 'Presumida'.

(iv) **Polyantha**: 'Anjani', 'Chatillon Rose', 'Echo', 'Orlean Rose', 'Mariposa', 'Mrs. Finch', 'Priti', 'Swati', 'Nartaki', 'Rashmi', 'Pink Showers', 'Emmeloord' (orange-red), 'Margo Koster' (Orange-Pink), 'May Wonder' (blood-red), 'Orange Triumph' (Salmon-red), 'Paul Crampbell' (orange-scarlet), 'Red Triumph', 'Sne Princes' (white).

(v) **Climbers**: ***Red*** – 'Fountain', 'Tempo', 'Alec's Red clg.', 'Sympathy', 'Lilli Marlene clg.', 'Compassion', 'Lawinia', 'Sonia Clg', 'Dublin Bay', 'Dynamite', 'Kassel'. ***Yellow*** – 'Golden Showers', 'Peace Clg.', 'High Field', 'Goldstern', 'Landora Clg', 'Gold Bunny', 'Easlea's Golden Rambler', 'Climbing Allgold', 'Lever Kusen'. ***White*** – 'Delhi White Pearl', 'Garden Party clg', 'Iceberg clg.', 'Swan Lake', 'John F. Kennedy clg', 'Climbing men', 'Mc Gredys Ivory'. ***Bicolor*** – 'Pinata', 'Vintage Wine'. ***Pink*** – 'Amarica', 'Leaping Salmon', 'Show Garden'. ***Orange/Vermillion/Apricot*** – 'Fragrant Cloud Clg', 'Apricot Prince clg.', 'Zenith', 'Bearth of Life', 'High Field', 'Auriel Dombasle', 'Danse du few'. ***Mauve*** – 'Anjel Face Clg', 'Paradise Clg', 'Blue Moon Clg'.

(vi) **Fragrant Roses**: 'Alec's Red', 'Amber Queen', 'America', 'Apricot Nectar', 'Arizona', 'Broad Way', 'Camelot', 'Chrysler Imperial', 'Color Magic', 'Compassion', 'Crimson Glory', 'Double Delight', 'Eiffel Tower', 'Forty-niner', 'Fragrant Cloud', 'French Lace', 'Granada', 'Intrigue', 'Jadis', 'Josephine Bruce', 'Margaret Merrill', 'MEI domance', 'Mirandy', 'Mister Lincoln', 'Nocturne', 'Papa Meilland', 'Perfume Delight', 'Prima Ballerina', 'Rose Sherbet', 'Royal Hifhness', 'Sheer Bliss', 'Sugandha', 'Sundowner', 'Sunsprite', 'Sutter's Gold', 'Sweet Surrender', 'Tiffany', 'Typhou Tea', 'Voodoo', 'Wendy Cussons', 'Whisky Mac', 'White Lightin', 'Ze-phirine Drouhin'.

(vii) **Indian Roses**: ***Hybrid Teas*** – 'ABA Saheb', 'Aphisarika', 'Achantha', 'Aditya', 'Adora', 'Agni', 'Agnihotri', 'Ahimsa', 'Akash', 'Akash Sundari', 'Andromeda', 'Annapurna', 'Anna Saheb', 'Anurag', 'Aravali Princess', 'Asha', 'Arhwini 89', 'Balaji', 'Balwant', 'Belle of Punjab', 'Bhanu', 'Bhargav', 'Bhavani', 'Black Delight', 'Blue Delight', 'Blue Ocean', 'Bodisattwa', 'Brahm Datt', 'Calcutta 300', 'Cauvery', 'Cecarcee', 'Chitra', 'Chitranjini', 'City of Panjim', 'Classic', 'Colour Harmony', 'Courageous Indira', 'Decan Delux', 'Die Della', 'Dil-ki-Rani', 'Diversity', 'Dr. B.P. Pal', 'Dr. Benjamin Pal', 'Dr. Kidwai', 'Dr. M.S. Randhawa', 'Dr. Noshir Wadia', 'Durgapur Delight', 'Durgapur Jubilee', 'First Rose Convention', 'Fragrant Beauty', 'Girija', 'G.K. Rose', 'Glamour Girl', 'Godavari', 'Golden Afternoon', 'Golden Biotech', 'Gomathi', 'Gulzar', 'Homage', 'Ichalkaranji 100', 'ICO Ambassador', 'ICO Beauty', 'ICO Delight', 'ICO Delux', 'ICO Trimurthi', 'Indu Singhal', 'Invention', 'Jaslok', 'Jawani', 'Jayant', 'Jayatsen', 'Jaypee Rose', 'Jogan', 'Kaladi', 'Kalyani', 'Kamaladevi Chttopadhyaya', 'Kanchi', 'Kanava', 'Kasturi Rangan', 'Kirang', 'Komala 189', 'Lemon Time', 'Madhosh', 'Mahalaxmi', 'Maharshi', 'Malak', 'Malkarsiddha', 'Malwa 194', 'Marie Palit', 'Manas', 'Mehar', 'Melody Queen', 'Memory of D.M. Roy', 'Mohak', 'Mother Teresa', 'Mrinalini', 'Nefertiti', 'Nehru Centinary', 'Our Indira', 'Padmavathi 95', 'Pahadi Dhun', 'Painted Melody', 'Pampa', 'Panchaganga', 'Papa Pirosha', 'Pastel Delight', 'Pink Fantasy', 'Pink Melody', 'Piroja', 'Polybag Joshi', 'Prof. Madhob Chandra Nath', 'Preyasi', 'Pride of Chalkaranji', 'Pride of Nagpur', 'Priyadarshini', 'Priyatama', 'Raja of Surendra Singh of Nalagarh', 'Rajni',

'Raktima', 'Rampa Pal', 'Ratan', 'Ratnaar', 'Red Recker', 'Rose Anil', 'Rose Bansal', 'Rose City of Nasik', 'Rose KSG Centinary', 'Sahasradhara', 'Santa Claus', 'Satrika', 'Savkar', 'Shalimar', 'Shankar Jaikishan', 'Shantaraj', 'Shanti Pal', 'Sheer Grace', 'Shimsa', 'Shoba', 'Shreyasi', 'Shri Swami Samarth', 'Siddhartha', 'Silky Petal', 'Sir C.V. Raman', 'Somasila', 'Speckled Delight', 'Srinivasa', 'Sugandha', 'Sugandha Raj', 'Suman', 'Sunanda', 'Subravat', 'Supriya', 'Surkhab', 'Surja Shikha', 'Swagatham', 'Sweet India', 'Tambrabarani', 'Tapti', 'Tata Centinary', 'Tenth Rose Convention', 'Thungabhadra', 'Tippu's Flame', 'Uma Rao', 'Vaishnavi 92', 'Vasavi', 'Viola', 'Week End', 'Yeshwant'. ***Floribunda*** – 'Ahalya', 'Akashdeep', 'Banjaran', 'Devi Gayatri', 'Dr. S.S. Bhatnagar', 'Hemavathy', 'ICO-B.K. Pati', 'ICO Pearl', 'ICO Talk', 'Indraman', 'Kanak', 'Kumaradhara', 'Kumari', 'Kusum', 'Lahar', 'Manasi', 'Manmantha', 'Mohini', 'Narmada', 'Neelakanti', 'Netravathy', 'Pushkarini', 'Rare Edition', 'Salmon Splash', 'Saroja', 'Shatadhara', 'Sushma', 'Thornless Beauty', 'Varsha'. ***Climbers*** – 'Climbing Kanaya Kumari', 'Climbing Dr. Homi Bhabha', 'Climbing Matangi', 'Climbing Sadabahar', 'Climbing Show Biz', 'Climbing Tata Centinary', 'Delhi Pink Pearl', 'Delhi White Pearl'. ***Polyanthus and Miniature*** – 'Anjani', 'Bharani', 'Chandrika', 'Dazzler', 'Delhi Starlet', 'Jimmy', 'Nartaki', 'Pink Showers', 'Pink Spray', 'Pushkala', 'Rashmi', 'Rosy Twinkler', 'Swati'.

(viii) **Export Rose Varieties**: 'Grand Gala', 'First Red', 'Ravel', 'Versillea', 'Femma', 'Konfetti', 'Noblesse', 'Skyline', 'Osiana', 'Suplesse', 'Lambada', 'Ambience', 'Cr. Prophyta', 'Saphir', 'Starlight', 'Corvetti', 'Escada', 'Papillion', 'Black Magic', 'Sundance', 'Capri', 'Toplesse', 'Ilseta', 'Timeless', 'Cappucino', 'Movie Star', 'White Nebeles', 'Sweet Red', 'Black Magic', 'Red Devil', 'Bianco', 'Golden eye', 'Sandy', 'Femma', 'Susanne', 'Maasai', 'Renali', 'Tineke', 'Sangria', 'Soledo', 'Cora', 'Diplomat', 'Escada', 'Laser', 'Texas', 'Sacha', 'Nicole', 'Pareo', 'Pavarotte', 'Samourai', 'Rodeo', 'Rossini', 'Vivaldi', 'Dallas', 'Vega', 'Saphire', 'Livia', 'Jacaranda', 'Tennesse', 'Marjan', 'Monica', 'Laminuette', 'Oriana', 'Pareo', 'Astra', 'Vera Lynn Kiss', 'Lambada', 'Nicole', 'Sandy', 'Sangaria', 'Soledo', 'Susanne'.

6.7 Cultivation

(i) **Preparation of Beds**: Beds are to be prepared at least 40–45 days before the planting. The beds are normally prepared during April–May. The preparation of beds depends upon the number of varieties. For individual plant about 55–75 cm. circular and 55–65 cm. deep pits are prepared. Approx 2 kg cow dung manure, 200 gm bone meal, and 1 kg pit soil are mixed separately and poured into the pit and then rest of the pit is covered with other pit soil. No chemical fertilizer should be applied to newly planted roses up to 1 month. Roses are planted during September/October but it may vary at other places depending upon the climatic condition. Before planting about 25–30 cm soil from each pit is removed and the plant along with the earth-ball is placed in such a way that the joint of stock-scion remains just above the soil level. Extra soil is added, and the plant is fixed firmly and watered immediately after planting. Planting distance (between plants and rows) depends upon the varieties. Conventional planting density of 60 × 60 cm distance is maintained. However, for commercial purpose high planting densities, that is, 60 × 30 cm; 30 × 30 cm; 150 × 30 cm; 100 × 30 cm; 30 × 20 have been recommended by different research institutions. Amateur growers can allow much space on two sides of plants for proper care. The rose bed should be wide enough that one can easily reach the plants from both sides. Hybrid Teas and grandifloras should be planted 70–80 cm apart. Floribundas do well 60 cm apart. The rose beds should get proper sunshine at least 5–6 h a day. Proper drainage system is vital because stagnant water at the roots will damage plants. It is necessary to water roses regularly during dry weather otherwise twice a day. The organic manures should be well rotted. Chemical fertilizers should be avoided at the early stage of plant growth. Plant health should be regularly checked, and periodical spraying of insecticide or fungicide is advisable.

(ii) **Pot Cultivation**: Roses can be grown in pots. Pots size should be minimum 12″ (30 cm.) diameter. For potted plants the soil mixture should also be prepared well in advance. Proper care should be taken of the bottom hole of the pot before filling pots. Bottom hole should be filled with corks, gravel, broken pot pieces, or coir fiber to ensure proper drainage. Generally, potted plants require daily watering.

(iii) **Greenhouse Cultivation**: Greenhouse cultivation is required not only for cooler regions but also for any climatic condition to grow a better crop. Roses for the international market are mostly cultivated under greenhouse. Green houses are framed structures covered with transparent or translucent material and large enough to grow crops under partial or fully controlled environment conditions to get maximum productivity and quality bloom. The main advantages of growing roses under greenhouse are high productivity per unit area, roses can be grown in any season of the year depending on the demand and market, blooms are of high quality and free from blemishes. The micro-climatic environment within green house is more suitable for crop development than the outside environment. The quality and quantity of production under greenhouse is better because it protects the crop damage caused by birds, rodents, pests, diseases, extreme temperature, heavy rain, hail storms, high wind, etc. Plants are mostly grown in growing media and therefore soil-borne disease can be avoided. Green house cultivation takes care of resources like effective water management, optimum use of pesticide and fertilizers, year-round growing of crops, etc. The flower export in India is increasing because of increased use of greenhouses in the country. The agro-technology adopted in greenhouse is quite variable.

6.8 Propagation and Cultural Practices

Roses can be propagated by seeds, cuttings, layering and by budding, but for commercial purpose it is mostly propagated by budding. True-to-parent type can be maintained through budding. The seed propagation is generally adopted by breeders for producing new cultivars. Grafting is an old method which, at present, is not commercially acceptable due to many practical disadvantages like non-synchronous nature of wood of both root stock and scion. Propagation of disease free plants through tissue culture is also being encouraged recently.

(i) **Seeds**: Seeds are formed within fruits. In rose the lower portion below the calyx swells and develops into fruit which is called "hip." Seed production varies from variety to variety. Some varieties produce good seeds. Unfortunately, all roses are not capable of producing seed or supplying functional pollen. Others may produce seed of poor germinative quality or have a positive tendency to pass on undesirable characteristics rather than good ones. Plants developed from seeds show variability in characters due to cross pollination and heterozygous nature. Fully ripened fruits are harvested and dried properly. It takes minimum of 3 months or more for the maturation of fruits. The seeds are air dried in a cool place. Mature seeds usually germinate after a resting period. Seeds are sown in earthen pots or "thalis" minimum after 1 month of collection. Seedlings are transplanted in nursery beds at 2–3 leaf stage.

(ii) **Cuttings**: Roses can be multiplied by cuttings but all the modern varieties do not respond to cuttings. Rootstocks are propagated by cuttings mostly during July–August. Rose varieties belonging to climbers, Rambler, Polyantha, and Miniature can be propagated by cuttings.

(iii) **Ground Layering**: Shoots of climbing/creeper varieties are bent to the ground or earthen pots and covered with soil leaving the terminal portion exposed. Sometimes small horizontal cutting or notching is done and rooting hormones are applied for improving root formation. Rooting starts normally after 25–30 days. Layering branch is separated when the roots are healthy.

(iv) **Air Layering**: Roses can also be multiplied through layering, but all the rose varieties do not respond to layering. For layering a small portion of stem bark is removed with a sharp knife and the portion of branch is covered with Sphagnum moss and covered with polythene after applying a small amount of root hormone (climbing and rambling roses are suitable for this method). After root formation, the branch is cut and planted.

(v) **Budding**: Budding is most suitable and commonly used method for the propagation of roses. The technique is simple but requires proper practice, precaution, and care. The first step is the selection of root stock (already described separately) and variety to be budded. Different steps followed in budding are as follows: selection of variety to be budded (scion) and stock variety; – selection of budding eyes (before sprouting situated in the axils of the leaves); selected branch with 3–4 eyes is cut and thorns are removed carefully; the upper portion of leaves of the branch is cut off leaving 1–2 cm. stalk; the eyes are removed with a sharp knife (budding-knife) and the eyes are removed along with a piece of bark 1.5 cm above and below the eye; the lower portion of stock (3–6″ above soil level) is cleaned by removing thorns and leaves; either I-shaped or T-shaped sharp cut is made at the lower portion of stock. For T-cut, first a horizontal cut and then a vertical cut below in the stock. For I-cut a vertical cut is made. Only the bark of the stem is cut. The size of the cut should be just sufficient to accommodate the eye of the scion; the removed eye along with the shield-shaped tissue attached to it very carefully kept into the T-shaped or I-shaped cut; – the eyes are tied with narrow tape-like alkathene fiber; under Lucknow condition budding is mostly performed during January–March. Union of scion and stock and sprouting time of budded eyes vary from variety to variety. After sprouting when the height of sprouted branch reaches approx. 6–8″ (15–20 cm.), the upper portion (above budding) of the stock is removed.

(vi) **Root Stocks**: Selection of proper root stock variety is most important for rose budding. The variety which is well acclimatized at our climatic and soil condition is selected. The root variety is called stock and the bud eye of selected variety is called scion. The success of budding depends upon the selection of proper root stock variety and its acceptance of scion bud and its sprouting. The success of budding depends upon the compatibility of stock and scion. The stock must be selected which will grow best in the particular area, should produce fibrous root, vigorous growth habit, thick bark, etc. Rooted rose stocks are planted in beds 25 × 25 cm apart. Different promising root stock varieties have been selected suitable for different agro-climatic zones of India. Five root stocks, viz., *Rosa indica* var. odorata, *R. multiflora*, *R. borboniana* (Edouard rose), *R. canina* (Dog rose) and *R. laxa* are mostly used. *R. indica* var. *odorata* has been found to be most promising under Lucknow condition. In normal conditions, only one variety is budded on each root stock. But more than one variety can also be budded one above the other on the same root stock in a zig-zag pattern.

(vii) **Pruning**: Pruning of roses is done for following reasons: to remove the weak branches; – to remove old woody branches; – to shorten the healthy branches; to stimulate new growth, to make the branches in a well-balanced arrangement; to stimulate quality blooming (although the quality of bloom is primarily a varietal character). Pruning has also great role in bloom quality); – to remove all dead and infected branches; to maintain floriferousness of plant. Pruning time is decided on the basis of climatic conditions of the region. Mostly pruning is done after rainy season. In Lucknow, India, pruning is normally done during the first week of October. Pruning operation should be done once a year. But in some places roses are pruned twice (November and March). Pruning should be done after one year age of the plant. In temperate climate, pruning is done in spring though some growers do it at the end of autumn. Pruning is done in the Indo Gangetic plains after cessation of rain when the cold season is approaching. The first week of October is suitable for pruning under North Indian conditions. Pruning operations are performed twice that is, November and June in Karnataka and Maharashtra. Pruning can be done with the help of secateurs or knives. The edges of both tools should be very sharp so that stems can be given a clean cut. The cut should be slightly slanting. The cut end should be immediately painted with any fungicide paste to avoid fungal attack. Pruning may be light or drastic. During light pruning, only the tips of branches are removed uniformly. In drastic pruning, maximum upper portion of each branch is removed leaving 4–5 eyes. The grower, by virtue of his experience, can decide the type of pruning required for his rose (type, variety, climate, spacing, and flower quality).

(viii) **De-suckering**: Suckers of wild stock plant appear below the budded portion. The growth of the sucker is very vigorous. It should be removed immediately otherwise the entire budded plant is ruined.

6.9 Micropropagation

Micropropagation is the true to type propagation of a selected genotype using in vitro culture techniques. Regeneration protocol has been standardized from different types of explants. Shoot regeneration has been reported from stem node-derived callus, immature embryos, somatic embryogenesis, and stem-derived callus. In addition to regeneration through callus, direct shoot regeneration from leaf midrib and petioles has been studied. Work has been done for large-scale multiplication of different types of roses starting from either shoot tip, axillary buds, or node with different degrees of success. Different species and cultivars of rose have shown to have different nutritional and hormonal requirements and also have different regeneration potentiality. In most of the cases, MS medium is used as a basal medium (0.8% agar, 3% sucrose, and pH 5.5 ± 1) and different hormone supplement used are IAA and BAP in different concentrations and combinations for culture initiation as well as multiplication, however, many have suggested modified MS as superior. Among different growth regulators used, NAA and BA are reported to be the best. A combination of cytokinin with a low dose of auxin, kinetin (0.21–8.9 mg/l), 2iP, application of GA_3 in the multiplication medium, etc. reported to have good results. Initially, rose micropropagation faced serious setbacks from yellowing of leaflets and vitrification, but this problem was averted by the addition of $AgNO_3$ in the culture media which resulted in the reduction of leaflet yellowing as well as enhanced growth of shoots (Chakrabarty and Datta 2006, c.f. Datta 2015). Different auxins (NAA and IBA) are used for root induction. The auxin requirement for root induction in rose is cultivar specific. Besides auxin, successful rooting and field transfer of rooted shoots, has been reported, to some extent, to be dependent on the level of cytokinin, vitamins, phenylalanin, and tyrosin used in the multiplication stage. Author and his colleagues standardized the micropropagation protocol for ten rose cultivars (namely Mrinalini, Queen Elizabeth, Windy City, Chandrama, Raktima, Doris Tysterman, Sylvia, Dr. B.P. Pal, and Contempo) (Chakrabarty et al. 2000). Following protocol has been followed for routine multiplication of rose. Nodal explants of 1.5–3 cm length are surface sterilized in 70% ethanol for 30 s followed by 0.1% $HgCl_2$ for 2 min and rinsed thoroughly in sterile distilled water. Cultures are inoculated under cool, white florescent light (36 μmol m^{-2} s^{-1}, 16 h photoperiod) in a culture room maintained at 22 °C ± 2 °C. For culture initiation MS medium supplemented with BA (2 mg/l) and IAA (0.2 mg/l) are used. For shoot multiplication, the same media composition along with 10 mg/l $AgNo_3$ is used. The addition of $AgNO_3$ did not reduce the shoot multiplication rate but it enhanced the shoot growth and reduced leaflet yellowing. Roots are initiated in one fourth strength of MS media supplemented with IAA (0.1 mg/l) and IBA (0.1 mg/l). Rooted well established regenerated plants were transferred to pots containing a mixture of sand: soil:manure (1:1:1 v/v) and were kept in a moist chamber with 80–90% relative humidity for 15 days before their transfer to the glasshouse. Depending upon the type of explant, one to five primary shoots developed prior to transfer to shoot multiplication medium. The bud size in the initial nodal explant had no effect on shoot numbers after 3 weeks, however, it was generally observed that larger buds regenerated more shoots than the smaller ones (c.f. Datta 2015).

6.10 Diseases and Pests and Plant Protection

(i) **Black Spot**: Black spot is one of the most dreaded of all rose diseases. It is a fungal disease caused by *D. rosae* (Deuteromycetes). As its name implies, it manifests itself by the appearance of black spots on the leaves. Black spots can develop rapidly on the leaves after a new infection begins, but ordinarily it is not seen or detected until about 2 or 3 weeks following initial infection. The symptom of the disease appears on both sides of the leaves like dark colored spots with fringed margins. The size of the black spot varies from a small dot to a centimeter in diameter. A large number of small spots fuse to form a large spot and almost cover the entire leaf. Some varieties shed off infected leaves while in others the infected leaves remain attached to the plants. The disease is more severe in the wet weather. It can be spread by rain and can be transmitted from one to another variety. It does not attack young leaves. The pathogen requires a temperature of 70 °F. Thousands and millions of spores formed from these spots are separated from each other and scattered to other bushes. ***Control measures***: Affected leaves should be picked off and burnt. Young leaves should be protected by periodical spraying (once in 10 days) Maneb (0.15%). A wide range of fungicides have been recommended which breeders can selectively apply to control black spot disease – Bavistin, Benlate, Captan, Dithane M-45, Folpet, Maneb, Farbam, Karathane, Thiram, Wettablesulfur, Zineb, etc.

(ii) **Powdery Mildew**: Mildew is a common disease of roses caused by *Sphaerotheca pannosa* member of the Ascomycetes group of fungi. The disease is popularly called Powdery Mildew, because of the powdery appearance on infected plants. All aerial parts of the plants are affected but initially more serious on lower side of the foliage with circular to irregular powdery growth. Younger leaves get curled exposing the lower surface on which are found raised blister-like areas, which may become coated with the white powdery growth of the mycelium and conidia of the fungus. ***Control measures***: It can be controlled by foliar spray with wettable sulfur at the rate of 1 Kg in 500 l at 10 days intervals depending upon the severity of the disease. A good amount of fungicides have also been recommended for satisfactory control, viz., Bavistin (0.05%), Benlate (0.02%), Captan (0.15%), DDT, Karathane (0.05%), Malathion, Maneb (0.2%), Saprol (0.15%), Zineb (0.4%), Magnesium sulphate (15 gm in 5 l of H_2O), Dithane Z 78, wettable sulphur (0.05%), etc. The dose may vary from cultivar to cultivar and intensity of disease.

(iii) **Downy Mildew**: It is caused by the fungus *Peronospora sparsa*, belonging to the family Peronosporacae of class Oomycetes. Red and brown or whitish-grey spots appear on leaves with downy fungus growth on the undersides. It looks more like black spots than mildew with purplish black spot on foliage. It begins at the top of the bush and infected leaves rapidly turn yellow. The fungus attacks the junction between the stem and the leaflets. As symptoms, leaves and buds drop and stem and canes begin to develop purple lesion which spoils the beauty of even the most perfect bloom. Detection of this disease is sometimes difficult as it is not superficial and is often embedded in the epidermis layer. Eventually, the entire plant may die.

(iv) **Rose Mosaic Viruses (RMV)**: The symptoms appear in different forms as yellow-white stripes around the veins, chlorotic mottles, blotches, and small leaves. Most of the diseased plants are less vigorous and the number of blooms is reduced. Two mosaic viruses, that is, Prunus Necrotic Ringspot Virus (PNRSV) and Apple Mosaic Virus (ApMV), have been identified. Viruses are transmitted through propagation. Viruses may even be transmitted through root stock and bud-wood. ELISA (Enzyme-Linked Immuno-sorbent Assay) is considered to be rapid, accurate, reproducible, and sensitive serological test for the identification of rose viruses. Cut flowers must be virus free to maintain maximum shelf life. Diseased flowers do not meet the quality standards demanded by the consumer. ***Control measures***: Different fungicides like Benlate Captan, Diethane Z-78, Diethane M 45, Karathane, etc. are used for its control.

(v) **Dieback**: This appears after pruning. It starts from pruned end and slowly gets downward. Paste of copper fungicide should be applied immediately on the cut surface after pruning. Sometimes cow dung and soil paste are also applied on the cut surface.

(vi) **Aphides**: Aphides (Plant lice) are very small light-green insect which attacks in clusters the tender shoots, buds, and flowers. The most common aphid is *Macrosiphum rosae*. They damage the plant by sucking sap. Mild insecticides are used for its control. Pyrodust (0.2%), Metacid (0.1%), rogor (0.2%), and malathion (0.1%) are most suitable for its control.

(vii) **White Ants**: White ants (Termites) remain in the soil and they mostly destroy the cuttings at the early stage of planting. Soil can be treated with 4% endo sulphan at the rate of 30 gm per bush. This can be checked only by treating soil with suitable insecticide before planting. BHC (5%), Aldrin (5%), or Chloro pyriphas (0.1%), Heptachlor or Chlordane are mostly used for its protection. Application of neem cake is also advisable as a repellant.

(viii) **Chafer Beetles**: It attacks roses during monsoon and mostly during night. It eats rose leaves by irregular cuts. Their larvae in the soil destroy the roots. For its control soil should be treated with soil insecticide (BHC or Aldrin) and plants should be sprayed with DDT (0.2%), endosulfon, or malathion, or rogor (1 ml in 1 l water).

(ix) **Jassids**: These are very small insects which suck cell-sap from leaves. The infected leaves turn yellow or whitish. It can be controlled by spraying parathion (0.02%), metacid (0.2%), etc.

(x) **Mites**: Red Spider Mite is one of the smallest insects which lives on the lower side of the leaves. They are well protected by a silky web and they suck plant sap and characteristic patches develop on infected leaves. In extreme cases, the entire plant is defoliated. Spraying of parathion (0.03–0.05%), Kelthane (0.2%), sulmite-75 (2 gm inl 1 water), and metacid (0.1–0.3%) are recommended for its control.

(xi) **Red Scales**: Sometimes stems of roses especially young shoots are severely attacked by these and scale like patches cover the stem. The plants are badly damaged due to continuous sucking of sap. For amateur growers, it can be checked by mechanical cleaning or otherwise it can be controlled by spraying folidol (0.2%) or folithion/eythion/metasystox (2 ml in 1 l of water).

6.11 Characterization

Characterization is very important for correct identification in addition to other requirements. It helps understand the genetic diversity, trace out the phylogenetic relationship, taxonomical status, preparation of catalogue, variation patterns, identification of desirable/novel genes, hybridization, registration, plant variety protection, farmer's right, etc. Cultivar identification and cultivar relatedness are important issues for horticultural breeders. A number of classical and modern techniques have been used in rose for testing hybridity and correct identification of varieties with diagnostic characters. The rich genetic resources of existing varieties can be utilized only after proper characterization and meaningful documentation. During characterization, special attention should be paid to important horticultural characteristics. The ultimate aim of characterization is their utilization in breeding. A number of morphological, cytological, anatomical, palynological, physiological, biochemical, phenolic compounds, pigment composition, and molecular (RAPD) parameters have been applied for characterization of germplasm and new varieties developed through conventional breeding, sport, and induced mutagenesis (Datta and Chakrabarty 2015; Datta and Singh 1999). Author characterized about 150 rose cultivars and characters like stem, young leaf, and flower color; prickles per unit area and prickle shape; petals per flower; leaf and petal size; number of leaflets; pollen grain size and fertility; phenolic compounds in leaves and petals; chlorophyll content in leaves; carotenoids in petals; RAPD markers; etc. were considered as parameters. Characterization identified a variety of specific desirable characters. Studies showed that visible colors of roses is influenced by carotenoids and selection of proper genotype with carotenoids combined with anthocyanidins and flavonols will help breeders to develop the selective combination of flower color as desired by trade (Datta 1999, 2015, 2018). Pollination mechanism, seed maturation, and germination have been studied for better hybridization (Gudin and Mouchotte 1996). Singh et al. (2003) characterized and identified parent varieties ('Sweet Afton', 'Pink Parfait', 'Crioson Glory', 'Charles Mallerin', 'Golden Splendour', 'Buccaneer', 'Swati', 'Anna Wheatcraft', 'Charleston', 'First Prize', 'Orangeade', etc.) having high female and male fertility for breeding. Molecular characters have been used for the characterization and identification of roses (Chakrabarty and Datta 2010, c.f. Datta 2018).

6.12 Development of New Variety

Among all the flowers, the modern roses have hypnotized mankind most and have, therefore, attained a unique status in human hearts with their revolutionary colors, tints, and shades. For modern and industrialized floriculture there is always a demand and necessity for new cultivars. Nowadays roses are available in numerous attractive colors and shades. All the present-day colorful roses are the result of extensive hybridization, spontaneous and induced mutations, and selections. The genetic constitution of the garden roses is very complex and the mode of inheritance of many characters has not been worked out. It is practically impossible to predict the results of any cross as modern hybrids carry the genes of many ancestral varieties. Following methods have played the most important role in developing new varieties of roses:

(i) **Spontaneous Mutation (Bud Sport)**

Spontaneous mutations have played very important roles in the evolution of many new varieties in roses. Spontaneous bud sports occur commonly in the existing cultivated rose varieties and account for considerable genetic variation. Such genetic changes have been noted from time to time by keen gardeners, horticulturists, and researchers in their germplasm collections giving rise to new varieties. Such mutations lead to the formation of chimera which becomes perpetual when propagated asexually. Stable chimeric sports have created attractive varieties over the years for floriculture trade. Bud mutation may arise through gene mutation or chromosomal variation. The contribution of such techniques in developing new rose varieties is very rich and the list of new varieties is very big. Bud sport has created interesting development of climbing habit of Hybrid Tea roses ('Crimson Glory', 'Mrs Sam McGredy', 'Climbing Blue Moon', 'Climbing Cinderella', 'Climbing Fragrant Cloud', 'Climbing Kronenbourg', 'Climbing Mr. Lincoln', 'Climbing Over the Rainbow', 'Climbing Peace', 'Climbing Queen Elizabeth', 'Climbing Ladies

Choice', 'Climbing Miss Harp', 'Climbing Guitare', 'Climbing High Field', 'Climbing Rina Herholdt', 'Climbing Sterling Silver', 'Climbing Yellow Doll', 'Climbing Zambra', etc.). About 18% of the varieties in the Hybrid Tea group have originated as sports. In the Dwarf polyanthas group, about 54% of varieties have been developed through bud sports. A number of striped roses have been developed via sports ('Careless Love', 'Candy Stripe', 'Banhar', etc.). Bud sports have contributed in miniature roses in the development of new varieties ('Climbing Baby Darling', 'Climbing Over the Rainbow', 'Climbing Mary Marshall', 'Climbing Yellow Doll', 'Climbing Peace', etc.). There are many roses that have sported and remained stable for hundreds of years – 'Winchester Cathedral', 'Redoute', 'Rose Marie', many sports of 'Peace' ('Climbing Peace' and 'Chicago Peace'), 'Prairie Snowdrift', etc.

(ii) **Hybridization**

The basic objective of rose hybridization is to incorporate and combine the desirable characteristics of both the parents in order to evolve new flower colors and forms, high fragrance, floriferousness, and disease resistance. Best roses can be developed only through rigorous selection. Desirable characters are scattered in different species and varieties and it is the task of breeders to unite these characters through selective breeding. The ancestry of most parents is very complex and different characters are governed by thousands of genes. Maximum breeding work has been done on roses. At an early stage, diversities developed through natural crossing among wild roses and their derivatives like damasks (*R. damascene*), albas (*R. alba*), centifolia (*R. centifolia*), gallicas (*R. gallica*), muska (*R. moschata*), and a few others. All rose species have contributed their specific traits at early breeding stages. Some significant notable contributions are – cold hardiness from *R. gallica*; recurrent blooming from *R. chinensis*; yellow from *R. foetida* (De Vries and Dubois 1978). Afterward, meaningful improvement in breeding took place through introduction of the Far Eastern rose species *R. chinensis* and *R. gigantean* into Europe. They were hybridized with the European species, *R. damascene*, *R. galica*, and *R. moschata* resulting important roses like Bourbon, Noisette, Portland, Hybrid Perpetuals, and Tea (Swarup 1988). Maximum breeding work is going on in different countries like France, Germany, the Netherlands, UK, USA, Canada, and other developed countries. Modern roses such as Portland, Bourbon, Noisette, Hybrid Perpetuals, etc. have been developed through crosses between Chinese and European roses.

Rose breeding stands on a very recognizable status as huge literature and varieties have been generated through worldwide research. Some basic important information have already been generated on the pattern of inheritance of important characters – maternal inheritance of plant vigor; single or two complementary dominant genes for the inheritance of prickles in the diploid; single recessive gene control of recurrent flowering; dominant monogenic inheritance of double flowers; probable single gene control of blossom color; leaf shape, plant stature, disease resistance, etc. Modern hybrids possess genes for many colors either in dominant or recessive form

(Pal 1982; Buck 1960; De Vries and Dubois 1984). A particular characteristic or group of characteristics may be influenced by a single gene or a particular characteristic may be influenced by several genes together. Desirable characters are present in elemental species and varieties. Target-oriented results can be achieved by breeders having knowledge of some fundamental facts about genetics.

For hybridization, crosses are made between two different known parents. The first step of hybridization is to choose the male and female parents. For crossing following steps are followed: •Selection of female parent and removal of anthers or stamens before anthesis with the help of fine scissors or forceps •Selection of male parent and removal of petals •Anthers at bursting stage are rubbed gently to the pistil of the female flower •Pollen may also be applied with the help of finger or soft bush •Pollination may be carried out from December to March • The female flower after crossing is protected by covering with cellophene bag/butter paper bag/cloth-bag •The female flower is properly labeled mentioning details about the crossing • Fruits are collected from successful crosses •Fruits on maturity become either brownish/yellow/orange/red or even remain green •Selections for new types are made from populations raised from seedlings of these cross-fertilized seeds. The selected desired new type is propagated and multiplied vegetatively. New types can also be selected from seedlings of naturally cross pollinated seeds.

Although there is meaningful progress in rose breeding, there is illimitable scope for the improvement of garden roses. There are many attributes which rose lovers would like to see in the garden roses. Knowledge on rose breeding is now very rich and more scientific. Desirable characters have been identified in different elemental species and varieties which are scattered at different places. There is every possibility for directive breeding for desired objectives as highlighted (cf. Datta 2018):

(a) ***Disease Resistance***: Varieties have already been identified which can be used as resistant parents in the breeding program (resistant to black spot and powdery mildew – 'Spotless Gold', 'Spotless Yellow', 'Spotless Pink', 'Ballet', 'Ovation', 'Captain Thomas', 'Prairie Princess', 'Music Maker', 'Appllow', 'Dezant', 'Gabricab', 'Jaguar', 'Golden Showers', 'A Mackenzie', 'Charles Albart', 'Champlan', 'William Battin', etc.; black spot immune – *R. bracteata, R. clinophylla;* tetraploid *R. multiflora* seedlings; mildew resistance – 'Golden Showers', 'Slver Jubilee', 'Pristine', etc.) (Saunders 1970; Knight and Wheeler 1978).

(b) ***Cold Resistance:*** Varieties have been derived from *R. rugosa* and *R. wichuriana* (Svejda 1979).

(c) ***Heat Resistance***: A number of promising species (*R. gigantean, R. clinophylla*), heritage roses ('Archduke Charles', 'Parle d'Or', 'Cecile Brunner', 'M. Falcot', etc.) and varieties ('Delhi Princes') have been identified which grow well under warm conditions. These roses may be included in breeding to develop better heat resistance roses (Viraraghavan 2003).

(d) ***Tthornless Roses***: The development of thornless varieties is a desirable objective in rose breeding. A number of species and varieties have been identified having less thorn or are relatively thornless. These have been included in the breeding program, and the resultants offer as most promising candidates for further breeding to develop thornless varieties. Present status of less thorn or no thorn roses are as follows: **Species** – *R. pendulina, R. blanda, R. Carolina, R. multiflora, R. penduliana, R. wichuriana*; *R. banksie lutes* are almost thornless. The 'Smooth Rose' sometimes called the 'Hudson Bay Rose' or 'Labrador Rose' is thought to be derived from the species *R. pendulina* and *R. blanda.* Varieties have been identified having less thorn or are relatively thornless ('Adam Messerich', 'Allister Stella Gray', 'Belle de Crecy', 'Betty Bland', 'Betty Bugnet', 'Blush Noisette', 'Chloris', 'Celestial', 'Charles de Mills', 'Cardinal de Richelieu', 'City of London', 'Cramoisi Picote', 'Duchesse de Buccleugh', 'Elizabeth Arden', 'Empress Josephine', 'Harmonie', 'Honorie de Brabant', 'Hippdyte', 'Officinalis', 'Jacaranda', 'J.P. Connell', 'Kathleen Harrop', 'Louis Bugnet', 'Martin Frobisher', 'Metis', 'Mine Legras de St. Germain', 'Modern Fireglow', 'Nastarana', 'Old Soothie', 'Paul Neyron', 'Prairie Youth', 'Playgirl', 'Royal Edward', 'Sutter's Gold', 'Stryker', 'Zephirine Drouhin', etc.) and thornless ('Bella Multiflora', 'Brunner', 'Camaieux', 'Cecile', 'Grand Gala', 'Hermosa', 'La Reine Victoria', 'Mrs. John Laing', 'Mme Pierre Oger', 'Nevada', 'Zephirine Drouthine'). **Fragrant varieties** with no or few thorns have been identified specially for using in garden for the blind ('Camaieux', 'Cecile Brunner', 'Drouhine', 'Hermosa', 'La Reine Victoria', 'Mrs John Laing', 'Mme. Pierre Oger', 'Nevada', 'Zephirine', etc.). **Climbers**: thornless or few at base ('Amadis', 'Amethyster', Allister Stella Gray', 'Blush Noisette', 'Blush Rambler', 'Burgundiana Rose', 'Chloris', 'Celestial', 'Georg Arends', 'Mine Legras de St. Germain', 'Mrs. John Laing', 'Nastarana', 'Paul Neyron', 'Reve d'or', 'Souv. due Dr Jamain', 'Tourde Malakoff', 'Tausendsction', 'Ulrich Brunner Fils', 'Veilehenbleau', 'Zephirine Drouhin'). **Shrubs**: ('Bellinda', 'Ballerina', 'Cecile Brunner', 'Gestendirector Otto Linne', 'Lavender Lassie', 'Margo Koster', 'Marguerite Hilling', 'Nevada', 'Nyphenburg'). **Old modern roses**: ('Adam Messerich', 'American Beauty', 'Baroness Rothschild', 'Belle de Crecy', 'Bells Isis', 'Blush Noisette', 'Camaieux', 'Celestial', 'Champney's Pink Cluster', 'Chloris', 'Commendant Beaurepaire', 'Complicats', 'Ducherse de Montebello', 'Ferdinand Pichard', 'Frau Karl Druschki', 'George Arends', 'Hermosa', 'Katheleen Harrop', 'La Reine Victoria', 'Lady Hillingdon', 'Louise Odier', 'Madame Legras de St. Germain', 'Madome Pierre Ogre', 'Madame Plantier', 'Maman Cochet', 'Marchioness of Londonderry', 'Marie Pavie', 'Mary Washington', 'Mrs Dudley Cross', 'Mrs. John Laing', 'Petite Lisette', 'Paul Neyron', 'Prince Charles', 'Reine des Violettes', 'Rosa Galica Officinalis', 'Rosa Mundi', 'Rosette Delizy', 'Ulrich Brunner'). **Floribunda**: ('Apache Tears', 'Apricot Nectar', 'Dusky Maider', 'Gruss an Aachen'). **HT**: ('Gpsy', 'Medallion', 'Sterling Silver'). **Miniature**: ('Andrea', 'Audrey Hepburn', 'Angel Dust', 'Blue Moon', 'Cinderella', 'Cinderella Gold', 'Cool Dude', 'Elizabeth Arden', 'English Porcelain', 'Fortune Cookie', 'Halo Rainbow', 'Halo Today', 'Harmonie', 'Jack Horner', 'Jacaranda', 'Little Linda', 'Madelyn Lang', 'Melody Marshall', 'Mistee', 'Old Soothi', 'Pompon de Paris', 'Pretty Penny', 'Royal Ruby', 'Sweety Fairy',

'Sutter's Gold', 'Stryker', 'Sugar Palm'). **Greenhouse Roses**: ('Bella Multiflora', 'City of London', 'Charlotte', 'Grand Gala', 'Heritage', 'Pink Parfait', 'Playgirl', 'Smooth Melody', 'Smooth Angel', 'Smooth Lady', 'Smooth Perfume', 'Smooth Romance', 'Sir Walter Roleigh'). It is now possible to develop more thornless roses through selective breeding (Porter 1999; Padhye 1988).

(e) ***Fragrance***: Rose and fragrance are synonymous. The fragrance is due to the presence of volatile oils. Commercially available rose perfume is mainly derived from *R. damascena* and *R. centifolia*. Fragrant varieties are present in different rose groups. Genetics of rose perfume has not been properly worked out. Over 30 compounds are involved in rose fragrance. Le Grice (1969) described rose scent as following types – Rose, Nasturtium, Orris, Violet, Apple, Lemon and Clove – to about 40 recorded by S. C. Harlord (Almond, Black berry, Honey, Magnolia, Musk, Myrrh, Pineapple, Raspberry, Bugs, Turpentine, etc.). The most common in Indian varieties is lemon ('Radhanath') and other examples are apple, clove, nasturtium, orris, violet, musk ('Heart Throb', 'Week End', 'Tribute', 'Double Helix'); raspberry, Parsley, wine orange, pineapple, mixed fruits ('Lalima', 'Kum Kum', 'Anirban', 'Bhanu', 'Brahm Datta', 'Red Perfume', 'Kasturi Rangan'); citrus, myrrh, strawberry, dianthus, tea ('Haridra', 'Raja Ram Mohan Roy', 'Sunanda', 'Corn Sukumarda', 'Nefertiti', 'Ganges Mist', 'Manipur Magic', 'Climbing Kanyakumari', 'Bhargav', 'Shantaraj', 'Willian Carey', 'Bharati'); honey, spicy ('Kishori', 'Fragrant Mauve', 'Touch of Heart', 'Mrs. Davis', 'Sudhanshu', 'Sweet India', 'Stealthy Kiss', 'Rajni', 'Asha'); Rose 'Sugandha', 'Fragrant Beauty', 'Rose Bengal', 'Our Indira', 'Classic', 'Pride of Nagpur', 'Dr Kane'). Roses with dark color petals, more petals, thick petals, and velvety petals have been noted highly scented and color wise categorized as -red and pink ones are most likely to smell like a "rose," while white and yellow ones incline towards orris, nasturtium, violet, or lemon. Orange-shaded roses usually have scents of fruit, orris, nasturtium, violet, or clover (Bhowmick 2006; Viraghavan 1988). Fragrance (F) of rose varieties have been categorized on a scale 1–10 and few examples are – 'Gruss An Coburg' (F9), 'The Doctor' (F9), 'Lady Luck' (F8), 'Granada' (F8), 'Oklahoma' (F9), 'Inge Horstmann' (F9), 'Blue Moon' (F8), 'Lemon Spice' (F8), 'Whisky Mac' (F8), 'Perfume Delight' (F9), 'Double Delight' (F9), 'Sweet Sarrender' (F10), 'Blue River' (F9), 'BelAmi' (F9), 'Ranjana' (F7), 'Sunsprite' (F8), 'Shocking Blue' (F9), 'Magali' (F8), 'Climbing Crimson Glory' (F9) (C.f. Datta 2018).

Considering the importance of fragrance, scented varieties have been identified from different groups of roses – ***Bush Roses*** ('Scented Air', 'Ena Harness', 'Fragrant Cloud', 'Margaret Merrill', 'Fountain', 'Royal Gold', 'Radox Bouquet', 'Double Delight'); ***Climbers*** ('Compassion', 'Breath of Life', 'Rosy Mantle'); 'Papa Meilland', 'Oklahoma', 'Mr. Lincoln', 'Sutter's Gold', 'Super Star', 'Tiffany', 'Lemon Spice'; ***Old varieties*** ('Lady Mary Fitzwilliam', 'Devonienses', 'Victor Verdier', 'Mme. Croline Testout', 'Opelia', 'Catherine Kordes', 'Crimson Glory', 'Soleil d'Or', 'Sensation', 'Souvenir de Claudius Pernet', 'Julien Potin', 'Talisman', 'Souer Therese', 'Peace', 'Signora', 'Charlotte Armstrong', 'Ena Harkness', 'Fashion', 'Sutter's Gold', 'Lemon Spice', 'Fragrant Cloud', 'Prima Ballerina', 'Tenerife', 'Forgotton Dreams', 'Dolly Parton', 'Velvet Fragrance', 'Radox Bouquet',

'Mr Lincol', 'Rosy Mantle', 'Compassion', 'Rajni', 'Somasila', 'Breath of Life', 'Spartan', 'Little Darling', 'Elizabeth of Glamis', 'June Park', 'Avon', 'Josephine Bruce', 'Wendy Cussons', 'President Hoover', 'Eden Rose', 'Tahiti'). Some varieties require special mention for their most specific dependable fragrant – 'Chrysler Imperial' (rose-clove flavor), 'Queen Elizabeth' (wood-like fragrance), 'Mister Lincoln' (tea and Damask), 'Camelot' (spicy), 'Tiffany' (lemon), 'Granada' (spicy-tea), 'Polynesian Sunset' (fruity), 'Junior Miss' (tea rose), 'Angel Wings' (apple), 'Mirandy' (rose–lemon), 'Golden Showers'(orris), 'Sutter's Gold'(quince), 'Charlotte Armstrong' (lemon–nasturtium), 'Eiffel Tower', 'Crimson Glory', 'Papa Meilland', 'Avon' (Damask-type perfume), 'Deshi roses viz', 'Edouard', 'Chait' (delightful fragrance) (Padhye 1982; Viraghavan 1988; Sidhu 1990; McCann 1987). In 1951, W. E. Lammerts found that a few of the older rose varieties were either only moderately scented or had no scent at all. In 1956, J. A. Gamble found on the examination of 3900 rose varieties, both old and new, that 25% were scentless, 20% strongly scented, and the rest had some scent (c.f. Datta 2018).

(f) ***Brown Color***: Series of beautiful brown roses are available – 'Amberlight', 'Auguste Renoir', 'Cesper', 'Chocolate Prince', 'Colorbreak', 'Brown Velvel', 'Dark Moments', 'Hot Chocolate', 'Hot Cocoa', 'Mayflower rose', 'Tasman Bay', 'Tombrown', etc. Some varieties have also been identified to develop further brown roses like 'Jocelyn', 'Kirsty Jane', 'Mary Summer', 'Princesse', 'Tane', etc.

(g) ***Better Red Roses***: Cyanidin, chrysanthemin and paeonin impart red colour to roses. Varieties with high amounts of these pigments may be included in breeding program to develop more perfect and desired red roses. Few rose species (*R. foetida* bicolor, *R. rugosa*, *R. stellata*, etc.), climbing roses ('Francois juraiville', 'Dorothy Perkins', and 'Souvenir de la Malmaison') and floribunda varieties ('Piccolo','Red pinocchio', and 'Ruby lips') contain a large amount of paeonines and chrysanthemin.

(h) ***Better Yellow and Orange Roses***: Flavonols – kaempferol and quercetin are present in a number of rose species and varieties of yellow color in combination with carotenoid. Pelargonidin and carotenoid will produce brilliance of color. 'Louise de Funes' has been developed from a mixture of cyanidin with carotenoid (Morey 1961).

(i) ***Miniature roses***: Hybridizers have been motivated to develop mini varieties considering their increasing popularity. All miniatures do not produce good seeds but they produce good pollen. Such minis can be used as pollen parents and crossed with climbers, floribundas, and shrub roses. Mini gene is generally dominant, so one can expect 90% of the progeny to be miniatures.

(iii) **Mutation**

Mutation is a sudden permanent genetic change which in turn produces a new form. Induced mutation has been very successful in vegetative propagated plants including rose. A number of new varieties have been evolved in rose by induced mutations. Radiations like X-ray, gamma ray; different chemicals like ethyl methane sulphonate, methyl methane sulphonate, ethylene imine, n-nitroso-N-methyl urethane, and colchicine have been successfully used for evolving new cultivars (Datta 1997). For induced mutagenesis experiment entire plant at different developmental stages can be exposed to radiation (say gamma ray) if there is gamma garden facility. Otherwise, size of plant materials depends upon the size of radiation source. For X-irradiation Philips X-ray machine is available. For gamma irradiation, specially designed gamma chamber unit of different capacity is available. The gamma radiation source of these units is Cobalt-60. For gamma irradiation budwood of 13 cm height containing 2–4 auxiliary buds (eye) are treated with 3, 4, and 5 krad. The dose rate and the time required for treatment is calculated based on the activity of Cobalt-60 source. After irradiation individual eyes are removed and budded on *Rosa indica* var. *odorata* (or any other suitable root stock). For use of chemical mutagens any of the following three processes may be adopted: (a) about an inch of a budwood containing 2–4 auxiliary buds can be dipped in a desired concentration of chemical mutagens for a specific period; (b) auxiliary buds are removed and dipped in a desired concentration of chemical mutagen for desired time; (c) shoot apex or auxiliary buds are covered with cotton and chemical mutagen is applied periodically upto desired time and then cotton is removed. After chemical treatment eyes are removed and budded. After budding of treated eyes, proper care is taken for sprouting and subsequent growth of the plant and careful attention should be paid up to the flowering period. Each and every petals of a flower should be examined to find out any change in flower color/type. Normally the mutation (i.e., change in flower color/shape) in rose arises as chimera (i.e., normal and mutated cells side by side). Isolation of mutant tissue from chimeric branch is the most important operation in mutation breeding. If one to few petals of a flower or one to few flowers of a branch are mutated, it is difficult to isolate the mutant issue. Isolation becomes easier if one complete branch is mutated. Sometimes all the control branches are removed to encourage proper growth of the mutated branch. Auxiliary buds of mutated branches are very carefully removed and propagated by budding. If tissue culture technique can be standardized to develop plants from petals only, then mutants can be isolated more easily even from the mutated sector of a petal. The new flower color/type is isolated in pure form by repeated budding and multiplied. Proper assessment of mutant character over the original character is most essential before release as a new cultivar. More than 70 mutant varieties have been reported worldwide. Induced mutagenesis at its present status appears to be well standardized, efficient, and cost effective. Mutation breeding can play a very important role in producing novelties in flower color/shape in outstanding cultivars of roses without losing any of their desirable characters. Colchicine has been tested for successful induction of mutations in roses (Ahloowalia et al. 2004; Datta 2012, 2014, 2017, 2020; Maluszynski et al. 1992).

6.13 Post-Harvest Management

Post-harvest management of cut roses plays a very important role in floriculture trade. Proper management increases the life of cut-flowers and reduces their loss. Massive amount of literature have been accommodated on this topic. Rose flowers are cut while still in the bud stage. In large-flowered roses, flowers along with the stem of prescribed length are cut when the first one or two petals start to unfold. The size of stem varies from 60 to 90 cm for large flower roses and 40–50 cm for small flowered. The size of large-flowered bud is 3–3.5 cm and 2–2.5 cm for the small flowered. The flowers are harvested and graded according to stem length and grouped in bunches of 10, 12, or 25 flowers. Standard packaging material and system are used for transport. For increasing the life of cut-roses both classical and modern methods are being used. Different factors related to pre-harvest, harvest, and postharvest have been studied and recommended. Most of the experimental results and recommendations are cultivar specific. There is no uniform specific treatment system which can be blindly used for all. A wide range of cultural practices and chemicals have been recommended for increasing the life of cut roses. Extensive work on post-harvest biology and technology, that is, pulsing, holding solution, and bud opening solutions which play a major role in improving the post-harvest life and quality of rose cut flowers have been studied. D-fructose (4%), $Al_2(SO_4)$ (300 ppm), $MgSo_4$ (150 ppm), $NiCl_2$ (330 ppm), Kinetin (2.5 ppm), 8HQC (150 ppm), Daminozide (25–100 ppm), L-ascorbic acid (500 ppm), Chrysal (1%), D-fractose (3%) + Kinetin (2.5 ppm), D-fructose (3%) + L-ascorbic acid (500 ppm) + Kinetin (2.5 ppm) showed appreciable effect on cut flower quality and longevity. Pulsing with different chemical combinations has been found to extend vase life - sucrose (3%) 18–24 hrs. at 25 °C, STS (0.5 mM) 45 min, $AgNo_3$ (1 mM) 45 min., $CaCl_2$ (1%) 20 hrs., $Na_2S_2O_3$, $5H_2O$ (1000 ppm) 8 hrs., SADH (500 ppm) 8 hrs. Some chemical combinations were tested and found to be promising in bud opening like sucrose (4%) + $CoCl_2$ (250 mM) + 8 HQC (200 ppm), 8 HQC (250 ppm) + Acetyle Salicylic acid (100 ppm) + D-fructose (1%), D-fructose (1%) + boric acid (500 ppm) + $CoCl_2$ (250 ppm) (c.f. Datta 2015). Flower buyers are very choosy and prefer quality blooms with longer vase life. Rose flower selling is now not a seasonal business. It is available all the year round. If the vase life of cut roses are taken care, the present retail system will definitely boost floriculture industry. Therefore, there is a need to develop appropriate crop-specific post-harvest technology, suitable to specific agro-climatic zone, to avoid losses at the growers, florists, and consumers level. There is a need to find out the best combinations of holding solutions for commercial exploitation to increase the vase life of cut roses.

6.14 Molecular Breeding

Tissue culture is a very important step in molecular breeding and this technique has been standardized in different rose species and cultivars – *Rosa multiflora*, *R. hybrida*, *R. canina*, *R. chinensis*, *R. rugosa*, *R. wichurarina*, *R. setigera*, *R. laevigata*, *R. banksiae*, *R. roxburghii*, *R. odorata*, miniature rose, hybrid tea, etc. using different explants like apical meristem, axillary meristem, shoot tip, lateral bud, etc. Techniques have also been standardized to regenerate from induced callus from another culture (tetraploid *Rosa damascena* and *R. hybrida*), protoplast derived calli from suspension cultures (*R. persica x xanthina* and *R. wichuraina* and

R. persica x xanthina and cv. All Gold). Somatic embryogenesis has been reported by a number of workers using different explants like leaf, petiole, internode, filament of stamen, immature seed, root, and zygotic embryo of different cultivars of *R. hybrida, R. persica x xanthina, R. rugosa,* etc. Work on genetic transformation in rose is limited. Reports available on – *Agrobacterium*-mediated transformation in *R. hybrida* cv. Royalty; transgenic rose using an embryogenic callus; introduction of rice chitinase transgene to reduce blackspot disease and enhanced resistance of *R. hybrida* cv. Carefree beauty to powdery mildew by expression of an antimicrobial protein gene; improved rooting characteristics in Rosa hybrid cv. Moneyway through the introduction of rolA, B, and C genes; transfer of antisense chalcone synthase cDNA to reduce anthocyanin biosynthesis to modify flower color (c.f. Datta 2005, 2015, 2018, Datta and Chakrabarty 2005). The most significant and exciting contribution of molecular breeding is the development of blue rose (Holton and Tanaka 1994, Tanaka et al. 1995, Hennayake et al. 1995, Katsumoto et al. 1978).

References

Ahloowalia BS, Maluszynski M, Nichterlein K (2004) Global impact of mutation-derived varieties. Euphytica 135:187–204

Bhowmick D (2006) Fragrant Indian roses. Indian Rose Ann XVII:107–111

Buck GJ (1960) Progress report on breeding hardy ever blooming roses. Am. Rose Ann 45:95–99

Chakrabarty D, Datta SK (2006) In vitro meristem tip culture of *Rosa hybrida*. Hortic Environ Biotechnol 47(2):71–74

Chakrabarty D, Datta SK (2010) Application of RAPD markers for characterization of gamma-ray-induced rose mutants and assessment of genetic diversity. Plant Biotechnol Rep 4:237–242

Chakrabarty D, Mandal AKA, Datta SK (2000) In vitro propagation of rose cultivars. Indian Journal of Plant Physiology 5(2):194–197

Datta SK (1997) Mutation studies on garden roses : a review. Proc Indian Natl Sci Acad B63 (1&2):107–126

Datta SK (1999) Flower colour analysis in garden roses : Carotenoides. Sci Hortic 6:151–156

Datta SK (2002) Rose (NBRI bulletin no. 9) page 1–26 (with 14 colour photographs)

Datta SK (2005) Technological advances for development of new ornamental varieties. In: Trivedi PC (ed) Advances in biotechnology. Agrobios, Jodhpur, pp 285–304

Datta SK (2011) Miniature roses – a fascinating group of rose with high research and economic potential. J Ornamental Hortic 14(1&2):1–15

Datta SK (2012) Success story of induced mutagenesis for development of new ornamental varieties. In : Bioremediation, biodiversity and bioavailability 6 (Special Issue I) : 15–26, Global Science Books

Datta SK (2014) Induced mutagenesis: basic knowledge for technological success. In: Tomlekova NB, Kozgar MI, Wani MR (eds) Mutagenesis – exploring genetic diversity of crops, vol 5. Wageningen Academic Publishers, pp 97–139

Datta SK (2015) Indian floriculture: role of CSIR. Regency Publications, A Division of Astral International(P) Ltd, New Delhi

Datta SK (2017) Improvement through induced mutagenesis : ornamental crops. In: Malik CP, Wani SH, Bhati H, Kushwaha, Kaur R (eds) Advanced technologies for crop improvement and agricultural productivity, vol 4. Agrobios, Jodhpur/Chater, pp 49–86

Datta SK (2018) Breeding of new ornamental varieties: Rose. Curr Sci 114(6):1194–1206
Datta SK (2020) Induced mutations: technological advancement for development of new ornamental varieties. Nucleus 63:119–129
Datta SK, Chakrabarty D (2005) Classical mutation breeding and molecular methods for genetic improvement of ornamentals. In: Datta SK (ed) Role of classical mutation breeding in crop improvement. Daya Publishing House, Delhi
Datta SK, Chakrabarty D (2015) Classical and modern methods for characterization of ornamental crops. Int J Horticult 5(21):1–45
Datta SK, Singh MS (1999) Survey of phenolic compounds in leaves of garden roses: miniature cultivars. Sci Hortic 6:157–164
De Vries DP, Dubois LAM (1978) On the transmission of the yellow flower colour from *Rosa foetidato* recurrent flowering hybrid tea-roses. Euphytica 27:205–210
De Vries DP, Dubois LAM (1984) Inheritance of the recurrent flowering and moss characters in F1 and F2 hybrid tea x *Rosa centifolia* muscosa (Aiton) Seringe populations. Gartenbauwissenschaft 49:97–100
Duha RS (1999) Rose. In: Bose TK, Maiti RG, Dhua RS, Das P (eds) Floriculture and landscaping. Naya Prokash, Calcutta, pp 368–405
Gudin S, Mouchotte J (1996) Integrated research in rose improvement – a breeder's experience. Acta Hortic 424:285–292
Hennayake CK, Kanechi M, Yasuda N, Uno Y, Holton TA, Cornish EC (1995) Genetic and biochemistry of anthocyanin biosynthesis. Plant Cell 7:1071–1083
Holton TA, Tanaka Y (1994) Blue roses – a pigment of our imagination? Trends Biotechnol 12: 40–42
Jian H et al (2010) Decaploidy in *Rosa praelucens* Byhouwer (Rosaceae) endemic to Zhongdian plateau, Yunnan, China. Caryologia 63:162–167
Kaicker US (1995) Rose breeding. In: Chadha KL, Bhattacharjee SK (eds) Advances in horticulture vol 12 – ornamental plants. Malhotra Publishing House, New Delhi, pp 335–349
Katsumoto Y, Fukuchi-Mizutani M, Fukui Y, Bruqliera F, Holton TA, Karan M, Nakamura N, Yonekura-Sakakibara K, Togami J, Pigeaire A, Tao GQ, Nehra NS, Lu CY, Dyson BK, Tsuda S, Ashikari T, Kusumi T, Mason JG, Tanaka Y (1978) Engineering of the rose flavonoid biosynthetic pathway successfully generated blue-hued flowers accumulating delphinidin. Plant Cell Physiol 48 (11):1589–1600. https://doi.org/10.1093/pcp/pcm131. First published online: 9 October 2007
Knight C, Wheeler BEJ (1978) Evaluating resistance of roses to blackspot. Phytopathol Z 91: 218–229
Le Grice EB (1969) Breeding blue and brown roses. RNRS Rose Annu:124–127
Maluszynski M, Sigurbjrnsson B, Amano E, Sitch L, Kamra O (1992) Mutant varieties – data bank FAO/IAEA data-base. Part 11. Mutat Breed Newsl 39:14–33
Matthews QL (2004) The moguls and the rose. Indian Rose Annual XX:86–88
McCann (1987) 'McCann on Roses', 'The Rose'. R.N.R.S. UK
Morey D (1961) Observations on the inheritance of yellow pigments in roses. Am Rose Annu 46: 120–124
Padhye VS (1982) Fragrance in roses. Indian Rose Ann:100–112
Padhye YS (1988) Thornless roses. Indian Rose Ann VII:84–94
Pal BP (1982) Rose breeding some thoughts and experiences. Indian Rose Ann 2:10–16
Porter B (1999) Thornless roses – fact or fiction? Indian Rose Ann XV:45–48
Saunders PJW (1970) The resistance of some cultivars and species of *Rosa to Diplocarpon rosae* wolf causing blackspot disease. Rose Ann 11:8–128
Sidhu GS (1990) Fragrant beauties past and present. Indian Rose Ann VIII:70–75
Singh AP, Prasad KV, Choudhary ML (2003) Breeding. In: Panorama of rose research, book published by A. P. Singh, All India coordinated research project on floriculture. Public Printing Service, New Delhi, pp 70–79
Svejda F (1979) Inheritance of winter hardiness in roses. Euphytica 28:309–314

Swarup V (1988) Rose breeding – an overview. Indian Rose Ann VII:43–53
Tanaka Y, Fukui Y, Fukuchi-Mizutani M, Holton TA, Higgins E, Kusumi T (1995) Molecular cloning and characterization of *Rosa hybrida* dihydroflavonol 4-reductase gene. Plant Cell Physiol 36:1023–1031
Vamosi JC, Dickinson TA (2006) Polyploidy and diversification: a phylogenetic investigation in Rosaceae. Int J Plant Sci 167:349–358
Viraghavan MS (1988) On the scent of 'scent'. Indian Rose Ann VII:95–100
Viraraghavan MS (2003) Rose breeding for warm climates. Indian Rose Ann XIX:47–50
Wylie AP (1955) The history of garden roses, part 2. J R Hortic Soc 79:8–24

Agro-Biodiversity: Conservation and Use of Plant Genetic Resources

7

Tulip

Puja Sharma, Bhavya Bhargava, Panchal Sangmesh, and Ujala

Contents

P. Sharma (✉) · P. Sangmesh
Dr. YSPUHF, Nauni-Solan, Himachal Pradesh, India

B. Bhargava
CSIR, IHBT, Palampur, Himachal Pradesh, India
e-mail: bhavya@ihbt.res.in

Ujala
Academy of Scientific and Innovative Research, New Delhi, India

S. K. Datta, Y. C. Gupta (eds.), *Floriculture and Ornamental Plants*, Handbooks of Crop Diversity: Conservation and Use of Plant Genetic Resources,
https://doi.org/10.1007/978-981-15-3518-5_9

Abstract

Tulips are the most popular bulbous ornamentals worldwide. The unique flower shapes, various colors, forms, and size have made them popular in cut flower and landscape industry. This bulbous geophyte is currently occupying the third position in most traded flowers in the world. Modern cultivars (predominantly *Tulipa gesneriana*) are grown for bulb production, cut flowers, flowering potted plants, and landscaping. Breeding objectives and approaches to achieve the desired characters are discussed on the basis of tulip biology. Besides breeding for aesthetics, the current trend is focused on breeding for environmental reasons, including breeding for resistance to pathogenic fungi and viruses. Precise thoughtfulness is paid to breeding methods: hybridization, mutation, and use of cytological and molecular markers that can reduce the selection method. Tissue culture techniques are being harnessed to overcome extended breeding cycle. The unending popularity of tulips opens up new vistas for exploitation of different traditional and molecular techniques for the development of novel cultivars.

Keywords

Tulipa · Hybridization · Breeding objectives · Cytological and molecular markers · Mutagenesis · Polyploidization · Micropropagation

7.1 Introduction

Tulip, a collective name given to a large number of tulip species and cultivars, is currently one of the most important ornamental bulbous crops in the world. These were introduced to Europe from Turkey in the middle of the sixteenth century. However, the primary gene center of the genus *Tulipa* L. is located in the Pamir-Alai and Tien Shan mountain ranges in Central Asia (Hoog 1973). The name "tulip" probably originates from the word "turban" (tulpana in Turkish). Tulip bulbs were brought to the imperial gardens in Vienna by Augerius Gislenius Busbequius, the ambassador for the Holy Roman Emperor Ferdinand I to the court of Suleyman the Magnificent of the Ottoman Empire. These were further taken via Carolus Clusius to the gardens of Leiden University, the Netherlands, where their cultivation and breeding began (Pavord 1999).

Diversification occurred from this region, resulting in a distribution from Morocco to Western Europe and eastward to Western China. A secondary gene center has been found in the Caucasus. They flowered for the first time in the

Netherlands somewhere in 1594. Around 1630, tulips were extremely popular. Currently, the total world area under tulip bulbs is about 13,000 hectares out of which about 88% area is located in the Netherlands. In other countries, Japan, France, Poland, Germany, and New Zealand, tulips are cultivated in an area of 122–300 ha (Buschman 2005). During the 2016–2017, more than 1800 cultivars were commercially cultivated on a total area of 11,843.91 ha, but only 20 cultivars were grown on more than 100 ha (BKD 2017a). The five most popular cultivars ("Strong Gold," "Leen van der Mark," "Purple Prince," "Purple Flag," and "Ile deFrance") occupy more than 19% of the acreage in the Netherlands (2257.5 ha) (BKD 2017a). The Netherlands produces more than 4 billion tulip bulbs, of which 2.3 billion (53%) are used for cut flower production in the Netherlands and other countries (Benschop et al. 2010).

The introduced tulips have been grown and bred for a long time. This has resulted in a wide diversity of flowering, growth, vigor, and flower shapes in the genus. These tulips, whose original species have not been determined, are grouped together and are called *T. gesneriana* L. The current commercial assortment still consists mainly of cultivars from *T. gesneriana*. The second group of cultivars, the Darwin hybrids, has been obtained from interspecific crosses between cultivars of *T. gesneriana* and genotypes of *T. fosteriana* Hoog ex W. Irving. Tulips are grown either for bulb production, forcing as cut flower and potted plant, and landscaping. The *T. gesneriana* and Darwin hybrids consist of more than 1100 cultivars (Van Scheepen 1996). The 10 most popular cultivars, however, occupy more than 35% of the planted acreage. Only 7% (643 ha) of the total area under tulip consist of species, of which *T. fosteriana*, *T. greigii* Regel., and *T. kaufmanniana* Regel. are the primary ones.

Tulip has been widely used as cut flowers, potted, and garden plants (Van Tuyl and Creij 2006; Ramzan et al. 2014). The genus *Tulipa* L., which belongs to the Liliaceae family, is native to the Pamir-Alai and Tien Shan mountain ranges in Central Asia (Hoog 1973) and comprises more than one hundred species (Diana 2013; Tang et al. 2015). Most species are distributed in Central Asia and the Caucasus (65 species), Iran and adjoining regions (36 species), and Turkey (18 species) (Eker et al. 2014). The history reveals that tulips were in cultivation in Turkey, Persia, and adjoining areas as early as twelfth century. From these places, tulips were introduced to Vienna in Austria in Europe in mid-sixteenth century. The first illustration of a tulip from Austrian garden was printed by Conrad Gesner in 1561 (Rees 1992). From Austria, tulip gradually spread to the Netherlands, UK, and other parts of Europe. In the Netherlands, it became a great favorite to the flower lovers. This love for tulips encouraged the development of a great bulb production industry in the Netherlands. Since then, the cultivation of garden tulips has become a craze throughout the world. In India also tulips have become very popular and recently it is being grown commercially in some parts especially in the foot hills of Himalayas in Kashmir, Himachal Pradesh, and Uttaranchal.

Gardeners and collectors are looking for genotypes with new shapes and colors, as well as with changed functional features, which enables the extended use of tulips in the landscape and in the floristic compositions. The tulip is one of Dutch national symbols and Dutch producers together with Dutch scientists work jointly on new breeding creations. The attractiveness of tulips on the international market stimulates other breeders in the world to actively breed this species. This can be especially prospective in countries possessing rich wildlife resources. For example, some tulip cultivars were recently released in China (Qu et al. 2016). Japan is the country where since the beginning of the twentieth century about 100 tulip cultivars were released. Some of them are from double haploid origin.

7.2 Botany

The genus *Tulipa* belongs family Liliaceae, of monocots. A tulip flower stalk is 7–30 cm in length, is stiff, straight, smooth, or hairy and in only a few species (*T. biflora*, *T. tarda*, *T. praestans*, and *T. turkestanica*) does the plant produce several flowers. Leaves are broad – oval or elliptical or equilateral – lanceolate, without a tail. In some species and varieties, the leaves have dark purple streaks or interrupted irregular rectangles on the leaf blades between the nerves. In most species, however, the leaves are gray-green and covered with a waxy layer. The number of leaves is usually three to five, with the more prominent at the bottom of the plant. The fruits are trivalent and have trivalent bags in which there may be more than 100 seeds. If their harvest takes place in the phase of adequate maturity, the seeds are easy to germinate and give rise to new plants. However, tulips are propagated by seeds only for breeding purposes due to a high heterozygosity. Since seedlings do not repeat the traits of the mother plant, tulips are reproduced only vegetatively by bulb proliferation.

The bulbs are modified, shortened underground stems. They consist of a basal plate (hill) and from two to six concentric fleshy scales covered by a brown dry tunic. The number of scales and bulb size determine its ability to flower. Usually, a bulb of four or five scales with a minimum circumference of 6–9 cm is able to produce a flower. In autumn, when bulbs are planted, a central bud already has fully developed miniature leaves, stem, and flower, including the pistil with visible stigma comprising three distinct lobes (stage G), which is the last stage of flower differentiation. From a practical point of view, tulip bulbs are considered annual. However, the whole life cycle of a single bulb from its initiation to flowering is about 29 months or an additional 12 months if the bulb is too small to initiate a flower in its second year (Rees 1992). One flowering-sized bulb, at planting, contains three generations: mother bulb, buds of daughter bulbs which will replace the mother bulb in the next season, and a meristem of each granddaughter bulbs, between the fleshy scales of the daughter bulbs. In tulip production, bulbs are dug up after the leaves dry, dried, and stored until autumn, when they are replanted.

Perianths are not differentiated on the calyx and the crown. They consist of three inner and three outer tepals. The length of tepals varies from 1.5 to a few centimeters. The sheets are usually glossy on the inside and nonglossy on the outside. Six large stamens are arranged in two circles surrounding the prominent pistil. On the elongated and triangular ovary, there is a stigma, but either the neck is short or does not exist at all. The flowers can be single or filled, and their shape can vary – from cups, bowls, calyx, bells, or lily-shaped to the twisted or rounded tepals. The range of tepal colors is very rich, with flowers displaying white, yellow, pink, red, and violet in different intensities – with the exception of either pure blue or black. There are a range of variation of colors, from being a single color to being different on the outer and the inner side of tepals or having different color at the edges. The color and shape of the tepals, as well as the inner base of the perianths, vary significantly among cultivars. A descriptor list of genus *Tulipa* (Petrová and Faberová 2000) shows ten flower shapes and nine inner base views. According to the UPOV guidelines for conducting tests for distinctness, uniformity, and stability of *Tulipa* (UPOV 2006), 16 main color groups of flowers are specified (cultivar example in brackets): (1) white ("Snow parrot"), (2) off-white (no example), (3) light yellow ("Yellow Purissima"), (4) medium yellow ("Yellow Flight"), (5) dark yellow ("Lady Margot"), (6) orange ("Orange Monarch"), (7) orange-red ("Temple of Beauty"), (8) medium red ("Lefeber's Memory"), (9) dark red ("Prominence"), (10) purple-red ("Blenda"), (11) light pink ("Bright Pink Lady"), (12) medium pink ("Angélique"), (13) dark pink ("Pink Impression"), (14) medium purple ("Attila"), (15) dark purple ("Queen of Night"), and (16) brown ("Cairo").

The Eriostemones are distinguished by possessing a little boss clothed with a tuft of hairs at the base of the filaments; the lower part of the three outer tepals is fringed with similar hairs, the leaves are long, narrow, and channeled, all springing from ground level; in well-grown specimens two or more flowers are carried on separate pedicels, also springing from ground level. The flowers generally show color at an early stage, even as the buds emerge from the ground; they are nodding and show a distinctive urn-shaped profile and usually open widely to a star. Leiostemones with large flowers and broad leaves are one extensive group which is characterized by scarlet or scarlet-crimson flowers possessing an olive or black blotch at the base often margined with yellow and more specifically a woolly coating of felted hairs between the tunic and the bulb.

In Autumn, the mother bulb produces only nonbranching roots (Kawa and De Hertogh 1992). Winter cooling then initiates hormonal changes, which are crucial for the proper subsequent growth and flowering of tulips, and additionally decreases the abscisic acid content and increases endogenous plant hormones stimulating the growth of plants (Saniewski and Kawa-Miszczak 1992). As the temperature increases in spring, a rapid growth of all above ground parts and flowering occurs. Tulips as other members of Liliaceae produce several secondary metabolites (alkaloids, flavonoids, glycosides, saponins, and tannins) and for that reason are used locally for therapy of different diseases, i.e., as an antibacterial agent (Ibrahim et al. 2016).

7.3 Origin, Domestication, and Distribution

Tulips originated in the Caucasus and Central Asia (Hoog 1973). They are now spread throughout Southern Europe, around the Black Sea coast south to Anatolia, in the Southern Ukraine to Central Siberia, the Caucasus, the Southern Balkans, North Africa through the Levant to Egypt and Iran, Saudi Arabia, Iraq, and Central Asia, and eastward to Western China, the Himalayas, and Mongolia. They are adapted to mountainous areas (sea level up to ~3000 m) and steppes in their natural habitats (Christenhusz et al. 2013). They were very popular around 1630 in the Netherlands. Breeding practices were carried out for a long time in history which lead to the development of ample diversity of flowers with different colors, shapes, and vigor. Original species of tulip are not known until today. They are grouped together under *T. gesneriana* L. Cultivars from *T. gesneriana* are the current commercial assortment. Interspecific crosses between cultivars of genotypes of *T. fosteriana* Hoog ex W. Irvingand *T. gesneriana* have resulted in the second group of cultivars, the Darwin hybrids. The classification and distribution of wild *Tulipa* species have been revealed universally. There are 18 *Tulipa* species in Anatolia and Turkey (Eker and Babaç 2010; Eker et al. 2014), 36 in the Iranian Plateau (Ghahreman et al. 2007), and 15–22 in the Balkan Peninsula (Govaerts 2015). According to Tang and Wang (1980), 13 wild tulip species were native to China. Among these species, 11 species were distributed in Xinjiang province. These include *T. altaica* Pall. ex Spreng., *T. schrenkii* Regel, *T. sinkiangensis* Z. M. Mao, *T. dasystemon* Regel, *T. kolpakowskiana* Regel, *T. iliensis* Regel, *T. patens* C. Agardh ex Schult, *T.edulis* (Miq.) Baker, *T. buhseana* Boiss., *T. heteroprtala* Ledeb, *T. thianschanica* Regel, and *T. heterophylla* Baker. Two species were distributed in areas from Northeast and Central China including the provinces of Liaoning, Shandong, Hebei, Zhejiang, Anhui, Hubei, Henan, and Shaanxi. Recently four new *Tulipa* species have been reported in China. These included *T. anhuiensis* X. S. Shen (2001), *T. tarbagataica* D. Y. Tan and X. Wei, sp. nov. (Tan et al. 2000), *T. wanzhensis* L. Q. Huang, B. X. Han and K. Zhang, sp. nov. (Han 2014), and *T. kuocangshanica* D Y Tan and D. Y. Hong (Tan et al. 2007).

7.4 Plant Genetic Resources

7.4.1 Geographic Distribution

Tulipa Linnaeus (1753: 305) is a large genus with 100 (Hall 1940) or 113 species as recently accepted by Govaerts (2010). The species are widely distributed from Northeastern China and Japan to Central and Southwest Asia, North Africa, and Europe (Botschantzeva 1982). The triangle between the Tien Shan and Pamir-Alai mountain ranges in Central Asia is considered to be the main center of diversity and the Caucasus region a secondary center (Zonneveld 2009). The Iranian Plateau is home to at least 36 species of wild tulips (Ghahreman et al. 2007). The number of native tulip species found in the Balkan Peninsula is much less, varying from c. 15 (Hayek 1933) to 22 (Govaerts 2010). Eighteen species have been recorded from European Turkey and Anatolia (Coşkunçelebi et al. 2008; Eker and Babaç 2010).

The number of species included in the genus depends on the species concept used and varies from more than a hundred (Hall 1940; Botschantzeva 1982) to about forty (Stork 1984). Section *Tulipa* (Marais 1984; originally described as sect. *Leiostemones* B.) is partially circumscribed in an artificial way and has been subdivided into several not clearly separated subsections. The other much smaller section (*Eriostemones* B.) consists of about 20 species which are divided into three subsections (Hall 1940). Some subsections are exclusively found outside the main gene center of the genus. The species of subsect. Australes occur in Europe, the Aegean Sea region, and Asia Minor. Species of subsect. Saxatiles are found in the Near East and at Crete. The area of subsect. Biflores ranges from Eastern Europe to Central Asia and Western China (Hoog 1973). Both diploids and polyploids occur in the three subsections of sect. *Eriostemones* (Kroon and Jongerius 1986).

7.4.2 Primary Gene Pool

Different authors consider that the genus is represented by 45–100 species. Hall (1940) and Botschantzeva (1962) reported approximately 100 species, Stork (1984) 45 species (after Van Eijk et al. 1991), Van Raamsdonk and De Vries (1992, 1995) 50–60 species, Zonneveld (2009) 87 species, and more recently Christenhusz et al. (2013) reported 76 species. In the World Checklist of Selected Plant Families, 554 names have been listed for genus *Tulipa*, but only 99 taxa have been accepted (Govaerts 2017). According to the taxonomic classification by Van Raamsdonk and De Vries (1992, 1995), the genus is divided into two subgenera: *Tulipa* and *Eriostemones* (Boissier). These subgenera are classified into eight sections (Table 1).

An intermediate garden classification of tulip has been introduced due to diversity in their forms (Table 2).

7.4.3 Wild Genetics Resources and Others

Among the *Eriostemones* Newton found two tetraploids, *T. silvestris* and *T. whittallii.* Out of these, *T. silvestris* is presumably derived from the diploid *T. australis* which is truly wild in the Appenines, Southern France, Spain, and Portugal, the only tulip indigenous to Western Europe. *T. silvestris* is a denizen along the Mediterranean, often as a weed of the vineyards in several English localities. One or two forms can be distinguished on close examination; particularly one from the neighborhood of Tabriz and N.W. Persia, where it is truly wild. *T. silvestris* shows relatively low fertility; evidence shows that it is an autotetraploid that has arisen by the doubling of a somatic cell and not by hybridization. *T. whittallii* probably bears the same relation to the very variable Grecian species, *T. orphanidea*, as *silvestris* bears to *australis*. Another presumed association exists between *T. bijora* (diploid) from Southern Russia and *T. turkestanica* (tetraploid) from further east. *T. turkestanica* was described by Newton as diploid, an accident due to the fact that the stocks in commerce are mixed and the two species closely resemble one another (Hall 1940).

The Eriostemones also include a highly polymorphic species to which many names have been given as *T. humilis*, *pulchella*, *violucea*. The Cretan tulip, *T. saxatilis*, which has been in cultivation for three centuries and was described by Parkinson, is a triploid belonging to this group. Its diploid analogue is the Mount Ida tulip, *T. cretica* (Hall 1940).

Table 1 The taxonomic classification of the species of the genus *Tulipa* in the sections (bold names) of the two subgenera, *Tulipa* and *Eriostemones*, according to Van Raamsdonk and De Vries (1992, 1995)

Subgenus *Tulipa*		
Tulipa *T. gesneriana* L. *T. armena* Boiss. *T. hungarica* Borbas *T. suaveolens* Roth *T. didieri* Jord.	***Eichleres*** (Hall) Van Raamsdonk *T. ingens* Hoog *T. lanata* Regel *T. tubergeniana Hoog* *T. eichleri* Regel *T. fosteriana* Hoog ex W. Irving *T. greigii* Regel *T. albertii* Regel *T. sosnovskyi* Akhverdov et Mirzojeva *T. praestans* Hoog *T. kaufmanniana* Regel *T. tschimganica* Bochantzeva *T. dubia* Vvedensky *T. subpraestans* Vvedensky	***Tulipanum*** de Reboul *T. agenensis* DC. *T. systola*Stapf *T. kuschkensis* B. Fedtschenko *T. julia* C. Koch *T. aleppensis* Boiss. ex Regel *T. praecox* Tenore
Kolpakowskianae (Hall) Van Raamsdonk *T. altaica Pall.* Ex Sprengel *T. lehmanniana* Mercklin *T. tetraphylla* Regel	***Clusianae*** Baker *T. clusiana* DC. *T. montana* Lindley *T. linifolia* Regel	
Subgenus *Eriostemones* (Boissier) Van Raamsdonk		
Australes sensu Hall *T. turkestanica* Regel *T. primulina* Baker *T. biebersteiniana* Schultes *T. sylvestris* L. *T. whittalii* (Dykes) A.D.Hall *T. ophanidea* Boiss. Ex Heldr. *T. hageri* Heldr.	***Saxatiles*** sensu Hall *T. humilis* Herb. *T. pulchella* Fenzl. *T. saxatilis* Sieb. ex Sprengel *T. bakeri* A.D. Hall *T. aucheriana* Baker	***Biflores*** sensu Hall *T. australis* Link *T. polychrome* Stapf *T. biflora* Pallas *T. sogdiana* Bunge *T. neustrueva* Pob. *T. tarda* Stapf *T. dasystemon* Regel

T. stellata is a hill species which ranges from Afghanistan to Kashmir and Himachal Pradesh in India. It is associated with a form differing only in possessing a yellow instead of a white basic color, known in gardens as *T. chrysantha*. This species is tetraploid, and both in external and chromosome morphology it approaches the tetraploid *Clusiana* (Hall 1940).

T. gumusanica Terzioglu is a new species occurring in the Vilayet Gumushane region of Turkey possessing chief character of pure yellow stamens and anthers and distinctly undulate leaves. *Tulipa albanica* is from a serpentine area in Kukësi district, Northeastern Albania; *T. koyuncui* from East Anatolia, the Irano Turanian (mountain) element; and Triploid ($2n = 3x = 36$) cytotype of *T. clusiana* var. *stellata* has been recovered from Kishtwar plateau located southeast of Srinagar.

Among the large-flowered species from Central Asia, *T. fosteriana, T. griegii*, and *T. kaufmanniana,* only diploids have been observed. Finally, at the extremity of the

Table 2 Horticultural classification of tulips (Orlikowska et al. 2018)

No.	Group (abbreviation)	Description	Example cultivars	Remarks
Early flowering				
1	Single early (SE)	Cup-shaped single flowers, stalk 15–45 cm	"Christmas Marvel," "Purple Prince," "Flair"	
2	Double early (DE)	Bowl-shaped double flowers, stalk to 40 cm	"Monte Carlo," "Viking"	
Mid-season flowering				
3	Triumph (T)	Cup-shaped single flowers, stalk 45–60 cm, stiff, and strong; the most popular for forcing; Triumph tulips are the result of hybridization between cultivars of SE and SL	"Apricot Beauty," "Leen Van Der Mark," "Strong Gold"	
4	Darwin hybrids (DH)	Ovoid-shaped big single flowers, stalk up to 70 cm; usually the result of hybridization between *T. gesneriana* and *T. fosteriana* and the result of hybridization between other cultivars and botanical tulips	"Ad Rem," "Apeldoorn," "Van Eijk"	
Late flowering				
5	Single late (SL)	Cup- or goblet-shaped flowers, stalk 45–75 cm. This group includes, e.g., the former Darwin Group and cottage group	"Queen of Night," "Menton"	
6	Lily flowered (L)	Flower form resembles that of a lily, with very pointed, reflexed perianth, stalk of variable length	"Claudia," "Pretty Woman," "West Point"	
7	Fringed (Crispa)	Single-flowered cultivars with tepals edged with crystal-shaped fringes	"Arma," "Fabio," "Davenport"	
8	Viridiflora (V)	Single-flowered cultivars with partly greenish tepals, stalk 36–60 cm	"Greenland"	
9	Rembrandt (R)	The "broken tulips," the colors of perianth are striped or blotched. The original Rembrandt tulips are no longer grown commercially, because unusual marbled coloring was caused by viruses and is protected only in historical collections	"Princess Irene" "Flaming Parrot"	Virus-free and genetically stable cultivars

(continued)

Table 2 (continued)

No.	Group (abbreviation)	Description	Example cultivars	Remarks
10	Parrot	Single-flowered cultivars with laciniate curled and twisted tepals, stalks 50–60 cm	"Bright Parrot," "Rococco," "Topparrot"	
11	Double late (DL)	Peony-flowered cultivars, stalks 40–60 cm	"Angelique," "Wirosa," "Yellow Pompenette"	
Species and their hybrids				
12	Kaufmanniana (K)	*T. kaufmanniana* and their subspecies, varieties, and hybrids; very early flowering; the foliage is sometimes mottled. Flowers with multicolored base open fully. The height is ca. 30 cm	"Showwinner," "Giuseppe Verdi," "The First"	For containers and greenery
13	Fosteriana (F)	*T. fosteriana* and their hybrids; the flowers are large. Some cultivars have mottled or striped foliage	"Purissima," "Red Emperor," "Yellow Emperor"	For containers and greenery
14	Greigii (G)	*T. greigii* and their hybrids; always with mottled or stripped foliage, flowering later than Kaufmanniana	"Red Riding Hood," "Toronto"	For containers and greenery
15	Miscellaneous (M)	Not a cultivar group but all species, varieties, and their cultivars in which the wild species are evident, not belonging to any of the above groups		For greenery, mainly for naturalization and for rock gardens

range of the genus one species *T. eduli* with a variety distinguished as *T. latifolia* is found in Japan and appears to be identical with a species from the Chinese mainland described by Baker as *T. wythronoides*. This minute tulip, the smallest of the genus, possesses 48 chromosomes and on account of certain morphological peculiarities notably two linear bracts a little below the flower, and should probably be assigned to a special genus (Figs. 1 and 2).

7.5 Collections

The basic sources of genetic variations are commercially grown cultivars and genotypes from different groups. The genus *Tulipa* is protected in situ and as ex situ collections in several countries, particularly those where natural gene

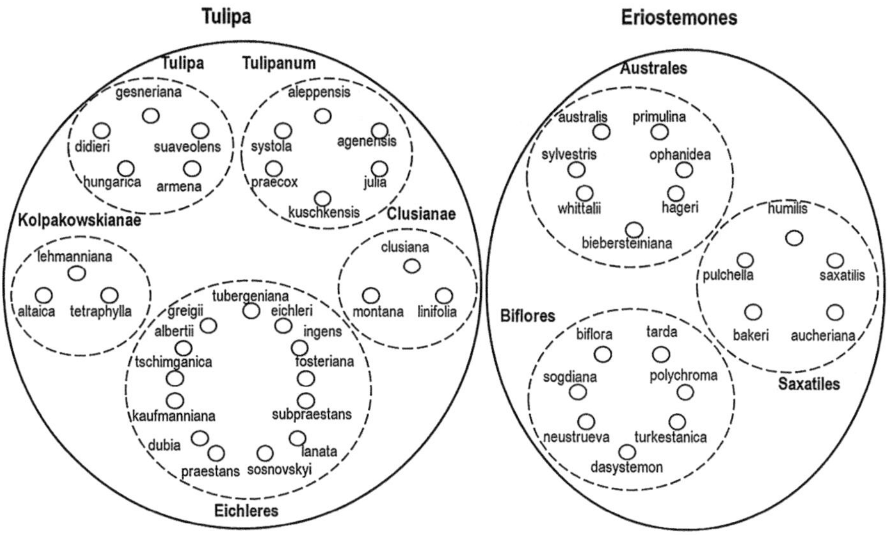

Fig. 1 The systematic of the genus *Tulipa*.L (Van Raamsdonk et al. 1997)

resources exist or where breeding programs are realized. Gene banks and ex situ field tulip collections are located in some research institutions and at universities, in botanical gardens, and in working collections of breeding companies and bulb-producing farms.

7.5.1 Gaps in Collection Both Geographical and Genetic

A huge genetic diversity present in germplasm of tulip across the planet. Yet, there exists a gap between genetic diversity present and the one which is utilized in plant breeding programs. Identification of gaps in the germplasm collections and their further utilization is necessary to achieve completeness of the collections and direct exploration of additional accessions. Practically the process involves a comparison between the range of natural diversity and the diversity already existing. To encourage and facilitate the use of germplasm collections, concept of core collection was introduced in 1984. A core collection comprises a representative subset of approximate 10% of the entire germplasm collection and is used for in-depth phenotypic and genotypic analysis. The technique can be used for improvement of several commercial traits in the tulips. The wild relatives of crops represent a major source of valuable traits for crop improvement. These precious resources are threatened by habitat destruction, land use changes, and other factors. Thus, there is a need for desperate collection and long-term requirement for research and breeding.

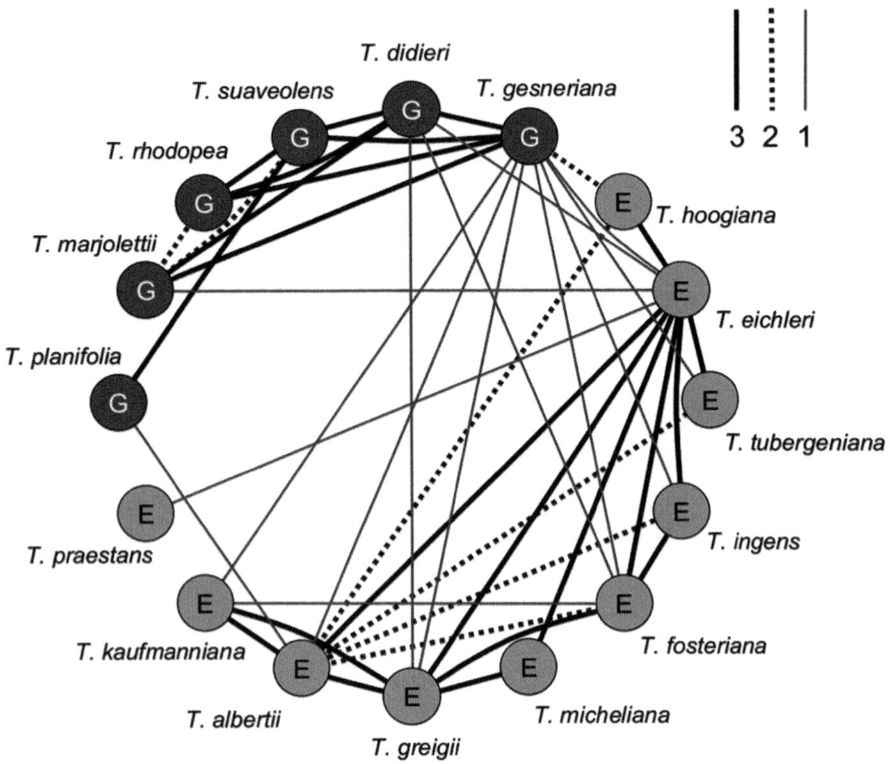

Fig. 2 Crossing polygon of species of section *Eichleres* ('E') and *Tulipa* ('G'). Meaning of lines: 1: several successful attempts, effectivity low; 2: one successful attempt, effectivity high; 3: several successful attempts, effectivity high. Low effectivity: less than 5F_1 bulbs per seed pod; high effectivity: more than 15 F_1 bulbs per seed. The data shown was pooled results of all crosses carried out per combination. Modified from Van Raamsdonk et al. (1995)

7.6 Conservation

7.6.1 Methods

7.6.1.1 Cold Storage

Collections of tulip genotypes are maintained by yearly planting, harvesting, and storage of the bulbs. Tulips require warm (17–20 °C)-cool (2–9 °C)-warm (17–20 °C) annual temperature sequence for its growth and development. The precise temperatures required are, however, dictated by bulb size, cultivar, usage, and the desired date of flowering, e.g., early, mid-season, or late. Under normal conditions of bulb harvesting (June to early July), a flower initiation and organogenesis take place at 17–20 °C. The farther, the temperature away from this range, the lower is the rate of floral differentiation. Once organogenesis has occurred, the bulbs require a period of 14–22 weeks of cold (2–9 °C) to complete flower initiation. The final stage of growth, flowering, and bulbing occurs at warm (17–20 °C) temperatures (John et al. 2006). Several studies have, however, been conducted to regulate flowering in tulips. To optimize the maintenance tulip bulbs were stored after flower development at −2 °C. By using this method, bulbs of ten genotypes could be stored for one growing season without losing their ability to form daughter bulbs. In the other method, bulbs were stored before the start of flower development at −0.5 °C, followed by a temperature regime of 4 weeks at 25 °C, 6 weeks 20 °C, and 16 weeks 5 °C, for development of sprouts. With the second method, bulbs of three genotypes could be stored for two seasons, without losing their ability to form daughter bulbs (Bonnier et al. 1997).

7.6.1.2 In Vitro Conservation

In tulips, bulb production of new cultivars at a quantity of about 10,000 is long lasting and expensive. By using in vitro reproduction, this period can be shortened. This technique is widely used in the rapid multiplication of the most promising breeding lines of many crops. However, the in vitro tulip propagation based on the direct regeneration of shoots or microbulbs on initial explants, scales or flower stem segments isolated from cooled bulbs described in accessible literature in the 1990s, was characterized by moderate efficiency (Nishiuchi 1980; Rice et al. 1983; Le Nard et al. 1987; Alderson and Taeb 1990; Baker et al. 1990; Hulscher et al. 1992; Famelaer et al. 1996; Kuijpers and Langens-Gerrits 1997). No effective secondary regeneration was achieved enabling the cyclic propagation of shoots or bulbs. In all the above reports, microbulbs induced on shoots were the final product of tulip propagation in vitro. They were transferred from in vitro conditions to 5–6 weeks long storage at room temperature, then rooted ex vitro at 9 °C, and finally grown at standard conditions for tulip active growth (15–20 °C). The above methods allowed a maximum of about 200 microbulbs to be obtained from one mother bulb. Recently, high regeneration potential has been demonstrated for explants derived from *T. tarda* seedlings that developed in vitro (Maślanka and Bach 2014). It seems that this method can be successfully used to propagate wild tulips, e.g., endangered species. An in vitro propagation method with relatively high efficiency, allowing the production of 400–1000 microbulbs from one bulb during 2–3 years, has been developed (Podwyszyńska and Marasek 2003; Podwyszyńska and Sochacki 2010; Podwyszyńska et al. 2014). In this method adventitious shoots are multiplied in vitro on an MS medium containing TDZ, N6-(-Isopentenyl) adenine (2iP), and NAA during several cycles over the course of 2 years after which the bulbing process was induced. The bulb development takes 8–10 months and comprises three phases: (1) 10-week special culture at 18 °C for dormancy induction in shoots, (2) 12–14 weeks long shoot cooling at 5 °C for dormancy release, and (3) bulb formation at 18 °C on a sucrose-rich medium during 12–14 weeks (Podwyszyńska 2006). The cyclic multiplication of adventitious shoots has been utilized successfully for maintaining breeding, producing virus-free plants, and inducing mitotic tetraploids (Podwyszyńska 2011; Sochacki and Podwyszyńska 2012). However, cyclic shoot multiplication should not take too long due to the possibility of somaclonal changes in some cultivars and on some medium compositions (Podwyszyńska et al. 2006, 2010a, b).

Research has also been conducted on tulip in vitro propagation by inducing the development of somatic embryos directly on initial explants or via callus culture. Gude and Dijkema (1997) induced somatic embryos on flower stem explants on a medium with 5 or 50 μM 2,4-dichlorophenoxyacetic acid (2,4-D) or picloram. Tulip propagation by inducing the development of somatic embryos from embryogenic callus cyclically, which multiplied up to 5 years, was reported by Podwyszyńska and Marasek (1999). The callus formation was induced on fragments of flower stems isolated from cooled bulbs and incubated on modified MS medium, in the presence of auxin alone (picloram, 2,4-D and NAA) or combined with TDZ. Embryogenic callus was obtained only with 2.5 mg l^{-1} 2,4-D in combination with 0.1–0.5 mg l^{-1} TDZ. Embryo-like structures developed on medium with a lower content of these growth regulators. The yield at the embryo formation step was very high (90 embryos per 100 mg callus), but finally only 30% of the embryos developed into shoots and 4% produced bulbs. As a consequence, work on the tulip propagation via somatic embryogenesis was not continued further. Bach and Ptak (2001) and Ptak and Bach (2007) reported the possibility of using the method based on somatic embryogenesis for tulip propagation by formation of numerous embryos on ovary segments isolated from bulbs and incubated on an MS medium supplemented with 25 μM of picloram and 0.5 μM 6-Benzylaminopurine (BAP). Further development of embryos in plants was conducted in the presence of 5 μM BAP and 0.5 μM NAA. The development of bulbs was achieved after the embryos were cultured in a sucrose-rich medium. It seems that the continuation of research on the optimization of individual stages of tulip propagation in vitro via ES can be highly effective.

7.6.2 Conservation in Gene Banks

The genebanks are created with the core purpose of conservation and management of genes or plant genotypes, from wild and cultivated species outside of their natural habitat, for current or future use. Although the germplasm of all species (that we can multiply) can theoretically be conserved in genebanks, traditionally genebanks have served as a repository for the germplasm of endangered species facing the threat of possible extinction, and for those forming the basis of human sustenance as a food source and for agricultural purposes.

Nowadays, genebanks accommodate a range of materials including varieties from traditional agriculture, i.e., landraces, primitive cultivars, and species that are harvested for use in sacred rituals as part of certain cultural practices, superior cultivars, advanced lines, mutants and synthetic materials emerging from scientific improvement programs, and even genetic fragments, cloned genes, marker genes, and transgenic plants resulting from biotechnology and genetic engineering initiatives.

7.6.3 Cryopreservation

Cryopreservation offers a viable and economical method for the long-term conservation of genetic resources of vegetatively propagated plants. Cryopreserved pollen can be a major assess point for prebreeding germplasm lines, hybrid seed production and biotechnology, and other basic studies. Malgorzata Maslanka and Anna Bach (2020) cryopreserved *T. tarda* Stapf. apical meristems by droplet vitrification. Storage below temperatures such as −130 °C has been argued to be an effective and safe mode for preservation.

7.7 Characterization and Evaluation

A broad exclusivity can be projected following hybridization between distant forms and different species. Despite difficulties, some distant hybrids were obtained from the progeny of *T. gesneriana* × *T. fosteriana*, *T. gesneriana* × *T. eichleri*, *T. gesneriana* × *T. kaufmanniana*, *T. gesneriana* × *T. albertii* and *T. greigii* × *T. gesneriana*, *T. praestans* × *T. gesneriana*, *T. tubergeniana* × *T. gesneriana*, and *T. ingens* × *T. gesneriana* (Custers et al. 1995; Van Creij 1997; Van Creij et al. 1999; Okazaki 2005; Marasek-Ciolakowska et al. 2018) (Table 3).

Embryo rescue is an in vitro technique that plays an important role in plant breeding allowing the development of many interspecific and intergeneric plant crop hybrids, including interploid ones (Sharma et al. 1996; Reed 2004). Interspecific or intergeneric incompatibility in plants can occur for many reasons, but embryo abortion occurs most often. Embryo rescue techniques have been widely used to overcome crossing barriers in intra- and interspecific hybridization of tulips (Okazaki 2005; Van Tuyl and Van Creij 2007). Depending on organs isolated from the flower after pollination, embryo rescue techniques include embryo, ovule, and ovary cultures. Developing of rescued embryo into a complete plant in vitro is influenced by several factors, such as the genotypes, the type of cultured organ (embryo, ovule, or ovary), the time of isolation after pollination, as well as in vitro culture conditions (medium composition, light, and temperature). Custers et al. (1992, 1995) reported successful embryo rescue in the interspecific cross *T. gesneriana* × *T. kaufmanniana* which was hindered by embryo breakdown. A low number of germinating seeds was obtained from ripening seed pods but using ovule or embryo culture higher efficiency of interspecific hybrid seedling production was enabled. Custers et al. (1995) proved that the optimal rescue time was 7–9 weeks after pollination using the culture of embryos or ovules on the medium containing half-strength Murashige and Skoog salts (1962) (MS), with 4% sucrose, 500 mg l^{-1} tryptone, and 4 nM 1-naphthaleneacetic acid (NAA) and with 0.75% agar. The cultures were initially maintained at 15 °C in darkness until 15 weeks after pollination. In order to induce germination, the embryos underwent the specific temperature treatment required for normal tulip seedling development as described by Niimi (1978). Thus, the embryos were cooled for 12 weeks at 5 °C, and subsequently seedlings formed bulbs at 15 °C within 12–18 weeks. Comparable efficiency of

Table 3 Interspecific crosses between species of section Tulipa and section Eichleres with modification from Van Raamsdonk et al. (1995) and Marasek-Ciolakowska et al. (2018)

	Female				
Subgenus or section	Male	*T. gesneriana*	*T. fosteriana*	*T. kaufmanniana*	*T. greigii*
Tulipa	*T. gesneriana*	a	c	c	c
	T. didieri	a	c	–	c
	T. suaveolens	a	–	d	d
	T. rhodopea	a	–	–	–
	T. marjolettii	a	–	c	d
	T. planifolia	–	–	–	–
Eichleres	*T. praestans*	–	d	b	–
	T. micheliana	–	d	d	a
	T. kaufmanniana	c	d	a	a
	T. vvedenskyi	c	b	a	a
	T. tubergeniana	c	a	d	c
	T. greigii	c	a	–	a
	T. hoogiana	b	–	b	–
	T. fosteriana	c	a	c	a
	T. ingens	c	a	–	–
	T. eichleri	c	a	c	a

[a]Several successful attempts, effectivity high among species of the same section (intrasectional crosses)
[b]One successful attempt, effectivity high
[c]Several successful attempts, effectivity low
[d]No seeds obtained
– Not determined

interspecific hybrid production using embryo or ovule cultures was obtained by Okazaki et al. (2005) in crosses of *T. gesneriana* × *T. fosteriana*, *T. gesneriana* × *T. eichleri*, and *T. gesneriana* × *T. greigii*. Embryos and ovules were excised 6–8 weeks after pollination and cultured them on a half-strength MS medium containing 3% sucrose and 0.2% gellan gum using temperature sequence treatment. Embryo and ovule cultures yielded a higher number of hybrid bulblets than those obtained from in situ seed production. Unique hybrids were also obtained by crossing *T. gesneriana* × *T. praestans* and from an incongruent cross *T. gesneriana* × *T. agenensis* by using ovule- and ovary-slice cultures (Van Creij et al. 1999, 2000). The efficiency of ovule culture initiated directly after pollination with ovary-slice culture and ovule culture initiated at 2–9 weeks after pollination was compared. In most cases, the percentage of germinated embryos increased with the advancement of the embryo developmental stage, regardless of the technique used. The efficiency of embryo rescue was improved by enhancing the sucrose concentration to 9% in the medium used at the stage of ovary-slice culture.

In some cases, when crossing two species is not possible, an intermediate cross with a third species, which is compatible with both parental species, can be used to overcome incongruity (Van de Wiel et al. 2010). The bridge cross with *T. greigii* as an intermediate parent was successfully used to overcome incongruity between *T. gesneriana* and *T. kaufmanniana* with the following hybridization sequence [*T. gesneriana* × (*T. kaufmanniana* × *T. greigii*)] (Van Eijk et al. 1991).

In conclusion, interspecific hybridization enables to combine desirable traits from different species into new cultivars, including disease resistance to tulip breaking virus (TBV), *Botrytis tulipae*, and *Fusarium oxysporum* f. sp. *tulipae*, shortening of the forcing period, extending flower longevity, and new flower shapes and colors (Van Eijk et al. 1991; Van Creij et al. 1997b, 1999; Van Raamsdonk et al. 1995; Van Tuyl and Van Creij 2007; Van Tuyl et al. 2012).

7.7.1 Genetics of Tulip

Chromosome complement in tulips was first determined by Guignard in 1900. De Mol in 1925 discovered polyploidy in the genus. Later he stated that after subjecting the bulbs to heat, diploid male gametes could be found; apparently the rare triploids had arisen through the entry of one of these unreduced cells into the sexual process.

The basic number of chromosomes in *Tulipa* is 12, the only apparent exception being *T. galatica,* in which Newton found 32 as the somatic number. The extra chromosomes have now been found to be supernumerary small chromosomes which do not form part of the regularly inherited complement, and vary in number in the same plant. Certain differences in the morphology and size of the chromosomes of different species exist which is of taxonomic value. Numerous trials have failed to affect any cross between species of the *Eriostemones* and of the *Leiostemones*.

The majority of tulip species and cultivars are diploid ($2n = 2\times = 24$) (Kroon and Jongerius 1986), and the somatic DNA 2C value ranges from 32–69 pg for the diploids (Zonneveld 2009). However, triploids ($2n = 3\times = 36$) have been found in different genotypes of *T. clusiana* and *T. kaufmanniana*; tetraploids (2n = 4× = 48) in *T. bifloriformis*, *T. sylvestris*, *T. kolpakowskiana*, and *T. tetraphylla*; pentaploids (2n = 5× = 60) in *T. clusiana*; and hexaploids in *T. polychrome* (Kroon and Jongerius 1986), which is the highest number of chromosomes thus far determined in either the wild species or the garden varieties of the genus *Tulipa*.

For the greater part, the chromosome numbers agreed with those reported in former cytotaxonomic research, but some deviations were found. In the section *Eriostemones*, *T. biebersteiniana* showed two ploidy levels: no. 67357 being diploid and no. 68012 tetraploid. Two origins of *T. bifloriformis* appeared to be tetraploid, whereas earlier counts of Newton (1927) and Botschanceva (1962) only showed diploid and triploid forms of this species. In *T. clusiana*, earlier determined as a pentaploid (Newton 1927), both a tetraploid and a pentaploid type were found and in *T. clusiana* var. *chrysantha* diploid, tetraploid and pentaploid forms were observed which was also done by others. *T. patens* was tetraploid, whereas Botschanceva (1962) discovered that this species was a diploid. Earlier reports stated that *T. polychroma* was a diploid (Upcott and La Cour 1936; Gabrielijan and Pogosjan 1971), but the material included tetraploid and hexaploid forms as well. In the section *Leiostemones*, the species *T. kolpakowskiana* and *T. ostrowskiana* appeared to include tetraploid forms not reported before. The chromosome numbers were determined in different *Tulipa spp.,* out of a total of 63 species 13 were found to occur at different ploidy levels. This Table 4 also includes the chromosome counts of Botschanceva (1962), Gabrieljan and Pogosjan (1971), Hall (1937), Upcott and La cour (1936), and Newton (1927).

Except for the cultivated tulips of garden origin, the genus is in Albania represented by only one native species: *T. australis* Link (Schrader 1799: 317) that was previously known as *T. grisebachiana* Pantocsek (1873: 265) and was first collected on the limestone mountain Mt Tomorri by Markgraf in 1928 (Markgraf 1931). *T. grisebachiana* was originally described from Mt Gliva (near Trebinje, Hercegovina). *T. australis* was also reported as *T. sylvestris* subsp. *grisebachiana* and subsp. *celsiana* by Hayek (1932–1933: 71, syn.: *T. celsiana* DC. in Redouté 1803: t. 38). It is not apparent whether Hayek was referring to plants from limestone or from serpentine substrate. They represent two different morphotypes with slight differences in perianth shape, size, and color. *T. grisebachiana* and *T. celsiana* have been maintained in synonymy of *T. australis* and have not formally recognized the two morphotypes from differing substrates. There is a report of the discovery of a second and undescribed tulip from Albania belonging to subg. *Tulipa* sect. The tulip has both red and yellow flowering forms.

7.7.2 Molecular Markers

The availability of markers associated with desirable traits can greatly facilitate and accelerate the breeding process. With their help it is possible to make a preliminary selection among the 1- to 2-year-old seedlings, thus limiting their number. This tool is especially important for plants as tulip, with a long vegetative (juvenile) phase,

Table 4 Chromosome numbers of tulip species (Kroon and Jongerius 1986)

Species	IVT No.	Kroon	Others	Origin
***Section Eriostemones* BOISS.**				
T. aucheriana BAKER	66242	24	24[5]	Van Tubergen, Haarlem
T. australis LINK	76121	24	24[4]	Elliott, Ashford
T. bakeri A. D. HALL	70604	24		Kooiman, Enkhuizen
T. biebersteiniana SCHULT.	67357	24	24[1]	hortus Leningrad
T. biebersteiniana SCHULT.	68012	48		hortus Moscow
T. biflora PALL.	74227	24	24[1,4]	Kooiman, Enkhuizen
T. bifloriformis VVED	65250	48	24,36[1]	hortus Leningrad
T. bifloriformis VVED	66090	48		hortus Leningrad
T. celsiana D.C.	65217	24		L.B.O., Lisse
T. clusiana D.C.	75306-A	48	60[4]	Kashmir, India
T. clusiana D.C.	75306-B	60		Kashmir, India
T. clusiana D.C. var. *chrysantha* (A. D. HALL)				
Sealy	71328	24	24,48,60[3],	Afghanistan, Ghazni
Sealy	71328	60	48[4]	Afghanistan, Ghazni
	65218	48		L.B.O., Lisse
Sealy	66330	48		Van Tubergen, Haarlem
T. clusiana D.C. var. *stellata* (HOOK.) REGEL	66328	48	48[4]	Van Tubergen, Haarlem
T. cretica BOISS.	73155	24	24[5]	Crete, Akrotiri
T. dasystemon REGEL	68013	24	24,48[1]	hortus Moscow
T. hageri HELDR.	65473	24		Thoolen, Overveen
T. humilis HERB.	64172	24	24[4]	Van Tubergen, Haarlem
T. neustrueva POB.	77343	24	241	Uzbekistan
T. orphanidea Bolss. et HELDR.	64179	24	24[4]	Van Tubergen, Haarlem
T. patens AGARDH. ex SCHULT.	67330	48	24[1]	hortus Copenhagen
T. polychroma STAPF.	71310	24	24[2,5]	South East Iran
T. polychroma STAPF.	71311	24		South-East Iran, Kuh-Taftan
T. polychroma STAPF.	76154	24		Iran, Shah-Dasht.
T. polychroma STAPF.	73113	48		Iran, Zagros mts.
T. polychroma STAPF.	71324	72		Afghanistan, Herat
T. saxatilis SIEB. ex SPRENGEL	64175	36		Van Tubergen, Haarlem
T. sylvestris L.	72106-B	48	48[4]	Iran, Hamadan
T. tarda STAPF.	67042	24	24,48[1],24[4]	hortus Tashkent
T. urumiensis STAPF.	64177	24	24[5]	Van Tubergen, Haarlem

(continued)

Table 4 (continued)

Species	IVT No.	Kroon	Others	Origin
T. whittallii (DYKES) ELWES	64180	48	48^{4}	Van Tubergen, Haarlem
***Section Leiostemones* BoISS.**				
T. acuminata VAHL	67747	24		Thoolen, Overveen
T. aitchisonii A. D. HALL	71327	24		Afghanistan, Faizabad
var. *aitchisonii*	75307	24		Kashmir, Jammu
var. aitchisonii	71337	24		Afghanistan, Band-e-Amir
T. aitchisonii A. D. HALL				
var. *clusianoides* WENDELBO	71329	24		Afghanistan, Kabul
T. albertii REGEL.	68061	24		hortus Tashkent
T. aleppensis BOISS. ex. REGEL	77159	24	36^{5}	Turkey, Antakya
T. anadroma Z. BOTSCH.	68068	24		hortus Tashkent
T. anadroma Z. BOTSCH.	76126	24		Elliott, Ashford
T. armena BOISS.	74223	24	24^{4}	Van Tubergen, Haarlem
T. banuensis GREY-WILSON	71334	24		Afghanistan, Pul-e-Isar
T. batalinii REGEL	64181	24	24^{4}	Van Tubergen, Haarlem
T. butkovii Z. BOTSCH.	76124	24		Eliott, Ashford
T. didieri JORD.	66243	24		Van Tubergen, Haarlem
T. ferganica VVED.	76128	24		Eliott, Ashford
T. greigii REGEL	77291	24	24^{4}	Western Tien Shan
T. grengiolensis THOMMEN	67738	24		Ruttier Lanche, Grenoble
T. hungarica BORB.	74197	24		Rumania, Danube valley
T. karabachensis GROSSH.	77161	24		Armenia, Shusha
T. karabachensis GROSSH.	77293	24		Armenia
T. kolpakowskiana REGEL	64183	24	24^{4}	Van Tubergen, Haarlem
T. kolpakowskiana REGEL	68064	48		hortus Tashkent
T. lanata REGEL	65369	36	$36^{1,5}$	Van Tubergen, Haarlem
T. linifolia REGEL	70596	24	24^{4}	Kooiman, Enkhuizen
T. linifolia REGEL	70662	24		Kooiman, Enkhuizen
T. marjoletti PERR. et SONG	65299	24		Inst. hort. Pruhonice
T. micheliana HOOG	71315	24	24^{5}	Iran, Birjand
T. montana LINDL	46156	24	24^{2}	Iran, MoravehTeppeh

(continued)

Table 4 (continued)

Species	IVT No.	Kroon	Others	Origin
T. montana LINDL. yellow form	73116	24	24 + 2B[2]	Iran, Elbruz mts.
T. ostrowskiana REGEL	65380	48	24[5]	Van Tubergen, Haarlem
T. planifolia JORD.	67737	24	24[5]	Ruttier Lanche, Grenoble
T. praecox TENORE	77101	36		Korcula, DonjeBlato
T. schrenkii REGEL	65150	24	24[2]	Van Eeden, Noordwijk
T. scardica BORNM.	75107	24		hortus Skopje
T. stapfii TURRILL	71340-A	24	24[5]	Iran, Gulestan Forest
T. systola STAPF	76155	24		Iran, MoravehTeppeh
T. tetraphylla REGEL	68040	48	24,48[1]	hortus Stockholm
T. ulophylla WENDELSO	77164	24		Wendelbo, Goteborg
T. vvdensky Z. BOTSCH	77289	36		Western Tien Shan
T. vvdensky Z. BOTSCH	76127	24		Eliott, Ashford
T. zenaidae VVEO.	77299	24		Kirgizkiy, Alatau
***Section Tulipanum* REB.**				
T. julia K. KOCH	68075	24	24[2]	hortus Tashkent
T. julia K. KOCH	72119	24		Armenia, Erevan
T. kuschkensis B. FEDTSCH	71320	24	24[1,5]	Afghanistan, Herat
T. tubergeniana HOOG	65391	24	24[1,5]	Van Tubergen, Haarlem
***Section Spiranthera* VVED.**				
T. kaufmanniana REGEL	77294	24		Western Tien Shan
T. tschimganica Z. BOTSCH	76123	24		Elliott, Ashford

during which spatial isolation of the clones is obligatory. The complexity of functional traits means that morphological and physiological markers for the selection of tulip are scarce. Nevertheless, from studies conducted on various levels (physiological, morphological, and molecular), some useful markers were discovered. The morphological and physiological markers resulted mainly from early genetic studies on this plant (Van Eijk and Toxopeus 1968; Van Eijk and Eikelboom 1975, 1976). They include the resistance to *Fusarium oxysporum f. sp. tulipae* that can be performed on one-year bulbils (Van Eijk and Eikelboom 1990; Romanow et al. 1991; Straathof et al. 1996a), similarly as ability for forcing and high bulb productivity (Van Eijk and Toxopeus 1968). Flower longevity can be evaluated at the first flowering (Van der Meulen-Muisers et al. 1997). The thickness of the wax layer on the leaves is related to susceptibility to *B. tulipae* (Reyes et al. 2005).

Cytological markers based on the construction of chromosomes can be used to observe the ploidy level (Abedi et al. 2015), for discrimination between genomes of closely related species in order to design a crossing program or to identify parents of hybrid plants or ancestors of allopolyploid species, which is especially important in incompatible and distant hybridization (Marasek et al. 2006) and to evidence introgressive fragments (Marasek-Ciolakowska et al. 2012).

In tulips, a number of studies have been published referring to chromosome number, chromosome identification, and karyotype analysis (Sayama et al. 1982; Kroon and Jongerius 1986; Van Raamsdonk and De Vries 1995; Marasek et al. 2006; Abedi et al. 2015). The basic chromosome number in the genus *Tulipa* is $x = 12$. They are large, but in the most species important for tulip breeding, *T. gesneriana* and *T. fosteriana* are similar in shape consisting of median, submedian, and subterminal chromosomes with no remarkable chromosome landmarks, such as secondary constrictions (Marasek et al. 2006). Chromosome identification possibility has been broadened by employing Giemsa staining of heterochromatin sections (C-banding) (Filion 1974; Blakey and Vosa 1982; Van Raamsdonk and De Vries 1995). Blakey and Vosa (1981, 1982) used conventional staining and C-banding techniques to establish a phylogenetic relationship between *Tulipa* species belonging to the subgenera *Eriostemones* and *Leiostemones*. In their studies several chromosome types were recognized with respect to their morphology and heterochromatin distribution, and groups of species with chromosome characteristics were described.

Fluorescence in situ hybridization (FISH) with 5S rDNA and 45S rDNA probes has provided molecular cytogenetic markers for the identification of most chromosomes, enabling to distinguish between *T. gesneriana* and *T. fosteriana* cultivars on the basis of variation in size, number, and chromosomal distribution of the FISH signals (Mizuochi et al. 2007). FISH karyotyping has been also efficient in detecting somaclonal variation in the population of long-term micropropagated *T. Gesneriana* "Prominence," revealing karyotype rearrangements and variations in the number of hybridized loci and in the size of hybridization signals between standard "Prominence" and somaclones (Marasek-Ciolakowska and Podwyszyńska 2008).

In tulips, genomic in situ hybridization (GISH) was useful in identifying the genome constitution in diploid, triploid, and tetraploid Darwin hybrids resulting from hybridization between *T. gesneriana* and *T. fosteriana* at the diploid level (Marasek et al. 2006; Marasek and Okazaki 2007). Darwin hybrids are becoming increasingly interesting since they combine the desirable horticultural traits from two sections, viz., *Tulipa* and *Eichleres*, such as good forcing quality, resistance to *Fusarium oxysporum*, and resistance or partial resistance to TBV (Van Tuyl and Van Creij 2007). Analysis of the genomic constitution of triploid cultivars of Darwin hybrids confirmed their origin via diploid gametes produced by diploid cultivars (Kroon and Eijk 1977; Marasek et al. 2006) with two genomes of the *T. gesneriana* (G) and one genome of the *T. fosteriana* and no recombinant chromosomes. Similarly, unreduced gametes took part in the origin of the tetraploid cultivars "Ollioules" which resulted from the cross between the diploid cultivar "Caravelle" and Darwin hybrids. The GISH assessment showed three genomes of *T. gesneriana* chromosomes and one *T. fosteriana*. According to Marasek-Ciolakowska et al. (Marasek-Ciolakowska et al. 2012), diploid Darwin hybrids can produce functional 2n gametes and also n gametes. This provides the opportunity to generate progenies of different ploidy levels from backcrossing hybrids to *T. gesneriana* parents (Marasek-Ciolakowska et al. 2014).

Southern hybridization and molecular cytogenetic analysis using GISH and subsequent FISH with 45S rDNA and 5S rDNA probes have been used to elucidate the genome composition of the diploid cultivar "Purissima," which is classified to the Fosteriana group based on its horticultural traits (Van Scheepen 1996) and shows a high crossability with *T. gesneriana.* Marasek and Okazaki (2008) found that "Purissima" is an interspecific hybrid with a genome comprised of one genome of *T. gesneriana* and one genome of *T. fosteriana*. "Purissima" was then used as pollen donor and diploids were obtained except for one triploid cultivar "Kouki." In diploid progenies of "Purissima," the presence of recombinant chromosomes was shown, whereas no recombinant chromosomes were observed in the triploid "Kouki" representing genome GGF. The total number of translocations ranged from one to six. A considerable amount of intergenomic recombinations between the parental chromosomes of the two species were recorded in both diploid BC1 and BC2 progenies of Darwin hybrids (Marasek-Ciolakowska et al. 2009, 2011, 2012), which is a breakthrough in introgression breeding in tulips.

Molecular markers will be increasingly important in tulip breeding but the road to mass application is a matter of the future (Arens et al. 2012). They can be obtained by different techniques of DNA analysis. Markers when applied in plant breeding can provide information about genome organization, gene introgression, and their linkage. They help in choosing suitable crossing partners and enable early selection (Moose and Mumm 2008). DNA markers were used, for example, in the identification of tulip cultivars (Bondrea et al. 2007) and species (Kiani et al. 2012) and in phylogenetic analysis of species (Nan Tang et al. 2013; Turktas et al. 2013).

The genomes of some plant species are sequenced, but the study of genes is difficult due to large size of genome and the presence of a multitude of noncoding introns and repetitive sequences. It is especially difficult in such plants as tulip possessing the extreme large genomes (Shahin et al. 2012). Moreno-Pachon et al. (2016) chose the method of sequencing of transcriptomes only, isolated from different tulip tissues. The sequencing and reading were directed toward the TCP transcription factor family of developmental regulators, specific for plants. Miao et al. (2016) used transcriptome analysis to study the molecular mechanism involved in stolon formation of *T. edulis* with the hope that it would enable genetic studies on the propagation of this endangered species. In order to study anthocyanin biosynthesis in red-flowered tulips, Yuan et al. (2013) isolated six cDNA clones from petals and studied their quantitative expression in comparison with anthocyanin accumulation with the aim of broadening the range of colors.

Some important genes that have been sequenced in tulips can serve as markers. Momonoi et al. (2007) isolated five cDNA clones encoding 1-aminocyclopropane-1-carboxylic acid oxidase (ACO), which were expressed in wilting tepals, leaves, and stems. ACO is a key enzyme that catalyzes the last step in ethylene biosynthesis. Ethylene takes part in several physiological processes, including leaf and flower senescence. In tulips, senescence does not have a climacteric character and it is not directly related to ethylene. ACO is coded in tulip by a gene family consisting of five genes, which are differentially regulated in both vegetative and generative tissues. Yamagami et al. (2000) isolated and sequenced tulip bulb chitinase from. *T. bakeri* and tested its activity in hydrolysis of glicolchitin. This knowledge can be eventually used in designing of resistance markers or in transformation of tulip. Kanno et al. (2003) studied the expression of floral homeotic genes in *T. gesneriana*, for which Van Tunen et al. (1993) proposed a modification of the ABC model describing flower morphology. His study on class B genes corroborated modified ABC model in tulip. The result could be important in transformation directed on floral morphology.

7.8 Information Documentation

The richest tulip field collection of historical cultivars from the sixteenth to nineteenth century is "Hortus Bulborum" in Limmen, the Netherlands, with a total number of accessions preserved there of about 2400 from which about 90% are no longer commercially grown. The collection was started in 1924, some old and legendary cultivars such as "Duc van Tol Red" and "Duc van Tol Yellow" (1595), "Zomerschoon" (1620), or a parrot of "Perfecta" (1750) are still preserved (Hortus Bulborum 2017b). In Turkey, which is one of the most important gene diversity centers of the genus *Tulipa*, large collections are located at the Ataturk Horticultural Central Research Institute in Yalova and at the Black Sea Agricultural Research Institute in Samsun. In the latter, the National Tulip Breeding Project was initiated in 2006.

7.9 Use of Plant Genetic Resources

7.9.1 Common Sources Used to Overcome Constraints

Distant crossing may not be successful due to different barriers, manifested before and after fertilization. To the first type of barriers can be included: lack of pollen germination on stigma, retarding the pollen tube germination in the stigma or in the pistil, and lack of ovule penetration, retarding the pollen tubes in the ovules. In tulip, to overcome the above barriers, cutting off the stigmas or inactivation of stigma enzymes in hot water (50 °C for 1–3 min) can enable fertilization (Okazaki and Murakami 1992). Shortening of style (Van Creij et al. 2000), placental pollination, and ovary grafting (Van Creij et al. 1997a, b) are the other techniques used to overcome prefertilization barriers. Postfertilization barriers concern endosperm dieback or its dysfunction, dieback of seedlings (chlorosis or albinism), inability to form flowers, or underdevelopment of generative organs. Application of amino acids or plant growth regulators to the ovary (Van Creij et al. 1999) embryo, ovule, or ovary culture in vitro (Van Tuyl et al. 1990; Custers et al. 1995; Van Creij et al. 1999, 2000; Okazaki 2005) are successfully utilized to overcome postfertilization barriers.

7.9.2 Breeding Options

There are over 14,000 tulip cultivars, and the latest edition of the Classified List and International Register of Tulip Names (Van Scheepen 1996) contains more than 8000 cultivars. Every year 100–150 new tulip cultivars from around the world are registered by the Royal General Bulb Growers' Association (KAVB) in the Netherlands. Tulip cultivars are available in a variety of colors, sizes, flower shapes, patterns of growth, and with numerous possibilities. High yield, easy forcing, easy transport, and attractive flowers are the prerequisite for novelty in tulip breeding.

Such new creations have potential for consolidation in floriculture, as well as in the world market. Additional traits, however, are also increasingly important, including resistance to biotic and abiotic stresses and length of vase and pot life. These traits, such as the tolerance to warm winters and lower dormancy level, are important for some regions (Schmidt 2015). An increasing mean temperature on our globe and the resulting higher temperature during the time of flower formation within the bulb often results in bud blasting due to dehydration of the flower. The question of how ambient temperature can modify morphological changes was partly answered by Leeggangers et al. (2017) who analyzed the expression of flowering repressor and activator genes in tulips. They stated that long before morphological changes in the shoot apical meristem, the expression of floral repressors is suppressed by increased ambient temperatures leading to the activation of flowering activators shortly before commencement of the phase change. Identification of these genes may be used in the future to preselect desired genotypes or in genetic transformation. To become more attractive for greeneries, the range of cultivars differing in flowering periods should be wider, including later and longer flowering. Another objective for greeneries is a perennialization – growing tulips without annual digging and planting, which has to be combined with their higher resistance for virus and fungal pathogen infections as well as with longer foliage retention and lesser bulb proliferation (Schmidt 2015).

7.9.3 Novel Colors

A dream of tulip breeders is to have black and blue tulips. Information pertinent to the breeding of blue tulips are the findings of Shoji et al. (2007, 2010) and Momonoi et al. (2009) who stated that the blue color, which is seen on the bottom of flowers of some cultivars, is likely the result of complexation of anthocyanin, and Fe^{3+} which concentrates in bottom parts of tepals is 25 times higher than in red parts while having a similar anthocyanin content in the tepals as a whole. It was also reported that for Fe^{3+}content increase, the responsible gene is a vacuolar iron transporter (*TgVit1*) that is active in blue cells, while the gene responsible for Fe storage protein, *TgFer1*, is expressed only in the red cells (Shoji et al. 2010). Thus, the expression of *TgVit1* and the suppression of *TgFer1* is critical for the accumulation of blue color in the bottom of tulip perianths.

7.9.4 Flower Longevity

Flower longevity is also a very important trait for producers of flowers, for stock and retail trade, and for consumers. It was a subject of studies by Van Eijk and Eikelboom (1976, 1986), Van Eijk et al. (1977), and Van der Meulen-Muisers et al. (1997). In their experiments, cultivars varied in flower longevity from 8–16 days and hybrid seedlings from 6–22 days, which indicates an additive effect of genes (Van der Meulen-Muisers et al. 1997). The authors received important information, leading to a better understanding of inheritance of flower longevity and an effective selection of plants possessing this trait. They found correlation between longevity of cut flowers in a vessel and noncut flowers grown in the field. This finding enables preselection of plants with high flower longevity at the first bloom. The most effective criterion of selection turned out to be a change of tepals' color. One important marker is the length of the last internode – the longer the internode, the shorter the longevity. Another marker is the intensity in water uptake after harvesting. Also, those cultivars requiring a longer period for forcing are usually more durable after cutting (Van Eijk and Eikelboom 1976), which interferes with a goal to select for the short period of forcing. According to Van der Meulen-Muisers et al. (1997) the effect of male parent on shoot elongation is greater than female.

7.9.5 Disease Resistance

The production potential of any crop depends upon its resistance to diseases. Tulips are also infected by a number of diseases and introduction of resistant genes into the cultivated varieties becomes important for commercial cultivation. Resistant screening is however another important method that is exploited for screening the cultivars for commercial purposes.

7.9.6 Botrytis Blight [GrayMold; *Botrytis tulipae* (Lib.) Lind]

Botrytis tulipae can infect tulip bulbs, leaves, and flowers. Infection of tulip leaves early in the growing season can result in rapid destruction of leaves and severe reduction of bulb yields. Resistance screening is described as an effective technique for screening tulips for resistance to Botrytis blight. Tulip plants were inoculated at the moment of flowering by rubbing leaves by hand, followed by spraying the leaves with a conidial suspension of *B. tulipae*. Inoculated plants were covered with plastic and incubated in a greenhouse at 18 °C. Disease severity was rated 8 dpi. The key steps of this technique were using plants at the moment of flowering for inoculation, removing the wax layer on tulip leaves and providing continuous leaf wetness for 5 days after inoculation.

Sources of resistance: Absolute resistance has been found in *T. tarda* (Straathof et al. 2002). However, this species cannot be crossed with the commercial tulip cultivars. Partial resistance has been observed in cultivars "Bellona" (with the minimal resistance level), "Flair," "Generaal de Wet," and "Johann Strauss," "Flair," and "Johann Strauss" and have been crossed with commercial cultivars and new hybrids were obtained. Cultivar "Princeps" showed a good level of disease resistance before flowering but has shown susceptibility during and after flowering.

Mode of inheritance: Chang-min et al. (2006) observed a continuous distribution among progeny of 20 crosses in resistance to Botrytis and concluded that the resistance trait was of a quantitative nature.

7.9.7 Bulb Rot or Basal Rot (*Fusarium oxysporum* f. sp. *tulipae*)

Bulb rot is the most serious fungal disease in tulips. Rotten bulbs produce stunted plants with yellowing of leaves and plants frequently die before flowering. Crop losses often reach 30–50% in susceptible tulip cultivars. Resistance screening was used. Reliable screening tests have been developed and standardized for tulip clonal materials and juvenile seedlings (Van Eijk et al. 1978, 1979). Tulip bulbs were planted and grown in Fusarium-infested soil for an entire season (November through June). Temperature of the soil was carefully controlled between 12 °C and 22 °C to promote Fusarium infection at the proper time. After harvest, bulbs were examined, and the percentage of healthy clusters was used as a criterion for disease resistance. To shorten this process, Tang et al. (2015) modified the test by using bulbs in their normal storage period and incubating the bulbs in a Fusarium-infested substrate at 20–24 °C for 8 weeks. A spot inoculation was described recently (Tang et al. 2015). The most aggressive Fusarium isolate Tu67 was tagged with the gfp gene. Tulip bulbs were wounded by punctures and inoculated by dipping on a cushion soaked with a spore suspension of the engineered isolate. Fusarium-infected areas in the bulbs emitted green fluorescence, and these areas were captured by an imaging system and quantified using the Fusarium Screen Analysis Software.

Sources of resistance and breeding: Almost absolute resistance has been identified in cultivars "Black Parrot," "Lucky Stride," "Mirjoran," and "Red Matador," and moderate levels of resistance in "Aristocrat," "Christmas Fire," "Mad. Lefeber," and "Rose Copland" (Van Eijk et al. 1978, 1979). In progeny tests, "Black Parrot," "Lucky Stride," and "Aristocrat" proved to be the most resistant cultivars and produced progeny with high levels of Fusarium resistance (Van Eijk and Eikelboom 1983). Cultivars "Kees Nelis," "Christmas Dream," and "Ile de France" also exhibited high to moderate levels of resistance (Tang et al. 2015). The resistance in some of these cultivars was inherited from old or older cultivars such as "Centenaire," "Panorama," "Van der Eerden," and "Philippe de Comines" (Van Eijk et al. 1985). High levels of Fusarium resistance have been also found in species and interspecific hybrids, such as *T. eichleri* hybrid "Clare Benedict," *T. eichleri* "Excelsa," *T. fosteriana* "Mad. Lefeber," *T. fosteriana* "Princeps," *T. kaufmanniana* "Guiseppe Verdi," and *T. vvedenskyi* (Van Eijk et al. 1985). These sources of Fusarium resistance have been used in crosses and new resistant cultivars, including "Pink Impression" and "Early Surprise," and "Furand," "Furanel," "Fusarino," and "Fusor" have been developed (Straathof and Eikelboom 1997; Van Eijk et al. 1985).

Crossing with “Don Quichotte” (resistant) has resulted in the development of “Arisa” with resistance against bulb rot and tulip mild mottle mosaic virus (Tsuji et al. 2005).

Mode of inheritance: Fusarium resistance in tulip cultivars, such as “Kees Nelis” and “Cantata,” is a quantitative trait mainly based on additive gene action (Van Eijk and Eikelboom 1983; Tang et al. 2015). GCA values were high and significant in reported tulip crosses. SCA values were low and its significance depended on tulip crosses. The Fusarium resistance of some progeny was higher than the resistant parent “Kees Nelis.” Thus, transgressive segregation exists for Fusarium resistance in tulip.

Molecular markers: Six putative QTL have been identified for the Fusarium resistance in cultivars “Kees Nelis” and “Cantata” and tagged with SNP markers. Five of the QTL each explained 13.6–20.7% of the phenotypic resistance variation. Three QTL (Fusarium2, Fusarium 3, and Fusarium6) were significant in all disease tests. Both the resistant parent “Kees Nelis” and the susceptible parent “Cantata” contributed resistance QTL to their progeny (Tang et al. 2015).

7.9.8 Mild Mottle Mosaic Disease (Tulip Mild Mottle Mosaic Virus, TMMMV)

TMMMV, transmitted by the obligate parasitic soil-inhabiting fungus *Olpidium brassicae*, has become a serious disease in some tulip production areas in Japan since 1979 (Morikawa et al. 2005).

Resistance screening: Tulip plants were exposed to the virus in the infected field for 2 years and then cultivated in containers with noninfected soil for one year. Peduncles were harvested from plants at flowering, and the presence of TMMMV was detected by tissue blot immunoassay (TBIA). The percentage of infected plants was used as the criterion for classifying disease resistance (Morikawa et al. 2005).

Sources of resistance: Morikawa et al. (2005) reported that 24 tulip cultivars could be considered highly resistant. Among these resistant cultivars are “Arisa,” “Ballade,” “Ballerina,” “Feu Superbe,” “Gander,” “Koiakane,” “Juan,” “Leo Visser,” “Orange Emperor,” “Princeps,” and “Snowstar.”

7.9.9 Tulip Breaking Virus (TBV)

TBV is a serious problem in the production of tulip bulbs and flowers in many countries. The virus can cause a significant reduction in tulip bulb number, weight, and quality. Reliable screening for TBV resistance has been performed at the clonal and seedling level by aphid or mechanical inoculation (Romanow et al. 1986, 1991). Aphid inoculation was achieved by placing viruliferous aphids on the leaves of tulip plants at flowering for one day. Tulip plants could be mechanically inoculated by pipetting droplets of virus inoculum onto tulip leaves followed by rubbing the leaves with carborundum powder by fingers. The inoculum was prepared by grinding

TBV-infected leaf tissue in water. The inoculated plants were observed for flower breaking one year after inoculation. TBV titers in the plants were determined by ELISA. Absolute resistance has been found in "Cantata" and "Princeps" (Eikelboom et al. 1992; Romanow et al. 1991; Straathof et al. 1996a). Cultivars "Juan" and "Mad. Lefeber" showed moderate levels of resistance. Cultivar "AdRem" was the only representative of *T. gesneriana* tested and could be classified as TBV resistant. This cultivar is a Darwin hybrid, resulting from a cross between "Denbola" and "Princeps." The resistance in "AdRem" should have come from "Princeps." TBV resistance in tulip is a quantitative trait. GCA effects explained the largest portion of the variation in tulip crosses (Straathof et al. 1996b). "Cantana" and "Princeps" showed significantly better GCA values than other cultivars, and they could be desirable parents for introducing TBV resistance from *T. fosteriana* into the *T. gesneriana* assortment.

7.9.10 Present Status of Use or Incorporation of Desired Traits

An effective regeneration of tulips from single cells has opened breeding perspectives for using transformation to introduce single desired genes to already known cultivars. The most desirable genes are those increasing the level of resistance to fungal and viral diseases that generate serious losses in tulip production. In numerous plant species, various techniques such as electroporation, microinjection, and *Agrobacterium*-mediated gene transfer have been developed to add or activate/deactivate desired genes (Moose and Mumm 2008). More recently, CRISPR (clustered regularly interspaced short palindromic repeats) and TALEN (transcription activator-like effector nucleases) techniques offer much precision (Gaj et al. 2013). Particle bombardment is the most common method applied successfully for monocot species, e.g., maize, for which transformation using *Agrobacterium tumefaciens* has been less successful (Wilmink and Dons 1993). In tulip, to date, only preliminary studies have been performed on genetic transformation using particle bombardment or *Agrobacterium*-mediated gene engineering techniques (Wilmink et al. 1995a; Wilmink 1996; Chauvin et al. 1997, 1999). The authors sought to find suitable selective agents, reporter genes, and promoters. As the majority of monocots, tulip explants revealed a very low susceptibility to kanamycin, a selective agent mostly used in the past time for plant transformation. Wilmink et al. (1995b) showed that herbicide Basta (PPT) is more useful for the selection, and the rice actin promoter is most active in the determination of GUS expression in transformation using particle bombardment of floral stem segments which possesses relatively high ability to adventitious shoot regeneration. Molecular analysis performed 12–15 months after the transformation of initial explants confirmed the presence of *bar* and *gus*genes in DNA of the regenerated tulip plants. However, expression at protein level was low or absent. Wilmink (1996) suggested that the introduced genes were present only in a part of the tissue indicating a chimeric structure of transformed plants which pointed to a multicellular origin of regenerates. In order to avoid the formation of chimeras, it was postulated to use the application of a selection system with a higher concentration of the selecting agent Basta used for a longer time during plant culture. However, the application of gene transformation in tulip breeding seems to be problematic due to the low in vitro regeneration ability of tulip explants and very long and complicated developmental cycle in this genus. It takes at least 8 years from transformation, through selection of transformed plant to flowering.

7.10 Future Perspective

The unending popularity of tulips around the world creates the need to breed new cultivars that can be commercially utilized in cut flower and landscape industry. Besides beauty, the new cultivars should possess the features necessary for effective bulb production and for forcing for a period from early autumn to late spring. Trends to protect the production environment are causing ever-stricter regulations in the use of crop protection products. This raises the need for cultivars that are more resistant to diseases and pests, less demanding in the field and during forcing. It is expected that with time, tulip breeding will be easier due to the increasing knowledge of genetic determinations related to the most important functional features. Simultaneously, more molecular biology tools need to be introduced to breeding programs; therefore, new molecular markers making tulip breeding more effective in a reduced time frame will be needed (Krens and Kamo 2013). Due to the fact that tulips are grown on large plantations producing large quantity of waste flowers, the idea was arisen to use anthocyanin-rich extracts as a new source of food colorant (Arici et al. 2016) and as a source of bioactive compounds (Sagdic et al. 2013). Above all, the demand for tulips as cut flower or landscape plant opens vistas for breeding new cultivars with novel characters.

References

Abedi R, Babaci A, Karimzadeh G (2015) Karyological and flow cytometric studies of *Tulipa* (Liliaceae) species from Iran. Plant Syst Evol 301(5):1473–1484

Alderson PG, Taeb AG (1990) Effect of bulb storage on shoot regeneration from floral stems of tulip *in vitro*. J Hortic Sci 65:71–74

Anonymous (2006) UPOV. Tulip. Guidelines for the conduct of tests for distinctness, uniformity and stability. TG/115/4. International Union for the Protection of New Varieties of Plants, Geneva

Anonymous (2017a) BKD. Voorlipigestatistiekvoorjaarsbloeiers 2016–2017

Anonymous (2017b) Hortus Bulborum. Stichting Hortus Bulborum. https://hortus-bulborum.nl/?lang=en. Access date 7 September 2021

Arens P, Bijman B, Shahin A, van Tuyl JM (2012) Mapping of disease resistance in ornamentals: a long haul. Acta Hortic 953:231–237

Arici M, Karasu S, Baslar M, Toker OS, Sagdic O (2016) Tulip petal as a novel natural food colorant source: extraction optimization and stability studies. Ind Crop Prod 91:215–222

Bach A, Ptak A (2001) Somatic embryogenesis and plant regeneration from ovaries of *Tulipa gesneriana* L. *in vitro* cultures. Acta Hortic 560:391–394

Baker CM, Wilkins HF, Asher PD (1990) Comparisons of precultural treatments and cultural conditions on *in vitro* response of tulip. Acta Hortic 66:83–90

Benschop M, Kamenetsky R, Le Nard M, Okubo H, De Hertogh A (2010) 1 The Global Flower Bulb Industry: production, utilization, research. Hortic Rev 36(1):1–115

Blakey DH, Vosa CG (1981) Heterochromatin and chromosome variation in cultivated species of *Tulipa,* subg. *Eriostemones* (Liliaceae). Plant Syst Evol 139:47–55

Blakey DH, Vosa CG (1982) Heterochromatin and chromosome variation in cultivated species of *Tulipa*subg. *Leiostemones* (Liliaceae). Plant Syst Evol 139(3):163–178

Bondrea OI, Pamfil D, van Heusden S, van Tuyl J, Meijer-Dekens F, Bondrea M, Sestras R, Rusu AR, Lucaci M, Patrascu BI, Balteanu VA (2007) AFLP as a modern technique for DNA fingerprinting and identification *Tulipa* cultivars. Bull USAMV-CN 63–64:317–319

Bonnier FJM, Eikelboom W, van Tuyl JM (1997) Storage methods to maintain a tulip collection. Acta Hortic 430:130

Botschantzeva ZP (1962) Tulips. In: Taxonomy, morphology, cytology, phytogeography and physiology (Russian edn), English translation: Varekamp HQ. Balkema, Rotterdam

Botschantzeva ZP (1982) Tulips: taxonomy, morphology, cytology, phytogeography and physiology. Balkema, Rotterdam, 227 pp

Buschman JCM (2005) Globalisation – flower – flower bulbs – bulb flowers. Acta Hortic 673:27–33

Chang-min PANG, van Empel P, van Tuyl JM (2006) Inheritance and resistance to Botrytis disease in F_1 generation of tulips. J Northwest Sci-Tech Univ Agric For 34:105–108

Chauvin JE, Hamann H, Cohat J, Le Nard M (1997) Selective agents and marker genes for use in genetic transformation of *Gladiolus grandiflorus*and *Tulipagesneriana*. Acta Hortic 430:291–298

Chauvin JE, Marhadour S, Cohat J, Le Nard M (1999) Effects of gelling agents on *in vitro* regeneration and kanamycin efficiency as a selective agent in plant transformation procedures. Plant Cell Tissue Organ Cult 58(3):213–217

Christenhusz MJM, Govaerts R, David JC, Hall T, Borland K, Roberts PS, Tuomisto A, Buerki S, Chase MW, Fay MF (2013) Tiptoe through the tulips – cultural history, molecular phylogenetics and classification of *Tulipa* (*Liliaceae*). Bot J Linn Soc 172(3):280–328

Coşkunçelebi K, Terzioğlu S, Türkmen Z, Makbul S, Usta A (2008) A comparative study on two closely relative Tulipa L. taxa from NE Anatolia. Plant Syst Evol *276*(3–4):191

Custers JBM, Eikelboom W, Bergervoet JHW, Van Eijk JP (1992) *In ovulo* embryo culture of tulip (*Tulipa* L.); effects of culture conditions on seedling and bulblet formation. Sci Hortic 51(1–2):111–122

Custers JBM, Eikelboom W, Bergervoet JHW, Eijk JV (1995) Embryo-rescue in the genus *Tulipa* L.; successful direct transfer of *T. kaufmanniana* regel germplasm into *T. gesneriana* L. Euphytica 82(3):253–261

Diana E (2013) The genus *Tulipa*. Royal Botanic Gardens, Kewdiana

Eikelboom W, van Eijk JP, Peters D, van Tuyl JM (1992) Resistance to tulip breaking virus (TBV) in tulip. Acta Hortic 325:631–636

Eker İ, Babaç MT (2010) *Tulipa koyuncui* Eker & Babaç sp. nov. (Liliaceae) from east Anatolia, Turkey. Nord J Bot 28(3):324–328

Eker I, Tekin Babac M, Koyuncu M (2014) Revision of the genus *Tulipa* L. (Liliaceae) in Turkey. Phytotaxa 157:001–112

Famelaer I, Ennik E, Eikelboom W, Van Tuyl JM, Creemers-Molenaar J (1996) The initiation of callus and regeneration from callus culture of *Tulipa gesneriana*. Plant Cell Tissue Organ Cult 47:51–58

Filion WG (1974) Differential Giemsa staining in plants. I Banding patterns in three cultivars of *Tulipa*. Chromosome 49:51–60

Gabrielijan EC, Pogosjan AI (1971) Taxanomic& karyologic investigations of Armenian tulips. Iz Akad Nauk Arm SSr 24(5):51–60

Gaj T, Gersbach CA, Barbas CF (2013) ZFN, TALEN and CRISPR/Cas-based methods for genome engineering. Trends Biotechnol 31(7):397–405

Ghahreman A, Attar F, Ghahremaninejad F (2007) A new species of Tulipa (Liliaceae) from western Iran. Novon: J Bot Nomencl 17(4):437–439

Govaerts R (2010) World checklist of selected plant families: the genus Tulipa L. Royal Botanic Gardens, Kew, UK: genus Tulipa. http://apps.kew.org/wcsp/qsearch.do

Govaerts R (2015) World checklist of selected plant families: the genus Tulipa L. Royal Botanic Gardens, Kew

Govaerts R (2017) World checklist of *Liliaceae*. Facilitated by the Royal Botanic Gardens, Kew. Published on the Internet; http://apps.kew.org/wcsp/

Gude H, Dijkema MHGE (1997) Somatic embryogenesis in tulip. Acta Hortic 430:275–280

Hall AD (1937) Polyploidy in Tulipa. J Linn Soc Bot 50(335):481–489

Hall AD (1940) The genus *Tulipa*. The Royal Horticultural Society, London

Han B (2014) Amana wanzhensis (Liliaceae), a new species from Anhui, China. Phytotaxa 177(2): 118–124. https://doi.org/10.11646/phytotaxa.177.2.3

Hayek A (1932–1933) Prodromus Florae peninsulae Balcanicae, Band 3 (Monocotyledoneae). Repertorium specierum Novarum Regni Vegetabilis, Beiheft 30, 472 pp

Hoog MH (1973) Lilies and other Liliaceae. The Royal Horticultural Society, pp 47–64

Hulscher M, Krijksheld HT, Van der Linde PCG (1992) Propagation of shoots and bulb growth of tulip *in vitro*. Acta Hortic 325:441–446

Ibrahim MF, Hussain FHS, Zanoni G, Vidari G (2016) Phytochemical screening and antibacterial activity of *Tulipa systole* stapf, collected in Kurdistan region-Iraq. ZANCO J Pure Appl Sci 28(4):44–49

John AQ, Neelofar, Khan FU (2006) The Tulip. Souvenir; National Symposium on Ornamental Bulbous Crops. Indian society of Orn Horticulture, pp 61–64

Kanno A, Saeki H, Kameya T, Saedler H, Theissen G (2003) Heterotopic expression of class B floral homeotic genes supports a modified ABC model for tulip (*Tulipa gesneriana*). Plan Mol Biol 52(4):831–841

Kawa L, De Hertogh AA (1992) Root physiology of ornamental flowering bulbs. Hortic Rev 14: 57–88

Kiani M, Memariani F, Zarghami H (2012) Molecular analysis of species of *Tulipa* L. from Iran based on ISSR markers. Plant Syst Evol 298(8):1515–1522

Krens FA, Kamo K (2013) Genomic tools and prospects for new breeding techniques in flower bulb crops. Acta Hortic 974:139–147

Kroon GH, Eijk V (1977) Polyploidyin tulips (Tulipa L.). The occurance of diploid gametes. Euphytica 26:63–66

Kroon GH, Jongerius MC (1986) Chromosome numbers of *Tulipa* species and the occurrence of hexaploidy. Euphytica 35:73–76

Kuijpers AM, Langens-Gerrits M (1997) Propagation of tulip *in vitro*. Acta Hortic 430:321–324

Le Nard M, Ducommun C, Weber G, Dorion N, Bigot C (1987) Obsevationssur la multiplication *in vitro* de la tulipe (*Tulipa gesneriana* L.) à partir de hampesfloralesprélevées chez des bulbesencours de conservation. Agronomie 7(5):321–329

Leeggangers HA, Nijveen H, Bigas JN, Hilhorst HWM, Immink RGH (2017) Molecular regulation of temperature-dependent floral induction in *Tulipa gesneriana*. Am Soc Plant Biol 173(3):1904–1919

Marais W (1984) Tulipa L. In: Davis PH (ed) Flora of Turkey and the East Aegean Islands 8. Edinburgh University Press, Edinburgh, pp 302–311

Marasek A, Okazaki K (2007) GISH analysis of hybrids produced by interspecific hybridization between *Tulipa gesneriana* and *T. fosteriana*. Acta Hortic 743:133–137

Marasek A, Okazaki K (2008) Analysis of introgression of the *T. fosteriana* genome into *Tulipa gesneriana* using GISH & FISH. Euphytica 160(2):217–230

Marasek A, Mizuochi H, Okazaki K (2006) The origin of Darwin hybrid tulips analyzed by flow cytometry, karyotype analyses and genomic in situ hybridization. Euphytica 151(3):279–290

Marasek-Ciolakowska A, Podwyszyńska M (2008) Somaclonal variation in long-term micropropagated tulips (*Tulipa gesneriana* L.) determined by FISH analysis. Floricultural and ornamental biotechnology. Global Sci Books 2:65–72

Marasek-Ciolakowska A, Ramanna MS, Van Tuyl JM (2009) Introgression breeding in genus Tulipa analysed by GISH. Acta Hortic 836:105–110

Marasek-Ciolakowska A, Ramanna MS, Van Tuyl JM (2011) Introgression of chromosome segments of *Tulipa fosteriana* into *T. gesneriana* detected through GISH and its implications forbreeding virus resistant tulips. Acta Hortic 886:175–182

Marasek-Ciolakowska A, He H, Bijman P, Ramanna MS, Arens P, van Tuyl JM (2012) Assessment of intergenomic recombination through GISH analysis of F1, BC1 and BC2 progenies of *Tulipa gesneriana* and *T. fosteriana*. Plant Syst Evol 298(5):887–899

Marasek-Ciolakowska A, Xie S, Arens P, Van Tuyl JM (2014) Ploidy manipulation & introgression breeding in Darwin hybrid tulips. Euphytica 198(3):389–400

Marasek-Ciolakowska A, Nishikawa T, Shea DJ, Okazaki K (2018) Breeding of lilies and tulips Interspecific hybridization and genetic background. Breed Sci 68:25–52

Markgraf F (1931) Pflanzenaus Albanien 1928. Hölder-Pichler-Tempsky

Maślanka M, Bach A (2014) Induction of bulb organogenesis in *in vitro* cultures of Tarda tulip (*Tulipatarda* Stapf.) from seed-derived explants. In Vitro Cell Dev Biol Plant 50(6):712–721

Maslanka M, Bach A (2020) Droplet-vitritcation cryopreservation of Tulipa tarda Stapf. apical meristems. Plant Cell Tissue Organ Cult (PCTOC) 144:91–95

Miao Y, Zhu Z, Guo Q, Zhu Y, Yang X, Sun Y (2016) Transcriptome analysis of differentially expressed genes provides insight into stolon formation in *Tulipaedulis*. Front Plant Sci 7:409

Mizuochi H, Marasek A, Okazaki K (2007) Molecular cloning of *Tulipafosterianar* DNA and subsequent FISH analysis yields cytogenetic organization of 5S rDNA and 45S rDNA in *T. gesneriana* and *T. fosteriana*. Euphytica 155:235–248

Momonoi K, Shoji K, Yoshida K (2007) Cloning and characterization of ACC oxidase genes from tulip. Plant Biotechnol 24(2):241–246

Momonoi K, Yoshida K, Shoji M, Takahashi H, Nakaqmori C, Shoji K, Nitta A, Nishimura M (2009) A vacuolar iron transporter in tulip, TgVit1, is responsible for blue coloration in petal cells through iron accumulation. Plant J 59(3):437–447

Moose SP, Mumm RH (2008) Molecular plant breeding as the foundation for 21st century crop improvement. Plant Physiol 147(3):969–977

Moreno-Pachon NM, Leeggangers HACF, Nijven H, Severing E, Hilhorst H, Immink RGH (2016) Elucidating and mining the *Tulipa* and *Lilium* transcriptomes. Plant Mol Biol 92(3):249–261

Morikawa T, Taga Y, Morii T (2005) Resistance of tulip cultivars to mild mottle mosaic disease. Acta Hortic 673:549–553

Newton (1927) J Linn Soc, Bot. xl vii, p 339

Niimi Y (1978) Influence of low and high temperatures on the initiation and the development of a bulb primordium in isolated tulip embryos. Sci Hortic 9:61–69

Nishiuchi Y (1980) Studies on vegetative propagation of tulip. 4. Regeneration of bulblets in bulb scale segments cultured *in vitro*. J Jpn Soc Hortic Sci 49(2):235–239

Okazaki K (2005) New aspects of tulip breeding: embryo culture and polyploidy. Acta Hortic 673: 127–140

Okazaki K, Murakami K (1992) Effects of flowering time (in forcing culture), stigma excision, and high temperature on overcoming of self-incompatibility in tulip. J Jpn Soc Hortic Sci 61:405–411

Okazaki K, Kurimoto K, Miyajima I, Enami A, Mizuochi H, Matsumoto Y, Ohya H (2005) Induction of 2n pollen in tulips by arresting meiotic process with nitrous oxide gas. Euphytica 143:101–114

Orlikowska T, Podwyszyńska M, Marasek-Ciołakowska A, Sochacki D, Szymański R (2018) Tulip. In: Ornamental Crops. Handbook of Plant Breeding. Springer Cham (Van Huylenbroeck J. ed.) 11:769–802

Pavord A (1999) The tulip. Bloomsbury, London

Petrová E, Faberová I (2000) Klasifikátor. Descriptor list. Genus *Tulipa* L. Výzkumný ústav okrasnéhozahradnictví, Prùhonice

Podwyszyńska M (2006) Improvement of bulb formation in micropropagated tulips by treatment with NAA and paclobutrazol or ancymidol. Acta Hortic 725:679–684

Podwyszyńska M (2011) *In vitro* tetraploid induction in tulip. Acta Hortic 961:391–396

Podwyszyńska M, Marasek A (1999) Somatic embryogenesis in tulip. 3rd international symposium in the series recent advances in plant biotechnology, from cells to crops, September 4–10, Stara Lesna. Slovakia Biologia 54(Suppl 7):25–26

Podwyszyńska M, Marasek A (2003) Effect of thidiazuron and paclobutrazol on regeneration potential of flower stalk explants *in vitro* and subsequent shoot multiplication. Acta Soc Bot Pol 72(3):181–190

Podwyszyńska M, Sochacki D (2010) Micropropagation of tulip: production of virus-free stock plants. In: Protocols for In Vitro Propagation of Ornamental Plants 589:243–256

Podwyszyńska M, Niedoba K, Korbin M, Marasek A (2006) Somaclonal variation in micropropagated tulips determined by phenotype and DNA markers. In: XXII International Eucarpia Symposium, Section Ornamentals, Breeding for Beauty 714:211–220

Podwyszyńska M, Kuras A, Korbin M (2010) ISSR evaluation of genetic stability of Polish tulip cultivars propagated in vitro. Biotechnologia 2(89):105–113. (In Polish, English abstract)

Podwyszyńska M, Novák O, Doležal K, Strnad M (2014) Endogenous cytokinin dynamics in micropropagated tulips during bulb formation process influenced by TDZ and iP pre-treatment. Plant Cell Tissue Organ Cult 119(2):331–346

Ptak A, Bach A (2007) Somatic embryogenesis in tulip (*Tulipagesneriana* L.) flower stem cultures. In Vitro Cell Dev Biol Plant 43:35–39

Qu LW, Xing GM, Zhang YQ, Su JW, Zhao Z, Wang WD, Lei JJ (2016) Native species of the genus *Tulipa* and tulip breeding in China. XII international symposium on Flower Bulbs and Herbaceous Perennials, Kunming, China, June 28–July 2, 2016. Book of Abstract 109

Ramzan F, Younis A, Riaz A, Ali S, Siddique MI, Lim KB (2014) Pre-planting exogenous application of gibberellic acid influences sprouting, vegetative growth, flowering, and subsequent bulb characteristics of 'Ad-Rem' tulip. Hortic Environ Biotechnol 55(6):479–488. https://doi.org/10.1007/s13580-014-0113-7

Reed S (2004) Embryo rescue. In: Trigiano RN, Gray DJ (eds) Plant development and biotechnology. CRC Press, Boca Raton, pp 235–239

Rees AR (1992) Ornamental bulbs, corms and tubers. CAB International, Wallingford, p 68

Reyes AL, Prins TP, van Empel J-P, van Tuyl JM (2005) Differences in epicuticular wax layer in tulip can influence resistance to *Botrytis tulipae*. Acta Hortic 673:457–461

Rice RD, Alderson PG, Wright NA (1983) Induction of bulbing of tulip shoots *in vitro*. Sci Hortic 20(4):377–390

Romanow LR, van Eijk JP, Eikelboom W (1986) Investigating resistance in tulip to tulip breaking virus and its transmission. Acta Hortic 177:235–239

Romanow LR, van Eijk JP, Eikelboom W, van Schadewijk AR, Peters D (1991) Determining levels of resistance to *Tulip breaking virus* (TBV) in tulip (*Tulipa* L.) cultivars. Euphytica 51(3):273–280

Sagdic O, Ekici L, Ozturk I, Tekinay T, Polat B, Tastemur B, Bayram O, Senturk B (2013) Cytotoxic and bioactive properties of different color tulip flowers and degradatioin kinetic of tulip flower anthocyanins. Food Chem Toxicol 58:432–439

Saniewski M, Kawa-Miszczak L (1992) Hormonal control of growth and development of tulips. Acta Hortic 325:43–54

Sayama H, Moue T, Nishimura Y (1982) Cytological study in *Tulipa gesneriana* and *T. fosteriana*. Jpn J Breed 32:26–34

Schmidt T (2015) Interspecific breeding for warm-winter tolerance in Tulipagesneriana L. Univ of Minnesota Aquaponics Report, 1–22, http://hdl.handle.net/11299/17843

Schrader HA (1799) Journal für die Botanik, vol 2. Joh. Christ. Dieterich, Göttingen, 502 pp. (dated 1799 on title page, published April 1800)

Shahin A, van Kaauwen M, Esselink D, Bargsten JW, van Tuyl JM, Visser RGF, Arens P (2012) Generation and analysis of expressed sequence tags in the extreme large genomes *Lilium*and *Tulipa*. BMC Genomics 13:640

Sharma DR, Kaur R, Kumar K (1996) Embryo rescue in plants-a review. Euphytica 89(3):325–337

Shen XS (2001) A new species of Tulipa (Liliaceae) from China. Acta Bot Yunnanica 23:39–40

Shoji K, Miki N, Nakajima N, Momonoi K, Kato C, Yoshida K (2007) Perianth bottom-specific blue color development in tulip cv. Murasakizuisho requires ferric ions. Plant Cell Physiol 48(2):243–251

Shoji K, Momonoi K, Tsuji T (2010) Alternative expression of vacuolar iron transporter and ferritin genes leads to blue/purple coloration of flowers in tulip cv. Murasakizuisho. Plant Cell Physiol 51(2):215–224

Sochacki D, Podwyszyńska M (2012) Virus eradication in narcissus and tulip by chemotherapy. In: Van Tuyl J, Arens P (eds) Floriculture and ornamental biotechnology, Special issue Bulbous Ornamentals, 2, vol 6. Global Science Book, Isleworth, pp 114–121

Stork A (1984) Tulipessauvagesetcultivees. Conservatoire et Jardinbotaniques, Seriedocumentaire 13, Geneve

Straathof TP, Eikelboom W (1997) Tulip breeding at CPRO-DLO. Daffodil and Tulip yearbook 8. Royal Horticultural Society, London, pp 27–33

Straathof TP, van Tuyl JM, Dekker B, van Winden MJM, Sandbrink JM (1994) Genetic analysis of inheritance of partial resistance to *Fusarium oxysporum* in Asiatic hybrids of lily using RAPD markers. Acta Hortic 414:209–218

Straathof TP, Eikelboom W, Van Tuyl JM, Peters D (1996a) Screening for TBV-resistance in seedling populations of Tulipa L. In: IX International Symposium on Virus Diseases of Ornamental Plants 432:392–399

Straathof TP, Löffler HJM, Roebroeck EJA, Linfield CA (1996b) Breeding for resistance to *Fusarium oxysporum* in flower bulbs. In: VII International Symposium on Flower bulbs 430:477–486

Straathof TP, Mes J, Eikelboom W, Van Tuyl JM (2002) A greenhouse screening assay for *Botrytis tulipae* resistance in tulips. Acta Hortic 570:415–421

Suji T, Imai T, Ikegawa S, Kanamori M, Umada Y, Koizumi M, Kizu M, Limura N, Urashima O (2004) The production and characteristics of the new tulip cultivar, 'Arisa'. In: IX International Symposium on Flower Bulbs 673:555–558

Tan DY, Wei X, Fang J (2000) New taxa of Tulipa L. from Xinjiang. Acta Phytotaxon Sin 38(3):302–304

Tan DY, Li XR, Hong DY (2007) Amana kuocangshanica (Liliaceae), a new species from south-east China. Bot J Linn Soc 154(3):435–442. https://doi.org/10.1111/j.1095-8339.2007.00660.x

Tang J, Wang FZ (1980) The flora of China, vol 14. Science Publishing House, Beijing, pp 81–97

Tang N, Shahin A, Bijman P, Liu J, van Tuyl J, Arens P (2013) Genetic diversity and structure in a collection of tulip cultivars assessed by SNP markers. Sci Hortic 161:286–292

Tang N, van der Lee T, Shahin A, Holdinga M, Bijman P, Caser M, Visser RGF, van Tuyl JM, Arens P (2015) Genetic mapping of resistance to *Fusarium oxysporum* f. sp. tulipae in tulip. Mol Breed 35:122

Turktas M, Metin ÖK, Baştuğ B, Ertuğrul F, Saraç YI, Kaya E (2013) Molecular phytogenetic analysis of *Tulipa* (Liliaceae) based on noncoding plastid and nuclear DNA sequences with an emphasis on Turkey. Bot J Linnean Soc 172(3):270–279

Upcott M, La Cour L (1936) Solms-laubach. 1899. Weizen und tulipe. Leipzig

Van Creij MGM (1997) Interspecific hybridization in the genus *Tulipa* L. PhD thesis, Wageningen University, The Netherlands

Van Creij MGM, Kerckhoffs DMFJ, Van Tuyl JM (1997a) Interspecific crosses in the genus *Tulipa* L.: identification of pre-fertilization barriers. Sex Plant Reprod 10(2):116–123

Van Creij MV, Van Went JL, Kerckhoffs DMFJ (1997b) The progamic phase, embryo and endosperm development in an intraspecific *Tulipa gesneriana* L. cross and in the incongruent interspecific cross *T. gesneriana* × *T. agenensis* DC. Sex Plant Reprod 10(4):241–249

Van Creij MGM, Kerckhoffs DMF, Van Tuyl JM (1999) The effect of ovule age on ovary-slice culture and ovule culture in intraspecific and interspecific crosses with *Tulipa gesneriana* L. Euphytica 108:21–28

Van Creij MGM, Kerckhoffs DMFJ, De Bruijn SM, Vreugdenhil D, Van Tuyl JM (2000) The effect of medium composition on ovary-slice culture and ovule culture in intraspecific *Tulipa gesneriana* crosses. Plant Cell Tissue Organ Cult 60:61–67

Van de Wiel C, Schaart J, Niks R, Visser R (2010) Traditional plant breeding methods. Wageningen UR Plant Breeding, Wageningen, May 2010. Report 338

Van der Meulen-Muisers JJM, Van Oeveren JC, Van Tuyl JM (1997) Breeding as a tool for improving postharvest quality characters of lily and tulip flowers. Acta Hortic 430:569–575

Van Eijk JP, Eikelboom W (1975) Criteria for early selection in tulip breeding. Acta Hortic 47:179–186

Van Eijk JP, Eikelboom W (1976) Possibilities of selection for keeping quality in tulip breeding. Euphytica 25:353–359

Van Eijk JP, Eikelboom W (1983) Breeding for resistance to *Fusarium oxysporum* f. sp. *tulipae* in tulip (*Tulipa* L.). 3. Genotypic evaluation of cultivars and effectiveness of pre-selection. Euphytica 32(2):505–510

Van Eijk JP, Eikelboom W (1986) Aspects of breeding for keeping quality in *Tulipa*. Acta Hortic 181:237–243

Van Eijk JP, Eikelboom W (1990) Evaluation of breeding research on resistance to *Fusarium oxysporum* in tulip. Acta Hortic 266:357–364

Van Eijk JP, Toxopeus SJ (1968) The possibilities of early selection for forcing ability and productivity in tulips. Euphytica 17(2):277–283

Van Eijk JP, Eikelboom W, Sparnaij L (1977) Possibilities of selection for keeping quality in tulip breeding. 2. Euphytica 26:825–828

Van Eijk JP, Bergman BHH, Eikelboom W (1978) Breeding for resistance to *Fusarium oxysporum* f. Sp. *tulipae* in tulip (*Tulipa* L.). 1. Development and screening test for selection. Euphytica 27 (2):441–446

Van Eijk JP, Garretsen F, Eikelboom W (1979) Breeding for resistance to *Fusarium oxysporum* f. sp. *tulipae* in tulip (*Tulipa* L.).2. Phenotypic and genotypic evaluation of cultivars. Euphytica 28: 67–71

Van Eijk JP, Eikelboom W, Hogenboom NG (1985) The importance of wild species and old cultivars for the breeding of flowerbulbs. Acta Hortic 177:399–403

Van Eijk JP, van Raamsdonk LWD, Eikelboom W, Bino RJ (1991) Interspecific crosses between *Tulipa gesneriana* cultivars and wild *Tulipa* species: a survey. Sex Plant Reprod 4:1–5

Van Raamsdonk LWD, De Vries T (1992) Biosystematic studies in *Tulipa* L. sect. *Eriostemones* (Liliaceae). Plant Syst Evol 179:27–41

Van Raamsdonk LWD, De Vries T (1995) Species relationships and taxonomy in *Tulipa* subg. Tulipa (Liliaceae). Plant Syst Evol 195:13–44

Van Raamsdonk LW, Van Eijk JP, Eikelboom W (1995) Crossability analysis in subgenus Tulipa of the genus *Tulipa* L. Botanical Journal of the Linnean Society 117(2):147–158

Van Raamsdonk LWD, Eikelboom W, De Vries T, Straathof TP (1997) The systematics of the genus *Tulipa* L. Acta Hortic 430:821–827

Van Scheepen J (1996) Classified list and international register of tulip names. Royal General Bulbgrowers' Association, KAVB

Van Tunen AJ, Eikelboom W, Angenent GC (1993) Floral organogenesis in Tulipa. Flowering Newsletter 16:33–38

Van Tuyl JM, Creij MGM (2006) Tulipa gesneriana and Tulipa hybrids. In: Flower breeding & genetics: Issues, challenges and opportunities for the 21st century (Springer Verlag.), pp 512–532

Van Tuyl JM, van Creij MG (2007) Tulip. In: Anderson NO (ed) Flower breeding and genetics. Springer, Dordrecht, pp 623–641

Van Tuyl JM, Bino RJ, Custers JBM (1990) Application of in vitro pollination, ovary culture, ovule culture and embryo rescue in breeding of Lilium, Tulipaand Nerine. In: De Jong J (ed) Integration of in vitro techniques in ornamental plant breeding, Proceedings of the Eucarpia Symposium, Section Ornamentals, November 10–14, 1990, Wageningen, pp 86–97

Van Tuyl JM, Arens P, Marasek-Ciolakowska A (2012) Breeding and genetics of ornamental geophytes. In: Kamenetsky R, Okubo H (eds) Ornamental geophytes: from basic science to sustainable production. CRS Press, London, pp 131–158

Wilmink A (1996) Genetic modification of tulip by means of particle bombardment. PhD thesis, University Nijmegen, The Netherlands

Wilmink A, Dons JJM (1993) Selective agents and marker genes for use in transformation of monocotyledonous plants. Plant Mol Biol Rep 11(2):165–185

Wilmink A, Van de Ven BCE, Custers JBM, Van Tuyl JM, Eikelboom W, Dons JJM (1995a) Genetic transformation in *Tulipa*species (tulips). In: Bajaj YPS (ed) Plant protoplasts and genetic engineering VI. Springer, Berlin/Heidelberg, pp 289–298

Wilmink A, Van de Ven BCE, Custers JBM, Nollen Y, Dons JJM (1995b) Adventitious shoot formation in tulip: histological analysis and response to selective agents. Plant Sci 110:155–164

Yamagami T, Tsutsumi K, Ishiguro M (2000) Cloning, sequencin, and expression of the tulip bulb chitinase-1 cDNA. Biosci Biotechnol Biochem 64(7):1394–1401

Yuan Y, Ma X, Tang D (2013) Isolation and expression analysis of six putative structural genes involved in anthocyanin biosynthesis in *Tulipafosteriana*. Sci Hortic 153:93–102

Zonneveld BJM (2009) The systematic value of nuclear genome size for 'all' species of *Tulipa* L. (Liliaceae). Plant Syst Evol 281:217–245

Value Addition: Dehydration of Flowers and Foliage and Floral Craft

8

Suhrita Chakrabarty Das and S. K. Datta

Contents

S. K. Datta has retired.

S. Chakrabarty Das (✉)
AICRP on Floriculture, Department of Post Harvest Technology, F/Horticulture, Bidhan Chandra Krishi Viswavidyalaya (BCKV), Kalyani, West Bengal, India

S. K. Datta
Floriculture Division, CSIR-National Botanical Research Institute, Lucknow, India

Bose Institute, Kolkata, India

Department of Plant Science and Biotechnology, Assam University, Silchar, Assam, India

S. K. Datta, Y. C. Gupta (eds.), *Floriculture and Ornamental Plants*, Handbooks of Crop Diversity: Conservation and Use of Plant Genetic Resources,
https://doi.org/10.1007/978-981-15-3518-5_21

Abstract

More than three fourth of the export basket of Indian floricultural products comprises of dry flowers and different handmade items made from botanical specimens, presented in a dried and colored form. With the increasing awareness for natural ecofriendly products, dried flowers have attained prime importance in the floriculture industry. It constitutes nearly 15% of the global floriculture business, and considering the present COVID-19-related pandemic situation, dry flower industry is going to become more relevant. At present, the industry relies substantially on gathering of flowers from the wild and drying those using conventional methods. However, some fresh flowers are also converted into dry flowers for better returns, including chrysanthemum, dahlias, marigold, jute flowers, wood roses, lotus pods, and lilies among others. Over 70 lakh people, mostly in rural areas, earn their livelihood from production of handicrafts and related activities through low capital investment. Dry flowers and plant materials have tremendous potential as substitute for fresh flowers and foliage for interior decoration as well as for a variety of other aesthetic and commercial uses. This chapter provides all relevant and latest information, which could be helpful in drawing the attention of the researchers and scientists to work on it. Besides the entrepreneurs would be directly benefitted by utilizing the knowledge reviewed in this chapter and expected to contribute a lot to the country's economy.

Keywords

Dry flowers · Botanicals · Drying · Preservation · Glycerinization · Skeletonization · Bleaching · Dying · Packaging · Value added products

8.1 Introduction

Holding on the piece of choice is human nature. This is very natural and this basic instinct has taught us the art and science of "preservation." The beauty of flowers and their serenity tempt us to wish that the flowers should never ever fade. In spite of using best chemicals for improvement of keeping quality and enhancement of vase life, the cut flowers cannot be stored for a long time. This greatly limits the use of flowers, to overcome which different methods for preserving flowers and other plant materials by drying and dehydrating have been tried since a very long period. History depicts that preserved garlands were placed in the tombs of ancient Egyptians. Centuries later, medieval monks harvested and dried flowers and herbs for medicinal purpose. Later an Italian Monk named Giovanni Batista Ferrari discussed about drying of flowers in his book *Flora-ouer o Cultura di Fiori*. During Victorian age, the skills in floral handicrafts and displaying dried flowers were highly esteemed.

In India, the dry flower industry took off more than 50 years back (Chakrabarty et al. 2011). India shares 10% of total global dried flower by exporting more than 10,000 tons of dried flowers, annually. India's total export of floriculture was Rs. 571.38 Crores/81.94 million USD in 2018–2019 (APEDA). The export basket of the Indian floricultural products comprises of dry flowers (77.1%), cut flowers (6.1%), bulbs and rhizomes (0.8%), cut foliage (0.02%), and others (15.9%).

There is no separate code for exporting dry flowers. Mostly all the dry flowers and handmade items fit into the HS code-06042000 (Fresh Foliage, Branches and Plants, Not Having Flowers/Buds, And Grasses, Mosses and Lichens Fresh, Dried, Dyed) and HS code-06049000 (Other (Excluding Fresh) Foliage, Branches and Plants, Without Flowers Buds And Grasses, Mosses and lichens). The major importing countries are the United States, the Netherlands, the United Kingdom, Germany, and the United Arab Emirates. At present there are more than 300 export-oriented units in India, 50% of which are based in Karnataka, Andhra Pradesh, and Tamil Nadu.

The shelf life of dehydrated floral material may be reasonably long if they are protected from moisture and dust by covering in glass or plastic jars. Dehydrated flowers and foliage can be used for designing distinctive and artistic greeting cards, sweet-smelling potpourris, collages, flower pitchers, flower balls, festive decorations, wall plates, bouquets, wreaths, landscapes, etc. Any plant part including inflorescence, pods, bract, peduncle, fruits anything over the earth, which is having moisture percentage between 15% and 20% or less (after drying) is suitable for this item. Stems, twigs, branches, bark, leaves/foliage, flowers, thorns/spines, fruits, cones, seeds, roots, lichens, fleshy fungi, mosses, selaginellas, ferns, etc. can be utilized for making various value-added floral crafts and flower arrangements which are nonperishable and have longer shelf life (Raj 2001).

The main advantage of dry flower industry is that it can be operated from a rural base, using Indian floral wealth to earn valuable foreign exchange. A number of supporting industries like terracotta, basket weaving, paper, textile, metal, glass, ribbon, and packaging have an ample scope for growth as a subsidiary to dry flower industry which as a whole can contribute immensely to the development of any area. Plants may be domesticated, semi-domesticated, or non-domesticated. We use only small number of crop plants for our basic requirements. Unutilized vegetation means whose potential has not been properly realized. There is an increasing interest in neglected and unutilized vegetation to develop new products for export and domestic markets. There are a wide range of wild/unutilized/underutilized plant species which have the potential for commercial exploitation in different forms. Road sides of rural and hilly areas are covered with different types of colorful flowers and foliage at different seasons round the year, and all these are wasted under natural process. The entire seasonal unutilized vegetations can be converted into value added products by using biotechnology-based dehydration technique.

Murugan et al. (2007) classified dry flower items (finished products) under eight categories: (1) dried flowers and plant parts in bulk, (2) potpourri, (3) arrangements, (4) floral handicrafts, (5) main blooms, (6) fillers, (7) liners, and (8) exotics.

Kamal Kant (2018) suggested drying techniques as one of the best methods to preserve ornamental parts of plant especially flowers for their year-round availability, longevity, quality, novelty, easy handling, low transportation cost, as well as eco-friendly nature. The demand of dry flower items has increased in such a way that India constitutes two thirds of dry flower export of the total floriculture exports and it offers a lot of opportunities for the floriculturist, entrepreneurs, industrialists, etc. to enter in the global floricultural trade. Moreover, preparation of dry flowers has been given top priority to cope with the pandemic situation due to COVID-19. Floriculture sector being badly affected with religious places closed, weddings postponed, major events in public sector and private sector deferred, no major social and religious functions lined up, farmers are facing hardship to market their produce, and flower consumption has come to a standstill. Many farmers started discarding the flowers not only in India but also in global auction houses located in the Netherlands. As a general advisory for all the Indian farmers, ICAR recommended that it would be a good idea to dry flowers that are colorful (rose, marigold, chrysanthemum, and China aster) instead of discarding the flowers. The dried petals could be used to make eco-friendly *gulal* and other marketable products for later use (Anon 2020).

8.2 Why Dry Flower Lasts Long?

Fresh flowers, after detaching from their mother plant, act as an individual living entity undergoing all biological processes starting from cell division to photosynthesis, respiration, etc. which strongly interpret that those are living entity. As applicable for all living entities, their senescence and death also come as a natural phenomenon; that is again triggered by microbial activity and several stress conditions like lack of water and food supply (due to clogging). On the other hand, removal of the water (75–90%) present in fresh flowers through drying and dehydration results in lesser water activity and increases resistance against most of the deteriorative agents. On the course of dehydration, moisture content of flowers is reduced to a point where growth of microorganisms is prevented and chemical changes are brought, almost, to a stand still. Reduction of moisture content in the dried flowers is the main cause of increased longevity, and it is inversely proportional to the durability. Reduced moisture content is related to uniform cell contraction in the flowers resulting in rigidity, while comparatively higher moisture content after drying leads to flaccid flowers. Thus the moisture content after drying influences flower shape, but there is a chance of rapid tissue desiccation, breakage, and petal shedding during handling in dried flowers. To obtain a good quality product with reasonable firmness and keeping quality for more than 6 months, a range of 8.0–11.5% moisture has to be maintained in dried flowers. Drying below 8.0% moisture content resulted in petal shedding which might be attributed to excessive loss in moisture that might have resulted in weakened adhesion and cohesion forces in flower tissue and might have caused softening of the middle lamella leading to abscission (Singh et al. 2004).

8.3 Collection of Materials for Drying

Almost all plant materials can be dried starting from flowers, foliage, and branches to grains, cones, nuts, berries, thorns, barks, and other fruits. Materials for drying (which may be under "flower" or "non-flower" category) are collected from nature as fresh or in semi-dried form. Collecting plants when they are wet or moist from dew should be avoided. Plant materials that are without insect or disease problems are to be avoided as any damage in the bloom will be magnified in drying process. After collection, the drying process of plant materials must be started as early as possible. Selected species are either commercially grown for obtaining good quality raw material or sometimes collected from the forests also. When the raw materials are fetched from commercially grown plants, we call it "harvesting" and otherwise termed as "collection."

8.3.1 Time and Season of Harvest

Plants for drying and preserving can be collected throughout the year. Flowers at different stages of development can be picked for drying purpose. Collecting plant materials when they are wet is always to be avoided. Stage of harvesting or collection is an important criteria for dry flower production. Harvesting prior optimum stage results in deformation upon drying. On the other hand, over-matured stage causes shattering of petals after drying. The stage of harvesting for different flowers varies according to the species and the form of flower desired. Depending upon the type of products, we can classify certain parameters of harvesting.

A) **For herbs**: Herb flowers should be picked just before they are fully open. Factors that can severely reduce the quality of herbs cause discoloration are sun rays, dew moisture, and frost.
B) **For flowers and foliages**: Datta (1999) reported that flowers and foliage should be collected from the fields, 1 or 2 days after irrigation. It is found that harvesting during dry or summer months gives excellent result as most of the water gets evaporated easily. Brightest colors are produced during winter season, but the plants are very much susceptible to pests and diseases in monsoon. Flowers should be picked on dry sunny days, mid-morning to mid-day (after the dew has dried off) being the best time. Picking in too late hours should be avoided because the strong sun can cause blooms to fade. It is best to cut the flowers in the morning hours after the dew has evaporated from the plants.

The flowers should be picked when they attain their peak bloom period. If picked past their prime, when they start turning brown, flowers will continue the browning process, and no amount of after care would prevent this. After cutting, leaves from the stem are stripped, since foliage on the stems do not dry properly. Everlasting flowers such as *Helichrysum* and *Helipterum* should be cut when they feel crisp and papery and start to dry on the plant. But flowers will disintegrate in storage if they are picked too late. *Helichrysum* harvested at fully open stage took lesser time for drying than those harvested at tight bud and half-open stage (Sangama 2004). For drying purpose, roses may be harvested when the buds are just beginning to unfurl. Mostly all flowers should be picked before they are in full bloom since they will open slightly while drying. Avoid harvesting too much matured flowers as they will generally shed upon drying and will not hold up well in arrangements. Foliage should be removed from the stem when those are fully matured. Lourduswamy et al. (2001) reported that for gomphrena full-bloom stage and for French marigold and zinnia half- and full-bloom stages are ideal for production of dry flower. Bhattacharjee and De (2003) found that half-bloom and full-bloom stages of chrysanthemum, rose, and celosia are best suited for drying purpose. Safeena et al. (2006) reported that faster dehydration of flowers harvested at half-bloom stage may be due to sensitivity of the flower tissues to ethylene or other hydrolyzing enzymes and senescence also. Hemant et al. (2016a, b) reported that harvesting at 50% open stage in microwave oven embedded with silica gel resulted best quality in terms of overall dry flower quality considering shape, color, flower diameter, intactness, pigment retention, and dry flower shelf life.

C) **For dry grasses, seeds, pine cones, and most seed heads:** Materials such as dry grasses, seeds, pine cones, and most seed heads should be harvested at the end of their growing season when they have reached the full maturity, but withering has not initiated.

The following conditions/criteria are normally followed during harvesting or collection of plant materials:

- Plant materials should be collected during dry season and on sunny days. Collecting plants when they are wet or moist (from dew/rain) must be avoided.
- Flowers are to be harvested when surface moisture (dew) is evaporated but before attaining the stage of wilting.
- Plants and flowers should be free from insect and disease damage as the damage becomes more obvious after drying.
- If the field is irrigated, then harvesting for dry flowers should be postponed for a day or two.
- Sharp knives or pruning shears must be used to cut flowers and plant materials.
- Time gap between harvesting and arranging the harvesting materials for drying should be least. Under unavoidable situation, plant materials may be placed in water to prevent wilting. Some flowers may hold color better if allowed to stand in water for a few hours.
- At least 10–15% more plant materials must be collected than needed to accommodate some unavoidable loss.
- While collecting plant materials, unlawful or endangered plants should be avoided.
- Flowers or plant parts selected at any stage for drying may be sprayed with Dithane Z-78 or neem-based pesticide (0.5%) for disinfection.

8.4 Techniques for Drying Flowers

The range of dried flowers and other attractive plant parts is quite extensive, namely, stems, roots, shoots, buds, flowers, inflorescences, fruits, fruiting shoots, cones, seeds, foliage, bracts, thorns, barks, lichens, fleshy fungi, mosses, sellaginellas, etc. A number of flowers respond well to drying such as anemone, zinnia, allium, sweet William, carnation, stock, freesia, narcissus, chrysanthemum, pansy, daffodils, marigold, rose, lilies, etc. and foliages like ferns, aspidistra, eucalyptus, ivy, laurel, magnolia and mahonia, etc.

Selection of a suitable crop for drying purpose is very important for success of the business. Some flowers loose its ornamental value after drying, like how sweet pea flowers loses their color and becomes dark brown when pressed dried which is not suitable for further use. Also, quality of dry flower may vary with cultivar of a particular crop. In *Helichrysum*, especially in yellow cultivars, in spite its inherent crisp texture, petals reflex downward, and center disc florets may shed after drying. This characteristic is encountered less with varieties of other colors (Sangama 2004).

Method of drying immensely affects the quality and appearance of dried flowers and other ornamental plant parts. Different techniques used for drying ornamental plant material include air drying, press drying, embedded drying, oven drying, freeze drying, etc. The National Botanical Research Institute (NBRI), Lucknow, India, is a pioneer institute where various dehydration techniques have been developed by which fresh look of flowers, twigs, branches, foliage, etc. can be retained for several months or years.

Commercially, flowers and other botanicals are normally processed by traditional means of sun drying. However, other drying techniques like air drying, oven drying, embedded drying with desiccants, microwave oven drying, freeze drying, press drying, and glycerinization treatment have also become useful. The dried flowers and botanicals can be used for making decorative floral crafts items like cards, floral arrangements, wall hangings, landscapes, calendars, potpourris, etc. for various purposes with potpourris being the major segment of dry flower industry (around 70% or more).

Several methods are practiced for drying or dehydration of different flowers or its plant parts. In these methods, removal of moisture is done artificially either by using desiccants or controlled temperature, humidity, and airflow. The principle involved in all the methods or techniques is common in which the plant material is exposed to a vapor pressure deficit ($v.p.d. = v.p._{source} - v.p._{sink}$), which induces water vapor to move by transpiration or evaporation from the plant material (source) into the surrounding environment (sink).

The flux of water vapor (J) is proportional to the vapor pressure deficit, i.e., $J = k\ x(v.p.d.)$, where k is a constant that depends upon water vapor transfer properties of the particular product (Joycee 1998).

Some of the important methods are described below.

8.4.1 Air Drying

Air drying is nothing but hanging of plant materials after tying up with a rope/wire, in a warm, clean, dark, and well-ventilated area with low humidity (for quick drying). Humid rooms with more than 75% relative humidity (like basements) should be avoided for air drying because of mold growth which may spoil the flower. Also drying in direct sunlight should be avoided as it causes discoloration or bleaching. Air drying in dark places (like closed cellar) help the flowers to retain its natural color. While drying, plant materials are bunched together in groups of maximum ten stems, and each bunch should be of one plant variety. Of course, large flowers can be hung individually rather than in bunches. The flowers without strong stems should be wired before drying. The flowers can also be spread over blotting sheets or newspapers and kept in dark or in sun for drying (Datta 2015). Short stem and small leaves like thyme may be placed in trays in thin layer for better drying. Flowers hung in a dark area took 8–10 days for drying when there is sufficient ventilation (Champoux 1999).

Air drying is simple and cheap, without use of any chemicals or desiccants, shrinkage of petals being the major disadvantage. Also, drying period is more and such flowers naturally retain straight stems upon drying. It takes 1–2 weeks for drying depending upon the moisture content, temperature, and humidity. Drying time is more for fleshy flowers and foliages. The stage of harvest is also important for getting superior quality of dry flower in this method. Flowers of good quality at bud stage, slightly immature stage, or partially opened should be selected in this method as they continue to open while drying and foliages must be removed from the flowers before drying.

Flower color should also be kept in mind before selecting the flowers for air drying. Bright red roses attain darker shade (like dried blood color) after drying; white roses are converted to yellow parchment color. However, blue- and yellow-colored flowers normally retain their color after air drying, but pink color fades away (Dilta et al. 2011).

Dubois (2005) reported that one way of increasing the drying rate is to raise the air temperature. When temperature is more than 60 °C, several enzymes responsible for browning of tissues within the plant tissue are destroyed.

Thus to sum up, we can put that time required for air drying depends on (1) the type of plant material, (2) harvesting stage and time, (3) relative humidity of the drying chamber, (4) air circulation of the room, and (5) temperature of the air.

Weather dependence and low quality of the product due to shrinkage and drooping of petals are the major drawbacks.

Flowers like *Helipterum* (acroclinium), *Helichrysum* (straw flower), goldenrod (*Solidago*), *Gypsophila* (baby's breath), *Limonium* (statice), *Achillea* (yarrow), *Gomphrena* (globe amaranth), *Anaphalis* (pearly everlasting), *Celosia* (cockscomb), *Centaurea cyanus* (bachelor button), *Consolida ajacis* (larkspur), *Cassia fistula* (golden rain tree), *Nigella* (fennel), *Bougainvillea*, *Setaria verticillata* (bristly fox-tail), *Miscanthus sinensis* (eulalia grass), *Pennisetum setaceum* (fountain grass), *Distichlis spicata* (spike grass), *Chasmanthium latifolium* (northern sea oats), *Callistemon lanceolatus* (bottlebrush), *Amaranthus caudatus* (love-lies-bleeding), *Jacobaea maritima* (dusty miller), *Physalis* (Chinese lantern), *Stachys* (lamb's ear) and *Alchemilla mollis* (lady's mantle), *Craspedia globosa*, *Anaphalis*, *Holmskioldia*, hydrangeas, xeranthemums, *Astilbe*, *Baptisia*, *Gaillardia*, *Larkspur*, lilac, marigold, milkweed, okra, paulownia, *Polygonum*, poppy, rose, sages, *Santolina*, *Acacia dealbata*, *Anthemis nobilis*, *Delphinium ajacis*, *Gaillardia pulchella*, *Protea* sp.,

Peltophorum ferrugineum, *Tagetes* sp., *Zinnia elegans*, *Salvia*, *Artemisia*, *Chrysanthemum*, *Delphinium*, *Oregano*, *Rumex* thistles, etc. could be easily dried by air drying (Datta 2015; Geetha et al. 2004, Raj 2006; Singh and Kumar 2008).

8.4.2 Press Drying

First reported in 1820, press drying is another common, easy, and inexpensive method of drying. In this method, the flowers and foliage are kept in blotting sheet/newspaper and pressed dried with the help of "plant press" or any heavy object. The plant press is made up of two wooden boards fixed with nuts and bolts at four corners. The size of plant press may vary (6 "X 12" to any desired size). Sheets could also be put one above the other, and corrugated boards of the same size are placed in between the folded sheets to allow the water vapor to escape. Collected leaves and flowers are kept between blotting sheets, and one type of leaves/flowers is always pressed in one sheet. All blotting sheets containing leaves/flowers are kept between two ply boards and tightened with nut and bolt. The materials may be kept at room temperature for dehydration. Drying room should be warm and dry. Blotting sheets are changed every third and fifth day to avoid fungal effect/contamination. This helps maintenance of original color of flowers and leaves.

After 24 h the bundle may be removed to an electric hot air oven for 24 h at 40–45 °C, for quicker drying. Press drying in oven at 35–39 °C for 48 h is optimum for pansy whereas 24 h for the leaves of silver oak, thuja, adiantum, and nephrolepis.

Placing the foliage between two pieces of wax paper and pressing medium hot iron easily preserve the flexibility and colors of foliage. New piece of wax paper must be used for each pressing.

Most flowers and leaves are suitable for pressing except those with bulky centers such as succulents and odd-shaped flowers such as daffodils. In that case it may be cut in half and opened out before pressing. Thick flowers like chrysanthemum need to have the calyx reduced in thickness; single petals can also be dried and reassembled when making craft. Time required for pressing varies with type of flowers and water content of tissue; however, it should be completed within 4 weeks, mostly requiring 1–3 weeks under ambient condition. Foliages (ivy) dry within 1 week and flowers like gerbera and chrysanthemum dry in a maximum of 2 weeks.

The drawbacks are that the finished material is rather more two dimensional than three dimensional and very brittle. The shapes of the flowers cannot be maintained as it becomes flattened, because the fresh material after pressing within the iron or wooden frame tends to stick to the paper. Pressing works best with leaves that are naturally flat. Press dried samples are used in card making, wall hangings, batches, stationary candles, etc.

Dehydration through press drying has already been standardized both for wide range of seasonal cultivated flowers/flower petals/foliages and unutilized rural, road side, and hill side flora: *Acalypha*, *Crocus*, pansy, *Alyssum*, daffodil, *Phlox*, anemone, daisy, *Primula*, azaleas, *Delphinium*, heather, bleeding heart, butterfly weed, heath, *Celosia*, *Bougainvillea*, *Ixora*, *Jarul*, *Caesalpinia*, *Lantana camara*, *Panicum*, *Mussaenda*, *Radhachura*, *Euphorbia hirta*, *Triumfetta rhomboidea*, *Polygonum*, *Oxalis*, *Cycas*, *Cleome viscose* and *Cleome rutidosperma*, *Desmodium gyrans*, *Mikania cordata*, *Atalantia* sp., *Oplismenus compositus*, *Hemigraphis hirta*, *Ipomea tridentate*, *Hemidesmus indicus*, *Vitex negundo*, *Teramnus labialis*, *Ziziphus oenoplia*, *Limonia acidissima*, *Cleome rutidosperma*, *Peperomia pellucida*, *Sida rhomboidea*, *Morus alba*, *Tephrosia purpurea*, *Scoparia dulcis*, *Phyllanthus simplex*, *Sapium sebiferum*, *Vitis* sp., *Merremia tridentata*, *Phoenix paludosa*, *Triumfetta rhomboidea*, *Boerhavia repens*, *Pouzolzia hirta*, *Prosopis juliflora*, *Ageratum conyzoides*, *Commelina benghalensis*, *Alysicarpus bupleurifolius*, *Urena lobata*, *Spilanthes calva*, *Cestrum fasciculatum*), *Adiantum*, *Selaginella*, *Candytuft*, *Chrysanthemum*, *Lantana*, *Rose*, *Statice*, *Zinnia*, *Verbena*, *Euphorbia*, *aster*, butter cup, geranium, marigolds, Queen Anne's lace, coral bells, lily, hardy geranium, bell flower, African violets, *larkspur*, *Hibiscus*, *Ixora*, nettle leaf, velvetberry, *Aster*, *Pentas*, *Bougainvillea*, *Plumeria rubra*, *Melia*, *Cesalpinia*; leaves like thuja, ferns, silver oak, blue gulmohar, thuja, and cockscomb; spiky leaves in iris, *Montbretia*, etc.; and a large number of unidentified materials (Geetha et al. 2004; Datta 2015; Mir et al. 2009).

8.4.3 Embedding/Drying Flowers by Using Desiccants

To maintain the three-dimensional structure, plant materials are dried in a desiccant by embedding. Embedded drying is one of the methods of flower dehydration useful for delicate flowers with high moisture content that shatter or misshapen when air dried. Also, this technique is advantageous to produce exquisite life to flowers in both form and color. But it is a costly method, and desiccated flowers are more fragile and vulnerable to atmospheric moisture.

Containers such as earthen pots, dark trays, microwave safe open containers, candy tins, plastic containers, large mouth jars, or any other with a tight-fitting lid are being used for embedding. These containers should be deep enough to accommodate the plant material without disturbing its shape and form. For embedding, the bottom of the containers is filled up to 5 cm. with the embedding materials, upon which stalks of the flowers are inserted upright. The entire material is covered with desiccants, and during the process sufficient care should be taken to protect the floral structure. After embedding the containers can be kept in room or can be exposed to sun on a regular basis or can be kept in oven (electric oven or microwave oven) for faster or rapid drying. Thus we can divide the process of "embedding" under the following categories:

1. **Sun** drying after embedding
2. **Oven/hot air oven** drying after embedding
3. Microwave drying after embedding

Desiccant is the material which removes moisture quickly from ornamental flowers or plant parts embedded without reacting with water vapor released during drying. Material which is used for embedding and drying flowers and foliage should be fine. It should not chemically react with floral parts. Sand, borax, alum, silica gel, yellow corn meal, etc., are such desiccants which may remove water from the flowers more rapidly than air drying and also retain their natural forms. Besides, borax + sand mixtures, fresh kitty litter, white cornmeal and borax mixture, silica gel + sand mixture, etc. can also be used as desiccants for drying flowers. Among these two most satisfactory desiccants used are sand-borax mixture and silica gel. Others such as kitty litter, perlite, saw dust, corn starch, and corn meal can be used but those are not reliable. This method of drying is usually preferred over air or oven drying as it reduces the problem of petal shrinkage. The desiccants can be reused provided they are free from any particle of dried flowers and thoroughly dried. The ideal size of desiccant should be 0.02–0.2 mm or 20–200 mesh (Raj 2014).

During desiccation the moisture of the plant material is absorbed by the surrounding desiccants that also support the flowers/foliage from all around and, thus, maintains original shape, color, and size of flowers for a long time. Drying with sand provides smooth petal texture, silica gel provides slight roughness, and borax causes fading of color and rough texture of petals. The flower quality is very well maintained with silica gel and sand regarding post-drying parameters and longevity (Malakar et al. 2019). Regarding pretreating the flowers with some chemicals, Sindhuja et al. (2016) reported pre-drying treatment of magnesium chloride (10%) for 5 h. could improve the quality of carnation flowers after drying by embedded in borax + silica gel (1: 1, v/v) mixture and dried in hot air oven at 55 ± 1 °C considering color, shape, texture, brittleness, overall acceptability, and total carotenoid content.

Drying of Plant Materials in Sand as Embedding Material

Drying with sand is one of the least expensive methods of dehydration of flowers. Flowers or leaves may be embedded in white silver sand in metallic, plastic, or earthen container at room temperature and allowed to dry naturally in shade and well-ventilated place. Fine white sand found on the seashore (river sand), clean, dry, and preferably salt-free, can be used for embedding because of its easy handling, availability, and no reaction with water vapor. Sand from the river or beach should be washed and dried. The fine sand does not react with the water vapor released during the process of drying like silica gel and borax. It allows the water vapor to escape into the air freely, thereby causing minimum loss in the size of flowers which is otherwise maximum in silica gel-embedded flowers. Since sand is heavier, it takes a longer time for drying than the other desiccants. It usually takes 4 days to 2 weeks for drying.

Flowers and leaves like anemone, black-eyed Susan, butterfly weed, chrysanthemum, corn flower, delphinium, gladiolus, rose, and pansy are suitable for embedded drying in sand along with asters, carnations, chrysanthemums, gladiolus, geranium, coreopsis, cosmos, tulips, and zinnias. Color retention is better in the flowers embedded in sand and dried under shade compared to oven-dried flowers of *Nerium oleander*. Among the various desiccants used to dry Indian blue water lily flowers, fine white sand was the best (Geetha et al. 2004). Color and structure of floral parts show no change after dehydration with sand and hot air oven but need careful handling. Total chlorophyll stability shows significant negative role. Enzyme stability examined in dehydrated floral parts was also significantly negative. But injury index shows significant positive association with per gram dry weight. Physiological parameters like diameter, length, color, and texture varied with each plant part.

Although sand is cheap and results better quality of dried flowers, it takes longer time for drying due to its heavy weight which sometimes destroy the specimens or tends to flatten flowers unless used very carefully. However, if necessary, it should be used in combination with cornmeal or borax for embedding purpose.

Drying of Plant Materials in Borax as Embedding Material

Borax is the least expensive and best-suited embedding material for drying delicate flowers which are less stiff than those preserved with air drying. Due to hygroscopic nature, borax shows slight fading of color (bleaching) and rough petal texture (brittleness) and sometimes burns the flower petals if embedded for a long time. To prevent this, washed, sifted fine sand mixed with two parts of borax can be used as desiccant to preserve flowers.

In borax mixtures flowers take longer time to dry than in silica gel, and it is difficult to dry delicate flowers with high water content such as rose buds. The use of borax for preserving flowers has an advantage in that the flowers hold their shape with minimum shrinkage, well-acceptable color, smooth petal texture, and less mechanical damage during handling. Generally, color of the flowers is assured except pinks and reds which may vary. But if the flowers remain in borax too long, they become brittle and may lose their petals. Rose, aster, carnation, marigold, dahlia, larkspur, geranium, zinnia, chrysanthemum, and delphinium are considered suitable for drying in borax.

Drying of Plant Materials in Silica Gel as Embedding Material

Among the different embedding materials tried, silica gel (60–120 mesh) is the best absorbent for removing moisture from flowers, and it prevents shrinkage of flowers and degradation of coloring pigments that could take place when petal tissues are exposed to high temperature. It is reported as the best desiccant for getting excellent quality of dry flowers that retain color and shape as it does not cause bleaching or brittleness to flower petals even if embedded for a long time. It can be used with all sizes of flowers; crystals can be ground down in a liquidizer or coffee grinder to a very fine powder and will dry even the smallest flower, enabling to keep their shape. The crystals leave less residue on the flowers and produce less air contamination. Self-indicating silica gel is blue in color when dry and becomes pink/white after absorbing moisture. It can be reused by warming up in oven (for 30 min at 250 °F) till the crystals turn blue in color again.

Silica gel, a xerogel of silicic acid, granular in shape, is composed of a network of interconnecting microscopic pores which attract and hold moisture by physical adsorption and capillary condensation, thus, acting as a dehydrating agent. It is suggested that silica gel can absorb large quantities of moisture (up to 40%).

Most flowers kept in silica gel dry within 36 to 72 h at 250°-300 °F (baby's breath, celosia, rose buds, rose, anemone, aster, dahlia, salvia and larkspur, aster, balsam, carnation, chrysanthemum, gladiolus, geranium, coreopsis, cosmos, tulip, and zinnia). Generally thin-textured flower or leaf takes 2 days, medium-textured takes 3–4 days, heavy-textured type takes 5–7 days, whereas in other desiccants thin-textured flower or leaf takes 4–5 days, medium-textured type takes 6–9 days, and heavy-textured varieties take 10–14 days.

Most of the flowers appeared almost fresh when dried in silica gel, although the color may be darkened to some extend. Colors that come out close to the original when dried in silica gel are white, yellow, lavender, and blue (non-roses). Darker colors such as red, deep pink, and orange tend to turn even darker (Safeena 2005; Dilta et al. 2011; Shankari and Anand 2014; Datta 2015; Shailza et al. 2018). Dried flowers readily absorb moisture from the air which can be avoided by placing in a closed container with the silica gel.

Mixture of More than One Embedding Materials

The proportion of the borax with sand mixture can vary with the color of flowers, ranging from a ratio of 1:50 to 1:1. For drying of rose, carnation, and gerbera, it is found that sea sand with a low proportion of borax for 10–15 days resulted in better drying. It is reported that flowers preserved in a mixture of borax and dry sand in the ratio of 1:1 hold their natural shapes better, shrinkage is minimal, and color retention is generally better. It is suggested that a mixture of borax and sand (2:1) will take 10–12 days to make the flowers papery and dry but less stiff than those preserved with "hang and dry" method, but the particles tend to cling to some flowers. In some cases, the sand because of its rough edges may produce small holes in the petals (Pertuit 2002a).

Borax and alum being light in weight could also be used for dehydration of flowers in combination with sand or corn meal for embedding, and borax combinations can be placed directly in a warm dry place without covering. Trinklein (2006) suggested that mixture of borax and cornmeal is better than mixture with sand as cornmeal is lightweight that makes flower boxes easier to handle and has less tendency to flatten flowers. Flowers that dry well in sand and borax include asters, cornflower, cosmos, liatris, daffodils, lupine, dahlias, daisy, delphinium, pansies, candytuft, dandelion, peony, rose, carnations, geranium, chrysanthemum, gladiolus, stock, coleus leaves, gloriosa, snap dragons, tulip, zinnias, etc. Drying will normally take 2–3 weeks. Malakar et al. (2016) reported that for foliages like *Araucaria cunninghamii*, *Thuja orientalis*, and *Juniperus chinensis*, silica gel and white sand+ microwave oven combination for 30 and 20 seconds was the best.

8.4.3.1 Sun Drying After Embedding

Sun drying is the most common method practiced in India. Flowers like small zinnias, marigold, pansies, and chrysanthemum embedded in sand in an upside down fashion and kept in the sun dry in 2 days. *Gomphrena*, zinnia, and French marigold take 3–4 days for sun drying. Flowers like gerbera, zinnia, and chrysanthemum dried well with minimum shrinkage when sun dried after placing them in a box containing sand which takes 4–5 days. A layer of desiccant (dried silver sand) at the bottom of the container is placed, and flower stems and wires are pushed into it and placed in sun for drying. Marigold, poppy, zinnia, chrysanthemum, acroclinium, and globe amaranths can be sun dried easily (Lourduswamy et al. 2001; Sujatha et al. 2001; Ranjan and Misra 2002).

The greatest advantage of solar drying is that it is cheap. All that is needed is a black painted tin shed or a black plastic tunnel. An efficient solar dryer can reduce drying time to less than 3 weeks. Even drying in solar dryer may be at par with hot air oven at 45 °C or microwave oven (450 Wt), thus an economically cheaper method of dehydration. In this context solar cooker and solar dryer have become very effective. Flowers can be directly embedded in the container of solar cooker and dried under the sun. The time of exposure varies according to day temperature. The solar cooker can also be operated by using electricity, and it could be the most suitable technique for rural women as they can cook food in a solar cooker and utilize it for dehydration for rest of the time. Drying in a solar dryer after being embedded in sterilized sand is found better as compared to other methods of drying like air drying, sun drying, and mechanical dehydration (Wilson et al. 2015). A hybrid solar dryer is developed by Padmapani et al. (2019) which successfully dried rose and marigold flowers in terms of anthocyanin and carotenoid content, shape, texture, brittleness, intactness, color, and percent reduction in diameter.

8.4.3.2 Oven/Hot Air Oven Drying After Embedding

Oven drying or hot air oven drying is also an important method of drying flowers and foliages. Drying is faster and quality of product is superior in oven drying. NBRI standardized the method of oven drying on *Helipterum roseum*, chrysanthemums, small flowered perennial, candytuft, *Dombeya* spp., gerbera, *Gomphrena globosa*, strawflower, *Limonium* spp. China aster, larkspur, rose, *Zinnia linearis*, bougainvillea, narcissus and dahlia, Gladiolus and large-flowered rose cvs, marigold and *Nymphaea* spp. (water lily), and foliage plants (Datta 2015).

From the reports of different scientists, some basic facts can be jotted:

- It takes lesser time for drying in oven as compared to shade drying for the same flower because at higher temperature the rate of transpiration is comparatively much higher. Diffusion pressure deficit of air increases with increase in temperature that stimulates diffusion and vaporization of internal moisture leading to high moisture loss (Mayak and Halevy 1980).
- The best temperature range is 45–50 °C for 24–72 h in oven depending upon the type of flower. The drying temperature varies from species to species and plant to plant.
- Drying temperature and duration vary with plant size, structure, and moisture content and stage of harvesting of the material.

- Half-opened flowers or flowers at bud stage are suitable for drying in hot air oven, while fully opened flowers are not suitable as their petals lose elasticity and peel off easily on drying.
- After dehydration, yellow flowers retain their color properly, but white becomes off-white, and red, blue, and other bright ones become considerably dark. The higher the temperature, the faster is the dehydration as well as degradation of pigments, viz., chlorophylls, carotenes, and xanthophylls. The anthocyanin content of flowers increases with increase in temperature which results in darkening of flower petals.
- The best advantage of oven drying method is that the process of drying is comparatively faster than air, water, and embedded drying, but if the temperature of hot air oven is more than 50 °C, the flowers will shrink.

Geetha et al. (2004), Deepthi and Kumar (2008), and Datta (2015) listed the flowers suitable for oven drying technique among which gerbera (*Gerbera jamesonii*), dahlia (*Dahlia variabilis*), chrysanthemum (*Dendranthema grandiflora*), China aster (*Callistephus chinensis*), statice, bougainvilleas (*Bougainvillea* sp.), marigold (*Tagetes* sp.), *Larkspur* (*Delphinium ajacis*), acrolinium (*Helipterum roseum*), *Gladiolus* sp., *Gomphrena globosa*, *Ixora coccinea*, *Narcissus* sp., *Nymphaea* sp., *Rosa* sp., *Zinnia linearis*, silver fern, golden fern, *Adiantum*, etc. are most prominent.

8.4.3.3 Microwave Drying After Embedding

Microwave drying takes only a few minutes and provides material that looks fresher and more colorful than that obtained by other methods. In microwave oven, the water molecules present in the sample are agitated by electronically produced microwaves which help in liberation of moisture, finally leading to drying. Microwave energy has the peculiar attribute of being preferentially absorbed by water and hence is a particularly efficient energy source for the process of drying. Drying is exceptionally fast in microwave oven, as it gets completed within a few minutes and generates little heat.

Only glass, paper, or special microwave containers should be used to hold the flowers and desiccants. Also placing cardboard box on an upturned saucer is recommended to raise the height so that the moisture can escape from its base. It is better if only one flower is dried at a time because the less heat the flower subjected to, the better color retention will result. Flowers can be kept into a plastic container filled with silica or just lay in between sheets of paper towel.

A cup of water in the microwave before drying is useful to prevent excessive drying which is a major threat in this case. After drying in the microwave oven, flowers must be left in the drying agent for a few hours for getting good appearance and color to the flowers. Thus, containers with flowers after taking out from microwave should be kept for a particular period of time at room temperature so that the moisture evaporates and the plant material is fully dried. This is called as "setting time" or "standing period" which varies from species to species. Also petals of dried flowers in microwave oven may be sprayed with hair spray or lacquer to prevent the absorption of air moisture. If the microwave oven has about 10 settings, using the setting of 4 (that's about 300 watts) is suggested. If the microwave oven has a defrost setting, we may use that (about 200 watts). It takes about two and a half minutes to dry flowers in a half pound of silica gel. The best way to determine the length of time required to dry flowers is to use a microwavable thermometer that is placed into the silica gel about a half inch from the covered plant material. When the temperature of silica gel reaches about 160 °F, the drying process is completed (Pertuit 2002a, b).

Microwave oven drying is not suitable for all flowers as some dried flowers are susceptible to breakage. It is the best for flowers with many petals such as marigold, rose, carnation, China aster, chrysanthemum, and zinnia and flowers with cluster of florets such as goldenrod, gypsophila, corn flower snapdragon, larkspur, acroclinium, ixora, candytuft, etc. Among nontraditional flowers, *Wedelia trilobata*, *Mussaenda luteola*, and *Cassia glauca* dried by microwave were also successful. Flowers with thin, delicate petals or those with hairy and sticky surfaces are not much suitable for drying in microwave. Also flowers with thick petals or high water content such as magnolia, hyacinth, and orchids do not dry well in microwave. Based on the studies conducted at NBRI, Lucknow, the time taken for drying a variety of flowers and leaves in a microwave oven ranges within 30 seconds to 10 min. Heating time in microwave oven for carnation is 2.5–3 min, for daffodil and violet 1.5–2 min, for pansy and rose 1.5 min, for sunflower 1.75 min, and for zinnia 2–2.5 min. Drying in microwave oven depends on the thickness of the flower. Roses take 2.5 min heating time and overnight standing time; zinnia, chrysanthemum, and marigold take 1.5 min heating time and 10 hrs standing time; large chrysanthemum takes 3 min heating time and 36 h standing time; and orchids take 1.5 min heating time and 24 h standing time. For *Hibiscus* and *Bougainvillea*, embedding in borax is suitable for drying at 2.5 min with flexibility and without brittleness. On an average, time required for dehydration in microwave oven varies from 1 min to 4 min and setting time from 24 h to 36 h. Among different combinations of embedding materials and drying methods, flowers dried in microwave oven embedded with silica gel scored highest for retention of color, shape, texture, and overall appearance. Microwave oven drying of rose flowers at 50 °C with silica gel was found the best from the reduction in size point of view. Microwave oven drying is the best method for attaining highest dry weight, lowest moisture loss, and least change in pigment levels, enzyme levels, epicuticular wax, and percentage of leachates retaining its total anthocyanin content, total phenolic content, and antioxidant activities (Biswas and Dhua 2010; Safeena and Patil 2013; Datta 2015; Acharyya et al. 2017; Arunkumar and Mangaiyarkarasi 2019; Ullas et al. 2018 and Preethi et al. 2019).

8.4.4 Freeze Drying

Freeze drying is the most advanced and effective method of flower preservation. The technique was originally introduced in 1813 by William Hyde Wollaston to the Royal Society in London. Procedures for freeze drying have been standardized for very few flowers till date. It relies on the principle of sublimation, where ice held under conditions of partial vacuum (less than 4.58 torr) and low temperature (less than 0 °C) evaporates on heating without going through a liquid phase. Absence of liquid water during the dehydration process discards many undesirable chemical reactions and helps in better retention of shape, natural color, and even fragrance (Dubois and Joyce 1989). It requires a special freeze drying machine where flowers are first frozen at −10 °C by placing in a refrigerated chamber for at least 12 h. Then vacuum is created in the chamber, leading to sublimation (transformation of solid to gaseous state bypassing the liquid phase) of moisture in the flowers. A vacuum pump slowly pulls the water out of the flowers as a vapor in one chamber which condenses as ice in another chamber. Freeze-dried flowers are allowed to warm up slowly to room temperature. The full drying cycle takes 5–9 days.

Chen et al. (2000) reported that vacuum-drying temperatures had more effect on the flowers than freezing time. Lower vacuum-drying temperatures resulted in flowers with color closer to fresh and control flowers, while higher vacuum-drying temperatures resulted in lower moisture contents and stronger/stiffer petals but more changes in color.

Major flowers dried by this method are roses, carnation, bridal bouquets, etc. For carnation freeze drying at (−)20 °C for 7 days (Bhattacharjee and De 2003) and for rose (cvs. Tineke, Golden Gate, Saphir, and Rote Rose) freeze drying for 14 days (Sohn et al. 2003; Behera 2009) are recommended. Shrinkage was observed in some cases, but the color remained similar to fresh rose. Freeze drying was standardized in China rose flowers (Liang et al. 2005) and pansy also (Deepthi and Kumar 2008).

Sireesha and Reddy (2016) reported that for freeze drying technique in orchid, application of preservative (a blend of Sprite and bleach) in lukewarm water (43–45 °C) during hydration process followed by pretreatment with base composition III and posttreatment by application of dried material preservative (DMP) as sealant was found to be effective for retaining inherent qualities of the flowers.

Fernandes et al. (2018) reported that freeze drying may be applied to produce dried borage flowers for infusions, while alginate coating is a promising treatment to increase shelf-life subject to further development.

Freeze-dried flowers are used to make open baskets, open wreaths, open bouquets, etc. Bridal bouquets could be preserved by this technique of drying without any damage. High cost involvement and lack of precise processing techniques are the major disadvantages of freeze drying. The initial costs of equipment investment, electrical energy consumption, and maintenance are relatively higher than those for other drying techniques; however, the quality of the finished floral product is almost doubled. Freeze-dried flowers are fragile, and care should be taken while handling and in their storage. The cartons containing dry flowers should be strong enough to protect them during transportation.

8.5 Factors Affecting the Rate of Moisture Loss During Drying

The rate of moisture loss during drying is primarily dependent upon the following factors:

- **Type of plant materials**: Selection of plant materials is the most important point for production of successful dried flowers to be kept for long duration, because the basic botany of the plant material, whether flower or foliage or any other plant parts, and the type of flower, its color, volume, petal types, thickness, moisture content, type of tissue (the more lignified is the tissue, the lesser is the distortion after drying), and finally stage of maturity – everything – matter. Scientists have worked on many items and standardized the protocol, but there are lots to be covered. Degradation of pigments as well as loss of water from the cells during drying results in change of color of flowers into brown which is not suitable for future use. It is reported that the red flowers turned to dark purple or bluish, pure blue acquired lavender or purplish color, pink color changed to red while magenta turned to lavender, but yellow and orange colors were usually well preserved and also intensified well after drying. Usually, the fading flower color was less in embedding method. Flower color after dehydration not only depends on category of pigment but also on internal or ultrastructural factors. Therefore, flower color as well as method of drying should be taken into consideration for getting excellent quality of dry flowers.
- **Temperature for drying**: According to thermodynamics, amount of heat transferred (H) from any surface is directly proportional to the temperature ($H\alpha t$). It indicates that the higher the temperature, the more rapid is the moisture loss. However, for living materials, though time taken for drying is definitely less in higher temperature, there is often a chance of charring of cells in high temperature which invariably reduces the quality of finished product. Therefore, it is advised that plant materials must be dried in a temperature range of 40–50 °C for obtaining a quality product.
- **Humidity**: Increased humidity in the air surrounding the item placed for drying always hinders the drying process. Wet air or the air containing higher moisture definitely has less power to absorb further moisture from the product. Besides, humidity affects the drying process by reducing the air temperature and preventing crispness of the surface of the product.
- **Airflow**: Absorbance of moisture from a product is primarily an evaporation process, which is directly proportional to the velocity of surrounding air. Actually it is not the air itself responsible for drying but the heat conducted by the air which helps in evaporation. That is why circulation of air is an absolute need for ready transmission of heat during dehydration. Hot air heats up the embedding material which in turn conducts the heat to the flower or foliage, liberating moisture in the form of vapor to be carried away by air again to the outside atmosphere.
- **Embedding material**: Embedding materials can be defined as the substances having a higher capacity of moisture absorbance which can act as a desiccant and is used for drying flowers and foliages in a better way. It can also be termed as drying material or desiccant. In general the embedding material should possess certain characteristic features like:

  Fineness – The finer the particles of embedding materials, the easier is to fill the crevices and cavities of flowers and foliages so that those are completely covered leaving no gaps. In this way the flowers and foliages are in closest contact with the desiccants, which ensures uniform escape of water from their surface and ensures preservation of the shape after drying. The ideal size of desiccant should be 0.02–0.2 mm or 20–200 mesh.

Reactivity – As a part of the process, it is obvious that water vapor will be liberated from the materials undergoing drying and get transferred to the embedding materials. It is very important that the embedding materials should never react either chemically or physically (lump formation) to this excreted water. Otherwise the entire drying process will be hampered. Besides the embedding material should inevitably be free from any salt or chemicals which can, by any chance, react with the petals and foliages and discolor them.

Heaviness – If the embedding material is too lightweight, it becomes difficult to envelop the plant material properly and leaves gap during embedding. To avoid certain circumstances, heavier materials are to be used. Again, if it is too heavy (like red sand), shape of the dried flowers may not be maintained and those may be disfigured.

8.6 Precautions for Drying

The following precautions may be followed at different steps of dry flower production to obtain quality products:

- All the materials should be collected after dew and surface moisture is evaporated.
- Harvesting of flowers and foliage should be done 1 or 2 days after irrigation, in case of cultivated crops.
- Dry seasons and sunny days are preferred for collecting plant materials.
- Faded or disease-/pest-infected materials should be discarded.
- Materials should be dried as early as possible after plucking.
- Soft brush should be used to clean any unwanted material sticking to the dry flowers.
- Dried flowers and foliage should be handled very carefully after dehydration.
- Dry flowers absorb moisture from the atmospheric very easily and tend to lose their shape. Therefore, they should be stored in moisture-proof containers immediately after drying. Different containers, like glass desiccators, tin boxes, and carton, wrapped with plastic sheet or wax paper, are used for storage.
- Small quality of silica gel should be placed inside the container to absorb the moisture if any.
- Dried items should be protected from light and direct sun light to preserve color.
- These should be handled very gently and carefully as these are very brittle and fragile.
- Dust particles not only spoil the beauty of flowers but also reduce storage longevity period. Hence storage containers used for dry flowers should always be free from dust particles.

8.7 Suitability of Technique

Dehydration technique has been standardized for a wide range of cultivated flowers, grasses, ferns, ornamental foliage, etc. The optimum stage, time of harvesting, and time required for dehydration vary from material to material. The technique has been extensively used for dehydration of several popular and common flowers, and several examples are given in Table 1. Wang (2019) reported that hot air drying (HD) of chrysanthemum flowers resulted in lowest total flavonoid values due to reaction between flavonoids and oxidase during drying. Microwave treatment for 30s combined with 75 °C hot air drying was the most effective treatment in preserving biologically active compounds, resulting in higher antioxidant capacity and greater inhibition of the enzyme acetylcholinesterase (AChE). FT-IR showed that the blanching-hot air drying, microwave treatment for 90s, and vacuum-hot air drying led to loss of more compounds than freeze drying. This study also found that 3,5-di-caffeoylquinic acid (3,5-DCQA), luteolin-7-O-glucoside (LuG), luteolin, and kaempferol were the key bioactive substances inhibiting AChE. Furthermore, molecular docking studies showed that a high inhibition of AChE by 3,5-DCQA and LuG could be attributed to the formation of strong hydrogen bonds. The study may be beneficial for understanding *C. morifolium*'s nutritional value and the effects of drying methods on the quality of the final product. On the other hand, vacuum air oven drying method was observed to be the best technique for retention of flower color and related traits (Pinder and Namita 2018) (Tables 1 and 2).

8.8 Post-dehydration Care

Drying is complete when the petals are completely dry and crispness can be felt. Flowers may be removed from the embedding materials by gently pouring off the desiccant particles and air dried to complete the process. Any remaining drying medium is whisked away with a soft brush. The thickest parts are slowest to dry in all types of drying. After drying, white or clear glue may be placed at the base of some flower petals to prevent shattering. Post-drying longevity studies revealed that the microwave-dried buds embedded in silica gel exhibited a longer shelf life than buds treated with the other treatments.

Flower shape after drying is influenced by its moisture content. Lower moisture content provides rigidity, and higher moisture content results flaccid flowers. Chen et al. (2000) reported low moisture content resulted in stronger and stiffer petals in dried flowers. Mechanical support provided by embedding media, throughout the drying process, ensured well-maintained flower shape when moisture content remains below 11.55%. The moisture content of dried flowers is inversely proportional to their longevity (Pandey 2001). A range of 8.0–11.5% moisture content in the dried flowers ensures good quality, firmness, along with keeping quality of more than 6 months. Drying below 8% moisture content showed shedding effect. Excessive moisture loss might be the cause of weakened adhesion and cohesion forces in

Table 1 Suitability of plant materials for different types of drying

Method of drying	Suitability of plant materials
Air drying	*Helipterum* (acroclinium), *Helichrysum* (straw flower), goldenrod (*Solidago*), *Gypsophila* (baby's breath), *Limonium* (statice), *Achillea* (yarrow), *Gomphrena* (globe amaranth), *Anaphalis* (pearly everlasting), *Celosia* (cockscomb), *Centaurea cyanus* (bachelor button), *Consolida ajacis* (larkspur), *Cassia fistula* (golden rain tree), *Nigella* (fennel), *Bougainvillea*, *Setaria verticillata* (bristly foxtail), *Miscanthus sinensis* (eulalia grass), *Pennisetum setaceum* (fountain grass), *Distichlis spicata* (spike grass), *Chasmanthium latifolium* (northern sea oats), *Callistemon lanceolatus* (bottlebrush), *Amaranthus caudatus* (love-lies-bleeding), *Jacobaea maritima* (dusty miller), *Physalis* (Chinese lantern), *Stachys* (lamb's ear) and *Alchemilla mollis* (lady's mantle), *Craspedia globosa*, *Anaphalis*, *Holmskioldia*, hydrangeas, xeranthemums, *Astilbe*, *Baptisia*, *Gaillardia*, larkspur, lilac, marigold, milkweed, okra, *Paulownia*, *Polygonum*, poppy, rose, sages, *Santolina*, *Acacia dealbata*, *Anthemis nobilis*, *Delphinium ajacis*, *Gaillardia pulchella*, *Protea* sp., *Peltophorum ferrugineum*, *Tagetes* sp., *Zinnia elegans*, *Salvia*, *Artemisia*, *Chrysanthemum*, *Delphinium*, oregano, *Rumex*, and thistles
Press drying	*Acalypha*, *Crocus*, pansy, *Alyssum*, daffodil, *Phlox*, *Anemone*, daisy, *Primula*, azaleas, *Delphinium,* heather, bleeding heart, butterfly weed, heath, *Celosia*, *Bougainvillea*, *Ixora*, *Jarul*, *Caesalpinia*, *Lantana camara*, *Panicum*, *Mussaenda*, *Radhachura*, *Euphorbia hirta*, *Triumfetta rhomboidea*, *Polygonum*, *Oxalis*, *Cycas*, *Cleome viscose* and *Cleome rutidosperma*, *Desmodium gyrans*, *Mikania cordata*, *Atalantia* sp., *Oplismenus compositus*, *Hemigraphis hirta*, *Ipomea tridentate*, *Hemidesmus indicus*, *Vitex negundo*, *Teramnus labialis*, *Ziziphus oenoplia*, *Limonia acidissima*, *Cleome rutidosperma*, *Peperomia pellucida*, *Sida rhomboidea*, *Morus alba*, *Tephrosia purpurea*, *Scoparia dulcis*, *Phyllanthus simplex*, *Sapium sebiferum*, *Vitis* sp., *Merremia tridentata*, *Phoenix paludosa*, *Triumfetta rhomboidea*, *Boerhavia repens*, *Pouzolzia hirta*, *Prosopis juliflora*, *Ageratum conyzoides*, *Commelina benghalensis*, *Alysicarpus bupleurifolius*, *Urena lobata*, *Spilanthes calva*, *Cestrum fasciculatum*), *Adiantum*, *Selaginella*, candytuft, *Chrysanthemum*, *Lantana*, rose, statice, *Zinnia*, *Verbena*, *Euphorbia*, *aster*, butter cup, geranium, marigolds, Queen Anne's lace, coral bells, lily, hardy geranium, bell flower, African violets, larkspur, *Hibiscus*, *Ixora*, nettle leaf, velvetberry, *Pentas*, *Plumeria rubra*, and *Melia*; leaves like thuja, ferns, silver oak, blue gulmohar, thuja, and cockscomb; and spiky leaves in iris and *Montbretia*
Embedding/drying flowers by using desiccants	*Ageratum*, dahlia, lily of the valley, anemone, daisy, magnolia, bells of Ireland, *Delphinium*, marigold, black-eyed Susan, dogwood, pansy, butterfly weed, false dragon head, passion flower, carnation, fever few, peony, *Chrysanthemum*, *Gladiolus*, rose, *Coleus*, hollyhock, *Salvia*, cone flower,

(continued)

Table 1 (continued)

Method of drying	Suitability of plant materials
	Lantana, snapdragon, coral bells, larkspur, stock, daffodils, lilac, *Verbena*, water lily, yarrow, *Zinnia*
Oven drying	*Bougainvillea* sp., *Callistephus chinensis*, *Dahlia variabilis*, *Delphinium ajacis*, *Dendranthema grandiflora*, *Helipterum roseum*, *Gerbera jamesonii*, *Gladiolus* sp., *Gomphrena globosa*, *Ixora coccinea*, *Narcissus* sp., *Nymphaea* sp., *Rosa* sp., *Tagetes* sp., *Zinnia linearis*
Drying in microwave oven	African daisy, *Aster*, *Calendula*, carnations, *Chrysanthemum*, *Clematis*, daffodil, dahlia, *Dianthus*, dogwood, marigold, orchids, pansy, peony, poppy, rose, salvia, Mexican sunflower, tulip, *Zinnia*

Table 2 Drying period of flowers and foliage (Geetha et al. 2004)

Flower and foliage plants	Hot air oven drying (in hours)		Press drying (in days)
	35–40 °C	45–50 °C	
Acroclinium	–	48	–
Acalypha	–	–	14
Aster	–	48	–
Antigonal	–	48	–
Bamboo	–	–	14
Bougainvillea	48	–	8
Caesalpinia	–	–	13
Chrysanthemum	–	45–48	–
Dahlia (pompon)	72	–	–
Digera muricata	–	–	12
Digitaria setigera	–	–	12
Echinochloa colona L	–	–	12
Fern	–	–	9
Ixora	–	36	11
Marigold small	–	48	–
Marigold large	–	72	–
Mimusops	–	–	12
Mussaenda	–	–	18
Narcissus	75	–	–
Nymphaea	–	120	–
Pansy	60	–	–
Zinnia linearis	–	48	–
Zinnia lilliput	–	72	–

flower tissue and softening of the middle lamella, which ultimately resulted into abscission (Singh et al. 2004).

Packaging is a very important aspect regarding post-dehydration care. Quality reliability and continuity are major considerations when purchasing dried flower products because dried flowers and foliages are fragile and require careful handling. Before using dried materials for making decorative items, it is necessary to protect them from all possible hazards.

Dry flowers are fragile and require careful handling. Flowers dried using silica gel will sometimes reabsorb moisture and wilt; therefore it is recommended that the flowers should be stored and displayed in a closed container to keep out moisture and dust. Since dry flowers absorb atmospheric moisture and lose their shape, they should be stored in moisture-proof containers like glass desiccators, tin boxes, and cartons wrapped with plastic sheet or wax paper wherein silica gel crystals are kept at the bottom. Storage containers should be dust-free, and it should be protected from light and direct sunlight to preserve color. Also flowers dried with sand can be stored in a strong carton to protect the petals from breaking.

Placing a layer of tissue paper in flowers may reduce breakage. Spraying the dried flowers with a clear plastic spray prevents them from absorbing water during humid periods and also keeps away dust from sticking and discoloring the petals. Silica gel crystals should be kept at the bottom of the storage containers like desiccators, glass jars, or plastic jars. It helps to prevent spoilage and maintenance of better quality for their future utilization. The flowers such as larkspur, hydrangea, or sweet Annie could be made durable by using a hair spray over them and wrapped loosely with tissue paper or newspaper and laid flat in the container kept in a cool dry place. Selection of proper packaging, giving proper cushioning, and use of moisture barrier packaging materials are major factors for successful marketing in dry flower industry. Boxes should be free from insects since they chew the soft tissue and flower petals shatter making the material unsuitable. Wrapping of dried flowers in newspaper and placing them in a cardboard box also work. The box should not be stored in an unusually damp or very dry place. A few moth balls can be kept to protect from small rodents and insects. Trinklein (2006) recommended various control measures against the household insects which move into the boxes during storage. Occasional checking of the box for insects and using naphthalene flakes are suggested.

Packaging for delicate dried plant materials should be done manually during transportation and distribution. It is always advisable to purchase a superior-grade or standard cartons or boxes for packaging. Dried plant parts should be stored in dust-free area, and cartons or boxes used for storage should be cleaned from time to time.

Insect pests can afflict all types of dehydrated plant parts. Book lice, silver fish, and mice are the common pests infesting dried plant material. These could be controlled by insecticides applied in the solid pest strips (dichlorvos), liquid (synthetic pyrethroids, ethyl parathion 0.01%), or gas (methyl bromide, phenyl tablets). The most common genera of fungi, namely, *Aspergillus*, *Penicillium*, and *Rhizopus* infest dried plant material. To prevent this plant material before collection may be treated with Dithane M-45 (0.2%). Sulfur burning or sulfur dioxide fumigation also reduced these fungi during storage. Oulakh and Radha Rani (2018) reported that different display packaging materials can be used to enhance the appearance of the products and also to retain the overall quality of the dried flowers for longer period. Sharma et al. (2019) reported that maximum score was allotted to flowers which were dried in microwave oven and kept covered in paper envelopes up to 120 days in

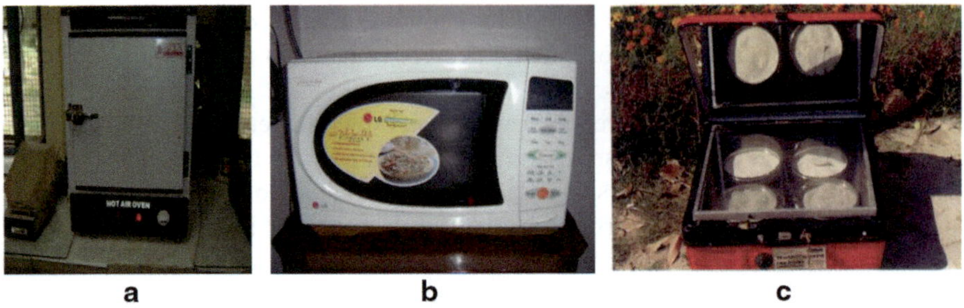

Fig. 1 Photographs of (**a**) cabinet dryer, (**b**) microwave oven dryer, (**c**) solar dryer

Fig. 2 Photographs of plant materials before and after bleaching: (**a**) *Celosia* sp., (**b**) *Gomphrena* sp., (C) *Inga dulcis*

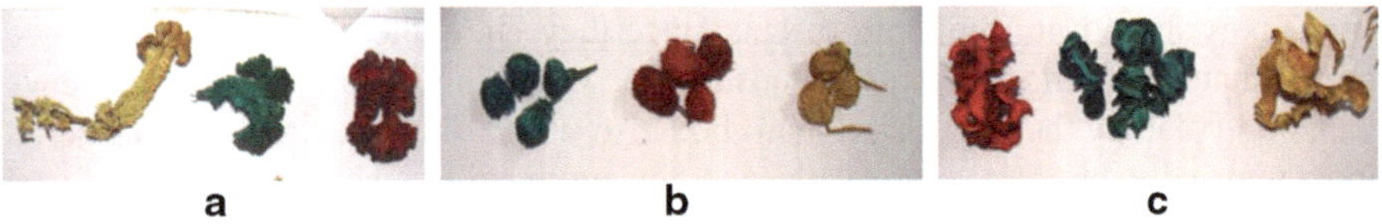

Fig. 3 Photographs of plant materials after dying: (**a**) *Celosia* sp., (**b**) *Gomphrena* sp., (**c**) *Inga dulcis*

storage. In case of dyed flowers, maximum presentability was found in flowers dyed with yellow fabric dye and stored in paper envelopes (Figs. 1, 2, and 3).

8.9 Utilization/Uses with Examples and Photographs

Both press-dried and embedded dry materials may be used for preparation of diversified value added products like bouquets, gift boxes, wall hanging, potpourris, artistic greeting cards, get well cards, wall plates, calendar, pictures, flower baskets, refrigerator magnets, mirror decoration, hats, embedding in gold/silver or resin to use as jewelry, landscape, table mats, coasters, three-dimensional arrangements of flowers for interior decoration, etc. Floral album may be prepared for identification of plants for taxonomic studies. Dehydrated flowers may be used as botanical specimens for demonstration and for teaching students. Pictures given Figs. 4 and 8.

Fig. 4 Photographs of plant materials after dying: (**a**) *Chrysanthemums* kept in glass vial after embedded drying. (**b**) Foliages pasted after press drying. (**c**) Bouquet with dried *Helichrysum* flowers and glycerinized leaves of *Grevillea robusta*

8.10 Combination of Other Botanicals with Dried Flowers and Its Potentiality

The dry flower industry deals with hundreds of plant materials for making different products like potpourri, arrangements, floral handicrafts, main blooms, fillers, liners, exotics, etc. for which other than natural flowers, lots of other botanicals are also used. Some techniques to produce those items are discussed below.

8.10.1 Preservation with Glycerine (Glycerinization)

Most of the plant materials contain more than 50% of their fresh weight as water. Upon drying, when the water content falls below 10%, brittleness becomes an issue for both handling and appearance. Glycerine drying has been used by several workers to preserve plant materials, especially foliage (whole bunches or single leaves) and retain their natural shape and flexibility. They lasted indefinitely and could be dusted or even wiped with damp cloth without risk owing to the leathery texture of leaves. Glycerine is reported as one of the best osmotic reagents, effective for dehydrating foliages as well as maintaining flexibility, shape, and texture. Thus, the preserved plant material is less brittle than dried material, making it less prone to shattering and mechanical damage (White 2007).

Many types of foliage have been successfully preserved by immersing leaves in diluted glycerol solution or placing crushed stems into it. Fresh and fairly matured foliage is ideal for glycerinizing. The resultant leaves are soft and flexible. Freshly cut statice stems may be preserved by soaking in 1:2 or 1:3 glycerol-water solution for 48 h followed by microwave drying for 1 min at medium high temperature (34 °C). A solution of 10–30% glycerol in water was found to be satisfactory for preserving most of the foliage by Dubois (2005). However, amount of glycerol accumulated by plant tissue increased with increasing concentration (from 10% to 30% v/v). Addition of 10 or 100 mM NaCI or KCI to the glycerol solution increased glycerol uptake by about 10%. Varying pH (2–8) of the glycerol solution had no differential effect on uptake. Glycerol uptake was not affected by soluble biocides like sodium dichloroisocyanurate, copper sulfate, and benzalkonium chloride. But poorly soluble fungicide mixtures like benomyl + iprodione + furalaxyl and propiconazole + propamocarb + procymidone may block the stems. Maintenance of a large vapor pressure deficit (vpo) greatly enhanced glycerol solution uptake. Thus pre-wilting the foliage for 12 h increased the initial rate of solution uptake. Fraying stem did not influence uptake of glycerol solution.

Also temperature of the solution had an impact on uptake. One part of glycerine mixed with two parts of hot water was the ideal mixture for twigs of a number of plant species like eucalyptus, hollyhock, hydrangea, etc. Addition of few drops of vegetable oil in the said solution intensified the color of immersed stems. Also mature leaves responded well to glycerine treatment as they translocated the solution readily to stems. Branches are allowed to absorb glycerine for 2–6 weeks depending on the texture and size of the leaves and branches.

Glycerine serves as a good source for microorganisms, so a pinch of antibiotic is necessary to prevent microbial growth in the dried specimens. Glycerine drying is actually replacement of cell moisture by glycerine which keeps the leaves soft and pliable for easier handling and less shedding. Thickness of the leaf is an influencing factor in glycerinization as it was found that thick magnolia leaves take longer time than a soft thin maple leaf. Though this method is most suitable for foliage than flowers, certain flowers like bells of Ireland, statice, hydrangeas, lady's mantle, narcissus, and rose hips can be used. Some plants take 30 h, while others may take 2–3 months. Actually glycerinization is the best for preserving small leafy tree branches where glycerin enters the leaves easily. Average time taken for drying is 2–3 weeks, and best results were reported during summer season when absorption is rapid and drying is completed within 2–6 days of immersion. In glycerine drying moisture in plant materials is replaced by a mixture of water and glycerine (Paul and Shylla 2002). Actively growing foliage gives the best results. Species which normally undergo cyclic growth patterns (like *Leucodendron*) may take up the glycerol solution better when picked during a growth "flush" (usually mid-summer). At other times soaking technique is better as slower rate of absorption takes place in soaking.

Also glycerinization can be effective to increase flexibility when flowers are placed in hot air oven with sand-embedded technique after pretreatment with glycerol and water mixed in the ratio of 1:5. This technique was effective in improving color, size, and flexibility in dry flowers of rose var. noblesse and *Chrysanthemum* var. yellow double. Similarly, air-dried helichrysum pretreated with ratio 1:4 glycerol was the best.

After glycerinization, though stem and leaves may turn brown in this process, they remain flexible and pliable almost indefinitely. There are two methods: glycerin uptake with the average time taken is 2–3 weeks for this treatment, and next is full-dip method, in which the plant material absorbs glycerin through the leaf surface and can be submerged in the solution done with ferns and single leaves of poplar and palmetto. The plant materials are not allowed to stand in the glycerine solution for a longer time as it results in glycerine bleeding. Sweating also occurs when there is a sudden drop in the humidity from a high level. Thus, there is drastic reduction in the water-holding capacity of dehydrating agent which then releases free water. This water cannot evaporate quickly enough and collects as droplets on the surface of the plant material where it provides an ideal environment for bacterial or fungal growth. Keeping the level of humectants in the plant tissues as low as practicable and by storing glycerine foliage at low humidity, sweating can be minimized. Marak et al. (2016) suggested that ideal concentration of glycerine best suited for preservation varied from species to species and method of treatment; however, best results were obtained in terms of texture, shape, brittleness, and overall acceptability for silver oak in 20% glycerine by uptake method. Yadav et al. (2018) reported that in *Buxus* leaves full-dip method of glycerine application was better as compared to uptake method and the best concentration of glycerine solution was 40% for drying of cut foliage. Karmakar et al. (2020) suggested that usable shelf life of *Nephrolepis exaltata* could be enhanced by processing with glycerin (40%) up to 87.33 days by full-dip method (Fig. 5).

The process of glycerinization as a whole can be jotted as:

- Normally the glycerine-water mixture is used to preserve foliage.
- This is done by placing the base of branches in a bucket or some other container of the glycerine-water mixture (normally containing one part glycerine to two parts water).
- The lower 2 inches of the branch may be crushed and placed in the jar for better absorption.
- The glycerine is drawn up the stem and moves into the leaves where it absorbs moisture.
- The process is complete when beads of glycerine begin to form on the edges of the leaves.
- Meanwhile, if the glycerine solution is depleted, supplementation with a solution of one part glycerine to four parts water is ideal.
- If the flowers seem to wilt after removing them from the solution, they need to be hung upside down to allow the glycerine to migrate to the leaf tips.
- Usually the process takes a couple of weeks. Excess glycerine is then washed off with soapy water. When dry, the leaves will be gray and pliable and can be sprayed with green floral paint.
- For best results, this method is used during the summer months when absorption is most rapid. Leaves that are thick and waxy will be dried with a soft and pliable texture, when immersed in glycerine solution for 2–6 days.

Geetha et al. (2004) suggested following plants for glycerinization:

Anthurium andraeanum, *Avena*, *Briza* sp., *Camellia japonica*, *Catharanthus* sp., *Citrus limon*, *Clematis*, *Codiaeum variegatum*, *Crotalaria selloana*, *Cyperus*

Fig. 5 Photographs of leaves preserved by glycerinization after 6 months (**a**) just after drying and (**b**) after 1 year of storage

alternifolius, *Digitalis purpurea*, *Dracaena*, *Eucalyptus* sp., *Fagus sylvatica*, *Grevillea robusta*, *Gypsophila elegans*, *Hordeum jubatum*, *Humulus lupulus*, *Hydrangea macrophylla*, *Ilex* sp., *Iris orientalis*, *Juniperus communis*, *Magnolia longiflora*, *Morina longifolia*, *Populus* sp. (Fig. 6).

8.10.2 Drying of Other Botanicals

Natural drying: The easiest and oldest method of drying used for leaves, flowers, pods, etc. is natural drying, drying in the sun. In this method, the flowers or plant parts are allowed to dry on the plant itself and collected when they are completely dried. Some of the important naturally dried plant parts are beautiful fruiting shoots of *Aegle marmelos*, *Bambusa* spp., *Cassia fistula*, *Caesalpinia sepiaria*, *Pinus roxburghii*, *Sapindus mukrossii*, etc. and seeds of *Abrus precatorius*/*Aesculus indica*, *Sapium sebiferum*, etc. were identified in the outer Himalayan regions. A number of botanicals (Table 3 and Fig. 7) were collected and dried by different drying methods by Anon (2018). The dried materials were tested for bleaching and dying, and several value added products were prepared from the said items like corner bouquet

Fig. 6 Photographs of *Helichrysum* after embedded dying in silica gel at 50 °C (**a**) just after drying and (**b**) after 1 year of storage

arrangement, stick materials, carry bags, jewelry box, greeting card, rakhi, etc. Products gave a profit margin around minimum 25–30%.

8.10.3 Skeletonization

When leaves or any other green tissues are left in the same water for quite some time (3–4 months), all their soft tissues decay, and the residual structure of leaf venation containing more lignified tissues may be recovered which is known as leaf skeleton. The network or skeleton of the original object can be bleached with a little lime and may form a nice decorative item. The home-scale process of skeletonization is:

- Leaves may be boiled for 40 min in a solution of 1 teaspoonful of baking soda or lye per quart of water.
- Then the boiled leaves are rinsed in cold water and spread on newspaper.
- The fleshy green pulps on both sides are carefully scraped off with a dull knife.
- If a lighter color is desired, the skeleton is immersed in 1 quart of water with 2 tablespoons of household bleach for 2 h.
- The venations are rinsed thoroughly and gently wiped with a clean cloth.
- Finally, those are placed between sheets of absorbent paper and pressed for minimum 2 h.

Table 3 Drying of different ornamental species and plant parts (Anon 2018)

Sl. no	Common name	Botanical name	Botanical parts being used	Method of drying	Bleaching	Coloring/painting
1.	Date palm	*Phoenix dactylifera*	Leaf and fruit bunches	Sun drying	Yes	Optional
2.	Atta fruit	*Annona squamosa*	Immature fruit	Sun drying	No	Painting
3.	Bakuli	*Lagerstroemia thorelli*	Fruit	Sun drying	Optional	Optional
4.	Broom grass	*Thysanolaena maxima*	Inflorescence	Air drying	Yes	Yes
5.	Tamarind hair	*Tamarindus indica*	Fruit hair	Sun drying	Yes	Optional
6.	Shola	*Aeschynomene aspera*	Plant stem	Sun drying	No	Optional
7.	Base of palmyra palm fruit	*Borassus flabellifer*	Fruit calyx	Sun drying	No	Painting
8.	Base of cauliflower	*Brassica oleracea* var. botrytis	Stem of cauliflower	Sun drying	Yes	Yes
9.	Siris	*Albizia lebbeck*	Pod	Sun drying	No	Optional
10.	Straw flower	*Helichrysum* sp.	Flowers	Embedding	No	No
11.	Marigold	*Tagetes* sp.	Flowers	Embedding	No	No
12.	Chrysanthemum	*Chrysanthemum*	Flowers	Embedding	No	No
13.	Bougainvillea	*Bougainvillea* sp.	Bracts	Embedding	No	No
14.	Boat fruit	*Delonix regia*	Mature fruit	Sun drying	No	Painting
15.	Small boat	*Peltophorum pterocarpum*	Mature fruit	Sun drying	No	Painting
16.	Amra pods	*Spondias pinnata*	Seed	Sun drying	Yes	Yes
17.	Arjun	*Terminalia arjuna*	Fruit	Sun drying	Yes	Yes
18.	Buddha nut	*Terminalia catappa*	Fruit coat	Sun drying	No	Painting
19.	Jarul	*Lagerstroemia indica*	Fruit	Sun drying	Yes	Yes
20.	Bel cup	*Aegle marmelos*	Fruit coat	Sun drying	Yes	Yes

21.	Casuarina	*Casuarina equisetifolia*	Cone	Sun drying	Yes	Yes
22.	Celosia	*Celosia* sp.	Flower	Silica embedded drying	No	No
23.	Coco petals	*Cocos nucifera*	Calyx	Sun drying	Yes	Yes
24.	Curly pods	*Inga dulcis*	Pod	Sun drying	Yes	Yes
25.	Curly tingting	*Cocos nucifera*	Mid-rib of leaf	Sun drying	Yes	Yes
26.	Eucalyptus pods	*Eucalyptus* sp.	Pod	Sun drying	Yes	Yes
27.	Grevillea	*Grevillea robusta*	Leaf	Press drying	Glycerine	Optional
28.	Hogla pencil	*Cyperus corymbosus*	Stem	Sun drying	Yes	Yes
29.	Jackfruit	*Artocarpus heterophyllus*	Immature fruit	Sun drying	No	Paint
30.	Cane	*Calamus* sp.	Stem	Sun drying	No	Painting
31.	BhuttaPata	*Zea mays*	Cover of the cob	Sun drying	Yes	Yes
32.	Pepal net	*Ficus religiosa*	Leaf skeleton	Press drying	Yes	Yes
33.	Mehogini spoon	*Swietenia mahagoni*	Petals	Sun drying	No	Painting
34.	Champa net	*Michelia champaca*	Leaf Skeleton	Press drying	Yes	Yes
35.	Mehogini center	*Swietenia mahagoni*	Flower Centre	Sun drying	Yes	Yes
36.	Garagara	*Coix lacryma-jobi L*	Fruit	Sun drying	No	Painting
37.	Cobra leaf	*Butea frondosa*	Leaf	Press drying	Yes	Yes
38.	Sun center	*Helianthus* sp.	Flower head	Silica embedded drying	No	No
39.	Jute stick	*Corchorus* sp.	Pith	Sun drying	Yes	Yes
40.	Coco coir/coco hair	*Cocos nucifera*	Fine coir	Sun drying	Yes	Yes
41.	Luffa	*Luffa acutangula*	Fully developed fruit, after removing the mesophyll tissue	Sun drying	Yes	Yes
42.	Nataraja	*Achras zapota*	Bark	Sun drying	No	Painting

(continued)

Table 3 (continued)

Sl. no	Common name	Botanical name	Botanical parts being used	Method of drying	Bleaching	Coloring/ painting
43.	Baramoti	*Litchi chinensis*	Immature fruits	Sun drying	No	Painting
44.	Supari	*Areca catechu*	Leaf base	Sun drying	Yes	Yes
45.	Shyama grass	*Echinochloa colona*	Inflorescence	Sun drying	Yes	Yes
46.	Lotus pod	*Nelumbo* sp.	Thalamus	Sun drying	Yes	Painting
47.	Moss	*Sphagnum moss*	Total plant body	Sun drying	Yes	Optional
48.	Banyan	*Ficus benghalensis*	Leaf	Press drying	No	No
49.	Base of palmyra palm fruit	*Borassus flabellifer*	Fruit calyx	Sun drying	No	No
50.	Base of cauliflower	*Brassica oleracea* var. botrytis	Stem of cauliflower	Sun drying	Yes	Yes
51.	Male inflorescence of Palmyra palm	*Borassus flabellifer*	Male inflorescence	Sun drying	No	No

Fig. 7 Photographs of different natural items: (**a**) *Coix*, (**b**) *Bauhinia*, (**c**) skeleton of *Michelia* leaf after dying, (**d**) lotus pod, (**e**) mehogini center, (**f**) skeleton of *Ficus* leaf after bleaching, (**g**) & (**h**) mixed items

Verma et al. (2012) reported that fully matured, healthy leaves of pipal immersed in NaOH (40%) solution could be skeletonized after 2 days. He also found out that heavy textured leaves are more suitable for this method. Also use of KOH (10%) in half a liter of water helps in removing the leaf tissues when immersed for 60–70 min. Marak and Chakrabarty (2014) reported that NaOH (40%) accelerated the process of skeletonization (removal of mesophyll layers) within 2 days with maximum visibility of veins. Heavy-textured leaves are the best choices for this method of preservation. The leaves of the ivy, the stink pod of the stramonium (which is now to be found exactly ripe for steeping), the oak leaf, *Ficus*, *Bauhinia*, *Michelia champaca*, etc. express greater beauty when skeletonized than when perfect (Fig. 8).

8.10.4 Bleaching and Coloring of Dehydrated/Skeletonized Botanicals

- **Bleaching**

Almost all the natural botanicals attain a darker shade during drying. Bleaching involves chemical process that makes the product to attain a visibly lighter shade and

Fig. 8 Plates containing different finished products prepared from dried botanicals. (**a**) Corner bouquet arrangement, (**b**) carry bag, (**c**) stick material, (**d**) pen stand, (**e**) jewelry box, (**f**) potpourri, (**g**) greeting cards, (**h**) greeting cards

can increase its ability to absorb light. Bleaching agents are chemicals that lighten or whiten a substrate through chemical action. The attractive plant material which is inherently colored by unwanted pigments or discolored due to oxidative browning can be bleached either by oxidative bleach or by reductive bleach or sulfured also. Both oxidative and reductive bleaches can be used for bleaching plant materials. Oxidative bleaches such as chlorites, hypochlorites, and peroxides or peroxy compounds tend to break down the colored compounds, while reductive bleaches such as borohydrides and sodium sulfide tend to modify them into colorless compounds. Sodium chlorite is relatively selective for lignin without damaging fiber, hence considered as an excellent bleaching agent. Among reductive bleaches, hydrosulfites (sodium or zinc hydrosulfite) are cheap and have maximum bleaching capacity. The bleached items can directly be used in various floral arrangements, or those can be dyed with different colors. Bleaching or discoloration allows the use of dyes for coloring of plant material. Without bleaching, the dying process cannot give the desired color and leads to uneven shades.

Sulfuring is also done to prevent enzymatic color change. Sulfur dioxide acts as bleaching material for colored plant material and, when used below a certain concentration, can help in fixing colors in some flowers also. Color fixation is related to acidification of the tissues. Sulfuring is produced by burning sublimed sulfur powder or by injecting sulfur dioxide gas (1–3%) into a sealed chamber. The plant materials are usually treated with sulfur dioxide overnight prior to ventilation of the chamber and subsequent completion of the drying process (Verma et al. 2012).

After bleaching with oxidative or reductive chemicals, often yellowing of the plant material may happen, which is the major problem. To avoid yellowing, multistep bleaching technique alternating with a reductive bleach results into less yellowing. A final wash in a 2% solution of barium hydroxide, calcium hydroxide, sodium bicarbonate, or aluminum sulfate may also prevent yellowing.

A number of preserved ornamental plant materials are bleached during preservation. Bleached plant materials may be recolored with dyes. Petal shades require almost total removal of color from the dried plant materials to avoid uneven dyeing. Profitability of bleaching depends upon attainment of bright white color and the efficient utilization of expensive bleaching chemicals. Bleached ornamental plant materials help in creating "contrast" effect when arranged with naturally dried or dyed items.

For bleaching dried flowers of celosia, rose, chrysanthemum, and *Plumeria alba* (small and large flowered types), 10–30% sodium chloride was very effective. Also sodium hypochlorite was the best chemical to bleach pink, globe amaranth, French marigold, and multicolored zinnia. Sujatha et al. (2001) reported that hydrogen peroxide was the best bleach at room temperature, caused the least damage to cell tissues, and was ideal for the bleaching. Dried pods of *Acacia auriculiformis, Sesamum indicum*, *Gossypium hirsutum*, and *Pongamia glabra* and cones of *Pinus* spp. exhibited minimum bleaching time of 6 h and lowest rate of damage with highest whiteness index and maximum score for shape retention when treated with 20% sodium chlorite +5% hydrochloric acid (cold water). Marak and Chakrabarty (2013) reported that bleaching with sodium hydroxide (10%) + sodium silicate (10%) + hydrogen peroxide-at 70 °C (hot) gave superior results for both pipal and champa with maximum sensory attributes. Mir and Jana (2015) reported effectivity of 20% hydrogen peroxide and "Ala"(15%) on bleaching of venation skeletons used for maximum of 2 h. Preethi et al. (2019) recommended 100% sodium chlorite for bleaching of dried plant materials like *Wedelia trilobata, Clitoria ternatea, Mussaenda luteola, Caesalpinia pulcherrima, Caesalpinia pulcherrima* 'flava', *Mussaenda luteola, Hamelia patens, Thryallis glauca, Ixora duffii, Ixora coccinea, Caesalpinia pulcherrima* 'rosea', *Saraca indica, Cordia sebestena*, and *Cassia glauca.*

- **Coloring**

The decorative value of dry flowers may be increased by external color, which may be in the form of dyes (water based) or paint (oil based). Coloring of dried flowers helps to retain their naturalness and adds more value to the product. Oil-based paints, being primarily a physical process, can be applied by aerosol spray or dipping. Dried foliage and flowers are painted by using brushes or dipping and spraying. Enamel paints used for interior decoration, poster paints, and tube paints can be employed for this purpose.

Lexically dyes are organic compounds which absorb light of 205–900 in ultra-frequencies, thus only reflecting a portion of the visible spectrum with the result that the eye sees color. Each pure dye compound has a unique color. A vast number of different colors may be developed by blending pure dye stuffs. The method of dying is actually a chemical process, and dyes are usually applied by immersion of ornamental plant parts which may be fresh, dried, or bleached. Resistance of a dye or pigment to chemical or phytochemical attack is an inherent property.

There has been significant development in organic color chemistry during the last few decades. The first ever dye (discovered in 1856) was named as Perkin's mauve, and following this, many color named indigosol, fire red, Hansa yellow color, sulfur black, etc. became popular. Basic dyes were the first of the synthetic color made out of coal tar derivatives. Although basic dyes produce brilliant colors, they have poor fastness Normally acid dyes are water solution anionic dyes, while the vat dyes are water-insoluble dyes. Acid dyes are effective for protein fibers such as silk, wool, nylon, etc. and fix to the fibers by hydrogen bonding, Van der Waals forces, and ionic linkages. In this case dyeing is generally carried out at boiling temperature for 30–60 min depending upon the depth of the shade and dyestuffs used. The wet and light fastness properties of acid dyes vary from poor to excellent, depending upon the molecular structure of the dyes. These dyes have very good leveling and migration properties along with low affinity for the fiber ultimately leading to poor fastness property, in general.

Tampion and Reynold (1971) explained three ways of dying flowers, viz., (i) by absorption (the cut stems are placed in a dye solution), (ii) by dusting the cut blooms (with powdered dye), and (iii) by dipping the cut blooms into a solute of dye. In case of dipping method, few drops of washing up liquid or surfactants may be added to the dye solution which can improve contact between dye bath solution and plant material and can increase the spread of dye molecules. Dyeing of carnation, *Chrysanthemum maximum*, star flowers, gypsophila, and hydrangea can be done by absorption method. Use of vat dyes is the best to dye celosia flowers at 0.2% concentration by cold method. To dye the dried plant parts, they also suggested the use of culinary dyes which are available in wide range of colors and nontoxic in nature. Dip dyeing and spraying are normally recommended to color the seeds and pods.

Sangama (2004) reported that by varying the dye concentration and combinations (food color and feulgen reagent), different shades of colors could be obtained in tuberose cultivars single and double. Yogita (2000) found that the basic group of dyes was the best at 3% and 0.3% for bleached roses and *Aerva* flowers, respectively. Among different dyes used, vat group was superior followed by direct and acid dyes which had low level of color fading on storage.

Dana and Lerner (2002) and White (2007) reported that fragile flowers should be dyed before drying especially when dried with a desiccant. They suggested three types of dyeing, viz., dip dyeing (grasses), spray dyeing (heavy-textured materials like pods, cone seeds), and absorption dyeing (fresh leaves with glycerine medium). In general, for good adsorption of a dye by the fiber, the later must contain acidic groups. The fastness of basic dyes after washing and rubbing is less than that for direct dyes. The products dyed with direct dyes are considered to be dull and less attractive to some extent. Preethi (2019) reported that *Clitoria ternatea* could be dyed with natural yellow dyes; *Mussaenda luteola* dyed with red, blue, and yellow natural dyes; and *Cassia glauca* dyed with natural yellow dyes where the color uptake was high and color fading low.

8.11 Making Different Products with Dried Flowers and Other Botanicals

All press-dried and embedded dried materials can be used for preparation of diversified value added products like bouquets, gift boxes, wall hanging, potpourris, artistic greeting cards, get well cards, wall plates, calendar, pictures, flower baskets, refrigerator magnets, mirror decoration, hats, embedding in gold/silver or resin to be used as jewelry, landscape, table mats, coasters, three-dimensional arrangements of flowers for interior decoration, etc. Preparation of some items are discussed below:

- **Pressed flower pictures, flowered trays, table top, and shadow boxes**: A piece of cardboard is covered with fabric or paper, and a design is sketched lightly on the front on which the pressed flowers are glued. When the glue is dried, it is covered with glass and framed as early as possible. The same for flowered trays or table tops can be prepared. Also shadow boxes can be prepared in this way with a specific planning for depth and skip covering with glass.
- **Wood panels/center piece/*pandal* decoration**: Plywood, wood, or bamboo structures can either be painted or rubbed with equal parts of turpentine and linseed oil followed by sketching a design on the surface. Then seeds, pods, dried branches, etc. are cemented upon which is finally covered with a coating of clear shellac. Centerpieces can be made in the same way.
- **Floral craft or arrangements**: Dehydrated flowers and foliage can be used for designing distinctive, fascinating, and artistic floral arrangements, bouquets, gift pack, festive decorations, collages, flower pitchers, floral balls, pomanders, wall sceneries, greeting cards, wedding cards, and sweet-smelling potpourris followed by items required for preparation of greeting cards, floral designs, pictures, landscapes, calendars, etc.
- **Corner bouquet arrangement**: These are front-facing arrangement, usually kept on side table. Size of decoration should be proportionate with the display table.
- **Stick materials**: Bleached and dyed foliages of different types may be used for making stick materials (normally made of bamboo stick of 10″, 12″, and 18″) by positioning the dried items around the stick in such a way that they form a beautiful shape. Accessories like ribbon and peeps can also be used to add beauty to the product.
- **Carry bags**: Elegant pieces of dried foliage bags can be made from dried foliages by stitching/adhering dry flower items of normal carry bags made up of jute, cloth, nylon, etc. Carry bag made with dried foliage is an art that stretches to antiquity and eternity.
- **Jewelry box**: Any leftover wooden or paper box can be painted with desirable colors, and then dried foliages were placed on them. The flowers are placed forming an exclusive design or pattern or some symbol. Writing names or messages with dried foliages can also be customized on these boxes.
- **Greeting card**: A piece of paper is cut and foliages/botanicals are arranged over the paper after applying adhesive so that when pressed, it does not come out of

flowers and leaves. This card/floral item is again placed under the glass table top for about an hour to dry, and these are kept away from moisture and dust. Some of the floral items can be framed or laminated. Various designs and patterns can be used along with different color combinations for making cards.

- **Potpourris**: Annon (2018) gave details about preparing **potpourri** containing lata ball, curly pod, coco chips, bakuli, and star flower (77.72 gram excluding package materials) with a **package** (H:13.5 cm X W:8 cm X L: 8 cm) **of** 34.78 gram. The total raw material cost for potpourri (in the year 2016) was Rs. 4.00/−. The total input cost per pack including packaging materials, labor, establishment, and others is Rs.15 per pack and sold at Rs. 40 per pack at wholesale price and with a retail price of Rs.s100/−.

Items	Quantity (pcs)	Total weight (gm)	Cost (Rs.) per pc/wt	Cost (Rs.) for measured volume in potpourri	Description	Collected from (places)
Lata ball	1	2.6	0.70/pc	0.70	Natural (size: 4 cm)	Baita, Midnapore
Curly pod	26	16.2	12.00 /kg	0.19	Natural (size: mix)	S.24 pgs., Midnapore
Coco chips	99	34.32	40.00/kg	1.37	Yellow colored	Bihar
Bakuli	2	4.03	15.00/kg	0.06	Natural, varnished	Midnapore
Star flower	59	20.57	60.00/kg	1.23	Natural	S.24 pgs., Midnapore

8.11.1 Some Novel Techniques

Processing

The principle of processing flowers is based upon replacement of internal moisture with 2-propyl alcohol or t-butyl alcohol (Romero-Sierra and Webb 1982). When these solvents are volatilized from the petals, the fresh texture of flowers is lost. Thus, a replacement technique has been developed with epoxy resin using an ascending series of acetone (von Hagens 1981), which results in complete discoloration of tissues. In "replacement technique" an ascending series of ethyl alcohol is used which causes pigments to leach out of petals during dehydration due to presence of water in the solvent. Markham et al. (2000) adopted a moisture replacement technique without using water-containing solvents. In this technique, pigments diffused but remained in the petals of dehydrated flowers. Ito et al. (2010) reported that processed flowers with natural pigments and texture can be made using normal monohydroxy alcohols and lateral chain diols. They developed a protocol where ethyl alcohol and polypropylene glycol were used as the primary and secondary soaking solvents, respectively. Primary soaking solvent like ethyl alcohol, which has a low viscosity, prevents petal shrinkage. When the petals were soaked preliminarily in some primary soaking solvents like ethyl alcohol, petal shrinkage induced by the secondary soaking solvents (polypropylene glycol) was greatly reduced. Polypropylene glycol was selected as the most appropriate secondary soaking solvent because it retained the petal shape and color of carnations (var. 'moondust velvet blue') for a long time. Such processed flowers retain natural pigmentation and textures almost similar to the fresh flowers. Considering cost, safety, and waste disposal issues, production of "processed flowers" almost similar to the natural one in texture and pigmentation seems to be a good choice.

Plant species where this protocol was successful are corn flower (*Centaurea cyanus*), dwarf delphinium (*Delphinium grandiflorum*), spiderwort (*Tradescantia ohiensis*), Asiatic dayflower (*Commelina communis*), pansy (*Viola wittrockiana*), lisianthus (*Eustoma grandiflorum*), Dutch iris (*Iris hollandica*), cockscomb (*Celosia cristata*), bougainvillea (*Bougainvillea spectabilis*), Easter cactus (*Rhipsalidopsis gaertneri*), portulaca (*Portulaca grandiflora*), four o'clock (*Mirabilis jalapa*), snapdragon (*Antirrhinum majus*), cooktown orchid (*Dendrobium phalaenopsis*), rose (*Rosa hybrida*), bigleaf hydrangea (*Hydrangea macrophylla*), daffodil (*Narcissus pseudonarcissus*), carnation (*Dianthus caryophyllus*), Iceland poppy (*Papaver nudicaule*), African violet (*Saintpaulia ionantha*), salvia (*Salvia guaranitica*), camellia (*Camellia japonica*), geranium (*Pelargonium incrassatum*), tulip (*Tulipa gesneriana*), sunflower (*Helianthus annuus*), pansy (*V. wittrockiana*), oilseed rape (*Brassica napus*), watermelon (*Citrullus lanatus*), and sweet William (*Dianthus barbatus*).

Polyset Drying It is a polymer preservation method, which is applied to the flowers and foliage about 45 min before drying. This method lessens the drying time and improves the intensity of flower color. It also minimizes shattering and wrinkling of the petals which may occur during drying. Drying of native flowering plants and their different parts with epoxy resin encapsulation is a novel approach. These are attractively embedded inside resin, avoiding dust, and beauty of native flowers can be cherished forever (Thakur et al. 2019).

8.12 Conclusion

Demand of dry flowers and floral craft is increasing day by day both in domestic and international market. Dry flower market has grown exponentially as consumers prefer it as the environmental-friendly and biodegradable alternative to fresh flowers. At present, India is a leading exporter of dried flowers with yearly exports averaging in excess of Rs.500 crores. India has rich source of plant materials year-round because of its diversity of topography and climate. Highly diversified agro-climatic conditions of India undoubtedly offer virtually countless varieties of wild flowers. Every state is very rich in natural resources of vegetations. A good amount of natural vegetations are wasted every year due to natural process. In spite of aesthetic beauty, all these natural resources are totally unutilized/wasted due to unawareness of their use. These entire seasonal colorful vegetations can be converted into value added products by using simple dehydration technique. Collection of these vegetations will not create any imbalance in nature. The technology will help proper judicial utilization of natural resources to prepare value added products. The technique has been simplified in such a way that one can learn it within 2 days. No sophisticated infrastructure is required. A cottage scale industry based on dehydrated floral craft can come up for self-employment. Such a creative occupation will help rural women to come out from their drudgery of daily life. There is large potential to develop the dry flower industry in India and to provide employment generation besides self-employment as the industry is labor-intensive. This is indeed what rural India needs. Lack of basic infrastructure and information is a stumbling block for this industry. With innovative training programs and awareness campaign, there is a lot which could be done for promotion of dry flowers' export.

References

Acharyya P, Majumder A, Malakar A, Biswas S (2017) Standardization of dehydration techniques of some selected flowers. Acta Hortic 1181:85–92. https://doi.org/10.17660/ActaHortic.2017.1181.11

Agricultural & Processed Food Products Export Development Authority (APEDA) Annual report (2018–19). In: http://apeda.gov.in

Anonymous (2018) Identification of ornamental species for preparation of value added dried products. In: Annual report of all India coordinated research project on floriculture. ICAR-Directorate of Floricultural Research, Shivajinagar, Pune, India

Anonymous (2020) Advisory to flower growers in the light of COVID 19 pandemic. In: National advisory floriculture. ICAR-Directorate of Floricultural Research, Shivajinagar, Pune, India

Arunkumar P, Mangaiyarkarasi R (2019) Microwave method for drying hibiscus and bougainvillea. Pharma Innov J 8(7):147–148

Behera TB (2009) Standardization of drying techniques of rose (Rosa hybrida L). (M.Sc.) Thesis, University of Horticulture and Forestry, Nauni, Solan, Himachal Pradesh, India

Bhattacharjee SK, De LC (2003) Advanced commercial floriculture. Avishkar Publishers, Jaipur

Biswas C, Dhua RS (2010) Microwave oven drying of cut carnation. J Ornam Hortic 13(1):45–49

Chakrabarty S, Roy S, Roy A (2011) An Entrepreneur's guide to dry flower: process and techniques in dry flower. Lambert Academic Publishing GmbH & Co. KG, Saarbrucken

Champoux J (1999) Tips and home remedies. http://www.kaepsmlin.com/tips. Accessed on 19 July 2020

Chen W, Gast KLB, Smithey S (2000) The effects of different freeze-drying processes on the moisture content, colour and physical strength of roses and carnations. Sci Hortic 84(3/4):321–332

Dana M, Lerner BR (2002) Preserving plant materials. http://www.agcom.purdue.edu/agcom/pubs/menu.htm. Accessed on 19 July 2020

Datta SK (1999) Dehydrated flowers and foliage and floral crafts. In: Bose TK, Maitti RG, Dhua RS, Das P (eds) Floriculture and landscaping. Naya Prokash, Kolkata, pp 696–703

Datta SK (2015) Dry flowers technology. Pointer Publishers, Jaipur

Deepthi S, Kumar S (2008) Dry flowers and natural splendor indoors. Floricul Today:42–48

Dilta BS, Sharma BP, Baweja HS, Kashyap B (2011) Flower drying techniques – a review. Int J Farm Sci 1(2):1–16

Dubois P (2005) Farnote 10/89: drying cut flowers and foliage. Website: www.agrc.wa.gov

Dubois P, Joyce D (1989) Drying cut flowers and foliage. Farm note No. 10/89, Western Australian Department of Agriculture, 3 pp

Fernandes L, Ramalhosa E, Casal S, José Pereira A (2018) The effect of different post-harvest treatments on the quality of borage (Borago Officinalis) petals. Acta Sci Pol Technol Aliment 17(1):5–10

Geetha CK, Rajeevan PK, Valsalakumari PK (2004) Technical bulletin on dry flowers. Kerala Agricultural University, Thrissur

Hemant U, Singh A, Ahlawat T (2016a) Standardization of dehydration technique for greenhouse cut rose var. Shakira. Indian J Hortic 73(1):99–103

Hemant U, Singh A, Ahlawat T, Palagani N (2016b) Influence of harvest stage and varieties on dry flower quality rose (Rosa & Times; hybrida). J Ornam Hortic 19(1–2):34–38

Ito H, Hayashi T, Hashimoto M, Miyagawa K, Nakamura S, Mizuta Y, Yazawa S (2010) A protocol for preparing preserved flowers with natural color and texture. HortTechnology 20(2):445–448

Joycee DC (1998) Dried and preserved ornamental plant material not new, but often overlooked and underrated. Acta Hortic 454:133–145

Kant K (2018) Drying techniques for preservation of ornamental parts of plant. Int J Sci Technol 7(5):1650–1654

Karmakar B, Chakrabarty S, Hayat A, Bagchi S (2020) Processing of Nephrolepis exaltata with glycerine to enhance shelf life by drying. Int J Curr Microbiol App Sci 9(3):348–356

Liang LY, Lai CY, Qing ZB (2005) Study on the application of freeze drying and microwave drying to cut flowers. Trans Chin Soc Agric Mach 36(1):71–74

Lourduswamy DK, Vadivel E, Azhakiamanavalan RS (2001) Research and development in dry flower technology. Floric Today 5(46):8–13

Malakar M, Acharya P, Biswas S (2016) Standardization of dehydration techniques of some Ornamental Foliages. Int J Agric Environ Biotechnol 9(4):555–562

Malakar M, Acharya P, Biswas S (2019) Effect of different Media for Dehydration of some ornamental flowers. Acta Hortic 1255:157–165. https://doi.org/10.17660/ActaHortic.2019.1255.25

Marak DW, Chakrabarty S (2013) Standardization of drying techniques for leaves of pipal (Ficus religiosa), Banyan (Ficus bengalensis) and silver oak (Grevillea robusta). Green Farming 6(6): 1388–1391

Marak DW, Chakrabarty S (2014) Value addition of Foliages for decoration. LAP Lambert Academic Publishing, Saarbrücken, Germany

Marak DW, Chakrabarty S, Singh B, Deepika (2016) Standardization of Glycerinisation for Leaves of Silver Oak (Grevillea robusta), Kanchan (Bauhinia purpurea) and Pipal (Ficus religiosa). Int J Agri Environ Biotechnol 9(5):815–820

Markham KR, Ryan KG, Gould KS, Rickards GK (2000) Cell wall sited flavonoids in lisianthus flower petals. Phytochemistry 54:681–687

Mayak S, Halevy AH (1980) Flower senescence. In: Thimann KV (ed) Senescence in plants. CRC Press, Boca Raton, p 132

Mir S, Jana MM (2015) Preparation of leaf venation skeletons of leaves for dry flower arrangement. Curr Hortic 3(1):53–56

Mir RS, Shinde BM, Chaskar MG, Jana MM (2009) Review of literature on dry flowers. Res Rev J Life Sci 9(2):52–55

Murugan PA, Thiyagarajan G, Ramesh K (2007) Dry flower technology. http://www.techno-preneur.net/information-desk/sciencetech-magazine/2007/Dec07/Dry-flower.pdf. Accessed on 19 July 2020

Oulakh B, Radha Rani P (2018) Quality characteristics of dried flowers as influenced by packaging materials. Indian J Hortic 75(3):498–499. https://doi.org/10.5958/0974-0112.2018.00084.1

Padmapani P, Sharma PK, Indra M (2019) Hybrid solar dryer for drying of high-value flowers. Curr Sci 116(9):1463–1466

Pandey PH (2001) Principles and practices of post harvest technology. Kalyani Publishers, Ludhiana

Paul D, Shylla B (2002) The art of flower drying. In: Mishra RL, Mishra S (eds) Floriculture research trends in India. Indian Society of Ornamental Horticulture, New Delhi, pp 41–46

Pertuit A (2002a) Drying flowers. http://hgic.clemson.edu/factsheets/hgic1151.html. Accessed on 19 July 2020
Pertuit A (2002b) Preserve flowers using the microwave. http://hgic/clemson.edu. Accessed on 19 July 2019. https://hgic.clemson.edu/factsheet/drying-flowers/
Pinder R, Namita (2018) Influence of dehydration techniques on colour retention and related traits of gerbera (Gerbera Hybrida) flowers. Indian J Agric Sci 88(5)
Preethi TL, Thangaselvabai T, Sankaranarayanan R (2019) Standardization of drying, bleaching and dyeing techniques for non-traditional flowers. Acta Hortic 1241:627–632. https://doi.org/10.17660/ActaHortic.2019.1241.92
Raj D (2001) Making floral crafts from forest product of the Himalayas. Indian Hortic 46(3):26–27
Raj D (2006) Advances in ornamental horticulture. Pointer Publishers, Jaipur
Raj D (2014) Floriculture at a glance. Kalyani publishers, Jaipur
Ranjan JK, Misra S (2002) Dried flowers: a way to enjoy their beauty for long period. Indian Hortic 47(2):32–33
Romero-Sierra C, Webb JC (1982) Flower preservation. United States Patent 4,349,459. U.S. Patent and Trademark Office, Washington, DC
Safeena SA (2005) Standardization of drying technology for Dutch roses. M.Sc Thesis University of Agricultural Sciences, Dharwad, Karnataka, India
Safeena SA, Patil VS (2013) Effect of hot air oven and microwave oven drying on production of quality dry flowers of Dutch roses. J Agric Sci 5:179–189
Safeena SA, Patil VS, Naik BH (2006) Standardization of stage of harvest for better quality of dry flowers of rose. J Ornam Hortic 9(3):224–226
Sangama (2004) Dehydration and product diversification of Helichrysum flowers. J Ornam Hortic 7(3–4):376–380
Shailza, Shalini J, Grewal HS (2018) Emerging prospective of floriculture industry: drying of ornamental plants and their parts. Int J Curr Microbiol Appl Sci 7(7):1619–1633. https://doi.org/10.20546/ijcmas.2018.707.190
Shankari A, Anand M (2014) Process of making waste into wealth – dry flower technology. Asian J Hortic 9(2):466–483
Sharma G, Kashyap B, Gupta YC, Sharma BP (2019) Studies on keeping quality of dried and dyed chincherinchee flowers. Indian J Hortic 76(3):552. https://doi.org/10.5958/0974-0112.2019.00089.6
Sindhuja S, Padmalatha T, Padmavathamma PS, Prasad ALN (2016) Effect of chemicals as pre-drying treatments on production of quality dry flowers in carnation. Progress Hortic 48(2):231–235
Singh D, Kumar S (2008) Dry flowers and natural splendor indoors. Floric Today 13(1):42–48
Singh A, Dhaduk BK, Shah RR (2004) Effect of different temperature and embedding media on flower dehydration of zinnia (Zinnia linearis Benth.). Indian J Hortic 61(3):249–252
Sireesha M, Reddy MV (2016) Effect of freeze drying process for retaining physical characteristics of orchid flower. J Res ANGRAU 44(1):44–52
Sohn K, Hwa J, Kwon H, Young KE (2003) Effects of drying methods on shape and colour of Rosa hybrida. Korean J Hortic Sci Technol 21(2):136–140
Sujatha AN, Damodaran T, Shiva KN (2001) Dry flower industry in Andamans. Kissan World 28:28
Tampion J, Reynolds J (1971) Botany of flower arrangers. Petham Books Ltd., London
Thakur P, Joshi AK, Chandra A (2019) Epoxy resin encapsulation: a novel drying technique for native flowers. Indian Hortic 64(4):40
Trinklein D (2006) Drying flowers and foliage for arrangements. http://extension.missouri.edu/publications/DisplayPub.aspx?P=G6540. Accessed on 19 July 2025
Ullas SP, Namita N, Singh KP, Panwar S, Kundu A, Krishanan GS, Kumar G (2018) Influence of drying techniques on retention of anthocyanin and their antioxidant activities in chrysanthemum (Chrysanthemum × Morifolium) flowers. Indian J Agric Sci 88(2):228–233

Verma AK, Dhiman MR, Kumar D, Gupta A (2012) Preserving flowers and plant parts. In: Post harvest technology for commercial floriculture. New India Publishing Agency, New Delhi, pp 143–171

von Hagens G (1981) Animal and vegetal tissues permanently preserved by synthetic resin impregnation. United States Patent 4,278,701. U.S. Patent and Trade- mark Office, Washington, DC

Wang Y, Sun J, Ma D, Li X, Gao X, Miao J, Gao W (2019) Improving the contents of the active components and bioactivities of Chrysanthemum Morifolium Ramat.: the effects of drying methods. Food Biosci 29:9–16. https://doi.org/10.1016/j.fbio.2019.03.003

White P (2007) Drying and preserving plant materials. Website: www.edis.ifas.ulf.edu/pdffles/ep/ep00400.pdf

Wilson D, Attri B, Sharma SK (2015) Evaluation of different methods for drying of chrysanthemum flowers. Prog Hortic 47(1):167–169. https://doi.org/10.5958/2249-5258.2015.00030.5

Yadav P, Kashyap B, Dhiman SR, Gupta YC, Sharma P (2018) Effect of glycerine on drying of cut foliage of *Buxus sempervirens*. J Ornam Hortic 21(1–2):17–23. https://doi.org/10.5958/2249-880X.2018.00004.X

Yogita C (2000) Studies on the standardization of bleaching and dyeing technology of flowers. M. Sc. thesis., Tamil Nadu Agricultural University, Coimbatore, Tamil Nadu.

Traditional Bulbous Plants 9

S. L. Chawla, Kalkame Ch. Momin, R. D. Pawar, Kiran Kumari, and Sudha Patil

Contents

S. L. Chawla (✉) · S. Patil
Navsari Agricultural University, Navsari, Gujarat, India

K. C. Momin
College of Horticulture & Forestry, Pasighat, Arunachal Pradesh, India

R. D. Pawar
RCSM College of Agriculture, MPKV, Kolhapur, Maharashtra, India

K. Kumari
College of Horticulture, SDAU, Jagudan, Gujarat, India

S. K. Datta, Y. C. Gupta (eds.), *Floriculture and Ornamental Plants*, Handbooks of Crop Diversity: Conservation and Use of Plant Genetic Resources,
https://doi.org/10.1007/978-981-15-3518-5_11

Abstract

Ornamental geophytes with hundreds of species from different genera as well as families grown as bulbous ornamentals hold an important position in world floriculture industry which are used for the production of cut flowers, potted flowering plants, landscaping, etc. For long-term survival, they bear the subterranean storage organs, viz., rhizomes, bulbs, corms, tubers, or tuberous roots, etc., to store food reserves, moisture, and nutrition which are very diverse in their structures. Market saturation with traditional plants and flowers has stimulated an increased interest in novelties through new crops and varieties. Researchers from many countries are thus evaluating and exploring their indigenous flora as a source of potential ornamentals. The rational exploration, evaluation, conservation of these natural resources, and their utility in breeding programs through traditional and advanced breeding options are essential. Exploration of natural resources and their utilization in commercial production is important for germplasm enhancement and breeding in search for new or advanced traits for major geophytes as well as introduction of new geophyte in the global market. Introduction and breeding may also facilitate the commercial production through new production centers with diverse climates in new regions as well as shaping market trends as well as development and accomplishment of consumer demand. These introductions will open doors for the advanced agrotechniques, cultivars, postharvest technologies, and marketing policies.

The globalization of the production and marketing chains in tune with the demand will benefit the researchers who have been engaged for years but also new researchers.

Keywords

Traditional · Bulbous · Breeding

Hundreds of species grown as bulbous ornamental crops commonly known as ornamental geophytes are used for the production of cut flowers, potted flowering plants, or in landscaping. Mostly they are monocots with few dicots and belong to different families as well as genera. They commonly have subterranean storage organs that are very diverse and may consist of bulbs, corms, rhizomes, tuberous roots or tubers, *etc.*, the main function of which is to store food reserves, moisture, and nutrition to ensure long-term survival of the species (De Hertogh and Le Nard 1993a). Out of total production of flower bulbs, 90% is accounted by only six genera, namely, *Tulipa, Lilium, Gladiolus, Narcissus, Iris,* and *Hyacinthus* (De Hertogh and Le Nard 1993a, b). These ornamental geophytes hold an important position in floriculture around the world. Few of them are discussed hereunder with their breeding perspectives.

9.1 Agapanthus

Agapanthus is herbaceous perennial that is cultivated as an ornamental plant because of its spectacular blue-violet to white spherical flowers. It is a very good cut flower and also mass planting of agapanthus flowers makes a wonderful display for mild climate, whereas the dwarf types are especially good as pot plant because restricted roots growth induces heavier flowering. The dried seed heads can be used in flower arrangements. *Agapanthus* not only consist ornamental value but medicinal too. *A. praecox* is an evergreen species with medicinal and ornamental uses. Some species of *Agapanthus* are commonly known as lily of the Nile.

9.1.1 Botany and Distribution

The genus *Agapanthus* was established by Charles Louis in 1788 in Sertum Anglicum, based on specimens he saw in London. With a long history of taxonomic confusion, finally the genus Agapanthus is placed in family Amaryllidaceae, as morphologically it is considered closely related to the Amaryllidaceae on account of its umbellate inflorescences, but is regarded as sufficiently different in the absence of amaryllid alkaloid compounds. It is the only genus in the subfamily Agapanthoideae. The strap type leaves of agapanthus are leathery and arranged in two opposite rows and having length up to 60 cm. The leaves are attractive even when the plant is not in flowering condition. As it is geophytes, it has underground rhizomes. It is a tender herbaceous perennial plant grows in upright clumps from fleshy rhizomes that produce short tuberous roots. Flower clusters are borne on sturdy, erect stems held well above the foliage. Each single terminal inflorescence consists of numerous tubular to bell-shaped flowers, each with six parts. There are 20 to 100 flowers in each rounded umbel, depending on variety and species. The individual flower looks like a miniature lily flower, that is, beautiful six tepals (three sepals and three petals) are present. The flower color varies from different shades of violet blue to white.

9.1.2 Origin, Domestication, and Spread

They are cultivated throughout warm areas of the world. They can especially be spotted throughout northern California. Boundaries of *Agapanthus* species are not clear in the genus, and in spite of having been intensively studied, the number of species recognized by different authorities varies from 6 to 10. Despite the common name of Lily of the Nile, they are not native to the Nile River basin of north-eastern Africa and all of the species are native to southern Africa (South Africa, Lesotho, Swaziland, Mozambique) though some have become naturalized in scattered places around the world (Australia, Great Britain, Mexico, Ethiopia, Jamaica, *etc.*). They range from the Cape Peninsula to the mountain ranges just south of the Limpopo River in Limpopo Province (Duncan 1998).

9.1.3 Plant Genetic Resources

Agapanthus is gaining more and more horticultural interest and large numbers of new cultivars are being introduced (Duncan 1985). All the species in *Agapanthus* have same chromosome number, that is, 2n=2x=30. When all species in a genus have the same chromosome number, differences in nuclear DNA content have proven to be very effective in delimiting intrageneric divisions in a number of taxa (Ohri 1998). *Agapanthus* is a genus of six species *viz. A. africanus, A. campanulatus, A. caulescens, A. coddii, A. inapertus,* and *A. praecox* (Zonneveld and Duncan 2003). On the basis of DNA content and pollen characters, Zonneveld and Duncan divided six species into two groups each with three species. The first group with lilac pollen, purple leaf base, deciduous growth, and DNA content below 24.5 pg contains *A. campanulatus*, *A. caulescens,* and *A. coddii*. The second group has yellow/brown pollen, green or purple leaf bases, a deciduous or evergreen growth habit, and a DNA content of 25.2–31.6 pg and includes *A. inapertus*, *A. praecox,* and *A. africanus* with *A. walshii* as a separate subspecies.

9.1.4 Conservation

An efficient *in vitro* technique for the rapid multiplication of *A. praecox* (medicinal plant) from shoot-tips cultured on MS medium containing BA (22.2 μM), IAA (2.9 μM), and TDZ (4.5 μM) combinations induced high number of shoots accompanied by *ex vitro* rooting (Baskaran and Van Staden 2013). The protocol developed is economically cheaper and can be applied for large-scale micropropagation for germplasm conservation and genetic transformation of *A. praecox*. Some studies also show successful regeneration of diploid and tetraploid from protoplast of *Agapanthus* (Nakano et al. 2003).

9.2 Canna

Canna is a plant of tropical or subtropical origin. *Canna*, the solitary genus, belongs to family Cannaceae which is one of the monocot families that is easy to recognize. The generic name has been derived from the Greek word "*Kanna*" meaning a reed, referring to its herbaceous stem (Everett 1980). It is commonly known by the names like Indian Shot, Achira, and African arrowroot, *etc.* Records that go back to 2500 B.C. in Peru show that the people were using the rhizomes of *Canna indica* (Gade 1966). *Canna* was also described in the writings of many botanists that came prior to Linnaeus and was listed in many gardens under different names. Although Cannas were previously considered as simple foliage plants, during the last two centuries, cultivation and improvement transformed them into attractive ornamental flowering plants. The flowers of *Canna* are mainly red, orange, and yellow and these majestic flowers provide a bold effect in the garden. It can beautify any wasteland or rugged part or any neglected look of garden, so it is versatile in nature.

Botany

Canna is perennial plant with rhizomes. Leaves are alternate and big, arranged spirally with a sheathing base and no ligulae. Leaves usually green and sometimes stained purple or bronze or variegated with clear midrib and numerous lateral veins. The inflorescence is terminal. Flowers are big and showy with range of colors like red, pink, yellow, or orange. The flower has three sepals and is variously interpreted as having staminodes functioning as petals or petals and staminodes (tepals), one is smaller than the other two. The ovary is inferior with three carpels. The fruit is hard and has a warty surface divided to three parts in which seeds are present. Seeds are hard and shiny black in color and globose to ovoid in shape. Plant height varies according to their species and cultivars. The fleshy, stubby, and horizontal rhizomes are rich in starch and edible. Various morphological, cytological, and taxonomical characteristics of family Cannaceae are closely related to other members of Zingiberales like Musaceae, Strelitziaceae, Lowiaceae, Heliconiaceae, Zingiberaceae, Costaceae, and Marantaceae (Cronquist 1981).

9.2.1 Origin, Domestication, and Spread

Canna species are native of South and North America, and with the course of time, they have been introduced in Asian paleotropics and subsequently evolved into native varieties. Charles de l'Ecluse, who first described and sketched *C. indica*, indicated this origin and stated that it was given the name *indica*, not because the plant is from India, in Asia, but because this species was originally transported from America and at that time, one described the tropical areas of that part of the globe as the Western Indies. From their original habitats, the species were introduced into Europe, beginning with *C. indica* in 1596 by Gerard. This was followed by *C. glauca*, whose exact year of introduction is not known but which was illustrated by Piso in 1648 (Baker 1893). The transformation of Canna from wild to cultivated condition is prevailed historically. The pan-tropical distribution of *Canna* species is most possibly the effect of human dispersal (Prince and Parks 2001). The transportation of Canna from their native place may have been the reason for occurrence of beautiful ornamental plant in Europe, Asia, and Africa (Maas-van de Kamer and Maas 2008). The first species of *Canna* introduced to Europe was *C. indica* L., which was imported from the East Indies, though the species originated from the America. Although canna has seen mostly in cultivation, it has naturalized in many parts and has been noted as weedy with the potential to be invasive and difficult to remove.

9.2.2 Plant Genetic Resources

Canna indica has become naturalized in many tropical areas around the world. The canna cultivars now in use are a tremendous improvement over the original botanical species. The most grown species of canna are *C. discolor*, *C. edulis*, *C. flaccida*, *C. glauca* var. rubra-lutea, *C. indica*, *C. iridiflora*, *C. langunose*, *C. liliflora*,

C. limbate, *C. lutea*, *C. nepalensis*, *C. pedunculata*, *C. speciose,* and *C. warscewiczii*. Some of the wild species of *Canna* are *C. glauca*, *C. indica*, *C. iridiflora*, *C. warscwiczii,* and *C. flaccid, etc.*, that involved in producing natural as well as manmade hybrids. The above five species are popularly known as elemental species of Canna. The entire cultivated garden Cannas are included under two artificial hybrid species *viz.*, *Canna* x *orchiodes* L. H. Bailey and *Canna* x *generalis* L. H. Bailey (Hannay 1936; Khoshoo and Mukherjee 1970). All the hybrid cultivars share some common features and bind themselves under the same horticultural species. Some phenotypic transformations were also taken place when they were shifted from wild to cultivated condition. Breeding for ornamental use started in 1848 when M. Annee brought some *Canna* species he had collected in South America to France. He is probably the first man to grow cannas through seeds sown in 1848 of the real species *Canna nepalensis* which with all probability was pollinated with *C. glauca*. He named the resultant race of tall cannas as *C. annaei*. Further in 1863, a new race *C. ehemanni* (syn. *C. iridiflora hybrida*) was developed through *C. iridiflora* and *C. warscewiczii* which was intermediate in stature and had showy foliage and more pleasing drooping flowers. Though *C. ehemanni* race is still in the trade, the original one is almost extinct. This race was further crossed with other species and races and a selection of dwarf but large-flowered type race was developed in France from a large population, and therefore, this race is known as French cannas or Crozy (one of the renowned breeder there) cannas. In the middle of twentieth century, in Italy, another race Italian or orchid-flowered cannas was developed by using *C. flaccida* with garden forms and with *C. iridiflora* and the resultant hybrids had iris-like outlines but the flowers were short-lived. The varieties under this race are America, Austria, Bavaria, Burbank, Burgundia, Italia, Pandora, *etc.*

9.2.3 Collection and Conservation

In India, hybridization has played a dominant and decisive role in the origin of ornamental cannas. This has been made possible by the ecospecific differentiation of the parental species, which implies lack of barriers and a good deal of recombination associated with reasonably high fertility. National Botanical Research Institute, Lucknow, is having very rich germplasm of *Canna*. Some of the introduced or indigenously developed varieties in NBRI, Lucknow, are Aida, After Glow, Ailsen, Ali Petzi, Alison, Anarkali, Aristocrat, Ariel, Arjun, Atom Bomb, Bardara, Bharat, Black Knight, Bridal Veil, Brocade, Carmine King, Charmion, Cherub, Masterpiece, Matchless, Morning Glow, Nerissa, Percy Lancaster, Perfection, Pink Satin, Plume, President, Queen Elizabeth, Raj Mahal, Rosamund Coles, Rose Queen, Sangrila, Sirius, Sir John Anderson, Sans Souci, Soldier Boy, Star of India, Striped Queen, Stromboli, Sun Set, Sweet Heart, The Queen, Yellow Gal, *etc.* Mukherjee and Khoshoo have deeply studied pollination mechanism, breeding systems, and variation in phenotype in canna (Mukherjee and Khoshoo 1970). Percy Lancaster also worked on canna and bred some cultivars at Agri-Horticulture Society of India, Alipore, Kolkata. Agri-Horticultural Society, Calcutta, introduced 51 canna cultivars from Italy in between 1895 and 1904 and 10 from USA after 1904, *viz.*, Africa,

Alemannia, America, Aphrodite, Asia, Atlanta, Attika, Australia, Austria, Bavaria, Borussia, Britannia, Burbank, Burgundia, Campania, Charles Naudin, Crown Prince of Italy, Edouard Andre, Emelia, Hellas, Heinrich Siedel, H. Wendland, Iberia, Indiana, Ischia, Italia, King Herbert, Kronos, La France, Mrs. Kate Grey, Oceanus, Pandora, Partenope, Pennsylvania, Pereus, Philadelphia, Phoebe, Pluto, Prof. Traub, Queen of Italy, Rhea, Roma, Romagna, Rossi, Sicilia, Suevia, Trinacria, Umbria, Wintzer's Colossal, Wm Beck, and Wyoming. There are so many species and cultivars of canna and the genus seems to be in no danger of genetic deterioration. However, it is important to conserve the vast genetic diversity.

9.2.4 Characterization and Evaluation

Various studies were carried out by researchers to characterize different species and cultivars of genus *Canna*. *Canna* x *generalis* (*C. glauca* x *C. indica* x *C. iridiflora* x *C. warscewiczii*) and *C.* x *orchioides* (*C. glauca* x *C. indica* x *C. iridiflora* x *C. warscewiczii* x *C. flaccida*) are horticultural species with a range of plant height from 50 to 160 cm under which all the ornamental cultivars of hybrid origin are included, whereas the height of the elemental species (*C. flaccida, C. glauca, C. indica, C. iridiflora,* and *C. warscewiczii*) ranges from 89 cm to 500 cm. The characteristic features of *C.* x *generalis* hybrids are short to tall and slender, leaves from glaucous grey and leathery to dark chocolate-red and thin, flower shape and color from small narrow segments to large and ruffled, color being from pale-yellow to orange or scarlet, and of *C.* x *orchioides* hybrids, the flowers are very large, tubular at base, petals reflexed, usually splashed, or mottled and three broad wavy staminodes exceed by the lip, but now both are so much interbred that these are now referred to only as *C.* x *generalis.*

In a comparative trial of the new cultivars and the old ones, the standard old cultivars such as President, Gartenfeuer (Liebesglut), Felix Ragout, and Garteninspektor Nessler were found better than the new ones; however, the yellow-flowered new cultivar Schwabenland though less attractive but was found more vigorous and highly tolerant to weather conditions (Bosse 1968). Khoshoo and Guha (1975) studied the existing cultivars of *Canna* and classified them on the basis of their height, morphology, foliage, and floral characters. Canna are categorized into selfs (without spots or margin, one color only), spotted (usually a shade of red on cream or yellow ground or red spots on orange of red ground), striped red on a cream or a yellow ground, margin yellow, margined with a darker shade than the ground color, flaked red or orange on a paler ground, and splashed orange on a deeper ground. Khoshoo and Guha (1975) also studied the five elemental species and other cultivars of canna and concluded that morphological transformation from the wild to cultivated form caused reduction in plant height, change in form and color of foliage, increase in hardiness, increase in flower size, free flowering erect flowers, intensification of color, and durability of flowers. It was found that next to hybridization, triploidy has been an important mechanism in the origin of cultivars with thicker, more durable, and larger flower parts. The two types of triploids, autotriploids and segmental allotriploids, are distinguishable by their morphological and cytogenetical properties (Khoshoo and Mukherjee 1970). The mutation breeding work on canna is limited to Chemarin et al. (1973) and a few breeders in India (Gupta 1966; Khoshoo 1968) and Thailand (Nakornthap 1965).

Sixty-six accessions of *Canna indica* from nine provinces in Indonesia are divided into two main clusters through molecular characterization: the green and red cultivar group. The green cultivar group is also divided into subcultivar green and green stripe purple based on color of sheaths, tip of bud, rachis inflorescence, petals, brachtea, and pattern of staminodea. The red cultivar is divided into subcultivar purple and subcultivar dark purple based on color of sheaths, rachis inflorescence, and petals (Sari et al. 2018). Edible Canna (*Canna edulis* Ker) was evaluated as an alternative starch source on the basis of genetic characteristics, agronomic traits, and starch properties (Piyachomkwan et al. 2002).

9.2.5 Future Perspective

Canna is an age-old plant with numerous economically important characters like presence of starch in rhizomes, antioxidant properties, medicinal properties, dye yielder, and ornamental value. So, selection among locally available material must be done for these characters, which can further be studied and improved by using breeding and molecular approaches. *C. indica* is also one of the several plant species that are used for waste water treatment as well as soil remediation. So, this aspect also needs further attention.

9.3 Crinum

9.3.1 Introduction

Crinum is an important and fascinating genus of the large and equally captivating family of Amaryllidaceae. Crinums are identified based on the lily-like flowers with underground bulbs. The name *Crinum* is derived from the Greek work "Krinos" meaning trailing hair or comet tail. There are about 130 species of *Crinum* widely distributed in and around the tropical and subtropical regions of the world. Larger in stature than most other Amaryllidaceae species, most *Crinums* are suitable as landscape plants. Furthermore, *Crinum* species have been used traditionally to cure ailments and diseases throughout the world and some of the most noted effects are analgesic, anticholinergic, antitumor, and antiviral.

9.3.2 Botany and Distribution

Crinums are bulbous, evergreen, perennial, herbaceous plants and assume a medium height. The plants have long, green, shiny, linear lanceolate, evergreen leaves around

1.0–1.5 meter long which emerges from large bulb. Flowers are white in color, which originates in clusters on thick and succulent stalk. Around 20–30 flowers are arranged in umbel. Corolla is around 8 cm long and white in color. Filament is about 5 cm long and anthers 2 cm in size. Ovary is three celled, with six stamens and one stigma. Roots are cylindrical which is around 25 cm long with a thickness of 1 cm. Fruit is globose nearly 5 cm across filled with large seed. Roots are adventitious which are below underground bulbs.

9.3.3 Origin, Domestication, and Spread

The representatives of the genus *Crinum* are found in the tropics of Africa, Asia, and America and in the temperate regions of the northern and southern hemispheres. In the south, crinums occur in South Africa, south-east Asia, and Australia, while in the north, it occurs mostly in Japan and the southern regions of the USA. The center of diversity is in Africa, south of the Sahara with a population of more than half the numbers of species found worldwide. In South Africa, 21 species are found, sparsely on mountain- or hill-slopes and more commonly in low lying areas, on river banks and at the coast.

Crinums were brought into general cultivation as early as the seventeenth century. The first species to be introduced to English gardeners was *C. asiaticum* from China in 1732. This was later followed by several species native to South Africa, including *C. bulbispermum* in 1752 and *C. macowanii* and *C. moorei* in 1874.

In India, *Crinum* is represented by 15 species, of which four are endemic to Western Ghats, viz. *C. brachynema* Herb., *C. malabaricum* Lekhak & S.R. Yadav, *C. wattii* Baker, *C. woodrowii* Baker ex W. Watson. *Crinum brachynema* and *C. woodrowii* are critically endangered species and are strictly confined to edges of lateritic plateaus and in semi-evergreen forests on hill slopes of Mahabaleshwar and adjoining areas of Satara district, Maharashtra (Yadav 1997; Gaikwad and Yadav 2004).

9.3.4 Plant Genetic Resources

Crinums are very desirable garden plants with their sturdy, decorative foliage, and large lily-like flowers. *Crinum asiaticum* L. and *C. latifolium* L. are commonly grown in gardens for their beautiful foliage and pure white to pink-tinged large elegant flowers. Some species, such as *C. moorei* and *C. neriodes*, are pleasantly scented. *C. bulbispermum* is a resilient variety as it can be grown in different types of soil, can survive long periods without water, and tolerant to cold.

Crinum brachynema is to endemic Mahabaleshwar (Kate's Point) in Satara district of Maharashtra and is listed as a critically threatened species. It is characterized by its fragrant night-blooming large, showy flowers and can be introduced into gardens as an ornamental plant. *Crinum malabaricum* is a recently described species which is so far known only from a fresh water stream bed at Periya region in Kasaragod district of Kerala. The species is represented by a population of about 1000 bulbs and restricted to about 0.5 sq. km area. The ribbon-shaped new leaves are reported to have the longest leaves in the genus that attains a remarkable length of 3.65–4.57 m in one month. Another indigenous species is *Crinum woodrowii* that is sporadically distributed in the main ranges of northern Western Ghats and good populations are found on hill slopes around Mahabaleshwar and Khandala range. The plant has glaucous leaves and fragrant white flowers that bloom in night.

There are certain species of *Crinum* which are known to be fragrant *viz.*, *C. moorei* is highly fragrant; flowers of *C. acaule* are delicately perfumed like carnations and those of *C. minimum* resemble the sweetly scented frangipani flowers. Many of the smaller South African crinums can be readily crossed and produce semi-dwarf hybrids that do not multiply vegetatively.

9.3.5 Collections and Conservation

The natural populations of *Crinum* species are dwindling day by day due to habitat encroachment and various other anthropogenic activities. These species need necessary conservation efforts for their survival. *Crinum* species are particularly vulnerable and are much sought after as ornamental plants and have a long history of ethno-botanical usage worldwide. They are also highly valued as medicinal plants and have attracted interest from medical science whose chemical analyses have confirmed the rationale for the plant's usage with the isolation of active compounds. The genus possesses several biological features that further exacerbate conservation efforts. Application of tissue culture technique may prove to be a worthwhile alternative to conventional propagation techniques for their *ex situ* conservation.

9.3.6 Breeding Options and Constraints

With a lot of variation in the *Crinum* species, hybridization is adopted to yield more colorful species. Though hybrids are commonly obtained, most are usually sterile. The hybrid pollen may be viable, but unless it is parental, is unlikely to be accepted. Success, therefore, is limited due to genetic conflicts. Despite these difficulties, enthusiastic breeders have persisted in their efforts with the result that there are, today, a number of popular hybrids, both intergeneric and interspecific. One such example is "Crinodonna" which is the result of crossing *Amaryllis belladonna* with *Crinum moorei* and was first described in Florence in 1921. Chittenden (1956) also reports that hybrids between *Crinum* and *Hymenocallis* have been produced. *C.* x *powellii* is an interspecific hybrid between *C. bulbispermum* and *C. moorei* which is grown for its heavy umbels of sweetly fragrant flowers and is one of the best for garden use.

9.3.7 Looking Forward for Future Perspective

Ornamental crop production has become a highly specialized industry and has resulted in an increased demand for new species and cultivars. The *Crinum* species as geophytes has been described as one with great potential to be developed as cut flowers and/or pot plants. This is because they are hardy and produce large numbers of attractive blooms for many months and good cut flowers with excellent vase life and some species are scented too. In addition, crinums have enormous phyto-constituents and pharmacological application showing its wide range of ethno-medicinal uses. Based on the potentiality of the crinums both for ornamental and medicinal values, efforts should be taken to conserve this plant species and should not be overexploited for maintaining their existence in the near future.

9.4 Dahlia

9.4.1 Botany and Distribution

Genus *Dahlia*, a member of family Asteraceae, is commonly grown as garden plants. Dahlia is a herbaceous perennial with hollow or solid, mostly erect, and branched stem growing up to 1.8–2.4 m and having fasciculated tuberous-roots. The tuber is having growing points in the crown where it is attached with the aerial stem. Leaves are opposite and simple or pinnatifide with serrations, and inflorescence is a solitary capitulum or head with long peduncle and consists of several hundred individual florets in a cyme. Heads are small to large, one per stem, having variable forms from open to ball and a diameter of 05 to 30 cm. Disc florets are yellow and fertile, actinomorphic, tubular, five lobed, and complete flowers, while ray florets are zygomorphic, pistillate, and spreading in outer whorls. They display bright colors. Fruit is an achene, oblong, or obovate with many flat seeds. Heads with varying ratios of disc and ray florets bloom in succession where ray florets (the only female flowers) open first from the outermost side towards the center. In single types, there are one to few whorls of ray florets while in double types, they for many whorls. Due to large, composite, and compact head, complete emasculation is a challenging task just like other members of Asteraceae.

The flower colors and their distribution in garden dahlia are a result of presence or absence of the two series of pigments viz., flavones and anthocynins as reported by W.J.C. Lawrence, 1929 with all types of color variations appearing only in ray florets with the disc florets remaining yellow. Lawrence (1942) also stated that the reasons for the profusion of different forms and colors are ~ (i) that dahlias being self-incompatible can never breed true, (ii) the factors producing flower color are in duplicate, two for flavone (ivory and yellow) and two for anthocyanin (pale and deep pigmentation), and (iii) each of these factors is represented four times. It was therefore evidently suggested that the plant arose as a hybrid between polyploid members of two color groups and that doubling of the chromosome number accompanied this hybridization.

Two basic different chromosome numbers are found in dahlia, *viz.* $x=8$ and $x=18$, the latter being derived from $x=9$ as reported by Gupta et al. (1972). Mehra and Ramanandan (1974) determined $2n=64$ chromosomes (Octaploid) in *D. coccinea* and *D. rosea* with basic chromosome number of $x=8$. The great variety results from dahlias being octaploids (they have eight sets of homologous chromosomes, whereas most plants have only two). They have haploid chromosome numbers of 16, 17, 18, and 32 with most of the garden cultivars including *Dahlia variabilis* being tetraploids having 2n=64 (Misra et al. 2017). Dahlia species with 2n=32 are allotetraploids, whereas the species, races, and varieties with 2n=64 are their autopolyploid (octaploid) derivatives. In situ hybridization using an rRNA gene probe indicated that the 2n=32 species have eight hybridization sites, while the 2n=64 species have 16 sites (Gatt et al. 1998).

9.4.2 Origin, Domestication, and Spread

The genus *Dahlia* is native to North America especially Mexico and is the national flower of Mexico. The distribution of Dahlia species is confined to the Central American region of Colombia, Mexico, and Guatemala, the majority of plants collected coming from Mexico which is a major center of diversification. Hence, the relative proximity of related species would increase the chances of interspecific hybridization, the probable result of which would be the evolution of forms showing considerable similarity (Lawrence 1929).

Antonio Jose Cavanilles, the staff of the Royal Botanic Garden in Madrid in the eighteenth century, gave the genus its Latin name "Dahlia" in the memory of a Swedish botanist and pupil of Linnaeus "Anders Dahl." Seeds of plants sent to Madrid from botanical garden of Mexico flowered for the first time in the botanical garden, Madrid, in October 1789, which was named *Dahlia pinnata* by Antonio Jose Cavanilles as per the records of his *Icones Plantarum* published in 1791 (Sørensen 1970). The dahlia was a cultivated plant much before its earliest scientific studies. In Mexico, it was named as Cocoxoehitl and was used as an ornamental, medicinal, and food plant. It was the part of the gardens of Aztecs, where it was domesticated and brought under cultivation even before the discovery of America. The records of Dahlia can be traced back in a book *Badianus Manuscript, An Aztec Herbal of 1552*, as the earliest illustration of a Dahlia written first in Nahuatl by Martinus de la Cruz and then translated into Latin by Juannes Badianus. After Spanish conquered the Mexico, King Phillip II of Spain commissioned his personal physician, Francisco Hernandez who was honored with the title "Protomedico of the Indies," to travel to Nueva Espana and prepare an account of the natural history of the land. Hernandez visited Mexico and Central America for the period of 1570 to 1577 where he noticed three spectacular dahlias, which he mentioned in his account of medicinal plants of New Spain published in 1651 entitled "Rerum Medicarum Novae Hispaniae Thesaurus seu Plantarum, Animalium, Mineralium Mexicanorum Historia." It had three sketches of dahlias along with their vernacular names in Nahuatl as cocotli, signifying word "syringa" meaning a hollow-stemmed plant; acocotli meaning "water-cane" or "water-pipe"; and cocoxochitl meaning "cane-flower" or "hollow-stem-flower." The French botanist Nicolas Joseph Thiery de Menonville who served King of France was sent on a secret mission "to secure (steal) living specimens of the jealously guarded cochineal insect (valued for its scarlet dye) and the Nopal cactus on which the insect lived." He narrated in his travel while collecting Nopal from the local merchants garden near Oaxaca city of Mexico he observed a double violet dahlia which he mentioned as a "double violet aster" produced on shrubs which described roughly the *D. tenuicaulis* (Sørensen 1970).

Late in the eighteenth century, Vincente Cervantes from Mexican Botanic Garden consigned a shipment of seeds of various Mexican plants that also included Dahlias to Antonio Jose Cavanilles, Madrid. Dahlia seeds flowered for the first time in the botanical garden in October 1789, which was used to describe the first three species of Dahlia by Antonio Jose Cavanilles, the head of the Madrid Botanical Garden, in his book *Icones et Descriptiones Plantarum* published in 1791 with first species as a double type *Dahlia pinnata*. The next volume published in 1796 described two other species as *Dahlia coccinea* and *Dahlia rosea* with single flowers. From Madrid, the dahlias were sent by Cavanilles to M. Thibaud and Alphonse de Candolle from France in 1802 as well as other scientists in England and other parts of Europe (Sørensen 1970).

Introduction of dahlia to the Netherlands florists from Mexico occurred when a box of its root-tubers was sent from Mexico there, where only one plant survived the trip which produced spectacular red flowers with pointed petals and this was named as *Dahlia juarezii*. European nurserymen used this in crossing with certain other dahlias already present there, as parents and these became the earliest progenitors of all modern dahlia hybrids. The *Jardin des Plantes* in Paris received *Dahlia variabilis* in 1802 from Madrid which was grown in 1804 by the gardener at Holland House, Kensington. Later it spread in the rest part of the world.

9.4.3 Plant Genetic Resources

a. Geographic Distribution

The distribution of Dahlia species is confined to the Central American region of Colombia, Mexico, and Guatemala, the majority of plants collected coming from Mexico which is a major center of diversification. Hence, the relative proximity of related species would increase the chances of interspecific hybridization, the probable result of which would be the evolution of forms showing considerable similarity (Lawrence 1929). Mexico represents the main source of germplasm with 35 endemic species of Dahlia. The genre is present in 26 states with the largest number of species in Hidalgo and Oaxaca followed by Guerrero. The state of Jalisco represents the greatest collection and efforts made. Dahlia species inhabit nine types of habitats with 35 species in coniferous and oak forests, deciduous tropical forest 20 and the xerophilous thicket 17 as the major habitats. The range of distribution is from 24 to 3,810 m elevation with the maximum number of species growing at 2,000 to 2,500 m. *D. coccinea* found growing in all the nine habitats and an elevation of 24–3033 m showing greatest ecological range. The species diversity of Dahlia is found in the Mexican Transition Zone extends towards the Cuenca del Balsas, the Pacific Coast, the Chihuahuan Desert, Tamaulipas, and Veracruz with the highest in the provinces of Sierra Madre del Sur and Sierra Madre Oriental with the Sierra Gorda in the state of Querétaro concentrating the greatest number of species (Carrasco-Ortiz et al. 2019).

b. Primary Gene Pool

Dahlia is a monophyletic group. Dahlias are well known in ornamental horticulture. They have been the subject of intense modification genetics that has produced more than 50,000 cultivated varieties. These have changed their characteristics by increasing the inflorescence, forms of ray flowers, diversity of colors, size of individuals, and flowering time (Carrasco-Ortiz et al. 2019). The current day varieties resulted through a series of hybridization, selection, and mutation breeding (Misra et al. 2017).

c. Wild Genetic Resources and Others

There are 42 accepted species of Dahlia recognized at present and are continue to be described. *D. pinnata, D. coccinea, D. ro*sea are the major species involved in the development of new age dahlias. The sectional classification of *Dahlia* was given by Sørensen (1969) with 27 species and was further updated with identification of new species. The updated sectional classification of *Dahlia* species as per "The American Dahlia Society" into four sections is as follows:

(I) **Section** *Pseudodendron* Sherff: *D. campanulata* Saar, Sørensen, & Hjerting, *D. excelsa* Bentham (uncertain), *D. imperialis* Rözl ex Ortgies, *D. tenuicaulis* Sørensen,

(II) **Section** *Entemophyllon* Sørensen, *D. congestifolia* Sørensen, *D. dissecta* S. Watson, *D. foeniculifolia* Sherff, *D. linearis* Sherff, *D. rupicola* Sørensen, *D. scapigeroides* Sherff, *D.sublignosa* (Sørensen) Saar & Sørensen,

(III) **Section** *Dahlia* Sherff: *D. apiculata* (Sherff) Sørensen, *D. atropurpurea* Sørensen, *D. australis* (Sherff) Sørensen, *D. barkerae* Knowles and Westcott, *D. brevis* Sørensen, *D. coccinea* Cavanilles, *D. cordifolia* (Sessé & Mociño) McVaugh syn. *D. cardiophylla*, *D. cuspidata* Saar, Sørensen, & Hjerting, *D. hintonii* Sherff, *D. hjertingii* Hansen and Sørensen, *D. mollis* Sørensen, *D. moorei* Sherff, *D. neglecta* Saar, *D. pugana* Rodriguez & Castro, *D. parvibracteata* Saar & Sørensen, *D. pteropoda* Sherff, *D. purpusii* Brandegee, *D. rudis* Sørensen, *D. sherffii* Sørensen, *D. scapigera* (A. Dietrich) Knowles & Westcott, *D. sorensenii* Hansen & Hjerting, *D. spectabilis* Saar, Sørensen, & Hjerting, *D. tamaulipana* Reyes, Islas, and Art. Castro, *D. tenuis* Robinson & Greenman, *D. tubulata* Sørensen, *D. wixarika* Art. Castro, Carr.-Ortiz & Aarón Rodriguez

Subsection Merckii Sørensen: *D. merckii* Lehmann (sometimes spelled Dahlia merkii)

(IV) **Section** Epiphytum Sherff: *D. macdougallii* Sherff

9.4.4 Conservation

In Mexico, there are 176 Natural Protected Areas (ANP) comprising six categories as Biosphere Reserves, National Parks, Natural Monuments, Resource Protection

Areas, Natural Areas of Protection of Flora, and Fauna and Sanctuaries. As mentioned by Carrasco-Ortiz et al. (2019) with the studies on geographical distribution and area of occupation by the taxon in Mexico, all dahlia species are at risk. *D. congestifolia, D. hjertingii, D. purpusii, D. spectabilis*, and *D. Tamaulipana* are critically endangered (CR), 31 species are endangered (EN), *D. merckii* is vulnerable (VU), and *D. coccinea* is almost endangered (NT). In Mexico, Oaxaca and Guerrero, the Sierra Madre Oriental, plus the coniferous and oak forests have the highest number of species in CR and EN. Finally, none of the species in CR is within Protected Natural Areas (ANP), while 21 species in EN are in ANP.

Along with the natural reserves of these species, cultivars and hybrids are conserved by different societies, institutes, as well as the breeder firms in field, seeds, and tubers in gene banks and in vitro cultures. The National Dahlia Collection, UK, established in 1983 in Oxfordshire and then transferred to The Duchy College near Rosewarne with its home at Varfell Farm since 1998 under the current custodianship of Greenyard Flowers, UK LTD. is having a collection of more than 1600 varieties in the Dahlia Garden with some old favorites and many heritage varieties.

9.4.5 Characterization and Evaluation

The efforts of E. E. Sherff, Prof. Paul D. Sorensen, Prof. Dayle E. Saar have systematized the taxonomy of dahlia along strictly botanical principles along with the valuable contributions by Danish professors Hans V. Hansen and J. P. Hjerting. And more recently, teams of Mexican botanists are finding new species in remote areas of the country (The Americ. Dahlia Soc. 2020).

An early breeder of dahlia was Comte Leon Charles LeLieur de Ville-sur-Arce who had four varieties to work with and by 1806 he had produced three double-flowered dahlias. Since 1813, many amateur breeders produced thousands of cultivars, usually chosen for their stunning and brightly colored waxy flowers (Misra et al. 2017).

9.4.6 Breeding Options

Dahlia breeding programs focused to enhance ornamental values, including flower color, size and form, and production quality. The conventional breeding programs *viz.,* introduction, hybridization, composite crossing, multiline, and backcross breeding have introduced variety of desirable characteristics. But classical breeding methods of dahlias are expensive and time-consuming methods. Also, there are many limitations such as limitations in distant hybridization due to incompatibility reaction and differences in ploidy level that is very common in dahlias, polygenic nature of traits such as uniform growth and synchronous flowering, and several viruses infecting dahlias being major hurdles and breakdown of resistance due to fast evolving pathogens. There is a need to use alternative breeding methods to provide faster developments of improved and novel types of cultivars through the use of recent developments in plant biotechnology such as directed mutation, genomics, and recombinant DNA technology (Dalda Şekerci and Gülşen 2016).

9.4.7 Looking Forward or Future Perspective

The future work on dahlia should include studies on fertility of different species and cultivars and generation of blue dahlias with the use of biotechnological tools with induction of delphinidins.

9.5 Eucharis

Eucharis is a genus of monocotyledonous geophytes, having horticultural importance due to their glossy broad leaves and striking white fragrant flowers. It is very good pot plant and adds elegant look to a shady corner in the garden. Its natural habitat is tropical rain forests thus adapted to low light conditions and plenty of moisture. It needs humidity and grows in pots and outside in shady places. Many species of this genus bloom throughout year and some bloom twice or thrice a year. *Eucharis* is popularly known as Amazon lily, but generally, the species *E. amazonica* and *E. grandiflora* are known by this name and these two species are grown for commercial purpose. There are some evidences which show that mucilage from *Eucharis* bulbs (*E. formosa*) used by the Jivaro Indians of Peru for treating facial blemishes and acne (Lewis 1986).

9.5.1 Botany and Distribution

Eucharis plant consists of large deep green glossy leaves which are 20–50 cm long and 10–20 cm broad and umbellate inflorescence which has 2–10 snow white flowers, generally on an erect 40–80 cm long scape. Flower of eucharis resembles little bit to narcissus flower in having a prominent central cup. The mature fruit is triloculicidal capsule. *Eucharis* is a perennial and propagated by removing the offsets or bulbs which are generally 2–6 cm (1–2 in) in diameter. The bulbs are globose or subglobose composed of concentric and modified leaf bases. The bulbs of Eucharis are sympodial in habit. The development of plant is directly related to size of bulb, nutrition, storage, and environmental factors. Bulb must go through juvenile stage and reach a size (>3.5 cm diameter) suitable for flowering (Schiappacasse 1996).

9.5.2 Origin, Domestication, and Spread

Most of the species of *Eucharis* genus are native to Central America and South America from Guetemala to Bolvia (Meerow 1989). Some species scattered to

Mexico, the West Indies, and tropical islands. The major center of distribution for *Eucharis* is in the western Amazon basins inclusive of major tributary system, that is, the Napo, Pastaza, and Huallaga and the adjoining lower slopes of the eastern Andean cordillera. The genus includes 17 species distributed throughout Central and South America, ten of which occur in Colombia *viz.E. candida* Planchon et Linden, *E. Formosa* Meerow, *E. bakeriana* N.E. Brown, *E. bonplandii* (Kunth) Traub, *E. caucana* Meerow, *E. ulei* Kranzlin, *E. lehmannii* Regel, *E. castelnaeana* (Baillon) Macbride, *E.* x *grandiflora* Planchon & Linden, and *E. sanderi* Baker (Yusti-Munoz and Velandia-Perilla 2013). *Eucharis x grandiflora* is natural hybrid between *E. moorie* and *E. sanderi* and widely commercialized. *E. sanderi* is endemic to western Columbia and is threatened extinction due to rapid loss of habitat. Some conservation measures are taken to protect the species at Isla Gorgona (Yusti-Munoz and Velandia-Perilla 2013).

9.5.3 Plant Genetic Resources and Conservation

A somatic chromosome number, that is, 2n=46 largely characterized in this genus. Two tetraplaoid species (2n=92) are *E. bonplandii* from Colombia and *E. bouchei* from Central America. *E. caucanafrom* Colombia is a hexaploid with the largest chromosome number in the genus (2n=138). *Eucharis amazonica*, 2n= 68, is the only known deviation from these 2x, 4x, or 6x karyotypes (Meerow 1987). On the basis of greenhouse pollination attempts, it was concluded that all species of *Eucharis* demonstrated some degree of self-incompatibility and only *E. castelnaeana* sets capsule with self-pollen. The phenomenon of protandry was also recorded and this further suggests that most species are predominantly outcrossing (Meerow 1989). A hybrid has been raised between *Eucharis* and the allied genus *Urceolina* and given the hybrid name *Urceocharis.* This genus have a great potential as a pot plant so further work for conservation of species, standardization of various multiplication methods and in the direction of hybridization and crop improvement is required.

9.6 *Hymenocallis*

The plants under the genus "*Hymenocallis*" are commonly called as Spider lilies. It was recognized as a distinct genus under family Amaryllidaceae since 1812. The generic name *Hymenocallis* is derived from the *Greek* words "*hymen*" (meaning a membrane) and "*callos*" (meaning "beauty") which literally means a beautiful membrane and refers to the membrane that unites the filaments and forms the staminal cup. Different species find their economic importance owing to their chemical constituents for medicinal components and their ornamental values. As stated by Singh and Saxena (2017), *Hymenocallis littoralis* an ornamental and medical plant which has been traditionally used for wound healing is found rich in many phyto-constituents that are useful in drug designing with antitumor, anticancer, antiviral, antimicrobial, antibacterial, antifungal cytotoxicity activity. It is also having allelochemical importance such as defensive compounds, insect repellents, attractants and also maintains the ecological balance.

9.6.1 Botany and Distribution

The 40 different species in the genus *Hymenocallis* are glabrous and perennial herbs having large, onion-like round to oblong tunicated bulbs and often underground stems (rhizomes) with thick white roots. The strap or sword shaped, narrow to broad, erect or arching, bright green to bluish green, deciduous or evergreen, often coriaceous, distichous, hairless, and somewhat fleshy leaves arise directly from the bulb in two ranks. Flowers are hermphrodite, sessile, bracteate, fragrant, and 1 to 16 in number borne terminally on leafless stalk called scape which may be two-edged or roundish. The flowers form an umbel. The ovary is three locular, with one or more ovules per locule. A flower bears a slender greenish white *floral tube which* divides at its tip into six long, reflexed and narrow, white or pale green membranous *tepals* that gives flower a spidery look. Above the tepals, a delicate white, saucer or funnel like membranous *staminal cup* joins the bases of the six long and thin stamens (Smith and Garland 2003; Tapia-Campos et al. 2012; Garland et al. 2013). The fruit is a fleshy capsule bearing 1–6 seeds (Smith and Garland 2003). Seeds are fleshy, heavy, exhibit no dormancy, and germinate in 3–4 weeks after dispersal (Meerow et al. 2002).

Jee and Vijayavalli (1999) studied and reported the karyotypes of eight taxa of *Hymenocallis* Salisb. viz., *H. harrisiana* Herb (2n=74), *H. speciosa* Salisb. (2n=54), *H. littoralis* Salisb., *H. rotata* Salisb., and *H. occidentalis* Kunth. (2n=46) with three different chromosome numbers (2n=42, 44, and 46) in three accessions of *H. daphe* Herb. They further suggested that polyploidy and aneuploidy have contributed significantly in the evolution of chromosomes in the genus with n=23 as the secondarily derived basic number for the genus. Haron (2003) has reported that the *H. littoralis* has 2n=44, 46, 48, and 68. Later 2n=68 in *H. littorallis* was also reported by Nwankiti (1985) and was recently recorded with 2n=52 from Thailand using the FISH technique (Tanee et al. 2018). Smith et al. (2001) identified a new species *H. frankliensis* from Florida with 2n=43, 44. The lowest chromosome number (2n=38) is reported in *H. henry* Traub (Smith and Flory 1990).

9.6.2 Origin, Domestication, and Spread

Hymenocallis Salisb. is an entirely American genus of the family Amaryllidaceae with species distributed along the warmer parts of North, South and Latin America, West Indies (Flory 1976; Jee and Vijayavalli 1999; Meerow et al. 2002; Grossi 2007), SE United states, Mexico, Central America. Mexico is home to the greatest number of species with a secondary area of diversity in the USA (Grossi 2007). Only three species are endemic to South America (*H. littoralis*, *H. pedalis,* and *H. tubiflora*) (Meerow et al. 2002; Grossi 2007). *Hymenocallis* sp. grow in diverse habitats. The North American species inhabit in and around the wetland habitats and less often in dry, flat woods or in disturbed sites while the Mexican species grow all around Mexico from xeric to aquatic conditions from tropical coastal areas to higher elevations and arid climates (Preuss 2002; Tapia-Campos et al. 2012). The genus ranges through the tropics, subtropics, and warm temperate regions of the New World and includes about 40 species (Garland et al. 2013). As indicated by Smith and his co-workers, 13 species are found in Florida (Garland et al. 2013).

9.6.3 Plant Genetic Resources

9.6.3.1 Geographic Distribution

H. littoralis originates from South and Central America, but it is cultivated and naturalized in tropical Africa, Asia, and the Pacific islands in the Malesian region, and it is recorded as naturalized in Java, the Philippines, and the Bismarck Archipelago (Haron 2003) Argentina, Brazil, China, Colombia, French, Guinea, Liberia, Mexico, Peru, South India, United States (Sarma and Debnath 2016) and in many other tropical regions as medicinal and an ornamental plant (Tanee et al. 2018). Jee and Vijayavalli (1999) collected eight taxa from South Indian states of Kerala and Tamil Nadu while *H. lottorolis* in Tripura by Sarma and Debnath (2016). Several species other than *H. littoralis* are cultivated as an ornamental in the Malesian region, for example, *H. caribea* (L.) Herbert, *H. Narcissiflora* (Jacq.) MacBr., and *H. speciosa* (Salisb.) Salisb., but only *H. littoralis* has become naturalized (Haron 2003).

9.6.4 Gene Pool

The different wild and cultivated species of *Hymenocallis* and closely related genera *Ismene, Elisena,* and *Leptochiton* previously classified as *Hymenocallis* and their complex hybrids and cultivars form the important gene pool for *Hymenocallis*.

9.6.5 Collections

The genetic diversity of cultivated and naturalized plants of *Hymenocallis* in South-East Asia may be very low. In Central America, several *Hymenocallis* species are classified as endangered. In Puebla (Mexico), a germplasm collection of ornamental geophytes is maintained including *Hymenocallis* (Haron 2003). The Centro de Investigación y Asistencia en Tecnología y Diseño del Estado de Jalisco (CIATEJ), Mexico, are also conducting collection, domestication, and breeding programs with different promising ornamental species native to Mexico along with *Hymenocallis* (Cruz-Duque et al. 2019). The institutes like USDA are emphasizing on conservation of natural habitats of *Hymenocallis* and the natural flora of America.

9.6.6 Characterization and Evaluation

Sealy (1954) narrated a history of the genus, taxonomic criteria, and brief review of 27 perfectly known and three other imperfectly known species of Hymenocallis while Traub (1962) has classified the species into six alliances based on the morphological characteristics as (I) Speciosa with species is having evergreen, petiolate, and broad leaves with tropical origin and distribution in Brazil, Mexico, and Carribean. (II) Caribaea is having tropical and subtropical species with evergreen, sessile, broad, oblong leaves and distribution in Central America, Florida, and Caribbean. (III) Littoralis is having species with slightly lanceolate and evergreen leaves and shortly joined perigonium segments with staminal corona and distributed in Colombia, Ecuador, and Mexico. (IV) Caroliniana with deciduous, South American species is having caduceus leaves, introrse anthers, and globular ovary with less than four ovules per locule. (V) Henryae distributed in Florida and Cuba is having cadudeus leaves, pale green tepals, introrse anthers, oblong ovaries with more than four ovules per locule and (VI) Mexicana with deciduous Mexican native species with sessile or subpetiolate, ensiform or caduceus elliptical leaves (Grossi 2007; Tapia-Campos et al. 2012).

Gerald L. Smith and co-workers *viz.*, Anderson, Garland, Flory, *etc.,* had a great work on identification and characterization of flora of *Hymenocallis* after the initial works by Tarub and others.

The scientific and common names of important species of spiderlilies as per the website of USDA are: *Hymenocallis caribaea* (L.) Herb. (Caribbean spiderlily), *Hymenocallis choctawensis* Traub (Choctaw spiderlily), *Hymenocallis coronaria* (Leconte) Kunth (Shoals spiderlily), *Hymenocallis crassifolia* Herb. (Coastal Carolina spiderlily), *Hymenocallis duvalensis* Traub (Dixie spiderlily), *Hymenocallis expansa* (Herb.) (West Indian spiderlily), *Hymenocallis franklinensis* G. Lom. Sm., L.C. Anderson & Flory (Franklin spiderlily), *Hymenocallis gholsonii* G. Lom. Sm. & Garland – (Gholson's spiderlily), *Hymenocallis godfreyi* G. Lom. Sm. & Darst (Godfrey's spiderlily), *Hymenocallis henryae* Traub (Henry's spiderlily), *Hymenocallis latifolia* (Mill.) M. Roem. (Perfumed spiderlily), *Hymenocallis liriosme* (Raf.) Shinners (spring spiderlily), *Hymenocallis littoralis* (Jacq.) Salisb. (Beach spiderlily), *Hymenocallis occidentalis* (Leconte) Kunth (Northern spiderlily), *Hymenocallis palmeri* S. Watson (Alligatorlily), *Hymenocallis puntagordensis* Traub (Punta Gorda spiderlily), *Hymenocallis pygmaea* Traub (Dwarf spiderlily), *Hymenocallis rotata* (Ker Gawl.) Herb. (Streambank spiderlily), *Hymenocallis speciosa* (L. f.) Salisb. (Greentinge spiderlily), *Hymenocallis tridentata* Small (Florida spiderlily).

9.6.7 Use of Plant Genetic Resources

These bulbous plants are suitable for tropical plains and also for hilly altitudes up to 3000 m. These plants are suitable for growing along swamps, water channels and pools, on the borders, along the paths, in the beds, in pots, and for mass effects.

Tropical Giant, Bellum, Sulphur Queen, Pax (yellow hybrid), Sofforthiae (most popular fragrant hybrid with bright golden yellow blooms most suitable as cut flowers), Zwanenburg (suitable for cut flowers or as pot plants), Zeylanicum, Carribea (a cut flower hybrid), *etc.*, are important genotypes developed in *Hymenocallis*.

9.6.8 Looking Forward or Future Perspective

The spider lilies are lacking in color variations which should be introduced through the genetic engineering tools, mutation breeding, and wide hybridization tools to create novelty in the genus. The types with more postharvest life should also be bred to enhance the status of the crop and growers around the world.

H. littoralis, an interesting medicinal plant, deserves more attention as a source of alkaloids and compounds with anticancer and antiviral (including anti-HIV) activities, but the reputed beneficial effects of bulb extracts internally on asthma and cough as well as externally on wounds, swellings, bruises, and boils also merit more attention and research (Haron 2003).

9.7 Iris

9.7.1 Introduction

The genus *Iris* is a large group of flowering plants comprising of 260–300 species with colorful showy flowers. It takes its name from the Greek goddess "Iris" word for a rainbow, referring to the wide variety of flower colors found among the many species. The genus *Iris* is separated into two major groups *viz.* the bulbous and the nonbulbous (rhizomatous) (De Munk and Schipper 1993). For flower production, the bulbous Irises are the most important and comprise of three major groups: Reticulata, Juno, and Xiphium. The nonbulbous (rhizomatous) group is important for flower bulb industry. Most of the commercial cultivars belong to the Dutch Iris (*Iris hollandica*) group, which was derived from crosses between two *Xiphium* species *viz., X. vulgare* and *X. tingitanum*.

9.7.2 Botany and Distribution

Species of the genus *Iris* are recognized by their basal fan of unifacial leaves; colorful perianth of three horizontal sepals and three upright petals that are basally fused into a tube; style branches that are fused at the base, petaloid distally and extend beyond the small flap-like, transverse stigma as a bifid crest; and three stamens that are opposite to the sepals and petaloid style. They have distichous leaves and the large spathe valves are either membranous or herbaceous. It is a terrestrial or lithophytic herb or undershrub to 1.5–2.0 m tall. Iris flowers round the year with its peak bloom during the months of March–May.

Various species of *Iris* grow abundantly in diverse habitats such as alpine and subalpine meadows, roadsides, stream banks, public gardens, orchards, saffron fields, graveyards, and cemeteries (Zeerak and Wani 2007). Plants of genus *Iris* comprises of over 3000 species in the world of which 12 species are reported in India (Bhattacharjee 1998) but most commonly grown *Iris* species found in India are *I. croceae, I. ensata, I. germanica, I. hookerian, I. kumaonensis,* and *I. kashmiriana* which are confined to Himalayan regions.

9.7.3 Origin, Domestication, and Spread

Irises are widespread and found growing wild in all parts of the northern temperate zone from California in the west right round the globe to China and Japan. To the north, they extend as far as Alaska, Kamchatka, and Siberia and to the south to Hong-Kong, Southern Arabia and Florida. They have a wide range of growth habit from a few yards of the sea as, for instance, in Portugal where *Iris subbiflora* grows on the coast near Coimbra, or in Maine where a form of *Iris setosa* flourishes within reach of the salt spray, while in China and Tibet they are found at elevations of more than 12,000 feet.

The **Xiphiums**, which comprises the so-called English and Spanish Irises, are confined to Spain and Portugal with the neighboring countries of Southern France and North Africa and are characterized by the shape of their flowers and by the bulbs.

The **Juno** species are found near the shores of the Mediterranean, in Asia Minor, Northern Mesopotamia, Turkestan, and in the Salt Gange on the Northwest frontier of India.

The third section **Reticulata** is confined to the region between the Caucasus and the south of Palestine, from a valley in Turkestan.

The nonbulbous Irises have a rhizome, which is a creeping stem. Based on the presence of rhizome, irises are classified as bearded and beardless species, which are known as Pogoniris and Apogons. Species of the Pogoniris section are found growing wild from the Atlantic coast of Portugal and the Atlas mountains of Morocco through Central and Southern Europe, and Asia Minor to Turkestan and the mountains of Manchuria, Tibet, and Western China. Further, the bearded irises are classified into three minor sections *viz.,* Oncocyclus, Regelias, and Pseudo-regelias. The Oncocyclus species ranges through Asia Minor, Syria, the Caucasus, and Western Persia. Another species Regelias are confined to the northern side of the great mountainous backbone of Central Asia, while Pseudoregelia are found on the southern side in northern India.

The Apogons or Beardless species are by far the most widely distributed of all the divisions of the genus and are found all over the temperate regions of Europe, Asia, and North America.

9.7.4 Collections and Conservation

The genus *Iris* comprises of 53 threatened taxa and out of which 29 belong to the section Oncocyclus (Walter and Gillet 1998). Oncocyclus have been of interest to the horticulturists due to its large showy flowers (Foster 1899). They are xerophytic plants growing naturally in the Caucasus, eastern Turkey, Syria, Lebanon, Jordan, and Israel, and further to the east in Iraq, Iran, and Afghanistan. Results of the field surveys by various workers have indicated that there has been a decrease in *Iris* populations worldwide and more specifically the Oncocyclus irises. Today the main factors affecting the potential extinction are the human activities which have shaped the number of populations to a critical stage during the past few years. In an effort to conserve the *Iris* populations in Lebanon, microreserves are being established. Rescue from possible extinction could be performed through the reinforcement of populations by introducing individuals from adjacent populations. In this regard, defining species and differentiating taxa is a crucial and a prerequisite in order to define conservation units.

Another species under threat is the Western blue flag (*Iris missourensis*) which is a native of Canada and is mostly confined to areas with high soil moisture in spring and dry conditions later in summer. The most significant threats to Western Blue Flag are the continuous habitat loss, fragmentation and degradation from trampling or overgrazing by livestock, and invasive alien plant species. Threats of a lower significance include fire suppression, excavation of soil, alteration of hydrology (sewer outlet), and encroachment of problematic native species (woody vegetation encroachment). In an effort to conserve the Western Blue Flag, a conservation program was established in 2002 to deliver and monitor the plant population. Management and conservation plan for *Iris missourensis* consists of inventory and monitoring; research as part of an adaptive management framework, communication, collaboration, and engagement; and habitat assessment, management, and conservation (Environment and Climate Change for Canada 2017).

Method of conservation	Place	Species conserved
Ex situ conservation	Millennium Seed Bank (Wakehurst Place, Botanic Gardens Kew, UK)	*I .cedretii* and *I. sofarana subsp. Kasruwana* seeds (Saad et al. 2009)
	Seed Bank at Herbarium of Hebrew University and University Botanical Garden of Jerusalem	Highly endangered *I. atropurpurea* and *I. hieruchamensis* (Cohen and Avishai 2000)
In situ conservation	Creation of two living collections outside the natural populations, but within the same ecological conditions; and relocation experiments from the Northern Negev, Israel	Critically endangered plant species, *I. atrofusca* (Volis et al. 2010)
	Creation of Nature Reserve in Israel	Highly endangered *I. atropurpurea* and *I. hieruchamensis* (Cohen and Avishai 2000)
	Wikemong Reserve in Canada	special concerned Dwarf Lake Iris (*I. lacustris*) (Cosewic 2010)

9.7.5 Characterization and Evaluation

It is very important to understand the genetic diversity and relationships among plant species and varieties from the breeding point of view and intellectual property rights (IPR) (Tay 2006; Wanjala et al. 2013). The genus *Iris* has a rich genetic diversity due to which there is relatively large genetic differentiation at the species level. Wild relative of *Iris* can be used for crossbreeding, as well as for ornamental and medicinal purposes. Morphological characteristics of 53 accessions were collected from Liaoning Province, which is a primary distribution area of *Iris* in China. *Iris tigridia* Bunge and *Iris ensata* Thunb. had shown to have better ornamental value than the other species under evaluation (Zheng et al. 2016).

Having a larger reservoir of variation leads to higher chances of finding particular characters, such as resistance genes for diseases and pests or for adaptation to wider ecological amplitudes and stress conditions. In recent years, more than 60 new genotypes of *Iris germanica* and *I. spuria* have been bred in Iran, which are used for hybridization and can provide new perspectives on the composition of wild types and other species. Important species of Iranian irises include *Iris acutiloba, Iris persica, Iris iberica*, and *Iris spuria*. From the medicinal point of view, *I. acutiloba* had the highest concentration of anthocyanins, flavonoids, and carotenoids (Azimi et al. 2019).

Iris species have beautiful linear foliage that makes them one of the most favored ornamental plants. Additionally, because of their high resistance to cold, drought, disease, and salinity, some species, such as *Iris lactea, Iris sanguinea, and Iris halophila*, can be successfully grown in coastal and saline-alkaline areas (Bai et al. 2008). In addition, *Iris tectorum, Belamcanda sinensis, Iris germanica,* and a few other species have medicinal value, containing flavonoids with good detoxification effects (Agarwal et al. 1984; Burcu et al. 2014). On the basis of soil and water requirements, *Iris* can be divided into three categories: the first group includes species that prefer weakly alkaline, calcareous, damp, fertile, and well-drained soil, such as *Iris tectorum* Maxim. and *I. germanica*; the second group includes the species that thrive in acidic and wet soil, such as *I. japonica* and *I. pseudacorus;* the third group includes species that can adapt to any type of soil-poor, dry or wet such as *I. lactea var. chinensis*. Among the different species, *Iris germanica* is also known to show the highest drought tolerance (Liu et al. 2005, Xu et al. 2019).

Many new cultivars with desirable traits have been bred through crossbreeding, by using the extensive collections of wild species and varieties. Embryo culture, somatic hybridization, and transgenic breeding are a few other successful methods used for *Iris* breeding (Shimizu et al. 1999). Till 2009, reports have shown that there were more than 30, 000 *Iris* cultivars in the world, as catalogued by the American Iris Association (Lin et al. 2010; Zhang 2010). In China, wild *Iris* resources with many good genes are abundant, and these may be used to improve, innovate, and preserve the *Iris* germplasm. However, *Iris* breeding started late and developed relatively slowly in China. Only a small part of the *Iris* resource was utilized directly without any modification, probably due to the lack of systematic research. In breeding programs, breeders typically select parents with good performance and a wide hereditary basis, according to their genetic diversity and relatedness to parental germplasm, which are very important criteria for crossbreeding (Hesham and Yan 2010; Matus and Hayes 2002).

There is little information available on the genetic diversity and relationships among the different species of iris considering its vast genetic diversity. Understanding the genetic variations within and between populations is essential for the establishment of effective and efficient methods of conservation of plants. In *Iris* particularly, where there is a great genetic diversity in species, the use of molecular markers is a powerful tool in the genetic study of such populations. The use of DNA markers, AFLP, SSR, RAPD, and ISSR represents an alternative method in detection of polymorphism. Genetic variations in some wild iris genotypes in Iran *viz. Iris kopetdaghensis, I. songarica, I. fosteriana* were evaluated using ISSR markers. These markers have proved to be an efficient tool in determining the high genetic diversity among these wild genotypes and have been successfully used in *Iris* breeding programs (Atari et al. 2017).

Amplified fragment length polymorphism (AFLP) markers in conjunction with a combination of EcoRI/MseI restriction enzymes have also been used to study the genetic diversity and relationships among 15 species of *Iris* collected in China. An assessment of genetic diversity parameters using AFLP markers showed that *Iris* has high genetic diversity at the species level. Clustering analysis and principal coordinate analysis showed that the 15 species of *Iris* were genetically similar and thus related. When the genetic similarity coefficient was 0.55, the 15 species could be divided into five distinct groups. These findings verify, replenish, and consummate the classical taxonomy and systematology of *Iris* and also provide references for the conservation, management, classification, identification, and breeding of *Iris* resources.

9.7.6 Breeding and Use of Plant Genetic Resources

Breeding and crop improvement of Iris had been started at the beginning of the twentieth century in the Netherlands (Dix 1974), and initially the "Spanish" irises (*X. vulgare* group) and many other varieties were released. The "German irises" were diploid bearded types that had 12 pairs of chromosomes and were descended from complex crosses involving two wild species-the lavender *Iris pallida* and the yellow and red *I. variegata*, both from southern Europe. During the initial phase of the year 1910, these garden diploids were crossed with a series of wild tetraploids (*I. ciipriana, I. ernesopotamica*, etc.) that had 24 pairs of chromosomes. These forms, all from the eastern Mediterranean region, were all purplish blue in color, taller, larger, as well as susceptible to cold and other unfavorable conditions than the older diploids. The modern tall bearded irises have been developed from these crosses.

Two very important breakthroughs took place in Iris breeding *viz.* the successful inter-specific crosses between small bulbing cultivars of the *X. vulgare* group and the crosses between the two different types of the *X. tingitanum* species. Subsequently, these crosses produced cultivars with large bulbs like Wedgwood, Prof. Blaauw, and Blue Magic which are adapted to year-round flower production (Dix 1974). "Blue Magic" is still the most widely grown cultivar in the Netherlands. Initially, breeding with these inter-specific hybrids was limited since they were sterile. However, a fertile spontaneous tetraploid of "Wedgwood" was observed during 1952 which made it possible to make crosses with cultivars of the *X. vulgare* types and several triploid cultivars, among which "Telstar" was one.

A few fertile plants were also found in "Prof. Blaauw" in different ecological conditions in western France (Brittany) (Le Nard, unpublished data). The seeds of these plants were collected and sown which also produced fertile plants having tetraploid chromosome. Inter-crossing and crossing of these plants with diploid cultivars were carried out and the resulting selections were released for cultivation in the year 1990. Systematic and scientific research devoted to *Iris* breeding is limited and is being carried out mostly by *Iris* growers and bulb-producing companies. In Netherlands, a breeding program using inter-specific hybridization was initiated to get different colored cultivars with the ability to flower year round (Eikelboom and Van Eijk 1990). The findings of the research showed some successful data on the transmission of flower colors; however, the use of colchicine for the production of fertile tetraploid plants was not successful. It has been suggested by Kim and De Hertogh (1997) to use a combination of mitotic substances and in vitro culture for successful breeding of the bulbous Iris.

9.7.7 Major Constraints in the Crop Production

Irises are very slow growing geophytes which take nearly two years or more for the plants to come into the flowering stage. This causes a major hindrance in the breeding program of Irises. For successful crosses, hybridization success rates vary from 25 to 65% and germination percentage of around 30 to 33%. The flowers of *Iris* are rarely pollinated naturally; however, it will set seed freely when hand-pollinated. Therefore, it is not necessary to remove the anthers and enclose the flowers in bags when making crosses. Irises are usually propagated through underground stems or rhizomes which perpetuate the genetic composition of the original plant. This makes it easy to keep parents indefinitely for comparison with or crossing with their descendants. Reticulata *Iris* breeding can be enhanced by developing polyploids either by tetraploids (4n) or possibly octaploids (8n). These polyploids have larger flowers, tolerate poor weather, and have a longer flowering period (Mc. Murtrie 2016).

Common sources used to overcome production constraints are listing of genetic resources, genetic stocks (including aneuploids series, substitution and translocation lines, recombinant inbred lines, *etc.*), inbred lines, released cultivars associated with desired traits, genes with gene symbols, mapping populations, *etc.*

9.8 Ixia

Ixia, a winter annual, is commonly known as "wand flower" that refers to the thin, wiry but strong stems that wave in the wind or "African cornflowers" or "corn lily" referring to the stem with its buds that resembles the stalk of corn. The derivation of the genus name "*Ixia*" has two different opinions. The one says that it came from the *Greek* word "*ixos*" meaning mistletoe (viscum), birdlime, referring to the viscous juicy sap, while the other says that Linnaeus derived the name from an old *Greek* name for a plant noted for the variability of its flower color (Lewis 1962). Since *Ixia* is a variable genus of many flower colors, the later explanation is more appropriate.

9.8.1 Origin, Domestication, and Spread

Ixia is native to the western and southern parts of South Africa, that is, Cape Province of South Africa. It is restricted to the winter-rainfall zone of South Africa. This African flower has been widely naturalized in Europe, south-western USA (*i.e.,* California), and parts of southern Australia (*i.e.,* in the coastal districts of central and southern New South Wales, Victoria, Tasmania, south-eastern South Australia and on Kangaroo Island, and in south-western Western Australia. However, in many parts, it has escaped the cultivation and became weed in woodlands, shaded moist sites, urban bushland, roadsides, disturbed sites, and waste areas.

9.8.2 Botany and Distribution

Ixias are tufted deciduous geophytic herbs characterized by sword like long and narrow leaves arising annually from perennial corm. Stems are 40 to 100 cm tall, slender, wiry, erect usually with 1 or more almost erect branches. Stems require support to be held upright. Leaves are flat or curved with narrowly elliptical blades 25 to 40 cm long and 3–10 mm broader and prominent veins. Roots are small, corms globular with fine roots and smaller cormels. Flower stems of almost equal length of 30–100 cm length borne during spring to summer. Flower heads are 4–20 flowered spikes having membranous floral bracts that are attached at the base of the ovary. Flowers are sessile, perianth 50–70 mm long of varying colors with a long narrow funnel shaped tube with six radiating tepals giving it a star shape outline. Outer three tepals are narrowly egg shaped, 20–25 mm long, and about 3–6 mm wide, while the inner three are slightly narrower. Ovary is inferior. Style is thread like, branching 2 mm long, spoon shaped, curved. Stamens are attached to the perianth tube, filaments are free, anthers 5–6 mm long and erect. Pollination is by insects *viz.*, bees, beetles, flies, and butterflies that visit the flowers for mating, to eat away the nectar and pollens. Fruits are botanically capsules with many brown seeds. The capsule dries and splits open and seeds are scattered about the parent as the stem weaves and bobs in the breeze. It is commonly reproduced vegetatively by corms. Spread of the species is by movement of corms in water flows and soil.

9.8.3 Plant Genetic Resources

The genus was established in late seventeenth century and currently genus Ixia consists of 99 species, each being a little different from others, especially in terms of its flower color and structure and are taxonomically classified into four sections *viz.*, *Ixia* (29 species), *Dichone* (17 species), *Hyalis* (19 species), *Morphixia* (34 species) with few circumscriptions between Lewis and Goldblatt & Manning. It belongs to the Iridaceae family and is placed in the largest subfamily Crocoidae (Goldblatt et al. 2015; Goldblatt and Manning 2012, 2016) in a tribe Ixieae with 14 genera.

Ixia* sect. *Dichone (Salisb. ex Baker) Goldblatt and J. C. Manning (Goldblatt and Manning 2011, 2016)

Dichone is one of four sections in the genus currently includes 17 species and three varieties. The meaning of a *Greek* word "*Dichone*" is two tubes that might refer to the incompletely dehiscent anther thecae. The section is vegetatively identical to section Ixia but is distinguished by the following floral characters: filiform lower part and short to vestigial upper part of the relatively short perianth tube; filaments not decurrent; stamens unilateral and reclinate in some species with horizontal to pendent anthers and incompletely dehiscent anthers in few species, opening from the base; style tightly clasping and branches involute-tubular and stigmatic only at the tips; anthers short, oblong to suborbicular, usually so-called subdidymous. Flowers are usually with shades of pink (rarely white), but blue-mauve in *I. brevituba*.

Ixia* sect. *Ixia (Goldblatt and Manning 2011, 2016)

Flowers with a filiform perianth tube, usually well developed and sometimes elongate; anthers linear, dehiscent along their entire length; style branches narrowly channeled (Lewis 1962, described the style branches as conduplicate, but they are not folded together, but rather form a narrow channel) and stigmatic along the margins, sometimes only toward the tips. The stamens are always symmetrically arranged with the filaments inserted at the base of the tepals and not decurrent. Leaf number is indeterminate, but usually at least four and up to 10. The upper leaves decrease in size progressively above, becoming partly to largely sheathing, but lack a sharp distinction between basal foliage and upper sheathing leaves. Flowers are usually unscented and often brightly colored and frequently have a dark central mark, either restricted to the tepal bases or including the filaments and sometimes the anthers and style branches. Nineteen species are currently recognized.

Ixia* sect. *Morphixia (Ker Gawl.) (Pax 1888; Goldblatt and Manning 2011, 2016)

Perianth tube is hollow; filaments inserted within the tube and decurrent, not connivent at the base. The style branches are usually fairly short and recurved, but of the same type as in sect. Ixia. Although usually central, the stamens are occasionally unilateral (a feature not known to Lewis): horizontal in *Ixia pauciflora*, declinate in *I. reclinata*, and ± arcuate in *I. stenophylla*. The tepals are seldom brightly colored, more often being muted shades of blue-grey, mauve pink, or white and only rarely with weakly developed markings. Floral scent is common and usually reminiscent of rose to violet but *I. rivulicola* has a rich floral scent with notes of passion fruit and sweet pea. Nectar is always present in the base of the perianth tube, sometimes in substantial quantities. An important feature of the section is that leaf number is determinate, usually three, occasionally more, and the basal leaves with expanded blades are sharply distinct from the uppermost one or two, which are entirely sheathing. There are 31 species recognized in the section.

Ixia sect. Hyalis (Baker) (Diels 1930; Goldblatt and Manning 2011, 2016)

A residual group includes species with style and anthers of the sect. Ixia type; an indeterminate leaf number, the upper leaves weakly differentiated from the lower, and usually with free blades; and tubular or funnel-shaped perianth tube ranging from hollow throughout to filiform in the lower two thirds and expanded above (thus, with decurrent filaments). Nectar is normally produced and the flowers are often scented (depending on the pollination system – those pollinated by long-proboscid flies are unscented). The outer bracts with usually only one prominent central vein are a feature of most species. The circumscription of sect. *Hyalis* adopted here thus differs from Lewis's; in that it is not restricted to species with an elongate perianth tube although the type species, *Ixia paniculata*, has the longest tube in the genus. There are 18 recognized species under this section.

Within sections Hyalis and Morphixia, there are recognized eight informal clusters of species sharing one or more derived features, for convenience called series, and most of these are monophyletic but some may be paraphyletic. The species of sect. Hyalis are distributed among three separate series and those of sect. Morphixia among five.

9.8.4 Cultivation

The Ixias are ideal ornamentals for the hot, dry climatic conditions and can be grown in warmer climates under protection from rains and water under partial or full sun. In landscaping, they are used for group planting in borders, containers, and bedding plants for their attractive and brilliantly blossoms with six petalled flowers colored from deep crimson to orange with contrasting spots and dots on them. The sites sheltered from high winds receiving full sun with well-drained sandy to loamy soils rich in organic matter and slightly acidic to slightly alkaline with lime are best. They are grown with the corms as are the members of Iridaceae family. Corms are planted in mid- to late autumn on beds at a depth of 3–4" and spacing of 3–4" apart. Apply irrigations @ 1 inch per week by moderate quality of water to just moisten and not to saturate the media with due care during active growth stages. Fertilizers are applied when sprouts come out within 1–3 weeks from planting that will bloom through spring and summer. After the harvest of cut flowers, keep the leaves intact for their corm development and multiplication. Later, when leaves turn yellow denoting the maturity of corms, the leaves are cut back and dig out corms and store in a cool location in a dry medium until the following spring or kept as it is for next season.

For pots, tubs, and urns, fill the containers with good quality, well-drained soil or any commercially available potting media mix with adequate drainage holes in container. Keep the containers where they will receive full sun. Plant the bulbs 2–3" apart for a good display at a depth of 4" inches into the media. After planting, water well to gently soak and settle the soil around the corms. The corms will give out roots earlier but shooting and flowering can be seen in spring and summer. After blooming has finished for the season, keep the leaves intact on plant for nourishing the developing corms. When leaves turn and dry out, remove them and let the corms remain in the container till next autumn in dormant stage. In dormant stage, protect them from mealybugs. In autumn, fertigate the corms for the nourishment and to encourage better flowering in the next spring.

A growth retardant, Paclobutrazol can be applied as a preplant corm soak, a postemergent drench, or a postemergent spray in combination with a 2- to 4-week preplant storage of corms at 7 °C and an 18 °C day/10 °C night forcing temperature to produce dwarfer, attractive, and marketable pot plants (Wulster and Ombrello 2000).

9.9 Nerine

The genus *Nerine* comprises of perennial bulbous ornamentals bearing colorful long lasting flowers of white pink to crimson shade in a spherical umbel of lily type flowers. Though they are known as Japanese spider lily and cape flower lily, these are not true lily, that is, liliaceae family, but more closely related to its relative *Amaryllis*. There are 20–30 species in the genus, but only *N. bowdenii*, *N. sarniensis*, and *N. undulate* (Syn.*N. flexuosa*) are commercialized. Nerine is gaining more and more horticultural interest for fresh cut flowers and garden plants but some species and cultivarshave gained popularity as potted plant too. The colorful, elegant flowers of Nerines can survive up to 14 days in a vase with water without showing any signs of wilting, senescence, or fading. It is an important cut flower but low flowering percentage limits its share in Dutch cut flower market.

9.9.1 Botany and Distribution

The genus *Nerine* is a member of class monocotyledonae, order Asparagales, family Amaryllidaceae, and tribe Amaryllidae. The geophytic organ is a true perennial bulb, comprised of fleshy leaf bases and covered with papery tunics. The bulbs of *Nerine* species need a minimum of two years growth and development in order to produce their first flowers. The largest bulbs can give rise to two stems or more if they have been grown under suitable conditions. Depending on the species, the leaves are long and slender and they can be evergreen or deciduous and sometime hysteranthous. In the case of deciduous species, the inflorescence may appear on naked stems before the leaves develop (hysteranthy), otherwise they appear together with the flowers (synanthy) or afterwards. Generally, flowers (8–20) are borne in umbels on leafless scapes and they can be red, rose, pink, purple, or white.

9.9.2 Origin, Domestication, and Spread

All species of the genus *Nerine* are native to South Africa and these species occupy wide spread range of habitat. This group is endemic to five southern African countries: Botswana, Lesotho, Namibia, South Africa, and Swaiziland (Duncan 2005). Concentrated in the summer rain fall zone in the eastern part of South Africa and comprises some 23 species (Zonneveld and Duncan 2006). The habitat diverse range starts from cool mountains, swamps, and desert plain ascending to mountain. They prefer rocky, arid habitats, and most species are found in the summer rainfall region (Snijman and Linder 1996).

9.9.3 Plant Genetic Resources

The basic chromosome number is 11 (2n=22) but plants with 2n=24 and triploids are also recorded (Traub 1967). During the early history of the crop improvement and hybridization of *Nerine*, the plant breeders contributed to great achievement using only a few species (Du Plessis and Duncan 1989). In nineteenth century, many hybrids appeared in England were developed using *N. sarnensis*, *N. undulate,* and *N. humilis*, but *N. bowdenii* was only introduced in the early 1900s providing breeders with the most valuable *Nerine* species of all (Duncan 2002). *N. bowdenii* is one of the hardiest species and thus used for breeding hardy varieties with wider color range. Some of the cultivars are *N. Bowdenii* "Alba," "Pallida," "Marnie Rogerson," "White Magic," "Stefanie," and "Blush White." Along with *N. bowdenii,* other species like *N. sarnensis* and different cultivars have been extensively used in plant breeding programs that have produced the majority of the commercially available hybrids. Consequently, commercial cultivation of *Nerine* now occurs throughout the world, particularly The Netherlands, Israel, South Africa, New Zealand, and Southern Australia. Its cultivation is also started in India. *N. sarniensis* is the main parent of 300 years of *Nerine* breeding. The world's largest collection of *N. sarniensis* hybrids is kept at Exbury Gardens in the United Kingdom.

9.9.4 Important Species of *Nerine*

Nerine bowdenii: *N. bowdenii* is named after Cornish Bowden who brought this variety from South Africa to England in 1903. This species is also known as Cornish lily. It is hardiest species and more robust than the other species of the genus *Nerine*. This may be the reason that it has been used as a parent in crosses to obtain hardy plants along with a wider color range and flower morphology.

Nerine sarniensis: *N. sarnensis*is also known as Guernsey lily as it is associated with the island of Guernsey. It has long lasting cut flowers. A wide range of color from white through pink, red orange, and mauve with well-formed inflorescence is present but multiplication rate is slow.

Nerine pancratiodes: It is a distinct species of the genus and cannot be easily confused with other members of the genus. It has pure white, funnel shaped, suberect flowers borne on densely pubescent suberect pedicels. The inflorescence is borne on a very long, erect, glabrous scape and the ovary is distinctive deep pinkish maroon.

9.9.5 Molecular Characterization and Evaluation

With the developing field of genome size measurement, it has been shown that when species in a genus have the same chromosome number, as in *Nerine* (2n=2x=22), differences in nuclear DNA content can be used as an additional character to delimit species and intrageneric divisions. Studies were carried out on genome size of Nerine and correlation was demonstrated with their nuclear DNA content, growth cycle, leaf width, and other morphological characters (Zonneveld and Duncan 2006).

9.9.6 Conservation

Some *Nerine* species in different parts of South Africa are rare or in danger of extinction due to degradation of their habitat due to summer rains, overgrazing, and erosion due to construction of roads. Some of these species are *N. pudica, N. marincowitzi, N. masoniorum, N. filamentosa, N. masoniorum, N. gibsonii, etc.* Graham Duncan has done great work on collection and hybridization and is well known for his expertise on Nerine. Different measures have been taken to conserve the species and to relieve the threat of extinction and one of them is *ex situ* conservation. Graham Duncan contributed a lot in collection and conservation of Nerine species at the Kirstenbosch National Botanical Garden. Some of these species are *N. filamentosa, N. gibsonni, N. huttoniae,* and *N. masoniorum*inthe. In Guernsey, the national flower is *Nerine sarniensis*, and the island collection of nerines is seeking recognition by National Council for the Conservation of Plants and Gardens as a national collection.

9.9.7 Future Perspective

This plant has great horticultural economic value. But the major problems in different species like low multiplication rate, short scape, long period to first flowering and comparatively less flowering limit its popularity in world flower trade. So, there is further need to work on its micripropagation and to bred early and profuse flowering cultivars with wide range of color.

9.10 Zephyranthes

9.10.1 Introduction

Zephyr lily is a bulbous perennial. Native to south-eastern North America, it has pale pink-tinged, white flowers. The name *Zephyranthes* is derived from word "Zephyrus" means the Greek God of west wind that reawakened nature each spring and "anthos" meaning flower. Common names for the species in this genus are fairy lily, rain flower, zephyr lily, magic lily and rain lily. *Zephyranthes* species grows from truncated bulbs and their active period of growth and flowering takes place in summer and a rest period during winter (John Peter Arulanandam et al. 2015). Many *Zephyranthes* species are appreciated as ornamentals and are traditionally known as "rain lilies" due to their tendency to flower shortly after rainy periods.

9.10.2 Botany and Distribution

The genus *Zephyranthes* vary in bulb, flower, and leaf characteristics in respect of size, color, *etc.* The species belonging to this genus are perennial bulbs which tolerate many natural habitats from wet soil to dry conditions. It is a glabrous perennial herb with bulbs, often clumped. Leaf blades are elongate and linear, grass like with overlapping sheath bases. Flowers are radially symmetrical, single, and terminal on scape, with a spathecous bract below, tubular in bud but splits at anthesis about half its length. The ovary is inferior and perianth is tubular with six segments. Flowers are yellow, pink, or white. Stamens are six in number, inserted on the throat of the perianth, anthers medially attached on the back, and stigmas three in number. Capsules are three lobed, locular with numerous, black and lustrous flat seeds.

9.10.3 Origin, Domestication, and Spread

The genus *Zephyranthes* is native to western hemisphere and to the higher altitudes like Mexico, Argentina, where the species possesses greatest cold hardiness potential. Several species have become naturalized and are cultivated as ornamental plants in other places like Hawaii, Indonesia, Thailand, *etc*. The species that is native to the higher altitudes in Mexico (*Z. lindleyana*) Central America, Costa Rica (*Z. carinata*), and parts of North America (*Z. longifolia*) or Argentina (*Z. candida*) represent the species having the greatest potential for cold hardiness. Broadly these plants are distributed in temperate to tropical areas of the world.

9.10.4 Plant Genetic Resources

Zephyranthes atamasca is native to Central Florida, North Florida, and the rest of the southeastern United States. This lily is commonly known as atamasco lily, rain lily, and blooms in March and April. Flowers are large – 3 inches wide or larger – white, and funnel shaped. The flowers are perched on 10-inch stems and will fade to pink. This rain lily is often found in rich, moist soils in swampy forests and coastal prairies and seen in roadside ditches. Clumps of broad, grassy leaves emerge in early winter and die down in late spring or early summer.

Zephyranthes candida, commonly known as fairy lily or white zephyr lily, is native to Argentina and Uruguay, where it is found along rivers and in marshes. This rain lily rapidly forms thick clumps of dark-green, upright, rush-like leaves. The clumps are up to 10 inches tall. In the late summer and fall, fairy lily produces 1–2 inch, white, crocus-like flowers.

Zephyranthes citrina, commonly known as yellow rain lily or citron zephyr lily, is native to the Yucatan peninsula of Mexico. During late summer and fall, these lilies produce small, deep-yellow flowers on stem which is 8–10 inches long. Leaves of this rain lily are up to 12 inches long.

Zephyranthes grandiflora, also known as pink rain lily or rose pink zephyr lily, is native to Central America and the West Indies islands. This rain lily produces bright-pink, funnel-shaped flowers all summer long. Individual flowers are up to 4 inches across and 3 inches long and perched on 10–12 inches tall stems. Each bulb produces clusters of dark-green, strap-like leaves up to 12 inches long. Bulbs rapidly form large clumps and can display up to 20 flowers at once. This lily is also known to be salt tolerant.

Zephyranthes treatiae is also known as Treat's zephyr lily and it flowers during the spring. It is found naturalized in wet pinelands and roadsides. It is native from Southern Georgia and Central peninsular Florida.

Zephyranthes simpsonii is known as red margin zephyr lily and is endemic to Central and Southern Peninsular Florida. It is found growing in wet pinelands and roadsides and flowers during spring and summer.

Zephyranthes rosea is native to Cuba and flowers during summer. Flowers are white to pink in color.

Zephyranthes tubispatha is native of Peru, tropical America, and the West Indies and has been naturalized all throughout the gardens. It is found in well-drained soils and grassy ground of hilly areas which flowers during September.

Zephyranthes susatana is found in semi-arid enclave of high plains of Bogota in Columbia. It is locally abundant in open dry grasslands, which are severely disturbed by man and cattle. It flowers twice in a year *viz.* February to March and November to December. Flowers are vinaceous yellow in the exterior and interior is dark yellow.

Zephyranthes albiella: This species is native to Central Columbia (Cundinamarca). The plants generally presenting leaves simultaneously with flowers. Flowers are white in color with greenish base.

Zephyranthes puertoricensis: It is a native of Puerto Rico and has flowers which are white, greenish in the throat and of intermediate size.

Zephyranthes robusta: This species is considered a native of Argentina, Uruguay, Rio de Grande, and Brazil.

9.10.5 Collections and Conservation

Zephyr lily is found in low woods and wet meadows. It readily colonizes road shoulders in areas of appropriate habitat, apparently benefiting from the extra light and decreased competition in this nonnatural setting. The main threat to this relatively common species is habitat destruction, fire suppression, conversion of habitat to pine plantations and agricultural fields, ditching, draining, and filling wetlands.

9.10.6 Characterization and Evaluation

Zephyranthes species can tolerate many ecological niches (periodically wet soil to desert conditions) and have many ornamental characteristics. The genus has been evaluated for possible medicinal properties and the biochemically toxic compounds are classed as alkaloids. Along with floral morphology, characteristics such as bulb size, bulb tunic, and leaf morphology have also been studied that help to identify individual species. Foliage in the wild is often ephemeral, but under cultivation it becomes more persistent. Leaf color ranges from the bright grassy green of *Z. candida* to rather broad glaucous colored foliage such as found in *Z. drummondii.* A few of the species have distinct bronze tints in the foliage when grown in bright light. Size of leaves in these species ranges from dark green and tiny grassy leaves in species like *Z. jonesi* or *Z. longifolia* to broader, glaucous leaves in species like *Z. drummondii.* Perhaps largest leaves of all are found on *Z. lindleyana* from Mexico, usually distributed as a cultivar called "Horsetail Falls"; this species has handsome broad leaves almost like a Hippeastrum. Flower color in the species ranges from white to yellow (various tints of this color from lemon to sulfur) and pink. *Zephyranthes* have erect flower stalks which support a flower that may be upward facing or slightly nodding. The funnel-shaped flowers with six petals can be crocus shaped, but may also open flat such as in *Z. jonesii* or even reflex slightly. The flowers of some species have a sweet, pleasant fragrance. Fragrance appears to be recessive in crosses, but there are a few species or hybrids, *Z. drummondii* (white), *Z. morrisclintae* (pink), and *Z. jonesii* (light yellow) that all carry the trait. Two of these open their flowers at night and are attractive to nocturnal insects. The flowers typically last only for a day or two; but new flowers may appear in a succession of blooms, especially during humid or rainy weather. Various members of the genus may bloom spring only or repeat and continue into autumn, often a few days after rainstorms thus one of the common names, rain lilies. Periods of synchronous bloom, which breeders have dubbed "blitzes," are part of their ornamental value and the breeders exploit for the purpose of producing new hybrids. Most species under cultivation will bloom without the naturally imposed drought and wet that occurs in nature. Greenhouse grown plants bloom very freely but cycle through periods of bloom. One of the longest blooming of all the species is *Z. primulina* which blooms from April until October. Some other species such as *Z. morrisclintae* appear to bloom only in the spring season. Most of these species are easily propagated vegetatively through offsets or twin scaling. A few of the species such as *Z. clintae* are slow to produce offsets. Sexual reproduction is via seed. Seed usually is best sown quickly after harvest, although short term storage can be successful. Maiden seedling can be brought into bloom for some of the hybrid in 8–12 months after sowing in ideal conditions. This makes it easy to carry out checks for apomixis. The apomictic species freely set seed and faithfully reproduce the maternal phenotype (Gangopadhyay et al. 2009).

9.10.7 Use of Plant Genetic Resources

In *Zephyranthes,* inter-specific and inter-generic hybridizations are complicated by the fact that some of the species are apomictic or pseudogamous, cross incompatible, or have widely variable 2n chromosome numbers. Chromosome number within *Zephyranthes* ranges from 2n=18-96, posing a barrier for crosses among the species (Raina and Khoshoo 1971). Among the different species of Zephyr lily, *Z. primulina,* although it is apomictic, it is a choice parent for crosses because of its rapid repeat flowering trait and long blooming season. In zephyr lilies, the stamens are often spatially separated with either exserted pistils or hidden pistils. This arrangement significantly influences pollination in the wild. Flowers with hidden stigmas are self-pollinated unless insect visited. Several species of *Zephyranthes*, such as *Z. jonesii* which have hidden pistils and are mildly fragrant, bloom at night and release pollen at dawn. This arrangement of stamens insures self-pollination unless the corolla with attached stamens is excised before pollen release when hand pollinating.

Although the presence of alkaloids has been documented in several species (Kojima et al. 1997), and other species have been evaluated for medicinal value (Katoch and Singh 2015), the primary interest in breeding programs is for the ornamental value of the flowers. Collections of wild naturally occurring hybrids such as *Zephyranthes tenexico* have apricot-colored flowers which can be used in breeding programs for extending the color availability (Fellers 1996). Although many of the species have small ephemeral flowers, some hybrids produce larger flowers that will remain open for up to 3 days. However, the potential for new cultivars via cross pollination is limited because of some reproductive barriers in these plants.

Pollination investigation reveals that all studied species of *Zephyranthes* are self-pollinated (Afroz et al. 2018). A summary of the barriers to cross pollination in *Zephyranthes* species and hybrids as well as other genera is as follows (Chowdhury and Hubstenberger 2006):

1. Plant structural morphologies: These include length of the floral tube, spatial arrangement of the stamens, and length of the pistils.

2. Chromosome number or ploidy level: A wide variety of 2n chromosome numbers is a deterrent to crosses.
3. Pollen production: Certain species and hybrids produce limited amounts of pollen.
4. Self or Cross incompatibility: Apparent sterility might actually be incompatibility.
5. Apomixis and/or Pseudogamy: These species often produce prolific seed which reproduces the maternal phenotype.
6. Flowering season: Some species are once-flowering; others repeat flowering throughout the growing season.
7. Receptivity of the stigma: Some species remain open for more than one day and may actually be receptive on day 2 of anthesis.

Identification of breeding lines is important for a successful breeding program. Some of the lines identified are hybrid F_1, *Z grandiflora* which are fertile and have the largest flowers which make it easier for pollination, tri-hybrid female parents [(*Z. candida* x *Z.citrina*) x *Z.macrosiphon*] which represents the cold tolerance of *Z. candida*, deep yellow color of *Z. citrina,* and large pink flowers of *Z. macrosiphon; Z. traubii* and *Z. labuffarosea* are nocturnal hybrids which are presumed to be insect pollinated and are fragrant.

9.10.8 Looking Forward or Future Perspective

The delimiting factors in *Zephyranthes* crop improvement lie in its reproductive biology. Taxonomic studies and classifications of either known or newly introduced species are important. Currently, studies using classical or molecular approaches are also limited and scanty. Clarifying the geographic distribution of wild geophyte species and their genetic diversity is important not only for breeding studies but also for the conservation of rare and endangered species. Re-evaluation of known geophytes is also necessary in their natural habitats both for conservation purposes and for obtaining information. Effective propagation systems, including *in vitro* propagation, need to be developed for many bulb crops.

References[1]

Afroz S, Rehman MO, Md H, A. (2018) Taxonomy and reproductive biology of the Genus *Zephyranthes* Herb. (Liliaceae) Bangladesh. Bangladesh J Plant Taxon 25(1):57–69

Agarwal VK, Thappa RK, Agarwal SG (1984) Isoflavones of two *Iris* species. Phytochemistry 23: 2703–2704

Atari SZ, Shoor M, Neghab MG, Tehranifar A, Shafaroudi SM (2017) Evaluation of genetic diversity of iris genotypes (*Iris spp.*) using ISSR. Majallah-i Ulum-i Bāghbānī 30(3):376–382

[1] *Originals not seen

Azimi MH, Jozghasemi S, Hasanzadeh Davarani F, Aliabadi HM (2019) A review of Iranian *Iris* germplasm. Acta Hortic 1240:51–56

Bai W, Li P-F, Li B-G, Fan F-C (2008) Some physiological responses of Chinese *Iris* to salt stress. Pedosphere 18(4):454–463. https://doi.org/10.1016/S1002-0160(08)60036-3

Baker JG (1893) On the species and garden forms of Canna. J Roy Hort Soc **2**:178–188

Baskaran P, Van Staden J (2013) Rapid *in vitro* micropropagation of *Agapanthus praecox*. S Afr J Bot 86:46–50

Bhattacharjee SK (1998) Handbook of medicinal plants. Pointer Publisher, Jaipur, pp 192–193

Bosse G (1968) Trials with new varieties of *Canna indica* (German). Neue Landschaft 13:99–102

Burcu B, Aysel U, Nurdan S (2014) Antimicrobial, antioxidant, antimutagenic activities, and phenolic compounds of *Iris germanica*. Ind Crop Prod 61:526–530

Carrasco-Ortiz M, Lino GM, Castro-Castro A, Vargas-Amado G, Harker M, Rodríguez A (Translation by Martin Kral) (2019) Species wealth, geographical distribution and conservation status of the genus *Dahlia* (Asteraceae) in Mexico. Acta Bot Mexicana 126:e1354

Chemarin NG, Glazurina AN, Abelin JA (1973) On uses of ionizing rays in *Canna* breeding. ByullGos Nikitsk Bot Sada 1:52–57

Chittenden FJ (1956) Dictionary of gardening: A practical and scientific encyclopaedia of horticulture, vol II. The Clarendon Press, Oxford

Chowdhury MR, Hubstenberger J (2006) Evaluation of cross pollination of *Zephyranthes* and *Habranthus* species and hybrids. J Arkansas Acad Sci 60 Article (18)

Cohen O, Avishai M (2000) The Irises still exist: The conservation status of species *Iris* section Onococyclus in Israel, a century after their description. Ann Bot LVIII:145–160

Cosewic (2010) Cosewic assessment and status report on the Dwarf Lake Iris (*Iris lacustris*) in Canada. Committee on the status of endangered wildlife in Canada Otawa, p 29

Cronquist A (1981) An integrated system of classification of flowering plants. Columbia University Press. 1262 pp

Cruz-Duque, A. A., Tapia-Campos, E., Rodriguez-Domínguez, J. M. and Barba-Gonzalez, R. (2019). Research on native ornamental species from Mexico. In: Acta Hortic. 1240. ISHS 2019. Proceedings of International Symposium on Wild Flowers and Native Ornamental Plants P. Azadi, L. Samiei and J. Van Huylenbroeck. DOI https://doi.org/10.17660/ActaHortic.2019.1240.1

Dalda Şekerci, A. and Gülşen, O. (2016). Overview of *Dahlia* breeding. Scientific Papers. Series B, Horticulture. Vol. LX: 199-204.

De Hertogh AA, Le Nard M (1993a) Botanical aspects of flower bulbs. In: Hertogh D, Nard L (eds) The physiology of flower bulbs. Elsevier, Amsterdam, pp 7–20

De Hertogh AA, Le Nard M (1993b) Botanical aspects of flower bulbs. In: Hertogh D, Nard L (eds) The physiology of flower bulbs. Elsevier, Amsterdam, pp 21–28

De Munk WJ, Schipper J (1993) *Iris*-Bulbous and rhizomatous. In: De Hertogh A, Le Nard M (eds) The physiology of flower bulbs. Elsevier Science Publication, Amsterdam, pp 349–376

*Diels L (1930) Iridaceae. In: A Engler and K Prantl, Die natürlichen Pflanzenfamilien 2, 15a: 463–505 . Leipzig.

Dix JF-C (1974) *De Veredeling van Tulpen, Hyacinten, Narcissen en Irissen. Koninklijke Algemeene Vereeniging voor Bloembollencultuur*, Hillegom

Du Plessis N, Duncan GD (1989) Bulbous plants of Southern Africa – A guide to their cultivation and propagation. Tafelberg Publishers, Cape Town, p 192

Duncan G (2002) Grow Nerines, Kirstenbosch Gardening Series. National Botanical Institute, Kirstenbosch

Duncan GD (1985) *Agapanthus* species-their potential, and the introduction of ten selected forms. Veld Flora 71(4):122–125

Duncan GD (1998) Grow Agapanthus, Kirstenbosch Gardening Series. National Botanical Institute, Kirstenbosch

Duncan GD (2005) *Nerine pussilla*. Curtis's Bot Mag 22(3):169–175

Eikelboom W, Van Eijk JP (1990) Prospects of inter-specific hybridization in Dutch *Iris*. Acta Hortic 266:353–356

Environment and Climate Change Canada (2017) Management plan for the Western Blue Flag (*Iris missouriensis*) in Canada [Proposed]. Species at Risk Act Management Plan Series. Environment and Climate Change Canada, Ottawa, pp 1–28

Everett TH (1980) The New York botanical garden illustrated encyclopedia of horticulture. Garland Publication Inc, New York

Fellers JH (1996) A Passion for rain lilies: *Cooperia, Habranthus* and *Zephyranthes*. Herbertia 51: 78–112

*Flory WS (1976). Distribution, chromosome numbers and types of various species and taxa of *Hymenocallis*. Nucleus, 19: 204-227.

Foster, M. (1899). *Iris sofarana* n sp *The Gardener Chronicle* II: 389.

Gade DW (1966) Achira, the edible *Canna*, its cultivation and use in the Peruvian Andes. Econ Bot 20:407–415

Gaikwad SP, Yadav SR (2004) Endemic flowering plant species of Maharashtra and their possible utilization. In: Pullaiah T (ed) Biodiversity in India, vol 3. Regency Publications, New Delhi, p 50

Gangopadhyay, Chakraborty M, Dewanjee S, Bhattacharya S (2009) Clonal propagation of *Zephyranthes grandiflora* using bulbs as explants. BiologiaPlantarum 54(4):793–797

Garland, M. A., Smith, G. L. and Anderson, P. J. (2013). The spider lilies (*Hymenocallis*) native to Florida. Botany Circular No. 39 (January, 2013), Florida Department of Agriculture and Consumer Services, Division of Plant Industry

Gatt M, Ding H, Hammett K, Murray B (1998) Polyploidy and evolution in wild and cultivated *Dahlia* species. Ann Bot 81:647–656

Goldblatt P, Manning JC (2011) Systematics of the southern African genus *Ixia* (Iridaceae). 3. Sections *Hyalis* and *Morphixia*. Bothalia 41(1):83–134

Goldblatt P, Manning JC (2012) Systematics of the southern African genus *Ixia* (Iridaceae: Crocoideae): 4. Revision of sect. Dichone. Bothalia 42(2):87–110

Goldblatt P, Manning JC (2016) Systematics of the southern African genus *Ixia* L. (Iridaceae): 5. Synopsis of section *Ixia*, including five new species. S Afr J Bot 104:175–198

Goldblatt P, Manning JC, Helme NA (2015) New species and taxonomic notes in *Ixia* (Iridaceae: Crocoideae) from western South Africa. S Afr J Bot 98:108–113

Grossi A (2007) *Hymenocallis* and other spider lilies. The Plantsman 2007:222–226

Gupta, M.N. (1966). Induction of somatic mutations in some ornamental plants. In: *Proc. All India Symp. Hort*., pp.: 107-114.

Gupta PK, Agarwal DK, Srivastava AK (1972) Further cytological investigations in Indian Compositeae. Cytologia 37:581–593

Hannay, F. (1936). The Canna: gorgeous tropical beauty. Pp 218. In: *Flower Grower.*

Haron NW (2003) *Hymenocallis littoralis* (Jacq.) Salisb. In: Lemmens RHMJ, Bunyapraphatsara N (eds) Plant resources of South-East Asia-Medicinal and poisonous plants 3. Backhuys Publishers, Leiden, pp 252–253

Hesham AA, Yan WG (2010) Genetic diversity and relatedness of rice cultivars resistant to straight head disorder. Plant Breed 129:304–312

Jee G, Vijayavalli B (1999) Karyomorphology of eight taxa of *Hymenocallis* from South India. Caryologia 52(1-2):59–64

John Peter Arulanandam L, Ghanthikumar S, Arunkumar T (2015) In vitro propagation of rain lily (*Zephyranthescitrine* L.). Bot Rep 4(4):5–7

Katoch D, Singh B (2015) Phytochemistry and pharmacology of Genus *Zephyranthes*. Med Aromat Plants 4:212

Khoshoo TN (1968) Genetic improvement of ornamentals in India. Indian J Genet 8A:187–198

Khoshoo TN, Guha I (1975) Origin and evolution of cultivated Cannas. In: Nair PKK (ed) Glimpses in plant research. Vikas Publishing House, New Delhi, pp 1–81

Khoshoo TN, Mukherjee I (1970) Genetic-evolutionary studies on cultivated Cannas: VI. Origin and evolution of ornamental Taxa. Theor Appl Genet 40(5):204–217

Kim KW, De Hertogh AA (1997) Tissue culture of ornamental flowering bulbs (geophytes). Hort Rev 18:87–169
Kojima K, Mutsuga M, Inoue M, Ogihara Y (1997) Two alkaloids from *Zephyranthes carinata*. Phytochemistry 48(7):1199–1202
Lawrence WJC (1929) The genetics and cytology of *Dahlia* species. J Genet XXI(02):125–159
Lawrence WJC (1942) The origin of the garden dahlia. John Innes Horticultural Institution, p 8
Lewis GJ (1962) South African Iridaceae: The genus *Ixia*. J South Afr Bot 28:45–195
Lewis WH (1986) The Jivaro Indians: note on an expedition. Discovery 17:2–6
Lin H, Tao QL, Zhang JJ (2010) *Iris*: introduction of the latest international classification. Garden Technol 118:1–3
Liu Q, Sun H, He D (2005) Plant responses to the high temperature and moisture stress. J China West Normal Univ Nat Sci Ed 26(4):364–368
Maas-van de Kamer H, Maas PJM (2008) The Cannaceae of the world. Blumea 53:247–318
Matus IA, Hayes PM (2002) Genetic diversity in three groups of barley germplasm assessed by simple sequence repeats. Genome 45:1095–1106
Mc Murtrie, A. (2016). Breeding new Reticulata irises. http://www.reticulata.com.
Meerow AW (1987) Chromosome cytology of *Eucaris, Caliphuria* and *Urceolina*. Am J Bot 74: 1559–1575
Meerow AW (1989) Systematics of Amazon lilies, *Eucaris* and *Caliphruria* (Amaryllidacea). Ann Mo Bot Gard 76(1):136–220
Meerow AW, Guy CL, Li QB, Clayton JR (2002) Phylogeny of the tribe Hymenocallideae (Amaryllidaceae) based on morphology and molecular characters. Ann Mo Bot Gard 89(3): 400–413
Mehra PN, Ramanandan P (1974) Cytological investigations on the Indian compositae, II Asteraceae, Heliantheae, Hebriaceae and Anthenudeas. Caryologia 27:255–284
Misra S, Misra RL, Banerji BK, Batra A, Dwivedi AK (2017) Dahlia. In: Misra RL, Misra S (eds) Commercial ornamental crops: Traditional and loose flowers. Kruger Brentt Publishers, pp 45–70
Mukherjee I, Khoshoo TN (1970) Genetic-evolutionary studies on cultivated Cannas. II.Pollination mechanism and breeding system. Proc Indian Natl Sci Acad 36(B):271–274
Nakano M, Tanaka S, Oota M, Ookawa E, Suzuki S, Saito H (2003) Regeneration of diploid and tetraploid plants from callus derived protoplasts of *Agapanthus praecox* sp. Orientalis Leighton. Plant Cell Tissue Organ Cult 72:63–69
Nakornthap A (1965) Radiation induced somatic mutations in the ornamental *Canna*: the use of induced mutations in plant breeding. Radiation Bot 5(Suppl):707–712
Nwankiti OC (1985) Cytotaxonomic survey of some tropical ornamental species V. Karyotype of two species of the genus *Crinum* and a related genus *Hymenocallis*. Cytologia 50:797–803
Ohri D (1998) Genome size variation and plant systematics. Ann Bot 82(Supplement A):75–83
*Pax F (1888) . Iridaceae . In: A Engler and K. Prantl, Die natürlichen Pflanzenfamilien 2, 5: 137 157 . Engelmann, Leipzig
Piyachomkwan K, Chotineeranat S, Kijkhunasatian C (2002) Edible canna (*Canna edulis*) as a complementary starch source to cassava for the starch industry. Ind Crop Prod 16(1):11–21
*Preuss KD (2002). Spider lilies for the south. Bulbs, 4: 27-31.
Prince LM, Parks CR (2001) Phylogenetic relationships of Theaceae inferred from chloroplast DNA sequence data. Am J Bot 88:2309–2320
Raina NS, Khoshoo TN (1971) Cytogenetics of tropical bulbous ornamentals IX.Breeding system in *Zephyranthes*. Euphytica 21:317–323
Saad L, Salma N, Talhouk, Mahy G (2009) Decline of endemic Oncocyclus irises (Iridaceae) of Lebanon: survey and conservation needs. Oryx 43(1):91–96
Sari A, Suryadiantina, Daryona BS, Purnomo (2018) Variability and interaspecific classification of Indonesian edible canna (*Canna indica* L.) based on RAPD marker analysis. SABRAO J Breed Genet 50(2):156–167

Sarma D, Debnath B (2016) Two species of Amaryllidaceae, new distributional record for Tripura, North East India. Ann Plant Sci 5(6):1360–1362

Schiappacasse, F. (1996). Tulipan. In: *Cultivo de Tulipan* (Tulip cultivation), ed. F. Schiappacasse, pp 11–29.

Sealy JR (1954) Review of the genus *Hymenocallis*. Kew Bull 9(2):201–240. https://doi.org/10.2307/4114384

Shimizu K, Miyabe Y, Nagaike H (1999) Production of somatic hybrid plants between *Iris ensata* Thunb. and *I. germanica*. Euphytica 107:105–113

Singh G, Saxena RK (2017) Chemistry and medicinal properties of *Hymenocallis littoralis*. Int J Sci Res 6(11):1327–1329

*Smith GL, Anderson, L. C. and Flory, W. S. (2001). A new species of *Hymenocallis* (Amaryllidaceaee) in Central Florida Panhandle. Novon, 11: 233-240.

Smith GL, Flory WS (1990) Studies on *Hymenocallis henryae* (Amaryllidaceae). Brittonia 42(3): 212–220. https://doi.org/10.2307/2807218

Smith GL, Garland MA (2003) Nomenclature of *Hymenocallis* Taxa (Amaryllidaceae) in Southeastern United States. Taxon 52(4):805–817

Snijman DA, Linder HP (1996) Phylogenetic relationship, seed characters and dispersal system evolution in Amaryllideae (Amaryllidaceae). Ann Mo Bot Gard 83(3):362–386

*Sørensen PD (1969). Revision of the genus *Dahlia* (Compositae, Heliantheae- Coreopsidinae). Rhodora, 71: 309-316.

Sørensen PD (1970) The Dahlia: an early history. Arnoldia 30(4):121–138

Tanee T, Sudmoon R, Siripiyasing P, Suwannakud KS, Monkheang P, Chaveerach A (2018) New Karyotype information of *Hymenocallis littoralis*, Amaryllidaceae. Cytologia 83(4):437–440

Tapia-Campos E, Rodriguez-Dominguez JM, Revuelta-Arreola MM, Tuyl JMV, Barba-Gonzalez R (2012) Mexican Geophytes II. The genera *Hymeocallis, Sprekelia* and *Zephyranthes*. Floric Ornam Biotech 6(Sp.I.: 1):129–139

Tay D (2006) Herbaceous ornamental plant germplasm conservation and use. In: Anderson NO (ed) Flower breeding and genetics. Springer, pp 113–175

The American Dahlia Society website.: https://www.dahlia.org/docsinfo/species-dahlias/list-of-species-dahlias/ dt. 09/08/2020.

*Traub HP (1962). Key to the subgenera, alliances and species of *Hymenocallis.* Plant Life, 18: 55-72.

Traub HP (1967) Review of the genus *Nerine* Herb. Plant Life 23:3–32

Volis S, Blecher M, Sapir Y (2010) Application of complex conservation strategy to *Iris atrofusca* of the Northern Negev, Israel. Biodivers Conserv 19:3157–3169

Walter KS, Gillet HD (eds) (1998) 1997 IUCN Red list of threatened plants. Compiled by the World Conservation Monitoring Centre, IUCN, Gland/Cambridge

Wanjala BW, Obonyo M, Wachira FN (2013) Genetic diversity in Napier grass (*Pennisetum purpureum*) cultivars: implications for breeding and conservation. AOB Plants 5:1–10

Wulster GJ, Ombrello TM (2000) Control of height and flowering of *Ixia* hybrids as container plants. HortScience 35(6):1087–1088

Yadav SR (1997) Endemic plants of Peninsular India with special reference to Maharashtra. In: Pokle DS, Kanir SP, Naik VN (eds) Proceedings, VIIIAAT annual meet and national conference, Aurangabad, pp 31–51

Yusti-Munoz AP, Velandia-Perilla JH (2013) First record of the Amazon lily *Eucaris sanderi* (Amaryllidacea) for Isla Gorgana, Colombia. Boletin Cientifico Museo de Hist Nat 17(1):46–49

Zeerak NA, Wani SA (2007) Diversity of Irises from Kashmir Himalaya. J Ornam Hort 10(2): 115–118

Zhang L (2010) The Iris application in American garden. Garden 8:38–41

Zheng Y, Meng T, Xiaoying B, Lei J (2016) Investigation and evaluation of wild *Iris* resources in Liaoning Province, China. Genet Resour Crop Evol 64:967–978

Zonneveld BJM, Duncan GD (2003) Taxonomic implications of genome size and pollen colour and vitality for species of *Agapanthus* L'Heritier (Agapanthaceae). Plant Syst Evol 241(1–2):115–112

Zonneveld BJM, Duncan GD (2006) Genome size for the species of *Nerine* Herb. (Amaryllidaceae) and its evident correlation with growth cycle, leaf width and other morphological characters. Plant Syst Evol 257(3–4):251–260

Xu YF, Hussain K, Yan XF, Chen XH, Shao MN, Guan P, Qu B (2019) Genetic diversity and relationships among 15 species of Iris based on amplified fragment length polymorphism markers. Appl Ecol Environ Res 17(3):5369–5381

Tuberose (*Polyanthes tuberosa* Linn./*Agave amica*) 10

R. Sadhukhan, T. K. Chowdhuri, and S. K. Datta

Contents

S. K. Datta has retired.

R. Sadhukhan (✉)
Department of Genetics and Plant Breeding and In-Charge, All India Coordinated Research Project on Floriculture, Directorate of Research, Bidhan Chandra Krishi Viswavidyalaya, Kalyani, West Bengal, India

T. K. Chowdhuri
Department of Floriculture and Land Scape Architecture, Bidhan Chandra Krishi Viswavidyalaya, Kalyani, West Bengal, India

S. K. Datta
Floriculture Division, CSIR-National Botanical Research Institute, Lucknow, India

Bose Institute, Kolkata, India

Department of Plant Science and Biotechnology, Assam University, Silchar, Assam, India

S. K. Datta, Y. C. Gupta (eds.), *Floriculture and Ornamental Plants*, Handbooks of Crop Diversity: Conservation and Use of Plant Genetic Resources,
https://doi.org/10.1007/978-981-15-3518-5_17

Abstract

Polianthes tuberosa Linn., commonly known as tuberose, is a monocot genus and a perennial, bulbous plant that belongs to the family Amaryllidaceae. It is native of Sonora desert of Mexico and successfully grown under commercial cultivation in tropical and subtropical areas for cut flower production, for floriculture trade, and as a source of oil. Tuberose has been classified into three types on the basis of flower, i.e., single-flowered, semidouble-flowered, and double-flowered. Considering its economic importance, extensive studies have enriched all basic scientific information on different aspects of tuberose. It is a unique flower – possibly the one which is used in every sphere of the Indian-Hindu life. The chapter will provide most information generated on different aspects related to geographic distribution, growth habit, physiology, agro-technology, techno-economics, ecological and flowering behavior, etc.

Keywords

Tuberose · Germplasm · Genetic diversity · Breeding · Characterization · Cytogenetics

10.1 Introduction

Polianthes tuberosa Linn., commonly known as tuberose, is a monocot genus and perennial, bulbous plant that belongs to the family Amaryllidaceae. It is known in different names in different countries like China (Wan 1Xiangyu, Ye Lai Xiang, Yue Xia Xiang), Cuba (Azucena, Guacamaya), the Philippines (Azucena), Czech Republic (Tuberóza), Denmark (tuberose), Poland (Tuberozo), France (Jacinthe Des Indes, Tubóreuse), Germany (Nachthyazinthe, tuberose), Hungary (Tubarózsa), and Indonesia (Sundel Melem). In India, it is known in different local names: English, tuberose; Bengali and Asamese, Rajanigandha; Hindi, Gulchari and Galshabbo; Telugu, Sukandaraji and Nelasanpenga; Tamil, Nilasompangi; Kannada, Sugandharaja; Manipuri, Kundaleiangouba; Urdu, Gul shabbo; and Marati, Gulcheri. It has also a number of synonyms like *Agave polianthes* (L.) Thiede & Eggli, *Agave tuberosa* (L.) Thiede & Eggli nom. Illeg., *Crinum angustifolium* Houtt., *Polianthes gracilis* Link, *Polianthes tuberosa* var. *gracilis* (Link) Beurl., *Polianthes tuberosa* f. *plena* Moldenke, and *Tuberosa amica* Medik.

Tuberose (*Polianthes tuberosa* Linn.) renamed as *Agave amica* stands out among the top in the list of fragrant ornamental flowers for its most delightful fragrance and essential oil and also for its beauty and elegance. It belongs to the family Amaryllidaceae as reported by Rose (1903–1905), whereas Hutchinson (1960) placed it under the family Agavaceae. It is a perennial, bulbous ornamental plant, native of Sonora desert of Mexico, and successfully grown under commercial cultivation in tropical and subtropical areas (Biswas et al. 2002). From Mexico, it has spread to different parts of the world. Though Aztecs grew it for ornamental and medicinal purposes for more than 400 years ago, in the course of time, it has secured a distinct position as one of the important floricultural crops and is cultivated widely across the globe at present. It is day-neutral in nature and can be cultivated round the year both in protected and open conditions. It is commercially grown in various countries, viz., India, Taiwan, China, Mexico, Bangladesh, Indonesia, Kenya, Italy, France, Morocco, the USA, Hawaii, and South Africa. In India, it is cultivated in

West Bengal, Karnataka, Maharashtra, Tamil Nadu, Haryana, Punjab, Gujarat, Andhra Pradesh, Assam, and Rajasthan (Biswas et al. 2002). The single-petalled varieties are commonly used for extraction of essential oil, making garland, loose flowers, bridal makeup, etc., whereas the double types are used as cut flowers, bouquets, garden display, and interior decoration. It has also used in landscaping due to their elegant, attractive appearance with sweet essence as a component of bed, borders, as well as potted plants. As a source of natural fragrance, the flowers are used in perfume industry due to the presence of essential oil as an important constituent in high-grade perfumes and are a source of volatile organic compounds and secondary metabolites. In addition to the perfumery industry, it has great importance in various folk medicinal uses as cosmetic, laxative, cooling, placebo, sexual disorder, hair color, emetic, diuretic, and gonorrhea. It is believed that the warm soothing effect of oil helps in releasing stress by uplifting mood, particularly in patients suffering from insomnia by enhancing emotional strength. Floral fragrance of tuberose has also useful effect on the brain and heart in improving emotional, psychological, and artistic inspirations. Tuberose flowers have also been recommended as edible (Barghout et al. 2020, Andrea et al. 2020).

Polyanthes means in Greek *many flowers*. *Polianthes tuberosa* L. was cherished and cultivated in Mexico before the conquest in 1522. It is one of the many flowers which has come to Mexican culture from the ancient culture of the Nahuatl-speaking people. It is also named from the Greek words *Plios* meaning white and *anthos* meaning flower (Datta 2006).

10.2 Botany

Tuberose is a semihard, herbaceous perennial plant commonly propagated through bulbs. Though originated in Mexico, it grows well in a variety of climates from tropical to temperate regions. Morphological features are to some extent similar to plants belonging to Amaryllidaceae and Liliaceae, both propagating through bulbs and to some extent with Agavaceae. The bulbs of tuberose, the modified underground stem, are made up of small scales and leaf bases. The base of the bulb is condensed to form a "basal disc" which gives rise to abundant, shallow, round, and adventitious roots at the base, and a floral stem emerges from its apex. Bulb shape is oval to conical and contains many buds which sprout into side suckers and form a cluster of plants. Bulb size may vary from very small to large having 5–6 cm in diameter. Small size bulbs are called bulblets. The leaves are sessile, lanceolate, fleshy, long, narrow, elongated and dense, grass-like and arranged in rosette near the base, and dark green to light green, sometimes reddish at the base, and grow up to 60 cm long and 1–3 cm wide depending on the genotype. Leaves are gathered at the base of the stem as flakes arched outward. Mutant tuberose genotypes with variegated leaves having yellow and silvery border have also been reported (Figs. 2 and 5). The stem is erect, round, and bright green with compact base, reaching up to 100 cm long even more depending on the genotype. The apical portion of the spike ends with a flower spike at its top called rachis. The flowers are hermaphrodite, regular, long, simple, and born on unbranched terminal racemose spike. They are basically funnel-shaped with highly fragrant waxy white tepals and larger at the base and smaller when approaching the apex of inflorescence. Inflorescence is a spike of about 30–60-cm-long rachis, with a pair of single or double flowers that are arranged toward the end and occupy more than a third portion of the stem. Stamens are six in number with dorsifixed anthers, and filaments are attached to the upper part of the perianth tube (corolla). The ovary is a tricarpellary and trilocular with three free stigmas at the apex in single-petalled varieties and tetralocular with four stigmas in double and in some single hybrids (Singh and Sadhukhan 2019) (Fig. 12). The ovary contains numerous ovules. Botanically fruit is a loculicidal capsule with three lobes. The capsule contains a number of black, deltoid or semicircular, and flat endospermic seeds (Fig. 6). Micro-morphological and anatomical studies revealed that the seeds lack starch but have exotesta with primary and secondary reticulation. The endotesta and linear, cylindrical embryo is surrounded by the oily endosperm. Seed is black in color which is due to the presence of phytomelanes in the epidermis. Seed dormancy is very low and remains viable for a very short period of time only. Floral aroma of tuberose is due to the presence of indole as the main component in its essential oils, which is also accompanied by benzyl acetate and methyl anthranilate. The fragrance is emitted at higher amount during evening hours. Analyses of emitted volatiles in tuberose flowers (Calcutta Single cultivar) revealed the presence of 24 compounds including benzenoids/phenylpropanoids (81.2%), terpenoids (13.64%), and fatty acid derivatives. The emitted volatilome of both in situ and plucked flowers has also been studied using different adsorbent matrices leading to the identification of 57 volatiles. Foliar anatomy of open leaf profile cross-section in tuberose is U-shaped in immature leaves and straight in mature leaves. Furthermore, the epidermis is made by a square, rectangular, or partially spherical cell layer (mono-stratified). The cell configuration of stomatal complexes is parasitic, with one or more attached cells and lining stomata, and arranged parallel to guard cells. Palisade parenchyma cells are poorly developed toward adaxial surface and, because of extremely poor development, virtually absent toward abaxial side. The normal development of the embryo sac follows monosporic pattern of the *Polygonum* type and starts its development from the chalazal megaspore. At maturity, the embryo sac is of pyriform shape with a chalazal haustorial tube that connects the embryo sac to the nucellus tissue of the ovule. Being located at the chalazal extreme of the embryo sac, the central cell nucleus shows a high polarity. Because of this particular characteristic, in order to fertilize such nucleus, the second sperm nucleus has to travel long distance. The chronological description of the events that happen from fertilization and early embryo development to the initial development of the endosperm was classified as the helobial type. Self-incompatibility of various degrees has also been reported in some cultivars (Datta 2017).

10.3 Systematics

Carl Linnaeus in 1753 described tuberose as *Polianthes tuberosa* belonging to the family Amaryllidaceae. According to the classification proposed by Hutchinson based on cytological studies of the chromosomes, the genus *Polianthes* was placed in the Agavaceae rather than in the Amaryllidaceae. However, the union of the

Amaryllidaceae and Liliaceae families and the recognition of Agavaceae as an independent family took place in 1988 (Cronquist 1988). The genus *Polianthes* includes all those species formerly placed in the genera *Bravoa*, *Pseudobravoa*, *Manfreda*, *Prochnyanthes*, and *Runyonia* and the herbaceous species of *Agave* (Shinners 1966). But with the recent phylogenetic studies based on morphological and molecular data, *Polianthes tuberosa* has been renamed as *Agave amica* and has been included in the family Asparagaceae and subfamily Agavoideae. Long before the botanical name of *Agave amica*, these intensely fragrant, white flowers were known to the Aztecs as *omixochitl* or "bone flower," derived from the words *omitl* meaning "bone" and *xochitl* meaning "flower," and in Spanish, it became "Flor de hueso." In Mexico, even today, one also hears the names "nardo," "azucena," "amole," and "amiga de noche" to call the white tuberose. It was called as *Hyacinthus indicus tuberosa radice* up to 1601, and after that, nard, as a species, was firstly classified by Linnaeus in 1737 in his book entitled *Genera Plantarum* with a generic name of *Polianthes*. The species was first described by Carl Linnaeus in 1753 as *Polianthes tuberosa* (Table 1). The word *Polianthes* is derived from the Greek words *polis* (gray) and *anthos* (flower). Similarly, the common name of the "tuberose" was probably derived from the French word *tuberuse*, meaning tuberous, but in English, the name also suggests the fragrance of the flower. In 1790, Friedrich Kasimir Medikus transferred this species into the genus *Tuberosa* as *Tuberosa amica*. The genus *Polianthes* is now included in a broad *Agave* after morphological and molecular phylogenetic studies.

When transferred to the genus *Agave*, two incorrect attempts were made to name the species. The published name *Agave tuberosa* was illegitimate as it was used previously by Philip Miller in 1768 for the species now called *Furcraea tuberosa*; hence, it cannot be used again. A new name *Agave polianthes* was published, but according to the *International Code of Nomenclature for algae, fungi, and plants*, it should be used when the older epithet is unavailable. The name for the species within *Agave* was correctly named as *Agave amica*. Thus, *Agave amica* is synonymous with *Polianthes tuberose* L., *Agave polianthes*, *Agave tuberose* (L.), or *Tuberosa amica* (Table 1).

Table 1 Classification of tuberose

Classification	Present classification (Thiede and Rafael Goverts, 2017)	Earlier classification (USDA)
Kingdom	Plantae	Plantae
Division	Angiosperms	Magnoliophyta – Flowering plants
Class	Monocotyledonae	Liliopsida – Monocots
Order	Asparagales	Liliales
Family	Asparagaceae	Agavaceae
Subfamily	Agavoideae	–
Genus	*Agave*	*Polianthes*
Species	*Amica*	*Tuberosa*
Cultivated species	*Agave amica*	*Polianthes tuberosa* L.

10.4 Cytology

Cytological studies of *P. tuberosa* revealed the chromosome number of single-flowered forms to be $2n = 60$, while in case of double types, a variation in chromosome numbers has been reported, i.e., $2n = 50, 54, 60$, and 120 (Sato 1938). However, Karihaloo (2019) in his investigations with single and double cultivars and a single x double hybrid concluded that the chromosome number in both single and double types was 2n = 60. He noticed regular meiosis with the chromosomes paired into 30 bivalents at metaphase I (M I) from which he inferred that flower doubling and leaf variegation in tuberose have arisen due to gene mutations and do not necessarily involve changes in chromosome number or structure. In single types, the haploid chromosome number is 30, of which 5 long and 25 small (Karihaloo 2019). Tuberose have a close genetic relationship and identical chromosome constitution with that of the *Yucca-Agave* group based on studies by Sato (1938) in tuberose, Watkins (1936) in *Yucca*, and Granick (1944) in *Agave*, and it was found that these all genera have 5 large L-shaped and 25 small haploid chromosomes. Studies have further suggested that tuberose is hybrid in nature and is believed to have originated from two different prototypes, one with medium, short chromosomes and the other one with long chromosomes (Sinha 1984 c.f. Datta and Banerji 1995). The cytomorphology of two gamma ray-induced mutants (Rajat Rekha and Swana Rekha) of *P. tuberosa* and their respective mother line have been examined (c.f. Datta and Banerji 1995). Cytological investigations showed chromosome number of 60 in single cultivars and 50 in double cultivars. It has been suggested that the double forms are derived from the single by loss of ten chromosomes due to abnormal mitotic or meiotic division. The karyotype of tuberose was similar to that of genera *Bravoa*, *Agave*, *Furcraea*, and *Beschorneria*. Multiple perianth varieties are autotetraploid with a compliment of 120 chromosomes, of which 20 were long and 100 were short chromosomes. The interrelationship, sterility, and occurrence of biotypes in two types of *P. tuberosa* have been studied. Autopolyploidy is often cited as one of the causes of partial sterility (c.f. Datta and Banerji 1995). Cytomorphological characters of gamma ray-induced mutants of tuberose were analyzed and detected in different types of chromosomal aberrations. Different types of chromosomal configurations were detected – single (2n = 60) and double (2n = 50) (Laxmi et al. 1984 c.f. Datta and Banerji 1995). In the diploid set, two pairs of chromosomes are associated with organization of the nucleolus and small bivalent and secondary pairing during I and II metaphase with two unequal-sized nucleoli formed during telophase. Further karyotype analysis in the genus *Polianthes* with several species reported a bimodal karyotype, consisting of 10 big chromosomes and 50 small ones (2n = 2x = 60).

10.5 Distribution

Though originated in the Sonora desert of Mexico, tuberose has spread to different parts of the world in the sixteenth century. Aztecs were found to be growing it nearly 600 years ago, who used the essential oil from the florets to flavor chocolate. It was among the first plants taken back to the Old World and was cherished and cultivated in Mexico before the conquest in 1522. In 1519, the Spanish found the Aztecs were growing it and took it to the Old World with them. Similarly, in 1500s, a French missionary returning from the Indies did so as well, and Simon Tovar (Spain physician) included it in his garden in 1596. This crop was introduced into Europe by the travellers from the Middle Ages in 1629, who generally passed on this planting material appropriately adapted under Mediterranean climate to royal gardens. After introduced into Europe, it became part of the moon gardens (a collection of white or pastel flowers, which release an intense fragrance after dusk) which were popular among the sun-shunning Victorian ladies, and gradually, the plant did fall out of favor when it became much overused at funerals. It is believed that tuberose was brought to India via Europe in the sixteenth century. William Bartram mentioned about tuberose in his early accounts of horticulture in the Gulf states. It was reported as useful and interesting exotics at a plantation near Baton Rouge, on the Mississippi, in 1777, "which grew from five to seven feet high in the open ground, the flowers being very large and abundant" (William Bartram in Bartram, William, Travels Through North & South Carolina, Georgia, East and West Florida, etc., Philadelphia, 1791. pp. 421–427). Pearly white tuberose has been reported to be growing in botanical garden at Birmingham and on the rock garden in the Tottenham nurseries in severe winter.

Due to its wide adoptability in various agro-climatic zones in the world, it is commercially cultivated in India, Bangladesh, Kenya, Mexico, Morocco, France, Italy, Hawaii, South Africa, Taiwan, North California, the USA, Egypt, China, and others countries. In India, the major area that cultivated tuberose is West Bengal (more than 5000 ha, in the districts of Nadia, Paschim Medinipur, Purba Medinipur, Howrah, and North 24 Parganas). Other major contributing states are Tamil Nadu (Coimbatore and Madurai), Karnataka (Mysore, Tumkur, Kolar, Belgaum, and Devanahalli Taluk), Maharashtra (Pune, Nashik, Ahmednagar, Thane, Sangli), Andhra Pradesh (East Godavari, Guntur, Chittoor, Krishna), Gujarat (Navsari and Valsad), Rajasthan (Ajmer, Udaipur, Jaipur), Orissa (Cuttack, Bhubaneswar, and Jaleswar), and Bihar (Samastipur and Muzaffarpur). Besides these states, recently other states have also taken the initiative to promote and produce tuberose like Chhattisgarh, Haryana, Madhya Pradesh, Telangana, Uttarakhand, etc. through AICRP on floriculture, ICAR, DFR, and other government activities. During the last decade, tuberose cultivation is widely promoted in village areas. The National Horticulture Board reported during 2013 that the total area under tuberose cultivation in this country is around 7.95 lakh hectare and it is estimated the quantity of production in terms of loose and cut flowers is about 27.71 '000 MT and 1560.70 lakh, respectively.

10.6 Origin, Domestication, and Spread

Polianthes tuberosa L. was cherished and cultivated in Mexico before the conquest in 1522. It is one of the many flowers which have come to Mexican culture from the ancient culture of the Nahuatl-speaking people. These intensely fragrant, cream-white flowers were known as *omixochitl* or "bone flower" from the words *omitl* which means "bone" and *xochitl* which means "flower." It is also named from the Greek word *Plios* meaning white and *anthos* meaning flower. In Mexico today, one also hears the names "nardo," "azucena," "amole," and "amiga de noche" applied to the white tuberose sold in the Indian markets of Mexico City. Tuberoses were among the first plants taken back to the Old World. They were revered by the Spanish and often used in their gardens. *P. tuberosa* was among the plants included in the garden of a well-known Sevilian physician, Simon Tovar, who died in 1596. Tuberoses were also mentioned in early accounts of horticulture in the Gulf states written by William Bartram. They were among many useful and interesting exotics at a plantation near Baton Rouge, on the Mississippi, in 1777, "which grew from five to seven feet high in the open ground, the flowers being very large and abundant" (William Bartram in Bartram, William, Travels Through North & South Carolina, Georgia, East and West Florida, etc., Philadelphia, 1791. pp. 421–427). Eleven or 12 different species of *Polianthes* have been discovered in the cool mountain valleys of western and northwestern Mexico. These species range in color from white, orange red, and red to stripe. All were found growing wild with the exception of *P. tuberosa*, which has never been found anywhere except under cultivation. The natural flower oil of *P. tuberosa* remains today one of the most expensive of the perfumer's raw materials (Emmart Trueblood 1973).

How and when the tuberose found its way to India and later to Ceylon and elsewhere in the Orient is probably an unanswerable question, but it was apparently taken to Europe toward the end of the sixteenth century. Originally, Linnaeus had used the name *Polianthes floribus alternis* in his *Hortus Cliffortianus* (1738); later in the species *Plantarum*, he grouped this and *Hyacinthus indicus tuberosus* of Gaspard Bauhin and Carls Clusius under the genus *Polianthes* and the species *Tuberosa*.

Fish (1881) reported tuberose variety 'Pearl' which grows dwarf and sturdy under cool and moist conditions. Reports are available on period of planting, selection of bulb, growth, cultural practices for cutting of bulbs and rooting, watering, storage temperature, flowering, application of stimulants, etc. for tuberose (W.P.R. 1883). An average of 20–30 flowers on a spike and sometime 30 flowers develop from a single bulb grown in 5″ pots. Ullrich (1993) attempted to compile a limited bibliography in which illustrations of *P. tuberosa* appeared. The *Index Londinensis*, an invaluable bibliography, cites three illustrations of *P. tuberosa* for the eighteenth century. The oldest known figures in the *Florentine Codex* were reported in detail by Emmart Trueblood (1973). Ullrich (1993) in his article has mentioned the early illustrations of tuberose in the European literature for the nineteenth century and reported his experience on tuberose. A number of quite rare species of *Polianthes* were collected from Central, Western, and Southern Mexico and mentioned all with proper accession numbers.

10.7 Plant Genetic Resources

Polianthes tuberosa Linn. was earlier categorized into the family Amaryllidaceae (Rose 1903–1905). However, Hutchinson (1960) placed it under the family Agavaceae. The most extensive contributions to new species of *Polianthes* were made by Rose (1903–1905). These species range in color from white, orange red, and red to stripe. Among those botanists whose studies were concerned with the genus *Polianthes* are Cyrus G. Pringle (1838–1928), Edward Palmer (1831–1911), Cristopher C. Pary (1823–1890), and J.N. Rose (1862–1928) (c.f. Emmart Trueblood 1973). All the species are wild with the exception of *P. tuberosa* which has never been found anywhere except under cultivation. It is a monotypic genus closely related to *Bravoa*. The genus *Polianthes*, apart from *Tuberosa*, contains several other species, but they are not clearly distinct and are not commercially exploited. 15 species native to Mexico were mentioned. *Polianthes tuberosa* L. is the only species cultivated as an ornamental cut flower in tropical and subtropical areas. However, according to Solano (2000), the genus *Polianthes* includes 14 species. The wild species mentioned were *P. palustris* (white), *P. durangensis* (purplish), *P. montana* (white), *P. longiflora* (whitish purple), *P. platyphylla* (white tinged with red), *P. graminifolia* (deep red), *P. geminiflora* (light orange red), *P. gracilis* (white), *P. blissii*, *P. pringlei* (white), *P. sessiliflora* (white), *P. nelsonill* (white), and *P. graminifolia* Rose.

10.7.1 Description of the Species

Solano (2000) described the different species of *Polianthes*. He reported 15 wild species of the genus *Polianthes*. Out of which, eight species belong to the subgenus *Bravoa* and seven species belong to subgenus *Polianthes*. *Polianthes tuberosa* is the only commercial exploited cultivated species having several varieties of single-, semidouble-, or double-petalled types. *Polianthes tuberosa* is highly fragrant and completely white in color with some tinge of pink colors in some cases at bud stage. Among the wild species, *P. montana*, *P. nelsonii*, and *P. sessiliflora* are also used for ornamental purpose locally. *P. montana* and *P. longiflora* have strong fragrance like *P. tuberosa*. Five species, viz., *P. platyphylla*, *P. venustuliflora*, *P. palustris*, *P. nelsonii*, and *P. sessiliflora*, have mild fragrance, and seven species, viz., *P. howardii*, *P. bicolor*, *P. graminifolia*, *P. geminiflora*, *P. oaxacana*, *P. zapopanensis*, *P. multicolor*, and *P. densiflora*, have no fragrance at all.

10.7.2 *Polianthes* Subgenus *Bravoa*

(a) *Polianthes howardii* Verh.-Will.: It is one of the endangered species of these genera which are mainly distributed in the Colima and Jalisco region of Mexico. It is a non-fragnant type. Flower color is reddish-purple, red in the base, and gradually green in the lobe. Flowering occurs mainly during July to September.
(b) *Polianthes bicolor* E. Solano & García-Mend.: Distributed mainly in the Oaxaca region. Flower color is orange-greenish with green lobe. This is quite similar to *P. geminiflora*. Flowering time is June to August. This is also a non-fragrant type.

(c) *Polianthes* Montana Rose: It is a strongly fragrant species. It is distributed in the Jalisco and Nayarit regions of Mexico. Flowering time is July to August. It is locally used as ornamental crop.
(d) *Polianthes graminifolia* Rose: Distributed in the Aguascalientes, Jalisco, and Zacatecas regions of Mexico. Flower color is red, orange, or pink. It is a non-fragrant type. It seems like *P. geminiflora*. Flowering time is July to September.
(e) *P. geminiflora*: It has been reported as a new species. It is widely distributed from Durango to Nayarit in the northwest and Puebla to Guerrero in the south of Mexico. Flower is non-fragrant and color is red, orange, or coral. Flowering time is July to August. Four intraspecific taxa have been reported – var. *geminiflora*, var. *clivicola*, var. *graminifolia*, and var. *pueblensis*. *P. geminiflora* (Lex.) Rose was described by Juan José Martínez de Lexarza as *Bravoa geminiflora* and was published in *Novorum Vegetabilium Descriptiones* in 1824. According to Lexarza, the plant was collected in the mountains of Valladolid, nowadays Morelia, Michoacán. J. N. Rose (1903–1905) transferred *Bravoa geminiflora* to *P. geminiflora*. *Polianthes geminiflora* has the widest geographical distribution and the highest morphological variation.
(f) *Polianthes oaxacana* García-Mend. & E. Solano: Distributed in the Oaxaca region. Flowers are tubular and bilaterally symmetrical and non-fragrant having pink outside and yellow inside. Flowers occur during September–October.
(g) *P. zapopanensis*: Distributed in Jalisco region. Flowers have thick infundibuliform and are orange or pink in color. Flowering occurs during July to September.
(h) *Polianthes multicolor* E. Solano & Davila: Distributed in the Guanajuato region. Flowers have many variations in color – almost white, orange, pink, orange-light yellow. This is non-fragrant type, and flower occurs during June–August.

10.7.3 *Polianthes* Subgenus *Polianthes*

(a) *Polianthes densiflora* (B.L. Rob. & Fernald) Shinners: Restricted in the Chihuahua region only and facing danger of extinction. Flower in non-fragrant and yellow in color. Blooming time is June to August.
(b) *P. platyphylla* Rose: This species has restricted distribution in South Durango, Jalisco, Nayarit, and Zacatecas of Mexico and is an endangered one. It bears fragrant flowers with white tinge and red. Flowering time is June to September.
(c) *P. venustuliflora* E. Solano & Castillejos: It is quite similar to *P. sessiliflora* and is distributed in Michoacán region. Flower is white tinged with pink and fragrant. Flowering duration is July to September.
(d) *P. palustris* (Rose) Thiede & Eggli: This species is similar to *P. nelsonii*. It is restricted only in Nayarit but was first identified in the swamps of western foothills of Sierra Madre. Flower is white and fragrant. Blooming period is August. Bulbs are oval to oblong with erect stem of length that reaches up to 40 cm and basal leaves with parallel venation. Flowers arise from single bract in

3–5 pairs with slightly curved perianth, short segments, and spreading filaments, short near the top of the tube without extended anthers; ovary is free at tip.

(e) *P. longiflora* Rose: This has been reported from Jalisco and Michoácan. It is also under the danger of extinction. The flower has strong fragrance and is used for ornamental purpose locally. Flower color is white tinged with purple. Blooming period is July to August.

(f) *P. nelsonii* Rose: Distributed in Durango, Aguascalientes and Chihuahua. Flower is fragrant, white, pink or sometimes red. Blooming takes place during June to August. It is used as ornamental locally.

(g) *P. sessiliflora* (Hemsl.) Rose: This is widely distributed in several parts of Mexico like Aguascalientes, Durango, Jalisco, Nayarit, San Luis Potosi, and Zacatecas. The flower is fragrant, white, pink, and sometimes red. Blooming takes place during July to September. This is also used as ornamental locally.

Solano and Feria (2007) compared three different approaches to measure the *Polianthes* species, area of distribution to assess the risk of species extinction, area of occupancy, extent of occurrence, and ecological modeling. They also compared the species distributions with terrestrial protected regions (TPR) and natural protected areas (NPA). They opined that most of the *Polianthes* species are distributed in the Sierra Madre Occidental and Transvolcanic Belt of Mexico. They concluded that one species, *P. palustris*, is likely to be extinct. *Polianthes geminiflora* (Lex.) Rose (Agavaceae) was neotypified, and its detailed description was presented by Solano and Feria (2007). This species was earlier described by Juan Jose Martinez de Lexarza as *Bravoa geminiflora* and published in *Novorum Vegetabilium Descriptiones* in 1824. The plant was collected in the mountains of Valladolid, nowadays Morelia, Michoácan. Rose (1903–1905) transferred *Bravoa geminiflora* to *P. geminiflora*. *Polianthes geminiflora* has the widest geographical distribution and the highest morphological variation of the genus, and they reported three recognized intraspecific taxa, *P. geminiflora* var. *clivicola*, *P. geminiflora* var. *geminiflora*, and *P. geminiflora* var. *pueblensis* within *P. geminiflora*. *Polianthes quilae* (*Polianthes* subgen. *Bravoa*) was reported as a new species from western Mexico. The new species *P. quilae is* related to *P. cernua* and *P. geminiflora* var. *clivicola* but was distinguished by the erect lanceolate leaves, glaucous inflorescence with 4–21 floral nodes, 0.9–2.1-cm-long pedicels, tubular ventricose and bicolorous perigone with ascending lobes, and filament insertion site of 3–5 mm above the ovary apex. Data on geographic distribution and ecology, phenology, and conservation status were presented by them.

10.8 Collection and Conservation

The wild genetic resources of *Polianthes* are endemic to Mexico and are distributed in over 18 states (Solano, 2000). These are found in grasslands, pine and oak forest, and tropical dry or semi-deciduous forests and have hardly been exploited for commercial purposes. The genus *Polianthes* has the highest concentration of species in the southern Sierra Madre Occidental and the northwest side of the Trans-Mexican

Volcanic Belt (Solano and Feria 2007). *P. tuberosa* is the only *taxon* in the genus *Polianthes* which is cultivated across the globe and used for medicinal, ornamental, and ceremonial practices since pre-Hispanic times. Several species like *P. bicolor*, *P. geminiflora*, and *P. longiflora* are important for their medicinal properties (Solano 2000, Solano and Feria 2007). Several species like *P. sessiliflora*, *P. nelsonii*, *P. longiflora*, and *P. montana* are used for ornamental purposes locally. Several species like *P. longiflora*, *P. densiflora*, *P. howardii*, *P. palustris*, and *P. platyphylla* are under the danger of extinction and are listed as rare by the IUCN. Solano and Feria (2007) opined that habitat destruction is the major reason affecting populations of *Polianthes* species. They compared three different approaches to measure the *Polianthes* species area of distribution to assess the risk of species extinction applying the MER (Method of Evaluation of Risk extinction of wild species for Mexico): area of occupancy, extent of occurrence, and ecological modeling. They also found the richness of areas of distribution of this genus, and they compared the species distributions with terrestrial protected regions (TPR) and natural protected areas (NPA). *P. zapopanensis* was described like a new species in Jalisco, Mexico. They concluded that most of the *Polianthes* species are distributed in the Sierra Madre Occidental and Transvolcanic Belt.

10.9 Genetic Diversity, Molecular Characterization, and Molecular Markers

Diversity in several tuberose genotypes was assessed by Kameswari et al. (2014), Khandagle et al. (2014), Bharti et al. (2016), and Chauhan et al. (2017). Several accessions of different species are collected from the wild and are characterized by amplified fragment length polymorphism (AFLP) with the aim to develop a breeding program. The compatibility among species was tested, and interspecific and intersectional hybridization was done. The results of the study revealed the molecular variability among species, the possibility of creation of interspecific hybrids of *Polianthes*, and the possibility of combining important horticultural traits from wild species into the cultivated one. Kameswari et al. (2014) employed six ISSR primers, viz., 808, 836, 840, 842, 855, and 857, to assess the genetic diversity in several tuberose genotypes. They noted 62 bands across 7 genotypes, of which 53 bands were polymorphic, accounting for 85.48% polymorphism. The genotypes were grouped into two major clusters. They observed that the genotypes 'Calcutta Single' and 'Prajwal' were closely related. 'Hyderabad Single' and 'Phule Rajani' were divergent from other genotypes. Khandagle et al. (2014) used 15 RAPD and 20 ISSR primers to assess the genetic diversity among ten tuberose varieties. In their study, both RAPD and ISSR primers revealed 53% and 73% polymorphism and generated 59 and 95 polymorphic markers, respectively. They observed two major clusters among the genotypes in the dendrogram generated by UPGMA methodology using Jaccard coefficient as distance matrix for RAPD and ISSR and three major clusters in the combined RAPD-ISSR cluster analysis, which was further grouped into sub-clusters based on their genetic relatedness/variation. Bharti et al. (2016) studied the genetic diversity and interrelationship among the single- and double-petalled cultivars using 16 RAPD and 40 ISSR markers. They observed 2 clusters among the 12 genotypes in the RAPD-based study, whereas 3 clusters were observed in their ISSR marker-based analysis. Chauhan et al. (2017) studied the genetic diversity among the 21 genotypes using 7 RAPD primers. They clearly differentiated all the genotypes by the RAPD fingerprints which produced reproducible brands. The dendrogram generated in their study by RAPD markers revealed considerable amount of genetic variation among the genotypes which was divided into four major groups. Mexican Double and Swarna Rekha cultivar were found to be diverse in their study. Six inter-simple sequence repeat (ISSR) markers were used for the evaluation of the tuberose genotypes. Studies reported three major groups based on the dendrogram generated by ISSR markers. Genotypes, namely, Hyderabad Single, Kalyani Single, and Pearl Double cultivar, were found to be more diverse in their study. Several ISSR markers were applied to study genetic divergences among tuberose populations and to study variability between control and variant lines. The primers UBC 825 and UBC 835 were found to be more polymorphic, while primers such as UBC 856 and UBC 841 recorded the least polymorphism between control and variants. The primer sets UBC 855, UBC 827, and UBC 808 produced the higher PIC, MI, and R_p values. 14 species of *Polianthes* from Mexico having colored flowers ranging from scarlet red to yellow were reported. All accessions from the wild for the breeding program and different genotypes were characterized by AFLP. Compatibility studies among species were successful in interspecific and inter-sectional hybridization. Results indicated molecular variability among species, creation of interspecific hybrids, and the possibility to combine important horticultural traits from wild species into novel tuberose cultivars.

Fan et al. (2018) reported several candidate genes associated with terpene and benzenoid biosynthesis, which are responsible for the floral scent emission in tuberose. They concluded that among the candidate genes, *PtTPS1*, *PtDAHPSs*, *PtPAL1*, and *PtBCMT2* might play important roles in regulating the formation of floral volatiles. Out of 228,706,703 high-quality reads that were obtained through transcriptome sequencing and de novo assembly, they identified 96,906 unigenes (SRA Accession Number SRP126470, TSA Acc. No. GGEA00000000). They functionally annotated approximately 41.85% of these unigenes using public databases, and a total of 4694 differentially expressed genes (DEGs) were identified during flowering. Through gas chromatography-mass spectrometry analysis, they concluded that the majority of the volatiles were benzenoids and small amounts of terpenes. From homology analysis, they concluded that 13 and 17 candidate genes were associated with terpene and benzenoid biosynthesis, respectively.

10.10 Use of Tuberose Genetic Resources

10.10.1 Major Constraints

Tuberose is mostly propagated by vegetative means through bubs. The seed setting in tuberose is quite erratic in the single-flowered cultivar and is not observed in the double flower. The variegated type, however, has high degree of seed setting compared to others (Datta 2006). There was no fruit set in single tuberose due to self-incompatibility but 63.78% fruit set when cross-pollinated with variegated cultivar. The variegated cultivar was both a male and female fertile variety with sufficient pollen production. Certain other single genotypes are also good seed setter on natural cross-pollination (Singh and Sadhukhan 2019). Honey bee is the chief pollinating agent. However, many single genotypes including 'Arka Prajwal' and 'Bidhan Snigdha' do not set seed. The exact cause of sterility is not known. Sterility is not due to any defects or deformation in the formation of the pollen grains or development of the embryo sac. Histological changes were studied during ovule development after crossing Mexican Single and Pearl Double and reported that reproductive compatibility stimulates the metabolic activities in the associated tissues leading to the normal development of fruit seed formation. Pollen viability, stigma receptivity, pollen germination, seed setting behavior, self- and cross-compatibility, and percentage of seed germination in different hybrids and varieties of tuberose have been studied extensively (Figs. 11 and 14). Studies revealed that single genotypes of tuberose provide an evidence of a gametophytic self-incompatibility system and are of cross-compatible nature. In interspecific and intergeneric crosses with *Polianthes*, it was found that *P. geminiflora* var. *clivicola* McVaugh was suitable both as pollen receiver and donor, with different species and with *Prochnyanthes*. *P. howardii* was identified as the better pollen receptor, because the pollen in these plants was sterile.

10.10.2 Overcoming the Constraints

Several techniques were applied to overcome the fertility barrier including bud pollination. Application of IBA and IAA at the time of pollination and use of gamma irradiated (0.5kR) pollen was found to be effective. Bud pollination failed to induce fruiting. Irradiated pollen that pollinated 2 days after anthesis resulted in fruit set up to 4.78 percent, and the seed viability was as high as 48.23 per cent. Datta (2006, 2017) elaborately compiled the plant genetic resource utilization for improvement of tuberose. Tuberose is commercially propagated by bulbs, and very few forms have been reported in both single- and double-petalled category. Lack of genetic variability in tuberose is the major constraint in conventional breeding. There are several cultivars of tuberose: a double called 'The Pearl or Dwarf Pearl Excelsior,' a single form usually called 'Mexican Single' but sometimes 'Mexican Ever Blooming' and one or more variegated forms. *Polianthes* hybridizing efforts mainly concentrated with "tuberose" (*Polianthes tuberosa* L.) due to its ready availability, large flowers, and outstanding fragrance. Few reports on early hybridization work are available. Bundrant (1985) reported very narrow genetic base of tuberose. He collected 'Mexican Single' tuberose from a local nurseryman in San Antonio, Texas, in about 1972. During that period, that variety only was in commerce. Three distinct cultivars known at that period assumed to have originated through mutation. Hybridization work started to develop further genetic variability in commercial cultivars. The genus *Polianthes* includes not only those species originally included in *Polianthes* but also all those formerly placed in the genera *Bravoa*, *Pseudobravoa*, *Manfreda*, *Prochnyanthes*, and *Runyonia* and the herbaceous species of *Agave*

(Shinners 1966). Hybrids between *Polianthes*, *Prochnyanthes*, and *Pseudobravoa* are believed to be possible. The first hybrid in this group was produced using *Polianthes* (*Bravoa*) *geminiflora* and *P.* (*Prochnyanthes*) *bulliana* in 1899, but the first cross involving the tuberose was reported in 1911 as *Polianthes* x *blissii*, a cross between *P. geminiflora* and *P. tuberosa*. Bundrant (1985) was successful in developing three hybrids: *P.* x *blissii*, *P.* x *bundrantii* (*P. tuberosa* x *P. howardii*), and *P. tuberosa* x *P.* (*Manfreda*) *maculosa*. *P. tuberosa* has the characters of dichogamy and self-incompatibility. Cross between single and double cultivars produced fruits and seeds when the female parents were fertilized 2–3 days after anthesis. Reciprocal crosses produced many single and few double plants in the progenies, and 12 seedlings with improved characteristics were selected. Howard (1978) reported successful cross among various *Polianthes* species. His objective was to develop colored flowers having the tuberose fragrance. He reported that *Polianthes tuberosa* crosses freely not only with other *Polianthes* but also with *Manfreda* and *Prochnyanthes*. Howard (1978) did experiment to combine with color and fragrance in hybrids with characteristics of the tuberose. He reported a new hybrid *P.* x *bundrantii* (*P. tuberosa* x *P. howardii*) which is similar to tuberose. The hybrid flowers had maroon interiors and rose pink exteriors tipped green with fragrance. Crosses and backcrosses among *P. tuberosa* 'Single,' *P. tuberosa* 'Double,' *P.* x *howardii*, and *P.* x *blissii* were made, and several hybrids showing pink, reddish-purple, purple, orange, and yellow flower colors were selected. However, the long spikes of flowers in these colored hybrids were only suitable for cut flowers. Four hybrids showing a dwarf plant type were reselected for use as pot and/or bedding plants. In interspecific and intergeneric crosses with *Polianthes*, it was found that *P. geminiflora* var. *clivicola* McVaugh was suitable both as pollen receiver and donor, with different species and with *Prochnyanthes*. *P. howardii* was better pollen receptor as the pollen in these plants was sterile. Double tuberose, which no longer exists in the wild gene pool, is sterile and cannot be used as pollen parents. In double cultivars, the pistil and stamens have become petaloid segments or staminode (Fig. 9). Artificial selection and continuous vegetative propagation have favored this transformation, and most individuals in cultivation are sterile (Solano 2000). Double cultivars are fertile in early-flowering stage, when the female parent in 2–3 days after anthesis can be used as both pollen and seed parents. There was no seed production in both selfed single varieties and selfed double varieties due to self-incompatibility.

10.11 Present Status of Breeding for Desired Traits

Howard (1978) collected several *Polianthes* species from different parts of Mexico and attempted hybridization. He described three interspecific hybrids with novel variants in flower colors: *P. blissii* (*P. geminiflora* x *P. tuberosa*) with orange red flowers, *P. bundrantii* 'Mexican Firecracker' like a modern hybrid between *P. howardii*, and *P. tuberosa* with flowers marked internally in shades of wine or purple and externally in red or pink and green. He attempted to develop colored flowers having the tuberose fragrance combing various *Polianthes* species with the cultivated one. He was motivated to initiate interspecific hybridization from the *Polianthes* x *blissii* Worsley hybrid. The hybrids developed by Howard were intermediate between its parents, having the delicious fragrance of male parent *P. tuberosa* combined with the rich rose pink color of female parent *P. geminiflora*. He developed many interspecific hybrids and reported that *Polianthes tuberosa* crosses freely not only with other *Polianthes* species but also with *Manfreda* and *Prochnyanthes* . Howard (1978) attempted to combine the color and fragrance. He reported a new hybrid *P.* x *Bundrantii* (*P. tuberosa* x *P. howardii*) which is similar to tuberose. The hybrid flowers had maroon interiors and rose pink exteriors that were tipped green and with fragrance. Crosses and backcrosses among *P. tuberosa* 'Single,' *P. tuberosa* 'Double,' *P.* x *howardii*, and *P.* x *blissii* were made, and several hybrids showing pink, reddish-purple, purple, orange, and yellow flower colors were selected. Huang et al. (2002) extensively studied breeding for colored tuberose and pigment composition of *P. tuberosa* x *P. howardii* hybrids. They bred and selected several hybrids having pink, reddish-purple, purple, orange, and yellow flower colors. Intergeneric crosses between *Manfreda* and *Polianthes* were attempted. The University of Arkansas started breeding work between *Polianthes* and *Manfreda* in 2003 in order to obtain cultivars more tolerant to hot and dry conditions and reported that the flower color of *Polianthes* is dominant over that of *Manfreda virginica* (L.) Salisb. and *M. maculosa* (Hooker) Rose. and these hybrids showed hybrid vigor characterized by larger plant size and extended blooming time. Some hybrids were successfully overwintered suggesting that *M. virginica* may confer additional cold hardiness.

The use of wild *Polianthes* species in tuberose breeding programs was also reported. Interspecific crosses were made mainly using single and double cultivars of *P. tuberosa* and *P. howardii* (reddish-purple flowers) in order to bring the flower color of *P. howardii* into the cultivated background. Several hybrids were reported showing various shades of color suitable as cut flowers. Huang et al. (2002), after analyzing the pigments in the hybrids, concluded that the use of *P. howardii* in tuberose breeding can contribute to the extension of the diversity of flower colors. The various flower colors in hybrids developed through crossings between *P. tuberosa* and *P. howardii* were developed from various pigment combinations like the yellow due to carotenoids; pink, reddish-purple, and purple colors due to anthocyanins; and orange colors due to the coexistence of anthocyanins and carotenoids. *P. howardii* contained all these pigments. Datta (2017) opined that the ratio of anthocyanins and carotenoids contents seems to play an important role in determining flower colors and this ratio can be changed through further crossing and selection which can generate various shades of orange colors. Huang et al. (2002) in their investigation reported that cyanidins contribute to red appearance and delphinidin to purple color in *P. howardii* and the hybrids. Thus, introduction of anthocyanins and carotenoids from *P. howardii* into *P. tuberosa* through interspecific hybridization created diversity of flower colors. Some interspecific and intergeneric hybridization have also been attempted in Japan, Taiwan, and the USA to develop orange, yellow, pink, and lavender flowered tuberose for the cut flower market as well as for garden use (Datta 2017). Huang et al. (2000) conducted experiments to study the stability of colors of tuberose of reddish-purple tuberose (*P. tuberosa*) hybrid line '77A05' after cultivation in open fields at four different altitudes (25 (Pingtung), 75 (Chiayi), 500 (Hsinshe), and 1200 m (Tapan)) in Taiwan in 1999. He reported that flowers cultivated at lower elevations were pale purple but reddish-purple at higher elevations (Hsinshe and Tapan). The anthocyanin content of the flowers was approximately two and three times higher at higher elevations. He concluded that the low temperature favored higher accumulation of anthocyanin. Changes in the accumulation of anthocyanins and the resultant color in tuberose after application of soil amendments have been studied.

Solano (2000) reported flower colors of different wild species – *P. howardii* (reddish-purple, red in the base, and gradually green in the lobes), *P. bicolor* (orange-greenish, green lobes), *P. montana* (white, pink), *P. graminifolia* (red, orange, coral), *P. oaxacana* (pink outside, yellow inside), *P. zapopanensis* (orange, pink), *P. multicolor* (almost white, pink, orange, orange-light yellow), *P. densiflora* (yellow), *P. platyphylla* (white tinged with red), *P. venustuliflora* (white tinged with pink), *P. palustris* (white), *P. tuberosa* (white, buds may have a light pink), *P. longiflora* (white tinged with purple), *P. nelsonii* (white, pink, and sometimes red), *P. densiflora* (white, pink, and sometimes red).

Colors of wild *Polianthes* species: *P. geminiflora* var. *clivicola*, *P. geminiflora* var. *graminifolia*, *P. graminifolia*, *P. howardii*, *P. palustris*, *P. platyphylla*, *P. pringlei*, *P. sessiliflora*, and *P. Montana*, which were white, yellow, pink red, etc. Taipei University (Taiwan) was deeply engaged in developing more prolific and higher-quality *Polianthes* varieties combining available vibrant colors, double flowers, and disease resistance strains (Datta 2017). Interspecific hybridization has been performed in the genus *Polianthes* in order to generate genetic variation, resulting in hundreds of hybrids, which are being selected for further breeding programs.

Induced mutation is a well-standardized technique for crop improvement, and a large number of new ornamental varieties have been developed. Working dose and LD50 of different mutagens (X-ray, gamma rays, fast neutrons, etc.) have been determined in both single and double cultivars. But no flower mutations could be induced except chlorophyll variegations in leaves (c.f. Datta 2006). Singh and Sadhukhan (2019) reported other mutants in addition to the leaf variegation due to induction of mutation by chemical or physical means. They reported a long stalk mutant and one branch mutant when Calcutta Double genotype was irradiated with gamma ray.

10.11.1 Breeding Breakthrough

A breeding breakthrough was the development of pink tuberose "pink sensation" (Datta, 2017). Dutch-based breeding company, Ludwig and Co., a licensee of new *Polianthes* breeding genetics, released the variety in 2014. The company has introduced a series of new *Polianthes*, both single and double flowering in soft yellow, dark yellow, soft pink, and lavender pink. "Pink sensation" is a single-flowered, soft pink, sweet-scented, shorter stem length and suitable for cut flower industry, pot culture, and garden display. Their stunning fragrance remains as good as ever (c.f. You Garden). From the above hybridization, it is clear that pigment composition plays an important role in developing new flower color variety.

10.11.2 Varieties Developed in India

Original cultivated tuberose genotypes were Mexican Single and Pearl Double or Mexican Double both being white in color belong to *Polianthes tuberosa* L. With time, several genotypes in both single and double category have been reported. In India, there are four types of cultivated tuberoses based on the number of rows of petals, viz., single, semidouble, double, and variegated. Single-flowered genotypes are often referred to as 'Mexicano' (Spanish for "Mexican") or 'Sencillo' (Spanish for "simple"), while the cultivar with paired flowers is commonly known as 'Perla' (Spanish for "pearl") or 'Doble' (Spanish for "double") or Dwarf Pearl Excelsior. Concrete content of the flowers of tuberose (single and double types) varies from season to season. Single types generally have higher concrete content (0.134–0.136%) than the double types (0.107–0.108%). They also observed that the flowers harvested during October yielded a higher concrete content than those harvested during March.

At present, total germplasm including new varieties available in India are as follows:

Single Types: Calcutta Single (Calcutta), Hyderabad Single (Hyderabad), Kahikuchi Single (Assam), Local Single, Pune Local Single (Pune), Mexican Single (Mexico), Nilakottai Local (Coimbatore), Sikkim Selection, Rajat Rekha (Gamma ray induced mutant, NBRI), Prajwal (Shringar x Mexican Single, IIHR), Phule Rajani, Shringar (Single x Double, IIHR), Arka Nirantara, GKTC-4, STR-501, Bidhan Snigdha (BR-1), Bidhan Ujjwal (Br-2), Bidhan Jyoti (Br-3), and Arka Sugandhi.

Semidouble or Double Types: Calcutta Double, Pune Local Double, Hyderabad Double, Pearl Double, Suvasini (Single x Double, IIHR), Vaibhav (Mexican Single x IIHR-2, IIHR), Swarna Rekha (Gamma ray induced mutant, NBRI), STR-505, Arka Sugandhi, BRH-19, and BRH-24.

10.11.3 Description of some Important Varieties Cultivated in India

Single

Single types bear one whorl of corolla which is generally highly scented (six tepals per flower) (Figs. 1 and 3). The floral buds are greenish white to white. The fully opened flowers are white in color. Seed setting is observed in several single genotypes. Loose flowers are extensively used for garland-making, flower arrangement, rangoli, extraction of essential oil, and concrete. Single-flowered varieties are more fragrant than double and are usually preferred for gajara and garlands, while double varieties are preferably used as cut flower for vase decoration. Mexican Single, Calcutta Single, Shringar, Prajwal, Arka Nirantara, Rajat Rekha (variegated mutant type), Hyderabad Single, Phule Rajani, Sikkim Selection, Bidhan Snigdha, Bidhan Ujjwal, Bidhan Jyoti, and Arka Sugandhi are the main single varieties (Figs. 7 and 15). Several genotypes similar to Mexican Single have been named after the locality they are grown, viz., Kahikuchi Single, Pune Single, Kalyani Single, etc.

Calcutta Single: It is extensively grown single genotypes in India used for both loose flower and cut flower. Spike length is around 80–100 cm. Flowers are white in color, lighter in weight (0.8–1.0 gram per floret), and highly fragrant. This is highly susceptible to foliar nematode (*Aphelenchoides besseyi*) prevalent in eastern India.

Shringar: This variety was released by the Indian Institute of Horticultural Research (IIHR), Bangalore, which is a cross between 'Single and Double.' Flower is highly fragrant medium spikes and attractive due to pinkish tinge. Spikes have more number and large-sized florets (floret weight 1.3–1.4 g). This is tolerant to root knot nematodes (*Meloidogyne incognita*).

Prajwal: This hybrid was released by the Indian Institute of Horticultural Research (IIHR), Bangalore, which is a cross between 'Shringar' and 'Mexican Single.' Florets are large (1.8–2.0 gram per floret) and moderately scented. Flower buds are slightly pinkish, and flowers are white when fully opened. It is recommended for loose flower and cut flower purpose. It is tolerant to foliar nematode. It is being widely cultivated by the growers and has replaced the Calcutta Single because of its higher yield and disease tolerance.

Arka Nirantara: The flowers are lighter in weight (1.0–1.2 gram per floret), white in color, and moderately fragrant. It was released by the Indian Institute of Horticultural Research (IIHR), Bangalore.

Phule Rajani: Also called Pune Single. Developed by MPKV. Moderate spike length (60–70 cm). Florets are white at bud stage and individual floret size 1.0–1.2 gram. The number of florets varies from 40 to 50 per spike. Good for loose flower as well as cut flower.

Bidhan Snigdha: This hybrid was developed by BCKV which is a cross between Arka Nirantara and Prajwal. Also named as Bidhan Rajani-1. Large floret (2.0–2.2 g) and high floret yield.

Bidhan Ujjwal: It is a dwarf genotype suitable for loose flower production, garden decoration, and pot culture. It has been developed at BCKV through hybridization between Sikkim Local x Phule Rajani followed by clonal selection. It is also named as Bidhan Rajani-2. Florets are bright white, lighter in weight (0.8–1.0 g per floret), and arranged on a compact rachis. Spike length is 40–50 cm.

Arka Sugandhi: Developed at the IIHR, this genotype is the most dwarf one reported so far. Spike length is 30–35 cm. Suitable for pot growing.

Semidouble

It bears two to three whorls of petals or corolla segments on straight spikes (Fig. 8). In this type of tuberose, the flower spike is straight, and the flower color is white also with pinkish red tinge.

Vaibhav is the most popular variety in this category. This variety, developed at the IIHR, Bangaluru, bears profusely round the year and is generally recommended for cut flower purpose.

Double

The varieties belonging to this group bear more than three whorls (Fig. 8). The main varieties are Swarna Rekha, Mexican Double or Pearl Double, Calcutta Double, Hyderabad Double, and Suvasini. No seed setting is observed in semidouble or double varieties. Double varieties are generally reported as less fragrant as the florets fail to open completely. Double cultivars generally bear stout spikes bearing around 30–40 bright florets on a compact rachis of 30–40 cm which makes it as an attractive cut flower. Though spike emergence takes place round the year in moderate climate, proper blooming, however, occurs during autumn or winter with the decrease in temperature. The quality of flower and fragrance are better during winter months than the summer months.

Calcutta Double: It is the most popular double genotype cultivated in India as cut flower for many years. The origin of the variety is unknown and may be a clonal selection from the original Pearl Double or Mexican Double. Spike is long (50–60 cm) with compact rachis. Florets are bright white in color and moderately fragrant. Profuse production takes place during early rainy season and early winter.

Suvasini: It is a double-type flower with multi-whorled variety released from the Indian Institute of Horticulture Research (IIHR), Bangalore, which is a cross between 'Single' and 'Double.' This variety produces more number of bold flowers per spike with uniformity. Spike is long (60–70 cm) with pinkish tinge on the young florets. During winter, the intensity of pinkish tinge is more on the tips and back side of the florets. Spike yield is 25% higher than Pearl Double cultivar.

Hyderabad Double: The variety, similar to Calcutta Double, is suitable for cut flower production. Rachis is compact and florets are moderately scented. Origin is unknown and might be a clonal selection from Calcutta Double or Mexican Double.

Variegated

The leaves are variegated, i.e., yellow on the margin. In these varieties, silvery white or golden yellow streaks are visible on leaves. The National Botanical Research Institute, Lucknow, has developed two variegated varieties, namely, Rajat Rekha and Swarna Rekha, through induced mutation by gamma irradiation from Mexican Single and

Mexican Double, respectively. In Rajat Rekha, leaves are silvery white with streaks along the middle of the leaf blade. Spikes are also silvery white. Floret size is small (0.6–0.8 g per floret). Swarna Rekha is a double genotype with golden yellow streaks along the margins of the leaf. Both Rajat Rekha and Swarna Rekha are very low in spike yield and are not recommended for commercial cut or loose flower production.

Another variegated single-petalled genotype named Sikkim Local or Sikkim Selection is maintained in the germplasm collection at BCKV and few other centers. The spike is the longest one (120 cm) reported so far. Sikkim local is a very good seed setter.

Colored Tuberose: Recently thorough intensive breeding effort through interspecific hybridization of several colored tuberose genotypes in both single and double categories has been made available for the growers. The genotypes are generally pink or yellow. Pink sensation and pink sapphire are available in public domain.

10.12 Cultivation Practices

10.12.1 Propagation

Tuberose is commercially propagated by vegetative methods (bulb, bulblet, and division of bulb) but rarely by seed. Recently, tuberose is also propagated by tissue culture to get virus-free planting material.

10.12.2 Seed Propagation

Success of intervarietal or interspecific hybrid production depends on the effective seed setting and seedling development. Seed and seedling performance in tuberose varies with genotype, prevalent temperatures during seed development, and the level of maturation at harvest. Seed setting of tuberose is observed only in single-type cultivar under suitable climatic conditions and known to exhibit low percentage of germination. Double cultivars are generally sterile with a rudimentary reproductive structure. The pistils and stamens have become a petaloid segments or staminodes. This structure has been favored due to continuous vegetative propagation and artificial selection for a long time (Solano, 2000). No double genotype exists in the wild. Seeds are sown in well-prepared growing medium containing leaf mold, vermicompost, and garden soil in equal proportion under portray nursery. Moisture and temperature have a marked effect on germination, and an ideal soil temperature of 25 °C is fully effective for increase seed germination. Before transplanting, bed should be prepared by digging, and sufficient quantity of FYM is to be mixed before sowing. The seedling is sown in rows 10–12 cm apart and 5 cm deep in heavy soil and 2.0 cm in light soil. Seed reproduction is mainly used to get new varieties or for very specific studies. It takes a long time before the plant blooms compared to vegetatively propagated plants. The offspring are dispersed due to genetic recombination, often leading to lose desirable characteristics present in the donor material.

10.12.3 Vegetative Propagation

In general, this species is multiplied by bulbs or bulblets surrounding the mother bulb. Division of suckers or micropropagation can also be done but not so popular in tuberose.

Propagation by Bulb This is the most common method practiced for the commercial multiplication of tuberose. The bulbs remain dormant during the winter months in places where the temperature is low. The dormancy of the bulbs can be successfully broken by treating the bulbs with 4% thiourea solution for 1 hour if early planting is desired, and ethylene chlorohydrin can also be used for breaking the dormancy of bulbs. Before sowing, the scale should be removed from the bulb for easily sprouting. Selection of ideal size bulbs is very important for quality production. Well-developed spindle-shaped bulbs free from diseases, with diameter 2.0–3.0 cm and above forming at the outer periphery of the clump, are considered ideal for planting. Experiments using different bulb size (1.5–2.5 or. 2.6–3.0 cm) and spacing (15x20, 20x20, or 25x20 cm) were conducted. Small bulbs planted at 15x20 cm gave highest yield of top-quality flowers. There are one important criterion to be kept in mind during planting, that is, depth of planting, here bulbs to be placed 4–5 and 5–6 cm depth in clay loam and sandy loam soil, respectively. Optimum bulb size for tuberose should be 2.1–3.0 cm. Similarly, the weight of bulbs plays a vital role for quality and quantity production of tuberose. It was observed that approximately 20–25 gm weight of bulb is the optimum level for planting of tuberose. Sharga (1982) graded rhizomes into six sizes ranging in weight from 3.0 to 49.5 g. He found rhizomes of 19.4 g gave satisfactory results with regard to plant growth and flowering. Cooling treatment with 10 °C temperature for a month before planting of tuberose bulb found to be very effective in terms of vegetative growth as well as better yield. There are some growth regulators (GA_3, ethrel, thiourea), when treated the bulbs to enhance sprouting and improve growth of tuberose plant. Dormant tuberose were treated with ethrel, cycocel, and GA for 1 hour and found that GA at 100 ppm or cycocel at 500 ppm advanced flowering and flower yield. It has been recommended to dip the bulbs in 5000 ppm CCC (5 g/lit) before planting to increase the yield. Improvement in plant growth and increased spike and flower yield were found when bulbs were stored at 4^0–10 °C temperature for 10–30 days and bulbs were soaked in GA_3 (200 mg/l) and thiourea (200 mg/l) solution for 6 hours (Dhua et al. 1987).

It is estimated that nearly about 1.20–1.25 lakh bulbs are required for planting 1 hectare area, the approximate weight of which should be 2.5–3 tons. Bulbs are matured after 2 years of growing in the field and harvested during the onset of winter season. Before lifting bulb from field, it is suggested to not use any water for irrigation 1 month before harvesting. After harvesting the bulbs, bulblets should be separated and cleaned properly with water followed by dipping fungicides (copper oxychloride at 0.2%) for half an hour and dried in shady place.

Propagation by Division Other propagation method of tuberose is division of bulb. Sprouting depends on the size of the bulbs; generally, the segments from large bulbs (2 cm or more in diameter) regenerate well. Bulbs are normally cut into two to three vertical sections; each segment must contain bud and a part of the basal plate. Each of these sections is treated with fungicide and planted vertically in a rooting medium with their tips just showing above the surface. New bulblets along with roots develop from the basal plate. At this stage, they are transferred to the field to continue to grow.

Micro-Propagation Several explants like root, bracts, petals, bulbs, scales, leaves, leaf callus cultures, and terminal and axillary stem scale of single and double types of tuberose have been cultured on various media. Embryoids were devoid of shoot apical meristems but developed primary root. Callus was successfully induced and subsequently regenerated plantlets from scale stem sections of tuberose cv. Single. Plantlets have been regenerated from in vitro anther culture with microspores at the uninucleate stage of single flower tuberose. Somaclonal variations were detected in inflorescence length, number of flower buds, and flower size and color. Liquid culture of tuberose has been standardized and optimized high production of polysaccharides by reducing the viscosity of culture medium. Direct shoot organogenesis has also been reported in tuberose. Disease-free plantlets have been produced using in vitro techniques (Khan et al. 2000, Krishnamurthy et al. 2001, Datta et al. 2002, Singh et al. 2020).

Planting Planting density is crucial for yield and quality flower production. The planting distance varies with the soil and climatic conditions. Both high and low planting densities adversely affect the quality flower production. About 45,000–55,000 bulbs are required per acre which is the optimum plant density for high plant growth and yield. For good economic returns, bulbs are planted at an optimum spacing of 30 x 20 cm or 20 x 20 cm or 30 x 30 cm with 5.0 to 7.0 cm depth. In tuberose, the spacing has a great importance for manipulating flower quality and quantity characteristics. Therefore, inter- and intra-row spacing is important for obtaining higher tuberose flower quality and quantity. The depth of planting depends upon the size of the bulb (large bulb more depth and small bulb less depth). The depth of planting varies from 3.0 to 7.0 cm depending upon the diameter of the bulb and the soil type. Generally, it should be 2.5 times more than the diameter of bulbs. While planting, the bulbs are planted at the recommended plant spacing, 4–6 cm deep on the sides of the ridges. Planting is deeper in sandy soil as compared to clay soil at the depth of 6.0 cm. In general, planting is done in such a way that the growing portion of the bulb is kept at the ground level.

10.12.4 Land Preparation and Planting

Well-leveled pulverized soil is suitable for this crop with proper drainage system. Plant requires huge quantity of water for cultivation, because soil is to be kept moist round the year, but standing water in the field is harmful to the crop, especially during winter and rainy season. Waterlogging for a short period damages the root system and affects growth and flowering. After three to four times of plowing, raised beds are prepared up to 6 inches height with 3 feet width, length depends on field size, and between beds should have space 1 foot for intercultural operation which is used as irrigation and drainage channel.

Before planting, bulbs are treated by dipping into the water for 30 minutes with fungicides (Blitox at 2gm/lit. of water). After fungicidal treatment, bulbs are dried in room temperature for a day followed by planting. Generally, tuberose bulbs are planted in different times at various parts of India. Planting can be done round the year depending on the availability of bulbs; March–April is the best time of planting in the subtropical region. In the hilly areas, western and southern part of India planting is done in the month of May–June. There are some pockets where tuberose is planted just after rainy season during middle of September to middle of October for getting winter flowering. Generally plant spacing is maintained based on plant-to-plant (20 cm) and row-to-row (30 cm) distance, but now high-density planting with large size of bulbs (15x15 or 15x20cm) is followed by the farmers for 1 year cultivation.

10.13 Nutrient Management

Tuberose plant is a heavy feeder crop because it has fibrous and shallow root system with huge canopy area, grows round the year in the field, and continues for 2 years. So, higher quantities of organic and inorganics are to be applied for obtaining optimum growth. In addition to that, dose and time of application at different growth stages are most important. But before application of any nutrients, status of growing media should be evaluated, because the requirements of nutrients are different for various categories of soil irrespective of varietal character. The requirements of various nutrients for the tuberose as recommended by the researchers are mentioned below.

10.13.1 Nitrogen

It is the most important primary nutrient for tuberose, which acts in the plant for synthesis of chlorophyll, helps in the photosynthesis, and improves vegetative growth. Plants are able to take nitrogen from either organic (cowdung manure, FYM, vermicompost, cakes and meals, etc.) or inorganic sources (different straight fertilizers like urea, DAP, ammonium sulfate, or NPK 10–26-26, etc.). The water-soluble fertilizers like NPK19–19-19, NPK 20–20-20, NPK 10–10-10, NPK 12–61-0, and NPK 13–0-45 were found to have good response through drip irrigation. Its requirement throughout the cropping period is about 200–300 kg/ha and usually is applied in three split doses starting from basal application at the time of planting, and another doses are applied 60 and 90 days at the age of plant. Increasing levels of nitrogen increased the number of leaves per plant, plant height, length of spike, length of rachis, yield of spike, and number of bulbs. Dhar et al. (2008) reported that the vegetative growth and flowering were influenced by application of N at 20 g/m2. Application of nitrogen at 220 kg ha^{-1} markedly improved vegetative growth and flowering of tuberose single varieties. Meena et al. (2018), however, reported that application of 250/200/200 kg ha^{-1} NPK through inorganic fertilizers had significant influence on plant height (72.10 cm), spike length (66.07 cm), floret spike^{-1} (39.43), floret diameter (3.49 cm), flower duration (32.63 days), durability of spike in the field (11.00 days), spike plant^{-1} (4.73), spike weight (75.12 g), bulb plant^{-1} (16.93), gross return (Rs. 914,450 ha^{-1}), net return (Rs. 678,646 ha^{-1}), and benefit-cost ratio (3.72).

10.13.2 Phosphorus

Root growth of tuberose plants fully depended on the availability of phosphorus in the growing media, but the use of more quantity badly affects root growth as well as enlargement of bulbs due to the increased hardiness of the soil. It should be applied at the beginning stage during field preparation at 200 kg ha^{-1}, while plant can develop a strong root system. Higher dose of phosphorus is beneficial to tuberose plant which was manifested in maximum number of leaves of plant^{-1} (33.73), number of tiller clump^{-1} (3.52), number of spike clump^{-1} (5.56), yield of spike ha^{-1} (749.98), diameter of largest bulb (5.47 cm), number of bulb clump^{-1} (17.83 g), weight of larger bulb (82.96 g), and diameter of largest bulb (5.47 cm) when plant were treated with N/P/K in a ratio of 180:360:180 kg ha^{-1}. During standing crop, it has good effect when used as water-soluble fertilizers. Organic sources of nutrients may be used in more quantity to fulfill the phosphorus requirement, because it is a bulbous plant. Its deficiency in soil is reflected in plant in such a way that it leads to dark green in color in upper leaves, pale green color in lower leaves, stunted growth, and reduced flower yield. Dhar et al. (2008) revealed that P increased the number of spikes, rachis length, and longevity of flower of tuberose. There was significant increase in the height of plant, number of leaves, and leaf area per plant. It was found that 20 g N/m^2 was optimum for vigorous growth of the plant.

10.13.3 Potassium

Similar to nitrogen and phosphorus, potassium has a prime role for growth and development of tuberose plant. It acts to increase enzyme activity, protein synthesis, and photosynthesis and cell growth. It also gives the plant good strength and improves flower quality. It's deficiency in the soil causes hindrance of cell division and plant growth.

10.13.4 Mode of Action of Nutrients

The N, P, and K contents of leaves significantly increased with the increase in rate of N, P, and K fertilizers, respectively. Leaf P and K concentrations decreased with increasing N fertilizer rate. N, P, and K contents in leaves were higher than those in bulbs (rhizomes). Bulb N increased with increasing rates of all fertilizers. Bulb P content was affected by N and P fertilizers, but not by K fertilizer. Bulb K content also increased with increasing rates of all fertilizers. A dressing of 120 kg of nitrogen increased the yield of flowers, with 60 kg of phosphorus and 40 kg potash per ha, but good plant growth and high flower yield were obtained with N/P_2O_5 at 300:200 kg/ha with N applied in 2 split doses, at planting and 40 days later. Application of split dose of one-third of the N dose before planting and one-third at 30 days and one-third at 60 days after planting showed greatest plant height and increased number of leaves and spikes plant^{-1}. N (5, 10, 15, or 20 g/m^2) improved vegetative growth, flowering, and bulb production in the first year, whereas P and K (20 or 40 g/m^2) increased spike number, rachis length, and duration of flowering in the second year. The optimum fertilizer application rates were determined as 15 g N + 40 g of P_2O_5 + 40 g K_2O/m^2, whereas fertilization of tuberose with N/P_2O_5/K_2O at 20:20:20 gm/m2 has been recommended. 80gN + 40 g P + 60gK/m^2 produced maximum spike yield and increased floret size and shelf and vase life of tuberose. Treatment combination of N3, P2, and K2 (200/150/200 kg/ha) was found significantly superior in respect of plant height (76.00 cm), flower stalk per plant (11.70), and number of bulbs per plant (43.70). In addition to the flower yield, essential oil is also one of the products of single tuberose which depends on nutrient management. Application of 80 kg N ha^{-1} increased essential oil. It was observed that application of graded levels of nitrogen (200 kg ha^{-1}) and sulfur (80 kg ha^{-1}) showed significant desirable improvement of vegetative and floral characters and also oil content in tuberose cv. Mexican Single. For achieving increased essential oil content in flowers and for the maximum recovery of concrete, a fertilizer dose of 80 kg N, 60 kg P_2O_5, and 40 kg K_2O ha^{-1} has been recommended.

Recently, an experiment was conducted on phenophase-based nutrient scheduling on flower yield, quality, and bulb production in tuberose by AICRP on Floriculture, BCKV, during 2019—2020 (unpublished). Application of NPK at 200/200/200 kg/ha in four split doses – at final land preparation (25% of recommended dose of fertilizer applied as basal dose using straight fertilizers along with 200 kg of neem cake/ha + 1 kg of *Trichoderma harzianum*), at vegetative stage after 30 days of planting using water-soluble fertilizer through drip (45:30:30% NPK of remaining recommended dose of fertilizers), at flowering stage after 90 days of planting using water-soluble fertilizer (45:60:60% NPK of remaining recommended dose of fertilizers), and at dormancy period using water-soluble fertilizer (10:10:10% NPK of remaining recommended dose of fertilizers), was found better for loose flower and bulb production.

10.13.5 Micronutrients

Apart from NPK, other macronutrients like calcium, magnesium, and some micronutrients, namely, iron, zinc, manganese, and boron, have also been found to influence the growth and flowering in tuberoses. Their effect on growth and importance on yield in tuberose plant describe below.

Calcium is one of the most important nutrients for tuberose; it gives strength to the plant and also prevents lodging of the spike. Bud rot of flower is due to severe deficiency of calcium in the soil. It may be corrected by application of lime or gypsum in the soil based on soil type. Zinc deficiency in tuberose leads to stunted growth, little leaf, and few other deformities in tuberose. It acts as balancing intake of P and K from soil. Soil application of zinc at 20 kg per hectare increased the spike length, rachis length, number of florets per spike, and number of bulbs per plant in tuberose cv. Double. In acidic soil, there is deficiency of boron, and in high rainfall area, it is found to be severe, resulting in deformed leaves and stunted growth of plant as well as smaller size spike. It may be met up with application of borax at 15 kg/ha or foliar spray on standing crop with 0.2–0.25%. Foliar application of boron 100 ppm twice at monthly interval produced the maximum height of plant and increased the number of leaves per clump resulting in improved yield of spikes of tuberose. Iron plays a vital role on growth of tuberose plant in terms of nitrogen assimilation and synthesis of chlorophyll. It's deficiency in tuberose plant leads to interveinal chlorosis in new leaves and very poor quality of flower produce. Magnesium deficiency causes interveinal chlorosis of older leaves in tuberose, and it may be corrected by application of $MgSO_4$ at 0.2%. Whereas interveinal chlorosis in new leaves are found due to the deficiency of manganese in growing media and sometimes its deficiency lead to yellowish flower spike, MnSO4 at 0.1% as foliar spray may correct the deficiency.

10.13.6 Integrated Nutrient Management

Integrated nutrient management of tuberose approaches recently come in the field with objectives to minimize the doses of fertilizer application, environmental chemical pollution avoidance, improve soil health, and reduce cost of cultivation. There are several research work done in this aspect and documented under mentioned with organic manures as a supplement of inorganic fertilizers and incorporation of bio-fertilizers.

Kabir et al. (2011) reported that in Bangladesh, recommended dose of fertilizer @ 400,300,300 and 100 kg ha^{-1} of urea, TSP, MP, and Gypsum of 50% along with 20 tons of poultry litter found highest bulb and flower yield followed by 20 tons cowdung manure over 100% recommended dose of fertilizers, whereas Tripathi et al. (2012) observed that maximum shoots $clump^{-1}$ (18.95), number of leaves $shoot^{-1}$ (19.44) and highest number of spikes (2,05,030/ha) with 75% of NPK (240: 160:100 kg ha^{-1}) + 50 tons FYM + 25 tons vermicompost. Highest NPK content in plant showed maximum number of spike $plant^{-1}$ and number of floret $spike^{-1}$ by application of 50% N through vermicompost and the rest 50% from urea, whereas phosphorus and potassium were used as per recommendation. There is some integration of micronutrients along with recommended dose of fertilizers. Fertigation with 80% RDF as water-soluble fertilizer along with boron at 0.1% or zinc at 0.5% through sprinkler irrigation improves vegetative growth and flowering of tuberose plant. Application of nitrogen at 150 to 200 ppm + 7.5 ppm zinc had improved the growth parameters and number of florets $spike^{-1}$ in tuberose cv. Double. Increased spike length and increase in chlorophyll content were also observed. Karuppaiah (2019) reported from his experiments the maximum vegetative growth and improved flowering of tuberose plant, when plants were treated with 25 tons ha^{-1} FYM + recommended dose of 200:200:200 kg ha^{-1} NPK + zinc sulfate at 0.50% + Borax at 0.50% on 30, 60, and 90 DAP. Highest flower yield was recorded in terms of number of spikes per plant or fresh weight (q/ha) with RDF which was at par with Comlizer applied as vermicompost 250 g + 75% RDF, or vermicompost 200 g + 50% RDF, or vermicompost 250 g + 50% RDF per square meter. Application of Comlizer as vermicompost 250 g + 35% RDF per square meter significantly reduced the number of spikes/plant and fresh flower yield (q/ha). Application of 75% RDF in integration with farm yard manure (FYM), vermicompost, *Azospirillum*, and phosphate-solubilizing bacteria (PSB) significantly yielded maximum number of spikes per plant with increased spike length, rachis length, number of florets per spike, and also maximum number of bulbs per plant (Fig. 2). Elisheba and Sudhagar (2019) observed that the tuberose crop receiving a combination of 75% RDF + vermicompost at 5 t ha-1 + humic acid at 0.2% was found to be best in all the growth characters.

10.13.7 Biofertilizer

Application of *Azotobacter* and FYM along with vermicompost recorded maximum vegetative growth, improvement in flower quality, days taken to lowest floret wither, and also in lowering down the number of unopened florets. The use of *B. subtilis* plus *P. fluorescence* mixture was remarkably advantageous to tuberose vegetative growth and flower production traits. Inoculation of tuberose plants with the K-solubilizing bacteria (*B. circulans*) greatly enhanced vegetative growth, cut flower yield, and flowering quality characteristics and increased nutrient contents in leaves, in comparison to the untreated plants (control), in both seasons. Tuberose plants received the combined mixture of *B. subtilis* + *P. fluorescence* amplified with *B. circulans* inoculation, at any level of potassin-P, and immensely exhibited noticeable performances. Application of PSB and *Azospirillum* along with various combinations of NPK was found to influence different growth and flowering characters. The effect of bio-fertilizers and vermicompost in tuberose treatments which comprised *Azospirillum* + vermicompost (10 t ha^{-1}) recorded maximum value for the parameters, namely, number of leaves on 90 days after planting, length of spike, length of rachis, number of florets per spike, number of spike per plant, and yield of spike/ha. Chemical fertilizer NPK at 15, 11.2, and 9.3 g/sq.mt. along with bio-fertilizer (*Azotobacter* and PSB) increased plant height, number of leaves, early emergence of flower, and bulb production (Srivastava and Govil 2006).

10.14 Growth Regulators

Growth regulators have good effect on tuberose cultivation starting from bulb treatment to foliar spray on standing crop. Chlormequat (CCC), maleic hydrazide (MH), ethrel, gibberellic acid (GA3), kinetin, thiourea, cyclocel, and paclobutrazol have positive effects on vegetative growth and flowering behavior in tuberose. CCC is an important growth regulator used in tuberose cultivation. Different doses of CCC are effective for flower production, to enhance flowering period, and to increase leaf dry weight, number of stalks and florets, fresh and dry weight of stalks, maximum number of florets per spike, and floret turgidity (Biswas et al. 1983).

Maleic hydrazide (MH) plays a vital role on growth and flowering of tuberose. Application of MH at 500 ppm significantly increased flower production, and application of ethrel at 100 ppm improves flower yield (Biswas et al. 1983). But advanced flowering at higher dose has been reported. Similar to CCC, MH, and ethrel, gibberellic acid (GA^3) is a popular growth regulator for tuberose flower production. Application of GA^3 at 50 ppm and at 100 ppm was found to be effective for higher flower production. Dipping of bulbs in GA3 at 300 ppm for 24 hours markedly increased flower yield and bulb production. Bigger sized bulb of Prajwal variety (4.0–4.5 cm diameter) when planted at wider spacing (40 × 20 cm) and sprayed with GA^3 at 200 ppm increased flower production. Bulb treatment of tuberose with growth regulators is better than spraying on foliage. Bulb soaking for 24 hours with GA^3 at 160 ppm had very positive response in growth and flowering of tuberose var. Phule Rajani rather than spraying with the same chemical with equal dose. GA^3 at 250 ppm concentration proved to be ideal to realize higher spike yield in tuberose. Foliar application of GA^3 at 150 ppm markedly increased vegetative growth and flowering of tuberose. Treatment with GA^3 at 30 and 60 days after sprouting of bulb had better production in terms of quality as well as quantity. GA^3 also enhances postharvest life of flower. GA^3 at 100 ppm enhanced postharvest life of flower by increasing uptake of water and also opened maximum number of flowers/spike. Increased water uptake, vase life, fresh weight, and percentage of opened florets, floret diameter, and length of florets with 100 ppm and 150 ppm GA^3 applied with sucrose at 2 and 4% concentrations have been reported. Kinetin at 100 ppm and at 150 ppm increased bulb qualities and maximum days to withering of first opened floret and flowering duration in tuberose var. Shringer and Kalyani Double. Thiourea (at 1000 ppm) and paclobutrazol (at 1500 ppm) showed influence on flower production in tuberose (Biswas et al. 1983).

10.15 Water Management

After planting of bulbs, soil is to be kept moist until bulbs have sprouted. After planting, water is to be provided to the plant through flood irrigation, or any other means twice in a week depending on the soil types and climatic condition and then once in a week is sufficient during summer. Plants irrigated every 7 days were larger, flowered earlier, and produced larger rhizomes. The water management practices and irrigation at different IW/CPE ratios had significant effect, and IW/CPE at 1.20 and nitrogen fertilization at 400 kg ha^{-1} recorded desirable effect on growth and floral characters. Irrigation applied at 0.8 IW/CPE ratio increased the number of florets per spike, and bulb yield increased over 23 and 22% over flood irrigation. Pal et al. (2019) conducted an experiment on three varieties of tuberose (Prajwal, Calcutta Single, Calcutta Double) along with three irrigation treatments on IW/CPE 0.4, 0.8, and 1.0. The experiment showed that the total water requirement of three varieties of tuberose for the period of March 2009 to March 2010 were 626.06 mm, 695.62 mm, and 751.27 mm for the Prajwal, Calcutta Single, and Calcutta Double, respectively. The irrigation requirements were 212.97 lit, 247.15 lit, and 278.32 lit for the Prajwal, Calcutta Single, and Calcutta Double, respectively. The different irrigation schedules regardless of the crop varieties on the number of spike per plot were significant. The maximum spike per plot was recorded at 1.0 IW/CPE which gave about 33.15 number of spike per plot, which was superior to 0.8 IW/CPE (32.25) and 0.4 IW/CPE (30.57).

10.16 Intercultural Operation

10.16.1 Earthing Up

Such type of operations is very much essential for tuberose cultivation for encouraging the bulbs to sprout more. It should be done manually and very carefully; earthing up is done in the tuberose field during weeding followed by fertilizer application 30, 60, and 90 days after planting.

10.16.2 Weed Management

Weed is a major problem of tuberose cultivation, because it grows in subtropical climate with high rainfall, high temperature, and high humidity where different kinds of weeds of both monocots and dicots grow vigorously round the year. About 40% of the total cost of cultivation goes for weed management. So weed management in tuberose field is a challenging job in present-day tuberose cultivation, because plant spread quickly toward horizontally and covers the space between row to row and plant to plant. After planting, manual weeding is best once in a month, but weed population in the tuberose field may be reduced by application of preemergence herbicides. Chemicals for weed control have been tested in tuberose. Pre-planted application of atrazine 3 kg/ha^{-1} caused the maximum reduction of weed population and resulted in good quality flowers. Under Gangetic alluvial soil, pre-planting application of glyphosate 71SG + oxyfluorfen 23.5 EC at PP 2–3 weeks before final land preparation at 2 g/lit of water followed by pendimethalin 30 EC (preemergence) at 1.0 kg a.i./ha within 24 hours of planting was found effective. No recommendation is available on suitable postemergence herbicide.

10.17 Harvesting and Yield

Harvesting depends on several factors mainly varietal potentiality, season of planting, size of bulb, and management of crop. Tuberose can be harvested as individual florets (loose flower) or as cut spikes. For loose flower production, fully matured flower buds are harvested preferably early in the morning or afternoon day before opening of flower. Depending on time of planting, size of bulbs, and several other factors, it starts flowering 80–100 days after planting. It has been found that maximum flower appears during summer and rainy season. Single varieties produce approximately 12–20 tons of loose flower/ha^{-1}/year^{-1}. Single, semidouble, and double varieties are suitable for cut flower production; spikes are harvested when the first pair of flower bud opens. Flower spikes are separated from the plant just 5–6 cm above the surface of the field. Nearly about 2.5–4 lakh spike can be obtained from 1 hectare of field per year.

10.18 Ratooning of Tuberose

After first year of flowering, plant goes to dormancy period during winter season. The crop regrows with proper management which is called rooting. Here, top dressing with fertilizers both organic and inorganic is to be repeated two times, just after winter season and before onset of rainy season. The production and quality will be less in comparison to the first year, because spike length, number of florets, size of florets, and weight of florets will reduce but the number of spike per plant will increase.

10.19 Lifting and Storage of Bulbs

After completion of flowering when vegetative growth almost ceases and leaves become yellowish followed by drying, it indicates that bulbs are matured to harvest. Irrigation is to be stopped into the field 1 month before lifting of bulbs, while new growth of bulblets will not appear and soil will dry up. Generally bulbs are ready to harvest just after the winter season in subtropical climate. After lifting of bulbs, bulblets are separated manually and graded based on their size. Very small bulblets are to be rejected for further use. Harvested bulblets are cleaned and treated with fungicides (copper oxychloride at 2 gm/lit of water) and are kept in cool, dry, airy, shady, and dark place for 3 months. Bulbs are stored at 10 °C and 15 °C and also at room temperature (25–30 °C) for 30 and 60 days. Storage of bulbs at low temperature delayed flowering and showed no appreciable beneficial effect. On an average, 5–8 number of good size bulblets (1.5–3 cm diameter) can be obtained from a single plant, and it is estimated that an average of 20–22 tons of bulblets are produced ha^{-1}, which may be used for planting in 6–7 ha area for the next season. Rhizomes of 2.6–3.0-cm-diameter and 1.5–2.0-cm-diameter bulbs when planted gave the highest yield of spikes and flowers. A good correlation between bulb size and growth and shoot emergence were observed with increasing bulb size, whereas other characters were enhanced with increasing bulb size of single variety. 2.6–3 cm diameter of single-type tuberose bulb is suitable for obtaining highest yields (17.79 ton/ha). Experiments were conducted on the control of flowering through storage of bulbs forcing with temperatures. The bulbs were stored at different temperatures (4, 12, and 27 °C) and weeks (4, 5, 6, and 7) under greenhouse condition. Early flowering observed in the treatment of 7 weeks of storage at 27 °C and the latest flowering happened with bulbs stored at 12 °C for 6 weeks (Dhua et al. 1987).

10.20 Postharvest Management

Tuberose is generally harvested as cut flower, i.e., spikes are cut at the base portion at one or two floret open stage. Double cultivars are generally used as cut flower, but few robust single cultivars are also used as cut flower. Majority of single cultivars are grown for loose flower production. Florets are picked at unopened bud stage day before actual opening. Loose flower are packed in polythene or bamboo basket and send to the local market, but for distance market ice-packed are used. Loose flower can be stored up to 4–5 days (at 10 °C temperature) in polyethylene. Tuberose flowers are very sensitive to the stresses of storage and transportation, particularly at warm temperatures. It is important to place the flowers in a proper vase preservative also. Vase deterioration of tuberose is mainly due to rapid respiration. Temperature management and carbohydrate supply are important to prolong vase life. Under normal display conditions, buds aborted early because of carbohydrate stress. A sugar-containing vase preservative (1.5%) and pretreatment with 20% sucrose for 15–20 hours improve vase life of the stems before or after storage. Tuberose florets produce very little ethylene, and are not generally affected by exposure to ethylene. Generally, harvesting tuberose inflorescence with several florets open improves their percent opening and vase life.

Effects of precooling and preservative chemicals on postharvest longevity of the florets of tuberose cv. Prajwal have been reported. Studies on the effect of pre-cooling of tuberose florets in combination with four chemical treatments reported that pre-cooling of flower buds improved the longevity (shelf life) of flower buds. Soaking of florets in 4% boric acid solution for 2 hours and air drying them before packing increased the shelf life up to 6 days. Loose florets when packed in polyethylene bag and stored in cool chamber exhibited better quality for a longer period of time. Katwate et al. (1995) studied storage of cut spikes of tuberose and reported that low temperature influences on longevity. The cut spikes after pre-cooling and pulsing with 20% sucrose solution were packed in five wrapping materials (LDPE-25 μ, 50 μ, PP-25 μ, 50 μ, and newspaper wrapping served as control) and stored on five levels of storage duration (0, 3, 6, 9, and 12 days) under refrigerated condition at 3–4 °C and 85–90% relative humidity. The vase life was maximum in polypropylene 25 μ (11.64 days) and minimum in low-density polyethylene 50 μ sleeves. Spikes wrapped with banana leaf exhibited maximum vase life with maximum floret opening and least amount of floret wilting due to the modified atmosphere having reduced oxygen and elevated carbon dioxide (CO_2). Spikes stored for 24 hours, especially at 10 °C, exhibited least floret wilting with better turgidity throughout the vase life. Retarded rate of respiration, transpiration, and ethylene production were observed, as well as the entire metabolism of the flower tissue as a whole which was facilitated by low temperature (during storage) that ultimately exhibited better performance of flowers in the vase.

Many antiseptic chemicals have been reported to enhance the vase life. 8HQC has long been known to have bacteriostatic and fungistatic properties and is extensively used as an antiseptic (cf. Datta 2006). Combined effect of antioxidant and α-lipoic acid along with sugars as pre-storage pulsing treatment and seal packaging with polyfilms, viz., HDPE (40 μ), LDPE (40 μ), PP (40 μ), and PP (30 μ), at low temperature (2 °C) storage for 15 days on post storage physiology and life of tuberose cut spikes has been worked out. Singh et al. (2018) observed that pre-storage pulsing of tuberose cut spikes with solution containing 50 mg 1-1 α-lipoic acid +15% sucrose for 6 hours and seal packaging with HDPE 40 μ polyfilm significantly influenced spike fresh weight, petal total soluble sugar level, and PAI (%) in floret tissue in low temperature stored cut spikes. They observed that untreated low-temperature stored tuberose cut spikes displayed drastic chilling injury and reduced vase life to 0–3 days after low-temperature storage which suggested that pre-storage pulse treatment of tuberose cut spikes with α-lipoic acid and packaging with HDPE during low temperature storage at 2 °C temperature retains higher spike fresh weight and petal sugar levels and stabilizes cell membrane integrity in petal tissue leading to a delay in petal senescence with 7 days of vase life even after 15 days of cold storage.

Various studies have found that bacterial contamination is one of the most important factors in reducing postharvest life of cut flowers with the negative impact on respiration, photosynthesis, and water uptake which causes water imbalance, indirectly stimulates ethylene production, and shortens postharvest life of cut flowers like tuberose. Therefore, the use of antimicrobial compounds, such as aluminum sulfate, to increase postharvest life of cut flowers like tuberose is recommended. Various reviews have shown that the use of calcium, aluminum, boron, copper, nickel, and zinc salts extends the vase life of cut flowers. In addition to the inhibitory effect of aluminum sulfate on reducing microorganism's activities, it reduces the transpiration rate and improves water absorption and fresh weight. Effects of different metal salts and chemicals on tuberose cut flowers have been studied. Flowers pulsed with 5 mM nickel for 12 hours showed highest water uptake and enhanced the floret opening as well as vase life. Zinc sulfate improved postharvest life. When spikes were held in GA 50 ppm + sucrose 8-HQS for 8 days, there was an increase in fresh weight and improved opening of florets and vase life. 8-HQS and combinations of 8-HQS with nickel or citric acid maintained higher rates of fresh weight of spikes. Spikes held in these solutions had higher number of opened and fully opened florets. The vase life of tuberose spikes was maximum (12.33 days) in 4% sucrose containing 8-HQS + nickel, while it was least (6.5 days) in spikes of distilled water. Boric acid (250 ppm), Al_2 (SO_4) (50 ppm), $CaCl_2$ (1000 ppm), $AgNO_3$ (50 ppm), $COCl_2$ (25 ppm), and citric acid (400 ppm) were beneficial for improving floret opening, flower diameters, and vase life. Different growth regulators (benzyladenine, GA, BA, NAA, and MH) at concentration of 50 or 100 ppm were tested their effects on vase life of cut flowers. Benzyladenine and GA at 100 ppm increased vase life and number of florets. BA at 100 ppm delayed petal senescence and freshness of flower for long time in comparison to 100 ppm GA. Tuberose florets produced very little ethylene, and were not affected by exposure to ethylene. Pulse pretreating the flowers with preservation solution containing 20% sucrose helped for long vase life. $AgNO_3$, $CaCl_2.2H_2O$, and Trimiltox Forte were found to delay flower opening as compared to ascorbic acid and standard preservative. $CaCl_2.2H_2O$ at concentrations of 750 to 1250 ppm and Trimiltox Forte at 1500 ppm resulted in minimum flower wilting after 6 days. $AgNO_3$ was found to have adverse effects on fragrance of the flowers that resulted in higher vase life of cut flowers (cf. Data 2006). Percentages of flowers opened and wilted were significantly negatively correlated with the vase life. Vase life was not correlated with fragrance of the flowers and net water uptake. Pulsing with gibberellic acid at 10 or 20 mg/l plus 8-hydroxyquinoline sulfate (200 mg/L) for 24 hours followed by continuous sucrose treatments (4 or 8%) plus 8-hydroxyquinoline sulfate extended the vase life and significantly promoted flower bud opening as compared with the 8-hydroxyquinoline sulfate controls. It has been suggested that a gibberellic acid pulse at 10 mg/L followed by continuous sucrose treatment at 4% be recommended to growers for extending the vase life and enhancing flower bud opening (Su and Huang 2001). Endogenous gibberellins (GAs) in corms of cv. Double were isolated and identified at the vegetative, early floral initiation and flower development stages. Vase life of flowers treated with 25, 50, and 75 mg/l accel. Increased up to 7.33, 6.67, and 5.67 days, respectively. Concentrations of 75, 100, and 50 mg/l of thyme oil increased vase life up to 4.33, 3.67, and 3.33 days, respectively. The influence of pre-storage pulsing with TDZ with sucrose and low-temperature storage with polyfilm packaging on post storage quality of tuberose cut spikes was studied, and it was concluded pre-storage pulsing of tuberose spikes with solution containing 0.5 mM TDZ +15% sucrose or for 6 hours, followed by 2 °C low-temperature storage under seal packaging with HDPE 40 μ for 15 days, significantly improved physiological parameters like water uptake (ml), total soluble sugars, and electrolyte leakage (%) in floret tissue, improved percent floret opening, and decreased abscission during post storage life.

10.21 Grading and Packaging

Cut spikes are graded based on varieties, spike length, rachis length, and spike thickness, as well as condition of flower spike in terms of freshness, straightness, or slight bendedness. Graded spikes should have equal length by cutting of cut ends and make 1378 round bundles with 25, 50, and 100 numbers of spikes. Then the whole bundles are wrapped with rubber band and covered with white soft tissue paper. After that, bundles are placed inside the cardboard boxes (90 x 40 x 20 cm), and its size is based on spike length and number of bundles. For distance market, perforated boxes are used for proper aeration. In case of loose flower, AICRP Floriculture, Kalyani Centre, BCKV, recommended thermocol box (1x 0.5 feet) with ice packing is the best packaging material for tuberose strings. Flowers stored at 30.5–33 °C and relative humidity – 65–75.5% – in thermocol boxes for 1 day exhibited the highest shelf life 1.94 days with least weight loss and more freshness (67.25%). The effects of different packaging materials [polypropylene bags and LDPE bags at different grades of thickness (100 and 200 gauge)] and storage conditions (ambient conditions, 40 °C and 60 °C) on keeping the quality of loose flowers and fully developed unopened buds of tuberose loose flowers cv. Bidhan Rajini-1 were studied and observed that packaging significantly influenced physiological loss in weight (PLW%), maximum relative water content (86.15%), and maximum days for 50 percent wilting (17.6 days) and the flowers remained fresh for 21 days, whereas flowers without packaging and kept at ambient temperatures lost their shelf life within 2 days. It has been mentioned that the combination of packaging and low-temperature storage helped to create the modified atmospheric condition (low temperature and high relative humidity) which resulted to maintain a better quality flower for a long time. The interaction and effect of packaging materials, i.e., containers (bamboo basket with newspaper lining, thermocol box, and corrugated fiberboard (CFB) box) and storage conditions (0, 1, 2, and 3 days), in tuberose flower strings of size 3 feet have been studied. Packaging significantly influenced weight loss, fresh weight, and freshness of flowers throughout the storage period. Thermocol boxes instead of CFB boxes have been recommended as better packaging material for tuberose strings where the strings could be stored for 1 day at ambient condition. Cut spikes harvested at 1–2 florets were packed after precooling and pulsing with 20% sucrose solution in five different wrapping materials as mentioned above and observed that vase life was maximum in polypropylene 25 μ (11.64 days) and minimum in low-density polyethylene 50 μ sleeves. Several eco-friendly alternative packaging materials were tested for tuberose and reported that areca nut sheath cup was suitable for retail packaging of tuberose with higher freshness (82.26%), fragrance (71.21%), and shelf life of up to 2 days in ambient storage condition when compared to flowers packaged in banana sheath cup and peepal cup. In low-temperature storage, also tuberose flower packaged in areca nut sheath cup had higher freshness index (87.84), color index (79.63), and fragrance index (71.21), as compared to flowers packaged in banana sheath cup and peepal leaf cup, and shelf life of 7 days. Loose flowers of tuberose were treated with gamma (γ) irradiation and packed in low-density polyethylene bags, heat sealed, and stored at 23 ± 2 °C, 80% relative humidity (RH) and 4 ± 1 °C, 40% RH. From sensory evaluation and biochemical analyses, they concluded that the longest shelf life of tuberose flowers could be 8 days at 23 ± 2 °C, 80% RH (compared to 4 days for control) and 24 days at 4 ± 1 °C, 40% RH (compared to 8 days for control) using combination treatment of low dose γ-irradiation (0.02 kGy) and preservative solutions (4% sucrose and 0.02% $CaCl_2$) (Archana et al. 2019, Baidya and Chakraborty 2020, Bhuvaneswari and Sangama 2017).

10.22 Extraction of Absolute and Concrete

The loose flower market in India is very unpredictable and frequently changing its market price. Its domestic market fully depends on local festival, but farmers are acquainted about this matter; therefore, sometimes they throw these flowers in the dustbin. However, it is very much essential to develop a processing unit for the benefit of the farmers at tuberose-growing area, while farmers can avail minimum market price. There are several research work done on value addition of tuberose flower. Safeena et al. (2015) reported that out of the approximate total yield of 30,000 kg of loose flowers from 1 hectare, in 3 years, 27.5 kg of "concrete" could be obtained. This concrete in turn will yield about 5.50 kg of absolute. One hectare of tuberose plantation may yield up to 12 kg of concrete (cost of 1 kg concrete is US$1350–1450 per Kg or Around Rs. 100,000.00 (0.1 million)/Kg).

Different essential oil extraction methods from double-flower variety of tuberose (*Polianthes tuberosa L.*) have been tested. Cold or hot enfleurage and solvent extraction with hexane or petroleum ether were tried. The chemical composition of the tuberose absolutes was analyzed by gas chromatography-mass spectrometry (GC-MS). The results showed that percentage yields of tuberose oil from cold enfleurage, hot enfleurage, hexane, and petroleum ether extractions were 0.3137%, 6.5808%, 0.0279%, and 0.0182%, respectively. The main chemical component in both enfleurage absolutes was methyl benzoate, while benzyl benzoate and pentacosane were found to be the main chemical components in hexane and petroleum ether absolutes, respectively. Extraction of essential oil by maceration method was studied and tested the antioxidant activities of produced essential oil. Hexane and petroleum ether were used as solvents to extract volatile compound from tuberose flower. The solvent and flower petals were soaked with a ratio of 1:2. Maceration process was carried out at ambient temperature for 24 hours. The solvent and flower petals were separated by filtration. The solvent was evaporated to obtain the extract and then measured the yield and checked the characteristics of the oil. The yield of maceration with hexane (0.12%) was higher than petroleum ether (0.08%).

10.23 Diseases

Disease problems include *Botrytis*, *Erwinia*, *Fusarium*, and anthracnose caused by *Colletotrichum*. Anthracnose is usually a problem during periods of high humidity. *Polianthes* is also susceptible to aphid, mite, and thrip infestations. Basal rot (causal organism *Sclerotium rolfsii*) is a very common fungal disease of tuberose during monsoon. Leaves slowly become yellow and dry up. Roots are affected first, and then the rot spreads upward through the bulb and collar portion of the stem. Dark brown lesions are found on infected bulbs and roots. Before the onset of monsoon, all old and dried leaves should be removed, and infected plants must be removed along with the soil. Carbendazim, copper oxychloride, or difenoconazole 250 EC at 0.5 liter per ha have been recommended. A new species of *Alternaria* which caused leaf spot disease of tuberose has been reported. Another disease is called mosaic caused by *Amorphophallus mosaie* virus. It causes mottling and distortion of the leaf lamina and reduces chlorophyll content. Removal of infected plants from field is recommended. Flower blight disease (caused by *Botrytis elliptica*) develops dark brown spots on flowers, and slowly the entire inflorescence dries up. *Potyvirus* causing mild mosaic on tuberose has been characterized. Different symptoms and management of fungal and bacterial diseases have been described. Root knot nematode and wilt disease of tuberose can be significantly decreased when neem-based formulation of *Pochonia chlamydosporia* and *Trichoderma harzianum* is applied before planting and subsequently once in 4 months. Bio-management of root-knot nematode has been worked out. The effect of bulb treatment/dressing with a combination of antagonistic fungus, *Pochonia chlamydosporia*, and plant growth-promoting fungus, *Glomus mosseae*, at higher concentrations has been studied in order to rationalize the use of these bioagents in tuberose against root-knot nematodes. Various bioformulations like *Paecilomyces lilacinus*, *Trichoderma harzianum,* neem cake, *Pseudomonas fluorescens*, and *Pochonia chlamydosporia* were effective and economic for managing *M. incognita* in tuberose. Among the bioformulations, application of *P. lilacinus* at 5 kg enriched with FYM 5 t/ha was found to be the most effective and economic for reduction of root knot nematode infestation, nematode population, and enhancement of cut flower yield of tuberose. They further reported that application of *T. harzianum* at 5 kg with FYM 5 t/ha was also comparable to carbofuran at 1 kg a.i./ha for root knot nematode management in tuberose (Sharma and Bhattacharjee 2002).

However, major diseases of tuberose are stem rot, leaf blight, wilt, leaf spot, and flower bud rot which are described below.

10.23.1 Stem Rot (*Sclerotium rolfsii*)

It is a serious disease of tuberose; initially, light green spots appear on the leaves; later on, it spreads on whole leaves, and leaves turned brown in color. Infected plant showing stunted growth and no flower spike appears. It happened due to high moisture in the soil with high temperature. As such type of symptoms appears in the field, soil moisture should be reduced. The disease can be managed initially by application of Neemcake @ 50 gm/sq.mt + *Trichoderma harzianum* @ 5 gm/sq.mt at an interval of 10–12 days (two times). Treatment of bulbs with *Trichoderma harzianum* at the time of sowing is effective. In case of severe symptoms chemical options of disease control should be explored.

10.23.2 Leaf Blight (*Botrytis elliptica*)

It is one of the major diseases of tuberose during rainy season in low-humidity and high-temperature area. Initially small brown spot appears on leaves which spreads of flower spike also, latter on dark brown spot appears on florets and gradually spreads over entire spike resulting in a dry look appearance. Affected portion should be collected and burnt. It may be controlled by burning of affected portion. As the symptoms appear, plants should be treated by foliar spay with Carbendazim @ 0.1% + Mancozeb @ 0.1% at an interval of 10–12 days.

10.23.3 Basal Rot and Wilt (*Sclerotium rolfsii*)

This is a common disease in tuberose during monsoon. It happened due to infection of root by the fungus, and leaves turn into yellow color and finally dry up. Initially infection started in the roots, and it then spread upward through the tuber to the collar region of the stem and ultimately to the bulb, and roots rot. White color cotton-like growth appears on the stem. Before the onset of monsoon, all old and dried leaves should be removed, and infected plants must be removed along with the soil. Infected field will not be used for tuberose cultivation for the next 3 years, and it can be controlled by drenching with copper oxychloride (Blitox at 0.4%) along with chlorothalonil 75% WP (Kabaz at 0.2%) at 10–12 days interval (two times).

10.23.4 Leaf Spot (*Alternaria polyantha*)

It is a common problem all over India due to fungus infestation; the symptom initially showing circular spots appears on the mid rib of the leaf. Later on, the number of spots as well as size increased and turned brown color. Finally, the whole leaves become yellow in color and defoliate, while the plant looks deciduous. Sometimes this symptom may appear on sepals, petals, and flower bud. It may be controlled by several ways; the first step is to collect all defoliated leaves, prune affected branches, and burn. As soon as the symptom has appeared, the plant is to be treated by foliar spay with carbendazim (Bavistin at 0.1%) + mancozeb (Dithane M-45 at 0.1%) at an interval of 10–12 days (two times).

10.23.5 Bud Rot (*Erwinia* Sp.)

It is a serious problem of tuberose found in high-humidity and high-temperature areas, where dry rotting appears on the buds and peduncle showing brown necrotic discoloration and inferior quality of floret produced. It can be controlled by foliar spraying with streptomycin or tetracycline at 50 mg/lit. of water at 10–12 days interval.

10.23.6 Flower Blight (*Botrytis elliptica*)

Initially small brown to black spots appear on the flower petals, and later on, it spreads to the whole flower looking burned. It can be controlled by foliar application of difenoconazole (0.1%) or iprodione + carbendazim @ 0.2% may be used to combat the situation.

10.23.7 Peduncle Blight (*Lasiodiplodia theobromae*)

Peduncle blight has been reported as a major limiting factor to the cultivation of tuberose, as the disease incidence was noticed up to 42.60% in pockets of Madurai District. Although the casual organism *Lasiodiplodia theobromae* was a ubiquitous pathogen, its occurrence on tuberose was a new record. The fungus included confounding symptoms which included blossom blight, peduncle blight, and leaf blight at tips as well.

10.24 Pest

There are several pests like thrips, aphids, bud borer, grass hoppers, weevils, nematodes, red spider mites, rodents, etc. that damaged the plants in several ways all most round the year. The most serious pests which are very much critical to manage are listed below.

10.24.1 Thrips

This insect sucks sap from leaves, spikes, and flower buds. The infected portion becomes deformed especially florets, resulting in the reduction of flower size as well as stalk length. There are some brown or silvery color patches developed on the leaves and severely attacked the leaf margin looking burned. As reported by Safeena et al. (2015), sometimes, these are associated with a contagious disease known as "bunchy top," where the inflorescence is malformed. The insect may be controlled by foliar spraying with imidacloprid 17.8% S.L (confider or jumbo at 0.15%) at an interval of 15 days.

10.24.2 Aphid

The insect is looking brown, green, and black, which sucks sap from the tender parts of the plant like new leaves and flower buds. The infected portion becomes deformed and the plant has stunted growth. It can be controlled initially by foliar spraying with neem oil at 2% followed by dimethoate 30% EC (Rogar at 0.2%) at an interval of 15 days or by application of thiamethoxam (25% WG) at 200 g per ha.

10.24.3 Bud Borer

It will enter into the flower bud and eat all the flower parts: It can be controlled by spraying of cartap hydrochloride (Kitap 1 gm/l), abamectin (Vertimec at 0.4 ml/l), or acephate (Acephate at 1.5 gm/l).

10.24.4 Red Spider Mites

It is a serious insect, which sucks sap from under the surface of leaves, and initially white specks developed on the leaves. Later on, leaves lost its shape, turned yellow in color, and fall down. The infestation on leaves by these insects will be found when temperature and humidity increased in maximum level. It can be controlled by spraying of propargite 57% EC (Simba or Omite at 0.2%) or flufenoxuron 10% EC (Cascade at 0.1%) or dicofol 18.5% EC (Kelthane at 0.15%) or diafenthiuron 50% WP (Pegasus or Polo at 0.1%) at weekly interval.

10.24.5 Nematodes

There are two serious root-knot nematodes found in the tuberose field, namely, *Meloidogyne incognita* and *Meloidogyne javanica*. The entire plant becomes dwarf, loses its fresh color, and lastly experiences wilting, and its flower will not open. Application of neem cake at 300 kg ha^{-1} or carbofuran (66 kg ha^{-1}, Furadon 3% granules/ha) or phorate 3G at 2.5–3.0 kg ha^{-1} and intercropping with marigold at the time of planting may reduce infestation of nematode. Root knot nematode and wilt disease of tuberose can be significantly decreased when neem-based formulation of *Pochonia chlamydosporia* and *Trichoderma harzianum* are applied before planting and subsequently once in 4 months.

10.24.6 Foliar Nematode

Foliar nematode caused by *Aphelenchoides besseyi* can lead to the loss of up to 38% spike yield and 59% loose flower yield. Khan et al. (2006) reported that treatment combinations, presoaking of bulbs in plain water for overnight followed by hot water treatment at 50 °C for 20 min. + two sprayings with monocrotophos 36SL at 0.15% at 30 days interval and the second and the third year crop receiving three sprays with monocrotophos 36SL at 0.15% at 30 days intervals, were found more effective for managing foliar nematode problem in tuberose. They suggested an integration of botanicals (NeemAzal, pongamia oil) with chemicals as well as hot water treatment for effective control of this nematode. *A. besseyi* maintained maximum population during July month of the rainy season that coincided with the start of heavy flush of tuberose. The least population was observed during December to February when average temperature, total rainfall, and relative humidity remained quite low. Among the weather factors, the nematode population had a positive correlation with maximum temperature, minimum temperature, rainfall, and relative humidity in infected flowers (Figs. 1, 2, 3, 4, 5, 6, 7, 8, 9, 10, 11, 12, 13, 14, and 15).

Fig. 1 Variability in flower sticks of tuberose. From left to right: Sikkim Selection, Calcutta Single, Prajwal, Hyderabad Single, Shringar, Phule Rajani, Hyderabad Double, Vaibhav, Calcutta Double

Fig. 2 Chlorophyll mutant of Calcutta Double (right). Normal plant (left)

Fig. 3 Single type with variable tepal number (seven to eight per floret) (left). Variegated single-type Sikkim Selection (right)

Fig. 4 Germplasm block of tuberose, National Active Germplasm Site at Horticulture Research Station, Bidhan Chandra Krishi Viswavidyalaya (Agriculture University), West Bengal, India

Fig. 5 Variegated double-type mutant Swarna Rekha (developed at NBRI) (left). Variegated single-mutant type Rajat Rekha with unique leaves and stems (developed at NBRI) (right)

Fig. 6 Fruits and seeds of tuberose: (left) fruit setting after natural cross-pollination in Bidhan Ujjwal; (middle) fruit setting after artificial cross-pollination, (right) seeds of tuberose

Fig. 7 Flower-type variation in tuberose: (right) single type (one whorl, 6 tepals), (middle) semidouble type (two whorls, 12 tepals), (left) double type (three or more whorls with 18 or more tepals)

Fig. 8 Flower-type variation in double types

Fig. 9 Left, gynoecium of double-type carpels fused together and rudimentary; middle, gynoecium of single-type tricarpellary and trilocular ovary; right, semidouble-type tetracarpellary and tetralocular ovary

Fig. 10 Anthesis and stigma receptivity: right, 48 hours after anthesis (receptive stigma); right middle, 24 hours after anthesis (non-receptive stigma); left middle, 6 hours after anthesis (non-receptive stigma); left, at the time of anthesis (non-receptive stigma)

Fig. 11 Stigma receptivity: left, at the time of anthesis (non-receptive stigma); left middle, 24 hours after anthesis (non-receptive stigma); right middle, 48 hours after anthesis (receptive stigma); right, 72 hours after anthesis (receptivity lost)

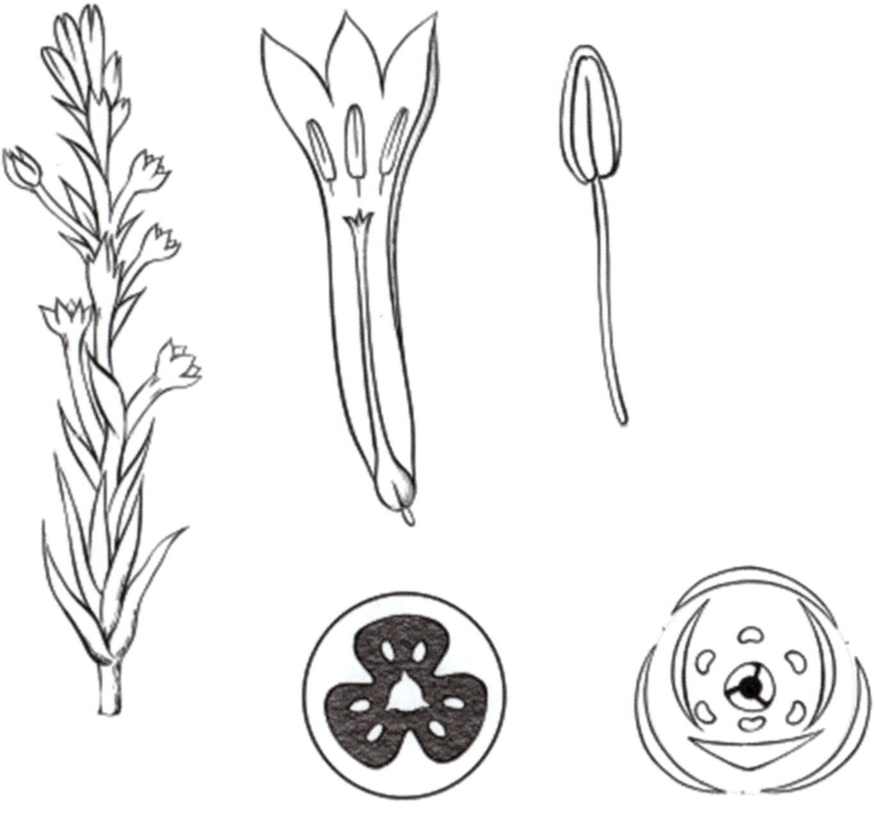

Fig. 12 Floral diagram of single tuberose. Left, spike; middle, cross-section of floret and ovary (trilocular); right, anther with filament and floral diagram of single type

Fig. 13 Stigma receptivity and floret appearance. From left to right, fully grown bud stage, anthesis (0 hour), 48 hour (receptive), and 72 hour (receptivity lost)

Fig. 14 Protandry in tuberose anthers shreds pollen before the pistil becomes receptive and thus prevents self-fertility

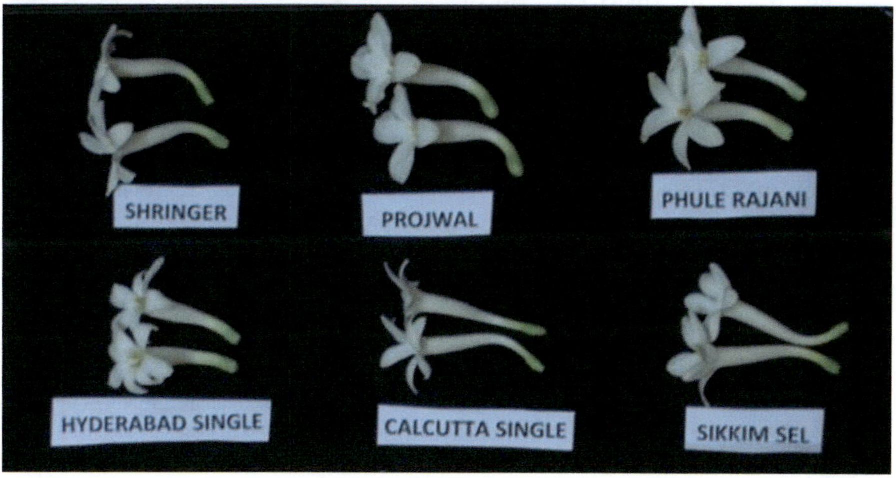

Fig. 15 Floret-type variation in single types

References

Andrea C, Ilaria M, CarloMascarello LP, Laurent C, Rosanna D, Barbara R (2020) Polianthes Tuberosa as edible flower: in vitro propagation and nutritional properties. Int J Food Eng 6(2): 57–62

Archana J, Girwani A, Reddy DVV, Goud CHR (2019) Effect of different packaging materials and storage temperatures on storage life of tuberose (Polianthes Tuberosa L.) cv. Bidhan Rajini-1. Int J Curr Microbiol Appl Sci 8(7) 2019, PP-2375-2385

Baidya BK, Chakrabarty S (2020) Influence of packaging materials and storage duration on quality of flower strings of *Polianthes Tuberosa* Linn. Cv. Prajwal. Int J Curr Microbiol Appl Sci 9(3): 664–674

Barghout N, Chebata N, Moumene SM, El-Hadi D (2020) Antioxidant and antimicrobial effect of alkaloid bulbs extract of Polianthes tuberosa L. (Amaryllidaceae) cultivated in Algeria. J Drug Delivery Therapeutics 10(4):44–48

Bharti H, Singh KP, Singh R, Kumar R, Singh MC (2016) Genetic diversity and relationship study of single and double petal tuberose (*Polianthes tuberosa* L.) cultivars based on RAPD and ISSR markers. Indian J Horticulture 73(2) 2016, PP-238-244

Bhuvaneswari S; Sangama, S. 2017. Eco friendly packages for freshness retention and shelf life extension of tuberose flower. Journal of horticultural Sciences; 12 (2):2017, 198–200

Biswas J, Bose TK, Maiti RG (1983) Effect of growth substances on growth and flowering of tuberose (*Polianthes tuberosa* L.). South Indian Horticulture 31:129–132

Biswas B, Kumar PN, Bhattacharjee SK (2002) Tuberose, AICRP on floriculture, technical bulletin no. 21. AICRPF, ICAR, New Delhi

Bundrant LA (1985) *Polianthes tuberose* and its hybrids. Herbertia:55–60

Chauhan P, Prakash S, Kumar M, Kumar N, Naresh RK, Singh MK, Rakesh SV, Chaudhary V, Sirohi U, Chand P (2017) Genetic diversity in single and double type of tuberose (*Polianthes Tuberosa* L.) germplasm using randomly amplified polymorphic DNA (RAPD) makers. Int J Agric Statistical Sci 13(2):519–526

Cronquist, A. 1988. The evolution and classification of flowering plants. 2nd ed., New York Botanical Garden Pr Dept, Bronx, 555 p., ISBN 978-0-89327-332-3

Datta SK, Banerji BK (1995) Cytogenetical studies in ornamental plants. In: Chadha KL, Bhattacharjee SK (eds) Advances in horticulture, vol. 12, ornamental plants, part I, vol 267. Malhotra Publishing House, New Delhi, p 334

Datta SK, Mishra P, Mandal AKAD, Chakrabortory D (2002) Direct shoot organogenesis from different explants of Chrysanthemum, marigold and tuberose. Israel J Plant Sci 50:287–291

Datta SK (2006) *Polianthes tuberosa* – an up-to-date research round up. Appl Botany Abstract 26(3):258–279

Datta SK (2017) Breeding of ornamentals: tuberose (*Polianthes tuberosa* L.). Curr Sci 113(7): 1255–1263

Dhar LN, Dubey RK, Kumar R, Kaur P (2008) Effect of nitrogen and phosphorus on plant growth, flowering and bulb production of tuberose (*Polianthes tuberosa* L). J Ornamental Horticulture 11(4):302–305

Dhua RS, Ghosh SK, Mitra SS, Yadav LP, Bose TK (1987) Effect of bulb size, temperature treatment of bulbs and chemicals on growth and flower production in tuberose (*Polianthes tuberosa* L.). Acta Horticulture 205:21–28

Elisheba BP, Sudhagar R (2019) Effect of integrated nutrient management on the growth of tuberose (*Polianthes tuberosa* L.) cv. Prajwal. Plant Archives 19(1):196–198

Emmart Trueblood EW (1973) "Omixochitl"- the tuberose (*Polianthes tuberosa*). Econ Bot 27(2): 157–173

Fan R, Chen Y, Ye X, Wu J, Lin B, Zhong H (2018) Transcriptome analysis of *Polianthes tuberosa* during floral scent formation. PLoS One 13(9):e0199261. https://doi.org/10.1371/journal.pone.0199261

Fish DT (1881) Cool treatment of tuberose. The Garden xx:127

Granick EB (1944) A Karyosystematic study of the genus Agave. Am J Bot 31:283–289

Howard TM (1978) Polianthes hybrids. Plant Life:126–129

Huang KL, Miyajima I, Okubo HJ (2000) Effects of temperature and shade treatment on flower colours and characteristics in newly established reddish-purple tuberose (Polianthes). J Faculty Agric Kyusku Univ 45(1):57–63

Huang, K. L., Miyajima, I., Okubo, H., Shen, T. M. and Huang, T. S. 2002. Breeding of coloured Tuberose (*Polianthes*) and cultural experiments in Taiwan. Proceedings of 8th Int. Symp. on Flower bulbs (eds Littlejohn, G. et al.), Acta Hortic., 570, ISHS, 367–371

Hutchinson J (1960) The families of flowering plants. Clarendon Press, Oxford

Kabir A, Iman M, Mondal M, Chowdhury S (2011) Response of tuberose to integrated nutrients management. J Environ Sci Nat Resour 4(2):55–59

Kameswari PL, Girwani A, Rani KR (2014) Genetic diversity in tuberose (*Polianthes tuberose* L.) using morphological and ISSR markers. Electron J Plant Breeding 5(1):52–57

Karihaloo JL (2019) Cytological and hybridization studies in three cultivars of tuberose (Polianthes tuberosa L.). Cytologia 84(1):47–52

Karuppaiah P (2019) Effect of zinc and boron on growth, yield and quality of tuberose (*Polianthes tuberosa* L.) cv. Prajwal. Horticulture Int J 3(1):7–11

Katwate SM, Patil MT et al (1995) Influence of low temperature storage on longevity of cut spikes of tuberose. J Maharashtra Agric Univ 20(2):289–290

Khan NH, Zaidi N et al (2000) Micropropagation potential of *Polianthes tuberosa* L bulbs, scales and leaves. Pak J Sci Ind Res 43(2):118–122

Khan, M.R., Shit, S., Pal, A.K. and Biswas, B. 2006. Management of Foliar Nematode, Aphelenchoides Besseyi infecting tuberose in West Bengal, India. Indian J Nematol. Vo. 36. No.1

Khandagale K, Padmakar B, Lakshmana RDC, Anuradha S, Aswath C (2014) Genetic diversity analysis and barcoding in tuberose (*Polianthes tuberosa* L.) cultivars using RAPD and ISSR markers. J Hortl Sci 9(1):5–11

Krishnamurthy KB, Mythili JB, Meenakshi S (2001) Micropropagation studies in 'single' vs 'double' types of tuberose (*Polianthes tuberose*). J Appl Hortic 3(2):82–84

Meena RK, Mahawer LN, Bairwa HL (2018) Standardization of NPK dose for growth, floral, bulb and economic parameters on Phule Rajani tuberose. Indian J Horticulture 75(3):486–491

Pal A, Adhikary R, Bera M, Garnayak R, Roy R (2019) Growing of tube rose with different irrigation treatment under alluvial soil of West Bengal. J Pharmacognosy Phytochem 8(3): 4534–4538

Rose, J.N. 1903-1905. Studies of Mexican and central American plants. Contribution United States National Herbarium, 8 : 1–55

Safena, S.A., Thangam, M., Priyadevi, S., Desai, A.R. and Singh, N.P. 2015. Ready Reckoner on cultivation of tuberose, Technical Bulletin No.50, Pub: ICAR-Central Coastal Agricultural Research Institute, Goa, India, pp.14

Sato D (1938) Karyotype alteration and phylogeny IV. Karyotype in Amaryllidaceae with special reference to the SAT chromosome. Cytologia 9:203–242

Sharga AN (1982) Effect of bulb size on vegetative growth and floral characters of tuberose (*Polianthes tuberosa* L). Progress Hortic 14(4):258–260

Sharma, P. and Bhattacharjee, S.K. 2002. Plant protection in ornamental crops. AICRP on Floriculture, Technical Bulletin No. 19

Shinners LH (1966) Texas, Polianthes, including Manfreda Agave subgenus Manfreda and Runyonia (Agavaceae). SIDA 2(4):333–338

Singh, A., Sahare, H., Sharma, G. S and Kumari P. 2018. Effect of Thidiazuron with sucrose pulsing and low temperature storage with Polyfilm packaging on floret opening and abscission of tuberose cut spikes J Pure Appl Microbiol; Vol.12 No.3, 2018., PP-1385-1392

Singh PK, Sadhukhan R (2019) Identification of variants induced by physical and chemical mutagens in tuberose (*Polianthes tuberosa* L.). J Crop Weed 15(2):40–45

Singh KBM, Madhavan J, Sadhukhan R, Chandra S, Rao U, Mandal PK (2020) Production of nematode free plantlets in *Polianthes tuberosa* using in vitro culture techniques. Hortic Environ Biotechnol IF 1.585. https://doi.org/10.1007/s13580-020-00263-5

Solano, C. E. 2000. Systematics of the genus *Polianthes* L. (*Agavaceae*). Ph.D. Thesis, National Autonomous University of Mexico, Mexico, 291 pp.

Solano E, Feria TP (2007) Ecological niche modeling and geographic distribution of the genus *Polianthes* L. (Agavaceae) in Mexico: using niche modelling to improve assessments of risk status. Biodivers Conserv 16:1885–1900

Srivastava, R. and Govil, M. 2006. Effect of Bio-fertilizers on vase life and corm production in gladiolus cv. American Beauty, "*National Symposium on Ornamental Bulbous Crops*". Held on 5–6 December, 2006 at S.V.B.P.U. of Ag. & T., Modipuram, Meerut (U.P.): 139

Su W-R, Huang K-L, Chang P-C, Chen W-C (2001) Improvement of postharvest vase life and flower bud opening in Polianthes tuberose using gibberellic acid and sucrose. Aust J Exp Agric 41(8):1227–1230

Tripathi SK, Malik S, Singh IP, Dhyani BP, Kumar V, Dhaka SS, Singh SP (2012) Effect of integrated nutrients management on cut flower production of tuberose (*Polianthes tuberosa* L), Var. Subhasini. Ann Horticulture 5(1):108–115

Ullirich (1993) Early illustrations of *Polianthes tuberosa* L. (agavaceae). Herbertia 49(1&2):50–57

Watkins GM (1936) Chromosome numbers and species characters. Am J Bot 23:328–333

Indian Orchid Germplasm

11

Conservation and Utilization

Ram Pal, P. Kiran Babu, and M. Dayamma

Contents

R. Pal (✉) · M. Dayamma
ICAR– National Research Center for Orchids, Darjeeling Campus, Darjeeling, India
e-mail: rampal@icar.gov.in

P. K. Babu
ICAR– National Research Center for Orchids, Pakyong, India

S. K. Datta, Y. C. Gupta (eds.), *Floriculture and Ornamental Plants*, Handbooks of Crop Diversity: Conservation and Use of Plant Genetic Resources,
https://doi.org/10.1007/978-981-15-3518-5_13

Abstract

Orchidaceae is the second largest family of flowering plants and valued for their ornamental and therapeutic importance. A number of species of this family are threatened with extinction due to specialized life cycle and diverse mode of living. The preservation of orchid germplasm traditionally accomplished by private nurseries for sale and breeding of orchids. The public-funded organization also preserved orchid germplasm for botanical and educational interest. ICAR-National Research Center for Orchids, Sikkim, was established in 1996 to provide research support to orchid growers in India. In collaboration with the National Bureau of Plant Genetic Resources (NBPGR), New Delhi, the center is engaged in preserving and sustainable utilization of orchid biodiversity. Preservation of orchid germplasm is challenging; nevertheless, concerted scientific efforts will ensure future conservation and sustainable use of orchid germplasm. There are various challenges in germplasm collection, conservation methodologies, gene pool creation, gene banking procedures and adherence to national and international laws, treaties and conventions for germplasm collection, as well as preservation and sustainable utilization. ICAR-NRC for Orchids, Sikkim, collected and conserved 3150 accessions of 352 native species of orchids. Global and regional networking are necessary for sustained collection, preservation, and exchange of orchid germplasm.

Keywords

Collection · Conservation · Gene banks · Gene pools · Orchidaceae · Genetic diversity

11.1 Introduction

Germplasm is the genetic materials used by plant breeders for the genetic improvement of crops. The genetic diversity includes wild species, weedy companions, subspecies, botanical varieties, landraces, extant varieties, genetic stocks, inbred lines, and modern cultivars. In Orchidaceae family, the reproductive barriers are feeble, and crossing among the genera are possible; therefore, the genera are also component of the genetic diversity. The genetic material also includes gene and its alleles, quantitative trait loci (QTLs), polyploid and aneuploid, different genome combinations, etc. The plant breeders use these elementary units to develop marketable varieties. These genetic resources provide genetic variability without which the success of any crop improvement program is incomplete. Plant germplasm conservation deals with preserving the genetic diversity of a targeted crop and its relatives in the form of seeds, meristem, or living plants for subsequent use. A standard germplasm conservation and management program consist of activities related to plant exploration, collection, conservation, characterization, multiplication, evaluation, documentation, and distribution. There are several crop-based national and international institutions for managing the germplasm of the food and fiber crop plants. The establishment of ICAR-National Research Center for Orchids under the aegis of Indian Council of Agricultural Research at Pakyong, Sikkim, in 1996 gave rise to systematic collection and conservation of orchid germplasm in the country. ICAR-NRC for Orchids, Sikkim, collected and conserved 3150 accessions of 352 native species of orchids.

11.2 Botany of the Orchids

Orchidaceae is one of the largest family of flowering plants accounting for ~7% of the total flowering plants. They are the most significant in the evolutionary form among the monocotyledonous families. Orchids are known for their pretty beautiful flowers of different sizes, fascinating shapes, and beautiful colors which have outnumbered other families of flowering plants by evolving higher levels of specialization in its vegetative and reproductive traits. The first record of an orchid in the world dates back to tenth to sixth century BC when *Ni,* a Chinese name for orchids, was first mentioned in *the Book of Songs* (in Chinese). The Greek philosopher, Theophrastus (370–285 BC), known as the "Father of Botany," referred these curious plants as orchis in his writing *Enquiry into Plants* and coined the name "Orchid." Taxonomic work on orchids started after Linnaeus (1707–1778), A.L. de Jussieu delineated a family for orchids for the first time in 1789. The name of orchids derived from the root tubers resembling testicle. Orchids have differentiated habit as epiphytes, lithophytes, terrestrial, and saprophytes. Epiphytic orchids grow on the main stem or branches of the trees without any internal connections. The roots of these orchids have specialized absorbent tissue called "velamen" to absorb moisture from the atmosphere. However, the roots of terrestrial orchids have a symbiotic mycorrhizal association with specific fungal species. The most visible manifestation of the diversity of orchids is their floral complexity: Orchids uniquely fuse their gynoecia and androecia, and one of the three petals is modified into a landing platform or attracting structure, the lip or labellum. It has long been assumed that in orchids the primary controls on floral morphology that operate in other monocots must have been modified. Orchid's flower shows several variations in shape, size, color, and posture with showy labellum or lip formed by posterior petal, the formation of gynostegium or column, and pollen united as pollinia and nonendospermic microseeds. Orchids are used as cut flowers, bouquet, loose flowers, dried flowers, a single flower mounting, potted plant display, by-products in handicrafts, and perfumery industries.

Orchids are one such group of plants which grow in a variety of habitats throughout the globe, but they are susceptible to habitat change. Orchids grow in a particular habitat because that is where they thrive. Removing and transplanting them elsewhere forces the orchids to adapt to an entirely new environment where the plant might not be as successful. Understanding the prospects, horticultural, and medicinal value, the family is gaining much attention throughout the world to unfold the biology, evolution, taxonomy, cytology, chemistry, hybridization and cultivation. In Northeast India, many orchids are used for different purposes, such as ornamentals, medicine, food, and sociocultural events (Deb 2009; Medhi and Chakrabarti 2009). Conferring to the molecular clock revealed by whole-genome sequencing of *Phalaenopsis equestris* (Cai et al. 2015), the emergence of orchids has occurred in the Late Cretaceous (76 Mya), and allowed them to cross the border of Jurassic mass extinction (66 Mya).

11.3 Origin, Domestication, and Spread

Orchids are herbaceous perennial plants with a wide range of growth habit and habitat. Nearly two-thirds of the known orchid species are epiphytes, and the remaining are lithophytes. As per the report of the IUCN, (1999), half of the extinct orchid species constitute terrestrial herbaceous perennials (IUCN 1999). With their distributional range, the orchids are broadly grouped into tropical, subtropical, and temperate. The distribution of orchids occurs throughout all the continents, except Antarctica. Orchids predominantly distributed in humid tropical forests of South and Central America, Southeast Asia especially India, Ceylon, Burma, Nepal, Bhutan, Laos, Philippines, and South China, Japan, Europe, Brazil, New Guinea, and Australia.

11.4 Plant Genetic Resources

The family Orchidaceae grouped into five subfamilies: Apostasioideae, Vanilloideae, Cypripedioideae, Orchidoideae, and Epidendroideae. The world estimate of orchid species usually assessed between 17,000 and 35,000 (Dressler 1993). Recent estimates suggest approximately 26,567 species and represented by nearly 25,000–35,000 species belonging to ~1000 genera (Chase et al. 2015; Willis 2017; Michael 2018). The largest genera are *Bulbophyllum* (2000 species), *Epidendrum* (1500 species), *Dendrobium* (1400 species), and *Pleurothallis* (1000 species) (IPNI 2012; World Checklist of Selected Families 2013).

11.4.1 Orchid's Distribution in India

Orchids have a wide range of distribution in India, and the distribution range begins with low-level plains to an elevation of 4300 m. Authors reported ~1300 orchid species distributed across the country (Mishra 2007; Medhi and Chakraborthi 2009; Singh et al. 2019b). Meitei et al. (2019a) compiled and reported the state-wise distribution of the orchids in India. The state Arunachal Pradesh represents the highest number of orchid species (558) followed by Sikkim (543), Meghalaya (532), and West Bengal (467) followed by Nagaland (396), Manipur (251), Uttarakhand (237), Kerala (186), Karnataka (175), Andaman and Nicobar (143), Odisha (128), Maharashtra (122), Madhya Pradesh (89), Andhra Pradesh (83), Himachal Pradesh (76), Tamil Nadu (72), Tripura (66), Jharkhand (Yin and Hong 2009), Jammu and Kashmir (Pamarthi et al. 2019), and least reported from newly formed state Telangana (Meitei et al. 2019a). Among them, eight Northeastern states of India are the paradise of the Indian orchids. These states hold about 876 orchid species in 151 genera and contributes 70% of the country's orchid wealth. Of which, many species are endemic and rare species shows high ornamental value. *Anoectochilus sikkimensis, Cymbidium eburneum, Dendrobium hookerianum, D. densiflorum, D. devonianum, D. thrysiflorum, P. fairrieanum, P. insigne, P. villosum, P. spicerianum, P. hirsutissimum, P. venustum, Papilionanthe teres, Pleione humilis, P. maculata, P. praecox, Renanthera imschootiana, Rhynchostylis retusa, Thunia marshalliana*, and *Vanda coerulea* are some of the promising orchids of these regions (Pal and Singh 2016).

11.5 Collections

Priorities and strategies for the collection followed by their conservation defined based on the economic value of cultivated species, distribution of wild species and its potential use in crop improvement program for sustainable utilization, and conservation of the genetic resources.

11.5.1 Methods

The primary plant genetic resources are landraces, wild species or relatives, and weedy races. These are classified into self-pollinated/autogamous, cross-pollinated/ allogamous, and asexually/vegetatively propagated based on their breeding behavior or pollination mechanism. In India, ICAR-NRC for Orchids, Botanical Survey of India, and various state and regional universities and institutes have carried out several explorations to collect and conserve the valuable orchid resources across the country. These collections used for taxonomic studies, breeding behavior, ecological studies, species distribution, plant-microbial-fungal studies, and development of varieties/hybrids. Most of the researchers carried their exploration programs based on the (i) published literature, (ii) preserved herbarium specimens, (iii) information on the spatial distribution of the species, (iv) RET status of orchids, etc. The orchid's collection site may be mountains, valleys, river beds, deep forests, farmer's backyards, etc. Based on some consideration in collecting valuable genetic resources, some procedures/strategies are proposed.

(i) *Appropriate site:* A site depicting the natural habitat or a cultivation field, preferably having a minimum of 50 random mating productive individuals (defining a minimal optimum size to qualify a population), should be appropriate. However, an explorer is the best judge to select a site/population for sampling, looking into all the factors involved/reflected. A unique type must always be sampled.

(ii) *Number of sites:* When no precise information on the distribution of variation in nature is available, the optimum number of sites for the sampling of a given landrace/species could be 50.

(iii) *Optimum sampling size:* A 95% confidence limit of sampling all the random alleles occurring in the target population, with a frequency more than 0.05, would be essential. Hawkes (1976) suggested that a bulked seed sample from 50–100 individuals shall be crucial to meet this requirement.

(iv) *Biased sampling:* Arguments favoring deliberate sampling for the desired phenotypes (rare or elites) have been agreed to in literature when the main objective is to collect the observable diversity in a population at a given time.

(v) *Coarse and fine grid sampling:* The preliminary survey and collection generally refer to the coarse grid sampling wherein the collection is made at longer intervals (25–50 km, depending upon the terrain), with a few samples taken per site. Further, depending on the time and resource, it is advisable to repeat an intensive sampling at sites where earlier collected samples have shown an exciting variation during a field evaluation of the collections made.

(vi) *Ecogeographic study*: It is a process of gathering and synthesizing taxonomic, geographic, and ecological data. It has three major components: (i) distribution of particular species in particular regions and ecosystems; (ii) pattern of inter-specific diversity; and (iii) relationship between ecological conditions and the survival of variants' frequency.

(vii) *GIS mapping:* Traits is to use a GPS/GIS map of particular germplasm or genotype collection sites which can be overlaid with climate data corresponding to vegetative and reproductive growth stages and to identify plant corresponding to sites with severe abiotic stresses during the previous 25 years (Petr et al. 2015).

(viii) *Documentation:* Data gathering is an integral part of the collection. Absolute minimum information to be recorded are: a. collectors' name and collection number, b. date and site of collection, c. geographical coordinates of collection sites, d. status of a sample (wild, weedy, and cultivated), e. source of the collection (field, market sample, or farm store), and f. label the collection bags both within and outside.

(ix) *Other considerations:* Besides breeding behavior, the botany of the target crop shall also affect the sampling decisions in the field. For example, in case of species with small capsules and limited seeds per capsule such as *Bulbophyllum* sp. a sampling of 5 or more ripe seeds from each of the three adjacent plants, every 3–4 paces, may be done to sample a total of 50 seeds. In case of species producing capsules with large amounts of seeds per capsule such as *Cymbidium* may be sampled on 1–2 plants at 2–3 paces apart from each other in all directions to constitute a bulk seed sample.

India is a mega-diversity center and a global hotspot for orchid biodiversity. One thousand three hundred fifty species of orchids belonging to 186 genera occur in eight orchid habitats of India (Singh et al. 2019a). In his monumental work *Hortus Malabaricus,* the then Dutch Governor of Malabar, Von Rheede (1678–1703) provided the first scientific account of Indian orchids. William Roxburgh (1832), the "Father of Indian Botany," published a treatment of 57 species in his *Flora Indica, vol. III.* But Sir J.D. Hooker (1888, 1890) made the most significant contribution to Indian orchids in *the Flora of British India* (Vol. 5 & 6), who described 1600 species of orchids from erstwhile British India.

After the establishment of Botanical Survey of India in 1890 at Royal Botanic Garden, Calcutta (now Acharya Jagadish Chandra Bose Indian Botanic Garden, Shibpur, Howrah), Sir George King as its director initiated a massive study on Indian flora including orchids. He published a floristic account linking to several classical works dealing with Indian orchids of which *The Orchids of Sikkim Himalayas* by King and Pantling (1898) and *The Orchids of North-Western Himalaya* by Duthie (1906) are worth to mention. Later on, more researchers contributed their significant studies on flora of Presidency of Bombay (Cooke 1908), Travancore (Rao 1914), Nilgiri and Pulney hilltops (Fyson 1924), Bihar and Orissa (Haines 1924), and Flora of Presidency of Madras (Fischer 1928) which were made during the pre-independence period. After the independence, the explorations and surveys focused on flora of states and districts fragile ecosystems and protected areas. In result, several publications on orchids, like of *Orchids of Meghalaya* (Kataki 1986); *Orchids of North-West Himalaya* (Deva and Naithani 1986); *Orchids of Nilgiri* (Joseph 1987); *Orchids of Arunachal Pradesh* (Chowdhery 1998); *Orchids of Nagaland* (Hynniewata et al. 2000); *Orchids of Kamrup* (Baruah 2001); *Orchis of Orissa* (Misra 2004); *Orchids of Kerala* (Kumar and Manilal 2004); *Orchids of Manipur* (Kumar and Kumar 2005); *Orchids of Sikkim and North-East India* (Lucksom 2007), *Orchids of Andhra Pradesh* (Mishra et al. 2008), and *Orchids of India* (Singh et al. 2019a) enriched the national repository of the Indian orchids.

Herbarium records are useful for exploring changes in orchid distributions and pollination rates (Robbirt et al. 2011), and evidence of phenological cues that track climate and the consequences for this under future climate change. Gaskett and Gallagher (2018) performed linear regressions for the number of orchid species (log10 transformed) and area in km^2 for (a) continents and (b) known orchid diversity hotspots. Some studies address orchid ecological preferences at regional scales, often with a conservation perspective (Phillips et al. 2010). A single, landscape-scale study is available, addressing orchid diversity, habitat, and climate in countries like China (Zhang et al. 2015).

11.5.2 Status of Collections

ICAR-NRC for Orchids, Sikkim, carried out several explorations to different parts of the country for collection and conservation of valuable orchid germplasm since 1996. Structured and well-planned explorations were conducted in the orchid-rich biodiversity hotspots. The germplasm collections were made as plantlets, tubers, capsules fruits, seeds, and floral parts. They were acclimatized and conserved in orchidariums and used in breeding programs. Till date, ICAR-NRC for Orchids has collected and preserved the germplasm of ~400 species from across the country. Among the collections, 83 species are rare, endangered, and threatened (RET) and 52 species are of medicinal interest. The RET collections include *Dendrobium draconis, D. ruckeri, D. praecinctum, Diplomeris hirsuta, Ornithochilus difformis,*

Paphiopedilum fairrieanum, *P. venustum*, *P. villosum*, *P. hirsutissimum*, *P. spicerianum*, *Renanthera imscootiana* (red vanda), *Satyrium nepalense*, *Taeniophyllum retrospiculatum*, *Vanda coerulea* (blue Vanda), *Zeuxine flava*, and *Z. reflexa*. The collected orchid germplasm is being utilized successfully in the breeding program to develop new varieties and hybrids. The other research institutes, universities, and regional research organizations are maintaining their collections in their respective locations. Collection of the precious orchid germplasm will be more focused on rare, endangered, and threatened (RET) species which could be helpful for identification of trait-specific germplasm and can be saved/conserve for future needs.

11.5.2.1 Threatened Orchid Species

Barik et al. (2018) compiled a list of 2704 threatened plant species belonging to 1031 genera and 217 families. The species include 2641 angiosperms, 23 gymnosperms, and 44 pteridophyte species representing ~13% of the estimated vascular plant diversity in India. Furthermore, they concluded that orchids are the most vulnerable group of plants comprising (624 species) 23% of the total threatened species. Several species are rare and threatened throughout India, owing to habitat degradation and fragmentation as a result of various anthropogenic influences such as land development activities, the building of dams, constructions of roads, commercial exploitation of the species, overgrazing, and sometimes natural calamities like frequent forest fires, landslides, heavy rainfall causing floods, etc. Some orchid species are particular to their requirements and confined to specific elevations and forest types. Some are naturally rare; others are so because of geographic distribution, narrow habitat requirements, and low-density populations. The red data book of Indian Plants published by BSI listed the plant species facing various threats requiring immediate attention. Recent estimates state that nearly 250 species of native orchids are under the risks of multiple categories. Certain species like *Aphyllorchis gollanii, Coelogyne truetleri, Anoectochilus rotandifolius, Paphiopedilum charlsworthii, Paphiopedilum wardii, Vanda wightiana*, and *Pleione lagenaria* are no longer found in natural habitats of India. Out of 352 orchids endemic to the country, 40 are "endangered," and 72 are "vulnerable" (Ram et al. 2011).

11.5.2.2 Endemism

Endemism in the flora of a country or geographical region provides an essential insight into that region's biogeography. It also contributes to the information on centers of diversity and adaptive evolution of the floristic components of that region. The complete endemic status of orchids in India is not available. According to a report published in 1983 (Das and Deori 1983), 85 species are endemic to this region. Of these, 20 species occur in Sikkim, 18 species in Meghalaya, 6 species in Assam, and 2 species in Nagaland. One hundred and thirty-five species, four subspecies, and three varieties belonging to 38 genera are endemic to Peninsular India, and 195 species are endemic to the Himalayan region (Kumar and Manilal 1994). Jalal and Jayanthi (2012) reported that 130 species belonging to 38 genera are endemic to peninsular India. Of 1300 orchid species reported from India, nearly 400 are endemic to India (Pal and Singh 2016).

Endemic Genera

There are 40 genera endemic to India (Irwin and Narasimhan 2011), of which four, namely, India A. N. Rao, *Aenhenrya* Gopalan, *Smithsonia* C. J. Saldanha, and *Xenikophyton* Garay belong to Orchidaceae family. Genus India A. N. Rao is distributed in northeast India (Arunachal Pradesh), whereas the remaining three found in the Western Ghats. There are three species in genus *Smithsonia* C. J. Saldanha, two in *Xenikophyton* Garay, and one in genus India A. N. Rao and *Aenhenrya* Gopalan. The six genera viz. *Cleisocentron* Bru¨hl, *Cryptochilus* Wall, *Diplocentrum* Lindl., *Diplomeris* D. Don, *Eparmatostigma* Garay, and *Penkimia* Phukan and Odyuo were earlier considered as endemic and have lost their endemic status due to new distributional record. Three genera *Jejosephia* A.N. Rao, *Proteroceras* J. Joseph and Vajr., and *Eparmatostigma* Garay have been merged with parent/allied genera (Pal and Singh 2016).

Endemic Species

In 2016, Pal and Singh compiled a list of 127 species; 6 varieties belonging to 52 genera are endemic to northeast India. Thus, 14.29% species are endemic to this region. Of these, 79 (62.2%) species are epiphytic, 40 (31.49%) are terrestrial, and 7 (5.51%) species are holomycotrophic (saprophytic). The analysis showed that 47.24% endemic species comprising genus *Bulbophyllum* (11 sp.), *Dendrobium* (10 sp.), *Oberonia* (9 sp.), *Eria* (7 sp.), *Liparis* (4 sp.), *Biermannia* (4 sp.), *Chierostyllis* (4 sp.), *Coelogyne* (4 sp.), *Herminium* (4 sp.), Octochilus (4sp.), and *Nervilia* (3 sp.) [60 all] are concentrated in these genera. The statewide analysis of endemic taxa showed that the highest numbers of endemic orchids are found in Arunachal Pradesh, followed by Sikkim, Meghalaya, Assam, Manipur, and Nagaland. The states of Tripura and Mizoram show a poor representation of endemic species. The highest numbers of strictly endemic species, 41, are found in Arunachal Pradesh followed by 25 in Sikkim, 18 in Meghalaya, 7 in Manipur, and 5 in Assam. No strictly endemic species recorded from Tripura and Mizoram (Pal and Singh 2016). In India, the Himalayan region has a high degree of endemism, making it the richest endemic center. There are about ~300 orchid species belonging to the endemic category. Endemic species are particular to habitat and sensitive to the microenvironment.

11.5.2.3 Identification of Gaps in the Germplasm Collections

The habitat loss, deterioration, and fragmentation of natural habitats, introduction of exotic species, overexploitation, environmental pollution, global warming, and commercialization of agriculture and forestry and *jhum* cultivation are the primary causes for the loss of diversity. India has strengthened diversity conservation by implementing a series of act, rules, laws, regulations, agreements, and developing network of protected areas. Though a large population of orchid species occur in their natural habitat, in many parts of the world, many orchid species have threatened due to habitat destruction, climate change, as well as illegal collection and trade. Illegal trade of orchids from Nepal is continuing over the last 25 years (Subedi et al. 2013). Identifying gaps in the germplasm collections is necessary to assess the completeness of the collection and exploration for further collection. Use of FloraMap, a GIS tool, creating the probability distribution map for each species in different states/districts of the country, is useful. Arc GIS tool is beneficial to check the accuracy of the coordinates by plotting all accessions on the latest political boundary map of each state/district. Diva-GIS is handy to assess the diversity in the assembled germplasm for each trait. Land cover maps of FAO provide information about vegetation and land cover in the targeted areas. Overlay the probability map, collection sites, and the diversity index of assembled germplasm and identify the gaps in trait-wise diversity. Overlay the collection sites or sampled sites on the probability map and identify the districts without and/or with few collection sites and high probability (>70%). Consult local government officials and extension officers working in the targeted area for crop cultivation and cropping pattern and then finalize the exploration area. Consulting regional institutes/universities that are working on the same region would help in collecting this precious germplasm.

11.6 Conservation

Orchids are one of the most threatened plants on this globe due to their specialized life cycle and mode of living. Hence, there is a need to investigate species diversity in orchids and their conservation. NRCO in coordination with National Bureau of Plant Genetic Resources, New Delhi, and other institutions plays an active role in the identification of orchid-rich habitats and finding out causes of loss of orchid biodiversity and suggesting appropriate measures for their conservation for in situ conservation. National Research Center for Orchids, Sikkim, has been designated as Active Germplasm Site for Orchids under National Active Germplasm Systems (NAGS) for sustainable conservation and use of orchid germplasm.

Increasing human interventions in natural habitats have increased threats for the survival of many orchid species in their homes. The risks are overexploitation, habitat loss, fragmentation of habitats, breakdown in ecological connections (pollinators, mycorrhiza, and, for holomycotrophs, loss of the chlorophyllous hosts), changed abiotic conditions (e.g., soil and hydrology), weeds, and introduction of pests and disease (Ibama 2008). Scientific preservation measures are necessary to guarantee their conservation and future use without losing their genetic variation. Plant specimens can be conserved in situ, i.e., in their natural habitat; however, this has a high maintenance cost. It requires large areas of land, and the plants remain vulnerable to abiotic factors such as bad weather, pests, and diseases (Santos 2000). Conservation through establishing botanic gardens, orchid biosphere reserves, orchid corridors, and cryopreservation will help for sustainable utilization for future generations (Dixon et al. 2003). Conservation of plant genetic resources broadly categorized into (A) in situ conservation and (B) ex situ conservation.

11.6.1 In Situ Conservation

Conservation of orchids involves protecting the natural habitat of the species, encouraging orchid biodiversity, and ensuring the survival of the rare species of orchids. Anthropogenic activities and natural calamities, disappearing of habitats, and large-scale illegal trade are the causes that may lead to risk for the extinction of thousands of rare and endemic species in the protected areas and sanctuaries. Hence, in situ conservation should also be supplemented with appropriate ex situ conservation measures. In situ conservation of species is most desirable for orchids as it ensures their natural growth, proliferation, and perpetuation, which allow the process of evolution to continue as part of the natural ecosystem. Protection of natural habitats by establishing sanctuaries, biosphere reserves, and forest reserves; the salvation of plants from degraded and threatened habitats and their culture in orchidaria, botanical gardens, and other rescue centers; and propagation of threatened plants through in vivo/in vitro and their re-introduction into well-protected habitats are among other measures suggested for orchid conservation.

India has an intricate protective area network (PAN) comprising 86 national parks and 480 wildlife sanctuaries covering about 4.66% of the total geographical area of the country and expected to further expand in the future (Ram et al. 2011). This PAN protects the species that are present in those forests. Unfortunately, many important and endangered orchid species lie outside the PAN (viz., *Paphiopedilum druryi* in Aghasthymalai hills of Kerala, *Vanda coerulea* in Meghalaya, *Paphiopedilum wardii* and *P. spicerianum* in Assam, and *Renanthera imscootiana* in Arunachal Pradesh). At present, the orchids also figure prominently in the Red Data Book prepared by the International Union for Conservation of Nature (IUCN). The entire family has been included in Appendix-II of Convention on International Trade in Endangered Species of Wild Fauna and Flora (CITES), where the international trade is strictly controlled and monitored (Pant 2013). There are some species such as *Liparis olivacea*, which have already been extinct from the wild (Subedi 2011). A few state governments like Arunachal Pradesh, Sikkim, Karnataka, and West Bengal have designated the orchid-rich habitats as "Orchid Sanctuaries" under the Wildlife Protection Act, 1972 (amended in 1992). Any attempt to remove orchids from their natural habitat should be considered poaching or smuggling.

11.6.2 Ex Situ Conservation

The ex situ conservation refers to off-site preservation of live plants, storage of genetically representative seeds and somatic tissues for the regeneration of plants from the stored material, and also storage of ecologically competent orchid mycorrhizas. Orchid germplasm can be preserved in botanical gardens, orchidaria, field gene banks, in vitro preservation, cryopreservation, DNA banking, etc. Ex situ conservation strategies such as propagation and seed banking are fundamental components in any integrated conservation approach (Cribb et al. 2003), providing long-term security ("extinction proofing").

11.6.2.1 Botanical Gardens

Establishment of botanical gardens are beneficial for collection, cultivation, and conservation of a wide range of orchid diversity at a single place. These species are labeled with botanical names which may help in identification of the plant species and provide educational information to the visitors/researchers.

11.6.2.2 Orchidarium

Orchids are often conserved in polyhouse/glasshouse having suitable climate conditions for their growth and flowering. These structures are termed as orchidarium, where a large number of orchid species are accommodated under a single roof. These structures need to be constructed, keeping in view the climatic conditions required for the species to be conserved. ICAR-NRC for Orchids is maintaining such orchidariums at its headquarters and regional station. Botanical Survey of India and forest departments of various states are also maintaining such orchidariums to promote ecotourism and conservation of orchid genetic resources. The orchids genetic resources conserved in such structures are also at high risk of pest and disease incidents and require well-trained human resources for proper care and monitoring of the plants. Simulated natural habitat: In India, orchid species are also conserved as live plants in various botanical gardens, scientific institutions, and departments of forests as an ex situ conservation measure to create artificial natural habitats. This method attempts to simulate the condition of nature. This method of conserving orchid germplasm reduces the cost of maintenance and prevalence of diseases and pests. Epiphytic orchids are tied on the suitable host plant, and terrestrials are planted in pots or on the ground. The Botanical Survey of India has established three field gene banks of orchids at Shillong, Yurcaud, and Howrah for conservation and multiplication of orchids. Similarly, states like Arunachal Pradesh, Assam, Mizoram, Karnataka, Nagaland, West Bengal, Sikkim, Himachal Pradesh, and Odisha have also collected and conserved the wild orchids. National Research Center for Orchids, Sikkim; Tropical Botanical Garden Research Institute (TBGRI), Kerala; Regional Plant Resource Center (RPRC), Odisha; and several other organizations including the state forest departments are also engaged in conserving orchids. The live specimens preserved ex situ in orchidaria are always at the risk of genetic erosion due to poor adaptation of local conditions and pests and diseases, and improper management. The collections need to be systematically duplicated using multiple conservation strategies.

11.6.2.3 Field Gene Banks

Field gene banks or simulated natural habitats of orchids are established in the vicinity of natural habitats. In this type of conservation method, the epiphytic orchids are mounted on suitable host trees whereas the terrestrial orchids are grown under the shades of trees. Maintaining orchid germplasm in this way not only reduces the cost of cultivation but also lessens the incidence of diseases and pests. However, it limits to the species occurring in that natural habitats or species having wider climatic adaptability. Field gene banks address the problems of maintaining plants in the orchidarium that requires huge investment on construction and maintenance of orchidria.

11.6.2.4 In vitro Conservation

In vitro germplasm banks of whole plants, cells isolated from plant tissues and organs through the use of tissue culture techniques, constitute a very efficient manner of conservation for orchids, and germplasm collections that have assumed a high-priority role in the preservation of species threatened with extinction. In vitro conservation is important because it makes possible the maintenance of cultures in active growth through the periodic subculture of buds and nodal segments. However, this technique allows for the conservation of plant germplasm for only a short period: 9–12 months, depending on the procedure and the plant species. In vitro conservation of orchid germplasm in slow-growth cultures requires urgent attention. This preservation technique can also be used to revitalize orchid germplasm affected by the virus and virus-like diseases through apical meristem culture. Orchids were the first plants to be tissue cultured. In vitro conservation of Indian orchids has yet not been attempted. There is a need for genetic stability studies to avoid somaclonal variants and slow-growth cultures for longer storage duration to avoid frequent transfers (Chang 2007). Development of tissue culture protocols for rare and endangered orchids helps rehabilitate the species and strengthen in situ conservation measures. ICAR-NRC for Orchids Indian Council of Agricultural Research – National Research Centre for Orchids, ICAR-NBPGR Indian Council of Agricultural Research – National Bureau of Plant Genetic Resources, state universities, regional research organizations/institutes, and state forest departments may collaborate to rehabilitate the orchid species facing extinction threats in their natural habitat.

11.6.2.5 Cryopreservation

Conservation measures need to be systematically redid using multiple conservation strategies like the seed, in vitro conservation in slow-growth cultures, and cryopreservation techniques, which have found increasing use in the preservation of plant genetic resources of rare, endangered, and threatened (RET) species. Cryopreservation at ultralow temperature (−196 °C) is a comprehensive alternative for long-term storage of orchids genetic resources and achieved by freezing of tissue at the temperature of liquid nitrogen (−196 °C) or gaseous phase (−140 °C) that suspends all metabolic activities. The critical point of all the methods available is to avoid intracellular ice crystal formation. In literature, there are four types of cryopreservation protocols – (i) conventional slow freezing; (ii) simple freezing; (iii) vitrification; and (iv) desiccation. Under these conditions, biochemical and physiological processes are entirely arrested, and the plant material can be stored for an unlimited period. The chances of genetic instability and the risk of contamination during subculturing are minimized. Cryopreservation works on the principle of avoidance of intracellular ice crystal formation during rapid cooling in liquid nitrogen, as they cause irreversible damage to cell membranes, thereby destroying their semipermeability. The concept revolves around dehydrating the cells followed by vitrifying the intercellular water directly to an amorphous state. For this, a concentrated cellular solution is desired, followed by rapid freezing, which is achieved by air drying, freeze dehydration, application of penetrating or nonpenetrating substances, or acclimation (Malhotra et al. 2019). The vitrification step completes by immersing the cells in LN after exposing them to concentrated cryoprotectant solutions (7 to 8 M). Tissues/explants of orchids may also be cryopreserved in liquid nitrogen cylinders as a long-term storage procedure after proper treatment of cryoprotectants like sucrose, glucose, proline, mannitol, glycerol, sorbitol, trehalose, polyethylene glycol, glycerol, and ethylene glycol applied in combination with dimethyl sulfoxide or DMSO and/or three-component mixtures, within the concentration range of 5–15% (w/v or v/v or 0.5–1.0 M) applied either at low (~0 °C) or at ambient (~25 °C) conditions and varying concentrations. Two critical advances in methodology have made this possible: the development of vitrification using PVS2 solution and the encapsulation– dehydration technique (Pal et al. 2020).

Tissue culture techniques emerged in the nineteenth century and got boosted in 1960. However, the cryopreservation techniques began in the 1980s, and this unique technique strengthens existing conservation techniques and is necessary for orchid conservation programs, offering adequate conservation and preservation potential as well as safe and good-quality genetic material for both research and commercial purposes. This unique technique has the potential to guarantee the long-term conservation of the orchid germplasm of species threatened with extinction and represents a valuable method for the preservation of the genetic resources of many orchid specimens. A single capsule can produce more than one million seeds, many times, and it is impossible to sow all the seeds and grow all of the plants simultaneously. It requires the development of preservation methods for maintaining the viability of the seeds for long periods. Likewise, the collection and storage of pollen must permit its use in crosses between plants that flower at specific times or places.

Scientists studied cryopreservation of orchids using seeds, pollen, shoot tips, floral parts, protocorms, zygotic embryos, etc. Vendrame et al. (2014) reviewed the cryopreservation strategies of many orchid species. Many researchers carried excellent research on conservation of orchids using with the cryopreservation strategies with various plant parts like seeds of the species of *Anoectochilus, Bletilla, Brassolaeliocattleya, Bratonia, Calanthe, Cattleya Dactylorhiza, Dendobium, Doritis, Encyclia, Grobya, Laeliocattleya, Oncidium, Phalaenopsis, Rhynchostylis,* and *Vanda* (Thammasiri 2008); immature seeds of *Bletilla striata* (Hongthongkham and Bunnag 2014); seeds and protocorms of *Bletilla striata, Dendrobium candidium,* and rare orchids (Nikishina et al. 2007); the protocroms of the species of *Doritis, Dendrobium, Rhynchostylis,* and *Seidenfadenia* (Thammasiri 2008; Yin and Hong 2009; Antony et al. 2010; Mohanty et al. 2012); protocorm-like bodies of *Cymbidium*, *Cleistostoma areitinum* (Maneerattanarungroj et al. 2007), *Dendrobium* Sonia 28 (Hooi et al. 2010), and *Phalaenopsis bellina* (Khoddamzadeh et al. 2011); shoot and shoot tips, shoot primordial, zygotic embryos, and pollen of the *Dendrobium* species (Vendrame et al. 2014); cell suspension cultures of *Doritaenopsis,* and leaf segments of *Aerides*.

11.6.2.6 DNA Barcoding of Indian Orchids

The conservation of orchids is carried out considering their status in the habitat. The basis of conservation is laid based on specific objectives such as conservation of threatened species, and molecular approaches, such as DNA-based methods, have transformed understanding and appreciation of conservation issues associated with orchids. Molecular data provide an empirical framework through which conservation practitioners are in a more informed position to define priorities, reduce costs, and optimize management decisions. Molecular data enables conservationists to address questions of genetic variation within and between populations, species, or provenance delimitation and the maintenance of evolutionary processes (Fay and Krauss 2003; Fay and Chase 2009).

11.6.2.7 Orchids DNA Banking

This comprises conservation of orchid's DNA in the form of leaf samples at -80oC in a deep freezer for a more extended period. Till date, the ICAR-NRC for Orchids did tremendous work on conservation of the samples through this method. Below are the detailed information:

DNA Bank and NCBI Deposits: Genomic DNA has been isolated from 260 species and stored. 65 DNA barcode sequences (using ITS, matK, rbcL, and trnH-psbA primers) were submitted to NCBI.

DNA repository of orchids: The DNA of native orchids are being preserved under −80 °C. Nearly 250 species samples are preserved carefully.

11.7 Characterization and Evaluation

The germplasm accessions of conserved species are evaluated for various horticultural traits, and accessions with unique traits are registered with the National Bureau of Plant Genetic Resources (NBPGR). The NRC for Orchids institute has conserved ~350 species of orchids collected across the country at its headquarters and regional station. For protecting the rare, endangered, and threatened (RET) species, the center is developing production and propagation protocols to bring them under cultivation to reduce the pressure of collection on natural habitats. Recently, Pamarthi et al. (2019) collected and characterized 351 species from 94 genera. Among the collected species, 205 species are threatened in their natural habitats, 90 species have breeding value, 87 species used in traditional medicine, 77 species have fragrance, and 11 species used in traditional dietary. Among the collections, the genus *Dendrobium* represents the highest number of species (68), followed by *Bulbophyllum* (Khoddamzadeh et al. 2011), *Cymbidium* (Gaskett and Gallagher 2018), *Coelogyne* (Fay and Krauss 2003), *Calanthe* (Deb 2013), *Liparis* (Cribb et al. 2003), and *Vanda* (Chowdhery 2009), as well as the followed genera *Eria*, *Pinalia*, *Paphiopedilum*, *Aerides*, *Gastrochilus*, *Oberonia*, *Pholidota*, *Cleistostoma*, *Goodyera*, *Luisia*, *Papilonanthe*, *Phalaenopsis*, *Pleione*, *Sunipia*, *Agrostophyllum*, *Crepidium*, *Epidendrum*, *Micropera*, *Otochilus*, *Phaius*, *Thunia*, *Zeuxine*, *Acampe*, *Ascocentrum*,

Callostylis, Ceratostylis, Cryptochilus, Esmeralda, Habenaria, Herminium, Panisea, Phreatia, Thelasis, which showed below 10 species. Whereas 45 genera hold single species. However, crop improvement programs also utilize wild species to a great extent in pre-breeding activities. Some important orchid species conserved at ICAR-NRC for Orchids, Sikkim, and Darjeeling Campus are shown in Plate 1.

After India became a signatory to the Trade-Related Aspects of Intellectual Property Rights Agreement (TRIPs) in 1994, legislation was required to be formulated. Article 27.3 (b) of this agreement calls for the member countries to protect plant varieties either by a patent or by an effective sui generis system or by any combination thereof. The Government of India enacted The Protection of Plant Varieties and Farmers' Rights Act (PPVFR) in 2001. Protection of Plant Varieties and Farmers Right Act (PPVFR Act, 2001) has been enacted to protect newly developed and extant varieties of crops as a follow-up act for TRIPS agreement (1995). The National Biodiversity Act, 2002, deals mostly with conservation and sustainable utilization of species diversity. Whereas PPVFRA takes into consideration mainly the commercial cultivars, farmer's varieties, extant varieties, and essentially derived varieties of a majority of the crop species which mostly utilize the genetic diversity of a particular crop species.

The PPVFR authority has notified 164 crop/species for registration in India. These include floriculture crops like Rose, Damask Rose, Chrysanthemum, Periwinkle, Bougainvillea, Canna, Gladiolus, Tuberose, Marigold, Crossandra, Jasmine, Carnation, and very recently seven genera of orchids have also been notified. Among the seven notified orchids, the genus–/species-wise observation characteristics varied from one to another. For example, *Cattleya* (Reddy et al. 2005), *Cymbidium* (Vendrame et al. 2014), *Dendrobium* (Pal et al. 2020), *Oncidium* (Tsai et al. 2017), *Paphiopedilum* (77), *Phalaenopsis* (Su et al. 2011), and *Vanda* (Robbirt et al. 2011) characters respectively. The DUS guidelines of the orchid's species are available on www.plantauthority.gov.in. Some other areas of PPVFR act where ICAR-NRC for Orchids can make a significant contribution are:

1. Establishment of distinct, uniform, and stable (DUS) criteria for various other orchid species and their hybrids which is presently constrained by lack of a sufficient number of varieties of different orchid species.
2. Developing complete passport data of the parental lines and making necessary arrangement for ensuring that genetic or parental material acquired for breeding or evolving the new variety has been lawfully acquired. This activity would enhance the flow of genetic material from other countries and improve the orchid breeding capabilities of the country.
3. Help the authorities in the protection of farmers rights. Many species of orchids are under cultivation by the forest-dwelling communities for their showy flowers or medicinal uses. These forest-dwelling communities conserve orchid genetic resources of wild relatives that contribute to their improvement through selection. Such efforts need to be identified, recognized, and identified as farmers' variety for registration under the PPVFR Act after consultation with the National Biodiversity Authority and State Forest Departments.

Plate 1 Some of the important orchid species conserved at ICAR-NRC for orchids **(a)** *Paphiopedilum insigne* (Wall. ex Lindl.) Pfitzer, **(b)** *P. fairrieanum* (Lindl.) Stein, **(c)** *P. hirsutissimum* (Lindl. ex Hook.) Stein, **(d)** *Diplomeris hirsuta* (Lindl.) Lindl., **(e)** *Dendrobium kingianum* Bidwill ex Lindl., **(f)** *Papilionanthe teres* (Roxb.) Schltr. and **(g)** *Spiranthes sinensis* (Pers.) Ames

11.7.1 Molecular Characterization, Identifying Genomic Resources

Cameron (2009) examined the utility of markers from different genomes in addressing phylogenetic questions and found that nuclear ribosomal genes provide a source of reliable phylogenetic data on the genera of subfamily Vanilloideae. Bateman et al. (2009) worked on phylogenetic relationships in subtribe Orchidinae, particularly Galearis and Platanthera, and demonstrated some of the problematic species best treated in these genera. An outgrowth of molecular data application to address orchid systematics has been the landmark Genera Orchidacearum series (Pridgeon et al. 2009). It also promoted the study of all aspects of orchid biology by providing a phylogenetic framework and summaries of previous research as a basis to enable yet further study. The molecular techniques play a significant role in protecting the genotypes from biopiracy. A program on orchid species' DNA fingerprinting is underway to protect the orchid biodiversity from illegal infringement.

The genome size of the family Orchidaceae with the ranging 168-fold (1C = 0.33–55.4 pg) is perhaps the most diverse among angiosperm families (Leitch et al. 2009). Genome studies of the orchids in India is in the nascent stage. Tsai et al. (2017) reviewed the genome sizes of various Orchidaceae subfamilies and some genera and species of this family. Being species-richest subfamily, Epidendroideae with genome contents ranging over 60-fold (1C = 0.3–19.8 pg) anchorage the most variable genome size in this family. Orchidoideae, where the largest descending/offspringing from species in subtribe Orchidinae, is depicted by a more restricted range of genomes (1C = 2.9–16.4 pg). And the subfamily Cypripedioideae show genome sizes ranging 1C = 4.1–43.1 pg which means tenfold only. Cypripedioideae contain the largest mean genome size (1C = 25.8 pg) among all the subfamilies. In Vanilloideae it ranged from 1C = 7.3 to 55.4 pg. In this subfamily, *Pogonia ophioglossoides* represents the largest genome size (1C = 55.4 pg) (Leitch et al. 2009). Apostasioideae, the primitive subfamilies, having 1C values range from 0.38 pg and *Apostasia nuda* to 5.96 pg in *Neuwiedia zollingeri* var. *javanica*, a close to the 16-fold range (Jersáková et al. 2013). Among the species which are used as parents for breeding in Taiwan, *P. equestris* and *P. aphrodite* subsp. *formosana*, the two native *Phalaenopsis equestris* species have a small genome size of 1.6 and 1.4 pg/1C (Chen et al. 2013, 2014).

The genetic linkage maps have been constructed for the medicinal orchids, *Dendrobium* (Lu et al. 2012), but not for the ornamental orchids. A total of 950 SSR markers from *P. equestris* were identified, and among them, 206 SSRs primer sequences were developed for *Phalaenopsis* genetic mapping (Hsu et al. 2011). *P. equestris* contains 38 chromosomes (2 N = 2X = 38) and its genome size is 1600 Mb per haploid genome. There are no physical maps constructed in the orchids because of the optical/physical mapping that has mainly applied for small genome organisms, for example, bacteria. Genome-wide association studies (GWAS) and genotyping-by-sequencing (GBS) studies were carried out in Phalaenopsis (Tsai et al. 2017). The orchid floral scent trait could be regulated by QTL. Recently, GBS has been applied to study the deceptive orchid *Orphis*, and found several SNPs linked to odor-related genes (Sedeek et al. 2014). The flower odor chemistry may fundamentally cause the reproductive hurdles, GBS showed common polymorphism all over the *Ophrys* genome but highly distinguished polymorphisms in genes involved in floral odor biosynthesis (Sedeek et al. 2014). Comparative genomics like receptor-like kinases (RLK) and terpene synthases (TPSs) in orchid genomes are carried out in *Dendrobium* and *Phalaenopsis* (Tsai et al. 2017). Recently, genome editing has been reported in *D. officinale* to knock out the expression of several lignocellulose biosynthesis genes, including C3H, C4H, 4CL, CCR, and IRX (Kui et al. 2016).

11.8 Information Documentation

OrchidoPedia: ICAR-NRC for Orchids developed "OrchidoPedia" android application which is an offline app covering 56 genera and 172 species native to Northeast regions of India. Presently, this application is in the English language, but it will also include local languages. The institute is regularly updating the orchid records and enhance its features and quantity of information about the orchids. The mobile android application can freely be downloaded and used to identify the orchid species using the following link. https://play.google.com/store/apps/details?id=nrco.orchidopedia&hl=en

Orchid Farming: The Orchid Farming mobile app of National Research Center for Orchids, Pakyong, Sikkim, imparts scientific knowledge and skills to orchid growers, entrepreneurs, and orchid lovers those who wish to grow orchids. It is an educational app providing information on different aspects of orchids management by choosing selective measures. The mobile android application is freely available for download and use on the following link: https://play.google.com/store/apps/details?id=nrco.cymbidiumorchid&hl=en

The EST dataset is valuable for the identification of specificity of orchids, annotation of genes for genomic sequencing, and assisting orchid genome organization. For storing and managing the vast expressed gene sequences from orchid, OrchidBase was designed to collect the transcriptomic sequences from 11 different tissues/organs of *Phalaenopsis* spp. and tissues of flowers of 10 species belonging to five subfamilies of Orchidaceae (Fu et al. 2011; Tsai et al. 2013; Niu et al. 2016). Both deep sequencings with ABI 3730, Roche 454, and Illumina/Solexa were applied to generate EST sequences collected in OrchidBase. OrchidBase is accessible at http://orchidbase.itps.ncku.edu.tw/. The database delivers a prominent feature of genetic resource for both data mining and experimental researches of orchid biology and biotechnology. Orchidstra (http://orchidstra.abrc.sinica.edu.tw), another orchid transcriptomic database, was developed to collect 233,924 unique contigs of *P. aphrodite* transcriptomic sequences by use of a Illumina/Solexa and Roche 454 platform. Profiling analysis with RNA-Seq was applied to categorize the genes with tissue-tropism expression patterns (Su et al. 2011). Besides, 50,908 contigs of sequences generated by using the Roche 454 platform from various organs of *Oncidium* were assembled into the OncidiumOrchidGenomeBase (http://predictor.nchu.edu.tw/oogb/) (Chang et al. 2011).

11.9 Use of Plant Genetic Resources

11.9.1 Major Constraints in the Crop Production

Orchids are herbaceous perennial, and juvenile phase varies for 2–4 years. The information on the inheritance pattern of economically important traits is minimal. In China, natural variations such as alba forms, inter- and intraspecific variations, and plants with variegated leaves and fragrance have the highest regard. Orchid growers in the Netherlands, Australia, New Zealand, and America cultivate *Cymbidium* varieties that are cool growing and bear large attractive flowers. However, growers in China, Japan, Taiwan, and Korea cultivate hybrids only for pot plants and appreciate it for flower, foliage, and fragrance (Pal et al. 2020). In many orchids, genetic incompatibility reinforces floral morphology to ensure that only outcrossing can occur; such a case was documented by Cheng et al. (2009). Peter and Johnson (2009) demonstrated that although geitonogamy occurred in some species, the outcrossing was predominant. They concluded that floral morphology influences and facilitates outcrossing, even if bee behavior favors geitonogamy.

In addition to native orchids, orchid breeders in India should also make use of advanced materials developed in other countries in their breeding program to obtain commercially viable hybrids. Intellectual property laws prevent any unauthorized use of planting materials/parental stocks/hybrids for developing new varieties/hybrids and their commercial utilization. With the enforcement of National and International treaties like ITPGRFA, CBD, and BDA, the procurement and introduction of breeding materials with appropriate denomination and parentage is desperately needed under relevant international laws with Standard Material Transfer Agreement (SMTA) and Mutually Agreed Term (MAT). Prior Informed Consent (PIC) concerned breeders and benefit-sharing agreement.

11.9.2 Common Sources Use to Overcome Production Constraints

A total of 90 species having a potential breeding value which were conserved at the ICAR-NRC for Orchids belonging to following genera viz. *Calanthe, Cymbidium, Dendrobium, Phalaenopsis, Cattleya, Oncidium, Paphiopedilum,* and *Vanda* were utilized in breeding programs to develop hybrids or improve lines. Subsequent milestones have been marked in hybrid development throughout the globe using modern technologies and approaches. Extensive researches viz. species compatibility, apomixis, genetic engineering, mutation breeding, ploidy breeding, etc. have been done by utilizing the native species and hybrids in the orchid improvement programs.

11.9.3 Breeding Strategies for Orchids

History of orchid breeding is nearly 150 years old. More than 200,000 hybrids have been registered with International Registration Authority, Royal Horticultural Society, Kew London. Most of these hybrids have been registered by private nurseries and private orchid breeders of different countries. The utmost vital traits for orchid breeding include flower self-life or longevity, attractive colors, variants including pure color without sustaining, multiple flower spikes, the orientation of flowers on the flower spike, upright round flower shape, free-flowering, compact in growth habit, fast growing, and self-supporting flower spikes. The flowers should have vibrant color, and the traits like fragrance, off-season flowering, the extension of the flowering season, and earliness would be of added advantage. A significant advantage of orchid breeding is that it may combine conventional breeding methods (through hybridization), coupled with clonal selection. Orchids cross easily due to the weak crossability barrier. Orchid breeding in India is still in the nascent stage. The current commercial cultivation of orchids in India is based on cultivars developed elsewhere in other countries like the Netherland, the USA, Australia, New Zealand, Thailand, Japan, etc. Many of these hybrids involve several species in their background. Hence, it is challenging to obtain commercially viable orchid hybrids with primary crosses involving different species. NRC for Orchids initiated its breeding program during 1998 and has developed few crosses over the last 22 years, e.g., *Cymbidium lowianum* x *Cym.* Showgirl and its reciprocal cross; *Cym* Oriental Legend' x *Cym* Showgirl "Cooksbridge" and it's reciprocal cross; *Cym* Sleeping Nymph' x *Cym* Goldengirl and it's reciprocal, etc. (Medhi et al. 2012). ICAR-NRC for Orchids registered the two crosses, namely Darjeeling Nymph (*Cym* Sleeping Nymph x *Cymbidium lowianum*) and Darjeeling's Delight (*Cymbidium lowianum x Cym Showgirl)* with International Orchid Registration Authority, RHS, London, in 2014. The selected clones of these crosses are shown in Plate 2. Protocols for in vitro seed culture and multiplication through tissue culture in various *Cymbidium* species and hybrids have also been developed successfully (Pal et al. 2020). The wild species have little value in the international floriculture trade due to inferior flower shape, size, and color in comparison to modern-day orchid hybrids. A total of 90 species with potential breeding value have been identified. Extensive researches viz. species compatibility, apomixis, genetic engineering, mutation breeding, ploidy breeding, etc. have been done by utilizing the native species and hybrids in the orchid improvement programs. Orchids are slow-growing plants with a long juvenile period, requiring 4–5 years on average to evaluate flower quality of the offspring and the attainment of new seeds. Some species exhibit complex reproductive processes (cross-pollination and specific physiology for seed germination), and these factors hinder their propagation and preservation (Nikishina et al. 2007).

11.9.3.1 Hybridization and Selection

Majority of orchids are highly heterozygous, and crossing between the same species or same genera usually leads to the production of fertile seeds that provides a range

Plate 2 (**a**) F1 population of Darjeeling Nymph (*Cym* Sleeping Nymph x *Cymbidium lowianum*) and (**b**–**e**) are selected clones from F1 population. (**g**) Selected clone of Darjeeling's Delight (*Cymbidium lowianum* x *Cym* Showgirl. Both the crosses were bred at ICAR-NRC for Orchids and registered with International Orchid Registration Authority, RHS in 2014

of genotypes. The segregating F1 population in these crosses gives useful gene recombinants for selection and cloning of new, desired individuals. Seed germination and raising of seedlings in tissue culture, acclimation of seedling and growing them to flowering size in greenhouses, evaluation of seedlings for desired characters, selection, multiplication of selected clones, testing the elite clones in genotype x environment under trials for testing their stability, and uniformity are the significant steps required for the development of the superior variety through hybridization and selection. It takes nearly about 12–14 years to release a commercial variety of *Cymbidium* orchid (Pal et al. 2020). The superior diploid and tetraploid clones are used further to improve desirable traits of the cultivar. Pal et al. (2020) listed intergeneric hybrids involving two, three, and four genera in crossing with *Cymbidium*.

11.9.3.2 Polyploidy Breeding

Colchicine played a significant role in developing many beautiful orchid hybrids by doubling the chromosome numbers. The change in ploidy level is also associated with a change in morphological and physiological characteristics of the plant viz. increase in number and size of stomata, increase in cell size and flower size, etc. (Pal et al. 2020). The importance of polyploidy in *Cymbidium* orchids was recognized soon after the flowering of *Cymbidium* Alexanderi "Westonbirt," a clone of *Cymbidium insigne* x *Cymbidium eburno-lowianum*, registered in 1911. It was a chance tetraploid with a large flower, heavy substance, superior color, and full shaped. This clone was repeatedly used to develop excellent varieties of *Cymbidium*: *Cymbidium* Rosanna, a cross of *Cymbidium* Alexanderi x *Cymbidium* Kittiwake, registered in 1927. *Cymbidium* Alexander was a tetraploid, while *Cymbidium* Kittiwake was a diploid, but a clone "Pinkie" was a tetraploid. Likely, *Cymbidium* Alexanderi "Westonbirt" and *Cymbidium* Rosanna "Pinkie" have had developed through the fertilization of unreduced gametes of their parents. Attempts made in *Cymbidium* on shoots, protocorms, and in PBLs (Pal et al. 2020) tetraploid flowers have a stronger scent than diploid flowers in *Cymbidium* Golden Elf "Sundust." The micro-propagation characteristics, such as rate of proliferation shoot bud and root differentiation, change after chromosome conversion. The importance of polyploidy in the breeding of *Cymbidium* was realized, and a number of superior hybrids were converted into tetraploids for using in the breeding program. The gene pool was expanded further by introducing miniature species, viz. *C. pumilum, C. devonianum, C. madidum*, and *C. ensifolium* in the breeding program, and many desirable traits such as aroma, cascading inflorescence, color, and size were improved. The breeders are now utilizing less-known species/wild species and related genera for introducing variability in the cultivated hybrids. Over the years, about 16,000 hybrids of *Cymbidium* have been registered with International Orchid Registration Authority, Royal Horticultural Society, London (Pal et al. 2020). The recent concept of pure lines in orchid has also been proposed based on haploid breeding approach suggesting the possibility to develop seed propagated orchids (Ichihashi 2008).

Present Status of Use or Incorporation of Desired Traits

The orchid cultivation and hybridization is prevalent in the worldwide orchid market. The well-designed elegant appearance, some even with charming fragrance, and prolonged long life for orchid flowers have promoted attractiveness of economically viable orchids among breeders, nurseries, and traders. Devadas et al. (2016) reported intra- and intersectional compatibility among Asiatic *Dendrobium* species. The high compatibility among Indian *Cymbidium* species and their hybrids and *Vanda* species have been recorded. High levels of incompatibility were observed among the *Phaius* genera with other orchid genera, like *Calanthe, Coelogyne, Phalaenopsis, Lycaste, Dienia, Cymbidium, Thunia, Paphiopedilum, Coelogyne, Eria*, and monopodial orchids like *Paplionanthe, Dendrobium, Arundina*, and *Vanda*. Primary species hybrid was made with *Phaius* using native species in both direct (PBX-11-22) and reciprocal combinations (PBX-11-25) (Devadas et al. 2019).

11.9.4 Other Uses of Orchids

11.9.4.1 Sociocultural and Religious Importance

In Nagaland, *Dendrobium hookerianum* and *Dendrobium nobile* symbolize purity and holiness. The headhunting community wears D*endrobium acinaforme* plant with the belief giving courage and good luck in their hunt. The beautiful foxtail orchid (*Rhynchostylis retusa*) locally called "Kopou Phul" in Assam is worn by ladies on their head as an ornament during different festival especially during "Bihu" festival in Assam. It symbolizes youthfulness during springtime, a symbol of love by the youth of the Ahom community (Medhi and Chakrabarti 2009). In Manipur (erstwhile *Kangleipak*), orchids are defined in many historic ceremonies. The flowers of orchids such as *Vanda tessellata* and *Coelogyne nitida* are used during local festivals in Assam and Arunachal Pradesh, and *Papilionanthe teres* flowers are used by the *Tai* ethnics of Assam and Arunachal Pradesh for offerings to Lord Budha and spirits (Medhi and Chakrabarti 2009; Meitei et al. 2019a). Various attractive eco-friendly products are made out of *Cymbidium* dried leaves, such as *Lepcha* hats, fruit and vegetable baskets, tea trays, containers, sitting mats, hanging pots, trash bins plant growing pots, etc. In Sikkim state, local people collect the raw material of *Cymbidium* dried leaves from their orchards, backyards of their houses, and make few traditional artifacts used in both for traditional and religious rituals as modern lifestyle accessories. These artifacts have a unique intricate style, design, and long life for which people appreciate it since the ancient period. It became the tradition to use specific *Cymbidium* artifacts in particular socio-religious rituals performed in the locality which has indirectly helped in the survival of this craft as the traditional knowledge concerning making those items (Singh et al. 2019).

11.9.4.2 Orchids Used in the Traditional Healthcare System

Orchids have been used in different systems medicine since the Vedic period (Deb et al. 2009). The medicinal value of orchids is recorded in the ancient scriptures as early as 250–300 BC by *Susruta* and *Vagbhata*, respectively, from ancient Sanskrit. Numerous orchids are used in traditional medical treatment as a remedy for several ailments since ancient times. The orchid genera used in the conventional health care system are *Calanthe, Coelogyne, Cymbidium, Cypripedium, Dendrobium, Ephemerantha, Eria, Galeola, Gastrodia, Gymnadenia, Habenaria, Ludisia, Luisia, Nevilia*, and *Thunia* (Gutierrez 2010). In India, some orchids like *Eulophia campestris*, *Orchis latifolia*, and *Vanda roxburghii* have drawn the attention of the scientific community because of their medicinal properties (Singh et al., 2009). In Ayurveda, a revitalizing herbal formulation “Astavarga” (Chyavanprash) derived from a group of eight herbs, four of them are orchids namely Jivak (*Malaxis muscifera*), Rishbhaka (*Malaxis acuminata*), Riddhi (*H. intermedia*), and Vriddhi (*H. edgeworthii*). *Dendrobium macraei* and *D. nobile* are another important orchids from an Ayurvedic point of view as it is reported to be a source of *Jivanti* (Meitei et al. 2019b). Reddy et al. (2005) reported the ethnobotanical and ethnoveterinary uses of orchids species of Andhra Pradesh, India.

11.9.4.3 Orchids in the Dietary System

Orchid’s importance comes into account in traditional food as side dishes or a supplement in many parts of the world. There are many wild orchid species used for food by the tribal people of Northeast India. Among many of the orchid species, the plant parts like pseudobulb, root, and rhizome are consumed as food. *Habenaria acuminata, H. susannae, Orchis latifolia, Pholidata articulate,* and *Satyrium species* are used as foods which play an important role in the nutrition of the people of Nagaland region (Deb 2013). Many tribes of the Nagaland state used leaves of

Cymbidium species as food. The new shoots of *Cymbidiums* are used with cereals to make a sauce, as well as the pseudobulbs of their orchids in a combination of common vegetables such as potato, tapioca etc. (Medhi and Chakrabarti 2009). The popular beverage called "Faham" or "Madagascar Tea" on the islands of Mauritius and Madagascar is prepared from the orchid *Jumellea fragrans*. Commercially used the vanilla flavor or vanillin extracted from *Vanilla planifolia. Anoectochilus* leaves are used as vegetables in Indonesia and Malaysia. Pseudobulbs of *Cymbidium maladimum* and *Dendrobium speciosum,* and tubers of *Microtis uniflora* and *Caladenia carnea* are eaten. The tubers from orchid genera such as *Acianthus, Dipodium, Glossodia, Lyperanthus, Prasophyllum*, and *Thelymitra* have been used as food by the inhabitants of Australia. In Africa, the tubers of *Cynorchis, Eulophia, Disa, Habenaria*, and *Satyrium* are used as food or juice is extracted from them. Roots, tubers, or rhizomes of *Eulophia, Gastrodia, Habenaria, Orchis, Pholidota, Platanthera*, and *Spiranthes* are used as food in Asia. Tubers of *Disa engleriana, D. robusta* and *D. zambica, Habenaria clavata, Satyrium ambylosacco, S. buchananii*, and *S. carsonii* are used as foods in Malaysia. In Bhutan, the inflorescence or the flowers and pseudobulbs of *Cymbidium* spp. are eaten (Bhattacharjee and Das 2008).

11.10 Looking Forward or Future Perspective

Orchids are one of the premier groups of flowering plants for evolutionary studies, and the massive amounts of DNA data now accumulating are revolutionizing our ideas about these beautiful plants. Strengthening the conservation practices through in situ/ex situ or on-farm conservation with the involvement of local communities will help to conserve these precious genetic resources. Therefore, under acceptable policy and guidelines, these resources can be more effectively utilized for horticultural crop improvement programs, sustainable utilization, and conservation strategies. The special incentives to farmers/local people for growing difficult or uneconomical material in private land or domestic gardens will help conserve wild relatives. The village communities may benefit through watershed management, wildlife habitats, and environmental stabilization. Scientists, individuals, departments, and institutions in particular State Agricultural Board and State Biodiversity Board should come forward and work together to protect these natural resources.

References

Antony JJJ, Keng CL, Rathinam X, Sinniah UR, Subramaniam S (2010) Preliminary study on cryopreservation of *Dendrobium Bobby* Messina protocorm-like bodies by vitrification. Afr J Biotechnol 9(42):7063–7070

Barik SK, Tiwari ON, Adhikari D, Singh PP, Tiwary R, Barua S (2018) Geographic distribution pattern of threatened plants of India and steps taken for their conservation. Curr Sci 114(3):470–595

Baruah (2001) Orchid flora of kamrup district, assam: with illustrations. Bishen Singh Mahendra Pal Singh, Dehradun, India, p 207

Bateman RM, James KE, Luo Y-B, Lauri RK, Fulcher T, Cribb PJ, Chase MW (2009) Molecular phylogenetics and morphological reappraisal of the Platanthera clade (Orchidaceae: Orchidinae) prompts expansion of the generic limits of Galearis and Platanthera. Ann Bot 104:431–445
Bhattacharjee SK, Das SP (2008) In: Orchids- botany, breeding, cultivation, uses and post-harvest management. Aavishkar Publishers, Distributors, Jaipur, India, p 396
Cai J, Liu X, Vanneste K, Proost S, Tsai WC, Liu KW, Chen LJ, He Y, Xu Q, Bian C, Zheng Z, Sun F, Liu W, Hsiao YY, Pan ZJ, Hsu CC, Yang YP, Hsu YC, Chuang YC, Dievart A, Dufayard JF, Xu X, Wang JY, Wang J, Xiao XJ, Zhao XM, Du R, Zhang GQ, Wang M, Su YY, Xie GC, Liu GH, Li LQ, Huang LQ, Luo YB, Chen HH, Van de Peer Y, Liu ZJ (2015) The genome sequence of the orchid Phalaenopsis equestris. Nat Genet 47:65–72
Cameron KM (2009) On the value of nuclear and mitochondrial gene sequences for reconstructing the phylogeny of vanilloid orchids (Vanilloideae, Orchidaceae). Ann Bot 104:377–385
Chang WC (2007) In vitro morphogenesis and micro-propagation of orchids. In: H-H Chen and WH Chen (eds) Orchid biotechnology. World Scientific Publishing Co. Pte. Ltd., Singapore
Chang YY, Chu YW, Chen CW, Leu WM, Hsu HF, Yang CH (2011) Characterization of Oncidium 'Gower Ramsey' transcriptomes using 454 GS-FLX pyrosequencing and their application to the identification of genes associated with flowering time. Plant Cell Physiol 52:1532–1545
Chase MW, Cameron KM, Freudenstein JV, Pridgeon AM, Salazar G, Berg CVD, Schuiteman A (2015) An updated classification of Orchidaceae. Bot J Linn Soc 177:151–174
Chen WH, Kao YK, Tang CY, Tsai CC, Lin TY (2013) Estimating nuclear DNA content within 50 species of the genus Phalaenopsis Blume (Orchidaceae). Sci Hortic (Amsterdam) 161:70–75
Chen WH, Kao YK, Tang CY (2014) Variation of the genome size among Phalaenospsis species using DAPI. J Taiwan Soc Hort Sci 60:115–123
Cheng J, Shi J, Shangguan F-Z, Dafni A, Deng Z-H, Luo Y-B (2009) The pollination of a self-incompatible, food-mimic orchid, Coelogyne fimbriata (Orchidaceae), by female Vespula wasps. Ann Bot 104:565–571
Chowdhery HJ (2009) Orchid diversity in North-Eastern States of India. J Orchid Soc India 23:19–42
Cooke T (1908) The flora of the Presidency of Bombay. C.I.E., formerly principal of the College of Science at Poona and director of the Botanical Survey of Western India. Published under the Authority of the Secretary of State for India in Council. Taylor & Francis, London
Cribb PJ, Kell SP, Dixon KW, Barrett RL (2003) Orchid conservation: a global perspective. In: Dixon KW, Kell SP, Barrett RL, Cribb PJ (eds) Orchid conservation. Natural History Pub, Kota Kinabalu, pp 1–4
Das S, Deori NC (1983) A census of endemic orchids of north-eastern India. In: Jain SK, Rao RR (eds) An assessment of threatened plants of India. Botanical Survey of India, Calcutta, pp 104–109
Deb CR (2013) Orchids of Nagaland, propagation, conservation and sustainable utilization: a review. Pleione 7(1):52–58
Deb CR, Deb MS, Jamir NS, Imchen T (2009) Orchids in indigenous system of medicine in Nagaland, India. Pleione 3(2):209–211
Deva S, Naithani HB (1986) The orchid flora of North-West Himalaya. Print & Media Associates, New Delhi, p 459
Devadas R, Pattanayak SL, Singh DR (2016) Studies on cross compatibility in Dendrobium species and hybrids. Indian J Genet 76:344–355
Devadas R, Pamarthi RK, Meitei AL, Pattanayak SL, Sherpa R (2019) Morphological description of Novel Phaius primary hybrid (Orchidaceae). J Exp Biol Agric Sci 7(2):138–147
Dixon KW, Cribb PJ, Kell SP, Barrett RL (2003) Orchid conservation. Natural History Pub, Kota Kinabalu, pp 1–24
Dressler RL (1993) Phylogeny and classification of the orchid family. Miss. Bot. Card. Cambridge University 47: 25–67
Duthie JF (1906) The orchids of the North-Western Himalaya. Ann Roy Bot Gard, Calcutta 9(2):81–211

Fay MF, Chase MW (2009) Orchid biology: from Linnaeus via Darwin to the 21st century. Ann Bot 104:359–364

Fay MF, Krauss SL (2003) Orchid conservation genetics in the molecular age. In: Dixon KW, Kell SP, Barrett RL, Cribb PJ (eds) Orchid conservation. Natural History Publications, Kota Kinabalu, Sabah, pp 91–112

Fischer CEC (1928) Orchidaceae. In: Gamble JS (ed) Flora of presidency of Madras, vol 8. West, Newman and Adlard, London, pp 1399–1478

Fu CH, Chen YW, Hsiao YY, Pan ZJ, Liu ZJ, Huang YM, Tsai WC, Chen HH (2011) OrchidBase: a collection of sequences of the transcriptome derived from orchids. Plant Cell Physiol 52:238–243

Fyson PF (1924) The flora of the Nilgiri and Pulney hill-tops (above 6,500 feet): being the wild and commoner introduced flowering plants round the hill-stations of Ootacamund, Kotagiri and Kodaikanal. Printed at Government Press, Madras, India

Gaskett AC, Gallagher RV (2018) Orchid diversity: spatial and climatic patterns from herbarium records. Ecol Evol 00:1–11

Gutierrez RMP (2010) Orchids: a review of uses in traditional medicine, its phytochemistry and pharmacology. J Med Plant Res 4(8):592–638

Haines HH (1924) The botany of Bihar and Orissa, vol 3. Adlard & Sons and West Newman Ltd., London, pp 1235–1280

Hongthongkham J, Bunnag S (2014) In vitro propagation and cryopreservation of Aerides odorata Lour. (Orchidaceae). Pakistan J Biol Sci 17:608–618

Hooi TH, James J, Julkiflee A, Poobathy R, Gnasekaran P, Subramaniam S (2010) Novel approach for preliminary PVS2 vitrification optimization parameters of Dendrobium Sonia-28 orchid with Evan Blue staining. Adv Environ Biol 4:284–290

Hooker JD (1888–1890) Flora of British India. Vols. 5 & 6. L. Reeve & Co., London. 667–858 & 1–198 pp

Hooker JD (1890) Flora of British India, vol 6. L. Reeve, London, p 184

Hsu CC, Chung YL, Chen TC, Lee YL, Kuo YT, Tsai WC, Hsiao YY, Chen YW, Wu WL, Chen HH (2011) An overview of the Phalaenopsis orchid genome through BAC end sequence analysis. BMC Plant Biol 11:3

Ibama (2008) Instituto Brasileiro do Meio Ambiente e dos Recursos Naturais Renováveis. Lista Oficial das Espécies da Flora Brasileira Ameaçadas de Extinção [Official List of Endangered Species of the Brazilian Flora]. Accessed on 12/05/2020. Available at: http://www.ibama.gov.br/documentos/lista-de-especies-ameacadas-de-extincao

Ichihashi S (2008) Method of constructing orchid haploid by treating unfertilized orchid flower with auxin and method of groing orchid. US patent no. US 7,442,549 B2

IPNI (2012) The International Plant name index. http://www.ipni.org

IUCN (1999) IUCN guidelines for the prevention of biodiversity loss due to biological invasion. Species 31–32:28–42

Jalal JS, Jayanthi J (2012) Endemic orchids of peninsular India: a review. J Threatened Taxa 4(15):3415–3425

Jersáková J, Trávníček P, Kubátová B, Krejčíková J, Urfus T, Liu ZJ, Lamb A, Ponert J, Schulte K, Čurn V, Vrána J, Leitch IJ, Suda J (2013) Genome size variation in Orchidaceae subfamily Apostasioideae: filling the phylogenetic gap. Bot J Linn Soc 172:95–105

Joseph J (1987) Orchids of Nilgiris. Botanical Survey of India, Howrah, p 66

Kataki SK (1986) Orchids of Meghalaya. Government of Meghalaya, Shillong, p 258

Khoddamzadeh AA, Sinniah UR, Lynch P, Kadir MA, Kadzimin SB, Mahmood M (2011) Cryopreservation of protocorm-like bodies (PLBs) of Phalaenopsis bellina (Rchb. f.) Christenson by encapsulation-dehydration. Plant Cell Tissue Organ Cult (PCTOC) 107(3):471–481

King G, Pantling R (1898) The. orchids of Sikkim Himalaya. Ann R Bot Gard. Calcutta 8:1–342, T. 1–448

Kui L, Chen H, Zhang W, He S, Xiong Z, Zhang Y, Yan L, Zhong C, He F, Chen J, Zeng P, Zhang G, Yang S, Dong Y, Wang W, Cai J (2016) Building a genetic manipulation tool box for

orchid biology: identification of constitutive promoters and application of CRISPR/Cas9 in the orchid, Dendrobium officinale. Front Plant Sci 7:2036
Kumar CS, Kumar PCS (2005) An orchid digest of Manipur, Northeast India. Rheedea 15(1):1–70
Kumar CS, Manilal KS (1994) A catalogue of India orchids. Bishen Singh Mahendra Pal Singh, Dehradun, iii+162 pp
Kumar CS, Manilal KS (2004) Orchids of Kerala, India. In: Manilal KS, Sathish Kumar C (eds) Orchid memories: a tribute to gunnar seidenfaden. Mentor Books, & IAAT, Calicut, pp 155–252
Leitch IJ, Kahandawala I, Suda J, Hanson L, Ingrouille MJ, Chase MW, Fay MF (2009) Genome size diversity in orchids: consequences and evolution. Ann Bot 104:469–481
Lu JJ, Zhao HY, Suo NN, Wang S, Shen B, Wang HZ, Liu JJ (2012) Genetic linkage maps of Dendrobium moniliforme and *D. officinale* based on EST-SSR, SRAP, ISSR and RAPD markers. Sci Hortic (Amsterdam) 137:1–10
Lucksom SZ (2007) The orchids of Sikkim and North East Himalaya. Spectrum House, Siliguri, pp 1–984
Malhotra EV, Gowthami R, Nerwal DK, Chandra R (2019) Preparation of solutions and media for cryopreservation, pp 1–4. In: Agrawal A, Gowthami R, Srivastava V, Malhotra EV, Pandey R, Sharma N, Gupta S, Bansal S, Chaudhury R, Rana JC, Tyagi RK, Singh K (eds) Laboratory manual for eighth international training course on in vitro and cryopreservation approaches for conservation of plant genetic resources. ICAR-National Bureau of Plant Genetic Resources (NBPGR), New Delhi, Bioversity International-India, Delhi, Asia Pacific Association for Agricultural Research Institutions (APAARI)/Asia-Pacific Consortium on Agricultural Biotechnology and Bio-resources (APCoAB), Bangkok, pp 5–19, xviii+83
Maneerattanarungroj P, Bunnag S, Monthatong M (2007) In vitro conservation of Cleisostoma areitinum (Rchb.f.) Garay, rare Thai orchid species by an encapsulation-dehydration method. Asian J. of. Plant Sci 6:1235–1240
Medhi RP, Chakrabarti S (2009) Traditional knowledge of NE people on conservation of wild orchids. Indian J Tradit Knowl 8:11–16
Medhi RP, Chakraborti M, Rampal (2012) Orchid biodiversity in India: conservation and utilization. Indian J Genet 72(2):148–156
Meitei AL, Babu PK, Singh DR (2019a) Orchids of India: the silent treasure distribution, utilization and conservation. In: Khuraijam JS (ed) Wild ornamental plants of India. Astral International Publishing Company, New Delhi, pp 277–292
Meitei AL, Pamarthi RK, Kumar R, Bhutia NT, Rai D, Babu PK, Singh AK, Gazmer R, Singh DR (2019b) Dendrobium nobile Lindl. in Indian Traditional medicine A phytochemical approach. Indian J Horticult 76:557–560
Michael FF (2018) Orchid conservation: how can we meet the challenges in the twenty-first century. Fay Bot Stud 59(16):1–6
Misra S (2004) Orchids of Orissa. Bishen Singh Mahendra Pal Singh, Dehra Dun, India
Misra S (2007) Orchids of India: a glimpse. Bishen Singh Mahendra Pal Singh, Dehradun
Mishra S, Panda SP, Sahoo D (2008) Orchid flora of Andhra Pradesh. Bull Bot Surv India 50:129–146
Mohanty P, Das MC, Kumaria S, Tandon P (2012) High-efficiency cryopreservation of the medicinal orchid Dendrobium nobile Lindl. Plant Cell Tissue Organ Cult 109(2):297–305
Nikishina TV, Popova EV, Vakhrameeva MG, Varlygina TI, Kolomeitseva GL, Burov AV, Popovich EA, Shirokov AI, Shumilov VY, Popov AS (2007) Cryopreservation of seed and protocorms of rare temperate orchids. Russ J Plant Physiol 54:121–127
Niu SC, Xu Q, Zhang GQ, Zhang YQ, Tsai WC, Hsu JL, Liang CK, Luo YB, Liu ZJ (2016) De novo transcriptome assembly databases for the butterfly orchid Phalaenopsis equestris. Sci Data 3:160083
Pal R, Singh DR (2016) Endemic orchids of North East India. In: Bioprospecting of Indigenous bioresources of North-East India. Springer, Singapore, pp 31–44
Pal R, Meena NK, Pant RP, Dayamma M (2020) Cymbidium: botany, production and uses. In: Mérillon J-M, Kodja H (eds) Orchids phytochemistry, biology and horticulture, reference series in phytochemistry. https://doi.org/10.1007/978-3-030-11257-8_8-1

Pamarthi RK, Devadas R, Kumar R, Rai D, Babu PK, Meitei AL, De LC, Chakrabarth S, Barman D, Singh DR (2019) PGR diversity and economic utilization of orchids. Int J Curr Microbiol App Sci 8:1865–1887

Pant B (2013) Medicinal orchids and their uses: tissue culture a potential alternative for conservation. Afr J Plant Sci 7:448–467

Peter CI, Johnson SD (2009) Reproductive biology of Acrolophia cochlearis (Orchidaceae): estimating rates of cross-pollination in epidendroid orchids. Ann Bot 104:573–581

Phillips RD, Hopper SD, Dixon KW (2010) Pollination ecology and the possible impacts of environmental change in the Southwest Australian biodiversity hotspot. Philos Trans R Soc B: Biol Sci 365:517–528

Pridgeon AM, Cribb PJ, Chase MW, Rasmussen FN (eds) (2009) Genera Orchidacearum Volume 5. Epidendroideae (Part II). Oxford University Press, Oxford

Ram RB, Lata R, Meena ML (2011) Conservation of floral biodiversity of Himalayan mountain regions with special reference to orchids. Asian Agri-History 15:231–241

Rao RM (1914) Flowering plants of Travencore. Printed at Government Press

Reddy KN, Subbaraju GV, Reddy CS, Raju VS (2005) Ethnobotany of certain orchids of Eastern Ghats of Andhra Pradesh (Revised). EPRTI Newsletter Vol. 11 No.3

Robbirt KM, Davy AJ, Hutchings MJ, Roberts DL (2011) Validation of biological collections as a source of phonological data for use in climate change studies: a case study with the orchid Ophrys sphegodes. J Ecol 99:235–241

Roxburgh W (1832) Flora Indica; or, descriptions of Indian Plants. by the late William Roxburgh, vol 2. Serampore: W. Thacker & Co. Calcutta; Parbury, Allen & Co., London, p 691

Santos IRI (2000) Criopreservação: potencial e perpectivas para a conservação de germoplasma vegetal [Cryopreservation: potential and perspectives for plant germplasm conservation]. Rev Bras Fisiol Veg 12:70–84

Sedeek KEM, Scopece G, Staedler YM, Schönenberger J, Cozzolino S, Schiestl FP, Schlüter PM (2014) Genic rather than genome-wide differences between sexually deceptive Ophrys orchids with different pollinators. Mol Ecol 23:6192–6205

Singh DR, Pamarthi RK, Kumar R, Rai D, Meitei AL, Babu PK (2019a) Traditional artifacts form dried leaves of Cymbidium species (Orchidaceae) in Sikkim, India. Indian J Tradit Knowl 18(2): 390–394

Singh SK, Agarwala DK, Jalal JS, Dash SS, Mao AA, Singh P (2019b) Orchids of India, a pictorial guide, botanical survey of India, Kolkata, p 548

Su CL, Chao YT, Alex Chang YC, Chen WC, Chen CY, Lee AY, Hwa KT, Shih MC (2011) De novo assembly of expressed transcripts and global analysis of the Phalaenopsis aphrodite transcriptome. Plant Cell Physiol 52:1501–1514

Subedi A (2011) New species, pollinator interactions and pharmaceutical potential of Himalayan orchids. Ph.D. Thesis, Leiden University, The Netherlands

Subedi A, Kunwar B, Choi Y, Dai Y, van Andel T, Chaudhary RP, de Boer HJ, Gravendeel B (2013) Collection and trade of wild-harvested orchids in Nepal. J Ethnobiol Ethnomed 9:64–74

Thammasiri K (2008) Cryopreservation of some Thai Orchids. Acta Hort (ISHS) 788:53–62. http://www.actahort.org/books/788/788_5.htm

Tsai WC, Fu CH, Hsiao YY, Huang YM, Chen LJ, Wang M, Liu ZJ, Chen HH (2013) OrchidBase 2.0: comprehensive collection of Orchidaceae floral transcriptomes. Plant Cell Physiol 54:e7

Tsai WC, Dievart A, Hsu CC, Hsaio YY, Chiou SY, Huang H, Chen HH (2017) Post genomics era for orchid research. Bot Stud 58:61. https://doi.org/10.1186/s40529-017-0213-7

Vendrame W, Faria RT, Sorace M, Sahyun SA (2014) Orchid cryopreservation. Ciênc Agrotec Lavras 38:213–229

Willis KJ (ed) (2017) State of the Worlds plants 2017. Report. Rayal Botanic Gardens, Kew, England

Yin M, Hong S (2009) Cryopreservation of Dendrobium candidum Wall. ex Lindl. protocorm-like bodies by encapsulation-vitrification. Plant Cell Tissue Organ Cult 98:179–185

Zhang Z, Yan Y, Tian Y, Li J, He JS, Tang Z (2015) Distribution and conservation of orchid species richness in China. Biol Conserv 181:64–72

Dahlia

12

Ranjan Srivastava and Himanshu Trivedi

Contents

Abstract

Dahlias (*Dahlia variabilis*) are popular Asteraceae ornamental plants cultivated in many countries due to huge variation in flower shapes, sizes and colors. This wide variation is based onto complicated genetic background, namely dahlia is an autoallooctaploid with the chromosome number ($2n = 8x = 64$) having a large genome size. Pigments contributing to wide range of flower color in dahlia are flavonoids, mainly anthocyanin, butein, and flavone derivatives. A number of F_1 hybrids, spontaneous and induced mutants have been developed leading to diversity in flower colour, form, size befitting different groups of dahlia. There is great scope for the improvement of dahlia for a variety of traits for overcoming biotic and abiotic stresses, improved vase life, fragrance etc. using proven technologies.

R. Srivastava (✉)
Department of Horticulture, G.B. Pant University of Agriculture & Technology, Pantnagar, India

H. Trivedi
School of Agricultural Sciences and Engineering, IFTM University, Moradabad, India

S. K. Datta, Y. C. Gupta (eds.), *Floriculture and Ornamental Plants*, Handbooks of Crop Diversity: Conservation and Use of Plant Genetic Resources,
https://doi.org/10.1007/978-981-15-3518-5_24

Keywords

Dahlia · Autoallooctaploid · Anthocyanin · Hybrids · Mutants

12.1 Introduction

Dahlia is one of the popular beautiful bulbous flowers. Wide spectrum of colors, variation in shape, form and size ranging from less than 2.5 cm across to more than 40 cm in the diameter made this flower immensely popular and loved all over the world. The Dahlia was named by Abbe Cavanilles in 1791 (Smith 1963).

Dahlias are extensively used for garden display, exhibitions, indoor landscaping, floral arrangements etc. in several ways. Dahlia has significant medicinal and nutritional properties. Tubers of dahlia are rich in inulin and fructose. They also have medicinal compound like phytin and benzoic acid (Whitey 1985). In cyanic strains of dahlia, an enzyme Flavonone-3 hydroxylase was detected in flower extracts (Forkmann and Stotz 1984). It is now commercially exploited for production of flowers in many countries like Japan, France, South Africa, U.K., Italy, Germany, and USA on a large scale (De Hertogh and Le Nard 1993).

12.2 Botany

Dahlias are half-hardy, bushy, herbaceous perennials, dicotyledonous plant with tuberous roots. Stem is straight, branched and feathery. Plant height ranges from as low as 30 cm (12 in) to more than 1.8–2.4 m (6–8 ft). Pinnate leaves are toothed, alternately arranged in various shapes and glandular in most species and mid-dark green in colour. Dahlias bear flower on long, stiff stem well above the foliage in various forms. The majority of species do not produce scented flowers. The inflorescence is capitate which comprises ray and disc florets. Hermaphrodite flowers open centripetally in succession. Ray florets are pistiliate which contain only stigma in the inner base of the petals. Disc florets are bisexual flowers and contain both female and male organs. Stigma is protogynous and become receptive when petal fully opens and remains receptive for 72 h. A receptive stigma looks like the '**Y**' in shape. Centre of the disc florets normally accommodates pollens. As a member of the Asteraceae family, the dahlia has a flower head that is actually a composite (hence the older name Compositae) with both central disc florets and surrounding ray florets. Each floret is a flower in its own right, but is often incorrectly described as a petal. The modern name Asteraceae refers to the appearance of a star with surrounding rays.

12.3 Origin, Domestication and Spread

The center of origin of dahlia is Mexico. The species introduced into the old world were *Dahlia imperialis*, *D. coccinea*, *D. merckii* and *D. juarezii*. The present day hybrids came from a hybrid *D. variabilis* as a result of continuous crossing among species and selection.

Dahlia is found predominantly in Mexico, but some species are found ranging as far south as northern South America. *D. australis* occurs at least as far south as southwestern Guatemala, while *D. coccinea* and *D. imperialis* also occur in parts of Central America and northern South America. Dahlia is a genus of the uplands and mountains, being found at elevations between 1500 and 3700 meters, in what has been described as a "pine-oak woodland" vegetative zone. Most species have limited ranges scattered throughout many mountain ranges in Mexico. A physician of Philip II Francisco Hernandez first mentioned dahlia in 1651. Dahlia was brought out from Mexico for the first time by Vincente Carvantes, Director of the Mexico City Botanic Garden in 1789. For the first time dahlia bloomed in the Europe in 1791 with the name 'Dahlia' in honour of Dr. Andreas Dahl (1751–1789). Then, dahlia became popular and spread to other countries of Europe. With in 15 years of its introduction, England developed 300 distinct cultivars from a few types. In 1841, one English dealer had over 1200 cultivars, the Caledonian Horticulture Society offered 50,000 francs in 1864 to the breeder of a true dahlia which does not exist now (Pizetti and Cocker 1975). Dahlia was first introduced into India during 1857 by the Agri-Horticultural Society of India (formerly Royal Agri-Horticultural Society of India), Kolkata. The dahlias are cultivated all over the world, but there are few places other than native land where they thrive well even if left to grow wild.

12.4 Plant Genetic Resources

(a) **Primary gene pool:** The most comprehensive taxonomy till date is of who recognized 36 species and four varieties on the basis of chromosomal count, extensive field collections and observations. The important species are *D. coccinea* Cavanilles, *D. pinnata* Cav., *D. imperialis* Rozel ex Ortgies (syn. *D. arborea, D. excelsa*), *D. tenuicaulis* Sorensen, *D. merckii* Lehm., *D. australis* (Sherff) Sorensen, *D. rudis* Sorensen, *D. tenuis* Robinson & Greenman, *D. Macdougallii* Sherff, *D. exelsa* Benth, *D. juarezii Berg.* (Cactus dahlia) and *D. rosea* Cav. (*D. barkeriae*). The wide variation in flower shape, size and colour is based onto complicated genetic background, namely dahlia is believed to be an autoallooctaploid with chromosome number $2n = 8x = 64$ (Lawrence and Scott-Honcrieff 1935; Gatt et al. 1998) having a large genome size (2C value = 8.27–9.62 pg; Temsch et al. 2008), and more than 50,000 cultivars have been bred during the last century (McClaren 2009). Polyploids are very common among plants. Polyploidy can be evolutionally advantageous in three points including heterosis, asexual reproduction and gene redundancy (Comai 2005).

(b) **Wild genetic resources and others:** The Mexican originated dahlia has its wild species like *Dahlia imparialis*, *Dahlia coccinea*, *Dahlia merckii* and *Dahlia juarejii*. Dahlia has more than 85 species out of which 25 are reported from the wild and remainders appeared in the gardens of Europe. They were considered as hybrids. In Mexico, dahlia grows wild in sandy uplands and its cultivation in that country was as early as 1570 under the name *Acoctli* or *Cocoxochitl*.

Cocoxochitl was discovered by Aztecs and it was used to grow as a medicinal plant as well as a flower for worship in the 1570s.

12.5 Collections

(a) **Methods:** In 1789, Vicentes Cervantes, Director of the Botanical Garden at Mexico City, sent "plant parts" to Abbe Antonio José Cavanilles, Director of the Royal Gardens of Madrid. Cavanilles flowered one plant that same year, then the second one a year later. In 1791 he called the new growths "Dahlia" for Anders (Andreas) Dahl. The first plant was called *Dahlia pinnata* after its pinnate foliage; the second, *Dahlia rosea* for its rose-purple color. In 1796 Cavanilles flowered a third plant from the parts sent by Cervantes, which he named *Dahlia coccinea* for its scarlet color. In 1798, Cavanilles sent *D. Pinnata* seeds to Parma, Italy. That year, the Marchioness of Bute, wife of The Earl of Bute, the English Ambassador to Spain, obtained a few seeds from Cavanilles and sent them to Kew Gardens, where they flowered but were lost after 2–3 years. Verma et al. (2002) collected 18 germplasm of dahlia from local habitat surrounding Nainital and Kumaon region and planted at NBPGR regional station, Bhowali (India) and reported wide range of variability amongst the accessions in different characters.

(b) **Status of Collections:** Mishra et al. (1990) collected and evaluated 23 varieties of dahlia (20 released varieties and 3 strains) of which 18 were exotic and 5 indigenous in nature namely, Decorative Yellow Mutant, My Beauty, Silkin Sheen, Elizabeth Margarette, Dark Red, Fojal Mutant, Variegated, Bharat Laxmi, Potridgar, Creamy Violet, Kitty, Avalanche, Garden Glory, Annapurna, Decorative Yellow, Seleetion-8, Powder Puff, Eggiestion, Duston Stone, Tam Tam, Arthur Godfrey, African Queen and Pioneer. They reported that plant spread, flower durability, flowers/plant, and flower diameter had maximum variation. All material showed maximum variation, while multivariate analysis also indicates that the varieties displayed high genetic diversity for all the characters.

(c) **Gaps in collections, both geographical and genetic:** Mishra et al. (1990) advocated that improvement in dahlia crop would be possible if promising varieties are chosen irrespective of origin for individual characters to create best recombinants.

12.6 Conservation

(a) **Methods:** The conservation of dahlia germ plasm resource adopts field cultivation to conserve mostly, but field cultivation faces limitation of light, temperature, water and Pest management leading to loss of germplasm resources. It requires more floor space and is labour intensive. Also there is huge requirement of material and financial resources. There are easy chances of mixing and planting sexual involution etc. In vitro conservation is gaining popularity. It also facilitates promotion of commercialization, and both at long-distance transportation good seed, also can preserve high quality seedling for a long time, and prevent kind of matter from degenerating.

(b) **Status of Plant Genetic Resources:** Since 1789 when Cavanilles first flowered the dahlia in Europe, there has been an ongoing effort by many growers, botanists and taxonomists, to determine the development of the dahlia to modern times. At least 85 species have been reported: approximately 25 of these were first reported from the wild, the remainder appeared in gardens in Europe. Morphological variation is highly pronounced in the dahlia. William John Cooper Lawrence, who hybridized hundreds of families of dahlias in the 1920s, stated: "I have not yet seen any two plants in the families I have raised which were not to be distinguished one from the other. Constant reclassification of the 85 reported species has resulted in a considerably smaller number of distinct species, as there is a great deal of disagreement today between systematists over classification (Weland 2015). The genus dahlia comprises about 20 species. A large number of cultivars are available to suit the requirement of the growers.

(c) **Gaps in available diversity:** Plant breeders require genetic diversity to develop cultivars that are productive, tolerant of biotic and abiotic stresses, and make efficient use of water and fertilizer. In the past dahlia has remained neglected, and improvement through breeding was not very effective due to limited genetic diversity. The recent germplasm accessions provide scope for Vigorous improvement through hybridization. The clusters comprising only one variety, i.e. Arthur Godfrey, African Queen and Pioneer, with specific valuable traits can also be used in hybridization to exploit hybrid vigour. The highly divergent groups are also likely to produce new genotypes with hitherto unknown combinations (Mishra et al. 1990). Further, the exotic genotypes had poor performance for some of the characters, these are a good source for increasing flower diameter, flower weight, plant spread, and number of flowers/plant. These can directly help in increasing flower number/plant and flower diameter. Thus, by increasing the total plant height and shoots/plant more remunerative varieties with more flowers of larger size and weight can be produced. Mishra et al. (1990) suggested that var. Arthur Godfrey for flower diameter, Kitty for early flowering and flower durability, Bharat Laxmi for plant height, Garden Glory for plant spread, flowers and shoots/plant, African Queen for length of floral stalk, and Annapurna for flower weight should be crossed, in single or double crosses to obtain the best hybrid and to create further variability for these characters. They further reported that varieties Garden Glory and Annapurna are the best parents for hybridization. There is need for blue dahlia, black (actually deep red but looking like black) dahlia or green dahlia which can add up to the colour diversity of dahlia. Also, the fragrant dahlia is another gap in the diversity.

12.7 Classification and Characterization

(a) **Classification**: All advanced dahlia-growing countries have their own system of classification. The scheme of classification in a country also changes from time to time. There is, however, a conscious attempt in different countries to have an identical method of classification. National Dahlia Society UK., Royal Horticultural Society, U.K. and the American Dahlia Society, USAhas recommended following groupsbased on flower shape, size, type colour, stem colour, leaflet number, etc.

i-. **Single flowered:** this group of dahlias originated from D. coccinea. Height 45-75 cm, flowers upto 10 cm diameter across with only a single row of petals. Disc apparent. Suitable for bedding purposes, e.g., Bambino, Little Dorrit, Inflammation, Margaret, Geerling, Mignon Single types, Mount Noddy, Orchid, Reddy, Sion, Pinnochio and Yellow Hammer.

ii-. **Anemone flowered:** Anemone flowered dahlias are developed from *D. pinnata, D. crocea, D. coronata and D. juarezii.* Flowers with one or more rows of petals surrounding a dense group of long tubular disc-florets. Fully double. Hieght 60–105 cm, producing a pin cushion effect showing no disc. Good for flower arrangement, e.g., Guinea, Honey, Lucy, Scarlet Comet, Stillwater pearl.

iii-. **Collerette:** Height 75–120 cm, large petals forming an outer collerette around a central disc which is apparent, bloom 12–15 cm across, having a ring of flate ray florets surrounding the disc floretes and inner rings of narrower florets which are about half of the length of outer florets, usually contrast in colour. Important ones are Can Can, Comet, Lilian Alice, Magic Night, Mrs. H. Brown, Nonsense, President Viger, Suntan, Starsister, etc.

iv-. **Water lily:** This is also introduced as Nymphaea flowered or Double camellia-flowered group. Height is 90–120 cm, flowers fully double, bloom 15 cm diameter, petals edges involute and look like widely open tubes giving saucer shaped appearance, bloom in August. Gelena, Peace Pact, Pearl of Heemstede are some examples.

v-. **Decorative:** This is one of the most popular groups, fully double flowers showing no central disc, cover wide range of colours and size from globular bloom with florets slightly flatter or curly, tips of petals either round or pointed, flower size 10–15 cm, 15–20 cm, 20–25 cm and > 25 cm diameter are available. African Queen, Arthur Godfrey, Avalanche, Bhola Baba, Black Out, Chinese Lantern, Colonel, Croydon, Apricot, Dutch triumph, Elma, Elizabeth, Kelvin Rose, Peter, etc. are popular cultivars.

vi-. **Ball:** Ball dahlias have fully double flowers. They are ball shaped or slightly flattened; petals are blunt or rounded at the tips and they are involute for more than half their length. Many balls look very similar to pompons but when attain size they are larger than the latter group Balls are subdivided into small ball (10–15 cm) and miniature ball dahlias

(5–10 cm). eg, Alltamy Cherry, Crichton Honey, Nijinsky and Risca Miner are small ball dahlias and Connoisscur's Choice, Nettie, Rothsay Superb and Wootton Cupid are miniature ball dahlias.

vii-. **Pompon:** Pompon dahlias have fully double blooms, globular in appearance. Petals are involute for whole of their length and in this formation an individual petal looks somewhat like the cell of a honeycomb. The size limit for pompon is 50 mm. These are extremely popular for flower arrangernent and have the longest vase life among all the dahlias, e.g., Albino, Barbara, Deepest, Glow, Diana Gregory, Hallmark, Honeycomb, Little Willo, Mark Lockwood, Noreen, Small World, William John and Yellow Baby.

viii-. **Cactus:** Fully doubled flower having no central disc and with long pointed ray florets, Straight or incurving petals are partially inturned along their length and tend to narrow and pointed which give the flower a star like appearance. Tips of petals of several varieties are lascinated or split. Large, medium, small and miniature types are available. Annapurna, Cardinal, Carnival, Danny, Gens, Garden Glory, Happy boy, Manali, Margaret, Preference, Pride of Holland, Orgeo, Royal Highness, White Pearl, Center Field, Chatnoir.

ix-. **Semi-cactus:** Bloom fully double with fully double flowers similar to cactus group. Important ones are Autumn Fire, Center Field, Hamari Boy, Latest Flame, Little Princess, Park Flame, Rottardam, Royal Wedding, Yellow Mood.

x-. **Miscellaneous:** Contains small, dissimilar classes eg. Akita, Fasciation.

xi-. **Paeony:** This group originated drom D. coccinea. Flowers posses 2–3 rows of florets which are generally flat with a central disk apparent. Flower size up to 18 cm.

xii-. **Star flowered/Stellar:** Small flowers having 1 or 2 rows pointed petals which overlap very slightly, recurved at the edges and form a shallow cup shaped flower around a central disk, eg. Tahoma Hope.

xiii-. **Climbing dahlia:** The only dahlia climbing species is *Dahlia macdougallii* and can climb about 10 m. The breeding work in this direction may lead to a series of beautiful climbing dahlias.

(b) **Characterization:** Darlington (1973) suggested that octaploid D. *variabilis* appears to be strictly self-incompatible. The incompatibility accelerated the very process which was necessary to produce recombination at a time when breeders would use natural open-pollinated seeds. The great variability in dahlia plants was found to be due to the prevailing self-incompatibility which maintained in all progeny, a rather high degree of heterozygosity. Swami Vinayananda (1986), however, stated that all dahlias are not self-incompatible and more than 25% cultivars are self-compatible, Hybrids raised from *Dahlia scapigera* x D. *coccinea* were also found to be self-compatible (Sorensen 1969). The self-fertilized seeds or resultant plants also retain the heterozygous nature which is common to all double dahlias.

From the pollen-fertility point of view, there are fertile and sterile cultivars. Easy seed-bearing but pollen-sterile cultivars are very useful for induced bee pollination. Both pollen-sterile and self-incompatible cultivars are much in use for hand pollination in the Western countries. According to seed-bearing habit, dahlia cultivars can be divided into three groups:

(i) easy seed-bearing, which forms seeds both in the disc and ray florets,
(ii) those which form seed only in the disc floret and the ray florets are sterile, and,
(iii) cultivars in which both disc and ray florets are sterile.

Characters like plant height, growth habit, appearance, position and colour of flowers, etc., are governed by genes. Only a few genes may be required to determine the length of the stem, while there may be several genes which are collectively responsible for the colour of a specific cultivar (Paul 1983). The salient characters of Dahlia species are as below (Bailey and Bailey 1977).

Dahlia species	Characters
D. Coccinea Cavanilles	Highly variable taxon possesses both diploid (n = 16) and tetraploid (n = 32) races. Plant slender, tall with glaucous stem and bipinnate leave. It bears single flowers of orange, red or yellow colour, disc florets are yellow or yellow tipped with scarlet
D. pinnata Cav.	Herbaceous perennial, tetraploid species, 90 cm height, leaves round, five foliate and large (25 cm long), single/double flowers of bluish red colour
D. imperialis Rozel ex Ortgies (syn. *D. arborea, D. excelsa*)	Known as Paenoy-flowered dahlia/tom thumb dahlia/tree dahlia, diploid species. Plants 3-6 m tall, four or six angled stem, bear bi-tripinnate foliage. Flowers single, large, long, stalked, funnel-shaped and white with red tinged colour
D. tenuicaulis sorensen	Tree like species, endemic in Mexico
D. merckii Lehm.	Possesses alternate pinnules on rachilla of its compound leaves and hollow leaf petioles, flowers small, lilac-yellow coloured
D. australis (Sherff) Sorensen	Variable in leaf profiles
D. rudis Sorensen	Tall, hollow and brittle stem, flowers are purple coloured
D. tenuis Robinson & Greenman	Plant tall with opposite pinnate leaves and small flowers with yellow ray florets with few disc florets
D. macdougallii Sherff	Epiphytic, perennial and only climbing species, plants 10 m tall, leaves pinnate, opposite
D. exelsa Berg.	Plants erect and tall, leaves bipinnate and purple flower heads

(continued)

Dahlia species	Characters
[a]*D. juarezii* Berg. (Cactus dahlia)	Scarlet flowers
[a]*D. rosea* cav. (syn. *D. barkeriae, D. coronatata, D. crocata, D. nana, D. purpurea, D. variabilis, D. superflual, D. royleana*)	Perennial, leaves typically one pinnate sometimes bipinnate, Single rose flowers, parent of old fashioned dahlia viz., Single, Pompon, Show and Fancy types

[a]Sometimes mentioned as species, but are in fact varieties

(c) **Evaluation of genetic diversity for desired traits, available sources of breeding value:** Precise information on the degree and nature of genetic divergence is useful for choosing the diverse parents for purposeful breeding programme. For that, surveying of the variations present in the germplasm is very important. Progenies expected to show a broad spectrum of genetic variability, derived from the diverse crosses. These progenies provide a greater scope for isolating transgressive segregants in the advance generations. In the past, dahlia remained neglected and improvement through breeding was not very effective due to limited genetic diversity. Mishra et al. (1987) studied variability and correlation among 23 varieties of dahlia. It was indicated that higher number of flowers and plant spread could be easily combined due to significant positive correlation among these two characters and cv. Garden Glory possesses both these characters. Mishra et al. (1990) investigated genetic divergence of 23 dahlia cultivars for nine developmental characters. The genotypes were grouped in to 10 clusters on the basis of multivariate analysis of divergence. The cluster I and II had four varieties, the cluster III and IV had three varieties each, cluster V, VI and VII comprised two varieties each and cluster VIII (Arthur Godfrey), IX (African Queen) and X (Pioneer) include one variety each, indicating these three varieties were more diverse tan others and can be used in hybridization to exploit hybrid vigour. Based on cluster means, characters like flower weight, flower diameter, number of flowers and shoots per plants and length of floral stalk were the major factors of differentiation among these 23 genotypes. The inner cluster D^2 values ranged from 74 to 3033, suggesting very little domestication in the crop. Improvement in this crop would be possible if promising varieties are chosen irrespective of origin for individual characters to create best recombinations. In this regard, variety Arthur Godfrey for flower diameter, Kitty for early flowering and flower durability, Bharat Laxmi for plant height, Garden Glory for plant spread, flower and shoots per plant, African Queen for length of floral stalk and Annapurna for flower weight should be crossed in single or double crosses to obtain the best hybrid and to create further variability for these characters. This study also suggested that exotic genotypes were good source for increasing flower diameter, flower weight, plant spread and number of flowers per plant. By increasing total plant height and shoots per plant, more remunerative varieties with more flowers of larger size and weight can be produced. Beura and Maharana (1992) reported that cultivars Croydon Apricot and Tenzing

> Norgey showed rapid flower initiation and best suited for late planting. Cultivars Swami Lokeshwaran and and Wallen were best for cut flower production having longer flower stalk, White Kenya produced largest flowers followed by Yellow Kenya. Cultivar Arthur Hambly had longest flowering span.

Mishra and Mohanty (2003) assessed genetic divergence among 18 dahlia varieties through multivariate analysis following Mahalanobis D^2 statistics, canonical variate analysis and numerical taxonomical approach. Significant differences among dahlia genotype were found for all the characters except for bloom life of flower and tuber number per plant. Based on D^2 values, the genotypes were grouped into six gene constellation. Three single variety clusters were distinguished for their high number of small flowers as compared to three multivariate clusters possessing less number of large sized flowers. Stalk diameter of secondary flower, number of ray florets in main flower, petal length and number of ray florets in secondary flower have major contribution to genetic divergence. In numerical taxonomic approach, the dendrogram obtained from cluster analysis of 18 genotypes divided all the accessions in to four clusters.

Feng et al. (2010) studied the genetic diversity of phenotype characteristics for 84 elite dahlia populations. Difference of stem diameter, leaf length and width, flower diameter and pedicle among different flower type populations were significant while plant height, petiole length were non significant. Negative and highly significant correlation between flower diameter and plant height, stem diameter in large flower type population was evident. Flower diameter and petiole length in small flower type population were correlated significantly. Leaf length was positively correlated with leaf width in four flower type populations and correlation degree of different phenotype characteristics was different in the same flower type population.

Shyamal and Kumar (2002) indicated that numbers of branches, days for flowering, number of flowers/plant, diameter of flower and vase life were the primary characters for economic yield in dahlia. Flower yield had strong positive correlation with diameter of flower and length of stalks at both phenotypic and genotypic levels.

12.8 Information Documentation

Previously, dahlia has remained neglected and owing to limited genetic pool, improvement though breeding was not very effective. The cataloguing of germplasm accessions has provided scope for vital improvement through hybridization. Mishra et al. (1987) studied variability and correlation among 23 varieties of dahlia and reported that high flower number and plant spread can be easily combined due to significant positive correlation between these two characters.

Bhattacharjee and Wahi (1982) indicated that in dahlia characters like flower diameter, flower longevity, plant height, and number of branches are important characters for flower yield per plant. Mishra et al. (1990) investigated genetic divergence of 23 cultivars of dahlia for nine characters. On the basis of multivariate analysis of divergence, the genotypes were grouped into 10 clusters. The clusters I and II had four varieties, clusters III and IV had three varieties, cluster V, VI and VII comprised two varieties each and cluster VIII, IX and X possessed one variety each indicating that these three varieties namely, Arthur Godfrey, African Queen and Pioneer were more diverse than others and can be used in breeding programme via hybridization to exploit hybrid vigour. Dhane et al. (2002) determined the magnitude of heritability, variability and genetic advance in yield related attributes in 25 dahlia varieties and reported that the flower peduncle length, flower diameter, flower fresh weight, days to freshness of flower from bud break to withering, number of disc florets per flower and flower number per plant offers a greater scope for improvement through selection because these attributes exhibited moderate to greater genetic advance coupled with high heritability and variability estimates.

Misra and Saini (1997) advocated additive gene action in dahlia for dry as well as green flower weight. Verma and Arya (2004) reported additive gene action for average flower weight, flower diameter, number of flowers per plant, days to 50% flowering whereas non-additive gene action for number of branches per plant, leaf width, flower disk diameter and number of seeds per flower.

12.9 Use of Plant Genetic Resources

(a) **Major constraints in the crop production:** Unknown genetic history and complicated genetics of modern cultivars, high ploidy level, high degree of heterozygosity self-incompatibility amongst large flowered varieties, poor seed set, make it difficult to select the varieties for breeding. Multiple genes with their strong interactions posed difficulty in improvement through hybridization. In the crop production, lack of area specific varieties/hybrid, climatic problems, lack of the knowledge of scientific production technologies, unavailability of market or export platforms are the factors which pose difficulties make it difficult for commercial production.

(b) **Common sources used to overcome production constraints:** Genus Dahlia has varius species of different ploidy levels. The cultivated dahlias are tetraploid with 2n = 32 having eight sets of homologous chromosomes. It also contains many transposons which contribute to manifestation of great diversity. Three basic numbers X = 16, 17 and 18 have been found in 29 species and species with X = 16 can have 2n = 32 or 2n 64 and also that both of these chromosomes are found with in several species.

Chromosomes counts of different species of dahlia (Sorensen 1969; Strother and Panero 2001; Temsch et al. 2008).

Species	Chromosome number (2n)	Species	Chromosome number (2n)
D. apiculata (Sherff) Sorensen	32	*D. variabilis*	32,64
D. australis (Sherff) Sorensen	32,64	*D. merckii*	36
D. coccinea Cav.	32,64	*D.macdougallii*	32
D. imperialis Roezl	32	*D. linearis*	34
D. rudis Sorensen	32	*D. coronata*	32
D. pinnata Cav.	64	*D.dissecta*	34–36
D. sherffii Sorensen	32, 64	*D. jaurezii*	64
D. tenuicaulis Sorensen	32	*D. maxionii*	32
D. Sorensenii	64	*D. scapigeroides*	34
D. brevis	32	*D. pteropoda*	64
D. tubulata	32	*D. rupicola*	34
D. companulata	32	*D. excelsa*	32
D. atropurpurea	64	*D. barkeriae*	64
D. parvibracteata	32	*D. mollis*	32

(c) **Breeding options:**

Dahlia was first introduced in India in 1857 under the Royal Agri-Horticultural Society of India formerly known as Royal Agri-Horticultural Society of India, Kolkata. The flower pot dahlia forms a major contribution of this country to the world of dahlia and these must have been cultivated even before 1935. Other notable contributions are the late cutting method of dahlia preservation developed in 1960, planned self pollination breeding in 1981 and planned breeding with only firstyear seedlings in 1982. The last two experiments gave a new theory called "Mutation theory of Dahlia evolution." In India, the contribution of Swami Vinayananda is very remarkable. He attempted planned crosses through hand pollination techniques raised open pollinated seedlings and developed few bi-coloured varieties. Some of his promising varieties are Lord Buddha, Prabodh, Sarda Devi, Swami Gauriswarananda. Swami Lokeshwarananda, Swami Madhavananda, Bhikkus Mother, Bhikkus Vivek, Jyotsna, Basudev, Blue Monarch, Chitchore, Disco, Jayanti and Shri Bhabani. Today's dahlias are not just hybrids but hybrids of hybrids. Origin of those cultivars can be traced back to a sizeable number of wild species which are still available in Mexico, Central America and northernmost South America. There are different opinions about the combinations of species that influenced the present dahlias most. Darlington (1973) considered it to be D. *imperialis* x D *coccinea* while Sorensen (1969) suggested it to be D. *coccinea* × D. *pinnate*.

Beside these three species, there are other species also which have contributed to the evolution of the present-day dahlia. Presently, there are now more than 57,000 registered cultivars which are officially registered through the Royal Horticultural Society (RHS).

(i) **Introduction:** In earlier days, introduction was the main method of the improvement in dahlia. Large number of decorative types of dahlias were introduced and became promising and highly adaptable viz. African Jaster, Croydon Monarch, Davidson, Kenya, Magistrate, Nearest blue, Prime minister, Thelma, The master, Uchhu, Welcome, Waverly, and Victor.

(ii) **Selection:** The first dahlia selected from planned hybridization in India was Bhikkus Mother in 1973. National Botanical Research institute, Lucknow released NBRI'S Pinki in 2001as a seedling selection from seeds procured from Australia. It is pompon, decorative, type with excellent growth and blooming quality. It was registered with Royal Horticultural Society, an International registration Authority for dahlia.

(iii) **Hybridization:** Breeding of dahlia through hybridization is a difficult and challenging task. Normally it takes a long time to cross-pollinate as the different rows of petals open in succession. Hybridization is restored to create recombinants and transgressive segregants.

Selection of parent is very difficult and important also. The quality of the parental line should be moderate depending upon the locality. A quality cultivar may not be a good parent, but a moderate one may prove to be an outstanding parent. Number of superior parents can be identified and used in subsequent combinations by progeny test. Breeder's own experience is very helpful in choice of parents. A potential parent should be observed for several years before test cross. Planting parents and combinations of parents in isolation from other dahlia can effect cross-pollination. For development of giant decorative dahlias through conventional breeding, Swami Vinayanand (1986) suggested parental lines Kenya Yellow and Yellow Monarch which may be selected.

The ray and disc florets of a dahlia flower open up in a succession. The inflorescence is capitate. The opening starts from the outer petals and takes couple of days to open up all the petals for a double bloom of dahlia. Dahlia is a bisexual flower in which ray florets are female and disc florets are complete flower. Stigma is considered to become receptive from the day the petal fully opens. The pollens are normally found at the center of the disc. Both, the pollens and the stigma of the same bloom usually mature at the same time. At the ripe condition of the pollen, it adheres to the finger tip if touched. In general, any plant to be used in breeding is usually grown to several blooms, a lateral may have only one bloom on its terminal bud. The same bloom may be used both as seed parent and as a pollen parent, the petal stigmas of the bloom are cross-pollinated with the pollen of some other cultivar and the pollen of the said bloom is used to cross pollinate some other cultivars. The same bloom can also be both parents of an offspring if the dahlia cultivar is self-compatible. Such seeds normally form in the disc florets. Schie and Debener (2013) made crosses between garden dahlias ($2n = 64$) and the epiphytic species *Dahlia macdougallii* ($2n = 32$) from the section Epiphytum in order to transfer new traits into the gene pool of garden dahlias (*Dahlia variabilis*) and obtained six plants whose hybris status was verified using three SSR markers. The hybrids exhibited indeterminate vegetative growth and the formation of flowers from axillary buds, similar to the father D. *macdougallii*. This result is of interest for breeding new varieties of dahlia with traits that are not present in the current gene pool. Some of the popular F_1 hybrids are as follows:

F_1 hybrid	Parents and/or group	Characteristics
Swami Vinayananda	Croydon monarch X Croydon masterpiece	Large floriferous decorative bicolour dahlia of amber yellow with ivory, erect & strong stalk, long lasting bloom, long vase life, vigorous & disease resistant
Trio	Shirly Wright X Croydon masterpiece	Bicolour dahlia of crimson white tips, produce three types of flowers- full crimson, bicolour crimson with white tip and crimson with whitish tips
Disco	Television X Terry	Small bicolour, informal decorative which changes to semi-cactus form on ageing, light red with white tips. Petals are narrow & slender, suitable for exhibition, garden display and cut flower
Julita, Krynica, Lucilla, Violetta, Orietta, Melba	Pompon X decorative	Earliness, abundance of flowering, wide range of flower colours, semi-dwarf/ miniature, inflorescence 6–7 cm, suitable for flower beds and cut flower
Aida, Kiker, Aga, Etna, Dafnis, Kaprys, Ozyrys, Korona, Wrzos	Cactus X decorative	Semi-cactus and cactus types, medium-tall, suitable for cut flower
Piko, Pepi, Noris, Syria	Dwarf X dwarf	Dwarf, suitable for bedding purpose

Source: Patil and Karale (2017)

(iv) **Mutation breeding:** Mutation is a natural or artificially induced change of the genetic information contained in a cell. This method has been found to be an effective and efficient breeding tool in dahlia. In this, one or two genes can be manipulated without disturbing the rest of genotype due to its heterozygous nature blended with its polyploid characters. High polyploidy and great number of flower colour genes gives a very wide variability. Singh and Roy (1970) mentioned that wide array of horticultural varieties in dahlia appeared to have arisen by gene mutation only. The possibility of mutation induction in octaploid garden dahlia strongly depends upon their genetic constitution. This has been successfully utilized to evolve greater number of mutations for flower colour and shape from garden dahlia cultivars Salmon Rays, Arthur Godfrey and Eldorado. Tubers are irradiated with 2–3 krads X-rays (Broertjes and Bellego 1967). Four mutants of Salmon Rays were awarded, named and registered as new varieties and brought in the market. Series of cuttings from the crown of the irradiated tubers, often result in mutated tissues taking part in the formation of new tubers, thus enabling the mutated character to be transmitted to the next vegetative generation. Soft wood cuttings, after over wintering under protected condition, were also tried to propagate the mutants. High percentage of blind tubers was observed in this case.

Natural colour sports which are available among the most popular cultivars in our country are discussed by Swami Vinayananda (1984). Swami Vinayananda (1986, 1991) described a new form (spoon flowered) of dahlia in the giant decorative cultivars Kelvin Rose and Kelvin. Spontaneous mutations have been isolated from different cultivars. Chimera is common source of new mutation. Manali, a deep magenta streaked flower located a sectorial chimera in White Pearl has been reported (Dohare et al. 1974). Similarly, Juanita, Pink Symbol, Padmaja Naidu, Kailaspati, Dandapani Sport and Prime Minister Sport are other sports.

Improvement of dahlia for form and quality of flowers has been found suitable through using physical mutagens. It is known as induced mutation. Broertjes and Bellego (1967) irradiated number of garden dahlia cultivars with various doses of X-rays. The optimum dose ranges 2–3 Krad. The great number of mutation for flower colour and shape were observed in the irradiated varieties Salmon Rays, Arthur Godfrey and Eldorado. Dube et al. (1980) reported that mutation frequency varied with the dose as well as cultivar and maximum number of mutants was found at 2Kr dose. A mutant of the 'Master' variety showed fimbriation of the petal tips. This is the first case of fimbriated Giant decorative. Das et al. (1978) attempted an improvement of dahlia through gamma irradiation of dahlia tubers at a dose up to 8Kr. They isolated 19 mutants, of which 18 were for flower colour and one for flower form. Kenya cultivar thrown out eight mutants followed by Eagle Stone-3, Black Out 2, The Master-2 and other cultivars with one mutant each. From them 11 mutants were named and released for commercial cultivation. Some of the induced and spontaneous mutants documented have been compiled below:

(a) **Induced Mutants**

Mutant	Parent cultivar	Characteristics
Pride of Sindri	Kenya	Primrose yellow flower, more compact, bloom 35 cm dia., flowers more freely
Bichitra		Mimosa yellow flower, bloom 30 cm dia., ray florets narrow, compact arrangement
Jyoti		Bloom 37 cm dia., marrow purple flowers, good form and habit
Twilight		Bloom 35 cm dia., purplish red flower
Jubilee		Bloom 35 cm dia., orange yellow flower colour with pink strips
Netaji	Eagle stone	Bloom 15 cm dia., attractive bloom of light primrose colour
Pearl		Bloom 15 cm dia., pearl white in colour

(continued)

Mutant	Parent cultivar	Characteristics
Black beauty	Black out	Bloom 28 cm dia., bloom darkest crimson colour with neyron rose strips, very attractive, size of bloom like parent
Vivekanand	The master	Bloom 28 cm dia., spirea red colour, ray florets are divided at tip
Happiness	Croydon monarch	Bloom size 30 cm dia., and as that of parent, deep ruby red coloured
Jayaprakash	Croydon apricot	Bloom 25 cm, phlox pink in colour, compact form & attractive shape
Autumn harmony	Arthur Godfrey	Cadmium orange with scarlet Centre, similar bloom size as of Arthur Godfrey
Holland jubilee		Light orange throughout, bloom firmer and more regular than original cultivar
Progression		Brick red throughout and which have similar size bloom as of Arthur Godfrey
Rosy mist		Empire rose and which have similar size bloom as of Arthur Godfrey
Selection	Salmon rays	Bloom 12 cm dia., colour vivid salmon with slight yellow suffusion at Centre, flower stalk longer than the parent, free flowering
Ornament		Bloom 12 cm dia., colour distinct apricot throughout
Rotonde		Bloom 12 cm dia., colour deep and vivid,, true pink colour, longer flower stalk
Gracieuse		Bloom 12 cm dia., colour violet mauve

(b) Spontaneous Mutants.

Parental cultivar with colour	Mutant Variety with colour
Alva's Supreme and Rusting (cream) Bhikkus Mother Bronze Symbol	White Alva's and White Rusting (white) Zail Singh, Bhikku & Swamiji Pink Symbol (dark pink) & Lavender Symbol (lavender and yellow blend)
Croydon Monarch	Yellow Monarch
Daleko Jupiter (orange and yellow blend)	Pink Jupiter (pink)
Donald van de Mark	Manjushri
Frank Hornsey (bronze and yellow blend)	Pink Frank Hornsey (pink), Yellow Hornsey (yellow), Lemon Hornsey (yellow), Pearl Hornsey (pink and white blend), Rose Hornsey (yellow with rose flush)
Comet (maroon)	Scarlet comet (light red)
Kenya	Kenya Blue, Kenya White, Kenya Yellow
Klankstad Kerkrade (cream flowered)	White Kerkrade (white), Majestic Kerkrade (cream and pink blend)
Paul Chester (bronze)	Lemon Chester and Yellow Chester (yellow)
Reginald Keene (bronze blend)	Candy Keene (pink blend) & Salmon Keene (salmon pink blend)
Shirley Alliance (orange)	Pink Shirley Alliance & Rosemary Clare (pink)

(continued)

Parental cultivar with colour	Mutant Variety with colour
Swamjji	Prabodh
Gallery Art Deco	Gallery Art Nouvean (strong red – purple)

Source: Patil and Karale 2017)

(v) **Molecular:** The AFLP marker analysis is a strong tool for genetic dissection of dahlia genome (Wegner and Debener 2008) Upon measuring the genetic distance between 19 dahlia cultivars, wild genotypes of *Dahlia* species and hybrids, it was concluded different clusters formed within cultivars did not show any relation to phenotypic characteristics like inflorescence morphology or breeding origin.

(d) **Present status of use or incorporation of desired traits:**

i-. **Giant cultivars:** Among different dahlia types, Giant decorative type dahlias are very popular in the world. In India, this type of dahlia is commonly found. Petal-stigmas located near the disc produce more seeds than other portion. Some Giant cultivars produce plenty of seeds in all petal-stigmas. They take the longest time to be fully opened. Only the thicker petalled cultivars stay fresh for a longer period. These are good material as seed parent. Cultivar Bhikkus Vivek was originated from such a giant seed parent. According to Swami Vinaynanda (1995) this is true also for smaller types, but for giant such cultivars are more useful.

ii-. **Flower colour:** A broad spectrum of flower colour exists ranging from white and yellow to hue or red, orange and magenta except blue. Apart from anthocyanins and 6'deoxychalacones flavones are the major pigments present in dahlia flower. Harborne (1967) explained that chalcones and aurones are the only colouring matter of yellow cultivars of dahlia. The enzyme dihydroflavonol 4- reductase (DFR) is involved in anthocyanin biosynthesis in dahlia (Fischer et al. 1988). Orange colour of flowers is mostly caused by the presence of common anthocyanins on a yellow background. Rose and lilac coloured cultivars are primarily based on lower chalcone synthase (CHS) activity. Yellow and white colours do not accumulate anthocyanins due to blockage of pathway. In yellow cultivars, absence of *DFR*, *FHP* and *ANS* gene expression is due to suppression of *bHLH* transcription factor. In white cultivars, *DFR* alone is affected. No difference between black and red cultivars is observed in the expression of transcription factors *IVS* and possible regulatory genes *WDR1, WDR2, MYB1, MYB2, 3RMYB* and *DEL* structural genes of flavonoid pathway. Despite suppression of *FHT* expression, F3'H enzyme activity is present in yellow and white cultivars. Black cultivars accumulate high amounts of anthocyanins, but show drastic reduction in flavones contents. The yellow pigments were characterized in the petald of *D. variabilis* (*D. pinnata*) cv. Rocqueen Court and *D. coccinea.*

They co-occur with malonylated anthocyanins in these flowers. Their presence in the Mexican *D. coccinea* supports the view that cultivated *D. variabilis* could have arisen from hybridization involving this wild species.

Crane and Lawrence (1956) elaborate that the great range of flower colours in dahlia was explained on the assumption that octaploid garden dahlia arose as a hybrid of two tetraploid species either by fusion or two unreduced gametes or by chromosome doubling. Either parent species have two different flower colour pigments. According to Swami Vinayananda (1990), the bicolour characteristics of seed parents of dahlias seem to be a dominant factor. Besides, if two different monochromes coloured cultivars are crossed, there may be some bicolour offspring. The chances are greater when one of the two is a white flower.

12.10 Looking Forward or Future Perspective

Dahlia has a broad scope for cultivation as well as breeding. It is a dire necessity to develop dahlia cultivars with desirable characteristics. The following desirable characteristics of idiotype dahlia are

i-. Development medium and strong stem with desired foliage
ii-. Large sized flowers
iii-. Earliness and longer crop duration
iv-. Deep and compact form of flower head
v-. Ideal flower pasture (about 45^0 angle)
vi-. Better and radically different colours which do not fade quickly
vii-. Good rooting ability and resistance to pest and diseases
viii-. Improving vase life
ix-. Breeding pure white, green, black and blue cultivars
x-. Ability to withstand extremes of climate
xi-. Adding fragrance to the flowers

References

Bailey LH, Bailey EZ (1977) Tagetus. In: Hortus third-A concise dictionary of plants cultivated in the United States and Canada. Macmillan Publishing Co. Inc, New York, p 1095
Beura S, Maharana T (1992) Effect of growth regulators on tuber production of dahlia (Dahlia variabilis D.). Orissa J Agric Res 5:174–178
Bhattacharjee SK, Wahi SD (1982) A correlation and path coefficient analysis of morphological traits in *Dahlia variabilis*. Hortic J 1:75–80
Broertjes C, Bellego JM (1967) Mutation breeding of *Dahlia variabilis*. Euphytica 16:171–176
Comai L (2005) The advantages and disadvantages of being polyploidy. Nat Rev Genet 6:836–846
Crane MB, Lawrence WJC (1956) The genetics of garden plants. Macmillan and Co, London

Darlington CD (1973) Chromosome botany and the origin of cultivated plants. George Allen & Unwin, London, pp 49–52
Das PK, Dube S, Dey AK (1978) *New* dahlias through irradiation. Indian Hortic 23(1):19–21
De Hertogh AA, Le Nard M (eds) (1993) The physiology of flower bulbs. Elsevier Science Publication, Amsterdam
Dhane AV, Nimbalkar CA, Sonawane PC (2002) Variability and heritability studies in dahlia. J Maharashtra Agric Univ 27(2):138–140
Dohare SR, Gill HS, Arora RS (1974) New dahlia mutant. Indian Hortic 19:21
Dube S, Das PK, Dey AK, Bid NN (1980) Varietal improvement of dahlia by gamma irradiation. Indian J Hortic 37(1):82–87
Feng LJ Yuan ZH Yin YL Zhao XQ (2010) Studies on the genetic diversity of phenotype characteristics for different dahlia populations. XXVIII Int Hortic Cong Sci Hort People (IHC2010): Int Symp Adv Ornamentals, Landscape and Urban Hort. 410
Fischer D, Stich K, Britsch L, Grisebach H (1988) Purification and characterization of (+) dihydroflavonol (3-hydroxyflavanone) 4-reductase from flowers of *Dahlia variabilis*. Arch Biochem Biophys 264:40–47
Forkmann G, Stotz G (1984) Selection and characterisation of flavanone 3-hydroxylase mutants of *Dahlia, Streptocarpus, Verbena* and *Zinnia*. Planta 161:261–254
Gatt M, Ding H, Hammett K, Murray B (1998) Polyploidy and evolution in wild and cultivated Dahlia species. Ann Bot 81(5):647–656
Harborne (1967) Comparative bio-chemistry of flavonoids. Academic, London, pp 56–70
Lawrence WJC, Scott-Honcrieff R (1935) The genetics and chemistry of flower colour in Dahlia: a new theory of specific pigmentation. J Genet 30:155–226
McClaren M (2009) Dahlia: history and species. In: McClaren B (ed) Encyclopedia of dahlias. Timber Press, Portland, pp 161–166
Mishra M, Mohanty CR (2003) Comparative estimates of genetic divergence in dahlia. Indian J Hortic 60(3):296–302
Mishra RL, Verma TS, Thakur PC, Singh B (1987) Variability and correlation studies in dahlia. Indian J Hortic 44:269–273
Mishra HP, Singh KP, Mishra GM, Prasad B (1990) Performance of some Dahlia variabilis varieties under late planted conditions in calcarious soil of plains Haryana. J Hortic Sci 19:284–290
Misra RL, Saini HC (1997) Genotypic and phenotypic variability in dahlia (*Dahlia variabilis*). Indian J Plant Genet Resour 10(2):269–271
Patil MS, Karale AR (2017) Flower breeding and genetics. New India Publishing Agency, New Delhi, pp 361–388
Paul H (1983) Dahlia of today. Puget Sound Association, pp 36–37
Pizetti I, Cocker H (1975) Dahlia in flower a guide to your garden. Harry N. Abrams Inc., New York
Schie S, Debener T (2013) The generation of novel species hybrids between garden dahlias and Dahlia macdougallii to increase the gene pool for variety breeding. Plant Breed 132(2):224–228
Shyamal MM, Kumar R (2002) Genetic variability and correlation and correlation studies in dahlia. J Ornam Hortic 5(1):40–42
Singh VK, Roy SK (1970) Cytology of Cuscuta Linn. Sci Cult 36:567–568
Smith AW (1963) A gardener's handbook of plant names. Harper and Roy Publishers, New York, p 428
Sorensen PD (1969) Revision of the genus Dahlia (Compositae, Heliantheae – Coreopsidinae). Rhodora 71:309–416
Strother JL, Panero JL (2001) Chromosome studies: Mexican compositae. Am J Bot 88:499–502
Temsch EM, Greilhuber J, Hammett KRW, Murray BG (2008) Genome size in Dahlia Cav (Asteraceae-Coreopsideae). Plant Syst Evol 276:157–166
Verma SK, Singh RK, Arya RR (2002) Inter-relationship among characters and path coefficient studies in dahlia. Ann Agric Res 23(2):271–274
Verma SK, Arya RR, Singh RK (2004) Genetic variability and correlation studies in chillies. Prog Hortic 36:113–117
Vinayananda S (1984) Dahlias of today. Puget Sound Dahlia Association, p 39

Vinayananda S (1986) In: Choudhary B, Chaddha KL (eds) Ornamental horticulture in India. ICAR, New Delhi, pp 72–85
Vinayananda S (1990) Dahlias. Flower show souvenir. Dahlia Society, New Delhi
Vinayananda S (1991) Dahlia musings. In: Show souvenir. Dahlia Society, New Delhi, pp 17–18
Vinayananda S (1995) Dahlia breeding. In: Chadha KL, Bhattacharjee SK (eds) Advance in horticulture 12. Malhotra Publishing House, New Delhi
Wegner H, Debener T (2008) Novel breeding strategies for ornamental Dahlias II: molecular analyses of genetic distances between Dahlia cultivars and wild species. Eur J Hortic Sci 73(3):97–103
Weland G (2015) The alpha- omega of Dahlis. American Dahlia Society, pp 1–47
Whitey GR (1985) The medicinal and nutritional properties of Dahlia spp. J Ethnopharmacol 14: 75–82

Canna

13

Ranjan Srivastava and Parul Punetha

Contents

R. Srivastava (✉)
Department of Horticulture, G.B. Pant University of Agriculture & Technology, Pantnagar, India

P. Punetha
Dept of Agriculture Sciences, Doon Group of Institutions, Rishikesh, India

S. K. Datta, Y. C. Gupta (eds.), *Floriculture and Ornamental Plants*, Handbooks of Crop Diversity: Conservation and Use of Plant Genetic Resources,
https://doi.org/10.1007/978-981-15-3518-5_25

Abstract

Canna, a member of the Cannaceae family, is an extensively cultivated garden plant for its decorative flowers. *Canna*, being the only genus in the family Cannaceae comprising of about 51 species, has almost universally been recognized by taxonomists. *Canna* includes diploid (2n = 18) and triploid (2n = 27) plants of *C. edulis*, and the latter includes autotriploids. Wild species of *Canna* are involved in producing natural as well as man-made hybrids. Despite the numerous diversity existence of canna in the world, still there is a need for varieties which can sustain long cold regimes, blue- or purple-flowered varieties. Crop improvement methods have been used extensively for inducing variability and conservation. Characteristics like high starch content, anthocyanin for industrial coloring of food material, and flower extract for dye-sensitized solar cells need to be exploited for commercialization.

Keywords

Canna · Ploidy · Diversity · Varieties · Novel characteristics

Canna, commonly known as Indian shot or canna lily, is a widely known ornamental plant having beautiful flowers. It is cultivated extensively around homes and public parks for its decorative and widely varying flower colors (Nakornthap 1965). It is extensively used in constructed wetland for removal of organic pollutants, nitrogen, phosphorous, and heavy metals (De Bust et al. 1995; Neralla et al. 1999). In addition, species of Canna are also edible (Tanaka 2004) and have medicinal properties (Choudhury et al. 2010; Mishra et al. 2012).

Canna belongs to the family Cannaceae, with a single genus *Canna*, and is a member of the Zingiberales, an order consisting of eight families: Musaceae, Strelitziaceae, Lowiaceae, Heliconiaceae, Zingiberaceae, Costaceae, Cannaceae, and Maranthaceae. The order is well-characterized and sharply defined, and its limits have occasioned no controversy (Cronquist 1981). The family Cannaceae, comprising of only the genus *Canna*, was placed on the first page of *Species Plantarum* by Linnaeus (1753) because the flowers have just one stamen and one style and therefore belonged in his monandria monogynia.

Canna, being the only genus in the family Cannaceae, has almost universally been recognized by taxonomists. The APG II system of 2003 (unchanged from the APG system, 1998) also recognizes the family and assigns it to the order Zingiberales in the clade commelinids, in the monocots. The generic name has been derived from the Greek word *Kanna*, a reed referring to its herbaceous stem (Singh et al. 2014).

Rogers (1984) further described the Cannaceae group as "a distinctive monogeneric family with a problem-free circumscription, in contrast with its internal taxonomy and nomenclatural disarray."

The genus *Canna* was earlier included in the family Scitamineae, later separated by Winkler (1930), and placed in monogeneric family Cannaceae because of the free sepals, single-chambered stamen adnate to lateral petaloid staminode, four petaloid staminodes, three-celled ovary with many ovules, style flattened, stigma terminal, and straight embryo. It was further supported by subsequent morphological and taxonomical work of Hutchinson (1959).

13.1 Botany and Distribution

The genus *Canna* comprises of about 51 species of flowering plants having rhizomatous rootstock (Kranzlin 1912). It is a coarse perennial herb, 90 centimeters to 3 meters tall. The plant grows from a large, thick, underground rootstock that is edible. Its large leaves resemble those of the banana plant but are not so large (Snafi 2015). The flower of *Canna* is mainly red, orange, yellow, purple, or white and cultivated as an ornamental plant around the globe. It is characterized by its asymmetric, showy flowers with a staminodial labellum, one to three relatively unspecialized petaloid sterile staminodes (Rogers 1984). In *Canna*, the ovary is inferior and has three locules with axillary placentation, each locule containing two alternating rows of ovules; three septal nectaries are filling the upper part of the ovary and opening at the top, the lower part of the ovary being the fertile part (Vogel 1969). Leaves are lanceolate or ovate 10–30 cm long and 10–20 cm wide, having large lamina up to 60 cm long. Inflorescence is waxy-glucose erect peduncle about 30 cm long. Leaves are dark green with purplish brown margins and veins. They are carline, simple, alternate, and spiral. The oblong leaves have their petioles extending downward to form a sheathing base around the stem. The lamina is pinnately, parallely veined. Leaf margins appear smooth and wavy with acute apex. The leaves are large and foliaceous reaching up to 65–70 cm in length and 30–35 cm in width (Launert 1989). Fruits are capsules, green, oblong, or aid, softly echinate (spiny), and 2 to 2.5 cm long. Capsules are about 40 x 25 mm in size with outer tepals (sepals) persistent at the apex. Seeds, initially white and when matured black with chestnut brown spots, are protected with a smooth coat (Ciciarelli 1995). The stem is pseudo which reaches up to 1.5–2.0 m in height. It is erect, herbaceous, sturdy, and cylindrical enveloped by the sheathing leaf bases. Rhizomes are yellowish white or pinkish on the outside and yellowish white within. At maturity, they turn brownish externally due to a thick outer covering. Rhizomes may be monopodial or sympodial, stoloniferous, or tuberous. Roots are thick, cylindrical, and creamy white in color with a diameter of 2–5 mm with numerous root hairs (Launert 1989).

Marantaceae show a certain resemblance to Cannaceae in floral morphology (Holttum 1951), and a recent molecular phylogenetic study indicates that Marantaceae is the sister family of Cannaceae (Kress 1990).

Although cannas were previously considered as simple foliage plants, during the last two centuries, cultivation and improvement transformed them into attractive ornamental flowering plants with reduction in plant height, increase in hardiness, variability in flower color, and such other positive attributes. Cannas are valued mostly for their large tropical foliage and flashy, brilliantly colored flowers. The foliage is as ornamental as the flowers. Canna foliage may be of various colors as pure green, ruby, coppery to purplish, and green with white stripes (Tjia and Black 1991). In addition to that, it is one of the world's richest starch sources. Some of the species like *Canna edulis* Ker Gawler and its cultivar varieties are described as edible and have been selectively grown for their rootstock, as a source of starch (Tanaka 2008).

Canna is distributed in the tropics and subtropics particularly of the Western Hemisphere. It is common in moist places along streams, springs, ditches, and margins of woods. It may also be found in wet temperate, mountainous regions. It is commonly cultivated in flower gardens (Bachheti et al. 2013; Snafi 2015). *Canna indica* was the first species of this genus introduced to Europe, which was imported from the East Indies, though the species originated from America (Indrayan et al. 2011).

13.2 Origin, Domestication, and Spread

Canna is a small genus of ten species that are originally native in (sub)tropical America (Maas-van de Kamer and Maas 2008), but it developed as an ornamental under different conditions (temperate) of Europe particularly in France and Italy (Mukherjee and Khoshoo 1969). Out of 51 species belonging to the genus (Kranzlin 1912), only 5 species, *C. glauca*, *C. indica*, *C. iridiflora*, *C. warscewiczii*, and *C. flaccid*, constitute the basal species responsible for the origin of the garden cannas included under an artificial hybrid species *C. generalis*.

Archaeological evidence suggests canna was one of the first plants to be cultivated during incipient civilization of Peru and Argentina in 2500 BC (Bird 1948). The first species of *Canna* introduced to Europe was *Canna indica* L., which was imported from the East Indies, though the species originated from the America. Charles de l'Ecluse, who first described and sketched *Canna indica*, indicates this origin and states that it was given the name of *indica* (Indrayan et al. 2011). Much later, in 1658, Pison made reference to another species which he documented under the vulgar or common name of "Albara" and "Pacivira," which resided in the shaded and damp places, between the tropics, and this species is *Canna angustifolia* L. (De l'Ecluse 1576). Without exception, all *Canna* species that have been introduced into Europe can be traced back to the America, and it can be asserted with confidence that *Canna* is solely an American genus. If Asia and Africa provided some of the early introductions, they were only varieties resulting from *Canna indica* and *Canna glauca* cultivars that have been grown for a long time in India and Africa, with both species imported from Central and South America.

Furthermore, *Canna* is an American genus, pointed by Lamarck (Pison 1658) where he argued that "Cannas were unknown to the ancients, and that it was only after the discovery of the New World, that they made their appearance in Europe." Since cannas have very hard and durable seed coverings (Tanaka et al. 2009), it is likely that seed remains would have survived in the right conditions and been found by archaeologists in the Old World. If the soils of India or Africa had produced some of them, they would have been imported before the 1860s into European gardens (Lerman and Cigliano 1971).

13.3 Plant Genetic Resources

13.3.1 Geographic Distribution

Mishra et al. (2013) reported that in America, wild species of *Canna* grow in the South of the United States, South America, from Venezuela to Argentina, and India. It is distributed across the world and is known by a variety of local names such as "Tous les mois" in West Indies, "Imbirg" in Brazil, "Chisgua" in Columbia, "Capacho" in Venezuela, "Queensland arrowroot" in Australia, "Zembu" in the Philippines, "Lotus tuber" in Taiwan, and "Sagu" in Thailand (Tonwitowat 1994).

It was introduced widely and is now cultivated pantropically and in many warmer regions of the world. In many regions, including Southeast Asia and the Pacific, it has also become naturalized. It is widespread and invasive on the Pacific Islands (PIER 2008) and is invasive in Queensland (Batianoff and Butler 2002) and South Africa (Henderson 2001). It is likely to be present in almost every tropical country in the world. It is also likely to be present in many subtropical and Mediterranean regions. Protected over winter, it is found in many temperate countries as a garden ornamental, and even though it can tolerate light frosts, it cannot survive persistent cold temperatures, and it could not naturalize in such climates. It is either introduced in African countries like Burundi, Kenya, Cameroon, Ethiopia, Ghana, South Africa, Uganda, and Zambia to Asian countries like China, India, Japan, Taiwan, the Philippines, and Thailand to European countries like Italy and Portugal to North American countries like the United States to Oceania like Australia, Fiji, and New Zealand or native of North American countries like Costa Rica, Cuba, Haiti, Jamaica, Mexico, Puerto Rico, etc. (Fig.1).

13.3.2 Primary Gene Pool

Genus *Canna* is comprised of five elemental species (Khoshoo and Mukherjee 1970), but its systematics seems imperfect to obtain a consensus. *Canna* includes diploid (2n = 18) and triploid (2n = 27) plants of *C. edulis*, and the later includes autotriploids (Khoshoo and Mukherjee 1970; Ishiki et al. 1997) and segmental allopolyploids (Khoshoo and Mukherjee 1970; Mukherjee and Khoshoo 1969; Koyama 1984). Herbarium specimens and living materials were examined by Segeren and Mass (1971), and they believed that *C. edulis* was synonymous to *C. indica*. Tanaka et al. (2009) in a study revealed that chromosome numbers in most taxa of the genus *Canna* were uniformly 2n = 18, with only a single instance of triploid, 2n = 27, in *C. discolor* Lindl. Venkatasubban (1946) also had previously reported *C. edulis* to be triploid.

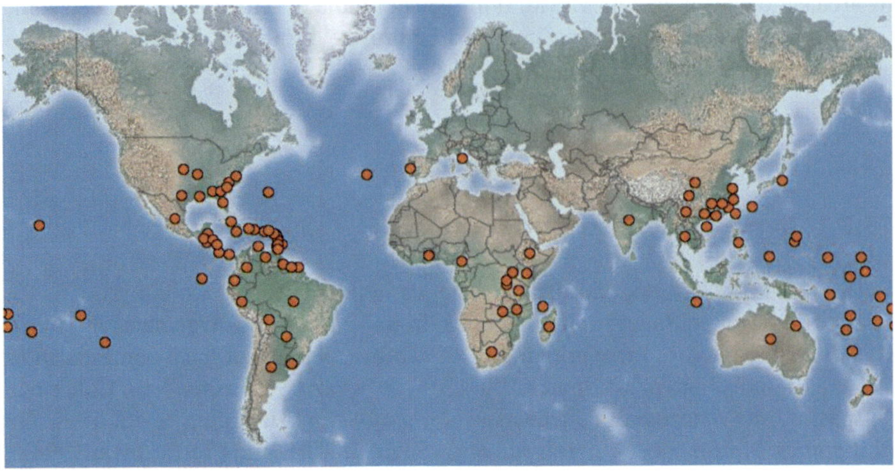

Fig. 1 Distribution of *Canna* in the world (Source: Invasive Species Compendium https://www.cabi.org/isc/datasheet)

13.3.3 Wild Genetic Resources and Others

Wild species of *Canna*, namely, *C. glauca*, *C. indica*, *C. iridiflora*, *C. warscewiczii*, *C. flaccid*, etc., are involved in producing natural as well as man-made hybrids. These five species are popularly known as elemental species of *Canna*. The entire cultivated garden cannas are included under two artificial hybrid species, i.e., *Canna* x *orchiodes* L. H. Bailey and *Canna* x *generalis* L. H. Bailey (Khoshoo and Mukherjee 1970). All these hybrid cultivars have some common features and bind themselves under the same horticultural species. Some phenotypic transformations were also taken place when they were shifted from wild to cultivated condition. The chief morphological factors involved are reduction in plant height, i.e., from tall to dwarf plant; variation in shape, size, and color of leaves; increase in flower size and diversity in flower color; durability of flowers; self-shedding pattern of flowers; etc. Along with these characters, physiological alteration like increase in cold hardiness and stress resistance has also been involved as the foliage plants transform into colorful ornamental plants (Khoshoo and Mukherjee 1970).

13.4 Collections

Diverse genetic resources are the essence of breeding programs, and new sources of genes from the collected and conserved germplasm are required for further breeding programs. Various characters have been used by taxonomists to discriminate between species, including the number of staminodes, the length of the corolla tube (including the adnate staminodes), flower color, the presence of spots on the flowers, rhizome diameter, etc. Species have been named based on flower color alone or for variation in the number of staminodia, a character that is plastic and controlled by only a few "Mendelian factors" (Honing 1939; Rogers 1984).

There are 51 species of this genus (Kranzlin 1912), out of which only 5 species, *C. glauca*, *C. indica*, *C. iridiflora*, *C. warscewiczii*, and *C. flaccida*, constitute the elemental or basal species responsible for the origin of the garden cannas included under an artificial hybrid species *C. generalis*. All these species belong to the subgenus *Eucanna*, Section Trialatae, and the four subsections, Glaucae, Coccineae (or Indicae), Elatae, and Achrida (Mukherjee and Khoshoo 1969).

Nineteen *Canna* species have been reported true species, and there have been two recent revisions of the genus *Canna* by botanists in recent years, firstly by Maas (in the Netherlands) and secondly by Tanaka (in Japan). The taxonomy presented below is based on the Tanaka et al.'s (2009) revision.

Source of *Canna* Species

Species	Source
C. edulis Ker Gawl.	Mr. J.W. Donahue., Gainesville, Florida, National Botanic Gardens, Lucknow; Botanischer Garten, Giessen
C. flaccid Salisb.	Mr. J.W. Donahue., Gainesville, Florida, National Botanic Gardens, Lucknow; Botanischer Garten, Giessen
C. glauca Linn.	Royal Botanic Gardens, Kew
C. humilis Bouché.	Prof. R. Prakken (Wageningen)
C. indica Linn.	National Botanic Gardens, Lucknow
C. iridiflora Ruiz & Pav.	Royal Botanic garden, Kew
C. lagunensis Lindl.	Mr. Donahue; Hortus Botanicus, Denmark
C. lutea miller	Mr. Donahue; National Botanic Gardens, Lucknow; Hortus Botanicus, Denmark; Botanischer Garten, Giessen
C. pallida roscoe	Mr. Donahue; NBG., Lucknow
C. warscewiczii E. Dietr.	Mr. Donahue; NBG., Lucknow

Kranzlin (1912) revised the group in 1912 and recognized ~59 species plus a list of incertae sedis; however, many species were based on a single herbarium sheet.

Kubitzki (1998) provides a general review of the group, but more modern, synthetic treatments are available. Tanaka recognized 19 species plus a number of subspecific taxa within *C. discolor* and *C. indica* in his 2001 revision, again recognizing several species based on a single specimen.

Additional newly described species include *C. ascendens* (Ciciarelli 2007), *C. tulianensis* (Tanaka 2008), and *C. variegate* (Ciciarelli 1995). Further, Maas-van de Kamer and Maas (2008) published the most recent monograph in which they provide an extensive review of the nomenclatural and taxonomic history, vegetative and floral morphology, and fruit, seed, and seedling morphology, pollination biology, distribution, and ecology. They recognize ten species including a large *Canna indica* complex. Several fossil taxa have also been described (e.g., Knowlton 1923; Becker 1969; and Daghlian 1982), but few can be assigned to the family with confidence with the exception of a leaf from the Eocene of Texas (Daghlian 1982).

13.5 Conservation

13.5.1 Methods

There are so many species and cultivars of *C. indica*, and the genus seems to be in no danger of genetic deterioration. However, it is important to conserve older and less popular cultivars and clones, to conserve the vast genetic diversity. No comprehensive germplasm collections exist at present. However, field gene bank conservation of about 150 *Canna* varieties have been reported at CSIR-National Botanical Research Institute, Lucknow, India (Chowdhuri and Deka 2019).

13.5.2 Status of Plant Genetic Resources

Hybridization has also been responsible for transgressive segregation, particularly in length and breadth of staminodia and luxuriance, affecting not only plant height but also flower size. Perhaps the most important single factor responsible for the evolution of ornamental cannas has been the repeated cycles of hybridization which have led to the breakage of size and other barriers; this seems to have been exploited continuously until very large flower size was built up and combined with other useful vegetative and floral characters such as color and number of flowers per inflorescence, extended blooming period, cold resistance, etc. The efficient vegetative propagation made fixing of the useful genotypes no problem, although they may contain a high degree of heterozygosity and sexual sterility.

Along these lines, Année (hybrids between *C. indica* and *C. glauca*) and Ehemann (hybrids between *C. iridiflora* and *C. warscewiczii*) cannas came into being in 1848 and 1863, respectively. Although both were a distinct improvement over the original species, they were still relatively small-flowered, and major improvements came round about 1868, when Crozy, Gladiolus, or French Dwarf cannas (*C. X generalis* Bailey) were released. This group arose from hybrids and backcrosses of the first two groups and contains diploids and interchanged heterozygotes and autotriploids. When further intercrossing, inbreeding, and selection yielded no significant improvement, "new blood" in the form of *C. flaccida* was introduced. The result was the release of Italian, Iris, orchid, or giant-flowered cannas (*C.* x *orchiodes* Bailey) in 1872. These were asynaptic seedless diploids and allotriploids or segmental allotriploids. By and large, Crozy cannas are the result of exploiting new genetic diversity and transgression, while Italian cannas owe their excellence to the luxuriance accompanying the introduction of *C. flaccida*.

Next to hybridization, triploidy (14%) has been an important mechanism in the origin of cultivars with thicker, more durable, and larger flower parts. The two types of triploids, autotriploids and segmental allotriploids, are distinguishable by their morphological and cytogenetical properties.

It is evident that during the 44 years (1848–1892), the speed of evolution was rapid and its direction governed by the following principles of selection: increase in hardiness, reduction in height, spikes well above foliage, free flowering, erect flowers, increase in flower size, color diversity, circular form of flowers, increase in thickness of flower parts and durability of flower, self-shedding flowers, etc. The result has been the transformation of cannas from simple foliage plants to attractive ornamental flowers (Khoshoo and Mukherjee 1970).

13.5.3 Gaps in Available (Useful) Diversity

Despite the numerous diversity existence of canna in the world, still there is a need for varieties which can sustain long cold regimes. Also, for canna to be a commercial variety for cut flower trade, there is a need for prolonged vase life. Fragrance is another trait that may add value to canna trade. There is lack of blue or purple color in canna.

13.6 Characterization and Evaluation

13.6.1 Characterization for Essential Features and Classification

In India, hybridization has been the major tool for canna improvement, and the pioneer work has been taken up by Agri-Horticultural Society, Kolkata. Numerous varieties have been developed as a result of hybridization and selection which are commonly grown in gardens. Depending on the size and shape of the flowers, cannas have been classified as follows:

Crozy and gladiolus flower: These hybrids were raised by Anne in about 1850 and Vilmorin in 1880.

Alipore hybrids: The selections are the results of years of hybridization and are a great improvement on the Crozy types from which they are derived. The size of the flowers markedly increased with a wide range of colors.

Giant or orchid flower: Originated in Italy and was very popular for many years. The large blooms of silky appearance resemble the flag iris, but not very hardy.

Dreadnaught: A great advancement on the ordinary Crozy or gladiolus-flowered canna both for the individual flowers and bunches.

Dwarf: This type includes varieties which do not exceed the height of 70–80 cm and are very effective for bedding purpose.

Bouquet: In this class, an ideal variety 'Cupid' has flowers on closely branched spikes. The plants of this variety are also dwarf.

Candelabra: This is a distinct break. The main flower stalk has branches, and as many as 8–12 spikes are produced instead of two or three.

Miniature: This is a small-flowering type derived by crossing a society's dwarf hybrid with *Canna indica*. The spikes are neat and compact.

Cannas are also divided into various classes as per the color of the flowers. These are as follows:

(i) Self – without spots or margin, one color only.
(ii) Spotted – usually a shade of red on cream or yellow ground or red spots on orange or red ground.
(iii) Striped red on a cream or a yellow ground.
(iv) Margin yellow.
(v) Margined with a shade darker than the ground color.
(vi) Flaked red or orange on a paler ground.
(vii) Splashed orange on a deeper ground.

The shades of color cover a wide range from creamy white through yellow, orange, and pink to an intense maroon red. There is no blue or purple color in canna. In one variety, the leaves are striped yellow and very attractive. The table below describes some good cultivars with their flower color.

Cultivar	**Flower color**
American beauty	Handsome orange-scarlet flowers
Apricot	Buff-yellow, base overspread with salmon pink
Assault	Flowers orchid-like, cardinal red
Aurora borealis	Canary yellow with rose-pink center, beautifully rayed
Carmine king	Bright carmine red, with yellow center
Cleopatra	Terracotta orange, light purple leaves
Dazzler	Flowers orchid-like, brilliant cardinal red, leaves bronze
Dorris	Pale salmon pink, flowers very pretty
Golden wedding	Flowers yellow, dwarf plants with long-lasting flowers
Louis cayeux	Large flowers of bright rosy scarlet color
Lucifer	Leaves purple, flowers yellow-edged, red
Mrs. Herbert hoover	Beautiful flowers with deep watermelon-pink color
Mrs. Pierre S. du Pont	Very charming flowers with crinkled edges and light watermelon-pink color
Orange perfection	Plants dwarf, orange flowers
Rosamond Coles	Dark reddish orange with deep orange-yellow border
Rosea gigantea	Very large flowers with soft rose to carmine pink color
Statue of liberty	Plants with bronze-colored leaves and blazing flame-red flowers
The president	Beautiful large flowers with rich glowing scarlet color
Yellow king Humbert	Bicolored flowers, bright yellow petals, marked with crimson dots

Source: Bose et al. (1998)

13.6.2 Development/Identification of Gene Pools and Core Collections

Hybridization has played a major part in the evolution of present-day cultivated *Canna* × *generalis* Bailey, which was initially released in 1868 by conventional breeding. Most of the present cultivars are polyploids and have been bred for increased hardiness, reduction in plant height, increase in flower size, and diversity in flower color for the selection of *Canna* (Khoshoo and Mukherjee 1970).

13.6.3 Evaluation of Genetic Diversity for Desired Traits

Morphological evaluation and characterization are needed to determine characters of *Canna indica* for genetic variability to improve canna varieties. Quite considerable genetic variability does exist among different species, varieties, and hybrids of *Canna* cultivated in Indian gardens. Prince and Kress (2001) used nuclear ITS and chloroplast rpl16 intron DNA sequence data for 22 plants representing 7 broadly defined species of *Canna*, and the molecular data confirmed the recognition of a limited number of species including a broadly defined *Canna indica*. Sari et al. (2016) conducted a research to determine diversity and phenetic relationship of *Canna indica* based on morphological characters. Sixty-six of *Canna indica* accession were used in this research. Morphological data were analyzed by description for characterization to construct identification key. The results revealed that *Canna indica* has many diversities, which appear on the characters of rhizomes, leaf, flower, and fruit. Sixty-six accessions of *Canna indica* from 9 provinces in Indonesia were divided into 2 main clusters. There were the green and red cultivar group with 67 percent similarity, and the green cultivar group also divided into green and green purple based on the color of sheaths, tip of bud, rachis inflorescence, petals, brachtea, and color pattern of staminodia with 85 percent similarity. Red cultivar is divided into red and red purplish based on the color of sheaths, rachis inflorescence, and petals with similarity index of 87 percent. Singh et al. (2014) compared the growth and flowering characteristics of ten *Canna* cultivars, viz., Allegheny, Angel Pink, Apricot Dream, Golden Lucifer, King City Gold, Latifolia, Lucifer, Orange Punch, Pink Sunrise, and Tropical Sunrise. Vegetative growth parameters and flower characteristics were analyzed and evaluated. The study indicates that the cultivars, viz., Tropical Sunrise, Pink Sunrise, Orange Punch, Golden Lucifer, and Allegheny, were better in respect of growth, rhizome, and floral characteristics and recommended for bedding purpose in landscaping. Piyachomkwan et al. (2002) evaluated edible canna (*Canna edulis* Ker) as an alternative starch source on the basis of genetic characteristics, agronomic traits, and starch properties. Four Canna varieties indigenous to Thailand were examined including Thai-green, Japanese-green, Thai-purple, and Chinese-purple and compared with cassava (*Manihot esculenta* Crantz). Using the random amplified polymorphic DNA (RAPD) technique employing ten base primers, four primers implied that at least three types of canna including Thai-green, Japanese-green, and Thai-/Chinese-purple existed and corresponded to plant characteristics as identified by flower, stem, leaf, and rhizome colors. Despite genetic diversification, starch properties were not variable. All four varieties produced 30.4–38.4 tons/ha of rhizomes with starch content about 13% (wet basis). Starch yields of canna (4.1–4.9 tons/ha) were comparatively lower than cassava (6.5 tons/ha). The starches were characterized by giant granules (10–80 μm) and compared with cassava starch pastes had a higher peak viscosity (930–1060 BU for canna starches and 815 BU for cassava starch), occurring at a higher temperature. Pastes of canna starch were more stable, and when cooled, viscosity increased to 1800 BU. Gelatinized pastes of canna starches also rapidly formed good gels on cooling.

13.6.4 Available Sources of Breeding Value

There are a number of problems related to the production of *Canna* and breeding of *Canna* for the creation of improved varieties. Germination of seed in *Canna* is very difficult and not practiced very often. Therefore, production of new variety with mating and cross-pollination is limited.

The common practice for canna propagation is asexual, mainly through multiplication of rhizomes. However, this poses the problem of genetic stagnancy, so genetic variation is limited. Varietal improvement of canna lies in genetic manipulation (Mishra et al. 2015a) and protoplast fusion (Mishra et al. 2015b). In both cases, a preliminary requirement is a robust micropropagation protocol. Only two groups have attempted this. Sakai and Imai (2007) regenerated *Canna edulis* from shoot tip culture by optimizing the combination of cytokinin BA, anti-auxin TIBA, and auxins NAA and IBA, but with limited success. Hosoki and Sasaki (1991) produced *C. edulis* with the longitudinal shoot split method.

Protoplast fusion technique was tried by Mishra et al. (2015b) to generate genetically modified hybrid varieties of *Canna*. In vitro generated leaves and shoots of *Canna indica* and *Canna edulis* were used as the source for isolation of protoplasts. Viable protoplasts (viability range 60–75%) were generated in enzymatic combination of cellulase (1%) and pectinase (0.5%) with an incubation temperature of 24 ± 3 °C for 16–18 hours in dark. They standardized protoplast fusion in *Canna by* using polyethylene glycol (PEG). The fused protoplasts were cultured in a medium consisted of banana micropropagation medium, supplemented with nutrients and growth regulators for regeneration. In both plants, good quality of protoplasts was produced from in vitro leaf tissue. They concluded that of all the combinations of enzymes, 1% cellulose +0.5% pectinase generated good quality protoplasts in both plants.

Mishra et al. (2015a) conducted a study on genetic fidelity and somatic embryogenesis of *Canna indica*. It is the first report of micropropagation of *C. indica*. They reported that it took about 12 weeks to obtain plantlets via somatic embryogenesis, and the regenerated plantlets were healthy. RAPD and ISSR analysis showed that they were genetically stable and identical to their parental counterpart. The protocol developed for the rapid regeneration and multiplication of *C. indica* through in vitro callus and somatic embryogenesis may be used for the mass propagation of medicinally important ornamental species.

Kromer and Kukulczanka (1985) have demonstrated the culture of shoot tips in *Canna indica*, on 2-iP supplemented in either half-strength MS liquid medium or semisolid medium, to obtain virus-free plants. The half-strength MS liquid medium was more effective in promoting regeneration in *C. indica*, presumably due to its semiaquatic nature of growth. Kromer (1979) studied the influence of different growth regulators in vitro, on the formation of callus, shoot buds, and roots from rhizome explants in *Canna indica* and obtained best results on media supplemented with 2 mg $\cdot$ L^{-1} IAA and 1 mg $\cdot$ L^{-1} kinetin leading to the development of shoot buds under in vitro conditions.

The physiological responses of three edible canna cultivars (*Canna edulis* Ker. cv. 'PLFR', 'Xingyu-1', and 'Xingyu-2') to continuous drought stress for 35 days were investigated by characterizing the water saturation deficit (WSD), relative electrical conductivity (REC), superoxidative radical content (SRC), ascorbic acid (AsA) content, glutathione (GSH) content, and protein content. It was observed that WSD, REC, and SRC progressively increased in the upper leaves of three cultivars under both control and drought treatments. The content changes of AsA, GSH, and water-soluble protein were lower in 'Xingyu-2' than in 'Xingyu-1' and 'PLRF' in upper leaves than in lower leaves. Compared with control, drought stress aggravated these physiological changes in all three cultivars. The correlation analysis showed that there were significant correlations between indexes except for protein content, which significantly correlated only in SRC. These indicated that drought stress directly led to water loss, and then the REC increased, while GSH and AsA played major roles in removing the SRC. These results revealed that 'Xingyu-2' was more tolerant to drought stress than 'PLFR' and 'Xingyu-1', and the lower leaves were more sensitive than the upper leaves. This study not only provides new insights into mechanisms of acclimation and tolerance to drought stress in edible canna but also provides clues for improving drought tolerance of edible canna through breeding or genetic engineering (Zhang et al. 2013).

13.6.5 Molecular

In nature, *Canna* seems to have been evolved essentially by gene mutation and re-patterning of chromosomes. Man's interference led to recombination emanating from interspecific hybridization, somatic mutations, and triploidy and has speeded up the process of evolution (Khoshoo 1979; Khoshoo and Mukherjee 1970).

Reports of the utility of molecular data to resolve relationships among cannas are also conflicting. Tanaka (2001) found support for several segregate species of *C. indica* using RFLP markers, and Hermann et al. (1999) was able to distinguish *C. edulis* (=*C. indica*) cultivars using RAPDs, yet Patra et al. (2008) were unable to distinguish cultivars of *C. indica* using either ISSR or RAPD markers.

Genetic relationships among 42 cultivars of *Canna* were determined by using amplified fragment length polymorphism (AFLP) marker. A total of 1607 DNA fragments was produced with 25 AFLP primer combinations, out of which 1491 (92.78%) were found to be polymorphic and 116 (7.22%) monomorphic. The number of polymorphic fragments varied from 33 (E-ACA/M-CAA) to 86 (E-ACT/M-CAA) with an average of 59.6 per primer combination, and percent polymorphism varied from 81.7% (E-AAG/M-CAA) to 100% (E-ACC/M-CTT) with an average of 92.8% per primer combination. The polymorphism information content (PIC) value ranged from 0.24 to 0.35 with an average of 0.30 per fragment, and the highest PIC value (0.35) was noticed for primer combination E-AAG/M-CAC followed by E-ACT/M-CTA, E-ACA/M-CAA, and E-AAG/M-CAA (0.34). Marker index (MI) and resolving power (RP) varied from 11.22 to 26.66 and 17.10 to 40.00, respectively. Jaccard's similarity coefficient varied from 0.33 to 0.72 with an average of 0.49 ± 0.03. The maximum genetic similarities (72%) were noticed between the cultivar NBC_1 and NBC_2 and NBC_16 and NBC_19 followed by NBC_19 and NBC_30 (70%). Based upon genetic similarity coefficient, the cultivars NBC_43, NBC_24, NBC_38, NBC_22, NBC_29, and NBC_36 were found to be most divergent among all the cultivars. The UPGMA clustering revealed four major groups accommodating 93% cultivars, and three cultivars each of different species, i.e., *C. argentina* (NBC_24), *C. latifolia* (NBC_43), and *C. generalis* (NBC_13), did not group with any clusters. Model-based clustering obtained from structure analysis also reveals similar pattern of grouping (Gupta et al. 2013).

Molecular structures of starches isolated from Japanese-green, Thai-green, and Thai-purple cultivars of edible canna (*Canna edulis* Ker) were also investigated by Thitipraphunkul et al. (2003). The absolute amylose content ranged from 19 to 25%. Degrees of polymerization (DPn) values of amylose determined by fluorescence-labeling method were 1590 for Thai-purple, 1620 for Japanese-green, and 1650 for Thai-green cultivars. The mole percent of branched fraction of amyloses from edible canna starches examined by an HPLC system after β-amylolysis of labeled amyloses was 13–16%. Branch chain-length distributions of amylopectin analyzed by HPSEC after debranching with isoamylase, followed by fluorescence labeling of unit chain, showed bimodal distribution with the DPn range of 25–28. The amylopectin of edible canna starches contained high amounts of organic phosphorus (391–420 ppm).

13.7 Information Documentation

The varieties of ornamental cannas offered in the trade today represent types that show great improvement over the original botanical species. Belling (1921) studied the behavior of homologous chromosomes in a triploid canna. The variety was secured under the name Gladiator which was a noteworthy variety as it was sterile, and the flowers, instead of setting seeds, dropped upon maturity. In nearly all cases, triploid varieties are partially or fully sterile and hence incapable of hybridization. Morphological characterization has been done by recording key and diagnostic characters of each variety. DUS guidelines are prepared and approved by PPV and FRA. Registration of new varieties is ongoing (NBRI, Annual Report, 2014–15).

In a pioneer project supported by the IAEA, Kasetsart University, Thailand, in collaboration with the local growers, had successfully generated 37 new mutant varieties. Some of the new color mutant varieties of canna induced by chronic gamma irradiation have unique flower colors (Fig. 2) that appealed to the florists and growers as decorative, and landscaping had been propagated for commercial purposes. The Malaysian Nuclear Agency has also developed new mutant varieties of canna with new flower colors and forms (Ibrahim et al. 2018).

Scanning electron microscopy investigations showed that the starch granules from all cultivars of *Canna* were oval-shaped granules with smooth surfaces and were around 10–100 μm in sizes. Proximate composition studies showed that the

Fig. 2 New mutant varieties of *Canna generalis* with unique floral colors by chronic gamma irradiation. The first columns on the left are the original patent varieties, and the second and third columns are mutant plants. (a) is parent GISC 1 with mutant plants 'Narippawaj' and 'Pink Peeranuch'; (b) is parent GISC 2 with mutant plants 'Red Ridthee' and 'Sumin'; (c) is parent GISC 24 with mutant plants 'Mattana' and 'Orange Siranut'; (d) is parent GISC 12 and mutant plant 'Yellow Arunee'

protein content in the *Canna* samples varied between 0.069 and 0.078%, lipid between 0.014 and 0.019%, and ash between 0.25 and 0.33%. All canna starches contained considerably high phosphorus (371–399 ppm), followed by calcium (113–154 ppm) and potassium (35–61 ppm). The absolute amylose content ranged from 19 to 25%. All three starches displayed a B-type X-ray diffraction pattern. The viscograms of canna starches determined by Rapid Visco Analyzer showed that Thai-green and Japanese-green starch pastes were quite stable during cooking and had high setback (Vankar and Srivastava 2018).

13.8 Use of Plant Genetic Resources

13.8.1 Major Constraints in the Crop Production

High-level heterozygosity, sterility, and the rather long duration between sexual generations all restrict improvement of canna through hybridization and selection. Thus, natural and induced somatic mutations may offer an easier method of producing new variation (Mukherjee and Khoshoo 1970). The availability of named varieties in commercial nurseries offers a major problem in the canna production. As the flowers do not last long as cut flowers and are very common garden plants, visitors do not show much interest, and hence, it is not regularly a choice plant for gardens as well. The edible canna has also been so neglected by scientists that everything remains to be done, from assessment to genetic improvement (Tanaka 1998). Recent social and economic changes are causing the society to be less dependent on its environment, leading to a rapid and irreversible loss of the minor crops which have potential for the plant resources in the future. The world food production is not enough to sustain the present population (Inatsu 1985), and there will be huge pressure in the future; therefore, immediate conservation of plant genetic resources such as edible canna plants is essential.

13.8.2 Common Sources Used to Overcome Production Constraints

Percy Lancaster (1967) reported that 51 canna cultivars were introduced from Italy and the United States into the Agri-Horticultural Society of India, Kolkata, which is considered as a major introduction in India. The cultivars introduced are presented in Table 1. The varieties suitable for production of starch from canna have opened up commercial prospects in many countries like Vietnam. The starch of canna plant has the largest granular size and enzymatic digestibility. Moreover, the leaves and stems can be fed to livestock. The canna starch has high digestibility, water solubility, and excellent infant food (Burkill 1985) which may attract the production of canna at large scale. The soil conservative property due to rapid leaf growth to cover ground surface can promote production. The exploitation of a natural coloring material from canna for cotton textile dyeing operations has been reported by Ghorpade et al. (2000). Samples dyed with canna have good fastness properties, in a range from

Table 1 Cultivars of *Canna* introduced into Agri-Horticultural Society of India, Kolkata

S. no.	Country	Year of collection	Introduced *Canna* cultivar
1.	United States	1904	Australia, Burbank, Indiana, king Herbert, Mrs. Kate Grey, Pennsylvania, Philadelphia, queen of Italy, Wintzer's colossal, Wyoming.
2.	Italy	1895	Austria, Italia, professor Traub
		1886	Alemannia, America, Burgundia
		1897	Africa, Asia, Attika, Britannia, crown Prince of Italy, Perseus, Pluto, Rhea, Suevia
		1901	Roma
		1902	Ischia, Rossi, Sicilia
		1904	Emilia, Romagna, Umbria

pinkish purple to dark purple (premordanting in stannic chloride), sap green (alum), dark green (ferrous sulfate), and mustard yellow (postmordanting in ferrous sulfate). The red flower petals of canna can be used for edible dye purpose.

Evaluation of *Canna* Sp. in Sodic Soils

The growth performance of *Canna* 'Red Dazzler' was evaluated under three different fertilization trials: (i) sole application of chemical fertilizers, (ii) FYM + PSB, + *Trichoderma*, and (iii) PSB + *Trichoderma*. The response of biofertilizers on the growth of *Canna* was not significant at the initial stage of investigation. However, response of FYM supplemented with PSB and *Trichoderma* and combination of both (PSB + *Trichoderma*) were found to be better than application of chemical fertilizers. The *Canna* varieties in CSIR-NBRI collections were screened to identify the best suited cultivars for sodic soil. Five varieties 'Caltolen', 'Pink Sunrise', 'Facceda', 'Black Night', and 'Bengal Tiger' were found as suitable for sodic lands. Significant variation was recorded in terms of their growth performance, although significant higher plant height was recorded in 'Black Night' throughout the year. This was followed by 'Bengal Tiger' and 'Caltolen', which were at par with 'Pink Sunrise'. However, the number of tillers per plant was significantly higher in later stage of growth at 300 DAP (days of planting). The varietal evaluation was further extended by adding 3 species and 24 varieties (NBRI, Annual Report, 2014–15).

13.8.3 Breeding Options

Hundreds of hybrids have evolved from complex crosses between various species of *Canna*; they are often grouped under the names *C. x generalis* L. H. Bailey and *C. x orchioides* L. H. Bailey. The distinction between these hybrid groups has been blurred by further interbreeding involving parents from one or more than one species, varieties, and hybrids. *Canna* x *generalis* (*C. glauca* x *C. indica* x *C. iridiflora* x *C. warscewiczii*) and *C.* x *orchiodes* (*C. glauca x C. indica x C. iridiflora x C. warscewiczii x C. flaccida*) are horticultural species under which all the ornamental cultivars and hybrids were included (Patra et al. 2008). Various techniques have been used as breeding options to improve the quality of flower.

Till date, there are five reports on in vitro regeneration from aseptically cultured explants of *Canna indica* and *C. edulis* (Kromer 1979; Kromer and Kukulczanka 1985; Hosoki and Sasaki 1991; Sakai and Imai 2007; Mishra et al. 2015a, b).

Agrobacterium-mediated transformation is a frequently used technique for quality improvement in plants, which has been optimized from time to time. Recently, this technique has been optimized in tea plants for its genetic improvement (Lv et al. 2017).

13.8.4 Present Status of the Use or Incorporation of Desired Traits

There have been a lot of research work on various quality parameters, but less focus is on the parameter vase life. *Canna* has a limited vase life due to the rapid loss of moisture from its perianth. For improving its market value, cuticularization of the perianth was done by the expression of a heterologous cut in producing gene by using the tissue culture and transformation by Singh et al. (2019). In this study, efficient, rapid, and direct adventitious shoot regeneration was successfully established in *Canna* × *generalis* using recalcitrant rhizome explants. The explants were cultured on MS medium supplemented with 6-benzylaminopurine (6-BA), thidiazuron (TDZ), and kinetin. Among the four genotypes taken for tissue culture, the 'Trinacria variegata' was the best responding cultivar. And 2 $mg \cdot L^{-1}$ 6-BA or 1.5 $mg \cdot L^{-1}$ TDZ along with 0.1 $mg \cdot L^{-1}$ IAA was optimum for their regeneration. The highest regeneration was achieved in 'Trinacria variegata' (36%) on 6-BA, 33% on TDZ, while kinetin failed to evoke any regenerative responses. Regeneration was enhanced by supplementation of 100 $mg \cdot L^{-1}$ ascorbic acid (AsA), while 100 $mg \cdot L^{-1}$ of L-cysteine or 100 $mg \cdot L^{-1}$ dithiothreitol (DTT) inhibited regeneration. Shoots were observed to develop 3–5 fibrous roots on MS medium supplemented with 0.5 $mg \cdot L^{-1}$ indole-3-butyric acid (IBA). The plantlets were transplanted into pots and acclimatized in glasshouse with 100% survival. For transformation of *Canna*, rhizome explants were co-cultivated for 60 minutes in *Agrobacterium* suspension. The explants were washed with 500 mg L^{-1} cefotaxime solution, subjected to 100 mg L^{-1} kanamycin selection followed by excision of the shoots and culturing them on IBA-supplemented media for root development. Transgene integration in the putative transformants was confirmed by PCR assay and copy number by Southern blot hybridization analysis.

13.9 Looking Forward or Future Perspective

Canna has high economic value due to high starch content in the rhizomes of edible canna, and the flowers are used as a natural source of anthocyanin pigment for industrial coloring of food material (Vankar and Srivastava 2010). *Canna indica* rhizomes have been found to possess inhibitory activity against reverse transcription of HIV-1 in vitro and are prescribed in Thai folk medicine for the treatment of AIDS (Woradulayapinij et al. 2005). Arabinoxylan, isolated from the dried rhizome, is a functional ingredient of bakery products and pharmaceutical preparations (Zhang and Wang 2011). The exploitation of a natural coloring material from canna for cotton textile dyeing operations and for edible dye purposes are other potential areas. *Canna indica* flower extract has been used for dye-sensitized solar cells (DSSC), and its efficiency in energy conversion has been 0.29% (Vankar and Srivastava 2018). It was an efficient method of conversion of visible light to electrical energy. These aspects of *Canna* are less exploited, so these should be further focused in the future.

References

Bachheti RK, Rawat GS, Joshi A, Pandey DP (2013) Phytochemical investigation of aerial parts of Canna indica collected from Uttarakhand India. Int J Pharm Tech Res 5(2):294–300

Batianoff GN, Butler DW (2002) Assessment of invasive naturalized plants in south-East Queensland. Plant Protection Quarterly 17(1):27–34

Becker HF (1969) Fossil plants of the tertiary Beaverhead basins in southwestern Montana. Paleontographica B 127:1–142

Belling J (1921) The behavior of homologous chromosomes in a triploid Canna. Genetics 7:197–201

Bird J (1948) Preceramic cultures in Chicama and Viru. In: Bennett WC (ed) A reappraisal of Peruvian archaeology Memoirs of the Society for American Archaeology. Botanical Society of America Miscellaneous Series Publications 12: pp 21–28

Bose TK, Maiti RG, Dhua RS, Das P (1998) Floriculture and Landscaping Naya Prokash Calcutta 757p

Burkill HM (1985) The useful plants of west tropical Africa ed 2. Royal Botanic Gardens, Kew 1: 313–315

Choudhury MD, Bawari M, Singha LS (2010) Some antipyretic ethno-medicinal plants of Manipuri Community of Barak Valley, Assam, India. Ethnobot Leaflets 1:21–28

Chowdhuri TK, Deka K (2019) Biodiversity and conservation of ornamental crops. In: Rajasekharan PE, Ramanatha Rao V (eds) Conservation and utilization of horticultural genetic resources. Springer, Singapore, pp 139–217

Ciciarelli M de las Mercedes (1995) Canna variegatifolia Ciciarelli sp. nov. (Cannaceae: Zingiberales) una especie nueva para Argentina. Revista del Museo de La Plata. Sección botánica 14: 333–341

Ciciarelli M de las Mercedes (2007) Canna ascendens (Cannaceae), una nueva especie de la provincia de Buenos Aires y comentarios sobre otras especies Argentinas de este género. Darwin 45: 188–200

Cronquist A (1981) An integrated system of classification of flowering plants. Columbia University Press, New York, pp 45–49

Daghlian CP (1982) Zingiberalean leaves from the lower Eocene of Texas. Botanical Society of America Miscellaneous Series Publications 162:57

De Bust TA, Peterson JE, Reddy KR (1995) Use of aquatic and terrestrial plants for removing phosphorus from dairy wastewaters. Ecol Eng 5:371–390

De l'Ecluse C (1576) Histoire des plantes rare observées en Espagne

Ghorpade B, Tiwari V, Vankar PS (2000) Dyeing of cotton fabric with canna flower extract using ecofriendly mordants. Colourage 47(1):27

Gupta A, Maurya R, Roy RK, Sawant SV, Yadav HK (2013) AFLP based genetic relationship and population structure analysis of Canna- an ornamental plant. Sci Hortic 154:1–7

Henderson L (2001) Alien weeds and invasive plants. Plant Protection Research Institute Handbook No 12. Roodeplaat, South Africa: ARC, Plant Protection Research Inst
Hermann M Quynh NK Peters D (1999) Reappraisal of edible canna as a high- value starch crop in Vietnam. CIP Program Report 1997–1998. 415–424
Holttum RE (1951) The Maranthaceae of Malaya. Gard Bull Singapore 13:254–296
Honing JA (1939) Canna Crosses IV. Genetica 15:23–47
Hosoki T, Sasaki H (1991) In vitro propagation of Canna edulis Ker. By longitudinal shoot-split method. Plant Tissue Cult Lett 8:175–178
Hutchinson J (1959) The families of flowering plants. Monocotyledons 2:588
Ibrahim R, Ahmed Z, Salleh S, Hassan AA, Ariffin S (2018) Mutation breeding in ornamentals. In: Huylenbroeck JV (ed) Handbook of plant breeding: ornamental crops. Springer, Singapore, pp 175–212
Indrayan AK, Bhojak NK, Kumar N, Shatru A, Gaura A (2011) Chemical composition and antimicrobial activity of the essential oil from the rhizome of Canna indica Linn. Indian J Chem 50B:1136–1139
Inatsu O (1985) Edible canna starch I. J Japanese Soc Starch Sci 30:38–47
Ishiki K, Arellano J, Salgado V (1997) Karyotype of achira (Canna edulis) cultivated in South America. Japanese J Trop Agri 41(Extra 2):3–4
Khoshoo TN (1979) Cytogenetics in relation to plant evolution and improvement. In: Khoshoo TN and Nair PKK (eds) Progress in Plant Research, NBRI Silver Jubilee Publication, Today and Tomorrow Printers and Publishers, New Delhi 2: p1–74
Khoshoo TN, Mukherjee I (1970) Genetic-evolutionary studies on cultivated cannas- VI. Origin and evolution of ornamental taxa. TAG 40:204–217
Knowlton FH (1923) Fossil plants from the Tertiary lake beds of South central Colorado. United States Geological Survey Professional Papers 131: 183–197
Koyama T (1984) Plant Resources Studies. Kodanshya, Tokyo. 152–155
Kranzlin FR (1912) Cannaceae. In: Engler A (ed) Das Pflanzenreich 47: 1–77
Kress JW (1990) The phylogeny and classification of the Zingiberales. Ann Mo Bot Gard 77:698–721
Kromer K (1979) Biological activity of endogenous and influence of exogenous growth regulators on Canna indica regeneration in vitro. Acta Hortic 91:295–300
Kromer K, Kukulczanka K (1985) In vitro cultures of meristem tips of Canna indica (L.). Acta Hortic 167:279–285
Kubitzki K (1998) Flowering Plants, Monocotyledons: Alismatanae and Commelinanae (except Gramineae). Berlin, Springer pp. 165–171
Launert E (1989) The Hamlyn guide to edible and medicinal plants of Britain and northern Europe. The Hamlyn Publishing Group Ltd, London
Lerman JC, Cigliano EM (1971) New carbon-14 evidence for six hundred years old Canna compacta seed. Nature 252:568–570
Lv Q, Chen C, Xu Y, Hu S, Le W, Sun K, Xuan C, Li X (2017) Optimization of agrobacterium tumefaciens-mediated transformation systems in tea plant (Camellia sinensis). Horticultural Plant J 17:30143–30147
Maas-van de Kamer H, Maas PJM (2008) The Cannaceae of the world. Blumea 53:247–318. [available online at: http://repository.naturalis.nl/document/564893]
Mishra T, Das AP, Sen A (2012) Phytochemical screening and in vitro antioxidant profiling of solvent fractions of Canna edulis Ker Gawler. Free Radical Antioxid 2:13–20
Mishra T, Goyal AK, Bhattacharya M, Kar P, Sen A (2015a) Polyethylene glycol mediated protoplast fusion of medicinally important Canna. Res Plant Biol 5(1):20–24
Mishra T, Goyal AK, Sen A (2015b) Somatic embryogenesis and genetic Fidelity study of the micropropagated medicinal species, Canna indica. Horticulturae 1(1):3–13
Mishra S, Yadav A, Singh SK (2013) A review on Canna indica Linn: Pharmacognostic and pharmacological profile. J Harmonized Res Pharmacy 2(2):131–144
Mukherjee I, Khoshoo TN (1969) Genetic evolutionary studies in cultivated cannas I. Variation Phenotype 36B(4):254–270

Mukherjee I, Khoshoo TN (1970) Genetic-evolutionary studies on cultivated cannas-IV: parallelism between natural and induced somatic mutations. Radiation Bot 10:351–364

Nakornthap A (1965) Radiation-induced somatic mutations in the ornamental Canna. In: Rome meeting on the uses of induced mutations in plant. Breeding Pergamon Press, New York, pp 707–721

Neralla S, Weaver RW, Varvel TW, Lesikar BJ (1999) Phytoremediation and on-site treatment of septic effluents in sub-surface flow constructed wetlands. Environ Technol 20:1139–1146

Patra B, Acharya L, Mukherjee AK, Panda MK, Panda PC (2008) Molecular characterization of ten cultivars of canna lilies (Canna Linn.) using PCR based molecular markers (RAPDs and ISSRs). Int J Integrative Biol 2(2):129–135

Percy Lancaster S (1967) Canna. Bull. Nat. Bot. Gdns. Lucknow No. 3

PIER (2008) Pacific Islands Ecosystems at Risk. USA: Institute of Pacific Islands Forestry. http://www.hear.org/pier/index.html

Pison (1658) Histoire naturelle du Brésil

Piyachomkwan K, Chotineeranat S, Kijkhunasatian C (2002) Edible canna (Canna edulis) as a complementary starch source to cassava for the starch industry. Ind Crop Prod 16(1):11–21

Prince LM Kress WJ (2001) Species boundaries in Canna (Cannaceae), evidence from nuclear ITS DNA sequence data. In: Proceedings of the conference "Botany 2001, Plants and People", August 12–16, 2001, Albuquerque, New Mexico

Rogers GK (1984) The Zingiberales (Cannaceae, Maranthaceae and Zingiberaceae) in the south-eastern United States. J Arnold Arbor 65:50–55

Sakai T, Imai K (2007) The influences of growth regulators and culture medium composition on shoot-tip cultures of edible Canna. Environ Control Biol 45:155–163

Sari N Purnomo Daryono BA Suryadiantina Setyowati M (2016) Variation and Intraspecies classification of Edible Canna (Canna indica L.) based on morphological characters. AIP Conference Proceedings 1744. https://doi.org/10.1063/1.4953515

Segeren W, Mass PJM (1971) The genus Canna in northern South America. Acta Botanica Neerlandica 20:663–680

Singh R, Dubey AK, Sanyal I (2019) Optimization of adventitious shoot regeneration and agrobacterium-mediated transformation in Canna X generalis (Canna lily). Horticultural Plant J 5(1):39–46

Singh SK Prasad R Roy RK Kumar S Goel AK (2014). Comparative analysis of growth and flowering characteristics of Canna cultivars in agro-climatic condition of Lucknow, Uttar Pradesh, India. 14(2): 935–938

Snafi AEA (2015) Bioactive components and pharmacological effects of Canna indica-an overview. Int J Pharm Toxicol 5(2):71–75

Tanaka N (1998) Economic botanical notes on edible canna (Cannaceae) in South Vietnam. J Japanese Bot 73:319–324

Tanaka N (2001) Taxonomic revision of the family Cannaceae in the New World and Asia. Makinoa ser 2(1):34–43

Tanaka N (2004) The utilization of edible canna plants in southeast Asia and southern China. Econ Bot 58(1):112–114

Tanaka N (2008) A new species of the genus Canna (Cannaceae) from eastern Honduras. J Japanese Bot 88:7–10

Tanaka N, Uchiyama EH, Matoba EH, Koyama ET (2009) Karyological analysis of the genus Canna (Cannaceae). Plant Syst Evol 280:45–51

Tjia B Black RJ (1991) Cannas for Florida landscape. Univ. of Fla. Coop. Ext. Serv., Circ. Univ. of Florida, Gainesville, FL. 424

Thitipraphunkul K, Uttapap D, Piyachomkwan K (2003) A comparative study of edible canna (Canna edulis) starch from different cultivars. Part II. Molecular structure of amylose and amylopectin. Carbohydr Polym 54(4):489–498

Tonwitowat R (1994) Yield and some agronomic characteristics of edible canna (Canna edulis Kerr.) as affected by varieties, planting dates, fertilizer rates and plant densities. [Thesis]. Kasetsart University. Bangkok

Vankar PS, Srivastava J (2010) Evaluation of anthocyanin content in red and blue flowers. Int J Food Engg 6(4):1–11
Vankar PS, Srivastava J (2018) A review- Canna the wonder plant. J Textile Engg Fashion Tech 4(2):159–163
Venkatasubban KR (1946) A preliminary survey of chromosome numbers in Scitamineae of Bentham and hooker. Proc Indian Acad Sci (Sect B) 23:281–300
Vogel S (1969) Chiropterophilie in der neotropischen Flora, B. Neue mitteilungen 3. Flora 158: 289–323
Winkler H (1930) Cannaceae. In: Engler HGA Prantl KAE (eds) Die natürlichen Pflanzenfamilien, 15a: 640–654
Woradulayapinij W, Soonthornchareonnon N, Wiwat C (2005) In vitro HIV type 1 reverse transcriptase inhibitory activities of Thai medicinal plants and Canna indica L. rhizomes. J Ethnopharmacol 101:84–89
Zhang J, Wang Z-W (2011) Arabinoxylan from Canna edulis Ker by-product and its enzymatic activities. Carbohydr Polym 84:656–661
Zhang W, Tian Z, Pan X (2013) Oxidative stress and non-enzymatic antioxidants in leaves of three edible canna cultivars under drought stress. Hortic Environ Biotechnol 54:1–8

Landscape Gardening Research 14

Diversity and Conservation

S. S. Sindhu and M. R. Dhiman

Contents

S. S. Sindhu (✉)
Division of Floriculture and Landscaping (School of Horticulture), ICAR- Indian Agricultural Research Institute, New Delhi, India

M. R. Dhiman
ICAR-INARI, Regional Station, Katrain, Kullu-Valley, Himachal Pradesh, India

S. K. Datta, Y. C. Gupta (eds.), *Floriculture and Ornamental Plants*, Handbooks of Crop Diversity: Conservation and Use of Plant Genetic Resources,
https://doi.org/10.1007/978-981-15-3518-5_14

Abstract

Research in landscape gardening has never been a priority although it has been a heritage and cultural needs. Floriculture is an important multibillion-dollar industry dealing with thousands of species and varieties of ornamental plants in both cultivation and the wild. In addition, it also provides an important livelihood option for several poor, especially in per-urban areas. Current global conservation efforts are focused mainly of food, forage, and industrial crops. The world's germplasm collection exceeds six million accessions, stored in about 1400 genebanks worldwide. However, barely less than 1% is ornamental plant species. It is timely in the twenty-first century the progressive world, besides conserving food crops, should also recognize the wealth of ornamental plant diversity to bring happiness and health to humankind. To date only one specialized ornamental plant genebank, the Ornamental Plant Germplasm Center, exists in Ohio, USA, although there could be small collections with different public and private sector organizations. More such centers are required to collect and conserve the disappearing genetic materials including new species not only for their aesthetic value but also as food, medicine, and beverage and for improvement. Presently ornamental germplasm are conserved through in situ and ex situ methods in forests, natural parks, botanical gardens, and arboreta. Various techniques ranging from the simple drying of seeds to cryopreservation of embryos are in practice. In this chapter, the diversity of ornamental plant species and their importance and potentials in landscape gardening are described. The methods of conservation using various approaches are described. It is hoped that the collecting, multiplying, and distributing of these invaluable ornamental germplasm will benefit one and all.

Keywords

Landscape gardening · Ornamental plants · Conservation · Utilization

14.1 Introduction

With the advancement of science and economic development, man, society, the environment, and the world are changing. Whether these changes are beneficial or catastrophic is the basic question we face today. The landscaping profession conveys and directs change so that human life will be more elegant, more satisfying, and more productive. There is no broader field in the world than landscape gardening (Bhattacharjee 2006). Landscaping is the integration of an art and science of using plants for beautification of a place for the purpose of aesthetic and environmental improvement. The use of plants and flowers in our daily life has increased manifolds and is regarded as absolute necessity. Landscape gardening is always a source of pleasure and enjoyment. Its functional use, aesthetic value, and ability to create an ambient environment are well accepted by one and all. Where there is human habitation, there is work for the landscape gardener. Every year, the amount of business in landscape gardening is increasing since there are increasing demands for more attractive parks, more beautiful schools and institutional grounds, pleasant residential grounds, hospitals, and environment-friendly cities and towns. Now, various types of gardens, such as home gardens, institutional gardens, industrial gardens, window gardens, roof gardens, rock gardens, water gardens, cemetery gardens, moon gardens, sunken gardens, marsh gardens, island gardens, boulevard gardens, and wall gardens, are established in our society. A landscape designer is looked upon as one of the most important men in the locality and admired by all on account of his capacity to execute important beautification programs. Large estates, parks, cementries and institutional grounds must have someone to look after their care and upkeep. The landscape gardener is a necessity because very few people are capable of selecting the right kind of plants for their grounds; they cannot imagine a landscape picture in terms of varieties and groups the way an expert can.

Today, most of us are familiar with the problems of our environment. We are destroying our natural gift by spoiling our forestland, ripping out hills and exposing them to erosion, and polluting our rivers, streams, and lakes with little understanding of the natural environment which supports human life. Unless we become conscious of developing and improving nature, we will soon live in barren communities. Most of us now realize that if we destroy the environment, we will destroy life. If we destroy the quality of our environment, we destroy the quality of life. The landscape architect enjoys a position of fundamental and special relevance regarding the relationship between people and the natural environment. Landscape architecture serves as a bridge between man-made and natural elements of the environment.

14.2 Perspective of Landscape Gardening

Changing scenario of gardens and landscape with changing culture is evident in historical perspective, and there is a great garden culture to share with the world. In 1200 to 1000 B.C., India had developed art of town planning and irrigation systems which can be seen in many legendary monuments. In the nineteenth century, urban planning became important, and this gave rise to the modern focus on landscaping, blend modern planning with the traditional knowledge which had been previously enjoyed. Now landscape architecture is a widely practiced with innovation in designs and plans for gardens and public spaces occurring around the world. The idea of garden is changing with focus on environment. Urbanization taking place calls for better landscape gardening with the choice of plants which can sequester more carbon and toxic gases. The challenge therefore is to have appropriate design and development of both exterior and interior garden which ensures the mitigation efforts to climate change besides aesthetic value for the people at large.

14.3 Planning of Landscape Garden

A great landscape is the result of planning and vision, and without proper planning, a new landscape can be more work than it has to be and can actually take away from the beauty of one's home and property. The first thing the landscape architect should do is to put on one sheet of paper a rough and ready plan of the garden to be made from scratch or altered. Before any plan can be studied and drawn, a map or survey of the ground must be secured. As a rule, however, all buildings, building entrances, windows, all walks, roads, paths, all physical features such as vegetations, topography, soil and climate, water sources, rocks, all terraces or other decided irregularities in the ground level, all unnatural features, or other features of importance should be located, named, and shown on the map. With the survey and contour maps completed, work on preliminary plan should start. It is the first plan made. In design it represents a thoroughly worked out scheme or solution of the problem in hand. The essentials of a preliminary plan are as follows: it should offer a practical workable scheme, it should be attractively drawn, and it should present enough information to explain the idea. A final plan represents a finished scheme, an accepted program of work. It may or may not be preceded by a preliminary plan. It differs from a preliminary plan in that it should be complete in all respects and presented in a form suitable for its use. In the final plan, one can show the adopted scheme, general plan, the planting arrangement, planting plan, and the dimension and construction details, working plan, grading plan, and drainage plans.

Before preparing the plans for landscape work, the first thing to do is to use our eyes and put our mind to work on the designs of gardens, which we can see around us. From these gardens, we may reach an idea of what sort of layout we would like to have ourselves. Whatever may be the ideas, important considerations for planning a new garden should be the personal taste, practical requirements, size and type of the ground, and financial resources. For such interesting ventures, many ideas will naturally crop up in the mind, but by no means would all these be used in any one garden. For the openness, the lawn and paved area is thought of, while for foundation planting, climbers and shrubs which will adopt themselves to walls may be planted against the house together with groups of shrubs. For a riot of colors within a short span of time, herbaceous borders are considered. Beds of roses, individual annuals, and bulbs are also very pleasing additions. Water gardens for aquatic plants; rock gardens for cultivation of xerophytic plants or alpine plants; arbors, screens, arches, and pergolas for growing many beautiful climbing plants; hedges for screening and decorative purposes; and position of birdbath, sundial, or a seat are some of the ideas, among many, which will no doubt appear in the mind of the designers. Gardens can be laid out in a severe formal manner or quite informally with no planned balance of line and form, or it may be a combination of these two extremes.

14.4 General Principles

A garden is primarily laid out for the enjoyment of human beings. In all gardens, it should be the endeavor of the designer to secure a reasonable amount of comfort and convenience for those for whom their pleasures are intended. Due respect for nature's teaching, dictates of artistic feeling, and ordinary good taste must be priority. The pleasure can be attributed to the following points.

All different parts of the garden or various features should exhibit equally in all respects and must be in agreement. Utility is another priority aspect where convenience, comfort, and neatness get enough consideration. The lines in the layout must be in order with all correctness and proper finishing. Whatever may be the style of the garden, uniform blend of all features and some sort of symmetry must be achieved in the finished garden. The garden must create a picturesque effect and bring in the balance of composition. A well-laid out garden excites and nourishes curiosity and gives a feeling of entanglement. The disposition of objects in the garden should, however, be without any confusion and must be simple in all respects. Abrupt breaks between the objects should be avoided in order to keep continuity and a link between the various parts. Each and every item of the garden must be set apart for a special purpose.

In landscape design, space is utilized for human use, convenience, and enjoyment. The basic materials used in the available space are arranged in such a way, so as to enhance the beauty as well as usefulness of each item. Land is the surface on which landscaping is done; this includes different types of soils, mineral materials, organic matter, water, air, rocks, vegetation, etc. Land projects and their designs are also dependent on the environment provided by the atmosphere and climate. Variation in grading in the land has a profound influence on the landscape scheme, so a contour map is made in order to study the plan intelligently and adopt a scheme that will fit with the ground.

14.5 Elements of Landscape Design

The components used in a landscape design have a number of characteristics which should be considered. These characteristics are sometimes referred to as design elements.

14.5.1 Line

A carefully planned group of lines in a landscape composition will direct the attention of the viewer to a focal point or a particular area of interest in the composition. Line has a crucial role in controlling movement, either visual or physical, in straight or curved directions. In a design, straight line indicates direct movement without hesitation. Interconnecting straight lines create points at the intersections for stopping, sitting, changes of views, and reflections back to the point of beginning. Sometimes, lines by their just a position make a sharp angle and suggest rigidity, while gentle curves suggest fluidity. Meandering curved lines invite slower movement and are useful in areas that should appear as natural as possible, such as paths through a wild garden.

14.5.2 Form

It is the degree of solidity of forms. In any design, form is very important in creating a particular design. In a tree, the trunk, branches, and leaves together create a form. Tall and slender trees have a vertical form. Low and spreading trees have a horizontal form. This form can be modified in proper landscape planning. A group of vertical plant forms, if grouped together in sufficient quantity so that the length of the group is greater than the height, makes the group appear to have a horizontal form.

14.5.3 Texture

It refers to the patterning of the components of the landscape. It may be fine or coarse, rough, or smooth.

14.5.4 Color

No garden can be really attractive that has color clashes. Human response to individual color varies. In general, reds, oranges, and yellows are considered warm colors and seen to advance toward the viewer. Greens and blues are cooler colors and tend to recede in a composition. Generally, contrasting colors are used.

14.5.5 Variety

It is a critical element in design. Too little variety leads to monotony, while too much brings confusion. While choosing plants to develop a landscape plan, it is never a good practice to choose one of everything. Using several of each kind will be much more apt to result in the unity of design. As a matter of fact, a variety of lines, forms, textures, and colors are needed to create an orderly, interesting landscape.

14.5.6 Repetition

It gives the element a variety of meanings and expression. Repetition reduces the confusion that may result from excessive variety. It also introduces a sense of order to the viewer of the landscape. Repetition can be achieved by planting individual plants in groups or masses of a single species or variety.

14.5.7 Balance

The visual equilibrium of different garden elements is known as balance in landscape design. Usually, it is possible to perceive a central axis in a landscape composition. It is on the basis of balance that landscapes are judged to be formal and informal and symmetrical and asymmetrical.

14.5.8 Emphasis

The eye is directed to one portion or object of a composition through the use of emphasis. There may, however, be primary or secondary points of emphasis.

14.5.9 Fragrance

One of the greatest delights of a garden is that the fragrance is recognized only in a chance manner. Gardens to be enjoyed primarily after dark could well be planned with this goal in mind. The scent of roses, jasmines, gardenia, *Magnolia grandiflora*, *Cestrum nocturnum*, lilies, and hyacinth are easiest to recognize and well-known to all. It is not only flowers; various kinds of grasses, aromatic herbs, shrubs, and trees induce an exhilarating aroma in the garden, which should not be ignored if there is scope to include them in the garden design.

14.5.10 Character

Many plants impart character to a garden setting. It is the pattern of the plant that is important. Pattern is a model that can be found in the natural arrangement of plant parts.

14.5.11 Harmony

The overall plan of the property should be harmonious and functional. One must seek harmony in the design of plant material too. While selecting plants, their ultimate role in the garden, eventual size, and the eye level should be thought of.

14.6 Garden Features

Thoughtful planning of garden features makes an invaluable contribution to garden design. It is possible to design a garden with a large number of man-made items. Indiscriminate and haphazard use of these items, however, will result in clutter. It should be kept in mind that whatever features may be chosen, they must add charm, create an atmosphere of naturalistic design, and add character to the quality of design. Some important garden features are water, rocks, roads, walks, pavements and steps, hedges and edges, arches, pergolas, screens and bridges, lawns, flower beds, borders, carpet bedding, and shrubbery. Many other features like stone lanterns, stone basins, statues, towers and wells, sundials, birdbaths, straddle stones, floral clocks, plant containers and raised beds, vertical gardens, and plant stands are also introduced in the garden as per the need of the landscape design. In some gardens, greenhouses, conservatories, and glasshouses are also constructed for comfortable accommodation of some special groups of plants. In modern gardens, artificial light plays an important role. Besides pleasing visual effects, sounds made by wind, water, insects, and birds in the garden create inquisitiveness and fill the mind with joy.

14.7 Center of Diversity of Ornamental Plants

Ornamental plant which includes wide range of plants from annuals to perennials accommodating seasonal annual flowers, climbers, shrubs, trees, foliage plants, ferns, grasses, cacti and succulents, palms, and other ornamental geophytes has become important and offer commercialization, as it provides high income per unit area and growing due to urbanization. In landscape gardening, plants are mainly used for social, spiritual, aesthetic, and decorative purposes. Mankind started appreciating the wide diversity available in nature and identified different genera/species for different uses (Singh and Malhotra 2017). Some of them are used for their attractive and colorful inflorescence and fragrance, while others are used for their foliage. However, with declining natural resources and a threat of climate change, there is a rising concern for meeting the growing requirement. These resources consist of the diversity of genetic material in the form of native and exotic plants and modern cultivars and other wild relatives occurring in nature. Over the years, hundreds of different plant species have been domesticated, and within each species, human and natural selection have combined to produce thousands of different varieties; in developed parts of the world, the "primitive cultivars" or "landraces" have given way to more productive, uniform, modern cultivars. Approximately 406,700 species of plants are available on our Earth, spread over the tropical, subtropical, and temperate zones of the world. There are 12 world mega-biodiversity centers, 17 mega-diverse nations, 8 centers of origin of crop plants, and 34 hot spots of biodiversity in the world. China, India with related center in Indo-Malaya, central Asia, the Near-East, the Mediterranean, Abyssinia (Ethiopia), southern Mexico, and Central America and South America (Peru, Ecuador, and Bolivia), with two lesser centers (the island of Chiloe off the coast of southern Chile and an eastern secondary center in Brazil and Paraguay), are center of diversities.

14.8 Genetic Resources of Ornamental Plants for Landscape Gardening

One of the secrets of successful landscape gardening is to choose the right plant for a particular place. Careful selection of the wealth of trees, shrubs, climbers, annuals, bulbs, foliage plants, grass cover, and aquatic ornamentals transforms a place into a haven of beauty. However, from the ornamental point of view, for the development of various types of gardens, parks, and roadside decoration with different plants, species having important flowers or good foliage or aesthetic value, etc. are collected, domesticated, and propagated by different botanical gardens, national institutes, central institutes, or State Agricultural Universities (SAUs) and some reputed nongovernmental organizations and nurseries. Important ornamental trees species planted in different parts of the world in tropical, subtropical, and temperate zones are listed in Table 1.

14.8.1 Diversity of Ornamental Trees in Tropical and Subtropical Garden for Beautification

Acacia auriculiformis, Acacia decurrens, Adansonia digitata, Albizia lebbeck, Alstonia scholaris, Alstonia macrophylla, Anthocephalus cadamba, Araucaria cookii, Araucaria cunninghamii, Araucaria heterophylla, Artocarpus altalis, Azadirachta indica, Barringtonia acutangula, Barringtonia racemosa, Baikiaea insignis, Bauhinia blakeana, Bauhinia purpurea, Bauhinia variegata, Bombax ceiba, Bombax ellipticum, Brassaia actinophylla, Brownea arisa, Brownea coccinea, Brownea grandiceps, Brownea hybrida, Brownea macrophylla, Butea monosperma, Cassia fistula, Cassia spectabilis, Casuarina equisetifolia, Ceiba pentandra, Chorisia speciosa, Couroupita guianensis, Callistemon citrinus, Callistemon polandii, Calophyllum inophyllum, Caesalpinia cacalaco, Cananga odorata, Careya arborea, Cassia grandis, Cassia lancasteri, Cassia marginata, Cassia moschata, Cassia nodosa, Cassia renigera, Cassia javanica, Castanospermum australe, Cochlospermum religiosum, Colvillea racemosa, Cordia sebestena, Crescentia cujete, Delonix regia, Diospyros peregrina, Duabanga sonneratioides, Erythrina indica, Eugenia operculata, Eugenia jambolana, Ficus benghalensis, Ficus benjamina var. *variegata, Ficus benghalensis* var. *krishnae, Ficus religiosa, Ficus rumphii, Ficus elastica, Filicium decipiens, Gardenia latifolia, Gustavia augusta, Grevillea robusta, Grewia columnaris, Heterophragma adenophyllum, Ixora parviflora, Jacaranda mimosifolia, Kigelia pinnata, Lagerstroemia flos-reginae, Lagerstroemia speciosa, Lagerstroemia thorelli, Mesua ferrea, Michelia champaca, Mimusops elengi, Mimusops elengi* var. *variegata, Magnolia grandiflora, Magnolia pterocarpa, Mimusops elengi* var.

Table 1 List of ornamental trees species biodiversity in the world

S.No.	Botanical name	Family	Origin
1.	*Acacia auriculiformis*	Mimosaceae	Australia
2.	*Acacia decurrens*	Mimosaceae	Australia
3.	*Acacia suma*	Mimosaceae	India
4.	*Acer caesium*	Aceraceae	Himalayan region
5.	*Adansonia digitata*	Bombacaceae	Africa and Australia
6.	*Ailanthus excelsa*	Simaroubaceae	East Asia and Australia
7.	*Albizia chinensis*	Mimosaceae	China
8.	*Albizia lebbeck*	Mimosaceae	Tropical Asia and Africa
9.	*Albizia procera*	Mimosaceae	Himalayan region
10.	*Alstonia macrophylla*	Apocynaceae	Malaysia
11.	*Alstonia scholaris*	Apocynaceae	India Sri Lanka Malaysia and China
12.	*Amherstia nobilis*	Caesalpiniaceae	Burma
13.	*Anogeissus acuminata*	Combretaceae	India and Burma
14.	*Anthocephalus indicus*	Rubiaceae	India, China, and Malaysia
15.	*Araucaria* spp.	Araucariaceae	South America and Australia
16.	*Artocarpus altilis*	Moraceae	East Asia
17.	*Azadirachta indica*	Meliaceae	India and Malaysia
18.	*Baikiaea insignis*	Caesalpiniaceae	West Africa
19.	*Barringtonia asiatica*	Lecythidaceae	Andamans of India, Australia, and Malaysia
20.	*Barringtonia acutangula*	Lecythidaceae	Western Ghat and Andamans of India, Australia, Malaysia, Burma
21.	*Barringtonia racemosa*	Lecythidaceae	India, Malaysia, Sri Lanka
22.	*Bauhinia acuminata*	Caesalpiniaceae	India
23.	*Bauhinia purpurea*	Caesalpiniaceae	Himalayan region
24.	*Bauhinia tomentosa*	Caesalpiniaceae	India
25.	*Berrya ammonilla* Tiliaceae India	Tiliaceae	India Malaysia and the Philippines
26.	*Bischofia javanica*	Euphorbiaceae	India, Burma, Bangladesh, and Malaysia
27.	*Bombax ceiba*	Bombacaceae	India and Malaysia
28.	*Brassaia actinophylla*	Araliaceae	Subtropical Australia
29.	*Brownea macrophylla*	Caesalpiniaceae	Tropical America
30.	*Butea monosperma*	Fabaceae	India
31.	*Caesalpinia coriaria*	Caesalpiniaceae	Central America
32.	*Callistemon lanceolatus*	Myrtaceae	Australia
33.	*Callistemon polandii*	Myrtaceae	Australia

(continued)

Table 1 (continued)

S.No.	Botanical name	Family	Origin
34.	*Careya arborea*	Lecythidaceae	India
35.	*Cassia excelsa*	Caesalpiniaceae	Brazil
36.	*Cassia grandis*	Caesalpiniaceae	South America and Caribbean Islands
37.	*Cassia javanica*	Caesalpiniaceae	Malaysia and Indonesia
38.	*Cassia lancasteri*	Caesalpiniaceae	India
39.	*Cassia marginata*	Caesalpiniaceae	Western Ghat of India
40.	*Cassia moschata*	Caesalpiniaceae	Tropical America
41.	*Cassia nodosa*	Caesalpiniaceae	India, Malaysia, Indonesia
42.	*Cassia spectabilis*	Caesalpiniaceae	India, Malaysia, and Burma
43.	*Casuarina equisetifolia*	Caesalpiniaceae	Malaysia, Australia, and Pacific Islands
44.	*Cedrela toona*	Meliaceae	Himalayan region, Burma, and Australia
45.	*Chorisia speciosa*	Bombacaceae	South America
46.	*Clitoria arborea*	Fabaceae	Tropical America
47.	*Cupressus sempervirens*	Cupressaceae	Italy
48.	*Cycas circinalis*	Cycadaceae	Tropical Africa, South India, and Sri Lanka
49.	*Delonix regia*	Caesalpiniaceae	Madagascar
50.	*Dillenia indica*	Dilleniaceae	India
51.	*Diospyros peregrina*	Ebenaceae	India, Malaysia, and Australia
52.	*Erythrina variegata*	Fabaceae	Tropical and regions of the world
53.	*Ficus benghalensis*	Moraceae	Himalayan Region
54.	*Ficus benjamina*	Moraceae	Western Ghat and Andamans of India
55.	*Ficus elastica*	Moraceae	Himalayan Region
56.	*Ficus hispida*	Moraceae	India, Sri Lanka, Burma, Australia, China
57.	*Ficus religiosa*	Moraceae	India and Burma
58.	*Gardenia latifolia*	Rubiaceae	Himalayan Region of India
59.	*Grevillea robusta*	Proteaceae	Australia
60.	*Ipomoea arborescens*	Convolvulaceae	Mexico
61.	*Ixora*	Rubiaceae	Western Ghat of India
62.	*Jacaranda mimosifolia*	Bignoniaceae	Tropical America
63.	*Juniperus chinensis*	Cupressaceae	China
64.	*Kigelia pinnata*	Bignoniaceae	Mozambique and Tropical Africa
65.	*Lagerstroemia parviflora*	Lythraceae	Himalayan Region and Western Ghat
66.	*Lagerstroemia thorelli*	Lythraceae	Western Ghat
67.	*Magnolia grandiflora*	Magnoliaceae	North America
68.	*Michelia champaca*	Magnoliaceae	Himalayan region of India and Malaysia
69.	*Millingtonia hortensis*	Bignoniaceae	Burma

(continued)

Table 1 (continued)

S.No.	Botanical name	Family	Origin
70.	*Murraya koenigii*	Rutaceae	Tropical Asia
71.	*Nyctanthes arbor-tristis*	Oleaceae	India
72.	*Parkia roxburghii*	Mimosaceae	Himalayan region, Malaysia Burma
73.	*Parkinsonia aculeata*	Caesalpiniaceae	Mexico and Central America
74.	*Pinus longifolia*	Pinaceae	Himalayan region
75.	*Plumeria alba*	Apocynaceae	India
76.	*Plumeria rubra*	Apocynaceae	Mexico
77.	*Podocarpus elongata*	Podocarpaceae	Western Africa
78.	*Podocarpus macrophylla*	Podocarpaceae	China and Japan
79.	*Podocarpus neriifolius*	Podocarpaceae	China to New Guinea
80.	*Polyalthia longifolia*	Annonaceae	Sri Lanka and Western Ghat of India
81.	*Polyalthia suberosa*	Annonaceae	Sri Lanka, Western Ghat of India and Burma
82.	*Pterospermum acerifolium*	Sterculiaceae	Tropical Asia
83.	*Putranjiva roxburghii*	Euphorbiaceae	India, Sri Lanka
84.	*Rhododendron arboretum*	Ericaceae	Himalayan region
85.	*Salix babylonica*	Salicaceae	China
86.	*Saraca indica*	Caesalpiniaceae	India, Burma, and Malaysia
87.	*Schima wallichii*	Theaceae	India and Indonesia
88.	*Spathodea campanulata*	Bignoniaceae	Tropical Africa
89.	*Sterculia foetida*	Sterculiaceae	Tropical Asia
90.	*Tabebuia argentea*	Bignoniaceae	South America
91.	*Tabebuia avellanedae*	Bignoniaceae	Tropical America
92.	*Tamarix chinensis*	Tamaricaceae	China
93	*Thespesia populnea*	Malvaceae	India and Burma
94.	*Thevetia peruviana*	Apocynaceae	West Indies
95.	*Thuja orientalis*	Cupressaceae	China
96.	*Wrightia tomentosa,*	Apocynaceae	India and Malaysia

variegata, *Millettia peguensis*, *Monodora grandiflora*, *Napoleona imperialis*, *Nauclea orientalis*, *Nyctanthes arbor-tristis*, *Parkia roxburghii*, *Plumeria lutea*, *Plumeria obtuse*, *Podocarpus macrophylla*, *Polyalthia pendula*, *Polyalthia longifolia*, *Pterospermum acerifolium*, *Putranjiva roxburghii*, *Pterygota alata* var. *diversifolia*, *Putranjiva roxburghii*, *Pachira cyathophora*, *Parkinsonia aculeate*, *Plumeria rubra*, *Ravenala madagascariensis*, *Salix babylonica*, *Samanea saman*,

Saraca thaipingensis, *Santalum album*, *Saraca asoca*, *Spathodea campanulata*, *Sterculia alata*, *Taxodium distichum*, *Tabebuia avellanedae*, *Tabebuia chrysantha*, *Tabebuia palmeri*, *Tabebuia rosea*, *Tecoma gaudichaudi*, *Thevetia peruviana*, *Terminalia catappa*, *Thespesia populnea*, *Thuja orientalis*, *Tipuana tipu*, *Wrightia tomentosa*.

14.8.2 Diversity of Ornamental Trees in the Temperate Garden for Beautification

Acer campbellii, *Araucaria cunninghamii*, *Araucaria heterophylla*, *Bauhinia* spp., *Citharexylum quadrangulare*, *Pinus longifolia*, *Pyrus persica*, *Rhododendron* spp., *Tamarix chinensis*, *Tamarix ramosissima*, *Taxus baccata*, *Ulmus alata*.

14.8.3 Biodiversity of Ornamental Shrubs

For garden beautification, ornamental shrubs have special characteristics such as attractive shape, colorful foliage, or various cheerful colors or fragrance. A huge number of shrubs have been diversified throughout the world for decoration of different gardens on the basis of such special characteristics as size and shape of plants, growth habit, growing locations, and color of foliage and flowers with or without fragrance. The major work on biodiversity of ornamental shrubs has been done by amateur garden lovers and nursery plant growers. Many clusters of nursery-based enterprises have been developed for ornamental plant production around urban areas. These entrepreneurs brought different varieties of rare and uncommon shrub species from abroad, collected for years as mother plant stock or genetic resources, followed by propagation for sale at high prices. Some important ornamental shrubs diversified in the tropical, subtropical, and temperate regions of the world are described here (Table 2).

14.8.3.1 Diversity of Ornamental Flowering Shrubs in Tropical and Subtropical Gardens for Beautification of Shrubbery Borders

Abelia chinensis, *Abutilon darwinii*, *Abutilon hybridum*, *Angelonia angustifolia*, *Angelonia grandiflora*, *Aphelandra sinclairiana*, *Asclepias curassavica*, *Asystasia gangetica*, *Barleria prionites*, *Barleria strigosa*, *Bauhinia acuminata*, *Bauhinia galpinii*, *Bauhinia tomentosa*, *Beloperone amherstiae*, *Beloperone guttata*, *Brunfelsia undulata*, *Buddleia asiatica*, *Buddleia davidii*, *Caesalpinia gilliesii*, *Caesalpinia pulcherrima*, *Calliandra brevipes*, *Calliandra haematocephala*, *Calliandra houstonii*, *Calliandra hybrida*, *Callistemon lanceolatus*, *Caryopteris mascanthus*, *Cassia alata*, *Cassia glauca*, *Cassia laevigata*, *Catesbaea spinosa*, *Cestrum aurantiacum*, *Cestrum diurnum*, *Cestrum elegans*, *Cestrum parqui*, *Cestrum nocturnum*, *Clerodendron fragrans*, *Clerodendron paniculatum*, *Crossandra undulaefolia*, *Cuphea hyssopifolia*, *Cuphea miniata*, *Datura fastuosa*,

Table 2 List of some ornamental shrub species biodiversity in the world

S.No.	Botanical name	Family	Origin
1.	*Abelia chinensis*	Caprifoliaceae	China
2.	*Abelia floribunda*	Caprifoliaceae	Mexico
3.	*Abutilon darwinii*	Malvaceae	Tropical and subtropical regions in the world
4.	*Acalypha* spp.	Euphorbiaceae	India, Africa, South America
5.	*Allamanda cathartica*	Apocynaceae	Central America
6.	*Allamanda neriifolia*	Apocynaceae	Brazil
7.	*Alternanthera* spp.	Amaranthaceae	Africa and Tropical America
8.	*Angelonia grandiflora*	Acanthaceae	Tropical America
9.	*Aphelandra fascinator*	Acanthaceae	Tropical America
10.	*Asclepias curassavica*	Asclepiadaceae	Tropical America
11.	*Asystasia gangetica*	Acanthaceae a	India
12.	*Azalea* spp.	Ericaceae	Himalayan region and Western Ghats of India
13.	*Barleria cristata*	Acanthaceae	India
14.	*Bauhinia acuminata*	Caesalpiniaceae	Central India
15.	*Bauhinia malabarica*	Caesalpiniaceae	Western Ghats of India
16.	*Beloperone guttata*	Acanthaceae	Mexico
17.	*Buddleia asiatica*	Loganiaceae	India and China
18.	*Buddleia lindleyana*	Loganiaceae	China
19.	*Caesalpinia pulcherrima*	Caesalpiniaceae	Barbados
20.	*Calliandra* spp.	Mimosaceae	Tropical America, subtropical America, and India
21.	*Callicarpa cana*	Verbenaceae	India
22.	*Camellia reticulata*	Theaceae	Himalayan region
23.	*Camellia japonica*	Theaceae	Japan and China
24.	*Campanula pyramidalis*	Campanulaceae	South Europe
25.	*Caryopteris mastacanthus*	Verbenaceae	Eastern Asia
26.	*Catharanthus roseus*	Apocynaceae	Madagascar
27.	*Cestrum diurnum*	Solanaceae	West Indies
28.	*Clerodendrum inerme*	Verbenaceae	India
29.	*Clerodendrum paniculatum*	Verbenaceae	Himalayan region
30.	*Coleus blumei*	Labiatae	Indonesia
31.	*Crossandra* spp.	Acanthaceae	India, Africa, and Sri Lanka
32.	*Cuphea hyssopifolia* Mexico	Lythraceae	Mexico
33.	*Datura fastuosa*	Solanaceae	India
34.	*Dombeya wallichii*	Sterculiaceae	Tropical Africa
35.	*Duranta repens*	Verbenaceae	West Indies, México, and Brazil
36.	*Eranthemum bicolor*	Acanthaceae	Africa

(continued)

Table 2 (continued)

S.No.	Botanical name	Family	Origin
37.	*Erythrina crista-galli*	Fabaceae	South America
38.	*Euphorbia pulcherrima*	Euphorbiaceae	Mexico
39.	*Excoecaria bicolor*	Euphorbiaceae	Tropical Asia and Africa
40.	*Gardenia jasminoides*	Rubiaceae	China
41.	*Hamelia patens*	Rubiaceae	West Indies and South Florida
42.	*Hibiscus mutabilis*	Malvaceae	China
43.	*Hibiscus rosa-sinensis*	Malvaceae	China
44.	*Hydrangea macrophylla*	Saxifragaceae	Japan
45.	*Iresine herbstii*	Amaranthaceae	Tropical and subtropical America
46.	*Ixora aliporensis*	Rubiaceae	India
47.	*Jacobinia carnea*	Acanthaceae	Brazil
48.	*Jasminum flexile*	Oleaceae	Himalayan Region and Western Ghats of India
49.	*Jasminum humile*	Oleaceae	Himalayan Region
50.	*Jasminum laurifolium*	Oleaceae	Himalayan Region
51.	*Jasminum primulinum*	Oleaceae	Western Ghats of India
52.	*Jasminum sambac*	Oleaceae	Tropical Asia
53.	*Jasminum syringifolia*	Oleaceae	Himalayan Region
54.	*Justicia furcata*	Acanthaceae	Mexico
55.	*Lagerstroemia indica*	Lythraceae	China
56.	*Lagerstroemia lancasteri*	Lythraceae	India
57.	*Lantana sellowiana*	Verbenaceae	South America
58.	*Lawsonia inermis*	Lythraceae	Arabia and Persia
59.	*Ligustrum robustum*	Oleaceae	Himalayan Region
60.	*Magnolia mutabilis*	Magnoliaceae	Tropical gardens
61.	*Malpighia glabra*	Malpighiaceae	Tropical and subtropical America
62.	*Malvaviscus arboreus*	Malvaceae	Tropical America
63.	*Murraya paniculata*	Rutaceae	India
64.	*Mussaenda luteola*	Rubiaceae	Tropical Africa
65.	*Nandina domestica*	Berberidaceae	China and Japan
66.	*Nerium indicum*	Apocynaceae	Mediterranean region
67.	*Ochna jabotapita*	Ochnaceae	Himalayan region
68.	*Pentas lanceolata*	Rubiaceae	Tropical Africa and Arab
69.	*Plumbago indica*	Plumbaginaceae	India
70.	*Reinwardtia trigyna*	Linaceae	India
71.	*Russelia coccinea*	Scrophulariaceae	Mexico
72.	*Sambucus hookeri*	Sambucaceae	Himalayan Region
73.	*Solanum macranthum*	Solanaceae	Brazil
74.	*Tecoma stans*	Bignoniaceae	West Indies and South America
75.	*Thunbergia* spp.	Acanthaceae	Tropical Africa
76.	*Woodfordia floribunda*	Lythraceae	India

Datura suaveolens, *Daedalacanthus macrophyllus*, *Dombeya mastersii*, *Dombeya natalensis*, *Dombeya wallichii*, *Erythrina* spp., *Gardenia jasminoides*, *Gardenia longistyla*, *Gardenia lucida*, *Galphimia gracilis*, *Gustavia insignis*, *Hamiltonia suaveolens*, *Hibiscus mutabilis*, *Hibiscus rosa-sinensis*, *Hibiscus syriacus*, *Hydrangea macrophylla*, *Ichroma tululosum*, *Ixora* spp., *Jacobinia carnea*, *Jasminum humile*, *Jasminum pubescens*, *Jatropha panduraefolia*, *Justicia aurea*, *Justicia ovate*, *Lagerstroemia lancasteri*, *Lantana camara*, *Lantana sellowiana*, *Ligustrum robustum*, *Magnolia mutabilis*, *Magnolia pumila*, *Malvaviscus arboreus*, *Melastoma malabathricum*, *Memecylon* spp., *Mussaenda erythrophylla*, *Nerium odoratum*, *Nerium oleander*, *Oncoba spinosa*, *Pachystachys lutea*, *Pentas karmesiana*, *Petrea arborea*, *Plumbago indica*, *Pootia grandiflora*, *Portlandia grandiflora*, *Quassia amara*, *Randia macrantha*, *Randia maculata*, *Reinwardtia trigyna*, *Rondeletia odorata*, *Ruellia* spp., *Ruttya fruticosa*, *Sambucus canadensis*, *Sarcocephalus cordatus*, *Stachytarpheta indica*, *Stachytarpheta mutabilis*, *Stemmadenia bella*, *Strobilanthus* spp., *Tecoma gaudichaudi*, *Tecoma stans*, *Tecomaria capensis*, *Thunbergia hybrida*, *Vitex agnuscastus*, *Woodfordia floribunda*.

14.8.3.2 Diversity of Ornamental Flowering Shrubs in Tropical and Subtropical Gardens for Beautification of Hedges and Edges

Acalypha hispida, *Allamanda neriifolia*, *Allamanda schottii*, *Barleria cristata*, *Clerodendron inerme*, *Hamelia patens*, *Ixora chinensis*, *Jasminum sambac*, *Lagerstroemia indica*, *Murraya exotica*, *Pentas lanceolata*, *Plumbago auriculata*, *Ravenia spectabilis*, *Thunbergia erecta*, *Vinca rosea*.

14.8.3.3 Diversity of Ornamental Foliage Shrubs in Tropical and Subtropical Gardens for Beautification of Shrubbery Borders

Acalypha godseffiana, *Alternanthera versicolor*, *Aralia* spp., *Cordium variegatum*, *Coleus blumei*, *Duranta repens*, *Eranthemum* spp., *Euphorbia cotinifolia*, *Excoecaria bicolour*, *Ficus triangularis*, *Graptophyllum pictum*, *Jacquinia ruscifolia*, *Jatropha podagrica*, *Lawsonia inermis*, *Malpighia coccigera*, *Malpighia glabra*, *Manihot esculenta* 'Variegata', *Muehlenbeckia platyclada*, *Murraya paniculata*, *Mussaenda* spp., *Nandina domestica*, *Nandina nivosus*, *Phyllanthus* spp., *Poinsettia heterophylla*, *Poinsettia leucocephala*, *Poinsettia pulcherrima*, *Polyscias* spp., *Pseuderanthemum* spp., *Sanchezia nobilis*, *Serissa foetida*, *Strobilanthes dyerianus*, *Thuja compacta*.

14.8.3.4 Diversity of Ornamental Foliage Shrubs in Tropical and Subtropical Gardens for Beautification of Hedges and Edges

Acalypha godseffiana, *Alternanthera amoena*, *Alternanthera bettzickiana*, *Alternanthera dentate*, *Alternanthera versicolor*, *Aralia* spp., *Duranta repens*,

Eranthemum bicolor, *Phyllanthus* spp., *Polyscias balfouriana*, *Polyscias filicifolia*, *Polyscias fruticosa*.

14.8.3.5 Diversity of Ornamental Flowering Shrubs in Temperate Gardens for Beautification of Shrubbery Borders and Hedges

Azalea spp., *Camellia* spp., *Hydrangea* spp., etc.

14.8.4 Biodiversity of Ornamental Climbers and Creepers

Climbers are grown throughout the world for the beautification of garden features such as covering arches, pergolas, bowers, walls, and topiary work. Also, some sweet-scented flowering climbers have been grown for commercial use in the perfumery industry in the private or public sector. Some important climbers grown in tropical, subtropical, and temperate gardens are listed in Table 3.

14.8.5 Biodiversity of Ornamental Palms

The ornamental palms of the world have a vital role in gardening, either outside or indoors, and most of these palms originated in tropical and subtropical zones. Tall palms are used for roadside beautification; small palms are grown in pots, and bushy palms are planted in gardens by a wall. The important palms used in gardening worldwide are summarized below.

14.8.5.1 Classification Based on Gardening

Indoor Gardening

Acoelorrhaphe wrightii, *Actinorhytis calapparia*, *Adonidia merrillii*, *Areca nagensis*, *Areca triandra*, *Arenga engleri*, *Arenga porphyrocarpa*, *Asterogyne martiana*, *Balaka seemannii*, *Bentinckia condapanna*, *Butia paraguayensis*, *Butia yatay*, *Chamaedorea adscendens*, *Chamaedorea arenbergiana*, *Chamaedorea brachypoda*, *Chamaedorea cataractarum*, *Chamaedorea costaricana*, *Chamaedorea elegans*, *Chamaedorea metallica*, *Chamaedorea seifrizii*, *Chambeyronia macrocarpa*, *Coccothrinax argentata*, *Cryosophila warscewiczii*, *Cyrtostachys renda*, *Dypsis albofarinosa*, *Dypsis lutescens*, *Dypsis madagascariensis*, *Elaeis guineensis*, *Howea forsteriana*, *Hyophorbe lagenicaulis*, *Latania lontaroides*, *Latania verschaffeltii*, *Licuala grandis*, *Licuala paludosa*, *Licuala peltata*, *Licuala ramsayi*, *Licuala spinosa*, *Livistona australis*, *Livistona chinensis*, *Livistona jenkinsiana*, *Livistona rotundifolia*, *Pelagodoxa henryana*, *Phoenicophorium borsigianum*, *Phoenix roebelenii*, *Phoenix rupicola*, *Pinanga adangensis*, *Pinanga coronate*, *Pinanga dicksonii*, *Pritchardia pacifica*, *Ptychosperma elegans*, *Ptychosperma furcatum*, *Ptychosperma lineare*,

Table 3 List of climbers and creepers species biodiversity in the world

S.No.	Botanical name	Family	Origin
1.	*Adenocalymma comosum*	Bignoniaceae	South America
2.	*Allamanda cathartica*	Apocynaceae	Brazil
3.	*Antigonon leptopus*	Polygonaceae	Tropical America
4.	*Aristolochia elegans*	Aristolochiaceae	Brazil
5.	*Aristolochia grandiflora*	Aristolochiaceae	South America
6.	*Aristolochia tomentosa*	Aristolochiaceae	North America
7.	*Beaumontia grandiflora*	Apocynaceae	India
8.	*Bougainvillea buttiana*	Nyctaginaceae	Colombia
9.	*Bougainvillea spectabilis*	Nyctaginaceae	Eastern and Central Brazil
10.	*Campsis grandiflora*	Bignoniaceae	China and Japan
11.	*Chonemorpha macrophylla*	Apocynaceae	India and Malaysia
12.	*Cissus discolor*	Vitaceae	Java
13.	*Clerodendrum splendens*	Verbenaceae	Tropical America
14.	*Clitoria ternatea*	Fabaceae	Tropical and subtropical gardens
15.	*Cryptostegia grandiflora*	Asclepiadeceae	Madagascar
16.	*Derris scandens*	Fabaceae	India
17.	*Ficus pumila*	Moraceae	Tropical and subtropical gardens
18.	*Gloriosa superba*	Liliaceae	Africa
19.	*Hiptage madablota*	Malphighiaceae	India
20.	*Ipomoea learii*	Convolvulaceae	Tropical America
21.	*Jasminum auriculatum*	Oleaceae	India
22.	*Lonicera japonica*	Caprifoliaceae	Japan
23.	*Passiflora laurifolia*	Passifloraceae	Tropical America
24.	*Passiflora edulis*	Passifloraceae	Brazil
25.	*Petrea volubilis*	Verbenaceae	Tropical America
26.	*Porana paniculata*	Convolvulaceae	India
27.	*Pyrostegia venusta*	Bignoniaceae	South America
28.	*Quamoclit lobata*	Convolvulaceae	Mexico
29.	*Quisqualis indica*	Combretaceae	Java and Malaysia
30.	*Senecio confusus*	Compositae	Mexico
31.	*Solanum jasminoides*	Solanaceae	South America
32.	*Strophanthus grandiflorus*	Apocynaceae	Tanzania and Mozambique
33.	*Thunbergia alata*	Acanthaceae	South Africa
34.	*Thunbergia fragrans*	Acanthaceae	Burma
35.	*Thunbergia laurifolia*	Acanthaceae	India and Malaysia
36.	*Trachelospermum jasminoides*	Apocynaceae	East Asia
37.	*Wisteria sinensis*	Fabaceae	China

Ptychosperma macarthurii, *Ptychosperma propinquum*, *Ravenea rivularis*, *Rhapis excelsa*, *Rhapis humilis*, *Reinhardtia gracilis*, *Roscheria melanochaetes*, *Sabal bermudana*, *Serenoa repens*, *Thrinax excelsa*, *Thrinax parviflora*, *Thrinax radiate*, *Wallichia caryotoides*, *Wallichia densiflora*.

Outdoor Gardening

Acrocomia aculeate, Aiphanes aculeate, Aiphanes minima, Archontophoenix alexandrae, Archontophoenix cunninghamiana, Archontophoenix myolensis, Areca catechu, Areca catechu cv. *alba, Areca catechu* cv. dwarf, *Areca concinna, Areca macrocalyx, Areca nagensis, Areca vestiaria, Arenga caudate, Arenga hookeriana, Arenga microcarpa, Arenga obtusifolia, Arenga pinnata, Arenga tremula, Arenga undulatifolia, Arenga westerhoutii, Arenga wightii, Astrocaryum alatum, Astrocaryum mexicanum, Attalea allenii, Attalea cohune, Bactris grasipaes, Bactris major, Bentinckia nicobarica, Brahea aculeate, Brahea edulis, Brahea aculeata, Calamus arborescens, Calyptrocalyx spicatus, Calyptrogyne ghiesbreghtiana, Calyptronoma plumeriana, Carpentaria acuminate, Caryota mitis, Chamaedorea arenbergiana, Chamaerops humilis, Cocos nucifera, Copernicia baileyana, Copernicia brittonorum, Copernicia gigas, Copernicia hospita, Copernicia macroglossa, Copernicia prunifera, Corypha utan, Corypha umbraculifera, Daemonorops jenkinsiana, Dictyosperma album, Drymophloeus hentyi, Dypsis cabadae, Dypsis decaryi, Dypsis lastelliana, Dypsis leptocheilos, Dypsis lutescens, Elaeis oleifera, Euterpe edulis, Gaussia maya, Heterospathe elata, Hydriastele microspadix, Hyophorbe lagenicaulis, Hyophorbe verschaffeltii, Hyphaene coriacea, Kerriodoxa elegans, Latania loddigesii, Livistona decora, Livistona muelleri, Loxococcus rupicola, Nannorrhops ritchiana, Normanbya normanbyi, Nypa fruticans, Oncosperma tigillarium, Phoenix paludosa, Pritchardia thurstonii, Pseudophoenix ekmanii, Rhopaloblaste augusta, Roystonea borinquena, Roystonea oleracea,*

Roystonea regia, Sabal "riverside," *Salacca zalacca, Satakentia liukiuensis, Schippia concolor, Syagrus cearensis, Trachycarpus fortunei, Trachycarpus martianus, Trachycarpus takil, Veitchia arecina, Veitchia joannis, Veitchia winin, Verschaffeltia splendida, Wallichia disticha, Washingtonia filifera, Washingtonia robusta, Wodyetia bifurcata.*

14.8.6 Biodiversity of Ornamental Ferns

Ferns, one of the most important graceful foliage ornamental plants, are grown in a wide range of climate, about 2120 species in shaded and semi-shaded areas on the ground worldwide. Almost all ferns like to grow in hilly areas in subtropical zones. Now, many ferns domesticated in the plains are used for house plants by growing them in pots or beds in the garden or cut foliage. Ferns look very beautiful when grown in a group in the garden. The most important ferns having ornamental value for garden beautification and indoor gardening include *Adiantum, Asplenium, Blechnum, Cyrtomium, Davallia, Dicksonia, Doryopteris, Nephrolepis, Pityrogramma, Platycerium, Polypodium, Polystichum,* and *Pteris* spp.

14.8.7 Biodiversity of Ornamental Grasses

Grasses have a vital role for beautification of gardens in the world, especially in lawn development, and now these are used commercially as turf in playgrounds. The many grasses diversified all over the world, as based on seasonal growing, are reported by Tiwari et al. (2015).

14.8.7.1 Warm Season Grasses

Axonopus affinis, *Axonopus aureus*, *Axonopus compressus*, *Axonopus fissifolius*, *Axonopus furcatus*, *Buchloe dactyloides*, *Cynodon dactylon*, *Eremochloa ciliaris*, *Pennisetum clandestinun*, *Paspalum notatum*, *Paspalum vaginatum*, *Stenotaphrum secundatum*, *Zoysia japonica*, *Zoysia tenuifolia.*

14.8.7.2 Cool Season Grasses

Agrostis palustris, *Agrostis tenuis*, *Festuca ovina*, *Festuca rubra*, *Lolium multiflorum*, *Poa annua*, *Poa perenne*, *Poa pratensis*, *Poa trivialis*.

14.8.8 Biodiversity of Ornamental House Plants

Most of the house plants of the world are grown under the forest canopy in tropical and subtropical zones in shaded and semi-shaded locations. It is important to realize that people in cities want to feel closer to nature by establishment of indoor gardening, and now the demand is increasing day by day in response to rapid urbanization. Some of the species available worldwide that are used for beautification of home gardens, indoor gardens, public gardens, etc. are classified below.

14.8.8.1 Classification Based on Light Requirement

Moderate Light to Full Filtered Sunlight

Aechmea spp., *Chlorophytum bichetii*, *Dracaena deremensis*, *Dracaena fragrans*, *Dracaena godseffiana*, *Ficus benjamina*, *Ficus cyathistipula*, *Ficus diversifolia*, *Ficus elastica*, *Ficus nitida*, *Ficus triangularis*, *Ficus triangularis* 'Variegata', etc.

Full Filtered Sunlight

Aeschynanthus spp., *Anthurium crystallinum*, *Anthurium forgetii*, *Anthurium magnificum*, *Anthurium pedatoradiatum*, *Anthurium scherzerianum*, *Asparagus densiflorus*,

Asparagus plumosus, *Asparagus setaceus*, *Fittonia verschaffeltii*, *Guzmania lingulata* 'Major', *Guzmania lingulata* 'Minor', *Leea rubra*, *Oxalis hedysaroides rubra*, etc.

Moderate Filtered Sunlight

Aglaonema spp., *Alocasia* spp., *Begonia* spp., *Billbergia* spp., *Bromelia balansae*, *Bromelia serra*, *Chrysothemis pulchella*, *Dianella tasmanica*, *Hypocyrta glabra*, *Impatiens linearifolia*, *Medinilla magnifica*, *Tillandsia* spp., etc.

Direct Sunlight

Alpinia zerumbet, Ananas bracteatus, Costus speciosus 'variegata', *Pandanus baptistii, Pleomele reflexa, Sansevieria cylindrica, Sansevieria guineensis, Sansevieria trifasciata, Stenotaphrum secundatum* 'Variegatum,' etc.

Partial Shade

Aphelandra squarrosa, Aspidistra elatior, Calathea spp., *Callisia repens, Cordyline terminalis, Cryptanthus bivittatus, Cryptanthus bromelioides, Cryptanthus* 'Golden green', *Episcia* spp., *Heliconia* spp., *Hemigraphis colorata, Hoffmannia refulgens, Pellonia daveauana, Peperomia* spp., *Phalaris arundinaceae, Philodendron* 'Black Cardinal', *Philodendron* 'Black Cardinal spot', *Philodendron* 'Blue mist', *Philodendron* 'Ceylon gold', *Philodendron* 'Charm', *Philodendron elegans, Philodendron erubescens* 'Gold', *Philodendron* 'Goldiana Spot', *Philodendron* 'Pink Princess', *Philodendron selloum, Philodendron* 'Serratum', *Ruellia devosiana, Saintpaulia ionantha, Schefflera arboricola* variegata, *Spathiphyllum clevelandii, Tradescantia albiflora, Tradescantia fluminensis, Vriesea* spp., *Xanthosoma lindenii, Zebrina pendula* 'Rubra', etc.

Shade

Ctenanthe lubbersiana, Cyclanthus bipartitus, Dieffenbachia spp., *Euonymus japonicus, Hedera helix, Homalomena* spp., *Kaempferia pulchra, Maranta arundinacea, Monstera deliciosa, Monstera oblique, Neoregelia carolinae, Nidularium innocentii, Pelargonium* spp., *Pilea serpyllacea, Pleomele angustifolia, Schefflera venulosa, Scindapsus* spp., *Stromanthe sanguine, Syngonium podophyllum* variegatum, *Tacca chantrieri*, etc.

14.8.8.2 Classification Based on Importance of Plant Parts

Foliage Beauty

Aglaonema spp., *Alocasia* spp., *Alpinia zerumbet, Ananas bracteatus, Anthurium crystallinum, Anthurium forgetii, Anthurium magnificum, Anthurium pedatoradiatum, Anthurium scherzerianum, Asparagus densiflorus, Asparagus plumosus, Asparagus setaceus, Aspidistra elatior, Billbergia* spp., *Calathea* spp., *Chlorophytum bichetii, Cordyline terminalis, Cryptanthus bivittatus, Cryptanthus bromelioides, Cryptanthus* 'Golden green', *Ctenanthe lubbersiana, Cyclanthus bipartitus, Dianella tasmanica, Dieffenbachia* spp., *Dracaena deremensis, Dracaena fragrans, Dracaena godseffiana, Euonymus japonica, Ficus benjamina, Ficus cyathistipula, Ficus diversifolia, Ficus elastica, Ficus nitida, Ficus triangularis, Ficus triangularis* 'Variegata', *Fittonia verschaffeltii, Hedera helix, Hemigraphis colorata, Homalomena* spp., *Leea rubra, Maranta arundinacea, Monstera deliciosa, Monstera oblique, Neoregelia carolinae, Neoregelia carolinae, Neoregelia carolinae, Nidularium innocentii, Oxalis hedysaroides rubra, Pandanus baptistii, Pelargonium* spp., *Pellionia daveauana, Peperomia* spp., *Phalaris arundinaceae, Philodendron 'Black Cardinal', Philodendron* 'Black Cardinal spot', *Philodendron* 'Blue mist', *Philodendron* 'Ceylon gold', *Philodendron* 'Charm', *Philodendron elegans, Philodendron erubescens* 'Gold', *Philodendron*

'Goldiana Spot', *Philodendron* 'Pink Princess', *Philodendron selloum*, *Philodendron* 'Serratum', *Pilea serpyllacea*, *Pleomele angustifolia*, *Pleomele reflexa*, *Ruellia devosiana*, *Sansevieria cylindrica*, *Sansevieria guineensis*, *Sansevieria trifasciata*, *Schefflera arboricola variegata*, *Schefflera venulosa*, *Scindapsus* spp., *Stenotaphrum secundatum* variegatum, *Stromanthe sanguine*, *Syngonium podophyllum* Variegatum, *Tacca chantrieri*, *Tradescantia albiflora*, *Tradescantia fluminensis*, *Xanthosoma lindenii*, *Zebrina pendula* 'Rubra', etc.

Flower and Foliage Beauty

Aechmea spp., *Aeschynanthus* spp., *Aphelandra squarrosa*, *Begonia* spp., *Bromelia balansae*, *Bromelia serra*, *Chrysothemis pulchella*, *Costus speciosus* 'variegata', *Episcia* spp., *Guzmania lingulata* 'Major', *Guzmania lingulata* 'Minor', *Heliconia* spp., *Hoffmannia refulgens*, *Hypocyrta glabra*, *Impatiens linearifolia*, *Kaempferia pulchra*, *Medinilla magnifica*, *Saintpaulia ionantha*, *Spathiphyllum clevelandii*, *Tillandsia* spp., *Vriesea* spp., etc.

14.8.9 Biodiversity of Bulbous Plants

Bulbous plants in horticulture include underground modified stems such as bulbs, tubers, corms, and rhizomes having tuberous roots. There are many bulbous plants growing in the world, and the rich sources are in the plains of tropical and subtropical zones as well as in some hilly areas. The major bulbous plants commercially exploited in the world are classified here.

14.8.9.1 Cut Flower Production

Alpinia, *Alstroemeria*, *Dahlia*, *Gladiolus*, *Heliconia*, *Lilium*, *Narcissus*, *Nelumbo*, *Polianthes*, *Solidago*, *Strelitzia*, *Tulipa*, etc.

14.8.9.2 Other Bulbous Plants for Garden Beautifications

Achimene, *Acidenthera*, *Agapanthus*, *Allium*, *Alpinia*, *Amaryllis*, *Arisaema*, *Begonia*, *Caladium*, *Canna*, *Cooperia*, *Costus*, *Crinum*, *Curcuma*, *Dahlia*, *Eucharis*, *Gloriosa*, *Heliconia*, *Haemanthus*, *Hedychium*, *Hemerocallis*, *Hippeastrum*, *Hymenocallis*, *Nymphaea*, *Strelitzia*, *Zantedeschia*, *Zephyranthes*, ornamental ginger (*Alpinia*, *Etingera*, *Tapeinochilos*, *Zingiber*), etc.

14.9 Wild Genetic Resources of Ornamentals in the Hills

The important hilly areas in India are located in the temperate North West Himalayas (Jammu and Kashmir, Himachal Pradesh, Uttarakhand) and subtropical East Himalayas (West Bengal, Sikkim, Meghalaya, Arunachal Pradesh); while some are scattered in the Aravali, Vindhya, Satpura ranges (West India), as well as Western Ghats, Nilgiris, and Eastern Ghats. Although the plant wealth of the hills (especially the Himalayas) has been well documented, precious little has been contributed specially on ornamentals. Collette (1921) described the flora of Shimla hills and surrounding areas and mentioned several trees like *Bombax*, *Cedrela*, *Aesculus*, *Acer*, *Erythrina*, *Prunus*, *Rhododendron*, *Putranjiva*, *Cupressus*, etc. Lele and Misra (1961) added several naturalized ornamentals to the above list, for example, *Tagetes minuta*, *Agave Americana*, *Yucca*, *Gloriosa*, *Oenothera*, *Cytisus*, *Sambucus*, *Solanum jasminoides*, *Opuntia*, *Passiflora*, *Lonicera*, *Wisteria*, *Lagerstroemia*, *Callistemon*, etc. Rathore (1981) evaluated 150 introduced ornamentals at Shimla, the promising ones being *Acer palmatum*, *Chaenomeles*, *Cotoneaster*, *Philadelphus*, *Viburnum*, and *Weigela*. Polunin and Stainton (1984) described the flora from Nepal to Western Himalaya, with color photographs of the important native and naturalized attractive species, thus supplementing the earlier work covering this area. Some of these species having beautiful flowers are *Erythrina stricta*, *Paeonia emodi*, *Hypericum*, *Gaultheria*, *Woodfordia*, *Daphne*, *Calanthe*, *Iris*, *Clematis*, and *Trichosanthes*. Kohli (1979) described several wild or naturalized bulbous flowers of Kashmir including *Anemone*, *Fritillaria*, *Hyacinthus*, *Narcissus*, *Tulipa*, *Iris*, *Paeonia*, *Lilium*, and *Gagea*.

Gupta (1967) compiled a list of plants flowering during different seasons in Missouri hills. Out of these *Asclepias*, *Tecoma*, *Osyris*, *Woodfordia*, *Magnolia*, *Hypericum*, *Reinwardtia*, and *Caryopteris* among the spring flowering; *Berberis*, *Viburnum*, *Robinia*, *Cocculus*, *Vitex*, and *Deutzia* among the summer flowering; *Cryptolepis*, *Verbascum*, *Barleria*, *Inula*, *Habenaria*, *Smilax*, and *Crassula* among the rainy season flowering; and *Spermadactylon* and *Primula* among the winter flowering species are worth mentioning. Ohashi (1975) enumerated several beautiful plants of Eastern Himalaya, including *Clematis alternata*, *Meconopsis horridula*, *Passiflora nepalensis*, *Calenthe trulliformis*, and *Anoectochilus lanceolatus*. Jana and Mukherjee (1979) listed promising species of bulbous flowers of Shillong area including *Agapanthus*, *Crocosmia*, *Hedychium*, *Hippeastrum*, *Iris*, *Lilium*, *Tigridia*, and *Zephyranthes*. Fyson (1974) described the plants of the Nilgiris (including Ooty), the important ones being *Hypericum humifusum*, *Ilex*, *Osyris*, *Prinsepisa*, *Thunbergia*, *Cassis tomentosa*, *Fudchsia fulgens*, and *Sarcococca*. Manilal (1988) recorded the flora of Silent Valley (Pal Ghat, Kerala), including *Rhododendron*, *Ligustrum*, *Euonymous*, *Lagerstroemia microcarpa*, *Mussaenda*, *Dendrobium pandburatum*, *Oberonia*, and *Habenaria*.

14.10 Characterization and Evaluation of Plant Diversity

The utilization of genetic resources in crop improvement programs rests on identification of promising accessions through characterization and evaluation and is the most important activity in germplasm management. Germplasm evaluation involves a whole range of activities including seed multiplication and collections, characterization, preliminary and detailed evaluation, regeneration, maintenance, and documentation. Characterization and preliminary evaluation involves recording of highly heritable (oligogenic) morphological characters, which describe the accession and enable any contamination or mix-up at later stage to be identified. It also includes scoring of a few more characters that are of interest to plant breeders. The detailed evaluation involves field trials of accessions over the years and locations and recording various characters in multi-disciplinary mode. Typically, these include yield, other agronomic characters, biotic and abiotic stress tolerance/resistance, and biochemical traits. The characterization has, generally, been based on morphological markers and agronomic traits. Molecular characterization has gained great importance and needs to be strengthened. The cultivars and promising germplasm need to be characterized using molecular markers to establish the identity and analyze genetic diversity and their utilization in crop improvement programs. The core collections are required to be made, which shall be part of an entire collection that represents the genetic diversity in a crop species and its wild relatives with minimum repetition. This core set can be evaluated intensively by breeders for economic traits.

14.11 Conservation of Biodiversity in Landscape Gardening

Conservation means protecting existing plants in a particular area from any natural calamities and providing scientific management for betterment of the plants for growth and development as well as improvement. The market for ornamental plants is constantly increasing with each passing day but at the same time is subjected to periodic trend-driven changes. However, changes in consumer preferences mean that cultivars unfashionable today may in the future once again be attractive for potential buyers. Furthermore, very often these constitute a great breeding material source. For this reason, the protection and storage of those valuable genetic resources is of great importance to be always able to meet market demands. Nevertheless, it is difficult for breeders and horticulturalists to provide enough space and funds for traditional cultivation of such numerous cultivars, which is laborious and threatened with biotic and abiotic stresses (Sekizawa et al. 2011). Traditional genetic conservation in the field or greenhouse requires intensive care of pot cultures or carefully separated field plots (Reed 2006). Additionally, many ornamental species are on the brink of extinction. Fast and easy access to high-quality gene banks of large material variety is the key for ornamental plant producers, and thus an efficient method for long-time conservation of the plant material may be extremely valuable for breeding and horticultural production (Halmagyi et al. 2004). Today, cryopreservation is believed to be the most promising and valuable long-term storage method. During the past decade, conservation of plants at international and national levels received excellent momentum, which is reflected in the establishment of different plant genetic resource centers; for example, in India NBPGR established its 10 Regional Stations and 59 National Active Germplasm Sites (NAGS), comprising ICAR Institutes, Project Directorates, NRCs, AICRPs, SAUs, KVKs, etc. Gene banks/germplasm banks refer to a place or organization where germplasm is conserved in the living state and the germplasm can be stored in the form of seeds, pollens, and in vitro cultures or as plants growing in the field.

14.12 Methods of Conservation of Ornamental Plants

There are two important methods of genetic resources conservation – in situ and ex situ. In ex situ conservation, the conventional methods are orthodox seeds or seed gene banks, in vitro culture (tissue culture/cryopreservation/DNA libraries), and field gene banks (plant conservation in a botanical garden/arboretum/greenhouse). For in situ conservation, the two important systems are natural habitats (an ecosystem with bioreserves/heritage sites/wildlife sanctuaries) and on-farm collections (farmers' fields/tribal areas).

14.12.1 Field Gene Banks

Field gene banks, also called plant gene banks, are areas of land in which germplasm collections of growing plants are assembled. This approach is also ex situ conservation of germplasm. Those plant species that have recalcitrant seeds or do not produce seeds readily are conserved in a field gene bank. In field gene banks, germplasm is maintained in the form of plants as a permanent living collection. Most of the trees, shrubs, climbers, palms, and grasses are conserved in open-field conditions, whereas house plants and ferns are conserved in a shaded structure. Bulbs are stored in cool, dry, and airy places. The agro-ecological conditions of the field gene bank should be as similar as possible to the environment where the collected plant materials are normally grown or collected, as the requirements of climate and soil of the different ornamental plants are varied. The field gene bank should be sited so as to minimize risks from natural and man-made disasters and hazards such as pests, diseases, animal damage, floods, droughts, fires, snow and freezing damage, volcanoes, hail, thefts, or vandals. Isolation distance should be adequate for the production of seeds for distribution (to minimize risks of gene flow from crops or wild populations) to maintain genetic integrity.

All germplasm accessions added to the gene bank should be legally acquired, with relevant technical documentation. Plants and soil should be regularly monitored for pests and diseases. Each accession in the field collection should be regenerated, when the vigor or plant numbers have declined to critical levels, to bring them to original levels and ensure that diversity and genetic integrity are maintained. True-to-type healthy plant material should be used for propagation. All accessions should be characterized. For each accession, a representative number of plants should be used for characterization. Accessions should be characterized morphologically using internationally used descriptor lists where available. Molecular tools are also important to confirm accession identity and trueness to type. Passport data for all accessions should be documented using the FAO/Bioversity multi-crop passport descriptors. All germplasm should be distributed in compliance with national laws and relevant international treaties and conventions. A risk management strategy should be implemented and updated as required that addresses physical and biological risks identified in standards. A gene bank should follow the local Occupational Safety and Health (OSH) requirements and protocols. There are many national and international institutes, in either the public or private sector, with a vital role in conserving ornamental plants, in addition to natural resources. Some institutes working under the government and private sectors are listed below in Table 4.

Table 4 Field gene bank conservation of different species or varieties of ornamental plants at different institutes in the world

Institutes	Name of the crop and its species/varieties/ genotypes
Rio de Janeiro Botanical Garden, Brazil	6500 species of tropical and subtropical trees, shrubs, climbers, palms, and other plants, with 900 varieties of palm
Brooklyn Botanic Garden, New York, North America	10,000 taxa of different kinds of plants, 200 cherry trees,5000 bushes, and 1500 kinds of roses
Royal Botanic Gardens, Kew, England	50,000 species of plants, specializing in roses, azalea, camellia, house plants, annuals, cacti, and succulents
Longwood Gardens, United States	11,000 species of different kinds of plants
Kyoto Botanical Garden, Japan	12,000 species of different kinds of plants including bamboos, *Camellia*, cherry trees, and lotus and bonsai plants
Nikko Botanical Garden, Japan	2200 species of plants with *Rhododendron*, *Prunus*, *Pinus*, and other rockery plants
National Botanical garden, Cape Town, South Africa	7000 plant species with trees, shrubs, climbers, lilies, and other plants
Fairchild Tropical Botanic Garden, Florida, USA	20,000 plants with different ferns, cycads, flowering trees, and vines
The German Genebank for Ornamentals, EUROPA-Rosarium, Sangerhausen, Germany	Wild roses (500 species), roses (7500 cultural tribes), *Rhododendron* (350 species and 3200 cultural tribes)
Acharya Jagadish Chandra Bose Indian Botanical Garden, Shibpur, Howrah, India	2350 species of plants with flowering trees, shrubs, climbers, cactus, succulents, palms, ferns, bulbs, etc.
Tropical Botanic Garden and Research Institute, Palode, India	40 species of ornamental plants, 45 species of jasmine, 100 species of cacti and succulents, 130 species of fern, 130 species of palms
Lal Bagh Garden, Bangalore, Karnataka, India	1000 species of flora with flowering trees, shrubs, climbers, cacti and succulents, water lilies, etc.
Lioyd's Botanical Garden, Darjeeling, West Bengal, India	More than 500 species of plants with cacti and succulents, orchids, *Rhododendron*, *Azalea*, bulbous plants, etc.
The Agri-Horticultural Society of India, 1-Alipore Rd., Cal-27, India	More than 1000 species of plants, with flowering trees, shrubs, climbers, cacti and succulents, palms, ferns, house plants, flowering plants

14.12.2 Cryopreservation

Cryopreservation refers to the storage of explants from tissue culture at the ultra-low temperature of liquid nitrogen (−196 °C). At such temperature, all the biological reactions within the cells are hampered; hence the technique makes available the storage of plant material for theoretically unlimited periods of time. At this temperature, biochemical, metabolic, and cell division activities are arrested, allowing for long-term storage. The main advantage of this method is the reduction of in vitro culture costs, required space, contamination, and somaclonal variation risk. The long-term cryoconservation of embryogenic cell lines could be a valuable tool for genetic transformation. Storage in liquid nitrogen would also help in preserving genetic diversity by storing wild species (e.g., for the purpose of breeding), some of which are already endangered.

14.13 Utilization of Plant Diversity in Landscape Gardening

The importance of the plant genetic resources is not only in the number of the available species but also in the genetic diversity existing in each species, which could provide the precious genetic bases for the flower improvement by breeding and for flexibility in conservation. Use of natural/underutilized plant species in landscape architecture creates a landscape harmonious with nature, and it is an application compatible with economic conditions. In this way, selection of plant materials that meet environmental conditions with minimal care should be preferred as an economic approach to planning. There are many wild species found all over the world, e.g., wild kangaroo paws of various colors now are found growing in home gardens. In Malaysia, there are many wild species growing in wastelands and roadsides like the *Dillenia* and *Melastoma* which can be easily introduced into home gardens and for landscaping (Chin and Tay 2017). The flowers of many ornamental species are edible either as food, vegetables, herbs, aromatic plants, or medicinal plants are just as important. Looking ahead it can be envisaged the great diversity of exiting ornamental species and their wild relatives have a great potential and a bright future for utilization by mankind and the future generation.

14.14 Future Perspective

Increasing human activity results in negative effects for the natural areas, threatening the existence of plants, animals, and other types of species. It is generally admitted that the extinctions of plant species are occurring at an unprecedented pace in recent years. Thus, agro-diversity conservation is an issue gaining greater importance in today's world. There are different ways and methods used for preserving plant diversity.

One method is to protect the endemic species and prevent their extinction by making use of these plants in gardens and other living environments of people. By doing so, there would be an important step taken for increasing people's consciousness. As people spend more time with natural plants, their protection motives and knowledge would certainly increase. However, in order for providing these environments to people and increase their access to natural plants, more works and projects for improving cultivation and production of natural plants are needed.

There is a scope for new varieties of landscape plants which must be tapped. Pollution mitigation and carbon sequestration by individual plant species have not been explored to the required level and need attention. The richness of species and genetic diversity in ornamental crops provides many opportunities, which can be achieved with adoption of more rational, science-based, and pragmatic approaches. This would facilitate conservation and sustainable use of genetic resources. Biodiversity and climate change are closely linked and each influences the other. Biodiversity is threatened by human-induced climate change, but adequate and efficient efforts to conserve biodiversity resources can be reduce the impacts of climate change on population and ecosystems. Therefore, landscaping deserves due importance in research priorities and policy planning for providing better living conditions to the people.

References

Bhattacharjee SK (2006) Landscape gardening. In: Advances in ornamental horticulture, vol 5. Pointer Publishers, Jaipur, pp 5–24

Chin HF, Tay D (2017) Conservation and utilization of ornamental germplasm. Acta Hortic 760: 581–587

Collette H (1921) Flora simlensis. Thacker Spink and Co., London

Fyson PK (1974) Flora of the Nilgiri and Pulney Hill-Tops. Bishan Singh and Mahinder Pal Singh, Dehradun

Gupta RK (1967) Seasonal flowers of the Indian summer resorts-Missouri hills. Nav Yug Traders, New Delhi

Halmagyi A, Fischer-Kluver G, Mix-Wagner G, Schumacher HM (2004) Cryopres-ervation of *Chrysanthemum morifolium* (*Dendranthema grandiflora* Ramat) using different approaches. Plant Cell Rep 22:371–375

Jana VK, Mukherjee A (1979) Shillong- a Treasure house of bulbous plants. Indian Hortic 24:17–22

Kohli PN (1979) Beautiful bulbous wild flowers of Kashmir. Indian Hortic 24:23–27

Lele VC, Misra DP (1961) Some unreported wild naturalized flowering plants of the Shimla Hills. Himachal Hortic 2:5–8

Manilal KS (1988) Flora of Silent valley. The Mathurbhumi Press, Calicut

Ohashi H (1975) The flora of Eastern Himalaya. University of Tokyo Press, Tokyo

Polunin O, Stainton A (1984) Flowers of the Himalaya. Oxford University Press, New Delhi, 580 pp

Rathore DS (1981) Advances in ornamentals in hills. Indian Hortic 26:19–26

Reed BM (2006) Cryopreservation of bermuda grass germplasm by encapsulation dehydration. Crop Sci 46(1):6–11

Sekizawa K, Yamamoto S, Rafique T, Fukui K, Niino T (2011) Cryopreservation of in vitro-grown shoot tips of carnation (*Dianthus caryophyllus* L.) by vitryfication method using aluminium cryo-plates. Plant Biotechnol 28:401–405

Singh HP, Malhotra SK (2017) Plant biodiversity for aesthetic value and landscape gardening. In: Malhotra SK, Lallan R (eds) Advances in floriculture and landscape gardening. Central Institute of Horticulture, Medziphema, Dimapur, pp 11–17

Tiwari AK, Singh KP, Amrapali S, Girish KS, Singh PJ (2015) Lawn management, ICAR-DFR Bulletin 14. ICAR-DFR, Pune

Jasmine

15

M. Ganga, M. Jawaharlal, and S. P. Thamaraiselvi

Contents

Abstract

Jasmine (*Jasminum* spp.) belonging to the family Oleaceae is esteemed for its attractive and fragrant flowers. The genus *Jasminum* comprises of around 89 species. *J. grandiflorum, J. sambac, J. auriculatum,* and *J. multiflorum* are commercially cultivated in India. Other species, namely, *J. nitidum*, *J. calophyllum,* and

M. Ganga (✉) · M. Jawaharlal · S. P. Thamaraiselvi
TNAU, Coimbatore, Tamil Nadu, India

S. K. Datta, Y. C. Gupta (eds.), *Floriculture and Ornamental Plants*, Handbooks of Crop Diversity: Conservation and Use of Plant Genetic Resources,
https://doi.org/10.1007/978-981-15-3518-5_16

J. flexile, also possess potential for use as loose flowers. Intellectual property rights in jasmine have been addressed through formulation of guidelines for DUS (Distinctness, Uniformity and Stability) testing. Genetic diversity of jasmine has been assessed with morphological and molecular markers. Various constraints in genetic improvement of jasmine crop through hybridization have been reported which include nocturnal flowering habit, insect interference due to fragrance, difficulty in emasculation and pollination due to long and delicate style, non-fruitfulness, development of a single seed per fruit, and slow growth of seedling. Breeding methods for jasmine crop improvement are clonal selection, open pollinated seedling selection, hybridization, mutation breeding, and ploidy breeding. Jasmine crop improvement work in India carried out by the Tamil Nadu Agricultural University (TNAU), Coimbatore, and the Indian Institute of Horticultural Research (IIHR), Bangalore, has resulted in development of nine improved varieties of various *Jasminum* species. *Jasminum* species are highly valuable genetic material which can be involved in jasmine breeding programs. Biotechnological tools such as embryo rescue can pave way for breaking the hybridization barriers in jasmine posed by factors such as endosperm antagonism and hybrid inviability.

Keywords

Jasminum spp. · Oleaceae · Ecotypes · DUS characterization · Clonal selection · Mutation

15.1 Introduction

Jasmine (*Jasminum* spp.) belonging to the family Oleaceae is one of the fragrant flowers cultivated since ancient times. Jasmine is esteemed for its attractive and fragrant flowers. From prehistoric times, jasmines have adorned the gardens of central Asia, Afghanistan, Iran, Nepal, India, and many other tropical and subtropical countries.

Jasmine flowers are popularly used for making garlands, adorning hair of women, in religious and ceremonial functions, and also for extracting perfumery oil. Jasmine essential oil is also used in preparation of cosmetics, soaps, confectionery, perfumed tobacco, syrups, aerated water, ointments and disinfectants, detergents, in medicinal and pharmaceutical beverages, and for curing various ailments. The flowers of "Arabian Jasmine" (*J. sambac*) are reported to be used in China for flavoring tea.

Jasmine growing countries of the world are India, Thailand, China, Sri Lanka, Philippines, France, Italy, Morocco, Algeria, North Africa, Spain, and Egypt. Of these countries, India, Thailand, China, Sri Lanka, and Philippines grow jasmine commercially for usage as fresh flowers and the other countries grow it for use in perfumery.

15.2 Botany

The taxonomic classification of Jasmine is as detailed below.

Kingdom	:	Plantae
Division	:	Magnoliophyta
Class	:	Mangnoliopsida
Order	:	Lamiales
Family	:	Oleaceae
Genus	:	*Jasminum*

Taxonomic descriptions of four species, viz., *J. sambac, J. pubescens J. primulinum,* and *J. grandiflorum* were made by Bor and Raizada (1946). Bailey (1947) gave the taxonomic description of 23 species of jasmine as well as the keys for their identification. Brief descriptions of eight South Indian species of jasmine, most of them being wild, have been given by (Anonymous 1959). The descriptions, medicinal uses, and cultivation aspects of 13 species of jasmine are also elaborated in "The Wealth of India" (Anonymous 1959).

Cytological investigations in jasmine were made by Krishnaswamy and Raman (1948), Raman (1955), Alikhan et al. (1969), and Murthy and Khanna (1971). The genus with basic chromosome number of $n = 13$ generally consists of diploid forms with $2n = 26$. However, natural occurrence of forms with higher ploidy levels was also observed as follows.

15.3 Origin, Distribution, and Domestication

Jasmines are native to the tropical and warm temperate regions of Europe, Asia, and Africa. The centers of diversity of jasmine are South Asia and Southeast Asia. India is one of the centers of origin of the crop. Nearly 72 species of *Jasminum* are distributed in India, Malaysia, and China. Among these, about 40 species are reported to occur in India. A large number of species are centered to the region comprising of the Himalayas, China, and Malaysia (Veluswamy et al. 1975).

The Arabian or Tuscan jasmine (*J. sambac*) is considered as a native of the East Indies. However, there are also early reports (Anon. 1959) which indicate its original home as the region west of India. The Royal jasmine or common white jasmine or Poet's jasmine (*J. officinale)* is considered to be of Persian origin. Cooke (1905) reported that the Spanish jasmine or Catalonian jasmine (*J. grandiflorum*) which is sometimes referred to as synonymous with *J. officinale* or *J officinale* var. *grandiflorum* is native of Kashmir, Afghanistan, and Persia.

There are also reports which indicate that *J. officinale* is a native of Northern India and Persia and *J. officinale* var. *grandiflorum* is a native of the North-West Himalayas (Anon. 1972). However, mutation studies by Veluswamy et al. (1973) led to inferences that *J. officinale* could be a natural budsport of *J. grandiflorum*.

J. auriculatum is distributed in the western peninsula of India (Cooke 1905), more specifically in Circars, Deccan Carnatic areas, and in the southern districts of Travancore up to dry slopes of the Western Ghats (Gamble 1957). According to Haines (1961), *J. auriculatum* is native of South India and the Central Provinces.

15.4 Plant Genetic Resources

The genus *Jasminum* is reported to comprise of around 500 species (Bailey 1958). However, a critical analysis of these species has revealed the number of true species to be only 89, of which 40 inhabit the Indian sub-continent (Veluswamy et al. 1980). The jasmine species *J. grandiflorum, J. sambac, J. auriculatum,* and *J. multiflorum* are commercially cultivated in India. Apart from the commercially important species, three lesser known species, namely, *J. nitidum*, *J. calophyllum,* and *J. flexile* possess economic importance, since they produce flowers which are suitable for use as loose flower, besides being ideal garden plants (Raman et al. 1969; Ganga et al. 2015).

15.4.1 Commercial Species

15.4.1.1 *Jasminum sambac* (Arabian Jasmine, Tuscan Jasmine)

It is the most popular species in India. It is the national flower of the Philippines. It is an evergreen shrub. Leaves are simple, opposite, or in threes, cordate to oblong, almost sessile and dark green.

Flowers are white, highly fragrant, borne usually in small, three forked cymes. Its flower buds are white with single or multi-whorled petals, used for garland-making, adorning hair, religious worships, decoration, and extraction of perfume.

Cymes two- or tri-chotomous; bract linear and simple or ovate, sometimes petaloid; corolla salver shaped; flowers are hermaphrodite having two bilobed anthers; filaments very short.

15.4.1.2 *J. auriculatum*

It is a scandent shrub having shiny leaves with minute lateral leaflets. Leaves are opposite, ashy-velvety, sometimes hairless, simple, or trifoliolate. Flowers are fragrant, in many-flowered cymes. The flowers are white, scented, star shaped and are very good as loose flowers.

15.4.1.3 *J. grandiflorum* (Italian Jasmine, Royal Spanish Jasmine)

It is a climbing large shrub having shiny dark green pinnate leaves with 7–9 leaflets of equal size. Flower buds are often tinged with pink outside; opened flowers are white and highly scented.

15.4.2 Lesser Known Species

15.4.2.1 *J. arborescens* (Tree Jasmine)

Flowers are white and have strong fragrance.

15.4.2.2 *J. azoricum* (Lemon-Scented Jasmine)

It is an evergreen twining vine native to the Portuguese island of Madeira. Flower buds are deep pink; flowers are white, fragrant, and star-shaped, appearing in panicles from the leaf axils in summer. The species is critically endangered. This species which does not tolerate freezing temperatures has long been in cultivation in Europe as a greenhouse plant.

15.4.2.3 *J. beesianum* (Rosy Jasmine)

It is hardy and low climber growing up to about 2.4 m height, with slender grooved stems. Leaves are simple, opposite, ovate lanceolate to lanceolate, sharply pointed and dark dull green in color. Flowers are small, fragrant, pink to deep rose.

15.4.2.4 *J. bignoniaceum* (Syn: *J. revolutum*) (Trumpet Jasmine)

An erect shrub with angular branches. Leaves are compound, with elliptic leaflets with pointed tips and margins entire. Flowers are borne in cymes, opposite to the leaves, bright yellow, trumpet shaped.

15.4.2.5 *J. calophyllum*

A profuse flowering species producing scented white flowers; generally grown in home gardens. Plant blooms throughout the year. Flowers are white and scented; free from pest and disease attack.

15.4.2.6 *J. dichotomum* (Rose Bud Jasmine, Everblooming Jasmine, Gold Coast Jasmine)

It is a vigorous evergreen climber which grows as a rambling shrub or a woody vine, up to 26 ft tall, with climbing stems. Leaves are simple, large, and dark green which are usually opposite but may be single or in whorls of three. Flower buds are pink and somewhat resemble rose-buds. Opened flowers are white, fragrant, and appear in clusters in leaf axils. It is native to Africa, cultivated in tropical regions as a garden plant.

15.4.2.7 *J. dispermum*

It is a woody deciduous climber. The vine has pinnate leaves with 3–5 leathery leaflets. Leaves are opposite, leaflets lanceolate, and usually in two pairs with a much larger terminal one. Flowers with pink flower-tube and 5 shorter oblong-round lobes are fragrant. Pink Jasmine is found in the Himalayas, from Kashmir to South West China and South East Asia, at an altitude of 900–2500 m.

15.4.2.8 *J. favreri*

It is a wide spreading evergreen shrub. Shoots are angled, downy, and purplish. Leaves are trifoliate and ovulate-lanceolate. Flowers are borne in terminal clusters, corolla bright yellow in color.

15.4.2.9 *J. flexile* (Syn: *J. caudatum*)

A profuse flowering species grown widely in home gardens of India for its scented white flowers. Flowers are produced throughout the year. Plants are generally free from pest and diseases. It is a native to south India. It is found in Peninsular India and Sri Lanka.

15.4.2.10 *J. floridum* (Showy Jasmine, Florida Jasmine, Yellow Jasmine, Fruity Jasmine)

Plants are evergreen and of rambling habit with angled and glabrous shoots. Leaves are dark green, alternate; leaflets oval or obovate. Flowers are borne in terminal clusters. Flowers are bright yellow and 1 inch across, with the petals shaped into a shallow cup. The bright yellow flowers make an attractive contrast with the dark green leaves. Fragrance is very delicate and fruity. It is native of Mediterranean and Asia Minor.

15.4.2.11 *J. fluminense* (Brazilian Jasmine)

It is an evergreen, climbing, woody vine, with young stems densely hairy and mature stems hairless. Leaves are opposite, compound, and trifoliate. Flowers are white, fragrant, in broad, branched clusters at leaf axils; petals fused into a narrow, slightly curved tube with 5–7 petals shorter than the tube, spreading in star-shaped fashion and are borne in loose clusters.

15.4.2.12 *J. fruticans* (Wild Jasmine, Shrubby Jasmine)

It is an evergreen or semi-evergreen shrub with a domed habit. Shoots are angled and glabrous. Leaves alternate, trifoliate, leaflets narrow. Flowers are yellow, mildly fragrant, and appear in terminal clusters. It is native to the Mediterranean region. It is a drought tolerant shrub. It can be useful as a tough summer flowering shrub.

15.4.2.13 *J. humile* (Yellow Jasmine)

It is a small erect much-branched shrub with green, angular branches. Leaves are pinnate with 3–7 ovate to lance-like leathery leaflets. Inflorescences are lax clusters of yellow tubular flowers at the ends of branches. Flowers are borne in clusters; have a slender tube, with 5 rounded spreading petals, fragrant, and yellow. It is found in the Himalayas, from Afghanistan to North East India and China, at altitudes of 1800–4000 m.

15.4.2.14 *J. multiflorum* (Syn: *J. pubescens*) (Winter Jasmine, Furry Jasmine, Downy Jasmine, Kund, Kunda, Kundo, Indian Jasmine, Musk Jasmine)

It is an evergreen, strong growing, woody branching vine that can be trained as a shrub, or as a spreading, vine-like shrub and climbs to a height of 1.60 m or more. The stem and leaves are covered with a downy pubescence. Leaves are opposite, simple, acute, and ovate-lanceolate. Flowers are borne in congested clusters at branch-ends on small side shoots, many flowered, white or pink and borne in terminal umbels; flowers very mildly scented. Calyx is densely hairy. Flowers almost throughout the year with profuse flowering during the cooler months.

15.4.2.15 *J. nitidum* (Syn: *J. laurifolium*) (Star Jasmine)

A strong growing woody vine; leaves simple, opposite, glossy, dark-green and ovate-lanceolate. Flowers are very fragrant and produced in few to many flowered clusters. Buds are tinted pink on the outside; flowers are white when open and about 4.0 cm across. It is native of North America.

15.4.2.16 *J. nudiflorum*

It is a hardy member of the genus and is popular in the West. It is a deciduous shrub. Leaves are opposite; leaflets deep glossy green and margins minutely hairy. Flowers are yellow, solitary, and axillary.

15.4.2.17 *J. officinale* (Poet's Jasmine)

It is a deciduous climber bearing stalkless leaflets. Flowers are white and fragrant, borne in loose clusters.

15.4.2.18 *J. parkeri* (Dwarf Jasmine, Himalayan Jasmine, Parker Jasmine)

A rare and endangered dwarf, prostrate, dome-shaped evergreen shrub, 20–30 cm tall, slender, sprawling, or sagging on rocky-slopes or stone walls. Leaves are pinnately compound and thickly leathery. Flowers are bright yellow, tubular, 5-lobed, borne in cymes in leaf-axils and at branch-ends.

15.4.2.19 *J. polyanthum* (Pink Jasmine, Chinese Jasmine)

This dense growing evergreen to semi-deciduous vine grows twining, branches up to 20 ft long. It remains attractive all year round, bearing dark, glossy evergreen foliage which contrasts with masses of white flowers. Leaves are compound, opposite, and deep green. Inflorescence is a panicle and is axillary and many-fid; flower buds are rose colored; flowers are pure white, star shaped, and highly fragrant.

15.4.2.20 *J. primulinum* (Syn: *J. mesnyi*) (Primrose Jasmine, Japanese Jasmine)

It is a rambling, open evergreen shrub with long, slender, arching stems that will climb like a sprawling vine if given support. Without support, it grows in a fountain like mound 5–10 ft in height and spread. The stems are square in cross section, and green, becoming woody with age. The glossy dark green leaves are opposite and divided into three leaflets. Flowers are trumpet shaped, semi-double with 6–10 petals and lightly fragrant. It makes a fine specimen shrub. It is native to southwestern China.

15.4.2.21 *J. pubigerum*

It is an evergreen erect shrub. Leaves are alternate and leaflets sessile. Flowers are yellow in color.

15.4.2.22 *J. revolutum*

It is an evergreen shrub of spreading growth habit. Flowers are yellow, fragrant, and borne in clusters of 6 to 12 or more.

15.4.2.23 *J. rex* (Royal Jasmine, King's Jasmine)

It is a quick growing, strong climbing, evergreen vine, and is good for fences and trellises. Leaves are broad, oval, deep green, and are relatively large with 5–7 leaflets in a group. Flowers are large, about 5 cm in diameter, pure white, borne in profuse clusters, nearly scentless. It has the largest blooms among *Jasminum* species. With warm temperatures they can bloom at intervals throughout the year. It is an ideal container plant; also suitable for training into topiary forms.

15.4.2.24 *J. smilacifolium*

It is a large climbing shrub having simple and ovate leaves. Flowers are white with pink tinge.

15.4.2.25 *J. stephanense* (Stephan Jasmine)

It is a deciduous climbing vine. It climbs virtually any surface such as walls, fences, posts, pergolas or trees by using twining stems. The vine also grows well in a container or pot where it can be allowed to climb or dangle. It has green pinnate leaves that often bear cream markings. Flowers are soft pink and trumpet shaped with delicate fragrance.

15.4.2.26 *J. trinerve*

It is a climbing shrub with glabrous branches and ovate leaves. Flowers are white.

15.4.2.27 *Jasminum volubile* (Syn: *J. gracile*) (Australian or Wax Jasmine)

It is a scandent shrub similar in growth habit to *J. nitidum*. Leaves are simple opposite and dark green. Flowers are pure white, fragrant, and are borne in clusters at the branch terminals.

15.4.2.28 *J. wallichianum*

It is a hardy, evergreen shrub. Flowers are borne in terminal or axillary clusters; corolla yellow and triangularly lobed.

15.4.3 Ecotypes of Jasmine

The following is a brief description of some of the ecotypes of the commercial *Jasminum* species grown in India.

J. sambac		
Gundumalli	:	This is the most popular and widely cultivated type. It is popularly known as "Ramanathapuram Gundumalli." It is a triploid with chromosome number 2n = 3x = 39. The flowers are round with blunt petal tips. The flowers are highly fragrant. It is a high yielder and yields 7 to 8 t/ha of flower buds.
Ramabanam	:	It bears double type flower buds which are longer and rounded. It is a diploid with chromosome number 2n = 2x = 26. It is a high yielder.
Madanban	:	Flower buds are long and bold with short corolla tube. It is a diploid with chromosome number 2n = 2x = 26. It is a high yielder.
Single Mogra	:	Flowers have three or four whorls of petals. It is a diploid with chromosome number 2n = 2x = 26. Corolla tube length is medium. Leaf tips are blunt.
Double Mogra	:	Flowers have 8–10 whorls of petals (multi-whorled) and resemble rose; have good fragrance. It is a tetraploid with chromosome number 2n = 4x = 52. Plants are intermediate in growth habit with 6–9 forks/cyme. Flower buds are medium bold with short corolla tube length
Iruvatchi	:	Flowers have 3 to 4 whorls of petals (double type), with shorter corolla tube. Two levels of ploidy, namely, diploidy (2n = 2x = 26) and triploidy (2n = 3x = 39), are prevalent in this type. Plants have a spreading growth habit. Young leaves have anthocyanin content. Calyx is rudimentary.
Kasthurimalli	:	Flowers have medium long corolla tube. It is a diploid (2n = 2x = 26).
Oosimalli	:	Flower buds are long and slender. It is a diploid (2n = 2x = 26).
Soojimalli	:	Flower buds are long. It is a diploid (2n = 2x = 26). Plants are upright in growth habit. Leaf and petal tips are sharp. Flowers are borne in terminal clusters. Flower buds are pointed and long. Corolla is star shaped.
Khoya	:	Flower buds are short having less fragrance. It is a diploid with chromosome number 2n = 2x = 26. Plants are tall growing with 1–5 forks /cyme. Open flowers are multi-whorled.
J. grandiflorum		
Triploid	:	Large sized flowers with pinkish buds. It is highly sterile. Chromosome number is 39

15.4.4 Varieties of Jasmine

Jasmine crop improvement work in India is being carried out at Tamil Nadu Agricultural University (TNAU), Coimbatore, and the Indian Institute of Horticultural Research (IIHR), Bangalore. Till date there are nine improved varieties of the various *Jasminum* species evolved by these two institutions, six from TNAU and three from IIHR.

All these varieties are improved clones of *J. grandiflorum, J. auriculatum, J. multiflorum,* and *J. nitidum,* except for one induced mutant of *J. grandiflorum.*

The Indian varieties of jasmine are described in Table 1.
See Plate 1.

15.5 Characterization

DUS testing is a way of determining whether a newly bred variety differs from existing varieties within the same species (Distinctness), whether the characteristics used to establish distinctness are expressed uniformly (Uniformity) and that these characteristics do not change over subsequent generations (Stability). DUS tests are a form of intellectual property rights designed to safeguard the substantial economic investment involved in modern plant breeding.

The examination, or "DUS Test," is based mainly on growing tests, carried out by the authority competent for granting plant breeders' rights or by separate institutions, such as public research institutes, acting on behalf of that authority or, in some cases, on the basis of growing tests carried out by the breeder. The examination generates a description of the variety, using its relevant characteristics (e.g., plant height, leaf shape, time of flowering), by which it can be defined as a variety in terms of Article 1 (vi) of the 1991 Act of UPOV Convention.

DUS (Distinctness, Uniformity and Stability) Test Guidelines as per the Protection of Plant Varieties and Farmers' Rights Act (PPV&FR Act 2001) have been developed by the Indian Institute of Horticultural Research (IIHR), Bengaluru, and Tamil Nadu Agricultural University (TNAU), Tamil Nadu, India, for *Jasminum sambac, J. auriculatum,* and *J. multiflorum.* These guidelines find a place in the Plant Variety Journal of India which is the Official Journal of the Protection of Plant Varieties and Farmers' Rights Authority (PPV & FRA), New Delhi, India, as well as in the official website of PPV&FRA (http://www.plantauthority.gov.in/crop-guidelines.htm).

15.6 Molecular Approaches

Genetic relationships among eight varieties of *Jasminum sambac* and two varieties of *Jasminum grandiflorum* were compared by morphological and molecular (RAPD) profiles by Sampath et al. (2008) and the studies revealed that the morphological data was obtained for their vegetative and reproductive characters. PCR-amplifiable DNA was isolated using the CTAB method and 120 amplified fragments were obtained using eight random primers. The genetic dissimilarity matrix was calculated based on Squared Euclidian Distances, which revealed a maximum genetic distance of 83% between vars. 'Co-2 Pitchi' and 'Single Mohra', which belong to different species and the minimum genetic distance (21%) was between vars. 'Khoya' and 'Khoya Large' belonging to same species (*J. sambac*). The Ward's method of cluster analysis grouped all the individuals on the dendrogram into two major clusters 'A' and 'B' at 58 linkage distances with varieties of *J. sambac* and *J. grandiflorum*, respectively. The study showed moderate to high genetic diversity

Table 1 Varieties of jasmine developed by Indian institutions

(1) ***J. sambac***		
(1a) Arka Aradhana	:	Developed by IIHR, Bangalore in the year 2000 Ideal for loose flower and essential oil extraction Full yield starts from 3rd year onwards. It can be maintained up to 12 years Yield: 8 t/ha of flower buds per year Concrete yield is 14.95 kg/ha
(2) ***J. auriculatum***		
(2a) Pari Mullai	:	Developed by TNAU, Coimbatore in 1972 It is a clonal selection from a germplasm clone (Medium Point) The plants exhibit resistance to gall mite The yield is 7800 kg of flower buds/ha/year The buds are white with moderate corolla tube length (1.25 cm). The concrete recovery is 0.29%
(2b) **CO.1 Mullai**	:	Developed by TNAU, Coimbatore in 1980 It is a secondary clonal selection from a local type (Long Round) The flower buds are white and bold with longer corolla tube (1.50 cm) The yield is 8800 kg of flower buds/ha/year Concrete recovery is 0.34%
(2c) **CO.2 Mullai**	:	Developed by TNAU, Coimbatore, in 1988 It is a clonal selection from progenies with desirable flower bud characters like long corolla tube and flower bud The length of the corolla tube is 1.70 cm Average yield 11,198 kg of fresh flower buds/ha/year It exhibits complete field tolerance to the phyllody disease and gall mite infestation
(3) ***J. grandiflorum***		
(3a) **CO.1 Pitchi**	:	Developed by TNAU, Coimbatore in 1980 It is a secondary clonal selection from germplasm collection The average flower yield is 10.2 t/ha/year The flower buds are pink tinged with long corolla tube (4 cm) It is suitable for oil extraction with a concrete recovery of 0.29% The concrete yield is 29.4 kg per hectare
(3b) **CO.2** Pitchi	:	Developed by TNAU, Coimbatore, in 1991 It is an induced mutant developed by treating vegetative cuttings of CO.1 Pitchi with gamma rays (1.5 kR) This mutant is characterized by bold pink buds The flower buds are bold and pink with 4.18 cm length The 100 buds weight is 10 g (as against 9.4 g in CO.1) Concrete recovery is around 0.30%
(3c) **Arka Surabhi**	:	It is a variety evolved through clonal selection by IIHR, Bangalore, in 1993 Yield: 10 t/ha/year of flower buds Concrete recovery is 0.35% It is tolerant to drought conditions It gives good yield up to 12 years Average yield is 11.68 t of flower buds per hectare

(continued)

Table 1 (continued)

(4) ***J. multiflorum***		
(4a) Arka Arpan	:	Developed by IIHR, Bangalore in 1999 Flower buds are pink in color and mildly fragrant. Yield is around 6.1 t/ha It has attractive foliage and is used as ornamental plant in landscaping
(5) ***J. nitidum***		
(5a) CO.1 Star Jasmine	:	Developed by TNAU, Coimbatore, in 2019 Year-round flowering; flowers will be available during lean season/off-season (November–February) Attractive bold buds with bright purple pink color Opened flowers are pure white, star shaped and fragrant Easy to pluck and highly suitable for string-making due to bold buds with long corolla tube Flower bud yield is 7.41 t/ha/year Good keeping quality (buds remain unopened for 12 h under room temperature and for 60 h under refrigeration) Plants are relatively free of pests and diseases Ideal as decorative ornamental also owing to attractive plant architecture

among the both *Jasminum* spp. RAPD markers combined with morphological analysis proved to be a quick, simple, and significant testing method to assess genetic diversity among *Jasminum* spp.

Knowledge of genetic relationships in crops is important for genetic resource conservation, plant breeding, variety protection, and genetic evaluation. Morphological and pedigree estimates of genetic relatedness among genotypes are not precise due to use of agronomic traits as parameters and effects of environments in which plants are cultivated. Molecular markers not only are used to distinguish genotypes but also provide information about the genetic relatedness between genotypes and even phylogenetic relationships between species and related taxa. Marker profiling gives an indication that the closer the genetic relationship between a given pair of genotypes is, the larger will be the number of shared markers.

ISSR markers revealed abundant polymorphism at both interspecific and intraspecific levels, implying that all of them could be applied to germplasm identification and genetic diversity assessment in the genus *Jasminum*. Most informative primers identified from preliminary studies produced 1277 discrete amplified fragments (Yohanan et al. 2019). The amplified bands were in a size range of 250–2000 base pairs with an average of 17.5 fragments per primer combination. ISSR primer UBC-844 produced a maximum number of bands, that is, 22 and UBC-822 produced a minimum number of bands, that is, 13. For the analysis of genetic diversity, it is important to know the type of markers, how many of them represent scorable variation in the entire genome, and whether they should be used for diversity estimation. There have been several efforts to transfer agro-economically important genes from wild to cultivated ones through conventional breeding practices [24]. However, knowledge of genetic relationships among various wild species is necessary for successful and efficient exploitation of genetic diversity present in the wild

Plate 1. Jasmine varieties evolved by Tamil Nadu Agricultural University (TNAU), India

Jasminum auriculatum

Pari Mullai

CO.1 Mullai

CO.2 Mullai

Jasminum grandiflorum

CO.1 Pitchi

CO.2 Pitchi

Jasminum nitidum

CO.1 Star Jasmine

Plate 1 Jasmine varieties evolved by Tamil Nadu Agricultural University (TNAU), India

species and such information is not available in the genus *Jasminum*, especially using molecular markers. The authors used ISSR markers to determine genetic relationship within 40 accessions of 23 species and to determine whether the markers can be effectively used in assessing genetic diversity.

Assessment of molecular diversity in jasmine (*Jasminum* spp.) using RAPD and ISSR markers by Ghosh *et al.* (2018) revealed a total of 33 polymorphic primers (20 RAPD and 13 ISSR) were used. Amplification of genomic DNA of 18 genotypes, using RAPD analysis resulted in a total of 248 products out of which 234 (94.24%) products were polymorphic. The polymorphism percentage ranged from 80% to 100% with an average of 94.24% polymorphism per primer. The similarity matrix developed using the NTSYS-PC 2.02 software showed that Jaccard's similarity index ranged from 0.62 to 0.89. In ISSR analysis, among 13 tested primers, a total of 595 products were produced out of which 562 (94.07%) products were polymorphic. The polymorphism ranged from 82.60% to 100% with an average of 94.07% polymorphism per primer. Based on the similarity matrix data dendrogram were prepared using UPGMA method. Genotypes were also classified into groups and several subgroups. The dendrogram revealed that the genotypes varied distinctly, with *J. primulinum* forming a separate cluster away from the other 17 genotypes.

Molecular markers offer an excellent alternative in development of improved disease resistant cultivars that would lead to increase in crop yield. They are employed for tagging the important disease resistance genes and provide valuable assistance in increasing selection efficiency for valuable traits via marker assisted selection (MAS).

PCR-based detection assay was developed to detect and distinguish these two viruses in several Hawaii-grown plants. Star jasmine (*Jasminum multiflorum*) plants growing in Hawaii expressing a diverse array of virus-like foliar symptoms were examined for the presence of a causal agent. Symptomatic tissues collected from three locations on the island of Oahu, Hawaii, consistently harbored double-stranded (ds)RNAs approximately 4.2 and 1.7 kbp in size. Sanger and high-throughput sequencing approaches revealed these dsRNAs were from two distinct virus species co-infecting the same host plant. One of these two viruses was the recently characterized Jasmine virus H (JaVH), and the second was designated as Jasmine mosaic associated virus (JMaV). The study also demonstrates that the recently described JaHV also appears widespread in the USA, being found in California, Hawaii, Maryland, and Washington, DC. The sequencing of two isolates of JaVH increases our understanding of the genetic diversity of this virus, and the virome of jasmine.

Arabian jasmine (*J. sambac* L.) plants showing witches broom (WB) symptoms were found in two regions in the Sultanate of Oman. Polymerase chain reaction (PCR) amplification of the 16S rRNA gene and the 16S–23S spacer region utilizing phytoplasma-specific universal and designed primer pairs, and transmission electron microscopy of phytoplasma-like structures in phloem elements confirmed phytoplasma infection in the symptomatic plants. PCR products primed with the P1/P7 primer pair were 1804 bp for jasmine witches broom (JasWB) and 1805 bp for alfalfa (*Medicago sativa* L.) witches broom (AlfWB). Actual and putative restriction fragment length polymorphic analysis indicated that jasmine and AlfWB phytoplasmas were molecularly indistinguishable from each other and closely related to papaya yellow crinkle (PYC), as well as being distinct from lime WB (LWB) and Omani alfalfa WB (OmAlfWB) phytoplasmas.

Wide genetic diversity as detected in this study has provided a solid plat form for further improvement of jasmine flower as aromatic floricultural crop. Developing new ornamental cultivars with improved floral attributes is a major goal in floriculture. Biotechnological approaches such as tissue culture and micropropagation techniques, polyploidy induction, mutation breeding, and genetic engineering have been used to develop many varieties of ornamental plants. Plant with new characteristics like floral architecture, color, fragrance, resistance to abiotic stress, and post-harvest life can be produced through biotechnology.

15.7 Floral Volatiles Emission Mechanism in Jasmine Flower

Floral scent emission is a time-controlled process guided by an internal circadian regulator. The VOCs emitted as floral scents are biosynthesized and present in floral tissues mostly as liquids or in solution. The physiological regulation controls the endogenous concentration which determines the emission rate of the scented VOCs. Many fragrance precursors are stored as glycosidically bound volatile components, mostly in nonvolatile form and are enzymatically transformed into volatile compounds during flower opening.

VOCs mainly benzenoids and terpenoids were identified from both the jasmine species. Most of the emitted VOCs of *J. auriculatum* showed the maximum rate of emission within few hours of floral opening in the evening and gradually decreased after midnight. The emitted VOCs of *J. multiflorum* showed a maximum rate of emission in the afternoon. The free endogenous concentrations of all VOCs were higher when corresponding emitted concentrations were high in both the species and vice versa. The biosynthesis of scent within the flower tissue is maintained; however, some external factors like temperature, light, or humidity might be responsible for slow release of the compounds from flower tissues after a particular time. Enzymatic treatment of floral extracts revealed that several alcoholic volatile compounds are biosynthesized and stored in the flowers as water-soluble glycosides. The concentrations of bound volatiles were higher at late bud stage.

Composition of floral scent of *Jasminum sambac* (L.) Aiton (Oleaceae) includes three major benzenoid esters – benzyl acetate, methyl benzoate, and methyl salicylate, and three major terpene compounds, viz., (E)-β-ocimene, linalool, and α-farnesene (Bera et al. 2015). In addition to emitted volatiles, a number of endogenous (non-emitted) floral volatiles were detected. These include monoterpenes, aromatic, and aliphatic alcohols, which were present in the form of water-soluble glycosides in the floral tissue.

The comprehensive profile of emitted volatiles from *J. sambac* flower using different adsorbent materials revealed that no single adsorbent alone can trap all the volatiles emitted from flower. However, when compared with other polymeric adsorbents, the performance of Porapak Q was found satisfactory and thus chosen to study the developmental emission rates of the dominating compounds present in the emitted floral volatiles. The concentrations and emission rates of benzenoids and terpenoids during the developmental stages of *J. sambac* flower are also analyzed. In addition to spatial emission from different floral parts, time-course mRNA accumulations of phenylalanine ammonia-lyase (PAL) and the two representative genes of terpenoid pathway, namely, 1-deoxy-D-xylulose 5-phosphate reductoisomerase (DXR), and terpene synthase (TPS), were also studied.

Using degenerate primer approach, the core cDNA fragment of all the above genes including ODORANT1 (ODO1) were isolated and cloned. ODO1 is a members of R2R3-type MYB transcription factor, which plays a regulatory role in floral volatile benzenoids synthesis. Transcript levels of ODO1 were increased before the onset of volatile emission and decreased upon the declination of volatile emission, suggesting that the benzenoids emission in *J. sambac* flower is partially regulated by ODO1. Further, *in vitro* activities of several enzymes of phenylpropanoid/benzenoid pathway, viz., PAL and acetyl-coenzyme A: benzylalcohol acetyltransferase (BEAT), *S*-adenosyl-L-methionine: benzoic acid carboxyl methyltransferase (BAMT), and S-adenosyl-L-methionine: salicylic acid carboxyl methyltransferase (SAMT), were examined throughout the floral life span. All the above enzyme activities along with the *in vitro* activities of DXR and TPS were found to follow a certain rhythm as observed in the emission of different benzenoid and terpenoid compounds.

Linalool emission peaked after petal opening and coincided with maximal expression of *JsTPS* gene as evidenced from RT-PCR analyses (semi-quantitative). The maximum transcript accumulation of this gene was observed in flower petals, indicating that the petals of *J. sambac* flower serve as a major contributor of volatile precursors (Bera et al. 2017). The transcripts accumulation of JsDXR and JsTPS in different developmental stages and in different floral part showed that emissions of terpenoid volatiles in *J. sambac* flower are partially regulated at transcription levels.

15.8 Breeding Options

15.8.1 History of Jasmine Breeding

Tamil Nadu Agricultural University (Coimbatore) is the pioneer in jasmine breeding in India. Considerable jasmine crop improvement work has been carried out at Tamil Nadu Agricultural University and the Indian Institute of Horticultural Research (Bangalore).

15.8.2 Breeding Methods for Jasmine

The following are the breeding methods for jasmine crop improvement:

(i) Clonal selection
(ii) Open pollinated seedling selection
(iii) Hybridization
(iv) Mutation breeding
(v) Ploidy breeding

15.8.2.1 Clonal Selection

Five improved clones of jasmine from TNAU, Coimbatore, and three from IIHR, Bangalore, have been evolved and released for commercial cultivation (Table 1 and Plate 1).

15.8.2.2 Open Pollinated Seedling Selection

Screening of open pollinated seedlings is a potential method for crop improvement in jasmine. Superior open pollinated seedlings of *J. auriculatum* have been reported by many earlier workers (Alikhan et al. 1969; Muthuswamy et al. 1972; Thangaraj et al. 1981; Bhupal Rao 1980; Muthukrishnan and Muthuswamy 1982; Veluswamy 1981; Indiresh et al. 1989) and it was inferred that OP seedling selection offers considerable scope for genetic improvement of many of the characters by selection.

Veluswamy (1981) evaluated open pollinated seedlings of Coimbatore clone of *J. grandiflorum* and found wide range of variation. A spontaneous variant with high seed setting potential and an open pollinated seedling variant bearing petaloid anthers have also been reported in *J. grandiflorum*

15.8.2.3 Hybridization

Genetic improvement in jasmine through hybridization has remained to be complicated till date. The most important barrier is non-fruitfulness is most species and cross combinations.

Genetic improvement in jasmines through hybridization is complicated, owing to the various constraints which are listed below.

- Nocturnal flowering habit
- Insect interference due to fragrance
- Difficulty in emasculation and pollination due to long and delicate style
- Non fruitfulness
- Only one seed develops per fruit
- Growth of seedling is slow

Bud pollination can be a method to avoid insect interference in jasmine hybridization (Bhupal Rao et al. 1977).

15.8.3 Fruit Setting Potential

In hybridization, fruit setting potential is an essential requirement for a species to make an ideal female parent. Fruit setting potentials vary widely among the *Jasminum* species. *J. auriculatum* and *J. flexile* have been found to have high fruit set rates (Bhatnagar 1960). In *J. grandiflorum,* only the diploid type sets fruit (Khan et al. 1969). High fruit set was observed in *J. auriculatum* (cv. CO. 1 Mullai), *J. flexile, J. calophyllum,* and *J. multiflorum* (Pink) under natural open pollination and artificial self-pollination and cross-pollination (Lakshmi 2017).

15.8.4 Natural Seed Set

Under natural conditions, only a few species of jasmine set seeds. *J. arborescens*, *J. auriculatum*, *J. angustifolium*, *J. calophyllum*, *J. flexile*, *J. grandiflorum*, *J. mesnyi*, *J. pubescens*, *J. rigidum*, *J. suavissinum,* and certain varieties of *J. sambac* have been reported to set seeds (Bhattacharjee 1978). Negligible seed set has been recorded in the morphotypes "Pink Pin" and "Pink Thrum" of *J. grandiflorum* and "Gundumalli" type of *J. sambac*, while 30% seed set in *J. calophyllum* and 50% in *J. auriculatum* variety 'Parimullai' were recorded.

15.8.5 Seed Set Under Artificial Pollination

Among a large number of crosses attempted involving the species *J. arborescens*, *J. calophyllum*, *J. flexile*, *J. grandiflorum*, *J. humile* var. 'Wallichianum', *J. nitidum,* and *J. rigidum,* most cross combinations failed to set seed and certain combinations resulted in highly shriveled seeds (Anon 1974). Among several intervarietal crosses of *J. grandiflorum*, the combination Seed set selection x 'Pink Pin' yielded 6.8% seed set. In others seed set was 0 to 3.2%. Hybridization of *J. pubescens* with several other species failed to yield any seed set.

Lakshmi (2017) observed fruit set in interspecific cross combinations involving *J. flexile*, *J. calophyllum,* and *J. multiflorum* (Pink) as female parent. The fruit set rate was 7.21% in *J. flexile* x *J. sambac* cv. Ramanathapuram Gundumalli, 70.79% in *J. flexile* x *J. rigidum,* and 86.75% in *J. calophyllum* x *J. rigidum*. However, seed set was not recorded in any of the successful interspecific cross combinations which resulted in fruit set. The fruit was retained on the plant for 30–45 days after which they turned necrotic and dried or dropped off. Analysis of the ovules of *J. flexile* x *J. auriculatum* cv. CO. 1 Mullai with TTC staining technique indicated viability at 15 days after pollination (DAP) and loss of viability at 30 DAP. Histological sections revealed necrosis of ovules in the crosses *J. multiflorum* (Pink) x *J. grandiflorum* (White) and *J. flexile* x *J. auriculatum* cv. CO. 1 Mullai. In the crosses *J. calophyllum* x *J. sambac* cv. Ramanathapuram Gundumalli and *J. auriculatum* cv. CO. 1 Mullai x *J. rigidum*, ovule development was absent, indicating failure in fertilization.

15.8.6 Seed Germination

Freshly harvested seeds have higher germination percentage. Time taken for germination varies from 3 weeks to 3 months depending upon the species. Seeds retain viability up to 10 months after which there is a rapid decline of germination. Higher temperature and high humidity hasten germination. Considerable seed germination rates have been reported in *J. auriculatum* (84%), *J. grandiflorum* (74%), and *J. flexile* (55%) (Veluswamy et al. 1975). Maximum germination of *J. auriculatum* seeds occurred in saw dust and sphagnum moss media (Veluswamy et al. 1974).

15.8.6.1 Mutation Breeding in Jasmine

India has a long history of induced mutagenesis in different species of jasmine. Occurrence of one bud sport has been reported by JVR and Krishnan (1980) in *J. auriculatum* Vahl. The improved characters of the mutants were enhanced length and width of flower bud, length of corolla tube, diameter of open flower, number of petals, length and width of petal, and weight of 100 flower buds. Veluswamy et al. (1976) in their mutation studies observed three white flowering plants from the mutated seeds of *J. grandiflorum* (pink flower type) which paved way to claim that *J. officinale* could be a natural bud sport of *J. grandiflorum* type which has attained the species status as *J. officinale,* but there is very little difference between these two species except for flower bud color.

Inducing a large spectrum of variability with employment of wide range of physical and chemical mutagens followed by cycles of recombination and selection can result in improvement of desirable attributes which otherwise may not respond to selection (Bhupal Rao et al. 1977).

In a mutation breeding experiment with *J. grandiflorum*, Nambisan et al. (1980) obtained two induced mutants, one with dwarf stature and the other with tolerance to leaf spot disease (*Cercospora jasminicola*). The yield of flowers and recovery of concrete were however less than in the parent clone "Thimmapuram."

Subjecting terminal cuttings of *J. grandiflorum* var. CO-1 Pitchi to gamma radiation at dose range of 3 to 5 krad resulted in reduction of sprouting and rooting rates (Kumar et al. 1983). Lethal dose for gamma irradiation of *J. grandiflorum* var. 'Pink Pin' was optimized by Devaiah and Srivastava (1989). It was found that LD_{50} value was close to 2.5 krad, 0.5 krad for 'Pink Thrum', close to 2.5 krad for *J. flexile*, close to 1 krad for *J. calophyllum* and 2 krad for *J. sambac* cv. Gundumalli. Linear reduction in rooting percentage, number of roots per cuttings and length and thickness of roots was also recorded with increase in intensity of gamma irradiation.

One variegated mutant of *J. auriculatum* with dwarf stature resulting from treatment of seeds with gamma irradiation and EMS was reported to possess ornamental value (Chezhiyan et al. 1984).

LD_{50} of gamma radiation for five genotypes of *J. sambac* and *J. grandiflorum* and effect of such radiation on their rooting parameters were studied Devaiah and Srivastava (1989). The result indicated that the LD_{50} was close to 2.5 krad for *Jasminum grandiflorum* var. Pink Pin and close to 0.5 krad for var. Pink Thrum, close to 2.5 krad for *J. flexile* Vahl. close to 1 krad for *J. calophyllum* Wall and 2 krad for *J. sambac* Ait var. 'Gundumalli'. Percentage of rooting, number of roots per cutting, and length and thickness of roots decreased with increase in intensity of gamma irradiation.

Soft wood cuttings of variety CO.1 Pitchi were subjected to 0.5 kR which led to release of 'CO.2 Pitchi' in 1992. This mutant is characterized by bold pink buds, long flower bud (4.18 cm), and high yield (11,680 kg/ha) with a concrete recovery of 0.30%.

Mutagenic treatment of semi-hard wood cuttings of *J. sambac* cv. Gundumalli with 2.0 kR followed by 1.5 kR + 30 mM EMS manifested higher variability in terms of PCV and GCV (Mekala et al. 2010). The floral characters bud length and flower weight decreased with increase in the doses while the flower width did not follow a definite trend. Induced mutations produced higher amount of phenotypic and genotypic co-efficient of variation for number of flowers and flower weight, indicating the scope for improvement of Gundumalli for yield and novelty by selection. Variability assessment of M_1V_2 population revealed high GCV for the morphological traits plant height, number of secondary and tertiary branches, and the floral trait number of flower buds per plant (Saranraj and Kannan 2013). High heritability with high genetic advance was observed for number of flower buds and yield of flower buds per plant.

Induced mutagenesis in five jasmine genotypes, namely, cv. White Pitchi and CO.1 Pitchi of *J. grandiflorum*, cv. CO.1 Mullai of *J. auriculatum*, cv. CO.1 Star Jasmine of *J. nitidum,* and cv. Arka Arpan of *J. multiflorum,* indicated that LD_{50} values for mutagen treatment with gamma rays and EMS were found to be in range of 19–20 Gy and 35–40 mM, respectively, for cv. CO.1 Pitchi, 15–20 Gy and 30–35 mM, respectively, for cv. White Pitchi,15–20 Gy and 40–45 mM, respectively, for cv. CO.1 Mullai, 25–30 Gy and 35–40 mM, respectively, for cv. Arka Arpan and 20–25 Gy and 40–45 mM, respectively, for cv. CO.1 Star Jasmine (Ghosh *et al.*, 2018). The study also revealed that the mutation effectiveness and efficiency were higher in gamma ray treatments compared to EMS.

15.8.6.2 Polyploidy Breeding in Jasmine

Artificial induction of polyploidy has been attempted and found feasible in some of the jasmine genotypes. Generally, triploids have been found to be more vigorous and higher yielders than diploids while tetraploids are found to be robust, hardy, and produce bolder flowers (Veluswamy *et al.*, 1980).

In *J. sambac,* tetraploids were more vigorous but produced lower number of bold flowers per plant (Alikhan *et al.* 1969). Triploidy in *J. grandiflorum* was found to increase concrete content, thereby indicating that ploidy breeding was a potential strategy for improvement of this species (Srivastava 1995). In *J. grandiflorum,* seed treatment with 0.5% colchicine resulted in six tetraploids which were, however, not found to be economically useful.

In *J. auriculatum*, seed and seedling treatment with 1% colchicine recorded 11.4–23.1% polyploidy. One tetraploid was obtained from Parimullai seeds treated with 0.5% colchicine for 12 h (Alikhan *et al.* 1969). In general, tetraploids were found to be poor yielders with low concrete recovery.

Induction of mutation through colchicine in two diploid genotypes, namely, White Pitchi (*J. grandiflorum*) and CO.1 Star Jasmine (*J. nitidum*), revealed that higher concentrations of colchicine resulted in compact growth habit with thicker and darker green leaves, while lower concentrations induced early flowering and higher number of flowering cymes/branch (Ghosh *et al.*, 2018). The flower quality parameters, namely, flower diameter, length of petal, width of petal, flower bud length, corolla tube length, and bud girth, were superior with 0.07% colchicine in cv. White Pitchi and 0.2% in CO.1 Star Jasmine (*J. nitidum*).

15.9 Future Perspective

Jasminum species are highly valuable genetic material which can be involved in jasmine breeding programs. Biotechnological tools such as embryo rescue can pave way for breaking the hybridization barriers posed by endosperm antagonism and hybrid inviability.

References

Alikhan MW, Raman VS, Raman KR (1969) South Indian Hortic 17:79–85

Anonymous (1959) Wealth of India. V. CSIR, New Delhi

Anonymous (1972) Encyclopaedia Britannica. Encyclopaedia Britannica Ltd, Chicago

Anonymous (1974) Second annual progress report. scheme for intensification of research on the development of rose and jasmine varieties with high essential oil content, ICAR. Tamil Nadu Agricultural University, Coimbatore

Bailey LH (1947) Standard cyclopedia of horticulture, II. MacMillan, New York, pp 1718–1719

Bailey, L. H. 1958. Manual of cultivated plants. McMillan, New York, 5(6): 192–202

Bera P, Kotamreddy JNR, Samanta T, Maiti S, Mitra A (2015) Inter-specific variation in headspace scent volatiles composition of four commercially cultivated jasmine flowers. Nat Prod Res 29: 1328–1335

Bera P, Mukherjee C, Mitra A (2017) Enzymatic production and emission of floral scent volatiles in *Jasminum sambac*. Plant Sci 256:25–28

Bhatnagar GS (1960) Studies on the biology of *Jasminum multiflorum* (Burm. F.) Andr. (*Jasminum pubescens* Willd.). Sci Cult 26:192–195

Bhattacharjee SK (1978) Jasmine- a most rewarding plant. J Sci Club 32(3 & 4):77–83

Bhupal Rao JVR (1980) A spontaneous variant with large flowers in *Jasminum auriculatum*. Vahl Curr Sci 49(8):323

Bhupal Rao JVR, Divakar NG, Negi SS (1977) Occurrence of floral dimorphism in *Jasminum grandiflorum* L. Indian Perfumer 21(3):160–161

Bor NL, Raizada MB (1946) Some beautiful Indian climbers and shrubs. J Bom Nat Hist Soc 46: 205–215

Chezhiyan N, Ponnuswamy V, Thamburaj S, Khader JMA (1984) Evaluation of certain hybrid progenies of *Jasminum auriculatum*. South Indian Hort 32(1):223–228

Cooke T (1905) The flora of Presidency of Bombay. Taylor and Francis, London

Devaiah KA, Srivastava HC (1989) Sensitiveness of jasmine cuttings to gamma radiation. Herba Hungarica 28(3):11 14

Gamble JS (1957) Flora of the Presidency of Madras, vol 09. Botanical Survey of India, Calcutta

Ganga M, Ranchana P, Ganesh S, Kannan M (2015) *Jasminum nitidum* – a potential unexploited jasmine species. In: Third international symposium on underutilized plant species – exploration and conservation for future generation, Madurai, Tamil Nadu, India

Ghosh S, Ganga M, Joel J (2018) Differential response of *Jasminum* spp. to propagation environment. J Crop Weed 14(3):18–24

Haines HH (1961) The botany of Bihar and Orissa, I-VI, 3rd edn. Botanical Survey of India, pp 1350

Indiresh KM, Uthaiah BC, Bhupal Rao JVR (1989) Variations for essential oil contents in *J auriculatum* Vahl. Indian Perfumer 33(1):44–47

Jvr BR, Krishnan R (1980) A spontaneous variant with larger flower in *Jasminum auriclatum* vahl. Curr Sci 49(8):323

Khan MA, Muthuswamy S, Raman KR (1969) New chromosomal forms of superior ornamental value in *Jasminum*. S Indian Hort 17:79–85

Krishnaswamy N, Raman VS (1948) Taxonomy and chromosome numbers of four varieties of *J. sambac* and other species. J Indian Bot Soc 1–7

Kumar N, Sambandamurthy S, Khadar JMA (1983) Sensitiveness of *Jasminum grandiflorum* cuttings to gamma radiation. Sci Ind Hortic 3194(5):250–251

Lakshmi J (2017) Interspecific hybridization and grafting studies in jasmine for off-season flowering and flower quality. PhD thesis, Tami Nadu Agricultural University

Mekala P, Kumar N, Kannan M, Vijayakumar G (2010) Effect of physical and chemical mutagens on floral characters of *Jasminum sambac* cv. Gundumalli and isolation of important mutants. J Ornament Hortic 13(3):200–206

Murthy AS, Khanna KR (1971) Cytological investigations in Jasmines. Curr Sci 40:555–566

Muthukrishnan CR, Muthuswamy S (1982) Studies on the variability, genetic advance and heritability in flower characters of open pollinated seedlings of *Jasminum auriculatum* Vahl. Indian J Hortic 39(1 and 2):125–129

Muthuswamy S, Khan WMA, Sayed S (1972) The extraction of jasmine concrete from *Jasminum auriculatum*, Vahl. Curr Sci 41(5):194–194

Nambisan K, Krishnan B, Veeraragavathatham D, Ramasamy N (1980) Induced mutants in jasmine (*Jasminum grandiflorum* l.): leaf-spot resistant and dwarf mutants. Sci Cult 46(12):427–428

Raman VS (1955) Cytogenetics of Indian Jasmine, the somatic chromosomes. Cytologia 20:19–31

Raman KR, Shanmugam A, Ahmad Sha H (1969) Studies on flowering habits and flower yields of some *Jasminum* species. South Indian Hortic 17:18–27

Sampath M, Sathyanarayana BN, Suresh NS, Simon L (2008) Analysis of genetic diversity among *Jasminum sambac* and *Jasminum grandiflorum* and molecular markers. Floric Ornament Biotechnol 52:60–64

Saranraj N, Kannan M (2013) Evaluation of v2 m1 generation of *Jasminum sambac* cv. Gundumalli (Ait.). Asian J Hortic 8(1):75–80

Srivastava HC (1995) French Jasmine. In: Chanda KD, Gupta R (eds) Advances in horticulture, vol VII. Malhotra Publishing House, New Delhi, pp 805–583

Thangaraj T, Muthukrishnan CR, Muthuswami S (1981) Studies on the evaluation and selection from open pollinated progenies of *Jasminum auriculatum* Vahl. South Indian Hort 29:143–148

Veluswamy P (1981) Genetic studies in *Jasminum auriculatum* Vahl. and *J. grandiflorum* L. Doctoral dissertation, Tamil Nadu Agricultural University, Coimbatore

Veluswamy P, Muthuswami S, Thangaraj T (1973) Origin and distribution of jasmine. Indian J Ornament Hortic 4:45–55

Veluswamy P, Muthuswami S, Thangaraj T (1974) A study on germination of seeds of some *Jasminum* species and clones. South Indian Hortic 22:127–128

Veluswamy P, Thangaraj T, Muthuswamy S (1975) A study on germination of seeds of some *Jasminum* species and clones. South Indian Hortic 23:71–72

Veluswamy P, Thangaraj T, Muthuswamy S (1976) Studies on anthesis and pollination in some *Jasminum* species and varieties. South Indian Hortic 24:45–49

Veluswamy P, Viiayakumar M, Muthuswami S (1980) Grafting in Jasmines. South Indian Hortic 28:156–157

Yohanan R, Jeyarani NJ, Devipriya V, Rather SA, Kasana S, Thakur J, Dwivedi MD, Pandey AK (2019) Evaluating genetic diversity within genus *Jasminum* L. (Oleaceae) using intersimple sequence repeats (ISSR) marker. Proc Natl Acad Sci India Sect B Biol Sci 90:531

Hibiscus (*Hibiscus rosa-sinensis*): Importance and Classification 16

Pablito M. Magdalita and Alangelico O. San Pascual

Contents

P. M. Magdalita (✉)
Institute of Crop Science, College of Agriculture and Food Science, University of the Philippines Los Baηos, College, Los Banos, Philippines

Institute of Plant Breeding, College of Agriculture and Food Science, University of the Philippines Los Baηos, College, Los Banos, Philippines
e-mail: pmmagdalita@up.edu.ph

A. O. San Pascual
Institute of Plant Breeding, College of Agriculture and Food Science, University of the Philippines Los Baηos, College, Los Banos, Philippines
e-mail: aosanpascual@up.edu.ph

S. K. Datta, Y. C. Gupta (eds.), *Floriculture and Ornamental Plants*, Handbooks of Crop Diversity: Conservation and Use of Plant Genetic Resources,
https://doi.org/10.1007/978-981-15-3518-5_18

Abstract

Hibiscus comprises a large genus of flowering plants under the Malvaceae family. One important Hibiscus cultivated worldwide is *Hibiscus rosa-sinensis*. It is a renowned ornamental plant that comes with different flower forms and colors. It is considered as the queen of Filipino garden in the Philippines. It is considered as the national flower of countries like Malaysia and Hawaii.

Aside from ornamental importance, Hibiscus is also used for feed, food, industrial, and medicinal preparations. It is used as colorant, component of salads, developed into jams, among others. Its extracts were used to remedy hair fall and dandruff. Further, different plant parts were used to treat infections. It also has expectorant, diuretic, emollient effects, among others. More than its food, feed, industrial, and medical uses, Hibiscus also has cultural significance and symbolism especially in Haiti, Tahiti, and Hawaii. The origin of Hibiscus is still uncertain but believed to come from India, China, or Americas as stated by some studies.

Hibiscus rosa-sinensis is widely distributed in tropical and subtropical conditions with different cultivars with different flower color and form. Other related species of *Hibiscus rosa-sinensis* were found to be hardy in winter. Further, other related species of Hibiscus have economic importance. They also possess important traits such as resistance or tolerance to pests and diseases, hardiness, drought or waterlogging tolerance, etc. These species can be used for genetic improvement of Hibiscus. However, extensive intraspecific breeding were done in Hibiscus around the world producing thousands of cultivars.

In the Philippines, several cultivars were released as Series by the University of the Philippines Los Baños (UPLB). Collectively, more than 40 hybrids were released by UPLB. In addition, private breeders in the Philippines and in different parts of the world also produced and released Hibiscus hybrids. Also, a cultivar library for majority of the hybrids was developed by the International Hibiscus Society where a search engine and a genealogy search can be used by breeders and collectors. Further, the chapter presents some of the cultivars released in the Philippines by UPLB, by private breeders, and other international breeders.

Keywords

Breeding · Germplasm · Hibiscus · Hibiscus rosa-sinensis · Rose mallow

16.1 Introduction

The *Hibiscus*, or rose mallow, is a large genus of about 300 species of flowering plants in the family Malvaceae and tribe Hibisceae (Purseglove 1987). The *Hibiscus* genus is quite large, comprising several hundred species that are native to warm-temperate, subtropical, and tropical regions throughout the world. The genus

Hibiscus includes both annual and perennial herbaceous plants, as well as woody shrubs and small trees. Member species are renowned for their large, showy flowers and are commonly known simply as *Hibiscus*. The natural beauty of the *Hibiscus* makes it one of the most widely cultivated flowers. The flower is short-lived at 1–2 days in the plant, but continuing blooms occur each day.

They are commonly called "gumamelas" in the Philippines, and the many splendid varieties are closest to the Filipino gardeners' heart. They are popularly grown in front yards either as ornamental plants, hedge, or fence as well as bleaching platforms ("kulahan") for clothes. The common names are China rose, Chinese *Hibiscus*, Ambashthaki, Bissap, Congura, Groseille de guinee, Guinea sorrel, Hibisco, *Hibiscus* Calyx, *Hibiscus sabdariffa*, Jamaica sorrel, Karkade, Oseille de guinee, Oseille rouge, Pulicha keerai, red sorrel, red tea, Rosa de Jamaica, "antolañga"/"antolañgan" in "Tagalog" and "Bisaya," "arotañgan" in "Pampanga," roselle, and so. In different areas of the Philippines, the *Hibiscus* is known in various names like "gomamela" or "gumamela" in "Tagalog," "Bisaya," and "Pampanga"; "kayanga" in "Ilokano," "Bikol," and "Bisaya"; "saysaya" in "Bontoc"; "tapolanga" in "Tagalog" and "Pampanga"; "tapuranga" in "Bisaya"; "tarokanga" in "Bisaya" and "Pampanga"; and "taukangga" in "Sulu, Mindanao" (Almario 2015).

The Chinese traders are believed to have introduced *Hibiscus* into the Philippines, long before the coming of the Spaniards. Father Manuel Blanco, a Spanish priest, first described *Hibiscus* in the Philippines in his book *Flora de Filipinas* published in 1883. The American colonizers who came to the Philippines in the 1990's introduced new Hibiscus forms and cultivars. The *Hibiscus rosa-sinensis* is the national flower and symbol of several nations and states. It is the national flower of Korea and Hawaii. It has been cultivated for centuries in tropical Asia (Magdalita and Pimentel 2013). They are large shrubs or small trees that produce huge, colorful, trumpet-shaped flowers over a long season. The "gumamelas" are deciduous shrubs with dark green leaves, and the plants can grow to 15 feet tall in frost-free areas. The flowers may be up to 6 inches in diameter, with colors ranging from yellow to peach to red. They can be planted singly or grown as a hedge plant. They can be pruned into a single-stemmed small tree.

16.2 Importance and Uses

Aside from being an ornamental plant, the *Hibiscus* is used for food, feed, industrial, and medicinal preparations (Rummel 2005). In Ceylon, the juice of the flowers is extracted and used for blackening shoes, while the Chinese and the Hindus use it for blackening their eyebrows (Burkill n.d.). The fresh flowers are also used as food coloring and as component of vegetable salad, while in Mexico, the dried flowers are considered edible, and they are used as a special delicacy and for making jams. It contains a little amount of the compound hibiscetin that is being used as sources of extracts like benzene and alcohol which possess antiestrogenic or antifertility properties (Kholkute and Udupa 1976).

At present, gumamela tea is gaining a worldwide popularity because the gumamela has been associated with longevity. The natives of Southern India use the red *Hibiscus* flower for hair care purposes. The extracts from the flowers and the leaves can be applied to remedy hair-fall and dandruff by inducing the hair follicles to make protective oils. The juice extracted from the petals can be mixed with olive oil and then boiled. This is used to stimulate hair growth and improve hair color (Stuart 2019).

The different parts of the *Hibiscus* like the flower, leaves, and roots are used as herbal medicine to treat different illnesses. In general, it is used for treating the following diseases: bronchitis as an expectorant, coughs and sore throat and fever as a refrigerant drink, dysentery, urinary tract infection and bladder infections, high blood pressure, constipation, headaches, boils, swelling, abscesses, and mumps (Barraquia et al. 2017). The root decoction is considered to have medicinal properties against various ailments in the Indian traditional medicine. Similarly, in the Philippines, the *Hibiscus* is used as an expectorant, diuretic, emollient, anti-infectious, anti-inflammatory, antipyretic, anodyne, and refrigerant (Stuart 2019).

In the Dutch Indies, the midwives apply the mucilage of the *Hibiscus* flower while a mother is in labor and at the same time give the mother draughts of the juice from the leaves (Magdalita and Pimentel 2013). The juice of the leaves is used by midwives to stimulate the expulsion of the afterbirth after delivery of the baby (Magdalita et al. 2016). The red flowers are used to regulate menstruation and are purgative. In addition, the flowers are fried in clarified butter and administered for suppressing excessive menstruation. The Chinese use the flowers against paralysis and dysmenorrhea.

In folk medicine, mild blood pressure is treated with *Hibiscus* tea. Numerous *in vitro* experiments have evaluated the effects of anthocyanin extracts from the *Hibiscus* flower against various cancer cell lines. Proposed mechanisms of action of *Hibiscus* were focused on its antioxidant activity and the ability to induce apoptosis, loss of appetite, colds, constipation, irritated stomach, fluid retention, and heart and nerve diseases. *Hibiscus* can be used also to treat upper respiratory tract pain, inflammation, and disorders of circulation.

In terms of symbolism and culture, the *Hibiscus* is being used as the national flower of Haiti. The *Hibiscus* species also represents several other nations. The *Hibiscus syriacus* is the national flower of South Korea, and *Hibiscus rosa-sinensis* is the national flower of Malaysia.

In the Philippines, the *Hibiscus* is used by children as part of a bubble-making pastime. The flowers and leaves are crushed until the sticky juices come out. Together with soap, *Hibiscus* juices produce more bubbles.

Doring (2020) also added that in Hawaii and Tahiti, the flowers are worn by the girls. The position where it is worn signifies a civil status. If worn in the left ear, the women are in a relationship, while when worn in the right, the woman is single.

Hibiscus produces showy flowers that attract insects such as butterflies and hummingbirds. The plant is also a versatile and hardy shrub when exposed to tropical conditions. However, some are tolerant to winter conditions. Its versatility makes it a good potted plant. In addition to its versatility is its continuous flowering habit throughout the year. It is very responsive to fertilizers.

16.3 Origin and Botany

The *Hibiscus* is a genus of flowering plants in the mallow family called Malvaceae. It is uncertain if the *Hibiscus* is a native of China even its Latin name, *Hibiscus rosa-sinensis*, wherein *rosa-sinensis* equates to Chinese Rose. Many people believed that the *Hibiscus* came from India. The genus *Hibiscus* is quite large, containing several hundred species that are native to the warm-temperate, subtropical, and tropical regions throughout the world. They can be grown in large containers under full sun. The genus *Hibiscus* has almost 232 species described with all the varieties and forms known to grow in full sun. The *Hibiscus* shrub is bushy and has several stems with a coarse texture and has either an upright or broad-spreading growth habit.

The Australian *Hibiscus* Society (2004) classified the bush height of *Hibiscus* into four major types, namely, low type whose height is less than 1 m, medium type whose height is from 1 to 1.5 m tall, average type that stands from 1.5 to 2.15 m tall, and tall type whose height is over 2.15 m. It is often many-stemmed especially if pruned regularly. The *Hibiscus* possesses some of the largest flowers of any plant. For example, the rose mallow (*Hibiscus moscheutos*) produces the largest flowers of all *Hibiscus* from late spring until the first frost, with some forms reaching up to 1 foot across. The *Hibiscus* has leaves that are alternate, ovate to lanceolate, often with a toothed or lobed margin.

In general, the *Hibiscus* flowers are glorious and at their best up to 15.24 cm in diameter. Flowers of many *Hibiscus* species are fare flared and bell shaped and may be single or double, smooth or scalloped. The flowers are solitary, axillary, and very large compared with other flowers. The outermost series of bracteoles are six, lanceolate, green, and 8 mm long or less. They have a long central tube with stamens and pistils at the tip. They are large, conspicuous, with five or more soft petals and attractive large stamens. The calyx of the flower is green, 2 cm long, and the lobes are ovate. The stamens form a long staminal tube enclosing the entire style of the pistil and protruding out of the corolla. The ovary is five-celled, styles are five and fused below. The fruits are called capsules, loculicidally five-valved and rarely formed under cultivation, unless cross-pollination takes place (Rummel 2005).

The Australian *Hibiscus* Society (2004) classified the flowers or blooms of *Hibiscus* into seven main bloom types, namely, single regular, single windmill, cartwheel overlap single, single crested, single fringed, semi-double, and full double. In single regular type, the petals are separated for less than half the distance from edge giving a regular scalloped appearance to edge. Most single types fall into this type. In single windmill type, the petals are narrow and separated for nearly their entire length. In cartwheel overlap single type, the petals are completely overlapped to tips giving a regular circular appearance. Single crested type has the basic type similar to any of the above, but its normal bloom exhibits petaloid on the end of the staminal column forming a perfect crest. In single fringed type, the edges of the petals are split and fringed. The staminal column is long and pendulous. Semi-double type has loose petal formation with few petals that may be twisted or quilled, and all the petals form from the base of the bloom. The staminal column may be missing. Finally, in full double type, the many petals and petaloid present are in a tight formation giving a full ball-shaped appearance. The staminal column is usually missing. There are no flat areas under the petals that are standing out. Sometimes this type has five florets in the center. In general, the *Hibiscus* flowers are from 4 to 18 cm broad. The Australian *Hibiscus* Society (2004) classified the bloom sizes into four major size. These are miniature type whose bloom is less than 12 cm in diameter, medium type whose bloom is 12 to 15 cm diameter, large type whose bloom is 15 to 20 cm in diameter, and extra-large whose bloom is over 20 cm in diameter. These different bloom size come in a variety of colors ranging from white to pink, red, orange, peach, yellow, or purple. It is an example of a complete flower. The fruit is a dry five-lobed capsule, containing several seeds in each lobe, which are released when the capsule dehisces or when the fruit splits open at maturity.

16.4 Taxonomy and Ecology

The Malvaceae family is distributed all over the tropical and subtropical areas of the world where 100 genera with 2000 species are included and originated from South America (Fryxell 1965). Furthermore, 78 species in 19 genera were from the Southern Peninsular India (Sivarajan and Pradeep 1996).

Devi (2004) stated that the genus *Hibiscus* had the characteristic of possessing the capsular dehiscent fruits that are schizocarpic. Devi (2004) continuously discussed that another distinguishing characteristic is the absence of gossypol glands, five-toothed staminal column apex, apical branching of the style, terminal stigmata, and equality in the number of style branches to the carpel as stated also by Sivarajan and Pradeep (1996).

Sectional classifications were proposed in 1824 by De Candolle where 117 were classified, but only in the 1900s that a systematic classification to this genus was done (Devi 2004). In 1900, Hochreutiner in his monograph *Hibiscus* described 197 species, while in India, Rakshit and Kundu (1972) revised the genus *Hibiscus* (Devi 2004). Devi (2004) also stated that Rakshit and Kundu (1972) further described the genus having 300 species distributed worldwide. However in India they stated that there are 28 species belonging to 10 sections found in that country.

To further understand the *Hibiscus* genus, recently, Hoskins (2016) employed phylogenetic analysis using chloroplast and nuclear DNA regions of *Hibiscus* section *Furcaria*. This is to determine the maternal genetic relationships between the diploids and the Australian hexaploid lineage in order to reconstruct the origin. Also it was done to determine if any surviving diploid donors exist which are related to the unknown J and V genomes of the Australian hexaploid group. The same author reported that using four chloroplast regions and two nuclear regions, it was found that the Australian hexaploid species form a well-supported clade using chloroplast genes, namely, *ndhCtrnV*, *ndhF-rpl32F-trnL*, *rps16-tmK*, and ITS with *Hibiscus sudanensis* as the maternal donor of the G genome group belonging to the hexaploid Australian species. This suggests that the hexaploid Australian G genome group could be related or could have come from the African diploid species *H. sudanensis*.

16.5 Distribution and Domestication

While it has been known that the *Hibiscus rosa-sinensis* complex is distributed and domesticated in the tropical and subtropical areas of the world with over 1000 cultivars coming in different colors and forms, it is not known to date the real origin of the species. Several authors claim it to be a native of China, while others claimed that it originated from Africa (Devi 2004). Furthermore they stated that its nearest relatives *H. schizopetalus* and *H. liliflorus* are both natives of east Africa and Mascarene Islands. While another author van der Pijl (1937) suggested that it originated from America since its pollinators which are the humming birds are also found in America. Khan et al. (2017) stated that *Hibiscus rosa-sinensis* probably originated from India. The authors also stated that Arabs believed that *Hibiscus* came from Spain.

Devi (2004) discussed that even though *H. rosa-sinensis* has been known and grown widely very early in China and called "China rose," no wild forms have been discovered in China. Devi (2004) in her study found that three out of the four genetically compatible species of *H. rosa-sinensis* are native to the South Indian Ocean Islands and the African East Coast, indicating that these places are the ancestral home of the early forms that involved in the *H. rosa-sinensis*.

A recent study on analyzing the two chloroplasts DNA sequences, namely, a coding region known as *ndh F* and a non-coding region called *rplI* 6 intron, integrated with the polyphyletic theory of origin of the tribe Hibisceae was conducted. Thus the study showed that the center of origin of the *Hibiscus* could be the Eastern Gondwana (Pfeil et al. 2002).

Recently, Craven et al. (2011) reported the taxonomic re-evaluation of *Hibiscus trionum* existing in Australia and New Zealand. The same author reported that there are three indigenous species in this *Hibiscus trionum* complex including *H. richardsonii*, *H. tridactylites*, and *H. verdcourtii*. *H. richardsonii* is distributed in the coastal regions of New South Wales, Australia, and in the northeastern half of the North Island, New Zealand, while *H. tridactylites* Lindley is distributed in the inland southern and eastern Australia. *H. verdcourtii* is distributed widely in inland Australia especially north of latitude 28° south.

16.6 Germplasm of *Hibiscus*

There have been 300 known species of *Hibiscus* worldwide (Wang et al. 2012). Among the species, *H. rosa-sinensis* is one of the most famous ornamental crops being cultivated for their numerous uses. In addition to H. *rosa-sinensis*, *H. sabdariffa* commonly known as roselle has been in cultivation to be used as a functional food due to its high anthocyanin content (Wu et al. 2018). In tropical and subtropical countries, *H. abelmoschus* or *Abelmoschus esculentus* commonly known as "okra" or lady's finger is a popular annual crop classified as a vegetable. In the Malvaceae family, other important species include *H. syriacus* or the Rose of Sharon and *H. vitifolius* that were considered to have ornamental and other pharmacological uses. *H. syriacus*, like *H. rosa-sinensis*, has been extensively used in breeding to produce hybrids of inter- and intraspecific origin as there were successful crosses of *H. syriacus* with other species such as *H. paramutabilis* (Van Laere et al. 2007) and *H. synosyriacus*. *H. paramutabilis*, *H. syriacus*, and *H. sinosyriacus* were native plants of China (Bates 1965).

Kenaf (*H. cannabinus*) is another important *Hibiscus* species that is commonly grown in Eastern Africa (Cheng et al. 2004). There were reports stating successful crosses of Kenaf and Roselle (Wilson and Menzel 1967) and Kenaf and its wild relative *H. acetosella* (Satya et al. 2012). In the Philippines, *Abelmoschus manihot*/*H. manihot* commonly known as "Lagikway" is a leaf vegetable where young leaf shoots were incorporated in fish viands (Maghirang et al. 2018). Another reported *Hibiscus* species in the Philippines is *H. surattensis* which was used as a souring agent (Maghirang et al. 2018).

Among the different *Hibiscus* species, *H. kokio*, *H. arnottianus*, and *H. waimeae* had diploid chromosome number of 80 and *H. schizopetalus* a diploid chromosome number of 40 (Wilcox and Holt 1913). Studies have shown that *H. schizopetalus*, the one that has a fringed petal and pendulous flower type, is considered a different species from *H. rosa-sinensis* as recognized by several taxonomists (Mabberley 1997). This observation jibed with the report of Fryxell (1965) indicating *Hibiscus* is frequently having intergradations with var. *rosa-sinensis* presumably due to hybridization.

Jo et al. (2019) reported that the *Hibiscus* genus has approximately 300 species distributed worldwide. They cited some species and the places where they were found. Among them are *H. schizopetalus* found in the East-African Coast, *H. liliiflorus* Cav. found in Mauritius and Rodriguez Islands, and *H. fragilis* and *H. boryanus* Hook and Arn. in Reunion Islands. They also added that some species are found in the Pacific Islands, including *H. arnottianus* Gray and *H. kokio* Hillebrand. Both are from the Hawaiian Islands. *H. storckii* Seeman is found in Fiji. Jo et al. (2019) elaborated that some species such as *H. syriacus*, *H. hamabo*, and *H. mutabilis* are native to Korea and are commonly planted as ornamental plants in this country (Fig. 1).

16.7 Genetics and Breeding

Hibiscus rosa-sinensis is one of many plant species with genetic characteristic known as polyploidy, in which there are more than two complete sets of chromosomes. A side effect of polyploidy is a condition where the phenotype of the offspring may be quite different from the parent, or indeed any ancestor, essentially allowing possibly random expression of all or any of the characteristics of all the generations that they have gone before. Due to this characteristic, *H. rosa-sinensis* has become popular with hobbyists who cross and re-cross varieties, creating new named varieties and holding competitions to exhibit and judge the many resulting new seedlings and often resulting to strikingly unique flowers. There are now more than 3000 hybrids registered around the world (Rummel 2005). More than the floral

Fig. 1 Different species of *Hibiscus*: (**a**) *H. syriacus*, (**b**) *H. mutabilis*, (**c**) *H. trionium*, (**d**) *Abelmoschus manihot*, (**e**) *H. sabdariffa*, (**f**) *H. surattensis*, (**g**) *H. schizopetalus*, and (**h**) *Corchorus olitorius*. (Pictures from Fig. 1A is from Rita Knust and 1B-C and 1D are lifted from https://pixabay.com)

traits for ornamental purposes, native, related, and wild species of *Hibiscus* can be exploited to incorporate other traits such as new and rare floral color, disease resistance, frost tolerance, longer floral display, and even incorporation of fragrance. To add to the genetic opportunities, *H. sinensis* has been attempted to hybridize with *H. moscheutos* and several other North American *Hibiscus* species, producing cold-hardy interspecific hybrids (Kuligowska et al. 2012). Improved and prolonged flower life is also explored to open the possibility of making *Hibiscus* as a cut flower through investigations on morpho-anatomical and physiological characters of its certain breeds (Magdalita et al. 2019).

Table 1 shows the different *Hibiscus* species that show different interesting traits that can be exploited in breeding programs. An example is the one described by Jo et al. (2019) who reported that there is an East Asian group of *Hibiscus* spp. and this group is known to be tolerant to cold stress. From these species *H. syriacus* showed the highest cold tolerance among the *Hibiscus* species. Hence development of new breeds using these species can be done to develop cold-tolerant *Hibiscus* hybrids.

16.8 Development of *Hibiscus* Hybrids

For ornamental purposes, hybrids were developed with outstanding floral characteristics such as novelty of floral color and unique color combinations, new forms via intra- and interspecific hybridization. Studies were done to test genetic compatibility of species. Among the commonly hybridized species are *Hibiscus rosa-sinensis* and *H. syriacus*. However, incompatibility of H. *rosa-sinensis* with *H. syriacus* and *H. moscheutos* to produce interspecific hybrids was observed (Kuligowska et al.

Table 1 Different *Hibiscus* species/cultivars and their interesting character/trait, the authors who reported presence of trait/character to the specific cultivar/species, and the places where the species or cultivars can be found

Hibiscus species/ cultivar	Trait or character of interest	Regions where the species are found to grow but not limited to	Author reported
Hibiscus surattensis Linn	Sourness, biodiesel, essential oil	West Africa, the Philippines	Biriok and Yawas (2017), Maghirang et al. (2018)
H. trionium	Resistance to *Verticillium dahliae*	Southern Europe, and the USA	Golubenko et al. (2007)
H. schizopetalus	Resistance to *Colletotrichum* leaf spot	Pacific, China, East Africa, Malaysia, the Philippines, India	Pascual and Magdalita (2012)
H. cooperi	Resistance to *Colletotrichum* leaf spot	Pacific, China	Pascual and Magdalita (2012)
H. rosa-sinensis 'Reddy or Not'	Resistance to *Colletotrichum* leaf spot	The Philippines, Hawaii	Pascual and Magdalita (2012)
H. rosa-sinensis 'Ruth Wilcox'	Resistance to *Colletotrichum* leaf spot	The Philippines, Hawaii, Australia	Pascual and Magdalita (2012)
H. rosa-sinensis 'Petite Peach'	Resistance to *Colletotrichum* leaf spot	The Philippines	Pascual and Magdalita (2012)
H2 and H4 genotypes of *H. sabdariffa*	Drought stress tolerance	Sudan	Mohamed et al. (2015)
H. rosa-sinensis 'Fantasia' and *H. rosa-sinensis* 'Jay's Orange'	Drought tolerance	Egypt	Shanan and Moghaieb (2016)
H. syriacus (Rose of Sharon)	Salinity tolerance, *Verticillium* wilt tolerance	China, India, the Philippines, tropical and subtropical Asia, south of the USA	Chen et al. (2019), Gilman and Watson (1993)
H. mutabilis	Mild frost tolerance	Korea, China, the Philippines	Lawton (2004), Shang et al. (2020) and Raut et al. (2014)
	Contaminant elimination in the soil		
	Flower color mutability		
Corchorus olitorius	High fiber	The Philippines, Africa, China, Vietnam, Fiji, Australia, Saudi Arabia, France, Egypt, Israel	Palve and Sinha (2005)

(continued)

Table 1 (continued)

Hibiscus species/ cultivar	Trait or character of interest	Regions where the species are found to grow but not limited to	Author reported
H. coccineus	Tolerance to extended flooding	Southeastern USA	Gilman (2014)
	Moderate drought tolerance		
H. arnnotianus	Fragrance	Hawaii, Africa	Native Plants Hawaii (2009)
H. tiliaceus	Tolerance to swampy waterlogged soils	East Africa to Central Pacific, Hawaii, Brazil, India, Indonesia and Myanmar, Guam, Jamaica, the Philippines	FAO (n.d.)

2012). However, studies on interspecific hybridization among *H. syriacus*, *H. sinosyriacus*, and *H. paramutabilis* have been reported (Van Laere 2007).

Due to incompatibility barriers of *H. rosa-sinensis* with other *Hibiscus* species, intraspecific hybridization became more extensive in this species producing thousands of hybrids. Extensive breeding of *Hibiscus rosa-sinensis* has been done in different parts of the world including tropical and subtropical Australia, South America, the Philippines (Magdalita and Pimentel 2013), Denmark (Kuligowska et al. 2012), India, and Taiwan among others. First reports of hybridization were published in 1914 in Hawaii.

Several *Hibiscus* breeding societies have emerged to group private *Hibiscus* breeders in Australia, in America, and in different regions of the world like Oceania, the Philippines, Europe, Bolivia, Polynesia, and South Africa among others (Fig. 2).

16.9 Plant Genetic Resources of *Hibiscus rosa-sinensis*

Different *Hibiscus* hybrids had been developed by the University of the Philippines Los Baños, a public institution of higher learning, and some private breeders in the Philippines. In other countries, majority of *Hibiscus* breeders are private hobbyists. Some of them have collectively created organizations and societies that aimed to provide technical expertise on breeding, propagation, and maintenance of *Hibiscus*. Among the different institutions and societies that were made to convene *Hibiscus* breeders from around the world are the Australian *Hibiscus* Society, American *Hibiscus* Society, and the largest, the International *Hibiscus* Society.

An important initiative of the International *Hibiscus* Society is the establishment of registry for cultivars around the world which breeders can use to trace genealogy of cultivars and trace the breeder, the location, and the origin of the hybrids. The search engine also allows the public to check on the other cultivars that have been developed using the parents of the registered hybrid.

Fig. 2 The *Hibiscus* breeding process being followed in the *Hibiscus* Breeding Program at the Institute of Plant Breeding, College of Agriculture and Food Science, University of the Philippines Los Baños

Fig. 3 Cultivar search engine of the International *Hibiscus* Society (n.d.) that documented cultivars of around the world. (Source: http://www.internationalHibiscussociety.org/SEArchive/SEindex1.php)

This is an important engine for breeders as they can be able to trace the inheritance of traits that can aid the breeders to select which parents to be used to attain a certain breeding objective. An example of the genealogy tree can be seen in Fig. 3.

This registry incorporates 22,339 cultivars. More hybrids are still undocumented as those reported in the registry are only those hybrids released and registered and/or applied for registration. The search engine can be accessed through http://www.international*Hibiscus*society.org/SEArchive/SEindex1.php.

Fig. 4 Cultivar search results for Amber Gem, an example of a genealogy tree result from the cultivar search of the engine. (Courtesy of the International *Hibiscus* Society. http://www.internationalHibiscussociety.org/new/)

An example of an entry can be seen in Fig. 4, the genealogy tree of the hybrid "Amber Gem." The results show the genealogy tree/pedigree of the hybrid and the other hybrids that have been developed from the parentals of the hybrid. The International *Hibiscus* Society holds the official database of International Cultivation Registration Authority (ICRA) for the *Hibiscus rosa-sinensis* cultivars.

16.10 Characterization of Candidate Cultivars and Hybrids' Bloom

Characterization of plant and flower characters most especially their colors were done using several instruments. Huang (n.d.) reported that it is important to recognize the features of candidate cultivars. This will check on the uniqueness of the hybrid when compared with new hybrids and others already released with similar characteristics. Huang (n.d.) developed a compilation of descriptions for the bloom characters which is an important factor in describing the hybrid's flowers. Another instrument being used is the Royal Horticultural Society Color Chart and their color coordinates (San Pascual et al. 2017). The figure shows the different bloom types and guide used in characterization of the hybrids. Brief characteristics of each bloom

Table 2 The brief description by Huang (n.d.) of each bloom/flower form of *Hibiscus*

Bloom characteristic/ flower form	Brief characteristics/exact description by Huang (n.d.)
Single and regular	Petals separated for less than half the distance from the outer edge giving a regular scalloped appearance to edge
Cartwheel	Petals completely overlapped to the tips giving a regular, circular appearance
Crested single	Basic type may be any form of single but normal bloom and exhibits petaloides on end of the staminal column forming a perfect crest
Single windmill	Petals narrow and separated for nearly their entire length
Fringed single	Edges of petals split and fringed. Staminal column sometimes long and pendulous
Double	Many petals and petaloides in a tight formation, giving a ball-type appearance on top of a flat circle of petals which stand out, staminal column usually missing
Semi-double	Loose petal formation with a few petals that may be twisted or quilled. All petals form from base of bloom. Staminal column may be missing
Crested semi-double	Loose double appearance, with petaloides arising from staminal column, stigmas usually present
Full double	Many petals and petaloides in a tight formation, giving a ball-shaped appearance. Staminal column usually missing. No flat under petals standing out, sometimes with five florets in the center
Cup and saucer	Outside guard petals follow single form, center tuft of petaloides all arise from the center and are distinctly separated from guard petals
Fluted	Can be single or double with petals that are soft on edges, sort of wavy
Ruffled	Frilly, ruffled edges and/or foliage
Tufted	Small upstanding creases on inside edge of petals

Lifted from: http://international*Hibiscus*society.org/new/images/Data/Nomenclature/Bloom%20Characteristics.pdf

form were also indicated by Huang (n.d.) and are presented in Table 2 and also described by the Australian Hibiscus Society (2004).

Quantitative characters such as petal thickness can be measured using a micrometer caliper, while the bloom size can be measured using rulers and vernier calipers. Normally, 20 flowers are being used to characterize the hybrid. Other characters such as foliage type and form, phyllotaxy, and stigma type among others can be characterized using the book by Radford (1972). This is the standard reference being used by the *Hibiscus* breeding at the Institute of Plant Breeding, College of Agriculture and Food Science, UP Los Baños in the characterization of their hybrids (Fig. 5).

16.11 *Hibiscus* Breeding Program in the Philippines (UPLB)

The *Hibiscus* Breeding Program at the Institute of Plant Breeding (IPB), College of Agriculture and Food Science (CAFS), University of the Philippines Los Baños (UPLB), Los Baños, Laguna, was started in 1994 by Mr. Reynold B. Pimentel and

Fig. 5 Representation of the different *Hibiscus* flower forms: (**a**) single regular, (**b**) cartwheel single, (**c**) crested single, (**d**) single windmill, (**e**) fringed, (**f**) double, (**g**) semi-double, (**h**) crested semi-double, (**i**) full double, (**j**) cup and saucer, (**k**) fluted, (**l**) ruffled, and (**m**) tufted. Pictures lifted from Huang (n.d.)

continued by Dr. Pablito M. Magdalita starting 2002. The parentals are local *Hibiscus* cultivars and those that were introduced from Hawaii, the USA, and Queensland, Australia. The products of these hybrids were named after accomplished Filipina and were collectively grouped according to a particular theme and grouped into series. The *Hibiscus* Breeding Program has released varieties named after several groups of extraordinary Filipinas who carved their names in Philippine history including Centennial Series, heroines of the Philippine Revolution; Millennium Series, women scientists of UPLB; Celebrity Star Series, accomplished celebrities in the Philippine cinema; Oblation Series, pioneering and outstanding UP alumnae who served their institution in extraordinary ways; Women in Public Service, outstanding women public servants who prioritize the needs of before their own others; Women in Science, outstanding women scientists; and Women in Media and the Arts Series, remarkable women in the Philippine mass media and the arts.

Hybrids from other countries such as Australia and the USA were procured and then established in the Institute of Plant Breeding, College of Agriculture and Food Science, UP Los Baños. Some foreign varieties were observed to have thicker petals and larger flower sizes and have longer floral retention ranging from 2 to 3 days as observed in subtropical countries where they were developed. After having them established in the Philippines, their growth and blooming characteristics were assessed. This is to test if they still have prolonged floral retention even they are already in tropical countries like the Philippines, since environmental conditions such as temperature, sunlight, relative humidity, etc. affect floral retention on the plant. These foreign varieties have been used as maternal parents and cross-pollinated with local varieties to infuse new floral colors and forms. In addition, these foreign varieties were also used for reciprocal crosses as paternal parent or pollen source or donor for the local hibiscus varieties to improve their plant and floral traits (San Pascual et al. 2017).

In the Philippines, registration of new cultivars and varieties was done in the National Seed Industry Council (NSIC), Department of Agriculture. This council scrutinizes the new variety prior to approval in accordance with the guidelines set by the Ornamental Crops Technical Working Group (TWG). Also the council release and publish catalogues containing registered cultivars in the Philippines in a periodic basis.

In the Philippines, hybrids of *Hibiscus* are collectively released and grouped in particular series such as the Centennial Series released in 1998 during the centennial anniversary of the Philippine Independence; hence these *Hibiscus* hybrids were named after the heroines of the 1898 Philippine Revolution who fought for freedom. All hybrids developed by UPLB were presented in Figs. 6, 7, 8, 9, 10, 11, 12, 13, 14, 15 and 16. In addition, researchers from UPLB have been publishing their hybrids and their characters in scientific journals. The descriptions of the different *Hibiscus* hybrids developed and released by the breeders at the Institute of Plant Breeding, College of Agriculture and Food Science, UP Los Baños are as follows.

Fig. 6 *Hibiscus* hybrids under the Centennial Series: *Hibiscus rosa-sinensis* (**a**) 'Tandang Sora', (**b**) 'Oryang', (**c**) 'Lolay', (**d**) 'Gabriela', (**e**) 'Goria', (**f**) 'Agueda', (**g**) 'Nazaria', (**h**) Marcela, (**i**) Ningning, (**j**) Sentenarya, and (**h**) Nay Isa

Fig. 7 *Hibiscus* hybrids under the Millennium Series: *Hibiscus rosa-sinensis* (**a**) 'Dolores Ramirez', (**b**) 'Gelia Castillo' (**c**) 'Obdulia Sison', (**d**) 'Emerita de Guzman', (**e**) 'Helen Valmayor', (**f**) 'Sentenarya' and (**g**) 'Clare Baltazar'

Fig. 8 *Hibiscus* hybrids under the Celebrity Star Series: *Hibiscus rosa-sinensis* (**a**) 'Star for All Seasons', (**b**) "Mega Star', (**c**) 'Nova Star', (**d**) 'Diamond Star', and (**e**) 'Super Star'

16.11.1 Centennial Series

The "Centennial Series" was the first batch of *Hibiscus* hybrids released last 1998 to coincide with the Centennial celebration of the Philippine Independence; hence the name "Centennial Series" was given. These hybrids were named after the Philippine

Fig. 9 *Hibiscus* hybrids under the Oblation Series: *Hibiscus rosa-sinensis* (**a**) 'Betty Go-Belmonte', (**b**) 'Estrella F. Alabastro', (**c**) 'Perla Santos-Ocampo', (**d**) 'Emerlinda R. Roman', (**e**) 'Mercedes B. Concepcion, (**f**) 'Nelia T. Gonzales'

Fig. 10 *Hibiscus* hybrids under the Women in Public Service Series I: *Hibiscus rosa-sinensis* (**a**) 'Rosa Rosal', (**b**) 'Kristie Kenny', (**c1** and **c2**) 'Loren Legarda', (**d**) 'Lilia de Lima'

Fig. 11 *Hibiscus* hybrids under the Women in Public Service Series II: *Hibiscus rosa-sinensis* (**a**) 'Cynthia A. Villar', (**b**) 'Maria Rosario Montejo; (**c**) 'Sylvia Lina', (**d**) 'Connie S. Angeles', (**e**) 'Arlene B. Arcillas', (**f**) 'Che-che Lazaro', (**g**) 'Pia S. Cayetano', (**h**) 'Vilma Abaya-Dimacuha', (**i**) 'Marilyn D. Marañon', (**j**) 'Domini T. Torrevillas', (**k**) 'Emmeline Aglipay-Villar', and (**l**) 'Carmen Pascual'

Fig. 12 *Hibiscus* hybrids under the Women in Science Series, *Hibiscus rosa-sinensis* (**a**) 'Solita F. Camara-Besa', (**b**) 'Ledivina V. Cariño', (**c**) 'Fe V. Del Mundo', (**d**) 'Lourdes Cruz' (in chemistry); Women in Media and the Arts Series, (**e**) 'Araceli Dans'; Women Saints and institutions named after them series, (**g**) 'St. Bridgette'; Women in Education Series (**h**) 'Patricia Licuanan'; and Women in Social Entrepreneurship (**i**) 'Milagros O. How'

Fig. 13 *Hibiscus* hybrid selections: (**a**) 'Tandang Sora' x 'Connie S. Angeles', (**b**) 'Nay Isa' x 'Helen Valmayor; (**c**) 'Gelia Castillo x Marilyn D Marañon' x Arlene B. Arcillas and 'Loren Legarda' x 'Estrella F. Alabastro'

heroines who struggled and fought for the freedom and liberation of the country against three centuries of Spanish rule and colonization and four decades of American rule. There were 11 National Seed Industry Council (NSIC)-approved *Hibiscus* hybrids under this series (Fig. 6). Among the hybrids under this series are:

1. *Hibiscus rosa-sinensis* 'Tandang Sora' which was named after the heroine, Melchora Aquino, the Mother of Katipunan. Katipunan is a Philippine revolutionary society during the Spanish colonization. The hybrid has creamy white petals with prominent red eye, 10–12 cm in diameter, and classified as single petal and regular.
2. *Hibiscus rosa-sinensis* 'Oryang' was named after Gregoria de Jesus or Oryang, the princess of the Katipunan. The flower is peach orange with a deep maroon red eye zone, large to extra large (20–22 cm in diameter), single petal and regular.
3. *Hibiscus rosa-sinensis* 'Lolay' was named after Doña Teodora "Lolay" Alonso, the mother of the Philippine National Hero, Dr. Jose P. Rizal. The flower is

Fig. 14 *Hibiscus* hybrids developed by private breeders in the Philippines: (**a**) 'JB's My Maryjane' (Bautista Joseph G), (**b**) 'Philippine Princess' (Ganigan Boyet), (**c**) 'Lemon Drew Maningas' and (**d**) 'Loving Rovilla' (Concepcion Niron Brylle), (**e**) 'Maegan' (Josan Fronteras and Raymond), (**f**) 'Philippine Devil's Advocate' (Ganigan Boyet), (**g**) 'RGKL David' (Lattao Sony Grace), (**h**) 'SMV Jianne Louise' (Velasquez Sheila Marie), (**i**) 'Taiwan Formosa Frozen Heart' (Lee Aneela), (**j**) 'TJ Flambeaux' (De La Torre Alan and Vera), (**k**) 'Hrya Espie' (Jorge Dominguez and Ryan Tayobong), (**l**) 'Philippine Typhoon Haiyan' (Boyet Ganigan). (Images lifted from and courtesy of: http://www.internationalHibiscussociety.org/new/, Dexter Rose Fernando, Jorge Dominguez and Boyet Ganigan)

apricot-orange with pastel pink eye with rays extending to the zone area, medium to large (12–17 cm in diameter), regular, with partially overlapped petals.

4. *Hibiscus rosa-sinensis* 'Marcela' was named after Marcela Agoncillo, the mother of the Philippine Flag. She with two other Filipinas made the first Philippine flag. The flower is lemon with yellow eye and prominent white zone and halo, medium to large (12–17 cm in diameter), regular and single petal.
5. *Hibiscus rosa-sinensis* 'Nay Isa' was named after Teresa Magbanua, the Visayan Joan of Arc who led combat troops in Visayas against Spaniards. The flower is brilliant yellow with splashes of orange on the eye zone, medium to large (15–17 cm), single row of petals fully overlapped and set seeds.
6. *Hibiscus rosa-sinensis* 'Ningning' was named after Trinidad Tecson y Perez, the Mother of 'Biak-na-Bato'. She fearlessly fought in battles against the Spaniards in the province of Bulacan. She is also called Henerala Ningning. The flower is

Fig. 15 Some other *Hibiscus* hybrids developed by private breeders in the Philippines: (**a**) 'BLG Valentine Wine' (Concepcion Niron Brylle), (**b**) 'Blood Moon' (De La Torre Alan and Vera), (**c**) 'Butch Campos' (Ganigan Boyet), (**d**) 'Divine Thaleia Laquerie' (Zenarosa Divine Grace), (**e**) 'Diwata' (Ganigan Boyet), (**f**) 'Elisha Lovie' (Lattao Sony Grace), (**g**) 'BLG Lea's Moonstone' (BLG Lea's Moonstone), (**h**) 'BLG Unsung Heroes' (Bautista Ron Ron Pearl), (**i**) 'Divine Gabriel Zenarosa' (Zenarosa Divine Grace), (**j**) 'Florence Zoe Dane' (Lattao Sony Grace), (**k**) Fabros Mira Claire F. (Fabros Franklin), (**l**) 'Janica's Heart' (Concepcion Niron Brylle), (**m**) 'Igorotak Aurora' (Morales Jojette), and (**n**) Igorotak Sabong She Bahong (Morales Jojette). (Images lifted from and courtesy of: https://internationalhibiscussociety.org/searchive/index, Dexter Rose Fernando, Boyet Ganigan and Jojette Morales)

orange with yellow overlay and lemon vein marking with pale pink eye, medium to large (15–17 cm in diameter), and has regular single row of petals.

7. *Hibiscus rosa-sinensis* 'Gabriela' was named after Maria Josefa Gabriela Cariño Silang. She led the revolt in the Ilocos region in the Philippines. The flower is bright scarlet with an average flower diameter of 15 cm. The flowers are single and overlapped.
8. *Hibiscus rosa-sinensis* 'Goria' was named after Gregoria Montoya y Patricio, a Filipina revolutionary 'Heroine' who led a 30-member unit of the Katipunan. The flower between is deep rose with a striking star-like white eye zone and prominent white vein markings, medium (15–17 cm), single and overlapped.

Fig. 16 Some other *Hibiscus* hybrids developed by private breeders in St Vincent and Grenadines in the Oceania, (**a**) 'St Vincent Bleeding Heart' (Shallow Gordon); in Mexico, (**b**) 'Melita' (Mario Pe); in Panama, (**c**) 'Panama Happy Road' (Montenegro Onesimo Del Cid) and (**d**) 'Yeroicy Pasion Tropical' (Guerra Yeimy); in Puerto Rico, (**e**) 'Caribbean Beautiful Heart' (Demirboga Adil) and (**f**) 'Tropical Yellow Eye' (Cardona Ernesto); in Bolivia, (**g**) 'Bolivian Adil Demirboga' (Veizaga Vargas Edwin Hugo); in Brazil, (**h**) 'Tom Jobin' (Jordan Elizabeth) and (**i**) 'Vanessa Saba' (Fonseca Sergio), (**j**) 'Spectroflash' (Bentes William), (**k**) 'Peter Maximilian' (Jordan Elizabeth), (**l**) 'PM Miss Parana' (Moll Peter E.). (Images lifted from: https://internationalhibiscussociety.org/searchive/index)

9. *Hibiscus rosa-sinensis* 'Agueda' was named after Agueda Kahabagan y Iniquinto. She is the first women general of the Armed Forces of the Katipunan. The flower is apricot orange with deep red eye and pink halo and red vein markings radiating to the petals, medium to large (15 cm in diameter), with overlapped row of petals.
10. *Hibiscus rosa-sinensis* 'Nazaria' was named after Nazaria Lagos, the Florence Nightingale of Panay. She was the director of the hospital of the Revolutionary Army of the Katipunan. The flower is cerise pink with deep red eye and prominent pink vein markings radiating to the petals, medium to large (15–17 cm), and has single row of cartwheel and fully overlapped petals.
11. *Hibiscus rosa-sinensis* 'Sentenarya' was named based from the word Centennial. The flower is scarlet red with deep glossy red eye, large to extra large (20–25 cm in diameter), and has single and overlapped row of petals. 'Sentenarya' is the signature in the IPB Centennial Series of *Hibiscus*, hence the name. This was

to honor and celebrate all those valiant Filipinas who unselfishly devoted their lives to the cause of freedom and independence over a hundred-year (1898–1998) period.

16.11.2 Millennium Series

The second batch of *Hibiscus* hybrid series was collectively named the "Millennium Series." These hybrids are tribute to the outstanding woman scientists of UP Los Baños who contributed to the growth and advancement of the Philippine agriculture. These hybrids were released in the new millennium (year 2000), hence the name (Fig. 7). The seven *Hibiscus* hybrids that belong to this series are the following:

1. *Hibiscus rosa-sinensis* 'Dolores A. Ramirez' was named after Dr. Dolores A. Ramirez, the Philippine's National Scientist for biochemical and cytogenetics. The flower has yellow-orange petals upon opening in the early morning, but the petals became rich golden yellow at mid-day with lilac eye zone and white veins radiating from the center to the petals.
2. *Hibiscus rosa-sinensis* 'Clare R. Baltazar' was named after Dr. Clare R. Baltazar, the National Scientist in the field of entomology. The flower has pale yellow petals with deep maroon eye zone and distinct white halo.
3. *Hibiscus rosa-sinensis* 'Gelia T. Castillo' was named after Dr. Gelia T. Castillo, the National Scientist in the field of rural sociology. The flower has golden yellow petals with deep maroon eye zone and orange halo.
4. *Hibiscus rosa-sinensis* 'Emerita V. de Guzman' was named after Dr. Emerita V. de Guzman, a well-known plant physiologist who developed the tissue culture protocol for 'Macapuno' coconut. The flower has petals and eye zone that are creamy yellow with lemon yellow edges and pink veins radiating through the petals and distinct lavender pink halo.
5. *Hibiscus rosa-sinensis* 'Obdulia Sison' was named after Dr. Obdulia F. Sison, a renowned professor in agricultural education. The flower has soft pink rose petals with white edges and red eye zone. It is also a 2-day bloom variety during the cooler months of the year.
6. *Hibiscus rosa-sinensis* 'Helen L. Valmayor' was named after Dr. Helen L. Valmayor, the "Mother of Philippine Orchids" who wrote the two-volume book *Orchidiana Philippiniana*, a monumental book on Philippine orchids. The flower has bright orange petals with yellow edging and pronounced red eye zone.
7. *Hibiscus rosa-sinensis* 'Millennia' was named after the new millennium (2000) and collectively composed this series. The flower has cardinal red petal with red eye zone. The foliage is medium to large, dark green, and glossy. The plant has medium height, bushy and medium grower. Millennia is the hybrid that represents collectively all the seven *Hibiscus* hybrids under the series.

16.11.3 Celebrity Star Series

This batch of *Hibiscus* hybrids called the "Celebrity Star Series" was released in 2002. This series was named after veteran and accomplished Filipina actresses in the Philippine cinema. There are five *Hibiscus* hybrids belonging to this series (Fig. 8).

1. *Hibiscus rosa-sinensis* 'Star for All Seasons' was named after Vilma Santos-Recto, popularly known as Star for All Seasons. The flower has semi-double lavender mauve petals. The foliage is medium to large, dark green, ruffled with distinct serrations.
2. *Hibiscus rosa-sinensis* 'Superstar' was named after the only Superstar Nora Aunor. The flower has red petals with bright yellow-orange splashes on middle section of the petal and has dark red eye.
3. *Hibiscus rosa-sinensis* 'Nova Star' was named after the Philippine Nova Star Kris Aquino. The flower has apricot-orange petals with speckles of yellow near the edges of the petals.
4. *Hibiscus rosa-sinensis* 'Megastar' was named after the only Megastar of the Philippine cinema, Sharon Cuneta. The flower has soft orange-yellow petal with partial overlap of brown tint with apricot-orange edges.
5. *Hibiscus rosa-sinensis* 'Diamond Star' was named after the Diamond Star of the Philippine movies, Maricel Soriano. The flower has creamy white petal with a very distinct red eye zone. The foliage is medium to large, green, and glossy.

16.11.4 Oblation Series

To commemorate the UP Centennial Celebration in 2008, new *Hibiscus* hybrids called "Oblation Series" named after the UP Oblation were released and made available to the public from 2006 to 2008 to culminate the celebration of the UP Centenary. These hybrids were named after outstanding alumnae of the university who were the first women to have assumed the highest position in the academic, scientific, and professional institutions and organizations where they served. These *Hibiscus* hybrids were the modest contribution of the Institute of Plant Breeding, UP College of Agriculture and Food Science, to make the university's centennial celebration meaningful and memorable. There were six National Seed Industry Council (NSIC)-approved *Hibiscus* hybrids under this series as follows and presented in Fig. 9:

1. *Hibiscus-rosa-sinensis* 'Emerlinda R. Roman' was named after the first woman University of the Philippines President Dr. Emerlinda R. Roman. The flower has deep yellow-orange petals with red eye zone and brown or gray-orange halo. It is large with an average bloom size of 15.75 cm.

2. *Hibiscus rosa-sinensis* 'Estrella F. Alabastro' was named after the first woman Secretary of the Department of Science and Technology of the Philippines. The flower has pinkish petals in the early morning which became white with pinkish veins radiating from the center going to the petals by mid-morning. It is medium with average bloom size of 12.70 cm.
3. *Hibiscus rosa-sinensis* 'Perla D. Santos-Ocampo', the Philippine's National Scientist in the field of medicine by the National Academy of Science and Technology (NAST) and the first woman Chancellor of the University of the Philippines (UP) Manila, among others. The flower has apricot-orange petals with red eye and yellow edging. It is medium to large with an average bloom size of 15.24 cm.
4. *Hibiscus rosa-sinensis* 'Mercedes B. Concepcion' was named after the National Scientist for social sciences by NAST and first Filipina Demographer, Dr. Mercedes B. Concepcion. The flower has canary yellow petals with starry red eye. It is large with an average bloom size of 15.50 cm.
5. *Hibiscus rosa-sinensis* 'Nelia T. Gonzales' was named after the first UP Los Baños Alumni Association Woman President for Life, UP Regent Nelia T. Gonzales. The flower has dark orange petals with red eye zone and yellow edges including yellow vein markings radiating from the center going to the petals. It is medium to large with an average bloom size of 15.24 cm.
6. *Hibiscus rosa-sinensis* 'Betty Go-Belmonte' was named after the first Filipino woman publisher, former UP Regent Betty Go-Belmonte. The flower has lemon yellow petals with red eye zone surrounded by white halo. It is medium with an average bloom size of 13.97 cm.

16.11.5 Women in Public Service Series

This is a batch of *Hibiscus* hybrids collectively called the "Women in Public Service" Series which was released by the Crop Science Cluster-IPB, College of Agriculture now CAFS, UPLB, as another contribution to the celebration in 2009 of the UP Centennial Year. These hybrids were named after women who have untiring devotion to public service and embodying the spirit of volunteerism in meeting the needs of others before their own without material or financial rewards. There are four National Seed Industry Council (NSIC)-registered *Hibiscus* hybrids under this series (Figs. 10 and 11).

1. *Hibiscus rosa-sinensis* 'Rosa Rosal' was named after the first woman face of the Philippine National Red Cross, Rosa Rosal. The flower has bloody red petals with red eye zone. It is medium with an average bloom size of 9.6 cm.
2. *Hibiscus rosa-sinensis* 'Kristie Kenney' was named after the first female US Ambassador to the Philippines, Kristie A. Kenney. The flower has bluish petals with red eye zone. It is medium with an average bloom size of 8.0 cm.
3. *Hibiscus rosa-sinensis* 'Loren B. Legarda' was named after Senator Loren B. Legarda who is a strong advocate of environmental protection and preservation in the Philippines. The flower has vivid dark tangerine orange petals with

white eye and pastel pink halo upon opening in the morning and turning golden yellow-orange with orange veins radiating from the center going to the petals toward mid-day.

4. *Hibiscus rosa-sinensis* 'Lilia B. de Lima' was named after the angel-warrior Director-General of the Philippine Economic Export Zone Authority (PEZA), Atty Lilia B. de Lima. The flower has petals wherein two-third is strong orange-yellow, while one-third is strong orange in the morning and became golden yellow with orange veins radiating from the center going to the petals toward mid-day. The eye is dark red with blotches and dark pink halo. It is large with an average bloom size of 15.3 cm.
5. *Hibiscus rosa-sinensis* 'Cynthia A. Villar' was named after Senator Cynthia A. Villar, a public servant who has a strong advocacy on livelihood creation. The single, regular flowers are orange (RHCC 28B) with red eye zone (RHCC 45 B) surrounded by pinkish halo. It is a medium-sized flower with bloom size of 133.5 mm in diameter.
6. *Hibiscus rosa-sinensis* 'Maria Rosario O. Montejo' is named after Maria Rosario O. Montejo, an outstanding public servant in Pulilan, Bulacan, Philippines, who worked passionately in the management and governance of community affairs. Flowers of the hybrids are single and regular with red-orange (RHCC 45B) petals and pinkish-red eye surrounded by yellow edges. Flowers are large with an average bloom size of 145.9 mm in diameter.
7. *Hibiscus rosa-sinensis* 'Sylvia P. Lina' is named after Ms. Sylvia P. Lina who whole-heartedly supported the livelihood promotion projects in the rural areas of Laguna, Philippines, together with her husband, Joey Lina, the former Governor of Laguna, Philippines, and the former Department of Internal and Local Government (DILG) Secretary. The flowers are simple and regular with neyron rose (RHCC 56A) petals with magenta eye (RHCC 66A). Flowers have an average diameter of 132.5 mm.
8. *Hibiscus rosa-sinensis* 'Connie S. Angeles' is named after Ms. Connie Angeles, a renowned public servant and TV host. Ms. Angeles is also known for her integrity as a public service host and a broadcaster like in her Channel 7 TV program "Kapwa Ko Mahal Ko" (I love my Fellowmen) with Orly Mercado, a co-host of this public service program. Flowers are regular and single with orpiment orange (RHCC 25 A) petal and cardinal red eye (RHCC 53 A) surrounded by pinkish halo. The flowers are medium in size with a diameter of 12.7 cm.
9. *Hibiscus rosa-sinensis* 'Arlene B. Arcillas'. The hybrid is named after the first woman Mayor of the City of Sta. Rosa in Laguna, Philippines. She had an untiring dedication and advocacy in improving nutrition in her city. She consistently promoted proper nutrition and health, environmental sustainability, and poverty reduction among its constituency. The hybrid has flowers with carmine rose (RHCC 52C) petals and cardinal red eye (RHCC 53C). The flowers are medium sized with a diameter of 13.05 cm.
10. The *Hibiscus rosa-sinensis* 'Che-che Lazaro' was named after the fearless broadcast investigative journalist of the former television program "Probe Team." The flower has strawberry red (RHCC 45B) petals with claret rose eye

(RHCC 50 A) and semi-ruffled petal edges. The flower is simple, regular, single petal, and large with an average bloom diameter of 147 mm.

11. The *Hibiscus rosa-sinensis* 'Pia S. Cayetano' was named after Senator Pia S. Cayetano, Senator of the Philippines and UP Regent. She is also an advocate of women's protection and children's rights, the environment, and the health and well-being of Filipinos. The flower is fuchsia-purple (RHCC 67 B) with dark purple eye (Fig. 5). The flower is simple, regular, single petal, and large with an average bloom diameter of 14.0 cm.
12. The *Hibiscus rosa-sinensis* 'Vilma Abaya-Dimacuha' was named after the former lady mayor of Batangas City, Philippines, who is a dedicated environmentalist and community builder. The flower has lemon yellow (RHCC 13 A)

Table 3 Number of registered cultivar in the International *Hibiscus* society as of October 2020

Country/area	Cultivar registered	Country/area	Cultivar registered
Australia	2206	Malaysia	4
Austria	14	Martinique	1
Barbados	4	Mexico	6
Bolivia	24	Morocco	4
Bosnia and Herzegovina	1	New Zealand	121
Brazil	432	Norway	8
British Virgin Islands	8	Panama	40
Bulgaria	27	Papua New Guinea	9
Canadian British Columbia	3	Puerto Rico	137
Canadian Ontario	50	Reunion Islands	4
China	42	Romania	50
Costa Rica	12	Russian Federation	248
Croatia	27	South Africa	120
Cuba	1	Spain	14
Czech Republic	258	Sri Lanka	36
Denmark	10	St. Vincent and Grenadines	11
Ecuador	5	Sweden	5
Fiji	112	Taiwan	1145
France	3	The Netherlands	68
French Polynesia	4923	The Philippines	161
Germany	226	Turkey	2
India	259	Ukraine	29
Indonesia	5	UK	5
Israel	1	USA	10,207
Italy	9	Uruguay	8
Jamaica	1	Unspecified	1280
Kazakhstan	4		

Fig. 17 Some other *Hibiscus* hybrids developed by private breeders in Australia: (**a**) 'Alan Poole' (Little Allan and Elaine), (**b**) 'All That Jazz' (Little Allan and Elaine), (**c**) 'Vanilla Sundae' (Cornwell Reg), (**d**) 'Tarantella' (Lindsay Greg and Julie), (**e**) 'ALKIRA Casablanca' (Lynam Richard and Brenda), (**f**) 'Stella Maris' (Mansbridge Richard), (**g**) 'Satan's Gold' (McMullen Allan), (**h**) 'Akane' (Purdie Jim and Ruth), (**i**) 'Avryl' (Westerman Alfred T), (**j**) 'White Wonder' (Van Moolenbroek Bev), (**k**) 'Dorothy Olive' (Richardson Norm and Betty), (**l**) 'Missy' (Purdie Tim). (Images lifted from: https://internationalhibiscussociety.org/searchive/index)

petals with striking red eye (RHCC 53 C). The flower is simple, regular, single petal, and large with an average bloom diameter of 130.5 mm.

13. *Hibiscus rosa-sinensis* 'Marilyn D. Marañon'. The hybrid was named after Dr. Marilyn D. Marañon who dedicated her service to the public by conducting numerous medical missions to the needy citizens of the Province of Negros Occidental, Philippines. She even gave up her career and pharmaceutical business just to attend to the health needs of the less fortunate Negrenses, the local people of her province. The bloom type is simple, regular, with a diameter of 14.54 cm. The petal is lemon yellow (RHCC 8A) with white eye zone (RHCC 25 A) surrounded by pinkish halo.
14. *Hibiscus rosa-sinensis* 'Domini M. Torrevillas' was named after the columnist of the Philippine Star, Ms. Domini M. Torrevillas. Ms. Domini M. Torrevillas is a multi-awarded and well-respected journalist, a freedom fighter, and a consummate storyteller. She authored three books: *Sounds of Silence*, *Sounds of Fury*, and *Vendors of Manila*. The flower has dark orange petals with dark red eye zone and yellowish veins going through the petals with a petal diameter of 17.54 cm.

Fig. 18 Some other *Hibiscus* hybrids developed by private breeders in China, (**a**) 'Hawaii Volcano' (Huang Xuguang), (**b**) 'Butterfly Dream' (Huang Xuguang), (**c**) 'CATAS Frozen Crater' (Niu Junhai), (**d**) 'CATAS Bloodied Plantain' (Niu Junhai); in Indonesia, (**e**) 'Irreple Black Copper' (Khasanah Iswatun); in Kazakhstan: (**f**) 'Andromeda Galaxy' (Vanyuk Anna); in Malaysia, (**g**) 'Wukhoon's the Beginning' (Voon Wu Khoon); in Russian Federation, (**h**) 'V's Raspberry Sherbet Punch' (Popova Valeria), (**i**) 'St Vincent Full Moon' (Kiseleva Svetlana), (**j**) 'Russian Anna Pavlova' (Tabuntsova Elena); and in Sri Lanka, (**k**) 'Midnight Diminish' (Weesinghe Sumeda), (**l**) 'Suni Rosalina' (Weesinghe Sumeda). (Images lifted from: https://internationalhibiscussociety.org/searchive/index)

15. Hibiscus rosa-sinensis 'Emmeline Aglipay Villar', was named after Atty. Emmeline Aglipay-Villar. She is an undersecretary of the Department of Justice of the Philippines and also a strong advocate of women's rights, child protection, and marginalized sectors' rights. She is also a fervent advocate of health awareness most especially of lupus. The flower is a tricolor hybrid with a spinel red petals, yellow-orange edges, and a red eye. It has a floral diameter of 12.8 cm.
16. *Hibiscus rosa-sinensis* 'Carmen Martinez Pascual' was named after Ms. Carmen M. Pascual, an artist and a public servant. She is also the wife of the former UP president, Dr. Alfredo E. Pascual. The flower has bluish-purple petals with a rhodonite red eye and white vein markings radiating from the eye to the petal and has a floral diameter of 14.5 cm.

Fig. 19 Some other *Hibiscus* hybrids developed by private breeders in Africa, 'Hanini Nada' (Hanini Ali), (**b**) 'Hanini Magic' (Hanini Ali), (**c**) 'Jim's Yellow Black' (Laurent Jim), (**d**) 'Jim's Yellow Cream' (Laurent Jim); in South Africa, (**e**) 'Joymac AD White' (Van Graan Susanne & Chris), (**f**) 'Joymac Anna Long' (Van Graan Susanne & Chris), (**g**) 'Joymac African Sun' (Van Graan Susanne & Chris), and (**h**) 'Joymac Coral Charm' (Van Graan Susanne & Chris). (Images lifted from: https://internationalhibiscussociety.org/searchive/index)

16.11.6 Women in Science Series, Women in Media and Arts Series, Women Saints and Institutions Named After Them, Women in Education Series and Women in Social Entrepreneurship Series

This batch of *Hibiscus* hybrids collectively called the "Women in Science" Series was released as a continuing contribution of the Institute to the celebration in 2009 of the UP Centennial Year. These hybrids were named after outstanding women scientists. Some of them were elected to the National Academy of Science and Technology (NAST) of the Philippines. At present, there are only two National Seed Industry Council (NSIC)-registered *Hibiscus* hybrids under this series.

1. *Hibiscus rosa-sinensis* 'Solita F. Camara-Besa' was named after Dr. Solita F. Camara-Besa. Academician Camara-Besa is a medical educator and a researcher. She has worked on the sodium and potassium content of Philippine foods and established standards useful in the preparation of diets. The flower has old rose or red-purple petals with red eye zone. It is medium with an average bloom size of 6.5 cm.
2. *Hibiscus rosa-sinensis* 'Ledivina V. Cariño' was named after Dr. Ledivina V. Cariño. Academician Dr. Ledivina V. Cariño is an outstanding and accomplished academician of NAST in the fields of public administration, political

Fig. 20 Some other *Hibiscus* hybrids developed by private breeders in Oceania and the Caribbean. Hybrids in Barbados, (**a**) 'Zion Bim' (Springer Johan), (**b**) 'Zeak Elderena' (Farara Lillian), (**c**) 'Zeak Savannah Georgie' (Farara Lillian); in Fiji, (**d**) 'Simmond's Red' (Simmonds H W), (**e**) 'Rag Doll' (Perks Ken); in New Zealand, (**f**) 'Apricot Parade' (Clark Jack), (**g**) 'Lavender Lady' (Clark Jack), (**h**) 'Mrs Jack Clark' (Clark Jack); in Virgin Islands, (**i**) 'Lavender Glow' (Lawaetz Kai), (**j**) 'Icy Passion' (Sinanan Steve), (**k**) 'Danmark' (Lawaetz Kai), (**l**) 'Northside Peach' (Stuedell David). (Images lifted from: https://internationalhibiscussociety.org/searchive/index)

science, and sociology. The flower has orange petals with pink eye zone and semi-ruffled edges. It is medium with an average bloom size of 8.5 cm.

3. *Hibiscus rosa-sinensis* 'Fe V. del Mundo' was named after the National Academy of Science and Technology's (NAST) National Scientist for pediatrics Dr. Fe V. del Mundo. This *Hibiscus* hybrid honors the outstanding achievements of Dr. Fe V. del Mundo as a pediatrician who developed the Philippine version of the incubator for prematurely born babies. The flower has tangerine orange petals (RHCC 29 A) with distinct reddish pink eye surrounded by white halo that radiates to the petals. The flower is simple, regular, single petal and large with an average bloom diameter of 145.9 mm.
4. *Hibiscus rosa-sinensis* 'Lourdes J. Cruz' was named after the National Scientist for biochemistry, Dr. Lourdes J. Cruz of NAST Philippines. Her research work concentrated on the biochemistry of conotoxins that have medical uses most especially in the nervous system. The hybrid is a double petal form with bright persimmon orange petals and petalloids with yellowish edging and guardsman red eye radiating to the petals. The flowers have an average diameter of 8.4 cm.

Fig. 21 Some other *Hibiscus* hybrids developed by private breeders in Taiwan: (**a**) 'Asia Queen Anna Sui' (Lai Vicki), (**b**) 'Asia Rosette Nebula' (Lai Vicki), (**c**) 'BLG Baby Angelo' (Huang Pohan and Tang Peter), (**d**) 'Bo Den Blue Ice' (Lee Aneela), (**e**) 'Bo Den Blue Moon' (Wu Kun Zhu), (**f**) 'Bo Den The Eye of Apollo' (Wu Kun Zhu), (**g**) 'Caliburn Pastel Palette' (Lee Linda), (**h**) 'Chris' A Thousand Kisses' (Chang Chris), (**i**) 'Chris' Countess' (Chang Chris), (**j**) 'Divine My Sweet' (Huang Pohan and Tang Peter), (**k**) 'Taiwan Sunny Waltz' (Lee Linda), and (**l**) 'TW Blue Thrill' (Wang Kuei Sheng). (Images lifted from: https://internationalhibiscussociety.org/searchive/index)

The petals and petalloids of this hybrid are arranged in a tight formation giving a ball-shaped appearance with five orange florets in the center of the flower.

The "Women in Media and the Arts" Series was formally launched starting 2010. These hybrids were named after remarkable women who made significant contributions to the Philippine mass media and the arts.

Hibiscus rosa-sinensis 'Araceli A. Dans' was named after the famous artist for painting and drawing Araceli A. Dans. This *Hibiscus* hybrid honors the notable achievements of Araceli A. Dans as an artist particularly in painting and drawing such as the "Calado," series, "Burdado," and "Flowers and Lace," among others, and being an art educator. The flower has nasturtium orange (RHCC 25 B) petals with jasper red (RHCC 39 C) eye and semi-ruffled petal edges. The bloom type is simple, single petal, regular, and large with an average bloom diameter of 145 mm.

This batch of *Hibiscus* hybrids called the "Women Saints Series and Institutions named after them" was launched in 2014. Hybrids under this series were named after

Fig. 22 Some other *Hibiscus* hybrids developed by private breeders in Polynesia bred by Johnson Richard: (**a**) 'Alkira Our Eva', (**b**) 'Borneo Legacy', (**c**) 'Life Creation', (**d**) 'Magenta Wind', (**e**) 'Sudu Boale', (**f**) 'Tahitian Xenosphere', (**g**) 'Tahitian Yellow Spiral', (**h**) 'Tahitian Valentine Princess', (**i**) 'Tahitian Strawberry Moon', (**j**) 'Tahitian Passion Flower', (**k**) 'Tahitian Princess', and (**l**) 'Tahitian Mahogany Marble'. (Images lifted from: https://internationalhibiscussociety.org/searchive/index)

beloved and respected women saints of the Catholic Church and institutions named after them. At present, only one hybrid belong to this series.

Hibiscus rosa-sinensis 'St. Bridget College' was named after St. Bridget College of Batangas City. St. Bridget is the patron saint of culture and arts in Ireland that was brought by the Religious of the Good Shepherd Sisters to the Philippines and became the patron saint of St. Bridget College. This college has been noted to produce high-caliber graduates in the region. The flower has unique bluish (RHCC 85 A) petals with barium yellow edging (RHCC 10 A) and a gray-red (RHCC N182 C) eye zone (Fig. 12). It is simple, regular, single petal, and large in size with an average bloom diameter of 147 mm. This *Hibiscus* hybrid is known as the Centennial Flower of St. Bridget College.

The batch of *Hibiscus* hybrids called the "Women in Education" Series was launched in 2015. The hybrids were named after remarkable women who made significant contributions in the field of education in the Philippines. At present, there is only one hybrid released under this series.

The *Hibiscus rosa-sinensis* 'Patricia B. Licuanan' was named after the former Commissioner of Higher Education and former UP Regent Dr. Patricia B. Licuanan.

Fig. 23 Some other *Hibiscus* hybrids developed by private breeders in Polynesia bred by Charles Atiu: (**a**) 'WMMA Cold as Ice', (**b**) 'WMMA Te Here', (**c**) 'Moorea For Ever', (**d**) 'Moorea Milagro', (**e**) 'Moorea Dreamcatcher', (**f**) 'Moorea Senior Charles', (**g**) 'Moorea Pacific Pearl Star', (**h**) 'Moorea Merveille Intensity', (**i**) 'Moorea Spring Sun', (**j**) 'AMMA Gentle Heart', (**k**) 'RK Cassiopeia' (Atiu Marianne), (**l**) 'RK Baron Draco' (Charles Atiu). (Images lifted from: https://internationalhibiscussociety.org/searchive/index)

The flower has solferino purple or red-purple (RHCC 65 B) petals with dark ruby red (RHCC 61 A) eye surrounded by grey halo with semi-ruffled petal edges. It is single petal, simple, regular, and large with an average bloom diameter of 145 mm. This hybrid honors Commissioner Licuanan's strong leadership roles in the education sector and her accomplishments as a social psychologist, educator, and advocate of women's rights.

This batch of *Hibiscus* hybrids is called the "Women in Social Entrepreneurship Series." This was launched last 2018. Hybrids under these series were named after outstanding women entrepreneurs with social consciousness. At present, only one hybrid was under the series.

Hibiscus rosa-sinensis 'Dr. Milagros O. How' is a hybrid named after Dr. Milagros How. Bloom type is simple with a regular bloom size of 14.56 cm. Flowers have a two-day longevity as an intact flower during the cold weather and a one-day longevity during summer as intact and cut flower. It has pinkish-white (RHCC N155B) petals during the morning and has white (RHCC N155B) petals in the afternoon. It has a red-purple (RHCC 65A) eye. The petals have ruffled edges.

Fig. 24 Some other *Hibiscus* hybrids developed by private breeders in India by Pushpah Suresh: (**a**) 'Anuraag', (**b**) 'Inner Sanctum', (**c**) 'Alluring Diamond', (**d**) 'Antique Coin', (**e**) 'Java Brown Windmill'; by Madappa Shyamala, (**f**) 'Divine Grace', (**g**) 'Little Ballerina', (**h**) 'Nectar fresh', (**i**) 'Classical Dancer'; by Pushpah Suresh and Madappa Shyamala, (**j**) 'DP's Love Whisperer', (**k**) 'Zeak Jackson', and (**l**) 'Zeak Julio Arana'. (Images lifted from: https://internationalhibiscussociety.org/searchive/index)

Dr. Milagros How is a fervent advocate of the advancement of Philippine agriculture and giving recognition to the efforts of the farms through the program "ToFarm." She has spearheaded various projects that have truly improved the lives of eight million Filipino farmers through training programs, scholarships, and farmer communities' empowerment, among others (Tofarm.org).

16.12 Hybridization of Hibiscus in Private Gardens

Hybridization of *Hibiscus* in the Philippines has not been limited to public research institutions like UPLB. Due to the availability of genetic materials and straightforward hybridization process, hybridization has been done by private breeders too.

Private breeders in the Philippines released, propagate, and commercialize their own *Hibiscus* hybrids. For instance, Mr. Boyet Ganigan has also developed several *Hibiscus* hybrids. Other private breeders in the Philippines include Mr. Jorge

Fig. 25 Some other *Hibiscus* hybrids developed by private breeders in America: (**a**) 'Alelulia' (Coryell Jill), (**b**) 'Zellie Waegner' (Miller August), (**c**) 'A Blue Diamond' (Reynolds Nola and Carlos), (**d**) 'Alluring' (Carran Bob), (**e**) 'Almost Breathless' (Reynolds Nola And Carlos), (**f**) 'Almost Perfection' (Reynolds Nola and Carlos), (**g**) 'Angel's Kiss' (Scobey Elaine and Russ), (**h**) 'Noemi' (Entz Douglas), (**i**) 'Adam's Reef' (Schlueter Barry and Susan), (**j**) 'Apricot Glow' (Schlueter Barry and Susan), (**k**) 'Wishing Well' (Schlueter Barry and Susan), (**l**) 'Godiva' (Theall Sandra), (**m**) 'Vero's Choice' (Schlueter Barry and Susan), (**n**) 'Adrenaline Junkie' (Alvis Todd and Janelle), (**o**) 'Accent' (Schlueter Barry and Susan), (**p**) 'Trivial Pursuit' (Schlueter Barry and Susan), (**q**) 'St Vincent Baby Angel' (Entz Douglas), (**r**) 'Caribbean Dark Galaxy' (Sinanan Steve), (**s**) 'Abracadabra' (Black Charles), (**t**) 'Acapulco Gold' (Black Charles), (**u**) 'AJ Crazy Beautiful' (Eminian Darren), (**v**) 'Bridal Veil' Hybridizer: Unknown (**w**) 'Cornucopia' (Black Charles), (**x**) 'Dutch Windmill' (Black Charles), (**y**) 'Belle du Jour' (Black Charles), (**z**) 'Tara's Heart' (Short Buddy), AA. 'Candy Rose' (USA Seed Bank) and AB. 'Lambada' (Sinanan Steve). (Images lifted from: https://internationalhibiscussociety.org/searchive/index)

Dominguez and Mr. Ryan Tayobong who have developed their HRYa *Hibiscus*; others such as Mr. Joseph Bautista and Mr. Niron Concepcion among others have registered their hybrids in the IHS database.

16.13 Some International Cultivars

The database of the International *Hibiscus* Society (IHS) indicated the following number of cultivars registered in each country/area. The number is only limited to those applied to the IHS. It covers those that are registered, some reported but not yet applied, and those applied for this but are not yet officially registered.

Table 3 shows some of the *Hibiscus* hybrids found in different parts of the world. Among the countries, the USA had the highest number of *Hibiscus* hybrids registered (10,207), followed by French Polynesia (4923), and then by Australia (2206) and Taiwan (1145). Some of the hybrids are shown in the following figures. Figures showing different *Hibiscus* are grouped according to place where they were either grown or developed. The name of the hybrid, their developer, and the origin of the hybrid were indicated in the figure (Figs. 17, 18, 19, 20, 21, 22, 23, 24 and 25).

Some of these hybrids are being exchanged or being propagated asexually. Since they can be propagated asexually, through cuttings or via grafting, these hybrids can be transported from one country to another with proper phytosanitary permits to prevent pathogen spread. In addition, hybrid seed exchange is also being done. In Taiwan, for example, breeders sell their seeds and ship them to the requesting country. The breeder still owns the intellectual property of the hybrid.

The exchange of *Hibiscus* asexually propagated plants allows the multiplication of plants with unique and novel flower characteristics. Some color combinations also became more evident in the later years when blue and brown colorations and even black-colored flowers of hybrids were offered for commercialization.

Pollen conservation and preservation were also being done by some breeders. They collect pollen of their hybrid and then keep it in refrigerated conditions and then ship them to where it should be so that the breeder will have a broad base of genetic materials for use as pollen source for their hybrids. Since the nature of *Hibiscus* is highly heterozygous, parental combinations may lead to novel color combination or even unique flower form among others.

Acknowledgment The authors would like to thank Mr. Boyet Ganigan, Prof. Jorge Dominguez, Dexter Rose Fernando, Jojette Morales, Rita Castro Knust, Sarang Bishnu, Jo Rosenberry, and Frank Dilenzo of www.pixabay.com and most especially to the International Hibiscus Society through its president Darren Eminian, for giving us permission to use their pictures for this book chapter.

References

Almario V (ed) (2015) Sagisag Kultura, vol 1. National Commission for Culture and the Arts, Manila. Retrieved from https://philippineculturaleducation.com.ph/gumamela/

Australian Hibiscus Society (2004) Hibiscus growing guide, 2nd edn. The Australian Hibiscus Society Inc., Queensland, 40 pp

Barraquia ARJ, Cabuso APL, Narvacan CM (2017) Antibacterial Activity of Ethanolic Extract of *Hibiscus rosa-sinensis* flower against Staphylococcus epidermidis and Staphylococcus saprophyticus. Lyceum of the Philippines-St. Cabrini College of Allied Medicine 2(2):39–50

Bates DM (1965) Notes on the cultivated Malvaceae: Hibiscus

Biriok GS, Yawas DS (2017) Comparative performance evaluation of biodiesel from *Hibiscus* Sabdariffa and *Hibiscus* Surattensis seeds oils. Int J

Burkill (n.d.) Cited by Bureau of Plant Industry-Department of Agriculture. January 1, 2006. *Hibiscus rosa-sinensis* Linn. Gumamela. http://www.bpi.da.gov.ph/Publications/mp/html/gumamela.htm. Accessed 15 June 2008

Chen JJ, Wang Y, Paudel A, Sun Y (2019) Comparing the salt tolerance of three landscape plants using a near-continuous gradient dosing system. Hortechnology. https://doi.org/10.21273/HORTTECH04385-19

Cheng Z, Lu BR, Sameshima K, Fu DX, Chen JK (2004) Identification and genetic relationships of kenaf (*Hibiscus* cannabinus L.) germplasm revealed by AFLP analysis. Genet Resour Crop Evol 51(4):393–401

Craven LA, de Lange PJ, Lally TR, Murray BG, Johnson SB (2011) A taxonomic re-evaluation of *Hibiscus* trionum (Malvaceae) in Australasia. N Z J Bot 49(1):27–40

Devi P (2004) Biosystematic studies in Hibiscus rosa sinensis L morphological palynological and molecular. http://hdl.handle.net/10603/165815. Accessed in: https://shodhganga.inflibnet.ac.in/handle/10603/165815

Döring M (2020) English Wikipedia – species pages. Wikimedia Foundation. Checklist dataset https://doi.org/10.15468/c3kkgh accessed via GBIF.org on 2020-11-15

[FAO] Food and Agriculture Organization (n.d.) *Hibiscus* tilaceus. In: Trees and shrubs of the Maldives. http://www.fao.org/tempref/docrep/fao/010/ai387e/ai387e04.pdf

Fryxell PA (1965) Stages in the evolution of *Gossypium* L. Adv Front Plant Sci 10:31–56

Gilman EF (2014) *Hibiscus* coccineus Scarlet Rose Mallow, Texas Star *Hibiscus*, Swamp *Hibiscus*. University of Florida, UF/IFAS Extension Service

Gilman EF, Watson DG (1993) *Hibiscus* syriacus: Rose-of-Sharon1. In: Fact Sheet ST-295, a series of the Environmental Horticulture Department, Florida Cooperative Extension Service. Institute of Food and Agricultural Sciences, University of Florida

Golubenko Z, Akhunov A, Khashimova N, Beresneva Y, Mustakimova E, Ibragimov F et al (2007) Induction of peroxidase as a disease resistance response in resistant (*Hibiscus* trionum) and susceptible (Althea armeniaca) species in the family Malvaceae. Phytoparasitica 35(4):401–413

Hochreutiner BPG (1900) Revision du genre Hibiscus. Romet

Hoskins WM (2016) Using phylogenetics to understand the evolutionary relationships of *Hibiscus* section Furcaria

Huang WJ (n.d.) Bloom characteristics learn to describe your Hibiscus. http://internationalhibiscussociety.org/new/images/Data/Nomenclature/Bloom%20Characteristics.pdf

International *Hibiscus* Society (n.d.). Cultivar search engine. http://www.international*Hibiscus*society.org/SEArchive/SEindex1.php

Jo YK, Mazharu IM, Kim CK, Kim HY, Lim KB (2019) Morphological Characteristics and FISH Analysis of *Hibiscus* F1 hybrids and parental lines. 원예과 h기술지 37(5):630–639

Khan IM, Rahman R, Mushtaq A, Rezgui M (2017) Hibiscus rosa-sinensis L. (Malvaceae): distribution, chemistry and uses. Int J Chem Biochem Sci 12:147–151

Kholkute SD, Udupa KN (1976) Antiestrogenic activity of *Hibiscus rosa-sinensis* Linn. flowers. Indian J Exp Biol 14(2):175

Kuligowska K, Simonsen M, Lütken H, Müller R (2012) Breeding of *Hibiscus rosa-sinensis* for garden use in Denmark. In: Proceedings of 2nd IS on woody ornamentals of the moderate zone. Acta Hort 990:235–241

Lawton BP (2004) Hibiscus: Hardy and tropical plants for the garden. Timber Press

Mabberley DJ (1997) The plant-book: a portable dictionary of the vascular plants. Cambridge University Press

Magdalita PM, Pimentel RB (2013) Development of *Hibiscus* hybrids 'Women in Public Service Series II' and propagation studies on *Hibiscus rosa-sinensis* 'Cynthia A. Villar'. Philipp Sci Lett 6(1):39–56

Magdalita PM, Cayaban MFH, Gregorio MT, Silverio JV (2016) Development and characterization of nine new *Hibiscus* hybrids. Philipp J Crop Sci 41(2):31–45

Magdalita PM, Valdoz JC, San Pascual AO, Sotto RC (2019) Phenotypic evaluation of floral characteristics for predicting the components of longer floral retention in *Hibiscus rosa sinensis* L. hybrids. Philipp Sci Lett 12(2):167–175

Maghirang RG, Oraye CD, Antonio MA, Cacal MS, City B (2018) Ethnobotanical studies of some plants commonly used as vegetables in selected provinces of the Philippines. J Nat Stud 17(2): 30–43

Mohammed BB, Sarwar MB, Hassan S, Rashid B, Aftab B, Husnain T (2015) Tolerance of Roselle (*Hibiscus* sabdariffa L.) genotypes to drought stress at vegetative stage. Adv Life Sci 2(2):74–82

Native Plants Hawaii (2009) *Hibiscus* arnottianus subsp. punaluuensis. http://nativeplants.hawaii.edu/plant/view/*Hibiscus*_arnottianus_punaluuensis

Palve SM, Sinha MK (2005) Genetic variation and interrelationships among fibre yield attributes in secondary gene pool of Corchorus spp. SABRAO J Breed Genet 37(1):55–64

Pascual CB, Magdalita PM (2012) Disease identification and in situ screening of Gumamela (*Hibiscus rosa-sinensis* Linn) germplasm and hybrids to leaf anthracnose for disease resistance breeding. Philipp Sci Lett 5(2):158–168

Pfeil BE, Brubaker CL, Craven LA, Crisp MD (2002) Phylogeny of Hibiscus and the tribe Hibisceae (Malvaceae) using chloroplast DNA sequences of ndh F and the rp 116 intron. Syst Bot 27(2):333–350

Purseglove JW (1987) Tropical crops. Dicotyledons. Longman Singapore Publishers Pte. Ltd., Singapore, 719 pp

Raut DN, Patil TB, Chaudhari SR, Pal SC, Mandal SC (2014) Antimitotic effect of ethanol fraction of *Hibiscus* mutabilis leaf and flowers. Res Pharm 4(5):16

Rummel DJ (2005) Botanical beauty book. C & E Publishing

San Pascual AO, Magdalita PM, Medina NG, Apacionado BV (2017) Characterization, pollen behavior and propagation of five selected '*Hibiscus*' hybrids ('*Hibiscus rosa-sinensis*' Linn.). Aust J Crop Sci 11(12):1508

Satya P, Karan M, Sarkar D, Sinha MK, Mahapatra BS (2012) Morphological and molecular characterization of interspecific hybrids of kenaf (*Hibiscus* cannabinus L.) and false roselle (H. acetosella Welw. ex Hiern). Indian J Genet Plant Breed 72(2):234

Shanan NT, Moghaieb RE (2016) Genetic diversity among four genotypes from *Hibiscus* rosasinensis in relation to drought stress. J Hortic For 8(3):24–31

Shang K, Hu YH, Vincent G, Labrecque M (2020) Biomass and phytoextraction potential of three ornamental shrub species tested over three years on a large-scale experimental site in Shanghai, China. Int J Phytoremediation 22(1):10–19

Sivarajan VV, Pradeep AK (1996) Malvaceae of southern peninsular India: a taxonomic monograph. Daya Books

Stuart G (2019) Gumamela. *Hibiscus rosa-sinensis* Linn. China Rose. Zhu Jin. http://www.stuartxchange.org/Gumamela.html. Accessed 13 Nov 2020

van der Pijl L (1937) Disharmony between Asiatic flower-birds and American bird-flowers. Ann Jard Bot Buitenzorg 68:17–26

Van Laere K, Van Huylenbroeck JM, Van Bockstaele E (2007) Interspecific hybridisation between *Hibiscus* syriacus, *Hibiscus* sinosyriacus and *Hibiscus* paramutabilis. Euphytica 155(1–2):271–283

Wang ML, Morris B, Tonnis B, Davis J, Pederson GA (2012) Assessment of oil content and fatty acid composition variability in two economically important *Hibiscus* species. J Agric Food Chem 60(26):6620–6626

Wilcox EV, Holt VS (1913) Ornamental Hibiscus in Hawaii (No. 29). Paradise of the Pacific Press

Wilson FD, Menzel MY (1967) Interspecific hybrids between kenaf (*Hibiscus* cannabinus) and roselle (H. sabdariffa). Euphytica 16(1):33–44

Wu HY, Yang KM, Chiang PY (2018) Roselle anthocyanins: antioxidant properties and stability to heat and pH. Molecules 23(6):1357

Modern Techniques for Plant Breeding in Ornamentals

17

Stephen F. Chandler and David Tribe

Contents

Abstract

The new breeding technologies of genetic modification and gene editing potentially offer breeders of ornamental plants the option of creating varieties that could not be obtained using conventional hybridization and mutagenesis techniques. Through knockdown of endogenous genes, precise changes to gene

S. F. Chandler (✉)
School of Applied Sciences, RMIT University, Bundoora, VIC, Australia
e-mail: schandler@florigene.com.au

D. Tribe
School of Biosciences, University of Melbourne, Parkville, VIC, Australia
e-mail: detribe@unimelb.edu

S. K. Datta, Y. C. Gupta (eds.), *Floriculture and Ornamental Plants*, Handbooks of Crop Diversity: Conservation and Use of Plant Genetic Resources,
https://doi.org/10.1007/978-981-15-3518-5_19

function can be made using gene editing, and by inserting genetic material from far outside a species, gene pool novel genotypes and phenotypes can be generated using genetic modification. Though there are significant technical requirements associated with the new technologies – particularly developing suitable gene delivery and plant regeneration methods – there are examples of their commercial application, outlined in this chapter. Genetically modified, novel flower color varieties of carnation and rose have been on the marketplace for many years, and research groups have shown that flower color can also be modified in a number of species using gene-editing methods. An important consideration associated with use of these new technologies is a requirement to meet any legislated regulatory requirements before research and commercialization of a genetically modified or gene-edited plant can be carried out. This is a financial hurdle for breeders, as additional trials and molecular characterization work is required. Also, though many ornamental plants are traded internationally, there is no international harmonization of regulation of genetic modification, and so far there is limited coordination in the adaptation and implementation of new rules to regulate gene-edited plants. As well as technical issues, balancing an assessment of potential commercial value against the cost of development and the complexity of obtaining regulatory approval is a key consideration faced by plant breeders contemplating the use of these new technologies.

Keywords

New breeding technology · Gene editing · Regulation · Genetic modification · CRISPR/Cas9

17.1 Introduction

Ornamental plants, or "ornamentals," are horticultural plant species grown largely for their aesthetic value. Excluding grass species used for lawns and turf, ornamental plants are, broadly speaking, plants grown for cut flower harvest for use in floristry and home decoration and plants grown for the decorative value of their foliage, flowers, and/or fruit. Ornamental plant species comprise a multitude of flowering and non-flowering plant varieties from thousands of species and may be grown in flower farms, pots, baskets, garden beds, and landscaped amenities such as gardens and parks. The range of ornamental plant species encompasses species adapted to all types of hardiness zone, indoor and outdoor plants, and most plant families and includes both annual and perennial species. Within a species there may be, literally, thousands of cultivars and varieties. The cultivated rose serves as an example; in this species there have been hundreds of years of breeding from species within the genus *Rosa*, resulting in the multitude of plant and flower forms we now see in the horticultural species *Rosa* x *hybrida*. Similar domestication of other very familiar ornamental plants, such as the chrysanthemum (Fukai 2003), has a history spanning more than a thousand years (Ko et al. 2019).

Historically, the development of new ornamental plant varieties was achieved through a combination of collection of wild species from all over the world, breeding through intraspecific hybridization, where possible crossing by interspecific hybridization, the generation and selection of mutant lines, and exploitation of natural variants (e.g., periclinal chimeras). Since the advent of tissue culture methods from the 1960s, breeding efforts have been assisted in recent decades with the development of sophisticated techniques to overcome interspecific barriers to hybridization, increase ploidy level, or allow clonal propagation of species which were difficult to propagate either sexually or by using conventional cutting or grafting methods. Later, as tools for genetic analysis became available, molecular markers have been used to assist breeders identify, track, and introduce desirable genes. From the 1980s methods for direct genetic modification (gene isolation and plant transformation) were invented, and these too have now been used to create new varieties in ornamental plant species. A potential breakthrough technology, genetic modification provided breeders the possibility of using genes from any species, including from bacterial origin, for new variety development. Now, new opportunities for developing novel and/or improved ornamental plants have emerged as whole genome sequencing and bioinformatics have become routine and affordable and gene editing, a more specific genetic modification technology, has been routinely demonstrated to be possible in plants.

In this chapter, we undertake to update readers on the use of genetic modification, gene editing, and other molecular tools for ornamental plant breeding. We have placed a particular focus on the potential applications of these technologies commercially, in the context of the global state of play for regulation of genetically modified and gene-edited plants. Regulatory requirements and considerations are unique to new ornamental plants which have been produced by genetic modification and gene editing in the sense that similar rules largely do not apply if more traditional breeding methods are used. Understanding the costs and complexity of meeting regulatory requirements is therefore central to any commercialization strategy for a genetically modified or gene-edited ornamental plant.

We have excluded from our review genetically modified and gene-edited ornamental plants that are primarily grown to provide edible plant parts (e.g., artichoke, asparagus, ornamental pepper, sunflower). This is because these plants would be subject to a wider regulatory review than non-food ornamental plants, as they will be considered to be foods. Space does not permit a discussion of food safety evaluation for genetically engineered ornamental plants which are also foods.

17.2 Molecular Tools for Ornamental Plant Improvement

17.2.1 Genome and Transcriptome Sequences

As the cost of nucleic acid sequencing has fallen, significant sequence information has been generated for individual genes, chloroplast genomes, and whole genomes in a number of ornamental plant species (Ahn et al. 2020). This includes whole genome sequence information of the important species rose (Saint-Oyant et al. 2018), carnation (Yagi et al. 2014), petunia (Bombarely et al. 2016), the orchid *Apostasia* (Zhang et al. 2017), *Asparagus setaceus* (Li et al. 2020), *Rhododendron lapponicum* (Jia et al. 2020), and morning glory (Hoshino et al. 2016). Sequence information allows application of plant genomics to plant improvement programs through identification of selectable markers for marker-assisted genotype-based breeding (Kumar 2019; Yagi 2018). Readers are directed to recent reviews for the ornamental plant species rose (Smulders et al. 2019), chrysanthemum (Su et al. 2019), gentian (Nishihara et al. 2018b), and *Paeonia* (Fan et al. 2019) for examples of the use of such markers.

Whole-transcriptome sequencing is a method for identification of the population of mRNA transcripts and noncoding RNA in plant cells. By sequencing different tissues, different physiological stages or a range of developmental stages differentially expressed genes can be identified and gene structure mapped. The technique may therefore lead to both gene identification and a fuller understanding of the control of gene expression and is a useful adjunct in the search for genes which may be responsible for important traits (Law et al. 2020). In genetically manipulated plants, whole-transcriptome sequencing is useful for understanding the long-term stability of inserted gene expression, which may be affected by methylation, for example. In ornamental plants, relatively little work has been done on transcriptome analysis (Su et al. 2019), but there are recent examples where the technique has been used for characterization of genes involved in pigment biosynthesis (Huang et al. 2019; Ma et al. 2018; Wei et al. 2020).

17.2.2 Uses of Sequence Information in Genetic Manipulation

Aside from identification of molecular markers, generation of sequence information is a critical component of genetic modification and gene-editing strategies. Using bioinformatic analysis, access to full genome sequence enables the identification and confirmation of genes from within genome sequences. In genetic transformation for novel trait development, homologous or foreign genes with known function may then be used for vector design and construction in combination with suitable DNA regulatory elements such as promoters and terminators. Sequence knowledge is also the basis for the identification of the targeted genome sequences and guide sequences required for gene-editing vectors. Obtaining legal regulatory approval for genetically modified plants and gene-edited plants is another area where sequence information may be critical. In some countries, the complete sequence of inserted genes and flanking genome sequence is required for approval of genetically modified organisms. Sequence information for the transformed plant species is then used to identify where in the genome insertion may have occurred, whether endogenous coding regions may have been interrupted and whether new open reading frames may have been created. In gene-edited plants, the same knowledge of the sequence in the target region of the genome will also be essential for regulatory scrutiny. Another area where sequence information will be fundamentally important is in the area of synthetic biology and the use of sequence information for generation of oligonucleotides and whole artificial genes for use in genetic manipulation. As yet, synthetic biology is yet to be demonstrated in ornamental plants.

For more details of the value and use of sequencing information in plant genome manipulation, wider ranging reviews have recently been written (Aubry 2019; Kausch et al. 2019; Shelake et al. 2019), and these may be referred to for a wider discussion and more detail.

17.3 Genetic Modification in Ornamental Plants

17.3.1 Genetic Modification

Genetic modification is the technique of inserting foreign genes into plants. The process of gene delivery to plant cells and generation of plants from those cells is termed transformation. Transformed (or transgenic) plants are described as genetically engineered organisms, genetically modified organisms (GMO), or living modified organisms (LMOs). In this document we use the more widely adopted acronym GMO. The term LMO is used in European and Japan legislation and is the exclusive term within the 2000 Cartagena Protocol on Biosafety to the Convention on Biological Diversity. The transformation process relies on the successful and coordinated application of three processes, outlined below:

1. Construction of a transformation vector. A vector designed for delivery of the genes to be inserted into the target plant will contain (a) the gene(s) required for generation of the modified phenotype, (b) the DNA sequences necessary for incorporation of the foreign genes into the target plant genome, and (c) genes whose transient or stable expression in the genome of the target plant can be used to select the cells where gene transfer has occurred. The vector will take the form of a binary plasmid for *Agrobacterium*-mediated transformation or a plasmid to be coated onto the inert metal particles used in biolistics (see below). Typically, multiple genes are included in a transformation vector, to produce both the desired modified phenotype and expression of a selectable marker gene to facilitate plant regeneration from solely transformed cells by negative selection against untransformed cells. The selectable marker gene may allow chemical selection (typically using genes conferring herbicide or antibiotic resistance) or a visual selection for transformed cells, based on fluorescence, for example. In the design of transformation vectors, selection of suitable genetic elements regulating gene expression (i.e., promoters, terminators) is critical (Smirnova et al. 2019). As well as adding new traits, genetic modification can also be used to down-regulate endogenous gene functions using antisense or co-suppression RNA interference (RNAi) technology (Schiemann et al. 2019). These techniques involve introduction into the transformation vector of a DNA sequence designed to be complementary to the mRNA of the targeted endogenous genes mRNA or to target inhibition of translation of the target mRNA.

2. A method for delivery of the vector to cells of the plant to be transformed. One of the most common delivery methods relies on co-cultivation of target tissues with a disarmed, non-virulent, strain of the natural plant pathogens *Agrobacterium tumefaciens* or *Agrobacterium rhizogenes* modified to carry a binary plasmid. In nature, these bacteria cause disease as a result of the insertion into the plant genome of bacterial genes, which are subsequently expressed to induce uncontrolled plant cell division, causing a gall to form. By removal of the bacterial genes which cause disease and adding a binary plasmid carrying genes of interest, the natural transfer mechanism can be utilized to enable transfer into the plant genome of the genes on the binary plasmid. In nature, *Agrobacterium* does not infect all plants, and many monocotyledonous species are not susceptible to infection. In these species, direct methods for insertion of transformation vectors have been developed such as biolistics and electroporation of protoplasts. Biolistics, in which plasmid DNA is coated onto metal particles which are then mechanically inserted into target tissues by explosive force, has become the most widely used gene delivery technique in plant species where infection with *Agrobacterium* is difficult.
3. A method for the regeneration of plants from transformed cells. To generate a transgenic plant, those cells which have been transformed and which survive culture on selective medium must be capable of regeneration into somatic embryos or adventitious shoots. Using tissue culture methods, these embryos and shoots can then be regenerated into whole plants for further propagation through seed or by asexual propagation. Usually, not all plant cells are capable of regeneration. Therefore, a transformation protocol will require the need to optimize a gene delivery system to the plant tissues containing those cells most capable of regeneration. Plant species vary widely in their competence for regeneration, and development of a regeneration system may require extensive optimization of culture conditions and media components and selection of those varieties within a species which are most amenable to regeneration.

Because there is no gene recombination (as there would be in breeding), an advantage of genetic modification is that it allows introduction of specific traits into elite germplasm while retaining the phenotypic characteristics of those genotypes. Genetic modification also allows introduction of genes from plant species other than the target species, surmounting the incompatibility barriers faced by plant breeders. Non-plant genes may also be expressed in plants after genetic modification, and many genetic elements (such as plant virus-derived constitutive promoters) and selectable marker genes (e.g., antibiotic resistance genes) are of viral and/or bacterial origin. Though the process of genetic modification is more specific than plant breeding, which involves either large-scale recombination or random generation of mutations, there is still a degree of unpredictability. This is because in a population of transformed plants, there will most probably be variation between transgenic events in the pattern of gene integration (measured as differences in copy number and possible re-arrangement of the inserted genes) and variation in the level of expression of the inserted genes. In the latter case, this is partly because it is not possible to control exactly where in the plant genome foreign gene integration will occur. Through screening populations of transgenic events, selection can be imposed to restrict commercial development to events where it is likely there has been integration of a single copy of the introduced gene(s). Care can also be taken to ensure there is no unrequired plasmid DNA incorporated into the transgenic event. These precautions have the advantage of making the molecular characterization of the transformant easier. As the exact site of integration cannot be controlled, there is the possibility that gene integration may have occurred at a site in the plant genome which may interrupt endogenous gene function, through disrupting the endogenous genome structure. An example of such an event is the GR2-R1 transgenic rice event where transgene integration interrupted an endogenous gene encoding an auxin transmembrane transporter protein (Bollinedi et al. 2017).

Genetic modification technology has allowed the development of widely grown insect and herbicide-resistant transgenic crop plants such as cotton, canola, maize, and soybean (Schiemann et al. 2019) and plants with enhanced nutritional value or production of recombinant proteins (Schillberg et al. 2019). Genetically modified crops with stacked genes (e.g., insect resistance and herbicide resistance) have also been commercialized. In North America and some South American countries, the majority of varieties of corn, soybean, cotton, and canola are now transgenic.

17.3.2 Genetically Modified Ornamental Plants

As has already been demonstrated for broad acre crops, genetic modification offers opportunities to introduce valuable and novel traits into ornamental plants. Examples of these traits, and many reports describing how they have been generated by researchers, have been reviewed by Azadi et al. (2016), Chandler and Sanchez (2012), and Kishi-Kaboshi et al. (2018). In brief, potential modifications include changes to traits important to horticulturalists and professional growers such as disease resistance, reduced chemical usage, enhanced productivity, and improved keeping quality and traits that might be of value to the consumer such as the modification of flower and leaf color (Chin et al. 2018; Noda 2018), modification of plant and flower shape and form (Sasaki et al. 2016), and enhancement of flower scent. Though there are no examples of stacked trait phenotypes in ornamental plants so far, there is no reason multiple transgenic traits could not be introduced into ornamental plants also.

For consumers, a new and exciting idea is the concept of transferring bioluminescence to plants using genes of insect, bacteria, or fungal origin. Demonstrated at the research level years ago using long-time camera exposures for visualization, new reports suggest we are now getting closer to the point where luminescence can be expressed in transgenic plants at high enough levels to be immediately visible to the naked eye under low light conditions (Mitiouchkina et al. 2020). Non-food ornamental plants in home use, landscaping, and "green buildings" are ideally suited for delivery of this type of "glow in the dark" phenotype (Ardavani et al. 2020).

In addition to insertion of genes encoding novel proteins, petal-specific RNAi expression has been used in transgenic ornamental plants to knockdown endogenous gene function. An example is downregulation of phytoene synthase gene activity in the orchid *Oncidium*, where subsequent reduction in xanthophyll levels in flowers leads to generation of new white color varieties (Liu et al. 2019b). Some commercialized transgenic carnation varieties which have been transformed for modified flower color also carry downregulated pigment biosynthesis genes (Tanaka and Brugliera 2013).

At the point of compiling this chapter in mid-2020, the only genetically modified (GM) ornamental plants that have actually been offered to the commercial market are flower color-modified varieties of carnation (Tanaka and Brugliera 2013) and rose (Nakamura et al. 2015). In both cases, varieties used for cut flower production have been genetically modified. The GM carnation was first developed and sold more than 20 years, and there are now more than nine varieties available (www.florigene.com). All the varieties are modified for novel flower color as the result of accumulation in flowers of delphinidin-related anthocyanins. Delphinidin-related anthocyanins are pigments with a bluer visible spectrum than the pink and red anthocyanins naturally found in carnation. To engineer delphinidin biosynthesis in carnation and rose genes were introduced from pansy or petunia. The GM rose variety Applause™ has also been developed for novel flower color as the result of the accumulation of delphinidin-related anthocyanins. The GM carnation and GM rose plants were generated using *Agrobacterium*-mediated gene transfer methods. Functioning selectable marker genes (herbicide resistance in the case of carnation and antibiotic resistance in the case of rose) were used for selection during regeneration. The GM carnation is grown in Ecuador, Colombia, and Australia and primarily sold in North America, Japan, Europe, and Australia. The GM rose is grown in Japan and Colombia and sold in North America and Japan.

The Biosafety Clearing House, an online database of living modified organisms (https://bch.cbd.int/) set up as part of the 2000 Cartagena Protocol on Biosafety to the Convention on Biological Diversity, is a catalogue of species which have been genetically modified and the genes which have been used for modification. This registry is dominated by food crops, modified for enhanced nutritional value and resistance to herbicides and insect pests. Based on known field trials and current submission to regulatory agencies, the product pipeline for genetically modified ornamental plants suggest that the next products to be commercialized will also be flower color-modified varieties. These possible products include varieties of the cut flower species chrysanthemum and gypsophila. In the case of chrysanthemum, the novel flower color is also the result of accumulation of delphinidin-related anthocyanins (Noda et al. 2013).

The fact there are not more genetically modified ornamental plants available to the market is not due to the fact that carnation and rose are the only two species that can be transformed. While barriers to regeneration and transformation in some ornamental plants should not be underestimated, there are many dicotyledonous and monocotyledonous ornamental plant species where transformation and regeneration is possible (Azadi et al. 2016; Ahn et al. 2020), and there are many published transformation protocols available for ornamental plants.

17.3.3 Regulation of Genetically Modified Ornamental Plants

From the beginning of the development of methods for producing genetically modified organisms, guidelines and later legislation has been introduced regulating research and commercial release of GMOs, including plants. Compliance with legislation governing regulation of GMOs is required for commercialization, and unapproved transgenic events are likely to be discovered and barred from the market, wasting all the investment made on research and product development. In the ornamental plant field, there is an example where this has happened. It was recently discovered that some varieties of the pot plant petunia contained a transgene from maize and a bacterial selectable marker gene (Fraiture et al. 2019). Clearly at some point in the past, a transgenic petunia had been used for breeding, as the genes were found in many different petunia varieties. The presence of the genes meant the varieties were unapproved flower color-modified transgenic events of the pot plant petunia. These unapproved varieties were identified in the marketplace in several countries (Bashandy and Teeri 2017) and were subsequently ordered to be removed from sale.

For a genetically modified plant variety, data collected to support a regulatory submission focuses on environmental risk analysis and characterization of the modified variety. In both cases, a comparison of the performance of a transgenic line in comparison to the parental variety used for transformation is a fundamental data requirement, as is good baseline knowledge of the biology of the plant species in question, a complete understanding of the DNA sequence of any inserted genetic material, and an understanding of insertion patterns in the transgenic variety. As a whole, this information is used to determine whether the transgenic variety is likely to pose any greater risk than non-transgenic varieties as result of the genetic modification. From a practical point of view, trials can be designed to collect information that can also be used for plant breeders right or plant patent applications. Measurement of vegetative and reproductive fecundity, estimation of levels of any known toxins or allergens, an assessment of the possible impact of any unique metabolites, the potential for inserted genetic material to cause generation of open reading frames that could encode proteins related to toxins or allergens, whether the target plant species is a weed, and whether the transgenic has enhanced weediness attributes are all components of the risk assessment.

Legislation governing research and commercialization requires developers of GMOs to undertake trials, biochemical characterization, environmental risk assessment, and molecular characterization not required for conventionally bred ornamental plant varieties. In some jurisdictions, fees are also imposed by regulatory agencies. This additional cost of regulatory compliance is significant (Kalaitzandonakes et al. 2007; Lassoued et al. 2019b; Schiemann et al. 2019; Smyth et al. 2014) and has been recognized as a barrier to commercialization in plants with less market size (and so potential revenue) than major crops (Chandler and Sanchez 2012; Miller and Silva 2018). Costs are borne during product development, trials, molecular characterization for selection of events, and during the approval process. In some countries, post-release monitoring is also required. The most important ornamental plants such as the major cut flower species and dominant indoor plant species are traded internationally. This presents the need to meet regulatory requirements in all countries in which a product may be traded. Unfortunately, there is little international harmonization of the regulation of GMOs outside of the Biosafety Clearing House mechanism. This lack of coordinated regulation increases regulatory cost as approval is required in multiple countries, using different (though in some cases overlapping) laws, languages, and technical information requirements.

17.4 Other New Breeding Technologies (NBT)

Advanced technologies other than genetic modification have now become available to further assist plant breeders. From the late 2000s, these have been coined *N*ew *B*reeding *T*echnologies, or NBT. NBT has arisen as the result of advances in gene and genome sequencing (which is now possible at a much-reduced cost compared to previous decades) and advances in plant molecular biology that have made plasmid, vector, and DNA manipulation easier and faster. NBT technology now allows, in principle, more precise changes (targeted to very specific and pre-determined DNA sequences) to the plant genome than occurs with genetic modification, a technique which, as we have discussed, involves potential problems associated with random integration of transgenes in the host genome. New plant breeding technologies have been reviewed by Hartung and Schiemann (2013), Lassoued et al. (2018), and Schiemann et al. (2019). Based on those reviews, Table 1 summarizes types of NBT.

With reference to Table 1, though many ornamental plants are grafted onto a rootstock to provide vigor to the flowering scions, there are no examples of the use of transgenic rootstocks in ornamental plants, even at the field trial level. Though vigor and disease resistance in rootstocks are important traits that could be targeted for improvement using gene manipulation, a transgenic rootstock would need to undergo extensive testing and analysis for regulatory approval. This would include detailed environmental risk assessment of the weediness potential of the rootstock were it to escape from cultivation, which has occurred for non-transgenic rootstocks in some cases (e.g., *Rosa canina* rootstock). Trial of transgenic rootstock would require many years of trials because rootstocks are typically from woody, persistent species and are kept for many years (rose rootstocks are a common and widely grown example).

Further referencing Table 1, the use of cisgenesis and transgenesis in genetically modified ornamental plant has also not eventuated at a commercial level. This is due to lack of genes suitable to affect the desired phenotypic change, including use of a selectable marker during transformation. Some cisgenic regulatory components have been used in the GM carnation that has been commercialized (Tanaka and Brugliera 2013), but these only account for a small proportion of the inserted DNA. A difficulty of the cisgenesis approach, which was originally formulated as a way to circumvent the need for GMO regulation, is that in those countries that regulate the technology by the process used to make the transgenic plants, rather than the phenotypic change, cisgenic transgenics are regulated in any event in the same way as GMOs. As mentioned earlier

Table 1 Summary of new plant breeding technologies applicable to ornamental plants. Employment of SDN and ODM methods are generically considered genome editing and/or gene-editing techniques

Method	Description		
Site-directed nuclease (SDN) mutagenesis, including base editing (Chen et al. 2019)	A nuclease is guided to a specific DNA sequence, where cleavage of both DNA strands occurs. More details of mechanisms of re-assembly of cleaved strands are described under SDN-1, SDN-2, and SDN-3. (Arora and Narula 2017; Schiemann et al. 2019)	SDN1 (van de Wiel et al. 2017)	The double-strand DNA break is repaired by the cells endogenous repair mechanisms only (primarily by non-homologous end-joining). A point mutation may occur if repair is inaccurate, or deletion of DNA sequence may occur if nuclease activity causes breaks at two nearby positions. Resulting genome sequence change may lead to loss of function of any genes contained within the modified DNA sequence
		SDN2	In addition to the nuclease, a template oligo-DNA is also added in the vector, and repair is primarily due to homology-directed repair. The template DNA is designed to have homology to the target DNA but has a marginally different sequence. If the template DNA is incorporated during the repair process, an altered genomic sequence is generated altering or interrupting the function of the target gene. Typically, the original and replacement sequence at the target site will be one to up to 20 nucleotide(s)
		SDN3	A nuclease and template DNA are added, but the template DNA is a longer DNA sequence, equivalent to sequence lengths in the T-DNA of binary vectors. The template is therefore long enough to include regulatory genetic

(continued)

Table 1 (continued)

Method	Description
	elements and gene(s). The DNA template is designed such that the ends of the insert have homology to significant lengths of genome sequence at the target DNA site, and the DNA between these homologous regions carries the desired genetic change. A homologous recombination (HR) pathway for repair (Schiemann et al. 2019) may result in incorporation of the new DNA sequence at the target DNA site
Oligonucleotide-directed mutagenesis (ODM) (Beetham et al. 1999)	Oligo-DNA or oligo-RNA, designed to be identical to target genomic DNA aside from specific base(s) change targeted to a change in gene function are introduced to plant cells in culture. As the introduced oligo binds to the target sequence during strand duplication the endogenous replication machinery of the cell may change the target strand DNA to match the introduced oligo, changing the genome sequence in the affected cell. The method is used to introduce specific mutations in target genes
Cisgenesis and intragenesis (Cardi and Varshney 2016)	A refined technique of both genetic modification and gene editing (SDN-3) in which all genetic elements in the transferred DNA are from the same species as the target plant (cisgenesis) or are from a plant sexually compatible to the target plant (intragenesis)
Grafting on GM rootstock	Some ornamental plants, such as the rose, may comprise a rootstock and a scion. In a scenario in which a non-transgenic scion was grafted onto a GM rootstock, the product from the plants (i.e., the flowers) would not contain any inserted genes. The desirable traits (e.g., disease resistance) will be expressed in the root system
Synthetic genomics and artificial genes (Pixley et al. 2019)	As part of the approach to synthetic biology, whole genes may be synthesized based on bioinformatic information and knowledge of DNA and/or protein function. These novel proteins can then be expressed in plants using genetic modification or gene-editing techniques

in this chapter, the use of synthetic biology for manipulation of ornamental plants is not likely in the near future, and so potential applications of this NBT in ornamental plants will not be further discussed. However, for reference, an interesting recent review outlines the current state of synthetic biology in plants, emphasizing the chloroplast genome as a possible target (Pixley et al. 2019).

In the remainder of this chapter the focus will be on genome editing and gene editing using the SDM and ODM techniques described in Table 1. It is in this area that there has been most NBT work with ornamental plants. The genome and gene-editing techniques shown in Table 1 are site-directed nuclease mutagenesis (SDN) and oligonucleotide-directed mutagenesis (ODM). In the former case, the extent of SDN has been classified by both scientists and regulators into SDN-1, SDN-2, and SDN-3 (Agapito-Tenfen et al. 2018) to reflect the scope of gene editing that has occurred. These categorizations have important implications for regulation of gene-edited ornamental plants, a subject which will be discussed in more detail later in this chapter.

17.5 Gene Editing in Ornamental Plants

The terms genome editing, genome engineering, and gene editing are used somewhat interchangeably in the scientific literature, but in the context of the development of new varieties of ornamental plants, the technology referred to here as gene editing is manipulation using site-directed nuclease (SDN) mutagenesis (Table 1). Though the oligonucleotide-directed mutagenesis method has potential to generate phenotypes comparable to SDN-1, the ODM technique has not been used to generate any new ornamental plant varieties, and from a review of published papers, current research efforts are focused on gene editing using SDN.

17.5.1 Gene-Editing Methods by SDN

The principle tool in gene editing by site-directed mutagenesis is the use of meganucleases, or homing endonucleases, termed sequence-specific nucleases (Belhaj et al. 2015). These nucleases cut DNA at specific DNA sequences in the plant genome, creating double-strand breaks which are then repaired by either the plant cells endogenous non-homologous end joining repair system or by homology-directed DNA repair (Malzahn et al. 2017). As outlined in Table 1, non-homologous end joining repair is useful for disruption of gene function, while homology-directed DNA repair is a tool for gene addition. By combining information on target DNA sequences and specificity of nuclease to a target sequence, gene editing is possible by either "knockdown" of endogenous gene function, modification of existing gene function, or insertion of new genes at desirable locations in the plant genome. The possibility of causing a phenotypic change through gene editing of regulatory elements, by mutagenesis or enhancement of promoter and terminator function, has been outlined as an alternative approach to targeting of specific genes (Abdallah et al. 2015; Bortesi and Fischer 2015; Shelake et al. 2019).

Initially, gene editing was accomplished using zinc finger nuclease (ZFN) (Abdallah et al. 2015; Arora and Narula 2017), and this was followed by use of transcription activator-like effector nucleases (TALENs) (Podevin et al. 2013; Shelake et al. 2019). Use of both ZFN and TALEN is suited to targeted mutagenesis and modification of endogenous genes. Now, the most widely used system is the CRISPR/Cas9 nuclease method which uses RNA as guide, rather than a protein guide (Bao et al. 2019). Details of the mechanism and protocol of CRISPR/Cas9 are provided in the following section. Because of the ease and relatively low cost of use (Abdallah et al. 2015), CRISPR/Cas9 has been almost exclusively used in ornamental plants and will be the method referred to in more detail in the rest of this chapter.

More recently, base-editing and prime-editing (Anzalone et al. 2019) methods have been described, in which, for example, a single base exchange (e.g., replacement of a C with a T) can be made using gene editing (Shelake et al. 2019). An example of such gene editing is creation of herbicide-resistant plants through a single base change (Han and Kim 2019; Shelake et al. 2019) without the unrequired mutations that would arise if a conventional mutation breeding strategy had been employed to make the same mutation.

17.5.2 Gene Editing Using CRISPR/Cas9

CRISPR/Cas9 is now a widely used gene-editing technique in human, animal, bacterial, and plant systems. Accordingly, there is a wealth of detailed technical knowledge that has been accumulated, and the method can only be outlined in summary format in this chapter. More details of the mechanism of action of CRISPR/Cas9 are provided in many recent reviews which describe application in plant biotechnology (Ahn et al. 2020; Bao et al. 2019; Chen et al. 2019; Jaganathan et al. 2018; Kuluev et al. 2019; Lozano-Juste and Cutler 2014; Malzahn et al. 2017; Schindele et al. 2019; Shelake et al. 2019; Xu et al. 2019). Readers are particularly directed to the reviews of Shan et al. (2020) and Jansing et al. (2019) as these authors outline broad strategies for possible application of CRISP/Cas9 gene editing including phenotype screening, transformation, and regeneration in both crop and non-model plant species. Shan et al. (2020) include tabulation of promoter and terminator combinations for CRISPR vectors in 45 plant genera.

In nature, CRISPR/Cas9 exists in prokaryotes as *c*lustered *r*egularly *i*nterspaced *s*hort *p*alindromic *r*epeat (CRISPR) DNA sequences accompanied by *cas* (CRISPR associated, or Cas) nuclease genes located within the same area of the genome. There are dozens of Cas genes in prokaryotes; Cas9 is from the bacterium *Streptococcus pyogenes* and is often designated SpCas9 in the scientific literature. The biological purpose of CRISPR/Cas is to confer a degree of bacteriophage/virus immunity in prokaryotes. The DNA sequences in CRISPR are derived from previously infecting viruses, and, in essence, RNA sequences (guide sequences) are transcribed from the CRISPR sequences. These guide the Cas proteins to similar virus DNA sequences which are then cut apart, interfering with virus replication.

The CRISPR/Cas system has been adapted for gene editing in plants by using the Cas9 nuclease in combination with a guide RNA(sgRNA). This ribonucleoprotein complex binds to the plant DNA and makes a double-stranded DNA break of the genomic DNA at specific sites which have both a) a PAM (*p*rotospacer *a*djacent *m*otif) sequence of the nucleotides NGG and b) any genomic DNA sequence immediately upstream to PAM which matches the sequence of the sgRNA. As sgRNA can be designed to complement known genomic DNA sequences, cuts in the genomic DNA are therefore made at precise points in the plant genome.

The process of repair of the DNA at the sites which have been degraded by Cas finalizes the gene-editing process. Broadly speaking, the processes effect the changes designated SDN-1, SDN-2, and SDN-3 through the DNA repair and modification mechanisms outlined in Table 1. DNA repair exploiting the plants own DNA repair mechanisms may lead to insertion or deletion of bases, potentially resulting in knockdown of function of the targeted DNA. Alternatively, homology-directed repair can be used to add new DNA sequence.

17.5.3 CRISPR/Cas Delivery Systems

To use a CRISPR/Cas system, there needs to be a delivery system to the plant cells for the nuclease (as DNA, mRNA, or protein), guide RNA, and, if so designed, additional DNA. To generate an altered phenotype, there must obviously also be a method to allow for regeneration of plants from any cells where the desired gene editing has taken place. To this end, regeneration protocols developed for genetic modification may normally be used.

There are several methods for Cas gene delivery to the plant cells, reviewed by Chen et al. (2019) and Kuluev et al. (2019). *Agrobacterium* co-cultivation-based methods commonly used for plant transformation (including protocols using *Agrobacterium rhizogenes*) are widely used for cassette delivery. This technique involves genetic modification of the plant with a vector containing the sgRNA/Cas gene complex, a selectable marker gene, and any additional DNA required for gene editing, all driven by suitable regulatory elements (Erpen-Dalla Corte et al. 2019). Expression of the inserted complex may then result in stable gene editing, and the resulting modified phenotype can then be expressed in subsequent vegetatively or sexually reproduced generations. There are technical advantages in this system for species where transformation protocols already exist but disadvantages if the goal is development of a commercial product. This is because, unless Cas is only transiently expressed and not integrated into the plant genome (this is a possibility), the gene-edited plants are genetically modified because of the integration of the Cas gene. Presence of the Cas gene not only has significant legal regulatory implications and but potentially also longer-term biological effects such as off-target mutations (i.e., editing of DNA at unintended locations in the plant genome) associated with permanent expression of the inserted sgRNA/Cas. There is also the possibility of multiple sgRNA/Cas insertion sites, with possibly partial, non-functional integration at some of those sites (Bao et al. 2019). Assuming simple integration during transformation, one solution to overcome regulatory compliance and long-term expression problems is removal of the integration sites by breeding. This may be a good commercial strategy as it is possible then to incorporate a gene-edited plant event into a larger-scale breeding program which might eventually lead to the incorporation of the gene-edited genetic change into multiple varieties. For a trait with value across many varieties, such as disease resistance, this could be very advantageous. Unfortunately, many ornamental plants, including all major cut flowers, are vegetatively propagated, either negating the possibility of breeding out the CRISPR/Cas integration or requiring a long-term backcrossing breeding program.

Various options have been proposed to exploit transient expression of sgRNA/Cas while facilitating stable gene editing of the genome (Jansing et al. 2019; Kausch et al. 2019). Under these scenarios, there would be no integration of CRISPR/Cas DNA into the plant genome, but transient expression of CRISPR/Cas would be sufficiently long for a stable gene-edit event to occur. Another approach is to exploit the transient expression of introduced genes which occurs to some degree in most cases when genes are introduced by *Agrobacterium* co-cultivation or by biolistics. By removing selection early in tissue cultivation, or not imposing selection, a larger population of gene-edited cells which are not stably transformed but where gene editing has occurred as a result of transient expression, will participate in plant regeneration. These cells, or plants regenerated from them, can be selected by high-throughput sequencing and/or screening for modified phenotype.

An alternate strategy to overcome stable integration of Cas is to transform cells with a transiently functional Cas protein and sgRNA in a ribonucleoprotein complex (Malnoy et al. 2016; Woo et al. 2015; Xu et al. 2020). This technique has been proposed as the method of choice for CRISPR/Cas delivery (Bao et al. 2019; Park and Choe 2019) and has been demonstrated using protoplasts (Li et al. 2013) in several species including the ornamental plant petunia (Subburaj et al. 2016; Xu et al. 2020). Agroinfiltration is also a suitable technique for delivery of the ribonucleoprotein complex.

A refinement of the gene-editing process is multiplex gene editing (Kawall 2019; Shelake et al. 2019). This is the technique of targeting a single gene with multiple guide RNAs or multiple genes with one or several guides. This technique may be useful for targeting phenotypic change for traits which are multigenic.

Finally, methods for gene delivery have been developed in plants which do not rely on regeneration, such as floral dip. Zlobin et al. (2020) have reviewed the potential application of these methods to gene editing in plants.

17.5.4 Gene Editing in Ornamental Plants: Other Technical Considerations

A gene-editing strategy in an ornamental plant will require good DNA sequence and genome information for the target species. This will allow identification of target genes to knock out gene function or add new genetic information and also allow design of suitable guide sequences. As outlined earlier in this chapter, whole genome sequences have been compiled for some important ornamental plant species, and more are likely to follow given the proven advantages of access to such information.

Assuming *Agrobacterium*-based transformation will be used, appropriate promoters will be required to drive the CRISPR/Cas9 and selectable markers (Bao et al. 2019; Jaganathan et al. 2018; Xu et al. 2019). In ornamental plants, there are species-specific considerations relating to promoter choice (Azuma et al. 2016) and selectable marker choice. These decisions may include choice of promoter and terminator elements from endogenous genes, which will function more efficiently and can be chosen from genes which express strongly in the tissues chosen for transformation and regeneration. In some ornamental plant species, such as chrysanthemum, the activity of constitutive promoters such as the commonly used 35S gene from cauliflower mosaic virus is transient and efficiency declines with the number of vegetative cycles even after its stable integration of the transgene. Shan et al. (2020) recommend using a transient assay screen to first optimize the components of a CRISPR/Cas9 system.

In a gene-edited plant, undesirable off-target effects are possible. This is because gene editing may occur at an undesired locus – most probably this would be at a position where there is a high level of sequence identity to a gene similar to the target DNA (Shelake et al. 2019). Sometimes, phenotypes caused by off-target gene editing can be identified, and molecular analysis can be used to identify others (Xu et al. 2020). Off-target events can then be eliminated from the transgenic population during assessment. To reduce off-target frequency, good genome sequence and bioinformatic information is required to ensure sequence guides do not have high enough sequence identity that they might target unintended parts of the genome. A thorough analysis of the available gene sequence depositories is therefore recommended to ensure minimization of unintended off-target effects (Abdallah et al. 2015; Uniyal et al. 2019). As has been discussed immediately above, the choice of delivery method is important as this may determine whether the gene editing is achieved by transient or stable expression of the CRISPR/Cas9. With stable integration of the CRISPR/Cas9 machinery, there is an increased chance of off-target mutations due to the stable expression of the CRISPR/Cas9 cassette causing more off-target changes with time. As discussed, fewer off-target effects are likely when a transient Cas9/sgRNA ribonucleoprotein delivery method is employed. However, due to the technical difficult of regeneration from protoplasts (there are relatively few protocols for ornamental plants), development of a transient expression system in an ornamental plant would probably only makes commercial sense if an existing protocol for regeneration is available. As gene-editing systems are finessed, molecular biology strategies are being developed to reduce off-target effects such as modification of Cas9, use of Cas genes other than Cas9, optimization of sgRNA sequence through screening variants, alternative PAM motif specificity (Ding et al. 2016), use of truncated sgRNA, and deployment of aptazymes (Bortesi and Fischer 2015; Fu et al. 2014; Hajiahmadi et al. 2019; Jansing et al. 2019). Use of Cas12/Cpf1 as an alternative to Cas9 is discussed by Kuluev et al. (2019), Liu et al. (2019a), and Lowder et al. (2016).

Some ornamental plant species are polyploid. For example, rose varieties may be diploid or tetraploid, and cut flower chrysanthemum varieties are tetraploid, aneuploid, or hexaploid. Polyploidy adds complexity to the gene-editing process because of the presence of multiple target loci and so the need for several gene-editing events to occur independently in order to make the same gene edit at all alleles (Su et al. 2019). Where possible, use of diploid varieties for gene editing would be sensible as the efficiency of gene editing is likely to be higher than in polyploid varieties of the same species (Shan et al. 2020).

In some cases, the regeneration of plants after transformation occurs from multiple cells, and if these cells are comprised of a mix of transformed and non-transformed cells, this may result in the regeneration of transgenic or edited/non-transgenic or non-edited chimeric plants. In the case of gene editing, this may result in plants in which both unmodified and gene-edited cells are present in the regenerated meristem or somatic embryo that will eventually form the regenerated plant (Arora and Narula 2017; Kishi-Kaboshi et al. 2018). There are pros and cons should this occur, depending on the type of chimera. Some interesting and unique plant and flower morphologies are of chimeric origin and commercial value. It is also possible that gene flow risk is reduced in plants which have periclinal chimeras. This is because the transgene may not be present in the cell layers which contribute to pollen formation. A downside of exploiting phenotypes which are chimeric in nature is that it will be necessary to confirm though multiple trial cycles that the chimeric phenotype is stably expressed.

17.5.5 Legal Compliance of Gene-Edited Plants

As legislation was developed for the control of research and commercialization of genetically modified plants, a divergence of approach emerged between different countries. While some countries focused on the modified phenotype for regulation and potential risk, other countries have used definitions of the genetic modification process or changes to the genome as triggers for regulation (Ellens et al. 2019; Kuluev et al. 2019; Metje-Sprink et al. 2019). This disparity of opinion has continued as either existing GMO regulations are adapted or new regulations penned for gene-edited plants. The exact process and technique of generating the gene-edited plant, often using the NBT definitions outlined in Table 1, has therefore been a central focus within draft and implemented legislated regulations (Jansing et al. 2019; Sprink et al. 2016). The regulation of gene-edited plants depends on the extent of editing (as outlined in Table 1) and is also dependent on whether there has been stable integration of the CRISPR/Cas cassette in the plant genome during the gene-editing process.

In the same way that there is no international harmonization for regulation of GMOs, different countries have made different decisions on how gene-edited plants will be regulated. At the time of writing, there are also some countries that are still deciding an approach or have drafted, but not finalized, legislation. As explained above, the exact process of gene editing has been an important consideration, and development of regulatory legislation of gene-edited plants has also been driven by a recognition by regulators that a unique problem is posed in cases where there is no stable integration of CRISPR/Cas9. In such events, though a gene-edited SDN-1, or in some cases, SDN-2 gene-edited plant can be identified by phenotype or by using molecular analysis methods such as Southern analysis, sequence analysis, and/or PCR-based protocols, the transformation process by which the plant was generated cannot be determined, and the phenotype may not be distinguishable from naturally occurring or induced mutations (Custers et al. 2018). This leads to problems of definition when the process or phenotype is used as the trigger for regulation. As one possible solution to this issue in gene-edited plants which are foods, Fraser et al. (2020) have proposed the use of metabolomic analysis for characterization.

The current status of regulation of gene-edited plants is summarized in Table 2 for a selection of countries, chosen because either genetically modified organisms are widely grown or there is extensive research with gene-edited plants. For an overview on regulatory status of gene-edited plants globally, and information on regulatory status in countries not shown in Table 2, readers are directed to Schmidt et al. (2020) and Zhang et al. (2020).

As Table 2 shows, there is a variation in the way countries have decided to regulate NBT. In a recent development, the USA issued new rules in May 2020 for the regulation of genetically modified and gene-edited plants. Under current regulations, certain gene-edited plants had already been deregulated (Waltz 2018), and the change in rules provides exemption of gene-edited plants where a phenotypic change results from cellular repair of a targeted DNA break in the absence of an externally provided repair template; or is due to a targeted single base pair substitution; or is the result of introduction of a gene known to occur in the plant's gene pool or a change in a targeted sequence to correspond to a known allele of such a gene or to a known structural variation present in the gene pool (Animal and Plant Health Inspection Service, USDA 2020). As the USA remains the center of innovation in the commercialization of genetically modified plants, this development is important as it brings that country within the majority group of countries that consider plants modified without the insertion of foreign DNA to not require regulation as a GMO (Schmidt et al. 2020).

Even in countries with legislation covering gene-edited plants, there is an element of uncertainty as each country actually implements their unique rules for regulation of gene-edited organisms and cases are actually reviewed (Hamburger 2018; Lassoued et al. 2018; Metje-Sprink et al. 2019; Sprink et al. 2016). As there have been few gene-edited plants that have been submitted for commercial approval outside of North America, it is not yet clear to what extent regulators will require supporting evidence for SDN-1-edited plants (e.g., evidence of lack of off-target editing or proof of lack of ORF generation or disruption in cases where a base has been deleted). Given the uncertainty, it is therefore important for any developer to know whether DNA has been integrated into a gene-edited plant from external sources – such as plasmid vector sequence, residual Cas9, or guide sequence. This knowledge will require high coverage sequencing of the final gene-edited plant proposed for commercial release. As ornamental plants are typically not used as food and are often grown in contained facilities over relatively small areas, the environmental risk assessment process for GMOs is generally more straightforward for an ornamental plant species, and the same will apply for a gene-edited ornamental plant. It is important to note though that related legislation, such as Environmental Protection Agency (EPA) requirements in the USA, will not necessarily distinguish between a gene-edited and genetically modified organism (Wolt and Wolf 2018). A gene-edited ornamental plant which has an insect resistance trait, for example, will need to be assessed and approved in the USA by EPA.

Table 2 Regulatory status of gene-edited plants

Country	Situation as at June 2020
United States of America (Animal and Plant Health Inspection Service, USDA 2020)	Rules for regulation of GMOs have been modified such that gene-edited plants will not be regulated if the genetic modification is solely a deletion of any size, is a single base pair substitution, is solely introducing nucleic acid sequences from within the plant's natural gene pool or from editing nucleic acid sequences in a plant to correspond to a sequence known to occur in that plant's natural gene pool
Canada (Ellens et al. 2019)	Gene editing is covered under existing legislation, which focuses on novelty of phenotype rather than the tool used to generate the phenotypic change
European Union (Bruetschy 2019)	Gene-edited plants are subject to the existing regulations for genetically modified plants. This applies whatever the tools used for transformation
Australia (Thygesen 2019; Australian government federal register of legislation 2019)	Plants produced using site-directed nuclease (SDN) 1 techniques are no longer defined as GMOs (so will not be regulated) provided that the organism has no other traits from gene technology, i.e., is already a GMO. If methods used to generate SDN-1 organisms produce GMOs in intermediate steps, those will also continue to be regulated. SDN-2 and SDN-3 will continue to be regulated in the same way as GMOs. Currently there is ongoing consultation with stakeholders about modernizing regulations, including definitions that allow exemption of certain techniques from regulation
Japan (Metje-Sprink et al. 2019)	Case by case decisions will be taken, but if no inserted nucleic acid or its replicated product is present, the gene-edited organism will not be considered a genetically modified organism
China (Zhang et al. 2020)	Under review. No position has yet been taken
Brazil (Eriksson et al. 2019)	Regulations were amended to cover new breeding technology where there is no recombinant DNA or RNA in the final product. On a case by case review basis, gene-edited products will be exempt from the regulations applicable to GMOs if there is no inserted foreign DNA or RNA
Argentina (Lema 2019)	Case by case decisions are made under existing regulations, which already address gene-edited plants. To date, plants produced using SDN-1 techniques have been presented for review, and it has been determined by the regulator that the final product is not defined as a genetically modified organism
Chile (Sanchez 2020)	Case by case decisions are made. To date, products developed by CRISPR have been assessed as non-GMO as no foreign DNA has been incorporated

As is evident from Table 2, it is not likely that there will be international harmonization of regulation of gene-edited plants, despite appeals for there to be some (Jansing et al. 2019; Kleter et al. 2019). The European court has ruled that gene-edited plants are to be regulated in the same way as GMOs (Kok et al. 2019), and the current position in New Zealand is also that gene-edited plants should be regulated like GMOs (Fritsche et al. 2018; Schmidt et al. 2020). This contrasts to the relaxation of regulation for gene-edited plants as part of the new rules for GMOs in the USA (McCammon and Mendelsohn 2019) and exemption of SDN-1-edited plants in new legislation in Australia and Japan. The European court decision raises the prospect of a lack of uptake of gene-edited plants in Europe (Halford 2019; Hundleby and Harwood 2019). This is important as Europe is a center for innovation in conventional ornamental plant breeding and production and is also an important consumer market for flowers, pot plants, and ornamental shrubs. In Europe, socioeconomic considerations are also part of the decision-making process for regulation of GMOs (Bruetschy 2019; Chneiweiss et al. 2017; Edvardsson Björnberg et al. 2019), and some European countries seek an evaluation of benefit to society as part of evaluation of requests to market a genetically modified organism. In the case of ornamental plants, the ethereal benefits of plants largely grown for their aesthetic value are sometimes difficult to quantify and justify when there are already many ornamental plant products available in the marketplace.

17.5.6 Gene Editing in Ornamental Plants

One of the earliest demonstrations of gene-edited mutagenesis was proven using transient expression of zinc finger nuclease in the ornamental plant petunia (Marton et al. 2010). However, petunia is considered a model plant for transformation due to the ease with which this species can be regenerated from leaf discs and protoplasts, and subsequent developments on gene editing in plants have tended to focus on the important crop plants such as rice, soybean, maize, and wheat. The lack of attention to ornamental plants is not to say that there is not great potential to generate useful ornamental plant varieties using gene-editing methods. This is because targeted mutations (SDN-1 and SDN-2 modifications) can be used to knock down gene function or genetic elements can be added, as with genetic modification methods, to introduce novel traits. In this respect, gene-editing and genetic modification techniques share the advantage that genetic and phenotypic changes can be obtained that could not possibly be achieved by conventional breeding due to incompatibility issues and lack of the desired genes in the accessible gene pool.

In crop plants, the potential for improving disease resistance, yield, and resistance to stress through gene editing has been reviewed thoroughly (Abdallah et al. 2015; Arora and Narula 2017; Jaganathan et al. 2018; Nogue et al. 2019). Some of the trait improvements suggested in crop plants would also be beneficial to producers of ornamental plants, such as disease resistance, flowering time, and improved post-harvest keeping quality (Chandler and Sanchez 2012). Perhaps uniquely for ornamental plant species, gene-editing strategies can also be envisaged for modification of traits of perceived value to the final consumer such as plant architecture, flower form, leaf pigmentation, flower color, flower fragrance, plant size (Abdallah et al. 2015), and plant and cut flower longevity (Erpen-Dalla Corte et al. 2019).

Looking more precisely at gene edits that have been used in ornamental plant species, Table 3 lists the recent examples of where gene editing has been used to generate new phenotypes. As Table 3 shows, several reports have been on flower color modification. One reason for this is the biochemistry, genetics, and molecular biology of pigment biosynthesis are well understood (Tanaka and Brugliera 2013). Using that information, knockdown strategies can be planned using gene editing to target some of the genes encoding the key enzymes in anthocyanin pigment biosynthesis, such as chalcone synthase and anthocyanidin synthase. Key points on the carotenoid biosynthesis pathway, such as phytoene synthase or carotene desaturase, are also suitable targets for gene editing in ornamental plants. In carnation, regulatory genes on the anthocyanin biosynthesis pathway have been identified (Totsuka et al. 2018), and these genes could also be targets for gene editing. Current examples of gene editing in ornamental plants are research-based, and there are at present no gene-edited ornamental plants in the marketplace anywhere in the world (Metje-Sprink et al. 2019). At the time of writing this review, we were aware of one application to the US authorities for a regulatory opinion on a gene-edited petunia in which gene editing was used to knock out the anthocyanin pathway, generating white and pale pink flower color lines from an originally purple flowered variety. It is possible that the intention of the researchers is to release the gene-edited lines as new varieties into the petunia marketplace.

In all the examples shown in Table 3, the transformation method was co-cultivation with *Agrobacterium tumefaciens* (*Rhizobium radiobacter*) carrying binary plasmids with the gene-editing CRISPR/Cas cassette. Table 3 complements the earlier reviews of gene editing in ornamental plants provided by Kishi-Kaboshi et al. (2018), Erpen-Dalla Corte et al. (2019), and Ahn et al. (2020). In those three reviews, *Agrobacterium*-mediated stable CRISPR/Cas integration was also the predominant delivery system.

In the examples of gene-edited ornamental plants shown in Table 3, the editing largely describes single base change "knock-out" changes as a result of insertion, deletion, or substitution, but only one report describes some T1 progeny that have been obtained without T-DNA insertion (Shibuya et al. 2018). These progenies would possibly be considered not to be GMOs. The lack of other examples of removal of the T-DNA may reflect the stage that research is currently at where it is easier to carry out gene editing after stable gene integration. It may also be an indication that it will be difficult to generate gene-edited ornamental plants that do not also have stable integration of the CRISPR/Cas cassette.

17.6 Concluding Remarks

Aside from the uses of molecular tools and sequence information to assist breeding, the highest profile of the new breeding technologies being applied in ornamental plants is genetic modification and gene editing. This is because these techniques

Table 3 Recent example of gene editing in ornamental plants

References	Plant	Target gene(s)	Transformation method	Altered phenotype
Tong et al. (2019)	Pot *Phalaenopsis* (phalaenopsis)	MADS DNA-binding proteins	*Agrobacterium* co-cultivation with antibiotic as the selection agent	Not described
Yan et al. (2019)	*Lilium pumilum* and *L. longiflorum* (lily)	Phytoene desaturase	*Agrobacterium* co-cultivation with antibiotic as the selection agent	Albino and pale-yellow leaf phenotypes
Nishihara et al. (2018a)	*Torenia fournieri* (torenia)	Flavanone 3-hydroxylase	*Agrobacterium* co-cultivation with antibiotic as the selection agent	Altered (paler) flower color due to reduced delphinidin accumulation
Watanabe et al. (2018)	*Ipomoea nil* (Japanese morning glory)	Carotenoid cleavage dioxygenase	*Agrobacterium* co-cultivation with antibiotic as the selection agent	Carotenoid accumulation leading to flower color change
Nishihara et al. (2018b)	*Gentiana scabra* (Japanese gentian)	Phytoene desaturase	*Agrobacterium* co-cultivation (selection not specified)	Albino in vitro leaf phenotypes
Shibuya et al. (2018)	Japanese morning glory	NAC transcription factor	*Agrobacterium* co-cultivation with antibiotic as the selection agent	Delay in petal senescence
Tasaki et al. (2019)	Japanese gentian	Anthocyanin 5-O-glycosyltransferase, anthocyanin 3′-O-glycosyl transferase and anthocyanin 5/3′-aromatic acyltransferase	*Agrobacterium* co-cultivation with herbicide as the selection agent	Altered flower color, red, pink, and mauve lines were generated from a blue parental variety
Xu et al. (2020)	*Petunia* X *hybrida* (petunia)	1-Aminocyclopropane-1-carboxylate oxidase	Transient expression of ribonuclease protein (RNP) mixture and stable integration by *Agrobacterium* co-cultivation with herbicide as the selection agent	Reduced ethylene production, reduced flower senescence, and increased longevity

have the potential to modify the genome of a particular plant species in ways not possible by conventional breeding methods, including traditional mutagenesis methods. The potential has been demonstrated in research studies, using both genetic modification and gene editing, as documented in this chapter. However, commercial stakeholders (breeders, large producers, grower consortia) have failed to capture innovative opportunities from this research (Ahn et al. 2020; Xu et al. 2019). In the case of genetic modification techniques, one of the primary reasons for lack of deployment of new varieties is the cost of securing regulatory approval, which is higher than for conventionally bred ornamental plant varieties. This financial penalty is an important one for developers of new ornamental plant varieties who must weigh up the potential market potential and so potential revenue against budgetary requirements elsewhere and a possible negative reaction from some distributors and/or consumers to being involved with or purchasing GM or gene-edited plants. The market opportunity for most ornamental plant species is small for variety developers, whose return from innovation is largely in the form of some sort of combination of royalties, license income, and/or contract growing. A revenue stream only emerges after investing capital on research and product development, and for ornamental plant species which have a small market share, there is a limited potential revenue as there is a finite number of plants that can be sold. Considerations of these commercial imperatives are as important as technical questions when considering the potential benefits of gene-editing methods which may, at least in the case of SDN-1 based gene-editing changes, be subject to less regulatory scrutiny.

To what extent then will gene editing reduce development costs compared to genetic modification and speed up new variety development in ornamental plants? There are market considerations of course, as developers have to identify a phenotypic change which will be commercially valuable and for which either producers or consumers are likely to be willing to pay a premium for the modified variety. There is some guidance from the experience of the release of the novel flower color genetically modified varieties of carnation. These varieties have been well received, and because they are novel, similar-colored varieties cannot be bred in competition. This has helped establish the new varieties in the marketplace. Similar arguments may not be able to be mounted for a) gene-edited varieties which can also be obtained using traditional breeding, for example, knockdown generation of un-pigmented varieties from pigmented varieties, b) gene-edited products where there already exists in the marketplace similar non-transgenic phenotypes, c) gene-edited products which cannot be readily identified by their phenotype (e.g., longer vase life), and d) traits where there are alternate technologies, such as the use of chemicals for disease control or chemicals to enhance storage quality. In these cases, it may be difficult to prevent illegal propagation and/or product replacement, and it may be difficult to obtain a significant premium for traits directed primarily to producers, such as insect or disease resistance.

The molecular genetics of the required phenotype change will need to be well understood, to the extent that either a targeted mutation can be made to the host genome or a foreign DNA cassette can be constructed. Assuming that these commercial and genome design prerequisites can be met, there are four other considerations which together present serious challenges (Bao et al. 2019) to widespread adoption of gene editing in ornamental plants.

The first consideration is that the gene editing of ornamental plants, similarly to genetic modification, relies on genetic transformation of plant organs or tissues and subsequent regeneration of plants. In the case of gene editing via gene delivery into protoplasts to enable transient expression of introduced RNP, a protoplast-based regeneration system will be required, and this is technically still very difficult for species from some groups of plants, such as woody plants and monocots. The availability of suitable transient expression, gene transfer, and plant regeneration methods are by no means trivial technical barriers (Jansing et al. 2019; Kausch et al. 2019; Kuluev et al. 2019) and, in some difficult to transform or to regenerate ornamental plant species, may be insurmountable. For minor crops like ornamental plants, it does not make commercial sense to spend too much money developing such enabling technologies from an absent or low knowledge base.

The second consideration for gene-edited ornamental plants remains the existence of legislated regulatory requirements. A survey of academic experts indicated that though the benefits of gene-edited plants was perceived to be greater than for GMOs, regulatory hurdles were seen as the number one factor affecting adoption (Lassoued et al. 2019a). A similar view was also expressed by Scheben and Edwards (2018) who nominated government regulation as a major bottleneck to adoption of gene editing in crops. In the case of the North America market, recent changes to the way GMOs are regulated, away from a process focus and more toward the evaluation of the final product, are set to streamline regulation. This should reduce costs, shorten the time frame to commercialization, and encourage innovation in gene editing. Because of this, and the extent of research activity in the USA, it is reasonable therefore to expect that any commercialized gene-edited ornamental plant products will first appear in North America. In markets outside of North America, the developers of gene-edited ornamental plants will still face significant additional cost associated with regulatory approval, compared to conventional breeders. In Europe, for example, as things now stand, the cost for approval will be the same for a gene-edited ornamental plant as for a genetically modified plant because of a legal decision determining that the two types of plants must be regulated under current legislation for GMOs. A recent survey of Dutch plant breeders indicates that the decision to regulate gene-edited plants in the same way as GMOs is a disincentive to use gene editing (Wesseler et al. 2019). No GMO ornamental plant has been approved for production in Europe, and it is unlikely that the further development of gene-editing techniques will change that situation. In other markets, the costs associated with regulatory compliance will vary depending on the complexity of the gene editing. For SDN-3 type genetic changes, gene-edited ornamental plants are likely to be regulated in the same way as GMOs, but even for SDN-1 type genetic changes, there will almost certainly be a need to include the cost of genome or insert sequencing and molecular analysis to demonstrate to oversight authorities that the only genetic change which has occurred is that linked to the target of the gene editing (Agapito-Tenfen et al. 2018). In general, Lassoued et al. (2019b) have determined that the cost and speed to market of commercialization of a gene-edited organism will be less than a GMO. If this is so, commercial decisions will be made on a case by case basis balancing the cost of development against the market potential of the gene-edited ornamental plant varieties.

The third consideration, which applies equally to genetically modified ornamental plants and gene-edited ornamental plants, is that ornamental plants are typically widely traded in the international marketplace. Key distribution routes are cut flowers, pot plants, and nursery stock from Mexico, Central America, and Northern South America to North America and Europe, flowers and pot plants from Asia and Europe to Japan, and distribution of flowers and pot plants from Europe to Asia. Harmonization of regulations of gene-edited plants would therefore be an additional boost to the application of this new breeding technology to ornamental plants. This has been recognized by those countries who rely on trade in food crops and for ornamental plants has also been acknowledged by countries such as Colombia, where production and export of ornamental plants is an important part of the economy. On the other side of coin, a lack of harmonization or establishment of different rules in different countries is a threat to the adoption of gene-editing technology in the ornamental plant industry. In the 25 years in which genetically modified plants have been on the marketplace, limited harmonization has been achieved, emphasized by the limited production of GMO commodity crops in Europe compared to the dominance of GMO varieties in North American (and to some extent South American) production of crops such as maize, soybean, and canola. Given no global consensus could be reached in that time, it is probably unrealistic to expect consensus on gene-edited plants will be reached in the near term.

The fourth consideration is the extent to which gene-edited plants will be accepted, in comparison to genetically modified plants, both by the general community and by non-government organizations currently opposed to GMOs (Gao 2018) who are by and large also distrustful of genetically edited plants (Helliwell et al. 2017). Already, it is clear that these organizations, particularly in Europe (Aerni 2019), would like gene-edited plants regulated and labeled, whether or not they can be phenotypically distinguished from natural or induced mutations. It has been postulated this could lead to a GEO (gene edited organism) category of plant variety (Kuluev et al. 2019) which may inevitably lead to GEOs following the same path as GMOs in the sphere of public opinion (Shew et al. 2018). Political interference and lack of synchronization of regulation have been identified as the biggest threats to adoption of gene editing in agriculture (Lassoued et al. 2018), and these threats would also apply to the ornamental plant industry.

At the end of the day, compared to major crop plant species, ornamental plants are of relatively minor importance. Exceptionally novel phenotypic changes, not possible by any means other than by genetic modification or gene editing, would need to be identified to justify the costs associated with genome analysis, development of cassettes and/or plasmids, plant regeneration protocols, and regulatory compliance. We can therefore expect future developments to be focused on those ornamental plants for which robust regeneration protocols already exist and where the plant species are already of significant commercial value such as the cut flowers species rose and chrysanthemum and pot and bedding plants like petunia, begonia, pansy, and lily. Though gene editing offers more precise genetic improvement with just as wide a range of applications as genetic modification, adoption of the technology may face the same existential barriers to commercialization as those faced by a genetically modified ornamental plant variety.

References

Abdallah NA, Prakash CS, McHughen AG (2015) Genome editing for crop improvement: challenges and opportunities. GM Crops Food 6:183–205. https://doi.org/10.1080/21645698.2015.1129937

Aerni P (2019) Politicizing the precautionary principle: why disregarding facts should not pass for farsightedness. Front Plant Sci 10:1053. https://doi.org/10.3389/fpls.2019.01053

Agapito-Tenfen SZ, Okoli AS, Bernstein MJ, Wikmark OG, Myhr AI (2018) Revisiting risk governance of GM plants: the need to consider new and emerging gene-editing techniques. Front Plant Sci 9:1874. https://doi.org/10.3389/fpls.2018.01874

Ahn CH, Ramya M, An HR, Park PM, Kim YJ, Lee SY, Jang S (2020) Progress and challenges in the improvement of ornamental plants by genome editing. Plan Theory 9:687. https://doi.org/10.3390/plants9060687

Animal and Plant Health Inspection Service, USDA (2020) Animal and plant health inspection service 7 CFR Parts 330, 340, and 372 [Docket No. APHIS–2018–0034] RIN 0579–AE47 Movement of Certain Genetically engineered organisms. In: Federal Register 85:29790–29837. Aviable via https://www.aphis.usda.gov/brs/fedregister/BRS_20200518.pdf. Accessed June 2020

Anzalone AV, Randolph PB, Davis JR, Sousa AA, Koblan LW, Levy JM, Chen PJ, Wilson C, Newby GA, Raguram A, Liu DR (2019) Search-and-replace genome editing without double-strand breaks or donor DNA. Nature 576:149–157. https://doi.org/10.1038/s41586-019-1711-4

Ardavani O, Zerefos S, Doulos LT (2020) Redesigning the exterior lighting as part of the urban landscape: the role of transgenic bioluminescent plants in mediterranean urban and suburban lighting environments. J Clean Prod 242:118477. https://doi.org/10.1016/j.jclepro.2019.118477

Arora L, Narula A (2017) Gene editing and crop improvement using CRISPR-Cas9 system. Front Plant Sci 8:1932. https://doi.org/10.3389/fpls.2017.01932

Aubry S (2019) The future of digital sequence information for plant genetic resources for food and agriculture. Front Plant Sci 10:1046. https://doi.org/10.3389/fpls.2019.01046

Australian Government Federal Register of Legislation (2019) Gene technology amendment (2019 measures No. 1) regulations 2019. Available at https://www.legislation.gov.au/Details/F2019L00573/Explanatory%20Statement/Text. Accessed 20 June 2020

Azadi P, Bagheri H, Nalousi AM, Nazari F, Chandler SF (2016) Current status and biotechnological advances in genetic engineering of ornamental plants. Biotechnol Adv 34:1073–1090. https://doi.org/10.1016/j.biotechadv.2016.06.006

Azuma M, Morimoto R, Hirose M, Morita Y, Hoshino A, Iida S, Oshima Y, Mitsuda N, Ohme-Takagi M, Shiratake K (2016) A petal-specific InMYB1 promoter from Japanese morning glory: a useful tool for molecular breeding of floricultural crops. Plant Biotechnol J 14:354–363. https://doi.org/10.1111/pbi.12389

Bao A, Burritt DJ, Chen H, Zhou X, Cao D, Tran LP (2019) The CRISPR/Cas9 system and its applications in crop genome editing. Crit Rev Biotechnol 39:321–336. https://doi.org/10.1080/07388551.2018.1554621

Bashandy H, Teeri TH (2017) Genetically engineered orange petunias on the market. Planta 246: 277–280. https://doi.org/10.1007/s00425-017-2722-8

Beetham PR, Kipp PB, Sawycky XL, Arntzen CJ, May GD (1999) A tool for functional plant genomics: chimeric RNA/DNA oligonucleotides cause *in vivo* gene-specific mutations. Proc Natl Acad Sci U S A 96:8774–8778

Belhaj K, Chaparro-Garcia A, Kamoun S, Patron NJ, Nekrasov V (2015) Editing plant genomes with CRISPR/Cas9. Curr Opin Biotechnol 32:76–84. https://doi.org/10.1016/j.copbio.2014.11.007

Bollinedi H, Prabhu KV, Singh NK, Mishra S, Khurana JP, Singh AK (2017) Molecular and functional characterization of GR2-R1 event based backcross derived lines of Golden Rice in the genetic background of a mega rice variety Swarna. PLoS One 12:e0169600. https://doi.org/10.1371/journal.pone.0169600

Bombarely A, Moser M, Amrad A, Bao M, Bapaume L, Barry CS, Bliek M, Boersma MR, Borghi L, Bruggmann R, Bucher M (2016) Insight into the evolution of the Solanaceae from the parental genomes of *Petunia hybrida*. Nat Plants 2:1–9

Bortesi L, Fischer R (2015) The CRISPR/Cas9 system for plant genome editing and beyond. Biotechnol Adv 33:41–52. https://doi.org/10.1016/j.biotechadv.2014.12.006

Bruetschy C (2019) The EU regulatory framework on genetically modified organisms (GMOs). Transgenic Res 28:169–174. https://doi.org/10.1007/s11248-019-00149-y

Cardi T, Varshney R (2016) Cisgenesis and genome editing: combining concepts and efforts for a smarter use of genetic resources in crop breeding. Plant Breed 135:139–147. https://doi.org/10.1111/pbr.12345

Chandler SF, Sanchez C (2012) Genetic modification; the development of transgenic ornamental plant varieties. Plant Biotechnol J 10:891–903. https://doi.org/10.1111/j.1467-7652.2012.00693.x

Chen K, Wang Y, Zhang R, Zhang H, Gao C (2019) CRISPR/Cas genome editing and precision plant breeding in agriculture. Annu Rev Plant Biol 70:667–697. https://doi.org/10.1146/annurev-arplant-050718-100049

Chin DP, Shiratori I, Shimizu A, Kato K, Mii M, Waga I (2018) Generation of brilliant green fluorescent petunia plants by using a new and potent fluorescent protein transgene. Sci Rep 8:16556. https://doi.org/10.1038/s41598-018-34837-2

Chneiweiss H, Hirsch F, Montoliu L, Müller AM, Fenet S, Abecassis M, Merchant J, Baertschi B, Botbol-Baum M, Houghton JA, Kritikos M (2017) Fostering responsible research with genome editing technologies: a European perspective. Transgenic Res 26:709–713. https://doi.org/10.1007/s11248-017-0028-z

Custers R, Casacuberta JM, Eriksson D, Sagi L, Schiemann J (2018) Genetic alterations that do or do not occur naturally; consequences for genome edited organisms in the context of regulatory oversight. Front Bioeng Biotechnol 6:213. https://doi.org/10.3389/fbioe.2018.00213

Ding Y, Li H, Chen LL, Xie K (2016) Recent advances in genome editing using CRISPR/Cas9. Front Plant Sci 7:703. https://doi.org/10.3389/fpls.2016.00703

Edvardsson Björnberg K, Zetterberg C, Hansson SO, Andreasson E, Zhu LH (2019) Consistent risk regulation? Differences in the European regulation of food crops. J Risk Res 22:1561–1570. https://doi.org/10.1080/13669877.2018.1501594

Ellens KW, Levac D, Pearson C, Savoie A, Strand N, Louter J, Tibelius C (2019) Canadian regulatory aspects of gene editing technologies. Transgenic Res 28:165–168. https://doi.org/10.1007/s11248-019-00153-2

Eriksson D, Kershen D, Nepomuceno A, Pogson BJ, Prieto H, Purnhagen K, Smyth S, Wesseler J, Whelan A (2019) A comparison of the EU regulatory approach to directed mutagenesis with that of other jurisdictions, consequences for international trade and potential steps forward. New Phytol 222:1673–1684

Erpen-Dalla Corte L, Mahmoud ML, Moraes ST, Mou ZW, Grosser J, Dutt M (2019) Development of improved fruit, vegetable, and ornamental crops using the CRISPR/Cas9 genome editing technique. Plan Theory 8:601. https://doi.org/10.3390/plants8120601

Fan Y, Wang Q, Dong Z, Yin Y, Teixeira da Silva JA, Yu X (2019) Advances in molecular biology of *Paeonia* L. Planta 251:23. https://doi.org/10.1007/s00425-019-03299-9

Fraiture MA, Ujhelyi G, Ovesná J, Van Geel D, De Keersmaecker S, Saltykova A, Papazova N, Roosens NH (2019) MinION sequencing technology to characterize unauthorized GM petunia plants circulating on the European Union market. Sci Rep 9:7141. https://doi.org/10.1038/s41598-019-43463-5

Fraser PD, Aharoni A, Hall RD, Huang S, Giovannoni JJ, Sonnewald U, Fernie AR (2020) Metabolomics should be deployed in the identification and characterization of gene-edited crops. Plant J 102:897–902

Fritsche S, Poovaiah C, MacRae E, Thorlby G (2018) A New Zealand perspective on the application and regulation of gene editing. Front Plant Sci 9:1323. https://doi.org/10.3389/fpls.2018.01323

Fu Y, Sander JD, Reyon D, Cascio VM, Joung JK (2014) Improving CRISPR-Cas nuclease specificity using truncated guide RNAs. Nat Biotechnol 32:279–284. https://doi.org/10.1038/nbt.2808

Fukai S (2003) *Dendranthema* species as Chrysanthemum genetic resources. Acta Hortic 620:223–230

Gao C (2018) The future of CRISPR technologies in agriculture. Nat Rev Mol Cell Biol 19:275–276. https://doi.org/10.1038/nrm.2018.2

Hajiahmadi Z, Movahedi A, Wei H, Li D, Orooji Y, Ruan H, Zhuge Q (2019) Strategies to increase on-target and reduce off-target effects of the CRISPR/Cas9 system in plants. Int J Mol Sci 20:3719. https://doi.org/10.3390/ijms20153719

Halford NG (2019) Legislation governing genetically modified and genome-edited crops in Europe: the need for change. J Sci Food Agric 99:8–12. https://doi.org/10.1002/jsfa.9227

Hamburger DJS (2018) Normative criteria and their inclusion in a regulatory framework for new plant varieties derived from genome editing. Front Bioeng Biotechnol 6:176. https://doi.org/10.3389/fbioe.2018.00176

Han Y-J, Kim J-I (2019) Application of CRISPR/Cas9-mediated gene editing for the development of herbicide-resistant plants. Plant Biotechnol Rep 13:447–457. https://doi.org/10.1007/s11816-019-00575-8

Hartung F, Schiemann J (2013) Precise plant breeding using new genome editing techniques: opportunities, safety and regulation in the EU. Plant J 78:742–752. https://doi.org/10.1111/tpj.12413

Helliwell R, Hartley S, Pearce W, O'Neill L (2017) Why are NGOs Sceptical of genome editing? NGOs' opposition to agricultural biotechnologies is rooted in Scepticism about the framing of problems and solutions, rather than just emotion and dogma. EMBO Rep 18:2090–2093. https://doi.org/10.15252/embr.201744385

Hoshino A, Jayakumar V, Nitasaka E, Toyoda A, Noguchi H, Itoh T, Shin T, Minakuchi Y, Koda Y, Nagano AJ, Yasugi M (2016) Genome sequence and analysis of the Japanese morning glory *Ipomoea nil*. Nat Commun 7:13295. https://doi.org/10.1038/ncomms13295

Huang P, Lin F, Li B, Zheng Y (2019) Hybrid-transcriptome sequencing and associated metabolite analysis reveal putative genes involved in flower color difference in rose mutants. Plan Theory 8:267. https://doi.org/10.3390/plants8080267

Hundleby PAC, Harwood WA (2019) Impacts of the EU GMO regulatory framework for plant genome editing. Food Energy Secur 8:e00161. https://doi.org/10.1002/fes3.161

Jaganathan D, Ramasamy K, Sellamuthu G, Jayabalan S, Venkataraman G (2018) CRISPR for crop improvement: an update review. Front Plant Sci 9:985. https://doi.org/10.3389/fpls.2018.00985

Jansing J, Schiermeyer A, Schillberg S, Fischer R, Bortesi L (2019) Genome editing in agriculture: technical and practical considerations. Int J Mol Sci 20:2888. https://doi.org/10.3390/ijms20122888

Jia X, Tang L, Mei X, Liu H, Luo H, Deng Y, Su J (2020) Single-molecule long-read sequencing of the full-length transcriptome of *Rhododendron lapponicum* L. Sci Rep 10:6755. https://doi.org/10.1038/s41598-020-63814-x

Kalaitzandonakes N, Alston JM, Bradford KJ (2007) Compliance costs for regulatory approval of new biotech crops. Nat Biotechnol 25:509–511

Kausch AP, Nelson-Vasilchik K, Hague J, Mookkan M, Quemada H, Dellaporta S, Fragoso C, Zhang ZJ (2019) Edit at will: genotype independent plant transformation in the era of advanced

genomics and genome editing. Plant Sci 281:186–205. https://doi.org/10.1016/j.plantsci.2019.01.006
Kawall K (2019) New possibilities on the horizon: genome editing makes the whole genome accessible for changes. Front Plant Sci 10:525. https://doi.org/10.3389/fpls.2019.00525
Kishi-Kaboshi M, Aida R, Sasaki K (2018) Genome engineering in ornamental plants: current status and future prospects. Plant Physiol Biochem 131:47–52. https://doi.org/10.1016/j.plaphy.2018.03.015
Kleter GA, Kuiper HA, Kok EJ (2019) Gene-edited crops: towards a harmonized safety assessment. Trends Biotechnol 37:443–447. https://doi.org/10.1016/j.tibtech.2018.11.014
Ko SS, Kanno A, Sanchez-Perez R, Yeh HH, Hohe A, Mondragon-Palomino M (2019) Editorial: from functional genomics to biotechnology in ornamental plants. Front Plant Sci 10:463. https://doi.org/10.3389/fpls.2019.00463
Kok EJ, Glandorf DCM, Prins TW, Visser RGF (2019) Food and environmental safety assessment of new plant varieties after the European Court decision: process-triggered or product-based? Trends Food Sci Technol 88:24–32. https://doi.org/10.1016/j.tifs.2019.03.007
Kuluev BR, Gumerova GR, Mikhaylova EV, Gerashchenkov GA, Rozhnova NA, Vershinina ZR, Khyazev AV, Matniyazov RT, Baymiev AK, Baymiev AK, Chemeris AV (2019) Delivery of CRISPR/Cas components into higher plant cells for genome editing. Russ J Plant Physiol 66: 694–706. https://doi.org/10.1134/s102144371905011x
Kumar M (2019) Application of molecular markers and their utility in genetic studies of floricultural crops: a review. Int J Agric Environ Biotechnol 12:229–247. https://doi.org/10.30954/0974-1712.08.2019.7
Lassoued R, Smyth SJ, Phillips PWB, Hesseln H (2018) Regulatory uncertainty around new breeding techniques. Front Plant Sci 9:1291. https://doi.org/10.3389/fpls.2018.01291
Lassoued R, Macall DM, Hesseln H, Phillips PWB, Smyth SJ (2019a) Benefits of genome-edited crops: expert opinion. Transgenic Res 28:247–256. https://doi.org/10.1007/s11248-019-00118-5
Lassoued R, Phillips PWB, Smyth SJ, Hesseln H (2019b) Estimating the cost of regulating genome edited crops: expert judgment and overconfidence. GM Crops Food 10:44–62. https://doi.org/10.1080/21645698.2019.1612689
Law J, Ng K, Windram OPF (2020) The phenotype paradox: lessons from natural transcriptome evolution on how to engineer plants. Front Plant Sci 11:75. https://doi.org/10.3389/fpls.2020.00075
Lema MA (2019) Regulatory aspects of gene editing in Argentina. Transgenic Res 28:147–150. https://doi.org/10.1007/s11248-019-00145-2
Li W, Teng F, Li T, Zhou Q (2013) Targeted genome modification of crop plants using a CRISPR-Cas system. Nat Biotechnol 31:684–686. https://doi.org/10.1038/nbt.2652
Li SF, Wang J, Dong R, Zhu HW, Lan LN, Zhang YL, Li N, Deng CL, Gao WJ (2020) Chromosome-level genome assembly, annotation and evolutionary analysis of the ornamental plant *Asparagus setaceus*. Hortic Res 7:48. https://doi.org/10.1038/s41438-020-0271-y
Liu Y, Zeng J, Yuan C, Guo Y, Yu H, Li Y, Huang C (2019a) Cas9-PF, an early flowering and visual selection marker system, enhances the frequency of editing event occurrence and expedites the isolation of genome-edited and transgene-free plants. Plant Biotechnol J 17:1191–1193. https://doi.org/10.1111/pbi.13118
Liu YC, Yeh CW, Chung JD, Tsai CY, Chiou CY, Yeh KW (2019b) Petal-specific RNAi-mediated silencing of the phytoene synthase gene reduces xanthophyll levels to generate new *Oncidium* orchid varieties with white-colour blooms. Plant Biotechnol J 17:2035–2037. https://doi.org/10.1111/pbi.13179
Lowder L, Malzahn A, Qi Y (2016) Rapid evolution of manifold crispr systems for plant genome editing. Front Plant Sci 7:1683. https://doi.org/10.3389/fpls.2016.01683
Lozano-Juste J, Cutler SR (2014) Plant genome engineering in full bloom. Trends Plant Sci 19:284–287. https://doi.org/10.1016/j.tplants.2014.02.014
Ma KF, Zhang QX, Cheng TR, Yan XL, Pan HT, Wang J (2018) Substantial epigenetic variation causing flower color chimerism in the ornamental tree prunus mume revealed by single base resolution methylome detection and transcriptome sequencing. Int J Mol Sci 19:2315. https://doi.org/10.3390/ijms19082315

Malnoy M, Viola R, Jung MH, Koo OJ, Kim S, Kim JS, Velasco R, Nagamangala Kanchiswamy C (2016) DNA-free genetically edited grapevine and apple protoplast using CRISPR/Cas9 ribonucleoproteins. Front Plant Sci 7:1904. https://doi.org/10.3389/fpls.2016.01904

Malzahn A, Lowder L, Qi Y (2017) Plant genome editing with TALEN and CRISPR. Cell Biosci 7: 21. https://doi.org/10.1186/s13578-017-0148-4

Marton I, Zuker A, Shklarman E, Zeevi V, Tovkach A, Roffe S, Ovadis M, Tzfira T, Vainstein A (2010) Nontransgenic genome modification in plant cells. Plant Physiol 154:1079–1087. https://doi.org/10.1104/pp.110.164806

McCammon SL, Mendelsohn M (2019) Innovation and the regulation of products of agricultural biotechnology in the United States of America. Transgenic Res 28:183–186. https://doi.org/10.1007/s11248-019-00150-5

Metje-Sprink J, Sprink T, Hartung F (2019) Genome-edited plants in the field. Curr Opin Biotechnol 61:1–6. https://doi.org/10.1016/j.copbio.2019.08.007

Miller HI, Silva B (2018) The flower industry gets the genetic engineering blues. GM Crops Food 9: 49–52. https://doi.org/10.1080/21645698.2018.1471962

Mitiouchkina T, Mishin AS, Somermeyer LG, Markina NM, Chepurnyh TV, Guglya EB, Karataeva TA, Palkina KA, Shakhova ES, Fakhranurova LI, Chekova SV (2020) Plants with genetically encoded autoluminescence. Nat Biotechnol. https://doi.org/10.1038/s41587-020-0500-9

Nakamura N, Katsumoto Y, Brugliera F, Demelis L, Nakajima D, Suzuki H, Tanaka Y (2015) Flower color modification in *Rosa hybrida* by expressing the S-adenosylmethionine: anthocyanin 3′, 5′-O-methyltransferase gene from *Torenia hybrida*. Plant Biotechnol 32:109–117. https://doi.org/10.5511/plantbiotechnology.15.0205a

Nishihara M, Higuchi A, Watanabe A, Tasaki K (2018a) Application of the CRISPR/Cas9 system for modification of flower color in *Torenia fournieri*. BMC Plant Biol 18:331. https://doi.org/10.1186/s12870-018-1539-3

Nishihara M, Tasaki K, Sasaki N, Takahashi H (2018b) Development of basic technologies for improvement of breeding and cultivation of Japanese gentian. Breed Sci 68:14–24. https://doi.org/10.1270/jsbbs.17074

Noda N (2018) Recent advances in the research and development of blue flowers. Breed Sci 68:79–87. https://doi.org/10.1270/jsbbs.17132

Noda N, Aida R, Kishimoto S, Ishiguro K, Fukuchi-Mizutani M, Tanaka Y, Ohmiya A (2013) Genetic engineering of novel bluer-colored chrysanthemums produced by accumulation of delphinidin-based anthocyanins. Plant Cell Physiol 54:1684–1695. https://doi.org/10.1093/pcp/pct111

Nogué F, Vergne P, Chèvre AM, Chauvin JE, Bouchabke-Coussa O, Déjardin A, Chevreau E, Hibrand-Saint Oyant L, Mazier M, Barret P, Guiderdoni E (2019) Crop plants with improved culture and quality traits for food, feed and other uses. Transgenic Res 28:65–73. https://doi.org/10.1007/s11248-019-00135-4

Park J, Choe S (2019) DNA-free genome editing with preassembled CRISPR/Cas9 ribonucleoproteins in plants. Transgenic Res 28:61–64. https://doi.org/10.1007/s11248-019-00136-3

Pixley KV, Falck-Zepeda JB, Giller KE, Glenna LL, Gould F, Mallory-Smith CA, Stelly DM, Stewart CN Jr (2019) Genome editing, gene drives, and synthetic biology: will they contribute to disease-resistant crops, and who will benefit? Annu Rev Phytopathol 57:165–188. https://doi.org/10.1146/annurev-phyto-080417-045954

Podevin N, Davies HV, Hartung F, Nogue F, Casacuberta JM (2013) Site-directed nucleases: a paradigm shift in predictable, knowledge-based plant breeding. Trends Biotechnol 31:375–383. https://doi.org/10.1016/j.tibtech.2013.03.004

Saint-Oyant LH, Ruttink T, Hamama L, Kirov I, Lakhwani D, Zhou NN, Bourke PM, Daccord N, Leus L, Schulz D, Van de Geest H (2018) A high-quality genome sequence of *Rosa chinensis* to elucidate ornamental traits. Nat Plants 4:473–484. https://doi.org/10.1038/s41477-018-0166-1

Sánchez MA (2020) Chile as a key enabler country for global plant breeding, agricultural innovation, and biotechnology. GM Crops Food. https://doi.org/10.1080/21645698.2020.1761757

Sasaki K, Yamaguchi H, Kasajima I, Narumi T, Ohtsubo N (2016) Generation of novel floral traits using a combination of floral organ-specific promoters and a chimeric repressor in *Torenia fournieri* Lind. Plant Cell Physiol 57:1319–1331. https://doi.org/10.1093/pcp/pcw081

Scheben A, Edwards D (2018) Bottlenecks for genome-edited crops on the road from lab to farm. Genome Biol 19:178. https://doi.org/10.1186/s13059-018-1555-5
Schiemann J, Dietz-Pfeilstetter A, Hartung F, Kohl C, Romeis J, Sprink T (2019) Risk assessment and regulation of plants modified by modern biotechniques: current status and future challenges. Annu Rev Plant Biol 70:699–726. https://doi.org/10.1146/annurev-arplant-050718-100025
Schillberg S, Raven N, Spiegel H, Rasche S, Buntru M (2019) Critical analysis of the commercial potential of plants for the production of recombinant proteins. Front Plant Sci 10:720. https://doi.org/10.3389/fpls.2019.00720
Schindele A, Dorn A, Puchta H (2019) CRISPR/Cas brings plant biology and breeding into the fast lane. Curr Opin Biotechnol 61:7–14. https://doi.org/10.1016/j.copbio.2019.08.006
Schmidt SM, Belisle M, Frommer WB (2020) The evolving landscape around genome editing in agriculture. EMBO Rep 19:e50680
Shan S, Soltis PS, Soltis DE, Yang B (2020) Considerations in adapting CRISPR/Cas9 in non-genetic model plant systems. Appl Plant Sci 8:e11314. https://doi.org/10.1002/aps3.11314
Shelake RM, Pramanik D, Kim J-Y (2019) Evolution of plant mutagenesis tools: a shifting paradigm from random to targeted genome editing. Plant Biotechnol Rep 13:423–445. https://doi.org/10.1007/s11816-019-00562-z
Shew AM, Nalley LL, Snell HA, Nayga RM, Dixon BL (2018) CRISPR versus GMOs: public acceptance and valuation. Glob Food Sec 19:71–80. https://doi.org/10.1016/j.gfs.2018.10.005
Shibuya K, Watanabe K, Ono M (2018) CRISPR/Cas9-mediated mutagenesis of the EPHEMERAL1 locus that regulates petal senescence in Japanese morning glory. Plant Physiol Biochem 131:53–57. https://doi.org/10.1016/j.plaphy.2018.04.036
Smirnova OG, Shumny VK, Kochetov AV (2019) Regulatory sequences for constitutive, tissue-specific, and induced expression of transgenes in ornamental plants. Russ J Plant Physiol 66: 679–693. https://doi.org/10.1134/s1021443719050182
Smulders MJ, Arens P, Bourke PM, Debener T, Linde M, De Riek J, Leus L, Ruttink T, Baudino S, Saint-Oyant LH, Clotault J (2019) In the name of the rose: a roadmap for rose research in the genome era. Hortic Res 6:65. https://doi.org/10.1038/s41438-019-0156-0
Smyth SJ, McDonald J, Falck-Zepeda J (2014) Investment, regulation, and uncertainty: managing new plant breeding techniques. GM Crops Food 5:44–57. https://doi.org/10.4161/gmcr.27465
Sprink T, Eriksson D, Schiemann J, Hartung F (2016) Regulatory hurdles for genome editing: process- vs. product-based approaches in different regulatory contexts. Plant Cell Rep 35:1493–1506. https://doi.org/10.1007/s00299-016-1990-2
Su J, Jiang J, Zhang F, Liu Y, Ding L, Chen S, Chen F (2019) Current achievements and future prospects in the genetic breeding of chrysanthemum: a review. Hortic Res 6:109. https://doi.org/10.1038/s41438-019-0193-8
Subburaj S, Chung SJ, Lee C, Ryu SM, Kim DH, Kim JS, Bae S, Lee GJ (2016) Site-directed mutagenesis in *Petunia* x *hybrida* protoplast system using direct delivery of purified recombinant Cas9 ribonucleoproteins. Plant Cell Rep 35:1535–1544. https://doi.org/10.1007/s00299-016-1937-7
Tanaka Y, Brugliera F (2013) Flower colour and cytochromes P450. Philos Trans R Soc Ser B Biol Sci 368:20120432. https://doi.org/10.1098/rstb.2012.0432
Tasaki K, Higuchi A, Watanabe A, Sasaki N, Nishihara M (2019) Effects of knocking out three anthocyanin modification genes on the blue pigmentation of gentian flowers. Sci Rep 9:15831. https://doi.org/10.1038/s41598-019-51808-3
Thygesen P (2019) Clarifying the regulation of genome editing in Australia: situation for genetically modified organisms. Transgenic Res 28:151–159. https://doi.org/10.1007/s11248-019-00151-4
Tong CG, Wu FH, Yuan YH, Chen YR, Lin CS (2019) High-efficiency CRISPR/Cas-based editing of *Phalaenopsis* orchid MADS genes. Plant Biotechnol J 18:889–891. https://doi.org/10.1111/pbi.13264
Totsuka A, Okamoto E, Miyahara T, Kouno T, Cano EA, Sasaki N, Watanabe A, Tasaki K, Nishihara M, Ozeki Y (2018) Repressed expression of a gene for a basic helix-loop-helix protein causes a white flower phenotype in carnation. Breed Sci 68:139–143. https://doi.org/10.1270/jsbbs.17072

Uniyal AP, Mansotra K, Yadav SK, Kumar V (2019) An overview of designing and selection of sgRNAs for precise genome editing by the CRISPR-Cas9 system in plants. 3 Biotech 9:223. https://doi.org/10.1007/s13205-019-1760-2

van de Wiel CCM, Schaart JG, Lotz LAP, Smulders MJM (2017) New traits in crops produced by genome editing techniques based on deletions. Plant Biotechnol Rep 11:1–8. https://doi.org/10.1007/s11816-017-0425-z

Waltz E (2018) With a free pass, CRISPR-edited plants reach market in record time. Nat Biotechnol 36:6–7

Watanabe K, Oda-Yamamizo C, Sage-Ono K, Ohmiya A, Ono M (2018) Alteration of flower colour in *Ipomoea nil* through CRISPR/Cas9-mediated mutagenesis of carotenoid cleavage dioxygenase 4. Transgenic Res 27:25–38. https://doi.org/10.1007/s11248-017-0051-0

Wei Z, Arazi T, Hod N, Zohar M, Isaacson T, Doron-Faigenboim A, Reznik N, Yedidia I (2020) Transcriptome profiling of *Ornithogalum dubium* leaves and flowers to identify key carotenoid genes for CRISPR gene editing. Plan Theory 9:540. https://doi.org/10.3390/plants9040540

Wesseler J, Politiek H, Zilberman D (2019) The economics of regulating new plant breeding technologies – implications for the bioeconomy illustrated by a survey among Dutch plant breeders. Front Plant Sci 10:1597. https://doi.org/10.3389/fpls.2019.01597

Wolt JD, Wolf C (2018) Policy and governance perspectives for regulation of genome edited crops in the United States. Front Plant Sci 9:1606. https://doi.org/10.3389/fpls.2018.01606

Woo JW, Kim J, Kwon SI, Corvalán C, Cho SW, Kim H, Kim SG, Kim ST, Choe S, Kim JS (2015) DNA-free genome editing in plants with preassembled CRISPR-Cas9 ribonucleoproteins. Nat Biotechnol 33:1162–1164. https://doi.org/10.1038/nbt.3389

Xu J, Hua K, Lang Z (2019) Genome editing for horticultural crop improvement. Hortic Res 6:113. https://doi.org/10.1038/s41438-019-0196-5

Xu J, Kang BC, Naing AH, Bae SJ, Kim JS, Kim H, Kim CK (2020) CRISPR/Cas9-mediated editing of 1-aminocyclopropane-1-carboxylate oxidase1 enhances petunia flower longevity. Plant Biotechnol J 18:287–297. https://doi.org/10.1111/pbi.13197

Yagi M (2018) Recent progress in whole genome sequencing, high-density linkage maps, and genomic databases of ornamental plants. Breed Sci 68:62–70. https://doi.org/10.1270/jsbbs.17080

Yagi MA, Kosugi SH, Hirakawa HI, Ohmiya AK, Tanase KO, Harada TA, Kishimoto KY, Nakayama MA, Ichimura KA, Onozaki TA, Yamaguchi HI (2014) Sequence analysis of the genome of carnation (*Dianthus caryophyllus* L.). DNA Res 21:231–241. https://doi.org/10.1093/dnares/dst053

Yan R, Wang Z, Ren Y, Li H, Liu N, Sun H (2019) Establishment of efficient genetic transformation systems and application of CRISPR/Cas9 genome editing technology in *Lilium pumilum* DC. Fisch. and *Lilium longiflorum* White Heaven. Int J Mol Sci 21:231–241. https://doi.org/10.3390/ijms20122920

Zhang GQ, Liu KW, Li Z, Lohaus R, Hsiao YY, Niu SC, Wang JY, Lin YC, Xu Q, Chen LJ, Yoshida K (2017) The *Apostasia* genome and the evolution of orchids. Nature 549:379–383. https://doi.org/10.1038/nature23897

Zhang D, Hussain A, Manghwar H, Xie K, Xie S, Zhao S, Larkin RM, Qing P, Jin S, Ding F (2020) Genome editing with the CRISPR-Cas system: an art, ethics and global regulatory perspective. Plant Biotechnol J. https://doi.org/10.1111/pbi.13383

Zlobin NE, Lebedeva MV, Taranov VV (2020) CRISPR/Cas9 genome editing through *in planta* transformation. Crit Rev Biotechnol 40:153–168

The Wild and Cultivated *Mussaenda* 18

Fernando B. Aurigue

Contents

Abstract

A brief history on the naming of the genus and general description of the different plant parts of *Mussaenda* are presented together with the list of species, subspecies or variety, author, year, and name of publication. In another table, the name of

F. B. Aurigue (✉)
Department of Science and Technology-Philippine Nuclear Research Institute, Quezon City, Metro Manila, Philippines
e-mail: fbaurigue@pnri.dost.gov.ph

S. K. Datta, Y. C. Gupta (eds.), *Floriculture and Ornamental Plants*, Handbooks of Crop Diversity: Conservation and Use of Plant Genetic Resources,
https://doi.org/10.1007/978-981-15-3518-5_20

varieties, breeder, year of registration or patenting, and the registration or patent number are enlisted.

The basic chromosome number x = 11 is consistent in all species studied to date. The inheritance of number and color of calycophylls, flower size, color and shape, and seed color are provided. For propagation, the procedures for the different asexual methods are presented step by step. For tissue culture techniques, the plant tissue culture medium formulation and results obtained are presented in a table. The cultural requirements are made available.

Lastly, the traditional uses for medicine of different plant parts, phytochemicals isolated through different extraction methods and identified, and their biological activities are tabulated. Information on germplasm collection, database, and conservation status are included.

Keywords

Bioactivity · Breeding · Calicophyll · Cytology · Genetics · Hybrid · Mussaenda · Phytochemistry · Propagation · Tissue culture

18.1 Introduction

The genus *Mussaenda* has become very popular as ornamental plant for landscaping or flowering pot plants because of its long-lasting and attractive expanded calyx lobes or calycophylls or pleasing visual effect when planted en masse. Many hybrids have been registered and even patented as new or improved varieties, some of which are already traded internationally.

With thriving nursery business or industry for mussaenda plant production, conventional means of vegetative propagation, such as by cutting, marcotting or air layering, inarching or approach grafting, grafting and budding, or tissue culture techniques (direct plantlet production from isolated planting materials and somatic embryogenesis), is necessary to keep up with the increasing demand. The use of rooting hormones and other plant growth regulators is necessary to effect desired results, especially in tissue cultures. Provision of the cultural requirements, like potting medium and fertilizers, for successful production of plants may vary from one nursery to another as it depends on the availability and economy of use of such material or supply. The major insect pests and diseases of mussaendas can be easily controlled or checked with biological or synthetic pesticide or sanitary practices in the production area.

Due to diversity and unique features of the species and varieties, researches on morphology, pigmentation, cytology, genetics, and isozyme polymorphism are conducted. However, the most important undertaking done is the verification of traditional use of plant parts for medicine. Secondary metabolites or phytochemicals extracted from different parts of the plant by different means are subjected to different tests or assay to determine their biological activities and application for treatment of certain disease or condition. Many new natural compounds have been isolated, identified, and tested in recent years which may result in drug production or confirmed medicinal value of the plant part.

There are more than 180 species of *Mussaenda*, and most of them are narrow endemics, hence rarely encountered in their natural habitat. The potential of these plant genetic resources for the development of new plant cultivars and/or new drugs necessitates their conservation and continuous utilization for basic and applied research activities.

18.2 Botany and Distribution

Mussaenda is a genus of flowering plants established by Carl Linnaeus in 1753 with *M. frondosa* from India as the type species. However, the first plant from the genus was first collected in Sri Lanka sometime in 1672–1677 by Paul Hermann of the Dutch East India Company, who is attributed for the generic name mussaenda later adopted by Linnaeus (Cramer and Bridgen 1997a). Mussaenda could be a Latinization of the Sinhala "bussanda."

Like other genera belonging to the family Rubiaceae, it is comprised of small- to medium-sized trees (rarely exceeding 9 m tall), scandent or climbing shrubs and subshrubs (suffrutices), and woody climbers (lianas). With at least 186 species distributed in paleotropics, it is the most species-rich genus under the tribe Mussaendeae, sub-family Ixoroideae.

A list of currently accepted taxa is given on Table 1.

The general description of the different parts of the plant is taken from Alejandro et al. (2016).

Stem – usually with dense fine, soft, short hairs when young, becoming hairless or nearly so as it matures; mature stems are cylindrical, with lenticels and bark that can be peeled easily. In young branches, the stem is somewhat flattened and covered with white or brownish to red hairs. Lateral branches are more or less horizontal or at an angle from the vertical (plagiotropic) except in suffrutices and lianas.

Leaves – opposite but alternate pairs are at right angle to each other (decussate), generally thin (membranaceous) to tough and leathery (coriaceous), generally with petiole (petiolate) to 8 cm, occasionally with a very short stalk (subsessile); leaf stalks are typically hairy; blades vary from ovate, or elliptic, rarely obovate to orbicular with size ranging from small (3.5–12.0 cm × 1.8–4.0 cm) to large with length of 34.0 cm, although most species are medium-sized, usually with hairs on both sides, especially along the midrib and secondary nerves on the lower surface, but those nearest the inflorescence are smaller, with more hairs, and often subsessile. The leaf tip could be acute or acuminate to subcaudate with a few having conduplicate apex. Meanwhile, the leaf base could be cuneate or attenuate but may vary from obtuse to acute. The midrib and secondary veins are conspicuous on the undersurface and mostly only slightly prominent on the upper surface. A few species have blade midrib and secondary veins in 5–18 pairs that are rarely sunken adaxially. Venation is loop-veined (brochidodromous)

Table 1 Species of *Mussaenda* accepted by the WCSP as of 31 May 2020 (IPNI 2020)

	Specific Epithet	Author	Year	Publication
1.	*acuminata*	Blume	1826–1827	Bijdr. Fl. Ned. Ind. 16: 986
2.	*acuminatissima*	Merr.	1921	Philipp. J. Sci. 17: 436
3.	*aestuarii*	K.Schum.	1905	Nachtr. Fl. Schutzgeb. Südsee [Schumann & Lauterbach] 394
4.	*afzelii*	G.Don	1834	Gen. Hist. 3: 490
5.	*afzelioides*	Wernham	1913	Cat. Pl. Oban 40
6.	*albiflora*	Merr.	1910	Philipp. J. Sci., C 5: 241
7.	*angolensis*	Wernham	1913	J. Bot. 51: 239
8.	*angustisepala*	Ridl.	1923	Fl. Malay Penins. ii. 60
9.	*anisophylla*	S. Vidal	1885	Phan. Cuming. Philipp. 178
10.	*antiloga*	Chun & W.C.Ko	1974	Fl. Hainan. 3: 582
11.	*aptera*	Pit.	1923	Fl. Indo-Chine [P.H. Lecomte et al.] 3: 185
12.	*arcuata*	Poir.	1797	Encycl. [J. Lamarck et al.] 4(1): 392
13.	*attenuifolia*	Elmer	1913	Leafl. Philipp. Bot. v: 1874
14.	*bammleri*	Valeton	1925	Bot. Jahrb. Syst. 60(1–2): 64
15.	*benguetensis*	Elmer	1906	Leafl. Philipp. Bot. i. 13
16.	*bevanii*	F.Muell.	1887	Proc. Linn. Soc. New South Wales Ser. II, ii. 419, t. 6
17.	*bityensis*	Wernham	1919	J. Bot. 57: 276
18.	*bodenii*	Wernham	1916	Trans. Linn. Soc. London, Bot. 9(1): 70
19.	*bonii*	Pit.	1923	Fl. Indo-Chine [P.H. Lecomte et al.] 3: 181
20.	*borbonica*	Lapeyrere	1888	Rev. Agric. Maurice, Reun. & Madag. ii. 85
21.	*brachygyna*	Merr. & L.M. Perry	1944	J. Arnold Arbor. 25: 193
22.	*breviloba*	S.Moore	1905	J. Bot. 43: 137
23.	*cambodiana*	Pierre ex Pit.	1923	Fl. Indo-Chine [P.H. Lecomte et al.] 3: 188
24.	*caudatiloba*	D.Fang	2002	Acta Phytotax. Sin. 40(2): 156
25.	*celebica*	Ridl.	1940	Bull. Misc. Inform. Kew 1939(10): 598
26.	*chevalieri*	Pit.	1923	Fl. Indo-Chine [P.H. Lecomte et al.] 3: 183
27.	*chinensis*	Lour.	1790	Fl. Cochinch. 1: 152
28.	*chingii*	C.Y.Wu ex H.H. Hsue & H.Wu	1986	Acta Phytotax. Sin. 24(3): 236
29.	*chippii*	Wernham	1913	J. Bot. 51: 237
30.	*chlorantha*	Merr.	1913	Philipp. J. Sci., C 8: 47
31.	*chrysotricha*	Valeton	1925	Bot. Jahrb. Syst. 60(1–2): 65
32.	*conopharyngiifolia*	Stapf	1905	J. Linn. Soc., Bot. 37: 104
33.	*cordifolia*	Wall. ex G.Don	1834	Gen. Hist. 3: 491, descr.
34.	*corymbosa*	Roxb.	1814	Hort. Bengal. 15

(continued)

Table 1 (continued)

	Specific Epithet	Author	Year	Publication
35.	*cuspidata*	E.T.Geddes	1927	Bull. Misc. Inform. Kew 1927(4): 170
36.	*cylindrocarpa*	Burck	1883	Ann. Jard. Bot. Buitenzorg 3: 112, 118
37.	*dasyphylla*	Miq.	1869	Ann. Mus. Bot. Lugduno-Batavi 4: 187
38.	*dawei*	Hutch.	1922	Bull. Misc. Inform. Kew 1922(6): 196
39.	*debeauxii*	Wernham	1916	J. Bot. 54: 300
40.	*decipiens*	H.Li	1980	Acta Phytotax. Sin. 18(1): 117
41.	*densiflora*	H.L.Li	1943	J. Arnold Arbor. 24: 455
42.	*dinhensis*	Pierre ex Pit.	1923	Fl. Indo-Chine [P.H. Lecomte et al.] 3: 173
43.	*divaricata*	Hutch.	1916	Pl. Wilson. (Sargent) 3(2): 394
44.	*dolichocarpa*	(Lauterb. & K.Schum.) Rechinger	1913	In Denkschr. Akad. Wiss. Wien, Math.-Nat. lxxxix. (Bot. & Zool. Ergebn. Samoa, Pt. 5, 166) 608
45.	*dranensis*	Wernham	1921	J. Nat. Hist. Soc. Siam iv. 136
46.	*elegans*	Schumach. & Thon.	1827	Beskr. Guin. Pl. 117
47.	*elliptica*	Hutch.	1916	Pl. Wilson. (Sargent) 3(2): 395
48.	*elmeri*	Merr.	1929	Univ. Calif. Publ. Bot. xv: 279
49	*emeiensis*	Z.Y.Zhu & S.J.Zhu	2008	Bull. Bot. Res., Harbin 28(3): 257 (-258; fig. 1)
50.	*epiphytica*	Cheek	2009	Nordic J. Bot. 27(6): 456 (-459; fig.1)
51.	*erosa*	Champ. ex Benth.	1852	Hooker's J. Bot. Kew Gard. Misc. 4: 193
52.	*erythrophylla*	Schumach. & Thonn.	1827	Beskr. Guin. Pl. 116
53.	*ferrea*	E.T.Geddes	1927	Bull. Misc. Inform. Kew 1927(4): 170
54.	*ferruginea*	K.Schum.	1889	Fl. Kais. Wilh. Land [K.M. Schumann & M.U. Hollrung] 129
55.	*fissibractea*	Merr.	1937	Mitt. Inst. Bot. Hamburg vii. 282
56.	*forbesii*	Wernham ex S.Moore	1923	J. Bot. 61 (Suppl.): 24
57.	*forsteniana*	Miq.	1869	Ann. Mus. Bot. Lugduno-Batavi 4: 188
58.	*frondosa*	L.	1753	Sp. Pl. 1: 177
59.	*garrettii*	Craib	1931	Bull. Misc. Inform. Kew 1931(9): 444
60.	*glabra*	Vahl	1832	Numer. List [Wallich] n. 6251
61.	*glabrata*	(Hook. f.) Hutch. ex Gamble	1921	Fl. Madras 610
62.	*gossweileri*	Wernham	1916	J. Bot. 54: 300
63.	*grandibractea*	Alejandro	2016	Ann. Missouri Bot. Gard. 101(3): 480
64.	*grandiflora*	Benth.	1849	Niger Fl. [W. J. Hooker]. 392
65.	*grandifolia*	Elmer	1906	Leafl. Philipp. Bot. i. 12
66.	*griffithii*	Wight ex Hook. f.	1880	Fl. Brit. India [J. D. Hooker] 3(7): 88
67.	*hainanensis*	Merr.	1935	Lingnan Sci. J. 14: 58
68.	*heinsioides*	Hiern	1877	Fl. Trop. Afr. [Oliver et al.] 3: 70

(continued)

Table 1 (continued)

	Specific Epithet	Author	Year	Publication
69.	*herderscheeana*	Valeton	1925	Nova Guinea 14: 261
70.	*hilaris*	Pierre ex Pit.	1923	Fl. Indo-Chine [P.H. Lecomte et al.] 3: 177
71.	*hirsuta*	Ridl.	1923	Journ. As. Soc. Mal. i. 68
72.	*hirsutissima*	(Hook. f.) Hutch. ex Gamble	1921	Fl. Madras 2: 610
73.	*hirsutula*	Miq.	1861	Journ. Bot. Neerl. i. (1861) 109
74.	*hoaensis*	Pierre ex Pit.	1923	Fl. Indo-Chine [P.H. Lecomte et al.] 3: 190
75.	*hossei*	Craib ex Hosseus	1911	Beih. Bot. Centralbl., Abt. 2. 28(2): 444, 457
76.	*incana*	Wall.	1824	Fl. Ind. (Carey & Wallich ed.) 2: 229
77.	*intuspilosa*	Jayaw.	1963	J. Arnold Arbor. 44: 257
78.	*isertiana*	Dc.	1830	Prodr. [A. P. de Candolle] 4: 371
79.	*johannis-winkleri*	Merr.	1937	Mitt. Inst. Bot. Hamburg vii. 283
80.	*kajewskii*	Merr. & L.M.Perry	1944	J. Arnold Arbor. 25: 194
81.	*kanehirae*	Merr. & L.M.Perry	1944	J. Arnold Arbor. 25: 195
82.	*keenanii*	Hook. f.	1880	Fl. Brit. India [J. D. Hooker] 3(7): 87
83.	*kerrii*	Craib	1911	Bull. Misc. Inform. Kew 1911(10): 389
84.	*kingdon-wardii*	Jayaw.	1965	J. Arnold Arbor. 46: 366
85.	*kintaensis*	King ex Stapf	1894	Trans. Linn. Soc. London, Bot. 4(2): 172, in obs.
86.	*kwangsiensis*	H.L.Li	1943	J. Arnold Arbor. 24: 455
87.	*kwangtungensis*	H.L.Li	1944	J. Arnold Arbor. 25: 427
88.	*lanata*	C.B.rob.	1911	Philipp. J. Sci., C 6: 357
89.	*lancifolia*	K.Krause	1920	Bot. Jahrb. Syst. 57(1): 28
90.	*lancipetala*	X.F.Deng & D.X. Zhang	2008	J. Syst. Evol. 46(2): 223 (220-225; figs. 1-2)
91.	*landolphioides*	Wernham	1913	J. Bot. 51: 238
92.	*lanuginosa*	Ridl.	1940	Bull. Misc. Inform. Kew 1939(10): 597
93.	*laxa*	(Hook.f.) Hutch. ex Gamble	1921	Fl. Madras 610
94.	*leptantha*	Wernham	1919	J. Bot. 57: 277
95.	*leucophylla*	E.M.A.Petit	1955	Bull. Jard. Bot. État Bruxelles 25:156
96.	*leucova*	Gilli	1980	Ann. Naturhist. Mus. Wien 83: 461
97.	*liedeae*	Alejandro	2016	Ann. Missouri Bot. Gard. 101(3): 485
98.	*linderi*	Hutch. & Dalziel	1931	Fl. W. Trop. Afr. [Hutchinson & Dalziel] ii. 101, in clavi, 103
99.	*longipetala*	H.L.Li	1943	J. Arnold Arbor. 24: 373
100.	*longisepala*	E.T.Geddes	1927	Bull. Misc. Inform. Kew 1927(4): 171
101.	*longituba*	Valeton	1907	Bull. Dépt. Agric. Indes Néerl. 10: 63
102.	*lotungensis*	Chun & W.C.Ko	1974	Fl. Hainan. 3: 581

(continued)

Table 1 (continued)

	Specific Epithet	Author	Year	Publication
103.	*macrantha*	Valeton	1911	Nova Guinea 8: 456
104.	*macrophylla*	Wall.	1820	Fl. Ind. (Carey & Wallich ed.) 1: 228
	macrophylla var. *grandisepala*	(Jayaw.) Alejandro	2016	Ann. Missouri Bot. Gard. 101(3): 489
105.	*magallanensis*	Elmer	1911	Leafl. Philipp. Bot. Iii. 996
106.	*maingayi*	(Hook.f.) Hemsl. ex T.Durand & B.D.Jacks.	1903	Index Kew. Suppl. 1(3): 284
107.	*malaccensis*	Ridl.	1923	Fl. Malay Penins. ii. 59
108.	*malacotricha*	Merr. & L.M. Perry	1944	J. Arnold Arbor. 25: 195
109.	*membranacea*	King	1904	J. Asiat. Soc. Bengal, Pt. 2, Nat. Hist. 72(4): 187
110.	*membranifolia*	Merr.	1923	Philipp. J. Sci. 23: 267
111.	*microdonta*	Wernham	1913	J. Bot. 51: 239
	microdonta subsp. *odorata*	(Hutch.) Bridson	1976	Kew Bull. 30(4): 696
112.	*milleri*	Elmer ex Alejandro	2016	Ann. Missouri Bot. Gard. 101(3): 494
113.	*mollis*	E.T.Geddes	1927	Bull. Misc. Inform. Kew 1927(4): 171
114.	*mollissima*	C.Y.Wu ex H.H. Hsue & H.Wu	1986	Acta Phytotax. Sin. 24(3): 235
115.	*monticola*	K.Krause	1912	Bot. Jahrb. Syst. 48(3–4): 406
116.	*motleyi*	Ridl.	1940	Bull. Misc. Inform. Kew 1939(10): 599
117.	*multibracteata*	Merr.	1916	Philipp. J. Sci., C 11: 34
118.	*multinervis*	C.Y.Wu ex H.H. Hsue & H.Wu	1986	Acta Phytotax. Sin. 24(3): 237
119.	*nannanii*	Wernham	1918	J. Bot. 56: 309
120.	*nervosa*	Elmer	1911	Leafl. Philipp. Bot. Iii. 994
121.	*nicobarica*	Shimpale, S.R. Yadav & Babu	2009	Rheedea 19(1–2): 54 (–56; Fig. 1)
122.	*nijensis*	R.D.Good	1926	J. Bot. 64 (Suppl. 2): 3
123.	*nivea*	A.Chev. ex Hutch. & Dalziel	1931	Fl. W. Trop. Afr. [Hutchinson & Dalziel] 2: 100, in clavi, 101
124.	*oblonga*	King	1904	J. Asiat. Soc. Bengal, Pt. 2, Nat. Hist. 72(4): 186
125.	*oreadum*	Wernham	1916	Trans. Linn. Soc. London, Bot. 9(1): 70
126.	*ornata*	S.Moore	1927	J. Bot. 65: 243
127.	*ovata*	Merr. & L.M. Perry	1944	J. Arnold Arbor. 25: 194
128.	*palawanensis*	Merr.	1915	Philipp. J. Sci., C 10: 103
129.	*paludosa*	E.M.A.Petit	1955	Bull. Jard. Bot. État Bruxelles 25: 164
130.	*parryorum*	C.E.C.Fisch.	1928	Bull. Misc. Inform. Kew 1928(7): 274
131.	*parviflora*	Miq.	1867	Ann. Mus. Bot. Lugduno-Batavi 3: 110
132.	*parvifolia*	Valeton	1907	Bull. Dépt. Agric. Indes Néerl. 10: 63

(continued)

Table 1 (continued)

	Specific Epithet	Author	Year	Publication
133.	*philippica* (Fig. 1)	A.Rich.	1830	Mem. Fam. Rubiac. (Mem. Soc. Hist. Nat. Paris, v. 245: 1834) 165
	philippica var. *pubescens*	Alejandro	2016	Ann. Missouri Bot. Gard. 101(3): 507
134.	*philippinensis*	Merr.	1908	Philipp. J. Sci., C 3: 264
135.	*pilosissima*	Valeton	1925	Bot. Jahrb. Syst. 60(1–2): 62
136.	*pinatubensis*	Elmer	1934	Leafl. Philipp. Bot. ix. 3210
137.	*pingbianensis*	C.Y.Wu ex H.H. Hsue & H.Wu	1986	Acta Phytotax. Sin. 24(3): 233
138.	*pluviatilis*	S.Moore	1927	J. Bot. 65: 244
139.	*polita*	Hiern	1877	Fl. Trop. Afr. [Oliver et al.] 3: 67
140.	*polyneura*	King	1904	J. Asiat. Soc. Bengal, Pt. 2, Nat. Hist. 72(4): 185
141.	*procera*	F.M.Bailey	1900	Queensland Agric. J. iii. 155
142.	*pubescens*	Dryand.	1810	Hortus Kew 1: 372*
143.	*pullei*	Valeton	1911	Nova Guinea 8: 459
144.	*purpurascens*	Ridl.	1912	J. Straits Branch Roy. Asiat. Soc. 61: 18
145.	*raiateensis*	J.W.Moore	1933	Bull. Bernice P. Bishop Mus. 102: 44
146.	*reinwardtiana*	Miq.	1857	Fl. Ned. Ind. 2: 211
147.	*ridleyana*	Wernham	1916	Trans. Linn. Soc. London, Bot. ix. 70
148.	*rivularis*	Welw. x Hiern	1898	Cat. Afr. Pl. (Hiern) i. 452
	rivularis var. *redheadii*	(E.M.A.Petit) Figueiredo	2008	Bot. J. Linn. Soc. 156(4): 569
149.	*roxburghii*	Hook.f.	1880	Fl. Brit. India [J. D. Hooker] 3(7): 87
150.	*rufa*	A.Rich.	1830	Mem. Fam. Rubiac. (Mem. Soc. Hist. Nat. Paris, v. 246: 1834) 166
151.	*rufescens*	Valeton	1911	Nova Guinea 8:458
152.	*rufinervia*	Miq,	1857	Fl. Ned. Ind. 2: 212
153.	*saigonensis*	Pierre ex Pit.	1923	Fl. Indo-Chine [P.H. Lecomte et al.] 3: 186
154.	*samana*	Jayaw.	1963	J. Arnold Arbor. 44: 259
155.	*sandakana*	Govaerts	2008	J. Arnold Arbor. 157(1): 120
156.	*sanderiana*	Ridl.	1909	Gard. Chron. ser. 3, 46: 34
157.	*scandens*	Elmer	1911	Leafl. Philipp. Bot. iii. 992
158.	*scratchleyi*	Wernham	1918	J. Bot. 56: 70
159.	*sericea*	Blume	1826	Bijdr. Fl. Ned. Ind. 16: 986
160.	*sessilifolia*	Hutch.	1916	Pl. Wilson. (Sargent) 3(2): 397
161.	*setosa*	Merr.	1915	Philipp. J. Sci., C 10: 104
162.	*shikokiana*	Makino	1904	Bot. Mag. (Tokyo) 18: 44
163.	*simpliciloba*	Hand.-Mazz.	1925	Anz. Akad. Wiss. Wien, Math.-Naturwiss. Kl. 1925, lxii. 147
164.	*soyauxii*	Büttner	1889	Verh. Bot. Vereins Prov. Brandenburg xxxi (1889): 81
165.	*spectabilis*	Ridl.	1918	J. Straits Branch Roy. Asiat. Soc. 79: 78

(continued)

Table 1 (continued)

	Specific Epithet	Author	Year	Publication
166.	*squiresii*	Merr.	1938	J. Arnold Arbor. 19: 68
167.	*subsessilis*	Pierre ex Pit.	1923	Fl. Indo-Chine [P.H. Lecomte et al.] 3: 187
168.	*sutepensis*	Hosseus	1911	Repert. Spec. Nov. Regni Veg. 10: 62
169.	*tenuiflora*	Benth.	1849	Niger Fl. [W. J. Hooker]. 392
170.	*teysmanniana*	Miq.	1857	Fl. Ned. Ind. 2: 213
171.	*theifera*	Pierre ex Carr.	1883	Rev. Hort. [Paris]. 1883: 93
172.	*thorelii*	Pit.	1923	Fl. Indo-Chine [P.H. Lecomte et al.] 3: 188
173.	*tomentosa*	Wall. ex G.Don	1834	Gen. Hist. 3: 491
174.	*treutleri*	Stapf	1909	Bot. Mag. 35: t. 8254
175.	*tristigmatica*	Cummins	1898	Bull. Misc. Inform. Kew 1898(135): 74
176.	*ustii*	Alejandro	2008	Bot. J. Linn. Soc. 158(1): 88 (-91; fig.1)
177.	*utakwae*	Wernham	1916	Trans. Linn. Soc. London, Bot. 9(1): 71
178.	*variolosa*	Wall. ex G.Don	1834	Gen. Hist. 3: 490, descr.
179.	*vidalii*	Elmer	1911	Leafl. Philipp. Bot. iii. 993
180.	*villosa*	Wall. ex G.Don	1834	Gen. Hist. 3: 489, descr.
181.	*viridiflora*	Alejandro	2008	Bot. J. Linn. Soc. 158(1): 90 (-91; fig. 2)
182.	*wallichii*	G.Don	1834	Gen. Hist. 3: 490
183.	*whitei*	S.Moore	1922	Proc. Roy. Soc. Queensland 34: 54
184.	*wrayi*	King	1904	J. Asiat. Soc. Bengal, Pt. 2, Nat. Hist. 72(4): 182
185.	*zenkeri*	Wernham	1913	J. Bot. 51: 276
186.	*zollingeriana*	Klotzsch	1853	Ber. Verh. Berl. Acad. (1853) 500

wherein the main veins emerging from the midrib at regular intervals, at the margin turning toward the apex and looping to join the next vein upward.

Stipules – interpetiolar, usually triangular or ovate, sometimes basally fused forming a ring around the node, persistent or deciduous, apices bifid 1/8 to 3/4 of their length, the lobes erect and rarely spreading away from each other, abaxially with indument, adaxially with indument all over or only at the base and/or apex.

Colleters – few or numerous, in continuous rows and/or in groups of two at the base of the stipule. Calyx colleters are small and usually found between sinuses of the calyx lobes or at the base of the calyx lobe typically solitary or in groups of two to five. The number of colleter is not a reliable character as it varies even among individual lobes of the same flower.

Inflorescence - typically terminal but may be displaced to pseudoaxillary which are cymose corymbs, congested or spreading, with few to many flowers, or rarely just one flower. The inflorescence axes are characterized by a regular, sympodial-dichasial branching of their ultimate elements and may vary from being densely covered with trichomes to sparsely so, or almost glabrous. Calycophyll, the enlarged, showy calyx lobe that make the inflorescence attractive, may be present as one or five to a flower or none at all. Its shape could be ovate to elliptic, or orbicular varying in sizes and colors, but typically white or yellowish white for

Fig. 1 *Mussaenda philippica* plants in their natural habitat (Photography by Mr. Benjamin Mabanta)

majority of the species and usually having five longitudinal nerves. The calycophyll of some African species are yellow-orange or carmine-red. Small bracts are generally present on the inflorescence axes and flower pedicels. They are opposite in arrangement and could be linear or lanceolate to ovate in shape with acuminate or gradually pointed tips and normally caducous. Bracts are usually entire or three- or five-lobed with the lateral lobes always shorter than the others and vary in the degree of indumentum and size.

Fruit - indehiscent and berry-like, with pedicel and generally warty, ellipsoid to obovoid, or rarely globose, green when young, and brown to blackish at maturity. Length rarely exceeds 2 cm. Conspicuous annular disks typically crown the fruits, and calyces may persist.

Seed – usually small, numerous, more or less angular to rounded, laterally flattened, and brown to black. Size is usually less than 1 mm but could be more than 1 mm in some African species. Size and shape of seeds may vary even from a single fruit.

18.3 Origin, Domestication, and Spread

Alejandro et al. (2005) inferred from available ITS and trnT-F data that Asian *Mussaenda* species originated from Africa and that African ancestor must have reached Asia through a long-distance dispersion event and then major radiations seem to have occurred only after colonization of Asia. The widespread occurrence of *M. arcuata* from mainland Africa to Comoro Islands, Madagascar, and the Mascarenes is probably by stepping-stone dispersal.

With regard to domestication, it is believed that *M. erythrophylla*, commonly known as Ashanti blood, has been cultivated due to its attractive inflorescence, particularly the bright red calycophyll and dark yellow, star-shaped flower (Fig. 2). From West Africa, it was brought to cultivation in different countries including Singapore and the Philippines, where only male forms are grown (Price 1974).

Also from West Africa, *M. elegans* was first recorded to be cultivated in the Philippines by Price in 1974. It has flowers without calycophyll. However, a sport of *M. philippica* with all five calyx lobes expanded into large, showy white structure that last for a long time was discovered, isolated, and successfully multiplied. This specimen, originally published as *M. philippica* var. *aurorae* and brought to other countries as greenhouse specimen, has been proven to be a natural mutant by Rosario (2007). In the same manner, it can be deduced that *M. garrettii, M. macrophylla* var. *grandisepala*, and an herbarium material of *M. whitei* from Dutch New Guinea (Brass 11,682 [A]) were also natural mutants. Being known from the type collection only, no living specimens of the three species exist.

Researches leading to the development of a female form of *M. philippica* var. *aurorae* and its cross-pollination by *M. erythrophylla* produced an interspecific hybrid which inspired the production of more hybrids (F_1s), back-crosses, and

Fig. 2 *Mussaenda erythrophylla* as cultivated plants used for ornamental purposes (Photography by Mr. Fernando Aurigue)

selfed-progenies (F_2s). Selection of new traits resulted in the development of different cultivated varieties that are traded internationally. Table 2 provides a list of the different cultivars duly registered or patented. However, there are several other hybrids which have not been registered yet but are already being commercialized.

18.4 Plant Genetic Resources

(a) Geographic distribution – *Mussaenda* naturally occurs in tropical Africa, eastern and southeastern Asia (including southern Japan), Caroline Group of Islands, Solomon and Fiji Islands, New Guinea, and Australia.
(b) Primary gene pool – many botanic gardens around the world have living specimens of *Mussaenda* in their greenhouses or conservatories. However, these are composed of just a few popular species and readily available cultivars. Breeding institutions in Australia, India, the Philippines, and Thailand have collections to create new varieties or for research purposes. In the Philippines, the National Plant Genetic Resources Laboratory (NPGRL) at the Institute of Plant Breeding (IPB), University of the Philippines Los Baños (UPLB), serves as repository of *Mussaenda* cultivars with duplicates at the Institute of Crop Science also at UPLB.
(c) Wild genetic resources and others – records show that many species are narrow endemics. For example, 23 out of the 25 taxa recorded from the Philippines can be found only in certain habitat or island (Alejandro et al. 2016). In Thailand, 17 species are naturally occurring in the wild, and one species, *M. garrettii*, is endemic to the country (Chantaranothai 2015). A genebank of wild *Mussaenda* species is not known to exist, but some botanic or private gardens may have a few in their collection.

18.5 Collections

(a) Methods – in the Philippines, a Gratuitous Permit (GP) should be secured from the Department of Environment and Natural Resources (DENR) office responsible in the collection area. It is a permit issued to an individual or entity engaged in noncommercial scientific or educational undertaking to collect wildlife. With the Wildlife Act (Republic Act No. 9147), collection may be allowed as long as it is not detrimental to the survival of the species or subspecies involved and/or their habitat provided that appropriate and acceptable techniques with least or no detrimental effects to the existing wildlife populations and their habitat be required. Collection of the threatened species, including their by-products and derivatives, from the wild shall be permitted only for scientific or propagation purposes by individuals, business, research, educational, or scientific entities accredited by DENR-Biodiversity Management Bureau. However, if the habitat is under the jurisdiction of indigenous people or a protected area, Prior Informed Consent from the concerned indigenous people in accordance with Indigenous

Table 2 Cultivars of *Mussaenda* that are registered or patented as of 31 March 2020

	Cultivar name	Breeder	Year of naming	Registration number
1.	Doña Aurora (natural mutant) [Fig. 3]	UPLB, College, Laguna, Philippines	1938	NSIC 1999 Or 14
2.	Doña Trining (selected *M. erythrophylla* introduced)	UPLB, College, Laguna, Philippines	1946	NSIC 1999 Or 15
3.	Doña Alicia (*M. philippica* x *M.* 'Doña Trining')	UPLB, College, Laguna, Philippines	1952	NSIC 1999 Or 16
4.	Doña Luz (*M.* 'Doña Hilaria' x *M.* 'Doña Aurora')	UPLB, College, Laguna, Philippines	1958	NSIC 1999 Or 17
5.	Doña Esperanza (F_2 of [*M. philippica* x *M.* 'Doña Aurora'] x *M.* 'Doña Trining')	UPLB, College, Laguna, Philippines	1962	NSIC 1999 Or 18
7.	Doña Evangelina (F_2 of *M.* 'Doña Hilaria' x *M.* 'Doña Trining') [Fig. 4]	UPLB, College, Laguna, Philippines	1962	NSIC 1999 Or 19
8.	Queen Sirikit (*M.* 'Doña Hilaria' x *M.* 'Doña Aurora')	UPLB, College, Laguna, Philippines	1963	NSIC 1999 Or 24
9.	Gining Imelda (*M.* 'Doña Hilaria' x *M.* 'Doña Aurora')	UPLB, College, Laguna, Philippines	1967	NSIC 1999 Or 20
10.	Doña Hilaria (F_2 of [*M. philippica* x *M.* 'Doña Aurora'] x *M.* 'Doña Trining')	UPLB, College, Laguna, Philippines	1974	NSIC 1999 Or 21
11.	Doña Paciencia (F_2 of *M.* 'Doña Hilaria' x *M.* 'Doña Trining')	UPLB, College, Laguna, Philippines	1974	NSIC 1999 Or 22
12.	Maria Clara (*M.* 'Maria Makiling' x similar to *M.* 'Doña Luz')	UPLB, College, Laguna, Philippines	1982	NSIC 1999 Or 25
13.	Maria Makiling (F_2 of [*M. philippica* x *M.* 'Doña Aurora'] x *M. philippica*)	UPLB, College, Laguna, Philippines	1982	NSIC 1999 Or 26
14.	Diwata (F_2 of [*M. philippica* x *M.* 'Doña Aurora'] x *M. philippica*)	UPLB, College, Laguna, Philippines	1982	NSIC 1999 Or 27
15.	Mutya (F_2 of [*M. philippica* x *M.* 'Doña Aurora'] x *M. philippica*)	UPLB, College, Laguna, Philippines	1982	NSIC 1999 Or 28
16.	Paraluman (F_2 of [*M. philippica* x *M.* 'Doña Aurora'] x *M.* 'Doña Trining')	UPLB, College, Laguna, Philippines	1982	NSIC 1999 Or 29
17.	Lakambini (F_2 of *M.* 'Doña Hilaria' x *M.* 'Doña Trining')	UPLB, College, Laguna, Philippines	1982	NSIC 1999 Or 30
18.	Diyosa (open-pollinated *M.* 'Queen Sirikit')	UPLB, College, Laguna, Philippines	1998	NSIC 1999 Or 31
19.	Doña Amelita (open-pollinated *M.* 'Queen Sirikit') [Fig. 5]	UPLB, College, Laguna, Philippines	1998	NSIC 1999 Or 23
20.	Corazon Aquino (*M.* 'Doña Evangelina' x *M.* 'Doña Trining')	UPLB, College, Laguna, Philippines	2001	NSIC 2001 Or 48
21.	Zenaida Umali ([*M. philippica* x *M.* 'Doña Aurora'] x *M.* 'Paraluman')	UPLB, College, Laguna, Philippines	2001	NSIC 2001 Or 49

(continued)

Table 2 (continued)

	Cultivar name	Breeder	Year of naming	Registration number
22.	Gloria Macapagal-Arroyo (*M.* 'Doña Evangelina' x *M.* 'Doña Aurora')	UPLB, College, Laguna, Philippines	2001	NSIC 2001 Or 50
23.	Capricorn Dream (*M.* 'Doña Evangelina' x *M. erythrophylla*)	Oram's Nurseries, Rockhampton, Queensland, Australia	2003	2003/021 PVJ 16(2)
24.	Capricorn Ice (*M.* un-named x *M.* 'Snow Queen')	Oram's Nurseries, Rockhampton, Queensland, Australia	2003	2003/108 PVJ 16(2)
25.	Emerlinda R. Roman (open-pollinated *M.* 'Doña Esperanza')	UPLB, College, Laguna, Philippines	2007	NSIC 2007 Or 63
26.	Clara L. Davide (*M.* 'Doña Esperanza' x *M.* 'Doña Aurora')	UPLB, College, Laguna, Philippines	2008	NSIC 2008 Or 68
27.	Marmalade (*M.* 'Doña Luz' x *Pseudomussaenda flava*) [Fig. 6]	Alipore, India	2009	USPP20221P3
28.	Teresita Lantin-Rosario (*M.* 'Maria Makiling' x *M.* 'Diwata')	UPLB, College, Laguna, Philippines	2016	NSIC 2016 Or 86

Fig. 3 *Mussaenda* 'Doña Aurora' as a natural mutant widely cultivated around the world (Photography by Mr. Fernando Aurigue)

Fig. 4 *Mussaenda* 'Doña Evangelina' is one of the most attractive interspecific hybrid developed in the Philippines (Photography by Mr. Fernando Aurigue)

Peoples Rights Act of 1997 (Republic Act No. 8371) or prior clearance from the concerned local government unit and the Protected Area Management Board must also be secured first, respectively. Other policies like the Expanded National Integrated Protected Areas System Act of 2018 (Republic Act No. 11038) and existing procedures of other relevant agencies, bodies, or institutions that exercise authority over the collection area should also be considered.

(b) Status of Collections (national, regional, and global with appropriate listing) – herbarium materials are well-represented and properly labeled and documented. Besides IPB and ICropS in UPLB, living collection of *Mussaenda* cultivars can be found in private gardens, while some commercial nurseries in the Philippines and other countries specialize in selling specific or selected cultivars. Unfortunately, the accessions are often mislabeled/misidentified. The NPGRL-IPB has passport data for each accession which is based on the catalogue of the National Seed Industry Council (NSIC), Bureau of Plant Industry (BPI), Department of Agriculture. There is no known active collection of *Mussaenda* species although several species may be found in botanic gardens.

(c) Gaps in collections, both geographical and genetic – as there is no active collection of *Mussaenda* species, representation of geographical and genetic

Fig. 5 *Mussaenda* 'Doña Amelita' is an open-pollinated hybrid of *M.* 'Queen Sirikit' (Photography by Mr. Fernando Aurigue)

variations within a species is wanting. Collection of a single species from various populations or in different areas as well as the variations within a population or area is governed by certain rules and regulations in the locality and the laws being implemented in the country as a whole.

18.6 Conservation

(a) Methods – the preservation, protection, and sustainable utilization of wild flora depend on the status of the species. Under the International Union for the Conservation of Nature (IUCN) Red List, three species of *Mussaenda* are of least concern (LC), but ex situ conservation is already being recommended. These species are *M. erythrophylla*, *M. mollissima*, and *M. philippica*. However, *M. philippica* is not mentioned in the Updated National List of Threatened Philippine Plants (DAO 2017). Instead, *M. acuminatissima*, *M. attenuifolia*, *M. chlorantha*, *M. grandifolia*, *M. lanata*, *M. magallanensis*, *M. milleri*, *M. nervosa*, *M. scandens*, *M. setosa*, and *M. vidalii* are classified under the vulnerable category (Vu), while *M. palawanensis* is included in the other threatened species (OTS) category.

Fig. 6 *Mussaenda* 'Marmalade' is a beautiful intergeneric hybrid developed in India (Photography by Mr. Fernando Aurigue)

Cultivars are propagated according to the need, and stock plants are usually maintained, so conservation of the materials does not appear to be a problem.

(b) Status of Plant Genetic Resources (general germplasm, base collection, active collection, breeder's collection, genetic stocks, pre-breeding material [including interspecific derivatives, etc.]) – collection of *Mussaenda* germplasm materials follows the needs or programs on a per country basis. Some genebanks or nurseries would maintain valuable or promising species for use as ornamental or medicinal plants. The NPGRL of the Philippines has genetic stocks of most of the different cultivars bred in the Philippines but none of the native species, except *M. philippica*.

(c) Gaps in available (useful) diversity – no gap is identified. The collection and actual use of different populations of a particular species or its samples from different areas are done only research purposes which require a Gratuitous Permit from the authority (e. g., DENR in the Philippines). Collection of materials from the wild for commercial use is unknown, but it would also require permission from more authorities. Harvesting of limited amount of materials from the wild for traditional use usually by indigenous people is allowed but has nothing to do with diversity.

18.7 Characterization and Evaluation

(a) Characterization for essential features and classification – the attractive calycophylls are the most important part of a *Mussaenda* plant, and it is characterized by its number, degree of enlargement, size (length x width), and color based on Royal Horticultural Society (RHS) Colour Chart. An essential classification in breeding is the functional reproductive part of the flowers: whether the plant can be used as a female parent, male parent, or both or neither a seed parent nor a pollen source (completely sterile).

18.7.1 Pigmentation

The orange flowers of *M. hirsutissima* contain aureusin, cernuoside, and the 4,6-diglucoside of aureusidin (Harborne et al. 1983). Tiambeng et al. (1969) determined that the anthocyanin pigments chrysanthemin and antirrinin and the anthocyanidin pigment cyanidin are present in the calycophylls of *Mussaenda erythrophylla* but none in *M.* 'Doña Aurora,' except for chlorophyll which is present in both upper and lower epidermal peels of the calycophyll. The hybrid cultivar, *M.* 'Doña Luz' contains 3.56 Klett units/100 mL HCl/g fresh weight of calycophylls. The red pigments of *M.* 'Doña Trining' and *M.* 'Lakambini' and the pink pigments of *M.* 'Doña Luz' and *M.* 'Doña Hilaria' are found distributed in the spongy parenchyma. Pigment particles are also present in vacuoles on the upper epidermal peels of the calycophylls of the said cultivars, being more concentrated for *M.* 'Doña Trining' and *M.* 'Lakambini.'

18.7.2 Cytology

Mussaenda erythrophylla and *M. philippica* have normal meiosis and chromosome counts show 11 bivalents. *M.* 'Doña Aurora,' the spontaneous mutant of *M. philippica*, is also a diploid, with a basic chromosome number $x = 11$. The hybrids also have 11 bivalents at diakinesis. Meanwhile, *M.* 'Doña Luz,' which is completely sterile, showed aberrations like bridges and laggards (Rosario, 1998). Over 30 taxa of *Mussaenda* from Africa and Asia have been studied for their chromosome numbers, and the basic count of $x = 11$ has been consistent (Table 3). No polyploid has been reported. Alejandro et al. (2016) counted $2n = 22$ from somatic metaphases on root tips of *M. philippica* var. *philippica*. The chromosomes measure about 1.15 μm in length and exhibit fairly uniform size.

(b) Development/identification of gene pools and core collections – there is so much to be desired for a germplasm collection, but it is obviously not a priority so funding for the development and maintenance of a *Mussaenda* gene bank seems elusive.
(c) Evaluation of genetic diversity for desired traits – studies on the genetic diversity of even a species that has larger geographical occurrence are wanting. The

Table 3 Chromosome number in *Mussaenda*

Species or cultivar	n	2n	Reference
arcuata	11	–	CCDB (2015)
chippie	11	–	CCDB (2015)
corymbosa	11	–	CCDB (2015)
elegans	11	–	CCDB (2015)
glabra	11	–	CCDB (2015)
incana	11	–	CCDB (2015)
laxa	11	–	CCDB (2015)
macrophylla	11	–	CCDB (2015)
maingayi	11	–	CCDB (2015)
philippica var. *philippica*	–	22	Alejandro et al. (2016)
pilosissima	11	–	CCDB (2015)
raiateensis	11	–	CCDB (2015)
roxburghii	11	–	CCDB (2015)
tristigmatica	11	–	CCDB (2015)

presence of an attractive calycophyll is the most desired trait, especially if there are five, fully enlarged calycophylls such as those in hybrids. Perhaps the other important characteristic that needs further evaluation is the presence of compounds with proven biological activities.

18.7.3 Genetics

Color of calycophyll – the cross between *M. philippica*, with white calycophyll, and *M. erythrophylla*, with red calycophyll, produced *M.* 'Doña Alicia,' with pink calycophyll. Similarly, the offsprings of *M.* 'Doña Aurora,' with white calycophylls, and *M.* 'Doña Trining,' with red calycophyll, have different shades of pink no matter what the number and size of the calycophylls are. Backcrossing one of the hybrids, *M.* 'Doña Hilaria,' with pink calycophyll, to the male parent, *M.* 'Doña Trining,' with red calycophyll, produced progenies with pink and red calycophyll. However, when *M.* 'Doña Hilaria' was crossed to *M.* 'Doña Aurora,' with white calycophylls, all the offsprings have pink calycophylls although the intensity and distribution pattern on the calycophyll varied. The results suggest that there is incomplete or partial dominance in inheritance of calycophyll color. *M. philippica* has recessive genes for white calycophyll, while *M. erythrophylla* has homozygous dominant genes for red calycophyll.

Number of calycophylls – *M.* 'Doña Alicia,' with one calycophyll, is the hybrid of *M. philippica* and *M. erythrophylla* both of which have only one calycophyll and considered homozygous. The second generation segregants from the cross between *M.* 'Doña Aurora,' with five calycophylls, and *M. philippica*, with only one calycophyll, are *M.* 'Maria Makiling' and *M.* 'Diwata,' both with five calycophylls of the same size, and *M.* 'Mutya,' also with five calycophylls but one or two are larger than the rest.

Meanwhile, the second-generation segregants from the cross between *M.* 'Doña Aurora,' with five calycophylls, and *M. erythrophylla*, with only one calycophyll, are *M.* 'Doña Hilaria,' *M.* 'Doña Esperanza,' and *M.* 'Paraluman' that have five calycophylls but one or two are larger than the others. The size of the calycophyll varies. The second-generation segregants from *M.* 'Doña Hilaria,' with five calycophylls of different sizes, backcrossed with *M.* 'Doña Trining,' with only one calycophyll, are *M.* 'Doña Evangelina' and *M.* 'Doña Pacencia,' with five calycophylls of similar size, and *M.* 'Lakambini,' with five calycophylls but one is larger than the rest.

On the other hand, the second-generation segregants from *M.* 'Doña Hilaria,' with five calycophylls of different sizes, backcrossed with *M.* 'Doña Aurora,' with five calycophylls of the same size, are *M.* 'Queen Sirikit,' *M.* 'Gining Imelda,' *M.* 'Doña Luz,' and *M.* 'Doña Leonila,' with five fully enlarged calycophylls. Likewise, hybrids between cultivars with five fully enlarged calycophylls such as *M.* 'Maria Makiling' x male segregant similar to *M.* 'Doña Luz' and *M.* 'Maria Makiling' x *M.* 'Diwata' also have five fully enlarged calycophylls, namely, *M.* 'Maria Clara' and *M.* 'Teresita Lantin-Rosario,' respectively. The number of calycophyll is quantitatively inherited. The hybrids can exhibit five calycophylls that are all fully enlarged or only one or two are fully enlarged.

Flower size, color, and shape – generally, cultivars with colored calycophylls have relatively larger flowers with darker-colored center which are apparently inherited from *M. erythrophylla* or *M.* 'Doña Trining.' Similarly, the flower shape has been transmitted to its offsprings like *M.* 'Lakambini,' *M.* 'Paraluman,' *M.* 'Doña Hilaria,' and *M.* 'Queen Sirikit.' However, the calycophyll color is not associated with the flower shape as the cultivars *M.* 'Doña Evangelina' and *M.* 'Lakambini' have completely different flower size and shape although both have red calycophylls. For progenies of parents with white calycophyll, whether only one or five and fully enlarged, the flower have shades of yellow at the center, unlike those of other cultivars. It appears that the color of hairs at the throat of flowers is associated with the calycophyll color as it is always observed in resulting offsprings.

Seed color – seeds of *M. philippica* are brown, while those of the hybrids between *M. philippica* and *M. erythrophylla* are black. This indicates that the seeds of *M. erythrophylla* are black, and the black color is dominant (Price 1974).

(d) Available sources of breeding value (listing of genetic resources available for various biotic and abiotic stresses and nutritional traits, other desirable traits, quantitative traits, particularly related with yield, etc.) – there is no known source of breeding value except for those *Mussaenda* species used in hybridization and whose desirable characteristics are known only to the breeders.
(e) Molecular characterization, identifying genomic resources, if any (?) such as molecular maps, molecular markers, tagged genes, marker associated maps, gene constructs, etc.

18.7.4 Isozyme Polymorphism

Isozyme polymorphism through electrophoresis is a useful tool to identify species, interspecific hybrids, and cultivars and to screen genetic variability in a population. Four enzyme systems have been used to analyze the isozymes in *Mussaenda* cultivars (Espada and Rosario 1992). Malic enzyme was found to be non-polymorphic. Two putative loci of esterase, designated as EST 1 and EST 2, with two alleles each (F and S) were found. A single zone of activity consisting of up to three bands was observed in acid phosphatase system, designated the ACP locus. For isocitrate dehydrogenase system, the zone of activity consisted of five bands and designated as the IDH locus. FM, MM, and SM have been assigned for the three banding genotypes observed in the IDH locus. Table 4 shows the characterization of some *Mussaenda* based on the esterase, acid phosphatase, and isocitrate dehydrogenase systems.

18.7.5 Microsatellite Markers

Microsatellite sequences are repetitive DNA sequences composed of several base pairs in length. They are composed of non-coding DNA and are not part of the genes but are useful as genetic markers. Microsatellites have proven to be efficient in species delimitation, phylogenetic reconstruction, and hybrid detection. For *M. pubescens*, 19 microsatellite markers were amplified successfully, and 17 primer pairs showed polymorphism for *M. pubescens*, *M. pubescens* var. *alba*, *M. esquirolii* (now *M. shikokiana*), *M. kwangtungensis*, *M. hirsutula*, and *Schizomussaenda henryi*. A maximum of eight alleles were detected per locus in 68 individuals at population level (Duan et al. 2012). However, these could not provide sufficient

Table 4 Genotypic designations of some *Mussaenda* based on four-locus combinations

Species/cultivar	(EST1/EST2)/ACP/IDH
philippica (male)	(–)/SS/MM
philippica (female)	(–)/SS/SM
Doña Aurora (male)	(FF/FF)/SS/FM
Doña Aurora (female)	(FF/FF)/SS/MM
Maria Makiling	(FF/FF)/SS/FM
Mutya	(FF/FF)/SS/FM
Diwata	(SS/FF)/FS/FM
Doña Alicia	(SS/SS)/FS/FM
Doña Hilaria	(FS/FF)/FS/FM
Doña Luz	(SS/FF)/FS/MM
Queen Sirikit	(SS/FF)/SS/MM
Gining Imelda	(FF/SS)/FS/MM
Doña Evangelina	(FS/SS)/SS/FM
Doña Esperanza	(SS/SS)/FS/SM
Paraluman	(FF/SS)/FS/SM

variation in population genetic structure analyses and in detecting interspecific hybridization in *Mussaenda* due to low polymorphisms. With the detection of 14 additional SSR loci that are highly polymorphic, these microsatellites are expected to provide additional tools for evaluating the intraspecific and interspecific population genetic structure and detecting the potential hybrids. Assessment of polymorphism was done using 169 individuals from three *M. pubescens* populations, two *M. pubescens* var. *alba* populations, and one population each of *M. caudatiloba*, *M. kwangtungensis*, and *M. hirsutula* (Duan and Zhang 2014).

18.8 Information Documentation

(a) Catalogues and databases – information are readily available and freely downloadable.
(b) Dissemination and exchange – distribution of information and exchange of knowledge about the plant or the breeding history of cultivated varieties are openly done.

18.9 Use of Plant Genetic Resources

(a) Major constraints in the crop production – besides the well-tested cultivars, only selected clones of a limited number of *Mussaenda* species are cultivated, and very few species have been used for varietal development.

18.9.1 Propagation

Mussaendas are propagated by cutting, marcotting or air layering, inarching or approach grafting, and grafting in the Philippines. Semi-hardwood cuttings about 15 cm long and 1–1.5 cm in diameter are used as propagules in India. Gopalaswamiengar (1991) wrote that *Mussaenda* is propagated by layering or cuttings. In Bangladesh, submerging the basal end of 20 cm stem cuttings in 0.3% NAA solution for 5 min significantly increased production of shoot buds and roots in *M. philippica* (pink flag bush), which is actually *M.* 'Doña Luz,' and *M. erythrophylla* (Rashid and Nahar 2019).

Protacio et al. (1999) reported that two-node cuttings of the cultivars 'Doña Luz,' 'Queen Sirikit,' 'Paraluman,' 'Doña Hilaria,' and 'Doña Aurora' do not require auxin application to form roots as long as cuttings were taken from current growth with green, hard, or partially matured stem. However, for similar cuttings of 'Mutya,' 'Gining Imelda,' 'Doña Paciencia,' 'Lakambini,' 'Diwata,' and 'Doña Evangelina,' a 15–30 min dip in 100 ppm indolebutyric acid (IBA) prior to sticking to the rooting bed is necessary to enhance and accelerate rooting. For 'Lakambini,' IBA combined with either dopamine or paclobutrazol enhanced rooting of two-node cuttings better than IBA, paclobutrazol, and dopamine applied alone (Protacio et al. 1998).

Commercial plant propagators use a mist system. Water is sprayed over the cuttings in the rooting beds or raised flatforms (benches or frames) until the cuttings form roots. Nozzles that break water into a fine spray of mist droplets measuring 50–100 m at 40 psi are ideal. The distance could be 1–1.25 m or as long as the mist completely covers or it is evenly distributed to the cuttings. The mist could be continuous or intermittent to maintain a film of water on the leaves during the day. The latter would require an installation of cyclic timer switches usually set at 3–15 s misting every 3–10 min.

To mass propagate by **cutting using a mist system**, the procedure provided by Protacio et al. (2000) is recommended as listed below (with editing):

1. Collect healthy cuttings early in the morning when stems are turgid. Cut diagonally just below the node using a clean (preferably surface sterilized) cutting instrument. Choose tip cuttings that are 3.5–5 inches long (comprised of 3–5 nodes) from partially mature stems. Newly emerged, succulent tip cuttings should not be collected; the stem should be approximately 1 month old from shoot emergence after light pruning.
2. Place the cuttings inside polyethylene plastic bag with small amount of water immediately after harvesting to prevent wilting.
3. Remove 2–4 older leaves. To allow closer spacing in the rooting bed, large leaves should be cut crosswise to reduce its size to half.
4. Treat the cuttings with auxin (either powder form containing 0.1% indole-butyric acid (IBA) or naphthalene-acetic acid (NAA) or 100 ppm IBA or NAA solution) to enhance and accelerate rooting by 1–2 weeks and to have uniform rooting.
5. Water the rooting medium thoroughly before sticking the cuttings. Insert the base of the cuttings to a depth of about 1.5–2.0 cm with a minimum of 10 cm × 10 cm spacing. A 4 m × 1 m bed can accommodate 400 cuttings per batch. An alternative is to directly stick a cutting in a No. 2 pot or seedling bag with a mixture of equal parts river sand and coconut coir dust as medium. The containers are placed on raised benches or frames under a mist system.
6. After 2 weeks in the mist, when roots have formed, a mild dose of complete fertilizer (14–14-14) at 1/4 teaspoon per gallon or 4 liters of water can be applied as foliar spray to enhance both root and shoot development. Gradually decrease the duration of the mist period to harden the cuttings for transplanting.
7. Three to 4 weeks after sticking into the rooting medium, when the length of the longest root is 2–3 cm, transplant the rooted cuttings into pots or seedling bags. Directly sticking the cuttings into individual containers with the growing medium prevents disturbance of the roots and transplanting shock.

In the absence of a mist system, mass propagation by **shoot tip cuttings** enclosed in large polypropelene plastic bags can be done. The steps prepared by Pimentel et al. (1999) are given below (with editing):

1. Collect healthy shoot tip cuttings from new flushes of stock plants. Choose shoot tips with 3–5 nodes. The shot tip cutting should be about 4–5 inches in length.

2. Remove the older leaves leaving only one fully expanded pair. Trim the retained leaves in half by cutting crosswise.
3. Drench the rooting medium with water or fungicide solution to disinfect fungal contaminants prior to sticking of the shoot tip cuttings. The rooting medium may be composed of any of the following:

(a) Pure coir dust
(b) Pure river sand
(c) Decomposed rice hull and garden soil (1:1)
(d) Coir dust and garden soil (1:1)
(e) Coir dust, sand, and decomposed rice hull (1:1/2:1/2)
(f) Coir dust and sand (1:1/2)
(g) Allow excess water to drain off

4. It is best to soak the shoot tip cuttings in fungicide solution for 5–10 minutes. Insert the base of the cuttings to a depth of 1.0–1.5 cm for faster/easier rooting. The number of cuttings to be inserted depends on the size of the container of the rooting medium. For a size 6 clay or plastic pot, 15 shoot tip cuttings can be accommodated. For a 5 in x 5 in x 7 in black plastic, 20–25 shoot tip cuttings may be inserted.
5. Pierce the center of the rooting medium with a bent wire (to be immersed also in fungicide solution before using) to prop the polypropelene plastic bag to be used as enclosure.
6. Enclose the entire system (cuttings, container with rooting medium and wire) with the polypropelene plastic and seal with a rubber band or a string.
7. Place the system setup in a partially shaded area.
8. Open the plastic bag at least once a week to:

(a) Aerate the cuttings.
(b) Remove any fallen leaf.
(c) Check if roots have started to develop by lightly pulling the cutting. If it resists, then roots may have already formed. If it is pulled up easily, check the base to see if roots have formed or if there is white mass formation which means that in another week, there will be roots formed already. Another indicator that the cutting has rooted is the production of new set of leaves.

9. Allow the cuttings to develop profuse roots in 1 or 2 more weeks.
10. After 2–3 weeks, loosen the rooting medium and separate rooted cuttings one by one by soaking in water.
11. Immediately transplant the rooted cuttings into individual pots or seedling bags with appropriate potting medium. Water the new plants and harden them in a partially shaded area.

In **marcotting or air layering**, the stem is encouraged to form roots while still intact. The process involves girdling or removal of the bark measuring 1.5–2.5 cm in width and scraping the phloem and cambium layers at the exposed part, on the aerial part of the plant. Hormones may be applied to promote faster root formation. The girdled part is covered with a ball moist sphagnum moss or coir dust and wrapped in polyethylene plastic sheet. Both ends are firmly tied with string. To prevent the girdled stem from breaking, especially during heavy rains and strong winds, the upper portion may be pruned to make it lighter. When a profuse root system has formed, as can be seen through the plastic film, the marcot is cut just below the girdled part of the stem. The plastic wrapper is untied and carefully removed, and then the marcot is planted immediately.

In **inarching or approach grafting**, two separate plants are brought together: one, which is in a smaller container, to serve as the rootstock and the other, which is difficult to propagate by cuttings, as the scion source. The stems to be joined together should be more or less of the same diameter. Rosario (1998) provided the steps in inarching (with editing).

1. Cut off a slice of bark and wood about 5.0–7.5 cm long from the stems of both the scion and the stock. The cut portion must be almost identical in size, smooth, and nearly flat as possible to ensure close contact of the cambium layers of the two stems.
2. Hold tightly the severed portions embracing each other with plastic straw or string.
3. When the cut parts became united, usually 2–3 months after inarching, gradually severe the stock above and the scion below the union.
4. Remove any shoot that will grow below the union area.

To do **cleft grafting** for cultivars that are difficult to propagate, Rosario (1998) provided the following steps below (with editing):

1. Cut off the shoot and make a 2–3 cm slit at the center of the cut end of the plant that will serve as root stock.
2. Harvest a scion stick from the plant to be propagated with the same diameter as the stem of the decapitated stock. Cut on both sides the base of the scion to the shape of a wedge.
3. Carefully insert the wedge-shaped scion into the slit made on the stock. Tightly tie the grafted area with a strip of plastic continuously from the base to the tip of the scion and covered with an appropriate plastic slip to prevent drying up of the cut ends of both the scion and stock.
4. Remove the plastic slip and gradually untie the plastic strip starting from the tip of the scion when union took place as indicated by the tip of the scion turning green and sprouting of new shoot. Completely remove the plastic strip when grafting is successful.

Budding is done for hard-to-propagate cultivars with very limited planting materials available for harvesting. Either patch budding method or T-budding method may be employed. The procedure below is also provided by Rosario (1998) with some editing:

1. Make a vertical cut about 1 inch long through the bark on the stem of a plant that will serve as root stock. Then make a horizontal cut about 1/3 of the stem diameter on top of the vertical cut to form the T-shape.
2. On the plant to be propagated, make a slicing cut through the bark 1/2 inch under and 1/2 inch beyond a bud. Harvest the bud piece by making a horizontal cut through the bark and into the wood such that the bud is at the center of a shield-shaped piece.
3. Insert the bud piece into the stock by pushing the lower part of the shield downward the T-shaped cut under the two flaps of bark. The horizontal cut on the stock should fit evenly on the stock.
4. Carefully wrap the bud piece on the stock by tightly tying with plastic straw or strip. The union is successful when the bud becomes active by turning green and enlarging to develop into shoot.

Tissue culture techniques have been tried to propagate selected *Mussaenda* cultivars. Table 5 summarizes the results on selected cultivars using different treatments. The procedure below was adopted from Cramer and Bridgen (1997b) with some modifications:

1. Recently matured, healthy, and turgid leaves are harvested from stock plants that are disease-free, not stressed from nutritional deficiencies, toxicities, or water excesses or deficiencies.
2. Excised leaves are surface sterilized with 2% commercial bleach solution (5.25% sodium hypochlorite) and 0.2% Tween 20 for 5 min with constant agitation. Rinse with several changes of sterile deionized water.
3. Cut out leaf midrib segments measuring 3–5 m long and inoculate into callus medium.

   Incubate at a 25-29 °C with cool white fluorescent lights (210–480 μmol/m^2/sec) at 16–24 hours for callus formation and embryogenesis.
4. After 2 weeks, when white, light green, and green calli have developed, subculture to fresh media every 4–5 weeks to prevent browning. Actively growing calli are divided with every subculture.
5. For shoot proliferation, incubate cultures at a 18-22 °C with cool white fluorescent lights (54 μmol/m^2/sec) at 16 hours.
6. Shoots are rooted in moist soilless potting mix in the growth chamber with high humidity condition.
7. Slowly acclimatize plantlets to normal greenhouse conditions.

In vitro germination of seeds from mature but unripe fruits of open-pollinated *M.* 'Queen Sirikit' produced new varieties (Rosario et al. 1990).

Propagation **by seeds** is done for hybrid production. In nature, seeds are disseminated by birds that eat the ripe fruits of several *Mussaenda* species as observed by Alejandro et al. (2016) in the Philippines.

Natural hybrids have been documented and studied by Chen et al. (2014). Possible pollinators for *M. shikokiana* are the most common floral visitors, namely,

Table 5 Summary of tissue culture of different *Mussaenda* species and cultivars

Species/cultivar	Explant/ material used	Tissue culture medium	Result	Reference
erythrophylla 'Scarlet'	Axillary buds	MS basal medium + 40 mg/L adenine sulfate + 2.25 mg/L BA	Shoots	Maity et al. (2001)
	In vitro derived shoots	1/2 MS + 0.5 mg/L IBA + 800 mg/L thiamin-HCl	Plantlets	
erythrophylla 'Rosea' → Doña Luz	Young leaf explants	MS + 0.5–1.0 mg/l BA + 2.0–3.0 mg/l IAA + 10 mg/l ascorbic acid	Embryogenic calli	Das (2010)
	Somatic embryos	1/2 MS + 0.25–0.5 mg/l BA + 0.1 mg/l GA_3 + 5.0 mg/l adenine sulfate + 2% sucrose	Plantlets	
Doña Luz	Shoot tips (5 mm)	MS basal medium + 0.6 mM myo-inositol + 1.2 μM thiamine-HCl + 10–20 μM BA + 87.7 mM sucrose + 7 g agar	Axillary and adventitious shoots after 6 weeks	Cramer and Bridgen (1997a)
Queen Sirikit	Midrib sections	MS basal medium + 5–10 μM BA + 5–20 μM IAA + 87.7 mM sucrose + 5 g agar	Calli after 2 weeks	Cramer and Bridgen (1997a)
	Calli	MS basal medium + 5–10 μM BA + 87.7 mM sucrose + 5 g agar	Somatic embryos after 8 weeks	

butterflies like swallowtails (Papilionidae), satyrids (Satyridae), skippers (Hesperiidae), and pierid butterflies (Pieridae); bumblebees; and hawkmoths. Except for swallowtails and hawkmoths, the same kinds of insects visit the flowers of *M. pubescens* var. *alba* and could possibly cause cross-pollination where the two species occur together and flowering time overlap. Borges et al. (2003) reported that *M. frondosa* is pollinated by the birdwing butterfly *Troides minos*.

18.9.2 Cultural Requirements

Medium – *Mussaenda* grows in a wide range of soil, whether containerized or directly planted on the ground as long as it is well-drained especially for the hybrids. For pot plant production, a potting mixture consisted of compost or coir dust, garden soil, and river sand in equal proportions is recommended.

Nutrients – complete inorganic fertilizer helps young plants grow vigorously when provided every month. The fertilizer in liquid form is usually applied as drench. Older plants, particularly those planted on the ground, may be fertilized in dry form just once a year, usually at the onset of the rainy season or when new buds appear. Spraying with foliar fertilizer or application of controlled released fertilizer granules and organic matter will promote better growth and flowering.

Moisture – best growth is attained when humidity is high and there is plenty of water for irrigation. Waterlogging or too much water in poorly drained media is detrimental to the plant, especially during rainy season. Nevertheless, flowering is promoted when there is plenty of rainfall. On the other hand, insufficient moisture in potting medium or irregular watering of containerized plants will cause defoliation, resulting in leggy plants, or premature dormancy or inactive growth.

Light – full and direct sunlight is required for optimum growth of plants and production of attractive calycophylls. Under filtered sunlight or partial shade, growth is lanky, and the color of calycophylls is not as intense.

Rest period and pruning – *Mussaenda* plants normally defoliate and become inactive at the start of the dry season or when the temperature becomes colder and the daylength shorter. Growth of internodes stops, shoot production is prevented, and development of inflorescence becomes scarce. The calycophylls wilts and turn brown or drop off. During the period of dormancy, pruning or judicious removal of the branches and cutting of the main stem is done to maintain the desired height and proportional size of the plant, especially those in large containers. Pruning should be done every year for more vigorous growth and better inflorescence production. However, even without pruning, the dormant shoots will be activated, and active vegetative growth followed by inflorescence production will take place sooner or later.

Temperature – as tropical plants, mussaendas grow best in warm condition. However, its interaction with light and moisture has not been studied with regard to growth and dormancy.

18.9.3 Paclobutrazol Treatment

Paclobutrazol is a plant growth retardant that has been tested for potted flowering plant production as drench for the potting medium. In *M.* 'Doña Luz,' drenching with 250–500 mg/L paclobutrazol solution arrested growth and even resulted in death of the plant. If there was growth, leaves formed are small, crinkled, or distorted and eventually fall off. For *M.* 'Lakambini,' a concentration of 20 mg/L has been found to be satisfactory, while for *M.* 'Mutya,' 25 mg/L resulted in the production of more curly and compact inflorescences (Protacio et al. 1997).

18.9.4 Pest and Diseases

(a) The most common insect pests attacking mussaenda plants are aphids and mealybugs. Various kinds of caterpillars, such as caseworms or bagworms, cutworms (larvae of tussock moth) and leafrollers, scale insects, and termites can also damage plant parts if not the whole plant.

Meanwhile, *Cercospora* leaf spot and sooty mold are important diseases of mussaendas. A possible threat is the virus as the plants are vegetatively propagated.

(b) Common sources used to overcome production constraints [listing of genetic resources (landraces, released cultivars/varieties), genetic stocks (including aneuploids series, substitution and translocation lines, recombinant inbred lines, etc.), inbred lines, released cultivars associated with desired traits, genes with gene symbols, mapping populations, etc.] – source of planting materials are usually the nurseries accredited by the government or recognized by the breeder or breeding institution.

(c) Breeding options – interspecific hybridization has been done in the Philippines for a very limited number of species. Crossing between hybrids or cultivars, back-crossing, sibbing, and selfing have been done successfully. In India, the cross between *Mussaenda* and *Pseudomussaenda* resulted in an intergeneric hybrid, called *M.* 'Calcutta Sunset' as the pollen source used to be known as *M. flava*. The procedure for artificial pollination has been tested and the problems often encountered were heterostyly, dioecy, pollen sterility, incompatibility, and low viability of resulting seeds or embryos. Nevertheless, embryo rescue has been tried (Rosario and Aurigue 2006).

Heterostyly – refers to the unequal length of style in flowers of different plants of the same species. The condition is related to the functional parts of the flowers. Alejandro et al. (2016) stated that all populations of Philippine *Mussaenda* species could be classified into short-styled and long-styled flower morphs.

Dioecy – refers to the presence of staminate flowers and pistillate flowers on different individual plants. The former, belonging to the short-styled flower (S-) morph, has functional stamens, producing and dehiscing pollen grains, but its gynoecium ceases to grow and remains sterile. The latter, of the long-styled flower (L-) morph, has sterile anthers but develops fruits when pollinated. Li et al. (2010) used the term cryptic dioecy for such conditions. Based on their studies of floral organogenesis in the L- and S-morphs of *M. pubescens*, male sterility in the L-morph results from earlier degradation of the tapetum, while female sterility of the S-morph appears to be mainly due to the failure of functional megaspore establishment. The condition is applicable only to the species because the cultivars or hybrids could be staminate, pistillate, or both (hermaphroditic condition). Problem occurs when only staminate or pistillate plants are in bloom or available at hand. This situation necessitates prior collection and storage of pollen to be used in hybridization. For hermaphrodite plants, emasculation or removal of the anthers before flower opening is necessary for artificial pollination.

Pollen Sterility – wild species that produce pollen have not been studied to determine their pollen fertility. The pollen grains are usually tetracolporate, but occasionally tricolporate in Philippine species. The size ranges from 10.8 μm to 19.8 μm. For cultivated varieties or hybrids, those that shed pollen are not necessarily 100% fertile. Not all short-styled flower morphs could be used as pollen source for hybridization because the pollen grains have low pollen fertility if not completely sterile. For example, the clone of *M.* 'Doña Luz' in cultivation or being propagated does not produce any pollen.

Incompatibility – germination of pollen grains of *M. philippica*, *M.* 'Doña Aurora,' and *M.* 'Doña Alicia' in Kwack's medium is normal, but when extracts from styles of *M.* 'Doña Evangelina,' *M.* 'Doña Hilaria',' and *M.* 'Doña Esperanza' are incorporated in Kwack's medium, there is a marked decrease in percentage of pollen germination, pollen tube elongation is inhibited, and pollen tubes become distorted or bursted. Similar observations were noted in other cultivars. The results indicate the presence of a factor in the style that inhibits pollen germination and pollen tube growth. This unknown stylar factor affects cross-compatibility and is one reason for the very low percentage of fruit setting for crosses made among hybrids. However, the identity and quantity of the inhibitory substance(s) present in the pistil of *Mussaenda* have not been determined.

Seed Viability – no studies on seed technology for *Mussaenda* is available. It is a common practice to harvest the fruits when already ripe and to process the seeds as soon as possible. Seeds are sown immediately as it may be impractical to store them. The viability of seeds, or its ability to germinate, could be affected by maturity and length of storage as well as storage conditions. After sowing, environmental factors, particularly moisture level of the germination medium and temperature, could affect germination rate and percentage.

(d) Present status of use or incorporation of desired traits – selected species and many cultivars developed by breeding are valuable as ornamental plants used mainly for landscaping. However, several species are also used in ethnic culture and traditional medicine as listed in Table 6.

Phytochemicals identified from different extracts from various parts of the plant of several species and their known biological activities are presented in Table 7.

It is interesting to note that shoots and flowers of *M. glabra* are occasionally eaten as vegetable, raw or steamed, in Indonesia (Lemmens 2003). Meanwhile, leaves (possibly dried) of *M. magallanensis* and *M. philippica* are used as a substitute for tobacco by some indigenous people in the Philippines. Moreover, the whole plant of *M. philippica* is used for agricultural rites in Mindanao, Philippines (Jayaweera 1964).

18.10 Looking Forward or Future Perspective

One of the objectives for future cultivars is to develop dwarf hybrids or mutants suitable as flowering pot plants. An ideal cultivar can be mass produced and induced to flower in due time and are tolerant to general shipping conditions such as fluctuating temperatures, darkness, and high ethylene levels.

Lemmens (2003) noted the potential of mussaendas as medicinal crop and promising results of pharmacological studies indicate the need for further research. Additionally, taxonomic studies for all published species are needed to clarify or update status of many taxa. DNA analysis or phylogenetic studies would be useful for this endeavor.

Table 6 Varied uses of some preparations of the different parts of certain species in ethnic culture and traditional medicine

Species	Part used	Preparation	Use	Reference
anisophylla	Fresh leaves	In the Philippines – decoction	Asthma	Jayaweera (1964)
elmeri (Sarawak)	Leaves	Boiled and taken orally	Toothache, headache and diabetes	Lemmens (2003)
ferruginea	Leaves	Papua New Guinea: decoction applied in a bath	Fever	Lemmens (2003)
		External application	Headache	
frondosa	Roots	Unspecified	India – white leprosy	Jayaweera (1963)
	Petaloid sepal	Unspecified	India – jaundice	
	Flowers	Unspecified	India – asthma, intermittent fever and dropsy	
	Leaves	External use as detergent	India – ulcers	
	Leaves and flowers	External use	Sri Lanka – inflammations	
	Leaves	Unspecified	Jaundice, asthma, hyperacidity, fever, cough, ulcer, leprosy, diuretic, wounds, and swells	Gopalakrishnan and Vadivel (2011)
		In India – taken along with milk	Jaundice	Sambrekar et al. (2010)
glabra	Sap	In Java – wash	Infected eye	Lemmens (2003)
	Flowers or leaves in combination with leaves of other medicinal plants	In Sumatra – infusion taken orally	Jaundice and headache	Lemmens (2003)
	Leaves	In Peninsular Malaysia – tea taken orally	Cough	Lemmens (2003)
		In Peninsular Malaysia – poultice	Headache	Lemmens (2003)
	Root	In Peninsular Malaysia – decoction	Cough, childbirth	Lemmens (2003)
laxiflora	Whole plant	Decoction and alcohol steeping	Numbness of limbs, injuries from falls,	Long and Li (2004)

(continued)

Table 6 (continued)

Species	Part used	Preparation	Use	Reference
			rheumatoid arthritis, arthritis, hemiplegia, and bellyache	
philippica	Roots and leaves	Decoction	Chest and lung infection	Lemmens (2003)
	Roots and enlarged calyx lobes	Unspecified	Jaundice	Lemmens (2003)
	Leaves	Decoction	Emollient	Lemmens (2003)
	Unspecified	Unspecified	Snakebites and dysentery	Lemmens (2003)
	Bark	In Mindoro, Philippines – juice	Headache	Jayaweera (1964)
	Plant	In Fiji – unspecified	Stomachache	
pubescens				
	Flowers	In Vietnam – decoction	Cough, asthma, and intermittent fever	Lemmens (2003)
		Chinese folk medicine	Common cold, laryngopharyngitis, acute gastroenteritis, and diarrhea	Zhao et al. (1994)
		Fujian Province folk medicine	Contraceptive	
	Leaves and roots	Decoction as analgesic	Rheumatism	Lemmens (2003)
raiateensis	Leaves	Fresh paste applied topically	Cuts and wounds	Kichu et al. (2015)
vidalii	Leaves	In the Philippines – soaked in water for a few minutes and solution is used as eye wash	Sore eyes	Lemmens (2003) and Jayaweera (1964)
		Macerated then applied externally	Headache and drunkenness	
Mussaenda spp. (Africa)	Leaf and bark	Sap	Eye infection	Lemmens (2003)
	Leaves and roots	Poultice	Wounds and sores	
		Unspecified	Elephantiasis	

Table 7 Phytochemical contents of some *Mussaenda* species and their biological activities

Species/ cultivar	Method of extraction	Compound	Biological activity	Reference
erythrophylla	Acetone-water and ethanol-water extract of leaves	Steroid, terpenoid, alkaloid, and flavonoid	In vitro anti-arthritic activity	Palit et al. (2013)
	Ethyl acetate extract of the stems	Steroids, triterpenoids, and flavonoids	Superoxide radical scavenging activity, hydroxyl radical scavenging activity, DPPH radical scavenging activity, inhibition of in vitro lipid peroxidation	Eswaraiah and Satyanarayana (2010)
	Methanolic extract of the stems	Glycosides, tannins, and saponins		
frondosa	Ethanol extract of whole plant	(−)-Quinic acid, 4-((1E)-3-hydroxy-1-propenyl)-2-methoxyphenol, napthalene, decahydro-2-methoxy-, and 1, 2, 3-benzenetriol, among others	Antiseptic, antioxidant, antidermatitic, fungicide, insecticide, antitumor, etc.	Gopalakrishnan and Vadivel (2011)
	Alcohol extract of leaves	Flavonoids, steroids, glycosides, saponin, resins, and mucilage	Antibacterial activity against *Escherichia coli* and *Pseudomonas aeruginosa*	Patil et al. (2011)
	Alcoholic extract	Flavonoids, sterols, proteins, and glycosides	Hepatoprotective activity against paracetamol-induced hepaticinjury in Wistar rats	Sambrekar et al. (2010)
	Aqueous extract	Flavonoids, proteins, and glycosides		
	Ethanolic extract of roots	Flavonoids, phenolic compound, tannins, and anthocyanins	Free radical scavenging and antistress activity	Koul and Chaudhary (2011)
glabra	Methanolic extract of leaves and stems	None stated	Slight in vitro antimicrobial activity against *Staphylococcus aureus*	Lemmens 2003
hainanensis	Ethanol extract of the plant	(1) pinostrobin; (2) quercetin; (3) 8-O-acetyl-shanzhiside methyl ester; (4) 8[E]-N-[2′-hydroxyl-tetracosan-cosoyl]-1-O-*β*-D-glucopyranosyl-8-en-octadecasphingenine;	DPPH radical scavenging ability of 1–3, 5–7, 9–12	Li et al. (2011)

(continued)

Table 7 (continued)

Species/ cultivar	Method of extraction	Compound	Biological activity	Reference
		(5) caffeic acid; (6) quercetin-3-O-β-D-glucoside (7) quercetin-7-O-β-D-glucoside; (8) 3-O-β-D-gluco-pyranosyl pomolic acid 28-O-β-D-glucopyranosyl ester; (9) 3,4-di-*o*-caffeoylquinic acid; (10) chlorogenic acid; (11) lamalbid; and (12) shanzhiside methyl ester		
macrophylla	Crude extract of root bark	(**1**) 3-*O*-β-d-glucopyranosyl-28-*O*-α-l-rhamnopyranosyl-16α-hydroxy-23-deoxyprotobassic acid, (**2**) 28-*O*-β-d-glucopyranosyl-16α-hydroxy-23-deoxyprotobassic acid, (**3**) 3-*O*-β-d-glucopyranosyl-28-*O*-α-l-rhamnopyranosyl-16α-hydroxyprotobassic acid, (**4**) 3-*O*-{[β-d-glucopyranosyl-(1 → 6)]-*O*-α-l-rhamnopyranosyl-(1 → 2)-*O*-β-d-glucopyranosyl-(1 → 2)}-*O*-β-d-glucopyranosyl-(1 → 3)-*O*-β-d-glucopyranosyl-cycloarta-22,24-dien-27-oic acid (mussaendoside W), (**5**) [3-*O*-acetyloleanolic acid, (**6**) 3-*O*-acetyldaturadiol (**7**) rotundic acid, and (**8**)16α-hydroxyprotobassic acid	Inhibitory activity of **1**–**6** against the periodontopathic bacterium *Porphyromonas gingivalis* **7** has cytotoxic activity	Kim et al. (1999)
	Methanolic extract of leaves	Phenolic compounds from dichloromethane soluble fraction	Moderate antioxidant activity	Islam et al. (2012)

philippica	Aqueous extract of leaves	Tannins, triterpenoids, saponins, flavonoids, phenols, and glycosides	Antiulcer activity in experimental rat models	Jena et al. (2019)
pubescens	Unspecified	Mussaendosides O, P, and Q	None stated	Zhao et al. (1994)
Queen Sirikit	70% methanolic extract of fresh flowers	Flavonol glycosides	Cytotoxic activity against fibroblast cultured from skin	Vidyalakshmi et al. (2007)
roxburghii	Chloroform soluble fraction from methanolic extract	Shanzhiol	Moderate antimicrobial activity against *Bacillus megaterium*	Islam et al. (2013)
	Petroleum-ether and carbon tetrachloride soluble fraction from methanolic extract	None stated	Strong cytotoxic activity	
tomentosa	Methanolic extract of leaves	Phenols, flavonoids, glycosides, saponins, terpenoids, tannins, reducing sugars, and proteins	Strong antioxidant activities	Muruganandam et al. (2016)

References

Alejandro GD, Razafimandimbison SG, Liede-Schumann S (2005) Polyphyly of *Mussaenda* inferred from ITS and trnT-F data and its implication for generic limits in Mussaendeae (Rubiaceae). Am J Bot 92(3):544–557

Alejandro GJD, Meve U, Liede-Schumann S (2016) A taxonomic revision of Philippine *Mussaenda* (Rubiaceae, Mussaendeae). Ann Missouri Bot Gard 101(3):457–524

Borges RM, Gowda V, Zacharias M (2003) Butterfly pollination and high-contrast visual signals in a low-density distylous plant. Oecologia 136:571–573

CCDB (The Chromosome Counts Database) Version 1.47 (2015) Published on the internet; http://ccdb.tau.ac.il/search/Mussaenda/. Accessed 21 Sept 2020

Chantaranothai P (2015) A synopsis of *Mussaenda* L. (Rubiaceae) in Thailand. Thai For Bull (Bot) 43:51–65

Chen S, Luo Z, Zhang D (2014) Pre- and post-zygotic reproductive isolation between co-occurring *Mussaenda pubescens* var. *alba* and *M. shikokiana* (Rubiaceae). J Integr Plant Biol 56(4): 411–419

Cramer CS, Bridgen MP (1997a) Somatic embryogenesis and shoot proliferation of Mussaenda cultivars. Plant Cell Tissue Organ Cult 50:135–138

Cramer CS, Bridgen MP (1997b) Micropropagation of *Mussaenda* cultivars. In: Bajaj YPS (ed) Biotechnology in agriculture and forestry 40. High-tech and micropropagation VI. Springer, Berlin/Heidelberg

DAO (DENR Administrative Order) (2017) Updated National List of Threatened Philippine Plants and Their Categories

Das P (2010) Mass cloning of rose and mussaenda, popular garden plants, via somatic embryogenesis. Hort Sci 37(2):70–78

Duan T, Zhang D (2014) Fourteen additional microsatellite markers for *Mussaenda pubescens* and cross-species amplification. J Genet 93:e44–e47

Duan T-T, Gong W, Zhang D-X (2012) Development of microsatellite markers from *Mussaenda pubescens* (Rubiaceae). Am J Bot 99:e437–e439

Espada LT, Rosario TL (1992) Isozyme polymorphisms in some ornamental mussaendas of the Philippines. Phil Agric 25(1 and 2):107–111

Eswaraiah MC, Satyanarayana T (2010) In vitro antioxidant and free radical scavenging activity of stem of *Mussaenda erythrophylla*. IJPSR 1(9):20–26

Gopalakrishnan S, Vadivel E (2011) GC-MS analysis of some bioactive constituents of *Mussaenda frondosa* Linn. Int J Pharm Bio Sci 2(1):313–320

Gopalaswamiengar KS (1991) Complete gardening in India (Revised and enlarged edition). Gopalaswamy Parthasarathy, Bangalore. 1035 pp.

Harborne JB, Girija AR, Devi HM, Lakshmi NKM (1983) Anthochlor pigments from the petals of *Mussaenda hirsutissima* and *Zinnia linearis*. Phytochemistry 22(12):2741–2742

IPNI (International Plant Names Index) (2020) Published on the Internet http://www.ipni.org, The Royal Botanic Gardens, Kew, Harvard University Herbaria & Libraries and Australian National Botanic Gardens. Retrieved 7 August 2020

Islam F, Quadery TM, Chowdhury SR, Kaisar MA, Uddin MG, Rashid MA (2012) Antioxidant and cytotoxic activities of *Mussaenda macrophylla*. Bangladesh Pharm J 15(1):69–71

Islam F, Kuddus MR, Latif F, Hossain MK (2013) Preliminary antimicrobial activity and cytotoxicity of leaf extracts of *Mussaenda roxburghii* Hook. f. Bol Latinoam Caribe Plant Med Aromat 12(6):612–617

Jayaweera DMA (1963) The rubiaceous genus Mussaenda: the species of India and Ceylon. J Arnold Arbor 44:232–267

Jayaweera DMA (1964) The rubiaceous genus Mussaenda: the species of the Philippine islands. J Arnold Arbor 45:101–139

Jena R, Kar DM, Rath D, Roy KS, Ghosh G (2019) Antiulcer property of *Mussaenda philippica*. Pharm J 11(3):603–607

Kichu M, Malewska T, Akter K, Imchen I, Harrington D, Kohen J, Vemulpad SR, Jamie JF (2015) An ethnobotanical study of medicinal plants of Chungtia village, Nagaland, India. Journal of Ethnopharmacology 166:5–17

Kim NC, Desjardins AE, Wu CD, Kinghorn AD (1999) Activity of triterpenoid glycosides from the root bark of *Mussaenda macrophylla* against oral pathogens. J Nat Prods 62(10):1379–1384

Koul S, Chaudhary A (2011) Radical scavenging and antistress activity of *Mussaenda frondosa* roots (Rubiaceae). Pharmacologyonline 1:1091–1097

Lemmens RHMJ (2003) Mussaenda L. In: Lemmens RHMJ, Bunyapraphatsara N (eds) Plant resources of South-East Asia no. 12(3), Medicinal and poisonous plants, vol 3. Prosea Foundation, Bogor, pp 311–314

Li A-M, Wu X-Q, Zhang D-X, Barrett SCH (2010) Cryptic dioecy in *Mussaenda pubescens* (Rubiaceae): a species with stigma-height dimorphism. Ann Bot 106:521–531

Li Y, Qiao W, Yuan K (2011) Isolation and structural elucidation of chemical constituents of *Mussaenda hainanensis* Merr. J Med Plants Res 5(8):1459–1465

Long CL, Li R (2004) Ethnobotanical studies on medicinal plants used by the Red-headed Yao People in Jinping, Yunnan Province, China. Journal of Ethnopharmacology 90(2–3):389–395

Maity SK, De KK, Kundu AK (2001) In vitro propagation of *Mussaenda erythrophylla* Schum and Thom cv. Scarlet through multiple shoot regeneration. Indian J Exp Biol 39(11):1188–1190

Muruganandam S, Rajesh V, Dhamodharan R, Arumugam P, Ravikumar S (2016) Antioxidant and antidiabetic potentials of *Mussaenda tomentosa* Wight Ex Hook. F. Int J Pharm Phytochem Res 8(1):64–70

Palit P, Talukdar S, Kundu A, Mandal SC, Mandal T, Dasgupta D (2013) Phytochemical investigations and pharmacological evaluation of the different bioactive fractions obtained from the leaves of *Mussaenda erythrophylla* and *Datura metel* Linn. and biodistribution studies of fluorescent based phospholipid vesicles of their extract. Biochem Pharmacol 2(4):122

Patil SA, Joshi VG, Sambrekar SN (2011) Evaluation of wound healing activity of isolated compound quercetin and alcoholic extract of leaves of *Mussaenda frondosa* Linn. Res J Pharm Phytochem 3(6):266–271

Pimentel R, Ledesma NA, Villanueva LL (1999) Mussaenda propagation with the 'Kulob' system. Integrated Ornamental Horticulture Research and Development Program brochure. Philippine Council for Agriculture, Forestry and Natural Resources Research and Development-Department of Science and Technology

Price GR (1974) Cultivated mussaendas in the Philippines. Kalikasan. Philipp J Biol 3:37–55

Protacio CM, Obmerga LR, Siar SV (1997) Production of dwarf compact plants of mussaenda by paclobutrazol treatment. Phil Jour Crop Sci 22(Supplement 1)

Protacio CM, Obmerga LR, Siar SV (1998) Propagation of Mussaenda by stem cuttings. Phil Jour Crop Sci 23(Supplement 1):28

Protacio CM, Obmerga LR, Siar SV (1999) Propagation of Mussaenda by stem cuttings. Philippine Agricult Sci 83(3):251–259

Protacio CM, Obmerga LR, Siar SV (2000) Mass propagation of mussaenda using mist system. Integrated Ornamental Horticulture Research and Development Program brochure. Philippine Council for Agriculture, Forestry and Natural Resources Research and Development-Department of Science and Technology

Rashid MHA, Nahar A (2019) Effect of naphthalene acetic acid on sprouting and rooting of stem cutting in *Mussaenda* species. Progress Agric 30(1):48–54

Rosario TL (1998) Ornamental Mussaendas of the Philippines (Revised edition). College of Agriculture, University of the Philippines Los Baños, College, Laguna, Philippines. 70 pp

Rosario TL (2007) Saga of a spontaneous mutant: *Mussaenda* 'Doña Aurora.'. Phil Jour Crop Sci 32(1):89–102

Rosario TL, Aurigue FB (2006) The origin and characterization of two new mussaenda hybrids. Philippine Agricult Sci 89(1):85–90

Rosario TL, Siar SV, Aurigue FB (1990) Morphology and cytology of the progenies of an open-pollinated *Mussaenda* 'Queen Sirikit.'. Philippine Agricult Sci 73(2):221–226

Sambrekar SN, Patil PA, Kangralkar VA (2010) Protective activity of *Mussaenda frondosa* leaf extracts against paracetamol induced hepatic damage in wistar rats. J Pharm Res 3(4):711–713
Tiambeng BJ, Ramirez DA, Espiritu LG (1969) Philippine Agriculturist 52:547
Vidyalakshmi KS, Dorni AIC, Vasanthi HR (2007) Antimitotic and cytotoxic activity of *Mussaenda queensirkit*. J Pharm Toxic 2(7):660–665
Zhao W, Xu J, Qin G, Xu R, Wu H, Weng G (1994) New triterpenoid saponins from *Mussaenda pubescens*. J Nat Prod 57(12):1613–1618

Chrysanthemum

19

S. K. Datta

Contents

S. K. Datta has retired.

S. K. Datta (✉)
Floriculture Division, CSIR-National Botanical Research Institute, Lucknow, India

Bose Institute, Kolkata, India

Department of Plant Science and Biotechnology, Assam University, Silchar, Assam, India

S. K. Datta, Y. C. Gupta (eds.), *Floriculture and Ornamental Plants*, Handbooks of Crop Diversity: Conservation and Use of Plant Genetic Resources,
https://doi.org/10.1007/978-981-15-3518-5_4

Abstract

The autumn-flowering garden *Chrysanthemum* (*C. morifolium* Ramat.) is an important flower crop throughout the world. It has earned tremendous popularity as a flower for the garden and as a cut flower for interior decoration or for greenhouse or conservatory display. The genus constitutes a large polyploid complex ranging from 2x to 22x, besides a number of aneuploids. The genetic diversity of chrysanthemum plays a very important role in developing new and novel desired forms through breeding and selection. All the present-day colorful varieties have been developed through complex interspecific crosses among elemental species, open pollination, indiscriminate intervarietal hybridization, spontaneous and induced mutation, and selection and management of chimera. The possibilities for creating different forms and improving chrysanthemum are infinite, and a breeder will always have future goals to work toward. This chapter will provide maximum information generated in India on different aspects along with important publications.

Keywords

Chrysanthemum · Germplasm · Genetic diversity · Breeding · Characterization · New varieties

19.1 Introduction

Chrysanthemum (*chrysos*, golden; *anthos*, flower) is one of the most interesting ornamental groups of plant in the world. The history of origin of chrysanthemum is very interesting and debatable among scientists. The history of the spread of chrysanthemum from its native place to the rest of the world and its development from wild daisy-like flowers to present-day magnificent cultivars make a fascinating reading. It is spread over temperate regions and in many parts of the globe. The present number of varieties in the world is reported to be above 2000, and in India, there are about 1000 varieties. Chrysanthemum has earned tremendous popularity as an ornamental flower for the garden and as cut flower for interior decoration or for the greenhouse or conservatory display. Its wide popularity is due to its large number of cultivars with respect to growth habit, size, color, and shape of bloom. As a short-day plant, it naturally flowers in the autumn and winter. Now, programmed blooming of chrysanthemum all year round is possible by environmental manipulations. Chrysanthemum has its admirers and enthusiasts all over the world for its use both as a commercial flower crop and as a popular exhibition flower. Chrysanthemum flowers vary greatly in shape, size, and color. All these variations have occurred due to the interplay of genetic factors, meaning thereby that the genetic resources have played a key role in bringing to the chrysanthemums their present fame and glory. All the present-day colorful chrysanthemum varieties have been developed through complex interspecific crosses among elemental species, open pollination, indiscriminate intervarietal hybridization, spontaneous and induced mutation, and selection and management of chimera. The main agencies responsible for varietal improvement are individual enthusiasts, nurserymen, and breeders working in research institutes and agricultural universities. This chapter will cover the information on research activities related to the history of origin, geographical distribution, enrichment and characterization of germplasm, their utilization in breeding program, and development of new and novel varieties through hybridization, selection, in vivo and in vitro mutagenesis, standardization of agro-technology, disease management, post-harvest physiology, programmed blooming, tissue culture, molecular techniques, etc.

19.2 Botany

The *Manual of Cultivated Plants* (Bailey 1949) and the *Dictionary of Gardening* (Chittenden 1951) list *Chrysanthemum morifolium* as the valid name for the varieties of chrysanthemum we grow today. *Chrysanthemum* belongs to the family Compositae, the most phylogenetically advanced dicotyledonous family (Hemsley 1889; Popham and Chan 1950), which constitutes a large polyploid complex ranging from 2x to 22x, besides a number of aneuploids. The recognizable feature of the family is that a large number of flowers are arranged on flattened axis to form a compact floral head which looks like a single bloom. The blooms of Compositae appear on capitulum inflorescence. The capitulum or head consists of a large number of small florets in very close formation. Two kinds of florets, i.e., ray florets and disc florets, are present in a bloom. Ray florets are large, attractive, colorful, and of various shapes which give beauty to the head, whereas disc florets are smaller and centrally placed. The difference between the two kinds of florets at a glance is generally that the disc florets are short while the ray florets are usually quite long. The ray florets are unisexual with only female parts, while the inner disc florets contain both male and female reproductive elements. The chrysanthemum bloom type depends mainly upon the relative number of the two kinds of florets, their shape, and the direction of growth. At present, there are about 200 species in the genus *Chrysanthemum*, among which the modern autumn-flowering perennial *Chrysanthemum morifolium* Ramat. is the most important (Carter 1980). The name *Chrysanthemum morifolium* has been changed to *Dendranthema grandiflora* (Heywood and Humphries 1977; Kitamura 1978; Anderson 1987). Reclassification of the *Chrysanthemum* by botanists has been based on cytology, cypsela anatomy and morphology, embryology, hybridization, and phytochemistry.

It is native to the Northern Hemisphere, chiefly Europe, and Asia, with a few in other areas. Many authorities claim that it originated in China (Carter 1980). Many species believed to have taken part in its development have been mentioned by various workers as *Chrysanthemum morifolium*, *C. sinense*, *C. indicum*, *C. japonicum*, *C. ornatum*, *C. satsumense*, *C. boreale*, *C. nipponicum*, *C. arcticum*, *C. sibiricum*, *C. rubellum*, *C. parthenium*, and *C. coronarium* (Ramatuelle 1792; Sabine 1822; Hemsley 1889; Niwa 1936).

19.3 Geographical Distribution

Chrysanthemum is known to have been cultivated for over 3050 years, probably originating about 500 BC. Confucius, a great Chinese philosopher, knew its blooms and described it in his work *Li-Ki* or *Ninth Moon* as "the chrysanthemum with its yellow glory"; other references referred them as "the golden flower." T'ao-Yuan-

Ming (AD365–427) was said to be mainly responsible for many early improvements in cultural methods. Chinese poets were writing in praise of the chrysanthemum, while replicas of chrysanthemums in their original form have been discovered on some fifteenth-century Chinese pottery, emphasizing their long-standing popularity (c.f. Jim Smith, NCS Yearbook 1984, Rosse 1964; Kyle 1952). The available records on history of chrysanthemum cultivation and their spread in different countries are as follows:

China – Chrysanthemums were first cultivated in China as a flowering herb as far back as the fifteenth century BC (c.f. History of the Chrysanthemum. National Chrysanthemum Society, USA). Those chrysanthemums were flowering herbs which have medicinal and aesthetic uses. Over 500 cultivars had been recorded by 1630 (c.f. *Chrysanthemum*. Flora of China. eFloras). Confucius described its yellow glory as early as in 550 BC. The credit for early remolding of the wild ancestors into much superior diverse forms also goes to the Chinese. The first white blooming chrysanthemum is also claimed by Liew Chieh Yuan of Peking to have been raised by T'ao Hungching during 452–532 AD (Cumming 1964). Tao Yuanming (AD365–427), a Chinese chrysanthemum specialist, is responsible for many early improvements in cultural methods. The ancient Chinese name for chrysanthemum is "Chu." The Chinese City of Chu-Hsien (which means Chrysanthemum City) was so named to honor the flower (Woolman 1957).

Japan – Seeds of chrysanthemum reached Japan from China via Korea as early as in 386 AD (Cumming 1964). It is believed that the cultivated chrysanthemum must have been brought from China to Japan after they had attained some degree of development in the era of Tempyo which is also called as the Nara period (710–794 AD) (Niwa 1936). Chrysanthemum was declared as the National Flower by Emperor Uda in the year 910 AD, and a 16-petaled single large chrysanthemum called "Ichimonji" was adopted as the imperial crest during the twelfth century. However, at present, chrysanthemum is the symbol of Royalty in Japan. Chrysanthemum underwent wonderful transformation representing widest range of variability (Nakajima and Young 1965).

Europe – The first mention of the chrysanthemum in Europe was made by botanist Bregnius in 1689, though it was 100 years later before the plant began to be cultivated in this continent, when Frenchman Pierre Louis Blanchard brought from Macao what was for a long time known as "The Old Purple." Eventually plants of this stock reached Kew, and a description of the new flower was featured in the *Curtis's Botanical Magazine* in 1796 (Kyle 1952). One Japanese cultivar was introduced in Holland in 1688. Jacob Bregnius, a Dutch botanist, described six Chinese cultivars being grown in Holland in 1690 (Niwa 1936; Cumming 1964). It was not much popular in Holland. In 1690, H. van Rheede described a Holland-grown type from India called "Gool-doodi." A French merchant from Marseilles named Pierre Louis Blanchard brought three cultivars home from China in 1789. Only one of these survived and was named "Old Purple," the first named cultivar to grow in the western world. Eventually this cultivar reached Kew Gardens, and its description was featured in the *Curtis's Botanical Magazine* of 1796. In 1827, seed was successfully produced in Europe by a retired French officer, Captain Bernet, and as many previous attempts by both English and French gardeners had failed, this date is of great historical importance in the chrysanthemum world. It is also on record that the first English raiser was Mr. Wheeler of Oxford in 1832 (cf. Jim Smith, History of the Chrysanthemum, NCS Yearbook 1984). In France, though its value was appreciated, it primarily remained a cemetery flower with a preference for white and purple. In England, the emphasis was on exhibitions, the first being held in 1843 at Norwich. Here, its popularity was steadily increased, and the National Register of Names of Chrysanthemum issued by the National Chrysanthemum Society of England listed more than 5000 cultivars.

France – French breeders accepted and appreciated chrysanthemum for commercial exploitation. Favorable climate helped to develop late-blooming types. White, violet, and purple varieties were imported by merchant A. Blanchard from China to Marseilles, and purple became very popular for gardens in 1789. Eight more frilled-type varieties were imported from Chinese during 1799–1808. Early and more better varieties were developed by Captain Bernet during 1826–1836. Several interesting varieties with new color and form were raised by Dominique Perfuse during 1836 from seeds given to him by Captain Bernet. Robert Fortune brought "Chusan Daisies" and large reflexing type from China in 1864 and popularized among the growers. Extremely early blooming group was introduced in 1890 by Simon Deleaux from crosses of Japanese importations (c.f. Kyle 1952).

USA – The first "dark purple" chrysanthemum was imported from England by scientist John Stevens in the United States in 1798, and it was followed by many more cultivars from China. The growing of chrysanthemum became commercialized in America in the shortest period. Different forms, such as pompon, decorative, anemone, and spider, received much attention here. Instead of big size and novelty in color, the main qualities sought-after were sturdy growth, long vase life, and higher yield. The National Chrysanthemum Society of America was founded in 1890, consisting exclusively of commercial growers, at New York. The society organized the first exhibition in 1902 at Chicago.

India – It is difficult to say since when its cultivation has begun in India as no historical literature is available on chrysanthemum. In Marathi literature, its first written mention seems to be in Dnyaneshwari (1290 AD) (Chap. 15, shloka 20). As already mentioned earlier, a cultivar from India has been reported to be grown in Holland in 1690 under the name 'Gool-doodi' which points toward the fact that chrysanthemum was in cultivation in North India in the seventeenth century. It is interesting to know that in Hindi, chrysanthemum is still called "Gul-daudi." In North India, chrysanthemum is primarily grown for decorating the landscape either in the ground or in pots. Contrarily, in southern part of the country, it is mostly grown in farmer's fields for supply to the market as loose flowers for garland, for hair decoration by the ladies, and for offering to god. While yellow-colored flowers are preferred in the South, in the North, various hues of red, purple, yellow, and white are found to be grown in abundance.

19.4 Origin

The origin of chrysanthemum is at the best hypothetical as there is no proper experimental proof and/or genetic evidence. Nonetheless, it is now generally accepted that modern, large, double, and exquisitely flowered cultivars owe their origin to relatively small, single, and not so attractive types. This great transformation is the result of centuries of natural cross-pollination and selection by keen growers and also by indiscriminate intervarietal hybridization by a galaxy of breeders and natural and induced mutations.

The chief mechanism in the cytogenetical evolution of garden chrysanthemum has been enumerated by Nazeer and Khoshoo (1982) as outbreeding, spontaneous, and intentional hybridization coupled with mutation, chromosomal differentiation, and repatterning and polyploidy. According to available literature, the modern chrysanthemums are the result of some 2500 years of continuous repeated cycles of hybridization and selection from several species of chrysanthemum that grew wild in China and Japan. After over 1000 years of development of cultivated chrysanthemum, these were exported from China to Japan about AD 750. Japanese started improvement of Chinese variety by cross-fertilization with the then existing wild Chinese varieties. Japanese hailed it as their National Flower in AD 797. Slowly the flower was first introduced to Holland in about 1668, to France in about 1789, and to England in 1795. Robert Fortune brought chrysanthemum at the Royal Horticultural Society from China in 1846 and from Japan in 1862.

The first English seedlings were raised in 1835 by Short and Freestone. John Salter established a nursery at Versailles in France in 1838 where he produced a number of seedlings, the two most notable being 'Annie Salter' (medium reflexed decorative, yellow) and 'Queen of England' (large pink incurve) introduced in 1847. John Salter started his improvement work on chrysanthemum at a nursery in Hammersmith, London, and isolated quite a number of sports from 'Queen of England'. A new seedling ('William Penn') was grown at the Pennsylvania Horticultural Society in 1841. Pottery dating back to the fifteenth century BC depicts the flower as we know it today. Respect for this flower ran so deep; a city was named after it: Chu-Hsien or Chrysanthemum City. In ancient China, almost all parts of the chrysanthemum had medicinal use. The chrysanthemum made its way to Japan around the eighth century AD, and the emperor adopted a single chrysanthemum as his crest and official seal. National Chrysanthemum Day is celebrated in Japan since 910 AD. Chrysanthemum is a hybrid species which has developed as a result of complex interspecific crossing among the elemental species for over and over a period of more than 2500 years. Literature survey indicates that the following elemental species have played very important role to build up the present status of chrysanthemum: *C. boreale*, *C. carinatum* (tricolored blooms), *C. coronarium* (native to South Europe, yellow and white blooms), *C. cinerariifolium*, *C. coccineum* (white, pink, and red), *C. frutescens* (white and soft yellow flowers), *C. indicum* (native of China and Japan, supposed to be one of the ancestral species involved in the evolution of modern florist's chrysanthemum, yellow flowers), *C. japonicum*, *C. maximum* (white and yellowish blooms), *C. ornatum*, *C. satsumense*, *C. sibiricum*, *C. sinense* (native of China, blooms white),

C. parthenium (flower white or pale yellow), *C. balsamita* (flower yellowish with some white rays), *C. maximum* (white perennials), *C. nipponicum* (native of Japan, white daisies), *C. rubellum* (England, pink to rose red), *C. cliginoseem* (white large daisies), *C. zawadskii* (species from Galicia and Siberia, rose pink daisies), *C. alpinum* (native to high Alps, daisies of glossy white), *C. arcticum* (Arctic regions, white or pink), *C. mawii* (pink daisies, white with pink reverse), *C. weyrich* (pink flowers), *C. carinatum* (tricolor Chrysanthemum, native to Morocco, purple or reddish rings with yellow and white base), *C. segetum* (corn chrysanthemum or corn marigold; native to Europe, Africa, and Asia; deep yellow or whitish blooms), *C. frutescens* (white or soft yellow daisies), *C. indicum* (native to India/China, tiny yellow blooms), *C. hortorum* (not a valid species, but all garden chrysanthemums have an occasion been grouped here), and *C. sibiricum* (this acted like a blood transfusion on the worn-out strains of *Chrysanthemum* when it was used for breeding in the early 1930s, single flowers, white aging to carmine pink). Exploitation of genetic resources of some more wild species like *C. oreastrum*, *C. hydrargyrum*, *C. zawadskii*, *C. chanetii*, *C. naktongense*, *C. mongolicum*, *C. argyrophyllum*, *C. rhombifolium*, *C. vestitum*, *C. dichrum*, *C. glabriusculum*, *C. lavandulifolium*, *C. foliaceum*, *C. nankingense*, *C. potentilloides*, and *C. maximowiczii* has been reported (Zhao et al. 2009; Datta 2013).

19.4.1 Domestication

Garden chrysanthemum has a long history of domestication. China is the primary center of origin, and Chinese have been growing improved forms for the last 3000 years. Two elemental species, *C. indicum* L. and *C. morifolium* Ramat. (*C. sinense* Sabine), played the major role in the development of garden forms. The original cultivars were likely single- and many-flowered. There has been a transformation of the corolla of individual florets into numerous forms during domestication for a long time. The first experimental hybrid was recorded in 1827, and the first bud sport arose in 1832 (Darlington 1973). The main improvement of chrysanthemum cultivars has been achieved in the gardens of China, Japan, Europe, and America through conscious and unconscious selection. Garden chrysanthemum reached various places in the form of seedlings or seeds. New cultivars have arisen from these stocks through further hybridization and bud sports. Indiscriminate intervarietal hybridization followed by selection was the main evolutionary factor (Nazeer and Khoshoo 1982).

19.5 Classification

The International Chrysanthemum Society and different National Chrysanthemum Societies have classified chrysanthemum based upon bloom structure. The commonly accepted classification of garden chrysanthemum is based on bloom shape and size and relative number of the two kinds of florets (ray florets and disc florets), their shape, arrangement, and direction of growth. They are mainly classified under two categories: large-flowered and small-flowered. Large-flowered chrysanthemums are further classified into 13 classes and small-flowered ones into 10 classes (Datta 1996). Examples for each are as follows:

(i) Large-flowered: Class 1. *Regular Incurve* – Ray florets are narrow to broad and very smoothly incurved in a regular pattern to form a perfect ball (Fig. 1a). The average bloom size varies from 10 to 15 cm (4″–6″). Disc not visible. *Examples* include 'Gairik', 'Shin Mei Getsu', 'Casa Grande', 'Kokka Soun', 'Snow Ball', 'S.S. Arnold', 'Autumn King', 'Chandrama', 'Sonar Bangla', 'R Venkatraman', 'Dream Castle', 'Bharat Ratna', 'Queen of Tamluk', 'Belur Math', 'Super Giant', and 'Adventure'. Class 2. *Irregular Incurve* – Ray florets are usually broad, smooth, incurved, and arranged in an irregular manner (Fig. 1b). The bloom size is very large, and the breadth and depth of the blooms are almost equal. The disc florets are entirely covered by the upper florets. The bloom size varies from 15 to 20 cm (6″–8″). *Examples* include 'Audrey Shoesmith', 'Hommand Philips', 'J.S. Salesbury', 'Kiku Biyori', and 'Mountaineer'. Class 3. *Skirted Incurve* –The lower florets, mostly the basal florets, bend downward in an irregular fashion to give a skirted shape. The bloom size varies from 15 to 20 cm (6″–8″). *Example* includes 'Dream Castle'. Class 4. *Incurving* – Ray florets are incurved upward in an indefinite manner. The bloom is not a compact ball due to

Fig. 1 Chrysanthemum flower types. Figure 1**a**. Regular incurve (Sonar Bangla). Fig. 1**b**. Irregular incurve (S.L. Andre Faurd). Fig. 1**c**. Incurving (Pink Cloud). Fig. 1**d**. Reflex (Alfred Wilson). Fig. 1**e**. Irregular reflex (Miss Maud Jeferi). Fig. 1**f**. Reflexing (Dee). Fig. 1**g**. Intermediate (Thaiching Queen). Fig. 1**h**. Quilled (Red Quill). Fig. 1**i**. Spider (Icicles)

loose fitting arrangements of florets (Fig. 1c). *Examples* include 'Classic Beauty', 'Pink Cloud', 'White Cloud', 'Scarlet Waltz', 'Gusman Red', 'Hanayome Sugata', 'Dr. S. Mukherjee', 'Leading Lady', and 'Hommand Philips'. Class 5. *Reflex* – Ray florets are narrow to broad and bent backward and downward. Inner florets remain incurved at the early stage concealing the disc florets of the bloom. Outer florets turn outward away from the central tuft. Blooms look globular but may be somewhat flattened. The average bloom size is 15–20 cm (6″–8″). The center or disc of the flower is not visible. On the basis of arrangements of ray florets, this class is further subclassified into (a) *regular reflex* (Ray florets are bent back and downward in a regular arrangement (Fig. 1d).), (b) *irregular reflex* (Ray florets are bent downward in a twisted and irregular way (Fig. 1e).), and (c) *reflexing* (Ray florets are like aster and have a tendency to reflex. The bloom appears flat shape (Fig. 1f).). *Examples* include 'Imperial', 'Dee', 'Mrs. A.I. Miller', 'John Webber', 'Julios Brinas', 'President Viger', 'Rose Day, 'Star of India', 'Hope', 'Mrs. Roager Thompson', 'Ashok Pillar', 'Mrs. Eleston', 'Kansya', 'Kasturba Gandhi', and 'Beatrice May'. Class 6. *Intermediate* – This class represents blooms intermediate between "incurved" and "reflex." Ray florets are narrow to broad and may be short. Few outer ray florets are partially incurved, but lower ray florets are reflex. Inner florets are incurved. Disc concealed. Centers may be slightly flattened or depressed. The bloom shape gives a globular effect (Fig. 1g). The bloom size is 15 cm (6″) or more. *Examples* include 'General Petain', 'Mrs. W.A. Reid', 'T-1', 'Cloth of Gold', and 'Sun Flight'. Class 7. *Ball* – The ray florets are straight and densely packed. These radiate uniformly in all directions to give the bloom a ball/ovoid/roundish shape. *Examples* include 'W-23', 'Pride of Madford', 'Nigeria', and 'Red Jack'. Class 8. *Quilled* – The ray florets are tubular and elongated with tips open or closed (Fig. 1h). The thickness of the tube varies from thin to medium to thick. *Examples* include 'W-11', 'Red Quill', 'Green Sensation', 'Tribhuban', and 'Pradhan's Pride'. Class 9. *Spider* – The florets (same as Class 8) are large, tubular, and elongated. Tips may be open or closed, but in either case, they are coiled or hooked. The rays may either fall or spread (Fig. 1i). *Examples* include 'Sunder Calcutta', 'Geetanjali', 'Achievement', 'Innocence', 'Carnation Gold', 'Icicles', 'Tokyo', 'Senkyo Emaki', 'Florida', 'Manick', 'Senkyo No Rya', 'Miss Universe', 'Valiant', 'M-30', 'Flirtation', 'Kogen No Hoshii', 'Diamond Jubilee', and 'T-8'. Class 10 – *Spoon*. The ray florets are tubular with spatula-like open tips. The size of open portion varies. Disc visible (Fig. 2a). *Examples* include 'Pink Casket', 'M-24', 'Crimson Tide', 'Carnation', and 'Puspahanas'. Class 11. *Anemone* – The ray florets are ligulate or quilled. Here, the disc is noticeably developed with florets. Disc usually hemispherical and raised (Fig. 2b). *Examples* include '007', 'Cloud Bank', and 'Red Admiral'. Class 12. *Single* – Ray florets are long, elongated, and strap-like (Fig. 2c). The number of whorl of florets is restricted up to four. The disc is conspicuously visible. *Examples* include 'Potomac', 'Joan Helen' and 'Surja'. Class 13. *Semidouble* – Ray florets are long, elongated, and strap-like. The number of whorl of florets is more than five. Disc conspicuous. *Examples* include 'Ronald' and 'Crimson Tide'.

(ii) Small-flowered: Class 1. *Anemone* – Disc florets are well developed and prominent. Ray florets may be flat, twisted, and quilled (Fig. 2d). *Examples* include 'Nirmod', 'Venus', 'Pink Cushion', 'Violet Cushion', 'Season', 'Ace',

'Perfect', 'Gaity', 'Dainty Maid', 'Mercury', 'Rosa', 'Maise', 'Gem', 'Modella', 'Harbinger' and 'Smita'. Class 2. *Button* – Florets are short rayonate-like and hemispherical. Florets radiate in all directions. Blooms are small and compact. 2–3 cm (about 1 inch) in diameter. *Examples* include 'Liliput', 'Bull Finch', and 'King Fisher'. Class 3. *Single Korean* – Ray florets are strap-like. Bloom flat and disc well visible. Ray florets are arranged in five or less whorls. *Examples* include 'Tune Full', 'Pat', 'Alpana', 'Dolore', 'Sharad Bahar', 'Sharad Shobha', 'Sharada', 'Sharad Seema', ' Prabha', 'Sharad Singar', 'Sunset', 'Kirti', 'Ragini', 'Luoy', 'Margery', 'Pilgrim', 'Pat', 'Fantasy', and 'Vinaya'. Class 4. *Double Korean* – Florets are the same as Class 3. The number of whorls of ray florets is more than five (Fig. 2e). Disc visible. *Examples* include 'Jyotsna', 'Khushrru', 'Flirt', 'Lalkila', 'Tara', 'Lilith', 'Priya', 'Purity', 'Criterion', 'Man Bhawan', 'Red Gold', 'Fatima', 'Lalpari', and 'Sonali'. Class 5. *Decorative* – Florets are the same as Class 4. Here, the disc is not visible due to developed ray florets. Ray florets regular or irregularly reflexed (Fig. 2f). *Examples* include 'Megami', 'Sharad Mala', 'Jayanti', 'Sujata', 'Jubilee', 'Nilima', 'Sonali Tara', 'Puja', 'Jaya', 'Suneel', 'Ajoy', 'Illini Cascade', 'Pink Gin', 'Jawra', 'Ratna', and 'Shabnam'. Class 6. *Pompon* – Ray florets are short, broad, and very systematically and uniformly arranged to

Fig. 2 Chrysanthemum flower types. Figure 2**a**. Spoon (M-24). Fig. 2**b**. Anemone. Fig. 2**c**. Single (Potomac). Fig. 2**d**. Anemone (Nimrod). Fig. 2**e**. Double Korean (Lilith). Fig. 2**f**. Decorative (Jayanti). Fig. 2**g**. Quilled. Fig. 2**h**. Stellate

give bloom a compact hemispherical shape. Width and breadth almost equal. Ray florets may be incurved or reflexed. Disc normally covered or inconspicuously open. *Examples* include 'Horizon', 'Apsara', 'Nanako', 'Cotton Ball', 'Birbal Sahni', 'Purple Star', and 'Maharaja'. Class 7. *Semi-quilled* – The ray florets are tubular up to certain length of the floret from the base and then open at the tip. Open-tip portion may be flat, reflexed, or incurved. Disc open. Examples include 'Jean', 'Alison', and 'Garnet'. Class 8. *Quilled* – Ray florets are elongated and tubular like a quill. The tips of florets may be open but not developed (Fig. 2g). *Examples* include - 'Fraiar', 'Munchausen', 'Q-3', 'Donald', 'Rita', 'Snow Crystal', 'Space in 83', and 'Green Nightingale'. Class 9. *Stellate* – Florets are like Class 3, but both sides of ray florets are reflexed downward. Florets may or may not be twisted. Disc flat with short florets (Fig. 2h). Examples include 'Laura', 'Heloise', 'Red Star', 'Harvest Home', 'Stella', 'Morning Star', and 'Gordon Tailor'. Class 10. *Cineraria* – Blooms are flat Korean type with diameter not more than 3 cm (about 1.2″). Examples include 'Phyllis', 'Jessie', 'Kashturi', 'Bindya', and 'Charmis'.

19.6 Species and Cultivars

(i) *Species*: The number of species under the genus *Chrysanthemum* varies from 100 to 200 (Niwa 1936; Carter 1980). Some important species of chrysanthemum are *Chrysanthemum boreale*, *C. carinatum* (tricolor), *C. coronarium* (garland chrysanthemum, yellow and white blooms), *C. cinerariifolium* (grown in temperate regions for making an insecticide called pyrethrum), *C. coccineum* (grown in temperate countries from seeds, blooms with white, pink, and red colors and look like anemone chrysanthemum, perennial), *C. frutescens* (white and soft yellow flowers), *C. indicum* (it is supposed to be one of the ancestral species involved in the evolution of modern florist's chrysanthemum, yellow flowers), *C. japonicum* (found to be growing in the Pacific coastal region of Japan), *C. maximum* (popular cut flower bearing white and yellowish blooms, perennial), *C. morifolium* (florist's chrysanthemum), *C. ornatum* (syn. *C. marginatum*, allied to *C. indicum* and *C. sinensis*), *C. rubellum* (it is exceptionally sturdy and used by breeders for its hardiness), *C. satsumense*, *C. sibiricum* (this is said to be one of the parents of Korean hybrids evolved in early 1930), *C. sinense* (this is supposed to be one of the sources (with *indicum*) of today's florist's chrysanthemum, native of China, and bears blooms with white ray florets), *C. maximum*, *C. pacificum*, *C. segetum*, *C. aphrodite* Kitam., *C. arcticum* L., *C. argyrophyllum* Ling, *C. arisanense* Hayata, *C. chalchingolicum* Grubov, *C. chanetii* H. Lév., *C. crassum* (Kitam.) Kitam., *C. cuneifolium* Kitam., *C. daucifolium* Pers., *C. dichrum* (C.Shih) H. Ohashi & Yonek., *C. foliaceum* (G.F.Peng, C.Shih & S. Q.Zhang) J.M.Wang & Y.T.Hou, *C. glabriusculum* (W. W. Sm.) Hand.-Mazz., *C. horaimontanum* Masam., *C. hypargyreum* Diels, *C. integrifolium* Richardson, *C.* × *konoanum* Makino, *C. lavandulifolium* Makino, *C. leucanthum* (Makino) Makino, *C. longibracteatum* (C.Shih, G.F.Peng & S.Y.Jin) J.M.Wang & Y.T.Hou,

C. maximoviczii Kom., *C. miyatojimense* Kitam., *Chrysanthemum* × *morifolium* (Ramat.) Hemsl., *C. morii* Hayata, *C. naktongense* Nakai, *C. ogawae* Kitam., *C. okiense* Kitam., *C. oreastrum* Hance, *C. parvifolium* C.C.Chang, *C. potentilloides* Hand.-Mazz., *C. rhombifolium* (Y.Ling & C.Shih) H.Ohashi & Yonek., *Chrysanthemum* × *shimotomaii* Makino, *C. sinuatum* Ledeb., *C. vestitum* (Hemsl.) Kitam., *C. yantaiense* M.Sun & J.T.Chen, *C. yoshinaganthum* Makino, *C. zawadskii* Herbich, and *C. zhuozishanense* L.Q. Zhao & Jie Yang.

Cultivars: The number of chrysanthemum cultivars is incredibly large with more than 15,000 listed in Japan only. The National Chrysanthemum Society of Britain lists over 6000 cultivars (Machine and Scopes 1978). In India, too, the number would easily cross 500 marks. The exact number will vary because every year new varieties are being developed throughout the world, but they are not documented at one place. Different countries have started more and more breeding work and developing new varieties regularly. Lists of some selected cultivars grown widely throughout the world color-wise are the following:

Spray Cultivars: *white*, 'Artic', 'Bonnie Jean', 'Cloudbank', 'Divinity', 'Elegance', 'Hurricane', 'Japanerin', 'Memento', 'Nimbo', 'Polaris', 'Schnesstern', 'Snow-don', 'Super White', 'White Illini Springtime', 'White Marble', 'White Sands', 'White Spider', 'White Taffeta'; *yellow*, 'Agenta', 'Celebrate', 'Golden Crystal', 'Golden Hurricane', 'Golden Polaris', 'Golden Sands', 'Golden Vedova', 'Golden Winner', 'Jubilee', 'Souvenir', 'Sunbeam', 'Super Yellow', 'Yellow Agenta', 'Yellow Bonnie Jean', 'Yellow Divinity', 'Yellow Galaxy', 'Yellow Horim', 'Yellow Illiini Springtime', 'Yellow Marble', 'Yellow Nimbo', 'Yellow Snowstar', 'Yellow Spider', 'Yellow Tuneful'; *pink*, 'Belair', 'Bluechip', 'Blue Winner', 'Dolly', 'Illini', 'Springtime', 'Riviera Spider', 'Taffeta', 'Vedova'; *pale pink*, 'Pink Marble', 'Pollyanne', 'Snapper'; *red*, 'Crackerjack', 'Red Fandango', 'Red Galaxy', 'Red Nero', 'Red Tuneful'; *bronze*, 'Belreef', 'Bronze Nero', 'Bronze Rosado', 'Flame Belari', 'Galaxy', 'Tuneful'; *light bronze*, 'Apricot Marble', 'Apricot Winner', 'Dramatic', 'Orange Aglow'; *purple*, 'Fandango', 'Flamenco'; *salmon*, 'Coral Marble'.

Standard Cultivars: *white*, 'Giant Indianapolis White', 'Improved Mefo', 'May Shoesmith', 'Beauty', 'Snow Ball', 'William Turner', 'Innocence', 'Gen. Petain', 'Valiant', 'Green Goddess', 'Ajina White', 'Premier'; *yellow*, 'Bright Golden Anne', 'Bright Yellow May Shoesmith', 'Rivalry', 'Yellow Fred Shoesmith', 'Chandrama', 'Kikubiori', 'Mountaineer', 'Super Giant', 'J.S. Lloyd', 'Triumphant', 'Evening Star', 'Melody Len'; *pink*, 'Cassandra', 'Deep Champagne', 'Pink Champagne', 'Promenade', 'Regal Anne', 'Ajina Purple', 'Pink Cloud', 'Pink Turner'; *red*, 'Crimson Anne', 'Red Anne', 'Red Resilient', 'Working Scarlet', 'Alfred Wilson', 'Alfred Simpson', 'The Dragon'; *bronze*, 'Bronze Princess Anne', 'Resilient'; *light bronze*, 'Gay Anne'; *purple*, 'Purple Anne'

Pot Cultivars: *white*, – 'Altis', 'Bonnie Jean', 'Mountain Snow', 'Neptune', 'Snow Crystal', 'White Anne', 'White Popsie', 'Windsong'; *yellow*,

'Armelle', 'Golden Crystal', 'Mountain Peak', 'Pride', 'Reaper', 'Spic', 'Stargold', 'Yellow Bonnie Jean', 'Yellow Delaware', 'Yellow Hector', 'Yellow Illini Spinwheel', 'Yellow Mandalay', 'Yellow Paragon', 'Yellow Popsie', 'Yellow Tuneful'; *pink*, – 'Always Pink', 'Dark Maritime', 'Deep Louise', 'Deep Popsie', 'Illini Trophy', 'Judith', 'Proud Princess Anne', 'Princess Anne Superb', 'Regal Anne', 'Rose Hostess', 'Royal Trophy'; *pale pink*, – 'Distinctive', 'Maritime', 'Wedgewood'; *red*, – 'Crimson Anne', 'Cromson Torch', 'Red Torch', 'Rory, Rufus', 'Working Scarlet'; *bronze*, 'Bronze Princess Anne', 'Copper Hostess', 'Gay Louise', 'Glowing Mandale', 'Mandalay', 'Rascal', 'Red Anne', 'Sparking Mandale', 'Tuneful'; *light bronze*, 'Bronze Popie', 'Dramatic', 'Gay Anne', 'Orange Aglow', 'Orange Bowl'; *purple*, 'Cerise Magnum', 'Purple Anne', 'Royal Purple'

It is difficult to mention the exact number of varieties available in India. New varieties are being developed by different research institutions, universities, amateur growers, etc. through selections, breeding. and induced mutations. A tentative list of varieties is mentioned.

CSIR-National Botanical Research Institute, Lucknow, India, is maintaining the following chrysanthemum cultivars (Datta 1998):

Large-flowered chrysanthemum – White: 'Beatrice May', 'Beauty', 'Bharat Mata', 'Casa Grande', 'Dee', 'Dorrige Queen', 'Frosty Whisker', 'General Petain', 'Green Goddess', 'Green Sleeves', 'Gypsy Queen', 'Icicles', 'Imperial', 'Jet Snow', 'John Webber', 'June Bride', 'Kasturba Gandhi', 'Kokka Soun', 'Maudjafferies', 'Mrs C Tolly', 'Nightingale', 'Pennylane', 'Purnima', 'Shamrock', 'S S Arnold', 'Snow Ball', 'Snow Don', 'Tokyo', 'Valiant', 'White Cloud', 'White Snow', 'White Sport of Pink Cloud', 'Willium Turner', 'Woolman Century', 'White Sport of Pride of Madford'. *Yellow*: 'Autumn King', 'Betty Barnes', 'Bhima', 'Bob Pulling', 'Chandrama', 'Cossak', 'Diamond Jubilee', 'Duskey Queen', 'Ella Dalby', 'Mahabi', 'Evening Star', 'Florida', 'Garden State', 'J S Salisbury', 'Kiku Biori', 'Kokka Yamata', 'L C Philips', 'Mountaineer', 'Mrs J A Miller', 'Mr Roger', 'Thompson', 'Mrs Nancy', 'Ferneaux', 'Pitamber', 'Queen of Tamluk', 'Rohinhood', 'R Venkatraman', 'Senyo No Rya', 'Sheila Morghan', 'Shin Mei Getsu', 'Sonar Bangla', 'Super Giant', 'Surya', 'Tamra', 'Thiokinga', 'Yellow Reflex', 'Yellow Rayonette', *Red*: 'Alfred Wilson', 'Arjuna', 'Black Hawk', 'Bicolour Incurved', 'Crimson Tide', 'Dorrige Velvet', 'Dragon', 'Gusman Red', 'Party Time', 'R.M. Quittenton', 'Red reflex', 'Leviathan', 'Mrs W A Reid', *Mauve*: 'Ajina Purple', 'Allahabad Reflex', 'Angeles Belle', 'Belur Math', 'Cover Girl', 'Coronation Pink', 'Edith Cavel', 'Fish Tail', 'Hope', 'H Townsend', 'Incurve Dwarf', 'Julius Brinas', 'Kenroku Kangiku', 'Kingford Smith', 'K N Modi', 'Kunchit', 'Mahatma Gandhi', 'Otome Zakura', 'Pink Brocade', 'Peacock', 'Pink Cloud', 'Pink Casekt', 'Pink Intermediate', 'Pink Rayonette', 'Pink Turner', 'Potamac', 'President Viger', 'Pride of Jamshedpur Raja', 'Royal Pinch', 'Royal Purple', 'Satish Modi', 'Scater's Waltz', 'Senkyo Emaki', 'Shefali', 'Spoon', 'Sport of H, Towensend', 'Tata Century',

'Taiho Tozan', 'Violent Queen', '(M45)', '(M-61)'. *Terracota*: 'Achievement', 'Alfred Simpson', 'Appart', 'Autum Blaze', 'Bhai- Bhai', '(T-10)', 'T-1', 'Captain Kettle', 'Chengis Khan', 'Dignity', 'Distinction', 'Gambit', 'Gen-Carpenter', 'Goliath', 'Heather James', 'Jane Sharp', 'Miss Universe', 'Mrs Helmipot', 'Orange-Fair Lady', 'Paul, Ronaldo', 'Sancho', 'S L Andre', 'Spider Bruno', 'Thiching Queen', 'Red Fair Lady', 'Red Quill'

Small-flowered chrysanthemum – Summer season cultivars: 'Himanshu', 'SU-1', 'Jwala', 'Jyoti', 'Su-3', 'Su-4', 'Phuhar'. *September–October blooming cultivars*: 'Ajay', 'Sharda', 'Sharad Kiran', 'Sharad Shobha', 'Vijay', 'Vijay Seedling'. *October blooming cultivars*: 'Arunima', 'Sharad Kanti', 'Sharad Mukta', 'Sharad Sandhya', 'White Dwarf (OO-8)'. *October–November blooming cultivars*: 'Chakra', 'Double Korean', 'Hemanti', 'Lalpari', 'Makhmal', 'Megami', 'Mohini', 'Nanako', 'Sharad Har', 'Sharad Mala', 'Tricolour', 'White Prolific', 'Yellow Prolific (NN-14)'. *November-December blooming cultivars*: 'Archana', 'Apsara', 'Birbal Deep Pink', 'Cotton Ball', 'Jayanti', 'Jubilee', 'Kundan', 'Ping Pong', 'Ratna', 'Yellow mutant of Ratna'. *December–January blooming cultivars*: 'Ratna', 'Button', 'Gauri', 'Gulal', 'Jaya', 'Khumaini', 'Lalima', 'Lilith', 'Mauve Spoon', 'Nilima', 'Puja', 'Purplish Red', 'Sunayana', 'Sunil', 'Vasantika', '(X-1)'. *Dwarf (no pinch no stake mini cultivars)*: 'Akita', 'Appu', 'Apurva', 'Arun Kumar', 'Arun Singar', 'Bindiya', 'Bronze', 'Cameo', 'Haldighati', 'Hemant Singar', 'Mahendra Singar', 'Mini Queen', 'Minihar', Orange, Pancho, Peet Singar, Pink Princess, 'Rangoli', 'Red', 'Red Anemone', 'Sengoku Ban', 'Sharad Singar', 'Shizuka', 'Suhag Singar', 'Shveta Singar', 'Swarn Singar', 'White Dwarf', 'White Pincushion', 'Yellow Charm'. *Decorative*: 'Alankar', 'Astral', 'Iiar', 'Jwara', 'Kalyani', 'Kanpur Yellow', 'Navneet', 'Pink', 'Renukoot', 'Seedling', 'Shyamal', 'Sonalitara'. *Stripped* 'S': 'Countees stripes', 'Duke', 'Karanfool', 'Kiran', Surekha. *Spoons* 'S': '(T-1)', '(T-3)', '(T-4)', '(T-5)', '(T-6)', '(T-7)', '(T8)', '(T-14)'. *Cultivars a*: 'Dainty Maid', 'Executive', 'Gaity', 'Perfecta', 'Venus', '(AA-4)', '(AA-9)'. *Cultivar b*: 'Angela', 'Anjali', 'Aura', 'Blaze', 'Coy', 'Gem', 'Lady-Roberts', 'Lord-Roberts', 'Marble', 'Marshal', 'Modella', 'Murcury', 'Rosa', 'Sukhai', 'Topaz', 'Vandana', '(A-8)'. *Cineraria* 'c': 'Bronze', 'Charmis', 'Philips', 'White Seedlings', '(C-5)'. *Cushion* 'E': 'Basanti', 'Fairy', 'Freedom', 'Himani', 'IIHR Selection', 'Kaumuduni', 'Kumkum', 'Processor', 'Harris', 'Seedling', 'Shanti', 'Snow White'. *Cultivars* 'F': 'Harvest Home', 'Laura', 'Stella'. *Cultivar* 'I': 'Molly', 'Fanny', '(I-3)', '(I-4)'. *Cultivar* 'Y': 'Pink', 'Rani', 'Sindoori', 'Sport of Y-1'. *Single Korean* 'N': '(N-1)', '(N-2)', '(N-3)', '(N-4)', '(N-6)', '(N-7)', '(N-8)', '(N-9)', '(N-10)', '(N-11)', '(N-12)', '(N-14)', '(N15)'. *Quilled* 'Q': '(Q-1)', '(Q-2)', '(Q-3)', '(Q-4)'. *Double Korean*: 'Aparajita', 'Batik', 'Cissie', 'Fatima', 'Flirt', 'Hindalco', 'Juno', 'Jyotsna', 'Khurso', 'Lalpari', 'Lalquila', 'Man Bhawan', 'Priya', 'Red Gold', 'Shabnam', 'Tara', 'White' (Korean Double), '(O-6)', '(O-21)', '(O-2)'. *Pin Cushion*: 'Malika', 'Mayur'.

The following new varieties of chrysanthemum have been developed through conventional breeding at CSIR-NBRI, Lucknow, and in some other ICAR institutions and universities in India: *CSIR-NBRI*: 'Ajay', 'Appu', 'Apsara', 'Apurva',

'Apurva Singar', 'Arun Kumar', 'Arun Singar', 'Bindiya', 'Birbal Sahani', 'Dhawal', 'Diana', 'Gauri', 'Gulal', 'Guldasta', 'Haldighati', 'Hemant Singar', 'Himanshu', 'Jaya', 'Jayanti', 'Jubilee', 'Jwala', 'Jyoti', 'Jyotsna', 'Kargil 99', 'Kaumudi', 'Kiran', 'Kirti', 'Kundan', 'Lal Kila', 'Lalima', 'Lalpari', 'Lilith', 'Maghi', 'May-Day', 'Mayur', 'Meghdoot', 'Mini-Queen', 'Mohini', 'Mother-Teresa', 'NBRI Pushpangadan', 'NBRI Khoshoo', 'NBRI Kaul', 'NBRI Himanshu', 'NBRI Little Orange', 'NBRI Little Hemant', 'NBRI Little Kusum', 'NBRI Little Pink', 'NBRI Yellow Bud Sport', 'Neelima', 'Niharika', 'Nirmal', 'Pancho', 'Peet Singar', 'Phuhar', 'Priya', 'Prof. Harris', 'Puja', 'Ragini', 'Rangoli', 'Sadbhavna', 'Shanti', 'Ratna', 'Sharda', 'Sharad Kanti', 'Sharad Kumar', 'Sharad Mala', 'Sharad Mukta', 'Sharad Sandhya', 'Sharad Shobha', 'Sharad Singar', 'Shizuka', 'Shyamal', 'Suhag Singar', 'Sujata', 'Suneel', 'Sunayana', 'Suparna', 'Surekha Yellow', 'Surya', 'Swarn Singar', 'Sweta Singar', 'Tushar', 'Vandana', 'Vasantika', 'Vijay', 'Vijay Kiran', 'Vinaya', 'White Charm', 'White Profile', 'Y2K', 'Yellow Charm', 'Yellow Prolific', 'NBRI Yellow Bud Sport'. *Indian Institute of Horticultural Research, Bengaluru*: 'Arka Ganga', 'Arka Pink Star', 'Arka Ravi', 'Arka Swarna', 'Chandrakant', 'Chandrika', 'Indira', 'Kirti', 'Nilima', 'Pankaj', 'Rakhee', 'Ravikiran', 'Red Gold', 'Yellow Star', 'Yellow Gold', 'Usha Kiran'. *Punjab Agricultural University, Ludhiana*: 'Anmol', 'Baggi', 'Gul-E-Sahir' [yellow], 'Royal Purple', 'Yellow Delight', 'Autumn Joy', 'Garden Beauty', 'Winter Queen'. *Tamil Nadu Agricultural University, Coimbatore*: 'CO.1', 'CO.2', 'MDU'. *Dr. YSPUHF, Nauni, Solan*: 'Solan Mangla'

Some outstanding chrysanthemum cultivars have originated in India through spontaneous mutations. The most notable varieties are 'Kasturba Gandhi' (white) developed from 'Mahatma Gandhi' (mauve), 'Sonar Bangla' (yellow) from 'Snow Ball' (white), 'White Cloud' (white) form 'Pink Cloud' (pink), 'Sharda' (yellow) from 'Sharad Shobha' (white), 'Queen of Tamluk' (yellow) from 'Casa Grandi' (white), 'R Venkatraman' (yellow) from 'S S Arnold' (white), 'William Turner' (white) from 'Pink Turner' (pink), 'J S Lloyd' (yellow) from 'William Turner' (white), 'White Ball' from 'Pride of Madford', etc. In addition to these promising varieties, a large number of new varieties have been developed in India through sports.

Mutant varieties developed in India: *CSIR-NBRI*, Lucknow – 'Agnishikha', 'Alankar', 'Anamika', 'Aruna', 'Asha', 'Ashankit', 'Basant', 'Basanti', 'Batik' (Fig. 6d), 'Colchi Bahar' (Fig. 6i), 'Cosmonaut' (Fig. 6f), 'Gairik', 'Hemanti', 'Himani', 'Jhalar', 'Jugnu', 'Kanak', 'Kansya', 'Kapish', 'Kumkum', 'Kunchita', 'Lalima Head Shape', 'Lalima Tubular Mutant', 'Lohita', 'Man Bhawan', 'Navneet' (Fig. 6g), 'Navneet Yellow', 'Nirbhaya', 'Nirbhik', 'Pingal', 'Pitika', 'Pitamber', 'Purnima' (Fig. 6c), 'Raktima', 'Rohit', 'Shabnam' (Fig. 6e), 'Shafali', 'Sharad Har', 'Sheela', 'Shweta', 'Surekha Yellow', 'Sonali' (Fig. 6h), 'Subarna', 'Tamra', 'Taruni', 'Tulika'

19.7 Cultural Practices

(i) *Pot culture*: Well-drained sandy loam of good texture and aeration, with a neutral or slightly acidic pH (6.5–7.0) and a high organic content, is the ideal soil for chrysanthemum culture (Datta 1996).

For pot culture of small-flowered chrysanthemum soil/F.Y.M./leaf mold (1: 2:2 v/v) is very suitable. Small amount of bone meal/super phosphate (2 table-spoon) is added sometimes with the compost. Proper vegetative growth at the early stage is most important. Feeding mixture, rich in nitrogen and potash content, should be applied regularly. Phosphorus is used as basal dressing. Small amount of oil cake is added over the soil in the pot at the early stage, and it is allowed to dissolve slowly by normal irrigation process. Application of liquid manure is started when the root system is well established. Fresh cowdung and oil cake are allowed to rot in a container in water. This decanted solution is applied to plants once a week as watering. Liquid manure (5 g potassium nitrate and 5 g ammonium nitrate dissolved in 10 liters of water) is applied twice at fortnight interval during September. The top portion (about 2 cm) of each pot is filled during the end of September by a compost mixture made up of neem cake/F.Y.M./soil/wood ash (1:4:8:4 part). Two doses (at fortnight interval) of liquid manure (as mentioned above) are applied at the time of flower bud initiation. Immediately after that, two doses (once a week) of another liquid fertilizer mixture (potassium nitrate 30 g + urea 5 g dissolved in 10 liters of water) are applied.

(ii) *Sterilization*: It is one of the most important operations for chrysanthemum culture. Old earthen pots are first cleaned with the help of a piece of cloth or jute and then dipped in 2% formalin solution and sun dried. Coarse sand is a very good medium for planting cuttings for rooting. The sand is sterilized by heating over an iron pan for half an hour. It may also be sterilized by using 2% solution of formalin. The compost containing of a mixture of leaf mold, cowdung manure, and soil is properly sprayed 2% formalin, mixed properly, and covered with alkathene sheet for about a week.

(iii) *Field preparation*: The field is ploughed twice or thrice, and 15–20 tons of F.Y.M. per acre are applied at the time of preparation of beds. Rooted suckers are planted at 30 cm (12″) distance. Fertilizers at the rate of 25 Kg nitrogen and 40 Kg each of P_2O_5 and K_2O per acre are applied for better performances. Application of urea at the rate of 30 kg per acre at color showing stage is recommended.

(iv) *Preparation of pots*: For maintaining proper drainage system, the drainage hole at the bottom of pots should be loosely covered first with broken pieces of earthen pots before filling compost. The drainage hole should never be clogged with compost and should be checked frequently.

(v) *Irrigation*: Chrysanthemum requires frequent and thorough watering before monsoon. Proper drainage system should be maintained both in beds and in pots as these plants are very sensitive to excessive water. There should not be waterlogging in beds and also on the pots during rainy season. If the water accumulation is due to clogging of drainage hole and faulty potting mixture, checking of drainage hole and changing of old potting mixture by new potting mixture are recommended. Excess water accumulation causes serious damage to the plant roots.

(vi) *Pinching*: Pinching is one of the most important operations in chrysanthemum culture. Soft vegetative shoot tips half to 1 inch long are removed. It is done with thumb and forefinger, although knives, scissors, etc. can also be used. Pinching gives the plant an appreciable shape and most essential for small-flowered chrysanthemum. Pinching is performed both in suckers and in cuttings. Two types of pinching, i.e., soft pinching (top soft tips of the shoot with 2–3 open leaves) and hard pinching (longer portion up to hard shoot), are performed.

(vii) *Suckers*: Rooted suckers are planted in field during January for stock plants. For vigorous and profuse branching, regular pinching is performed in these plants. These stock plants may be used both for preparation of cuttings and as potted plants for flower show and other displays. The first pinching is performed in April, second pinching in May, and third pinching in June. After third pinching, cuttings are taken from these mother plants. Fourth pinching is performed during August, and the final pinching of stock plant is completed by the middle of September.

(viii) *Cuttings*: Suckers are planted in open beds in January, and all care is taken for their vigorous growth and profuse branching by frequent pinching. Terminal shoot cuttings (7–13 cm or 3–4″ long) are taken from these stock plants in June. 15–20 cuttings may be available from a healthy well-grown plant. Cuttings are prepared during June to August. The cuttings may be planted in 10–25 cm (4″–10″) pots, flat thalis, or rectangular wooden box (10–13 cm or 4″–5″ deep) filled with sterilized coarse sand. The basal portion (10 mm or less than half inch) of cuttings are dipped either in Seradex/Keradex (rooting hormone) or in any other rooting hormone (1000 ppm solution of indolebutyric acid) for better rooting. Before planting, 3–4 basal leaves of each cutting are removed, and the lower portion of cuttings are treated with some copper fungicide to avoid fungal growth. The container, after planting cuttings, should be kept in semi-shade for about 2 weeks to avoid hot sun. Watering is done once or even twice a day as per weather condition.

(ix) *Disbudding and Dis-shooting*: These operations are mostly performed for large-flowered and decorative-type chrysanthemums. Disbudding operation is an important factor in the maintenance of high-quality product. Many of the varieties are disbud or standard types, in which the largest terminal bud is reserved and all axillary buds are removed. For disbudding of spray varieties, only the large apical bud is removed, and the axillary buds are allowed to develop. When growers want to develop three blooms per plant or one bloom per plant, these operations are most essential. For taking three blooms per plant (June-planted cuttings), the first pinching is done in August. Three lateral strong shoots are allowed to grow and others are removed. Disbudding starts in October when all but the central buds on each lateral shoot are removed. Lateral buds and side shoots are removed at their early stage of growth from time to time. For taking one bloom per plant (June-/July-planted cuttings), no pinching is done. Only the main stem is allowed to grow.

Disbudding and dis-shooting of undesirable lateral buds and shoots are done as in the case of three-bloom type.

(x) *De-suckering*: During the vegetative growth phase, plants grow upward. New suckers continue to develop from the base of plants. For proper and vigorous growth of plants, suckers are removed from time to time.

(xi) *Staking*: Staking is necessary to keep plants erect and to maintain proper shape of plants and bloom. Growers need only one stake for single bloom and three stakes for three blooms per plant. In small-flowered, 5–8 stakes are used for profuse blooming.

19.8 The Art of Training

(i) *Standard*: Large-flowered chrysanthemums are trained as standard for better shape of the plants and attractive extra-large flower producing 1–3 blooms per plant. Suckers are planted in January. The plants are transplanted several times in bigger pots and finally into 25 cm (10″) pot in August. These plants bloom in November–December. These plants are normally very tall and need care throughout the year. According to recent cultural methods for developing better standards, plants are developed from cuttings in July. It avoids unnecessary care of plants for about 6 months.

(ii) *Sen Rin Tsukuri*: It is a Japanese style of chrysanthemum culture whose literal meaning is "growing thousand blooms." The plant is designed to a geometrical shape (6–10 concentric circles in stepped manner), and it is trained in such a way that about 200–300 blooms, and even more, are formed per plant having an approximate height of 153–183 cm (56 ft.) and a diameter of 183–244 cm (6–8 ft.). Varieties with vigorous growth habit in all directions, long internodes, flexible stems, etc. are suitable for this. Suckers are planted in December in 15 cm (6″) pots filled with the compost made of leaf mold/light clay/charcoal powder (10:2:1). Second potting is done into 20 cm (8″) pots during the end of February in a potting mixture of cowdung manure/leaf mold/light clay + bone meal (4:2:2 + 1 tablespoon). Two tablespoonful of oil cake is added on the top of each pot for vigorous growth. The first pinching is performed when the plant attains a height of 20–25 cm (8″–10″). Plants are shifted in bed in March for more vigorous growth. The bed is well manured with rotten farm yard manure, bone meal, and oil cake. The planting distance is 92 cm (3 ft.). The main stem is made vertical with the help of a strong bamboo stake. The lateral shoots which come out after first pinching are made horizontal with the help of hooked wire inserted in the pot. Pinching is continued for profuse branching till the end of June. The new branches are trained in desired direction. Liquid cowdung manure and oil cake are added to accelerate the growth rate. Disbudding is done from October to maintain only one terminal bud. The plant is finally shifted into the container. The lifting of plant from bed to container is a very important operation and should be performed with great care to maintain proper beauty of plant. The plant should be lifted without damaging the ball. The final pot size may be 31 cm (12″). When the plant is well established in pot after lifting, the plant is given the final shape. Now the structure is made by splitting bamboo around the plant. The shape of the structure may be given according to choice, but the most popular shape is hemisphere or dome-shaped. Then branches and individual buds are tied at definite places so as to give the plant a particular uniform shape. Before lifting of plant from bed and the time of final tying, the branch irrigation is stopped for 2–3 days so that the branches become soft and more flexible.

Suitable varieties: 'Beauty', 'Maud Jefferies', 'John Webber', 'Evening Star', 'Shin Mei Getsu', 'Allhabad Reflex', and 'Raja'

(iii) *Bush Form*: It is a specific cultural practice for small-flowered chrysanthemum where the plant is given a bush appearance by specific pinching and training. Blooms are arranged compactly to give an effect of a floral carpet. Korean, anemone, button, charm, stellate, decorative, quilled cultivars having profuse branching habit are the most suitable. First soft pinching is started in March when the plant attains about 20 cm (8″) height. The lateral primary branches are again soft pinched, and the process is continued till September. For maintaining uniform spreading, bamboo stakes are used around the periphery and tied the plants with a ring of wire.

(iv) *Pot Mums*: One cutting is planted in one pot which grows tall, and the lower portion of stem looks naked. 5–7 cuttings are also planted at equal distance around the periphery of a pot (20–25 cm or 8″–10″) during June/July. Optimum conditions are provided for proper vegetative growth of plants for 2 months till the initiation of flower bud. The compost mixture of clay/F.Y.M./leaf mold (1:2:2) has been found to be very good for proper growth of plants. Top dressing with neem cake about a month after potting is very useful. Liquid manuring with fertilizer mixture during early September is recommended for vigorous growth. The height of plants of pot mums is mostly uniform. In no-pinch pot mums, the number of flowers are almost as many as there are plants, i.e., 5–7, and the flowers are bigger in size. If larger number of flowers are desired, the branching is encouraged by soft-pinching.

Suitable varieties: 'Beatrice May', 'Kasturba Gandhi', 'General Petain', 'Otome Zakura', 'Pink Cloud', 'Pink Casket', 'Fish Tail', 'Jack Straw', 'Evening Star', 'Goldie', 'John Reid', etc.

(v) *Cascade Form*: Cascade form plants give the excellent look effect of a water fall in blooming stage. Cascades may be small, medium, and large size. Varieties with both long and short internodes, thick flexible stem, profuse branching, and prolific blooming habit and specially anemone and Korean types are suitable. Selected varieties are planted in the bed during March in a slanting position (60° angle). A strong bamboo stake is inserted in the soil at the same angle. The plants are dug out from bed with large balls in July and planted in a large pot at 45° angle. Bamboo frames of desired shape and design are prepared. The main stem and branches are tied to the frame at several places. The frame is bent gradually downward by applying pressure taking care that the main stem is not broken or cracked. The operation should be done very carefully and slowly step-by-step so that the plant acquires a horizontal shape by the end of August. The bending process is continued for the next 2 months (September–October). Pinching is the most critical technique in the formation of cascade. Pinching is started at the height of 15–25 cm (6″–10″) from the ground, and the process is continued till September. Both soft and hard pinching are performed. October is the most crucial period when bud initiation starts. One should keep regular vigilance to plants, and buds should be arranged systematically by bending and trying to cover the entire structure.

Suitable varieties: 'Perfecta', 'Modella', 'Jaya', 'Aparajita', 'Mayur', and 'Flirt'

(vi) *Coniform*: The plant shape is made conical by special training. The most suitable varieties are those which produce profuse lateral branching from the base of the main branch upward. Staking and pinching are most important for giving a perfect coniform shape. A strong vertical bamboo stake is used from the very beginning to keep the main stem erect. The first pinching is performed during late March, and the last pinching is most important step which is performed in four stages during September. The plant is divided into four regions – lower, middle, upper, and tip. The branches at lower portion (1/third height) are pinched first. The middle portion is pinched after interval of another 3–4 days. The tip is pinched at the end after another 3 days. For support to the branches, additional bamboo stakes are used to maintain a perfect "coniform" plant. The first lateral shoots from the top are removed. The second lateral shoots are allowed to grow upward. Subsequently other lateral shoots which develop late are pinched selectively. The longer shoots are at the base and shorter at the upper level.

(vii) *Fan Form*: It looks like a hand fan and small-flowered chrysanthemums are suitable. It looks flat, round, and vertical made of split bamboo. The bamboo structure is fixed in between the two identical plants planted close to each other in 25 cm (10″) pot. All the branches of both plants are tied to the bamboo structure to give the plants a flat shape. Both soft and hard pinching start during the end of February and continues up to September. The central portion of the plant is made first by hard pinching. The area surrounding the central portion is pinched after 3–4 days. The pinching date should be calculated in such a way that the last pinching is performed by the middle of September (Datta 1996).

19.9 Protected Cultivation

Protected cultivation has been taken up extensively for the year-round production of chrysanthemum. Greenhouse cultivation has been adopted in many countries for commercial exploitation of cut flowers. Low-cost polyhouse model has been developed in some countries using bamboo poles or GI pipes covered with white/yellow HDFE sheets. Low-cost greenhouse model has also been created by using mild-steel tubes and ultraviolet stabilized polythene sheet. Cooling and exhaust systems do not cost much. It may be mentioned that highly sophisticated greenhouses are used in case of commercial production of high-quality flowers, but in case of nursery, low-cost greenhouse is ideal.

19.10 Postharvest Managements (Harvesting, Grading, and Packaging)

Harvesting is done in the morning hours at appropriate stage, and hand cutters are used for this purpose. The correct stage of harvesting depends on the cultivar, marketing, and other facilities available to a grower. Single cultivars are harvested when the maximum number of flowers is open, but before the pollen grains are shed from the outer row of the disc florets. Anemone cultivars are harvested before the central cushion in the topmost flower is fully developed, whereas decorative types should be harvested when the petals in the center of the topmost flower are almost fully developed. In standards, harvesting is generally done when outer ray florets ceased to elongate. Pot mums are sent to the market when flowers are about half to full open (Machine and Scopes 1978). The stems after cutting are stripped off leaves (one-third) and collected in buckets containing cold water at 15–18 °C. The buckets are kept in cold rooms at 10 °C for cooling for a minimum duration of 2 h. The water contains a biocide to prevent the growth of microorganism. Silver nitrate at the concentration of 25 ppm has been found to be most effective biocide for chrysanthemums (Kofranek 1980). The bud opening solution is used at the marketplace, and 200 ppm 8-HQC and 2% sucrose are ideal for bud opening. Bud opening solution containing 2.5–3.0% sucrose, 25–30 ppm silver nitrate, and 75 ppm citric acid has been suggested. BAP has been reported to be one of the limiting factors in opening of immature buds. Sucrose (5%) in combination with biocide ($AgNO_3$, 25 ppm + citric acid, 75 ppm.) substantially increases bud size and vase life.

Cut flowers are graded into several grades depending on stem length, spray type, color and diameter of flowers, weight, number of flowers, etc. For pot mums, there is no standard grade. Plants should be bushy with good growth, 2.0–2.5 times as tall as their pots, having a minimum of 15 flowers and free from pests and diseases. A plant having 20–25 good-quality flowers would be more desirable. Normally, corrugate fiberboard (CFB) material is used for preparation of packing box. The dimensions of packing box vary on the number of stems. The CFB box is lined by spreading an adequately large sheet of tissue paper along the cartoon walls. The stems are placed horizontal layer by layer and in an alternating pattern using tissue paper between the layers. The top layer is covered with the tissue paper. The tissue paper is just moistened with water. The boxes are closed using self-adhesive BOPP tape. The boxes are stacked in cold rooms or air-conditioned vans as the case may be.

Chrysanthemum flowers show visible senescence symptoms after 12–15 days. Florets/petals primarily determine the commercial longevity of flowers. Knowledge on flower senescence is very important, and majority of work on flower senescence focused on ethylene sensitive plants. Chrysanthemum is an ethylene-insensitive plant, and experiment has been conducted to study the physiological, biochemical, and genetic processes that occur during floret senescence. Reactive oxygen species

(ROS) concentration and lipid peroxidation increase from young floret stage to the senescent stage. Activities of superoxide dismutase (SOD), ascorbate peroxidase (APX), peroxidase (POD), and catalase (CAT) show uniform increases from young floret through to the mature stage and thereafter decline. Lipid peroxidation and membrane damage are involved in flower deterioration. Variation in oxidative stress and the antioxidant enzyme activity in the florets of chrysanthemum during senescence have been studied. Among the SOD isoforms, Fe-SOD and Cu/Zn-SOD were induced during the onset of senescence. Similarly different isoforms of APX and glutathione reductase (GR) also appeared during the senescence process. The capacity of the antioxidative defense system increased during the onset of senescence, but the imbalance between ROS production and antioxidant defenses ultimately led to oxidative damage. It has been proposed that a decrease in the activity of a number of antioxidant enzymes that normally prevent the buildup of free radicals can at least partially account for the observed senescence and cell death in chrysanthemum florets (Chakrabarty et al. 2007).

19.11 Manipulation of Growth/Flowering and Vase Life

A wide range of growth substances and chemicals have been used to regulate growth and flowering in chrysanthemum. GA_3 shows marked variation in response to growth and development depending on the time of application and stage of growth. Considerable encouragement of floral initiation has been observed after spraying of β-sitosterol and lanosterol (100 ppm) and glycocholic acid (500 ppm). Branching and plant height can be increased by spraying BA (40 ppm) and PBA (200 ppm) and morphactin, respectively. Flowering can be delayed using ethephon and CEPA ((2-chloroethyl)phosphonic acid, 100 ppm). Vegetative growth and flower yield are increased by spraying GA_3, ascorbic acid, and IAA alone and in different combinations:

(i) **Growth retardation**

A number of growth retardants have been identified to restrict stem elongation and to develop well-shaped pot plants. *SADH* (Alar, B-nine 2000–4000 ppm) reduces stem length, and stems are thicker and stronger. Application at 250 ppm 3 weeks after the start of short days increases flower life. *PHOSFON* (300–750 g/m^3) is used mainly for potted plants, and the rates may vary depending on cultivar, potting compost material, and season. It reduces cell division which effect on growth due to shortening of internodes. Higher doses delay flowering and reduced flower size. It reduces apical dominance, to encourage branching, to improve petal color, and also to develop resistance to disease and drought (Machine and Scopes 1978; Sachs and Kofranek 1963). *Paclobutrazol* is a gibberellin biosynthesis inhibitor which suppresses internode and long-term growth, dwarfing, and delay in flowering properties (Kher 1973; Gilbertz 1992). A number of other chemicals like CCC (chlormequat) at 1.25%, Amo-1618, EI-531 (α-cyclopropyl α-4-methoxyphenol-5-

pyrimidin methanol) as a single spray at 150 ppm, ancymidol, pp. 333, etc. have been found effective to reduce stem length (Sachs and Kofranek 1963; Barrett and Bartuska 1982).

(ii) **Vase life**

The use of proper preservative solution throughout the period of postharvest handling is very important to prolong the life of cut flowers. Dipping of stem for a very short period (5 s) in 1200–4800 ppm silver nitrate has been suggested. Addition of 2% sucrose to silver nitrate is beneficial which increases the vase life from 12 days to 20 days. The type of preservative used is very critical because of 8-HQC that even at level as low as 150–200 ppm can cause stem discoloration and 1–3% sucrose may result to foliar chlorosis and/or necrosis. Sodium dichloro-S-triazinetrione (SDT) in combination with sucrose is equally effective as silver nitrate and sugar. The flower stems should be placed into the preservative solution as soon as possible after harvesting or removing them from storage.

(iii) **Effect on flowering and programmed blooming**

Blooming season of chrysanthemum could be extended slightly by traditional methods like late planting and delayed pinching. Being a short-day plant, flower bud formation in chrysanthemum can be prevented by reducing the length of continuous night below critical level through artificial light, and the plants remain vegetative till the long day.

Chrysanthemums are classic short-day plants with blooming controlled by the relative length of the day and night. Since chrysanthemums are considered a short-day (SD) plant, the plants show vegetative growth under long days and flowers under short days. Most cultivars begin to develop flower buds when days are less than 12 h long. When the night gets longer than 9 h, chrysanthemums begin to set flower buds. The blooming period under traditional culture is short, few weeks or month, depending on the geographical location of growing area. Each chrysanthemum variety has its specific photoinduction requirement, i.e., the number of lights and continuous long dark periods (short days) required by its plant for coming into bloom after it has attained maturity. A combination of day and night is termed as one photoperiod. Varieties require differential photoperiod ranging from 8 to 15 weeks for coming into bloom. A variety requiring 8 weeks for photoinduction period is said to belong to 8-week response group. The term "response group" refers to the number of weeks to flowering from the commencement of short days in a night temperature of 15.6 °C in light conditions near to those prevailing for natural flowering (October and November). The varieties have been thus classified into eight groups, i.e., 8, 9, 10, 11, 12, 13, 14, and 15 weeks' response group, depending on their respective photoinduction requirement. A succession of chrysanthemum of some choicest varieties is obtained by planting, pinching, lighting, and shading the plants according to a definite plan. By manipulating the planting date and the light inside the greenhouse, the grower can coordinate the response of several varieties as per specific flowering dates and marketing requirements. The success of year-round blooming has been based on sound business thinking. With planned scheduling, now salable chrysanthemum flowers can be produced every day of the year (Datta and Gupta 2012; Kher 1969, 1972, 1976).

19.12 Chrysanthemum Trade

Chrysanthemum stands third in the world cut flower trade and first in Japan and China, third position in Germany, and second position in the United Kingdom. Chrysanthemums are produced in almost all major flower-producing regions of the world, with major area being in Japan followed by the Netherlands, Colombia, Italy, and Spain. Standard (Inga, Lilac Eleonora, Boris Becker, Rivalry, Tom Pearce, Fred Shoesmith, Migoli, Astro, Pandion, Alee Bedser), Spray (Spider White, Spider Yellow, Super White, Bosshardt Majoor, Glance, Starmark, Green Peas, Gold Peas, Statesman), and Santinis (Fatima, Kermit, Stallion, Delilah) are the main commercial types for international market. Spray-type chrysanthemums are of much demand in the market. White color dominates the chrysanthemum sales followed by yellow, purple, and pink, and top-selling varieties belong to the Reagan family, followed by spiders.

19.13 Diseases

Chrysanthemum suffers from different diseases which severely damage the plant at various stages of growth. Researchers, nurserymen, gardeners, and amateurs have reported their views about the symptom and control of different diseases in different journals, books, bulletins, monographs, newspapers, etc. The literature on diseases and control measures of chrysanthemum are very rich:

(i) *Wilt* - Wilt is the most serious and destructive disease of chrysanthemum. The word "wilt" may be used in conjunction with other words as the name of specific diseases. It may occur due to a number of different diseases and disorders or even just a lack of water. Wilting may cause rot of the root or the base of the stem. Yellowing and browning of leaves is the first symptom, which causes the death of the base of the plant upward. Infected plants are stunted and often fail to produce flower.

Symptom: Different fungal species responsible for wilting are *Fusarium* sp., *Verticillium* sp., *Pythium* sp., *Phytophthora* sp., *Phoma* sp., and *Rhizoctonia* sp. *Verticillium dahliae* is mainly found in greenhouse soil and enters the plant through roots, later invading the vessels of the stem and cutting off the water supply. Fungus disease occurs in some specific soil when the vegetative growth of plant ceases and the bud development begin. Lower leaves turn yellow, and the lower portion of the stem turns black and deep brown due to *Fusarium* wilt. Collar rot or root rot disease causes damping off of cuttings in the rooting bench. In warm humid condition, root rot wilt will occur in beds and plants wilt suddenly. The foliage margin first turns yellow, then the whole leaf becomes brown, and gradually the entire plant wilts due to fungal disease "Seidwetz disease" caused by *Verticillium alboatrum*. The symptoms of *Verticillium* wilt develop in two ways. First, the plant may be stunted, with interveinal yellowing of lower leaves associated with browning of the petioles. Second, more severe symptom develops with general wilting leaves turning brown but often attached to the plants. Vascular tissues may be brown and discolored although wilting may not occur until the flower buds are developing. Unsterilized or poorly sterilized soil encourages *Phoma* root rot (*Phoma chrysanthemicola*). Plants become stunted, often with chlorotic lower leaves which may eventually show necrotic areas (spotting). The necrotic areas spread slowly and the entire leaf is affected; ultimately, the whole plant will wilt and die. *Phytophthora* and *Pythium* species can invade sterilized soil and infect root. *Pythium* species infect cuttings in the rooting bench causing damping off at the base of the stem. Spores of these fungi are readily produced in wet soils and rapidly invade other plants, especially those with damaged roots. *Rhizoctonia solani* may infect roots but is more frequently found attacking the base of stems of young plants. *Sclerotinia* rot caused by *Sclerotinia sclerotiorum* is uncommon and rarely causes serious losses. Light brown lesions develop on stems which become covered with fully fungal growth. *Verticillium albo-atrum* wilt disease is the most destructive which causes conspicuous yellowing and browning of the leaves. Infected plants are stunted and often fail to produce flowers. The fungus is soil-borne and enters through the roots later invading the vessels of the stem and cutting off the water supply. Other two wilt diseases are caused by *Fusarium oxysporum* var. *tracheiphila* and *Pythium aphanidermatum*, respectively. Infection results in plugging of xylem vessel elements with gum and pectinaccous materials, hypertrophy, and hyperplasia of xylem parenchyma cells, abnormal activity of the vascular cambium derivatives, formation of cavities within tissue, and eventual colonization of phloem and cortex parenchyma cells which result in their collapses. The collapse and necrosis of these cells result in black streak appearing on the stems of severely wilted plants.

Control: Disease can be greatly reduced by growing new plants developing from tip cuttings of apparently healthy plants. The soil must be properly sterilized. The affected plants should be immediately destroyed. Regular steam sterilization and use of thiram or dusting with PCNB are good measures for control. Avoidance of overwatering, steam sterilization of beds, and good soil condition for proper air circulation to keep the plants dry will help to avoid the disease. Drenches of carbendazim (25–30 gm/100 ltr) and thiram (150 gm/100 ltr.), dusting quintozene (13 gm/m^2), and spraying or dipping in benomyl (240 gm/100 ltr.) have been found to protect young plants. High lime and nitrate nitrogen fertilizer treatments, together with benomyl drenches for control of *Fusarium* wilt and soil drenches with

Dexon or Truban have been recommended for excellent control. The use of Basamid (dazomet) in different propagation media composed of two methyl bromide compounds can control *Fusarium* wilt disease. Faltap (0.2%) and Capfat (0.3%) reduce collar and root rot disease severity significantly caused by *Phoma chrysanthemicola*. A wild-type isolate of *Trichoderma viridis* (T-1) and a benomyl-resistant biotype (T-1-R9) alone or in combination with *Aspergillus ochraceus* reduces disease by at least 50%. Application of *Aspergillus*, *Paecilomyces*, *Penicillium*, *Trichoderma*, and *Bacillus* spp. cultures on solid media is found to protect chrysanthemum from infection by *Rhizoctonia solani*.

(ii) *Rust*

Symptom: Rust is a troublesome disease in chrysanthemum. The symptom of rust, caused by *Puccinia chrysanthemi*, develops as blister-like swelling, which breaks open and discharge masses of brown, powdery spores. Severely infected plants became very weak and fail to bloom properly. The rusts may be carried on stocks and thus introduced into greenhouses. Rust pustules from undersides of the leaves discharge brown spores. White rust of chrysanthemum due to *Puccinia horiana* and is observed, and its relation to temperature has been noted. Infection of leaf petioles, stem, petals, sepals, and bracts is usually associated with sori formation. Mass infection leads to total necrosis of leaves, absence of flowering, and premature plant death.

Control: Early removal of infected leaves, better ventilation, supply of water directly to the pots without wetting the plants, and dusting of plants with sulfur have been recommended for control. Spraying the plants thoroughly with potash sulfide at the rate of half ounce to a gallon of water or a wettable sulfur spray can be used. Any good fungicides such as zineb, ferbam, phalton, captan, thylate, etc. can be used for its remedy. Spraying of oxycarboxin on plants infected with *Puccinia horiana* has been recommended. Plantvax (oxycarboxin) has been recommended for good control of white rust. Spraying of fungicides (propiconazole, triadimefon, benodanil, triadimenol, etc.) before appearance of disease symptoms has been recommended for control. Systemic fungicide (triforine, bitertanol, oxycarboxin, and propiconazole) are effective for control of white rust of chrysanthemum. Phytosanitary and cultural measures have been recommended to protect against infection. Weekly foliar spray of hexaconazole (at 50 g/ml) alone and/or in combination with captan (1:17.5) and with mancozeb (1:28) has been recommended for control of white rust.

(iii) *Leaf Spot*

Symptom: Two types of leaf spots are found in chrysanthemum. The one type of leaf spot is caused by *Septoria chrysanthemella*. In this case, the spots are at first yellowish and then become dark brown and black. Serious infection may result in premature withering of the leaves; the dead leaves hang to the stem for some time. The leaves curl and fall prematurely. The lower leaves are infected first. Numerous spores are found in the diseased area which are long and slender and marked by crosswalls. There are other types of leaf spots in chrysanthemum, caused by *Cylindrosporium chrysanthemi*. The spores are dark brown with yellowish margins. Excessive nitrogenous feeding stimulates blotch. Several other fungi cause spots on this host. Most prevalent are *Alternaria chrysanthemi* and *Cercospora chrysanthemi*. This produces dark brown spots with yellowish margins.

Control: Handpicking and burning to destroy the infected leaves have been recommended as good practices for control. Spraying/dusting with sulfur compounds or Bordeaux mixture, lime sulfur or karathane, bavistin (carbendazim), benomyl (0.1%), chlorothalonil, fermate, captan, phaltan, maneb, tecto (thiabendazole), nanob, and zineb and avoiding wetting leaves while watering and mulching with peat moss are perfect for control. Acceptable control was achieved in the field with 80% difolatan (Captafol) W.P. and 50% Zincofol W.P. alternately or alternately with 50% Benlate (benomyl) or by alternatively 65% Neo Asozin with 80% Captafol.

(iv) *Viral Disease*

Virus-infected plants show varying amounts of distortion and loss of flower color and reduce flower size. Color bleaching occurs on red, bronze, and pink varieties producing streaks or flecks of white or yellow on the ray florets. The virus is mostly transmitted by aphids. The number of viral diseases affecting chrysanthemum is nearly 20, and some notable are *Chrysanthemum* stunt, tomato spotted wilt, tomato aspermy, *Chrysanthemum* flower distortion, *Chrysanthemum* mosaic, *Chrysanthemum* rosette, etc.

Plants should be destroyed as soon as the symptoms are detected. Heat therapy consisting of treating the plants at 37.2 °C for 4 weeks has been found effective in eliminating aspermy virus. Meristem culture after heat treatment produces virus-free cuttings. Aspermy, English stunt, and most mosaic viruses are controlled, while vein mottle, American stunt, and some mosaic viruses are not controlled by heat treatment. Control of thrips and weeds and use of cuttings from virus-free indexed stock can control viral diseases.

(v) *Pests* - The most common pests of chrysanthemum are aphids, hairy caterpillars, red spider mites, root cutting grub, thrips, and nematodes. Leaf miner is another pest very extensively reported from many countries.

(vi) *Aphids*: Aphids are greenish to black dot-like insects which suck the sap from growing stem tips and undersurface of laves, causing loss of vigor. They are also carriers of virus diseases, and they excrete honey dew which forms a substratum for fungal growth. *Myzus persicae*, *Macrosiphoniella sanborni*, *Brachycaudus helichrysi*, *Aphis gossypii*, *Aphis fabae*, and *Aulocorthum circumflexum* are the aphids attacking chrysanthemums.

The control measures include spraying with tobacco-soap decoction. Several insecticides, viz., malathion, parathion, pirimicarb, and aldicarb, smoking with 98% deltamethrin at 1 g/m^3 for 1 h and fumigation with phosphine for another

5 h, and spraying spores of a fungus *Verticillium lecanii* have been used to control aphids.

(vii) *Hairy Caterpillars*: It (*Diacrisia obliqua*) attacks the plants in rainy season and continues till winter. The pest is easily recognized by the presence of hair on their body. They multiply fast and have gregarious habit during early stage. As they eat up the leaves from the surface, papery skeletons are left which dry up. Manual collection and destruction in early stages can check heavy infestation. Spraying thiodon 35 EC or ecaulux 35 EC at 1.25 ml/l is recommended as a control measure.

(viii) *Red Spider Mites*: Mites (*Tetranychus urticae*) look red dot-like bodies on undersurface of leaves causing white specks in the early stages. They occur in hot season and damage leaves and buds which give a pale appearance.

Control: Soil application of systemic insecticide UC 21149; conventional acaricides, e.g., dicofol, tetradifon, and aldicarb; and sprays of demeton-S-methyl (22 g/100 l), diazinon (16 g/100 l) or quinomethionate (125 g/100 l), and cyhexatin (25 g/100 l) are recommended.

(ix) *Grubs*: Grub (*Holotrichia* sp.) is a troublesome pest which remains underground, particularly in shade under the trees. It cuts the underground portion of stem or roots causing sudden wilting of healthy plants during dry hot months. Soil application of aldrin, lindane, or thimet has been found effective against the grub. Tropical cut-worm (*Spodoptera littoralis*) attacks chrysanthemum cuttings which can be controlled by using 0.04% dieldrin emulsion.

(x) *Thrips*: Thrips are slender white to black insects which feed on growing point causing mottling and distortion of leaves and also leaf silvering due to separation of upper epidermal tissue from the rest of the leaf. The insects also damage flowers of summer blooming cultivars. Several insecticides, viz., malathion, lindane, metasystox and nicotine spray, deltamethrin, and phosphine, have been found effective. *Thrips tabaci*, *T. nigrophilosus*, *Frankliniella tritici*, and *Hercinothrips* sp. are thrips commonly attacking chrysanthemums.

As the insects pupate in soil, drenching with BHC (200 g/100 l) or diazinon (40 g/100 l) also helps in controlling their population.

(xi) *Leaf Miners*: *Phytomyza syngenesiae* and *Liriomyza trifolii* have been identified as major leaf miners in greenhouse grown chrysanthemum. The larvae of these insects make tunnels in lea and pupate at the end, leaving characteristic trails behind.

Liriomyza can be controlled by foliar treatment with methamidophos, cypermethrin, and permethrin but found methamidophos phytotoxic. Pyrazophos has been found to be effective in reducing the damage by *Phytomyza* leaf miner. Interplanting of *Vicia faba* (field bean) has been recommended as trap crop.

(xii) *Nematodes*: Nematodes travel up the plant in films of water, enter leaves through the stomata, and feed on them causing triangular brown and black patches between the veins, a characteristic of eelworm infestation. Lower leaves are attacked first. Nematodes damaging chrysanthemum are *Pratylenchus penetrans*, *Aphelenchoides ritzemabosi*, *Belonolaimus longicaudatus*, and *Meloidogyne incognita*.

The use of eelworm-free stock plants for propagation and sterilized compost, application of aldicarb (500 g/ha) before planting, and treating with parathion and mocap (O-ethyl S, S-dipropyl phosphorodithioate) can reduce the damage.

(xiii) *Biological Control*: Chrysanthemum is largely damaged by various pests like aphids, leaf-eating caterpillars of moths and butterflies, etc. Moths and butterflies lay their eggs on leaves, and newly emerging buds cause irreparable damage. The insecticides sprayed on the plants fail to kill them. The growth of the plant is also greatly affected by environmental pollutions like toxic gases, smoke, heavy dew drops, smog, dirt, etc. It is not practicable to spray chemicals and insecticides directly on flower and buds. The use of translucent paper bags or polythene bags to cover the buds and flowers has been suggested. It protects the buds and flowers from moths and butterflies and does not allow them to lay eggs on them. It also saves the buds and flowers from dirt, dust, smoke, dew, toxic gases, and other sources of pollution.

(xiv) *Variety selection*: Selection of resistant variety/strain through screening and breeding for resistance is very encouraging to combat diseases. Encouraging experimental results are available for screening varieties under greenhouse and natural condition resistance to *Puccinia horiana*, Japanese white rust, *S. chrysanthemella*, *Fusarium oxysporum* f. sp. Chrysanthemi, and *F. oxysporum* f. sp. *tracheiphilum* race 1, collar and root rot disease, etc.

19.14 Chromosome Status

The genus *Chrysanthemum* constitutes a large polyploid complex ranging from 2X to 22X, besides a number of aneuploids (Dowrick 1952, 1958; Endo 1969; Nazeer and Khoshoo 1982, 1983; Miyazaki et al. 1982; Maoxue et al. 1983). The basic chromosome number of the genus is 9 with varying numbers of chromosome in different cultivars of the species with 2n = 36, 45, 53–60, 62–65, 67, 68, 72, and 75. The model number is 2n = 54(6X) followed by 2n = 55(6n + 1) and 53(6X-1). Nazeer (1981) reported the presence of one B-chromosome in four cultivars. The chromosome complement in most of the cultivars is composed of metacentric, submetacentric, and sub-telocentric chromosome. Telocentric chromosome in few cultivars has also been reported (Nazeer and Khoshoo 1983). The size difference of chromosomes is not much prominent. In tetraploid cultivars, the chromosomes range from 3.5 to 5.0 μm and hexaploid cultivars from 3.2 to 4.8 μm, whereas a smaller chromosome size 2.1–3.1 μm is reported in octoploid. Karyotypic heteromorphicity is observed in the complement of almost all the taxa. The appearance of unusually long and/or small odd chromosomes in the complement may indicate unequal translocations or centric fusion. Nucleolar chromosomes are present in almost all the taxa and nucleolar or organizer appears as satellites. Meiotic analysis shows the presence of only a few multivalents accompanied by a very large number of bivalent which indicate segmental aneuploid nature of the taxa. Diploid-like meiotic behavior, characterized by predominant bivalent formation, is also noticed in some cultivars. There is a decrease in size of the chromosome with an increase in the grade of ploidy. DNA content among cultivars varies from 12.64 to 25.33 pg and shows a ratio of approximately 1:1.2:1.5:2 between tetraploid/pentaploid/hexaploid/octaploid (Nazeer and Khoshoo 1982). Chromosome pairing during meiosis has been studied in colchiploids *C. japonense* 6X and *C. boreale* 2X. In both species, chromosomes paired predominantly as bivalents, even though because of complete homology they possess the ability to pair as multivalents (Watanabe 1983). Detailed analysis of the chromosome compliment revealed that considerable reshuffling and structural alterations have taken place during the course of differentiation. The chief mechanism in the cytogenetical evolution of garden chrysanthemum has been enumerated as outbreeding, spontaneous, and intentional hybridization coupled with mutation, chromosomal differentiation, and repatterning and polyploidy.

19.15 Genetics

Ornamental plants are essentially cross-pollinated with considerable heterozygosity on account of attractive flowers with fragrance and/or nectaries. The variability in habit, height, vigor, period and quality of bloom, color, size and shape of flowers, and fertility is expressed under cultivation. Compared to the wild types, there have arisen a large number of shapes, colors, etc. in chrysanthemum. The useful aspect underlying the transformation in the genetic system accompanying the change in habitat, from wild to domesticated condition, has been one of the important factors contributing to the origin of new varieties.

Genetic studies showed that self-incompatibility was sporophytic and involved more than one locus (Fryxell 1957; Drewlow et al. 1973). Zagorski et al. (1983) detected on the basis of selfing and crossing in a complete diallel outcrosses in 11 garden chrysanthemums that at least three genes were responsible to govern self-incompatibility. Stewart and Derman (1970) worked out the complex nature of color inheritance on the basis of somatic-genetic analysis of the apical layers of 16 chimeral sports of the Indianapolis chrysanthemum cultivars. It has been clearly detected that a cultivar can be genetically one color in L1 and of another color in L2. As the sex cells arise from L2, a cultivar which is white in L1 and pink in L2 would have white color but breed as pink and vice versa. Genetic studies with special reference to both self-incompatibility and self-compatibility have been studied thoroughly in garden chrysanthemum (Drewlow et al. 1973; Mulford 1937; Ronald and Ascher 1975). Kawase and Tsukamoto (1977) made a thorough study with 57 cultivars and 2 wild species for investigating self-fertility. All the parental plants in this experiment were self-incompatible which, however, varied with cultivating year or the flower head even for the same cultivar. Some parental plants exhibited a high degree of self-fertility. The inheritance of characters, cell sap anthocyanin, and chromoplast carotene in petal cells in a wide range of progenies from crosses has been studied. The presence of one gene A ensured anthocyanin production, while gene I inhibited carotene production. Yellow flowers were formed in the absence of both A and I while white flowers in the absence of A alone. The combination of A and I resulted in pink, carmine, bluish red flowers, whereas A without I resulted in bronze and brownish red flowers. Segregation of flower color was suggested by them to help in the identification of parental genotype. An extensive genetic approach was made in a large number of population to study the inheritance pattern of flowering behavior and floral characteristics (Jong 1989). He found 70% heritability of characters, like number of days to flower and number of flowers per plant, in 79 F_1 population from 15 parents. A significant general combining ability effect for days to flowering was noted. The progenies of parents that flowered early at 1 temperature were generally early at all temperatures at which plants were studied (12°, 13°, 15°, and 17 °C at night). He also studied the inheritance of flower doubleness and floret corolla shape in 70 F_1 population from 16 parents. Singleness was partially dominant over doubleness of type double X single (29% single, 58% semidouble, and 12% double). The percentage varied with parents. The anemone type was inherited as if it was single. Crosses between ligulate, tubular corollas produced ligulate, intermediate, and tubular types in various ratios depending on parents used. Flower doubleness and corolla shape were not linked.

19.16 Improvement

(i) *Breeding*: Breeding is the art and science of changing the genetics of plants in order to produce desired characteristics. Breeding of chrysanthemum is carried out in many countries by individual enthusiasts, gardeners, and farmers or by professional plant breeders at government institutions and agricultural universities, crop-specific industry, or research centers. New varieties of chrysanthemum are now being developed methodically through efficient, intelligent, and systematic work. It will not be possible to highlight all the breeding work going on in different countries. In general, common techniques followed everywhere for developing new varieties have been highlighted. Closely or distantly related varieties are crossbred to introduce traits/genes from one variety or line into a new genetic background. In breeding with heterozygous material, the breeder does not know exactly what genes have been introduced to the new cultivars.

(a) *Hybridization/selection*: Chrysanthemum is a hybrid species which is the result of repeated cycles of complex interspecific crossing among elemental species extending over a period of more than 2600 years. Chrysanthemum has developed considerable heterozygosity and the variability in habit, height, vigor, period and quality of bloom, color, size and shape of flowers, and fertility. Cross-breeding is one of the main methods to increase the genetic variability, with which the plant breeder tries to combine the beneficial characters from different sources into one genotype. Cross-breeding has been utilized as one of the main methods to increase further genetic variability in chrysanthemum. Systematic efforts have been made at different research institutions and private nurseries to develop high-yielding variety, pot culture variety, cut flower variety, garland purpose, and exhibition type by selection, incorporating desirable genes through natural crossing or conscious selective crossing.

Under hybridization program, new varieties of chrysanthemum are developed by the following methods: (a) Seedlings, with promising desirable characters, are selected from natural cross-pollination that resulted in the development of many new varieties. (b) Parent varieties with desired characters are grown separately in the field for natural crossing. Seedlings from this crossing result into development of new varieties. (c) To avoid contamination, selected parent varieties are grown in pots, and they are kept separately under net. Seedlings with desirable character are selected from these crosses as new varieties. (d) Deliberate conscious/selective artificial cross-pollination between closely and distantly related individuals resulted in the development of many promising varieties.

Male and female parents with desirable characters are selected, and anthers of disc florets of female parent are clipped before anthesis. Disc florets are bagged with cellophane paper bag to avoid natural pollination. Long ray florets of female plants are cut to expose stigma. Pollen grains are collected from male parent and dusted on stigma of female parent. Seedling with superior characters over the existing parents are selected and multiplied as new variety. Promising varieties comprising novel commercial characters like attractive flower color and shape, no pinch no stake dwarfness, out-of-season blooming, cut flowers (attractive color, long erect stem, uniform bloom opening, tough florets, long vase life, and healthy leaves), pot culture (dwarf and compactness, profuse branching, uniform spreading of branches, simultaneous blooming habit, attractive color and good color retention quality, and healthy leaves), high yielding, garland purpose, exhibition type, chlorophyll variegation in leaves, showy decorative leaves, etc. have come out from systematic efforts of all abovementioned methods.

Selection through hybridization resulted in the development of more than 85 new promising varieties at CSIR-National Botanical Research Institute, Lucknow, India (mentioned above). Some promising varieties with attractive floral characteristics are mentioned:

(i) *Pompon Type*: Seedlings of Japanese pompon variety 'Nanako' were selected and further crossed among themselves resulting into development of promising high-quality pompon varieties most suitable for cut flowers by virtue of their attractive form, color, and good keeping quality ['Apsara' (Fig. 3a) (rosy flush on white background), 'Vasantika' (Fig. 3b) (yellow), 'Birbal Sahni' (white), 'Kundan' (Fig. 3c) (yellow), 'Jayanth' (Fig. 3d) (yellow), 'Jubilee' (orange), 'Maghi' (pink), 'Shanti' (Fig. 3e) (white), etc.].

(ii) *No Pinch No Stake Mini Chrysanthemum*: Year-round cultural operations and a lot of additional care like "pinching" and "staking" are required for all garden chrysanthemums which are expensive and time-consuming. Japanese varieties 'Akita' and 'Koben' were repeatedly crossed among themselves, and a series of dwarf varieties were selected and grouped under "mini chrysanthemum." These are unique genetic selections with dwarf, bushy, compact, round-shaped, profuse blooming habit which require neither "pinching" nor "staking." These can be grown in small container smaller than ice cream cup to 10–12″ pots. These have created a good awareness of "mini chrysanthemum culture" in the society. The mini varieties developed at CSIR-NBRI, Lucknow, India, are 'Apurva' (Fig. 4a), 'Appu', 'Arun Singer', 'Bindiya', 'Cameo', 'Haldighati' (Fig. 4b), 'White Charm' (Fig. 4c), 'Peet Singar' (Fig. 4d), 'Hemant Singar' (Fig. 4e), 'Pancho' (Fig. 4f), 'Mini Queen' (Fig. 4g), 'Diana' (Fig. 4h), 'Kusum' (Fig. 4i), 'Sweet Singar', 'Yellow Charm', 'Kargil'99' (Fig. 5a), 'Sadbhavna' (Fig. 5b), 'Y2K' (Fig. 5c), 'NBRI Mini Jessie' (Fig. 5d), 'NBRI Little Darling' (Fig. 5e), 'Mother Teresa' (Fig. 5f), 'Shanti', 'NBRI Iidiana', 'Orange Little Darling' etc. The variety 'Mother Teresa' got US patent (PP13,678). There are series of new selections for release as new varieties (Figs. 5g–k). These are the only varieties in chrysanthemum which can be utilized for multicolor culture in one pot and landscaping.

(iii) *Out-of-Season Blooming Varieties*: Normal blooming season of chrysanthemum in Northern India persists approximately 6 weeks. Conscious selective crossing among Japanese varieties 'Shin Fuzi', 'Bosetsue', 'Yuki Kaza', etc. and selections resulted in development of out-of-season varieties. Chrysanthemum can be grown almost round the year exploiting these genotypes by planting in the right time and right genotype ['Himanshu' (April–May/second flush in October); 'Jawala', 'May Day' (May–June/second flush in November); 'Tushar', 'Jyoti' (June–July/second flush in November); 'Meghdoot', 'Phuhar' (July–August); 'Sharad', 'Ajay' (September–October); 'Sharad Mala' (October); 'Sharad Singar' (October); 'Haldi Ghati' (October–November); 'Vasantika', 'Jaya' (December–January); 'Maghi' (January–February)].

(iv) *Development of chlorophyll variegated varieties*: Plants with chlorophyll variegated leaves look beautiful even when there is no flower. Chrysanthemum remains in vegetative growth almost 9–10 months in a year. Continuous efforts are being made at NBRI to develop suitable genetic strains with variegated leaves by conventional and induced mutation breeding methods. Six chlorophyll variegated varieties (DWS-2, DWS-12, DWS-15, B-16, B-17, and OO-2) have been selected through selections from conventional breeding and one from induced mutation. The intensity of variegation varies from variety to variety. Chlorophyll variegations of leaves of these varieties have become additional beauty which can be enjoyed throughout the year. All these chlorophyll variegated strains are being utilized in different hybridization and mutation breeding programs to develop more and more new attractive variegated varieties (Datta 2015).

Fig. 3 Cut flower varieties developed through breeding. Figure 3**a**. 'Apsara'. Fig. 3**b**. 'Vasantika'. Fig. 3**c**. 'Kundan'. Fig. 3**d**. 'Jayanti'. Fig. 3**e**. 'Shanti'

(b) *Induced Mutation*: Extensive work on induced mutations has been done on *C. morifolium* Ramat. by a number of workers in different countries, and a wide range of physical mutagens (X-ray, gamma rays, fast neutrons, thermal neutrons, radioactive phosphorous) and chemical mutagen (ethylene imine, ethyl methane sulfonate, and colchicine) have been used for its improvement. Appreciable knowledge and literature have been generated on practical experiments for crop improvement using classical and modern (in vitro and management of chimera) induced mutagenesis techniques on different aspects like LD_{50} dose, radiosensitivity (with respect to influence of various factors including flower type, shape and color, chromosome number, INV, ICV, 2c DNA content, and chromosome number and chromosomal aberrations), selection of material (suckers/cuttings), methods of exposure to mutagens, determination of suitable dose of mutagen, combined treatment, recurrent irradiation, split dose, colchicine mutation, mutant genotype, detection of mutation, mutation frequency and spectrum of mutations, nature of chimerism, classical and modern methods for management of chimera, in vitro mutagenesis, isolation of mutants, etc. (Datta 2015). The details of utilization of induced mutations and its prospects and released mutant varieties have already been reviewed (Datta 1988, 1997, 2015; Broertjes and Van Harten 1988; Ahloowalia et al. 2004; Datta and Chakrabarty 2005).

Fig. 4 No pinch no stake mini chrysanthemum. Figure 4**a**. 'Apurva'. Fig. 4**b**. 'Haldi Ghati'. Fig. 4**c**. 'White Charm'. Fig. 4**d**. 'Peet Singar'. Fig. 4**e**. 'Hemant Singar'. Fig. 4**f**. 'Pancho'. Fig. 4**g**. 'Mini Queen'. Fig. 4**h**. 'Diana'. Fig. 4**i**. 'Kusum'

Approximately 280 mutant varieties have been developed throughout the world (FAO/IAEA, Vienna, Mutant Variety Database) which are mainly color mutations. More than 75 gamma ray-induced mutant varieties (mentioned above) have been reported from CSIR-NBRI, Lucknow, India (Datta 2015). Radiation-induced phenotypic variations including several interesting changes in flower form for novelties have been developed ('Tulika', 'Shabnam' (Fig. 6e), 'Cosmonaut' (Fig. 6f), etc.). Induction of tubular florets is one of the interesting observations in chrysanthemum. Complete tubular floret mutant can be induced in those cultivars where there is small tube at the base of each floret. Plants with chlorophyll variegated leaves look beautiful even when there is no flower. Chrysanthemum remains in vegetative growth almost 9–10 months in a year. Conventional breeding/selection resulted in the development of six chlorophyll variegated traits (DWS-2, DWS-12, DWS-15, B-16, B-17, and OO-2). The intensity of variegation varies among the traits. Two chlorophyll variegated varieties ('Kargil'99' (Fig. 5a) and 'Niharika') have been developed. Perhaps these are the first chlorophyll leaf variegated varieties of *Chrysanthemum* developed at NBRI for floriculture trade reported so far.

Colchicine for the first time has been used for development of new flower color in chrysanthemum. *Chrysanthemum* cv. 'Sharad Bahar' was purple, whereas the mutant ('Colchi Bahar', Fig. 6i) color was terracotta red. Colchicine treatment can produce gene mutation (Colchi mutation (C-mutation)) instead of inducing polyploidy (Datta 2015).

(c) *Chimaera and Its Management*: Mutation is a single-cell event. Mutated cell is exposed to the so-called diplontic selection, that is, the competition between the mutated cell and the surrounding non-mutated ones. The mutated cell develops into a group of cell and finally into a cell layer. All the somatic mutations in flower color/shape induced after gamma irradiation appeared as chimera (Fig. 6a, b). The size of mutant sector varied from a narrow streak on a floret to entire floret, from few florets to whole flower head, and from a portion of a branch to entire branch. Isolation of mutant tissue is one of the most important operations in mutation breeding. Isolation of mutant tissue is not possible with the help of available conventional propagation technique when it is chimera of a flower and a huge number of such spontaneous and/or induced mutant tissues are lost. Standardization of in vitro technique has opened a new way for isolating new ornamental varieties through retrieval of chimeric tissues. A novel technique has been standardized for isolation of new chimeric flower color/shape mutants through in vitro direct shoot regeneration from ray florets and

Fig. 5 No pinch no stake mini chrysanthemum. Figure 5**a**. 'Kargil 99'. Fig. 5**b**. 'Sadbhavna'. Fig. 5**c**. 'Y2K'. Fig. 5**d**. 'Mini Jessie'. Fig. 5**e**. 'Little Darling'. Fig. 5**f**. 'Mother Teressa'. Figs. 5**g**–**k**. New Selections

Fig. 6 Chimera and induced mutants of *Chrysanthemum*. Figure 6**a**. Chrysanthemum flower bud chimera (original red and mutant yellow). Fig. 6**b**. Chrysanthemum flower chimera (original red and mutant yellow). Fig. 6**c**. 'Purnima'. Fig. 6**d**. 'Batik'. Fig. 6**e**. 'Shabnam'. Fig. 6**f**. 'Cosmonaut'. Fig. 6**g**. 'Navneet'. Fig. 6**h**. 'Sonali'. Fig. **i**. 'Colchi Bahar'

development of solid mutants through in vitro mutagenesis in chrysanthemum (Datta 2015).

(d) *Possibilities of Inducing Desired Flower Color Mutation*: Although mutation is a chance process, from the repeat experiments with the same and/or different cultivars of chrysanthemum, it has been determined that if white varieties are irradiated, the mutation will either be in flower shape or color (yellow). Red varieties, on the other hand, will produce either a completely yellow mutation or a mixture of red and yellow. If yellow varieties are irradiated, the mutation will be either different shades of yellow or white or mixture of yellow and white (Datta 2015).

19.17 Tissue Culture and Gene Transfer

Micropropagation is the practice of rapidly multiplying stock plant material to produce a large number of progeny plants. Most of the modern ornamental breeders follow the micropropagation techniques for large-scale propagation of ornamentals. Conventionally, chrysanthemums are propagated through suckers/cuttings. Some cultivars are easy to propagate, while others are very difficult to propagate mainly due to low number of suckers and low survival rate of plants. Any new variety developed can be made available in the market at the earliest due to high rate of in vitro multiplication. In vitro techniques have been well established for multi-purpose use specially for rapid multiplication for commercial exploitation from almost all plant parts like nodal explant, stem tips, shoot apices, petal epidermis, petal, bud, protoplast, achenes, leaf discs, etc. Tissue culture of different explants has also been done after treatment with X-rays and gamma rays for in vitro mutagenesis (Datta 2015; Misra et al. 2010).

Explant Culture Ray florets were collected and washed thoroughly under running tap water for 15 min and for another 5 min with 5% aqueous solution of Godrej Liquid Cleaner or aqueous solution of Teepol. The explants were then quickly dipped in 70% ethanol and surface-sterilized with 0.1% $HgCl_2$ for 2 min followed by repeated rinsing with sterile distilled water. Florets of the original cultivar were cultured on MS medium (Murashige and Skoog 1962) supplemented with different combinations of 1-naphthaleneacetic acid (NAA) and 6-benzylaminopurine (BAP) + sucrose (3%) and Bacto4 agar (0.8%). All cultures were kept under a 16-h photoperiod (36 lmol/m2 ás) at 25 °C and 55 ± 60% relative humidity. The pH of the medium was adjusted to 5.6 before autoclaving at 121 °C for 15 min. All the regenerated shoots (2–3 cm in length) were transferred to half-strength MS medium containing 1.5% sucrose and 0.8% BactoAgar for root induction. Rooted plantlets were transferred to plastic pots containing soil and leaf mold mixture (1:3) and placed under high humidity for 1 week for hardening. After another 2 weeks of hardening, plantlets were transplanted in the field. In this system, we obtained up to 79.2% response with 16.6 number of shoots per responding explants on MS medium containing 0.2 mg/l NAA and 2 mg/l BA. Shoots were excised and easily rooted on MS medium without any growth regulators (Datta 2015).

Mutated ray florets collected from flower heads (after approx. 105 days of planting) were washed thoroughly in running tap water for 15 min and for another 5 min with 5% aqueous solution of Teepol (a liquid detergent) and then washed again in tap water. Florets were disinfected with 70% ethanol for 20 sec followed by surface sterilization with $HgCl_2$ solution (0.1%, w/v) for 2 min and then washed thoroughly in sterile distilled water. Individual whole florets were used as explants. They were cultured on MS medium (Murashige and Skoog 1962) containing 30% sucrose, 0.8% BactoAgar, and growth regulators. For cv. 'Purnima', different combinations of NAA (0, 0.2, 0.5, and 1.0 mg/l) and BAP or kinetin (0, 0.2, 0.5, and 1.0 mg/l) were tested, while in the case of cv. 'Colchi Bahar' only one combination (NAA 0.2 mg/l BAP 0.5 mg/l) was tested. Medium pH was adjusted to 5.6 before autoclaving at 121 °C for 15 min.

Ray florets were cultured on MS medium (Murashige and Skoog 1962) supplemented with 0.2 mg/l NAA, 1 mg/L BAP, sucrose (3%), and BactoAgar (0.8%). The medium pH was adjusted to 5.6 prior to autoclaving at 121 °C for 15 min. For rooting, 2–3 cm shoots were excised from explants and placed in rooting medium (MS + 0.2 mg/liter, NAA + 3% sucrose +0.8% ~ gar). All the cultures were kept under a 16-h photoperiod (36 ~ molm2 S-I) at 25 ± 1 °C and 50–60% relative humidity. The rooted plantlets were then placed ·in potted soil [containing sand/soil (1:1)] and kept in humid chamber (with 80–90% relative humidity) for 15 days prior to their transfer to field.

19.18 Biotechnology

Biotechnology, the world's fastest-growing and most rapidly changing technology, has revolutionized research activities in the area of floriculture. Concentrated research in the molecular and cellular biology of important ornamental crops has resulted in the development of effective gene transfer procedures, with subsequent recovery of engineered plants. Isolation of desirable gene with floricultural properties, characterization, manipulation, and their transfer to desired variety have opened up novel routes to the development of improved ornamental plant varieties. Series of publications are available on standardization of regeneration protocol using different explants and *Agrobacterium*-mediated Bt II, GUS, PFG, hygromycin phosphotransferase, CHS, TSWV nucleocapsid protein, etc. Improvement of chrysanthemum through biotechnology has recently been reviewed (Teixeira da Silva 2004). Transgenic plants having genes of economic important characters have been regenerated (c.f. Datta and Chakrabarty 2005). White-flowered plants have been produced by successful isolation and introduction of CHS gene to a pink cv. Moneymaker (Courtney et al. 1994). Kunitake et al. (1993) studied the transformation in chrysanthemum through polyethylene glycol and electrophoretic transfection. Chrysanthemum is susceptible to infection by wild-type *Agrobacterium tumefaciens*. *Agrobacterium*-mediated Bt II gene has been transferred to leaf explants, and transgenic calluses have been regenerated. *Agrobacterium*-mediated NPT II as selectable marker gene and GUS as reporter gene have been successfully introduced into *D. morifolium* and *D. indicum* genotypes (Ledger et al. 1991). *Agrobacterium*-mediated transgenic (NPT II, GUS, and hygromycin phosphotransferase genes, delta endotoxin gene, nucleocapsid protein gene, flavonoid 3′5′hydroxylase gene) plants have been regenerated, and different problems related to transformation have been discussed by several workers (c.f. Datta and Chakrabarty 2005; Takatsu et al. 1999; Dolgov et al. 1995; Renou et al. 1993; Pavingerová et al. 1994; Fukai et al. 1995). *Agrobacterium*- and biolistic-mediated transformation procedure has been developed for nucleocapsid protein genes (Yepes et al. 1995).

19.19 Characterization

Characterization is the most important for correct identification of cultivars. It helps to understand the genetic diversity and to trace out the phylogenetic relationship, taxonomical status, preparation of catalogue, variation patterns, identification of desirable/novel genes, hybridization, registration, plant variety protection, farmer's right, etc. Different workers have developed their own characterization system for chrysanthemum. The author did an extensive work on the characterization using classical and modern techniques. Different parameters of cytology, morphology, physiology, chemical and biochemical, DNA markers, etc. have been utilized for characterization. Such analysis will provide important information for preparation of checklist of chrysanthemum. This will be the most helpful not only for identification of varieties but also for selection of desirable characters for intervarietal breeding program (Datta 2015).

Under characterization program, a wide range of morphological characters, viz., plant height, plant spread (N–S), plant spread (E–W), branch number, leaf length, leaf width, petiole length, number of flowers/branch, number of flowers/plant, size of flower, size of floret, number of florets/flower head, moisture content, weight of flower, time of flowering, and type of flower, were taken into consideration. All morphological characters with special reference to floral characters are of much commercial importance for assessment of cultivars. A good amount of variety-specific morphological characters and desirable genes have been identified through characterization (Datta 2015). Chlorophyll (Chl. a, Chl. b, and total) contents in leaves were estimated both at vegetative and flowering stages. Chlorophyll content varied from cultivar to cultivar. It was interesting to note that chlorophyll content in all the cultivars increased at flowering stage.

Voluminous works have been done on chrysanthemum for characterization of normal varieties and mutant (spontaneous and induced) varieties. Induced mutations have developed a large number of flower color/type and chlorophyll variegated mutants worldwide in chrysanthemum. The author did comparative cyto-palynological, anatomical, biochemical, and DNA-based fingerprinting of original and mutant cultivars for better and clear understanding of the exact mechanism involved in the origin and evolution of somatic flower color mutation.

(i) **Cytological studies**

The study of mitosis was done by preparing slides by squash technique. First of all, fresh roots were pretreated in paradichlorobenzene and a pinch of aesculin for 5 min in −20 °C followed by 15 min at 4 °C and then fixed in the fixative propionic acid/alcohol (1:3). Roots were hydrolyzed in 1(N) HCl for 13 min and stained by usual Feulgen staining procedure (Datta 1997).

Original and all mutant cultivars showed $2n = 6x = 54$ chromosome number. Different types of chromosomal aberrations (bridge, fragment, laggard, early separation, clumping, exclusion, micronucleus, etc.) were observed during root tip mitosis in few cells. No mutant-specific abnormality could be detected. Ideograms showed that the mutants did not differ from their respective original cultivars in number of types of chromosomes and number of each type represented in them. The karyotype in the analyzed original and mutant cultivars was reasonably symmetrical. Cytological analysis clearly indicate that changes in flower color may be considered to have taken place through gene mutation but neither through change in chromosome number and aberration nor due to change in karyomorphology (Datta 1994).

(ii) **Micromorphological studies**

Flower petals were fixed and then mounted on SEM stubs using double-sided adhesive tapes after critical point drying. Subsequently the materials were sputter-coated with gold (200 Ao thickness), and scanning photomicrographs were taken in JEOL-JSM 35C Scanning Electron Microscope at 10 kV (Datta and Shome 1994).

The original and their respective induced mutant cultivars showed considerable variations in petal epidermal micromorphology particularly in cell boundaries, cell surface, striations, and papillae. The petal micromorphological characters can be utilized for identifying mutants (Datta 2015).

(iii) **Palynological studies**

Pollen grains are collected soon after anthesis. These are dusted on glass slides, washed in absolute alcohol, and mounted in malachite green-glycerin jelly. The pollen grains which are regular in shape and full and had uniform stain are considered as fertile, while those which are irregular in shape, empty, and hyaline are regarded as sterile. For the study of pollen grain morphology, pollen slides are made using the revised acetolysis method of Erdtman (1960).

Most of the chrysanthemum cultivars show regular size pollen grains, but few cultivars and their mutants had dimorphic pollen grains. The shape of endocolpium is variable; lalongate types are the most common, but lolongate, circular, square, and indiscernible types are also found. Exine surface pattern shows conspicuous changes. The base and tip of exine spines are variable in shape. The original exine surface, i.e., fosso-reticulate pattern with narrow muri and irregularly shaped lumina, changed to reticulate exine with broad muri and uniformly circular lumina in the "mutants." The tips of the spines changed from straight to bent. The reticulate undulated exine surface changed to a scrobiculate wrinkled surface. The punctata exine surface pattern of the original transformed to a scrobiculate pattern in the mutant (Datta 2015).

(iv) **Pigment analysis**

Chrysanthemum pigments have been studied by a number of workers using thin-layer chromatographic and spectrophotometric methods and categorized into a large number of chrysanthemum cultivars according to the presence or absence of different pigments.

For spectrophotometric analysis of phenolic compounds, 200 mg of florets were extracted in 50-ml methanol containing 1% HCl. The extracts were scanned from 200 to 800 nm region of wave length in Utltroscope 2000, Pharmacia Biotech (Datta 2015).

For the study of phenolic compounds by thin-layer chromatographic methods (TLC), mature leaves and petals were extracted in methanol containing 1% HCl. The chromatograms were developed on glass plates (6.3 X 10 cm) coated with silica gel emulsion. The plates were run 8 cm in a mixture of benzene/propionic acid/H2O (20:

40:10 v/v). They were then dried in air, and the spots were observed and marked under the necked eye and under UV. The plates were then sprayed with flavone reagent (diphenylboric acid ethanolamine complex) and again marked under UV. The color reaction of each spot and their Rf values were determined from six good chromatograms. These were then transformed into hRf (Rf X 100) values (Datta 2015).

Floret pigment analysis by thin-layer chromatographic and spectrophotometric methods of large number of mutants and original cultivars indicated that somatic flower color changes were due to both qualitative and quantitative changes in pigments as a result of mutations induced by mutagens in pigment biosynthesis pathway.

Chrysanthemum pigments have been studied by a number of workers using thin-layer chromatographic and spectrophotometric methods and categorized into a large number of chrysanthemum cultivars according to the presence or absence of different pigments. Several pigment compounds have been qualitatively characterized by thin-layer (TLC) and paper (PC) chromatography. Carotenoid and anthocyanin pigments were extracted from the flowers and analyzed both qualitatively and quantitatively. Series of carotenoids have been reported, but only a few have been identified, including lutein or xanthophyll, tetraxanthin, and flavoxanthin. Cultivars were classified into four basic groups according to analysis: (1) white, containing flavonols; (2) pink (pink to red-purple), anthocyanins, and flavonols; (3) yellow, carotenoids; and (4) (i) orange and (ii) red, anthocyanin, flavonols, and carotenoids. Anthocyanin compounds include cyanidin-3-glucoside (chrysanthemin) and its derivatives, malvin and ensatin. Four flavonoids – apigenin, acacetin, quercetin, and luteolin – have been identified (Kawase and Tsukamoto 1974, 1976, 1977).

(v) **Molecular characterization**

DNA-based markers provide powerful and reliable tools for discerning variation within crop germplasm, cultivar identification, and pedigree analysis and to study evolutionary relationships. DNA fingerprinting is used to analyze the various aspects of plant genus such as taxonomy, phylogeny, ecology, genetics, and breeding in interspecific or intraspecific level. These techniques include RAPD, RFLP, SSR, STS, SNP, VNTR, STR, SFP, and AFLP. The genetics of chrysanthemum are very complex. RAPDs, however, are a powerful tool to detect different molecular markers in hybrid populations of *Chrysanthemum* cultivars. Huge amounts of work have been done on molecular characterization of chrysanthemum from different laboratories. RAPD markers have been successfully used to distinguish between the radiation-induced mutants from original group in an early vegetative stage, sports, commercial varieties, and nonrelated cultivars. Because of the high level of polymorphism and clonal stability, RAPD fragments are useful for cultivar identification. The genetic variability among related *Chrysanthemum* species was too high to study genetic distances either among cultivars within chrysanthemum or among species related to chrysanthemum (Wolff and Peters-Van Rijn 1993; Wolff et al. 1994, 1995; Rumin et al. 2004; Trigiano et al. 1998; Martin et al. 2002; Wolff and Peters-van Rijn 1993; Scott et al. 1996).

The author and his colleagues selected large number of large-flowered, small-flowered, mini varieties, mutant varieties of chrysanthemum, and the objective of the study was to estimate the genetic relationship among the different cultivars with relation to their morphological and biochemical characteristics and geographical distribution. Special attempt was to investigate the molecular differences for better understanding of the origin and evolution of somatic flower color mutation.

DNA Extraction The total genomic DNA was extracted from young leaves of rose cultivars by CTAB procedure (Saghai-Maroof et al. 1984) with some modifications. Extraction in chloroform/isoamyl alcohol (24:1) followed by centrifugation twice at 14,000 g helped to remove polysaccharides. RNA contaminants in all the samples were digested with 100 mg/mL RNase A for 30 min at 37 °C, extracted once with phenol/chloroform/isoamyl alcohol (25:24:1). After ethanol precipitation, DNA was resuspended in 100 mL of TE (10 mM Tris–Cl + 1 mM EDTA) buffer (pH 8.0). The average yield was calculated by a spectrophotometer (Ultraspec 2000, Pharmacia Biotech), and DNA samples were stored at −20 °C. PCR Conditions 20 arbitrary decamer primers (Bangalore Genei, India) were used for polymerase chain reaction (PCR). PCR was performed in 20 mL reaction mixture containing 5 ng template DNA, 1 unit of Taq DNA polymerase, 100 μM dNTPs, 1.0 μM primer, 2.5 mM MgCl2, 10 mM Tris-HCl (pH -9.0), 50 mM KCl, and 0.01% gelatin. PCR amplification was performed using a PTC-100 Peltier Thermal Cycler (MJ Research, USA) using the following conditions, preheating of 4 min at 94 °C and 45 cycles of 15 sec at 94 °C, 45 sec at 36 °C, and 1.5 min at 72 °C, and elongation was completed by a final extension of 4 min at 72 °C. The final reaction mixture was cooled down to 4 °C. After amplification PCR product was resolved by electrophoresis in 1% agarose gel with 1 × TAE buffer. Bands were visualized by staining with ethidium bromide (0.5 μg/mL-1) under UV light and protographed. Only distinct bands were counted for data analysis and faint was not considered. The size of the amplification products was estimated from a 100 bp DNA ladder (Sigma). All the reactions were repeated at least twice, and only those bands reproducible on all runs were considered for analysis. DNA fragment profiles were scored in binary fashion with “0” indicating absence and “1” indicating presence of band. Genetic distance was calculated by Jaccard’s coefficient (Jaccard 1908) which is as follows, Sij – Nij/(Nii + Nij + Njj), where Sij is the similarity index between the ith and jth genotype, Nij is the number of bands present in both genotypes, Nii is the number of bands present in the ith genotype but absent in the jth genotype, and Njj is the number of bands absent in the ith genotype and present in the jth genotype. The similarity matrix was converted to dissimilarity matrix (1-Sij), and a dendrogram was constructed using the neighbor joining tree method using RAPDistance Package version 2.0 (Armstrong et al. 1998).

Variation between cultivars was high. The intervarietal and intravarietal taxonomic relationship of C. x *grandiflorum* was obtained. Moreover, the selected unrelated cultivars and mutants of chrysanthemum were distinguishable from each other through banding patterns obtained from RAPD. Cultivars, whether parent or mutant, with different flower colors could be clearly distinguished. Some primer yielded extremely different banding patterns which help to identify mutant from its mother plants. It is, however, possible that some of the specific bands present for some of the mutants may code for flower color, but this can only be verified by using SCAR markers and cloning cDNA. The analysis of genetic similarity indices revealed low diversity within the radiomutants. High genetic distance among the different *C.* x *grandiflorum* parents and mutants provides the chance to introgress new and novel genes from the chrysanthemum gene pool (Chatterjee et al. 2005; Chatterjee et al. 2006a, b; Datta 2015). Data shows an estimation of genetic diversity which cannot be simply interpreted by classical morphological, cytological studies. Present experiment of estimation of genetic diversity is useful for the farmers for the breeding purpose to increase the heterosis of hybrids and introgress the new genes in the gene pool. These studies can be very effectively proof to be a technique for the development of mapping population for tagging of agronomically important traits. Difference in morphology is totally independent of geographical distance, negating a simple isolation by distance model. As all the selected cultivars are morphologically very similar, it is very difficult to identify them until the blooming season. DNA fingerprinting is the only solution to solve this problem. This RAPD profile serves as a way for identification of species-specific marker and also produces a reliable data to construct the phylogenetic relationship of the genus *Chrysanthemum* (Chatterjee et al. 2005).

References

Ahloowalia BS, Maluszynski M, Nichterlein K (2004) Global impact of mutation-derived varieties. Euphytica 135:187–204

Anderson YO (1987) Reclamation of the genus Chrysanthemum L. HortScience 22(2):313

Armstrong J, Gibbs AJ, Peakall B, Weiller G (1998) The RAPDistance package. http://life.anuedu.au/molecular/software/rapd.html

Bailey LH (1949) Manual of cultivated plants. MacMillan Co., New York

Barrett JE, Bartuska CA (1982) PP333 effects on stem elongation dependent on site of application. HortScience 17:737–738

Broertjes C, Van Harten AM (1988) Applied mutation breeding for vegetatively propagated crops. Elsevier, Amsterdam

Carter GD (1980) An introduction to floriculture. Academic Press, San Diego

Chakrabarty D, Chatterjee J, Datta SK (2007) Oxidative stress and antioxidant activity as the basis of senescence in chrysanthemum florets. Plant Growth Regul 53:107–115

Chatterjee J, Mandal AK, Ranade SA, Datta SK (2005) Estimation of genetic diversity of four Chrysanthemum mini cultivars using RAPD. Pak J Biol Sci 8(4):546–549

Chatterjee J, Mandal AK, Ranade SA, Datta SK (2006a) Intravarietal variation of miniature Chrysanthemum cultivars using random amplified polymorphic DNA. Hortic Environ Biotechnol 47(4):192–192

Chatterjee J, Mandal AKA, Ranade SA, Teixeira da Silva JA, Datta SK (2006b) Molecular systematics in chrysanthemum x grandiflorum (Ramat.) Kitamura. Sci Hortic 110:373–378

Chittenden FJ (1951) The Royal Horticultural Society. Dictionary of gardening. A practical and scientific cytopedia of Horticulture, Assisted by specialist vol 1, A-Co., Oxford, at the Clarenden Press

Courtney-Gutterson N, Napoli C, Lemieux C, Morgan A, Firoozabady E, Robinson KEP (1994) Modification of flower colour in florist's chrysanthemum: production of a white flowering variety through molecular genetics. Biotechnology 12:268–271

Cumming RW (1964) The chrysanthemum book. D. Van Nostrand, Princeton

Darlington CD (1973) Chromosoma botany and the origins of cultivated plants, 3rd edn. G. Allen and Unwin Ltd., London

Datta SK (1988) Chrysanthemum cultivars evolved by induced mutations at National Botanical Research Institute, Lucknow. Chrysanthemum 44(1):72–75

Datta, SK (1998) Chrysanthemum germplasm at NBRI, Lucknow and search for novel genes. Applied Botany Abstracts, 18(1): 45–72

Datta SK (1994) Cytological interpretation of induced somatic flower colour mutation in garden chrysanthemum. In: Plant Cytogenetics in India (edited by Sibdas Ghosh). Published by the Association for Cell and Chromosome Research, Dept. of Botany, University of Calcutta, Calcutta, pp 107–114

Datta SK (1996) Chrysanthemum, NBRI Bulletin No. 1. EBIS, NBRI, Lucknow, pp 1–26

Datta SK (1997) Ornamental plants- role of mutation, vol 219. Daya Publishing House, Delhi

Datta SK (2013) *Chrysanthemum morifolium* Ramat. – A unique genetic material for breeding. *Sci Cult* 79(7–8):307–313

Datta SK (2015) Indian floriculture; role of CSIR. Regency Publications, A Division of Astral International(P) Ltd., New Delhi 110 002. [432 pages]

Datta SK, Chakrabarty D (2005) Classical mutation breeding and molecular methods for genetic improvement of ornamentals. In: Datta SK (ed) Role of classical mutation breeding in crop improvement. Daya Publishing House, Delhi, pp 260–303

Datta SK, Gupta VN (2012) Year round cultivation of garden chrysanthemum (*Chrysanthemum morifolium* Ramat.) through photoperiodic response. Sci Cult 78(1,2):71–77

Datta SK, Shome U (1994) Micromorphological studies of original and mutant cultivars of ornamentals. Feddes Repertorium 105(3–4):167–174

De Jong J (1989) The flowering of *Chrysanthemum morifolium* seedlings and cuttings in relation to seasonal fluctuation in light. Sci Hortic 41(1–2):117–124

de Ramatuelle M (1792) Description de la Camomille á grandes fleurs. Lul à l'académie des sciences, le 5 décembre 1792. J d'Hist Nat 2:233–240

Dolgov SV, Mityshkina TU, Rukavtsova EB, Buryanov YI (1995) Production of transgenic plants of *Chrysanthemum morifolium* Ramat with the gene of *Bac. thuringiensis* 6-endotoxin. Acta Hortic 420:42–45

Dowrick GJ (1952) The chromosomes of chrysanthemum I. Heredity 6:365–376

Dowrick GJ (1958) Chromosome numbers and the origin and nature of sports in the garden chrysanthemum. Natl Chrysanth Soc:60–69

Drewlow LW, Ascher PD, Widmer RE (1973) Genetic studies of self incompatibility in the garden chrysanthemum *C. morifolium* Ramat. Theor Appl Genet 43:1–5

Endo N (1969) The chromosome survey of cultivated chrysanthemums, *Chrysanthemum morifolium* Ramat. I. On the chromosome numbers of cultivated chrysanthemum (part). J Jpn Soc Hortic Sci 38:611

Erdtman G (1960) The acetolysis method. A revised description. Sv Bot Tidskr 54:561–564

Fryxell PA (1957) Mode of reproduction in higher plants. Bot Rev 23(3):135–233

Fukai S, de Jong J, Rademaker W (1995) Efficient genetic transformation of chrysanthemum *(Dendranthema grandiflora* Ramat. Kitamura) using stem segments. Breed Sci 45:179–184

Gilbertz DA (1992) Chrysanthemum response to timing of paclobutrazol and uniconazole sprays. Am Soc Horticult Sci 27(4):322–323

Hemsley WB (1889) The history of the chrysanthemum: the principle modifications of the chrysanthemum. Gardeners Chronicle 6:586–587

Heywood VH, Humphries CJ (1977) Anthmideae systematic review. In: The biology and chemistry of compositae. Academic, London

Kawase K, Tsukamoto Y (1974) Studies on flower colour in *Chrysanthemum morifolium* II. Absorption spectra of intact flowers. J Jpn Soc Hortic Sci 43:165–171

Kawase K, Tsukamoto Y (1976) Studies on flower colour in *Chrysanthemum morifolium* III. Quantitative effects of major pigments on flower colour variation, and measurement of colour qualities of petals with a colour difference meter. J Jpn Soc Hortic Sci 45(1):65–75
Kawase K, Tsukamoto Y (1977) Studies on breeding in chrysanthemum. Chrysanthemum morifolium Ramat I. Self-fertility (In Jap.). J Jpn Soc Hortic Sci 46:101–112
Kher MA (1969) Photoperiodic response in Chrysanthemum. Plant Sci 1:200–206
Kher MA (1972) Precision blooming in Chrysanthemum. Indian J Ornam Hort 3:80–82
Kher MA (1973) Effect of B-9 and ethrel on the growth and flowering in chrysanthemum. Progress Hortic 5:77–83
Kher MA (1976) Planning a longer chrysanthemum season. Indian J Ornam Hortic VII:36–39
Kitamura S (1978) Dendranthema at Nipponanthemum. Acta Phytotaxon Geobot 29:165–170
Kofranek AM (1980) In: Larson RA (ed) Cut Chrysanthemum: introduction to floriculture. Academic, pp 5–43
Kunitake H, Imamizo H, Mii M (1993) Somatic embryogenesis and plant regeneration from immature seed-derived calli of rose (*Rosa rugosa* Thunb). Plant Sci 90:187–194
Kyle F (1952) *Chrysanthemums.* A complete guide to their culture. Ward, Lock & Co., Limited, London/Melbourne
Ledger SE, Deroles SC, Given NK (1991) Regeneration and *Agrobacterium*-mediated transformation of chrysanthemum. Plant Cell Rep 10:195–199
Machine J, Scopes N (1978) Chrysanthemum year round blooming. Blandford Press, Dorset
Maoxue L, Xiaofang Z, Junyu C (1983) Cytological studies on some Chinese wild Dendranthema species and Chrysanthemum cultivars. Acta Hortic Sin:1983–1993
Martin C, Uberhuaga E, Perez C (2002) Application of RAPD markers in the characterization of chrysanthemum varieties and assessment of somaclonal variation. Euphytica 127(2):247–253. https://doi.org/10.1023/A:1020215016347
Misra P, Chakrabarty D, Datta SK (2010) Role of micropropagation for commercial floriculture. In: Datta SK, Chakrabarty D (eds) Floriculture. Role of tissue culture and molecular techniques. Pointer Publishers, Jaipur, pp 165–188
Miyazaki S, Tashiro Y, Kanazawa K, Takeshita A, Oshima T (1982) On the flower characteristics and the chromosome numbers of Higo chrysanthemum (*Chrysanthemum morifolium* Ramat.). Bull Fac Agric Saga Univ 52:1–11
Mulford FL (1937) Results of selfing twenty-four early blooming chrysanthemums. Proc Am Soc Hortic Sci 35:818–821
Murashige T, Skoog F (1962) Physiol Plant 15:473–497
Nakajima T, Young HC (1965) The art of chrysanthemum. Harper & Row Publishers
Nazeer MA (1981) Karyological studies in two species of Chrysanthemum L. Cell Chrom Newslett 4(2):33–35
Nazeer MA, Khoshoo TN (1982) Cytological evolution of garden chrysanthemum. Curr Sci 51: 583–585
Nazeer MA, Khoshoo TN (1983) Variation in chromosome complement of *Chrysanthemum morifolium* complex. Nucleus 26(1):22–29
Niwa T (1936) Chrysanthemums of Japan. The Sanseido Company, Tpkyo
Pavingerová D, Dostál J, Bísková R, Benetka V (1994) Somatic embryogenesis and *Agrobacterium*-mediated transformation of chrysanthemum. Plant Sci 97:95–101
Popham RA, Chan AP (1950) Zonation in the vegetative stem-tip of *Chrysanthemum morifolium*. Am J Bot 37:476–484
Renou JP, Brochard P, Jalouzot R (1993) Recovery of transgenic chrysanthemum *(Dendranthema grandiflora* Tzvelev.) after hygromycin resistance selection. Plant Sci 89:185–197
Ronald WG, Ascher PD (1975) Self compatibility in garden chrysanthemum: occurrence, inheritance and breeding potential. Theor Appl Genet 46(1):45–54
Rosse A (1964) The Chrysanthemum book. D. Van Nostrand Company, Inc, Princeton/ Toronto/New York/London

Rumin LJ, Zalewska M, Sadoch Z (2004) Radiomutants of chrysanthemum (*Dendranthema grandiflora* Tzvelev) of the lady group: RAPD analysis of the genetic diversity. Plant Breed 123(3):290–293. https://doi.org/10.1111/j.1439-0523.2004.00996.x

Sabine J (1822) XXIV. Observations on the *Chrysanthemum Indicum* of Linnæus. Transactions of the Linnean Society of London. https://doi.org/10.1111/j.1095-8339.1821.tb00071.x

Sachs RM, Kofranek AM (1963) Comparative cytohistological studies on inhibition and promotion of stem growth in *Chrysanthemum morifolium*. Am J Bot 50(8):772–779

Saghai-Maroof MA, Soliman KM, Jorgensen RA, Allard RW (1984) Ribosomal DNA spacer length polymorphisms in barley: Mendelian inheritance, chromosomal location, and population dynamics. Proc Natl Acad Sci U S A 81(24):8014–8018. https://doi.org/10.1073/pnas.81.24.8014

Scott MC, Caetano-Anolles G, Trgiano RN (1996) DNA amplification fingerprinting identities closely related chrysanthemum cultivars. J Am Soc Hortic Sci 121(6):1043–1048

Stewart RN, Derman H (1970) Determination of number and mitotic activity of shoot apical initial cells by analysis of mericlinal chimeras. Am J Bot 57:816–826

Takatsu Y, Nishizawa Y, Hibi T, Akutsu K (1999) Transgenic chrysanthemum (*Dendranthema grandiflorum* (Ramat.) Kitamura) expressing a rice chitinase gene shows enhanced resistance to gray mold (*Botrytis cinerea*). Sci Hortic 82:113–123

Teixeira da Silva J (2004) Ornamental chrysanthemum: improvement by biotechnology. Plant Cell Tissue Organ Cult 79:1–18

Trigiano RN, Scott MC, Caetano-Anolle's G (1998) Genetic signatures from amplification profiles characterize DNA mutation in somatic and radiation-induced sports of Chrysanthemum. J Am Soc Hortic Sci 123:642–646

Watanabe K (1983) Studies on the control of diploid-like meiosis in polyploid taxa of Chrysanthemum. IV. Colchiploids and the process of cytogenetical diploidization. Theor Appl Genet 66: 9–14

Wolff K, Peters-van Rijn J (1993) Rapid detection of genetic variability in chrysanthemum (Dendranthema grandiflora Tzvelev.) using random primers. Heredity 71(4):335–341. https://doi.org/10.1038/hdy.1993.147

Wolff K, Rijn PV, Hofstra H (1994) RFLP analysis in chrysanthemum probe and primer development. Theor Appl Genet 88(3–4):472–478. https://doi.org/10.1007/BF00223663

Wolff K, Zietkiewicz E, Hofstra H (1995) Identification of chrysanthemum cultivars and stability of DNA fingerprint patterns. Theor Appl Genet 91:439–447

Woolman J (1957) Chrysanthemum in pictures. John Gifford Ltd., London

Yepes LC, Mittak V, Pang SZ, Gonsalves C, Slightom JL, Gonsalves D (1995) Biolistic transformation of chrysanthemum with the nucleocapsid gene of tomato spotted wilt virus. Plant Cell Rep 14:694–698

Zagorski JS, Ascher PD, Widmer RE (1983) Multigenic self incompatibility in hexaploid Chrysanthemum. Euphytica 32:1–7

Zhao H, Liu ZH, Hu X, Yin JL, Li W, Rao GY, Zhang XH, Huang CL, Anderson N, Zhang QX, Chen JY (2009) Chrysanthemum genetic resources and related genera of Chrysanthemum collected in China. Genet Resour Crop Evol 56(7):937–946. https://doi.org/10.1007/s10722-009-9412-8

Bougainvillea

20

S. K. Datta, R. Jayanthi, and T. Janakiram

Contents

S. K. Datta has retired.

S. K. Datta (✉)
Floriculture Division, CSIR-National Botanical Research Institute, Lucknow, India

Bose Institute, Kolkata, India

Department of Plant Science and Biotechnology, Assam University, Silchar, Assam, India

R. Jayanthi
Department of Horticulture UAS, GKVK, Bangalore, India

T. Janakiram
Horticultural Science Division, Indian Council of Agricultural Research, New Delhi, India

S. K. Datta, Y. C. Gupta (eds.), *Floriculture and Ornamental Plants*, Handbooks of Crop Diversity: Conservation and Use of Plant Genetic Resources,
https://doi.org/10.1007/978-981-15-3518-5_2

Abstract

Bougainvillea (Fam. Nyctaginaceae) is one of the most important perennial ornamental shrubs, sometimes a climber, in tropical and subtropical gardens. It is a very important floriculture crop for multipurpose use. The genus has approximately 18 species of which three (*B. spectabilis* Willdenow, *B. glabra* Choisy, and *B. peruviana* Humboldt and Bonpland) are horticulturally important. A good number of varieties have been developed in different parts of world/India through detection and isolation of bud sports, crosses among three basal species and varieties, selection and induced mutations. All basic scientific information is available on different aspects of *Bougainvillea*. Basic cytogenetic studies were directed towards determination of chromosome number, DNA content, mitotic and meiotic divisions, karyotypic analysis, colchi-ploidy, etc. Studies have opened a new way to synthesize new variety through chromosomal manipulations. Pollination mechanism, breeding system, and cross compatibility relationship have also been studied. Horto-taxonomical studies were initiated on important *Bougainvillea* species and cultivars. Extensive studies in this direction have enriched information on morphological descriptions, habit, growth, agrotechnology, techno-economics, flowering behavior, affinities with colored illustrations, and their usage. Extensive work has been carried out on induced mutagenesis and relative sensitivity of different single and double bracted varieties to mutagens have been determined very critically, and a number of promising mutant varieties have been developed and released. Mist propagation and micropropagation technique has been standardized for difficult-to-root and promising cultivars. A good amount of variety-specific morpho-chemical characters and desirable genes have been identified through characterization.

Keywords

Bougainvillea · Elemental species · Bud sport · Cytogenetics · Genetic diversity · Breeding

20.1 Introduction

Bougainvilleas are the most important ornamental flowering shrub as well as climber used for landscaping in various ways for landscaping. They are grown for the beautiful colorful bracts all over the world. It has a unique capacity of absorbing pollutants and in tolerating all environmental fluctuations. It flourishes well in all places even in polluted areas of cities and industrial towns also, that's why it is called as glory of the garden. Bougainvilleas are popular ornamental plants in most areas with warm climates for multipurpose use and very suitable plant for multidisciplinary research work. Numerous varieties and cultivars have enriched bougainvillea as one of the important components in floriculture. Most of the present day varieties have developed through bud sports, chance crossing, breeding, and induced mutation. Different aspects of Bougainvillea like distribution, classification, uses, propagation, pests and diseases, description of different varieties, flowering behavior, characterization, etc., have been reported from time to time by a number of researchers from different countries. Present chapter provides information on introduction, history, systematic position, species and cultivars of horticultural importance, taxonomical description, habit, morphological characters and flowering behavior, agrotechnology, propagation, horto-taxonomy, improvement, etc. Attempt has been made to put together all information to develop a complete documentation of the results of the research and demonstrations conducted by different researchers.

20.2 Uses

Bougainvillea is a very interesting ornamental which has multipurpose use due to its diversity in habit, growth, and color spectrum of varieties. It is also tolerant to a number of adverse conditions like dry soil, saline soil, pollution, etc. It can be grown as pot plants, standard, bushes, hedges, climber, bonsai, hanging baskets, landscape, as an espalier, factory gardens, institutional gardens, terrace garden, wind curtain, topiary, pergola, desert gardens, as ground cover on banks, as an excellent ground cover for difficult-to-maintain areas, to cover hill side, and various other multipurpose (Fig. 4). Varieties acclimatized and developed in India have been identified and categorized on the basis of their use as **Climber** ("Chitra," "Dr.R.R. Pal," "Elizabeth," "Roseville's Delight," "Los Banos Beauty," "Scarlet Queen," "Shweta," "Cherry Blossom" etc.), **Specimen plant** ("Blondie," "Begum Ali Yuwar Jung," "Mary Palmer," "Sonnet," "Mahara" etc.), **Group Planting** ("Tomato Red," "Chitra," "Cherry Blossom," "Arjuna," "Flame," "Purple Star" etc.), **Shrub/Bush** ("Asia," "Flame," "Spring Festival," "Summer Time," "Parthasarathy," "Glabra," "Blondie" etc.), **Topiary** ("New Red," "Lady Richard," "Happiness," "Perfection," "Purple Star," "Green Smith" etc.), **Hedge** ("Mary Palmer Special," "Mrs. H.C. Buck," "Shubhra," "Thimma," "Partha," "Parthasarathy" etc.), **Standard** ("Begum Sikander," "Glabra," "Jayalakshmi," "Rose Queen," "Spring Festival," "Speciosa" etc.), **Pot Plant** ("Begum Sikander," "Blondie," "Cherry Blossom," "Chitra," "Flame," "Wazid Ali Shah," "Shubhra," "Jaya Lakshmi," "Happiness," "Zakirina" etc.), **Ground Cover** ("Dr. H. B. Singh," "Dr. R.R. pal," "Mrs. H.C. Buck," "Shubhra," "Splendens" etc.), **Bonsai** ("Tomato Red," "Thmma," "Golden Glow," "Zakiriana," "Mrs. H.C. Buck," ""Shubhra," "Begum Sikander," etc.) (Datta et al. 2017).

20.3 Botany

It belongs to the family Nyctaginaceae (Four o clock plant family). On the basis of bract nature, Bougainvillea is classified into (1) Single bracted and (2) Double or multibracted Bougainvillea. Variegated foliage bougainvilleas are sometimes proposed as third group. But in reality, all variegated varieties are developed either through spontaneous or induced mutations from existing single or double bracted varieties. Majority of Bougainvillea belong to single bracted group with wide range of bract color spectrum. The wide range of adaptability along with diverging habitat and broad spectrum of bract color place this plant on a very important position among ornamentals. Double bracted group of Bougainvillea consists of initial four varieties ("Cherry Blossom," "Los Banos Beauty," "Mahara," and "Roseville's Delight") and few more varieties were developed through spontaneous and induced mutations. *B. glabra*, *B. spectabilis,* and *B. peruviana* are the only species of the Bougainvillea genus, possessing showy colorful bracts. *B. glabra* and *B. spectabilis* show natural variation in shape of leaves and bracts, base and apex of leaf, color of bracts and degree of hairiness, division of main short and flowering cymes, and shape of flower tube. Flowers are small, tubular, and surrounded by showy, colorful petaloid bracts. The fruit is an elongated achene less than 1 /2 inch long. It is rather inconspicuous, not showy, and has a dry, hard fruit cover. Seeds germinate readily and require no treatments to break dormancy. *B. peruviana* does not show any variability and is represented by only one cultivar. *B. glabra* and *B. spectabilis* are similar in essential morphological features. However, *B. peruviana* differs from two species in leaf shape, division of floral cyme, shape of floral tube, and unbranched main shoot. The cultivars which do not fall within the taxonomic limits of above species can be, by and large, divided on morphological grounds into three hybrid groups viz. *B.* x *buttiana* (*B. glabra peruviana*), *B.* x *specto-peruviana,* and *B.* x *specto-glabra*. The cultivars show a great range of shape, size, and color of bracts. About 80 per cent of garden Bougainvillea are found male and female sterile on the basis of pollination mechanism and breeding system. Fertile forms do not set seed on selfing but do so readily on crossing. Within the species, only the "*B. spectabilis*," the "*B. glabra*," and the "*B. peruviana*" are primary, whereas the rest are hybrids obtained from natural or artificial interbreeding of the above-mentioned species. The most diffused species are the "*B. x buttiana*" which derives from the interbreeding of the glabra and the peruviana plants, the "*B. x spectoperuviana*" which derives from the interbreeding of the spectabilis and the peruviana plants, and the "*B. spectoglabra*" which derives from the interbreeding of the spectabilis and the glabra plants (Hammad 2009).

20.4 Origin, Elemental Species, and Cultivars

The Bougainvillea is native to South America from Brazil west to Peru and South to Southern Argentina. Now bougainvillea is a popular plant in Southern California, Florida, the Caribbean, and other areas with tropical and warm climates. In 1789, a documented description of the bougainvillea as "Buginvillea" appeared in the "Genera Plantarium" by A.L. de Jusseau. An amendment to the spelling of the name to the correct version, bougainvillea, appeared in the Index Kewensis in the 1930s. The plant was discovered in Rio de Janeiro during an expedition which took place in 1768 by the French naturalist Dr. Philibert Commerson. Later on this plant took the name of Bougainvillea in honor of Louis Antoine de Bougainville, the French explorer who was in charge of that expedition. Only three species out of the eighteen of Bougainvillea

(*B. spectabilis* Willd., *B. glabra* Choisy, and *B. peruviana* H & B) are ornamental and by hybridization and mutation have given rise to the large number of modern-day garden cultivars differing in color, shape, and size of bracts. The "*spectabilis*" and the "*glabra*" kind were initially imported and cultivated in Europe, in particular in France and Great Britain. At a later stage, the Kew Gardens cultivated and distributed numerous specimens in Australia, New Zealand, East Africa, India, and in other British colonies of America. Originally, *B. spectabilis* and *B. glabra* were undifferentiated until the mid-1980s when botanists classified them as distinct species. Currently, there are over 300 varieties of bougainvillea around the world. Because many of the hybrids have been crossed over several generations, it's difficult to identify their respective origins. Natural mutations seem to occur spontaneously throughout the world. This had led to multiple names for the same cultivar (or variety) and has added to the confusion over the names of bougainvillea cultivars. In the early nineteenth century, these two species were the first to be introduced into Europe, and soon, nurseries in France and Britain sold these varieties in Australia. Meanwhile, Kew Gardens distributed plants it had propagated to British colonies throughout the world. Soon thereafter, a crimson specimen in Cartagena, Colombia, was added to the genus descriptions. Originally thought to be a distinct species, it was named *Bougainvillea* × *buttiana* honor of the European who first encountered it. However, later studies classified it as a natural hybrid of a variety of *B. glabra* and possibly *B. peruviana* – a "local pink bougainvillea" from Peru. Natural hybrids were soon found to be common occurrences all over the world. For instance, around the 1930s, when the three species were grown together, many hybrid crosses were produced almost spontaneously in East Africa, India, the Canary Islands, Australia, North America, and the Philippines (Datta et al. 2017, Salam et al. 2017).

Different species of bougainvillea are recognized as the official flowers of the islands of Grenada and of Guam; Lienchiang and Pingtung countries in Taiwan; lpoh in Malaysia; Tagbilaran in the Philippines; and Camarillo, Laguna Niguel, and San Clemente in California.

Short description of elemental species and important role played by Bougainvillea cv. "Mrs. Butt" for the evolution of many important cultivars of modern bougainvillea is very important. One significant incident in the history of bougainvillea took place with the discovery of a crimson bougainvillea in Cartagena, a Spanish port in the Mediterranean, by Mrs. R.V. Butt. Originally thought to be a distinct species and named *B. buttiana*; however, it was later discovered to be a natural hybrid of a variety of *B. glabra* and possibly *B. peruviana* – a "local pink bougainvillea" from Peru. The prominent feature of few important species on the basis of morphological characters is mentioned (Anonymous 1924, Kobayashi et al. 2007, Pancho and Bardenas 1959):

***B. spectabilis*:** German botanist Carl Ludwig Willde identified from Brazil in 1798. Leaves and stems hairy; leaves are large and ovate; thorns large curved; bracts are red, dark pink, or purple; flowers small flowers cream colored; perianth tube hirsute or villous, bloom cycle seasonal.

***B. glabra*:** Swiss botanist Jacques Denys Choisy identified in 1850. It is evergreen, climbing, leaves elliptical, green or variegated, glabrous, some puberulence, bracts are in many sizes and shapes, purple or mauve although white bracts are also present, thorns are small with curved tips, perianth tube puberulent or glabtate, blooms several times a year.

***B. peruviana*:** German naturalist and explorer Alexander von Humboldt identified from Peru in 1810. Most stable species; Bark green; leaves are long, thin, ovate and glabrous; Bracts rounded magenta to pink; flowers are yellow; thorns short and straight; perianth tube glabrous, perianth tube variously pubescent. The plants may bloom several times a year.

***Bougainvillea* hybrids:** B. glabra x B. peruviana is most common and known as Bougainvillea x buttiana. This hybrid was made by Mrs. R. Butt in Trinidad. Leaves are large and ovate/heart-shaped with slight hairiness on both sides; thorns are straight and short; bracts are usually rounded, red or dark pink; flowers small cream colored with pink tones; B. x buttiana hybrids bloom several times a year.

Bougainvillea x spectoperuviana is another common hybrid. Leaves are large, dark green, ovate, hairless; thorns straight; bracts are coppery red in the juvenile stage, turning to various shades of magenta or pink as they age; flowers are cream colored; bloom several times a year.

Bougainvillea x spectoglabra is another common bougainvillea hybrid. Leaves are small and dark green; thorns are numerous and curved; bracts mauve or purple; flowers small almost white; blooming is several times a year.

Mrs. Butt: It was first observed by Mrs. R.V. Butt in 1910 in Cartangena (in Columbia) from a priest's garden. "Mrs. Butt" was distributed to east Africa, Australia, and New Zealand and introduced in India in 1923. It was different from *B. glabra* and *B. spectabilis* and this was published as *B. x buttiana* "Mrs. Butt." "Mrs. Butt" is having the characters of both parents, that is, leaves are like *B. peruviana* and floral bracts are like *B. glabra*. "Mrs. Butt" played very important role for evolution of many important cultivars of modern bougainvillea. Series of new varieties with new bract colors (dianthus purple, phlox purple, burnt orange, Spanish orange, majolica yellow, etc.) have originated from "Mrs. Butt" through spontaneous mutations. Important varieties like "Scarlet Queen," "Surekha," "Scarlet Queen Variegata," "Versicolour," "Alick Lancaster," "Louise Wathen" (syn. Orange Glory), "Enid Lancaster," "Bhabha," "Louise Wathen Medipicta" (sy. "Louise Wathen variegata"), "Mrs Butt Variegata," "Magenta Queen" (syn. "Mrs Butt Magenta," "Purple Queen"), "Purple King," "Purple Prince," "Rao," "Kuvempu," "Vellayani" "Mrs McClean" (syn "Orange King" Practorius), "Yellow Queen," "Roseville's Delight" (syn "Dona Rosita Delight," "Doubloom"), "Archana," "Mary Baring," "Golden Glow," "Lady Mary Baring," "Gangaswamy," "Gangamma," etc., have either direct or indirect link with Mrs. Butt.

20.5 Introduction in India

India is one of the major repositories of wide range of genotypes of *Bougainvillea* and is perhaps the foremost among nations which have done appreciable work on *Bougainvillea.* Voluminous work has been done in India on different aspects like enrichment of germplasm, agro-technology, characterization, and improvement. All the present day Bougainvillea cultivars being grown in India are either exotics introduced from Africa, USA, UK, South Africa, Australia, Philippines, Sri Lanka, Jamaica, Canary Island, Singapore, Java, West Indies, etc., or developed by concentrated efforts of nurserymen, individuals, or scientific institutions. Bougainvilleas are now widely grown in Indian gardens. Interest in bougainvillea began in India in the mid-1800s and it became popular in 1920s with the arrival of the brilliant red cultivar "Scarlet Queen." A number of new cultivars including variegated forms developed through bud sports and the bract color expanded to orange, deep pink, rose, and blends of these colors. "*B. spectabilis*" was introduced in India from Europe in 1860 followed by "*B. glabra*" cv. "Splendens" (1969), "*B. x buttiana*" cvs. "Scarlet Queen" (1920), "Mrs. Butt" (1923), and "*B. peruviana*" cv. "Princes Margaret Rose" (1935). The first hybrid between "B. glabra" x "B. peruviana" was "Partha" released in 1942. "Mary Palmer," with both white and magenta pink bracts in the same inflorescence, was developed as sport from "Mahatma Gandhi" in the late 1940s. Voluminous work has been done in India on different aspects like enrichment of germplasm, agro-technology, characterization, and improvement at ICAR-Indian Agricultural research institute (IARI), New Delhi; CSIR-National Botanical Research Institute (NBRI), Lucknow; ICAR-Indian Institute of Horticultural Research (IIHR), Bangalore; Lalbaugh Botanical Gardens, Bangalore; Bhaba Atomic Research Centre (BARC), Trombay; Agri-Horticultural Society, Calcutta. Different Agricultural universities, amateur growers, and nurserymen are also maintaining huge cultivars due to its increasing demand and popularity. Division of Floriculture and Landscaping, Indian Agricultural research Institute, New Delhi, was recognized as the International Registration Authority in 1966 for *Bougainvillea* cultivars. Since then the Division is compiling, maintaining, publishing lists of names of cultivars, and registering new varieties. Voluminous compilation work resulted publication of first "The International Bougainvillea Check List" in 1981 and subsequently "The New International Bougainvillea Check List" in 1999 (Singh et al. 1999). These check lists are of immense importance to *Bougainvillea* lovers, researchers, floriculturists, landscape architects, town planners, and gardeners of the world.

20.6 Germplasm

Germplasm of bougainvillea represents assemblies of genotypes or populations representative of collection of cultivars, genetic stocks, wild species, etc., which are maintained in forms as plants. These are used as raw materials for future breeding program. India is one of the major repositories of wide range of genotypes of Bougainvillea and is perhaps the foremost among nations which have done appreciable good work on Bougainvillea. Germplasm collections in India is spread at Indian Agricultural research institute (IARI), New Delhi; National Botanical Research Institute (NBRI), Lucknow; Indian Institute of Horticultural Research (IIHR), Bangalore; Lalbaugh Gardens, Bangalore; Bhaba Atomic Research Centre

(BARC), Trombay; Agri-Horticultural Society, Calcutta. Different Agriculture Universities, amateur growers, and nurserymen are also maintaining good amount of cultivars due to its increasing demand and popularity. Number of germplasm collections at each center varies. There are more than 400 varieties growing in the Indian gardens. If we want to mention the names of varieties as reported by Institutions/Universities/Scientists, there will be repetition of varieties. At the same time, development of new varieties is a routine process to enrich the germplasm. Some important varieties maintained in India are "Amarnath," "Amarault," "Anindita," "Arjuna," "Autumn," "Baby Margaret rose," "Bangalore Variegata," "Beauty," "B.G.K. Hamilton," "Bougainvillea glabra," "Bonfire," "Brenda Ohare," "Cherry Blossom," "Cherry Ripe," "Chitra," "Chitravati," "Common Rose," "Conquest," "Crimson King," "Crimson Lake," "Crispa," "Daya," "Deep Cherry," "Dorothy Jivarajadasa," "Double Delight," "Dr. B.P. Pal," "Dr. Bhabha," "Dr. H.B. Singh," "Durga"s Delight," "Eclipse," "Ethiraj," "Eusga," "Exquisite," "Feathery Fantasy," "Feathery Fantasy Bi Coloured," "Frazer," "Gagarin," "Gangamma," "Gangaswamy," "Gem," "Geoffrey Nagpal," "Glabar var Cypheri," "Golden Queen," "Gopal Swamy," "Gulaby," "Isobel," "Happikiness," "Himani," "H.C. Buck," "Hiawatha," "Intermedia," "John Latin," "Jane Stanfeld," "Jasper Rose," "Jawaharlal Nehru," "Jayalakshmi Variegata," "Jennifer," "Jennifer Nagpal," "Kayata," "Kuvempu," "Lady Elizabeth," "Lady Mary Baring," "Lalbagh," "Lal Baugh Louis Wathen" "Lakshmi," "Laxminarayana," "Lazat of Mysore," "Leah Nagpal," "Lilac Beauty," "Lilacina," "Los Banos Variegata," "Louise Wathen Mediopicta," "Magenta Queen," "Mahara," "Mahara Variegata," "Maharaja of Mysore," "Mahatma Gandhi," "Mahatma Gandhi Variegated," "Marigowda," "Mary Palmer," "Mauve Queen," "Meera," "Meera Sport," "Midget," "Mrs. Butt Magenta," "Mrs. Butt Scarlet," "Mrs. Chico," Mrs. Eva Variegata "Light Purple", "Mrs. Fraser," "Mrs. G.S. Randhwa," "Mrs. Kay Malvenan," "Mrs. Marie Buck," "Mrs. R.B. Carrick," "Mrs. McClean Nirmal," "Mundanna," "New Red," "Nirmal Chandra," "Orange Glory," "Padmi," "Pearl," "Philoman," "Pink Beauty," "Pixie," "Poultoni Variegata," "Preeti," "Purple King," "Purple Mrs. Butt," "Purple Prince," "Purple Queen," "Purple Rose," "Purple Wonder," "Queen Elizabeth," "R.B. Singh," "R.S. Bhutt," "Raman," "Ratna," "Red Glory Improved," "Red September," "Refulgens," "Rose Fuschia," "Roseville"s Delight," "Sharma," "Shubhra," "Sholay," "Singapor Red," "Snow Queen," "Soundarya," "Spectabilis Variegated," "Specaible," "Spitfire," "Spring Festival," "Stanza," "Star Mauve," "Sweet Heart," "Tetra McClean," "Thimma," "Tyrian Rose," "Udai Chandra," "Usha," "Vesuvius," "Vijaya," "Winsome," "Yellow Queen," "Zakir Hussain," etc. (Figs. 1, 2, and 3).

20.7 Classification

The classification and nomenclature of the species, varieties, or cultivars of Bougainvillea are not very much well understood and still there are many discrepancies in the views of various workers. On the basis of bract nature, Bougainvillea is divided into single bracted and double or multibracted Bougainvillea. Variegated foliage bougainvilleas are sometimes proposed as third group. But in reality, all

Fig. 1 Some representative varieties of Bougainvillea

variegated varieties are developed either through spontaneous or induced mutations from existing single or double bracted varieties. Majority of Bougainvillea belong to single bracted group with wide range of bract color spectrum. Double bracted group of Bougainvillea consists of only four normal varieties viz. "Cherry Blossom," "Los Banos Beauty," "Mahara," and "Roseville's Delight." Two chlorophyll variegated mutants "Marietta" and "Archana" have developed through spontaneous mutations from "Mahara" and "Rosevilles delight," respectively. Four gamma ray induced chlorophyll variegated mutants have been developed – "Pallavi" from "Rosevilles delight"; "Mahara variegata" from "Mahara"; "Los Banos Variegata" and "Los Banos Variegata Jayanthi" from "Los Banos Beauty" (Datta 1992, Datta et al. 2017). Four more chlorophyll variegated induced mutant varieties have been reported: "Jaya" induced from original cv. "Jayalaxmi"; "Jayalaxmi variegated" induced from "Jayalaxmi"; "Lady Hudson of Ceylon variegata" induced from "Lady Hudson of Ceylon" and "Silvertop" induced from original cv. "Versi Colour" (Abraham and Desai 1977).

20.8 Characterization

Characterization is most important for correct identification of cultivars. It helps to understand the genetic diversity, to trace out the phylogenetic relationship, taxonomical status, preparation of catalogue, variation patterns, identification of

Fig. 2 Some representative varieties of Bougainvillea

desirable/novel genes, hybridization, registration, plant variety protection, farmer's right, etc. Different parameters of cytology, morphology, physiology, chemical and biochemical, DNA markers, etc., are utilized for characterization. Under characterization program of bougainvillea different characters like bract color and size; leaf color and size; branch and leaf number; leaf shape; spine number and size; stem color; stomatal index; moisture content of bract; pollen grain sterility; bract pigment and phenolic compounds in leaves and bracts and RAPD markers have been worked out in different varieties (Swarup and Singh 1964,. Datta et al. 2017).

(i) **Hortorium taxonomy:** CSIR-National Botanical Research Institute, India, is one of the pioneer institutions where Horto-taxonomical studies started from very beginning. Horto-taxonomical study deals with the fixity of names to the cultivars according to the rules of "International Code of Nomenclature for Cultivated Plant," their detailed morphological account including ancestry, habit, growth, flowering behavior, affinities with color illustrations, and their usage. Many hybridizers in different countries introduce and name seedlings without evaluation or registration. Lack of proper pedigree records and other relevant data before the cultivar is named or included in further breeding program, not only creates confusion but gives rise to a large number of cultivars of uncertain identity which would often not merit separate taxonomic status and entity. Therefore, all ornamental varieties need to be

Fig. 3 Some representative varieties of Bougainvillea

studied taxonomically in order to arrange all variations in a systematic manner. To overcome this problem, CSIR-NBRI was very conscious from beginning about characterization of cultivars to prepare a proper check list and maintenance of authentic germplasm collection (Datta et al. 2017).

(ii) **Bract color:** Bract color is one of the most important identifying characters of Bougainvillea. All the single bracted varieties have been classified into eight color groups viz., ***Red-Purple Group***, ***Purple-Violet Group***, ***Purple Group***, ***Red Group***, ***Violet Group***, ***Orange and Orange-Red Group***, ***White Group,*** and ***Bi-coloured Group.*** In the Multi-Bracted Group eight varieties viz., Archana (Bi-Colour), Cherry Blossom, Los Banos Beauty, Los Banos Variegata, Mahara, Mahara variegate, Pallavi (Bi-Colour), and Rooseville's Delight (Bi-Colour) have been studied. Color of matured bracts of Cherry Blossom, Mahara, and Mahara Variegata are Red-Purple Group, Los Banos Beauty is Purple Group, and Los Banos Variegated is Purple-Violet Group, Archana, Pallavi and Roseville's Delight are bi-colored. In some varieties, bract color of young and mature are different and in some cultivars bract color changes with age.

(iii) **Bract size:** The average bract size varies from variety to variety and these variations are observed in each color group. The smallest bract size (length) is observed in "Enid Lancaster" (1.75 cm) and the variety "Begum Sikander" (4.26 cm) showed the biggest size.

Fig. 4 Different uses of Bougainvillea. (**a**). Pot culture; (**b**) and (**c**). Bonsai; (**d**). Topiary; (**e**). Group planting; (**f**). Flat pot culture; (**g**). Hedge; (**h**). Arches and Pergolas; (**i**). Climber

(iv) **Flower size:** The length of flower tube and diameter of the star have been determined in all the single bracted varieties. In double bracted varieties, there is no flower.

(v) **Leaf color:** Leaf **c**olor of all the varieties have been determined at three stages (new leaf initiation stage, young, and matured) and the varieties have been categorized into different groups viz., Yellow-Green Group, Green Group, Brown Group, Greyed-Orange Group, Variegated, Greyed-Purple, Coppery-Red, Greenish-Brown, and Greyed-Brown on the basis of their color similarity to color chart of Royal Horticultural Society of London.

(vi) **Leaf size:** Leaf size has been measured and a wide range of variability in leaf length, leaf width, and petiole length are found among the varieties of each group.

(vii) **Number of branches:** The varieties can be categorized into three classes on the basis of branch number per 30 cm mature shoots area, that is, 1–5, 5–10, and 10–25.

(viii) **Number of leaf:** All the varieties have been categorized into five classes (**Class I 1–50; Class II 51–100; Class III 101–150; Class IV 151–200; and Class V > 200**) on the basis of total leaf number per 30 cm of mature shoot. Maximum varieties of each group have leaf number between 51 and 100.

(ix) **Shape of the leaf apex:** The shape of the leaf apex at full mature stage in different varieties has been identified under five groups, that is, acute, acuminate, acute to mucronate, acute to acuminate, and mucronate.

(x) **Number of spine and size:** A wide range of variation in spine number are observed per 30 cm length of mature stem in each group. Maximum varieties have spine number between 11 and –15 in almost all the groups. Size variation among the group and between the group are not very prominent.

(xi) **Stem color:** The color shades of mature stem of all the varieties of each group vary from Brown Group, Grey group, Greyed-Green, or Greyed-Brown Group.

(xii) **Stomatal Index:** Size and number of stomata per unit area vary both among the varieties of each group and also between the varieties of different groups. Anatomical studies were made from apical peelings and transverse sections of leaves after staining in 4 per cent aqueous solution of silver nitrate.

(xiii) **Moisture content of bracts:** Moisture content varies from 47 to 62 per cent among the varieties.

(xiv) **Pollen grain sterility:** A wide range of variation in pollen sterility are observed among the varieties and the sterility percentage varied from 10 per cent to above 90 per cent. Pollen grains of all the varieties are uniform, that is, "Normal" (almost same size). "Small" and "Big" sized pollen grains in addition to "Normal" pollen grains have been observed in some of the varieties. The smallest pollen grain of the Big category is larger than the largest pollen grain of the normal category and the largest grain of the small category is smaller than the smallest grain of the normal category. Basically pollen grains of Bougainvillea are 3(–4) colpate, prolate spheroidal to oblate spheroidal in shape with prominently reticulate exine. In most cultivars, two sets of bacula are present, the first and outer bigger set forms the muri and the second inner set supports the lumina. The shape and size of the lumina are variable among the cultivars.

(xv) **Phenolic compounds in leaves:** Young leaves and bracts were extracted separately in methanol containing 1 per cent concerted hydro chloric acid (approximately 0.1 g material in 0.8 ml solvent). The methanolic leaf extracts were tested for phenolic compounds with ferric chloride. The chromotograms were developed on glass plates (22 x 22 cms) coated with silica gel emulsion. Fixed amount of extract was spotted on the glass plates. The plates were run (18 cm) in a mixture of benzene:propionic acid:water (20:42:10 v/v). Plates were dried in air and the spots were observed and marked. Color of each spot was noted in visible light. The Rf values of each spot was determined from six good chromatograms These were transferred into hRf (Rf x 100) values. The distribution of the different compounds, their color, and hRf (Rf x 100) values have been determined. Totally 15 spots were observed in the varieties under investigation on the basis of Rf values and color. It was interesting to note that only one spot could be detected in one variety ("Krumbiegel") under Red Purple Group using present solvent system. The number of spots for phenolic compounds in all other varieties varies from two to five. The intensity of different spots varied on the basis of concentration of each compound. Color of different spots for spot 1 to Spot 15 were as follows: Yellow, Light Grey, Light Grey, Light Yellow, Light Pink, Dark Grey, Yellowish Brown, Grey, Pinkish Brow, Grayish Yellow, Dark Brown, Dark Grayish Green, Dark Brown, Greenish Grey, and Dark Yellowish Brown, respectively.

(xvi) **Phenolic compounds in bracts:** A total of 17 spots were detected among all the tested varieties. The spots were numbered serially 1–17 and hRf values of each spot of each variety have been determined. The intensity of different spots varied on the basis of concentration of each compound. The color of different spots observed for spot 1 to Spot 17 were Pink, Dark Grey, Dark Grey, Pink, Light Grey, Dark Grey, Dark Grey, Light Grey, Light Grey, Pink, Pink, Greyish Brown, Dark Brown, Dark Grey, Greyish Yellow, Yellowish brown, and Dark Grey, respectively.

(xvii) **Bract pigment:** For spectrophotometric analysis of bract pigments, 100 mg bracts were extracted in 25 ml methanol containing 1% HCl. The extract was scanned from 200 to 700 nm region of wave length in spectrophotometer scanning equipment (uvikon 930 Kontron Instruments). A total of thirty varieties, representing each color group, were selected for analysis of bract pigment by Spectrophotometer. Methanolic bract extracts of each variety were scanned in spectrophotometer covering 200–700 nm wave length. Total number of peaks found at different wave length and their O.D. values were recorded. Peaks were observed at different wave lengths like 276, 282, 290, 296, 320, 324, 326, 328, 334, 336, 352, 478, 524, 540, 542, and 544 nm. Intervarietal differences were observed and confirmed by number of peaks and their O.D. values.

Heuer et al. (1994) analyzed betacyanins from the bracts of *B. glabra* and characterized them by spectrophotometric techniques. Kochhar et al. (1979) reported photo control of betacyanin synthesis on the basis of studies on bougainvillea growing in normal sunlight and those enclosed in black polythene bags. Pigments in the magenta colored bracts of Bougainvillea CVs. Princess Marget Rose, Mrs. H.C. Buck, and Mary Palmer were resolved by paper electrophoresis into 5 bands, 3 pertaining to betacyanins and 2 to betaxanthim with different electrophoretic mobilities. The genes responsible for the synthesis of betacyanins have been suppressed in Mary Palmer, a mutant of Mrs. H.C. Buck. They are, however, partially activated by light and high temperature. His potentiality is almost lost in Shubra, in which no bands of betacyanins were described (Kochhar and Ohri 1977). Piattelli and Imperato (1970) and Imperato (1975) isolated seven betacyanin from purple bracts of Bougainvillea "Mrs. Butt." They also detected acylated betacyanin, a violet-red pigment isolated from bracts of *Bougainvillea glabra* var. "Sanderiana." Sabale and Bhosale (1986) reported the presence of C3 and C4 photosynthesis in the Bougainvillea Cv. Mary Palmer. Their (1986) further anatomical, photosynthetic, and photo respiratory observations on three samples of variegated leaves revealed a combination of characters of both C3 and C4 type of photosynthesis.

(xviii) **Molecular characterization of germplasm/hybrids and mutants:** Molecular characterization is essential for elucidating the genetic relationships among the different groups of this species and cultivars. DNA-based markers provide powerful and reliable tools for discerning variation within the germplasm, genotype identification, taxonomical studies, and to study evolutionary relationships. These techniques include RAPD (Randomly amplified polymorphic DNA), RFLP (Restriction Fragment Length Polymorphism), SSR (Simple Sequence Repeats) or microsatellites, STS (Sequence Tagged Sites), SNP (Single Nucleotide Polymorphism), VNTR (Variable number tandem repeat), STR (Short tandem repeat), SFP (Single feature polymorphism), and AFLP. These results opened new avenues for the development of useful markers for further improvement of crops. One can compare the genetic makeup between different species and even different varieties or cultivars of a species by using RAPD markers. The genetic variation in plants is considered to be an unlimited resource for introgression of foreign genes.

Techniques have been standardized at different institutions for molecular characterization of bougainvillea. One of the authors (S.K.Datta) along with his team members has done extensive work on characterization of different ornamentals (germplasm and new varieties developed through conventional breeding and induced mutagenesis) at Floriculture Laboratory, CSIR-National Botanical Research Institute (NBRI), India. Accurate characterization is necessary to trace out the parents of new varieties. Use of molecular markers in addition to the classical methods provides more positive identification of new cultivars. Genomic DNA was extracted from the very young leaves of bougainvillea by using DNA extraction procedure of Saghai-Maroof et al. (1984). After amplification, PCR product was resolved by electrophoresis in 1% agarose gel with 1X TAE buffer. Bands were visualized by staining with ethidium bromide (0.5 μg/ml) under UV light and photographed. Only distinctive polymorphic bands were counted for data analysis, and faint bands were not considered. Presence and absence of bands were indicated by 1 and 0, respectively. Genetic distance was calculated by Jaccard's coefficient (Jaccard 1908). Ninety-two most important cultivars were selected from germplasm collections of National Botanical Research Institute, Lucknow, India, to determine the genetic relationship among the varieties using RAPD technique. Ten random decamer primers were selected and each primer produced a unique set of amplification products ranging in size from 300 to 2500 bp. Each primer produced a unique set of amplification products ranging in size from 300 bp to 2500 bp. The number of bands for each primer varied from 10 in P2 to 18 in P10. These 10 primers used in this analysis yielded 167 scorable bands with an average of 11.3 bands per primer. Of the 167 fragments scored from these primers, 26 were monomorphic and 141 were polymorphic (84.4%). The highest number (87%) of polymorphic bands was obtained by primer P10 and lowest (50%) by primer P2. Mean genetic similarity GS_M was calculated for the total 92 cultivars which was 0.301 indicating a high genetic diversity in the bougainvillea gene pool. RAPD analysis was found to be very helpful for the identification of cultivars, documentation, and to trace out the molecular affinity of origin of unknown group of *Bougainvillea* cultivars. Genetic diversity of a large number of *Bougainvillea* cultivars has been estimated. Origin and interaction of different species/cultivars of the four major groups were studied successfully; origin and affinity of 41 cultivars of unknown origin could be traced out up to certain level. Such study is very helpful and necessary for assessment of genetic diversity of large germplasm collections of horticultural species and their further improvement through selective breeding program. The information obtained from this work may be useful for better management, identification of accessions, and also in avoiding duplications or mislabeling of the genotypes studied (Chatterjee et al. 2007). Genetic relatedness and molecular characterization of 50 bougainvillea cultivars that belong to four major species of bougainvillea (*B. glabra*, *B. spectabilis*, *B.peruviana,* and *Bougainvillea* × *buttiana*) have been studied. Five microsatellite (simple sequence repeat; SSR) markers with high PIC values were used to characterize these bougainvillea cultivars. A total of 18 rare alleles were identified. The study proved the efficiency of SSR markers in documentation, identification, and tracing out the molecular origin among unknown cultivars of bougainvillea (Kumar et al. 2014). The genetic relationships among the most important Bougainvillea cultivars grown at most of the gardens in Egypt were determined using random amplified Polymorphic DNA (RAPD) technology and isozymes patterns. Isozymes revealed higher polymorphism among bougainvillea cultivars with all enzymes used (Hammad 2009). Diversity among the bougainvillea cultivars have been analyzed using morphological markers. A total of 38 morphological descriptors of UPOV guidelines were recorded for hundred bougainvillea cultivars. Significant amount of variation was observed for all the morphological descriptors (Kumar et al. 2015).

Horto-taxonomical description of some leading Bougainvillea species/cultivars has been prepared according to their affinities to the species and hybrid seedlings:

Aida: *B. spectabilis*, seedling of *B. spectabilis*, Lancaster 1940, Calcutta, India; tall, growth vigorous; young stem greenish coppery, tomentose; leaf ovate, base acute, apex acute to acuminate, coppery, old dark green, puberulent; thorn straight; young bract carmine changing to rose madder, elliptic, apex obtus; profuse blooming; ideal as shrub, and pot culture.

Blondie: *Bougainvillea* x *buttiana*; seedling of *B.* x *buttiana*, Kenya, Africa and named in California, USA; tall, growth vigorous; greenish coppery young stem, tomentose; leaf elliptic, base acute, apex accuminate, youngleaf coppery, old leaf green; thorn curved; young bract majolica yellow changing to Saturn red and finally to amaranth rose, ovate, base cordate, apex acute, margin wavy, puberulent, nonpersistent; profuse blooming; ideal as bush, standard, ground cover, and pot culture.

Camarillo Fiesta: *Bougainvillea* x *buttiana*; USA; tall, growth vigorous; young stem greenish coppery, tomentose; leaf ovate, base acute, apex accuminate, green, puberulent; thorn curved; young bract burnt orange changing to magenta rose, ovate, base cordate, apex acute, puberulent, nonpersistent; profuse blooming; ideal as bush, climber, and pot culture.

Dr. Harbhajan Singh: *B. glabra*; seedling of "Trinidad" and "Formosa," IIHR (1979); dwarf, drooping; young stem green, veluntinus; leaf elliptic, acute, acuminate, green, puberulent; thorn curved; young bract mauve, apex acuminate; profuse blooming; ideal as bush, hedge, pot culture, bonsai, and hanging basket.

Enid Lancaster: *Bougainvillea* x *buttiana*; bud sport of "Louise Wathen," Lancaster, Delhi; tall, growth vigorous; young stem coppery, glabrate; leaf blade ovate, base acute, apex accuminate, young light coppery, old green; thorn slightly curved; young bract Spanish orange changing to brick red, ovate, base cordate, apex acute, puberulent, nonpersistent; profuse blooming; ideal as shrub, climber, standard, and pot culture.

Flame: *B. spectabilis*; hybrid seedling; tall, growth vigorous; young stem coppery, puberulent, leaf ovate, base acute, apex acuminate, coppery, old green, puberulent; thorn curved; young bract claret rose changing to solferino purple, ovate, apex acute; profuse blooming; ideal as shrub, climber, standard, and pot culture.

Glabra: *B. glabra*; Brazil (1860); young stem green, tomentose; leaf elliptic, base acute, apex acuminate, green, glabrate; thorn curved; young bract phlox purple; profuse blooming; ideal as bush, climber, standard, and for pot culture.

Isabel Greensmith: *B. peruviana*; hybrid seedling, Kenya; tall, growth vigorous; young stem greenish coppery, glabrate; leaf elliptic to ovate, base acute, apex acuminate; thorn slightly curved; young bract burnt orange changing to rose opal, elliptic to ovate, base cordate, apex acute to accuminate, glabrate, nonpersistent; profuse blooming; ideal as bush, climber, standard, cascade, pot culture, hanging basket, and bonsai.

Lady Mary Baring: *Bougainvillea* x *buttiana*; bud sport of "Golden Glow," Nairoby; tall, growth vigorous; young stem coppery, glabrate; leaf broadly ovate, base acute, apex accuminate, young coppery, old green, puberulent; thorn slightly curved; young bract Indian yellow changing to darker, ovate, base cordate, apex acute, puberulent, nonpersistent; profuse blooming; ideal as bush, climber, standard, espalier, and pot culture.

Lateritia: *B. spectabilis*; seedling of B. spectabilis, London 1865; tall, vigorous growth; young stem green, villous; leaf velutinous; thorn slightly curved; young bract brick red changing to darker, ovate, apex acute; profuse blooming; ideal as climber, on arch, up a tree and pot culture.

Mrs. H.C. Buck: *B. peruviana*; Madras 1930; seedling of "Princess Margeret Rose"; tall, vigorous growth; young stem coppery, villous; leaf ovate, base acute, apex acuminate, light green; thorn slightly curved; young bract magenta rose changing to darker, ovate, base cordate, apex acute; profuse blooming; ideal as shrub, climber, ground cover, hedge, topiary, standard, and pot culture.

Palekar: *B. peruviana*; selection, Bombay; tall, drooping, growth vigorous; young stem greenish coppery, puberulent; leaf elliptic, base acute, apex acuminate; thorn slightly curved; young bract red changing to salferino purple, elliptic to ovate, base cordate, apex acute, glabrate, nonpersistent; profuse blooming; ideal as bush, climber, standard, cascade, pot culture, hanging basket, and bonsai.

Poultoni Special: *Bougainvillea* x *buttiana*; seedling of *B.* x *buttianna*, Durban, S. Africa; tall, growth vigorous; young stem green, pilose; leaf ovate, base acute, apex accuminate, green, puberulent; thorn curved; bract largest in size (4.5–60 x 3.2–4.8 cm), young bract solferino purple changing to darker, ovate, base cordate, apex acute, puberulent, nonpersistent; profuse blooming; ideal as bush, standard, and pot culture.

Princess Margaret Rose: *B. peruviana*; Agri-Horticultural Society Madras 1935; tall, growth vigorous; young stem green, glabrate; leaf blade ovate, base acute, apex acute, green, glabrate; thorn slightly curved; young bract fuchsine pink changing to darker, ovate, apex acute; profuse blooming; ideal as bush, climber, and pot culture.

Shweta: *B. glabra*; bud sport of "Trinidad," CSIR- NBRI, India 1979; young stem green, tomentose; leaf blade elliptic, base acute, apex acuminate, green, puberulent; thorn curved; young bract sap green changing to white, elliptic, apex acuminate; profuse blooming; ideal as bush, standard, pot culture, and bonsai.

Splendens: *B. spectabilis*; seedling selection (London 1861); tall and vigorous growth; young stem green; young leaf coppery, elliptic, base acute, apex acuminate, mature leaf green; thorn straightly curved; young bract imperial purple darker, ovate, apex acute; profuse blooming; suitable as bush, climber, and pot culture.

Thimma: *B. peruviana*; bud sport of "Mary Palmer," Lal Baugh Garden, Bangalore, 1960; tall, growth vigorous; young stem pinkish yellow, glabrate; leaf elliptic to ovate, base acute, apex acute to accuminate, green with yellow blotches in the center, puberulent; thorn slightly curved; bracts three types – young magenta rose, parchment white with magenta sector, parchment white, parchment white with magenta sector, ovate, apex acute to obtuse, puberulent, profuse blooming; ideal as bush, climber, hedge, standard, espalier, topiary, pot culture, and bonsai.

Mahara: *Bougainvillea* x *buttiana*; bud sport of "Mrs. Butt" (Philippines 1963); tall, vigorous growth; young stem greenish coppery, puberulent; leaf ovate to elliptic, base acute, apex accuminate, green; thorn curved; multibracted, young bract rhodamine purple changing to lighter, ovate to elliptic, base cordate to oblique, apex acute to accuminate, puberulent, persistent; profuse blooming; ideal as bush, climber, bonsai, and pot culture.

Marieta: *Bougainvillea* x *buttiana*; bud sport of "Mahara" (Philippines, 1967); medium, growth restricted; young stem coppery, tomentose; leaf blade ovate, base acute, apex accuminate, young leaf with pink margin and grey green center, old leaf with cream variegation at margin and irregular grey green center; thorn point curved; multibracted, young bract rhodamine purple changing to lighter, ovate to elliptic, base cordate to oblique, apex acute to accuminate, puberulent, persistent; medium blooming; ideal as bush, climber, and pot culture.

Partha: *B. peruviana*; seedling of "Princess Margaret Rose," Bangalore, 1942; tall, growth vigorous; young stem greenish coppery, puberulent; leaf blade ovate to elliptic, base acute, apex accuminate, young light coppery, old green, puberulent; thorn slightly curved; bract brick red changing to fuschia purple, ovate, base cordate, apex acuminate, puberulent; profuse blooming; ideal as bush, climber, ground cover, hedge, standard, espalier, pot culture, and bonsai.

Roseville's Delight: *Bougainvillea* x *buttiana*; bud sport of "Mrs. McClean" (Philippines, 1962); tall, growth vigorous; young stem greenish coppery, glabrate; leaf ovate, base acute, apex accuminate, young leaf coppery and old green; thorn slightly curved; multibracted, young bract orpiment orange changing to rose madder, elliptic to ovate, base acute to cordate, apex acute to accuminate, puberulent, persistent; profuse blooming; ideal as bush, climber, bonsai, and pot culture.

Los Banos Beauty: *Bougainvillea* x *buttiana*; bud sport of "Pink Beauty" (Philippines, 1967); tall, growth vigorous; young stem greenish coppery, tomentose; leaf ovate, base acute, apex accuminate, green, puberulent; thorn slightly curved; multibracted, young bract mallow purple changing to lighter, elliptic to ovate, base cordate, apex acute, puberulent, persistent; profuse blooming; ideal as bush, climber, bonsai, and pot culture.

Cherry Blossom: *Bougainvillea* x *buttiana*; bud sport of "Los Banos Beauty" (Philippines, 1967); tall, growth vigorous; young stem coppery, puberulent; leaf ovate, base acute, apex accuminate, green, puberulent; thorn slightly curved; multibracted, young bract white suffused with mallow purple, elliptic to ovate, base acute to cordate, apex acute to accuminate, puberulent, persistent; profuse blooming; ideal as bush, climber, bonsai, and pot culture.

20.9 Genetics and Cytogenetics

Cytogenetics helps in solving intricate problems of evolution of economic plants and their relationship with relevant wild ancestors. Bougainvillea is a very interesting ornamental plant for evolutionary studies. Such study provides valuable information on the nature of evolutionary steps and the changes that the breeding system and chromosomes underwent. Such study also provides the genetic and cytogenetic reasons for the diversity and helps to chalk out a meaningful breeding methodology for genetic improvement of bougainvillea. Most of the garden forms are of bud sport and hybrid origin, raised over a number of years, and named differently by different workers and nurserymen based on their locality and bract color. Cytogenetic studies were mainly concentrated towards determination of chromosome number, mitotic and meiotic divisions, karyotypic analysis, colchi-ploidy, DNA content, etc. (Sen and Sen 1954). Aminotriazole pretreatment has been reported to be very helpful for chromosome studies in leaf and shoot tip squashes (George and Sobhana 1976). More often than not, there is no authentic record of their ancestry and they show a great deal of morphological and cytological polymorphism (Khoshoo 1990, Datta and Banerji 1995). $2n = 34$ chromosomes were recorded in large number of varieties belonging to "B. glabra," "B. buttiana" and "*B. spectabillis.*" $2n = 32$ chromosomes were detected in two horticultural varieties of Bougainvillea, one with scarlet bract and the other with brick red bract, which morphologically resembled to "*B. spectabillis.*" $2n = 51$ chromosomes were recorded in few varieties. $2n = 20$ chromosomes were detected in some buds from red sectorial part of variety "Mary Palmer." Natural triploidy, new "n" number ($n = 8$) and a good amount of meiotic irregularities due to variable pairing of chromosomes have been reported (Banerji and Banda 1967). DNA content has been estimated from three basal "B. species," hybrid groups, triploid hybrid cultivars, and induced tetraploids. DNA content of "B. peruviana" was found to be significantly different from "B. spectabilis" and "B. glabra" and synthetic hybrids were intermediate between parents. Direct correlation of DNA content and ploidy level has been recorded. Pollination mechanism, breeding system, and cross compatibility relationship have also been studied. Tetraploidy has been induced which restores fertility in sterile cultivars. Induced tetraploids are self-incompatible like the fertile diploid cultivars but set seeds readily on crossing. Diploid progenitors show irregular meiosis, while in tetraploid counterparts there is predominant bivalent pairing. Low frequency of quadrivalents at 4x level implies that pairing at diploid level is between chromosomes heterozygous for cryptic structural changes. Chromosome and DNA studies have solved many taxonomic problems, phylogenetic relationships, affinities of different species and varieties and also have opened a new way to synthesize new variety through chromosomal manipulations. There is strong evidence that morphological and horticultural diversity may be primarily due to variability within species "*B. spectabilis*" and "*B. glabra*" which were involved in indiscriminate hybridization among themselves and with "*B. peruviana.*" Somatic mutations seem to account for variability in color of bracts, flower tube shape, and leaf variegation. The new color combinations may have also arisen as a consequence of hybridization and interaction of alleles responsible for bract color in the basal types (Sharma and Bhattacharya 1960).

20.10 Breeding System in Bougainvillea

Normally bougainvillea cultivars are sexually sterile and are multiplied by vegetative propagation. Understanding the basic information on breeding system is very important to prepare most suitable breeding methodology for genetic improvement of bougainvillea. Studies clearly indicate that selection pressure along with genetic-evolutionary influence helped to develop new variations in bougainvillea. Khoshoo (1981) proposed such excellent knowledge in breeding system of ornamentals in general and bougainvillea in particular. Studied clearly indicate that most of the present day cultivars of bougainvillea have developed from three basal species, namely, *B. glabra* Choisy, *B. peruviana* Humboldt & Bonpland, and *B. spectabilis* Willdenow and one hybrid species *B. x buttiana* Holttum & Standley. The mechanisms involved were natural and selective hybridization, chromosomal manipulations, and spontaneous and induced mutations. Bougainvillea has comparatively simpler genetic system as all the three elemental species are diploid and the hybrids between them are either completely sterile or semi sterile. This interspecific hybridization has produced intermediate morphological and phenological characters and entirely novel bract colors in the three hybrid groups. The fertility in these hybrid groups could be restored by induced polyploidy. At the early stage, a good amount of variation was developed due to natural interspecific hybridization among the three species of bougainvillea. Further development of novel colours and interesting chimeric designs and double bracted cultivars developed through bud-sports from these hybrids. Pollen and/or seed sterility is noteworthy handicap for large-scale breeding of bougainvillea. This restricts in selection of male and female parents for breeding. Use of chromosome manipulation is worth mentioning as a good tool for development of new varieties. Colchicines-induced polyploidy restored the fertility and helped development of desirable, colorful and floriferous, often bicolored, hybrid bougainvillea (triploid, tetraploid, and aneuploid). Induced tetraploidy showed new perspective in bougainvillea breeding. Triploid varieties were developed by crossing tetraploid with diploid. Cultivars "Shubhra," "Mrs. McClean," "Mary Palmer," and "President Rosevelt" are diploid and sterile. Tetraploidy was induced in these varieties. Tetraploids are self-incompatible which after crossing with other fertile forms produce seeds and developed very promising cultivars at CSIR-NBRI viz. "Begum Sikander," "Wajid Ali Shah," "Mary Palmer Special," and "Chitra" by hybridizing "Dr.B.P. Pal" (tetraploid of "Shubhra") and "Tetra Mrs. McClean" (tetraploid of "Mrs. McClean") (Datta et al. 2017, Khoshoo and Zadoo 1969, Khoshoo 1990, Ohri 2013, Ohri and Khoshoo 1982, Ohri and Zadoo 1979, 1986, Zadoo et al. 1975a,b,c,d).

20.11 Breeding

A number of plant breeding methods are available for crop improvement and to develop new varieties. Creation of genetic variability is pre-requisite for development of new variety. For development of new varieties in bougainvillea, attention is paid on spontaneous mutations, selection, classical breeding, chromosome manipulations, induced mutagenesis, etc. There are more than 400 cultivars of Bougainvillea evolved from the three basal species, namely, "*B. glabra* Choisy," "*B. peruviana*" Humboldt & Bonpland, "*B. spectabilis*" Willdenow, and one hybrid species ""*B. x buttiana*" Holttum & Standley, through natural and planned hybridization and spontaneous and induced mutations. The main handicap in bougainvillea breeding is pollen and seed sterility, which limits the selection choice of male and parents. *Bougainvillea* has comparatively simpler genetic system as all the three elemental species are diploid and the hybrids between them are either completely sterile or semi sterile. This interspecific hybridization has produced intermediate morphological and phenological characters and entirely novel bract colors in the three hybrid groups. The fertility in these hybrid groups could be restored by induced polyploidy and this expanded germplasm was used to produce tetraploid and triploid hybrids with highly positive horticultural traits. All basic genetic information of bougainvillea in terms of breeding system, experimental hybridization, polyploidization, etc., has been worked out for cultivated and elemental species. This helps for developing meaningful methodology in the creation of new and novel cultivars of commercial importance.

Spontaneous bud sports have played very important role in developing new and novel characters of ornamental value in different ornamentals including bougainvillea. Such natural mutations have been noted from time to time by keen gardeners, horticulturists, researchers, etc., in their germplasm collections which are being maintained by vegetative means. Spontaneous mutations have been responsible for three types of changes in cultivated bougainvilleas viz. change in bract color, development of imperfect floral tubes, and leaf variegation (Zadoo et al. 1975a).

Conventional plant breeding can be used as a system for selection of superior genotypes from genetically variable populations derived from sexual recombination. Hybridization involves crossing two plants so as to produce a genetically and phenotypically superior offspring than either of the parents. Bougainvillea cultivars are sexually sterile and are multiplied by vegetative propagation. Some cultivars produce seeds infrequently. Desired hybridization between the cultivars producing seeds and those having fertile pollen is in practice for evolving new cultivars. Selection choice of male and parents is limited. Breeding program is hampered in bougainvillea due to widespread pollen and/or seed sterility. Studies indicate that total sterility in diploids is the result of recombination between homologous chromosomes, or segregational hybrid sterility. Studies proved that fertility can be restored by the colchicine-induced polyploidy, and for developing new varieties, chromosome manipulation has been used as a good toll. This method has given a wider choice for breeding bougainvillea and resulted development of a number of very promising, colorful and floriferous, often bicolored, hybrid bougainvillea (triploid, tetraploid, and aneuploid). Induced tetraploidy has opened a new perspective of Bougainvillea breeding. A number of triploids ($2n = 3x = 51$) have been raised by crossing tetraploids ($2n = 4x = 68$) with diploids. Tetraploidy was induced in diploid and sterile varieties ("Shubhra," "Mrs. McClean," "Mary Palmer," and "President Rosevelt"). Tetraploids are self-incompatible, but produce seeds readily on crossing with other fertile forms. By hybridizing "Dr.B.P. Pal" (tetraploid of "Shubhra") and "Tetra Mrs. McClean" (tetraploid of "Mrs. McClean") with other sterile, diploid cultivars, a number of very beautiful, bicolored cultivars have been evolved, viz. "Begum Sikander," "Wajid Ali Shah," "Mary Palmer Special," and "Chitra" (Khoshoo 1990). Such manipulation of fertility in sterile bougainvillea cultivars has widened the scope of bougainvillea breeding to evolve more promising cultivars through planned hybridization which could have not been possible otherwise due to pollen or seed sterility.

20.12 Induced Mutations

Induced mutagenesis is now an established method for crop improvement using physical and/or chemical mutagens. Induction of somatic mutation by mutagenesis is an important method in improving the specific characteristics. CSIR-National Botanical Research Institute, India, is one of the pioneer institutions where commendable work has been done on induced mutagenesis. Appreciable information has been accumulated on different aspects like radiosensitivity, selection of material, methods of exposure to gamma rays, suitable dose of gamma rays, colchicine treatment, recurrent irradiation, detection and isolation of mutants, commercial exploitation of mutant, etc. Mutation techniques by using ionizing radiations and other mutagens have successfully produced quite a large number of new promising varieties in bougainvillea (Datta 2004, Jayanthi et al. 1999, Swaroop et al. 2015). Relative sensitivity of different single and double bracted bougainvillea varieties to different mutagens have been determined very critically on the basis of several parameters like sprouting; survival; plant height; leaf, branch, and bract number and size; chromosomal aberrations and mutation frequency. Determination of suitable dose of gamma radiations for the induction of somatic mutation is most essential. Extensive work has been carried out to determine the radiosensitivity and LD_{50} dose. The suitable dose of gamma rays for irradiation of stem cuttings has been standardized from 250 to 1250 rads. Induction of somatic mutation by mutagenesis is an important method in improving the specific characteristics. The most promising and beautiful chlorophyll variegated mutant "Arjuna," induced after treatment with 250 rads of gamma rays of Bougainvillea cv. "Partha," has been commercialized (Gupta and Shukla 1974). Only four cultivars viz. "Cherry Blossom," "Los Banos Beauty," "Mahara," and "Rosevilles Delight" comprise the double bracted group. Two chlorophyll variegated mutants "Marietta" and "Archana" have developed through spontaneous mutations from "Mahara" and "Rosevilles delight," respectively. Mutation breeding is the only method for improvement of double bracted Bougainvillea as improvement by conventional cross-breeding is not possible due to absence of true flower. A total of four chlorophyll variegated mutants have been detected. One chlorophyll variegation detected in 10Gy (1Gy = 100rads) treated population of "Rosevilles Delight" has been isolated and named "Pallavi" and released as new cultivar (Banerjee et al. 1987). Chlorophyll variegation detected in "Mahara" has been named "Mahara Variegata" (Datta and Banerji 1994). Two chlorophyll variegations have been detected in "Los Banos Beauty" which have been released as "Los Banos Variegata" and "Los Banos Jayanthi" (Banerji et al.1987, Datta 1992, Datta and Banerji 1990, 1997). As far as bract color is concerned, still no new bract color could be induced. It was interesting to note that although there was formation of no new bract color, formation of chimera of existing bract color was recorded. Irradiated population of "Roseville's Delight" produced chimeric bract of "Mahara," "Cherry Blossom," and "Los Banos Beauty" type; "Cherry Blossom" produced "Mahara" and "Los Banos Beauty" type and "Los Banos Beauty" produced only "Mahara" type chimera. But surprisingly no change in bract color in irradiated population of "Mahara" was recorded (Banerji and Datta 1993, Datta and Banerji 1997, Datta et al. 2017). The chlorophyll variegated mutants look very attractive even during off season when plants are devoid of colorful bracts.

It was interesting to note that no mutation in bract color could be detected through classical mutation technique except in one cultivar "Palekar." Stem cuttings of five single bracted varieties were irradiated with 250, 750, and 1500 rads of gamma rays. Somatic mutations in leaves and bract color were recorded. Somatic mutation in bract color was observed in one branch in chimeric form in 750 rad gamma rays treated stem cuttings. Cuttings were taken from the mutant branch and mutant has been established in pure form. The original young bract color of "Palekar" is Current Red (821/3) changing to Salferino purple (26/1) on maturity. The mutant bract color at young stage is Orange (12/1) which on maturity becomes Azalia Pink (523/1).

Leaves in original cultivar are obtuse and flat while in mutant these are elliptic and incurved (Srivastava et al. 2002).

Four gamma ray-induced mutants have been released by Abraham and Desai (1977). The mutants are "Jaya" with ornamental novelty, induced from original cv. "Jayalaxmi"; "Jayalaxmi variegated" with variegated leaves, induced from "Jayalaxmi"; "Lady Hudson of Ceylon variegata" with variegated leaves, induced from "Lady Hudson of Ceylon" and "Silvertop" with ornamental novelty, induced from original cv. "Versi Colour."

Chemical mutagen has been applied to induce new genetic variability in multi-bracted bougainvillea. Mature stem cuttings of "Los Banos Variegata" were treated with different concentrations, that is, 0.01% to 0.3% of Ethyl Methane Sulphonate (EMS) for 6 hours. One plant from 0.02% EMS treated population exhibited one mericlinal chimeroid branch with variegated leaves in M_1V_1 generation. The size of chlorophyll variegation in leaves varied from a narrow streak on a leaf to whole leaf. The growth of the axillary bud associated with variegated leaf was arrested by the influence of the apical dominance of the main apex of the mericlinal shoot. The chimeric branch has been isolated in pure form by air layering method and subsequently multiplied through repeated cuttings. The variegated mutant branch has been isolated, multiplied and named "Los Banos Variegata-Jayanthi" to release as a new variety (Jayanti et al. 2000).

20.13 Propagation/Cultural Practices

Bougainvillea is mostly propagated by hard wood cuttings and through different asexual methods (cuttings, layering, grafting, budding) and micropropagation (Sindhu and Dagar 2015).

i. **Stem cuttings**: Propagation through stem cuttings is very simple, easy, and less expensive. Mature healthy (semi-hardwood or hardwood) branch is selected for preparation of cuttings. Cuttings are prepared with the help of disinfested knife or secateurs and each cutting should be 2 to 6 inches long with 2 to 6 leaves (5 to 9 nodes in length). Media like soil, sand, moss, vermiculite, perlite separately and combinations of some of these are used for planting cuttings. For better rooting, either rooting powder hormones (seradix, keradix, rootadex, etc.) or growth regulators (Indole Butyric acid –IBA; Napthalene Acetic Acid – NAA etc.) are used of different concentrations (200–6000 ppm) (Bhattacharjee and Balakrishna 1983).
ii. **Rooting:** Voluminous work has been done on rooting of Bougainvillea. Effect of the type of wood, length of cutting, seasonal effect, effect of growth regulators, moisture management, etc., on rooting of bougainvillea have been studied extensively by a number of workers and recommended – better rooting of soft wood cuttings in February, maximum rooting in cuttings during September pretreated with 0.2 per cent NAA, better rooting of tip cuttings than mature cuttings, plant growth regulators (IAA, IBA, and NAA) are superior to control,

best rooting of hard wood cuttings when dipped in IBA at 6000 ppm and NAA at 1000–2000 ppm, highest rooting with IAA at 4000 ppm, cuttings treated with 4000 ppm IBA stimulates maximum percentage of rooting and cent per cent survival under mist, best rooting when cuttings are soaked in water followed by a 10 minutes dip in 2500 ppm IBA + 2500 ppm fenulic acid, better rooting of hard wood cuttings than semi-hard wood cuttings, 100 ppm IBA and ascorbic acid 10 per cent sucrose, terminal cuttings for rooting in 18 hours treatment in solution of p-hydroxybenzoic acid for rooting, basal end of both hard wood and semi-hard wood treated with IBA at 2000 ppm are very effective for rooting, cuttings of largest diameter treated with IBA at 6000 ppm develop highest rooting percentage, more compact and well-branched root system when treated with copper hydroxide, use of coir dust, etc. Better rooting has been reported in "*Bougainvillea x buttiana*" "Mrs Butt" under intermittent mist after wounding the stem and /or treated with IBA as a quick-dip at 5000 ppm for 30 seconds. Wounding and dipping in IBA alone significantly increased number of roots per cutting compared with controls (Datta et al. 2017, Gupta and Kher 1991, Yadav et al. 1978, Singh et al. 2013).

iii. **Air layering:** This method is not used for large scale multiplication. Only selected varieties and selected branches are isolated and multiplied. Roots are developed in shoots that are still attached to the parent plant. The selected branch is brought to the level of an earthen pot by bending and buried in light soil and tip remains uncovered. When roots are developed from covered portion, the stem is cut very cautiously and planted in pots. Another method of air layering is called "gootie." Small ring of bark (about 1 cm) is removed from the selected mature shoot and rooting hormone is applied on exposed stem. Exposed portion is covered with moss and with a plastic and tied with rope. When root develops, the branch is cut and planted in pot after removing moss.

iv. **Budding:** Budding method is also used for multiplication of bougainvillea. This technique is simple but requires proper practice, precaution, and care. Vigorous growing and easily multiplied varieties are selected as root stock. The root stock plant with one branch is prepared in January/February. Budding eyes are selected before sprouting situated in the axils of leaves of the desired variety to be multiplied. The eye is removed with a sharp knife (budding knife) along with a piece of bark 1.5 cm above and below the eye. I-shaped or T-shaped sharp (bark of the stem) cut is made at the lower portion of stock (3–6″ above soil level). The size of the cut should be sufficient to accommodate the eye of the scion. The eye along with the shield-shaped tissue is carefully kept into the T-shaped or I-shaped cut. The eye is tied with narrow tape-like alkathene fiber. The budding is mostly performed during April under subtropical condition. Union of scion and stock and sprouting time of budded eyes vary from variety to variety (2–4 weeks). The eye sprouts and develops into a branch on root stock. The upper portion (above budding) of the stock is removed when the height of sprouted branch reaches approx. 6–8″ (15–20 cm). The stock pant along with sprouted scion is removed from the soil and planted at desirable place (pot/bed).

v. **Grafting:** This method is used for specific mode of multiplication. Like stalk selection in budding method, well-established vigorous growing variety is selected as understock variety for grafting. Branch of pot/soil growing desired variety to be grafted is brought near the branch of understock variety. A small portion of bark of both the branches are removed. Then root hormone is applied in both the branches and tied tightly with plastic after keeping moss. When the branches join properly, the upper portion of understock and lower portion of desired variety are removed to allow growth of shoot of desired variety.

vi. **Pruning:** This operation is very important to develop new branches and flowering and also to give a proper shape of the plant. Bougainvillea generally blooms on new branches. Normally branches are cut back after flowering is over. This helps to produce new branches and bloom. Pruning is performed to maintain good shape of the plant. For busy appearance, new growing tips of a new bougainvillea should be pinched every few weeks. Bougainvillea varieties can be grown as a bonsai even in ground if it is pruned timely by removing most of the growing branches.

20.14 Tissue Culture

One of the major constraints of floriculture industry is nonavailability of large scale genuine quality planting materials. Conventional methods of propagation cannot meet the increasing demand of propagating materials. Micropropagation is perhaps the most widely used biotechnology tool for large scale propagation of floricultural crops. The technology has now been commercialized globally and has contributed significantly towards the enhanced production of high quality planting materials. There are limitations of propagation rate of bougainvillea through conventional methods, further some of the promising varieties where root formation is difficult. A perusal of literature shows that only a limited attempt has been made to micro-propagate this plant (Cooper 1931). In vitro methods have been standardized for large scale development of planting materials. Shoot apices of *Bougainvillea buttinia* "Scarlet Queen Variegated" were induced to multiply in vitro. Shoots were initially grown in the basal medium for a minimum period of 120 days before their culture in a medium containing 1 mg/liter BAP plus 0.1 mg/liter IAA, in which ca. 7 off-shoots were produced in 60 days. 100% rooting was achieved in shoots cultured in 5 mg/liter NAA for 15 days followed by their transfer in 0.5 mg/liter NAA for a further period of 15 days, while the pH of the medium was 4.5. A maximum of 42 plantlets were produced from each culture of a shoot apex in one year. The in vitro raised plants grew normally in soil and flowered true-to-type under field conditions (Sharma and Chaturvedi 1988). Shoot apices of Bougainvillea glabra "Magnifica" were induced to regenerate an average of ten shoots from their base in response to BAP (0.5 mg/l) plus IAA (1.5 mg/l). All the isolated shoots from such cultures were rooted in a medium containing 0.1 mg/l each of IBA and 2,4,5-T and lacking BAP. Plantlets were then successfully grown in potted soil where they flowered normally. The results offer an opportunity to develop a practical method of large-scale clonal propagation of new cultivars of bougainvillea at a faster rate than is possible by using conventional method of propagation (Sharma et al. 1981). In case of *Bougainvillea glabra*, "magnifica" roots were induced when individual shoots were subculture either first in 0.75 mg/l NAA for 30 days and then transferring them on 0.5 mg/l 2,4,5 trichlorolphenoxyacetic acid (2,4,5-T) supplemented media (Chaturvedi et al., 1978) or by NAA or IBA (0.1 mg/l) and 2,4,5-T (0.1 mg/l; Sharma et al. 1981). In case of *Bougainvillea buttiana* "Scarlet Queen Variegated," roots were induced by subculturing the shoots on high NAA containing medium for 15 days followed by their transfer on low NAA containing medium for 15 days, while the medium pH was 4.5. In the present protocol, successful rooting was achieved in a single step. Well-rooted plantlets were transferred to soil and leaf mold mixture in plastic pots and after three weeks of hardening, they were finally transferred to 10 inch earthen pot and kept in field conditions. Plants were grew well in field and flowered true to type.

Two difficult-to-root varieties, one chlorophyll variegated mutant variety "Los Banos Variegata" and one hybrid seedling variety "Mary Palmer Special," were selected and standardized protocol for large scale multiplication. One-cm-long stem nodes with one axillary bud were collected from field grown plants of both the varieties. Explants were washed thoroughly in running tap water for 15 min and for another 5 min with 5% aqueous solution of liquid detergent and then washed again in distilled water. They were disinfected with 70% ethanol for 30 sec followed by surface sterilization with 0.1% $HgCl_2$ for 2 min and then washed thoroughly in sterile distilled water. They were cultured on MS medium with 3% sucrose, 0.8% agar, and different combinations of 1-naphthaleneacetic acid (NAA), 6 benzyladenine (BA), and gibberellic acid (GA_3). Medium pH was adjusted to 5.6 before autoclaving at 121^0 C for 15 min. Cultures were incubated at 25 ± 1 °C under cool white light with a 16th photoperiod (36 μ mol m^{-2} s^{-1}). For root induction, isolated shoots of 2–3 cm long were cultured on MS medium supplement with different concentrations of NAA, IBA or IAA either singly or in combination. Well-rooted plantlets were transferred to plastic pots containing mixture of soil and leaf mold (1:1) and kept covered to maintain high humidity for the first 10 days. After 3 weeks, plants were finally moved to field conditions.

Sprouting of axillary bud was observed within 2 weeks of culture initiation. Initially one or two shoots were formed from each axillary bud but multiple shoot formation was observed only after one subculture in the same medium. The best response was obtained with 0.5 mg/l NAA and 2 mg/l BA treatment in both the cvs. In this treatment, response of 60% explants (recorded after 4 weeks) with 7.4 number of shoots per responding explant (recorded after 8 weeks) was recorded in cv "Los Banos Variegata" and 7.5 explants (recorded after 4 weeks) with 7:2 number of shoots per responding explants (recorded after 8 weeks) was recorded in cv. "Mary Palmer Special." To further improve the regeneration frequency and shoot multiplication rate, different concentrations of GA_3 were incorporated into NAA 0.5 mg/l and BA 2 mg/l containing medium. Among different concentrations of GA_3, 0.5 mg/ l was found best suitable where hundred percent explants responded (recorded after 4 weeks) with 8.3 and 8.4 number of shoots per responding explant (recorded

8 weeks) in cv "Los Banos Variegata" and "Mary Palmer Special" respectively. In 0.2 and 1 mg/l GA_3, 100% explants responded with 6.8 ± 0.3 and 7.9 ± 0.3 number of shoots per responding explants respectively in case of cv. "Los Banos Variegata" while in case of cv. "Mary Palmer Special" 100% explants responded in these GA_3 treatments with 8.2 ± 2.2 and 6.6 ± 1.8 number of shoots per responding explant, respectively. Vigorous growth of shoots was observed in GA_3 containing medium. As addition of 0.5 mg/l GA_3 was found beneficial, therefore, shoot clumps formed in all other treatments were transferred to NAA 0.5 mg/l + BA 2.0 mg/l + GA_3 0.5 mg/l containing medium. Proliferating cultures were maintained in this medium by regular subculture at 4 weeks interval. Cultures were maintained in this medium for last six years without losing their multiplication rate. For root induction, 2–3 cm long shoots were excised from shoot clumps and individual shoots were transferred to rooting medium containing NAA, IBA, or IAA (1,2 or 4 mg/l), NAA (1 mg/l) + IBA (1 mg/l) or NAA (1 mg/l) + IAA (1 mg/l). Among all these treatments, roots were formed only in NAA 2 mg/l and 4 mg/l supplement medium within 3–4 weeks. In 2 mg/l NAA treatment, 40% and in 4 mg/l NAA treatment 83% rooting was obtained in the cv. "Los Banos Variegata" and in cv. "Mary Palmer Special" 56% and 82% rooting was obtained in these two treatments, respectively. Plantlets were ready to transfer in soil within another week (Datta and Mandal 2012, Datta et al. 2017).

20.15 Physiology and Postharvest Management

Application of TRIA (a saturated primary alcohol n-C30H61OH) improves the plant's physiological activities, stimulates flowering, enhances plant growth, and increases the quality of potted Bougainvillea plants. Flowering in many species can be induced by a variety of environmental techniques and the application of growth-promoting chemicals. The application of kinetin and the removal of young leaves enhance inflorescence development and improve the quality of Bougainvillea bracts. Application of sucrose during postharvest storage increases the longevity of Bougainvillea bracts (Datta et al. 2017, Hossain and Boyce 2008). *Effects of silver thiosulfate and naphthalene acetic acid (0.45% NAA + 1.2% NAA-amide at 500 mg L − 1) on flowering bud development, anthesis duration, bract longevity, and bract photosynthetic rate in Bougainvillea spectabilis have been studied. Gago and Monteiro (*2012*).* Effect of gibberellic acid (GA_3 100 and 150 ppm), phloemic stress, and combination of 100 ppm GA_3 and phloemic stress on *Bougainvillea* bract blooming, expansion, development and bract longevity have been studied and found that 100 ppm GA_3 increased the length of petiole, bract size, and shape by 40%. Bract blooming was three days earlier in 100 ppm GA_3 treated branches and 4 days earlier in 150 ppm GA_3 than in water control. The number of bracts per branch was higher in 100 ppm GA_3 + phloemic stress and phloemic stress than the other treatments. Petal size and petiole length were the highest in 100 ppm GA_3. The findings suggested that gibberellic acid played an important role to induce rapid bract blooming and expansion, whereas phloemic stress increased total number of bract and longevity (Saifuddin et al. (2009). Fresh flowers lose their freshness and quality both during travel and also during and after arrangements due to flower specific short vase life. Such deficiencies can be ameliorated through application of nutrient additives to vase water. Normally Bougainvillea is not used as cut-flowers, in spite of the fact that it blooms profusely in varied hues and shades in most part of the year in tropics/subtropics. This is due to the reason that its bracts start wilting within hours after the branch is severed from the parent plant. The flowering branches can be kept turgid and fresh with erect and natural-looking bracts up to 5–6 days if kept in solutions of citric acid (1000 ppm) and succinic acid that are supplemented with NAA (50 ppm) to prevent shedding of bracts (Kochhar et al. 1992, Shukla and Kher 1979).

20.15.1 Disease

Bougainvillea are relatively pest-free plants, but may suffer from worms, snails, and aphids. The larvae of some Lepidoptera species also use them as food plants, for example, the Giant Leopard Moth (*Hypercompe scribonia*). It gets attacked by spiders which can be controlled by using miticide. Attack of fungus *Phytophthora* can be checked by spraying Blitox (copper oxychloride) (Datta et al. 2017).

i. **Leaf spot**: Bougainvillea is susceptible to fungal and bacterial leaf spot. Reddish-brown spots develop on young foliage and spread distorting the growth of the plant. The infected leaves and branches should be removed and destroyed to check spreading. Dry condition of leaves and foliage, pruning and application of fungicide help to minimize the spread of the infection. "*B. glabra*" is reported to be resistant or slightly susceptible to the leaf spot disease compared to "*B. spectabilis.*" Leaf blight caused by *Phytophthora para-sitica, Cercosporidium bougainvillea,* Fusarium wilt (Fusarium oxysporum), Bacterial leaf spot (*Pseudomonas andropogonis)* have been reported (Kobayashi and Oniki 1994, Polizzi et al. 2010, Nema et al. 1999, Alfieri, Jr. 1970).
ii. **Root rot**: It is caused by fungus like *Rhizoctonia*, *Pythium,* or *Phytophthora.* Fungus infects roots and symptoms are stunted growth, plant dieback, wilting, and chlorosis. Water logging condition is susceptible to develop root rot. Infected plants should be removed and destroyed as soon as possible. Application of broad-spectrum fungicide (copper ammonium) during planting can reduce chances of infection.
iii. **Mildew**: Mildew infection can be identified by the symptom of white, powder-like appearance on the foliage. Excess water and poor soil drainage cause mildew infections. Removal of diseased foliage and application of mild fungicides may reduce the chances of mildew infection.
iv. **Scale disease:** Scale insects or mealybugs attack bougainvillea and suck the sap of plants. Insects are observed on the underside of leaves. Scale infections can be detected from canker development on branches and stunted plant growth. Pyrethrin-based pesticides may be sprayed by preparing a mixture with water.

v. **Pests**: Very few reports are available on incidence of pests on Bougainvillea. Mites (*Phylcoptes bougainvillea, Vittacus bougainvillea*), scales (*Coccus hysperidium*), and association of phytophagous nematodes with rhizosphere soil and roots have been reported on *Bougainvillea glabra* (Pathak et al. 1995).
vi. **Aphids:** Suck sap from leaves. Aphids may be controlled by spraying them with a direct stream of water.
vii. **Nematode:** Nematodes are parasitic roundworms that attack bougainvillea roots and plants infestations cannot absorb water and nutrients. Poor growth, wilting, leaf loss, yellow foliage, and deformed roots are the symptoms of nematode infestations. Nematode population can be reduced by incorporating compost, wood shavings, or manure into the soil before planting. It can be killed by solarizing the soil using plastic tarps before planting.

20.15.2 Marketing:

Readymade *Bougainvillea* plants are in great demand in market and sold in the market in different forms according to their uses. Well-grown plants in larger containers are sold for landscape. Large scale plants of similar size and different bract colors are also in great demand for selective landscaping. There is mass marketing of plants both in 3 to 4 inch pots and 6 to 8 inch pots for multipurpose use like standard, bedding plants, industrial plantation to combat air pollution, etc. Plants in 8 to 10 inch pots are sold for hanging baskets. Selective potted plants are sold for road side plantation.

20.16 Registration of Bougainvillea Cultivars

Indian Agricultural Research Institute, New Delhi –110,012 was appointed as the International Registration Authority for Bougainvillea cultivars at the 17th International Horticultural Congress held at Maryland, USA in 1966 (Choudhury and Singh 1981). The primary functions of International Registration Authority are as follows:

(a) To compile, maintain, and publish a list of names of cultivars and subsequently to publish such supplements and new edition of the list as circumstances may require
(b) To register names which conform to the rules and recommendations of the International Code of Nomenclature for Cultivated Plants
(c) To endeavor to get raiser, introducers, and others concerned with the distribution of plants to submit all new names to the Registration Authority and to use only names which confirm to the code

Division of Floriculture and Landscaping, Indian Agricultural Research Institute, New Delhi, and The Bougainvillea Society of India (BSI) register new bougainvillea varieties which serves as a documental proof of new variety, author, mode of origin, place, and date. A check list of all the existing important cultivars based on the available information from literature as well as from amateurs, professional traders, and other sources from different countries have been prepared (Anonymous 1981, Singh et al. 1999, Sindhu 2015).

The first Indian ornamental variety which has been patented is bougainvillea. Indian Institute of Horticultural Research, Bangalore, developed one variety "Dr.H. B. Singh" (Dr.R.N. Bhat and team) which got patent in Australia in the name of "Krishna" (Dr. H. B. Singh: "Krishna," Application No: 97/119 Accepted: 12 Jan 1998. Applicant: Jan Iredell, Moggil, QLD.). The variety is exclusively used as a potted plant and gained considerable popularity.

References

Abraham V, Desai BM (1977) Radiation induced variegated mutant in bougainvillea. Curr Sci 46(10):351–352

Alfieri SA Jr (1970) Bougainvillea blight caused by Phytophthora para-sitica. Phytopathology 60:581

Anonymous (1924) Mrs. Butt. Gardener's Chronicle 76:243

Anonymous (1981) The international Bougainvillea check list. Published by international registration authority, division of vegetable crops and floriculture. Indian Agricultural Research Institute, New Delhi, pp 1–106

Banerji SK, Banda GK (1967) A natural triploid in Bougainvillea. Indian J Hort 24:106–108

Banerji BK, Data SK (1993) 'Mrs butt' – bud sports and its near relatives. Sci Hortic 3:141–151

Banerji BK, Nath P, Datta SK (1987) Mutation breeding in double bracted Bougainvillea cv. 'Roseville's delight. J Nucl Agric Biol 16:45–47

Bhattacharjee SK, Balakrishna M (1983) Propagation of bougainvillea from stem cuttings I. Effect of growth regulators, rooting media, leaf number, length and woodiness of cuttings. Haryana J Hortic Sci 12(1–2):7–12

Chatterjee J, Mandal AKA, Chakrabarty D, Datta SK (2007) Use of RAPD analysis to determine genetic diversity and relationships among Bougainvillea cultivars at inter-specific levels. Hortic Environ Biotechnol 48(1):43–51

Chaturvadi HC, Sharma AK, Prasad RN (1978) Shoot apex culture of *Bougainvillea glabra* 'Magnifica'. Hortic Sci 13:36

Choudhury B, Singh B (1981) The international Bougainvillea check list. Indian Agricultural Research Institute, New Delhi

Cooper DC (1931) Micropropagation in *Bougainvillea glabra*. Amer J Bot 18:337–344

Datta SK (1992) Mutation studies on double bracted *Bougainvillea* at National Botanical Research Institute (NBRI), Lucknow, India. Mutation Breeding Newsletter, IAEA, Vienna, Issue No. 39: 8–9

Datta SK (2004) Bougainvillea research at National Botanical Research Institute, Lucknow. J Ornamental Hortic 7(1):1–14

Datta SK, Banerji BK (1990) 'Los Banos Variegata' – a new double bracted chlorophyll variegated Bougainvillea induced by gamma rays. J. Nucl Agric. Biol. 19:134–136

Datta SK, Banerji BK (1994) 'Mahara Variegata'- a new mutant of Bougainvillea. J Nucl Agric Biol 23(2):114–116

Datta SK, Banerji BK (1995) Cytological studies in ornamental plants. Advances in horticulture vol. 12 – ornamental plants part I. In: Chada KL, Bhattacharjee SK (eds) . Malhotra Publishing House, New Delhi, pp 267–334

Datta SK, Banerji BK (1997) Improvement of double bracted bougainvillea through gamma ray induced mutation. J Plant Sci 64(4):395–400

Datta SK, Mandal AKA (2012) Standardization of *in vitro* multiplication of two difficult- to-root bougainvillea cultivars for commercial exploitation. Sci Cult 78(5–6):251–254
Datta SK, Jayanthi R, Janakiram T (2017) Bougainvillea. New India Publishing Agency, New Delhi. 110 034
Gago CML, Monteiro JA (2012) NAA and STS effects on potted bougainvillea; early flower death allows delayed bract abscission. Postharvest Biol Technol 74:49–54
George K, Sobhana P (1976) Leaf and shoot tips squashes by aminotriazol pretreatment. Nucleus 19(1):29
Gupta VN, Kher MA (1991) A note on the influence of auxins in regeneration of roots in the tip cuttings of bougainvillea sp. var. garnet glory under intermittent mist. Haryana J Hortic Sci 20(1–2):85–87
Gupta MN, Shukla R (1974) Mutation breeding in bougainvillea. Indian J Genetics 34 (A):1295–1299
Hammad I (2009) Genetic variation among *Bougainvillea glabra* cultivars (Nyctaginaceae) detected by Rapd markers and isozymes patterns. J Agric Biol Sci 5(1):63–71
Heuer S, Richter S, Metzger JW, Wray V, Nimzt M, Strack D (1994) Betacyanins from bracts of *Bougainvillea glabra*. Phytochemistry 37(3):761–767
Hossain ABMS, Boyce AN (2008) Influence of ethanol on the longevity and delayed senescence of bougainvillea flower. J Appl Hortic 10(2):149–153
Imperato F (1975) A-branched tri-sacharide in the Beta-cyanins of *Bougainvillea glabra*. Phytochemistry 14:25–26
Jaccard P (1908) Nouvelles recherches sur la distribution florel. Bull Soc Vaud Sci Nat 44:223–270
Jayanti R, Datta SK, Verma JP (2000) Chemical induced mutation in double bracted bougainvillea cv. 'Los Banos Beauty'. Indian Bougainvillea Annual 15. April 9–11
Jayanyhi R, Datta SK, Verma JP (1999) Effact of gamma rays on single bracted Bougainvillea. J. Nucl Agric. Biol. 28(4):228–233
Khoshoo TN (1981) Breeding systems in relation to evolution in some ornamentals. Proc Indian Natl Sci Acad B47(2):279–301
Khoshoo TN (1990) Creativity in ornamentals. Indian Hortic 34:25–29,33
Khoshoo TN, Zadoo SN (1969) New perspectives in Bougainvillea breeding. J Hered 60(6): 357–360
Kobayashi T, Oniki M (1994) Circular leaf spot of Bougainvillea caused by Cercosporidium bougainvillea in Indonesia. Ann Phytopathol Soc Jpn 60(2):221–224
Kobayashi DK, McConnell J, John Griffis J, (2007) Bougainvillea. Ornamentals and Flowers: 1–12
Kochhar VK, Ohri D (1977) Bio-chemical analysis of bract mutations in bougainvilleas I. 'H.C. Buck' family. Zeitschrift fur pflanzenzuchtung 79(1):47–51
Kochhar VK, Kochhar S, Ohri D (1979) Bio-chemical analysis of bract mutation in bougainvillea cv. Mrs. H.C.Buck. Curr Sci 48(4):163–164
Kochhar VK, Shukla R, Murty AS, Kher MA (1992) Studies on prolonging vase life of Baugainvillea. Indian J Agric Res 26(1):15–24
Kumar PP, Janakiram T, Bhat KV, Jain R, Prasad KV, Prabhu KV (2014) Molecular characterization and cultivar identification in Bougainvillea spp. using SSR markers. Indian J Agric Sci 84(9): 1024–1030
Kumar PP, Janakiram T, Bhat KV, Prasad KV, Jain R (2015) Genetic divergence analysis of bougainvillea (Bougainvillea spp) cultivars using morphological markers. Indian J Agric Sci 85(5):661–665
Nema S, Sharma ND, Nema S (1999) *Phytophthora nicotianae*: incitant of blight of Bougainvillea. J Mycol Plant Pathol 29(2):262
Ohri D (2013) Cytogenetics of domestication and improvement of garden Gladiolus and Bougainvillea. Nucleus. https://doi.org/10.1007/s13237-013-0091-7
Ohri D, Khoshoo TN (1982) Cytogenetics of cultivated Bougainvillea. Z Pflanzenzuchtg 88: 168173
Ohri D, Zadoo SN (1979) Cytogenetics of cultivated Bougainvillea VIII. Cross compatibility relationship and origin of Bougainvillea 'H.C. Buck' family. Z. Pflanzenzuchtg 82:182–186

Ohri D, Zadoo SN (1986) Cytogenetics of cultivated Bougainvillea IX. Precocious centromere division and origin of polyploid taxa. Plant Breed 97:227–231

Pancho JV, Bardenas AE (1959) Bougainvilleas in the Philippines. Baileya 7(3):91–100

Pathak KN, Kumar P, Mishra HP, Kumar P (1995) Association of phytophagous nematodes with rhizosphere soil and roots of some ornamental plants. J Ornamental Hortic 3(1–2):55–58

Piattelli M, Imperato F (1970) Pigments of *Bougainvillea glabra*. Phytochemistry 9(12):557–2560

Polizzi G, Aiello D, Guarnaccia V, Vitale A, Perrone G, Stea G (2010) First report of Fusarium Wilt of paper flower (*Bougainvillea glabra*) caused by *Fusarium oxysporum* in Italy. Plant Dis 94(4): 483

Sabale AB, Bhosale LJ (1986) Photosynthesis in the variegated leaves of Bougainvillea cultivar Rao. Ind J Plant Physiol 29(3):196–206

Saghai Maroof MA, Solinan KM, Jorgensen RA, Allard RW (1984) Ribosomal DNA spacer-length polymorphisms in barley: Mendelian inheritance, chromosomal location, and population dynamics. Proc Natl Acad Sci U S A 81:8014–8018

Saifuddin M, Sharif Hossain ABM, Osman N, Moneruzzaman KM (2009) Bract size enlargement and longevity of *Bougainvillea spectabilis* as affected by GA_3 and Phloemic stress. Asian J Plant Sci 8:212–217

Salam P, Bhargav V, Gupta YC, Nimbolkar PK (2017) Evolution in Bougainvillea (Bougainvillea Commers.) – a review. J Appl Nat Sci 9(3):1489–1494

Sen SN, Sen NK (1954) Chromosome number and abnormalities in Bougainvillea. Bull Bot Sur Bengal 8:140–142

Sharma AK, Bhattacharya NK (1960) Cytological investigations on Bougainvillea as an aid in interpreting the evolution and affinities of different species and varieties. Nucleus 3:19–76

Sharma AK, Chaturvedi HC (1988) Micropropogration of Bougainvillea x Buttiana ';scarlet queen variegated' by shoot tip culture. Ind J Exp Biol 26:285–288

Sharma AK, Prasad RN, Chaturvedi HC (1981) Clonal propagation of Bougainvillea glabra 'Magnifica' through shoot apex culture. Plant Cell Tissue Organ Cult 1:33–38

Shukla R, Kher MA (1979) A note on prolonging vase life of bougainvillea. Indian J Hortic 36(1): 93–95

Sindhu SS (2015) Registration of new cultivars. Indian Bougainvillea Annuals 26:30

Sindhu SS, Dagar SS (2015) Propagation methods of bougainvillea. Indian Bougainvillea Annual 26:1–3

Singh B, Panwar RS, Voleti SR, Sharma VK, Thakur S (1999) The new international Bougainvillea check list. Indian Agric Res Inst, New Delhi

Singh KK, Singh T, Tomar YK (2013) Multiplication of Bougainvillea cv. Torch glory through shoot tip cutting under mist chamber. HortFlora Res Spectrum 2(2):139–144

Srivastava R, Datta SK, Sharma SC, Roy RK (2002) Gamma ray induced genetic variability in Bougainvillea. J Nucl Agric Biol 31(1):28–36

Swaroop K, Jain R, Janakiram T (2015) Effect of different doses of gamma rays for induction of mutation in bougainvillea cv. Mahatma Gandhi. Indian J Agric Sci 85(9):1245–1247

Swarup V, Singh B (1964) Pollen morphology and leaf harirs in classification of *Bougainvillea*. Ind J Hortic 21(2):156–164

Yadav LP, Bhattacharya AP, Pandey HS (1978) Effect of season on the rooting of bougainvillea cuttings. Progressive Hortic 9(4):72–73

Zadoo SN, Roy RP, Khoshoo TN (1975a) Cytogenetics of cultivated Bougainvillea III. Bud sports, Z. Pflanzenzuchtg 74:223–239

Zadoo SN, Roy RP, Khoshoo TN (1975b) Cytogenetics of cultivated Bougainvillea V. Tetraploidy and restoration of fertil;ity in sterile cultivars. Ehphytica 24:517–524

Zadoo SN, Roy RP, Khoshoo TN (1975c) Cytogenetics of cultivated Bougainvillea VI. Hybridization, Z. Pflanzenzuchtg 75:114–137

Zadoo SN, Roy RP, Khoshoo TN (1975d) Cytogenetics of cultivated Bougainvillea II. Pollination mechanism and breeding system. Proc Indian Acad Sci 41:498–502

GERBERA (*Gerbera jamesonii* Bolus ex. Hooker F.) 21

Rajiv Kumar, Tarak Nath Saha, and Swati Saha

Contents

Abstract

Plant genetic resources include primitive forms of cultivated plant species and landraces, modern cultivars, obsolete cultivars, breeding lines and genetic stocks, weedy types, and related wild species. Genetic resources provide basic material for selection and improvement through breeding. Gerbera is one of the most important cut and pot flowers grown worldwide owing to its wide range of colors, forms, and attractive geometrical shape. In this chapter, we have discussed the recent status of gerbera with respect to genetic diversity and conservation. The

R. Kumar (✉)
Division of Flower and Medicinal Crops, ICAR-Indian Institute of Horticultural Research, Bengaluru, Karnataka, India

T. N. Saha
ICAR-Directorate of Floricultural Research, College of Agriculture Campus, Pune, Maharashtra, India

S. Saha
ICAR-Indian Agricultural Research Institute, Pune, India

S. K. Datta, Y. C. Gupta (eds.), *Floriculture and Ornamental Plants*, Handbooks of Crop Diversity: Conservation and Use of Plant Genetic Resources,
https://doi.org/10.1007/978-981-15-3518-5_15

chapter deals with the updated information on its botany, origin, domestication and spread, geographical distribution, wild genetic resources, collection and conservation methods, status of plant genetic resources, characterization and evaluation including genetic diversity for desirable traits, molecular characterization, and uses of plant genetic recourses of gerbera. The strategy for better utilization of gerbera germplasm for development of novel, biotic, and abiotic resistant to new varieties has also been discussed.

Keywords

Gerbera · Origin · Botany · Collection · Conservation · Characterization · Evaluation · Genetic resources · Genetic stock

21.1 Introduction

Gerbera is one of the most important cut and pot flowers worldwide and occupies fifth position in the international flower trade after rose, carnation, chrysanthemum, and tulip in the global cut flower trade owing to its wide range of colors, forms, and attractive geometrical shape. It is perennial and reproduces asexually. It produces colorful flowers with white, red, yellow, pink, orange, or bicolored. It belongs to the family Asteraceae, the largest family of flowering plants. It is native to tropical regions of Africa, Asia, and South America. It is also known as Transvaal daisy or Barberton daisy. Gerbera is also commonly known as the African daisy. It is scattered from Africa to Madagascar into tropical Asia and South America.

Gerbera jamesonii was first described by Robert Jameson in 1889 while exploring the Barberton area in the Lowveld region of Mpumalanga Province, South Africa. It was the first species of gerbera to be the subject of a scientific description, studied by J.D. Hooker in *Curtis's Botanical Magazine* in 1889. The flag and coat of arms of the Province of Mpumalanga include a depiction of this flower.

Among the different species, *Gerbera jamesonii* is the only species under commercial cultivation. Modern cultivated gerbera arose from the hybridization between *G. jamesonii* and *G. viridifolia* and possibly other species. The genus *Gerbera* was named in honor of the German naturalist Traugott Gerber, while the species *jamesonii* was named in honor of Captain Jameson.

Its demand as cut flower and also as an ornamental potted plant is gaining importance in the world market and has a very good export potential because of its graceful appearance, hardiness, and ability to with stand during transportation and long shelf life. The tremendous variability in gerbera with reference to flower color, shape, and size makes it more useful for cut flowers, bouquet, and decoration in marriage and landscaping in gardening. Apart from domestic consumption, it has got export potential also. In India, gerbera is mainly grown under polyhouse in North Eastern States, Karnataka, Maharashtra, and Uttarakhand. However, higher altitudes such as Meghalaya, Uttarakhand, and Yercaud with temperate climate favor growing gerbera under open conditions.

The domesticated cultivars are mostly a result of a cross between *Gerbera jamesonii* and another South African species *Gerbera viridifolia.* The cross is known as *Gerbera hybrida.* They vary greatly in shape and size. Colors include white, yellow, orange, red, and pink. The center of the flower is sometimes black. Often the same flower can have petals of several different colors. It is also important commercially. With the increase in economic importance of ornamentals in many countries, the international demand for gerbera flowers has also rapidly increased and become one of the most important commercial cut flower for presentation and interior decoration.

There is a great demand for gerbera particularly in European markets during winter months and almost throughout the year in India. Since, India is situated comparatively closer to major flower-consuming countries than its Asian counterparts; it has very good scope and potential in the flower trade. Severe winter in major flower-producing European countries is also an advantageous factor to India, specially cities such as Bengaluru, Pune, Hyderabad, Nasik, etc. which has moderate climate all through the year which has got a great potential for producing gerbera on commercial scale for export.

21.2 Gerbera: Diversity and Plant Genetic Resources

21.2.1 Botany and Distribution

Gerbera, a genus of ±30 species of perennial herbs, is a member of tribe Mutisieae, which includes some 14 genera and more than 200 species (Katinas et al. 2009; Ortiz et al. 2009) and belongs to the family Asteraceae and sub-family Mutisioideae. Its flower head comprised three types of flowers, i.e., ray florets in outer ring, trans-florets in middle ring, and disc florets in inner ring. The marginal ray flowers are large, strongly ligulate, and zygomorphic with fused showy petals. The ray flowers have showy ventral ligule formed by three fused petals, while the two dorsal petals are rudimentary. The female ray flowers also develop anther but get aborted later in the development. The ray and disc flowers are packed tightly into a flower head called the capitulum. The trans-flowers in Gerbera are female, like ray flowers, but just smaller in size, and the length of the petals varies among varieties. Gerbera flowers also possess hair-like structures, called pappus. The central disc florets are hermaphrodite and bear carpels and stamens. The disc flowers are gradually changing from bilateral symmetry toward radial symmetry with more in central position of the capitulum. The capitula show spiral phyllotaxis with left and right turning rows of floral primordia. There is a developmental delay in marginal ray flowers compared to the adjacent trans or disc flowers, respectively. These structures are modified sepals that show flower-organ determination may be modified with MADS-box genes. The stamens of gerbera flowers are aborted in marginal flowers, the petals and anthers are fused into tubular structures, and the plant possesses inferior ovaries. Early in development, the three main flower types (ray, trans, and disc) are morphologically similar. Morphological differentiation of flower types occurs through the action of the TCP transcription factors on CYCLOIDEA-like TCP protein domains.

21.2.1.1 Vegetative Morphology

The species of gerbera are acaulescent, herbaceous perennials growing up to 80 cm high, with rosulate, petiolate foliage (Hansen 1985). The petioles are thin, 1–3 mm diameter, and glabrescent to woolly or rarely glabrous, brownish in color; however, green and maroon petioles are observed in *G. crocea*. The leaf blades vary in shape, degree of marginal toothing, or incision and in vestiture and are critical for species identification. Gerbera are simple, petiolate, and coriaceous, either glabrescent or variously felted with the degree of felting and the color varying among the species. Most species have lanceolate to ovate leaves, with the margins variously incised, dentate, retrorse-serrate, sinuate or rarely entire, and revolute or rarely involute (only *G. ovata*). *Gerbera linnaei* is unique in having distinctly pinnatisect or sometimes deeply pinnatifid leaves with 5–18 jugate lobes often flexed upward along the lower edge. The leaves in almost all species are discolorous, with the upper or adaxial surface green and either glabrescent and dull or shining or thinly felted or woolly. The lower or abaxial surface is usually densely felted (glabrous or glabrescent in *G. crocea*), with the color of the felting characteristic for most species: whitish-grayish felted in *G. crocea*, *G. sinuata*, and *G. wrightii* and yellowish-brownish felted in *G. grandis*, *G. linnaei*, *G. ovata*, *G. serrata*, and *G. tomentosa*.

21.2.1.2 Reproductive Morphology

The inflorescences in most Mutisieae are radiate capitula with the marginal florets specialized for increased pollination (Katinas et al. 2009). Gerbera is characterized by scapose, radiate capitula, often large with conspicuous rays. The scape is sparsely to more closely bracteate. The corolla in all florets is ± bilabiate, the two upper (inner) limbs forming a cleft inner lip, and the three lower (outer) limbs fused into a three-dentate limb that is greatly extended in the marginal florets. All species of *Gerbera* have dimorphic florets. The marginal florets are unequally bilabiate and radiate and are female fertile; the disc florets are weakly bilabiate and bisexual. The color of the marginal florets is variable within species, ranging from white to pale or deep pink in some and from white or cream to bright yellow in others. The under surface of the rays is almost invariably flushed dark red to purple. Yellow-flowered forms are found only in some of the species that develop yellowish indumentum on the leaf under surface, suggesting a possible genetic link in pigment production in the foliage and flowers. The color of the anther appendages is also a useful diagnostic character, varying from dark brown to blackish in *G. crocea*, *G. linnaei*, and *G. sinuata* but pale yellow in *G. grandis*, *G. ovata*, *G. serrata*, *G. tomentosa*, and *G. wrightii*. Most species flower in spring and summer.

21.2.1.3 Flower Forms of Gerbera

Broadly, gerbera flowers are classified into four groups:

- **Single flowers:** There is a row of nonoverlapping ray florets with green disc florets. These are the most common gerbera grown for bedding purpose.
- **Double or duplex:** There is a double row of overlapping ray florets with green/black disc florets.
- **Crested doubles:** These are doubles containing two rows of overlapping ray florets with one or more inner rows of shorter ray florets with green/black disc florets.
- **Full crested doubles:** These have solid overlapping rows of ray florets with an inner row diminishing in size, covering the disc florets entirely.

21.2.2 Origin, Domestication, and Spread

Manning et al. (2016) revised the taxonomy of *Gerbera*. They recognized eight species in the section. Most of the species are obligate pyrophytes, flowering only in the spring and summer following a burn. Populations from the Cederberg, Olifants River Mountains, and Piketberg previously included in *Gerbera crocea* but anomalous in that species in their sinuately incised leaves with whitish-gray felted under surface are segregated as *Gerbera sinuata*. The disjunct Eastern Cape populations of *Gerbera tomentosa* lack the imbricate involucral bracts diagnostic for that species and are segregated as *Gerbera ovata*, a new name for *G. tomentosa* var. *elliptica*, characterized by small, ovate leaves with involute margins. *Gerbera grandis* is a new species from the Kleinrivier Mountains recognized by its large size, lanceolate leaves, and long, subulate involucral bracts. They also correct the recorded distribution of *G. crocea* and *G. tomentosa* and transfer the name *G. sinuata* var. *undulata* from *G. crocea* to synonymy under *Gerbera viridifolia*.

21.2.2.1 Origin

Gerbera jamesonii was discovered by William Greenstock in the summer of 1875–76 in the Houtbosch area of the Transvaal, but Anton Rehmann, an Austrian, is often given credit for its discovery in the summer of 1879–1880. However, it was not described until Robert Jameson, a prominent manufacturer of jams and condiments in Durban, collected specimens while on a gold expedition to the Barberton area in 1884. He took the plants to the Durban Botanical Garden, where the director, John Medley Wood, sent material to Harry Bolus in Cape Town for identification. Bolus collected additional material in 1886 and sent it to J.D. Hooker of the Royal Botanical Garden at Kew, with the request that it be named after Jameson. Wood sent live material to Kew in 1888, and one plant survived to flower in the greenhouse in the spring of 1889. This plant is the source of an illustration that appeared in

Hooker (1889) and serves as the citation for the name *Gerbera jamesonii* H. Bolus ex Hooker (Codd 1979).

However, two earlier descriptions of *G. jamesonii* were published, one by R.W. Adlam of South Africa and the other by Justin Allen of Kew (Adlam 1888; Allen 1889; Codd 1979; Hansen 1985). Seeds were sent to England from South Africa by George Thorncroft in 1887, possibly to Mr. Tillett of Norwich, who grew and provided them to Kew. Richard Irwin Lynch, the curator of the Cambridge Botanic Garden, also received seeds from Adlam (Codd 1979; Lynch 1905). The seeds produced a plant with yellow flowers that Lynch named Sir Michael. Lynch maintained the plants at Cambridge, and with its inability to withstand severe frosts though, it was marketed by Hugh Low and Co. and Bush Hill Park Nursery as early as 1902 (Bowe et al. 1969; Lonsdale 1908; Nichols 1902). Dümmer lists 44 references to gerbera between 1888 and 1911 in his 1914 revision of the genus (Dümmer 1914). Early literature on gerbera appeared in both *The Gardeners' Chronicle* and the *Journal of the Royal Horticultural Society* from the late 1880s through 1910.

Gerbera is an herbaceous perennial with the source of cultivated germplasm being *G. jamesonii* and *G. viridifolia.* The genus was established in 1737 by Gronovius and commemorates Traug Gerber, an eighteenth-century German naturalist (Dümmer 1914). The first official description of the South African species *Gerbera jamesonii* was made by J. D. Hooker in 1889 in *Curtis's Botanical Magazine.* It bears a large capitulum with prominent, yellow, orange, white, pink, or various red colored ray florets. At the end of the nineteenth century in Cambridge, England, two South African species, *G. jamesonii* and *G. viridifolia*, were crossed by R.I. Lynch and named the hybrid as *Gerbera* × *cantebrigiensis*, known today also as *Gerbera hybrida*.

The development of *G. jamesonii* as a floricultural crop is traced from its cultivation as a novelty in South Africa to its establishment as a commercial crop in the 1930s. The relative contribution of *G. jamesonii* and *G. viridifolia* Sch. Bip. to the modern crop is unknown, but much of the cultivated germplasm can be traced back to material that passed through the Cambridge Botanic Garden, UK, and La Rosarie, Antibes, France. It is a diploid species with the somatic chromosome number $2n = 50$.

21.2.2.2 Domestication

Gerbera gossypina (Royle) Beauved ($2n = 46$) is a wild plant of Western Himalayan origin. *G. gossypina* may be used as an ornamental pot plant due to its compact plant size, small flowers of light pink color (RHS 76D), and green foliage with woolly fibers on the under surface. The wild gerbera species is free of diseases such as powdery mildew and gray mold. Morphologically the species exhibits adaptive features such as woolly fibers on the under surface of leaves and forms small light purple colored flowers. The species is not commonly distributed in Himalayas and is being domesticated for its conservation and sustainable utilization as an ornamental pot plant (Singh and Dhyani 2016). *Gerbera asplenifolia*, *G. aurantiaca*, *G. kunzeana*, and *G. viridifolia* are some other important species in the genus.

21.2.2.3 Spread

The genus *Gerbera* is distributed in the America from Mexico to South America, Asia from Yemen and countries east of and including the Himalayan plateau to Bali, and Africa including sub-Saharan Africa and Madagascar. Dispersal-variance analyses indicate South America as the ancestral area of the *Gerbera*-complex, consistent with the hypothesis that the basal clades of the Asteraceae arose in South America (Panero and Funk 2008).

The distribution of *G. jamesonii* is restricted to southern Africa (between latitude 20°S and 30°S, and east of longitude 25°E), where it is endemic to Transvaal and Swaziland. It is found in Bushveld and steep slopes, on dolomitic and stony clay soils, and on burnt ground and dry, shaded habitats (Hansen 1985). *G. viridifolia* has a broader distribution that ranges along eastern Africa (between lat. 5°S and 35°S, and east of long. 25°E), in open grassland with stony soil.

21.2.3 Plant Genetic Resources

The global demand of flower crops is increasing. To meet this demand innovative production and marketing efforts are needed. Searching of new sources of genes from the already collected and conserved germplasm is greatly needed with reference to the utilization of germplasm in breeding programs. The genotypes developed with significant promising traits from the valuable conserved germplasm play a major role in the varietal development. The diverse genetic resources are the essence of breeding programs and also more demand for tropical flowering plants.

21.2.3.1 Geographic Distribution

Eight species are endemic to the Cape Floristic Region of South Africa. Most specimens are readily assigned to species, but hybridization between species is not uncommon, resulting in a range of intermediate forms that will confuse species identification without field knowledge of the situation. All of the species have ± discrete areas of distribution, and this is a valuable adjunct to correct identification. Some species appear to replace one another along a geographical cline, with a blurring of morphological discontinuities in the zones of overlap. This is especially evident in *G. tomentosa G. serrata-G. ovata* and *G. linnaei-G. serrata*. The mainland African species of *Gerbera* were monographed by Hansen (1985), who recognized 13 species from continental Africa, mainly southern Africa which is a center of diversity, 8 species in Madagascar, and 9 species in Asia.

The species of *Gerbera* are segregated among six sections (Hansen 1985, 1990, 2006), with sect. *Gerbera* restricted to the Cape Floristic Region (Manning and Goldblatt 2012). As circumscribed by Hansen (1985), *G. tomentosa* has a disjunct distribution in the southwest and east of the Cape Floristic Region, with *G. serrata* occupying the gap between. Re-examination of the material of these two taxa showed that the eastern populations of *G. tomentosa*, corresponding to *G. tomentosa* var. *elliptica* DC., are not only morphologically distinct in leaf form from the western populations but also lack the imbricate involucral bracts that are diagnostic for the species. Populations in the north of the range, however, on the Cederberg, Olifants River Mountains, and Piketberg, have leaves with crenate or roughly incised margins and thickly whitish-gray felted under surface. Hansen (1985) associated the Cederberg plants with *G. crocea* despite their anomalous leaf indumentum but regarded the Piketberg collections as the hybrid G. × *G. wrightii* (a species with whitish-felted leaf under surface) on the mistaken assumption that they had been collected on the Cape Peninsula, where *G. wrightii* Harv. is endemic. The distribution in some *Gerbera* species is given hereunder:

***Gerbera sinuate*:** It is restricted to the western coastal mountains of Western Cape, from the Piketberg and central Cederberg along the Olifants River Mountains to the Witzenberg near Ceres and the Elandskloofberge near Gouda, occurring on stony sandstone slopes.

***Gerbera crocea*:** It is relatively common in the mountains in the southwestern part of Western Cape, from Groot Winterhoek above Tulbagh and the Witzenberg near Ceres to the Kogelberg and the Cape Peninsula thence eastward to the Potberg and inland to the Naudesberg near Koo with outlying populations in the southern foothills of the Swartberg near Oudshoorn occurring on sandstone slopes.

***Gerbera wrightii*:** It is endemic to the Cape Peninsula, from Constantiaberg to Scarborough occurring on sandstone slopes. *G. wrightii* hybridizes with *G. linnaei* where they co-occur on the Cape Peninsula.

***Gerbera linnaei*:** It is endemic to the extreme southwestern part of Western Cape, from the Hex River Mountains to the Cape Peninsula and the Kleinrivier Mountains.

***Gerbera serrata*:** It is restricted to the southern coastal plain and adjacent mountains of Western Cape, from Riviersonderend to Plettenberg Bay, extending inland onto the Gamkaberg and Rooiberg in the Little Karoo occurring on well-drained stony loam soils and on coastal terraces on hard packed sandy soils.

***Gerbera ovata*:** It is restricted to the eastern end of the Cape Floristic Region, from the Baviaanskloof in Eastern Cape through the Langkloof to the Vanstadensberg and Great Winterhoek Mountains, occurring on stony and rocky sandstone soils.

***Gerbera tomentosa*:** It is restricted to the mountains in the extreme southwestern part of Western Cape, from Bainskloof to the Riviersonderend and Kleinrivier Mountains, Bredasdorpberge, and western Langeberg above Swellendam. The species occurs on stony slopes on sandstone, granite, or clay, sometimes in marshy or peaty soils, and flowering takes place only after fire.

***Gerbera grandis*:** It is a local endemic of the Kleinrivier Mountains in Western Cape, where it is known so far only from Vogelgat above Hermanus in the west and Salmonsdam Nature Reserve near Stanford in the east, occurring on sandstone slopes at around 400 m.

***G. viridifolia*:** *Gerbera viridifolia* is widespread, particularly in the KwaZulu-Natal, Mpumalanga, and Swaziland regions. The species is found in grasslands, in stony soil, on mountain slopes, and even in damp areas throughout the eastern regions of Africa. It species is common and widespread in areas of moderate to high rainfall.

***Gerbera sylvicola*:** It is a new South African endemic species. The species appears to be closely related to *G. kraussii*, and prominently veined abaxial leaf surfaces and several ecological, reproductive, and vegetative features clearly distinguish it.

21.2.3.2 Primary Gene Pool

Species

The first scientific description of a *Gerbera* was made by J.D. Hooker in *Curtis's Botanical Magazine* in 1889 when he described *Gerbera jamesonii*, a South African species. Out of all the recorded species, only *Gerbera jamesonii* is under cultivation. Some important species are described below:

***Gerbera aurantiaca*:** It is an endangered species in South Africa and produces large flower heads with dark red, long ray florets, and a black center.

***Gerbera galpinii*:** It is endemic to moist habitats and has entire, glabrous leaves. These species may be useful germplasm for improving cultivated gerberas for aesthetic values, stress tolerance, and/or environmental adaptation.

***Gerbera asplenifolia*:** Leaves narrow, 10–15 cm long, more or less deeply lobed, leathery; flower heads purple on a hairy scape.

***G. aurantiaca*:** Leaves lanceolate to long, acute, 12.5–15 cm long, entire, or toothed. Flower heads orange and anthers yellow.

***G. jamesonii* (Barberton daisy):** Hairy throughout; base woody; leaves lobed; solitary orange-scarlet heads, 7.5–12.5 cm or more across, are borne from November to February in the plains; single or double-flowered cultivars and hybrids in attractive pastel colors are generally available.

***G. kunzeana*:** A Himalayan species whose flowers scarcely open.

***Gerbera ambigua*:** It is widely throughout Africa from the coast to about 1900 m. This species is found in grassland, open woodland, and damp areas.

***G. viridifolia*:** Its leaves are elliptic or oblong, obtuse, flower stalk short. The specific epithet *'viridifolia'*, means "with green leaves." *Gerbera viridifolia* is one of the parents of the commercially produced *Gerbera* hybrids.

21.2.3.3 Wild Genetic Resources and Others

Gerbera delavayi (fireweed) is a perennial herb native to southwest China and the neighboring Vietnam. In recent years, the species distribution and abundance have seriously declined, because its leaves have been excessively harvested for textile industry. Isolation barriers and human overexploitation have led to the moderate genetic diversity in the populations of *G. delavayi*. Environmental factors, floristic composition, and geographic barriers of mountain ranges affected the genetic structure of *G. delavayi*. When utilizing the current wild resources, protection of genetic diversity both in situ and ex situ need to be considered (Xu et al. 2017).

The phylogenetic analysis showed two clades inside the *Gerbera*-complex. Clade A contains only South American endemic genera, in which *Lulia* is sister to *Brachyclados* and *Trichocline*. Clade B mainly contains groups of taxa that colonized other continents including areas in the northern temperate latitudes. Clade B is further divided into two clades where *Gerbera* is shown to be non-monophyletic because the African *Gerbera* clade is sister to *Amblysperma* and the Asian *Gerbera* clade includes *Uechtritzia*. The biogeographic and molecular dating showed a South American origin for the early-divergent nodes of the sub-family with a node age of 47.52–49.67 Ma in the Eocene. The *Gerbera*-complex is likely to have originated in the Andes in the late Oligocene (mean node age of around 25.74 Ma) followed by long-distance dispersal events to North America and Asia and separate dispersal events to Africa and Australia. This is the first phylogenetic analysis to show the systematic positions of *Amblysperma*, *Lulia*, and *Uechtritzia* (Pasini et al. 2016).

Gerbera ambigua is a low-growing perennial daisy with white or yellow flower heads which is an excellent edging plant for a border or a contrast plant for grasses in a natural garden. It is a stemless perennial herb with a basal rosette of leaves emerging from a silky crown. It is also a useful rock garden. *G. kraussii* has subsequently been incorporated into *G. ambigua* by Hansen (1985). *Gerbera kraussii* was first collected "near Pietermaritzburg" by Christian Krauss in South Africa, between 1838 and 1840. Natural hybrids between *G. ambigua* and the Hilton daisy (*G. aurantiaca*) have been recorded (Hansen 1985).

21.2.4 Collections

21.2.4.1 Methods

Plant exploration and germplasm collection is the first and foremost activity in plant genetic resources (PGR) management. This activity is executed through planning, coordinating, and conducting collaborative explorations for collection of germplasm of different agri-horticultural crops and their crop wild relatives from various diversity-rich regions. The germplasm is a source of traits for crop improvement, resilience, and stability. Germplasm is collected from centers of diversity, gene banks, gene sanctuaries, farmer's fields, markers, and seed companies. The objective of PGR collecting may be achieved through collection of maximum genetic diversity, properly documented and to be used by scientific community. The crop gene pool includes its wild relatives, landraces, primitive cultivars, obsolete, and promising plant genetic resources.

The gerbera germplasm may be collected through the exploration of targeted areas, private breeding companies, farmers, and exchange of germplasm.

21.2.4.2 Status of Collections (National, Regional, and Global with Appropriate Listing) (Tables 1 and 2)

21.2.4.3 Gaps in Collections, Both Geographical and Genetic

- Worldwide, the lack of sufficient human capacity in collecting of genetic resources, e.g., taxonomy, genetics, and ecology.
- There is a lack of resistance germplasm for phytoplasma, which is one of the major diseases in gerbera cultivation.

Table 1 Potential novel germplasm of *Gerbera jamesonii* registered with ICAR-NBPGR, New Delhi, India, for unique traits

Sl. No.	Accession	Pedigree	Promising trait
1.	IC556977	*Gerbera jamesonii* (GJ4) x open pollinated lines of *Gerbera jamesonii*	Floriferous and double type
2.	IC556978	*Gerbera jamesonii* (GJ4) x open pollinated lines of *Gerbera jamesonii*	Floriferous and good for open cultivation
3.	IC0613966 Him peace	CSIR-IHBT-gr-1 x CSIRIHBT-gr-7	Double flower, medium size, and white (RHS 155D)
4.	IC0613967 Him glow	CSIR-IHBT-gr-2 x CSIRIHBT-gr-3	Double flower shape, standard size with yellow orange flower (RHS 16C)
5.	IC0621472 (IIHR 8–45)	Half-sib selection from line IIHR-1	Flower head color (50A, red group) and double-type flower head
6.	IC06214771 (IIHR 3–34)	Half-sib selection from line IIHR-3	Flower head color (68D, red purple group), and double-type flower head
7.	IC0632739 (IIHRGO-1)	IIHR99–5 x Savana	Bright red flower color (40A, red group), double-type flowers, and ability to grow under open condition

Table 2 Valuable gerbera germplasm for open-grown conditions

Sl. No.	Variety	Sources
1.	Arka Krishika	ICAR-IIHR, Bengaluru
2.	IIHRGO-1 (IC0632739): Flower color and flower form. Bright red (RHS color: 40A, red group) and double-type flowers. Ability to grow under open field condition.	
3.	RCGH-12 (IC-0633067): Flowers are strong red (53C), double form, flower diameter (9.18 cm), yielding 23.7 flower per plant and 5.04 days vase life	ICAR RC NEH region, Umiam, Meghalaya
4.	RCGH-22 (IC-0633068): Flowers are vivid reddish orange (32A), double form, flower diameter (10.64 cm), yielding 23.5 flower per plant and 5.12 days vase life	
5.	RCGH-114 (IC-0633069): Flowers are vivid reddish orange (40A), double form, flower diameter (11.26 cm), yielding 24.5 flower per plant and 5.07 days vase life	
6.	RCGH-117 (IC-0633070): Flowers are strong orange (24 A), semi-double form, flower diameter (11.70 cm), yielding 23.4 flowers per plant and 5.83 days vase life	
7.	RCGH-28 (IC-0633071): Flowers are vivid reddish orange (N30A), double form, flower diameter (10.87 cm), flowers (26.7 per plant) and dwarf stalk length (26.7 cm)	
8.	YCD-1, YCD-2	TNAU, Coimbatore

- There is restriction in the collection of wild relative in the protected areas.
- It is not always clear whether it is better to collect from farmers' fields or natural habitats.
- Inheritance pattern of most of the economic traits are not well defined.

21.2.5 Conservation

a. **Methods**

Collection of germplasm and the search for desirable traits are of utmost importance in flower crop breeding. The natural variation present in growth, yield, and inflorescence characters among germplasm can be utilized in breeding programs.

In situ conservation: The germplasm is conserved in natural environment by establishing biosphere reserves such as national parks and sanctuaries. This is used in the preservation of land plants in a near natural habitat along with several wild types. Germplasm is maintained in the form of plants as permanent living collection. Mostly gerbera plants are maintained under shaded structure like polyhouse/net house.

Ex situ conservation: This method is used for the preservation of germplasm obtained from cultivated and wild plant materials. The genetic material in the form of seeds or in vitro cultures are preserved and stored as gene banks for long-term use.

In vivo gene banks have been made to preserve the genetic resources by conventional methods, e.g., seeds, vegetative propagules, etc. In vitro gene banks have been made to preserve the genetic resources by nonconventional methods such as cell and tissue culture methods. This will ensure the availability of valuable germplasm to breeder to develop new and improved varieties. The methods involved in the in vitro conservation of germplasm are:

Cryopreservation In cryopreservation, the cells are preserved in the frozen state. The germplasm is stored at a very low temperature using solid carbon dioxide (at −79 °C), using low temperature deep freezers (at −80 °C), and using vapor nitrogen (at −150 °C) and liquid nitrogen (at −196 °C). The cells stay in completely inactive state and thus can be conserved for long periods. Any tissue from a plant can be used for cryopreservation, e.g., meristems, embryos, endosperms, ovules, seeds, cultured plant cells, protoplasts, and calluses. Certain compounds like DMSO (dimethyl sulfoxide), glycerol, ethylene, propylene, sucrose, mannose, glucose, praline, acetamide, etc. are added during the cryopreservation. These are called cryoprotectants and prevent the damage caused to cells (by freezing or thawing) by reducing the freezing point and super cooling point of water.

The method of slow growth storage consists in decelerating or suppressing the plants physiological metabolism. Differently of cryopreservation, in which for long periods, the plant material is stored at ultralow temperatures, suppressing

Table 3 Slow growth storage of ornamental plants

Species	Storage temperature (°C)	Storage light/ radiance	Storage period (months)	Recovery (%)	Explant type
Gerbera cv. Marleen	4	Darkness	3	100	Shoot cultures

Table 4 Availability status of important global ornamental germplasm

Crop	GENESYS	USDA-GRIN	ERISCO
Gerbera	18	–	16

Table 5 Priority ornamental crops and its species for further introductions

Crop	Species	Source
Gerbera	*Gerbera piloselloides*, *G. aurantiaca*, *G. wrightii*, *G. viridiflora*, *G. linnaei*, *G. cordata*, *G. aristata*, *G. ambigua*	Genetic Resources Unit, Institute of Biological, Environmental and Rural Sciences, Aberystwyth University, UK

growth, in order to avoid deterioration of plant (Grout 1995; Silva and Scherwinski-Pereira 2011). This method has been used for conservation in short- and medium-term, mainly for meristems and/or shoot tips of several species. It consists in reduction of the growth and increases the intervals between the subcultures, without affecting significantly the viability of the explants (Engelmann 2011) (Table 3).

In *Gerbera*, shoot tips are the commonly used explants, while adventitious shoot induction from the capitulum is also a popular method. Genotype is one of the most influential factors on the response of gerbera in vitro. Despite this, no successful universal protocol has yet been developed for multiple cultivars, limiting the usefulness of current protocols for commercial biotechnology labs. Epigenetic variations in micro-propagated gerbera are frequently observed only with high concentrations of cytokinins in the culture medium, but somaclonal variation is rare. The in vitro rapid shoot multiplication *Gerbera jamesonii* var. Pink Star offers ample scope to develop large number of clones from a stock for in vitro conservation.

b. **Status of plant genetic resources (general germplasm, base collection, active collection, breeder's collection, genetic stocks, pre-breeding material, including interspecific derivatives, etc.)** (Tables 4 and 5)
c. **Gaps in available (useful) diversity**
   - Evaluation of germplasm for phytoplasma incidence.
   - Evaluation of germplasm against biotic stress such as thrips, mites, and white flies.
   - Evaluation of germplasm for pot culture.

21.2.6 Characterization and Evaluation

(a) **Characterization for essential features and classification**

Characterization and evaluation is the key to assess the potential and actual value of germplasm. A number of techniques are available for characterization and evaluation depending upon the need, type, and nature of the plant/material available. Among these, a) agro-morphological traits, b) biochemical traits, and c) molecular or DNA-based markers are the prominent techniques for characterization and evaluation of PGR. The DNA-based markers or molecular markers are also gaining importance because of the lack of environmental influence on these molecular makers. Characterization of traits would be of great help to identify promising gerbera hybrids for domestic as well export market and assist in selection of potential parents for their further utilization in gerbera improvement programs.

Hybrids RCGH-12, RCGH-22, RCGH-114, and RCGH-117 were found superior for vegetative and flower quality traits for open-grown condition in Meghalaya conditions. However, genotypes Dana Ellen, Livia, Corona, and Fredi found suitable for flower quality traits under polyhouse.

The analysis of genetic variability in gerbera is a prerequisite for breeding programs because it can generate data on the genetic relationships existing among the genotypes of this genus. Among the strategies available for evaluating genetic variability, the use of molecular markers is the most widely applicable because these markers are best suited for understanding the genome and may be used in paternity testing, genetic variability characterization, the elucidation of genetic relationships between genotypes, developing methods for the maintenance of genetic variability existing in germplasm banks, and identifying genes or combinations of features related to key traits of biological and agronomic interest (Hayden et al. 2010).

GERB12 and GERB17 markers are correlated with ray floret width and stem length, respectively, which enabled to predict the best controlled crosses between contrasting genotypes and allow for molecular marker-assisted selection associated with morphological traits.

The genetic diversity of 12 accessions of gerbera was assessed through random amplified polymorphic DNA (RAPD) markers. Ten (10) decamer RAPD primers produced a total of 49 scorable bands from 12 genotypes of *G. jamesonii* Bolus, out of which 42 were polymorphic and seven were monomorphic. The percentage of polymorphism ranged from a maximum of 100.00% by OPE-02, OPE-14, OPF-18, OPG-18, OPG-16, and OPG-17 to a minimum of 50.00% by OPE-08 (Prajapati et al. 2014).

(b) **Development/identification of gene pools and core collections**

Gerbera aurantiaca (Hilton Daisy), an endangered species, populations exhibit considerable variation in flower color, and morphological and molecular techniques are being used to ascertain taxonomic boundaries and genetic variation within and between the populations.

Asen (1984) analyzed and isolated flavonoids through high pressure liquid chromatography (HPLC). The anthocyanins isolated from 18 cultivars, ranging in color from orange through lavender, were pelargonidin and cyanidin 3-malonylglucosides accompanied by smaller amounts of pelargonidin and cyanidin 3-glucosides. Related flavonoid copigments were apigenin and luteolin 4′-glucosides and 7-glucosides, apigenin 7-malonylglucoside, kaempferol and quercetin 3-glucosides, 4′-glucosides, and 3-malonylglucosides. Both qualitative and quantitative differences in these flavonoid chemical markers distinguished cultivars with very similar colors.

Gerbera gossypina (Royle) Beauv. (Asteraceae) is a perennial wild plant of Himalayan origin. It may be used as an ornamental pot plant due to its compact plant size, small flowers of distinct light pink color, and green foliage with woolly fibers on the under surface. *G. gossypina* has distinct and homogenous flower color (RHS 76D) and is free of diseases such as powdery mildew and gray mold. It can be utilized for improvement of cultivated gerbera through a backcross hybridization program (Singh and Dhyani 2016).

(c) **Evaluation of genetic diversity for desirable traits**

Genetic diversity is the key for genetic improvement in important ornamental species. DNA-based molecular marker RAPD is a powerful, less time-consuming, and cost-effective molecular technique for assessment of genetic diversity among different genotypes of *G. jamesonii.*

Singh et al. (2009) generated variability through hybridization and characterized floral traits in *Gerbera.* Considerable variations were observed for the floral traits, viz., flower color, shape, diameter, disc color, scape length, disc diameter, ray floret width, and ray floret number among the parental genotypes and 46 F_1 progeny of an inter-varietal cross IHBT Gr4 x IHBT Gr5. Flower shape and disc color were observed to be qualitative traits, while flower color, scape length, flower diameter, disc diameter, ray floret width, and ray floret number had quantitative inheritance.

Genetic diversity was evaluated by RAPD markers and morpho-agronomic characters for a total of 42 accessions of *Gerbera jamesonii.* Jaccard and Shannon indices indicated the presence of higher genetic variation among commercial accessions in comparison to the cluster representing non-commercial accessions (Mata et al. 2009).

Polymorphism was observed in markers OPE-02, OPE-14, OPF-18, OPG-18, OPG-16, OPG-17, and OPE-08. The availability of gerbera RAPD markers would facilitate the use of molecular markers in gerbera breeding and genetic studies (Prajapati et al. 2014).

(d) **Available sources of breeding value (listing of genetic resources available for various biotic and abiotic stresses and nutritional traits, other desirable traits, quantitative traits, particularly related with yield, etc.)**

Potential Germplasm Suitable for Resistance to Biotic Stress

A number of pathogens infect gerbera flowers, leaves, roots, and crowns and cause severe diseases symptoms that can limit gerbera plant development, lower flower

Table 6 Promising genotypes resistant to biotic stress

Sl. No.	Biotic stress	Resistant/tolerant genotypes/line
1.	Powdery mildew (*Podosphaera* (*Sphaerotheca*) *xanthii* (syn. *Podosphaera fusca*)	Cultivars Terra Fame (cut flower) and Festival Semi-Double Orange (pot type)
		Breeding lines, viz., plants 176 and 214, UFGE 31–19, UFGE 5–23, UFGE 4033
2.	*Alternaria* leaf spot (*Alternaria alternata*)	Cultivars mammoth and tiramisu

quality, reduce flower yield, and impact plant performance. Identifying sources of disease resistance and incorporating the resistance into new cultivars has become a main objective in recent gerbera breeding efforts and genetic studies (Table 6).

(e) **Molecular characterization, identifying genomic resources, if any (?) such as molecular maps, molecular markers, tagged genes, marker-associated maps, gene constructs, etc.**

Estimates of additive, dominance, maternal, and paternal components of variance were obtained for a sample of 18 traits, including yield, scapes, flowers, disk, ray and trans-florets, leaves, and branching in the Davis population of *Gerbera hybrida*. The results, based on the covariance of reciprocals, indicate that although heritability averaged 0.52, extra nuclear maternal or paternal effects are not important sources of variability. Therefore, reciprocal differences do not seriously affect estimates of additive variance or heritability in this population (Harding et al. 1991).

Song and Deng (2013) studied the inheritance of powdery mildew in gerbera. To determine the mode of inheritance for powdery mildew resistance in UFGE 31–19, one of its PM-resistant (PM-R) progeny, UFGE 4033, was crossed with PM-susceptible (PM-S) cultivar, Sunburst Snow White, and their progeny were evaluated for PM severity, suggesting that the PM resistance in UFGE 4033 and UFGE 31–19 is a quantitative trait, likely controlled by major genes. It was proposed that the two regions be named Rpx1 and Rpx2 (resistance to *P. xanthii*).

Sandigawad (2014) studied genetic diversity among seven genotypes of *Gerbera* using Random amplified polymorphic DNA (RAPD) markers. Out of 20 RAPD decamers tested, only 13 amplified genomic DNA across all the 7 genotypes of *Gerbera*.

Identifying Genomic Resources

Kloos et al. (2005) reported a single dominant gene *Pmr1* controlling the resistance of gerbera plants 176 and 214 to *P. xanthii* (causal agent of gerbera powdery mildew). The dominance or expression of *Pmr1* varied with genetic background as well as gene dosage and suggested that other unidentified genes might modify *Pmr1*'s effect on PM resistance phenotypes. This study represents the first effort to identify and tag QTLs for disease resistance traits in gerbera.

Song et al. (2012) represent the first effort to sample and characterize *R*-gene candidate sequences in gerbera. The obtained sequences may provide a valuable entry point to obtain full-length or additional sequences of gerbera NB-LRR genes. The SCAR, CAPS, and TRAP markers developed may be valuable for mapping of genes or quantitative trait loci responsible for disease resistance in gerbera.

The SNP (single nucleotide polymorphism) markers for the *2-PS* gene can be used to improve *Botrytis* resistance in gerbera by selecting for the favorable alleles of the *2-PS* gene in marker-assisted selection (Fu et al. 2015).

Genetic Transformation in Gerbera

Genetic modification of the flavonoid pathway has been used to produce novel colors and color patterns in ornamental plants. It has been suggested that coordinate control of multiple steps of the pathway with the help of regulatory genes would lead to a more predictable control of metabolic flux. Regulation of anthocyanin biosynthesis has been studied in *Gerbera hybrida*. An R2R3-type MYB factor, GMYB10, shares high sequence similarity and is phylogenetically grouped together with previously characterized regulators of anthocyanin pigmentation. Ectopic expression of GMYB10 leads to strongly enhanced accumulation of anthocyanin pigments as well as to an altered pigmentation pattern in transgenic gerbera plants. Anthocyanin analysis indicated that GMYB10 specifically induces cyanidin biosynthesis in undifferentiated callus and in vegetative tissues.

Furthermore, in floral tissues, enhanced pelargonidin production is detected. Microarray analysis using the gerbera 9 K cDNA array revealed a highly predicted set of putative target genes for GMYB10 including new gene family members of both early and late biosynthetic genes of the flavonoid pathway. However, completely new candidate targets, such as a serine carboxy peptidase-like gene as well, as two new MYB domain factors, GMYB11 and GMYB12, whose exact function in phenylpropanoid biosynthesis is not clear yet, were also identified.

Transformation of GMYB10, an anthocyanin regulator identified from an ornamental plant *Gerbera hybrida*, indicated that regulatory genes can be used to modify secondary metabolic pathways in a highly predicted manner. cDNA microarray analysis of transgenic tissues overexpressing GMYB10 revealed completely new candidate targets, such as a serine carboxy peptidase-like gene, as well as two new MYB domain factors whose exact function in phenylpropanoid biosynthesis is not yet clear. In addition, several new gene family members of the biosynthetic genes of the flavonoid pathway, which are likely to encode enzymes functioning in the same metabolon or enzyme complexes, could be identified (Laitinen et al. 2008).

The first genes isolated from gerbera belonged to the anthocyanin pathway. The dark color of the disk of *G. hybrida* is due to the presence of anthocyanin pigments in the pappus bristles of florets.

Gerbera contains a family of three CHS encoding genes showing different spatial and temporal regulation. *GCHS1* and *GCHS4* are the two *CHS* genes highly expressed in gerbera petals. *GCHS4* expression in gerbera petals is regulated post-transcriptionally, at the level of either translation elongation or protein stability. *GCHS4* is the only CHS encoding gene that is expressed in the cyaniding-pigmented vegetative tissues of gerbera cv. Terraregina. *GCHS3* expression is pronounced in the pappus bristles of the flowers. Expression of both *GCHS1* and *GCHS4* is high in the epidermal cells of gerbera petals, but only GCHS1 is contributing to flavonoid biosynthesis (Deng et al. 2014).

Elomaa et al. (1998), while examining the effect of the transacting regulator GMYC1 in the regulation of the dihydroflavonol-4-reductase (*dfr)* gene, a late gene of the anthocyanin pathway, incidentally observed that in crosses of gerbera cultivars. The expression of the *gmyc1* gene in different cultivars suggested that it regulates *dfr* activity in the corolla and carpel, but not in the pappus and stamen.

RNA expression in flower-type-specific MADS protein complexes may play a central role in differential development of ray and disc flowers across the gerbera capitulum and that some commonality is shared with known protein functions in floral organ determination. Gerbera flowering head is more than a mere floral analog at the level of gene regulation. Global gene expression analyses using the gerbera cDNA microarray indicate that rapid transcriptional changes correlate with morphological differentiation of individual flower types. Several genes were identified encoding MADS domain transcription factors that are differentially expressed in developing ray and disc flower primordial.

Plants of four gerbera cultivars transformed with nucleocapsid N-gene of tomato spotted wilt virus were evaluated in terms of resistance to the virus and several phenotypical traits. Sixteen out of 33 transformed genotypes (with transgenic plant status confirmed by PCR with specific primers for N and npt II genes) survived when transferred to the greenhouse (Korbin 2006).

Hussein et al. (2013) investigated a transient expression system for gerbera petals based on the *Agrobacterium* infiltration protocol using the reporter genes ß-glucuronidase (gus) and green florescence protein (gfp). The transient expression results showed change in the anthocyanin pigment in all infiltrated flowers with color genes. Additionally, blue color was detected in the stigma and pollen grains in the infiltrated flowers.

To improve the cold tolerance in commercially important ornamental plant gerbera cv. Gold Eye, introduction of the *Arabidopsis* Ca^{2+}/H^{+} antiporter gene (CAX1) via *Agrobacterium*- mediated transformation was carried out. Expressing the cold-tolerant gene CAX1was incorporated, and its functional role as further assessment improving cold tolerance was investigated. Introduction of the gene to the cultivar improved the cold tolerance in the winter months (Chung et al. 2016).

21.2.7 Information Documentation

(a) **Catalogues and databases**

The germplasm catalogues and database of gerbera germplasm may be maintained by the PGR department of the respective country.

(b) **Dissemination and exchange**

In India, the gerbera germplasm are being disseminated and exchanged through the AICRP on Floriculture for multi-location evaluation of germplasm. The germplasm exchange for breeding or characterization purpose is being facilitated by ICAR-NBPGR, New Delhi.

21.2.8 Use of Plant Genetic Resources

(a) **Major constraints in the crop production**

- High costs of planting material and protected structure for growing.
- Non-availability of skilled labors.
- Non-availability of planting material.
- Incidence of serious insect-pests such as thrips, mites, white flies, etc.
- Incidence of diseases such as phytoplasma, powdery mildew, roots rot, etc.
- Fluctuations in market price.
- Lack of advanced production technology.
- Lack of adequate infrastructure facilities for quick disposal of the produce in the market.
- Floricultural crops have not been included with the overall land use planning.
- Inadequate support to postharvest management including grading, storage, marketing, and processing.
- Poor extension and training efforts in the sector.
- Physiological disorders such as:

  Stem break: It is a common postharvest disorder in cut gerbera which is mainly caused by water imbalances. It could be ethylene controlled and associated with early senescence caused by water stress.

  Yellowing and purple margin: Nitrogen deficiency causes yellowing and early senescence of leaves. Phosphorus deficiency causes pale yellow color with purple margin.

  Flower bent: Loss of cell turgidity and undernutrition (lack of calcium).

  Pre-harvest stem break: High root pressure and high humidity in the air.

  Premature wilting of *Gerbera* flower: Cloudy weather followed by bright sun or carbohydrate depletion.

  Double-faced *Gerbera* flower: It is caused by imbalance of nutrients. Too much growth too little flower buds.

  Nonuniform flower blooming: Physical injury to flower stem/pest damage/phytotoxicity.

  Short stem length: High salinity level, moisture stress, low soil temperature.

(b) **Common sources used to overcome production constraints [listing of genetic resources, genetic stocks (including aneuploids series, substitution and translocation lines, recombinant inbred lines, etc.), inbred lines,**

released cultivars associated with desired traits, genes with gene symbols, mapping populations, etc.]

To meet this rising demand, floriculture becomes popular, and in recent days it has emerged as one of the most lucrative professions in all over the world. The future scope of *Gerbera* flowers is increasing demand for cut flowers, live plants, quality plant production, the sale of garden tools, pots, etc.

ICAR-IIHR, Bengaluru, have developed six indigenous gerbera varieties suitable for polyhouse and open-grown condition. The polyhouse grown varieties were found at par with the exotic varieties. The cost of planting of these varieties comes down to half of the exotic varieties of gerbera, thereby increasing in per unit income. The brief description of these varieties is given below.

Arka Krishika It has stalk length of 39 cm and produces 3.6 flowers per plant/month. Its yellow-colored flowers are double types. It is suited for open-grown condition and used for cut flower and flower decoration.

Arka Red It has bright red color (RHS group Red 40A) flower, double in nature, suitable for growing outside for both beds and cut flower. It yields 40 flowers/plant/year.

Arka Ashwa It has double-type flowers. It performs well under polyhouse with 50% shade net and at par with commercial varieties. It produces flowers with diameter (10.85 cm), flower stalk length (61.06 cm), flower stalk diameter (6.42 mm), and 3.23 numbers of flowers/month. It is suitable for cut flower and flower arrangement.

Arka Nesara It has double-type flowers. It performs well under polyhouse with 50% shade net and at par with commercial varieties. It produces flower diameter (10.43 cm), flower stalk length (61.11 cm), flower stalk diameter (5.63 mm), and 2.89 numbers of flower/month. It is suitable for cut flower and flower arrangement.

Arka White It has semi-double type of flowers with white color flower (White Group NN155A). It performs well under polyhouse with 50% shade net and at par with commercial varieties. It produces average flower diameter (11.89 cm), flower stalk length (61.39 cm), flower stalk diameter (5.79 mm), and 2.87 numbers of flowers per month. Its flower quality traits are at par with commercial checks and suitable for cut flower and flower arrangement.

Arka Pink It has double-type flowers with pink color flower (Red-Purple Group 65A). It performs well under polyhouse with 50% shade net and at par with commercial varieties. It produces average flower diameter (12.89 cm), flower stalk length (65.64 cm), flower stalk diameter (5.77 mm), and 2.85 numbers of flowers per month. Its flower quality traits are at par with commercial checks and suitable for cut flower and flower arrangement.

(c) **Breeding options**

The breeding of gerbera started at the end of the nineteenth century in Cambridge, England, when two South African species. *G. jamesonii* and *G. viridifolia*, were crossed by R.I. Lynch. He named the hybrid *Gerbera* × *cantebrigiensis*, known today also as *Gerbera hybrida*. The majority of the present commercially cultivated varieties originate from the crossing progenies of these two species. Natural hybrids of the two species have not been found (Hansen 1985). It is possible that also other wild gerbera species have been used in breeding.

Gerbera breeding is mostly based on sexual crossing and selection. The goals sought in breeding programs for gerbera are improvements in aesthetic properties (flower color and morphology) and in economic characteristics such as flower productivity, timing and synchrony of flowering, vase life, and resistance to insects and especially to fungal diseases such as *Phytophthora*, *Verticillium*, *Fusarium*, and *Botrytis*.

Conventional breeding has received great benefit from the large genetic variation among the *Gerbera* species and has already resulted in the production of elite genotypes rich in colors and color patterns within the capitulum, combined with good production characteristics. Still, combining all the properties wanted by consumers and by growers using traditional methods is very laborious and time-consuming, the other facet of the high heterozygosity in *Gerbera*.

One of the greatest weaknesses of using traditional vegetative propagation by splitting or division of rhizomes or clumps is the low rate of propagation, the long period of time required to obtain commercial quantities of new plants, around five plants from one per year and the increase in the frequency of plantlets with phytosanitary problems, mainly due to the use of non-sterile tools at the time of cutting and division of parts of rhizomes or clumps. In fact, micropropagation using terminal buds/apices or through organogenesis of somatic tissue is considered to be the only possible viable method for the rapid mass propagation of elite gerbera germplasm while maintaining the genetic fidelity. Moreover, plant tissue culture methods such as shoot tip culture result in disease-free plantlets.

Micropropagation is the main system used to clonally propagate gerbera in vitro resulting in the production of millions of plantlets each year. Numerous types of explants and protocols for micropropagation have been established and used for gerbera. Shoot tips are the commonly used explant, while adventitious shoot induction from the capitulum is also a popular method.

Efficient tissue culture, micropropagation, and gene transfer methods were developed for gerbera species. Consequently, modification of flowers colors in gerbera became one of the first examples of genetically modified traits in ornamentals. The possibility of producing genetically modified gerbera plants has made gerbera a well-recognized model species within flower developmental biology. The genes could be engineered to modify flower color. It also served as marker genes in studies of inflorescence development since pigmentation patterns in various gerbera varieties are strictly correlated with the anatomy of the complex inflorescence (Helariutta et al. 1995b). More recent studies have identified a number of regulatory genes that are required for organ- and tissue-specific activation of biosynthetic genes and anthocyanin pigmentation (Elomaa et al. 1998, 2003). These studies have provided intrinsic scientific information about transcriptional regulation of gene expression in plants. Tissues of gerbera plants are very rich in secondary compounds that are attractive targets for molecular breeding.

(d) **Present status of use or incorporation of desired traits**

The desirable traits (quality, yield, biotic and abiotic) into modern gerbera can be transferred through conventional breeding and transformation. Unique breeding lines of gerbera have been registered with ICAR-NBPGR, New Delhi (Table 1), and promising genotypes resistant to biotic stress (Table 5), which can be utilized for incorporation of these unique traits in new gerbera hybrids.

The studies on flavonoid pathway genes led to the discovery of a novel gene function responsible for synthesis of secondary compounds involved in pathogen and insect resistance in gerbera (Helariutta et al. 1995a, 1996; Eckermann et al. 1998).

Chung et al. (2016) improved the cold tolerance of the gerbera cv. Gold Eye by introduction of the *Arabidopsis* antiporter gene (CAX1) via *Agrobacterium*-mediated transformation. Gerbera expressing the *Arabidopsis* antiporter gene (CAX1) were obtained using the optimized concentrations.

Huang et al. (2020) reported that GhEIL1 forms part of the ethylene signaling pathway and activates GEG (a gerbera homolog of the gibberellins-stimulated transcript 1 [GAST1] from tomato) to regulate ray petal growth during the late developmental stage in *G. hybrida*.

21.2.9 Looking Forward or Future Perspective

As *Gerbera* is one of the important commercial flower crops mainly grown worldwide for its cut flower and potted plants. National Active Germplasm Sites for vegetatively propagated flower crops such as *Gerbera* should be strengthened for proper maintenance of germplasm under polyhouse and open-grown conditions. Focus is to be given on traits-specific exploration and collection from primary and secondary sources. The germplasm exchange among the countries may be increase through proper channel. The novel breeding lines/genetic stocks/varieties with flower quality traits, resistant to biotic and abiotic stress may be utilized in future breeding program of gerbera. Genetic transformation is one of the future tools for bringing novel gerbera. When utilizing the current wild resources, protection of genetic diversity both in situ and ex situ needs to be considered. The private sector should be encouraged by the government for developing and disseminating flower cultivation technology and market to the farmers' level. A strong coordination should be established between research organization and extension agencies for disseminating suitably advanced technology to farmer's level. Senior officials of Department of Agricultural Extension (DAE) should be designated for an extension, coordination, and monitoring of floricultural, especially for *Gerbera* flower activities.

References

Adlam RW (1888) Colonial notes. Gardeners' Chron 3(78):775

Allen J (1889) Gerbera jamesoni. Gardeners' Chron 5(130):772–773

Asen S (1984) High pressure liquid chromatographic analysis of flavonoid chemical markers in petals from *Gerbera* flowers as an adjunct for cultivar and *germplasm* identification. Phytochemistry 23(11):2523–2526

Bowe R, Danhardt W, Fritzsche W, Gerstner W, Junges W (1969) Gerbera. Verlag J. *Neumann-Neudamm*. Springer, Berlin

Chung M-Y, Kim MB, Chung YM, Nou I-S, Kil KC (2016) *In vitro* shoot regeneration and genetic *transformation* of the *Gerbera* (*Gerbera hybrida* Hort.) cultivar Gold Eye. J Plant Biotechnol 43(2):255–260

Codd LE (1979) The story of the Barberton Daisy, *Gerbera jamesonii*. Veld Flora 65(4):114–115. Contrib Bot 22:539–543

Deng X, Bashandy H, Ainasoja M, Kontturi J, Pietiäinen M, Laitinen RAE, Albert VA, Valkonen JPT, Elomaa P, Teeri TH (2014) Functional diversification of duplicated chalcone synthase genes in anthocyanin biosynthesis of *Gerbera hybrid*. New Phytologist 201:1469–1483

Dümmer RA (1914) The South African gerberas. J Royal Hort Soc 40:236–262

Eckermann S, Schroder G, Schmidt J, Strack D, Edrada RA, Helariutta Y, Elomaa P, Kotilainen M, Kilpelainen I, Proksch P (1998) New pathway to polyketides in plants. Nature 396:387–390

Elomaa P, Mehto M, Kotilainen M, Helariutta Y, Nevalainen L, Teeri TH (1998) abHLH transcription factor mediates organ, region and flower type specific signals on dihydroflavonol-4reductase *(dfr)* gene expression in the inflorescence of *Gerbera hybrida* (Asteraceae). Plant J 16:93–99

Elomaa P, Uimari A, Mehto M, Albert VA, Laitinen RAE, Teeri TH (2003) Activation of anthocyanin biosynthesis in *Gerbera hybrida* (Asteraceae) suggests conserved protein-protein and protein-promoter interactions between the anciently diverged monocots and eudicots. Plant Physiol 133:1831–1842

Engelmann F (2011) Use of biotechnologies for the conservation of plant biodiversity. In Vitro Cell Dev Biol Plant 47:5–16

Fu Y, Chen M, Van Tuyl J, Visser RGF, Arens P (2015) The use of a candidate gene approach to arrive at *botrytis* resistance in gerbera. Acta Hortic 1087:461–466

Grout BWW (1995) Introduction to the *in vitro* preservation of plant cells, tissue and organs. In: Grout BWW (ed) Genetic preservation of plant cells *in vitro*. Springer, Berlin Heidelberg/New York, pp 1–20

Hansen H (1985) A taxonomic revision of the genus Gerbera (Compositae, Mutisieae) sections Gerbera, Parva, Piloselloides (in Africa), and Lasiopus. Opera Botanica 78:5–36

Hansen HV (1990) Phylogenetic studies in the Gerbera-complex (Compositae, tribe Mutisieae, subtribe Mutisiinae). https://doi.org/10.1111/j.1756-1051.1990.tb00537.x

Hansen HV (2006) Comments on the Gerbera-complex (Asteraceae: Mutisieae). SIDA. SIDA Contrib Bot 22(1):539–543

Harding J, Huang H, Byrne T (1991) Maternal, paternal, additive and dominance components of variance in Gerbera. Theor Appl Genet 82:756–760

Hayden MJ, Tabone TL, Nguyen TM, Coventry S, Keiper FJ, Fox RL, Chalmers KJ, Mather DE, Eglinton JA (2010) An informative set of SNP markers for molecular characterization of Australian barley germplasm. Crop Plant Sci 61:70–83

Helariutta Y, Elomaa P, Kotilainen M, Griesbach RJ, Schröder J, Teeri TH (1995a) Chalcone synthase like genes active during corolla development are differ- entially expressed and encode enzymes with different catalytic properties in *Gerbera hybrida*. Plant Mol Biol 28:47–60

Helariutta Y, Kotilainen M, Elomaa P, Teeri TH (1995b) *Gerbera hybrida* (Asteraceae) imposes regulation at several anatomical levels during inflorescence development on the gene for dihydroflavonol-4-reductase. Plant Mol Biol 28:935–941

Helariutta Y, Kotilainen M, Elomaa P, Kalkkinen N, Bremer K, Teeri TH, Albert VA (1996) Duplication and functional divergence in the chalcone synthase gene family of Asteraceae:

evolution with substrate change and catalytic simplification. Proc Nat Acad Sci USA 93: 9033–9038

Huang G, Han M, Jian L, Chen Y, Sun S, Wang X, Wang Y (2020) An ethylene insensitive3-like1 protein directly targets the geg promoter and mediates ethylene-induced ray petal elongation in *Gerbera* Hybrida. Front Plant Sci. https://doi.org/10.3389/fpls.2019.01737

Hussein GM, Abu E-h GA, Abdou Sara M, Abdallah Naglaa A (2013) Optimization of transient gene expression system in *Gerbera* Jemosonii petals. GM Crops Food 4(1):50–57

Katinas L, Sancho G, Telleria MC, Crisci JV (2009) Mutisieae sensu strict (Mutisioideae sensu stricto) 2009. In: Funk VA, Susanna A, Stuessy TF, Bayer RJ (eds) Systematics, evolution and biogeography of compositae. International Association for Plant Taxonomy, Vienna, pp 229–248

Kloos WE, George CG, Sorge LK (2005) Inheritance of powdery mildew resistance and leaf macrohair density in *Gerbera hybrida*. HortScience 40:1246–1251

Korbin M (2006) Assessment of *gerbera* plants genetically modified with TSWV nucleocapsid gene. J Fruit Ornamen Plant Res 14(Suppl 1):85–93

Laitinen RAE, Ainasoja M, Broholm SK, Teeri TH, Elomaa P (2008) Identification of target genes for a MYB-type anthocyanin regulator in *Gerbera hybrid*. J Exp Bot 59(13):3691–3703

Lonsdale E (1908) Jottings for florists. Florists' Exch 26(8):236

Lynch RI (1905) Gerbera, with a coloured plate of the new hybrids. Flora Sylva 3:206–208

Manning JC, Goldblatt P (2012) Plants of the greater cape floristic region 1: the Core cape Flora. *Strelitzia* 29. South African National Biodiversity Institute, Pretoria

Manning JC, Simka B, Boatwright JS, Magee AR (2016) A revised taxonomy of Gerbera sect. Gerbera (Asteraceae: Mutisieae). S Afr J Bot 104:142–157

Mata TL, Segeren MI, Fonseca AS, Colombo CA (2009) Genetic divergence among *gerbera* accessions evaluated by RAPD. Sci Hortic 121(1):92–96

Nichols H (1902) Home correspondence. Gardeners' Chron 31(806):377

Ortiz S, Bonifacino JM, Crisci JV, Funk VA, Hansen HV, Nicholas Hind DJ, Katinas L, Roque N, Sancho G, Susanna A, Telleria MC (2009) The basal grade of compositae: Mutisieae (sensu Cabrera) and Carduoideae 2009. In: Funk VA, Susanna A, Stuessy TF, Bayer RJ (eds) Systematics, evolution and biogeography of compositae. International Association for Plant Taxonomy, Vienna, pp 193–213

Panero JL, Funk VA (2008) The value of sampling anomalous taxa in phylogenetic studies: major clades of the Asteraceae revealed. Mol Phylogenet Evol 47:757–782

Pasini E, Funk VA, de Souza-Chies TT, Miotto STS (2016) New insights into the phylogeny and biogeography of the Gerbera-Complex (Asteraceae: Mutisieae). Taxon 65(3):547–562

Prajapati P, Singh A, Patel NL, Singh D, Srivastav V (2014) Evaluation of genetic diversity in different genotypes of *Gerbera jamesonii* Bolus using random amplified polymorphic DNA (RAPD) markers. Afr J Biotechnol 13(10):1117–1122

Sandigawad AM (2014) Analysis of molecular diversity in *Gerbera* genotypes revealed by RAPD markers. Adv Biores 5(4):92–97

Silva TL, Scherwinski-Pereira JE (2011) *In vitro* conservation of *Piper aduncum* and *Piper hispidinervum* under slow-growth conditions. Pesq Agrop Brasileira 46:384–389

Singh S, Dhyani D (2016) Domestication of wild Gerbera (*Gerbera gossypina* (Royle) Beauved). Int J Plant Sci Ecol 2(1):10–14

Singh S, Dhyani D, Karthigeyan S (2009) Characterization of variability in F_1 progeny of inter-varietal cross of *gerbera* (*Gerbera jamesonii* Bolus Ex. Hooker F). J Ornam Hortic 12(4):225–231

Song X, Deng Z (2013) Powdery mildew resistance in *Gerbera*: mode of inheritance, quantitative trait locus identification, and resistance responses. J Am Soc Hortic Sci 138(6):470–478

Song X, Deng Z, Gong L, Hu J, Ma Q (2012) Cloning and characterization of resistance gene candidate sequences and molecular marker development in *Gerbera* (*Gerbera hybrida*). Sci Hortic 145(20):68–75

Xu X-D, Zheng W, Chen L-Q, Wen J (2017) Genetic diversity and population structure of *Gerbera delavayi* (Asteraceae) in Southwest China: implications for conservation. Ann Bot Fenn 54(4–6):409–422

Amaryllis/Hippeastrum

22

S. K. Datta

Contents

Abstract

The genus *Hippeastrum*, also referred to as *Amaryllis*, belonging to the family Amaryllidaceae is an ornamental bulbous flowering plant. It originated in the subtropical Americas, from Eastern Brazil to the Southern central Andes of Peru, Argentina, and Bolivia. Many species in this genus and their hybrids have large and showy colorful flowers. They are native to Central and South America, and the

S. K. Datta has retired.

S. K. Datta (✉)
Floriculture Division, CSIR-National Botanical Research Institute, Lucknow, India

Bose Institute, Kolkata, India

Department of Plant Science and Biotechnology, Assam University, Silchar, Assam, India

S. K. Datta, Y. C. Gupta (eds.), *Floriculture and Ornamental Plants*, Handbooks of Crop Diversity: Conservation and Use of Plant Genetic Resources,
https://doi.org/10.1007/978-981-15-3518-5_23

members of this genus are easily grown in the tropical and subtropical regions. There is a huge diversity of different hybrids of the *Hippeastrum* plant. The present-day *Hippeastrum* is the result of selections that have continued for over 200 years using whatever species were available. It can be propagated through seeds, offset bulblets/bulbils, basal cuttings of the bulbs, twin scaling, and tissue culture. Wide range of variations in flower colors, plant shape, time of flowering, etc. are found when multiplied by seeds. Different aspects related to growth, physiological, ecological, and flowering have been studied. Breeding potential of wild species and hybrids has been assessed in terms of cytological variations, embryology, molecular markers, etc. These provide valuable information about the constitution of germplasm, genetic compatibility of different taxa involved in hybridization.

Keywords

Amaryllis · *Hippeastrum* · Germplasm · Genetic diversity · Breeding · New variety · Characterization

22.1 Introduction

Hippeastrum is an excellent bulbous plant, bearing beautiful flowers. Earlier, many genera of Amaryllidaceae were under *Amaryllis*. Some confusion exists between the *Amaryllis* and *Hippeastrum* but all have been transferred to *Hippeastrum*. The popular name of *Amaryllis* still holds good but *Hippeastrum* is considered in most recent publications. Flowers are large and showy with various forms and colors and remain fresh for 4–5 days from the date of opening. The flowers are usually borne in an umbel on the flowering scape emerging from bulbs. It looks very beautiful when planted in clumps on raised slopes. It flowers from mid-February to April in the North Indian Gangetic plains when there is real scarcity of flowers. It is grown for garden beautification as in beds, borders, and is suitable for planting in pots, greenhouses, window gardens, rockery, shrubbery, and also in landscaping. It is also used as cut flowers because of their large size and attractive color. The long-lasting colorful cut flowers also serve as an excellent material for ikebana. Scientists are engaged in research on *Hippeastrum* on different multidisciplinary aspects like enrichment and characterization of germplasm, their utilization in breeding program, and development of new and novel varieties through hybridization and selection. The chapter gives an overview of the R&D activities on *Hippeastrum* summarizing the accumulated experience related to geographical distribution, agro-technology, cytogenetics, improvement, post-harvest management, diseases and pests, characterization, etc.

22.2 Botany

Plants of the genus *Amaryllis* are known as belladonna lily, Jersey lily, naked lady, amarillo, Easter lily in Southern Australia or, in South Africa, March lily due to its propensity to flower around March. It is a monotypic genus. *Amaryllis belladonna*

L. is a perennial bulbous plant that belongs to the family Amaryllidaceae. *Amaryllis* is the only genus in the subtribe Amaryllidinae, in the Amaryllideae tribe. The genus name *Amaryllis* comes from the Greek word "amarysso," which means "to sparkle." The taxonomy of the genus *Amaryllis* is complicated. For many years there was confusion among botanists over the generic names *Amaryllis* and *Hippeastrum*. The first issue is whether the name should more properly be *Amaryllis* L. In 1753 Carl Linnaeus created the name *Amaryllis belladonna*, the type species of the genus *Amaryllis*, in his Species Plantarum along with eight other *Amaryllis* species. The common name "Amaryllis" is mainly used for cultivars of the genus *Hippeastrum*, often sold as indoor flowering bulbs particularly at Christmas in the Northern hemisphere. By contrast the generic name Christmas applies to bulbs from South Africa, usually grown outdoors. The genus is native to tropical and subtropical regions of the Americas from Argentina north to Mexico and the Caribbean. South African and South American plants were placed in this same genus. *Amaryllis* became a polymorphic genus with about 50 species by the early nineteenth century, and a dozen genera from those are reflected today, and attempts were made to separate it into different genera. English botanist, Revd. William Herbert began this work in 1819 and reported in Curtis's Botanical Magazine which he expanded in 1821 in The Botanical Register, identifying 14 species of the new genus of *Hippeastrum*, and only leaving three species in *Amaryllis*. The rest of the *Amaryllis* species he transferred to other genera, several of which he created. Herbert further refined his descriptions of *Hippeastrum* in his work on the Amaryllidaceae in 1837. A number of subgenera have been proposed over the years. John Gilbert Baker in 1878 reorganized *Hippeastrum* and described nine sections of the genus. He included seven subgenera by 1888 like *Rhodophiala* (5 species), *Macropododastrum* (1 species), *Omphalissa* (6 species), *Aschamia* (10 species), and *Lais* (3 species), some of which have since been treated as separate genera (*Habranthus*, *Rhodophiala*). Currently these subgenera are not widely used due to indistinct boundaries of some of the divisions. *Hippeastrum* is a genus in the family Amaryllidaceae (subfamily Amaryllidoideae, tribe Hippeastreae, and subtribe Hippeastrineae. The name Amaryllis is confusing. *Amaryllis* is the common name for flowers of the genus *Hippeastrum*. This genus was separated from genus *Amaryllis* in the early nineteenth century. However, the genus *Amaryllis* still exists today. It is most common species is the naked lady. *Hippeastrum* and *Amaryllis* have similar shapes, though the *Hippeastrum* has a hollow stem. *Hippeastrum* is Greek for horseman's star or knight's star, as the flowers have a star-like shape. There had dispute in the taxonomy of the genus. Carl Linnaeus coined the name *Amaryllis belladonna*, the type species of the genus *Amaryllis* in 1753. At the time both South African and South American plants were placed in the same genus. Subsequently they were separated into two different genera. The decision was taken at the 14th International Botanical Congress in 1987 that *Amaryllis* L. should be a conserved name (i.e., correct regardless of priority) and ultimately based on a specimen of the South African *Amaryllis belladonna* from the Clifford Herbarium at the Natural History Museum in London (Meerow et al. 1997). The name *Amaryllis* is very popular but in most recent publications it is considered *Hippeastrum*. The genus is sometimes mistaken for the genus *Hippeastrum*, since plants belonging to this genus are commonly known as *Amaryllis*. Although dispute over the name continues (Dandy and Fosberg 1954, Traub 1983), the name *Amaryllis belladonna* has been internationally allocated to the cape belladonna staff (Dyer 1976). It is perennial herbaceous bulbous plants. They generally have large fleshy bulbs. Bulbs are tunicate and generally between 5 and 12 cm in diameter and produce two to seven long-lasting evergreen or deciduous leaves. The leaves are hysteranthous, sessile, rarely persistent, and subpetiolate. Flowers (2–14), which are more or less zygomorphic and hermaphrodite, are arranged in umbelliform inflorescences supported on an erect hollow scape. The perianth has six bright colored tepals and the tepals are united at the base to form a short tube. The androecium consists of six stamens with filiform filaments and the anthers are dorsifixed or versatile. The ovary is inferior and trilocular. Fruits form a trivalve capsule containing seeds. About 20 to 30 seeds are contained in each ovary or 80 seeds in each capsule (Okubo 1993). At the beginning of the growth season (March) the plant bears 6–12 white- to-pink flowers on a solid scape, hence the name March lily. *Amaryllis* is commonly known as *Belladonna* lily or March lily. The belladonna lily's specific flowering time is late summer, February and March. *Amaryllis* is Greek feminine and is named after a beautiful shepherdess. The specific epithet belladonna means beautiful lady.

Amaryllis is native of Central and South America, and is easily grown in the tropical and subtropical regions. The native land of *Hippeastrum* is tropical and subtropical America. *Hippeastrum* is considered synonym to *Amaryllis*. *Hippeastrum* is applied on South America and *Amaryllis* to South African species (Moore 1963). The genus *Hippeastrum* has about 90 species and over more than 600 hybrids and cultivars. 75 tropical and subtropical species have been mentioned in "Dictonary of flowering plants and ferns" (Willis 1973). Bailey Hortorium Staff (1976) also reported 75 species and all are originated in South America. Few more important species are *H. solandriforum* Herb., *H. reticulatum* Herb., *H. stylosum* Herb., *H. aulicum* Herb., *H. reginae* Herb., *H. leopoldii* Dombr., *H. vittatum* Herb., *H. johnsoni* Bury., *H. equestre* Herb. Meerow et al. (1997) as per discussion at the 14th International Botanical Congress in 1987, 42 species were transferred from *Amaryllis* to *Hippeastrum* which aid nomenclatural clarity and enable botanists and horticulturists to refer to these species by their correct names. A lot of controversy developed during the period 1938 to 1984 of the name *Amaryllis* and *Hippeastrum* and the history of controversy was critically discussed and reviewed by a number of researchers in the interest of nomenclatural clarity (Uphoff 1939, Sealy 1939, Traub and Moldenke, 1949; Dandy and Fosberg 1954; Traub 1958, Hunziker and Cocucci 1959, Moore 1963, Tjaden 1981, Ravenna 1981, Goldblatt 1984, Williams and Dudley 1984).

22.3 History and Geographical Distribution

Amaryllis L. is the type genus of the family Amarylliceae. The genus is native of tropical and subtropical America being distributed from Mexico to West Indies, Southward to Chile and Argentina. Only one species (*A. reginae*) crosses the ocean to the African continent where it grows in Princes' Island in estuary of Congo River in West Central Africa. The maximum concentration of species is in Amazon River Basin of Brazil, Bolivia and Peru, an area which may legitimately be looked upon as the center of diversity and dispersal of the genus (Traub 1958).

Like all cultivated plants, present-day garden amaryllis has been developed by the conscious and subconscious selection by man over the years from the wild species of the genus *Amaryllis*. The first species to be introduced from its native Maxican-Brazillian region was *A. belladonna* Linn., in 1689 by the Dutch medical man-botanist, Dr. Paul Hermann.

It is believed that the Portuguese brought the Amaryllis bulb to Europe in the early sixteenth century. Eduard Friedrich Poeppig (1798 to 1868), German botanist discovered the *Amaryllis* growing on a hillside in Chile. Portugal, Spain, and Italy were among the first European countries to introduce the *Amaryllis belladonna*. *Amaryllis belladonna* arrived in England around the early eighteenth century. *Amaryllis* plant bulbs were labeled as lilies during the eighteenth century. It is called March lily in South Africa, belladonna lily or the Jersey lily in the United Kingdom, Madonna lily in Italy. The Portuguese name of the *Amaryllis* means "St. Joseph's Staff." *Amaryllis belladonna* was introduced into cultivation at the beginning of the eighteenth century. Goldblatt (1984) mentioned that *Hippeastrum* belong to the Caribbean species cited by the Linneaus and *Amaryllis* to the South African plants. Amaryllis bulb was probably brought to Europe around the early sixteenth Century. Different species are distributed at several places like South America, Brazil, Guyana, Mexico, Peru, France, England, etc. *Hippeastrum* are widely grown in the USA, The Netherlands, South Africa, Swaziland, and Israel (Okubo 1993, Rees 1985). There are many recognized centers in America and European countries notably in Holland, where a large number of new hybrids have been developed. Holland is a major grower and exporter of the Amaryllis especially during Christmas time. A few species and hybrids are cultivated in Indian Gardens.

22.4 Distribution and Habitat

Hippeastrum is found in a wide range of habitats – tropical and subtropical and also temperate. It prefers both shade and full sun. There are epiphytic species such as *Hippeastrum aulicum*, *Hippeastrum calyptrum*, *Hippeastrum papilio*, and *Hippeastrum arboricola*. Eastern Brazil and central southern Andes of Peru, Bolivia, and Argentina are the main centers of diversity of *Hippeastrum*. About 34 of the species have been found in Brazil. Some species are reported from Mexico and the West Indies.

22.5 Genetic Diversity

The study of genetic divergence is a useful and effective tool for screening accessions in germplasm banks, studies of organism evolution, and identification of superior parents in breeding programs. Diversity in vegetative and floral characters has been studied at different institutions using classical and molecular methods. Considerable variations in the floral characters among the accessions have been detected. Diversity in different quantitative traits like bulb length and width, number of leaves, leaf length and width, petal length, flower length, number of flowers per bulb, flower and stem length, etc. have been recorded. The present-day *Amaryllis* is the result of selections using whatever species were available. There is a huge diversity of different hybrids of *Hippeastrum* plants on the market today (Robert et al., 2006). Extensive variation was observed in all the floral characters of the hybrid group. Hybrid flower size was either medium or large. Wide range of variations in flower color and striping has been reported. In addition to the amazing diversity of hybrids, a large number of cultivated wild species are in cultivation with great range of diversity. Indiscriminate use of hybrids and wild species in breeding resulted in considerable diversity in flower size and color. Every center has examined the diversity using different parameters. In Vietnam, genetic diversity and distribution of *Hippeastrum* were studied on 25 accessions collected from 17 provinces in the northern, central, and southern regions of Vietnam. All the cultivated *Hippeastrum* of Vietnam were classified into four distinct groups on the basis of floral observation and RAPD analysis. Wide variation in flower color, shape, and size was detected among accessions of the hybrid group. Variation was observed between northern, central, and southern *H. puniceum* populations. While *H. puniceum* is the most widely distributed species from the northern to the southern regions, *H.* x "Johnsonii" has been observed only in the northern region, but not in the central or southern regions, so the distribution of *Hippeastrum* within Vietnam might depend on adaptation to environmental conditions (Phuong et al. 2014).

22.6 Classification

Hippeastrum has been classified according to the form, shape, and size of flowers by American Amaryllis Society (1933). The wild species of *Amaryllis* have been separated into five divisions (subgenera) and the subgenera have botanical names, viz.: (a) Macropodastrum for Elegans, (b) Lais for Striata group, (c) *Amaryllis* for the Bellanonna group, (d) Omphalissa for Aulica group, and (e) Sealyan for Reticulata group.

A workable classification of cultivars has been proposed and cultivated *Amaryllis* has been classified in eight divisions on the basis of agro-horticultural use (Khoshoo 1971, Traub 1958, Narain 1977a, b, c, d, Narain and Khoshoo 1977).

Division 1: **Cultivated Wild Amaryllis (D-1)** – It includes all cultivated species of the genus as they are known in the wild, including subspecies, varieties, and forms. The wild species in cultivation are listed as *Amaryllis belladonna* var. major and *Amaryllis belladonna* var. plena, *Amaryllis striata* var. fulgida, *Amaryllis immaculate*, and *Amaryllis psittacina*.

Division 2: **Long-trumpet Amaryllis Hybrids (D-2)** – The hybrids developed through crossing of *Amaryllis elegans* with *Amaryllis stylosa*. *Amaryllis striata* and *Amaryllis vittata* are fragrant and long trumpet flowers are distinctly drooping. The present-day hybrids have relatively long and trumpet shaped pedicels. The tepal tube is very long (11–15 cm). The flowers are like Easter lilies, diploids (2n = 22).

Division 3: **Belladonna-Type Amaryllis Hybrids (D-3)** – *Amaryllis johnsonii*, the first *Amaryllis* hybrid was reported as early as 1799 belongs to this group. The pedicels are relatively long and the flowers usually drooping but not always so. The tepal tube is much shorter than those in Division 2, the flowers are much shorter, and variously shaped, showing the influence of species with an informal flower structure, such as *Amaryllis belladonna* and *Amaryllis striata*, the diploids (2n = 22).

Division 4: **Reginae-Type Amaryllis Hybrids (D-4)** – The pedicels are relatively shorter than the long-trumpet and belladonna-type hybrids. Flowers are drooping, horizontal or slightly upright. The tepal tube is short and the flowers are moderately open faced, showing the influence of *Amaryllis reginae*, *Amaryllis correiensis*, and similar species. The important cultivars are divided into two groups, markedly imbricated and less imbricated: ***Markedly imbricated type*** – The flowers are markedly imbricated, that is, the tepal segs overlap for about ¾ or more of their length and the tips are rounded, rarely somewhat pointed, for example, *Amaryllis reginaeoides* cv "Picotee." ***Less imbricated type*** – The flowers are less imbricated, that is, the tepal segs overlap for less than ¾ of their length, pointed and reflexed, and the tips are somewhat rounded or pointed, for example, *Amaryllis reginaeoides*.

Division 5: **Leopoldi-Type Amaryllis Hybrids (D-5)** – These are similar to Reginae-type hybrids, except that the flowers are flat and wide open-faced type with almost overlapping, regular, rounded tepal segments and held horizontally, showing the influence of *Amaryllis leopoldi* and *Amaryllis pardiana*. The important cultivars are divided into two groups, markedly imbricated and less imbricated: ***Markedly imbricated type*** – The flowers markedly imbricated, that is, the tepal segs overlapping for almost their entire length, and the tips are rounded, for example, *Amaryllis leopoldaeoides* cv "Doris Lillian." ***Less imbricated type*** – The flowers are similar to above subdivision except that the tepal segs are less imbricated, and the tips may be either rounded or slightly pointed, for example, *Amaryllis leopoldaeoides* cv. "McCulloch."

Division 6: **Orchid-Flowering Amaryllis Hybrids (D-6)** – This type may have originated from crosses involving *Amaryllis sybister.* The tepal segs are not arranged according to the usual *Amaryllis* flower pattern but are variously shaped, often twisted reflexed. The interesting example of appearance of orchid-flowering individuals among the progeny of the Leopoldii strain is Cannae Butterfly. Blooms large with strongly ridges and twisted tepal segments which begin their reflexing the rigid, thick, greenish white midrib, the general appearance is red with white ribs.

Division 7: **Double Amaryllis Hybrids (D-7)** – Flowers are double and semi-double. Different colors such as pink with purple tinge and with yellow center are available, for example, "Helen Hull" and "Mckean's Hybrid."

Division 8: Miniature Amaryllis Hybrids (D-8) – Miniature hybrids are distinguished with dwarf-statured forms, including various flower forms, showing the influence of *Amaryllis espiritensis*, *Amaryllis traubii*, *Amaryllis reticulate*.

22.7 Species and Cultivars

A large number of species (approx. 91) of *Amaryllis* have been reported by various workers in worldwide, but only few of them are under cultivation. The widely grown species in gardens of tropical and subtropical regions is *Amaryllis belladonna*. Some important species are: *Amaryllis angustifolia*, *A. barbata*, *A. calyptrate*, *A. correiensis*, *A. cybister*, *A. elegans*, *A. evansiae*, *A. forgetii*, *A. leopoldii*, *A. miniata, A. moreliana*, *A. reginae*, *A. papilio*, *A. psittacina*, *A. reginae*, *A. striata*, *A. traubii*, *A. viridiflora,* etc. There are 10 major ancestral species which seem to be involved in the origin of garden *Amaryllis* through rampant hybridization followed by selection. Subgenus – Lais: *A. vittata*, *A. striata*; Subgenus – Amaryllis: *A. leopoldii*, *A. reginae*, *A. espiritensis*, and *A. belladonna*; Subgenus – Omphalissa: *A. psittacina*, *A. aulica,* and *A. pardina*; Subgenus – Sealyana: *A. reticulate* (Narain 1974). Some more important species of *Hippeastrum* are: *Hippeastrum aulicum* (Ker Gawl.) Herb, *H. correiense* (Bury) Worsley, *H. evansiae* (yellow), *H. papilio* (Ravenna) Van Scheepen, *H. brasilianum*, *H. doraniae, H. equestre*, *H. flammigerum*, *H. hybridum*, *H. pardinum* (green, yellow, and scarlet) (Hook. F.) Dombrain, *H. pretense*, *H. procerum*, *H. reticulatum* (rose) (L'Her.) Herb. Syn. *H. striatifolium* (Sims), *H. rutilum*, *H. stylosum*, *H. vittatum* (crimson) (L'Her.) Herb, *H. angustifolium* Pax, *H. arboricola* (Ravenna) Meerow, *H. aviflorum* (Ravenna) Dutilh, *H. calyptratum* (Ker Gawl.) Herb., *H. canterai* Arechav., *H. cybister* (Herb.) Benth. ex Baker, *H. evansiae* (Traub & I.S.Nelson) H.E.Moore, *H. ferreyrae* (Traub) Gereau & Brako, *H. iguazuanum* (Ravenna) T.R.Dudley & M.Williams, *H. johnsonii*, *H. leopoldii* (crimson and white) T. Moore, *H. miniatum* (Ruiz & Pav.) Herb., *H. petiolatum* Pax, *H. psittacinum* (orange and scarlet) (Ker Gawl.) Herb., *H. puniceum* (orange-red or salmon-red) (Lam.) Voss. Syn. H. equestre (Aiton), *H. reginae* (L.) Herb., *H. striatum* (Lam.) H.E.Moore *syn. H. rutilum* (Ker Gawl.) Herb. *Hippeastrum petiolatum* is sterile and unable to produce seeds. *H. petiolatum* is a sterile triploid that reproduces asexually, producing many bulbils around the mother bulb. Other species such as *Hippeastrum reticulatum* are self-pollinating, reproducing by distributing seed.

According to the International Union for the Conservation of Nature (IUCN) Red List of Threatened Species following species were considered threatened: *Hippeastrum arboricolum* (Argentina), *Hippeastrum aviflorum* (Argentina), *Hippeastrum canterai* (Uruguay), *Hippeastrum ferreyrae* (Peru), *Hippeastrum petiolatum* (Argentina and Brazil).

Most modern cultivars lack of fragrance, although "Dancing Queen" represents an exception. Recessive nature and genetic relationship of fragrance to flower color (white, or pastel shades) have been reported. *Hippeastrum ramboi*, a new species endemic to the mountainous region of Rio Grande do Sul (Brazil) has been described its habitat, ecology, geographical distribution, and its threat status has been evaluated. The new species has morphological affinity with *H. sanctaecatharinae* and *H. breviflorum* (Novelties et al. 2017).

The cultivars/varieties present at CSIR-NBRI, Lucknow, India, have been classified following the system suggested by Traub (1958):

Long-trumpet Amaryllis hybrids: Pedicels relatively long, flowers drooping, tepal tube very long. Flowers like Easter lilies. Diploids (2n = 22): "Becon" and "Sneezy."

Belladona type Amaryllis hybrids: Pedicels relatively long, flowers usually drooping but not always so. Tepal-tube shorter than above.

Varieties with different chromosomal status have been categorized (Narain and Khoshoo 1968):

Diploid (2n = 22) – "Achilles", "Adonis", "Amazon", "Aphrodite", "Bashful", "Beauty", "Bride", "Bridesmaid", "Brilliant", "Buccancer", "Cardinal", "Cerberus", "Ceres", "Coquette", "Diana", "Deepali", "Dopey", "Firefly", "Fresta", "Flora", "Gloriosus", "Hysperian", "Ida", "Itene", "Jenny", "Jenus", "Leo", "Lucifer", "Melpomone", "Mercurius", "Neptune", "Nesta", "Olymphus", "Orion", "Orthello", "Percy Lancaster", "Prima Donna", "Prime Minister", "Sleepy", "Sneezy", "Star of India", "Sybil", "Tara", "Uranus", "Vesta".

Triploid (3n = 33) – "Apple Blossom", "Becchus", "Bleeding Heart", "Bouquet", "Candy Cane", "canary Bird", "Carmine Dove", "Carmine King", "Charm", "Christmas Gift", "Country Girl", "Cymbister", "daintiness", "Dutch Belle", "Equestre", "Fantastica", "Fantasy", "Fascination", "Flirt", "Formossima", "Happy Memory", "Happy", "Indira", "Jessica", "Jupiter", "Lyric", "Mckean's Hybrid", "Moon Magic", "Morning Star", "Mrs. Percy Lancaster", "My Love", "Orange King", "Orange Majesty", "Our Red", "Pecotee", "Phoenix", "Pink Pearl", "Pepperment", "President", "Red Coat", "Romeo", "Salmon Purple", "Scarlet Beauty", "Scarletto", "Silver Jubilee", "Sperlkelia Hybrid", "Star Dust", "Streaking Stripes", "Stylosus", "United nations", "Universal Double", "White Giant", "Zebrina".

Tetraploid (4n = 44) – "Admiral", "Admiration", "Aeneas", "Actna", "Alexander", "Andromeda", "Apollo", "Aries", "Aurora", "Autocrat", "Beautiful", "Begum Secunder", "Black Prince", "Bridegroom", "Bright Red", "Charming", "Charm", "Chaste", "Ciree", "Cordelia", "Dainty", "Day Break", "Definance", "Deepa Kaul", "Denslow", "Diamond", "DOC", "Edith", "Enchantress", "Fiery Belt", "Flame", "Ganymede", "Glory", y', "Gorgeous", "Gracilis", "Grumphy", "Hannibl", "Hayward", "Heba", "Heliose", "Invincible", "Ivy", "Juliet", "Juno", "Kadam Rasul", "Lalkilla", "Maharaja", "Mars", "Mary", "Mentor", "Meteor", "Minerva", "Mother's Day", "Mount Everest", "Nizam", "Peacefulness", "Perseus", "Picture", "Piligrim", "pinkie", "Plauto", "Princess", "Prof.

Kaul", "Rose Queen", "Salmon Beauty", "Saturn", "Shah Nazaf", "Sheba", "Shiela Kaul", "Silver Lining", "Sieren", "Snow White", "Spitfire", "Star", "Starway", "Stylx", "Sweet Heart", "Taurus", "Thora", "Unnamed", "Venus", "White Queen", "Wyndham".

Reginae-type Amaryllis hybrids: Pedicels usually relatively shorter than the above two. Flowers rather drooping, horizontal, or slightly upright. Tepal tube short.

(a) ***Markedly imbricated type: tetraploids (2n = 44)***: "Begum Secundra", "Black Prince", "Emperor", "Glory", "Gorgeous", "Prof Kaul", "Shah Nazaf", and "Taurus".
(b) ***Less marked imbricated type: Tetraploids (2n = 44)***: "Autora", "Charming", "Day Break", "Edith", "Enchantress", "Fiery Bett", "Grumpy", "Heba", "Picture", "Snow White", and "Thora".

Leopoldii-type Amaryllis hybrids: Flowers are wide open, apparently flatish, and held horizontally.

(a) Markedly imbricated type: Diploids (2n = 22): "Ida" and "Sybil". Tetraploid: "Doc".
(b) Less marked imbricated type: Tetraploid: Unnamed (c.v. 35).

Semidouble hybrids*: *Diploid: "Firefly".

22.8 Breeding/Hybridization

Before starting any breeding work, it is very prerequisite to work out blossom biology and pollination mechanism. This study has been done very carefully in *Amaryllis*. The floral bud development takes place while the buds together with the scape are hidden underground within the bulb. The flower is wide open normally a day prior to anthesis. Dehiscence occurs between 8.00 to 10.00 am. and the anther splits longitudinally, under subtropical conditions. The stigma becomes receptive 12 to 24 hours later and by 10.00 am or at the most 3.00 pm next day as indicated by bright shining secretion of sugary fluid on the stigma. Both style and stigma point upwards at maturity. Pollen grains are not carried by wind due to its heaviness. Insects are attracted by bright color of the flowers, presence of nectar and fragrance. Both self-fertile and self-sterile systems are present in *Amaryllis* (Narayan 1981). Pollen fertility in the elemental species is fairly high and ranges between 56 and 80% except in *Amaryllis* species which is totally male sterile but fully female fertile. The size of the pollen grains varies from 70.3 to 75.04u and number of seeds per capsule ranges between 40 and 60. In the garden cultivars, there is great variation of pollen fertility which varies from 20 to 60%. The number of seeds per capsule ranges between 15 and 45 but the germination is poor and very few seeds are viable (Gupta and Datta 2006).

A large number of breeders used a wide number of species for development of new varieties. A brief outline of the early major events in origin and evolution of garden amaryllis have been highlighted (Traub 1958). After announcement of first hybrid, cross breeding work expanded in Europe. The early hybrids were not significantly improved and a total of about 175 hybrids were described (Sweet 1839). Slowly hybridization started in different centers of Holland, Belgium, and England and beautiful hybrids with brilliant flowers in diverse forms were evolved. Two hybrids were developed by The Garroway Company (Bristol, England) – one hybrid *A.* x *acramannii* (more vigorous with large open faced flowers) from *A. aulica* var. platypetala X *A. psittacina* and another hybrid "Acramannii Pulcherrima" developed in 1850 from *A. aulica* and *A.* x *johnsonii*. Subsequently several hybrids were developed in Belgium (by Louis van Houtte) involving species like *A. aulica*, *A. psittacina*, and *A. vittata*. De Graaffs at Leyden successfully developed many hybrids which were responsible for producing the Reginae Group of hybrids. This began with the origin of "Gravenna." They produced small-flowered colorful very famous selection Amaryllis "Graveana" in 1850s by crossing *A. vittata* with *A. striata* vars. "Fulgida" and "Crocata." This hybrid was crossed to *A. psittacina* and *A. striata* by S.A. de Graaff resulting in the famous cultivar "Empress of India," having several flowers per spike and a number of "Empress of India" strains were evolved by crossing it with other taxa. Traub (1958) considered it the beginning of the important group hybrids (gh.) collectively called "Reginae hybrids" (A. x gh. Reginaeoides) developed through hybridization and selection involving "Empress of India" and several available older hybrids raised by S.A. de Graaff in 1960s. The Reginae group of hybrids bears some resemblance to *A. reginae* and they are large flowered. R.W. Pearce from Peruvian-Bolivian region introduced two important species (*A. pardina* and *A. leopoldii*) with widely open flowers during 1867 and 1870. These species played important role in the origin of the Leopoldii class of hybrids grouped under A. x gh. Leopoldaeoides. There were only two flowers per spike. Increase in number of flowers was obtained by incorporating "Empress of India." The Leopoldii class was further improved in England, Holland, and Belgium (Traub 1958). The first almost pure white Leopoldii hybrid "Anowdon" was produced in 1904 after much work by C.R. Fielder. But, unfortunately many such important hybrids were lost due to nonavailability of techniques for vegetative multiplication. Amaryllis breeding industry all over the world particularly Europe was seriously affected during the two world wars (1914–1918 and 1939–1945) and many cultivars were lost. But some important cultivars were developed between 1919 and 1938 and after 1945. Double type amaryllis was raised by crossing a fully double *A. belladonna* var. plena f. albertii from Cuba with Mead strains of Reginae hybrids (McCann 1937, 1950). Some hybrids were developed by Van Tubergen (Haarlem) by crossing *A. aulica* with *A. reticulate* which could be grown outdoors. One most important species (*A. espiritensis*) introduced from Brazil was used as male parent and crossed to *A. belladonna* var. haywordii from Bolivia by Mrs. M.G. Henry and developed a very floriferous dwarf type with carmine-pink reasonably large flowers known as *A.* x *henryae* or "Henry Miniature" hybrids. She further developed vigorous tall and floriferous hybrids of Belladonna type called

A. x *gladwynesis* by crossing *A. belladonna* var. haywardii with a hybrid very near to *A.* x *johnsonii.*

After analysis of hybrids and breeding procedure, it is very clear that there are vast improvements in the cultivated types over the ancestral species. The main principles in developing desired hybrids are selection pressure toward – number of flowers per spike, number of scapes per bulb, period of blooming, foliage, large flower size, quality of flower, fragrance, breeding for landscaping, dwarf habit, etc. The genetic system of a taxon controls its heredity and variation and one of its important components is the breeding system. For such type of crop like *Amaryllis* the process of selection begins as soon as a wild species is brought under cultivation for domestication. Usually, an elemental taxon is domesticated in environments far removed from its native place. Under such circumstances, geographical migration of species is easier if self-pollination is possible rather than when the species is dioecious or self-incompatible and thus cross-pollinated (Baker 1955). Under domestication, altogether different selection pressures as also different objectives of selection exert their own genetic-evolutionary influence in molding the course of future evolution under the influence of man. One of the important factors that helps in generating variability, including different characteristics of the breeding and mating systems, is through Mendelian variation and recombination followed by selection within the confines of a single taxonomic species.

Hippeastrum comprises of about 60–80 species and many hybrids (Gender 1973, Everett 1980; Rees 1985; Okubo 1993). Hybrids are complex and produced through crosses between many selected species and cultivars (Okubo 1993, Everett 1980). Proper knowledge of flowering physiology (flower initiation and development), reproductive biology (pollen viability, stigma receptivity, pollination processes, pollen-pistil interactions, incompatibility, etc.), and methods for overcoming incompatibility and environmental factors are very important for successful hybridization. The first successful hybrid between *Hippeastrum reginae* and *H. vittatum* was reported in 1799 by Mr. Arthur Johnson and named as *H.* x *johnsonii* (Everett 1980, Huxley et al. 1992). *H. johnsonii* generally acknowledged as the first amaryllis hybrid was a primary hybrid of *H. vittatum* and *H. reginae* (Traub 1934). The initial centers of amaryllis breeding are Holland, South Africa, and Florida (Barnhoorn 1976, Hayward 1934, Goedeet 1961, Ludwig and Co. 1948). At early stage there had no choice and any available materials were included in breeding. Few natural tetraploid species were incorporated in early breeding. Partial fertile triploid hybrids were produced by crossing diploids and tetraploids (Bell 1978). The first pink hybrid was developed through crossing between *A. evansiae* and a white-flowered Dutch hybrid (Nelson and Foret 1968). Hybrids under Reginae strain were produced by Jan de Graff in the Netherlands in the mid-nineteenth century by crossing *Hippeastrum vitatum* and *Hippeastrum striatum* with *Hippeastrum psittacinum* and some promising hybrids available in Europe. Few interesting hybrids were "Graveana" and "Empress of India." Richard Pearce, British explorer developed Leopoldii hybrids crossing *Hippeastrum leopoldii* and *Hippeastrum pardinum* with the best of the Reginae strains and produced a lineage of very large open flowered specimens, with up to 4–6 flowers on each scape. Henry Nehrling and Theodore Mead started breeding in the late nineteenth and early twentieth century in the USA and developed some modern hybrids by crossing their hybrids with Dutch stock. Dutch growers began cultivation and breeding in 1946 in South Africa. At early stage there had been challenges in *Amaryllis* breeding. All present-day *Amaryllis* in Florida are the results of extensive selections (Bell 1978). Different species had different desired characters which were utilized by breeders. For developing true yellow hybrids *Amaryllis evansiae*, *A. aglaiae*, and *A. parodii* were the most suitable. *A. angustifolia* and *A. cybister* were suitable for new floral forms. *A. reticulata* var. stratifolia was identified and utilized in hybridization for developing new foliage with special aesthetic merit, increasing length of flowering season and disease resistance. It was interesting to note that hybrids could grow easily than species (Bell 1978). Breeding started with species – *H. fragrantissium* (*Amaryllis fragrantissima cardenas*), *H. lapasence* (*A. lapacensis cardenas*), *H. cardenasianum* (*A. cardenasiana* Traubt Dorun), *H. papilio* (*A. papilio* Ravenna), *and H. reticulatum* Herbert var. Striatifolium Herbert) (Meerow 1988, Meerow et al. 1990). Sir William Macarthur (1800 to 1882), an Australian botanist, started breeding in Australia during the middle of the nineteenth century using *Amaryllis bellanonna* and brunsvigia and the resultant hybrids were *Brunsvigia multiflora* and *Amaryllis multiflora* which today are known as x Amarygia. Dutch breeders provided brilliant color for the gardeners. Size and quality increased in hybrids through further hybridization between Dutch and Florida strains. A number of large flowered, tetraploid hybrids were developed through more than 200 years breeding history (Traub 1958). Meerow (1988) focused his breeding on *H. papilio*. The main intension was to develop evergreen hybrids with attractive floral form and keeping qualities, but with an increase in floret number and variation in pigmentation. The main intergeneric breeding objectives were to develop cold resistance with diverse flower color and fragrance. The intergeneric hybridization was based on flowering physiology, reproductive biology, and in vitro plant growth techniques (Jirakiattikul 1999). Some other breeding objectives were disease resistance, extension of flowering season, etc. Most cultivars of *Hippeastrum* come from the Dutch and South African sources, and bulbs are now being developed in the USA, Japan, Israel, India, Brazil, and Australia. Double flowers from Japan are very attracting. Amaryllis hybrids are listed as "Dutch," "Israeli," "Peruvian," etc., depending on the country of origin. Most modern commercial hybrids are derived from the following species: *H. vittatum, H. leopoldii, H. pardinum, H. reginae, H. puniceum, H. aulicum.* Hippeastrum breeding program initiated to develop triploid (Meerow 1988, Meerow et al. 1992, Bell 1978). Triploid ($2n = 33$) hybrids ("Bahia"PPAF, "Rio"PPAF, and "Sampa"PPAF) were developed crossing *Hippeastrum papilio* (Rav.) Van Scheepen (diploid, attractive, compact, evergreen foliage; and long-lasting flowers with unusual color range (shades of purple) and zonation) with other commercial tetraploid varieties (Meerow 2000).

Next to hybridization, polyploidy has been an important factor and some of the species involved are tetraploid. Of particular interest are the triploid hybrids which are outstanding and surpass both the diploids and tetraploids in all floral characters. The polyploids (particularly 3x) have shorter and sturdier spikes with larger flowers which are longer lasting (c.f. Narain and Khoahoo 1977). Attempts were made to tetraploidize some of the progeny in order to overcome self-incompatibility. There are some evidence that tetraploidy can be induced with colchicine in plantlets of *Hippeastrum*. Colchicine treatment and embryo culture resulted development of encouraging hybrids (Bell 1978).

CSIR-NBRI, India is maintaining germplasm collections of *Hippeastrum* including NBRI developed hybrids through selective breeding, as well as hybrids developed with unknown parentage, local species and Dutch Hybrids. Only two species, *Hippeastrum belladonna* and *Hippeastrum gracillis*, formerly grew in plains at NBRI. A total of three varieties, two ("Snow White" and "Firefly") under *H. belladonna* and one ("Charm") under *H. gracillis*, which produce small-sized, bell-shaped flowers with a narrow color range, are available on the plains. A large number of Dutch hybrids, both imported and available from different regions of India, that produce giant size blooms in a much wider range of colors were collected. Using these germplasms as baseline materials for a hybridization program, CSIR-NBRI has successfully developed new varieties through selective hybridization and selection of natural hybrids for research purposes. No detailed information is available on the genetic aspects of the genus, and factors underlying its evolution are unclear. A total of 24 varieties were selected, of which 16 were Dutch hybrids, three were subtropical varieties, and five were hybrid varieties developed at NBRI. CSIR-NBRI, Lucknow, India, has been most successful in raising many desirable varieties with attractive and new color combinations. Some of the most promising hybrids are: "Apurb", "Begum Secundra", "Chitwan", "Deepali", "Garima", "Jwala", "Jyoti", "Kiran", "Kiran Rekha", "Miss Deepa Kaul", "Mrs. Percy Lancaster", "Niharika", "Poonam", "Prof. K N Kaul", "Raktamanjari", "Samrat", "Shah Najaf", "Smriti", "Prakash", "Dhruva", "Agni", "H.S. 1", "H.S.2".

Only one intergeneric hybrid, *Hippeaskelia* (*Hippeastrum* x *Sprekelia*), has been reported (Okubo 1993). Numerous studies have examined many barriers to intergeneric hybridization, and a large number of breeding strategies have been used to overcome these barriers. For example, in vitro pollination, ovule culture, ovary culture, and embryo culture have been used to produce interspecific hybrids in *Lilium* (Van Tuyl et al. 1991).

22.9 Cytology

Cytological studies on *Hippeastrum* are solitary. Different scientists studied the chromosomes of *Hippeastrum* as per their objectives. The majority of *Hippeastrum* species are diploid ($2n = 22$) (Arroyo 1982, Flory and Coutthard 1981, Naranjo and Andrada 1975). The chromosome counts of many cultivars of *Hippeastrum* are based on $x = 11$ but certain discrepancy from this number are also on record. The chromosomes of some of the species are based on $n = 8$ (Fernandez 1970), $x = 10$ (Arroyo 1982), and $x = 12$ (Williams and Dudley 1984). Various degree of heteromophy was reported by them in most of chromosomes. Normally the plants had the basic Karyotype of 2 metacentric, 2 submetacentric, 4 subacrocentric, and

3 acrocentric chromosomes. Narain and Khoshoo (1968) proposed a basic Karyotype of 2 metacentric, 5 submetacentric, and 4 acrocentric chromosomes in diploid and tetraploid cultivars. Naranjo and Andrada (1975) reported 4 metacentric, 4 submetacentric, and 3 acrocentric chromosomes in different diploid species. Khaleel et al. (1991) suggested that the most frequent basic karyotype comprised of 4 metacentric, 5 submetacentric, and 2 acrocentric chromosomes in majority of tetraploid cultivars. Such variations among the Karyotypes of allied species indicates prominent role of chromosomal structural changes operating in the evolution (Rees and Jones 1977). Presence of mismatched and heteromorphic chromosomes provides evidence of extensive hydridization in past and hybrid nature of genus. Presence of heteromorphic chromosomes is considered as the clue to the structural hybridity (Sharma and Bal, 1956). High degree of heteromorphy or mismatching have been reported by Narain and Khoshoo (1968) in cv. "Andromeda" contained seven sets with four chromosomes each, two sets with six chromosomes each and two pairs with chromosomes each. Shafiq and Vahidy (1998) studied karyomorphological variations of three cultivars ("Apple Blossom," "Cardinal" and "Minerva") of *H. vittatum* and observed heteromorphy of varying degree in several chromosomes of diploid and tetraploid plants. Khaleel et al. (1991) analyzed katyotype, genome and development of gametophytes of nine hybrids of garden *Amaryllis*. They reported the basic chromosome number x = 11 and all nine hybrids as tetraploid with altered karyotypes. Meiosis showed bivalents, trivalents, and quadrivalents. The fruiting behavior and low percentage of seed set are associated with pollen sterility, and degeneration of embryo sacs. Hang et al. (2015) determined chromosome number and karyotypes of 97 *Hippeastrum* accessions from different provinces of Vietnam. Most accessions showed diploid chromosomes (2n = 2x = 22), three accessions triploid (2n = 3x = 33), 32 accessions showed tetraploids (2n = 4x = 44). Karyotypes of three *Hippeastrum* species (*H.* x "Johnsonii," *H. puniceum*, and *H. reticulatum* var. *striatifolia* Herb.) showed differences in chromosome lengths, long arm to short arm ratios, centromere positions, and chromosomes types. These species carry one to three unequal pair(s) of chromosomes in their chromosome complements. CSIR-NBRI Lucknow, India, made elaborate studies on chromosomes of *Amaryllis*. Fifty cultivars covering five elemental species (*A. vittata, A. belladonna, A. stylosa, A. reticulate*, and *A. species*) were studied. Among the elemental species *A. stylosa, A. reticulate*, and *A. species* are diploid, whereas *A. vittata* was found to be both diploid and tetraploid. *A. belladonna* occurs in three races, namely diploid, triploid, and tetraploid. In cultivars, 30 are diploid, one is triploid and 14 are tetraploid. A cytogenetic survey of 136 cultivars were carried out and both diploid (2n = 22) and tetraploid (2n = 44) chromosome numbers were determined. Aneusomaty was found in root tip mitosis of a tetraploid cultivar of hybrid origin. Cytogenetic survey detected ten major ancestral species which seems to be involved in the origin of garden Amaryllis through hybridization and selection. *A. stylosa* was detected to be an interchange heterozygote. From chromosome survey it was found that *Amaryllis* is monobasic (x = 11), the polyploids ranging from 3x to 7x are all based on this number. Cytology was also used as an aid to understand its evolution (Mookerjea 1955, Khoshoo and Narain 1967,

Narain 1974, 1977a, Narayan 1981, Narain and Khoshoo 1968, Raina 1969). The basic chromosome number of *Hippeastrum* is x = 11. The basikaryotype consists of two metacentric (median), five submetacentric (submedian), and four acrocentric (subterminal) chromosomes. However, most of available hybrids are tetraploid (2n = 44) such as "Apple Blossom," "Lucky Strike," "Red Strike," "Basuto," "Bold Leader," "Cocktail," and "Desert Dawn" and all these hybrids have the basic chromosome number for the genus (Khaleel and Siemsen, 1989; Khaleel et al.. 1991). Also, there are some polyploids with 2n = 33, 66, 77 (Guha, 1979, cited in Khaleel et al., 1991).

22.10 Propagation

Hippeastrum can be propagated by seeds, offset bulblets, twin scaling, and in vitro culture. Due to heterozygosity, seeds are used for the development of new varieties (Okubo 1993). Commercial growers propagate through separation of offsets to maintain the purity of a cultivar or sterile hybrid clones. Seeds should be immediately sown after harvest (Okubo 1993) otherwise seed viability is lost during storage (Carpenter and Ostmark 1988). It is also propagated by offsets or daughter bulbs. Three or more small offsets are produced at the base of the mother bulb and it depends on the cultivar (Okubo 1993).

Twin-scaling is an important propagation method as some cultivars of *Hippeastrum* produce only a few or no offsets (Okubo 1993). For commercial flower production bulbs of circumference greater than 20 cm are planted (De Hertough 1994). Storage of seeds lose their viability. Therefore, seed should be sown immediately after harvest (Roger 1976 c.f. Carpenter and Ostmark 1988). From germination to develop a flowering size bulb takes almost 2 years. Three or more offsets or daughter bulbs (depending upon cultivar) are produced at the base of mother bulb. Twin-scaling is also an important method of propagation.

i. **Seeds:** Seeds are developed in the ripened capsule after pollination and fertilization of ovules which takes 3–4 weeks. The fully ripened capsules are harvested and air-dried for 24–48 hours at room temperature of 40 to 45 °F. Air-tight container is used for storing the seeds. The highest germination is usually obtained when the seeds are planted immediately after harvesting. Seeds are sown in shallow boxes or seed pans containing sterilized soil, covered with a layer of fine leaf mold and watered through a fine nose. Seeds normally germinate in 7–14 days under favorable conditions but it may take more time for the first leaflet to show above the soil surface. The seedlings should be fed regularly with a good complete mixed fertilizer and watered. Seedlings are ready for transplanting in 3–4 weeks. Each seedling is transplanted in 4″ pots at 2 to 3 leaf stage. It is again transplanted to bigger pots of 6 to 8″ when new growth starts. These are transplanted to the field after a fortnight. The seedlings are to be shaded for the first few days in the middle of the day, and watering should be sprinkled every day by a fine nose. It takes 3–5 years to bloom. Seeds do not breed true and are used for the development of new cultivars.

ii. **Vegetative propagation**: Propagation through separation of bulblets is a very slow process for obtaining a large number of plants in a short time. One plant produced 2–3 bulblets in a year of growth. Number of bulblets can be increased using a novel method "notching" (Gupta and Kher 1983a, b, 1987). Mature large sized bulbs are dug out from the bed during January, the roots are cut back to about an inch from the base and also the green leaves are removed by cutting. Bulbs are dipped in potassium permanganate (0.2%) for 5 minutes for disinfections. Several cuts, such as 1, 2, 4, and 8 cross cuts, are given with the help of a sharp knife by holding the bulbs upside down to divide the disc into 2, 4, 8, and 16 notches of equal size. It is possible to continue the cut to 32 by expert professionals. The cuts are extended only up to two-third depth of the bulbs, so as to keep the segments attached at the top. Each cut piece must contain a piece of the basal plate (stem tissue) or it will not root because roots develop from stem tissue, not leaf tissue. Sphagnum moss are used to wrap the notched bulbs and planted immediately in the 10″ earthen pots containing compost and coarse sand (1:2 ratio). Routine light watering is necessary to keep the medium moist during callusing period. Watering is increased gradually during April when new growth started. Liquid manuring along with a mixture of chemical fertilizers is applied at fortnight intervals for better growth of young bulblets from June onwards. The clumps are lifted from the pots during the last week of October and all the bulblets are separated very carefully with the help of a sharp knife resulting on an average 12 to 16 bulblets per bulb. The individual bulblets are planted in 6″ pots containing a mixture of soil:lead mold:sand:F.Y.M. (1:1:1:1). Each bulblet takes 2–3 years to reach standard sized bulb with potential to produce flowers (Gupta and Datta 2006).

iii. **In vitro propagation** Tissue culture techniques have been standardized for quick multiplication from different explants. Seabrook and Cumming (1977) standardized tissue culture technique for rapid propagation of amaryllis (*Hippeastrum* spp. Hybrids) using different explants like leaf bases, scapes, peduncles, inner bulb scales, and ovaries. In vitro propagation of *Hippeastrum* has been successful using twin scales, and floral stem or scape tissue. Mii et al. (1974) reported that the first appearance of a bud occurred 1 month after inoculation of twin scales on media containing 5.0 mg/L NAA and 10.0 mg/L kinetin. Huang et al. (1990) reported production of direct bulblets from twin scales in MS medium supplemented with 1.0 mg/L IAA and 5.0 mg/L.. When scape sections were cultured on Lindsmaier-Skoog medium supplemented with 0.3 mg/L NAA, callus formation, and numerous bulbs occurred 10 weeks after culture (Pajerslci and Ascher 1977). They also reported that adventitious shoots were regenerated from scape sections using MS medium containing 0.5–1.0 mg/L NAA and with or without 0.1 mg/L BAP. Hussey (1980) reported regeneration of adventitious shoots from leaf base and scape tissue on media supplemented with 0.5–0.8 mg/L NAA. Significant regeneration response occurred on media supplemented with 0.5–0.8 mg/L NAA and appreciable more shoots were regenerated from explants taken from the upper half of the stems (Alderson and Rice, 1986). Sultana et al. (2010) standardized in vitro bulb production of *Hippeastrum*. Tested MS medium supplemented with different hormone concentrations of BAP (0.0, 2.0, 4.0, 6.0, and 8.0 mg/L) and CCC (0.0, 125, 250, and 500 mg/L) and sucrose levels (30, 60, 80, 90, and 110 g/L) and reported sucrose level at 90 g/L produced the maximum average weight as well as the highest regeneration percentage. The increasing rate of CCC increased the number and average weight of bulb. The maximum bulb formation observed in media supplement with 6.0 mg/L BAP and 500 mg/L CCC fortified with 90 g/L sucrose. Chen et al. (2017) noted the highest bulblet proliferation rate (11.6 bulblets) from per 2.5 cm circumference in "Red Lion" cultivar under in vitro by culturing quarter-bulb explants on the MS media fortified with 0.1 mg L^{-1} thidiazuron (TDZ) and 0.2 mg L^{-1} α-naphthalene acetic acid (NAA) for 8 weeks.

22.11 Cultivation

i. **Climate and soil:** *Hippeastrum* can be grown under wide environmental conditions ranging from tropical to subtropical or temperate climate. It prefers rich sandy loamy porous soil with proper drainage and plenty of moisture. *Hippeastrum* cultivation is done better in well-drained sandy soil with organic manure and pH between 6 and 7 (Traub, 1958, Everett 1980, Okubo 1993). It grows well under shade of trees provided direct light falls on them for a portion of the day. There is poor flower formation in total shade. Plants grown in full sun will flower better but the foliage will discolor.
ii. **Preparation of beds**: Soil should be thoroughly dug or ploughed up to a depth of 40–45 cm and exposed to sun for 15 days to kill insect, pests, and pathogens. There should be proper drainage facilities. Well-rotted cow dung manure or compost should be mixed @ 20 kg/m^2 along with 300 g ammonium sulfate, 250 g superphosphate, and 50 g muriate of potash for better blooms and bulblets. For pot culture minimum 25 cm pot should be used for single bulb. Sand, farmyard manure, leaf mold, and soil in equal proportions with a tablespoonful of bone meal should be used as potting mixture.
iii. **Planting:** Planting time in the plains is during December–January and in hills February–March. Well-matured bulbs (5.1 to 5.5 cm diameter) should be planted at 8 to 10 cm deep and 25 cm apart and water should be added immediately. Routine cultural practices like weeding, hoeing, and irrigation are performed. The flower stalks are staked to keep them erect and stakes are removed once flowering is over.
iv. **Feeding**: For normal growth and development of plants proper nutrition is very important. It is wise to broadcast the manure (nitrogen, phosphorous, and potassium – 200 kg N, 400 kg P, and 200 kg K per hectare, Bhattacharjee et al. 1982) and plough it into the soil through spading. Manure should always

be fine and well-rotted for heavy clay or gravelly soil. Chemical fertilizers are used with manure. Full dose of phosphorus and potash and half dose of nitrogen are applied during final land preparation before planting and remaining half dose of nitrogen after 40 days of planting. For pot culture, liquid manure along with a mixture of chemical fertilizers containing 10 g ammonium sulfate and 30 g muriate of potash in every 5 liters of liquid are applied during bud opening at fortnightly intervals.

v. **Watering:** Optimum moisture level of soil should be maintained. In dry weather irrigation is done at an interval of 15–20 days. Watering should be done during vegetative growth and it is withheld after flowering till new growth starts.

vi. **Weeding and hoeing**: These are very important steps for preparation of beds and maintenance. Hand weeding is possible for small size plantation area. For commercial plantation, pre-emergence application of weedone (2,4,5-T) at 10 liters per hectare proved very effective in controlling the weeds.

vii. **Growth and development:** Effect of different day/night temperature (low, middle, and high) on the growth and development of the vegetative and reproductive parts have been studied and found that high temperature favored promotion of mother bulb enlargement, but the moderate and low temperatures were optimum for bulb and flower production (Ijiro and Ogata 1997). Post-harvest management of cut spikes is very important to increase the shelf life of cut flowers and to reduce their loss. The scapes are cut when the buds are fully elongated. The stalks are placed in water after harvesting in the early morning. Cut flowers may be stored in refrigerated showcases or storerooms to minimize the spoilage. Flower scapes are packed in proper cartons with adequate protection so that cut flowers reach the customer in good condition. Scapes packed in moist cloth remain fresh for long periods. The life of cut flowers reduces due to quick scape rotting. Chemical manipulations can increase the vase life of cut flowers. Three chemicals in different concentrations and combinations have been tested to increase the vase life – 8-hydroxyquinoline citrate (250, 500, 1000 ppm), citric acid (500, 1000, 1500 ppm), and sodium chloride (1.0, 1.5 and 2.0%). 1% sodium chloride with 3% sucrose is found to be most effective both for enhancing the vase life of cut flowers and preventing scape rotting.

Bulbs are stored when the flowering is over. The old flower along with faded stalks is cut just near the base of the ground and bulbs are allowed to mature. Abundant watering is required during active growth. Fully matured bulbs are removed from the soil during middle of December. The roots from the bulbs are cut back to about 1 inch in length from the base of the bulb's neck. The bulbs are kept in shade for few hours to dry and shade off soils. The dried and clean bulbs are treated with 0.2% "Captan" solution for few minutes to prevent fungal infection. After drying of bulbs, bulbs were further kept in perforated iron trays in cool dry place at room temperature for storage till middle of July. After storage they are planted in the soil at fortnightly intervals to obtain a succession of bloom from June to August (Gupta and Kher 1983a, b).

viii. **Flowering**: Depending upon the climate and situations *Hippeastrum belladonna* and Dutch hybrid cultivars of *Hippeastrum hybridum* start flowering during March–April in plains. *H. stylosum*, *H. cybister*, and Sprekelia hybrid flower in May while *H. reticulatum* cv. Mrs. Garfield flowers in July–August. One or two flower scape appear from the ground and produce 2 to 4 flowers on each scape.

22.12 Diseases and Pests

Several diseases and pests cause extensive damage to the plants and bulbs of *Hippeastrum* varieties.

i. **Fungal Diseases**: ***Red Leaf Spot*** – It is very common in India and caused by *Stagonospora curtisii*. Symptom develops in the form of spots on the leaves and flower stalks at early stage and slowly fungal spores enlarge forming cankers with red boarders at brown or gray center. Copper fungicide (Blitox 0.2%) spray or hot water treatment of bulb at 43.5 °C for 2 hours can manage the disease. Some other fungal diseases are "soft rot" of bulbs caused by *Rhizopus stolonifer*, "bulb decay" caused by *Pyhium debaryanum*, "black mildew" caused by *Asterinella hippeastri*, "leaf spot" caused by *Aecidium amaryllidisi*, and anthracnose caused by *Septogloeum amaryllis*. Copper fungicide (Blitox 0.2%) spray can control the disease.
ii. **Viral Diseases**: ***Cucumber Mosaic Virus*** (CMV) – Leaf surface develops large yellowish green patches at random. Malathion (0.2%) spray at 15 days interval can control the vector.

Tomato Spotted Wilt – Numerous yellow spots develops on the leaves of affected plants which ultimately die. The disease may spread by thrips which can be controlled by spraying 0.2% Metasystox or Malathion.

iii. **Pests: *Caterpillar*** – Black caterpillar (*Polytela* glorisae) and grasshoppers attack plants which eat soft leaves and buds within 4–5 days. Either 0.1% Monocil solution can be sprayed at fortnightly intervals for management of disease or it can be removed by hand picking (Percy Lancaster and Percy Lancaster 1958). The main pest problems of amaryllis are mosaic virus and "red blotch" which is caused by the fungus *Stagonospora curtisii*. It is very destructive and spreads rapidly. Captan, Bordeaux, and Benlate mixtures are recommended.
iv. **Thrips** – Thrips suck the juice from leaves, flower stalks, and flowers in both greenhouse and field grown plants. Weekly application of chlorodane dust can manage the thrips.
v. **Aphids, Mealy Bugs, and Scale insects** – Aphids may spread virus diseases. The insects puncture the tissue and suck the sap of the surface. About 15 g of nicotine sulfate and 25 g of soapchips in 1 gallon of water or 0.2% of Malathion spray at 10-days interval are used to control.

vi. **Mites** – Spider mites (*Tetraychus bimaculatus*), bulb scale mites (*Tarsonemas solani*), and the bulb mites (*Rhizoglyphus echinopsus* and *Rhizoglyphus solani*) damage tissues by sucking the sap from beneath the surface during warm, dry periods. Spider mites and bulb scale mites can be controlled by spraying Metasystox 30 E.C. @ 0.2% at weekly interval and dipping the bulbs in hot water at 110 °F for 60 minutes, respectively.

Nematodes, pocket gophers, rabbits, and squirrels can damage Amaryllis bulbs.

22.13 Manipulation of Flowering Period

The blooming period of *Amaryllis* is very short, that is, about 6 weeks only during March–April due to genetic phenomenon. CSIR-NBRI Lucknow, India, standardized methods for manipulation of flowering period for commercial purposes. **Early flowering** can be induced by raising the temperature using incandescent bulb during night. Flowering can be obtained as early as in December end by this method. Indoor potted plants are covered with black polythene sheet and an electric bulb is put inside the cover to raise the temperature during night for about a month before desired date of flowering. **Delayed flowering** is also possible by adopting some cultural practices. Bulbs are taken out from the ground during mid-December and green leaves and roots are removed and stored at room temperature for about 210 days till mid-July. These bulbs may even be planted in soil and started watering from May onwards at fortnight intervals. Flowering starts within 20 days and flowers may be available from June to August (Gupta and Kher 1983a, b).

Naturally, *Hippeastrum* flowers in the spring, from the middle of March until the end of April. World market demand for *Hippeastrum* flowers at the end of December at during the Christmas Holidays. Bose et al. (1980) have shown that there is no effect of day length on the flowering of the *Hippeastrum*. Earlier studies reported that the induction for flowering of the *Hippeastrum* is autonomous and primarily connected with the size of the bulb (Bose et al. 1980, Rees 1985). Ephrath et al. (2001) found that the exact flowering time can be manipulated by applying an appropriate thermal regime to large sized bulbs. Temperature had a strong effect on bulb and leaf development. They studied the effect of ambient temperature on the growth rate of bulbs, the effect of soil temperature on the growth rate of bulbs, and the susceptibility of several bulbs' sizes to various thermal regimes. A high linear temperature-dependent correlation was found between leaf area and bulb size. The distribution of the fresh matter among the roots, leaves, and bulbs was also found to be temperature dependent.

22.14 Characterization of Germplasm by DNA Markers

Characterization is most important technique for correct identification of plants. It helps to understand the genetic diversity, phylogenetic relationship among the genotypes, taxonomical status, registration, plant variety protection, farmer's right etc. Correct identification of new hybrids is extremely important to protect plant breeders' rights for commercial exploitation. A number of classical and advanced methods comprising different parameters of cytology, morphology, physiology, phenolic compounds, DNA markers, etc. are utilized. Use of molecular markers in addition to the classical methods provides more positive identification of new varieties. Random amplified polymorphic DNAs (RAPDs) analysis has been used to characterize genotypes of known and unknown origin and to measure genetic relationships among the hybrids of *Hippeastrum*.

Chakrabarty et al. (2007) examined the varietal identification and assessment of genetic relationships in 24 varieties of *Hippeastrum* at National Botanical Research Institute, Lucknow, India, by using RAPD markers. Out of 24, 16 were introduced Dutch Hybrids, three subtropical varieties, and five hybrid varieties developed at NBRI. The details of all the varieties are "Dhruva" (seedling selection, red, throat green yellow), "Prakash" (seedling selection, red markings on white back ground), "Firefly" (2n = 22, basal species, reddish pink, maroon eye), "Charm" (4n = 44, basal species, carmine red), "Pink Pearl" (3n = 33, Dutch Hybrid, pink, "Agni" (hybrid "Charm" x "Pink Pearl," red), "HS-1" (selfing- "Carmine Dove," carmine red), "Carmine Dove" (3n = 33, Dutch Hybrid, carmine), "Snow White" (4n = 44, basal species, white), "Canary Bird" (3n = 33, Dutch Hybrid, white with maroon stripes all over the petals), "HS-2" (hybrid "Snow White" x "Canary Bird," red stripe on the white base), "Mckean's Hybrid-1" (3n = 33, Dutch Hybrid, red, double form), "Dutch Belle" (3n = 33, Dutch Hybrid, gorgeous rose), "Country Girl" (3n = 33, Dutch Hybrid, orange red), "Bleeding" (3n = 33, Dutch Hybrid, red lines on white back ground), "Sperkelia Hybrid" (3n = 33, Dutch Hybrid, dark red), "Silver Lining" (4n = 44, Dutch Hybrid, orange red), "Happy Memory" (3n = 33, Dutch Hybrid, red and white), "Carmine pink" (3n = 33, Dutch Hybrid, carmine red), "Scarletto" (3n = 33, Dutch Hybrid, sparkling bright dark red), "Salmon Purple" (3n = 33, Dutch Hybrid, purple red), "Scarlet Beauty" (3n = 33, Dutch Hybrid, scarlet red), "Salmon Beauty" (4n = 44, Dutch Hybrid, ground color salmon red without marking) and "Mckean's Hybrid-2" (3n = 33, Dutch Hybrid, red stripes on the pink base of the petal) (Fig. 1). Primarily 20 primers were screened out of 7 primers produced bands in the ranged between 5 and 13 DNA bands per primer in 24 cultivars of *Hippeastrum*. A total of 59 bands were produced out of which 48 were polymorphic bands and generated 81.35% polymorphism. The similarity coefficients ranged between 0.2 and 1.0. The constructed dendrogram showed that 7 primers discriminated among all varieties into two major clusters. The first cluster contained 11 varieties and the second cluster comprised 13 varieties. The overall results of the cluster analysis fit with the available pedigree data of the species. Hybrids with common parents (one or more) clustered together indicating their relationship among the genotypes. A clear genetic association between *Hippeastrum* cultivars and hybrids with parents was found. They further suggested that RAPD profiles of *Hippeastrum* hybrids and/or varieties where the pedigree is not known can now be used as a tool to study ancestral relationships.

Genetic diversity of 25 accessions collected from 17 provinces in the northern, central, and southern regions of Vietnam was studied through RAPD analysis.

Fig. 1 Representative varieties of Amaryllis. **Basal species**: Charm, Fire Fly, Snow White; **Dutch hybrids**: McKean's Hybrid 1, McKean's Hybrid 2, Country Girl, Dutch belly, Bleeding heart, Pink pearl, Carmine Pink, Happy memory, Carmine Dove, Scarletto, Canary Bird, Silver lining; **NBRI Developed varieties**: Prakash, Dhruva, Agni, HS1, HS2

Twenty-five primers produced a total of 230 distinct bands, of which 167 were polymorphic (about 72.6%). Extensive variation in flower color, shape, and size was detected among accessions of the hybrid group (Phuong et al. 2014). Genetic diversity and population structure of 104 amaryllis accessions were screened using molecular markers (SSRS and SNPS). About 269 unique genes were determined that might regulate the flower's path development. Results indicated that all accessions can be grouped into three main clusters, which can be further divided into two subgroups. Results enable large-scale transcriptomics and classification of *Hippeastrum* genetic polymorphisms and will be useful in the future for resource conservation and production (Wang et al. 2018). Xiong et al. (2020) examined the genetic diversity in single-flowered amaryllis (*Hippeastrum hybridum*) germplasm using SRAP markers. Twenty-three screened sequence related amplified polymorphism (SRAP) produced 244 clearly amplified fragments, out of which 242 were polymorphic and 2 were monomorphic which generated (99.2%) polymorphism. Genetic distance and structure-based analyses mainly clustered all accessions into two or five subgroups, indicating the complexity of the hybrid pedigree of single-flowered amaryllis. To effectively explore the novel variation in these 82 genetic resources, an initial core collection with 15 accessions with the goal of maximizing the SRAP alleles. This preliminary core set retained 100% of the SRAP variation with greater genetic diversity and heterogeneity.

References

Alderson PG, Rice RD (1986) Propagation of bulbs from floral stem tissues. In: Withers LA, Alderson PG (eds) Plant tissue culture and its agricultural applications. University of Cambridge, Cambridge, pp 91–97

Arroyo S (1982) The chromosomes of *Hippeastrum*. *Amaryllis* and *Phycella* (Amaryllidaceae). Kew Bull 37:211–216

Bailey Hortorium Staff (1976) Hortus third. Macmillan, New York

Baker, J.G. 1878. Trimen, Henry (ed.). "An enumeration and classification of the species of *Hippeastrum*". J Bot British Foreign 16*:* 79–85

Baker HG (1955) Self-compatibility and establishment after "long-distance" dispersal. Evolution 9: 347–349

Barnhoorn F (1976) Breeding the 'Hadeco' amaryllis hybrids. Plant Life 32:59–63

Bell WD (1978) New potentials in *Amaryllis* breeding. Fla State Hortic Soc 86:462–466

Bhattacharjee SK, Yadav LP, Mukhopadhyay TP (1982) Effect of nitrogen, phosphorus and potash fertilization on Hippeastrum. Indian Agric 26:193–197

Bose TK, Jana BK, Mukhopadhyay TP (1980) Effects of growth regulators on growth and flowering in *Hippeastrum hybridum* hort. Sci Hortic 12(2):195–200. https://doi.org/10.1016/0304-4238(80)90127-2

Carpenter WJ, Ostmark ER (1988) Sensitivity of seed germination of *Amaryllis* to light and temperature. HortScience 23:1002–1004

Chakrabarty D, Gupta VN, Datta SK (2007) Varietal identification and assessment of genetic relationships in *Hippeastrum* using RAPD markers. Plant Biotechnol Rep 1:211–217

Chen UC, Din I, Wu ZY, Tsao JY, Hsia CN (2017) Effect of surface disinfection, explant type, and plant growth regulators on establishment of *in vitro* bulblet proliferation of Amaryllis. J Taiwan Agric Res 66(4):286–297

Dandy JE, Fosberg FR (1954) The type of *Amaryllis belladonna* L. Taxon 3:231–232

Dyer RA (1976) The genera of southern African flowering plants Vol. 2. Gymnosperm and Monocotyledons, vol 1. Department of Agricultural Technical Services, Pretoria, pp 35–66

Ephrath E, Ben-Asher J, Alekparov C, Silberbush M, Wolf S, Day ANE (2001) The effect of temperature on the development of *Hippeastrum*: a phytotron study. Biotronics 30:51–62

De Hertogh AA (1994) Home forcing of potted *Amaryllis (Hippeastrum).* Leaflet No. 8529. North Carolina Cooperative Extension Service

Everett TH (1980) The New York botanical garden illustrated encyclopedia of horticulture, vol 5. Garland Publishing, New York

Fernandez PE (1970) Cytotaxonomy of two species of the genus Hippeastrum (Amarillidaceae). Cytologia 35:431–433

Flory WS, Coutthard RF Jr (1981) New chromosome counts, number and types in green *Amaryllis*. Plant Life 37:43–56

Gender R (1973) Bulbs: A Complete Handbook of Bulbs, Corms and Tubers. Richard Clay Ltd, Bungay, Suffolk

Goedeet RD (1961) Hadeco *Amaryllis* hybrids grown in South Africa. Plant Life 17:85–86

Goldblatt P (1984) Proposal to conserve 1176 *Amaryllis* and typification of *A. belladonna* (Amaryllidaceae). Taxon 33:511–516

Guha S (1979) Cytological studies in the genus *Amaryllis*. Nelumbo Bull Bot Survey India 21:18–21

Gupta VN, Datta SK (2006) Amaryllis. NBRI Bulletin No.16, pp 1–26. Economic Botany Information Service, National Botanical Research Institute, Lucknow

Gupta VN, Kher MA (1983a) A note on the influence of different dates of digging and differential period of storage at room temperature on flowering in *Hippeastrum*. The Lal-Baugh 28 (3&4):74–76
Gupta VN, Kher MA (1983b) Rapid multiplication of *Hippeastrum* bulbs by notching. Indian J Hortic 40(1–2):98–101
Gupta VN, Kher MA (1987) Improved techniques for growing *Amaryllis*. Indian Hortic 33(1):12–24
Hang TTM, Phuong PTM, Dung TTM, Miyajima I (2015) Preliminary chromosome studies on *Hippeastrum* species from Vietnam. J Fac Agric, Kyushu Univ 60(1):51–55
Hayward W (1934) The Mead strain of the Nehrling *Amaryllis*. Yrbk Amer Soc 1:62–63
Huang CW, Okubo H, Uemoto S (1990) Comparison of bulblet formation twin scales and single scales in *Hippeastrum hybridum* cultured *in vitro*. Sci Hortic 42:151–160
Hunziker AT, Cocucci AE (1959) Estudioss obreA maryllidaceae. I. Una nueva es-pecie de Hippeastrum del centro de Argentina: H. parodii nov. sp. Bol Acad Nac Cienc 41:5–16
Hussey G (1980) Propagation of some members of the Liliaceae, Irisdaceae and Amaryllidaceae by tissue culture. In: Birkell CD, Cutler DF, Gregory M (eds) Petaloid monocotyledons. Academic, London, pp 33–42
Huxley A, Griffiths M, Levy M (1992) The new Royal Horticultural Society: dictionary of gardening, vol 2. The Macmillan Press, London
Ijiro Y, Ogata R (1997) Effect of ambient temperature on the growth and development of Amaryllis (Hippeastrum hybridum hort.) bulbs. J Japan Soc Hortic Sci 66(3.4):575–579
Jirakiattikul Y (1999) Study of intergeneric hybridization in *Hippeastrum*. Submitted in fulfilment of the requirements for the degree of doctor of philosophy. University of Tasmania, Agricultural Science, Hobart
Khaleel TF, Siemsin D (1989) Cytoembryology of *Amaryllis* hybrids. Can J Bot 67:839–847
Khaleel TF, Haven S, Gilg T (1991) Karyomorphology of *Amaryllis* hybrids. Cytologia 56:31–41
Khoshoo TN (1971) Hortorium taxonomy. Indian J Genet 31:305–315
Khoshoo TN, Narain P (1967) Aneusomaty in an *Amaryllis* hybrid. Indian J Genet Plant Breed 27: 322–333
Ludwig & Co. (1948) The breeding hybrid Amaryllis. Herbertia 15:69
McCann JJ (1937) New double hybrid *Amaryllis*. Herbertia 4:185–186
McCann JJ (1950) McCann double *Amaryllis*. Plant Life 6:107–108
Meerow AH (1988) New trends in *Amaryllis* (*Hippeastrum*) breeding. Proc Fla State Hortic Sci 101:285–288
Meerow AW (2000) 'Rio'PPAF, 'Sampa'PPAF, and 'Bahia'PPAF, three new triploid Amaryllis cultivars. HortScience 35(1):147–149
Meerow AW, Broschat TK, Kane ME (1992) Breeding of new *Hippeastrum* cultivars using diploid species. Acta Hortic 325:583–589
Meerow AW, Scheepen JV, Dutilh JHA (1997) Transfers from *Amaryllis* to *Hippeastrum* (Amaryllidaceae). Taxon 46:15–19
Meerow AW, Kane ME, Broschat TK (1990) Breeding of new Hippeastrum cultivars using diploid species. I. The F-1 evaluation. Proc Fla State Hortic Sci 103:168–170
Mii M, Mori T, Iwase N (1974) Organ formation from the excised bulb scales of *Hippeastrum hybridum in vitro*. J Hortic Sci 49:241–244
Moore HE (1963) *Amaryllis* and *Hippeastrum*. Baileya 11:15–16
Mookerjea A (1955) Cytology of *Amaryllids* as an aid to the understanding of evolution. Caryologia 7: 1–71
Narain P (1974) Hybridization and polyploidy in relation to *Amaryllis* species and cultivars. Plant Life:94–96
Narain P (1977a) Cytogenetics of garden Amaryllis. Plant Life 1:38–63
Narain P (1977b) Cytogenetics of garden Amaryllis I. Systematic position and taxonomic treatment of Indian cultivar. Plant Life 33:38–41
Narain P (1977c) Cytegenetics of garden Amaryllis II. Chromosome variation. Plant Life 33:48–64

Narain P (1977d) Cytogenetics of garden Amaryllis. III. Chromosome variation. Plant Life 33:43–64
Narain P, Khoshoo TN (1968) Cytogenetic survey of *Amaryllis* cultivars. J Cytol Genet 3:41–45
Narain P, Khoshoo TN (1977) Origin and evolution of garden *Amaryllis*. Indian J Hortic 34(1): 80–85
Naranjo CA, Andrada AB (1975) El cariotipa fundamental en el genero *Hippeastrum* Herb (Amaryllidaceae). Darwin 19:566–582
Narayan P (1981) Polllination mechanism and hybridization in *Amaryllis*. Plant Life 37:114–118
Nelson B, Foret JA (eds) (1968) Breeding records of amaryllis hybrids produced by Ira S. Nelson. 1956–1965. Bull Louisiana Soc Hortic Res 3:129–145
Novelties DT, Buneker HM, Bastian RE (2017) Taxonomic novelties in south Brazilian Amaryllidaceae – 1: *Hippeastrum ramboi* a new species from Rio Grande Do Sul and Lectotypification of H. breviflorum herb. Balduinia 60:1–10, 15- X-2017. https://doi.org/10.5902/2358198029456
Okubo H (1993) Hippeastrum. In: A. A. De Hertogh and M. Le Nard (eds.). The Physiology of Flower Bulbs. Elsevier Science Publisher, Amsterdam, pp. 321–334
Pajerslci L, Ascher PD (1977) Propagation of *Amaryllis (Hippeastrum)* from cultured scaped tissue. HortScience 12:25
Percy Lancaster S, Percy Lancaster A (1958) *Amaryllis*. Bull. No. 9. National Botanic Gardens, Lucknow
Phuong PTM, Isshiki S, Miyajima I (2014) Genetic variation of *Hippeastrum* accessions in Vietnam. J Fac Agric Kyushu Univ 59(2):235–241
Raina SN (1969) Cytogenetics of some amaryllids. Ph. D. Thesis, Agra University, Agra
Ravenna P (1981) Contributions to south American Amaryllidaceae VII.P I. Life 37:57–83
Rees AR (1985) Hippeastrum. In: Halevy AH (ed) CRC handbook of flowering. CRC Press, Florida, pp 294–296
Rees H, Jones N (1977) Genetics, principle and perspectives: a series of texts, vol 1. University Park Press, Baltimore, p 61
Robert B, Jordheim M, Kiremire B, Namukobe J, Andersen ØM (2006) Anthocyanins from flowers of Hippeastrum cultivars. Sci. Hortic. 109:262–266
Roger AL (1976) Storage of Amaryllis seed. Plant Life 32:95–96
Seabrook JA, Cumming BG (1977) The *in vitro* propagation of *Aamaryllis* (*Hippeastrum* spp. hybrids). In Vitro 13(12):831–836
Sealy JG (1939) *Amaryllis* and *Hippeastrum*. Bull Misc Inform Kew 1939:49–68
Shafiq FA, Vahidy AA (1998) Chromosomal heteromorphy in the karyotypes of three local cultivars of *Hippeastrum vittatum* (Amaryllidaceae). Pak J Biol Sci 1(1):5–10
Sharma AK, Bal AK (1956) A cytological study of a few genera of Amaryllidaceae with a view to find out the basis of their phylogeny. Cytologia 21:329–352
Sultana J, Sultana N, Siddique MNA, Islam AKMA, Hossain MM, Hossain T (2010) In vitro bulb production in *Hippeastrum* (*Hippeastrum hybridum*). J Cent Eur Agric 11(4):469–474
Sweet R (1839) Hortus Brittanicus or catalogur of plants etc, 3rd edn, London
Tjaden W (1981) *Amaryllis belladonna* Linn. – an up-to-date summary. PI. Life 37:21–26
Traub HP (1934) A preliminary Amaryllis (Hippeastrum) checklist. Year book of the Amer. Amaryllis Soc 1:45–51
Traub HP (1958) The *Amaryllis* Mannual. Mac Millian and Co., New York
Traub HP (1983) The lectotypification of Amaryllis belladonna L. (1753). Taxon 32:253–267
Traub HP, Moldenke HN (1949) Amaryllidaceae: tribe Amarylleae, Stanford
Uphoff JCT (1939) Critical review of Sealy's "*Amaryllis* and *Hippeastrum*". Herbertia 6:163–166
Van Tuyl JM, Van Dien MP, Van Creij MGM, Van Kleinwee TCM, Franken J (1991) Application of *in vitro* pollination, ovary culture, ovule culture and embryo rescue for overcoming incongruity barrier in interspecific *Lilium* crosses. Plant Sci 74:115–126

Wang Y, Chen D, He X, Shen J, Xiong M, Wang X, Zhou D, Wei Z (2018) Revealing the complex genetic structure of cultivated amaryllis (*Hippeastrum hybridum*) using transcriptome-derived microsatellite markers. Sci Rep 8(1):10645. https://doi.org/10.1038/s41598-018-28809-9

Xiong M, Wang Y, Chen D, Wang X, Zhou D, Wei Z (2020) Assessment of genetic diversity and identification of core germplasm in single-flowered amaryllis (Hippeastrum hybridum) using SRAP markers. Biotechnol Biotechnol Equip 34(1):966–974

Williams M, Dudley TR (1984) Chromosome count for *Hippeastrum iguazuanum*. Taxon 33:271–275

Willis JC (1973) A dictionary of the flowering plants and ferns, 8th edn. (Revised by Airy Shaw, HK), Cambridge University Press, Cambridge

Heliconias: Dramatic Flowers of the Tropics and Subtropics 23

Moumita Malakar and Sukanta Biswas

Contents

M. Malakar (✉)
Department of Horticulture & Floriculture, Central University of Tamil Nadu, Thiruvarur, Tamil Nadu, India

S. Biswas
Department of Horticulture, Institute of Agricultural Science, University of Calcutta, Kolkata, West Bengal, India

S. K. Datta, Y. C. Gupta (eds.), *Floriculture and Ornamental Plants*, Handbooks of Crop Diversity: Conservation and Use of Plant Genetic Resources,
https://doi.org/10.1007/978-981-15-3518-5_26

Abstract

Heliconias have long been popular horticulturally because of their showy inflorescences. Owing to their exotic shape and unusual posture, early explorers of the tropics introduced them to Europe with several species that became prized greenhouse specimens. They originally were classified as species of bananas because of their similar foliage. In 1771, Linnaeus established the new genus *Heliconia*, naming it after *Helicon*, a mountain in Greece, the home of Apollo and the muses. In tropical America, *Heliconias* are often called "wild bananas" or "tropical kings"; locally, they are often referred to as "false birds of paradise." They are monophyletic genus comprises of about 380 species with high ornamental potential. Some controversy exists over the taxonomic status of this crop. The identification of genotypes mainly based on morphological characteristics generated synonyms, but it's possible to differentiate genotypes and elucidate the origin of hybrids and triploids of *Heliconia* by means of chromosome counting, base-specific fluorochrome staining such as CMA (chromomycin A3) and DAPI (4′,6-diamidino-2-phenylindole), and FISH (fluorescence in situ hybridization). On the other hand, self-incompatibility (SI) is prominent in some populations of Central American *Heliconia*. The availability of long-distance pollinators (hummingbirds) and low daily flower output may promote outcrossing despite the scarcity of physiological self-incompatibility in these plants. Conclusively, based on variability and diversity study, few genotypes, viz., *H. psittacorum* × *H. spathocircinata* cv. 'Golden Torch,' *H. psittacorum* var. 'Choconiana,' *H. psittacorum* var. 'Lady Di,' and *H. rostrata*, found to be promising under Indian condition and may be considered as potential parents in future breeding program.

Keywords

Heliconia · Taxonomy · Origin · Self-incompatibility (SI) · Diversity · Breeding

23.1 Introduction

Flowers are an integral part of human life due to their diversity in beauty, form, texture, color, and fragrance (Urooj-Ul-Nissa et al. 2015). Production and consumption of tropical flowers are increasing in many countries around the world. Tropical flowers are perceived by many flower consumers as exotic and unusual, with a potential market in temperate countries. In this backdrop, *Heliconia* is gaining popularity in the world as well as Indian market as an emerging flower crop. *Heliconia* is derived from the Greek word *helikonios*. Interestingly, it refers to Mount Helicon in Greece, home to the muses, goddesses of the arts, and sciences in Greek mythology. The muses were said to be eternally young and beautiful; thus, the name *Heliconia* refers to the flowers' long-lasting and attractive qualities. Predominant constraints for successful thrive of *Heliconia*, the trendiest cut flower of the tropics, is the shortage of apposite genotypes under Bengal circumstance. *Heliconias* are tropical plants related to bananas, cannas, and gingers and have more than 100 species, mostly native to the New World tropics where they are found in open clearings. This plant belongs to Heliconiaceae family and is commonly known as “lobster’s claw,” “parrot flower,” “parrot plantain,” “false plantain,” etc. (Janakiram and Kumar 2011). This plant is also known as “The King of Tropicals.” It is well adapted to major agro-climatic conditions and can be grown up to a height of 3000–4000 m. *Heliconia* with about 250–300 species disseminated primarily in neotropical areas from the North of Mexico to the South of Brazil (Dahlgren et al. 1985). Only a small paleotropical group, with approximately six species, is endemic to the Pacific Islands (Silva et al. 2016; Maglianesi et al. 2015; Berry and Kress 1991; Andersson 1998). In Brazil, there are about 40 species distributed in two main areas, the Amazon Basin and the Atlantic Forest, which correspond to the primary areas of the distribution of the genus in the country (Berry and Kress 1991). Most commonly grown landscape *Heliconia* species are *Heliconia angusta*, *H. bihai*, *H. brasiliensis*, *H. caribaea*, *H. latispatha*, and *H. pendula*. The continuous increase in the production and commercialization of *Heliconia* as cut flower and ornamental plant in many countries around the world shows the importance of *Heliconia* cultivation to agribusiness. The *Heliconia* plants are herbaceous, perennial, and rhizomatous. The erect shoots are pseudostems formed by overlapping sheathing leaf blades. The vegetative growth is vigorous, yielding many shoots and forming large clumps. The inflorescence formation happens at the upper part of the pseudostem. It can be either upright or pendent, with distich or spiral bracts in several spacing arrangements, exposed or even hidden among the leaves in the clump. The bract’s color can be variations of green, yellow, orange, red, pink, and purple, which characterizes *Heliconia* as an exotic cut flower (Berry and Kress 1991). They flourish well in loamy soils rich in humus. They need sunlight, with temperatures not lower than 30 °C. Furthermore, rhizomes that might or might not have young shoots, should be planted and old shoots, if there are any, should be cut back to about 6 inches before planting, sometimes mandatory for its successful thrive. *Heliconia* cut flowers and potted plants are used for interior decoration. This can also be used in landscaping. Roots and seeds of some varieties are useful for medicinal purposes. It may be propagated through division of rhizomes or from seeds, although dividing rhizome is by far the most common method. An obvious problem in quest of new hybrids is the low rate of hybridization. Several causes may be liable to this difficulty, among which key is the low rates of cross-pollination by hummingbirds or other natural agents (Temeles et al. 2000; Maglianesi et al. 2015). Assuming that the natural pollination is the “bottleneck” in hybridization, then the obvious solution would be the “artificial pollination” by the growers.

There is a clear need to study the existing germplasm of wild and cultivated *Heliconia* species especially in India to better contribute to their conservation while advancing selection and multiplication programs to improve the genetic quality of planted materials. Several DNA markers, viz., amplified fragment length polymorphism (AFLP), random amplified polymorphic DNA (RAPD), and inter-simple sequence repeats (ISSRs), have been widely used in genetic variation studies and are becoming increasingly popular in low-level systematization (Bussell et al. 2005). Consequently, that may serve as input for future breeding programs and compose an approximation to the analysis of the phylogenetic relationships between *Heliconia* species, hybrids, and cultivars (Plate 1).

23.1.1 Traditional Uses

Heliconias are grown for the florist's trade and as landscape plants. These plants do not grow well in cold, dry conditions. They are very drought-intolerant but can endure some soil flooding. *Heliconias* need an abundance of water, sunlight, and soils that are rich in humus in order to grow well. These flowers are grown in tropical regions all over the world as ornamental plants. The flower of *H. psittacorum* (parrot heliconia) is especially distinctive; it has greenish-yellow flowers with black spots and red bracts reminiscent of the bright plumage of parrots.

As a cut flower: Brilliant color, exotic form, long straight/drooping peduncles, and outstanding postharvest durability eventually tagged them as "specialty cut flower."

In the landscape: *Heliconia psittacorum* × *H. psittacorum* cv. 'Golden Torch,' *H. psittacorum* var. 'Lady Di,' *H. stricta*, and *H. angusta* as potted plants and for interior landscape can be employed. Its exotic appearance and brilliant colors fetch premium price in the market. Leaves of some varieties of *Heliconia* are also sold as cut leaves for flower decoration. On an average, 600–700 plants of *Heliconia* can be planted in 1 hectare of land. The average price may range from Rs.5 to 20 but depending on its variety, it may vary significantly. Hence, a minimum income of around 25,000/ha to maximum of 90,000/ha can be realized depending upon the variety of *Heliconia* plants (source: Extension Folder No. 71, 2014, ICAR- Research Complex GOA, Links: https://www.ccari.res.in/Extension%20Folder%20No.%2071.pdf) (Figs. 1 and 2).

On the other hand, *H. bihai* is widely utilized in subtropical to tropical landscapes throughout the world as accents or focal points in the landscape; good shade tolerance in some species renders them suitable for understory use; smaller stature species work well in containers as well as in the ground; intermediate and taller species make good background or screening plants, particularly for areas with less than perfectly drained soils; their outdoor utilization is restricted to protected areas along the Gulf Coast and deep South Texas; *Heliconia* spp. are staples in conservatories and are sometimes included in larger-scale interiorscapes; several species are commonly grown for the cut flower trade.

Heliconia rostrata Ruiz & Pav. (Heliconiaceae) is a comparatively rare plant found in the Rema-Kalenga Wildlife Sanctuary in Sylhet Division in the northeast of Bangladesh and Bagerhat District in the southwest part of Bangladesh. In English, it is known as "lobster claw plant." The flower of this plant is considered as the national flower of Bolivia. To our knowledge, only one previous ethnomedicinal

H. psittacorum x H. spathocircinata cv. 'Golden Torch'

H. psiattacorum var. 'Lady Di'

H. psittaocorum var. 'Choconiana'

H. wagneriana

H. stricta var. 'Dwarf Jamaican Red'

H. stricta

H. rostrata

H. humilis

H. metallica

H. indica var. 'Indica'

Plate 1 Photoplates of commonly available *Heliconia* species and cultivars

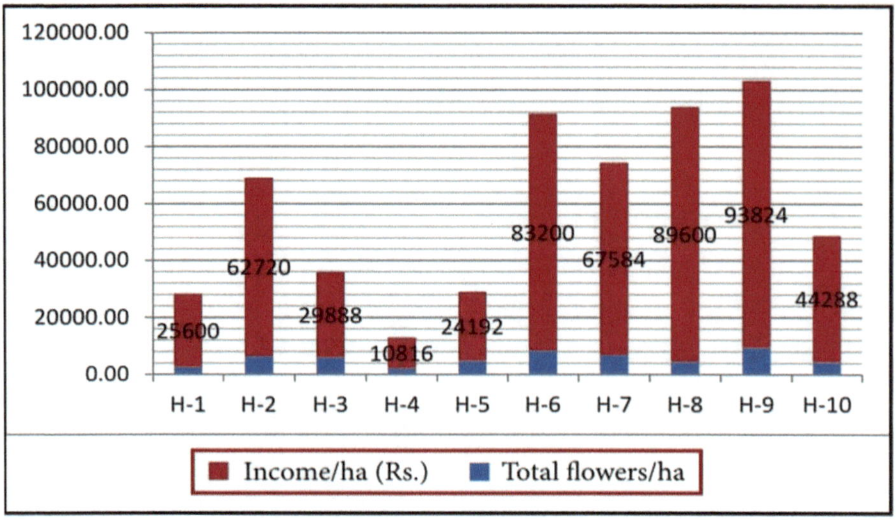

Fig. 1 Graph showing flower yield and total income from different *Heliconia* types

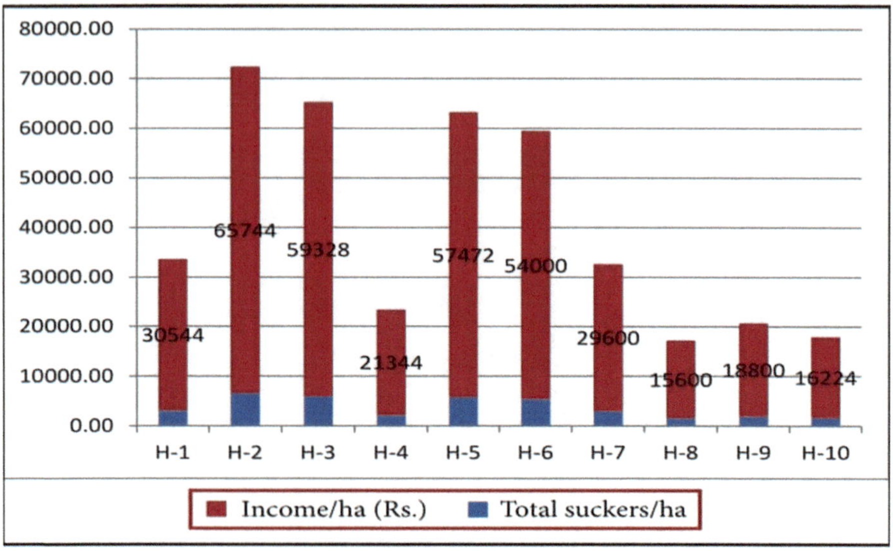

Fig. 2 Graph showing sucker yield and total income from different *Heliconia* types

use of this plant has been reported from Bangladesh. Folk medicinal practitioners in two villages by the Rupsha River in Bagerhat District of the country use leaves and seeds of the plant as a tonic and to treat headache, sprains, and pain (Mollik et al. 2010). Diabetes mellitus, characterized by high blood glucose levels, is becoming a serious problem worldwide because it cannot be cured with existing medicines and, moreover, with time, leads to further serious complications like disorders of the heart, kidney, and eyes. Kidney disorders due to diabetes (diabetic nephropathy) can lead to edema or swellings in various parts of the body (Hommel et al. 1990). Although oral medications and insulin injections are available to lower blood glucose, such medications are costly or cumbersome (like daily injections of insulin) and, as a result, not much of use among the common rural people of Bangladesh. As a result, there is a constant search going on for new antidiabetic drugs. Since plants have always proved to be good sources for lead compounds and new medicines, many antidiabetic research works have centered on plants (Kayarohanam and Kavimani 2015). The use of *Heliconia rostrata* Ruiz & Pav. by a folk medicinal practitioner (FMP) practicing at Chunarughat, Habiganj District, Sylhet Division, Bangladesh, to treat diabetes and diabetes-induced swellings on legs was quite popular (Shahriar et al. 2017). Diabetes was diagnosed by him on the basis of several symptoms in combination like fatigue, frequent urge to eat, frequent thirsts and urination, excessive sweating, and occasional swellings in different parts of the body. Also, on occasions, the FMP relied on diagnosis in a modern diagnostic clinic that the patient has diabetes. The FMP did not know how to distinguish between the various types of diabetes like diabetes type 1, 2 and gestational diabetes also. The FMP's mode of treatment consisted of collecting leaves of *Heliconia rostrata* Ruiz & Pav. and crushing the leaves between two pieces of stone to obtain juice. The juice was then filtered through a piece of clean cotton cloth. About 120 ml of the juice was advised by the FMP to be orally taken with 1 g of kali jeera seeds (*Nigella sativa* L.; family, Ranunculaceae) every evening for 3 consecutive months. *H. rostrata* Ruiz & Pav. specimen (as shown by the FMP) was photographed and identified at the Bangladesh National Herbarium. The FMP mentioned that he collected leaves of *H. rostrata* Ruiz & Pav. from Rema-Kalenga Wildlife Sanctuary. *H. rostrata* Ruiz & Pav. has not been studied thus far for its antidiabetic potential or active ingredients and so merits scientific research toward discovery of possible novel antidiabetic compound(s) (Shahriar et al. 2017).

23.1.2 Evaluation of Genotypes for Multifaceted Uses

In creation of lush, tropical settings, landscape designers are lured by the long slender foliage of *Heliconias*, gingers, and their relatives, but their showy inflorescences are an added bonus. Defining space, creating boundaries, screening for privacy, controlling erosion, and hiding foundations are appropriate functional uses for these plants, while aesthetic choices are made based on color, scale, texture, and attractiveness singly, grouped, or clumped. Selection criteria include both plant characteristics (size, floral display, foliage, growth habit) and site properties such as exposure in sun, salt, wind, soil (pH and drainage), and temperature, but its successful establishment whether outdoor, i.e., in ground, or indoor, i.e., in containers or in interiorscaping, demands knowledge of growth habits and environmental requirements. Maintenance considerations include grooming and pest-weed control, while problems include chlorosis, failure to flower and excessive growth.

Pinheiro et al. (2011) reported *Heliconia* × *nickeriensis* for ornamental characteristics and management for garden use. *H.* × *nickeriensis* is described as a short-height plant, with quick development, dispersed type of clump architecture, and open growth habit. The clump area reached 5.14 m^2 at 18 months after planting (MAP), demonstrating the necessity of a large space for development. The shoot emission in the internal part of the clump permitted complete soil coverage. The dark green leaves contrast with the yellow-orange inflorescences which are easily visualized above its foliage. The flowering period started at 9 MAP reaching an emission of more than 17 inflorescences per clump at 17 MAP. The inflorescences kept the quality for more than 25 days after its emission in the clump. This genotype represents a good option to be used either isolated for covering large open areas or grouped with other ornamentals to create compositions with different colors, forms, and textures. He again carried out work to characterize *Heliconia* genotypes for landscape use. Qualitative and quantitative characteristics related to clump and inflorescence aspects were analyzed in *H. psittacorum* × *H. spathocircinata* cv. 'Golden Torch Adrian' and 'Golden Torch,' *H. psittacorum* var. 'Suriname Sassy' and 'Red Opal,' *H. collinsiana*, *H. rostrata*, *H. forereoi*, and *H. bihai*. Except for 'Red Opal,' *H. psittacorum* cultivars and their interspecific hybrids were considered to be short plants (less than 1.51 m height) and adequate for flower bed as they would not interfere in the open views of gardens. The species *H. bihai*, *H. collinsiana*, *H. rostrata*, and *H. forereoi* formed inflorescences which last longer than 78 days in good conditions in the clump, a remarkable aspect in the landscape. *H. bihai* inflorescences were difficult to visualize, and its bracts accumulated water, attracted insects, and had repulsive odor. The evaluated genotypes have traits that cause acceptation and adequacy to landscape design, allowing landscape designers to diversify plant indications (Table 1).

Srinivas et al. (2012) studied 18 genotypes of *Heliconia* and observed wide variation for vegetative and floral characteristics. Varieties 'Lady Di,' 'Sassy,' 'Iris Banochi,' 'Kawauchi,' and 'Macas Pink' were observed to be ideal for cut flowers and recommended for commercial cultivation. Hybrid cultivar 'Golden Torch' was observed to flower profusely and exhibited prolonged blooming and hence ideal for the landscape (Table 2). Evaluation of 26 genotypes, i.e., *H. bihai*, *H. bihai* var. 'Kamehameha,' *H. bihai* var. 'Nappi Yellow,' *H. caribaea* × *H. bihai* cv. 'Carib Flame,' *H. collinsiana*, *H. episcopalism*, *H. latispatha* var. 'Distans,' *H. latispatha* var. 'Red-yellow Gyro,' *H. orthotricha* var. 'She,' *H. pendula*, *H. pseudoaemygdiana*, *H. psittacorum* × *H. spathocircinata* cv. 'Alan Carle,' *H. psittacorum* × *H. spathocircinata* cv. 'Golden Torch,' *H. psittacorum* × *H. spathocircinata* cv. 'Golden Torch Adrian,' *H. psittacorum* var. 'Red Opal,' *H. psittacorum* var. 'Red Gold,' *H. psittacorum* var. 'Strawberries' & 'Cream,' *H. psittacorum* var. 'Suriname Sassy,' *H. rauliniana*, *H. rostrata* (3 days/10 days), *H. stricta*, *H. stricta* var. 'Fire Bird,' *H.* × *nickeriensis*, and *H. wagneriana*, were made for main characteristics, viz., plant height, presence of wax and hairs, flowering period, precocity, clump area, growth habit, color of inflorescence, and leaves. All 26 genotypes evaluated proved that they can be used in landscape design according to the project if their particular characteristics are taken in consideration.

Table 1 Classification of *Heliconias* of Brazil

Subg. *Taeniostrobus Kuntze* (Griggs)
H. episcopalis Vell.
Subg. *Heliconia*
Sect. *Heliconia*
H. adeliana Emygdio & Santos
H. bihai (L.) L.
H. stricta Huber
Sect. *Tortex* Andersson
H. pabstii Emygdio & Santos
H. spathocircinata Aristeg.
Sect. Novo I *H. lourteigiae* Emygdio & Santos
Sect. Novo 11
H. farinosa Raddi
H. kautzkiana Emygdio & Santos
H. rivularis Emygdio & Santos
H. sampaioana Emygdio
H. velloziana Emygdio
Subg. *Griggsia* Andersson
H. chartacea lane ex Barreiros
H. juruana Loes.
H. marie-augustae Emygdio & Santos
H. pendula Wawra
Subg. *Stellochlamys* Baker
Sect. Stenochlamys (Baker) Schumann
H. acumillata L. C. Rich.
H. angusta Vell.
H. auriculata Barreiros
H. aurorea Emygdio & Santos
H. citrina Emygdio & Santos
H. fluminensis Emygdio & Santos
H. lacletteana Emygdio & Santos
H. laneana Barreiros
H. psittacorum L.f.
H. richardiana Miq.
H. timothei L. Anderss.
Sect. *Lanea* Andersson
H. aemygdiana Burle Marx
H. pseudoaemygdiana Emygdio & Santos
Sect. *Proxichlamys* Andersson
H. densiflora Verlot
Sect. *Lasia* Andersson
H. lasiorachis L. Anderss.
H. velutina L. Anderss.

(continued)

Table 1 (continued)

Sect. *Cannastrum* Andersson
H. metallica PI. & Lind. ex. Hook
H. subulata R. & P.
Sect. Zingiberastrum Andersson
H. apparicioi Barreiros
H. hirsuta L.f.
H. schumanniana Loes.

After Andersson (1981a, b)

In Hawaii, sales record from several commercial growers were collected over several years to identify the peak seasons of flowering as well as the duration of the flowering seasons by Criley (2000). Such information is helpful for the flower markets. The natural flowering season for *Heliconia* species in their natural habitats is sometimes indicated in the taxonomic literature, but it may be influenced locally by rainfall and drought periods as well as by photoperiod and may not be reliable in indicating production periods elsewhere. Wild collected plants are often introduced with little or no information on their season of flowering. With more than three-dozen species of *Heliconia* grown and exported in international trade, the seasonality of flowering is important to the supply and marketing of this bold tropical flower. *Heliconia* species of commercial interest with strong seasonal flowering periods are noted (*H. angusta*, *H. bihai*, *H. caribaea*, *H. caribaea* × *bihai*, *H. collinsiana*, *H. lingulata*, *H. rostrata*, *H. stricta*, and *H. wagneriana*) as well as species with longer flowering periods such as *H. psittacorum* cultivars and hybrids.

Pinheiro et al. (2011) conducted an experiment with one of the objectives of introducing the most competitive tropical floral cultivars for diversification of floral plant assortment. *Heliconia* flower compositions can be used alone or in combination with other exotic species as well as with few delicate and elegant flowers (*Anthurium*, *Cymbidium*, *Protea*, *Trachelium*, *Zantedeschia*, *Zingiber*, etc.) or can be an excellent choice for container plants that can be grown indoors for the winter and moved outdoors for the spring and summer.

Cut flower production of *Heliconia* is becoming an important agribusiness in Brazil. Various investigations on the selection of genotypes to support the production of inflorescences in quality and quantity were made strategically for expanding industry. The number of days from inflorescence emergence to harvesting (DIH), fresh weight of the stem (FWS), number of leaves in the stem at inflorescence emission (NLI), diameter stem 20 cm under the inflorescence (DI), stem length (SL), and inflorescence length (IL) were some important traits to be considered for genotype selection. Eighteen genotypes from the *Heliconia* Germplasm Collection of Federal Rural University of Pernambuco State (*H. psittacorum* L. f. × *H. spathocircinata* Aristeguieta cv. 'Golden Torch Adrian'; H. psittacorum L. f. × *H. spathocircinata* Aristeguieta cv. 'Alan Carle'; H. psittacorum L.f. var. 'Strawberries'; *H. psittacorum* var. 'Cream'; H. psittacorum L.f. var. 'Suriname Sassy; H. psittacorum var. 'Red Opal'; *H. pseudoaemygdiana* L. Em. & Em;

Table 2 Floral characteristics of selected varieties of *Heliconia* sp.

Genotypes	Bract	Rachis	Perianth	Ovary	Peduncle
Heliconia humilis	Vivid reddish orange bracts with dark green lip along the kneel in shape of lobster claws, upright flower stalk	Pale red with green tinge	Whitish green almost completely concealed in spathes	Yellow, proximally green	Light green mottled with red
Heliconia stricta var. 'Dwarf Jamaican Red'	Pale red bracts with light green lip along the kneel, boat-shaped bract, upright flower stalk	Pale green mottled with light red	Greenish yellow	Light yellow	Reddish green
Heliconia psittacorum × *H. spathocircinata* cv. 'Golden Torch'	Orange-yellow colored upright bract with light red patch at the base of cheeks, basal bract with green kneel	Golden, often with small red areas at the base	Golden with faint green tip	Golden on distal 1/3 and top yellow below	Yellow with green tint
Heliconia psittacorum var. 'Choconiana'	Strong orange-yellow with pale red toward the cheek	Pale red	Light orange-yellow and dark green spots at the tip	Yellow, proximally orange	Light yellow or cream colored
Heliconia psittacorum var. 'Lady Di'	Cherry red bracts with pink tinge	Deep reddish pink colored	Faded yellow with dark green spots at the tip	Faded yellow	Faded green in color
Heliconia rostrata	Hanging inflorescence of inverted, recurved, rose-red, yellow and green tipped, velvety texture	Red colored, scalloped	Yellow to whitish yellow	Orange-yellow	Red
Heliconia wagneriana	Bracts are somewhat variable, bright red areas over most of the cheek, and are surrounded by pale green along the lip of the	Light red in color	Dark green on distal 1/3 with light green tip, yellow on proximal part	Light green	Light green distally and cream proximally

(continued)

Table 2 (continued)

Genotypes	Bract	Rachis	Perianth	Ovary	Peduncle
	kneel and tip with light yellow				
Heliconia stricta	Large stout inflorescence with solid red color	Deep red	Orange-yellow	Orange	Red
Heliconia metallica	Small faded green colored, boat-shaped, not attractive	Light green	Rosy pink in color with white tip	Rosy pink in color with white tip	Light green
Heliconia indica	It is a foliage variety, reared for its exquisite colored foliages. The leaves are tinged with red color, mottling with moss green color. The plant is bigger with longer leaves				

H. psittacorum var. 'Red Gold'; *H.* × *nickeriensis* Maas & de Rooij; *H. latispatha* Benth (orange); *H. latispatha* Bentham cv. 'Yellow Gyro'; *H. rauliniana*; *H. latispatha* Benth cv. 'Distans'; *H. rostrata* R. & P. (10 days); *H. rostrata* R. & P. (3 days); *H. wagneriana* Peters; *H. bihai* (L.) L. cv. 'Kamehameha'; *H. psittacorum* × *H. spathocircinata* Aristeguieta cv. 'Golden Torch'; *H. bihai*) were evaluated. The inflorescences were harvested when they presented between two and four open bracts. Significant (5%) differences were observed for all the traits among genotypes. The DIH ranged from 14.85 to 29.53 days, and the plants have 4.69 to 6.29 leaves. This information is important to help the growers on planning the harvesting program according to the species cultivated. The FWS varied from 0.36 kg to 0.4 kg and the DI 27.34 to 5.18 mm. The SL ranged from 56.58 to 125.47 cm and the IL from 15.6 to 32.87 cm. It is important to notice that large and heavy inflorescences are not suitable for transportation. These results give indications of genotype selections for further studies and commercial use.

23.1.3 Common Names

Heliconia is commonly known as "lobster's claw," "parrot flower," "parrot plantain," "false plantain," "toucan beak," "The King of Tropicals," and "false bird of paradise."

23.1.4 Habitat and Distribution of *Heliconia*

Heliconias are native primarily in the American tropics from the Tropic of Cancer in Central Mexico to the Tropic of Capricorn in South America, including the Caribbean. A curious disjunction of group of six species of *Heliconia* which occurred thousands of miles from most other species is found in the Old World tropics (Berry and Kress 1991). The center of diversity of the genus is found along the northern Andes (Colombia and Ecuador) extending into southern Central America [Panama and Costa Rica]. Most species inhabit moist or wet regions, but some are found in seasonally dry areas. Although *Heliconia*s attain their most luxuriant vegetative growth in the humid lowland tropics at elevations below 500 m, the greatest numbers of species (many locally endemic) are found in middle-elevation (800–1500 m) rain and cloud forest habitats. Few species occur above 2000 m. The most conspicuous members of the genus inhabit open sites in secondary growth along roadsides, on river banks, and in forest light gaps. With increased destruction by man of the tropical rain forest, these species readily invade and colonize the newly opened areas. Other species never attain such extensive vegetative growth and are restricted in more shaded habitats of the primary forest. These latter species are often found to be locally endemic and are fast becoming extinct as destruction of the tropical forest accelerates.

23.1.5 Diversity and Distribution of *Heliconia* in Brazil

Heliconia bihai, a widespread neotropical species, was published by Linnaeus in 1771 and is therefore the earliest name for a species occurring in Brazil. However, the earliest name for an endemic Brazilian *Heliconia*, *H. angusta*, was published by Vellozo in 1825 in Flora Fluminensis (Palma and Vieira 2006). Since that time, many new taxa have been described, and many new names have been published for Brazilian *Heliconia*s. Brazilian botanists Dr. Luiz Emygdio de Mello Filho, Dr. Emilia Santos, and Dr. Humberto de Sóuza Barreiros had been instrumental in furthering our understanding of the native taxa. Sixty-five species names have been applied to the *Heliconia*s that occur in Brazil. Of these 65 names, 28 are generally recognized synonyms. Of the remaining 37 species of Brazilian *Heliconia*, some controversy exists over the taxonomic status of at least 8. For example, Andersson (1998) had synonymized with *H. angusta* five species described from the Atlantic coastal forests. Estimates of the number of *Heliconia* species in Brazil therefore range between 29 and 37.

Two primary areas of distribution of species of *Heliconia* exist in Brazil: the Amazon basin (21 species) and the Atlantic coastal forests (20 species). The species occurring in each of these general regions can be classified as endemic to one of these regions, disjunct between one of these regions and some other region (e.g., the Guianas, the Planalto), and are widespread (found in more than one region of South America). With respect to the species found in the Amazon basin, eight (38%) are endemic to the Amazonian region in the broad sense. Only two of these species (*H. adeliana* and *H. auriculata*) are endemic to Brazil (and both are taxonomically controversial). Eight species (38%) are widespread throughout South America, four (19%) are found in the Amazon region and the Guianas, and one is distributed in the southern Amazon region and parts of the Planalto.

23.1.6 Geographic Distribution of Heliconia Which Occur in Brazil

Amazonian region	Atlantic coastal region
***H. acuminata* + *H. aemygdiana*	
**H. adeliana* **H. angusta*	
+*H. aemygdiana* **H. aurorea*	
**H. apparicioi* **H. citrina*	
**H. auriculata* + *H. episcopalis*	
+*H. bihai* **H. farinosa*	
***H. chartacea* **H. fluminensis*	
***H. densiflora* **H. kautzkiana*	
+*H. episcopalis* **H. lacletteana*	
+*H. hirsuta*	**H. laneana*
**H. juruana* **H. marie-augustae*	
**H. lasiorachis*	**H. pabstii*
***H. lourteigiae*	***H. pendula*
+*H. metallica*	**H. pseudoaemygdiana*
+*H. psittacorum*	+*H. psittacorum*
**H. schumanniana*	***H. richardiana*
+*H. spathocircinata*	**H. rivularis*
+*H. stricta*	**H. sampaioana*
++*H. subulata*	+*H. spathocircinata*
**H. timothei*	**H. velloziana*
**H. velutina*	
21 species	20 species
Total: 41 species	

= endemic species
+ = widespread
++ = and Planalto
** = and Guianas

Of the 20 species found in the Atlantic coastal rain forests of Brazil, 14 (70%) are endemic to the region. Even if one accepts Andersson's many synonyms for *Heliconia angusta*, the region is exceptionally high in endemics. This high degree of endemism for *Heliconia*s in the coastal forests is similar to the patterns of distribution of forest trees reported by Andersson (1998). Of the remaining species of *Heliconia* in this region, four (20%) are widespread, and two (10%) are also found in the Guianas.

There is at least one species of Brazilian *Heliconia* in each taxonomic category in Andersson's classification of the genus. This distribution of species suggests that all of the major lines of evolution within the genus have radiated into Brazil. The broad taxonomic representation is also true if just the Amazonian species are considered. However, in the Atlantic coastal forests, only one-half of the taxonomic categories are represented.

In conclusion, Brazil is not exceptionally species rich for the genus *Heliconia*. Species that are endemic occur primarily in the Atlantic coastal rain forests and represent principally two taxonomic groups in the genus. It is doubtful that many new species are still to be found in Brazil relative to the great number of new taxa being discovered in the Andean regions of Ecuador, Colombia, and Peru.

However, for the Brazilian *Heliconias*, further systematic and evolutionary studies are certainly warranted. For example, investigations of reproductive biology and plant-pollinator relationships have not yet been thoroughly studied in Brazilian species. Moreover, studies on genetic diversity using isozyme electrophoresis and nucleic acid variation, especially in the Atlantic coastal species (e.g., the *H. farinosa* complex and *H. angusta* complex), should prove to be particularly interesting and informative for an understanding of species boundaries and the processes of speciation in these plants.

23.1.7 Cultivation of *Heliconia*

Heliconias are usually propagated by division of the roots or underground stems and planted 2 feet apart in rich, deep, porous soil. New upright stalks arise from the tips of lateral underground stems. Growth will be rapid if the watering gets limited until the aboveground shoots are developed. When shoots are 8 to 10 feet high, then abundant watering and heavy fertilizing become mandatory. Plenty of sunlight is recommended in general. The soil should be prepared by plowing followed by mixing one part of sand, one part of dry cow-dung manure, and three parts of vermicompost. After attaining the height of approximately 30 cm, transplanting should be made in the main field (ideal plot size $5 \times 2\ m^2$) at a distance of $1\ m \times 1\ m$ by the end of July to early of August. They perform best in semi-shaded condition (Malakar et al. 2015). Uniform cultural operations should be practiced. To harvest flower stalks, remove the whole stem by cutting it at the soil level to induce the production of new stems. After 4 or 5 years, the roots and underground stems become thick. Division and replanting will increase the yield of flowering stems. Rarely is injury from insects and diseases a serious problem. Flowering is most common in late spring and summer months.

Predominant constraints for successful thrive of *Heliconia* the trendiest cut flower of tropics are the shortage of apposite genotypes and growth medium under Bengal circumstance. In this backdrop, a field experiment was conducted by Malakar et al. (2019b) being employed *Heliconia psittacorum* × *spathocircinata* cv. 'Golden torch,' *Heliconia psittacorum* var. 'Choconiana,' *Heliconia psittacorum* var. 'Lady di,' *Heliconia rostrata*, *Heliconia humilis*, *Heliconia stricta* var. 'Dwarf Jamaican Red,' *Heliconia wagneriana*, *Heliconia stricta*, and *Heliconia metallica* in Agricultural Experimental Farm of the University of Calcutta at Baruipur, West Bengal, during 2013–2014 and 2014–2015 using six varied growth media, viz., top soil (control), river sand, top soil + river sand (1:1), top soil + dry cow-dung manure (1:1), top soil + vermicompost (1:1), and river sand + vermicompost (1:1). The finest outcomes for all parameters considered are obtained by top soil + vermicompost (1:1) and top soil + river sand + vermicompost (1:1) irrespective of species after 4 weeks of planting. Plant height increased manifold times in all species and varieties, but *H. stricta* attained the highest (395.02 cm), while significant plant spread of 674.22, 354.79, and 903.73 cm^2 are perceived in 'Golden torch,' 'Choconiana,' and *H. rostrata*, respectively. The latter two *Heliconia* plants yielded also maximum number of shoot $clump^{-1}$ (13.01 and 14.12), whereas leaf blade length of 102.72 cm is found in *H. stricta*. Noteworthy stem length is noticed in *H. wagneriana*. On the other hand, magnificent inflorescences – the most striking part of *Heliconia*, are produced by both *H. rostrata* and *H. stricta* (77.11 and 60.95 cm). On the contrary, utmost number of inflorescences/plant/year is produced by hybrid cultivar 'Golden torch' (102.23). River sand solely failed to prove its competency, while the interaction of top soil + cow-dung manure (1:1) bestowed lowest vegetative traits.

The use of shading screens of different colors can change the spectral quality of radiation and, as a consequence, the growth and production of crop plants. Silva et al. (2017) evaluated various aspects of the growth, yield, and quality of floral stems of *H. psittacorum* × *H. spathocircinata* cv. 'Golden Torch' grown under different light conditions. The treatments consisted of four conditions: blue photoconversion screens (35–40%), red photoconversion screens (18–21%), black shading screens (45–49%), and full sunlight. The experiment they conducted is divided into 2 parts. In the first part, growth was assessed, while in the second, aspects related to the production and quality of floral stems were examined. The experimental design was in randomized blocks with split plots in time. Each plot contained 6 blocks, with 11 replications per block, giving a total of 66 pots per plot. The evaluations were performed at 30-day intervals over a 6-month period. In the analysis of growth, red screens contributed to the growth of the plants, increasing the number of shoots and leaves and also plant height. The productivity and quality of floral stems were, however, highest in plants grown under blue and black screens. Silva et al. (2017) concluded that under the prevailing experimental conditions, the use of blue screens (35–40%) and black screens (45–49%) is suitable for production of the floral stems of 'Golden Torch' *Heliconia*.

Loges et al. (2012) studied *H. pogonantha* for ornamental characteristics and management for garden and cut flower use. The study carried out at Federal Rural University of Pernambuco (UFRPE) using the *Heliconia* germplasm collection. Characteristics that were identified as desirable for use as an ornamental plant were the growth habit of the clump and shoot emergence in the internal part of the clump. The plants present dark green leaves and dark red pendent inflorescence production mainly from May to November. There were 6 inflorescences per clump at 41 MAP (May 2010) and 194 cumulative shoots per clump at 54 MAP (June 2011). To assess the inflorescence for cut flower use, the flower stems were harvested with three or four open bracts, and the postharvest durability was 13 days. It looks 396 days from shoot to inflorescence emergence and 19 days from inflorescence emergence to harvesting. Its stem diameter was 37 mm and stem height 1.44 m, while its inflorescence length and width were 0.47 and 0.25 m, respectively. Its stem fresh weight was 0.46 kg with 0.80 m height. This species represents a good option as ornamental plant due to the pendant inflorescence with dark red bracts. Nevertheless, as a cut flower, the stem is too heavy especially for transport. Furthermore, during postharvest, the red bracts changed to black very quickly decreasing the visual quality of the inflorescence.

23.1.8 Agro-morphological Traits of *Heliconia*

The wide variation in vegetative growth, size, shape, and arrangement of bracts has been reported by different authors. Selection and introduction of ideal genotypes for supporting the production of cut flowers with quality are strategic for expanding the floral industry. Very few or no systematic work has been reported in this country on *Heliconia* with respect to evaluation and genetic amelioration. Also, its commercial potential has not been exploited. In this backdrop, Malakar et al. (2015) evaluated different *Heliconia* genotypes for vegetative and flowering traits for their suitability as cut flower or as landscape plant. According to them, maximum height was recorded in *Heliconia stricta* with 10.1 ft followed by *Heliconia wagneriana* with 8.93 ft. In the case of *Heliconia stricta* var. 'Dwarf Jamaican Red,' the lowest height was observed, i.e., 1.48 ft. *Heliconia indica* var. 'Red' exhibited maximum plant spreading with 63.41 ft^2 followed by *Heliconia wagneriana* with 34.36 ft^2. In *Heliconia metallica*, 30. 11 ft^2 plant spreading was noted which was on par with *Heliconia rostrata* (28.49 ft^2) and *Heliconia stricta* (28.19 ft^2). Minimum plant spread was observed in *Heliconia stricta* var. 'Dwarf Jamaican Red' (4.56 ft^2). Significant variation was also noted in leaf blade length among the genotypes. Leaf blade length ranged from 121.72 cm. and 27.80 cm. in *Heliconia indica* var. 'Red' and *Heliconia psittacorum* var. 'Lady di,' respectively. *Heliconia wagneriana* (106.56 cm.) and *Heliconia stricta* (101.36 cm.) were also at par with *Heliconia indica* var. 'Red' (Malakar et al. 2015). Maximum number of shoots per clump (10.33) was recorded in *Heliconia psittacorum* var. 'Choconiana' followed by *Heliconia stricta* var. 'Dwarf Jamaican Red' and *Heliconia rostrata* (10 each), *Heliconia humilis* (6.66), *Heliconia wagneriana*, *Heliconia metallica*, and *Heliconia psittacorum* × *H. spathocircinata* var. 'Golden torch' (6), whereas, *Heliconia psittacorum* var. 'Lady di' and *Heliconia stricta* recorded minimum number of shoots per clump (5). Genotype *Heliconia stricta* produced maximum number of leaves per stem (7) followed by *Heliconia indica* var. 'Red,' *Heliconia psittacorum* var. 'Lady di' (6), and *Heliconia humilis*, *Heliconia psittacorum* × *H. spathocircinata* var. 'Golden torch,' *Heliconia psittacorum* var. 'Choconiana,' and *Heliconia rostrata* (5) while minimum was recorded in *Heliconia wagneriana* (3). Maximum stem length was observed in *Heliconia wagneriana* (152.42 cm) which was similar to *Heliconia rostrata* (150.41 cm) and *Heliconia stricta* (147.49 cm). Shortest stem length was found in *Heliconia stricta* var. 'Dwarf Jamaican Red' with 20.39 cm. Variation was observed in number of flowering stems per clump. Highest number of flowering stems per clump was noted in *Heliconia metallica* (6.66) followed by *Heliconia stricta* (5.33), *Heliconia psittacorum* var. 'Lady di' (5), *Heliconia humilis* (4.66), and *Heliconia psittacorum* var. 'Choconiana' (4.66). Besides this, in *Heliconia rostrata*, the least number of flowering stems/clump was recorded (3). In *Heliconia indica* var. 'Red,' flowering stem (0) was absent; however, it is valued for its beautiful foliage (Malakar et al. 2015). This variability may be associated with adaptability to the climatic conditions. Significant variation was observed by Malakar et al. (2015) among genotypes for flowering traits such as inflorescence length, number of open bracts, flowers per bract, and number of bracts. Inflorescence length (with peduncle) was observed to be maximum in *Heliconia rostrata* (75.12 cm), categorized under very long length. Genotypes *Heliconia psittacorum* var. 'Golden torch' (30.12 cm), *Heliconia psittacorum* var. 'Choconiana' (48.34 cm), *Heliconia psittacorum* var. 'Lady di' (30.44 cm), *Heliconia wagneriana* (30.46 cm), and *Heliconia metallica* (40.56 cm) were grouped under long length inflorescence. Medium inflorescence length was recorded in genotypes *Heliconia stricta* var. 'Dwarf Jamaican Red' (16.59 cm) and *Heliconia humilis* (27.30 cm). Inflorescence of *Heliconia psittacorum* was having maximum length of 18.56 cm. A large number of open bracts was observed in *Heliconia stricta* (13) and *Heliconia rostrata* (9). In *Heliconia wagneriana* and *Heliconia psittacorum* var. 'Golden torch,' a number of bracts were 7.66 and 6, respectively, while the remaining genotypes recorded lesser number of open bracts (Malakar et al. 2015). Inflorescence with lesser number of open bracts at harvesting stage is preferred for their longer durability and ease in handling and packing. Lowest number of flowers per bract was observed for the genotypes *Heliconia humilis*, *Heliconia psittacorum* var. 'Golden torch,' and *Heliconia rostrata* (5), and highest was observed for the genotype *Heliconia stricta* (14.66). In case of *Heliconia wagneriana*, the number of opened bracts was quite high (12). Similarly, in *Heliconia psittacorum* var. 'Choconiana,' *Heliconia psittacorum* var. 'Lady di,' and *Heliconia metallica*, the number of flowers per bract were 8, 6.33, and 6, respectively. The number of bracts ranged from 3 to 7.66 in *Heliconia psittacorum* var. 'Lady di' and *Heliconia wagneriana*, respectively, and the highest was observed in genotype *Heliconia rostrata* (14.33) which was more or less at par with *Heliconia stricta* (12.66). The size of bract was highest in the genotype *Heliconia stricta* (62.26 cm^2) which significantly differed from others. In the case of *Heliconia humilis* (50.67 cm^2) and *Heliconia wagneriana* (45.41 cm^2), bract sizes were also larger. The lowest value for size of bract was for the genotype *Heliconia psittacorum* var. 'Choconiana' (5.39 cm^2) which was at par with *Heliconia metallica* (7.65 cm^2). In *Heliconia psittacorum* var. 'Lady di,' bract size was 12.87 cm^2, and in *Heliconia psittacorum* var. 'Golden torch,' *Heliconia stricta* var. 'Dwarf Jamaican Red' and *Heliconia rostrata* bract sizes ranged from 29.63 cm^2 to 38.70 cm^2. *Heliconia wagneriana* had broader bracts arranged in compact manner than in *Heliconia stricta* var. 'Dwarf Jamaican Red' where the distance between two bracts are quite less. In *Heliconia psittacorum* cv. 'Golden torch' (long, boat-shaped bract), *Heliconia stricta* (long, broad, boat-shaped bract) and *Heliconia rostrata* (lobster claw-shaped bract), the bracts were arranged at wider spacing in the inflorescence rachis. The study conducted by Malakar et al. (2015) revealed that under hot humid situation prevailing over here, there was no uniformity in flowering behavior of different varieties and species. Considering the flowering behavior, some of the varieties showed continuous flowering throughout the year, whereas some were significantly seasonal. Our results showed that *Heliconia psittacorum* cv. 'Golden torch' showed perpetual flowering and produced maximum number of inflorescence per plant per year (98). In *Heliconia humilis* and *Heliconia rostrata*, there was no flower production during winter season; however, peak flowering was noted during the month of April–July in both species. The yield of inflorescence/plant/year was moderately high in *Heliconia psittacorum* var. 'Choconiana' (54.6), *Heliconia psittacorum* var. 'Lady di' (52.6), and *Heliconia rostrata* (48.6), while lowest yield was recorded in *Heliconia metallica* (15.66) which was similar to *Heliconia wagneriana* (16.66). In *Heliconia stricta*, the number of flowers per inflorescence (186) was highest, while in *Heliconia humilis* lowest number of flowers per inflorescence (18.33) was noticed. In the remaining genotypes, number of flowers per inflorescence ranged from 19 to 92 which exhibited actually a wide variation. The least number of days from bud emergence to full unfurling of bracts was observed in the genotype *Heliconia psittacorum* var. 'Golden torch' (16.33 days) which was in abeyance to that of *Heliconia metallica* (19 days). The highest number of days were taken by the genotype *Heliconia wagneriana* (40.66 days) followed by *Heliconia stricta* (35.33 days) and *Heliconia humilis* (31.66 days). Chlorophyll content of the leaves were recorded to be high in genotype *Heliconia stricta* (4.76 mg/g of tissue), which was at par with *Heliconia humilis* (4.44 mg/g of tissue) and *Heliconia wagneriana* (4.22 mg/g of tissue). Anthocyanin content of leaves as measured showed higher values for the genotype *Heliconia indica* var. 'Red' (59.25 mg/100 g of tissue) followed by *Heliconia metallica* (9.84 mg/100 g of tissue) (Malakar et al. 2015).

23.1.9 Postharvest Life of *Heliconia* Inflorescences

Heliconia is the most flourishing cut flower in the tropics owing to its charismatic form and alluringly blended hues of its bracts. Most *Heliconias* will have a longer shelf life if the leaves are removed when they are harvested. To improve the appearance of the bracts, wash them in water with a little detergent, and rinse them with fresh water. Removal of the flowers will often improve the appearance of the bracts. To lengthen and keep their quality, immersing of bracts in fresh water for 30 min every 3 days are imperative. Cut *Heliconias* will respond to the use of chemical preservatives in the water and the presence of indirect sunlight. Refrigeration of cut stems is discouraged in this crop (Watson and Smith 1914). To label postharvest complications and to protract vase life for a substantial span, an experiment was designed by Malakar et al. (2019a) employing antithetical vase solutions of various concentrations (T_1 = control (DIW); T_2 = standard preservative (2 teaspoonful of fresh lemon juice, 1 teaspoonful of common sugar, 1/2 teaspoonful of household bleach in 1 L of water); T_3 = 8-HQC at 500 mg L^{-1}; T_4 = 8-HQC at 500 mg L^{-1} + sucrose 2%; T_5 = $AgNO_3$ at 1500 ppm + 8-HQC at 500 mg L^{-1} + sucrose 2%; T_6 = $CaCl_2$ at 750 mg L^{-1} + 8-HQC at 500 mg L^{-1} + sucrose 2%; T_7 = GA_3 at 80 ppm + 8-HQC at 500 mg L^{-1} + sucrose 2%; T_8 = NAA at 100 ppm + 8-HQC at 500 mg L^{-1} + sucrose 2%; T_9 = citric acid at 200 mg L^{-1} + 8-HQC at 500 mg L-1 + sucrose 2%; T_{10} = BAP at 50 ppm + 8-HQC at 500 mg L^{-1} + sucrose 2%) and fresh, mature, cut spikes of nine different

Heliconia species and varieties. T_5 and T_6 treatment combinations contain $AgNO_3$ at 1500 ppm and $Cacl_2$ at 750 mg/l along with 8-HQC at 500 mg/l and sucrose 2% outstandingly elevated solution uptake and flower opening inside bracts principally in all species and varieties, respectively. Untreated spikes of *Heliconia psittacorum* var. 'Lady di,' *Heliconia wagneriana*, *Heliconia stricta*, *Heliconia stricta* var. 'Dwarf Jamaican Red,' and *Heliconia metallica* exhibited noteworthy fresh weight retention on eighth day exceptionally, while in contrast T_4, T_5, T_7, and T_3 evinced themselves unrivalled for the rest of the genotypes on the same duration. To preserve bract and flower's carotene, anthocyanin, and chlorophyll pigments, all treatment's impact were truly species-specific apparently. T_7 and T_9 hold GA_3@80 ppm and citric acid@ 200 mg/l along with germicide and sucrose respectively, while silver nitrate@1500ppm (T_5) and calcium chloride@750mg/l (T_6) yielded utmost vase life by delaying senescence in all inflorescences. These identical treatments amplified the levels of catalase, peroxidase, and enzymatic activities and collaterally declined the lipid peroxidation on seventh day in the bract. Therefore, Malakar et al. (2019a) concluded that $AgNO_3$, $Cacl_2$, GA_3, and citric acid at mentioned concentration render magnificent beneficial effects on vase life attributes being subsisted with oxidative stress of *Heliconia* inflorescences.

Mangave et al. (2013) also studied the effects of postharvest spray application of plant growth regulators, gibberellic acid (GA) and benzyl adenine (BA), alar (daminozide), and chemicals like bovine serum albumin (BSA) and potassium permanganate (KMnO4) on postharvest quality of *Heliconia* inflorescence. Postharvest spray treatments significantly influenced postharvest quality and life of *Heliconia* inflorescence as compared to control. Spray treatments of GA (100 mg l^{-1}) and BSA (50 mg l^{1}) effectively increased water uptake and retained fresh weight of cut inflorescence. The same treatments also reduced the levels of catalase (CAT) and peroxidase (POD) enzymatic activity and decreased the lipid peroxidation (measured as TBARS) in the bract tissue. Percent absolute integrity of bract cell membrane (PAI) was also high in GA (100 mg l^{-1}) and BSA (50 mg l^{-1}) spray treated cut inflorescence on eighth, tenth, and twelfth day of vase life. Postharvest spray treatment of GA (100 mg l^{-1}) showed significant increase (by almost twofold) in the vase life of *Heliconia* inflorescence as compared to control. These results suggest that postharvest spray of GA (100 mg l^{-1}) or BSA (50 mg l^{-1}) maintains higher inflorescence fresh weight, improves water uptake and reactive oxygen species (ROS) scavenging capacity, and stabilizes absolute integrity of cell membrane leading to a delay in bract cell death in *Heliconia* inflorescence cv. 'Golden Torch.'

Bahubali et al. (2014) studied the effects of vase solutions comprising of 8-HQC, sucrose, calcium chloride, α-lipoic acid, sodium benzoate, spermine, citric acid, and commercially available surfactant on postharvest quality, and life of *Heliconia* inflorescence was investigated. Vase solution treatments, T_4 (α-lipoic acid 100 mg/l + 8-HQC 250 mg/l + sucrose 3%) and T_6 (spermine 100 mg/l + 8-HQC 250 mg/l + sucrose 3%) effectively increased water uptake and retained fresh weight of *Heliconia* inflorescence and maintained pigment (carotene) in the bracts. The same treatments also reduced the levels of catalase (CAT) and peroxidase (POD) enzymatic activities, decreased the lipid peroxidation (measured as TBARS), and improved percentage of absolute integrity (PAI) in the bract. Maximum vase life was recorded in *Heliconia* inflorescence held in vase solution comprising of α-lipoic acid 100 mg/l + 8-HQC 250 mg/l + sucrose 3% followed by spermine 100 mg/l and $CaCl_2$ 250 mg/l along with 8-HQC 250 mg/l + sucrose 3%.

23.1.10 *Heliconia*: A Potential Storehouse of Natural Colorants

Indigo is known to be the oldest natural colorant in India. The advent of synthetic dye during 1856–1900 jeopardized the market of natural colorants, as synthetic dyes were cheaper and gave excellent fastness and reproducible color shades. Since natural dyes derived from flora are safe because of their nontoxic, biodegradable, and noncarcinogenic characteristics, recently, the textile dye industry has been forced to stop the production of potentially dangerous dyes and subsequently reduce the toxic effluents. The colorants (from plant origin) used in dyeing various fabrics are mainly flavonoids, along with anthraquinones and indigoids. Most commonly found flavonoids are flavonols, flavones, and anthocyanins. These flavonoids give variety of yellow, brown, and green shades. The principal coloring components from *Heliconia* flower bracts are tannins and flavonols. However, the common drawbacks of natural dyes are their nonreproducible and nonuniform shades, poor to moderate color fastness, and lack of scientific information on the chemistry of dyeing and standardized dyeing methods (Kumar and Prabha 2018).

In India, some 500 varieties of plants are available which are potential for natural dye yield. In recent years, the name *Heliconia* (family, Heliconiaceae), a tropical "specialty cut flower," has gained fame as an emerging cut bloom for its meteoric and alluringly blended inflorescences' hues (like red, pink, orange, yellow, and different combinations), and exotic posture despite it is entirely unutilized in natural dyeing industry although the eco-friendly nature of them has been evidenced earlier. So, to bring this underexploited ornamental, which could be a representative source of natural dye, into notice of entrepreneurs by developing a new, simple, and affordable dye extraction method (may be employing microwave and biodegradable enzyme-based extraction principles) could be a noteworthy aim. The extracted natural dye could be characterized too by Fourier transform infrared spectroscopy (FTIR) and UV-VIS spectrophotometer since the former one identifies chemical bonds in a molecule by producing an infrared absorption spectrum, while the latter detects the presence of color. Since Egyptian era, natural dyes were used, but textile apparels witness several setbacks (such as poor adhesion, color uniformity, and fastness properties), so to eradicate these problems, dyeing of cotton fabrics with and without metallic mordants followed by evaluation of color fastness will be indispensable. The global natural dye market is anticipated to generate revenues of approximately $5 billion by 2024. The reason behind utilizing this new emerging specialty ornamental cut flower is to employ them for value-added production. Exploitation of its brilliantly multihued flowers for dye extraction would be quite significant since natural dyes appear to be the ideal choice under the current national and international awareness about environmental ecology and pollution controls. There is growing demand for developing suitable extraction techniques for more effective and efficient extraction from plant material.

23.2 Botany and Taxonomy

Taxonomic Tree

Kingdom: Plantae
Clade: Tracheophytes
Clade: Angiosperms
Clade: Monocots
Clade: Commelinids
Order: Zingiberales
Family: Heliconiaceae Vines
Genus: *Heliconia*

The taxonomy of *Heliconia* (family, Heliconiaceae; order, Zingiberales; Monocotyledonae) was investigated by a number of Europeans (Andersson 1981a, b) and North Americans (Daniels and Stiles 1979; Kress 1984, 1989, 1990) botanists, and a consensus on species boundaries and relationships failed to reach. For this reason, estimates on the number of species of *Heliconia* range from 120 to over 400. However, recent progress on enumerating and describing taxa indicates that 200 to 250 species is a good estimate for the genus. Sheela (2007), on the other hand, reported about 89 species under the genus and more than 350 varieties. Nascimento et al. (2014) remarked that the genus *Heliconia* is not much studied and the number of existing species in this genus is still uncertain. It is known that this number varies between 150 and 250 species.

In this genus, there are lot of taxonomic confusion and uncertainties about the number of species and the relationship among them (Marouelli et al. 2010). Significant variations exist within and between species of *Heliconia*. However, natural variation among individuals and populations of *Heliconia* has caused confusion among hobbyists and commercial growers (Berry and Kress 1991). Originally, *Heliconia*s were included in the family Musaceae, but the genus was always considered to be homogenous with its own characteristics such as inverted flowers, single staminode, and drupe-type fruits. *Heliconia* is the only genus in the plant family Heliconiaceae, which is a member of the order Zingiberales (earlier called the Scitamineae). In addition to the several cellular features (short root hair cells, sieve tube plastids with starch, silica bodies, inaperturate and exineless pollen) that distinguish the Zingiberales from other monocots, there are several conspicuous characters by which they can be recognized, including (1) large leaves with long petioles and blades possessing transverse venation; (2) large, usually colorful, bracteate inflorescences; and (3) arillate seeds. This order is most closely related to the family Bromeliaceae and their relatives in the superorder Bromeliiflorae (Dahlgren et al. 1985).

Most taxonomists (Berry and Kress 1991) recognize eight separate families in the Zingiberales: Musaceae, Strelitziaceae, Lowiaceae, Heliconiaceae, Zingiberaceae, Costaceae, Cannaceae, and Marantaceae. Most of the members of these eight families are native to the tropical regions of the Earth, and many are found in Brazil. *Heliconia* has been variously associated with the Musaceae or the Strelitziaceae but is now placed in its own family Heliconiaceae. The inverted flowers, the presence of a single staminode, and the drupaceous fruits are special features of *Heliconia*. Many species and varieties native to Brazil are now being grown as pot plants and as cut flowers.

23.2.1 Classification of *Heliconia*

Over the past 100 years, several intrageneric classifications of *Heliconia* have been proposed; each one has been more complex than its predecessors (Kress 1984). His classification consists of two subgenera and six subordinate taxa of unspecified rank. Plant habit and height, inflorescence orientation, cincinnal bract orientation, and the distance between adjacent bracts are characters used to define subgeneric groups. Andersson (1981a) had proposed a new, slightly different classification of the genus in which the ranks and arrangement of Griggs' taxa were changed (Table 1). Andersson's subgenera and sections are more carefully defined and circumscribed than those of Griggs'. Although he had not discussed the phylogenetic relationships of his taxa nor provided evidence that his subgenera and sections are monophyletic, Andersson's system is a great improvement on the system of Griggs. A detailed analysis of the classification of subgen. *Griggsia*, one of the largest taxa that are made up of species possessing pendent inflorescences, is in progress (Kress unpubl.).

23.2.2 Botanical Description

Cronquist (1981) stated that the leaves of *Heliconia*s are distichous with a long basal sheath and a long and expended petiole. Based on the leaf, they are classified into three groups, namely:

1. Musoid (banana-like leaves): e.g., *H. bihai*
2. Zingiberoid (ginger-like leaves): e.g., *H. hirsuta*
3. Cannoid (canna-like leaves): e.g., *H. metallica*

The most typical laminas of the three genera, viz., Heliconiaceae, Musaceae, and Zingiberaceae, are large and oblong and have an acute or irregular apex and an asymmetrical chordate base. The laminar veins are parallel and are oriented essentially perpendicular to the costa and the margin. However, there is considerable variation in lamina architecture in the Heliconiaceae. Thickness ranges considerably across the lamina with considerable variation between species. The only pattern common to all species is that the blades are thinner at the margin than at the costa.

The veins of the lamina occur in a number of size classes that are arranged into a repeating unit called a set. Set patterns are described for the species and families. Although the Heliconiaceae and Musaceae do not differ in general set pattern, variation in the Heliconiaceae tends toward simpler sets, whereas the Musaceae tends toward more complex sets. Lamina anatomy is similar to that reported in the literature for these families with several notable exceptions. The Heliconiaceae and Musaceae may be distinguished by characteristics of their lamina anatomy and by the fact that Musaceae blades have an irregular apex.

Morphological and anatomical features of roots, stems, leaves, and scapes were studied in *H. angusta* and *H. velloziana* from the Atlantic forest in the southeastern of Brazil. Morphologically, *H. angusta* and *H. velloziana* show differences in their sizes, blade shapes, and number and shape of inflorescence bracts. On the other hand, they have common anatomical characteristics such as roots with air canals in the cortex, rhizomes with isolated fiber bundles, collateral vascular bundles and uniseriate endodermis and pericycle, leaves presenting air canals and collateral vascular bundles forming arcs and thin-walled epidermal cells, and scapes with collateral vascular and fiber bundles in the cortex. The distribution of the fiber bundles in the leaves and in the scapes was different for each species, having a taxonomical value, *H. velloziana* presenting continuous fiber bundles. Air canals in roots and leaves with narrow mesophyll might be related to the moist understory of the Atlantic forest habitats (Simao and Lucia 2001).

Inflorescences are the most striking part of *Heliconia* plants. The inflorescence is terminal, erect, and pendant composed of bracts in one plane (distichous) or spirally arranged. Each bract constitutes and involves one cincinnus with many flowers. The bracts are modified leaves, cymbiform, or lanceolate to conduplicate with variable coloration, size, arrangement, texture, and number, and some of these features are used in the subgenus classification (Cronquist 1981; Berry and Kress 1991; Anderson 1992).

Guimarães et al. (2014) worked on phenotypic diversity of *Heliconia* based on qualitative descriptors. The aim of this study was to characterize *Heliconia* genotypes phenotypically using 26 qualitative descriptors (Table 3). The evaluations were conducted in 5 flowering stems per clump in 3 replicates of 22 *Heliconia* genotypes. From the output generated by the Mahalanobis dissimilarity matrix and the clusters formed among the *Heliconia* genotypes studied, the phenotypic characterizations that best differentiated the genotypes were the pseudostem and wax green tone (light or dark green), leaf wax petiole, petiole hair, cleft margin at the base of the petiole, midrib underside shade of green, wax midrib underside, color sheet (light or dark green), unequal lamina base, torn limb, inflorescence wax, position of inflorescence, bract leaf in the apex, twisting of the rachis, and type of bloom. These results will be applied in the preparation of a catalogue for *Heliconia* descriptors, in the selection of different genotypes with most promising characteristics for crosses, and for the characterization of new genotypes to be introduced in germplasm collections.

The ability to identify plants from their pollen has enabled botanists and ecologists to reconstruct past assemblages of plants and identify periods of environmental change (Faegri and Iversen 1950; Moore et al. 1991). Morphological characteristics

Table 3 Qualitative descriptors related to the pseudostem, leaf, and inflorescence of collected *Heliconia* genotypes

	Descriptors																								
	Pseudostem			Leaf																Inflorescence					
Genotypes	PDG	PW	PH	WP	HP	BPW	BMP	OEP	MUG	MUSG	WMU	MUH	MLP	LD	DGC	LH	WL	LUB	CLB	BLA	TSB	WI	IP	HB	TR
H. psittacorum × *H. spathocircinata* cv. 'Golden Torch'	+	0	0	0	0	+	0	0	0	+	+	0	+	+	+	0	+	0	0	0	0	0	0	0	+
H. psittacorum cv. 'Lady Di'	0	0	0	0	0	0	0	0	+	+	+	0	+	0	0	0	+	0	0	0	+	0	0	0	+
H. psittacorum cv. 'Choconiana'	0	0	0	+	0	+	0	0	+	+	+	0	0	0	0	0	+	0	0	0	+	+	0	0	+
H. rostrata	+	0	0	+	0	0	0	0	0	+	0	0	0	+	+	0	0	0	0	0	+	+	+	+	0
H. stricta	+	0	0	0	0	+	+	0	0	+	0	0	+	+	+	0	0	0	0	0	+	0	0	0	0
H. wagneriana	0	+	0	+	+	+	0	0	+	0	+	0	+	+	+	0	0	0	0	0	+	0	0	0	0
H. metallica	0	+	0	+	0	+	0	0	0	+	0	0	+	0	0	0	0	+	0	0	+	0	0	0	+
H. humilis	+	0	0	0	0	+	0	0	0	0	0	0	+	+	+	0	0	+	0	0	0	0	0	0	0
H. stricta var. 'Dwarf Jamaican Red'	+	0	0	0	0	0	0	0	0	0	0	0	+	0	0	0	0	0	0	0	+	0	0	0	+
H. indica var. 'Indica'	0	0	0	0	0	+	0	0	0	0	0	0	+	0	0	0	+	0	0						

"+" denotes present and "–" denotes absent

PDG pseudostem of dark green color, *PW* wax, *PH* hair, *WP* wax in the petiole, *HP* hair on petiole, *BPW* base of the petiole winged, *BMP* basal margin of the petiole cleft, *OEP* open edge of the petiole, *MUG* midrib underside shade of green, *MUSG* midrib upper shade of green, *WMU* wax midrib underside, *MUH* midrib underside hair, *LH* leaf hair, *WL* waxy leaves, *DGC* dark green leaf color, *MLP* margin of leaves shade of purple, *LD* leaf blade dark, *LUB* leaf blade uneven bases, *CLB* cut leaf blade, *WI* waxy inflorescence, *IP* inflorescence position pending, *BLA* bract leaf at the apex, *HB* hair in bracts, *TR* torsion of rachis, *SR* stiffness of rachis, *TSB* type of seasonal blooming of pollen grains also can be useful characters in studies of plant taxonomy because many pollen traits are influenced by the strong selective forces involved in various reproductive processes, including pollination, dispersal, and germination (Erdtman 1952; Moore et al. 1991; Nowicke and Skvarla 1977; Stuessy 1990). At the same time, characters subject to strong selection can be misleading, if they reflect convergent evolution with similar evolutionary responses by unrelated taxa to similar environmental conditions. Thus, the use of pollen morphology as a taxonomic character is challenging, and pollen characteristics must be considered in concert with other characteristics in evolutionary reconstructions.

23.2.3 Floral Biology and Genetic Variability in *Heliconia*

Exploitation of genetic variability in the available germplasm is prerequisite in the breeding program for effective selection of superior genotypes. Natural genetic variation for most of the yield attributes in *Heliconia* is high, but information on the nature and magnitude of variation available in the material and part played by environment in expression of different characters is needed. Yield being the most important and complex character is governed by many physiological processes within the plant and is also influenced by the environmental factors. A rational approach toward improvement of yield necessitates the selection of desirable components of yield and suitable genotypes. Inadequate information on nature and magnitude of genetic variability may limit the breeding processes for yield improvement.

Improvement of crop yield largely depends on the extent to which the yield-attributing and stress-resistant (both biotic and abiotic stresses) characters are heritable from generation to generation. Measure of heritability alone does not give an idea about the expected genetic gain in the next generation, but it has to be considered with genetic advance (Johnson et al. 1955). High heritability coupled with high genetic advance indicates the importance of additive gene effects for conditioning of the particular character (Panse 1957). High mean associated with high variability is considered as a better index for selection. High genotypic coefficient of variation (GCV) in the population could be exploited in the selection program. Moderate to high genetic variability along with higher heritability and genetic advance indicates the extent of scope for further improvement through phenotypic selection. Higher phenotypic coefficient of variation (PCV) than genotypic coefficient of variation (GCV) for a character indicates much environment effect in the expression of the character. Moderate to high GCV together with high heritability and genetic advance for a character indicates the effectiveness of selection for the character.

Sheela et al. (2006) assessed 12 *Heliconia* cultivars for morphological, flowering, and vase life parameters during 2003–2005 in Vellayani, Kerala, India. Analysis of variance revealed significant differences among cultivars for all characters studied indicating the high potential for crop improvement through selection. Estimation of phenotypic coefficients of variation was higher than the genotypic coefficients of variation for all characters studied indicating the role of the environment on the expression of the genotype. High heritability coupled with high genetic advance was recorded for number of flower shoots, stalk length, plant height, duration of flower, bract size, vase life, and days to first flowering indicating the preponderance of additive effects in expression of these characters.

Exploration on genetic variability and diversity of *Heliconia* genotypes among yield and floral biology traits were executed by Malakar et al. (2019c) being intended to investigate the prime floral (Table 1) parameters quite potential to enact vital role in breeding and also inflorescence productivity per annum. The study revealed noteworthy dissimilitude in flowering behavior of different varieties under West Bengal condition. Here, 'Golden Torch,' a hybrid cultivar, merely showcased perpetual flowering, whereas others are considerably seasonal. Highest genotypic and phenotypic variabilities were observed for anthocyanin content. Likewise, PCV of 61.43% for duration of male phase and GCV of 58.34% for number of flowers/inflorescence were apparent as second highest. Here characters days from first to last flower opening, days from emergence to male and female phase, and duration of male and female phase, recorded high heritability coupled with high genetic advance which manifests possibility of genetic improvement through selection. Duration of male and female phase duo showed significant positive phenotypic and genotypic correlation with days from first to last flower opening and days from bud emergence to full unfurling of bracts. Furthermore, inflorescence productivity year^{-1} was found to be positively correlated with inflorescence length and number of open bracts at both genotypic and phenotypic levels.

An investigation on "floral biology and compatibility studies in *Heliconia*" was carried out by Sanjeev et al. (2010) with the objective of studying the floral biology of 15 *Heliconia* species and varieties for 16 floral characters. The study revealed that under South Kerala condition, there was no uniformity in flowering behavior of different varieties. Considering the flowering pattern, some of the varieties showed continuous flowering throughout the year, whereas some others were significantly seasonal. Pollen fertility estimated using acetocarmine staining method revealed that 'Lady Di' had the highest pollen fertility of 89.39% which is on par with 'Parakeet.' The lowest pollen fertility was recorded for 'Guyana' (23.23%). Variability studies revealed that highest variability for phenotypic component was observed for the character pollen fertility (97.97%) followed by anthocyanin content (87.76). Genotypic component of variability was highest with 86.16% for anthocyanin content. The characters anthocyanin content and number of flowers per bract recorded high heritability coupled with high genetic advance, which shows that genetic improvement of these characters is possible through selection. Correlation studies revealed that days from first to last flower opening showed significant positive phenotypic correlation with days from emergence to male phase and pollen size.

23.2.4 Pollen Viability and Stigma Receptivity

Heliconia species are gaining importance due to their desirable horticultural properties and postharvest characteristics. Much, though, remains to be known about this plant. Thus, it has become doubly important to carry out investigations on its breeding behavior and other factors affecting its fertility. Since germinability is very important in determining fertility, germination tests were carried out by Salih (1994), investigating the effects of flower collection time, temperature conditions before and after harvest, and medium components. It was found that all the above played very important roles, with temperature probably being the most important one. Harvesting of flowers very early in the day at low-temperature conditions was found to enhance germinability. The components of the germination medium were also concluded to be important. Sucrose may have a role to play as an osmoticum and may also be involved in pollen metabolism. Boric acid may enhance pollen tube growth. Gibberellic acid and calcium enhance germination but not tube growth, with calcium being enhanced in its action by magnesium and potassium. Viability studies carried out gave consistent results for all tests except for the TTC test which gave no result for all taxa of *Heliconia* tested. This test was therefore concluded to be not reliable or valid for this genus. Stigma receptivity studies gave expected results for all taxa tested except for the hybrid cultivar 'Golden Torch,' which had a receptive stigma despite no pollen production, and the cultivar 'Andromeda,' which may possibly be exhibiting protogyny. In vivo germination and in vitro pollination studies gave no results, possibly because of the use of the wrong cultivar. *H. rostrata* was found to behave in a similar fashion to *H. psittacorum* cultivars. This suggests that the similarity between members of this genus may be more than the degree of disparity. Future prospects for research include the use of *H. psittacorum* cv. 'Black Cherry' and cv. 'Susi' as parents in hybridization, especially with 'Susi' as pollen parent as it has such a good growth rate of the pollen tube. Also, the use of growth regulators to retard senescence and therefore promote germination and fruit set may have some potential.

Since knowledge of floral biology is essential for crop improvement programs in any crop plant, a study on pollen viability was undertaken by Narkar et al. (2017). He studied the pollen grain fertility of few *Heliconia* sp. for breeding improvement. The studies were undertaken on 30 selected species and cultivars of *Heliconia* having good cut flower qualities and popularity in the market. A thorough understanding of the pollen characters is a fundamental requirement for any successful breeding. The interspecific hybrids recorded lower fertility percentage, while the other species and cultivars recorded higher fertility. Pollen shape varied within the same species in different cultivars. This indicates that for selection of suitable cultivars in further breeding programs, their compatibility and fertility characteristics should be assessed beforehand.

Sanjeev et al. (2010) also studied pollen fertility of 15 *Heliconia* species and cultivars. Pollen fertility as estimated using acetocarmine staining method revealed that 'Lady Di' had the highest pollen fertility of 89.39% which is on par with 'Parakeet' variety. The lowest pollen fertility was documented for 'Guyana' (23.24%).

Six cultivars of *H. psittacorum* were selected for studies on their natural fruit-bearing ability, pollen formation, and pollination under the tropical climatic conditions of Singapore by Lee et al. (1994). Three of them, namely, 'Tay,' 'Andromeda,' and 'Lady Di,' were partially fertile with a very low rate of fruit set ranging from 2.8% to 4.7%. They were found to be diploid with $2n = 24$ chromosomes. The process of pollen formation (microsporogenesis) was normal, and pollen grains were all uniform in size and appeared normal. The poor fruit set of these three cultivars were attributed to poor pollen germination on stigmas rather than poor pollination of self-incompatibility. The other three cultivars, namely, 'Petra,' 'Sassy,' and 'Iris,' were completely sterile. Their pollen grains were of variable sizes and appeared to be abnormally fragmented. Over 80% of the pollen grains aborted 1–2 days before pollination. These abnormal features were consistent with irregular distribution of chromosomes during meiosis in microsporocytes. All three cultivars were confirmed to be triploid ($2n = 3x = 36$).

In *Heliconia* sp., an obvious problem in quest of new hybrids is the low rate of hybridization. Several causes may be liable to this difficulty, among which key is the low rates of cross-pollination by hummingbirds or other natural agents (Temeles et al. 2000; Maglianesi et al. 2015). Assuming that the natural pollination is the 'Bottleneck' in hybridization, then the obvious solution would be the "artificial pollination" by the growers. However, nobody succeeded in artificially pollinating *Heliconia*s even within a species (Nelson 2000). Customarily, barriers for hybridization between plants can be placed into general categories: (1) "pre-zygotic" factors involving cross-pollination between species and (2) "post-zygotic" factors relating to the germination and growth of hybrid seeds and seedlings. There are several "pre-zygotic" hindrances preventing hybridization between *Heliconia* sp. such as ineptitude of pollens of some species to fertilize the ovules of other species due to biochemical incompatibility, and differences in flowering seasons may not be random but staggered (Berry and Kress 1991). Since stigma receptivity is a crucial stage in the maturation of a flower which may greatly influence the rate of pollination (Stone et al. 1995), scanning of the ability of pistils for capturing lodged pollens onto it is consistently significant like pollen grain study. Till date, published information of *Heliconia* sp. on pollen and stigma are very finite. Thus, an elaborate comprehension and appraisal of pollen grain fertility and stigma receptivity are indispensable for *Heliconia* breeding. So, investigation on pollen viability and stigma receptivity by Malakar (2020b) opened a vista for further development of this crop. Her investigation exhibited highest fertility of pollen by *H. psittacorum* var. 'Choconiana' (100%), *H. psittacorum* var. 'Lady di' (94%), *H. metallica* (93%), and *H. humilis* (91%) along with sizeable pollens, while pollen sterility was found in case of hybrid cultivar 'Golden torch' and *H. stricta* with significantly reduced pollen size. Therefore, pollen size is positively correlated with pollen fertility. Stigma receptivity studies yielded expected consequences for all taxa tested by Baker's and Peroxtesmo tests. On the contrary, unanticipated outcome was perceived for hybrid cultivar 'Golden Torch' and *H. stricta* since they had receptive stigmas despite of no viable pollen production. So, the utilization of 'Golden torch' and *H. stricta* as female parents with notable pollen parents can prosper *Heliconia* sp. hybridization (Table 4).

Table 4 Morphological characters of stigma and its receptivity using different test solutions of nine species and varieties of *Heliconia* sp.

Name of species	Stigma size and form	Stigma type	Stigma color	Baker's test	6% hydrogen peroxide test	Peroxtesmo test
Heliconia psittacorum x *Heliconia spathocircinata* cv. 'Golden torch'	Filiform with three distinct prominent lobes (each 1 mm long); both sides are gibbous, papillae developed throughout stigma and style, only the head/upper surface is receptive	Dry	Faded whitish green	The entire upper surface only stained dark brown-purple, sometimes heads of two side lobes get stained	Weak positive response (+) only at the tip	Only the tip of stigma turned blue
Heliconia psittacorum var. 'Choconiana'	Filiform, the tip and inner part of stigma is receptive (1–2 mm), multiseriate papillae throughout the receptive part and style, tips of papillae are also receptive	Dry	Do	The entire tip of stigma and the inner part (2 mm) stained purple	Strong positive response (++) at the tip of stigma and papillae	2 mm inner surface of stigma stained blue
Heliconia psittacorum var. 'Lady di'	Filiform (diameter 3 mm), entire tip, inner portion (1–2 mm in length) and tip of papillae are receptive	Dry	Greenish	Tip and 1-2 mm of the inner part of stigma stained light-deep purple	Weak positive response (+)	Whole surface stained blue
Heliconia rostrata	Filiform, the receptive part is entire tip (1–2 mm in length); extended down also, trilobed	Dry	Highly faded pinkish green	Only the tip of the stigma stained dark purple	Strong positive response (++)	Stigma center stained blue
Heliconia humilis	Capitate, seriate papillae at the receptive area, entire surface is receptive (2 mm length)	Dry	Yellowish-green	The whole surface stained orange-yellow	Weak positive response (+)	Only the tip of the stigma turned blue

Heliconia stricta var. 'Dwarf Jamaican Red'	Filiform, tip of the stigma is receptive (1 mm), multiseriate papillae noted	Dry	Whitish	Only tips of papillae turned dark purple	Weak positive response (+)	Only 1 mm of the tip stained blue
Heliconia wagneriana	Filiform, trilobed, whole surface is receptive (2 mm length)	Dry	Whitish yellow	Whole trilobed stigma was stained brownish orange	Strong positive response (++)	Positive and immediate reaction (blue) on whole surface of the stigma
Heliconia stricta	Filiform, trilobed, one lobe is larger (2–3 mm) than the other; only the head/upper surface is receptive, multiseriate papillae throughout the receptive part and style	Dry	Yellowish	Whole surface stained dark purple	Very strong positive response (+++) overall the stigma, especially in the tip	Strong reaction, deep blue
Heliconia metallica	Filiform, multiseriate papillae throughout the stigma and style, tips of papillae are also receptive	Dry	Greenish	Entire surface and tips of papillae turned brown-purple	Strong positive response (++)	Tips of papillae and stigma center stained blue

23.2.5 Key Features of Some Common *Heliconia* Species

In 1980, Smith named and described the middle-American species of *Heliconia*s of Florida. With the help of that, Watson and Smith (1914) described 13 species that commonly grow in Hawaiian gardens. They fall in the following four categories:

1. *Inflorescence erect and in one plane*
   Heliconia aurantiaca Ghiesbreght – orange bracts, yellow flowers
   H. aureo-striata Bull – green and yellow bracts, variegated leaves
   H. bourgaeana O. G. Petersen – rose-colored spathe with green margins.
   H. caribaea Lamarck – large yellow bracts
   H. humilis Jacquin – large red bracts
   H psittacorum Linnaeus f. – rosy-red bracts, conspicuous orange flowers
   H. wagneriana O. G. Petersen – bracts with red base, yellow streak on top
2. *Inflorescence erect and in more than one plane*
   Heliconia metallica Planchon and Linden ex Hooker – small green to yellow bracts
   H. latispatha Bentham – large orange bracts
3. *Inflorescence pendant and in one plane*
   Heliconia rostrata Ruiz et Pavon – rose-colored bracts with green tips
   H. catheta R. R. Smith sp. nov. var. 'Catheta' – red triangle at the base of bracts, green tips
4. *Inflorescence pendant and in more than one plane*
   Heliconia collinsiana Griggs var. 'Velutina' R. R. Smith var. 'Nov' – red triangle toward the base of bracts, yellow tips
   H. collinsiana Griggs var. 'Collinsiana' – red bracts covered with white bloom.

Heliconia aurantiaca Ghiesbreght: Plants rarely more than 3 ft tall, leaf blades oblong, elliptical. Bracts orange or green with orange base. Conspicuous cream to yellow flowers, 2 in long, attractive for cutting. Fruit dark blue. Often used in borders where it spreads quickly and makes an attractive mass effect. Leaves wilt when used as cut foliage.

Heliconia aureo-striata Bull: Bracts large, short, thick, non-touching, pale yellow, and 5 in long. Flowers pale green when fresh, protruding from bract. Leaves 6 to 12 ft long, dark green with yellow stripes. Midrib and occasionally the vein are rose-colored. Flower has a good keeping quality for use as a decoration.

Heliconia bourgaeana O. G. Petersen: Vigorous plant, often 15 ft high, with large green leaves, pale green on the underside. Bright red inflorescence, 1 to 2 ft high. Fat, boat-shaped bracts with green margins. Flowers green and white, fruit blue. This species is extremely ornamental and keeps well as a cut flower.

Heliconia caribaea Lamarck: Large, heavy plant, 12 to 15 ft high. Large, golden yellow, short, erect, overlapping bracts, 5 in long, especially broad and boat-shaped. Dull green flowers dry to an unsightly brown as the bracts age.

Heliconia humilis Jacquin: This species is native to Central America, the Caribbean Islands, South America, and some of the islands of the South Pacific. Commonly called "lobster claw red." Vigorous plant with leaves 8 to 10 ft high; bright red bracts in shape of lobster claws. Upright flower stalk, up to 3 ft long, composed of a series of flat, keel-shaped bracts, 5 to 7 in wide, with small green flowers almost completely concealed in spathes. As a cut flower, long lasting and extremely rigid. Makes a good garden plant where there is plenty of space for it to spread.

Heliconia psittacorum Linnaeus f.: Dwarf form with leaves 212 to 4 feet high. Many flower stalks, conspicuous among leaves. Large, rosy-orange bracts, 2 to 3 in long; three to seven conspicuous, orange, tubular flowers with black tips. Although the flowers drop as the inflorescence matures or is cut, plant spreads rapidly and makes a good garden plant for foundation plantings or to rapidly cover the soil in partially shaded, damp locations.

Heliconia wagneriana O. G. Petersen: Medium-sized plant up to 15 feet high. Bracts 5 in long, orange-pink with green, later yellow, strip along the border of each bract. Because flowers are concealed, makes a decorative cut specimen.

Heliconia metallica Planchon and Linden ex Hooker: Medium-sized plant. Lower surfaces of leaves tinged with purple. Erect inflorescence, bracts 6 in long, pale green with conspicuous orange flowers. Useful as a garden plant.

Heliconia latispatha Bentham: Plant 7 feet high, leaves pointed, large, erect inflorescence in more than one plane. Lowest bract is highly developed into a leaf-like blade. Pumpkin-orange bracts from 4 to 7 in long. Two lowest bracts with the leaves are longer. Flowers yellow to yellow-green. Rampant growth that does not suit small gardens, artistic design of the inflorescence unique and suitable for large tropical flower arrangements.

Heliconia rostrata Ruiz et Pavon: An herbaceous, perennial native to North Western region of South America, familiar as "hanging lobster claw." This species has downward-facing flowers and often used in landscaping. The inconspicuous, faded yellow flowers emerge from claw-shaped bracts magnificent up to 3 feet (0.9 m) long, pendent, zigzagged inflorescence. The leaves are simple, alternate, long-petiolate, and have green, lanceolate leaf blades that are easily shredded by the wind. The pseudostems (formed by the leaf sheaths) emerge from underground rhizomes. It prefers sunny to partly shady location at lower elevations. The flowers make long-lasting cut flowers in tropical flower arrangements. Typical pendent inflorescence of red and yellow bracts appears in late spring and early summer. The inflorescence lasts for several weeks on the plant but is generally very short lived as a cut flower.

Heliconia catheta R. R. Smith sp. nov. var. '*Catheta*': Plant 15 feet high, leaves green on both sides, pendant inflorescence. It's composed of rose-red bracts with green margins, 8 to 10 in long, widely spaced on stalk. Yellow-colored flowers are produced.

Heliconia collinsiana Griggs var. '*Velutina*' R. R. Smith var. 'Nov'.: Large plants 12 to 15 feet high. Pendant inflorescences with widely separated bracts, 6 to 12 in long and in more than one plane. Bracts have a red triangle toward the base, with chartreuse to yellow tips that usually hide the yellow to orange flowers. Cut stalks are used in flower arrangements. Plants are extremely large for small gardens.

Heliconia collinsiana Griggs var. '*Collinsiana*': Large, pendant, swinging inflorescence, borne on tall stalk among large leaves, often up to 12 feet high of red, narrow, pointed bracts covered with white bloom. Large bracts range from 2 to 4 in long with 2 to 3 in between the attachments to the flower stalk. Several orange flowers protrude beyond each bract. This species keeps well and is useful for large arrangements in tropical and contemporary settings. The plants spread rapidly and require large space as a garden plant.

H. psittacorum L.f. × *H. spathocircinata* cv. '*Golden Torch*': A very popular type of *Heliconia* commonly known as "parakeet flower" or "parrot's beak," *Heliconia* resembles most to the "bird of paradise." Striking inflorescence and bracts are yellow-pointed. These tough, leathery structures protect the actual flower which is yellow and quite delicate. They bloom all-round the year and are good cut flower, lasting for 2–3 weeks in vase. They also make great container plants.

H. psittacorum cv. '*Choconiana*': Another variety of *H. psittacorum*. It has orange-colored bracts and flowers, while flower's tip has dark green patch. The inflorescence as a whole is small and light. So, it is ideal as cut flower with its long stalk.

H. psittacorum L.f. cv. '*Lady Di*': Striking inflorescence, maroon-pointed bracts. These tough, leathery structures protect the yellow-colored delicate actual flower. They bloom in summer and are good cut flower, lasting for 2–3 weeks in vase. They also make great container plants.

H. stricta var. '*Dwarf Jamaican Red*': A variety of *H. stricta* which is very short in height. The faded red-colored flowers are of immense beauty and ideal as indoor pot plant.

H. stricta: It is a plant species native to Brazil, Colombia, Venezuela, Ecuador, Peru, Bolivia, Guyana, and Suriname, reproducing by seeds and by underground rhizomes. It is reportedly naturalized in Cuba and Puerto Rico and cultivated as an ornamental in many other warm regions. Deep red bracts having bright yellow flowers, inflorescence is bulky. The plants are long with long green leaves. It looks beautiful as landscape plant.

H. wagneriana: Medium-sized plant up to 15 feet high. Bracts are light red with green-bordered edge along with whitish base. They are about 5 in long. Looks like stoke of paint brush has been given with red over the whitish base color. Because flowers are concealed, it makes a decorative cut specimen.

H. metallica: It is a medium-sized plant. Ventral surface of soft leaves is tinged with purple, while dorsal surface is green. These bicolor leaves have huge ornamental value. Erect inflorescence, bract 6 in long, pale green with conspicuous brilliant pink flowers inside. It is useful as a garden plant.

H. indica var. '*Indica*': It is a foliage variety, reared for its exquisite colored foliages. The leaves are tinged with red color, mottling with moss green color. The plant is bigger with longer leaves.

H. bihai: Leaves are sympodial with the emerging stalks; blades are elliptic to paddle-shaped, simple, with entire margins and acute to acuminate tips; petiole bases wrapped around the stalk; depending upon the species, they can be held horizontally or slightly recurved to strongly erect; venation is pinnate or nearly parallel; size is highly variable from around 1′ to as much as 6′ long; the bold foliage can be medium green, bluish green to a dark lustrous green at maturity; leaves are usually lighter colored beneath, and most are not pubescent. Small perfect flowers are inconsequential aesthetically, but the showy bracts and bracteoles can be spectacular; few to 50 flowers are held in erect or pendent terminal racemes; flowers have 5 functional stamens and 1 staminode; the primary bract is lobster claw-shaped to spathe-like and frequently very showy; flowers have variations of bright rich dark green, bright, yellow, orange to red colors; secondary bracteoles may also be showy and come in similar colors but may also be white; the size of the flower parts within the bract vary among species and in some are essentially entirely hidden within the bract, whereas in others they rather resemble a bug upturned in the center of a boat-like bract; individual flowers bloom for a day, but the bracts remain effective for an extended period of time and are often used as cut flowers in the floral industry; individual stalks are polycarpic, dying back to the ground, while new stalks arise from the rhizome. Fruits are small knob-like drupes, not ornamental and frequently produced.

23.2.6 Genetic Relatedness of *Heliconia* Species Using DNA Markers

Enormous genetic diversity available for *Heliconia* breeding has facilitated the development of new varieties and hybrids. The search of superior hybrid parents in *Heliconia* breeding programs is commonly based on the estimation of the general combining ability (GCA) and specific combining ability (SCA) of inbred lines. However, the application of this procedure is expensive and time-consuming. The development of DNA-based molecular markers represents an alternative procedure for the identification of promising parental lines for high-performance hybrid production. The random amplified polymorphic DNA (RAPD) marker has been widely used for the estimation of genetic distance among closely related individuals. Thus, molecular markers, as RAPD, could be used for germplasm classification and clustering, producing valuable information for heterosis prediction.

Heliconia is a beautiful tropical flower which may help to increase profits of cut flower industry worldwide. The inflorescence durability, stock length, and vivid colors are supporting the commercialization on the international market. In comparison to the traditional cut flowers like rose, chrysanthemum, gladiolus, etc., *Heliconia* is a more recent introduction in floriculture trade. Different studies have been conducted in various parts of the world in developing *Heliconia* for commercial use, and those are mostly on cultural aspects. The genus *Heliconia* contains approximately 180 species of different tropical origin. In this genus, there are lot of taxonomic confusion and uncertainties about the number of species and the relationship among them (Marouelli et al. 2010). Molecular studies are therefore necessary for better understanding of the species boundaries of these plants for further development. A random amplified polymorphic DNA (RAPD) marker is one such powerful tool for genetic variability studies and clarification of the relationship between *Heliconia* species (Marouelli et al. 2010). However, little studies on this line have been reported so far on this species of plant.

Exploitation of molecular markers may aid in perceptive recognition of several species and varieties of them since taxonomic confusions and uncertainties subsist amid them. In this context, ten species and varieties were analyzed by Malakar et al. (2020a) using RAPD markers. Chosen 30 primers among 70 amplified 1281 polymorphic DNA fragments with each primer giving a mean of 42.7 polymorphic bands. The genetic similarity matrix constructed with Jaccard's coefficient using RAPD marker scores showed that the highest value was between *H. psittacorum* var. 'Choconiana' and *H. psittacorum* var. 'Lady Di' (0.384), while the lowest was between *H. psittacorum* var. 'Choconiana' and *H. stricta* var. 'Dwarf Jamaican Red' (0.244). Ten species and varieties of *Heliconia* formed three distinct clusters at similarity coefficient value of 0.33, implying a parallelism between genetic and morphologic or taxonomic variability of *Heliconia* genotypes. 'Golden Torch,' 'Choconiana,' 'Lady Di,' and *H. wagneriana* included in cluster I, while 'Choconiana' and 'Lady Di' formed a more cohesive entity. On the contrary, *H. stricta* var. 'Dwarf Jamaican Red' produced a separate cluster validating that taxonomically related entries clustered together and distant ones segregated.

Molecular marker is important from the genetic variability assessment point of view, because the rDNA multigenic family once subjected to a rapid evolution in concert event allows greater precision in the reconstruction process of the relationship between species based on sequencing, since this phenomenon increases the intra-genomic uniformity (Baldwin et al. 1995). These authors also affirmed that due to the biparental inheritance of the nuclear genome, it is possible to study the origin of hybrids and their parents. Moreover, chloroplast genes (cpDNA), such as the leucine and phenylalanine of RNA transporter (trnLtrnF), the threonine and leucine of RNA transporter (trnT-trnL), and the protein small 4 (rps4), have been used successfully to solve genetic diversity doubts in taxonomic lower levels. Johansen (2005), also studied the genetic diversity in Zingiberales order, using cpDNA, which had positioned all Heliconiaceae and Musaceae families within the same clade.

Goh et al. (1995) conducted an experiment to determine the genetic variations detected with RAPD markers in *Heliconia*. They reported many *Heliconia* sp. are polymorphic with large number of cultivars. Accordingly, they developed a method for DNA extraction from leaves and the subsequent RAPD analysis was also used to detect genetic variations and similarities among the different species, cultivars and hybrids utilized in their study. Significant differences in RAPD profiles occurred among different *Heliconia* sp. and some distantly related plants. *H. rostrata* had three prominent bands that were absent in *H. psittacorum* cultivars. All 19 cultivars of *H. psittacorum* tested had similar profiles using 3 different primers. Three cultivars of the hybrid *H. psittacorum* × *H. spathocircinata* cv. 'Golden Torch,' 'Red Torch,' and 'Alan Carle' had slightly different RAPD profiles which were distinct from that of *H. psittacorum* × *H. marginata*. Of the three triploid cultivars of *H. psittacorum* tested, 'Iris' and 'Petra' had identical profiles with all the primers. A prominent band identified in 'Iris' and 'Petra' was absent in 'Sassy,' thus being concluded that 'Iris' and 'Petra' are the same genotype.

Marouelli et al. (2010) conducted an experiment on genetic analysis of *Heliconia* species and cultivars with random amplified polymorphic DNA (RAPD) markers. Many *Heliconia* species are polymorphic with a large number of cultivars. Cultivar identification has been primarily based on morphological differences of the flowers and inflorescences. A protocol was developed to extract DNA from *Heliconia* leaves and to analyze genetic variation using RAPD. The percentage of genetic similarities among *Heliconia* species, cultivars, and hybrids were determined. Data from 11 primers suggested that the RAPD technique can be used to distinguish species and cultivars of *Heliconia*. Using a single 10-mer primer (OPA 18), distinct RAPD profiles were determined for 16 cultivars of *H. psittacorum* grown at the Jurong Bird Park, Singapore, a *Heliconia* Society International Depositary. It was suggested that the characteristic profiles generated by RAPD may be used as additional DNA markers for classifying different species and cultivars of *Heliconia*. The phylogenetic tree derived from the RAPD data showed that all of the 16 cultivars examined are closely related to each other providing the first genetic evidence that this large group of cultivars has a common genetic background. Moreover, two triploid cultivars of *H. psittacorum* ('Iris' and 'Petra') showed identical RAPD profiles with ten different primers in agarose and polyacrylamide gels, suggesting that they are of the same genotype.

Brown and Weir (1983) made a study on the genetic diversity of selected Panamanian *Heliconia* species in the Section *Barbatae* using RAPD markers. This research focused on a genetic analysis of a complex of Panamanian *Heliconia* species including *H. pogonantha* var. 'Pogonantha' and var. 'Verguasensis,' H. ramonensis var. 'Ramonensis,' *H. lanuginose*, *H. xanthotricha*, *H. glabra*, *H. magnifica*, and *H. xanthovillosa*.

In India, perhaps the molecular characterization of *Heliconia* by RAPD assay was first conducted by Sheela et al. (2006), and thereafter very little study has been reported on this relevant topic from India. They analyzed 17 *Heliconia* species and varieties were analyzed using RAPD markers. Eight primers, which produced the highest number of bands, were used for DNA amplification. The genetic similarity matrix constructed with Jaccard's coefficient using RAPD marker scores showed that the highest value was between 'Petra Orange' and 'Parakeet,' while the lowest was between 'Golden Torch' and *H. humilis*. The 17 species and varieties of *Heliconia* formed 9 distinct clusters at similarity coefficient value of 0.42, implying a strong parallelism between genetic and morphologic or taxonomic variability of *Heliconia* genotypes. 'Petra Orange,' 'Deep Orange,' 'Parakeet,' 'Pascal,' and 'Alan Carle' formed a big cluster within which 'Petra Orange' and 'Parakeet' formed a more cohesive entity.

Marouelli et al. (2010) concluded through an experiment of genetic relationships among the *Heliconia* (Heliconiaceae) species based on RAPD markers. In order to establish a relationship among the different species, molecular studies are therefore necessary for better understanding of the species boundaries of these monogenic species. They examined the genetic variability and the phylogenetic relationships of 124 accessions of the genus *Heliconia* based on RAPD markers. Phenetic and cladistic analyses, using 231 polymorphic RAPD markers, had shown that the genus *Heliconia* is monophyletic. Groupings corresponding to currently recognized species and some subgenera were found, and cultivars and hybrids were found to cluster with their parents. RAPD analysis generally agreed with morphological species classification, except for the position of the subgenus *Stenochlamys*, which was found to be polyphyletic.

Isaza et al. (2012) studied the genetic diversity and molecular characterization of several *Heliconia* species in Colombia. They characterized the genetic variability of 67 genotypes of cultivated *Heliconias* belonging to *H. caribaea* Lamarck, *H. bihai* (L.) L., *H. orthotricha* L. Anderson, *H. stricta* Huber, *H. wagneriana* Petersen, and *H. psittacorum* L.f., as well as that of several interspecific hybrids such as *H. psittacorum* L.f. × *H. spathocircinata* Aristeguieta and *H. caribaea* Lamarck × *H. bihai* (L.) L. An approximation to their phylogenetic analysis was also created. Molecular analysis using amplified fragment length polymorphism (AFLP) markers revealed a total of 170 bands. Two large, well-defined groups resulted: the first group was very closely related to *H. caribaea* and *H. bihai* species with those of *H. orthotricha* and *H. psittacorum* and the second group to *H. stricta* and *H. wagneriana* cultivars. The lowest percentage of polymorphism was found in *H. psittacorum* (17.65%), and the highest was in *H. stricta* (55.88%). Using AFLP, phylogenetic analysis of the species study revealed the monophyletic origin of the Heliconiaceae family and identified the *Heliconia* subgenus as monophyletic while providing evidence of the polyphyletic origin of several representatives of the *Stenochlamys* subgenus.

Filho et al. (2016) undertook an investigation on genetic diversity and morphological characterization of half-sib families of *H. bihai* L., *H. chartacea* Lane ex Barreiros, and *H. wagneriana* Peterson. Its main aim was to characterize 15 half-sib families originating from commercial cultivations, by morphological and molecular markers. The genetic diversity ($\hat{H}E$), considering all individuals of the three species, were 0.103. For *H. bihai* half-sib families, the value of $\hat{H}E$ was 0.242, showing high genetic diversity. The $\hat{H}E$ value for *H. chartacea* was 0.068, indicating low genetic diversity. All individuals of *H. wagneriana* showed the same band patterns, suggesting that the two parental plants were propagated vegetatively from the same plant and may have undergone some endogamic crossings. These results showed that molecular characterization can differentiate individuals closely related as half siblings for *H. bihai* and *H. chartacea*, despite the low variation observed with morphological descriptors. The high genetic diversity observed in *H. bihai* half-sibling genotypes can provide valuable resources for breeding programs.

23.2.7 Cytological Markers to Affirm the Genesis of *Heliconia* Species and Hybrids

The identification of *Heliconia* genotypes – mainly based on morphological characteristics – has generated synonyms. Therefore, it becomes necessary to apply novel techniques to differentiate genotypes and elucidate the origin of hybrids and triploids. Hence, Costa et al. (2016) characterized 18 *Heliconia* genotypes, including 3 hybrids, 2 triploids, and their possible parent plants cytogenetically, by means of chromosome counting, staining with the fluorochromes CMA (chromomycin A3) and DAPI (4′,6- diamidino-2-phenylindole), and FISH (fluorescence in situ hybridization) for 45S and 5S rDNA sites. The analyses revealed 16 diploid ($2n = 2x = 24$)

and 2 triploid genotypes ($2n = 3x = 36$), with new counts for 8 genotypes, including the triploid 'Suriname Sassy.' The karyotypes were symmetric, with chromosomes varying from 0.88 to 3.35 μm. Regarding the CMA/DAPI staining and FISH, one $CMA^+/DAPI^0/5S$ rDNA band was observed in the proximal region of two and three small chromosomes in the diploid and triploid genotypes, respectively. The $CMA^{++}/DAPI^-/45S$ rDNA markings varied from two to four sites per genotype, revealing some interesting heteromorphisms. The distribution of 5S and 45S rDNA sites, both CMA-positive, corroborated the origin of the triploid genotypes 'Sassy' and 'Suriname Sassy' from diploid genotypes of *H. psittacorum*, as well as the origin of the hybrids of *H. psittacorum* × *H. spathocircinata* ('Golden Torch' and 'Golden Torch Adrian') and *H. caribaea* × *H. bihai* cv. 'Jacquinii' from their respective parent species.

Cytogenetic studies of the genus *Heliconia* had revealed the basic number $x = 12$, with a predominance of diploid species ($2n = 2x = 24$), besides some triploids with $2n = 3x = 36$ (Lee et al. 1994; Kaemwong and Eksomtramage 1998; Criley 2000). However, all previous works have only been based on conventional staining and restricted to chromosome counting. Costa et al. (2016) confirmed the numbers determined for *H. psittacorum*, *H. spathocircinata*, *H. latispatha* cv. Distans, H. *pendula* (Anderson 1984), *H. bihai* (Omanakumari and Mathew 1976), *H. rostrata* (Hanson et al. 2001), *H. stricta* cv. 'Fire Bird,' and the hybrid *H. caribaea* × *H. bihai* cv. 'Jacquinii' (Guangsui et al. 2010), for the hybrids 'Golden Torch' and 'Golden Torch Adrian' (Kaemwong and Eksomtramage 1998), and for the triploid *H. psittacorum* cv. 'Sassy' (Lee et al. 1994). Data regarding chromosome numbers were described for the first time by authors (Costa et al. 2016) for *H. stricta*, *H. wagneriana*, *H. psittacorum* 'Paquevira,' *H. psittacorum* cv. 'Red Opal,' *H. psittacorum* 'Strawberries & Cream,' *H. psittacorum* cv. 'Suriname Sassy,' *H. collinsiana*, and *H. rauliniana*.

Despite conservation of the diploid number $2n = 24$, the number of 45S rDNA sites had undergone some marked variations, as the species *H. psittacorum* and *H. bihai* and most of the further analyzed diploid species presented two sites, whereas *H. spathocircinata* and *H. latispatha* cv. 'Distans' exhibited four sites (Costa et al. 2016). The latter two species belong to the subgenus *Heliconia*, despite being clustered in different clades according to phylogeny of Marouelli et al. (2010). On the other hand, *H. bihai* also belongs to the subgenus *Heliconia* but is clustered in one clade with *H. spathocircinata* (Marouelli et al. 2010), although, from the cytogenetic point of view, it has the most distinct karyotype, with a CMA and 45S rDNA pattern different from the two other species. Moreover, polymorphisms were observed in relation to size, evincing bands with distinct sizes both in the CMA and in the 45S rDNA sites, probably related to differences in the copy number of DNA repeats within the 45S rDNA sites (Costa et al. 2016). Additionally, it is important to highlight that subgenera are still under discussion, and despite *Heliconia* being considered monophyletic, at least subgenus *Heliconia* is probably polyphyletic (Marouelli et al. 2010), supported at least in part by FISH analyses.

The number and position of the 45S rDNA sites in the diploid *Heliconia* species may be used as further evidence with regard to the degree of relationship between species and their possible hybrids, as also observed in *Citrus* (Moraes et al. 2007) and *Lilium* (Wang et al. 2015). For instance, the *H. psittacorum* cultivars 'Red Opal,' 'Strawberries & Cream,' and 'Paquevira,' though presenting large morphological differences, such as in bract coloration and inflorescence type, were cytologically very similar, exhibiting slight differences in relation to the secondary constriction. Only 'Red Opal' exhibited size heteromorphism for the chromosome pair carrying the 45S rDNA site. In analyses applying molecular markers, Marouelli et al. (2010) suggested that the latter genotype might be a hybrid between *H. psittacorum* and *H. spathocircinata*. However, the present analyses by CMA/DAPI and FISH did not corroborate this suggestion, since 'Red Opal' presented two CMA^{++}/hybrid origin involving another still unidentified species (Costa et al. 2016).

Some genotypes mentioned in several literature as triploids, for instance, 'Petra,' 'Sassy,' 'Iris,' and 'Fire Flash' (Lee et al. 1994; Kaemwong and Eksomtramage1998), or as hybrids, such as 'Golden Torch,' 'Golden Torch Adrian,' 'Alan Carle,' 'Fire Opal,' 'Jacquinii,' and *H. psittacorum* × *H. marginata*, with putative parent species being indicated for some of them (Costa et al. 2007; Marouelli et al. 2010; Rocha et al. 2010; Isaza et al. 2012; Guimarães et al. 2012, 2014). For the genotypes 'Golden Torch' and 'Golden Torch Adrian,' which were reported as natural hybrids between *H. psittacorum* and *H. spathocircinata* based on morphological aspects and on research using molecular markers (Berry and Kress 1991; Marouelli et al. 2010), three chromosomes carrying CMA^{++}/45S rDNA sites were observed. Two sites were terminal in chromosomes of different size (one big chromosome and one smaller), similar to those carrying 45S rDNA in *H. spathocircinata*, and the third was present in a satellite-carrying chromosome with distended secondary constriction, typical of the carrier chromosome in *H. psittacorum*. These cytogenetic data support their hybrid origin.

In the putative parental species *H. bihai* and in the hybrid *H. caribaea* × *H. bihai* cv. 'Jacquinii,' two chromosomes were identified as possessing CMA^{++}/45S rDNA sites, both with size heteromorphism for the chromosome and the 45S rDNA site. One of the chromosome pairs of 'Jacquinii' showed similar size and morphology to the 45S rDNA carrier *H. bihai* chromosome, with variation only in the size of the 45S rDNA site. This observation noted by Costa et al. (2016) seems to corroborate the indication of *H. bihai* as a possible parent, as proposed by Marouelli et al. (2010) based on molecular analyses, considering that the rDNA loci patterns may present variations in hybrids, as a result of genome-dependent dynamics (Ksiazczyk et al. 2015). In turn, in the triploid genotypes 'Suriname Sassy' and 'Sassy,' three large chromosomes carrying a subterminal $CMA^{++}/DAPI^{-}$/45S rDNA site adjacent to a terminal $CMA^{0}/DAPI^{0}$ satellite were detected, a feature typical of diploid populations of *H. psittacorum*, corroborating their suggested origin.

For taxonomists, there are obstacles regarding the definition of the number of *Heliconia* species: 204 are accepted, and 176 are considered synonyms (The Plant List 2013). This difficulty arises from their description being mainly based on morphological aspects, although various molecular analyses already exist that emphasize the genetic relations between the species, demonstrating the adequate clustering of the hybrids and their possible parent plants, as well as the identification of each species' subgenera (Marouelli et al. 2010; Rocha et al. 2010; Gowda et al. 2012; Isaza et al. 2012; Guimarães et al. 2012). Taxonomic hurdles are also observed for the genus *Citrus*, for which the number of accepted species has varied from 16 to 162 (Swingle and Reece 1967). At present, only three species are accepted as true, the remaining being considered hybrids (Nicolosi et al. 2000). In this genus, the pattern of CMA^{+} bands associated with the localization of the 5S and 45S rDNA sites allowed suggesting the distinction between pure species and those of hybrid origin (Carvalho et al. 2005; Moraes et al. 2007). In *Heliconia*, the cytogenetic characterization using rDNA probes also proved an efficient approach for differentiating genotypes and identifying possible parental species, as well as hybrid and triploid species, by means of typical site patterns and chromosome types.

23.2.8 Self-Incompatibility (SI) in *Heliconia*

Physiological self-incompatibility (SI) is one of the most common mechanisms by which outcrossing is promoted in the angiosperms (East 1940; Whitehouse 1950; Nettancourt 1977). Considerable information on the taxonomic distribution of SI and the mechanisms of self-inhibition is available (Heslop-Harrison 1975; Nettancourt 1977). Yet we are far from a full understanding of SI. Most of our knowledge of SI in angiosperms comes from investigations of temperate zone plants, especially herbaceous annuals and perennials (Nettancourt 1977). Relatively little is known about the breeding systems of trees (Hagman 1975) and tropical species.

Information on breeding systems of plants is important to our understanding of the processes of speciation (Baker 1959). Several early speciation models that attempted to explain the great floristic diversity of the tropics were based on opposing assumptions about breeding systems: one assumed predominant self-pollination of tropical trees (Nettancourt 1977) and the other predominant cross-pollination (Baker 1959). These phyletic models of speciation have since been challenged by other models including quantum evolution (Endler 1977). Even though no consensus on speciation processes in the tropics has been reached, the controversy has focused attention on our lack of knowledge about breeding systems of tropical plants.

Earlier several workers had reported that the majority of tree species they tested in subtropical and tropical, wet, and semideciduous forests are obligate outcrossers (Arroyo 1979; Sobrevila and Arroyo 1982). Only a very small percentage of the total flora, however, has been sampled. In addition, the distinction between self-compatible and self-incompatible species is not always clear cut: self-incompatible plants occasionally set fruit following self-pollination (Bawa 1979).

Studies on breeding systems of tropical plants so far have been restricted to trees of mature and secondary forests; the prevalence of SI in the extensive herbaceous understory flora of these same forests is unknown. The most conspicuous and dominant elements of the lowermost strata of the wet neotropical forests are the large, herbaceous, broad-leaved monocots of the Heliconiaceae, Marantaceae, Zingiberaceae, Costaceae, Araceae, Arecaceae, and Cyclanthaceae. Flowers of all species of the first four families are hermaphroditic. Members of the latter three families tend to be dioecious or monoecious and strongly dichogamous (East 1940; Mora-U and Solis 1980).

Reportedly, in *H. mathiasiae*, no self-pollinations were successful that is why it appears to be fully self-incompatible (Kress 1983). Self-incompatibility is "the inability of a fertile hermaphrodite seed plant to produce zygotes after self-pollination" (De Nettancourt 1977). This is mainly due to the recognition and rejection of self-incompatible pollen, which leads to block self-fertilization. Morphologically self-incompatibility is either homomorphic or heteromorphic, whereas genetically it is grouped as gametophytic or sporophytic based on its determination by haploid genotype or by the genotype of the pollen parent. Sporophytic homomorphic self-incompatibility is not as widespread as gametophytic homomorphic self-incompatibility. A number of families employ gametophytic homomorphic self-incompatibility; however, only a few have been studied at the molecular level. Progress in understanding the molecular biology of heteromorphic self-incompatibility systems has been much slower, and not much is known about how self-incompatible pollen is rejected in these systems. Diverse mechanisms have been reported for the rejection of self-incompatible pollen due to the inherent capability to recognize their own pollen and subsequently prevent self-fertilization. Molecular studies are showing that there are in fact at least two tightly linked loci, one controlling self-incompatibility in the style and second controlling self-incompatibility in the pollen. The stylar S-gene product has been extensively characterized, while the pollen S-gene has not yet been identified. The S proteins are subsequently shown to possess ribonuclease activity and are then referred to as S-RNases. Despite the high sequence diversity, the S-RNases contain a number of conserved regions. There are five highly conserved regions designated C1–C5 (Ioerger et al. 1991). C1, C4, and C5 contain mostly hydrophobic amino acids and may be involved in forming the core structure of the S protein (Kao and McCubbin 1996). Ioerger et al. (1991) identified two areas exhibiting the highest levels of variability and were referred to as "hypervariable domains" HVa and HVb. Due to the high sequence diversity and the hydrophilic nature of these regions, they are thought to be involved in determining S allele specificity (Ioerger et al. 1991). Earlier, Ishimizu et al. (1998) identified four different regions, called PS1 to PS4 ("positive selection"), which may function as the determinant of S allele specificity. Gray et al. (1991) provided further evidence to support the role of S-RNase in pollen rejection when they showed that the S-RNase was able to enter pollen tubes grown in vitro. S-RNases are necessary and sufficient in the pistil for the recognition and rejection of incompatible pollen tubes. These genes, referred to as modifier genes (De Nettancourt 1977), have been placed into three groups depending on how they affect the determinants of allelic specificity (McClure et al. 2000). Group I factors would affect the expression of the S genes such as transcription factors. In recent years, exciting progress has been made toward understanding the molecular basis of several self-incompatibility systems. Several S locus genes have been identified, and the information is starting to emerge on how the encoded S proteins function to cause the rejection of self-incompatible pollen.

Self-incompatibility was studied in populations of 19 species of Central American *Heliconia*. Pollen tube growth and in some cases fruit set were used to determine the extent of physiological self-incompatibility in each species. Responses ranged from total self-rejection in one species to full self-compatibility in the majority of taxa studied. Partial self-incompatibility, as expressed in the number of styles accepting self-pollen and the number of pollen tubes penetrating these styles, was found in one species. Levels of autogamy in excess of 25% were detected in five species. The prevalence of self-compatibility in these herbaceous members of the understory of tropical wet forests contrasts with the common occurrence of obligate outcrossers in the canopy layers. The availability of long-distance pollinators and low daily flower output may promote outcrossing despite the scarcity of physiological self-incompatibility in these plants (Kress 1983).

With respect to *Heliconia*, the two species that show a significant degree of SI, *H. mathiasiae* and *H. tortuosa*, are not closely related to each other; the closest relatives of each are found in taxonomically unrelated groups (Kress 1981). A phylogenetic tree (cladogram) of 27 species of *Heliconia* that was previously constructed using morphological characters includes most of the species tested for SI (Kress 1981). If breeding system data are superimposed on the cladogram, both *H. mathiasiae* and *H. tortuosa* are derived from "hypothetical" ancestors that were self-compatible. According to this evidence, SI has evolved independently in self-compatible taxa twice in the genus. The evolutionary pattern in *Heliconia*, therefore, contradicts the one documented for most other flowering plant groups.

Comparisons of breeding systems of genera closely related to *Heliconia* provide no additional insight into the evolution of SI. *Musa* and *Ensete*, the only two genera in the Musaceae, are primarily monoecious and highly protogynous, both derived character states for the order. It is not known whether members of the three genera of the Strelitziaceae are self-incompatible (Kress 1983).

Potential for physiological self-incompatibility in *Heliconia bihai* as a mechanism to promote outcrossing was studied by Meléndez-Ackerman et al. (2008) in St. Lucia. Their results suggested that *H. bihai* is self-compatible. However, plants vary in their degree of compatibility from full to partial self-compatibility. They found only traces of physiological incompatibility, but conclusive determination of its mechanism would require further testing. In contrast to Central American *Heliconia* species, they found that more pollen tubes were able to grow inside *H. bihai* styles following artificial pollinations. Additional studies would be needed to test if other populations of *H. bihai* share this phenomenon.

23.2.9 Other Mechanisms Promoting Outcrossing

Lack of physiological SI does not preclude the functioning of other mechanisms that may promote cross-fertilization in *Heliconia*. Orientation of floral parts differs among species and results in differential placement of pollen on hummingbirds. Within any flower, however, separation of the stigma and anthers is not great and probably has little effect on stigmatic selection of cross-pollen over self-pollen.

Preliminary studies suggest that dichogamy (protandry and protogyny) does not play any part in the breeding systems of these plants. Anthers dehisce before anthesis and stigmatic surfaces are receptive at flower opening (Kress, unpubl.). Without appropriate genetic markers, it is very difficult to ascertain whether competition between cross and self-pollen in pistils leads to significant outcrossing (Bateman 1956; Levin 1975). In several species of *Heliconia*, outcross pollen and pollen tubes germinated and grew at roughly the same rate (6 mm per hour) as self-pollen in non-mixed pollen applications (Kress, unpubl.). However, undetected subtle differences in growth rates of self- and cross-pollen may favor cross-fertilization.

The amount of inbreeding within a population depends on pollinator foraging behavior as well as physiological SI (Linhart 1973; Levin 1975; Schmitt 1980). In species of *Heliconia* visited primarily by territorial non-trap lining pollinators, pollen flow between individual plants will be severely restricted (assuming that hummingbird territories are limited to a small number of clones). In contrast, interplant pollen flow will be significantly greater in species that are visited by trap lining pollinators. These predicted patterns of pollen flow have been demonstrated to a limited extent in *Heliconia* using histological marker dyes to track pollinator movement within and between plants of several species (Linhart 1973). Recent studies by Ray, Stiles, and Kress (unpubl.) provide additional evidence for limited interplant pollen flow in *H. imbricata*, a species pollinated by non-trap lining hummingbirds, even though it was observed that territoriality among hummingbirds eventually broke down and pollen was exchanged among territories.

23.2.10 Future Prospect

Heliconia plants are found in all tropical regions around the world. The economic importance of *Heliconia* genus relies on its cultivation as an ornamental plant which is used either as cut flowers or as garden and potted plants. In northeast Brazil, especially under the rain forest conditions in the State of Pernambuco, these plants have increased their production, and most genotypes bloom throughout the year (Costa et al. 2006). This genus has great potential to be exploited in the international flower market by adjusting the quality of its products to international requirements which can be achieved through genetic breeding programs.

References

Anderson GJ (1984) Chromosome number reports LXXXV. Taxon 33:756–760

Andersson L (1981a) Revision of *Heliconia* sect. *Heliconia* (Musaceae). Nord J Bot 1:759–784

Andersson L (1981b) The neotropical genera of Marantaceae: circumscription and relationships. Nord J Bot 1:218–245

Andersson L (1998) Heliconiaceae. In: Kubitzki K (ed) The families and genera of vascular plants. Flowering plants monocotyledons, alismatanae and commelinanae (except Gramineae), vol 4. Springer, Berlin, pp 226–230

Arroyo MTK (1979) Comments on breeding systems in neotropical forests. In: Larsen K, Holm-Nielson LB (eds) Tropical botany. Academic, New York, pp 371–380

Bahubali MD, Singh A, Mahatma MK (2014) Post harvest physiology and quality of *Heliconia* inflorescence cv. 'Golden Torch' as influenced by antioxidants. Indian J Hortic 69:259–264

Baker HG (1959) Reproductive methods as factors in speciation in flowering plants. Cold Spring Harb Symp Quant Biol 24:177–191

Baldwin BG, Sanderson MJ, Porter JM et al (1995) The ITS region of nuclear ribosomal DNA: a valuable source of evidence on angiosperm phylogeny. Ann Missouri Bot Gard 82:247–277

Bateman AJ (1956) Cryptic self-incompatibility in the wallflower: *Cheiranthus cheiri* L. Heredity 10:257–261

Bawa KS (1979) Breeding systems of tree species of a lowland tropical community. Evolution 28:85–92

Berry F, Kress J (1991) *Heliconia* Identification Guide. Smithsonian Institution Press, Washington, DC, p 344. Retrieved from: https://www.cambridge.org/core/journals/journal-of-tropical-ecology/article/f-berry-w-jkress-1991-heliconia-an-identification-guidesmithsonian-institution-press-washington-dc-usa344-pagesclothback-isbn-1560980060-price-35-00-paperback-isbn-1560980079-price-1695/3422D0B88965F25E6070FD74A1B13747

Brown AH, Weir BS (1983) Measuring genetic variability in plant populations. In: Tanskley SD, Orton TJ (eds) Isozymes in plant genetics and breeding: Part A. Elsevier, Amsterdam, pp 219–229

Bussell JD, Waycott M, Chappill JA (2005) Arbitrarily amplified DNA markers as characters for phylogenetic inference. Perspect Plant Ecol Evol Syst 7:3–26

Carvalho R, Soares FWS, Brasileiro-Vidal AC (2005) The relationships among lemons, limes and citron: a chromosomal comparison. Cytogenet Genome Res 109:276–282. https://doi.org/10.1159/000082410

Costa AS, Loges V, Castro ACR et al (2006) Perfilhamento e expansão de touceiras de helicônias. Hort Bras 24:460–463

Costa AS, Loges V, Castro ACR (2007) Variabilidade gene'tica e correlações entre caracteres de cultivares e hı'bridos de *Heliconia psittacorum*. Revista Brasileira de Ciências Agrárias 2:187–192

Costa MGS, Leite BSF, Loges V et al (2016) Chromosome markers confirm origin of *Heliconia* hybrids and triploids. Euphytica 25:1–14. https://doi.org/10.1007/s10681-016-1780-5

Criley RA (2000) Seasonal flowering patterns for Heliconia shown by grower records. Acta Hortic 541:159–165. https://doi.org/10.17660/ActaHortic.2000.541.22

Cronquist A (1981) An integrated system of classification of flowering plants. Columbia University Press, New York, pp 1–1262

Dahlgren RMT, Clifford TH, Yeo PF (1985) The families of monocotyledons. Springer, Berlin, pp 350–358

Daniels GS, Stiles FG (1979) The Heliconia taxa of Costa Rica.Keys and descriptions. Brenesia15:1–150

De Nettancourt D (1977) Incompatibility in angiosperms. Sex Plant Reprod 10:185–199

East EM (1940) The distribution of self-sterility in the flowering plants. Proc Am Phil Soc 82:449–518

Endler JA (1977) Geographic variation, speciation and clines. Princeton University Press, Princeton

Erdtman G (1952) Pollen morphology and plant taxonomy, angiosperms. Almquist & Wiksell, Stockholm

Faegri K, Iversen J (1950) Textbook of modern pollen analysis. Munksgaard, Copenhagen, p 169

Filho RMM, Pereira FRA, Martins RSS (2016) Genetic diversity and morphological characterization of half-sib families of *Heliconia bihai* L., *H. chartacea* Lane ex Barreiros, and *Heliconia wagneriana* Peterson. Genet Mol Res 15(2):1–9

Goh CJ, Kumar PP, Yau CK (1995) Genetic variations detected with RAPD markers in *Heliconia*. Acta Hortic 420:72–74

Gowda V, Erickson DL, Kress WJ (2012) Development and characterization of microsatellite loci for two Caribbean *Heliconia* (Heliconiaceae: *H. bihai* and *H. caribaea*). Am J Bot 99:81–83. https://doi.org/10.3732/ajb.1100386

Gray JE, McClure BA, Bonig I et al (1991) Action of the style product of the self-incompatibility gene of *Nicotiana alata* (S-RNase) on *in vitro* grown pollen tubes. Plant Cell 3:271–283

Guangsui Y, Junmei Y, Qingyun L, Yanxia Z, Zhiqun Z (2010) Karyotypes comparison of eight varieties of *Heliconia* spp. Chin J Trop Crops 12:2118–2123

Guimarães WNR, Ferraz GMG, Martins LSS et al (2012) Genetic diversity analysis of *Heliconia psittacorum* cultivars and interspecific hybrids using nuclear and chloroplast DNA regions. In: Aflakpui G (ed) Agricultural science. InTech, Rijeka, pp 11–22. https://doi.org/10.5772/48815

Guimarães WNR, Martins LSS, Castro CEF (2014) *Heliconia* phenotypic diversity based on qualitative descriptors. Genet Mol Res 13:3128–3142. https://doi.org/10.4238/2014.April.17.9

Hagman M (1975) Incompatibility in forest trees. Proc R Soc Lond B 188:313–326

Hanson L, McMahon KA, Johnson MA, Bennett MD (2001) First nuclear DNA C-values for another 25 angiosperm families. Ann Bot 88:851–858

Heslop-Harrison J (1975) Incompatibility and the pollen-stigma interaction. Annu Rev Plant Physiol 26:403–425

Hommel E, Mathiesen ER, Aukland K, Parving HH et al (1990) Pathophysiological aspects of edema formation in diabetic nephropathy. Kidney Int 38:1187–1192

Ioerger TR, Gohlke JR, Xu B et al (1991) Primary structural features of the self-incompatibility protein in Solanaceae. Sex Plant Reprod 4:81–87

Isaza L, Marulanda ML, López AM (2012) Genetic diversity and molecular characterization of several *Heliconia* species in Colombia. Genet Mol Res 11:4552–4563

Ishimizu T, Endo T, Yamaguchi-Kabata Y et al (1998) Identification regions in which positive selection may operate in S-RNase of Rosaceae: implication for S-allele-specific recognition sites in S-RNase. FEBS Lett 440:337–342

Janakiram T, Kumar PP (2011) Enhancing flower potential of Heliconia. Indian J Hortic 56(1):22–25

Johnson HW, Robinson HF, Comstock RE (1955) Estimate genetic and environmental variability in soyabean. Agron J 47:314–318

Johansen LB (2005) Phylogeny of Orchidantha (Lowiaceae) and the Zingiberales based on six DNA regions. Syst Bot 30:106–117

Kaemwong S, Eksomtramage L (1998) Chromosome numbers of genus *Heliconia*. Plant Genet Breed 20:489–495

Kao TH, McCubbin AG (1996) How flowering plants discriminate between self and non self-pollen to prevent inbreeding. Proc Natl Acad Sci U S A 93:12059–12065

Kayarohanam S, Kavimani S (2015) Current trends of plants having antidiabetic activity: a review. J Bioanal Biomed 7:55–65

Kress WJ (1981) Reproductive biology and systematics of central American *Heliconia* (Heliconiaceae). Ph.D. dissertation, Duke Univ, Durham

Kress WJ (1983) Self-incompatibility in central American *Heliconia*. Evolution 37(4):735–744

Kress WJ (1984) Systematics of central American *Heliconia* (Heliconiaceae) with pendent inflorescences. J Arnold Arbor 65:429–532

Kress WJ (1989) New taxa and notes on Heliconia (Heliconiaceae). Selbyana 11:49–53

Kress WJ (1990) The diversity and distribution of Heliconia (Heliconiaceae) in Brazil. Acta Bot Bras 4:159–167

Ksiażczyk T, Zwierzykowska E, Molik K (2015) Genome-dependent chromosome dynamics in three successive generations of the allotetraploid *Festuca pratensis* and *Lolium perenne* hybrid. Protoplasma 252:985–996. https://doi.org/10.1007/s00709014-0734-9

Kumar V, Prabha R (2018) Extraction and analysis of natural dye. J Nat Prod Plant Resour 8(2):32–38

Lee YH, Nig NY, Goh CJ (1994) Pollen formation and fruit set in some cultivars of *Heliconia psittacorum*. Sci Hortic 60:167–172

Levin DA (1975) Gametophytic selection in *Phlox*. In: Mulcahy DL (ed) Gamete competition in plants and animals. North Holland Publ. Co., Amsterdam, pp 207–218

Linhart YB (1973) Ecological and behavioural determinants of pollen dispersal in humming bird pollinated *Heliconia*. Am Nat 107:511–523

Loges V, Castro CEF, Guimarães WNR (2012) Agronomic traits of *Heliconia* for cut flowers use and molecular markers. Acta Hortic 937:535–543

Maglianesi MA, Boehning-Gaese K, Schleuning M (2015) Different foraging preferences of hummingbirds on artificial and natural flowers reveal mechanisms structuring plant-pollinator interactions. J Anim Ecol 84:655–664

Malakar M, Acharyya P, Biswas S (2015) Evaluation of *Heliconia* species based on agro-morphological traits. Int J Agric Environ Biotechnol 8(4):957–964. https://doi.org/10.5958/2230-732X.2015.00109.6

Malakar M, Acharyya P, Biswas S (2019a) Consequences of divergent vase solutions on post-harvest durability and quality of *Heliconia* inflorescences. Acta Hortic 1256:77–94. https://doi.org/10.17660/ActaHortic.2019.1256.12

Malakar M, Acharyya P, Biswas S (2019b) Rejoinder of diverse organic growing media on morphological attributes of nine *Heliconia* species and varieties under West Bengal condition. J Crop Weed 15(1):35–44

Malakar M, Acharyya P, Biswas S (2019c) Study on genetic variability and correlation for floral traits and yield in *Heliconia* genotypes. J Crop Weed 15(1):64–72

Malakar M, Acharyya P, Biswas S (2020a) Molecular characterization of ten *Heliconia* (Heliconiaceae) genotypes by means of RAPD markers. Acta Hortic 1288:95–102. https://doi.org/10.17660/ActaHortic.2020.1288.14

Malakar M, Acharyya P, Biswas S (2020b) Discernment of pollen grain fertility and stigma receptivity plying hasty-easy techniques of *Heliconia* species L. under microscope. Acta Hortic 1288:139–148. https://doi.org/10.17660/ActaHortic.2020.1288.21

Mangave BD, Singh A, Mahatma MK (2013) Effects of different plant growth regulators and chemicals spray on post-harvest physiology and vase life of *Heliconia* inflorescence cv. 'Golden Torch'. Plant Growth Regul 69:259–264

Marouelli LP, Inglis PW, Ferreira MA (2010) Genetic relationships among *Heliconia* (Heliconiaceae) species based on RAPD markers. Genet Mol Res 9(3):1377–1387

McClure BA, Cruz-Garcia F, Beecher BS et al (2000) Factors affecting inter and intra-specific pollens rejection in *Nicotiana*. Ann Bot 85:113–123

Meléndez-Ackerman E, Rojas-Sandoval J, Planas S (2008) Self-compatibility of micro-gametophytes in *Heliconia bihai* (Heliconiaceae) from St. Lucia. Caribb J Sci 44(2):145–149

Mollik MAH, Hassan AI, Paul TK (2010) A survey of medicinal plant usage by folk medicinal practitioners in two villages by the Rupsha River in Bagerhat District, Bangladesh. Am Eur J Sustain Agric 4(3):349–356

Moore PD, Webb GA, Collinson ME (1991) Pollen analysis, 2nd edn. Oxford Blackwell Scientific Publications, London

Moraes AP, Soares-Filho WS, Guerra M (2007) Karyotype diversity and the origin of grapefruit. Chromosom Res 15:115–121. https://doi.org/10.1007/s10577-006-1101-2

Mora-U J, Solis EM (1980) Polinizacion en Bactris gasipaes H. B. K. (Palmae). Rev Biol Trop 28: 153–174

Narkar ND, Sheela VL, Kadam DS (2017) Studies of pollen grains in *Heliconia* (*Heliconia* spp.) for improvement. Curr Hortic 4(1):48–51

Nascimento HRD, Gallo R, Karsburg IV (2014) Cytogenetic and identification of the nucleolus organizer region in *Heliconia bihai* (L.). Rev Ceres 61(4):451–457

Nelson J (2000) The HSPR News Letter, Published by *Heliconia* Society of Puerto Rico. 2:1–3

Nettancourt DDE (1977) Incompatibility in angiosperms. Springer, New York

Nicolosi E, Deng ZN, Gentile A, La Malfa S, Continella G, Tribulato E (2000) Citrus phylogeny and genetic origin of important species as investigated by molecular markers. Theor Appl Genet 100:1155–1166. https://doi.org/10.1007/s0012200.51419

Nowicke FW, Skvarla FF (1977) Pollen morphology and the relationship of the plumbaginaceae, polygonaceae, and primulaceae to the order centrospermae. Smithson Contrib Bot 37:1–23. Retrieved from: http://rcpol.org.br/wp-content/uploads/2018/03/95-Nowicke-_-Skvarla-1977-Pollen- Morphology-and-the-Relationship-of-the-Plumbaginaceae-Polygonaceae-and-Prirnulaceae-to-the-Order- Centrospermae.pdf

Omanakumari N, Mathew PM (1976) Developmental abnormalities in microspores of *Heliconia bihai* Linn. New Bot 3:33–37

Palma ART, Vieira IN (2006) Thylamys genus in Brazil: natural history and geographic distribution. In: Caceres NC, Emygdio L, Monteiro-Filho A (Org) The Marsupails of Brazil: biology, ecology and evolution. Campo Grande: Publisher of the Federal University of Mato Grosso do Sul, pp 271–286

Panse VG (1957) Genetics of quantitative characters in relation to plant breeding. Indian J Genet 17(3):10–328

Pinheiro CR, Amorim JAE, Diniz LEC et al (2011) Diversidade genética de isolados de Ralstonia solanacearum e caracterização molecular quanto à filotipos e sequevares. Pesq Agropec Bras 46: 593–602

Rocha FHA, Loges V, Costa ASD (2010) Genetic study with *Heliconia psittacorum* and interspecific hybrids. Crop Breed Appl Biotechnol 10:282–288. https://doi.org/10.1590/S1984-70332010000400001

Salih J (1994) Pollen studies on *Heliconia*. A dissertation submitted in partial fulfilment for the degree of Bachelor of Science with honours in botany. National University of Singapore, pp 85–86. Retrieved from: http://lkcnhm.nus.edu.sg/dna/docs/ed566648139cd5114a1ac01a07611274.pdf

Sanjeev SJ, Sheela S, Geetha Lekshmi PR (2010) Floral biology and correlation studies in *Heliconia*. J Ornam Hortic 13:8–19

Schmitt J (1980) Pollinator foraging behaviour and gene dispersal in *Senecio* (Compositae). Evolution 34:934–943

Shahriar S, Farzana B, Rahmatullah M (2017) *Heliconia rostrata* Ruiz & Pav. (Heliconiaceae) – a previously unreported plant for treatment of diabetes and diabetes-induced edema. Asian J Pharmacogn 1(4):51–54

Sheela VL, Lekshmi PRG, Nair CSJ (2006) Molecular characterization of *Heliconia* by RAPD assay. J Trop Agric 44(1/2):37–41

Sheela VL, George TS, Rakhi R (2007) Variability studies in cut flower varieties of *Heliconia*s. - Indian J Hortic 64(1):109–111

Silva MW, Ono EO, Santos MHLC et al (2016) Growth and production of *Heliconia* under different light conditions. Ciências Agrárias, Londrina 38(1):7–18

Silva MW, Ono EO, Santos MHLC et al (2017) Growth and production of *Heliconia* under different light conditions. Agronomia 38(1):7–18

Simao DG, Lucia SV (2001) Morphology and anatomy in Heliconia angusta Vell. and H. velloziana L. Emygd. (Zingiberales: Heliconiaceae) from the Atlantic forest of south-eastern Brazil. REvta Brasil Bot 24(4):415–424

Sobrevila C, Arroyo MTK (1982) Breeding systems in a montane tropical cloud forest in Venezuela. Plant Syst Evol 140:19–37

Srinivas M, Kumar R, Janakiram T (2012) Evaluation of *Heliconia* genotypes for vegetative and flowering traits. Indian J Genet 72(3):397–399

Stuessy TF (1990) Plant taxonomy: the systematic evaluation of comparative data. Columbia University Press, New York

Swingle WT, Reece PC (1967) The botany of Citrus and its wild relatives. In: Reuther W, Weber HJ, Batchelor LD (eds) The Citrus industry: history, world distribution, botany and varieties. University of California Press, Berkeley, pp 190–430

Temeles EJ, Pan IL, Brennan JL (2000) Evidence for ecological causation of sexual dimorphism in a hummingbird. Science 289:441–443

Urooj-Ul-Nissa B, Khan FU, Neelofar N et al (2015) Physiological and flowering response of Dahlia (*Dahlia variabilis* Desf.) cv. 'Pink Attraction' to growing media. J Plant Pest Sci 2(1): 33–42

Wang Q, Wang J, Zhang Y, Zhang Y, Xu S, Lu Y (2015) The application of fluorescence *in situ* hybridization in different ploidy levels cross-breeding of lily. PLoS One 10:1268–1299. https://doi.org/10.1371/journal.pone.0126899

Watson DP, Smith RR (1914) Ornamental *Heliconia*s. Published by Hawaii Cooperative Ex tension Service, Circular 482. Retrieved from: https://www.ctahr.hawaii.edu/oc/freepubs/pdf/C1-482.pdf

Whitehouse HK (1950) Multiple-allelomorph incompatibility of pollen and style in the evolution of the angiosperms. Ann Bot 14:199–216

Biology of Floral Scent Volatiles in Ornamental Plants

24

Upashana Ghissing and Adinpunya Mitra

Contents

U. Ghissing · A. Mitra (✉)
Natural Product Biotechnology Group, Agricultural and Food Engineering Department, Indian Institute of Technology Kharagpur, Kharagpur, West Bengal, India
e-mail: adin@iitkgp.ac.in

S. K. Datta, Y. C. Gupta (eds.), *Floriculture and Ornamental Plants*, Handbooks of Crop Diversity: Conservation and Use of Plant Genetic Resources,
https://doi.org/10.1007/978-981-15-3518-5_27

Abstract

Floral scent has fascinated humans since antiquity and since then played a major aesthetic and commercial role in our lives. Yet, the principal function of the plethora of volatile compounds is to promote ecological interactions between flowers and their specific pollinators. Floral fragrance is composed of low molecular weight volatile organic compounds (VOCs) and derived from the terpenoid, phenylpropanoid/benzenoid, and fatty acid biosynthetic pathways. These pathways are regulated by a network of complex endogenous and external factors that generate a fine-tuned temporal emission of floral scent. They are produced in different subcellular compartments of the floral tissue and rely on primary metabolic pathways for the supply of precursors for their biosynthesis. Recent advances in instrumentation, in association with our current ability to isolate and characterize genes and the enzymes they encode, have greatly improved our understanding of how plants synthesize and regulate the production of these specialized compounds. In this chapter, we provide an overview of the complex biological mechanisms governing the biogenesis of scent in flowers.

Keywords

Flower volatiles · Pollinators · Biosynthesis · Terpenoids · Phenylpropanoids/benzenoids regulation · Localization

24.1 Introduction

Intricate perianth shapes, spectacular color patterns, and a wide range of fragrance are the key features of ornamental flowers. Humans have since long made huge profits from these lucrative features of ornamentals in the form of perfumes, toiletries, cosmetics, flavorings, and so on. Despite their attractiveness to human minds, plants have not evolved their highly valued features for human benefit but as manipulative strategies to lure their valuable pollinator partners. In flowers, color is rarely a specific signal, since most animals have limited abilities for distinguishing subtle color tinges. Solely relying on visual signals is more difficult for use by nocturnal pollinators that forage on night-blooming flowers. Owing to its chemical complexity, scent on the other hand is surprisingly species specific, and its perception triggers innate behavioral responses in potential pollinators. The unique signature combinations of volatile molecules creating the small and not-so-small variances in fragrance spectra among flowers of different species can be differentiated by the olfactory receptors of insect antennae assisting them to find and visit their flower(s) of choice. Olfactory cues enhance attraction over long distances as fragrance signals are primarily dispersed by air currents away from their sources. Further it is the principal cue for foraging under crepuscular conditions such that the signature scent signals fabricated by the flowers pollinated by nocturnal visitors like moths have been described as "white olfactory image" (Paul et al. 2020). The presence of scent signals thus clearly affects pollinator behavior and, therefore, is a crucial part of floral signaling involved in plant reproductive success. Thus, thousands of plant species actively emit scent signals despite the fact that scent biosynthesis and emission are both metabolically very costly. The goal of this chapter is to provide a general survey of the literature of fragrance biology that covers diverse functions, chemistry, biosynthesis, regulation, detection, localization, and trafficking in floral organs.

24.2 Diverse Functions of Floral Scents

Flowers present pollinators with an almost unlimited range of species-specific odors. The number and diversity of these constituents make floral scent one of the most variable aspects of plant phenotype. Some compounds are more or less ubiquitous, while others are found uniquely only in certain species. Floral scents are labelled as "pleasant," "strong," "fruity," "perfumy," "of decay," etc. where each scent type is suggestive of attracting different pollinator types. These suites of VOCs constituting the floral scents of individual species are dynamic in space and time, thus displaying plasticity in response to pollinator abundance profiles.

Pollinators have excellent olfactory acuity and can learn to associate specific scents with floral rewards. This is highly beneficial as distinctive signals that directly increase the efficiency of pollen transfer. Diurnal butterflies utilize floral scent as an innate cue to identify nectar-providing flowers. Behavioral observations showed that the potential pollinator fauna can discriminate the promising targets based on qualitative as well as quantitative aspects of floral scent composition. It was observed that the bumblebee *Bombus terrestris* can even assess a correlation between the strength of an olfactory scent signal and the amount of reward that is being offered by a flower; such assessment may be particularly significant in plants where the amount of nectar is highly variable (Knauer and Schiestl 2015). The discriminatory visitation based on floral scent has important implications for plant reproductive success.

Flowers produce pollen and ovules for the next generation, so they have a high fitness value to the plant, and therefore the finding of chemicals that defend flowers against herbivores and pathogens is not surprising. Thus, the unique constitutive volatile blend produced by flowers in addition to attracting mutualists such as pollinators also deters antagonists such as overexploiting non-pollinating vectors, pathogens, floral herbivores, etc. Floral traits therefore represent a compromise among selection by mutualists and antagonists. Selection pressure attributed by florivores is particularly strong, as per damage to floral tissues may have a greater impact on plant fitness than damage to other plant tissues. Florivores directly impact plant fitness by consuming pistils and stamens and gametes or causing the abortion of flowers. They also cause collateral damage to plant-pollinator interactions by damaging petals which alters floral shape and size or simply by being present on a flower. Therefore, fragrance bouquets in addition to its attractiveness also displayed defensive functions too. 2-Phenylethanol emission and flower damage by ants were negatively correlated. Isoeugenol and benzyl benzoate have also been shown to act as feeding deterrents, thus reducing damage from florivores (Kessler et al. 2013). In the ephemeral flowers of *Murraya paniculata*, high emission of linalool observed particularly in the monsoon season is thought to be strategic in retarding bacterial and yeast growths in the floral phyllosphere that are often introduced by nectar-foraging bees (Paul et al. 2020).

Flowers in addition have also been demonstrated to produce deceptive mating signals of female insects, generally hymenopterans. An outstanding example in this area concerns the spectacular flowers of an orchid species *Ophrys arachnitiformis* that are visited by male solitary bees who appear to be deceived by the scent signals that are very similar to the female sex pheromones (Vereecken and Schiestl 2009). The chemical signals from the flowers elicit sexual behavior in males, which try to copulate with the flower labellum. Further copulatory attempts with another flower ensure that the pollinia is transferred to the other flower's stigmatic surface. This phenomenon known as pseudocopulation generates a high degree of mutual dependence between the plant and its pollinators. Thus, understanding the function of floral scent may be central to the evolution of floral signal complexity.

The gaining of comprehensive insights into the ecology of floral scent has been notoriously challenging because of its functional complexity. Knowledge of how insect pollinators respond to specific floral volatiles is still elementary, so very few individual substances have been assigned specific adaptive values. Hence based on their diverse functions, it can be assumed that floral scent has evolved as a mosaic, with different VOCs facing different selections and to some extent independent evolutionary fates.

24.3 The Chemical Diversity of Floral Scent Volatiles

Understanding of the chemistry of floral volatiles is critical in interpreting their evolution and biological function. Scent is a highly complex element of the floral phenotype, with dynamic patterns of emission and chemical composition bearing a substantial economic significance. Identification of floral scent requires highly advanced specifically designed equipment and interdisciplinary skills for chemical characterization. Further, while evaluating the chemical profile of a floral scent, the reports on the fragrance chemistry are never final as different methodological approaches may report different scent compositions in the same plant species, particularly in the relative proportions of individual volatiles. However, in the past decade, improvement of efficient technologies for collection, identification, and quantification has advanced our understanding of floral scents and their variation. To date, over 1700 compounds have been identified from the scent of flowers belonging to over 90 different plant families (Dudareva et al. 2013). In the atmosphere, the fragrant signal information becomes more complex due to extreme dilution and variation in chemical structures and properties of the VOCs. It consists of a composite blend of low molecular weight compounds (under 300 Da), with high vapor pressure, typically lipophilic in nature, and is mostly emitted from the floral organs into the external atmosphere. Floral volatiles mostly comprise a blend of the following classes of compounds: terpenoids, fatty acid derivatives including lipoxygenase pathway products, phenylpropanoids and benzenoids, C5-branched compounds, and various nitrogen- and sulfur-containing compounds. Chemistry of the fragrance differs among species on the basis of the number and relative proportion of constituent volatile compounds.

Significant progress has been made in determining the biochemical routes to floral volatile formation along with the molecular mechanisms that regulate their formation. The diversity in the fragrant molecules is mainly derived from specific enzymatic catalysis. Characterization of enzymes and genes involved in the biosynthesis of fragrance molecules in model plant species such as *Clarkia breweri*, snapdragon (*Antirrhinum majus*), *Petunia hybrida*, rose (*Rosa* spp.), and *Nicotiana* spp. has significantly contributed to our understanding of the genesis of floral volatiles and their role in the biology of plants.

24.3.1 Biosynthesis of Terpenoids

Among plant secondary metabolites, terpenoids are structurally the most diverse group and play an important role in plant-pollinator, plant-pathogen, and plant-plant interactions. They also function as phytoalexins in plant defense or as signals involving indirect defense responses against herbivores and their natural enemies, thus avoiding further damage. In *Arabidopsis* flowers, the expression of monoterpene and sesquiterpene synthases is limited to the stigma, ovule, nectaries, and sepals, signifying the role of volatile terpenoids in protecting vulnerable sites of flowers in addition to its role as an attractant to pollinators (Tholl and Lee 2011). They are also key components of membranes (sterols), photosynthetic pigments (carotenoids, chlorophylls, plastoquinone), phytohormones (cytokinins, abscisic acid, gibberellins, brassinosteroids), and ubiquinones functioning in mitochondrial electron transport. Terpenoids (isoprenes, monoterpenes, and sesquiterpenes) are found in almost all higher plants and constitute the largest class of floral volatile compounds. Terpenoids are also of immense significance in the pharmaceutical, food, and cosmetic industries. Hence, based on their broad distribution and functional resourcefulness, constant efforts are being made to understand the biosynthetic pathways and comprehend the regulatory mechanisms governing the production of terpenoids. Till date, more than 40,000 individual structures have been identified encompassing 556 scent compounds and constitute approximately 55% of secondary metabolites in plants (Abbas et al. 2017). They show an extremely variable chemical structure yet share a common pathway of biosynthesis. They occur in almost all plant organs, viz., leaves, roots, stems, fruits, and seeds; however the highest amounts are present predominantly in flowers. Here they serve as a major cue for attracting pollinators serving as vectors in pollen transfer especially in long-distance pollinator attractant for night-emitting plants. For example, *Petunia axillaris*, which is a moth-pollinated species, emits larger amounts of volatile terpenoids than day-emitting bee pollinated plants of the same genus such as *P. integrifolia* (Abbas et al. 2017).

Terpenoids are biosynthesized from two interconvertible C_5 units: isopentenyl diphosphate (IPP) and its allelic isomer dimethylallyl diphosphate (DMAPP). These C_5 units serve as substrates/precursors for the biosynthesis of various terpenoids. Two compartmentally detached pathways, the mevalonic acid (MVA) and methylerythritol phosphate (MEP) pathways, are involved in the biosynthesis of terpenoid volatiles.

24.3.1.1 The Mevalonic Acid Pathway

The MVA pathway provides the precursor molecules for the biosynthesis of sesquiterpenes, phytosterols, brassinosteroids, and triterpenes. In plants, it is thought to be localized mainly in the endoplasmic reticulum (ER)/cytosol compartment of the cell (Fig. 1). Recent evidences in *Arabidopsis* however prove the targeting site for several of its enzymes in the peroxisome (Simkin et al. 2011). This pathway comprising of six enzymatic reactions begins with the consecutive condensation of three molecules of acetyl-CoA to the C_6-compound 3-hydroxy-3-methylglutaryl-CoA (HMG-CoA). The condensation steps are catalyzed by two separate enzymes, acetoacetyl-CoA thiolase (AACT, EC 2.3.1.9), which is involved in the condensation of two acetyl-CoA molecules to form the intermediate acetoacetyl-CoA (AcAc-CoA), and HMG-CoA synthase (HMGS, EC 2.3.3.10) which is responsible for forming HMG-CoA by condensation of the intermediate molecule of AcAc-CoA with a third molecule of acetyl-CoA. Following this in an irreversible step catalyzed by the enzyme HMG-CoA reductase (HMGR, EC 1.1.1.34), HMG-CoA is converted into mevalonate by two sequential reduction steps, each utilizing NADPH as a reducing equivalent. It is generally accepted that the reaction catalyzed by HMGR is the main rate-limiting step in the MVA pathway. This has been drawn from experiments on *hmg* mutants of *Arabidopsis* with reduced sterol and triterpene levels, whereas an increase in sterol levels was found in transgenic *Arabidopsis* overexpressing HMGR. Positive correlations between *HMGR* gene expression and the induced production of sesquiterpene phytoalexins have also been reported in many plants (Tholl and Lee 2011). Further, mevalonate is transformed into IPP by two sequential ATP-dependent phosphorylation steps, catalyzed by the enzymes mevalonate kinase (MVK, EC 2.7.1.36) and phosphomevalonate kinase (PMK, EC 2.7.4.2). Finally in the pathway an ATP-driven decarboxylative elimination is catalyzed by mevalonate diphosphate decarboxylase (MPDC, EC 4.1.1.33). IPP derived from the MVA pathway in the cytosol is further acted upon by isopentenyl diphosphate isomerase (IDI, EC 5.3.3.2), a divalent, metal ion-requiring enzyme, to form dimethylallyl diphosphate (DMAPP).

24.3.1.2 The Methylerythritol Phosphate Pathway

In all plants, both IPP and DMAPP are also synthesized by a second alternative pathway, 2-C-methyl-D-erythritol 4-phosphate (MEP), which is located entirely in the plastid (Fig. 1). In floral tissue, this plastidial localization of the terpenoid biosynthesis was first seen in daffodil petals by demonstrating capabilities of the isolated chromoplasts in catalyzing the production of limonene, myrcene, ocimene, and linalool (Mettal et al. 1988). The MEP pathway involves seven enzymatic steps and begins with the Mg^{2+}-dependent condensation of D-glyceraldehyde 3-phosphate

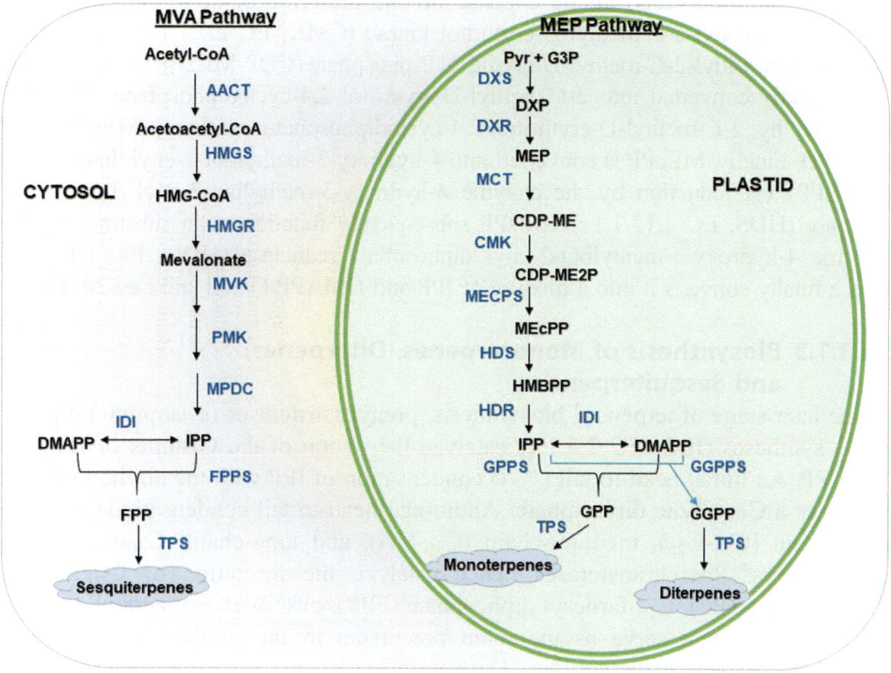

Fig. 1 Simplified representation of the biosynthetic pathways leading to volatile terpenoids in plants. Abbreviations: AACT, acetoacetyl-CoA thiolase; CDP-ME, 4-diphosphocytidyl-2-*C*-methyl-D-erythritol; CDP-ME2P, 4-diphosphocytidyl-2-*C*-methyl-D-erythritol 2-phosphate; CMK, 4-diphosphocytidyl-2-*C*-methyl-D-erythritol kinase; DMAPP, dimethylallyl diphosphate; DXP, 1-deoxy-D-xylulose 5-phosphate; DXR, 1-deoxy-D-xylulose 5-phosphate reductoisomerase; DXS, DXP synthase; FPP, farnesyl diphosphate; FPPS, farnesyl diphosphate synthase; G3P, glyceraldehyde-3-phosphate; GGPP, geranylgeranyl diphosphate; GGPPS, geranylgeranyl diphosphate synthase; GPP, geranyl diphosphate; GPPS, geranyl diphosphate synthase; HDR, 4-hydroxy-3-methylbut-2-enyl diphosphate reductase; HDS, 4-hydroxy-3-methylbut-2-enyl diphosphate synthase; HMBPP, 4-hydroxy-3-methylbut-2-enyl diphosphate; HMGR, HMG-CoA reductase; HMGS, HMG-CoA synthase; IDI, isopentenyl diphosphate isomerase; IPP, isopentenyl diphosphate; MCT, 4-diphosphocytidyl-2-*C*-methyl-D-erythritol synthase; MECPS, 2-*C*-methyl-D-erythritol 2,4-cyclodiphosphate synthase; MEcPP, 2-*C*-methyl-D-erythritol 2,4-cyclodiphosphate; MEP, 2-*C*-methyl-D-erythritol 4-phosphate; MPDC, mevalonate diphosphate decarboxylase; MVK, mevalonate kinase; PMK, phosphomevalonate kinase; TPS, terpene synthase

(GAP) and pyruvate (Pyr) to produce 1-deoxy-D–xylulose 5-phosphate (DXP) (Dudareva et al. 2013). This first committed step is catalyzed by the enzyme DXP synthase (DXS, EC 2.2.1.7). In the second step, DXP is transformed into 2-*C*-methyl-D-erythritol 4-phosphate (MEP) through an intramolecular rearrangement of DXP, followed by an NADPH-dependent reduction. The reaction is reversible and is catalyzed by the enzyme 1-deoxy-D-xylulose 5-phosphate reductoisomerase (DXR, EC 1.1.1.267). MEP is converted into 4-diphosphocytidyl-2-*C*-methyl-D-erythritol (CDP-ME) via a CTP-dependent reaction catalyzed by the enzyme 4-diphosphocytidyl-2-*C*-methyl-D-erythritol synthase (MCT, EC 2.7.7.60). CDP-ME is further phosphorylated at the expense of one molecule of ATP by the enzyme 4-diphosphocytidyl-2-*C*-methyl-D-erythritol kinase (CMK, EC 2.7.1.148) to produce 4-diphosphocytidyl-2-*C*-methyl-D-erythritol 2-phosphate (CDP-ME2P). CDP-ME2P is subsequently converted into 2-*C*-methyl-D-erythritol 2,4-cyclodiphosphate (MEcPP), catalyzed by 2-*C*-methyl-D-erythritol 2,4-cyclodiphosphate synthase (MECPS, EC 4.6.1.12). Finally, MEcPP is converted into 4-hydroxy-3-methylbut-2-enyl diphosphate (HMBPP) via reduction by the enzyme 4-hydroxy-3-methylbut-2-enyl diphosphate synthase (HDS, EC 1.17.7.1). HMBPP subsequently functions as a substrate for the enzyme 4-hydroxy-3-methylbut-2-enyl diphosphate reductase (HDR, EC 1.17.7.4), which finally converts it into a mixture of IPP and DMAPP (Tholl and Lee 2011).

24.3.1.3 Biosynthesis of Monoterpenes, Diterpenes, and Sesquiterpenes

In the later stage of terpenoid biosynthesis, prenyltransferases or isoprenyl diphosphate synthases (IDS, EC 2.5.1.1) catalyze the fusion of the C_5-units of IPP and DMAPP. An initial head-to tail (1′-4) condensation of IPP with the allylic DMAPP produces a C_{10}-allylic diphosphate. Additional head-to tail condensations result in short-chain (C_{15}-C_{25}), medium-chain (C_{30}-C_{35}), and long-chain (C_{40}-C_n) prenyl diphosphates. Prenyltransferases hence catalyze the formation of C_{10}-geranyl diphosphate (GPP), C_{15}-farnesyl diphosphate (FPP), or C_{20}-geranylgeranyl diphosphate (GGPP) that serve as the main precursors in the production of diverse specialized terpenoid metabolites. Despite the strict compartmentalization of the MEP and MVA pathways, studies on incorporation of stable-isotope precursors demonstrated a degree of exchange of isoprenoid intermediates between plastids and cytosol. Thus, IPP, DMAPP, and small prenyl diphosphates (GPP and FPP) act as connecting metabolites and facilitate a metabolic crosstalk among the compartmentally separated MVA and MEP pathways. Trafficking of these compounds across the inner envelope membrane of plastids is mediated by an unidentified metabolite transporter. This connection between the isoprenoid biosynthetic pathways allows the MEP pathway (often with a higher carbon flux than the MVA route) to support biosynthesis of terpenoids in the cytosol (Dudareva et al. 2013). For example, in snapdragon flowers, the MEP pathway provides precursors for cytosolic sesquiterpene formation.

The C_5- to C_{20}-prenyl diphosphate compounds produced by prenyltransferases are additionally altered by enzymes known as terpene synthases (TPS) into C_5-hemiterpenes such as isoprene, C_{10}-monoterpenes, C_{15}-sesquiterpenes, and C_{20}-diterpenes. The TPS gene family comprises of >100 members characterized from several plant species (Dudareva et al. 2013). These enzymes produce single or multiple cyclic and acyclic products and thus are fundamental to the diversity of terpene compounds found in nature. TPS enzymes localized in plastids generally produce monoterpenes or diterpenes from the predominantly plastidial pools of GPP and GGPP, respectively, while TPS enzymes targeted to the cytosol primarily convert FPP to sesquiterpenes (or squalene in the biosynthesis of C_{30} terpenes). The terpene synthase products can be further altered by secondary reactions such as hydroxylation, peroxidation, methylation, acylation, glycosylation, or cleavage, which further add to the compound diversity (Tholl and Lee 2011). Additionally, several TPS enzymes can accept more than one substrate, which also magnifies the diversity of produced terpenoids by targeting bifunctional enzymes to different compartments with a wide range of accessible substrates (Dudareva et al. 2013).

In addition to mono- and sesquiterpenes, few flowers also release irregular terpenoids derived from oxidative cleavage of carotenoids (C_8 to C_{18}) which constitute a small class (~7%) of floral terpenoids. They are synthesized through a three-step modification process, namely, (i) an initial dioxygenase cleavage catalyzed by the group of enzymes called carotenoid cleavage dioxygenases (CCDs), (ii) subsequent enzymatic transformation of the cleavage product giving rise to polar intermediates (aroma precursors), and (iii) acid-catalyzed conversion of the nonvolatile products into active volatile forms. An example exemplifying this three-step process is the formation of β-damascenone from neoxanthin. The initial oxidative cleavage product of neoxanthin has to be enzymatically reduced and acid-catalyzed before finally being converted into the volatile ketone (Muhlemann et al. 2014a). Interestingly, volatile products such as α- and β-ionone, geranylacetone, and pseudoionone found in petunia scent can result from the primary dioxygenase cleavage step itself, indicating the significant role of CCDs in floral volatile biosynthesis.

24.3.1.4 Regulation of Terpenoid Biosynthetic Pathway

Multiple factors including circadian clock, light, and transcription factors control the coordinated formation and emission of volatile terpenoids at molecular, metabolic, and biochemical levels. Besides, sucrose has been reported to increase the transcript levels of DXS, DXR, MCT, and CMK in dark grown plants (Tholl and Lee 2011). Terpenoids show a highly defined regulation of their biosynthetic pathways that are regulated mainly at the level of transcription. They are de novo synthesized and released from individual plant tissues only at particular times suggesting a spatio-temporal regulation of gene expression. Further, the amount of substrate molecules supplied to the cell largely governs the content of VOCs. For linalool synthase, under the control of the cauliflower mosaic virus 35S constitutive promoter when introduced into petunia W115, the amount of linalool or its glycoside synthesized depended on the availability of the geranyl pyrophosphate (GPP as substrate) in the tissue rather than the expression of the linalool synthase gene (Lücker et al. 2001).

Transcription factors are also known to play a key role in regulating the metabolic flux through terpenoid biosynthetic pathway. Production of *ANTHOCYANIN PIGMENT1* (*PAP1*), a *Myb* transcription factor, enhances production of phenylpropanoid and terpenoid scent compounds in rose flowers. *PAP1*-transgenic rose lines released approximately up to 6.5 times elevated levels of terpenoid volatile compounds. Olfactory assay also showed that bees and humans could tell apart the floral scents of *PAP1*-transgenic and control flowers (Zvi et al. 2012). MYC2, a basic helix-loop-helix transcription factor, was also lately identified in *Arabidopsis* inflorescences and was shown to stimulate the expression of two sesquiterpene synthases (TPS11 and TPS21) through the gibberellic and jasmonic acid signaling pathways (Muhlemann et al. 2014a).

24.3.2 Biosynthesis of Phenylpropanoids/Benzenoids

Phenylpropanoids and benzenoids compromise the second major class of plant VOCs and are derived exclusively from the aromatic amino acid phenylalanine. These groups of compounds are essentially reduced at the C_9 position (either to aldehyde, alcohol, or alkane/alkene) or contain alkyl additions to the hydroxyl groups of the phenyl ring or to the carboxyl group (i.e., ethers and esters). Phenylpropanoids and benzenoids constitute the key aroma blend of many flowers, viz., *Petunia hybrida*, *Clarkia breweri*, *Polianthes tuberosa*, *Jasminum* spp., and many more. On the basis of the structure of their carbon skeleton, this category of VOCs can be subdivided into three subclasses: phenylpropanoids (with a C_6-C_3 backbone), benzenoids (C_6-C_1), and phenylpropanoid-related compounds (C_6-C_2) (Muhlemann et al. 2014a). Biosynthesis of all plant phenylpropanoids and benzenoids ultimately starts from the shikimate pathway. Seven plastidial enzymatic reactions linked to the shikimate pathway and three of the arogenate pathway connect the central carbon metabolism to phenylalanine.

24.3.2.1 The Shikimate Pathway

Precursor molecules of the shikimate pathway, phosphoenol pyruvate (PEP) and D-erythrose 4-phosphate (E4P), are derived from glycolysis and the pentose phosphate pathway, respectively. These pathways also supply precursors to the MEP pathway. The latter has to compete for carbon allocation with the shikimate/phenylpropanoid pathway. Approximately 30% of photosynthetically fixed carbon is directed to phenylalanine biosynthesis (Dudareva et al. 2013). The first committed step of the shikimate pathway is governed by the first gene of the shikimate pathway, 3-deoxy-D-arabino-heptulosonate 7-phosphate synthase (DAHPS; EC 2.5.1.54), which plays a key role in controlling carbon flux into the pathway (Fig. 2). Plant DAHPS enzymes are likely to be subjected to feedback inhibition by the aromatic amino acids (phenylalanine, tryptophan, or tyrosine). The second enzyme of the shikimate pathway, 3-dehydroquinate synthase (DHQS; EC 4.2.3.4), converts DAHP into 3-dehydroquinate. The third and fourth steps are catalyzed by the bifunctional enzyme 3-dehydroquinate dehydratase/shikimate 5-dehydrogenase (DHD/SKDH; EC 4.2.1.10 and EC 1.1.1.25), leading to the formation of shikimate. The fifth step of the shikimate pathway is catalyzed by shikimate kinase (SK; EC 2.7.1.71), which converts shikimate to shikimate 3-phosphate. The sixth enzymatic step is catalyzed by 5-enolpyruvylshikimate 3-phosphate synthase (EPSPS) (EC 2.5.1.19), which leads to the synthesis of enolpyruvylshikimate 3-phosphate (EPSP). The final step in the shikimate pathway is catalyzed by chorismate synthase (CS; EC 4.2.3.5), which converts EPSP to chorismate. Chorismate, the terminal metabolite of the shikimate pathway, is a central branch point metabolite in the synthesis of aromatic acids. Chorismate is then converted to phenylalanine by a three-step pathway involving chorismate mutase (CM; EC 5.4.99.5), prephenate aminotransferase (PPA-AT; EC 2.6.1.78), and arogenate dehydratase (ADT; EC 4.2.1.91) (Widhalm and Dudareva 2015) (Fig. 2).

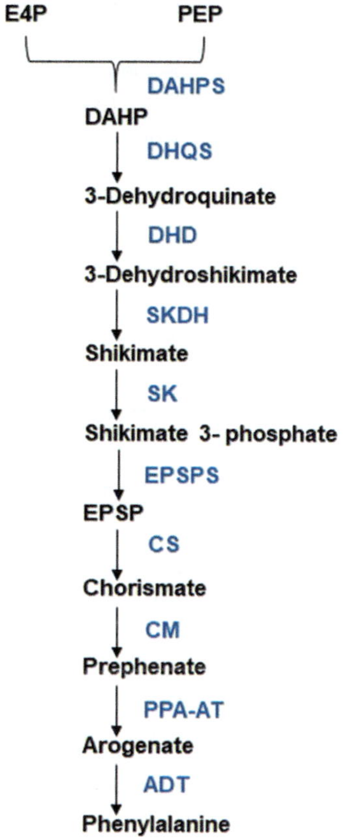

Fig. 2 Shikimate pathway showing the biosynthetic route of phenylalanine, an aromatic amino acid that is the precursor for volatile benzenoids and phenylpropanoids. Abbreviations: ADT, arogenate dehydratase; AS, anthranilate synthase; CM, chorismate mutase; CS, chorismate synthase; DAHP, 3-deoxy-D-arabino-heptulosonate-7-phosphate; DAHPS, 3-deoxy-D-arabino-heptulosonate synthase; DHD, 3-dehydroquinate dehydratase; DHQS, 3-dehydroquinate synthase; E4P, D-erythrose 4-phosphate; EPSP, 5-enolpyruvylshikimate-3-phosphate; EPSPS, 5-enolpyruvylshikimate-3-phosphate synthase; PDT, prephenate dehydratase; PEP, phosphoenolpyruvate; PPA-AT, prephenate aminotransferase; SK, shikimate kinase; SKDH, shikimate dehydrogenase

24.3.2.2 Phenylpropanoid Metabolism

The phenylpropanoids are a major group of volatile aromatic compounds derived from phenylalanine. They function beyond pollinator attraction and are generally toxic to animals and microorganisms. Hence, plants produce these specialized metabolites in their vegetative parts as defense against herbivores and pathogens. Further, they have been used in food flavoring and preservation, are important constituents in many spices used by humans, and have therefore played important roles in human nutrition. The floral scent bouquet of many plant species (e.g., *Clarkia breweri*, *Petunia hybrida*) constitutes volatile phenylpropenes (Koeduka 2014). For example, flowers of *Clarkia breweri* release a mixture of volatiles that include important phenylpropenes, viz., eugenol, isoeugenol, methyleugenol, and methyl-isoeugenol. Flowers of *Petunia hybrida* also emit intense levels of iso-eugenol, along with minor amounts of eugenol. In orchid flowers (*Bulbophyllum patens*), methyleugenol was identified as the major volatile that attracts male fruit flies (pollinators) (Koeduka 2014). Additionally, many herbs, such as basil (*Ocimum basilicum*), synthesize and store phenylpropenes (eugenol, chavicol, or their

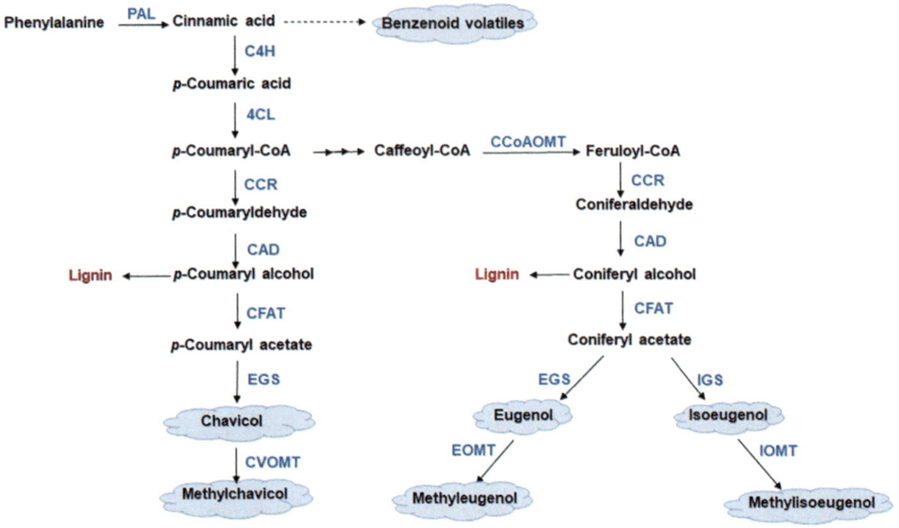

Fig. 3 Schematic representation of the pathway leading to the biosynthesis of eugenol, isoeugenol, and their derivatives. Volatile compounds are demonstrated by the clouds. Stacked and dashed arrows indicate multiple enzymatic reactions. Abbreviations: 4CL, 4-coumarate-CoA ligase; C4H, cinnamate-4-hydroxylase; CAD, cinnamyl alcohol dehydrogenase; CCoAOMT, caffeoyl-CoA 3-O-methyltransferase; CCR, cinnamoyl-CoA reductase; CFAT, coniferyl alcohol acetyltransferase; CVOMT, S-adenosyl-L-methionine:chavicol O-methyltransferase; EGS, eugenol synthase; EOMT, S-adenosyl-L-methionine: eugenol O-methyltransferase; IGS, isoeugenol synthase; IOMT, S-adenosyl-L-methionine: isoeugenol O-methyltransferase; PAL, phenylalanine ammonia lyase (Modified version of Muhlemann et al. 2014b)

methylated derivatives) in specialized glandular trichomes on the surface of their leaves (Gang et al. 2002).

The phenylpropanoid volatiles (C_6-C_3 compounds) share initial steps of their biosynthesis with the lignin biosynthetic pathway (Muhlemann et al. 2014b). This pathway begins with the deamination of phenylalanine to cinnamic acid (CA) catalyzed by the enzyme L-phenylalanine ammonia-lyase (PAL; EC 4.3.1.24) (Fig. 3). Hydroxylation of CA by the cytochrome P450-dependent monooxygenase enzyme, cinnamic acid-4-hydroxylase (C4H; EC 1.14.14.91), leads to formation of *p*-coumaric acid/4-coumaric acid (*p*-CA). *p*-CA is converted to *p*-coumaroyl-CoA by the 4-coumaroyl-CoA ligase (4CL; EC 6.2.1.12) family of enzymes (Fig. 3). The 4CL enzyme is vital in regulating the flow of hydroxycinnamic acids into successive biosynthetic pathways. The first committed step in the biosynthesis of lignin monomers comprises the reduction of the CoA thioesters to their respective aldehydes, which is catalyzed by cinnamoyl-CoA reductase (CCR, EC.1.2.1.44). CCRs have broad substrate specificity and are therefore capable of converting a range of cinnamoyl-CoA thioesters (feruloyl-CoA, *p*-coumaroyl-CoA, sinapoyl-CoA, caffeoyl-CoA, and 5-hydroxyferuloyl-CoA) to their corresponding aldehydes. Regardless of the broad substrate specificity, CCRs demonstrated the highest catalytic efficiency for feruloyl-CoA producing coniferaldehyde. Subsequently, coniferaldehyde is converted to coniferyl alcohol by coniferaldehyde dehydrogenase (CAD; EC 1.2.1.68) (Muhlemann et al. 2014b). The phenylpropene biosynthetic pathway diverges from the lignin pathway after the formation of monolignols such as *p*-coumaryl alcohol or coniferyl alcohol. The first committed step in phenylpropene biosynthesis is catalyzed by an acyltransferase, namely, coniferyl alcohol acetyltransferase (CFAT; EC 2.3.1.84), of the BAHD family that transfers the acetyl moiety of acetyl-CoA to the hydroxyl group of *p*-coumaryl alcohol or coniferyl alcohol, respectively. These resultant monolignol acetates are reduced by two distinctive NADPH-dependent reductases, eugenol synthase (EGS; EC 1.1.1.318) and isoeugenol synthase (IGS; EC 1.1. 1.319), to produce an allyl-phenylpropene (eugenol and chavicol) or propenyl-phenylpropene (isoeugenol) (Koeduka 2014) (Fig. 3). Further alterations of the benzene ring may also follow, including methylation, prenylation, or formation of a methylenedioxy bridge resulting in the structural diversity of volatile phenylpropenes. The enzyme S-adenosyl-L-methionine: eugenol/isoeugenol/chavicol O-methyltransferase (EC 2.1.1.146) that catalyzes the final methylation of eugenol, isoeugenol, and chavicol has been characterized. These structural differences affect the biological activity of the phenylpropenes. For example, phenylpropenes with a *p*-hydroxyl group and sometimes also a *p*-methoxyl group on the benzene ring, such as eugenol, isoeugenol, and chavicol, exhibit antagonistic activity against pathogens such as *Botrytis cinerea*, whereas prenylated phenylpropenes, where the hydroxyl group is masked with a dimethylallyl group, displayed a potent oviposition deterrent activity for mites (Koeduka 2014). Moreover, these structural differences also influence the aromatic property of phenylpropenes. Eugenol and isoeugenol bear a pungent and spicy aroma, whereas methylated compounds have a fresh herbaceous smell.

24.3.2.3 Benzenoid Metabolism

The C_6-C_1 benzenoid compounds are derived from the C_6-C_3 phenylpropanoids by shortening of the three-carbon chain attached to the phenyl ring of phenylpropanoids by two carbons. The mechanism by which this is achieved is not completely understood. The first committed step in benzenoid biosynthesis is also catalyzed by the well-characterized and most widely distributed enzyme, L-phenylalanine ammonia-lyase (PAL; EC 4.3.1.24), which deaminates phenylalanine to *trans*-cinnamic acid (CA) (Fig. 4). CA was then shown to proceed via either a *β*-oxidative or a non-β-oxidative pathway for production of benzenoid volatiles such as benzyl acetate, methyl salicylate, methyl benzoate, benzyl benzoate, etc. A major difference between β-oxidative pathway and non-β-oxidative pathway is that the former route starts with CoA esters instead of free acids. The *β*-oxidative pathway has been fully interpreted in petunia flowers and appears to be localized in peroxisomes (Muhlemann et al. 2014a). In the past decade, substantial advancement has been made in the characterization of enzymes and genes contributing to the final steps of benzenoid volatile formation. It begins with the conversion of CA to cinnamoyl-CoA that is subsequently accompanied by hydration, oxidation, and cleavage of the

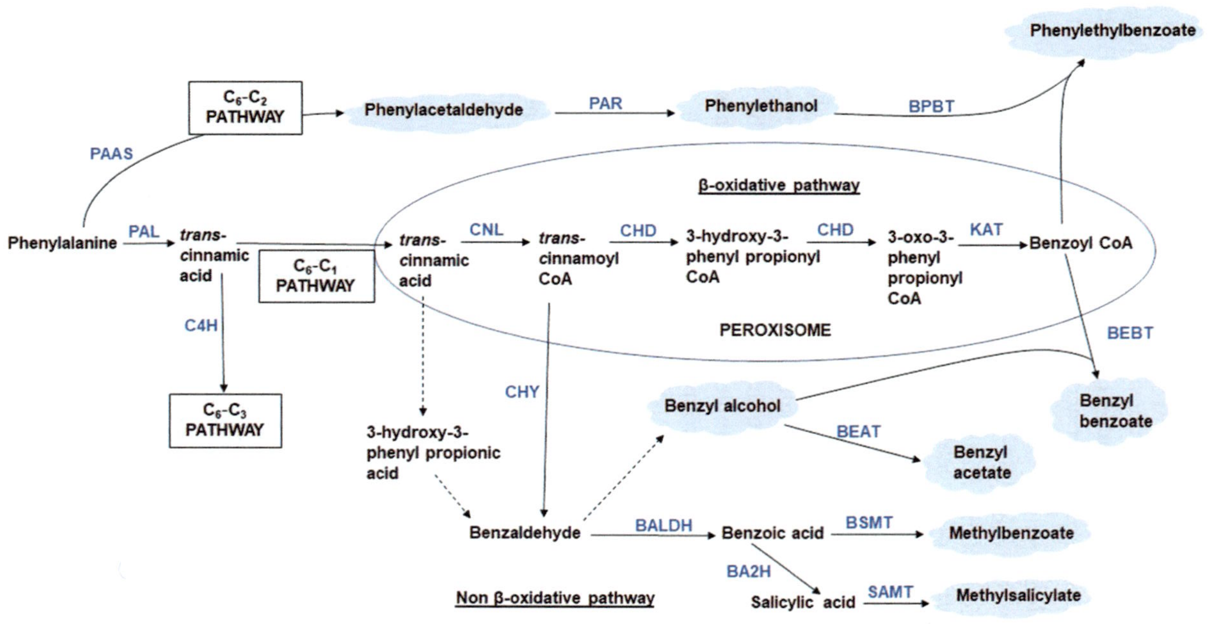

Fig. 4 **Schematic representation of VPB biosynthesis in plants demonstrating the C_6-C_1, C_6-C_2, and C_6-C_3 pathways**. The peroxisomal β-oxidative pathway and a non-oxidative route are responsible for the formation of various benzenoid volatiles. Volatile compounds are demonstrated by the clouds. Solid arrows indicate established biochemical steps, while hypothetical steps not yet described are represented by broken arrows. Abbreviations: BA2H, benzoic acid 2-hydroxylase; BALDH, benzaldehyde dehydrogenase; BEAT, acetyl-CoA:benzyl alcohol acyl transferase; BEBT, benzoyl-CoA:benzyl alcohol benzoyl transferase; BPBT, benzoyl-CoA:benzyl alcohol/2-phenylethanol benzoyl transferase; BSMT, benzoic acid/salicylic acid carboxyl methyltransferase; C4H, cinnamate 4-hydroxylase; CHD, cinnamoyl-CoA hydratase-dehydrogenase; CNL, cinnamate-CoA ligase; CHY, 3-hydroxyisobutyryl-CoA hydrolase; KAT, 3-ketoacyl-CoA thiolase; PAAS, phenylacetaldehyde synthase; PAL, phenylalanine ammonia lyase; PAR, phenylacetaldehyde reductase; SAMT, S-adenosyl-L-methionine: salicylic acid carboxyl methyltransferase (Modified version of Dudareva et al. 2013)

β-keto thioester via a reverse Claisen reaction resulting in the formation of benzoyl-CoA (Dudareva et al. 2013) (Fig. 4).

The alternative CoA-independent, non-β-oxidative pathway that occurs in the cytoplasm comprises hydration of the CA to 3-hydroxy-3-phenylpropionic acid (Fig. 4). This product when subjected to a reverse aldol reaction leads to the production of benzaldehyde. Alternatively, 3-hydroxyisobutyryl-CoA hydrolase (CHY; EC 3.1.2.4) is an important enzyme leading to formation of benzaldehyde from the β-oxidative pathway intermediates (Zhang et al. 2016). Benzaldehyde, the chief intermediate in the non-β-oxidative pathway, is oxidized to benzoic acid by a NAD+-dependent benzaldehyde dehydrogenase.

Additional alterations, viz., methylation and acylation of benzoic acids, are carried out by enzymes to catalyze the formation of volatile esters. BAHD superfamily of acyltransferases and the SABATH family of methyltransferases were found to greatly contribute to the final biosynthetic steps of volatile benzenoids. These include the most widely distributed enzymes in the plant kingdom, namely, acetyl-CoA:benzyl alcohol acetyltransferase (BEAT; EC 2.3.1.224), S-adenosyl-L-Met:salicylic acid carboxyl methyltransferase (SAMT; EC 2.1.1.274), S-adenosyl-L-Met:benzoic acid carboxyl methyltransferase (BAMT; EC 2.1.1.273), benzoyl-CoA: benzyl alcohol benzoyl transferase (BEBT; EC 2.3.1.196), and orcinol O-methyltransferase (OOMT), which are responsible for the formation of benzyl acetate, methyl salicylate, methyl benzoate, benzyl benzoate, and dimethoxytoluene, respectively (Muhlemann et al. 2014a).

24.3.2.4 Biosynthesis of Phenylpropanoid-Related Compounds

The biosynthesis of another branch of volatile phenylpropanoid-related (C_6-C_2) compounds, such as phenylacetaldehyde and 2-phenylethanol, does not transpire via cinnamic acid. In contradiction to benzenoids and phenylpropenes, the biosynthesis of this particular group of volatiles competes with PAL enzyme for phenylalanine utilization (Fig. 4). In petunia petals, phenylacetaldehyde is produced directly from phenylalanine via a combined decarboxylation-amine oxidation reaction catalyzed by phenylacetaldehyde synthase (PAAS; EC 4.1.1.109). Phenylacetaldehyde reductase (PAR) then catalyzes the conversion of phenylacetaldehyde into 2-phenylethanol (Dudareva et al. 2013). In roses, though, it is formed via two alternate routes: the first employs an aromatic amino acid decarboxylase (AADC; EC 4.4.1.28) to synthesize phenylacetaldehyde, while the second route (Fig. 5, dashed arrows) involves deamination of phenylalanine by an aromatic amino acid aminotransferase (AAAT; EC 2.6.1.57) followed by decarboxylation of the produced phenylpyruvate intermediate by phenylpyruvate decarboxylase (PPDC; EC 4.1.1.43). The synthesized phenylacetaldehyde is then subsequently transformed into 2-phenylethanol by PAR (Muhlemann et al. 2014a).

24.3.2.5 Regulation of Phenylpropanoid/Benzenoid Metabolism

Little is known about the mechanisms regulating the signaling at the molecular levels in floral volatile phenylpropanoid/benzenoid (VPB) synthesis. Multiple factors including circadian clock, light, temperature, humidity, and transcription factors

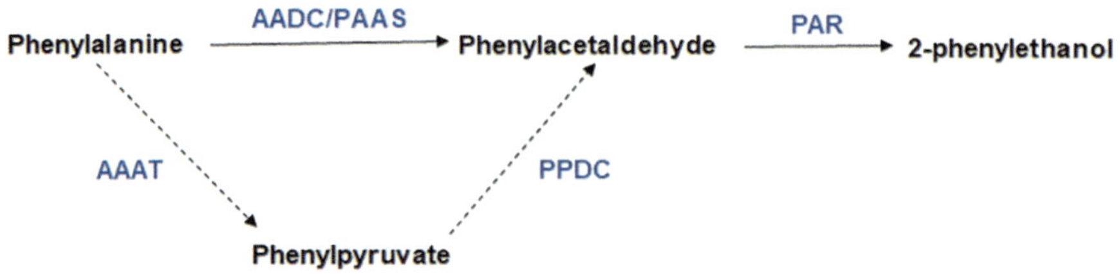

Fig. 5 Schematic representation of the pathways leading to the biosynthesis of 2-phenylethanol. Abbreviations: AADC, aromatic amino acid decarboxylase; AAAT, aromatic amino acid aminotransferase; PAAS, phenylacetaldehyde synthase; PAR, phenylacetaldehyde reductase; PPDC, phenylpyruvate decarboxylase

control the formation and emission of VPBs at molecular, metabolic, and biochemical levels. 2-Phenylethanol, for example, increased with a red and far-red light treatment in petunia flowers (Colquhoun et al. 2013). Another important component regulating the production and emission of VPBs is the substrate availability. Methyl benzoate, for example, is regulated primarily by the level of substrate availability (benzoic acid), which in turn could be regulated at the level of expression of genes encoding the key enzymes of its biosynthesis (Kolosova et al. 2001).

Comparative analysis on the regulation of VPBs in flowers revealed that the orchestrated emission of phenylpropanoid and benzenoid compounds is controlled upstream of individual metabolic pathways and consists of a coordinated and rhythmic expression of genes that encode enzymes involved in the final steps of scent biosynthesis. For example, rhythmic expression of BEAT gene in the ornamental flowers of *Jasminum sambac* was accompanied by a rhythmic release of benzyl acetate (Bera et al. 2017). Little is known about the mechanisms regulating the signaling at the molecular level of floral VPB synthesis. The transcription factors, ODORANT1 (ODO1) and EMISSION OF BENZENOIDS II (EOBII), have been recognized as the major molecular regulators of the VPB pathway in plants. Suppression of ODO1 belonging to the MYB transcription factor family leads to decreased levels of emitted volatile benzenoids and phenylpropanoids in *Petunia hybrida*. Downregulation of ODO1 also showed a dramatic effect on the genes from the S-adenosyl-methionine (SAM) cycle that supplies the methyl groups necessary for volatilization of some compounds such as benzoic acid. EOBII was found to be particularly flower specific and temporally and spatially associated with scent production/emission, suggesting its critical role as a regulator of the machinery responsible for floral fragrance. Suppression of EOBII expression led to substantial decrease in the amount of VPBs biosynthesized and emitted by flowers, such as benzaldehyde, phenylethyl alcohol, benzyl benzoate, and isoeugenol. Thus, up/downregulation of EOBII altered the transcript levels of numerous floral scent-related genes related to the synthesis of the VPBs including the primary shikimate pathway genes that determine substrate availability. Due to its coordinated effect on the production of floral volatiles, EOBII is proposed to have a central regulatory role in the biosynthesis of phenylpropanoid volatiles. In petunia a recently identified

EOBI was identified as another flower-specific transcription factor governing VPBs, which acts upstream of ODO1 and downstream of EOBII. Silencing of the EOBI expression led to downregulation of several genes regulating the shikimate pathway (5-enolpyruvylshikimate 3-phosphate synthase, 3-deoxy-D-arabino-heptulosonate 7-phosphate synthase, chorismate synthase, chorismate mutase, arogenate dehydratase, and prephenate aminotransferase) as well as downstream genes (phenylalanine ammonia-lyase, isoeugenol synthase, and benzoic acid/salicylic acid carboxyl methyltransferase) (Muhlemann et al. 2014a). MYB4 is another R2R3-MYB transcription factor that regulates the biosynthesis of VPBs. It negatively regulates cinnamate-4-hydroxylase in petunia and thus indirectly controls the precise emission of VPBs derived from *p*-coumaric acid, such as eugenol and isoeugenol. Therefore, MYB4 represses the level of C4H transcription to achieve a tightly controlled amount of VBPs in flowers and thus indirectly functions to "fine-tune" the bouquet of floral fragrance (Dudareva et al. 2013).

In numerous plants, ethylene is synthesized in a localized, specific, and reproducible manner after pollination, highlighting the importance of role of ethylene during pollination and fertilization. The hormone ethylene has also been shown to govern the biosynthesis of benzenoids in petunia flowers. Post-pollination (as a result of successful fertilization), ethylene is produced by the petals and ovaries that accelerates corolla senescence and also rapidly downregulates the biosynthesis and emission of all benzenoids. It directs the coordinated downregulation of the vital volatile biosynthetic genes, viz., BSMT, BPBT, and PAL. Consequently, it also regulates the synthesis of the rate-limiting substrates of benzenoid synthesis, namely, cinnamic acid and benzoic acid (Underwood et al. 2005).

24.3.3 Biosynthesis of Fatty Acid Derivatives

The third class of floral VOCs comprising compounds such as hexanal, hexenol, nonenal, methyl jasmonate, etc., is an important category of volatiles derived from unsaturated C_{18}, viz., linolenic and linoleic fatty acids. The precursor molecule responsible for the biosynthesis of these fatty acids is the plastidic pool of acetyl-CoA that is generated from pyruvate (the final glycolytic product). The linoleic and linolenic acids undergo stereospecific oxygenation by action of lipoxygenase at carbon atom 9 (9-LOX) or at 13 (13-LOX) of the hydrocarbon backbone of the fatty acid leading to the production of their respective 9- and 13-hydroperoxides (Dudareva et al. 2013). These intermediates then move into two separate branches of the LOX pathway, leading to the production of diverse volatile compounds. The allene oxide synthase (AOS; EC 4.2.1.92) branch utilizes the 13-hydroperoxy intermediate that leads to the synthesis of jasmonic acid (JA) (Fig. 6). JA is further acted upon by the enzyme JA carboxyl methyltransferase that catalyzes the production of methyl jasmonate. JA is nonvolatile; then again its methyl ester is an important volatile plant hormone that carries out vital plant functions as a cellular regulator and mediates various developmental processes and defense responses. Contrary to the AOS branch, the hydroperoxide lyase (HPL; EC 4.1.2.-) branch

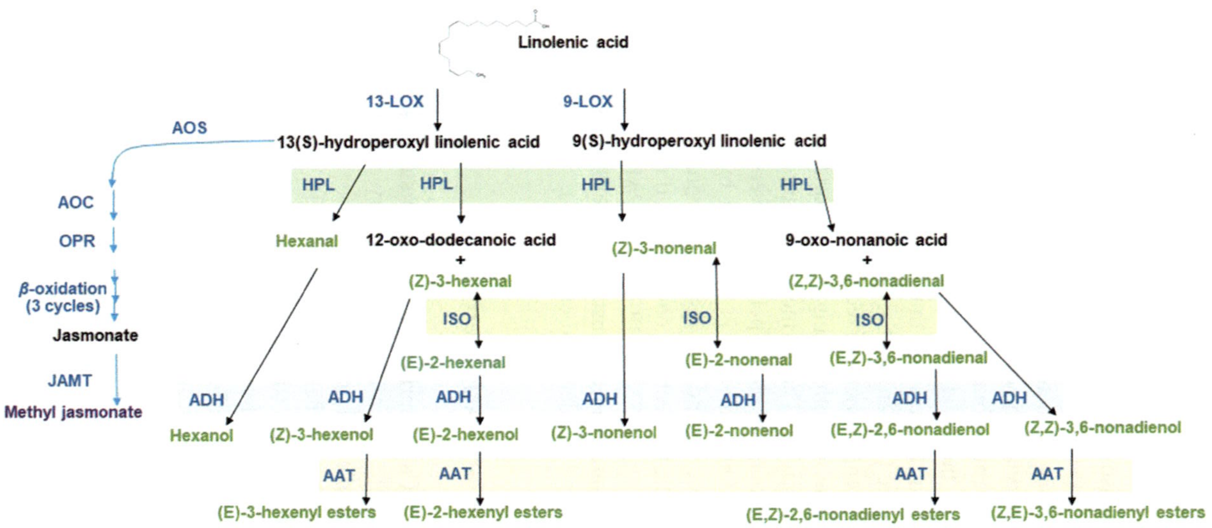

Fig. 6 **Schematic representation of the biosynthesis of fatty acid-derived VOCs and green leaf volatiles (GLVs)**. Both linoleic and linolenic acid are the precursors for the array of fatty acid-derived VOCs. These precursors are directed to the lipoxygenase (LOX) pathway by an oxidation producing 9-hydroperoxy and 13-hydroperoxy intermediates that are finally transformed to volatile derivatives by hydroperoxide lyases, isomerases, alcohol dehydrogenases, and alcohol acyltransferases (shown in boxes). Abbreviations: AAT, alcohol acyltransferase; ADH, alcohol dehydrogenase; AOC, allene oxide cyclase; AOS, allene oxide synthase; HPL, hydroperoxide lyase; ISO, isomerase; JAMT, jasmonic acid carboxyl methyl transferase; 9-LOX, 9-lipoxygenase; 13-LOX, 13-lipoxygenase; OPR, 12-oxophytodienoate reductase (Redrawn from Dudareva et al. 2013)

can convert both of the 9- and 13-hydroperoxide derivatives into C_6 and C_9 volatile aldehydes (hexanal, *(Z)*-3-hexenal and *(Z)*-3-nonenal) (Fig. 6). The aldehydes are formed with a (*Z*) configuration, though isomerization to the (*E*) isomer by an alkenal isomerase generally follows. These unsaturated or saturated C_6 and C_9 aldehydes act as substrates for alcohol dehydrogenases that produce volatile alcohols, which subsequently can be further altered to their esters. The C_6 and C_9 aldehydes and alcohols commonly denoted as the "green leaf volatiles" are mostly synthesized in vegetative tissues. However, recent studies demonstrate that they are also essential constituents of the floral volatile bouquet and contribute significantly to the full fragrance signature of numerous plant species such as carnation and wild snapdragon (Muhlemann et al. 2014a).

24.3.4 Modifications in the Volatile Backbone with Special Emphasis on Glycosylation

The volatiles that are produced in plants are not immediately released from the respective tissues by active or passive transportation. A major fraction of the floral scent volatiles are frequently found as reserves post their biosynthesis in special storage structures in sink organs, such as flower petals. Volatile phenylpropanoids and other specialized scent metabolites, i.e., terpenoids, flavonoids, fatty acid derivatives, etc., undergo various modifications, such as glycosylation, methylation, and acylation by the action of specific tailoring enzymes. These alterations elevates their stability, facilitates cellular transport, increases water solubility, and reduces toxicity by blocking reactive groups, thus enabling storage and trafficking in subcellular compartments (Cna'ani et al. 2017). Methyltransferases of the SABATH family and acyltransferases of the BAHD family are important enzymes catalyzing vital modifications for the production of volatile scent compounds in flowers (Muhlemann et al. 2014a).

Among all alterations, glycosylation is the most ubiquitous modification of plant secondary metabolites that are fundamental to several biological processes. Glycosylation reactions, in general, involve sugar conjunction or sugar polymerization with other biomolecules. The addition of sugar to the aglycones changes their solubility, biological activity, chemical properties, and membrane transportation. Glycosylation causes an increase in hydrophilicity and molecular weight making the glycosyl-bound volatiles (GBVs) more water soluble. In general the instability and reactivity of the aglycone is reduced (Song et al. 2018). In 1969, the first glycosylated volatiles (geranyl, neryl, and citronellyl glucoside) were identified in rose petals (Francis and Allcock 1969). Thus, GBVs are a relatively new class of plant secondary metabolites. They had escaped detection for a long time due to their frequent low abundance in plant tissues and missing chromophores (Song et al. 2018). Continuous progress in the development of novel biochemical techniques and highly sensitive and sophisticated analytical instruments has enabled the detection of this fraction of plant metabolites demonstrating their natural diversity.

Glycosyl conjugation of volatiles is a very interesting phenomenon as it ceases the volatility of the metabolites, rendering them nonvolatile so that they can serve as a readily hydrolysable pool for the release of aglycones, thus delivering an extra layer of complexity to the machinery regulating floral scent production. The occurrence of glycosides as the vital form of pooled volatiles has been validated in many plants, such as 2-phenylethanol glycosides in flowers of *Jasminum auriculatum* (Barman and Mitra 2021), phenylethyl alcohol glycosides in *Rosa damascena* flowers, eugenol glycosides in *Solanum lycopersicum* fruit, and benzyl alcohol and phenylethyl alcohol glycosides in leaves of *Camellia sinensis* (Cna'ani et al. 2017). Additionally, monoterpene alcohols, such as geraniol, linalool, and linalool oxides, or the phenolic aglycone of methyl salicylate and C_{13}-norisoprenoids, or the aliphatic alcohol aglycone of (*Z*)-3-hexenol is also capable of forming complex glycosides in plants (Cui et al. 2016).

Recently, several papers described the role of novel glycosyltransferase enzymes that are able to catalyze the last step of glycosyl-bound volatile formation. Uridine diphosphate sugar-dependent glycosyltransferases (UGTs, EC 2.4.x.y) are the key enzymes catalyzing glycosylation by transferring a carbohydrate from a nucleotide-activated monosaccharide, usually a UDP-sugar (Song et al. 2018). All recognized UGT enzymes bear a universal motif comprising 44 amino acids. The substrate promiscuity is a very interesting feature of GBVs-producing UGTs, i.e., UGTs show preference for one volatile; however, they have the ability to glycosylate additional substrates that adds to their function in detoxification of endogenous and exogenous substances (Song et al. 2018). Thus, UGTs aid to catalyze the reactions between sugar donors and a wide range of substrates. They can utilize a range of sugar donors, viz., UDP-glucose, UDP-xylose, UDP-rhamnose, and UDP-galactose. They frequently display a preference for some specific sugar donors. All plant GBVs show a direct association of the aglycone moiety to a β-D-glucopyranose. This glucose molecule is frequently esterified to a malonyl moiety or is substituted with one (disaccharide) or more added sugar units (such as trisaccharide).Thus, mono-, di-, and trisaccharide conjugates of scent volatiles with various types of associated sugars are reported in the various plant parts. Generally α-L-rhamnopyranose, α-L-arabinofuranose, α-L-arabinopyranose, β-D-xylopyranose, β-D-apiofuranose, and β-D-glucopyranose have been recognized as the terminal sugars in disaccharide glycosides, which are linked to the hydroxyl group at position 2, 3, or 6 of the first β-D-glucopyranose moiety (Song et al. 2018). Thus, the GBVs are classified into seven categories based on the diverse sugar moieties and are identified as β-D-glucopyranosides, 6-*O*-β-D-xylopyranosyl-β-D-glucopyranosides (primeverosides), 6-*O*-α-L-arabinopyranosyl-β-D-glucopyranosides (vicianosides), 6-*O*-α-L-arabinofuranosyl-β-D-glucopyranosides, 6-*O*-β-D-apiofuranosyl-β-D-glucopyranosides (acuminosides), 6-*O*-α-L-rhamnopyranosyl-β-D-glucopyranosides, and 6-*O*-β-L-rhamnofuranosyl-β-D-glucopyranosides (Cui et al. 2016). In addition, volatiles associated with trisaccharides have been found in tomato and tentatively in grape (Song et al. 2018).

Glycosylated metabolites derived from distinct branches of the volatile biosynthetic pathways share a common strategy for vacuolar sequestration (Fig. 10). The conjugated volatiles are reserved in the vacuoles not only as a surplus source or as detoxification mechanism but also as a station on their route to secretion. Vacuolar sequestration comprises vesicle trafficking, membrane transporters, and glutathione S-transferase-mediated transport whereby the blend between substrates and preferred transporter types might be different depending on plant species (Song et al. 2018). Recently, extraction of sequestered benzyl alcohol and phenylethyl alcohol glycosides has been illustrated from isolated intact vacuoles of *Petunia* petals (Cna'ani et al. 2017).

Glycosylated scent compounds are generally regarded as storage forms or precursors for the emission of aglycones at the appropriate time or stage of plant or organ development. The sugar moiety is cleaved off through the corresponding enzymatic action of β-glucosidase (EC 3.2.1.21). Feeding trials demonstrated that the stored pool of glycosylated compounds is metabolized during the period of maximum scent emission. This dynamic nature of the vacuolar pool supports its probability of being harnessed for scent production. This has been well demonstrated in *Rosa* flowers where the pool sizes of free volatiles are seen to be in direct correlation with those of their conjugates throughout the flower life span. Levels of phenylethyl alcohol conjugates are generally higher in early stages of bud development, as opposed to the free-form phenylethyl alcohol which is only present from anthesis onward. Phenylethyl alcohol-β-D-glucopyranoside is thought to be the chief precursor for aglycone emission after petal opening in rose flowers. Furthermore, the glucosidase activity was found to be elevated up to five times suggesting that glucosidase is partly responsible for controlling the rhythmical diurnal emission of 2-phenylethanol from *Rosa damascena* flowers (Baldermann et al. 2009). Similarly observation was reported in *Jasminum auriculatum* where levels of β-glucosidase were seen to increase significantly during flower opening which correlated straightaway with an increase in volatile emission (Barman and Mitra 2021). Thus, GBVs might act as the primary source of rhythmically emitted volatiles, being released periodically by the action of β-D-glucosidase throughout the photoperiod. Changes in rates of glycoside formation and hydrolysis may be responsible for rhythmic emission of volatiles. This indicates that although the odorless nonvolatile forms are less recognized compared to the volatiles, the glycosylation of aroma compounds impacts the flowers in significant ways.

24.4 Mechanisms Governing Scent Emission from Flowers

Plants naturally regulate floral fragrance emission to guard the cost-benefit balance. Oscillations of scent emission have been observed in numerous flower species. Understanding the role of the different physiological and biochemical factors governing the biosynthesis and emission of volatiles is pertaining to the study of floral biology. A few of the regulatory mechanisms governing floral scent are briefly discussed here.

24.4.1 Circadian Rhythms and Mechanisms of Circadian Oscillations in Floral Scent

The meaning of the Latin word circadian is "approximately one day." Circadian rhythms are thus self-sustained biological oscillators affecting patterns of behavioral, physiological, and molecular rhythms within a cycle length of approximately 24 h. Plant biological rhythms greatly impact the physiology of individuals and have evolved to boost plant fitness. These oscillations are endogenous in nature and not a reflection of response to a rhythmic environment. Studies conducted using normal light/dark or constant light conditions have determined the notion that an internal circadian clock regulates diurnal rhythms in flowers. Flower diurnal rhythms include floral opening/closing, nectar accumulation, and synchronized scent emission patterns. These physiological changes are likely to have been coevolved with the activity of potential pollinators to maximize outcrossing, thus allowing an efficient methodological resource utilization. Successful pollination of plants requires spatial and temporal coordination from both partners. For example, *Cestrum nocturnum* also called "night-blooming jasmine" opens and emits a strong, pleasant scent at night attracting nocturnal pollination vectors. In an experiment flowering plants of *C. nocturnum* were kept under continuous light/dark at a constant temperature. A clear 24 hour rhythm was observed even under the altered growing conditions with periodic scent emissions occurring during the time corresponding with what would have been night. This clearly demonstrated the regulation of scent emission by an endogenous clock (Overland 1960). Further, the cell autonomous nature of the rhythm was also demonstrated by using detached corolla lobes which emitted scent in the same circadian fashion as that of an intact flower. This temporal regulation of scent emission is tied to the activity of the specific metabolic pathways responsible for scent production. The control of internal clocks on the floral scent emission profiles has been studied in different flowers, viz., *Petunia axillaris*, *P. parodii*, *Nicotiana attenuata*, rose, and many others. Diel rhythmicity in modulating nectar secretion (volume and concentration) has also been observed in several plant species, suggesting a strong regulation of nectar secretion rhythms directing the temporal behavior of individual foragers.

Although contribution by the circadian clock has been recognized over a range of synchronous behavior in plants, the mechanisms by which the circadian clock is regulated still remains elusive. The basis of circadian petal movement can be differential growth or cell expansion and contraction devised by ion uptake. In the flowers of *Kalanchoe blossfeldiana*, an uptake of K^+ ions during the day increases cell osmolarity and turgor pressure affecting cell expansion in the upper epidermis resulting in petal opening. At night, decreasing ion levels and low cell pressure are associated with flower closure (Schrempf 1977). Additionally, hormones also play a role in circadian rhythm-controlled petal opening. For example, in waterlily, the central regulatory role of auxin in floral movement rhythm was observed, thus directing flower opening and closure (Ke et al. 2018).

A new study has led to the conclusion that plants adjust their circadian rhythms pertaining to the cycles of day and night by sensing the amount of sugar in their cells. This has been demonstrated in *Arabidopsis* where timing of the diel turnover of starch is correlated to the circadian system indicating that the circadian clocks in plants respond to metabolic rhythms. Photosynthesis, the key metabolic process, is timed accordingly to the rhythmic light-dark cycle. Consequently, the oscillation of the circadian clock contributes to the regulation of photosynthesis, and in turn the diurnal accumulation of sugars through photosynthesis feeds back to regulate the circadian oscillator (Haydon et al. 2017). Thus, it has been established that sugars affect the stability of circadian oscillator proteins indicating that sugars not only act as source of energy but also act as signals in plants. This unfolds routes to novel molecular pathways for input of sugar signaling to the circadian network in plants and establishes a strong link between vital plant metabolic pathways and clock mechanisms.

24.4.2 Understanding Cuticular Waxes and Their Role in Regulating Floral Scent

Understanding of the surface layer plays essential roles in plant development and physiology, as it is an interface between the plant and its external biotic and abiotic environment. The appearance of cuticle on outermost surface of the land plants is one of the most successful evolutionary products, as a result of their evolution to dry environments from relatively aquatic environments such as green algae. The plant cuticle is a thin extracellular hydrophobic film sealing the aerial epidermis of most plant parts that provides an efficient protection against invasion by microorganisms, regulation of gas exchange, chemical transport, prevention of organ fusions and attenuates the detrimental effects of ultraviolet radiation. It also plays a role in specific plant-insect interactions and imparts viscoelastic properties to the underlying polysaccharide cell wall, thus protecting the plant from mechanical damage (Yeats and Rose 2013).

The microscopic structure of cuticle can be divided into two parts based on histochemical staining and chemical composition: cutin and cuticular waxes. The former also called "cuticular layer" is a polymer of inter-esterified hydroxylated C_{16} and C_{18} fatty acid derivatives with embedded polysaccharides, which coats the cell wall of the epidermal cells. This layer is enclosed by and embedded within the cuticular waxes also referred to as the "cuticle proper" (Fig. 7). Cuticular waxes are

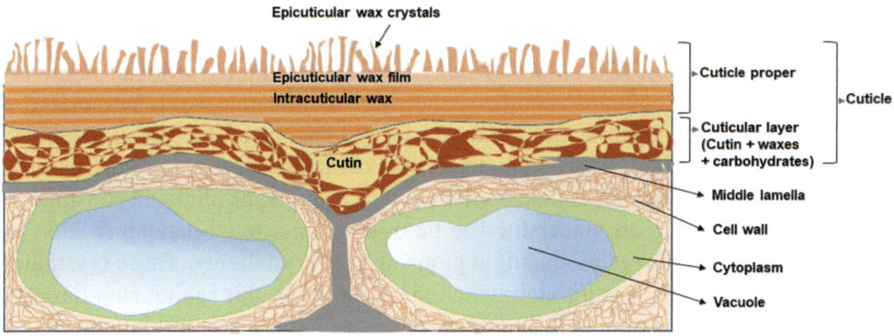

Fig. 7 Diagrammatic representation of plant cuticular structure highlighting the major structures of plant cuticle overlying the epidermal cell wall layer (Redrawn from Yeats and Rose 2013)

organic solvent extractable wax covering on the cutin surface composed of lipids derived primarily from very long-chain fatty acids (C_{20}-C_{32}). They are either deposited within the cutin matrix (intracuticular wax) or are found to accumulate on the surface as epicuticular wax crystals or films (Yeats and Rose 2013). The composition of cuticular waxes varies not only among species but also among different surfaces of the same plant. The basic wax compounds identified in all species are similar, consisting of hydrocarbons (n-alkanes), wax, esters, aldehydes, alcohols, and fatty acids.

Flowers are especially evolved and adapted to ensure reproductive success by attracting pollinators. Hence, they must also be able to resist unfavorable environmental conditions, for instance, a dehydrating atmosphere. It has been revealed that petals are also covered by cuticles, which are quite similar in composition to those present in other plant organs. The understanding of the effect of the film of cuticular layer covering the floral epidermal surfaces is somewhat controversial as scanty research has been done on the effect of cuticle on volatile emission. In the floral wax of some species, few distinctive compounds were identified including diverse triterpenoids (e.g., triterpene alcohols, acids, and esters), long-chain aliphatics with secondary functional groups (e.g., secondary alcohols, ketones, alkanediols, and γ-lactones), as well as other cyclic components like tocopherols. Special esters have also been identified including phytyl esters and cinnamyl esters. Additionally, a large variety of hydrocarbons other than the ubiquitous n-alkanes have been reported for specific petal waxes. These include series of iso-alkanes, anteiso-alkanes, and alkanes with methyl branches (Jetter 2006). This varying composition of cuticular waxes might promote diversity in the properties of the transpiration barrier in different plant organs.

Floral volatiles are de novo synthesized in the epidermal cell layers of flower organ. Hence, they must move from their intracellular sites of biosynthesis through the cuticle film before their release from the flower surface. Exact site of export through the cuticle is not defined, and moreover multiple pathways have been proposed. Few studies have documented the presence of cuticular pores. Other documentations show that wax-covered cuticles might permit the navigation of smaller volatiles. Still, petal cuticle might provide a level of regulation toward the volatile emission, perhaps through developmental control of cuticle permeability (Goodwin et al. 2003). In *Polianthes tuberosa* Calcutta double cultivar, relatively higher amount of endogenous volatiles with lower emission rate was hypothesized to be attributed to wax accumulation on the surface of the petaloid tepals in the flower (Kutty and Mitra 2019). Thus, epicuticular waxes appear to be a crucial element responsible for determining cuticle permeability. The n-alkane aliphatic constituents of the cuticular waxes are understood to be oriented in a fitted-parallel fashion by weak van der Walls forces producing regions of high crystallinity. Those crystalline regions act as barred zones through which the scent molecules cannot pass through. However, substitutions creating branched alkanes produce stearic hindrances that check the local packing of wax molecules in the cuticle film. Goodwin et al. (2003) suggested that branched alkanes, and possibly the hydroxyl esters, reduce the crystallinity of cuticle lipids and are responsible for loose molecular associations producing a more permeable cuticle in petals of *Antirrhinum majus*. Thus, physico-chemical properties of cuticle are suggested to be an essential feature facilitating effective release of volatiles.

Recently, a very remarkable observation was reported in petunia by Liao et al. (2021) that the cuticle is not just merely a final diffusion barrier for volatile compounds but also acts as a sink/concentrator for VOCs at the surface level. It was demonstrated to hold >50% of internal VOCs. This vital function of the cuticle provides protection from the potentially toxic internal accumulation of the hydrophobic volatile compounds inside the cell. Contradictory to the previous hypothesis that a thinner cuticle would permit more volatile release, decreasing the cuticle thickness by targeting the wax transporter *PhABCG12* via RNAi approach reduced the total emission of VOCs by 50–56% along with a 31–43% decrease in their internal pools in relation to controls. Evaluation of the total biosynthetic flux showed that biosynthesis was reduced by 52 $\pm$ 7.8% in the transgenic flowers. Additionally the rhythmicity in VOC release is also eliminated. Taken together, the cuticular layer adds an additional layer of complexity to the dynamics of volatile emission process and thus acts as an integral member of the VOC network.

24.4.3 Regulatory Role of Primary Metabolism and Its Intermediates in Modulating Floral Secondary Metabolism

The chemistry of flowers is thus exceptional, as it sustains very diverse physiological functions. Exquisite displays of the metabolic resources present in flowers in the form of fragrance and colors resulted from the biosynthesis and accumulation of tinted and scented specialized metabolites. The maintenance of flowers is exorbitant in terms of respiratory energy and nutrients. The sustenance of this secondary or specialized metabolism is dependent on less sensational but equally as important reactions and byproducts of the conserved primary metabolism. This has been well established in flowers of snapdragon (Muhlemann et al. 2012) and *Polianthes tuberosa* (Kutty and Mitra 2019). There is no defiance that carbon allocation must be a central player behind this huge network of secondary metabolite biosynthesis in plants. Thus, primary and specialized metabolic pathways are not independent; rather, specialized metabolites are mostly produced in the terminal branches of the network of primary metabolism. Furthermore, sugars and amino acids serve as a dietary reward for the pollinators feeding on nectar and pollen. Indeed, this central mechanism sustains the physiology of flowers, provides the precursor molecules, and meets the energy demand of specialized metabolism in flowers (Muhlemann et al. 2012). Biosynthesis of VOCs also depends on the energy as well as the availability of substrates (carbon, nitrogen, and sulfur) provided by primary metabolism. Consequently, the availability of these building blocks regulates the production of all secondary metabolites, including VOCs, demonstrating the high connectivity between primary and specialized metabolism (Fig. 8). Yet this fundamental process is still scarcely understood, so rational engineering is not yet practical and demands further research. Therefore, understanding how primary metabolites

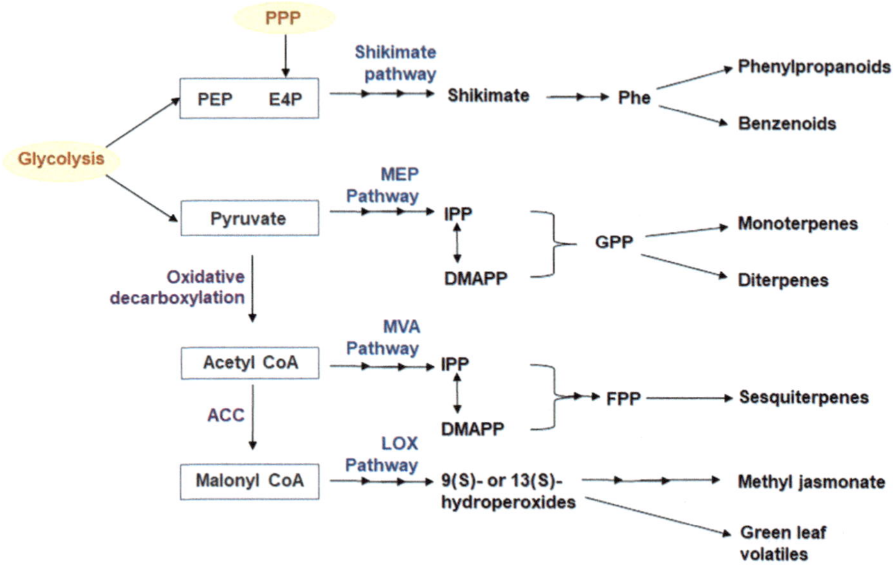

Fig. 8 **Overview of the diverse biosynthetic pathways leading to the production of plant VOCs. The precursors for all the different pathways** (shown in boxes) **originate from primary metabolism intermediates**. These pathways lead to the biosynthesis of major volatiles, namely, phenylpropanoids/benzenoids, terpenoids, and methyl jasmonate/green leaf volatiles. Stacked arrows demonstrate the participation of several enzymatic reactions. Abbreviations: ACC, acetyl-CoA carboxylase; DMAPP, dimethylallyl pyrophosphate; E4P, erythrose 4-phosphate; FPP, farnesyl pyrophosphate; GPP, geranyl pyrophosphate; IPP, isopentenyl pyrophosphate; PEP, phosphoenol pyruvate; Phe, phenylalanine; PPP, pentose phosphate pathway (Modified version of Dudareva et al. 2013)

are de novo synthesized or imported, and how they are secreted and catabolized, is the core of floral biology.

Flower parts are mostly green during early developmental stages apparently due to the presence of chlorophyll and are capable of fixing CO_2, thus providing an additional strategy of carbon acquisition. The presence of chlorophyll is not visibly evident in mature petals as chloroplasts were shown to deteriorate into chromoplasts during the later stages (Borghi and Fernie 2017), and it is generally assumed that mature white or colored corollas are devoid of chloroplasts. Reproductive structures have been shown to derive up to 60% of their total carbon requirement from their own CO_2 fixation. Light and dark carbon dioxide fixation occurs in young flower buds when petals are still green. This has been demonstrated in many flowers including *Nicotiana tabaccum*, Cymbidium, *Citrus sinensis* (*L.*) Osbeck, and *Petunia hybrida* (cv Hit Parade Rosa). Floral photosynthesis is further energetically favorable as it reduces the cost of flowering and its maintenance. However, as the flower matures they gradually lose their ability to fix carbon and eventually become fully heterotrophic. This shift from autotrophic to heterotropic metabolism is when specialized metabolite accumulation becomes predominant (Muhlemann et al. 2012).

Floral tissues rely heavily on the source-to-sink flow of carbon for the supply of photo-assimilates. The strong sink-like properties of flowers draw nutrients toward them. The generated gradient is maintained by a series of symplastic and apoplastic phloem unloading and distribution of photo-assimilates. The nonreducing disaccharide sucrose is the preferentially transported sugar from photosynthetically active tissues of the leaf and stem along active phloem transporters as was demonstrated in cut carnation flowers (Borghi and Fernie 2017). Over the past 20 years, many sugar transporters have been identified in plants. Sucrose is loaded into the phloem by the synchronized action of "SUGARS WILL EVENTUALLY BE EXPORTED TRANSPORTER" (SWEET) and "SUCROSE UPTAKE TRANSPORTER" (SUT) proteins. Approximately 20 SWEET paralogs are expressed in plant genomes, which are differentially expressed, associating them with a variety of sugar translocation steps. SWEETs are localized to different cellular compartments, particularly in the plasma membrane (SWEET1, 8, 9, 11, 12, and 15), tonoplast (SWEET16 and 17), and Golgi complex (SWEET9 and 15). Post-unloading from the phloem tissues, in the apoplast of the sink cells, carbon in the form of sucrose is hydrolyzed into fructose and glucose by enzymes of the β-fructofuranosidase family called the invertases (INVs; EC 3.2.1.26). Finally, plasma membrane SUGAR TRANSPORTER PROTEINs (STPs) mediate the uptake of the hexoses released in the apoplasmic space (Borghi and Fernie 2017). Changes in sugar metabolism and cell osmolarity are hence suggested to be the chief driving forces regulating petal movements. Rapid flower opening in many species, including roses, *Hemerocallis* hybrid cv. Cradle Song, Asiatic lilies, and *Campanula rapunculoides*, was related to the hydrolysis of reserve carbohydrates.

In flowers, amino acids also play highly diverse and essential roles. Apart from being the building blocks of vital enzymes and proteins, they offer essential constituents for floral metabolism and structure. They also serve as precursors as well as nitrogen donors for the biosynthesis of a diverse range of compounds central to plant growth and development, viz., chlorophyll, nucleotides, hormones, and various secondary metabolites. The incorporation of inorganic nitrogen into organic compounds occurs primarily in the root, where the absorbed nitrate or ammonium is primarily assimilated into glutamine, asparagine, glutamic acid, and aspartic acid. Following biosynthesis, amino acids are transported via the phloem to the sink organs in the developing vegetative and reproductive tissues. However recent evidences confirm that in addition to being transported exclusively from roots, the de novo synthesis of amino acids also occurs in floral tissues. The biosynthesis of phenylalanine in the floral parts has been demonstrated in many species. In addition the de novo biosynthesis of asparagine, γ-amino butyric acid (GABA), and proline has been confirmed in flowers (Borghi and Fernie 2017). Proline in particular has been linked to roles in flowering, as an osmotic protectant and as a signal molecule.

Numerous volatile compounds, exclusively those exceedingly abundant in plants, originate from the amino acid phenylalanine. However, other amino acids such as alanine, valine, leucine, isoleucine, and methionine also serve as precursors for a wealth of plant volatiles including aldehydes, alcohols, esters, acids, and nitrogen- and sulfur-containing volatiles. This general scheme of biosynthesis of amino acid-

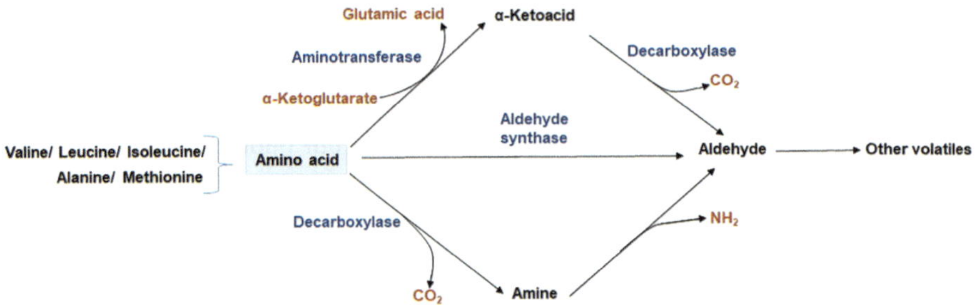

Fig. 9 Schematic representation demonstrating the biosynthesis of VOCs from branched-chain amino acids. (Modified version of Dudareva et al. 2013)

derived volatiles is thought to proceed similarly to that present in bacteria or yeast, where these pathways have been understood more extensively. The amino acids are subjected to an initial deamination or transamination catalyzed by aminotransferases, leading to the production of the corresponding α-ketoacid. These α-ketoacids can be additionally subjected to decarboxylation, followed by reductions, oxidations, and/or esterifications, forming aldehydes, acids, alcohols, and esters. A typical biochemical scheme for the biosynthesis of VOCs from amino acids is given in Fig. 9. Amino acids can further be the precursor molecules of acyl-CoAs, which are used in alcohol esterification reactions catalyzed by alcohol acyltransferases (AATs; EC 2.3.1.84) (Dudareva et al. 2013). Examples of such alcohol acyltransferases, catalyzing the formation of volatile esters from alcohols and acyl-CoAs derived from amino acids, include the well-known benzoyl-CoA:benzyl alcohol benzoyl transferase (BEBT) and benzoyl-CoA:benzyl alcohol/2-phenylethanol benzoyl transferase (BPBT).

Furthermore, high rates of respiration have been documented in floral organs. The exceptionally high rates of respiration back the process of thermogenesis, as seen in some exemplary plants such as the Araceae family including *Philodendron melinoni*, skunk cabbage (*Symplocarpus foetidus*), and voodoo lily (*Amorphophallus* spp.). Post-anthesis, proteins participating in mitochondrial electron transport chain are also increased along with the enormous respiration rates via the alternative oxidase pathway. This releases free energy in the form of heat, i.e., equivalent to a 10 °C to 25 °C increase in the tissue temperature (Borghi and Fernie 2017).This escalation in temperature assists in the production as well as volatilization of floral scents, which in addition to attracting insects is also involved in the generation of a warm environment for them.

Thus, the current knowledge on floral metabolism proves that the vital mechanisms of flower growth and development are based on and interlinked with the primary metabolism in plants. However, how the entire network operates from sucrose to secondary metabolites during flower development is yet to be determined. Numerous issues regarding floral central metabolism and the implications for flower development still remain challenging. Thus, further understanding of central metabolism in flowers will provide a firm foundation to understand the link between floral metabolism and function.

24.5 Tissue-Specific Localization of Volatiles in Floral Parts

Production and emission of volatiles in vegetative tissues are usually linked to certain cells, organs, and compartments like oil cavities, secretory cells, glandular hairs, scales, and oil or resin ducts. This compartmentalization has been intensively investigated and is well characterized in plant species. However, the mechanisms of volatile release from flowers have gained momentum in recent times demonstrating unique patterns in the localization and spatial arrangement of floral scent. Localization of volatiles is usually tissue specific in most species. The emission of fragrances can differ qualitatively and quantitatively in different parts of flowers. Generally, petals are the primary major biosynthetic source of floral volatiles; however, other parts of the flower (sepals, stamens, pistils, and nectaries) also contribute significantly to the floral bouquet in many plant species. This tissue specificity of volatile emission is regulated by an array of expression of the scent biosynthetic genes and enzyme activity. Maximum amount of scent biosynthetic genes is isolated from the scent-producing parts of the flower (Muhlemann et al. 2014a).

Volatile production and emission have been demonstrated to occur via specialized tissue areas called "osmophores" or scent glands. Many investigations were devoted to the understanding of fragrance emission via those specialized tissue structures. They vary in shape and size and are usually found toward the adaxial side of the perianth and display a bullate, rugose, pileate, conical, or papillate epidermis. This wrinkled or rugose epidermis facilitates an increase in the surface area of fragrance emission (Effmert et al. 2006). Osmophores consist of a multilayered glandular epithelium and is quite often observed in Orchidaceae. The subepidermal cell layers of the glandular tissue are modified into specially designed structures with huge storage and scent-producing units that are distinctly differentiated from the epidermal tissues. These cells of the glandular tissue contain enlarged nuclei with a dense cytoplasm. Transmission electron microscopy reveals the presence of a large amount of endoplasmic reticulum, dictyosomes, mitochondria, lipid droplets, and amyloplasts. The lipid droplets are surrounded by a layer of phospholipids and contain essential oils to be released during the emission process. Further, presence of abundant amyloplasts, i.e., starch deposits, might be the driving source of carbon and energy during scent emission as a significant depletion of these deposits was observed post-emission. A network of dispersed vacuoles also turns into an enlarged one post-anthesis, which probably plays a role in volatile emission. The epidermis is usually devoid of starch deposits, and thus based on these observations, the glandular cells are distinguished as the site of production, and the epidermis is distinguished as the site of emission (Effmert et al. 2006).

Osmophores are predetermined for scent production and emission, but there are examples where emission occurs even in their absence. For example, species such as *Clarkia breweri* (Onagraceae) and *Stephanotis floribunda* (Asclepiadaceae) lack the specialized cells in their floral tissues. Consequently, under such situations, the homogeneous epidermis cells are seen to be involved in both the biosynthesis and emission of scent volatiles. Thus the unique features of the osmophores, i.e., starch accumulations and lipid inclusions, are primarily found in epidermal cells itself instead of the underlying subepidermal cell layers (Effmert et al. 2006).

Unlike most osmophores, a few orchid species like *Cyclopogon elatus* are found to show a very unusual anatomical structure called "trichome osmophores" (Wiemer et al. 2009). Anatomical evidence suggests that two parallel oval-shaped patches of unicellular trichomes on the abaxial surface of the labellum are osmophores. Glandular nature of the trichome is supported by their immediate release of odor when rubbed and the presence of lipid droplets in their cytoplasm. Histochemical staining with neutral red staining was positive in flower buds and freshly opened flowers, showing high metabolic activity in two parallel oval-shaped areas located on the abaxial surface of the labellum. In withered flowers, a reduction of the cytoplasm along with the consumption of starch in those particular cells provides further proof of its glandular activity.

Other types of specialized cells typically dedicated to the volatile biosynthesis and localization are glandular and non-glandular trichomes which have frequently been reported on many flower surfaces such as *Angelonia salicariifolia*, *Vanda* Mimi Palmer, *Plectranthus ornatus*, *Lavandula angustifolia*, and *Helianthus annuus*. Glandular trichomes allow the sequestration (excretion and storage) of specialized metabolites at high concentrations which prevent their intracellular accumulation and toxicity. They show significant variation in morphology and the type of compound produced and have a definite structure dedicated to its storage. For example, the peltate trichome of the Lamiaceae including aromatic plants such as peppermint (*Mentha piperita*), basil (*Ocimum basilicum*), and lavender (*Lavandula* spp.) is positioned under the cuticle (subcuticular). A second type of trichome known as the biseriate, found in species of the Asteraceae, including *Artemisia annua* and sunflower (*Helianthus annuus*), bears the storage space on top of the glandular cells. During cavity development, the observation of a thickening of the cuticle in sunflower trichomes further suggested that the cuticle is strengthened to resist the pressure generated by metabolites accumulating in the storage space. Trichome-borne compounds include all major classes of secondary metabolites but are often dominated by volatile terpenoids and phenylpropanoids. The storage of VOCs in dedicated spaces of trichomes provides insights into novel polymer chemistries about cell wall and cuticle composition that selectively retain hydrophobic compounds, prevent re-entry of VOCs into the cells, as well as allow the storage cavities to expand as VOCs accumulate (Tissier et al. 2017). Presently it remains unclear how much trichome volatiles contribute to floral fragrance compositions.

24.5.1 Uncovering of Scent-Emitting Tissues in the Flower

Prior to the middle of the past century, the human olfactory system served as an important tool for understanding scent emission from floral tissue. However, with the advent of modern technologies, an unbiased technique for detecting osmophores is based on a dye called neutral red (Wiemer et al. 2009). Due to increased permeability of the cell wall in these specialized tissues, this vital stain is selectively taken up and retained in the cells. Intact osmophore tissue is able to selectively take up and retain this stain inside the vacuoles by an "ion-trap mechanism" such that once inside the vacuole it becomes impermeable across the tonoplast (Effmert et al. 2006). Therefore, the permeability of the cuticle and cell walls along with the presence of a vacuole which are all found in osmophores is pivotal for distinct staining with the neutral red dye. Availability of other commercial histochemical stains has helped in gaining additional understanding of the scent-synthesizing and scent-emitting structures. Sudan III, Sudan IV, Sudan black B, and Nile blue are effective dyes for the detection of lipids. Ruthenium red stains polysaccharides, nadi reagent indicates terpenoids, ferric trichloride targets polyphenols, and concentrated sulfuric acid effectively detects sesquiterpenes.

24.5.2 Collection and Analysis of Volatiles

In addition to morphological and anatomical inspection, analysis and identification of emitted volatiles is indispensable for studying the biogenesis of scent. For precise and unbiased chemical studies, a variety of methods have been used for collecting and exploring flower volatiles. New and improved approaches are being continually developed and implemented in parallel with more sensitive instrumentation. Earlier the most commonly used methods for collecting volatiles, i.e., solvent extraction and distillation, have now been largely replaced by the nondestructive headspace techniques, which are more reliable, accurate, and easily adaptable with current analytical instrumentation. Additionally, convenient collection of volatiles in situ from whole plants is an added significance of this method. In this technique, the sample of interest (a plant/plant part) is enclosed in a closed collection chamber, and the emitted volatiles in the airspace adjoining the sample (headspace) are trapped onto an adsorbent (Qualley and Dudareva 2009). Thus, this method is advantageous for sampling volatile compounds from low-emitting plants. The standard adsorbents presently used for trapping flower volatiles are activated charcoal and the porous polymers Tenax GC and Porapak Q. They differ to some extent in their adsorption properties and in the procedures used for volatile desorption, both of which influence the final volatile recovery. Porapak Q was shown to have a higher trapping efficiency for volatiles than either Tenax GC or charcoal. This almost certainly can be attributed to Porapak Q's greater surface area per unit mass. The trapped volatile samples are eluted with suitable pure solvents (dichloromethane, hexane, etc.). Volatile extracts can be further concentrated by solvent evaporation under a nitrogen stream before they are finally analyzed through GC-MS system. A vital advancement in headspace analysis was the development of solid phase micro-extraction (SPME) which is a cheap, fast, simple, and solvent-free sample preparation technique. SPME sampling requires no pump or hardware, thus making it ideal for use in field. It is based on adsorption and desorption of volatiles from an inert fiber coated with different types of adsorbents. Following equilibration between the fiber and the volatile sample (a few minutes to half an hour), the fiber can be directly subjected to thermal desorption onto a gas chromatograph for analysis. Thus, a careful selection of the polarity and thickness of the fiber coating allows sampling of a range of compounds across different polarity and volatility.

To date two types of headspace sampling methods have been devised for volatile metabolome investigations. The different sampling techniques are the static headspace sampling (without air circulation) with the use of SPME and the dynamic headspace sampling which comprises close-loop stripping (air is constantly recycled) and pull and push-pull systems (air is constantly taken up from the outside, passed over the plant sample, and through an absorbent trap). On comparison of the two methods, dynamic sampling method is the most frequently used technique in all areas of plant volatile analysis. This method collects a much larger and concentrated quantity of compounds as the constant flow of air allows the absorbent to act as a trap for the volatiles. Upscaling of sampling can be easy by increasing the quantity of sorbent, airflow rate, sample mass, and sampling time. Precise quantification of compounds is feasible as multiple columns can be connected in tandem to compensate for differences in affinity of adsorbents, a technique known as multiple layer adsorption. In addition, push and pull headspace sampling helps escape problems often encountered with the sealed systems of static headspace and closed-loop stripping including heat, water vapor, ethylene accumulation, etc. that can affect sampling efficiency (Qualley and Dudareva 2009).

24.6 Interpreting the Release and Emission of Floral Scent

Finally, understanding the mechanism behind the synchronized release of volatiles is vital to the regulation of scent biogenesis in plants. The boiling points of nearly all volatile compounds recognized as floral scents range from 150 °C to 350 °C. The VOCs are existent in floral tissues mostly as liquids or in solution. A negative correlation was found among the boiling points of the VOCs and the ratio of the emitted and endogenous concentrations, indicating that the emission of an individual compound is strictly dependent on the level of the endogenous concentration of the compound. Sagae et al. (2008) suggested that an increase in ambient temperature resulted in increased emission in *Petunia axillaris* floral volatiles. The floral scent emission and endogenous levels were investigated in P. *axillaris* at different temperatures (20°, 25°, 30°, and 35 °C) under the hypothesis that floral scent emission would be regulated by both metabolic and vaporization processes. The total endogenous content of scent components decreased with increasing temperature with the total emission showing a peak at 30 °C. Thus, the vaporization of scent was predicted to be influenced by ambient temperature. Recent studies by Barman and Mitra (2021) have also demonstrated similar correlation between the temporal accumulations of endogenous volatiles with the concentrations of emitted volatiles in *Jasminum auriculatum* at different temperatures. Additionally, the volatile production was negatively affected by high-temperature growth conditions due to inhibition of the expression of pathway genes responsible for the biosynthesis as well as emission of volatiles. The dynamic nature of floral scents thus highlights a greater need for understanding of physiological and ecological mechanisms driving the vital phenomenon of scent emission.

Volatiles have a small molecular size; hence, it was earlier reasoned that they could easily pass through pores in the cell wall and passively diffuse across cellular barriers into the external atmosphere, which has a low concentration of the scent compounds as compared with the interior of the cell. The process modeled on the principle of Fick's first law governs the steady-state flux of molecules according to the concentration gradients. This default assumption that VOCs simply diffuse out of cells may be true for small VOCs such as isoprene. However, numerous examples exist when rates of VOC emission cannot be explained by a concentration-dependent diffusion mechanism. Extremely high levels of accumulations (50 to 120 mM) of the volatile compounds are required at each cellular barrier to achieve the observed emitted flux (Widhalm et al. 2015). Hence, the prediction that emissions of most VOCs are driven solely by diffusion would also result in toxic VOC accumulation in membranes which could prove detrimental to membrane integrity and function. Such harmful accumulated levels could further cause the reorganization of cellular membranes leading to leakage of organelles or cellular content. This elucidates another hypothesis that VOC emission additionally requires other transport devices to attain the observed emission rates.

Floral organs are metamorphosed leaves, so epithelium modifications usually include stomata. Stomata as a channel of scent release have been mentioned from time to time. A plausible hypothesis of scent emission via floral stomata of the petaloid tepal in tuberose has been put forward recently. Scent emission in tuberose follows a nocturnal pattern further supporting the connection between scent release and stomatal opening (Maiti and Mitra 2017). Emission of volatiles through stomata has also previously been reported in leaves of *Pinus* sp.

Regardless of whether they are released via specialized structures or through stomata, considering at the cellular level, VOCs must travel from their site of production through the cytosol to the cell membrane and then subsequently traverse the hydrophilic cell wall and, occasionally, the waxy cuticle complex to finally get released into the atmosphere. The trafficking mechanisms play an important role in the overall emission of floral scents. However, they are comparatively difficult to investigate as the specific processes cannot be singled out for separate quantification. As a result, transport of scent molecules has long been neglected, and direct evidence is scarcely available till date. Mechanical disruption allows direct access for volatile compounds inside the cell to the atmosphere, however, it remains vague how from intact cells VOCs cross these cellular barriers and release directly to the environment.

A mechanism based on correlation to the intracellular trafficking of other lyophilic compounds, such as waxes and diterpenes, has been considered that could also be involved in shuttling VOCs out of undamaged cells (Jetter 2006). VOCs synthesized in internal organelles, such as plastids, can be effectively trafficked to the plasma membrane via endoplasmic reticulum (ER) through inter-organellar membrane hemi-fusion in association with the *trans*-Golgi network (TGN). Thus, direct contact between membranes of the ER and the plasma membrane creates a lipophilic pathway for intracellular trafficking of floral scent molecules (Fig. 10). Evidence supporting this mechanism was seen in the highly specialized scent-producing cells

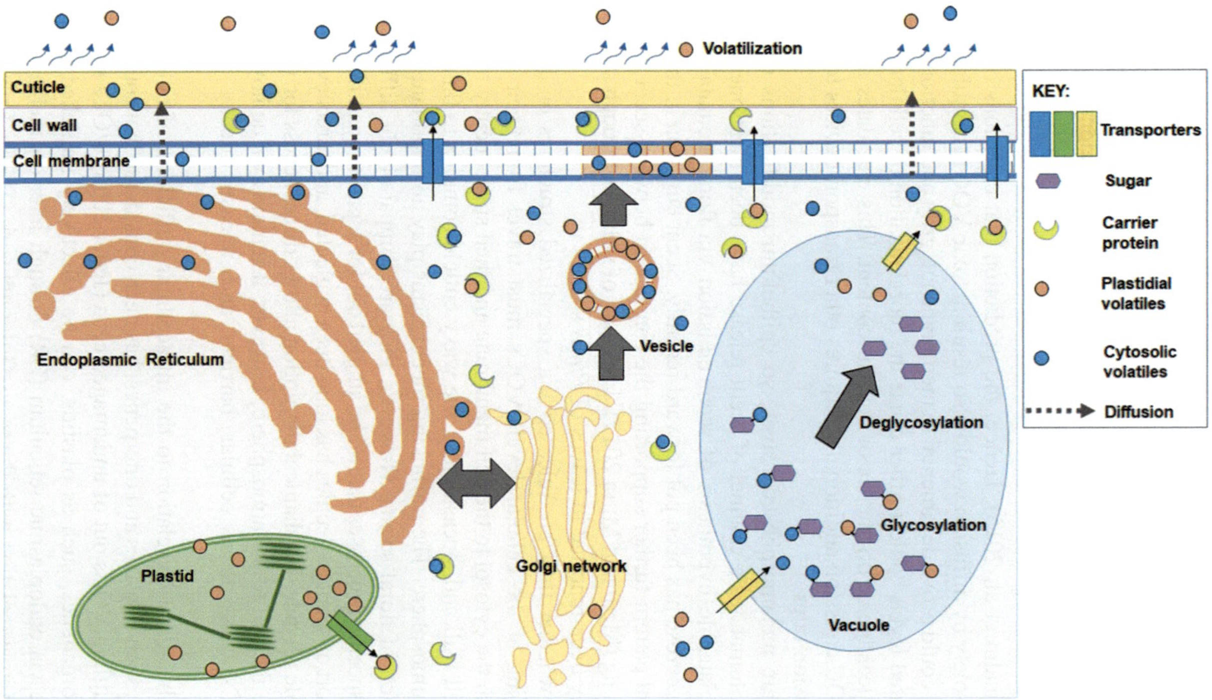

Fig. 10 **A simplified overview of the proposed trafficking model for VOCs in plant cells. VOCs may either be diffused through each cellular barrier or be trafficked via tracking routes through the cytosol as illustrated by block arrows**. Carrier proteins such as lipid transfer proteins may assist in shuttling the hydrophobic VOCs through the hydrophilic cytoplasm and cell wall. Predicted active transporters such as ABC transporters located in the plasma membrane might further be involved in exporting VOCs which may then diffuse through the cuticle into the external atmosphere. Some of the VOCs may be accumulated in the vacuoles as nonvolatile glycosylated forms where they act as storage units, and only upon action of hydrolytic enzymes (deglycosylation) the sugar moiety is released and the resultant free volatiles are emitted out of the cell (Modified version of Widhalm et al. 2015)

in the *Sauromatum guttatum* appendix, where membrane channels are created by fusion of the rough ER to the plasma membrane. This microscopic finding was suggested to be the transport mechanism for the emission of the sesquiterpenoids, viz., copaene and caryophyllene (Skubatz et al. 1995). This process could also efficiently reduce the accumulation of volatile compounds in the internal membranes. Alternatively, VOC trafficking to the plasma membrane could also be facilitated by carrier proteins bearing hydrophobic pockets which are capable of escorting lipophilic compounds across aqueous environments of the cell (Widhalm et al. 2015).

The non-polar nature of the scent molecules contributes to its very low solubility in any aqueous environment, which consequently and substantially also hampers their transport across the cell wall. Hence, transport across the hydrophilic cell wall may be facilitated energetically, probably by soluble carrier proteins with hydrophobic pockets (Fig. 10) (Widhalm et al. 2015).

Alternatively, lipid transfer proteins (LTPs) could also be involved in volatile transport across the epidermal cell wall. LTPs are small (<10 kDa) soluble cysteine-rich proteins that are capable of binding small hydrophobic molecules. LTPs reversibly bind lipids, in many cases nonspecifically, thus increasing their solubility in aqueous solution. The LTP-lipid complexes are presumed to be transported along the concentration gradients of bound molecules, thus effectively shuttling them from regions of relatively high concentration. LTPs have been detected in the epidermal apoplast of barley leaf tissue suggesting a role in the export of nonvolatile hydrophobic molecules, such as cuticle precursors, across the hydrophilic cytosol and cell wall to reach the cuticle. Hence, LTPs might as well be involved in the transport of scent molecules across the epidermal cell wall (Jetter 2006). Recently transcriptome analysis has revealed that genes encoding LTPs, as well as members of the ABCG subfamily, are highly expressed in glandular trichomes that produce VOCs (Tissier et al. 2017). Correspondingly a remarkable proposal of active biological transport for exporting cargo to the cell exterior via plasma membrane specialized adenosine triphosphate-binding cassette (ABC) transporters has been suggested in *Petunia hybrida*. The transporter of the subfamily G member 1, PhABCG1, is predicted to be a plasma-membrane transporter expressed exclusively in petals of open flowers and is regulated by the ODORANT1 transcription factor, which controls VOC biosynthesis in petunia flowers. Downregulation of PhABCG1 by RNA interference resulted in a decrease in emission profile with toxic accumulation levels of VOCs in the plasma membrane with subsequent detrimental effects on membrane integrity. RNA interference lines exhibited 52–62% decrease in total VOC emission and a corresponding 101–157% increase in the internal pool (Adebesin et al. 2017).

In conclusion, vital steps involved in the export of VOCs through the membrane are energy dependent. Intracellular trafficking, transport through the cell membrane, and export from the epidermal cell wall consequently levy transport barriers and small quantities of products accumulate in corresponding membrane-bound compartments. As the concentration increases, a gradient is built up from inside to outside. This gradient energies transport across the cell wall and cuticle. This transport may also be aided by proteins, especially in the transfer of the lipophilic molecules across the aqueous environment of the cell wall. In case of toxic accumulations in the cell, the compounds might be routed to the vacuoles and stored as nontoxic glycosylate forms (Fig. 10).

24.7 Practical Applications of Floral Scent Research

The commercial significance of scent metabolites has encouraged current interest in understanding their biosynthesis in planta. As discussed earlier, pollination is vital for plant survival, and pollinators utilize scent as a major cue in finding potential hosts. The genetic engineering of floral scent could thus effectively enhance the pollination by increasing the attractiveness of a host to potential vectors. The ability to manipulate floral scent offers a foundation for studying the effects of changes in scent and the roles of specific volatiles in individual pollinator attraction and, thus, possibly will allow us in the future to fill vital lacunae in our understanding of plant-insect interactions. Alternatively, tailoring floral scent for specialized pollinators will efficiently diminish the chance of pollen loss and ineffective interspecific pollination, thereby escalating plant reproductive success. Unique compositions of floral scents also draw interest from the cosmetic and fragrance industries, which are in a continuous search for novel and rare volatile compounds and scents. Nonetheless, all of these potential applications for modifying volatile emission in plants call for a genetic engineering approach, which can be accomplished only with an ample understanding of the control of the volatile biosynthesis machinery. Thus, recent findings of genes encoding enzymes catalyzing the biosynthesis of volatile compounds, and advancements in our comprehension of the regulation of scent biosynthetic genes, have made possible to engineer plants with altered aromas. Approaches for metabolic engineering of floral scent have used several strategies in the past, including introducing new scent genes and blocking the existing non-scent pathways to focus metabolite flux into scent production.

In the last decade, numerous efforts have been made to induce the biosynthesis of new floral scent components in flowers. Some of the pioneering experiments are discussed below which show that volatile compounds can be synthesized in heterologous systems through metabolic engineering. The linalool synthase (LIS) from *Clarkia breweri* (which converts GPP to (3*S*)-linalool) was the most repeatedly used gene in these attempts. The LIS gene was introduced into *Petunia hybrida* W115 (Lücker et al. 2001) and carnation (*Dianthus caryophyllus*) (Lavy et al. 2002), under the control of the cauliflower mosaic virus 35S (CaMV) constitutive promoter. Transgenic petunia expressing the gene produced linalool; however, the whole product was transformed into a nonvolatile compound glycoside by the action of endogenous glycosyltransferases. Thus, no alteration was found on the olfactory properties of the transformed plants. Correspondingly, in transgenic carnations, biosynthesis of linalool was achieved, but a substantial fraction of the synthesized linalool was converted to linalool oxides by an endogenous enzyme. Further, the transgenic tissue did not store the linalool but was rapidly emitted post-synthesis, so there was hardly any change in floral aroma in smell tests (Lavy et al. 2002). In another experiment, the LIS gene when introduced into tomato (*Lycopersicon esculentum* Mill.) under the control of the tomato late-ripening-specific *E8* promoter, the transgenic tomato fruits accumulated >50-fold more linalool than wild-type plants without any major alteration of other metabolic pathways. Some amount of the linalool was also oxidized to the volatile 8-hydroxylinalool, but the concentrations were sufficient enough for olfactory recognition (Lewinsohn et al. 2001). These studies revealed that unexpected problems can be encountered during the genetic engineering of floral fragrance.

In an alternative molecular genetic approach, metabolic attempts were made to modify the floral scent by redirecting the metabolic flux in carnations (Zuker et al. 2002). In their experiment, the antisense suppression of the flavanone 3-hydroxylase (encoding a key enzyme in the anthocyanin pathway) led to the disappearance of the original orange/reddish color of flowers and increased methyl benzoate production. As both anthocyanins and methyl benzoate originate from the phenylpropanoid pathway, these results suggest that redirecting the metabolic flux can result in enhanced scent production in transgenic flowers.

Arabidopsis *PAP1* coding for the *Myb* transcription factor was cloned into *Rosa hybrida* 'Pariser Charme' where enhanced levels (nearly 6.5 times higher) of phenylpropanoid and terpenoid compounds were observed. Olfactory assay revealed that bees and humans could discriminate between the floral scents of *PAP1* transgenic and control flowers. The escalation of volatile production in the *PAP1* transgenic plants was probably due to the transcriptional activation of the respective biosynthetic genes and enhanced metabolic flux in both the phenylpropanoid and terpenoid pathways. The mechanism(s) controlling the production of specialized metabolites however remains to be elucidated (Zvi et al. 2012).

More recently, penetrating radiations such as γ-rays have also been utilized for the alteration morphology, anatomy, biochemistry, and physiology of plants; hence, γ-irradiation technology has been substantially utilized for the improvement of ornamental plants. In *Polianthes tuberosa* flowers, an increase in the amount of emission of floral volatiles was observed in plants developed from tubers irradiated with a low-dose application of γ-irradiation. Benzaldehyde, methyl salicylate, and trans-isoeugenol increased significantly as compared to control, thus improving the floral scent profile of the irradiated plants (Kutty et al. 2020). Similar observations were seen in *Jasminum auriculatum* where it was demonstrated that low dose of γ-irradiation could enhance the contents of desired floral volatiles as compared to non-irradiated plants, indicating a hormetic effect (Barman et al. 2020). Thus, γ-rays can act as an environmentally friendly, low-cost technology, to successfully develop new floricultural crops with improved traits.

Therefore, although advancement in floral scent engineering has been substantial, floral scent fabrication in planta remains rather unpredictable. The complex networks of floral volatile biosynthesis are a set of overlapping and competing pathways, where the regulatory mechanisms are barely understood. Introduction of new gene or upregulation of genes might result in substrate shortage or accumulation of toxic gene products and can potentially cause cell death since many of the scent compounds are cytotoxic. On the contrary, downregulation of native genes may result in unexpected re-channeling of metabolites. Moreover transgene expression can be achieved more efficiently by the usage of flower-specific promoters instead of the CaMV promoter. Future attempts at scent genetic engineering will hence need to utilize specific floral gene promoters and be based on a stronger understanding of metabolic fluxes. Nonetheless, transgenic flowers emitting suitable amounts of engineered scents may be of immense delight to human senses. Additionally, considering how these deliberate attempts to alter the volatile signatures of plants will impact our ecosystem is important. Research in this direction is required to provide a deeper understanding on the effect of alterations in plant volatile compounds on a global scale.

24.8 Conclusion

Scent that emanates from flowers has been highly valued by humans since antiquity. The commercial importance of volatile secondary metabolites has prompted interest in understanding their formation and in engineering their biosynthesis in planta. Floral scent research has also become an enriched area of plant science based on their importance in ecology and evolution. Contemporary research has identified numerous floral volatile compounds in nearly 1000 plant species (Muhlemann et al. 2014a). However plant specialized metabolites including floral scents are produced in very low concentration and low quantity in tissues. Due to this reason, separation of compounds is generally expensive and also inefficient, and many aspects of their biosynthesis and function, including transcriptional regulation, are limited. Hence, future studies must address the composition and function of floral scent bouquets on a finer spatial scale within the flower. Additionally, the temporal variations in floral scent generated by floral rhythms clearly warrant further investigation.

Furthermore, petals display a very high chemo-diversity and undergo extensive metabolic changes during flower development, but the measure by which floral metabolism depends on de novo biosynthesis versus intake of sugars and amino acids remains largely uninvestigated. Moreover how the entire metabolic network from sucrose to secondary metabolites operates during flower development is yet to be determined. Lack of characterization of transcription factors that coordinately regulate gene expression controlling fragrance biosynthesis and an insufficient understanding of the intracellular metabolite trafficking are the key lacunae in understanding the floral metabolic pathways. Existing knowledge clearly indicates that engineering plants to synthesize and emit more volatiles from their flowers is now possible. In this sense, the use of genetically modified plants, overexpressing the genes responsible for the limiting steps in the interface between primary and secondary metabolism, might be an essential tool for allowing a better understanding of the regulatory network that controls metabolic flux in planta.

In conclusion, floral scent biology has become an intensive area of study for researchers with expertise in the most dissimilar disciplines including plant developmental biology, molecular biology, metabolomics, genomics, entomology, and behavioral biology. Thus, a more holistic and collaborative approach is demanded to shed light on the mechanisms that control the biology of floral scent and its implications on flower development and plant pollinator interactions.

Acknowledgment Research work on floral scent biology in the authors' laboratory was supported by research grants from the Science and Engineering Research Board (www.serb.gov.in), India [EMR/2015/000247/PS], Department of Science and Technology, Government of India [DST/INT/RUS/RFBR/P-329] and by the Council of Scientific and Industrial Research (www.csirhrdg.res.in), India [38(1336)/12/EMR-II and 38(1420)/16/EMR-II]. U Ghissing was a recipient of an individual research fellowship [09/081(1291)/2017-EMR-I] from CSIR.

References

Abbas F, Ke Y, Yu R, Yue Y, Amanullah S, Jahangir MM, Fan Y (2017) Volatile terpenoids: multiple functions, biosynthesis, modulation and manipulation by genetic engineering. Planta 246:803–816

Adebesin F, Widhalm JR, Boachon B, Lefèvre F, Pierman B, Lynch JH, Alam I, Junqueira B, Benke R, Ray S, Porter JA (2017) Emission of volatile organic compounds from petunia flowers is facilitated by an ABC transporter. Science 356:1386–1388

Baldermann S, Yang Z, Sakai M, Fleischmann P, Watanabe N (2009) Volatile constituents in the scent of roses. Floricul Ornament Biotech 3:89–97

Barman M, Mitra A (2021) Floral maturation and changing air temperature influence scent volatile biosynthesis and emission in *Jasminum auriculatum* Vahl. Environ Exp Bot 181:104296

Barman M, Kotamreddy JN, Agarwal A, Mitra A (2020) Enhanced emission of linalool from floral scent volatile bouquet in *Jasminum auriculatum* variants developed via gamma irradiation. Ind Crop Prod 152:112545

Bera P, Mukherjee C, Mitra A (2017) Enzymatic production and emission of floral scent volatiles in *Jasminum sambac*. Plant Sci 256:25–38

Borghi M, Fernie AR (2017) Floral metabolism of sugars and amino acids: implications for pollinators' preferences and seed and fruit set. Plant Physiol 175:1510–1524

Cna'ani A, Shavit R, Ravid J, Aravena-Calvo J, Skaliter O, Masci T, Vainstein A (2017) Phenylpropanoid scent compounds in petunia x hybrida are glycosylated and accumulate in vacuoles. Front Plant Sci 8:1898

Colquhoun TA, Schwieterman ML, Gilbert JL, Jaworski EA, Langer KM, Jones CR, Rushing GV, Hunter TM, Olmstead J, Clark DG, Folta KM (2013) Light modulation of volatile organic compounds from petunia flowers and select fruits. Postharvest Biol Technol 86:37–44

Cui J, Katsuno T, Totsuka K, Ohnishi T, Takemoto H, Mase N, Toda M, Narumi T, Sato K, Matsuo T, Mizutani K (2016) Characteristic fluctuations in glycosidically bound volatiles during tea processing and identification of their unstable derivatives. J Agric Food Chem 64:1151–1157

Dudareva N, Klempien A, Muhlemann JK, Kaplan I (2013) Biosynthesis, function and metabolic engineering of plant volatile organic compounds. New Phytol 198:16–32

Effmert U, Buss D, Rohrbeck D, Piechulla B (2006) Localization of the synthesis and emission of scent compounds within the flower. In: Dudareva N, Pichersky E (eds) Biology of floral scent. CRC Press, Boca Raton, pp 105–124

Francis MJ, Allcock C (1969) Geraniol β-D-glucoside; occurrence and synthesis in rose flowers. Phytochemistry 8:1339–1347

Gang DR, Simon J, Lewinsohn E, Pichersky E (2002) Peltate glandular trichomes of *Ocimum basilicum* L.(Sweet Basil) contain high levels of enzymes involved in the biosynthesis of phenylpropenes. Int J Geogr Inf Syst 9:189–195

Goodwin SM, Kolosova N, Kish CM, Wood KV, Dudareva N, Jenks MA (2003) Cuticle characteristics and volatile emissions of petals in *Antirrhinum majus*. Physiol Plant 117:435–443

Haydon MJ, Mielczarek O, Krank A, Roman A, Webb AA (2017) Sucrose and ethylene signaling interact to modulate the circadian clock. Plant Physiol 175:947–958

Jetter R (2006) Examination of the processes involved in the emission of scent volatiles from flowers. In: Dudareva N, Pichersky E (eds) Biology of floral scent. CRC Press, Boca Raton, pp 125–144

Ke M, Gao Z, Chen J, Qiu Y, Zhang L, Chen X (2018) Auxin controls circadian flower opening and closure in the waterlily. BMC Plant Biol 18:143

Kessler D, Diezel C, Clark DG, Colquhoun TA, Baldwin IT (2013) Petunia flowers solve the defence/apparency dilemma of pollinator attraction by deploying complex floral blends. Ecol Lett 16:299–306

Knauer AC, Schiestl FP (2015) Bees use honest floral signals as indicators of reward when visiting flowers. Ecol Lett 18:135–143

Koeduka T (2014) The phenylpropene synthase pathway and its applications in the engineering of volatile phenylpropanoids in plants. Plant Biotechnol 31:401–407

Kolosova N, Gorenstein N, Kish CM, Dudareva N (2001) Regulation of circadian methyl benzoate emission in diurnally and nocturnally emitting plants. Plant Cell 13:2333–2347

Kutty NN, Mitra A (2019) Profiling of volatile and non-volatile metabolites in *Polianthes tuberosa* L. flowers reveals intraspecific variation among cultivars. Phytochemistry 162:10–20

Kutty NN, Ghissing U, Kumar M, Maiti MK, Mitra A (2020) Intense floral scent emission in *Polianthes tuberosa* L.(Tuberose) variants sprouted from γ-irradiated tubers. J Plant Growth Regul 39:112–121

Lavy M, Zuker A, Lewinsohn E, Larkov O, Ravid U, Vainstein A, Weiss D (2002) Linalool and linalool oxide production in transgenic carnation flowers expressing the *Clarkia breweri* linalool synthase gene. Mol Breed 9:103–111

Lewinsohn E, Schalechet F, Wilkinson J, Matsui K, Tadmor Y, Nam KH, Amar O, Lastochkin E, Larkov O, Ravid U, Hiatt W (2001) Enhanced levels of the aroma and flavor compound S-linalool by metabolic engineering of the terpenoid pathway in tomato fruits. Plant Physiol 127:1256–1265

Liao P, Ray S, Boachon B, Lynch JH, Deshpande A, McAdam S, Morgan JA, Dudareva N (2021) Cuticle thickness affects dynamics of volatile emission from petunia flowers. Nat Chem Biol 17: 138–145

Lücker J, Bouwmeester HJ, Schwab W, Blaas J, Van Der Plas LH, Verhoeven HA (2001) Expression of Clarkia S-linalool synthase in transgenic petunia plants results in the accumulation of S-linalyl-β-d-glucopyranoside. Plant J 27:315–324

Maiti S, Mitra A (2017) Morphological, physiological and ultrastructural changes in flowers explain the spatio-temporal emission of scent volatiles in *Polianthes tuberosa* L. Plant Cell Physiol 58: 2095–2111

Mettal U, Boland W, Beyer P, Kleinig H (1988) Biosynthesis of monoterpene hydrocarbons by isolated chromoplasts from daffodil flowers. Eur J Biochem 170:613–616

Muhlemann JK, Maeda H, Chang CY, San Miguel P, Baxter I, Cooper B, Perera MA, Nikolau BJ, Vitek O, Morgan JA, Dudareva N (2012) Developmental changes in the metabolic network of snapdragon flowers. PLoS One 7:e40381

Muhlemann JK, Klempien A, Dudareva N (2014a) Floral volatiles: from biosynthesis to function. Plant Cell Environ 37:1936–1949

Muhlemann JK, Woodworth BD, Morgan JA, Dudareva N (2014b) The monolignol pathway contributes to the biosynthesis of volatile phenylpropenes in flowers. New Phytol 204:661–670

Overland L (1960) Endogenous rhythm in opening and odor of flowers of *Cestrum nocturnum*. Am J Bot 47:378–382

Paul I, Bhadoria PB, Mitra A (2020) Seasonal and diel variations in scent composition of ephemeral *Murraya paniculata* (Linn.) Jack flowers are contributed by separate volatile components. Biochem Syst Ecol 89:104004

Qualley AV, Dudareva N (2009) Metabolomics of plant volatiles. Methods Mol Biol 553:329–343

Sagae M, Oyama-Okubo N, Ando T, Marchesi E, Nakayama M (2008) Effect of temperature on the floral scent emission and endogenous volatile profile of *Petunia axillaris*. Biosci Biotechnol Biochem 72:110–115
Schrempf M (1977) Studies of the circadian rhythm of petal movement in *Kalanchoe blossfeldiana*. Biol Rhythm Res 8:396–400
Simkin AJ, Guirimand G, Papon N, Courdavault V, Thabet I, Ginis O, Bouzid S, Giglioli-Guivarc'h N, Clastre M (2011) Peroxisomal localisation of the final steps of the mevalonic acid pathway in planta. Planta 234:903–914
Skubatz H, Kunkel DD, Patt JM, Howald WN, Hartman TG, Meeuse BJ (1995) Pathway of terpene excretion by the appendix of *Sauromatum guttatum*. Proc Natl Acad Sci 92:10084–10088
Song C, Härtl K, McGraphery K, Hoffmann T, Schwab W (2018) Attractive but toxic: emerging roles of glycosidically bound volatiles and glycosyltransferases involved in their formation. Mol Plant 11:1225–1236
Tholl D, Lee S (2011) Terpene specialized metabolism in *Arabidopsis thaliana*. The Arabidopsis Book 9:e0143
Tissier A, Morgan JA, Dudareva N (2017) Plant volatiles: going 'in' but not 'out' of trichome cavities. Trends Plant Sci 22:930–938
Underwood BA, Tieman DM, Shibuya K, Dexter RJ, Loucas HM, Simkin AJ, Sims CA, Schmelz EA, Klee HJ, Clark DG (2005) Ethylene-regulated floral volatile synthesis in petunia corollas. Plant Physiol 138:255–266
Vereecken NJ, Schiestl FP (2009) On the roles of colour and scent in a specialized floral mimicry system. Ann Bot 104:1077–1084
Widhalm JR, Dudareva N (2015) A familiar ring to it: biosynthesis of plant benzoic acids. Mol Plant 8:83–97
Widhalm JR, Jaini R, Morgan JA, Dudareva N (2015) Rethinking how volatiles are released from plant cells. Trends Plant Sci 20:545–550
Wiemer AP, More M, Benitez-Vieyra S, Cocucci AA, Raguso RA, Sérsic AN (2009) A simple floral fragrance and unusual osmophore structure in *Cyclopogon elatus* (Orchidaceae). Plant Biol 11: 506–514
Yeats TH, Rose JK (2013) The formation and function of plant cuticles. Plant Physiol 163:5–20
Zhang GH, Jiang NH, Song WL, Ma CH, Yang SC, Chen JW (2016) De novo sequencing and transcriptome analysis of *Pinellia ternata* identify the candidate genes involved in the biosynthesis of benzoic acid and ephedrine. Front Plant Sci 7:1209
Zuker A, Tzfira T, Ben-Meir H, Ovadis M, Shklarman E, Itzhaki H, Forkmann G, Martens S, Neta-Sharir I, Weiss D, Vainstein A (2002) Modification of flower color and fragrance by antisense suppression of the flavanone 3-hydroxylase gene. Mol Breed 9:33–41
Zvi MM, Shklarman E, Masci T, Kalev H, Debener T, Shafir S, Ovadis M, Vainstein A (2012) PAP1 transcription factor enhances production of phenylpropanoid and terpenoid scent compounds in rose flowers. New Phytol 195:335–345

Mirabilis: Medicinal Uses and Conservation 25

Useful Tropical Plant

Moumita Malakar and Sukanta Biswas

Contents

M. Malakar (✉)
Department of Horticulture & Floriculture, Central University of Tamil Nadu, Thiruvarur, Tamil Nadu, India

S. Biswas
Department of Horticulture, Institute of Agricultural Science, University of Calcutta, Kolkata, West Bengal, India

S. K. Datta, Y. C. Gupta (eds.), *Floriculture and Ornamental Plants*, Handbooks of Crop Diversity: Conservation and Use of Plant Genetic Resources,
https://doi.org/10.1007/978-981-15-3518-5_28

Abstract

Among the estimated 35,000 to 70,000 medicinally important plant species, the genus *Mirabilis* is a significant perennial herb resides to Nyctaginaceae which contains more than 25 species but especially on *jalapa* and *longiflora* species much emphasize for several genetical and pharmaceutical investigations were given. The former one is familiar as 'Four O'clock and 'Sweet Marvel of Peru', the most well-known garden ornamental herb. The somatic chromosome number is 58 with three of the chromosome pairs large or rather large which remain attached to the nucleolus. Fertility is highly reduced in F_1 generation and the same trend too found in successive generations. Flower color inheritance shows a perfectly regular Mendelian segregation with anthocyanin versus without and lilac color versus pink. Possible causes of flower variegation are cytoplasmic inheritance via uniparental mode of inheritance neither reciprocal differences nor plastid. Additionally, it is traditionally used as purgative and emetic, for treatment of many gastrointestinal disorders, including dysentery, diarrhea, muscle pain, and abdominal colic. Besides this it also exhibits certain useful activities like antiviral activity, antimicrobial activity, antimalarial activity, anthelmintic activity, antioxidant activity, and many others.

In different plant parts like from *M. jalapa* flowers 8 betaxanthins (indicaxanthin, vulgaxanthin-I, miraxanthin), from roots rotenoids (mirabijalone A, B, C, and D, 9-O-methyl-4-hydroxyboeravinone B, boeravinone C and boeravinone F, and 1,2,3,4-tetrahydro-1-methylisoquinoline-7,8-diol), and from seed oil 8-hydroxyoctadeca-cis-11,14-dienoic acid, a minor fatty acid could be isolated. Methanolic extract of the aerial parts yields β-sitosterol, stigmasterol, ursolic acid, oleanolic acid, and brassicasterol. Besides pharmacologically active compounds, ether compound-3,3′-Methylenebis (4- hydroxycoumarin), N-D-alpha Phenylyglycinelaminaribiitol-3-(4-(dimethylamino) cinnamoyl) and 4-hydroxycoumarin are found also in alcoholic extract.

Keywords

Mirabilis · Inheritance · Pharmaceutical · Genetical · Ornamental

25.1 Introduction

Mirabilis is an important member of Nyctaginaceae family of herbs or shrubs, mainly native to tropical America and Northern Mexico and commonly known as "four o'clock" (The wealth of India 1949). About 60 species of the genus have been reported that grows worldwide (Hutchinson 1969) out of which *Mirabilis himalaica* is indigenous to India mainly in West Bengal, Manipur, and Western Himalayas cultivated for ornamental purpose (The wealth of India 1949). It had been reported that the plant was introduced into Europe by the Spaniards in 1596, but it was first recorded botanically in 1753 as the most commonly grown ornamental species of *Mirabilis* plant and is available in wide range of hues. The reason of poising one of its common name "4'o clock" since flowers usually open from late afternoon or at dusk. During anthesis blooms exhale strong, sweet-smelling fragrance throughout the night and close during morning time (Hutchinson 1969). Intense aroma released from blooms after sunset shoo away the mosquitoes. The name of *Mirabilis jalapa* given by Carl Von Linne in 1753 is formed from the scientific Latin *Mirabilis* meaning "admirable" by allusion to the remarkable colors of its flowers and the specific name *jalapa* that would refer to its origin in the Jalapa in Guatemala. But the epithet of *jalapa* could also refer to the city of "Xalapa" (Jalapa) in Mexico from which came a former purgative drug, named "jalap," taken from the tubers of the tuberous jalap (Hutchinson 1969). Tournefort identified medicinal jalap as *Mirabilis*, after being "deceived" by persons who claimed to know the plant (Felter and Lloyd 2010). Perhaps this was an honest mistake by Tournefort's sources as both plants have enlarged roots and large tubular flowers. Also, frequently specimens of *M. jalapa* and many Convolvulaceae taxa have cordate leaves. In addition, both *M. jalapa* and medicinal "jalap" have been reported as purgatives. Whether due to deliberate deception or innocent misidentification by others, Tournefort's mis-identification led to further confusion about the identity of medicinal jalap and more names. Balfour (1879) identified medicinal jalap as *Exogonium purga*, though no author was given (Felter and Lloyd 2010). The family Nyctaginaceae was established by Jussieu in 1789 (Zant 2016). In Indian context, Roxburgh (1832) had made first report and descriptions of *Mirabilis* genus of the family Nyctaginaceae. Based on that plants are usually herbaceous, shrubs, or trees. Leaves are simple, opposite, alternated or whorled, and exstipulate. Flowers are bisexual, rarely unisexual, usually actinomorphic, solitary, or fasciculate and arranged to cymes. Bracts are calyx-like, sometimes brightly colored and often connate or free. Perianths are adnate, petaloid, tubes campanulate, tubular, or salver form, apex 3–10 lobed. Stamens are 1 to numerous and hypogynous. Filaments are free or connate at base. Anthers are two-loculed, dehiscence, and longitudinal. Ovary is superior, 1- loculed, and carpel 1; ovule is 1. Style is solitary. Stigma is globose. Fruit is an achene-like anthocarp, enclosed by persistent perianth, ribbed or winged, often glandular. Seeds are one, endosperm copious. Embryo is erect or curved.

25.1.1 Traditional Uses

Traditional herbal medicines have improved the general health of most people. The primary benefits of using plant-derived medicines are that they are relatively safer than synthetic alternatives. According to the World Health Organization (WHO), about 65–80% of the world's population in developing countries depend essentially on plants for their primary health care due to poverty and lack of access to modern medicine. For centuries, the plant was used as anti-diabetic in China (Zant 2016). Decoction of the entire plant was given orally to treat kidney infections and for its diuretic action (Sharma et al. 2012). Stems were used as tonic (Chetty et al. 2008). In Latin America and South Africa, roots of *M. jalapa* L. were traditionally used for its purgative, emetic, and cathartic properties (Chetty et al. 2008; Breyer-Brendwijk 1962). Roots were used in the treatment of abnormal accumulation of pus or liquid in cavities, cellular tissues, and inflamed and enlarged lymph nodes. Bruised leaves were employed for poulticing abscesses and boils. Juices expressed from leaves were used in treatment of skin allergy by indigestion, earaches in children, tonic (Chetty et al. 2008), and external applicator to wounds and bruises and for allaying itching in urticaria (Chetty et al. 2008; Sharma et al. 2012; The wealth of India 1949). Leaves infusion was applied topically to reduce swelling in conditions like bone fractures or twisting (Sharma et al. 2001). In south of Brazil, leaves were also used in traditional folk medicine to treat inflammations and pain-related diseases and as laxative (Walker et al. 2008). In China, leaves and stems of the plant were cooked with pork and used as tonic (The wealth of India 1949). Seeds and flowers were used to remove freckles of the skin. In Malagasy, the plant was used to treat intestinal pains (Breyer-Breandwijk 1962).

It is used also for centuries for the treatment of various ailments and emetic, for treatment of many gastrointestinal disorders, including dysentery, diarrhea, muscle pain, and abdominal colic. Besides herb is considered to possess anti-diabetic and purgative and has certain useful activities like antiviral activity, antimicrobial activity, antimalarial activity, anthelmintic activity, antioxidant activity, cytotoxic activity, anti-tubercular drug-induced hepatotoxicity, antinociceptive activity, antifungal activity, anti-corrosion activity, antispasmodic activity, anti-inflammatory activity, and many others like pharmacologically active compounds including active alcoholic extract, ether compound-3,3′-Methylenebis (4- hydroxycoumarin) N-D-alpha-Phenylyglycinelaminaribiitol-3-[4-(dimethylamino) cinnamoyl] 4-hydroxycoumarin. Overall, the genus *Mirabilis* has the versatile potentiality that has been underestimated in a systematic way.

25.1.2 Vernacular Names

Assamese, Gophuligopal; **Bengal**, Krishnakeli, Sarpamani; **English**, four o' clock, Marvel of Peru; **Gujrati**, Gubbaji; **Hindi**, Gul-abbas; **Kannada**, Sanjamalligie, Chandramalligie; **Malayalam**, Antmalari; **Marathi**, Gulbas; **Oriya**, Rangai; **Persian**, Gul-i-abbasa; **Punjabi**, Gulabbas; **Sanskrit**, Krishnakeli; **Tamil**, Andhimalligai; **Telugu**, Chandrakanta (Khurian 2003; The wealth of India Raw materials 1998).

Synonyms
The synonyms of *Mirabilis jalapa* are *M. dechotoma* Linn. (In Brazil), *M. dechotoma* Linn and *M. longiflora* Linn. (In tropical America), *M. lindheimeri* Linn., and *M. odorata* Linn.

25.1.3 Cultural Practices

The plant does best in full sun. Often in the sun, the leaves wither and then return vigorously in the evening, when temperatures start to fall and the sun sets. The plant perennializes in warm, coastal environments. It cannot stand the cold as the aerial part, with the first frosts, deteriorates and can die, but the underground part that can return to vegetate in spring remains vital. The plant is self-seeded, often spreading rapidly if left unchecked in a garden. It is recommended that the seeds should be soaked before planting, but this is not totally necessary. The fragrance of the flower is more intense and noticeable during the warm period of the day. The plant is easy to grow, as long as it is sunny or partially shaded. Under these conditions, it grows very quickly.

It grows preferably in light soil, rich in humus and well-draining; it is neutral side acidity (pH). Pot cultivation is always possible with a mixture of 80% soil and 20% garden soil and a very deep container with the tubers being put at a depth of 10 cm. It is usually sown from mid-February to May. The seeds germinate rapidly at a temperature of 18 °C (Vivanco 1999).

Mirabilis usually propagates by seeds and tubers. Seeds remain viable for several years. The large elongated tubers make large specimen difficult to transplant.

25.2 Botany and Taxonomy

25.2.1 Taxonomic Tree

Domain: Eukaryota
Kingdom: Plantae
Phylum: Spermatophyta
Subphylum: Angiospermae
Class: Dicotyledonae
Order: Caryophyllales
Family: Nyctaginaceae
Genus: *Mirabilis*
Species: *Mirabilis jalapa*

25.2.2 Botanical Description

It is a perennial, herbaceous, bushy plant that reaches stature heights of mostly 1 m, rarely up to 2 m. It may be grown as an ornamental annual, especially in the temperate zone. The stems are thick, full, and quadrangular with many ramifications, swollen and rooting at the nodes. The posture is often prostrate. Roots are thick, tuberous, and 10 cm or more in diameter. Leaves are ovate and cordate. The single-seeded fruits are spherical, wrinkled, and black upon maturity, having started out greenish yellow. Flowers are shortly stalked funnel-shaped found in clusters (The wealth of India 1949), subtended by an involucre of 5, ovate, connate bracts (Chetty et al. 2008), striped or blotched, fragrant, white, yellow, purple, or red colored (The wealth of India 1949). Perianth is funnel-shaped and 5-lobed. Stamens are 3–6 in numbers and exerted. Anthocarps are globose and black when ripped (Chetty et al. 2008).

Elaborately, cotyledons are long stalked (2–3 cm), with broader than long blade, almost rectangular of 18–22 cm long and 28–36 mm wide. The base is slightly cordate and the apex slightly truncated. It is tri-nerved at the base, light green to yellowish green, glabrous. First leaves are simple, opposite, borne on a petiole 2–3 cm long. The blade is almost triangular in shape, base slightly cordate, 3–4 cm long and 2–3 cm wide, rather strongly veined, light green, glabrous. Habit is a bushy herbaceous because of strong stems branching by 2 or 3. Underground system consists of large tuberous roots, stem full square, fleshy, green to reddish, thickened at nodes. Leaves simple, opposite supported by a stalk from 1 to 4 cm long, without stipule. The lamina is fleshy, soft enough, oval, base truncate, rounded or slightly cordate, and acute apex from 4 to 12 cm long and 3–8 cm wide. The margin is entire, densely ciliated, hairs multi-cellular. Both sides of lamina are glabrous and the visible nerves are arched. Inflorescence is a contracted cyme from 1 to several flowers clustered in an involucre of bracts at the ends of branches. The calyx, long 7–12 mm, tubular, and topped by triangular lobes finely ciliated. The corolla consists of five petals united into a long narrow tube flaring widely at the end.

It measures 4–5 cm long and 3–4 cm wide, pink yellow, scarlet, purple or white in color. Stamens 5–6 in number, unequal in size emerging beyond the corolla tube. The style is even longer. The flowers are short-lives, hey bloom at night, fade in the next morning, and the corolla falls off very quickly, fruits are achene, subglobose to ellipsoidal from 6 to 8 mm long, green to black when mature, with 5–10 longitudinal ribs, mute between the ribs (Le Bourgeois et al. 2008).

25.2.3 Flower's Colors

A curious aspect of *M. jalapa* is that flowers with different colors grow simultaneously on the same plant. Additionally, an individual flower can be splashed with different colors. Flower patterns are referred to as sectors (whole sections of flower), flakes (stripes of varying length), and spots. A single flower can be plain yellow, red, magenta, pink, or white or have a combination of sectors, flakes, and spots (Le Bourgeois et al. 2008). Furthermore, different combinations of flowers and patterns can occur on different flowers of the same plant (Hassler 2012). Usually, the flowers are yellow, pink, and white, but a different combination of flowers growing on the same single four o'clock plant can be found. Another interesting point is a color-changing phenomenon. For example, in the yellow variety, as the plant matures, it can display flowers that gradually change to a dark pink color. Similarly, white flowers can change to light violet. Despite their appearance, the flowers are not formed from petals – rather they are a pigmented modification of the calyx. Similarly, the "calyx" is an involucre of bracts. The flowers are funnel-shaped and penta-lobed; they have no cup (replaced by bracteates leaves) but are made of a corolla (Hutchinson 1969). The inflorescences contain 3–7 unpopped flowers. Earning the name "4'o clock flower," the fragrant flowers open in the late afternoon or early evening, and also in overcast weather, and exhale a scent reminiscent of the tobacco flower and attract moths for pollination. The anthesis lasts from 16 to 20 h and thus remains visible part of the day. The flowers are pollinated by long-tongued moths of the family Sphingidae, such as the sphinx moths or hawk moths and other nocturnal pollinators attracted by the fragrance (Hassler 2012).

25.2.4 Habitat

This weedy species can be found scattered throughout the world disturbed sites including waste ground and old home sites. It is also cultivated in anthropogenic (man-made or disturbed habitats), meadows, and fields. *M. jalapa* is a herbaceous perennial plant with numerous branches that exist in southern and warm western regions, and an annual in cooler northern regions, of its native tropical South America. It has been naturalized in many parts of the world, including Israel (Retrieved from: https://www.cabi.org/isc/search/index?q=Mirabilis%20jalapa).

25.2.5 Pollen Morphology

The shape of the pollen grains of *M. jalapa* is spheroidal, oblate spheroidal, with a diameter ranging from 125 to 140 μm and thickness of 10–15 μm. Exine ornamentation is spinulose; spinules 0.5–1 μm high, randomly distributed, aperture type pantoporate with numbers ranging from 18 to 20 μm the diameter of aperture varies between 6.3 and 10 mm, while the membrane of the aperture is margin ornate, membrane provided with spinulose and granules. Pollen dimorphism is frequently found in this species (white-pink, mixed, and mixed radiated); occasional giant, dimorphic anomalous, deformed, and joint grains have been observed. All these anomalous pollen grains except giant pollen grains are sterile (Dutta et al. 2015).

25.2.6 Ecology and Phytogeography

Nyctaginaceae, family of 4'o clock flower, is mostly pantropical, particularly distributed in South America with the poor representations of most widespread species in the warm parts of the Old World. Although the family is predominantly tropical, its area reaches to 38°SL in New Zealand and to 45°SL in Argentina. The abovementioned family is a low land family, occurring up to about 2000 m. in not too dry climates, rather indifferent to soil. The genus *Mirabilis* L., with about 60 species, is mostly American, from California to Argentina. In India, only the species *M jalapa* L. is found, native to Peru, now cultivated as an ornamental or medicinal plant and occasionally escaped from the gardens in all tropical regions. Now it is abundantly growing throughout the tropical region of India as propagated through seeds. The species is reported cultivating up to 1200 m height (Rozina 2016) (Table 1).

Table 1 Morphological similarities between *Mirabilis* genus and *Mirabilis jalapa* L.

Characteristics	*Mirabilis* genus	*Mirabilis jalapa* L.
Nature	Perennial, subshrub	Perennial, subshrub, showy
Stem	Repeatedly forked, decumbent to erect	<1 m, glabrous to sparsely short-hairy
Leaf	Generally petioled	Blade 5–14 cm, ovate to cordate
Inflorescence	Forked, calyx-like involucres densely clustered or solitaty in axils, bell to saucer-shaped; flowers 1–16 per involucre, blooming sequentially	Clusters of involucres, short-stalked, terminal; involucres bell-shaped; bracts 5, 5–15 mm. 1/2–1/4 fused; flower 1 per involucre
Flower	Perianth funnel, to bell-shaped, lobes 5; stamens 3–5, generally exerted; stigma spheric, generally exerted	Perinath 30–50 mm, narrowly funnel-shaped, generally bright magenta (yellow, white, or variegated)
Fruit	Round to club-shaped, smooth to 5-ribbed; winged	8–10 mm, ovoid, glabrous to minutely short -hairy

25.2.7 Distribution, Collection, and Conservation

M. jalapa is probably native to Mexico, as it is one of the species reported in the chronicles of the conquers to Mexico, growing in Aztec gardens prior to 1521 (Douglas and Manos 2004). According to Vibrans (2009), the species is native to Tropical America, possibly only to Mexico, and introduced elsewhere. Some references cite the species as native to the Peruvian Andes, because it was exported from that region into Europe during the 1500s (Broome et al. 2007). The species type is from India, but described from a cultivated plant in 1753, which accounts for some references citing the species as native from that country (LeDuc 1995). The species now occurs in North America, Central America, the Caribbean, South America, Europe, Africa, Asia, and Oceania (Table 2). It is expected to occur in other countries due to its seeds being available over the Internet and its recommended uses as an ornamental species. Although not reported for some countries, the available common names in several languages suggest a wider distribution.

25.2.8 History of Introduction and Spreading

Based on literature, *M. jalapa* was already used in gardens by the Aztecs before the Conquest prior to 1521. It was also used as an ornamental in towns and cities established by Spain in the New World during colonization. The species was exported to Europe in the sixteenth century from Peru and cited as an ornamental in Spain before the 1600s (CABI 2019). It is reported as present in Europe by1596 in the Annals of the Museum of Natural History of Paris. Some botanical gardens in Europe had the species in their lists of seeds available for exchange, the Imperial Botanic Garden in St. Petersburg by the mid-1800s, in the late1800s at the Kew Royal Botanic Gardens, and by1895 at the Berlin Botanic Garden. Its popularity as an ornamental during the Victorian era contributed to its use and movement into European colonies on other continents (CABI 2019). It is reported in Africa since the 1700s, in Asia since the 1800s, and in North America (USA) and in the Caribbean since the 1800s (Missouri Botanical Garden 2017). In the tropics and subtropics, the plant will self-seed, contributing to its spread and escape from cultivation (Table 3).

25.2.9 Origin, Domestication, and Spreading

M. jalapa hails from tropical South America, but has become naturalized throughout tropical, subtropical, and temperate regions. In cooler subtropical and temperate regions, it will die back with the first frosts or as the weather starts to cool down (especially after it fully matures and finished self-seeding), re-growing in the following spring from the tuberous roots. *M. jalapa* is native to the dry tropical regions of Central and South America: Guatemala, Mexico, Chile, and Peru. Locations within which *M. jalapa* is naturalized include the warmer parts of Australia, the USA, and New Zealand and many Pacific islands. It is currently naturalized in many

Table 2 Distribution table

Continent/country/region	Distribution	Last reported	Origin	First reported	Invasive	References
Africa						
Benin (datasheet/108375)	Present		Introduced			PROTA (2016) (datasheet/34254#RE DDB-167308)
Burundi (datasheet/108374)	Present		Introduced			Missouri Botanical Garden (2016) (datasheet/34254#RE DDB-181421)
Cameroon (datasheet/108397)	Present		Introduced			Missouri Botanical Garden (2016) (datasheet/34254#RE DDB-181421)
Comoros (datasheet/108474)	Present		Introduced	1850		Missouri Botanical Garden (2016) (datasheet/34254#RE DDB-181421)
Congo, Democratic Republic of the (datasheet/108615)	Present		Introduced			PROTA (2016) (datasheet/34254#RE DDB-167308)
Congo, Republic of the (datasheet/108392)	Present		Introduced			CABI (Undated) (datasheet/34254#RE DDB--27)
Egypt (datasheet/108418)	Present, only in captivity/cultivation		Introduced	1750		Ladwig-Cooper (2012) (datasheet/34254#RE DDB-179830)
Ethiopia (datasheet/108422)	Present		Introduced		Invasive	Witt and Luke (2017) (datasheet/ 34254#REDDB-148565); CABI (Undated) (datasheet/34254#RE DDB--27)
Gabon (datasheet/108430)	Present		Introduced			Missouri Botanical Garden (2016) (datasheet/34254#RE DDB-181421)
Ghana (datasheet/108436)	Present		Introduced			Missouri Botanical Garden (2016) (datasheet/34254#RE DDB-181421)

Kenya (datasheet/108470)	Present		Introduced		Invasive	BioNET-EAFRINET (2016) (datasheet/34254#REDDB-181396); PROTA (2016) (datasheet/34254#RE DDB-167308)
Libya (datasheet/108492)	Present		Introduced			Uotila (2011) (datasheet/34254#RE DDB-177746)
Madagascar (datasheet/108498)	Present		Introduced			Missouri Botanical Garden (2016) (datasheet/34254#REDDB-181421); PROTA (2016) (datasheet/34254#RE DDB-167308)
Malawi (datasheet/108512)	Present		Introduced		Invasive	Witt and Luke (2017) (datasheet/ 34254#REDDB-148565); PROTA (2016) (datasheet/34254#RE DDB-167308)
Mauritius (datasheet/108510)	Present		Introduced			PROTA (2016) (datasheet/34254#RE DDB-167308)
Rodrigues (datasheet/108547)	Present		Introduced			PROTA (2016) (datasheet/34254#REDDB-167308); Balfour (1879) (datasheet/34254#RE DDB-168315)
Mozambique (datasheet/108515)	Present		Introduced			PROTA (2016) (datasheet/34254#RE DDB-167308)
Namibia (datasheet/108516)	Present		Introduced			Encyclopedia of Life (2016) (datasheet/34254#RE DDB-166304)
Rwanda (datasheet/108551)	Present		Introduced		Invasive	Witt and Luke (2017) (datasheet/34254#RE DDB-148565)
Seychelles (datasheet/108554)	Present		Introduced		Invasive	PIER (2016) (datasheet/34254#RE DDB-167278)
Sierra Leone (datasheet/108562)	Present		Introduced			Missouri Botanical Garden (2016) (datasheet/34254#RE DDB-181421)

(continued)

Table 2 (continued)

Continent/country/region	Distribution	Last reported	Origin	First reported	Invasive	References
South Africa (datasheet/108613)	Present		Introduced		Invasive	CABI (Undated) (datasheet/34254#RE DDB--27)
Tanzania (datasheet/108591)	Present		Introduced		Invasive	Witt and Luke (2017) (datasheet/ 34254#REDDB-148565); Missou Botanical Garden (2016) (datasheet/34254#RE DDB-181421)
Uganda (datasheet/108594)	Present		Introduced		Invasive	BioNET-EAFRINET (2016) (datasheet/34254#REDDB-181396); PROTA (2016) (datasheet/34254#RE DDB-167308)
Zambia (datasheet/108614)	Present		Introduced		Invasive	Witt and Luke (2017) (datasheet/ 34254#REDDB-148565); PROTA (2016) (datasheet/34254#RE DDB-167308)
Zimbabwe (datasheet/108616)	Present		Introduced			PROTA (2016) (datasheet/34254#RE DDB-167308)
Asia						
Afghanistan (datasheet/108351)	Present		Introduced			PROTA (2016) (datasheet/34254#RE DDB-167308)
Chagos Archipelago (datasheet/108365)	Present		Introduced			PIER (2016) (datasheet/34254#RE DDB-167278)
China (datasheet/108398)	Present		Introduced		Invasive	Zhou XiaoKui et al. (2008) (datasheet/34254#REDDB-119336); PIER (2016) (datasheet/34254#RE DDB-167278)
Guangdong (datasheet/108671)	Present		Introduced			PROTA (2016) (datasheet/34254#RE DDB-167308)

Hong Kong (datasheet/108678)	Present		Introduced			PIER (2016) (datasheet/34254#RE DDB-167278)
India (datasheet/108459)	Present		Introduced			PROTA (2016) (datasheet/34254#RE DDB-167308)
Andhra Pradesh (datasheet/108721)	Present, only in captivity/cultivation		Introduced			Flora of India (2015) (datasheet/34254#RE DDB-166479)
Assam (datasheet/108723)	Present		Introduced			Encyclopedia of Life (2016) (datasheet/34254#RE DDB-166304)
Gujarat (datasheet/108732)	Present		Introduced			Bedi (1978) (datasheet/34254#RE DDB-13974)
Karnataka (datasheet/108738)	Present, only in captivity/cultivation		Introduced			Flora of India (2015) (datasheet/34254#RE DDB-166479)
Kerala (datasheet/108737)	Present, only in captivity/cultivation		Introduced			Flora of India (2015) (datasheet/34254#RE DDB-166479)
Maharashtra (datasheet/108740)	Present		Introduced			Encyclopedia of Life (2016) (datasheet/34254#RE DDB-166304)
Punjab (datasheet/108748)	Present		Introduced			Encyclopedia of Life (2016) (datasheet/34254#RE DDB-166304)
Tamil Nadu (datasheet/108751)	Present, only in captivity/cultivation		Introduced			Flora of India (2015) (datasheet/34254#RE DDB-166479)
Indonesia (datasheet/108455)	Present		Introduced		Invasive	PIER (2016) (datasheet/34254#RE DDB-167278)
Israel (datasheet/108457)	Present		Introduced			Uotila (2011) (datasheet/34254#RE DDB-177746)
Japan (datasheet/108467)	Present		Introduced			PIER (2016) (datasheet/34254#RE DDB-167278)
Jordan (datasheet/108466)	Present		Introduced			Uotila (2011) (datasheet/34254#RE DDB-177746)

(continued)

Table 2 (continued)

Continent/country/region	Distribution	Last reported	Origin	First reported	Invasive	References
Lebanon (datasheet/108482)	Present		Introduced			Uotila (2011) (datasheet/34254#RE DDB-177746)
Malaysia (datasheet/108514)	Present, only in captivity/cultivation		Introduced			Hamilton and Holttu (1933) (datasheet/34254#RE DDB-171449)
Maldives (datasheet/108511)	Present		Introduced		Invasive	PIER (2016) (datasheet/34254#RE DDB-167278)
Nepal (datasheet/108524)	Present		Introduced			Manandhar (1991) (datasheet/34254#RE DDB-173814)
Pakistan (datasheet/108537)	Present		Introduced			PROTA (2016) (datasheet/34254#RE DDB-167308)
Philippines (datasheet/108535)	Present		Introduced		Invasive	PIER (2016) (datasheet/34254#RE DDB-167278)
Singapore (datasheet/108557)	Present		Introduced			PIER (2016) (datasheet/34254#REDDB-167278); Cantley and Dennys (1886) (datasheet/34254#RE DDB-169160)
Sri Lanka (datasheet/108485)	Present		Introduced			PROTA (2016) (datasheet/34254#RE DDB-167308)
Syria (datasheet/108572)	Present		Introduced			Uotila (2011) (datasheet/34254#RE DDB-177746)
Europe						
Austria (datasheet/108361)	Present, only in captivity/cultivation		Introduced			DAISIE (2016) (datasheet/34254#REDDB-150692); Antoin (1876) (datasheet/34254#RE DDB-168093)
Belgium (datasheet/108370)	Present, only in captivity/cultivation		Introduced			DAISIE (2016) (datasheet/34254#RE DDB-150692)

Croatia (datasheet/108452)	Present		Introduced			DAISIE (2016) (datasheet/34254#RE DDB-150692)
Cyprus (datasheet/108408)	Present		Introduced	1985		Della and Latrou (1995) (datasheet/34254#REDDB-169985); DAISIE (2016) (datasheet/34254#RE DDB-150692)
Czechia (datasheet/108409)	Present, only in captivity/cultivation		Introduced			DAISIE (2016) (datasheet/34254#RE DDB-150692)
Denmark (datasheet/108412)	Present		Introduced			DAISIE (2016) (datasheet/34254#RE DDB-150692)
France (datasheet/108429)	Present		Introduced			Uotila (2011) (datasheet/34254#RE DDB-177746)
Corsica (datasheet/108704)	Present		Introduced			DAISIE (2016) (datasheet/34254#RE DDB-150692)
Gibraltar (datasheet/108437)	Present		Introduced			Uotila (2011) (datasheet/34254#RE DDB-177746)
Greece (datasheet/108443)	Present		Introduced			DAISIE (2016) (datasheet/34254#RE DDB-150692); Heldreich (1863) (datasheet/34254#REDDB-179625); Uotila (2011) (datasheet/34254#RE DDB-177746)
Hungary (datasheet/108454)	Present, only in captivity/cultivation		Introduced			DAISIE (2016) (datasheet/34254#RE DDB-150692)
Italy (datasheet/108464)	Present		Introduced			Uotila (2011) (datasheet/34254#RE DDB-177746); Pampanini (1911) (datasheet/34254#RE DDB-175100)
Sardinia (datasheet/108758)	Present		Introduced			Uotila (2011) (datasheet/34254#RE DDB-177746)

(continued)

Table 2 (continued)

Continent/country/region	Distribution	Last reported	Origin	First reported	Invasive	References
Sicily (datasheet/108757)	Present		Introduced			Uotila (2011) (datasheet/34254#RE DDB-177746)
Malta (datasheet/108509)	Present		Introduced			DAISIE (2016) (datasheet/34254#RE DDB-150692)
Netherlands (datasheet/108522)	Present		Introduced			PROTA (2016) (datasheet/34254#RE DDB-167308)
Portugal (datasheet/108542)	Present, only in captivity/cultivation		Introduced			DAISIE (2016) (datasheet/34254#RE DDB-150692)
Azores (datasheet/108776)	Present		Introduced			DAISIE (2016) (datasheet/34254#REDDB-150692); Trelease (1897) (datasheet/34254#REDDB-177610); Uotila (2011) (datasheet/34254#RE DDB-177746)
Madeira (datasheet/108777)	Present		Introduced			DAISIE (2016) (datasheet/34254#RE DDB-150692)
Romania (datasheet/108548)	Present, only in captivity/cultivation		Introduced			DAISIE (2016) (datasheet/34254#RE DDB-150692)
Russia (datasheet/108550)	Present, only in captivity/cultivation		Introduced			DAISIE (2016) (datasheet/34254#RE DDB-150692)
Central Russia (datasheet/108782)	Present, only in captivity/cultivation		Introduced			Uotila (2011) (datasheet/34254#RE DDB-177746)
Slovenia (datasheet/108559)	Present		Introduced			Encyclopedia of Life (2016) (datasheet/34254#RE DDB-166304)
Spain (datasheet/108421)	Present		Introduced			DAISIE (2016) (datasheet/34254#RE DDB-150692)

Balearic Islands (datasheet/108701)	Present		Introduced			DAISIE (2016) (datasheet/34254#REDDB-150692); Uotila (2011) (datasheet/34254#RE DDB-177746)
Canary Islands (datasheet/108702)	Present		Introduced			DAISIE (2016) (datasheet/34254#REDDB-150692); Uotila (2011) (datasheet/34254#RE DDB-177746)
Sweden (datasheet/108556)	Present, only in captivity/cultivation		Introduced			DAISIE (2016) (datasheet/34254#RE DDB-150692)
United Kingdom (datasheet/108431)	Present, only in captivity/cultivation		Introduced			PFAF (2016) (datasheet/34254#REDDB-167246); Eyles Stiles (1765) (datasheet/34254#RE DDB-181893)
North America						
Antigua and Barbuda (datasheet/108352)	Present		Introduced			Acevedo-Rodríguez and Strong (2012) (datasheet/34254#REDDB-151259); Broome et al. (2007) (datasheet/34254#RE DDB-168959)
Bahamas (datasheet/108382)	Present		Introduced			PROTA (2016) (datasheet/34254#REDDB-167308); CABI (Undated) (datasheet/34254#RE DDB--27)
Barbados (datasheet/108368)	Present		Introduced			Acevedo-Rodríguez and Strong (2012) (datasheet/34254#REDDB-151259); CABI (Undated) (datasheet/34254#RE DDB--27)
Belize (datasheet/108387)	Present		Introduced			Missouri Botanical Garden (2016) (datasheet/34254#RE DDB-181421)

(continued)

Table 2 (continued)

Continent/country/region	Distribution	Last reported	Origin	First reported	Invasive	References
Bermuda (datasheet/108377)	Present		Introduced	1913		CABI (Undated) (datasheet/34254#RE DDB--27)
British Virgin Islands (datasheet/108602)	Present		Introduced			Acevedo-Rodríguez and Strong (2012) (datasheet/34254#REDDB-151259); CABI (Undated) (datasheet/34254#RE DDB--27)
Canada (datasheet/108388)	Present		Introduced			Missouri Botanical Garden (2016) (datasheet/34254#RE DDB-181421)
Cayman Islands (datasheet/108479)	Present		Introduced			Acevedo-Rodríguez and Strong (2012) (datasheet/34254#RE DDB-151259)
Costa Rica (datasheet/108402)	Present		Introduced			Missouri Botanical Garden (2016) (datasheet/34254#RE DDB-181421)
Cuba (datasheet/108405)	Present		Introduced		Naturalized	Oviedo Prieto et al. (2012) (datasheet/34254#REDDB-151278); Missouri Botanical Garden (2016) (datasheet/34254#RE DDB-181421)
Dominica (datasheet/108413)	Present		Introduced			Missouri Botanical Garden (2016) (datasheet/34254#RE DDB-181421)
Dominican Republic (datasheet/108414)	Present		Introduced			Missouri Botanical Garden (2016) (datasheet/34254#REDDB-181421); CABI (Undated) (datasheet/34254#RE DDB--27)
El Salvador (datasheet/108571)	Present		Introduced			Missouri Botanical Garden (2016) (datasheet/34254#RE DDB-181421)

Grenada (datasheet/108432)	Present		Introduced			Acevedo-Rodríguez and Strong (2012) (datasheet/34254#REDDB-151259); Broome et al. (2007) (datasheet/34254#REDDB-168959); CABI (Undated) (datasheet/34254#RE DDB--27)
Guadeloupe (datasheet/108441)	Present		Introduced	1895		CABI (Undated) (datasheet/34254#REDDB–27); Broome et al. (2007) (datasheet/34254#RE DDB-168959)
Guatemala (datasheet/108445)	Present		Introduced			Missouri Botanical Garden (2016) (datasheet/34254#RE DDB-181421)
Haiti (datasheet/108453)	Present		Introduced			Acevedo-Rodríguez and Strong (2012) (datasheet/34254#RE DDB-151259)
Honduras (datasheet/108451)	Present		Introduced			Acevedo-Rodríguez and Strong (2012) (datasheet/34254#RE DDB-151259)
Jamaica (datasheet/108465)	Present		Introduced	1890		Missouri Botanical Garden (2016) (datasheet/34254#RE DDB-181421)
Martinique (datasheet/108506)	Present		Introduced	1819		Missouri Botanical Garden (2016) (datasheet/34254#REDDB-181421); Broome et al. (2007) (datasheet/34254#RE DDB-168959)
Mexico (datasheet/108513)	Present, widespread		Native			CONABIO (2016) (datasheet/34254#REDDB-178825); Missouri Botanical Garden (2016) (datasheet/34254#RE DDB-181421)
Montserrat (datasheet/108508)	Present		Introduced			Acevedo-Rodríguez and Strong (2012) (datasheet/34254#REDDB-151259); Broome et al. (2007) (datasheet/34254#RE DDB-168959)

(continued)

Table 2 (continued)

Continent/country/region	Distribution	Last reported	Origin	First reported	Invasive	References
Netherlands Antilles (datasheet/108356)	Present		Introduced			Acevedo-Rodríguez and Strong (2012) (datasheet/34254#REDDB-151259); Broome et al. (2007) (datasheet/34254#RE DDB-168959)
Nicaragua (datasheet/108521)	Present		Introduced			Missouri Botanical Garden (2016) (datasheet/34254#RE DDB-181421)
Panama (datasheet/108530)	Present		Introduced			Missouri Botanical Garden (2016) (datasheet/34254#RE DDB-181421)
Puerto Rico (datasheet/108541)	Present		Introduced			Acevedo-Rodríguez and Strong (2012) (datasheet/34254#REDDB-151259); CABI (Undated) (datasheet/34254#RE DDB--27)
Saint Kitts and Nevis (datasheet/108475)	Present		Introduced	1901		CABI (Undated) (datasheet/34254#RE DDB--27)
Saint Lucia (datasheet/108483)	Present		Introduced			Acevedo-Rodríguez and Strong (2012) (datasheet/34254#REDDB-151259); Broome et al. (2007) (datasheet/34254#RE DDB-168959)
Saint Vincent and the Grenadines (datasheet/108600)	Present		Introduced			Acevedo-Rodríguez and Strong (2012) (datasheet/34254#REDDB-151259); Cleall e al. (1807) (datasheet/34254#REDDB-169546); Broome et al. (2007) (datasheet/34254#RE DDB-168959)
Turks and Caicos Islands (datasheet/108575)	Present		Introduced			PROTA (2016) (datasheet/34254#RE DDB-167308)
U.S. Virgin Islands (datasheet/108603)	Present		Introduced	1896		Missouri Botanical Garden (2016) (datasheet/34254#RE DDB-181421)

United States (datasheet/108597)	Present		Introduced			USDA-NRCS (2016) (datasheet/34254#RE DDB-150759)
Alabama (datasheet/108796)	Present		Introduced			USDA-NRCS (2016) (datasheet/34254#RE DDB-150759)
Arizona (datasheet/108798)	Present		Introduced			SEINet (2016) (datasheet/34254#RE DDB-167402)
Arkansas (datasheet/108797)	Present		Introduced			USDA-NRCS (2016) (datasheet/34254#RE DDB-150759)
California (datasheet/108799)	Present		Introduced			USDA-NRCS (2016) (datasheet/34254#RE DDB-150759)
Colorado (datasheet/108800)	Present, only in captivity/cultivation		Introduced			CABI (Undated) (datasheet/34254#RE DDB--27)
Connecticut (datasheet/108801)	Present		Introduced			USDA-NRCS (2016) (datasheet/34254#RE DDB-150759)
District of Columbia (datasheet/108802)	Present		Introduced			USDA-NRCS (2016) (datasheet/34254#RE DDB-150759)
Florida (datasheet/108804)	Present		Introduced			USDA-NRCS (2016) (datasheet/34254#RE DDB-150759);
Georgia (datasheet/108805)	Present		Introduced			USDA-NRCS (2016) (datasheet/34254#RE DDB-150759)
Hawaii (datasheet/108806)	Present		Introduced		Invasive	PIER (2016) (datasheet/34254#RE DDB-167278)
Idaho (datasheet/108808)	Present, only in captivity/cultivation		Introduced			CABI (Undated) (datasheet/34254#RE DDB--27)
Illinois (datasheet/108809)	Present		Introduced			SEINet (2016) (datasheet/34254#RE DDB-167402)
Indiana (datasheet/108810)	Present		Introduced			USDA-NRCS (2016) (datasheet/34254#RE DDB-150759)

(continued)

Table 2 (continued)

Continent/country/region	Distribution	Last reported	Origin	First reported	Invasive	References
Iowa (datasheet/108807)	Present, only in captivity/cultivation		Introduced			CABI (Undated) (datasheet/34254#RE DDB--27)
Kansas (datasheet/108811)	Present	1889	Introduced			Smyth (1889) (datasheet/34254#RE DDB-176812)
Kentucky (datasheet/108812)	Present		Introduced			USDA-NRCS (2016) (datasheet/34254#RE DDB-150759)
Louisiana (datasheet/108813)	Present		Introduced			USDA-NRCS (2016) (datasheet/34254#RE DDB-150759)
Maryland (datasheet/108815)	Present		Introduced			USDA-NRCS (2016) (datasheet/34254#RE DDB-150759)
Michigan (datasheet/108817)	Present, only in captivity/cultivation		Introduced			CABI (Undated) (datasheet/34254#RE DDB--27)
Minnesota (datasheet/108818)	Present, only in captivity/cultivation		Introduced			CABI (Undated) (datasheet/34254#RE DDB--27)
Mississippi (datasheet/108820)	Present, only in captivity/cultivation		Introduced			USDA-NRCS (2016) (datasheet/34254#RE DDB-150759)
Missouri (datasheet/108819)	Present, only in captivity/cultivation		Introduced			CABI (Undated) (datasheet/34254#RE DDB--27)
Montana (datasheet/108821)	Present, only in captivity/cultivation		Introduced			CABI (Undated) (datasheet/34254#RE DDB--27)
Nebraska (datasheet/108824)	Present, only in captivity/cultivation		Introduced			CABI (Undated) (datasheet/34254#RE DDB--27)
Nevada (datasheet/108828)	Present		Introduced			USDA-NRCS (2016) (datasheet/34254#RE DDB-150759)
New Hampshire (datasheet/108825)	Present, only in captivity/cultivation		Introduced			CABI (Undated) (datasheet/34254#RE DDB--27)

New Jersey (datasheet/108826)	Present		Introduced			USDA-NRCS (2016) (datasheet/34254#RE DDB-150759)
New Mexico (datasheet/108827)	Present, only in captivity/cultivation		Introduced			CABI (Undated) (datasheet/34254#RE DDB--27)
New York (datasheet/108829)	Present		Introduced			USDA-NRCS (2016) (datasheet/34254#RE DDB-150759)
North Carolina (datasheet/108822)	Present		Introduced			USDA-NRCS (2016) (datasheet/34254#RE DDB-150759)
Ohio (datasheet/108830)	Present		Introduced			USDA-NRCS (2016) (datasheet/34254#RE DDB-150759)
Oklahoma (datasheet/108831)	Present		Introduced			USDA-NRCS (2016) (datasheet/34254#RE DDB-150759)
Oregon (datasheet/108832)	Present		Introduced			SEINet (2016) (datasheet/34254#RE DDB-167402)
Pennsylvania (datasheet/108833)	Present		Introduced			USDA-NRCS (2016) (datasheet/34254#RE DDB-150759)
Rhode Island (datasheet/108834)	Present		Introduced			USDA-NRCS (2016) (datasheet/34254#RE DDB-150759)
South Carolina (datasheet/108835)	Present		Introduced			USDA-NRCS (2016) (datasheet/34254#RE DDB-150759)
Tennessee (datasheet/108837)	Present, only in captivity/cultivation		Introduced			CABI (Undated) (datasheet/34254#RE DDB--27)
Texas (datasheet/108838)	Present		Introduced			USDA-NRCS (2016) (datasheet/34254#RE DDB-150759); Reverchon (1886) (datasheet/34254#REDDB-175838); PROTA (2016) (datasheet/34254#RE DDB-167308)
Utah (datasheet/108839)	Present		Introduced			SEINet (2016) (datasheet/34254#RE DDB-167402)

(continued)

Table 2 (continued)

Continent/country/region	Distribution	Last reported	Origin	First reported	Invasive	References
Vermont (datasheet/108841)	Present		Introduced			USDA-NRCS (2016) (datasheet/34254#RE DDB-150759)
Virginia (datasheet/108840)	Present		Introduced			USDA-NRCS (2016) (datasheet/34254#RE DDB-150759)
Washington (datasheet/108842)	Present		Introduced			USDA-NRCS (2016) (datasheet/34254#RE DDB-150759)
West Virginia (datasheet/108844)	Present, only in captivity/cultivation		Introduced			CABI (Undated) (datasheet/34254#RE DDB--27)
Wisconsin (datasheet/108843)	Present, only in captivity/cultivation		Introduced			CABI (Undated) (datasheet/34254#RE DDB--27)
Oceania						
American Samoa (datasheet/108360)	Present		Introduced			PIER (2016) (datasheet/34254#RE DDB-167278)
Australia (datasheet/108362)	Present		Introduced			PIER (2016) (datasheet/34254#RE DDB-167278)
New South Wales (datasheet/108620)	Present		Introduced			Australian Tropical Rainforest Plants (2016) (datasheet/34254#RE DDB-181392)
Queensland (datasheet/108621)	Present		Introduced			Australian Tropical Rainforest Plants (2016) (datasheet/34254#RE DDB-181392)
Christmas Island (datasheet/108407)	Present		Introduced			PIER (2016) (datasheet/34254#RE DDB-167278)
Cook Islands (datasheet/108395)	Present		Introduced		Invasive	PIER (2016) (datasheet/34254#RE DDB-167278)
Federated States of Micronesia (datasheet/108427)	Present		Introduced		Invasive	PIER (2016) (datasheet/34254#RE DDB-167278)

Fiji (datasheet/108425)	Present		Introduced		Invasive	PIER (2016) (datasheet/34254#RE DDB-167278); Greenwood (1944) (datasheet/34254#RE DDB-1045)
French Polynesia (datasheet/108533)	Present		Introduced		Invasive	PIER (2016) (datasheet/34254#RE DDB-167278)
Guam (datasheet/108446)	Present		Introduced			PIER (2016) (datasheet/34254#RE DDB-167278)
Kiribati (datasheet/108473)	Present		Introduced		Invasive	PIER (2016) (datasheet/34254#RE DDB-167278)
Marshall Islands (datasheet/108499)	Present		Introduced			PIER (2016) (datasheet/34254#RE DDB-167278)
Nauru (datasheet/108526)	Present		Introduced		Invasive	PIER (2016) (datasheet/34254#RE DDB-167278)
New Caledonia (datasheet/108517)	Present		Introduced		Invasive	PIER (2016) (datasheet/34254#RE DDB-167278)
New Zealand (datasheet/108528)	Present		Introduced		Invasive	PIER (2016) (datasheet/34254#RE DDB-167278)
Niue (datasheet/108527)	Present		Introduced		Invasive	PIER (2016) (datasheet/34254#RE DDB-167278)
Palau (datasheet/108543)	Present		Introduced			PIER (2016) (datasheet/34254#RE DDB-167278)
Papua New Guinea (datasheet/108534)	Present		Introduced			PIER (2016) (datasheet/34254#RE DDB-167278)
Pitcairn (datasheet/108540)	Present		Introduced		Invasive	PIER (2016) (datasheet/34254#RE DDB-167278)
Solomon Islands (datasheet/108553)	Present		Introduced		Invasive	PIER (2016) (datasheet/34254#RE DDB-167278)
Tonga (datasheet/108585)	Present		Introduced		Invasive	PIER (2016) (datasheet/34254#RE DDB-167278)

(continued)

Table 2 (continued)

Continent/country/region	Distribution	Last reported	Origin	First reported	Invasive	References
Tuvalu (datasheet/108589)	Present		Introduced			PIER (2016) (datasheet/34254#RE DDB-167278)
U.S. Minor Outlying Islands (datasheet/108596)	Present		Introduced		Invasive	PIER (2016) (datasheet/34254#RE DDB-167278)
South America						
Argentina (datasheet/108359)	Present		Introduced			Missouri Botanical Garden (2016) (datasheet/34254#RE DDB-181421)
Bolivia (datasheet/108379)	Present		Introduced			Missouri Botanical Garden (2016) (datasheet/34254#REDDB-181421); Rusby (1895) (datasheet/34254#RE DDB-176088)
Brazil (datasheet/108381)	Present		Introduced			Missouri Botanical Garden (2016) (datasheet/34254#RE DDB-181421)
Acre (datasheet/108626)	Present		Introduced			Missouri Botanical Garden (2016) (datasheet/34254#RE DDB-181421)
Alagoas (datasheet/108627)	Present		Introduced			Flora do Brasil (2016) (datasheet/34254#RE DDB-166445)
Amazonas (datasheet/108628)	Present		Introduced			Flora do Brasil (2016) (datasheet/34254#RE DDB-166445)
Bahia (datasheet/108630)	Present		Introduced			Missouri Botanical Garden (2016) (datasheet/34254#RE DDB-181421)
Ceara (datasheet/108631)	Present		Introduced			Flora do Brasil (2016) (datasheet/34254#RE DDB-166445)
Espirito Santo (datasheet/108632)	Present		Introduced			Flora do Brasil (2016) (datasheet/34254#RE DDB-166445)

Goias (datasheet/108634)	Present		Introduced			Missouri Botanical Garden (2016) (datasheet/34254#RE DDB-181421)
Mato Grosso (datasheet/108638)	Present		Introduced			Flora do Brasil (2016) (datasheet/34254#RE DDB-166445)
Minas Gerais (datasheet/108636)	Present		Introduced			Flora do Brasil (2016) (datasheet/34254#RE DDB-166445)
Paraiba (datasheet/108640)	Present		Introduced			Flora do Brasil (2016) (datasheet/34254#RE DDB-166445)
Parana (datasheet/108643)	Present		Introduced			Flora do Brasil (2016) (datasheet/34254#RE DDB-166445)
Pernambuco (datasheet/108641)	Present		Introduced			Flora do Brasil (2016) (datasheet/34254#RE DDB-166445)
Rio de Janeiro (datasheet/108644)	Present		Introduced			Flora do Brasil (2016) (datasheet/34254#RE DDB-166445)
Rio Grande do Norte (datasheet/108645)	Present		Introduced			Flora do Brasil (2016) (datasheet/34254#RE DDB-166445)
Rio Grande do Sul (datasheet/108648)	Present		Introduced			Flora do Brasil (2016) (datasheet/34254#RE DDB-166445)
Santa Catarina (datasheet/108649)	Present		Introduced			Missouri Botanical Garden (2016) (datasheet/34254#RE DDB-181421)
Sao Paulo (datasheet/108651)	Present		Introduced			Missouri Botanical Garden (2016) (datasheet/34254#RE DDB-181421)
Sergipe (datasheet/108650)	Present		Introduced			Flora do Brasil (2016) (datasheet/34254#RE DDB-166445)
Chile (datasheet/108396)	Present		Introduced			Missouri Botanical Garden (2016) (datasheet/34254#REDDB-181421); PIER (2016) (datasheet/34254#RE DDB-167278)

(continued)

Table 2 (continued)

Continent/country/region	Distribution	Last reported	Origin	First reported	Invasive	References
Easter Island (datasheet/108666)	Present		Introduced		Invasive	PIER (2016) (datasheet/34254#RE DDB-167278)
Ecuador (datasheet/108416)	Present		Introduced			Missouri Botanical Garden (2016) (datasheet/34254#RE DDB-181421)
Galapagos Islands (datasheet/108700)	Present		Introduced		Invasive	PIER (2016) (datasheet/34254#RE DDB-167278)
French Guiana (datasheet/108434)	Present		Introduced			Missouri Botanical Garden (2016) (datasheet/34254#RE DDB-181421)
Guyana (datasheet/108448)	Present		Introduced			Missouri Botanical Garden (2016) (datasheet/34254#RE DDB-181421)
Paraguay (datasheet/108544)	Present		Introduced			Missouri Botanical Garden (2016) (datasheet/34254#RE DDB-181421)
Peru (datasheet/108532)	Present		Introduced			Missouri Botanical Garden (2016) (datasheet/34254#RE DDB-181421)
Suriname (datasheet/108568)	Present		Introduced			Missouri Botanical Garden (2016) (datasheet/34254#RE DDB-181421)
Uruguay (datasheet/108598)	Present		Introduced			Missouri Botanical Garden (2016) (datasheet/34254#RE DDB-181421)
Venezuela (datasheet/108601)	Present					Missouri Botanical Garden (2016) (datasheet/34254#RE DDB-181421)

Table 3 Introduction

Introduced to	Introduced from	Year	Reason	Introduced by	Established in wild through: Natural reproduction	Established in wild through: Continuous restocking	References
Singapore	Peru	1886	Ornamental purposes (pathway cause) (datasheet/109051)		Yes	No	Cantley and Dennys (1886) (datasheet/ 34254#dc3 37f2-490e-addd-aa2175bb5b8e)
Cyprus		1885	Ornamental purposes (pathway cause) (datasheet/109051)		Yes	No	Della and Latrou (1995 (datasheet/ 34254#036 370a-4f42-b060-279b35b18e5e)
Egypt		1750	Ornamental purposes (pathway cause) (datasheet/109051)		No	No	Ladwig-Cooper (2012) (datasheet/ 34254#5d6 3b92-46fe-8b46-e27cd2cff57b)
Rodriguez Island		1874			Yes	No	Balfour (1879) (datasheet/34254#035 ce1a-406d-b493-15f75ed6b39d)
USA		1886			Yes	No	Reverchon (1886) (datasheet/34254#a9a 29e5-4c30-9960-96853ebae104)
Bahamas		1889			Yes	No	Gardiner and Brace (18 (datasheet/ 34254#Non
Saint Vincent and the Grenadines		1806			Yes	No	Cleall et al. (1807) (datasheet/34254#c6e 5d32-41c2-9279-034f9c8684c7)
Bolivia		1895			No	No	Rusby (1895) (datasheet/34254#72c 83ec-43bd-8eb6-73f3e913856b)

(continued)

Table 3 (continued)

Introduced to	Introduced from	Year	Reason	Introduced by	Established in wild through		References
					Natural reproduction	Continuous restocking	
Austria		1873	Ornamental purposes (pathway cause) (datasheet/109051)		Yes	No	Antoine (1876) (datasheet/34254#3af dfbe-44bb-badf-06cfd8ee6b9b)
Greece		1863	Ornamental purposes (pathway cause) (datasheet/109051)		Yes	No	Heldreich (1863) (datasheet/34254#224 9bdb-4997-8ec4-2c14de9bea02)
Italy		1911	Ornamental purposes (pathway cause) (datasheet/109051)		Yes	No	Pampanini (1911) (datasheet/34254#079 4b44-4c24-b1b6-c9720baa1ce3)
UK		1765	Ornamental purposes (pathway cause) (datasheet/109051)		Yes	No	Eyles Stiles (1765) (datasheet/34254#af2 24fd-406a-a624-541b00f06c26)
Fiji		1943			Yes	No	Greenwood (1944) (datasheet/34254#f31 f64d-44b1-8a6d-fe6c6b377234)

countries in Asia, Africa, the USA, Middle East, and Europe. In Réunion, *M. jalapa* was initially an ornamental species; however it became naturalized on the west coast, between 400 and 700 m altitude, and on the south coast between 0 and 700 m. It occurs in a rural debris area and is relatively common in weedy sugarcane fields on the west and south coasts. Its high seed production and rapid growth allow it to cover up to 30–50% in cane plots. Furthermore, *M. jalapa* had also been introduced, naturalized, or invasive in East Africa including parts of Tanzania and invasive in Kenya and Uganda. In Kenya, *M. jalapa* is cultivated in areas around Nairobi and Naivasha; in Uganda in the Ankole and Mengo districts and in the Mabira Forest; and in Tanzania in the Rubondo, Lushoto, and Mpwapwa districts (Mirabilis jalapa 2011).

Kress et al. (2003) had provided a checklist of eight species under four genera of trees, shrubs, herbs, and climbers of Nyctaginaceae of Myanmar and had added their habit, distributions, and common or regional or vernacular names. They had described the members of the family Nyctaginaceae of Mustang, Nepal. They have retained the genus *Oxybaphus* and not included it under *Mirabilis* as synonym. Paria and Chattopadhyay (2000) and Kirtikar and Basu (1918) all had worked in different pockets of India including some districts of Murshidabad and described different members (like *Boerhavia* L. and *Mirabilis* L. Verma) of the family Nyctaginaceae in India.

25.2.10 Available Species and Cultivars of Genus *Mirabilis*

M. aggregate (Ortega) Cav., *M. albida* (Walter) Heimeri (white four o'clock), *M. alipes* (S. Watson) pilz (winged four o'clock), *M. austrotexana* B.L. Turner (Lonestar four o'clock), *M. coccinea* (Torr.) Benth. & Hook. F. (scarlet four o'clock), *M. comata* (Small) Standl. (hairy-tuft four o'clock), *M. decipens* (Standl.) Standl. (broadleaf four o'clock), *M. expansa* (Ruiz & Pav.) Standl., *M. gigantea* (Standl.) Shinners (giant four o'clock), *M. glabra* (S. Watson) Standl. (smooth four o'clock), *M. glabrifolia* (Ortega) I.M.Johnst. (flat-top four o'clock), *M. greenei* S. Watson (Greene's four o'clock), *M. hirsuta* (Pursh) Macmill (hairy four o'clock), *M. jalapa* L. (Marvel of Peru), *M. laevis* (Benth.) Curran (desert wishbone-bush), *M. laevis* (Benth.) Curran var. *cedrosensis* (Standl.) Munz (California four o'clock), *M. laevis* (Benth.) Curran var. *crassifolia* (Choisy) Spellenb (California four o'clock), *M. laevis* (Benth.) Curran var. *laevis*, *M. laevis* (Benth.) Curran var. *retrosa* (A. Heller) Jeps. (wishbone-bush), *M. laevis* (Benth.) Curran var. *villosa* (Kellogg) Spellenb, *M. linearis* (Pursh) Heimeri (narrowleaf four o'clock), *M. longiflora* L. (sweet four o'clock), *M. longiflora* L. var. *longiflora* L. [excluded], *M. longiflora* L. var. *wrightiana* (A. Gray ex britton & Kearney) Kearney & peebles, *M. macfarlanei* Constance & rollins (Macfarlane's four o'clock), *M. melanotricha* (Standl.) Spellenb., *M. multiflora* (Torr.) A. Gray (Colorado four o'clock). (Refer photoplates)

25.2.11 Plant Genetic Resources

Mirabilis is an ordinary ornamental genus over the world. Breeding of new pedigrees depends mainly on conventional crossing. Here the introduction of wild species and related genera of *Mirabilis* is to extend the genetic resources that can be used for genetic improvement of *Mirabilis* by hybridization. In the genus *Mirabilis* and closely related genera, there are few wild species especially *M. nyctaginea* (Michx.) MacMill. (Heart 4'o clock); those are able to cross successfully. These represent important genetic resources for introducing new characters into *Mirabilis.* It consists of more than 54 species distributed globally in India: Assam, Bihar, Maharastra, Gujrat; Tropical America, China, and locally in Bongaigaon, Kamrup (Ved et al. 2016). More than 20 species of *Mirabilis* are native to tropical America and Mexico. Here, the basic about wild species will aid to understand and exploit the genetic resources much better. These species include *M. coccinea*, *M. expansa*, *M. jalapa*, *M. longiflora*, and *M. multiflora.*

25.2.12 Characterization and Evaluation

Among *Mirabilis* genus, especially *M. jalapa* L. is a traditional medicine widely used in many parts of the world for the treatment of various diseases. Traditionally, the tuberous root of it has been used for treatment of carbuncles (a skin infection caused by *Staphylococcus aureus*). Phytochemical investigation of the extracts from this plant showed that it is rich in many active compounds including triterpenes, proteins, flavonoids, alkaloids, and steroids. Alanine, alpha-amyrins, arabinose, beta-amyrins, campesterol, daucosterol, and dopamine are the other compounds reported from extracts of this plant. Gas chromatography (GC)/mass spectral analysis of dichloromethane and methanol extracts of *M. jalapa* tubers indicated that oleic acid and beta-sitosterol, respectively, were the major compounds present. Liquid chromatography/mass spectroscopy analysis of the aqueous extract of the tuberous root of the plant showed a high content of flavonols. Phenolic acids such as ferulic acid and caffeic acid were also detected in the plant extracts. It has been also reported that the water extract of tuberous root of *Mirabilis* could be potentially used to heal many ailments (Nidavani and Mahalakshmi, 2014).

Gagoi et al. (2016) carried out an investigation to isolate and characterize bioactive components from *M. jalapa* L. by thin-layer chromatography for the separation of spots from fractions of the crude extract. They evaluated free-radical scavenging activity by spraying thin-layer chromatography plates (spotted with fractions) with 0.2% of 2,2-diphenyl-1picrylhydrazyl solution. Activities against human pathogens such as *Staphylococcus aureus* and *Candida albicans* were determined using the agar diffusion method. Potential spots were subjected to infrared (IR) analysis and gas chromatography for characterization. Two spots (5F1 and 1F3) showed free-radical scavenging activity. The 1F3 spot was active against both *S. aureus* and *C. albicans*, whereas the 5F1 spot was active against *S. aureus* only. IR spectral analysis indicated that 5F1 spot to be a triterpenoid. Using IR spectral analysis and an IR library search, the 1F3 spot was identified to be a flavone, which may have a hydroxyl group in ring "A" of the flavone nucleus. They concluded that the 1F3 and 5F1 spots are potential free-radical scavengers. Both 1F3 and 5F1 exhibited antimicrobial activity. IR spectral analysis coupled with an IR library search indicated 1F3 and 5F1 to be a flavone and a triterpenoid, respectively.

Recent advances in elucidating the synthetic pathways of plant pigments have been made in parallel with advances in biochemical and molecular biological techniques. In particular, the anthocyanin and carotenoid biosynthetic pathways have been well characterized biochemically and at the molecular level (Springob et al. 2003). Strack et al. (2003) have explored the presence of trogonelline, betaxaxanthin, and betacyanin as isolated products in their phytochemical study with different extracts of leaf of *M. jalapa*. Sasaki et al. (2004) also detected UDP–glucose, i.e., cyclo-DOPA 5-O-glucosyltransferase activity in four o'clocks (*M. jalapa* L.) since it had been reported earlier that cyclo-DOPA (dihydroxyphenylalanine) glucoside is a more efficient precursor in amaranthin biosynthesis than betanin or betanidin.

25.2.13 Active Constituents

Chemical analysis of different parts of *M. jalapa* manipulated the presences of alkaloids, flavonoids, phenols, steroids, triterpenes, glycosides, tannins, saponins, and lignins. The detailed study of these compound compounds from TLC visualized alanine, arabinose, campesterol, daucosterol and dopamine, d-glucan, hexacon-1-ol, indicaxanthin, isobetanin, 6-methoxyboeravinone, C-methylabronisoflavones, miraxanthins, n-dotriacontane, n nonacosane, npentacosane, and n-triacontane. Flowers mostly contain anthocyanins and flavonoids. A number of active compounds were extracted from different organs of *M. jalapa*, including anti-fungal phenolic compounds, ribosome in activating protein (RIP) which is associated with anti-viral activity, antimicrobial peptides, and rotenoids that are potent inhibitors of HIV-1 reverse transcriptase (Kaladhar and Nandikolla 2010). The alcoholic extract of *M. jalapa* is a possible source of active compounds against pathogenic enteric organisms (Enejii et al. 2011). About 20 different chemical constituents have been identified from the methanolic extract of the whole plant of *M. jalapa* by gas chromatogram mass spectrometry (GC-MS) analysis. The presence of various bioactive compounds justifies the use of whole plant for various ailments by traditional practitioners. GC-MS analysis of *M. jalapa* revealed the existence of the ether compound-3, 3′-methylenebis (4- hydroxycoumarin) (17.07), ND-alpha-phenylglycine (38.76), laminaribiitol (7.753), 3-(4-(dimethylamino) cinnamoyl) 4-hydroxycoumarin (16.89), unknown (5.284), and unknown (10.26) (Mahalingam et al. 2012). Four new rotenoids named mirabijalone A–D1 (Irena et al. 2007; Dutta et al. 2015; Kaladhar and Nandikolla 2010; Enejii et al. 2011), boeravinone C (Yi-Fen et al. 2002), and F (Pumacahua et al. 2015), together with 9-O-methyl-4-hydroxyboeravinone B (Mahalingam et al. 2012), were extracted from the roots of *M. jalapa*. Their structures were determined on the basis of their HR-EIMS, UV, IR,

1H- and 13C-NMR (DEPT), and 2D NMR (HMQC, HMBC, NOESY) data (Yi-Fen et al. 2002). Rozina (2016) had evaluated the presence of phytoconstituents in different parts of *M. jalapa* and found that the major phytoconstituents present in the plant are rotenoids (mirabiljalones A-D, boeravinones C and F), an isoquinoline derivative as well as terpenoids, steroids, phenolic compounds, alanine, arabinose, urolic acid, mirabilosic acid, mirabalisol, trigonellin, and antiviral protein. The root of the plant contains alkaloids, glycosides, flavonoids, trigonelin, saponin, and lignin. Mahalingam et al. (2012) had performed the phytochemical analysis of *M. jalapa* and found few bioactive constituents in the plant such as – 3, 3′-methylenebis (4-hydroxycoumarin), N – D – alphaphenylyglycine, laminaribiitol, and 3- (4 – (dimethylamino) cinnamoyl) – 4 – hydroxycoumarin. Jyotchna et al. (2016) reported through investigation that the extract of *M. jalapa* is rich in many active compounds including triterpenes, proteins, flavonoids, alkaloids, steroids, alanine, alpha-amyrins, arabinose, beta-amyrins, campesterol, and daucosterol. They also investigated that the gas chromatography (GC) or mass spectral analysis of dichloromethane and methanol extracts of *M. jalapa* tubers had shown the presence of oleic acid and beta-sitosterol.

25.2.14 Physicochemical Characteristics

Chemical analysis of *M. jalapa* seeds manifested that they contain 98.73% total carbohydrates, protein 0.8%, lipids 0.23%, ash 0.24%, and amylose 8.60%. Thermal analysis of *M. jalapa* showed typical starch behavior during mass loss (TG/DTG) and gelatinization (DSC), with gelatinization temperature and enthalpy around 80 °C and 5.62 J gl, respectively. The tiny (around 1 mm) diameter of starch from *M. jalapa* could be appropriate for use in the cosmetic and pharmaceutical industry credit to their high adsorption capacity. It is also suitable in the food industry for encasing flavors, essences, and other substances (Pumacahua et al. 2015). Kumar and Usha (2015) had explored that the root extract of *M. jalapa* (Nyctaginaceae) contains an antiviral protein which is type I ribosomal inactivating protein (RIP) named as Mirabilis antiviral protein (MAP). He had isolated, purified, identified, and studied its potential application as an anticancer agent. Asima et al. (2014) had isolated protein from seed of *M. jalapa* and reported that seed contains high amount of protein and amino acid and also confirm the presence 17 essential amino acid out of which 9 amino acids are essential.

25.2.15 Medicinal Uses

M. jalapa L. has a great importance in the field of ethnobotany. Purgative and emetic properties of its leaves are well-known. Decoction of leaves is used against genito-urinary system disorders, while poultice of subterranean parts is used to treat injuries (Weckerle et al. 2009). The indigenous people of Mexico use this plant for treatment of many gastrointestinal disorders, including dysentery, diarrhea, muscle pain, and abdominal colic. The extract of *M. jalapa* exhibits an inhibitory effect on digestive gut and smooth muscle contractility, whereas it stimulates the contraction of rabbit aortic muscle in a concentration-dependent manner (Aoki et al. 2008). Sometimes decoction is also used for constipation. The leaves are crushed and mixed with salt and use in sprain and bruise (Boulogne et al. 2011). It is also used for treating amenorrhea and dysmenorrhea in women (Srithi et al. 2012). Juice of leaves is mixed with water and used for treatment of jaundice (Sharma et al. 2012). Paste of leaves has emollient property (Mohammed et al. 2012) and is useful for skin eruption, while leaves juice is taken orally in hepatitis (Muhammad et al. 2013). Root extract has hypolipidemic and hypoglycemic activity and aphrodisiac, diuretic, and purgative properties. Stem with leaves are used for depigmentation (Kamagaju et al. 2013). Roots are used to arouse aphrodisiac activity. Leaves are fried in clarified butter and are fastened on the abscess. Leaf juice is used as eye drop to soothe eye inflamma-tion. Boiled leaves are eaten to reduce body pains. Tuber is administered in minute quantities to cure piles (Bhatia et al. 2014). Since ancient era, this plant is widely used as antidysenteric, antiparasitic, carminative, digestive stimulant, tonic, vermi-fuge, wound healer, and cathartic. It is also used as dropsy. Dried flowers are used as a snuff for headaches, fungal infection, and root decoction to wash wounds and treat skin afflictions as leprosy.

25.2.16 Biological Activities

The leaves of *M. jalapa* are used as traditional folk medicine in the south of Brazil to treat inflammatory and painful diseases and also used as a laxative. Leaf juice is used on boils, wound as external application, and bruises and for the treatment of urticarial (Kumar et al. 2010).

25.2.17 Antimicrobial Activity

It's evident that red flowered *M. jalapa* plant has strong antibacterial potential and is active against a wide range of microorganisms, but in contrary its aqueous ethanolic extracts didn't proved antibacterial activity. The potentials of its alcoholic extract are a possible source of active compounds against pathogenic enteric organisms. Further isolation of active components in their purest form is made possible that may serve as suitable candidates in the design of antibacterial drugs that are of plant origin (Eneji et al. 2011). It was noticed that neither aqueous nor methanolic extracts of *M. jalapa* were able to inhibit any of the tested bacterial strains (Nair et al. 2005). The antibacterial activity of various extracts of *M. jalapa* stem had been studied. Methanolic stem extract showed potent antibacterial activity against Gram (+) bacteria, while dichloromethane stem extract showed potent anti-bacterial activity against Gram (−) bacteria, which was less than that of the standard drug (Devi et al. 2010). However, the therapeutic potency for use in folklore medicine is due to the presence of some of the secondary metabolites like alkaloids, saponins, tannins, and flavonoids known for antimicrobial activity which was noticed (Akintobi et al. 2011). The antimicrobial activity of methanolic extract of *M. jalapa* is worthy of further investigation as a natural wide-spectrum antibacterial agent (Zachariah et al. 2012). Antibacterial activities of the ethanolic extract of the red-colored flower of *M. jalapa* had been examined in vitro against *Staphylococcus aureus*, *Salmonella typhi*, *Escherichia coli*, *Bacillus subtilis*, *Vibrio cholera*, *Serratia marcescens*, and *Pseudomonas aeruginosa*. The ethanolic extract of the flower showed highest inhibition against *Bacillus subtilis* (47%) followed by *Staphylococcus aureus*, *Salmonella typhi*, *Pseudomonas aeruginosa*, *Escherichia coli*, *Vibrio cholerae*, and *Serratia marcescens* (Sinha 2015). Moreover the qualitative phytochemical screening of ethanolic extract confirmed the presence of high content of tannins along with moderate amounts of alkaloids, carbohydrates, saponin, and terpenes. The existence of high amount of tannins in the extract of this plant indicated the potential antibacterial and antiviral properties of the plant. Most of the bioactive compounds responsible for bactericidal activity might be present in ethanolic extract, whereas the bacterial insensitivity to aqueous extract might be due to the antimicrobial peptides and other bioactive compounds which might not be soluble in water.

25.2.18 Antiviral Activity

Various antiviral compounds derived from plant are active against animal, plant, and human viruses. Such compounds are grouped together as furocoumarins, terpenoids, alkaloids, linens, and other specific proteins (Rozina 2016). One of the plant-derived antiviral proteins called ribosome-inactivating proteins (RIPs), which are widely distributed in higher plants (Zipf 1995). Root extracts of *M. jalapa* were sprayed on test plants 24 h before the virus or viroid inoculation which inhibited the infection by almost 100%, as corroborated by infectivity assays and the nucleic acid spot hybridization test (Vivanco et al. 1999). Noronha et al. (1993) noted that extracts of *M. jalapa* reduced the multiplication of Tobacco mosaic virus (TMV) by 50% when added to the inoculum. Because of its antiviral activity, it is not surprising that it hosts very few pathogens except *Phytophthora mirabilis* infecting only leaves and other aboveground plant parts (Goodwin and Fry 1994), while it is known to be host to two viruses: *Parietaria* mottle virus (Parrella 2002) and *Mirabilis* mosaic virus (family Caulimoviridae) (Brunt and Kitajima 1973). The existence of high amount of tannins in the extract of this plant indicated the potential antibacterial and antiviral properties of the plant (Satish et al. 2007).

25.2.19 Antifungal Activity

Plant extracts of different parts of many higher plants have been reported to exhibit antifungal properties under laboratory study. The organic extract of the cell mass form manipulated plant cell culture of *M. jalapa* L. resulted in the isolation of the three new phenolic compounds. Two of the phenolic compounds were found to show antifungal activity against *Candida albicans*. The methanol extracts of *M. jalapa* L. were tested for their antifungal activities against *Aspergillus niger*, *Candida albicans*, and *Daedalea flavida*. Results showed that *M. jalapa* has the potential inhibitory effect against *Aspergillus niger* and *Daedalea flavida* while having no effect on *Candida albicans* (Kakad et al. 2015).

25.2.20 Antioxidant Activity

The antioxidant activity of extracts of *M. jalapa* is confirmed through several studies. In vitro antioxidant potential of methanolic extracts of aerial parts of it revealed the immense potential and elucidates their tentative mechanisms of action. The total flavonoid content of the extract was discovered to be an active compound responsible for antioxidant activity and could serve as a free radical inhibitor or scavengers. Further study is necessary for isolation and characterization of the active antioxidants, which can be used to treat various oxidative stress-related diseases. Mahapatra and Bhaskar (2013) had confirmed the presence of flavonoid by performing phytochemical analysis from the methanolic extract of leaf of *M. jalapa* and reported that this active compound was responsible for antioxidant activity and could serve as a free radical inhibitor or scavengers. Shaik et al. (2012) reported the anti-lipidperoxidation activity of ethanolic extract. They used goat liver as lipid source, and in vitro evaluation was done by malondialdehyde of tissue homogenate.

25.2.21 Antispasmodic Activity

Antispasmodic activity is the activity of a drug to contract smooth muscles. The methanolic extract of *M. jalapa* L exhibits inhibitory effects on gut smooth muscle contractility as well as at the same time stimulated the contraction of rabbit aortic muscle in a concentration-dependent manner (Shaik et al. 2012). Methanolic extract of *M. jalapa* flowers contains several compounds showing spasmolytic activity such as some routioides (boeravinone F, at least), sitosterol-d-glucoside, and ursolic acid. However, additional studies are requiring explaining the mode of action of *M. jalapa* extract (Aoki et al. 2008).

25.2.22 Anti-inflammatory Activity

Anti-inflammatory refers to the property of a substance or treatment that reduces inflammation or swelling. The alcoholic, aqueous, ether extracts from the leaves of *M. jalapa*. L were studied for obtaining the anti-inflammatory activity by carrageenan-induced paw edema, formalin-induced paw edema, and cotton pellets-induced granuloma models in Wistar albino rats, which confirmed the anti-inflammatory activity of the extract of *M. jalapa* (Nath et al. 2010).

25.2.23 Antinociceptive Activity

The antinociceptive activity found to be present in extracts from leaves and stems of *M. jalapa* reduces sensitivity to painful stimuli. Ethyl acetate extracts made from leaves of *M. jalapa* produce antinociceptive in clinically related models of pain without the induction of tolerance, namely, postoperative, chronic inflammation, and neuropathic pain model. Moreover, this effect in the chronic inflammation model seems to be an intermediate agent in the activation of the cholinergic system, through pain inhibition. So, *M. jalapa* can be used successfully in various ethnopharmacological preparations to treat different painful diseases (Walker et al. 2013).

25.2.24 Anthelmintic Activity

The activity of drugs that expel parasitic worms (helminths) and other internal parasites from the body by either stunning or killing them and without causing significant damage to the host is called anthelmintic activity. The aerial parts of *M. jalapa* showed significant anthelmintic properties at higher concentration (Zachariah et al. 2012).

25.2.25 Anti-diabetic Activity Studies

Long-term hyperglycemia leads to the development of complications associated with diabetes. Diabetic complications are nowadays a global health problem without any effective therapeutic approach. The ethanol extract of the roots of *M. jalapa* has anti-diabetic activity. Oral administration of ethanolic extract of root of the *Mirabilis jalapa* L. (10 mg/kg and 20 mg/kg) significantly reduced serum, glucose, triglycerides, urea, creatinine, total cholesterol, LDL cholesterol, and the activity of gluconeogenic enzyme glucose-6-phosphate, but increased serum insulin, HDL cholesterol, protein, liver, and skeletal muscle (Zhou et al. 2012). Root extracts of *M. jalapa* L. could be used as an oral hypoglycemic agent or functional food for diabetic patients with hyperlipidemia and for persons with high risk of diabetes (Zhou et al. 2012).

25.2.26 Anti-tubercular Drug-Induced Hepatotoxicity

Studies were carried on the anti-tubercular activity of *M. jalapa*. It was deduced that leaves of it shows a protective effect on hepatotoxicity induced by anti-tubercular drugs (Jyothi et al. 2013).

25.2.27 Cytotoxic Activity

The leaf extract of *M. jalapa* L contains RIP (ribosome inactivating protein) which has cytotoxic activity on HeLa and Raji cell line with different levels (Ikawati et al. 2003).

25.2.28 Anti-corrosion Activity

Inhibitors protect metals from corrosion by adsorbing onto the surface by forming a thin adsorption layer. The efficiency of an inhibitor is largely dependent on the extent of adsorption of the inhibitor molecules on the metal surface. The nature of corrosion inhibitor has been deduced in terms of the adsorption characteristics of the inhibitor. *M. jalapa* flowers acted as an efficient corrosion inhibitor for mild-steel in 1 ml HCl. The inhibition efficiency of the extract of *M. jalapa* was maximum at 5% v/v concentration of the stock and at 49.89 °C (Thilagavathy and Saratha 2015).

25.2.29 Biofilm

Biofilms present significant therapeutic barriers for many antibiotics, and the discovery of agents which could prevent their formation or adherence would be of great use. Plant extracts of *M. jalapa* demonstrated limited bacteriostatic activity. It is inferred from different ethnobotanical usage for inhibition of growth and biofilms in methicillin-resistant *Staphylococcus aureus* that *M. jalapa* shows no considerable biofilm inhibition activity (Abdullah et al. 2016).

25.2.30 Endophytic Mycoflora

Endophytic mycoflora of *M. jalapa* was studied and isolated. Experiments showed the existence of about 17 endophytic fungal belonging to 10 genera. Their colonization frequency investigated found the fungal composition included 70.2% of hyphomycetes, 17.5% of coelomycetes, 11.6% of ascomycetes, and 11.6% of sterile mycelia (Rakshit and Sreedharamurthy 2011).

25.2.31 Poisonousness

Poisonous plants have always been part of our daily life, and some of them are so common that we do not even suspect their toxic nature. Although tubers are eaten in pickled form, intake of tubers as such is poisonous causing severe irritation in the mouth, chocking, and paralysis of the respiratory system.

25.2.32 Gold Nanoparticles

Extract of different parts of various plant species has been found to be environmentally friendly and cost effective for the production of nanoparticles on a large scale. Flowers of *M. jalapa* can be the cheapest source as registrant for the preparation of gold nanoparticles in just 1–2 h. The presence of pink colorant (anthocyanin) is an obvious choice for the preparation of gold nanoparticles (Vanker and Dhara 2010).

25.2.33 Phytoremediation

Phytoremediation of soils contaminated by organic chemicals is a challenging problem in environmental science and engineering. *M. jalapa* L is a widely spread species that can be effectively applied to phytoremediation of $\leq$10, 000 mg/kg petroleum-contaminated soil. Peng et al. (2009) investigated that the remediation capability of *M. jalapa* L. to treat petroleum contaminated soil efficiency of removing total petroleum hydrocarbons (TPHs) over the 127-day culture period was high, up to 41.61–63.20%, when the removal rate of natural attenuation was only 19.75–37.92%.

25.2.34 Molecular Characterization of Several Physiochemical Factors of *M. jalapa*

The paucity of cultivated land poses a problem toward the cultivation of conventional legumes. These factors inspired the many to search for non-conventional protein sources from non-conventional plant resources in our country. Keeping this view Ghosh et al. (2014) undertaken the investigation on seeds of *M. jalapa* as non-conventional source of protein. They concluded that the protein (TPI [Triose-phosphate isomerase]) isolated from the seeds of *M. jalapa* contains most of the essential amino acids including appreciable amounts of sulfur containing amino acids. The seed protein showed high solubility in aqueous solution. The proteins can be easily extracted from the seeds. There are also plenty of low molecular weight proteins present in the seed. Considering all these factors, it appears that the *M. jalapa* seed protein would prove to be a good source of edible protein after proper toxicological screening and would serve as an important source of unexploited protein from legume.

Wong et al. (1992) purified a protein from root tubers of *M. jalapa* to homogeneity by ion-exchange chromatography on CM-Sepharose CL-6B and FPLC on Mono-S column. The purified protein was confirmed to be *Mirabilis* antiviral protein (MAP). However, in addition to its antiviral property, the MAP was demonstrated to possess abortifacient activity in pregnant mice, inhibitory effect on cell-free protein synthesis, and antiproliferative effect on tumor cells. As judged from its biological and physiochemical properties, MAP is a type I ribosome-inactivating protein.

Isolation and characterization of a novel class of plant antimicrobial peptides from *Mirabilis jalapa* L. seeds result in isolation of two antimicrobial peptides, designated Mj-AMP1 and Mj-AMPS, respectively. These peptides are highly basic and consist of 37 and 36 residues for Mj-AMP1 and Mj-AMPB, respectively. Both peptides contain three disulfide bridges and differ from one another only by four amino acids. Sodium dodecyl sulfate-polyacrylamide gel electrophoresis (SDS-PAGE) analysis of the reduced and unreduced peptides suggests that the peptides associate with dimers in their native form. The Mj-AMPs exhibit a broad spectrum of antifungal activity since they are active against all plant pathogenic fungi tests. Concentrations required for 50% inhibition of fungal growth vary from

8 to 300 pg/ml for Mj-AMP1 and from 0.5 to 20 pg/ml for Mj-AMPB. These peptides were also active on two tested Gram-positive bacteria but were apparently nontoxic for Gram-negative bacteria and cultured human cells. Although the Mj-AMPs show sequence similarity to pagatoxins, a class of insecticidal neurotoxic peptides isolated from the venom of spiders, they do not affect pulse transmission in insect nerves. MJ1 and MJ2 both clones of cDNA encode two antimicrobial peptides (Mj-AMP1 and Mj-AMP2) of *M. jalapa*, but consequently reduced amino acid sequences reveal the presence of a putative signal sequence proceeding the mature peptide, indicating that the Mj-AMPs are expressed as preproteins. The Mj-AMP1 and Mj-AMP2 encoding genes are interrupted in their coding sequences by a single intron (380 bp and 900 bp for Mj-AMP1 and Mj-AMP2 genes, respectively). Southern blot analysis indicates that the Mj-AMP encoding genes belong to a gene family of low complexity. Northern blot analysis suggests seed-specific expression of Mj-AMPs since transcripts of the expected size could only be detected in near-mature and in mature seeds of *M. jalapa*.

Considering the range of brilliant colors of *M jalapa* L, Jiang et al (2017) identified and characterized it's novel genomes since genetic research on it is quite limited. Using fluorescent differential display (FDD) screening, they reported the identification of a novel Ty1-copia-like retrotransposon in the genome of the red flower of *M. jalapa* L. followed by naming it REC66 based on its sequence homology to the GAG protein from Ty1-copia retrotransposon. Using degenerate primers based on the DNA sequence of REC66, a total of 14 different variants in reverse transcriptase (RT) sequence were recovered from the genomic DNA. These RT sequences show a high degree of heterogeneity characterized mainly by deletion mutation; they can be divided into three subfamilies, of which the majority encode defective RT. This is the first report of a Ty1-copia retrotransposon in *Mirabilis jalapa* L. The finding could be helpful for the development of new molecular markers for genetic studies, particularly on the origin and evolutionary relationships of *M. jalapa* L, and the study of Ty1-copia retro-transposons and plant genome evolution in the genus *Mirabilis* or family Nyctaginaceae.

The isolation, identification, and characterization of different genes from various flowers (mainly from petals) associated with senescence have been discussed earlier by many authors. A large population of genes associated with flower senescence have been identified and isolated in *Mirabilis* (Channelière et al. 2002; Hunter et al. 2002; van Doorn et al. 2003; Breeze et al. 2004; Xu et al. 2007). Shari and Tahir (2014) reported that cysteine proteases or the genes encoding cysteine proteases (assigned a central role in protein degradation) had been identified from various flower systems, but no cysteine protease had been identified from senescing *M. jalapa* flowers. In senescing, *Mirabilis* flower's genes encoding MADS-domain transcription factors, MYB transcription factors, gibberellin-induced protein, and Cytochrome P_{450} – a homolog of "clock gene" (CCA 1) and aspartyl protease are involved in downregulation of genes (Shari and Tahir 2014). The degradation of proteins is one of the hallmarks of senescence which is brought about by a variety of proteases and ubiquitin-mediated proteasomes. In addition to cysteine endopeptidases, the genes encoding ubiquitin involved in proteasomal protein degradation have been identified from various flower systems (e.g., a gene encoding poly-eubiquitin [an essential element in ubiquitin pathway] from *M. jalapa*). The identification and upregulation of a "Ring Zinc Finger Ankyrin Protein" (*Mj XB3*) have also been reported from senescing *M. jalapa* flowers which share similarity to *XBAT31* and *XBAT32* of *Arabidopsis thaliana* and *glycine max*, respectively. These ankyrin repeat RING domain-containing proteins are reported to have ubiquitin ligase activity (for which the RING domains are essential) and have been found to share high homology to that of E3-type binding proteins or ubiquitin ligases that targets proteins for proteolysis via ubiquitin pathway (Lorick et al. 1999; Schnell and Hicke 2003; Stone et al. 2005; Wang et al. 2006). Of the different types of ubiquitin ligases, the *Mj XB3* (isolated from *Mirabilis* flowers) has been fully characterized containing an open reading frame (ORF) of 1341 bp. The transcripts encoding a range of transcription factors have been isolated and found to be differently regulated during development and senescence in various flower systems, e.g., MYC protein zinc finger protein from *Mirabilis*. Moreover, the homologs of b-Zip and HD-Zip proteins (plant-specific transcription factors) have been found to be upregulated in senescing *Mirabilis* flowers, thought to be induced in response to the changing osmotic and water relations of the opening and senescing flowers in *Mirabilis jalapa* (Xu et al. 2007) and that HD-Zip transcription factor isolated from *Mirabilis* has been found to be a member of HD-ZIP-I family that also includes *Athb-7* and *Athb-12* transcription factors from *Arabidopsis thaliana* (Sessa et al. 1994; Lee and Chun 1998).

Mathur (2018) also characterized the extracted seed oil of *M. jalapa* using petroleum ether as solvent. The seeds were found to contain 4.5% oil. Refractive index, specific gravity, saponification number, iodine value, and acid number were determined as main physicochemical characteristics. The density of extracted oil was found as 0.70 g/ml. The iodine value was found as 83 and saponification value 174. This saponification value indicates moderate cleansing ability, whereas the high iodine value suggests reactivity toward atmospheric oxidation on exposure to air. He concluded that it is non-drying oil, so it can be used in manufacturing of medicated cosmetics. In addition to its medicinal values, the oil can also be used as drug binder.

M. expansa constitutes a significant staple being a novel source of bioactive phytochemicals and forthcoming novel food products for the inhabitants of highlands of Peru, Bolivia, and Ecuador. Vivanco et al. (1999) found two novel type I ribosome-inactivating proteins (RIPs) in the storage roots of *Mirabilis expansa*, an underutilized Andean root crop. The two RIPs, named ME1 and ME2, were purified to homogeneity by ammonium sulfate precipitation, cation-exchange perfusion chromatography, and C_4 reverse-phase chromatography. The two proteins were found to be similar in size (27 and 27.5 kD) by sodium dodecyl sulfate-polyacrylamide gel electrophoresis, and their isoelectric points were determined to be greater than pH 10.0. Amino acid N-terminal sequencing revealed that both ME1 and ME2 had conserved residues characteristic of RIPs. Amino acid composition and Western blot analysis further suggested a structural similarity between ME1 and ME2. ME2 showed high similarity to the *M. jalapa* antiviral protein, a type I RIP. Depurination of yeast 26S rRNA by ME1 and ME2 demonstrated their ribosome-inactivating activity. Because these two proteins were isolated from roots, their antimicrobial activity was tested against root-rot microorganisms, among others. ME1 and ME2 were active against several fungi, including *Pythium irregulare, Fusarium oxysporum solani, Alternaria solani, Trichoderma reesei,* and *Trichoderma harzianum*, and an additive antifungal effect of ME1 and ME2 was observed. Antibacterial activity of both ME1 and ME2 was observed against *Pseudomonas syringae, Agrobacterium tumefaciens, Agrobacterium radiobacter*, and others.

25.2.35 Breeding Status

Mirabilis as a model organism was used to study the cytoplasmic inheritance. Variegated plant leaves were used to prove that certain factors outside the nucleus affected phenotype in a way not explained by Mendel's theories (Miko, 2008). Miko (2008) proposed that leaf color in *Mirabilis* was passed on via a uniparental mode of inheritance. Also, when plants with dark-pink flowers are crossed with white-flowered plants, light-pink-flowered offspring are produced. This is seen as an exception to Mendel's law of dominance because in this case, the dark-pink and white genes seem to be of equal strength, so neither completely dominates the other. The phenomenon is known as "incomplete dominance." However, the Mendelian principle of uniformity in the F_1-generation and the principle of segregation in the F_2-generation of genes do apply, which confirms the importance of Mendel's discoveries (Rudolph 2017).

Various analogous features of cross between *M. jalapa* L. and *M. longiflora* L. key are a) perfectly clear Mendelian segregation for some alternative flower color factors; b) polymeric segregation for quantitative characters such as intensity of flower color (probably three factor pairs), length of flower, and type of hairiness; and c) extremely complex segregation for growth habit, which is not at all surprising, as the very typical habits of both parent species depend upon numerous characters.

Though >1000 F_2 plants were studied, not a single one showing the pure paternal (*M. longiflora*) type as regards length of flower or type of hairiness was found. This may be explained by the highly polymeric character of the segregation, but it might also be caused by a plasmatic influence of *M. jalapa*. A study of the reciprocal cross would elucidate this point and helps to realize the cross at young stages of pollination and transplantation of a *M. jalapa* style. In this connection, it may be remembered that the length of the *M. longiflora* style is not the only factor which prevents the fertilization with *M. jalapa* pollen. In the *M. longiflora* style, the *M. jalapa* pollen tube ceases to grow before it reaches the length of the *M. jalapa* style.

Flowering and breeding characteristics are important for understanding plant population reproduction. Chen et al. (2008) studied the flowering dynamics and breeding system (synonymously known as facultative autogamy) of *M. jalapa* L. using the data of pollen-ovule ratio (P/O), pistil recipient phase, outcrossing index (OCI), pollen viability, numbers of pollen grains and their germination ratio on stigma, and emasculation. It had found that the flowering period of *M. jalapa* was about 4 months (from June to October), and the life span of individual flower was generally 2–3 days. For each flower, the flowering process could be divided into five periods, i.e., pre-dehiscence, initial dehiscence, full dehiscence, perianth closure, and withering, based on the changes in floral morphology and anther dehiscence. In the period of perianth closure, the percentage of stigma inside of closed perianth was 66.80% when it was sunny, 81.65% when it was cloudy, and 99.22% when it was rainy. The floral diameter was 22–28 mm. In the majority of *M. jalapa*, there was a temporal isolation between male and female organs within the same flower, i.e., the gynoecium ripened early, stigmas and anthers were arranged at the same height, and the outcrossing index (OCI) was 3. However, in a lesser percentage of flowers, there were temporal and spatial isolations between male and female organs within each flower. The gynoecium ripened early, with an outcrossing index (OCI) of 4. According to the criteria put forward by Dafni and Firmage (2000), most of *M. jalapa* individuals had the breeding system of self-compatible and only requiring pollinators sometimes, and fewer were of outcrossing, partially self-compatible, and requiring pollinators. The pollen-ovule ratio (P/O) was approximately 269. For about 6 h after flowering, when stigmas were at the highest receptivity, 56.29% of the pollen was viable. According to the results of emasculation, bagging, and artificial pollination, there was no agamospermy, and the breeding system of *M. jalapa* was determined to be largely self-compatible and only requiring pollinators sometimes. *M. jalapa* preserved a high percentage of stigma inside of closed perianth, which could be regarded as an adaptive strategy for reproduction under unfavorable environmental conditions.

25.2.36 Genetics

First hybrids of *Mirabilis* are *M. jalapa* and *M. longiflora*. Polyploidy status of *M. jalapa* is established by its karyo-morphologial study (i.e., chromosome number $n = 58$). Germplasm collections of the species are maintained at various institutions worldwide (Kew Royal Botanic Gardens 2016) as model for describing incomplete dominance inheritance.

A genetic model for flower variegation in *Mirabilis jalapa* was given by Spitters et al. (1975). Two loci effecting pigment production were assumed by authors. They summarized that the dominant alleles of each of these two loci may be repressed, and thus they didn't express themselves. In the presence of a dominant variegation gene, a repressed allele could become active at some phase of the ontogeny of the plant. Repression and later release of repression results in a pattern of variegation of a dominant color phenotype on the background of a recessive color phenotype.

25.2.37 Mechanism of Flower Color Inheritance

This species is cultivated for the brilliant color and pleasing odor of its flowers. Due to its simplicity, the breeding behavior of the self-flower colors has furnished classical material for illustrating the simple laws of inheritance. It may be shown, however, that the breeding behavior of these self-flower colors is not so simple as was first implied and that where formerly there were only a few flower color classes recognized, now there are many. To recognize and classify the flower colors of *Mirabilis* correctly, the interpretation from "yellow X white" varietal cross along with genetical investigation is needed. Generally red colors (red being used in a very general sense to cover a number of different shades) are obtained in the F_1 and F_2 generation, but many unsuccessful cases have been found for distinguishing between the F_2 red flower color types. The 'rose pink' homozygote and 'light pink' heterozygote, both were probably the first true 'pinks' known to genetic literature on M jalapa, consequenced due to a cross made between 'red x white' 'four-o-clock' plants. Earlier it's hue was assumed as 'magenta' instead 'pink'.

Crimson, yellow, and dominant white are the most common hues of the true-breeding varieties. In yellow, there is a soluble yellow pigment base. Many researchers thought that all the varieties of *M. jalapa* "sprang originally from the crimson, which apparently contains a mixture of magenta anthocyanin and soluble yellow pigment." By the loss of anthocyanin, the yellow variety was obtained, and by the loss of yellow pigment, the white varieties resulted. The action and interaction of factors at two loci results in whole range of flower color classes of *M. jalapa*. The one controls the color base, the other modifies the color base, but the latter locus is typified by an allelic series. The loci may be designated as the Y and R loci, respectively (Y-factor for yellow color or color base; y-allele of Y, absence of color base; R-factor which modifies Y to red, and in the absence of Y, white color results; Rp-allele of R, modifies Y to red, and when Y is replaced by y, it gives rose pink color; r-allele of R and Rp, gives non-modification with Y, and in its absence [in the presence of y], white color results) with additional symbols to indicate the other factors of the R series (rp-allele of the R series, gives non-modification with Y, and when Y is replaced by y, flesh pink color results; rp is dominant to r; Rpl-factor for light rosaline purple, also allele of the R series). It is incompletely dominant to R and Rp and presumably dominant to r and rp. In the presence of Y, it produces red color. A complete account of breeding behavior couldn't be given by Showalter (1933). For clearness the several crosses made during investigation by Showalter (1933) treated individually. Reciprocal crosses gave the same results. From all the data obtained as well as those of previous experimenters, it was evident that the factors Y and R were inherited independently of each other.

It is not uncommon to see a plant of *M. jalapa* bearing two distinctly different flower colors, in the same, or in different flowers. In cases where mutation occurs in self-colored plants, there is generally a preponderance of flowers of the same color, with a few of a different or mutant color. These few are borne on one (rarely more than one) terminal branch, indicating that somatic mutation took place late in the growth of the plant, giving a new flower color. In some cases, not all the flowers of such a branch are self-colored. A single blossom may show part mutant and part non-mutant color, indicating that the whole of the growing tip of the branch was not involved in the mutation. It's owing to "flaking" problem. With Showalter (1933) in his investigation of the 5000 plants (slightly more) of *M. jalapa*, 21 self-flower color mutations had been observed and recorded. Most of the observations had been confined to material with known breeding behavior. The somatic mutations he observed were as follows: (1) white with light pink branch (R to Rp), 6 cases; (2) white with light pink branch (R to rp), 2 cases; (3) white with pale rosaline purple branch (R to RpZ), 3 cases; (4) rose pink with light pink branch (Rp to R), 1 cases; (5) crimson with orange red branch (R to r), 4 cases; and (6) flesh pink with light pink branch (rp to R), 2 cases.

Segregation for length of flower and type of hairiness seem to be controlled by polymeric factors. The pure paternal (*M. longiflora*) type of any of these two characters never occurred. This may be due either to a high number of factors or to plasmatic influence.

25.2.38 Inheritance and Genetic System of Flower Variegation in *Mirabilis jalapa*

Variegated flowering plants often carry completely mutated flowers; sometimes whole branches show but completely mutated flowers. The progeny of these completely mutated flowers is identical to the progeny of variegated flowers of the same plant, which illustrates that mutation in the gametes producing L_2-tissue occurs independently of mutation in the flower color determining L_1-tissue of the same flower. Temperature appears to have an influence on the degree of variegation. At 28 °C the percentage of mutant tissue in the flower is smaller and with it the degree of variegation than at 20 °C (Spitters et al. 1975). The degree of variegation is not affected only by temperature but also by a shift in timing of the mutation. Mutation happens in the ontogeny of the plant at smaller percentage in the mutant tissue of the unfolding flowers. Unstable gene systems, causing an irregular pattern of variegation in pigment distribution, may be responsible for flower variegation in other flowering genera (e.g., *Antirrhinum majus*). Characteristic for the group of unstable genes is the very high mutation rate in somatic tissue which is often noticed in the gametes as well. Moreover, they appear only in restricted genetic backgrounds, such as in the presence of a certain variegation gene. In most cases, the mutations are dominant and resemble reversion to a wild type. A lower mutation rate of unstable genes at higher temperatures had been reported for kernel variegation in maize (Rhoades, 1941) and for flower variegation in *Portulaca* and *Nicotiana* (Sand 1957). Spitters et al. (1975) had found similar results in *M. jalapa* at temperatures of 20 °C and 28 °C, although the percentage of the mutated tissue and not the number of spots was estimated as an indication for the mutation rate. It had also been observed that the mean percentage of mutated tissue of the flowers decreases toward the end of the flowering period of the plant. What ontogenic factor was responsible for the decrease was undescribed by authors.

The research work on the genetics of flower variegation had been done by many authors together with an investigation of other possible causes of variegation, viz., cytoplasmic inheritance, variation in somatic chromosome number, and virus:

1. Cytoplasmic inheritance: Leaf variegation is often cytoplasmically inherited. Common criteria for this way of inheritance are non-Mendelian inheritance,

reciprocal differences, and somatic segregation during ontogeny of the plant. Spitters et al. (1975) found a genetic model for flower variegation based on Mendelian inheritance. Furthermore, no reciprocal differences were observed. Whatever may be the cause, plastids are not generally responsible for flower variegation in *M. jalapa* since there is strictly maternal inheritance as far as plastids are concerned. A somatic segregation during ontogeny was only observed in a dominant direction: white to yellow, white to red, and yellow to red. So, flower variegation seems not to be cytoplasmically inherited.

2. Variation in somatic chromosome number: It could be argued that variation in somatic chromosome number causes flower variegation. Since the case of *M. jalapa* variegation implies a change from the recessive color to the dominant color, variation in somatic chromosome number can only be the cause if a chromosome which carries an inhibitor gene for activation of Y^V and R^V alleles is eliminated. Shi et al. (2010) showed that in root tip cells of variegated plants, the somatic chromosome number was 58, and no variation in this number had been observed. Although the cells of the perianth had not been investigated, it was very likely that variation of somatic chromosome number was not the cause of flower variegation.
3. Virus: In several cases virus induces variegation like broken flowers (e.g., in tulip). It was attempted by Spitters et al. (1975) to demonstrate the presence of a virus by grafting and by hot water treatment. Grafting was carried out as a virus of the variegated part of the graft combination could be transmitted to a possible virus-free part (i.e., uniformly colored flowers) and induce flower variegation. If variegation results from the activity of a virus, the virus must be transmitted through the fruit. So, hot water treatment of the fruit was applied to inactivate the hypothetical virus. Treated fruits of true breeding variegated plants should therefore have produced plants with uniformly colored flowers.

The repressing and regulating systems of maize is a very likely explanation for the flower variegation in *M. jalapa*. The dominant genes for pigment production Y and R could be repressed in their activation, for which one suppressor for each allele is needed. The repressed alleles together with their suppressor are given the respective symbols Y^V and R^V.

25.2.39 Reproductive Biology

M. jalapa reproduces by seeds and vegetatively by tubers (Royal Horticultural Socicty, 2016). Its morphological adaption promotes outcrossing but, in many cases, it has been found as a selfer. The species will self-pollinate by the stamens rolling toward the stigma when the flower closes (Valla and Ancibor, 1978). *M. jalapa* is pollinated by sphinx moths or hawk moths (belong to Sphingidae family) at >13 °C temperature as temperature and nectar production affect the pollination (Martínez and Búrquez 1986).

25.2.40 Fertility

Study on fertility status of five *M. jalapa* and seven *M. longiflora* plants showed 97, 93, 92, 91, 81 and 97, 97, 96, 95, 94, 68, 59 percent of quality pollen by acetocarmine-glycerine staining method, and based on that, they were classified into > good > and > poor> grains (the former being colored and having a more or less normal appearance of the protoplasm). Seed setting in the pure species is very abundant. Generally, in *Mirabilis* F_2 generation shows a rather reduced fertility. After free pollination but spatial isolation, F_1 plants produced some hundreds of seeds yearly. In F_2 generation 10–15% and F_1 generation 38% good pollen grains were found in both species which unfortunately was too low. Some plants set seeds at an early stage, and in others the flower-buds do not develop. The majority of plants, however, flowered abundantly, but most of these normally flowering plants produced no seeds or less than 5, and plants with more than 20 seeds were very rare. Author deduced that the number of pollen mother cells in these loculi varied from 7 to 11. In almost all F_2 and F_1 plants, fertility seemed to be reduced still more, most of them producing no or very few seeds. An opposite relation was found in *Tragopogon pratensis* x *T. porrifolius* cross (Winge, 1938), which showed a rather reduced fertility in F_1 but quite normal fertility in F_2, and the following generations. However, as to *Mirabilis*, it must be considered that the fertility of first year F_2 plants was compared with that of old F_1 plants, and that fertility in the latter may had increased in the course of years, a phenomenon that has been observed in many hybrids.

25.2.41 Cytology

In agreement with Showalter (1935), the somatic chromosome number was determined to be 58. Among the 29 pairs of *M. longiflora* and *M. jalapa*, two large pairs and one rather large could be distinguished. Often one or two of the large pairs were connected to the nucleolus or were lying close to it, and in a few cases small satellites were seen in two of them. The reason of nucleolus forming chromosomes was not discussed by Tischler (1929). Showalter (1935), again, found that generally 2, and sometimes 3, chromosomes (or groups of chromosomes) were attached to the nucleolus, but he didn't always recognize very large chromosomes in them. According to him, at diakinesis chromosome pairs were attached end to end into groups, an arrangement not occurred own material. Now, it seems possible that he may had misinterpreted large chromosomes attached to the nucleolus as groups of small chromosomes. The number of chromosomes in *Mirabilis* and related groups and especially the presence of 3 pairs of large nucleolus chromosomes suggest the probability of (secondary) polyploidy in *Mirabilis* (Tischler 1928). Chromosome pairing in F_1 was rather regular: often 29 bivalents occur. Even the one highly heteromorphic large pair almost always forms a bivalent. In spite of the size difference in this and in some other pairs, indicating major structural differences between the parent species, no clear cases of mu1tivalent chromosome configurations were seen (Prakeen 1944). Cross of *Tragopogon pratensis* (n = 6) x *T. porrifolius* (n = 6) and *Nicotiana longiflora* (n = 10) X *N. alafa* (n = 9) could revalidate the above observation. In the first case, the structural differences seem to be small, and pairing was quite regular. In the *Nicotiana* cross, however, the differences were very striking and seem to affect every chromosome. Nevertheless, in 31 of the 52 cells studied, the configuration at first metaphase was $1_{111} + 8_{11}$ or $9_{11} + 1_{1}$, and only once was a cell with two trivalents found. Lack of multivalent configurations may depend upon various structural peculiarities: size of chromosomes, position of centromere, size of translocated or duplicated segments, and number and localization of chiasmata. But, analogous to polyploids, some other imperfectly known factors such as differential affinity and tendency to two-by-two pairing may possess some importance. The earlier meiotic stages in *Mirabilis* deserve a thorough reinvestigation for many reasons, viz., chiasma formation and the extreme terminalization, attachment of chromosomes to the nucleolus, and in F_1 the regular pairing of the heteromorphic pair and the non-pairing of one other large pair.

25.2.42 Future Prospect

"Unstable genes" are quite common in plants, but as they are "unreliable," their practical use in plant breeding is limited. In general, however, temperature effects on timing and frequency of mutation, i.e., extent of flower variegation in plants of *M. jalapa* and other species are constant. This means that plants of varieties with flower variegation grown in the tropics will express less variegation than sister plants grown in temperate regions. Furthermore, variegated plants multiplied in the tropics will produce in their progenies fewer plants with uniformly colored flowers than in temperate countries. The same holds for plants grown and multiplied in heated greenhouses and outdoors, respectively. So, it's a key point for plant breeders to pay attention on it. On the other hand, from the time immemorial, plants have been widely used as curative agents for variety of ailments. *M. jalapa* preparations were widely employed by practitioner using herbal-based natural therapies. Although the various scientific studies have proved its efficacy, most of the studies were conducted using crude preparations or extracts without mentioning its active phytoconstituents. The detailed research work on isolation of bioactive phytoconstituents followed by standardization is seriously demanded. Further, translational potential and possible novel bioactivities and novel targets are yet to be discovered with this amazing plant.

Photoplates of Available *Mirabilis* Species and Cultivars(See Table 4)

Table 4 Photoplates of available *Mirabilis* Species and Cultivars

 *Mirabilis expansa*	 *Mirabilis comata*	
 *Mirabilis coccinea*	 *Mirabilis alipes*	
 *Mirabilis albida*	 *Mirabilis gigantea*	 *Mirabilis glabra*

Mirabilis jalapa

Mirabilis macfarlanei

Mirabilis hirsuta

Mirabilis longiflora

Mirabilis greenei

Mirabilis laevis

Mirabilis linearis

References

Abdullah MA, El-Baky RMA, Hassan HA (2016) Fluoroquinolones as urease inhibitors: anti-*Proteus mirabilis* activity and molecular docking studies. American J Microbiol Res 4:81–84

Acevedo-Rodríguez P, Strong MT (2012) Catalogue of the Seed Plants of the West Indies. Smithsonian Contributions to Botany, Washington DC, USA. 98:11–92. http://botany.si.edu/Antilles/WestIndies/catalog.htm

Akintobi OA, Agunbiade SO, Okonko IO et al (2011) Antimicrobial evaluation and phytochemical analysis of leaf extracts of *Mirabilis jalapa* against some human pathogenic bacteria. Nat Sci 9(10):45–53

Antoine F (1876) Das Pflanzenreich auf der Wiener Weltausstellung im jahre 1873. Notizen uber die exponirten pflanzen, pflanzenrohstoffe und produkte, sowie uber ihre bildlichen Darstelllugen. Osterr Bot Z 26(11):376–383

Aoki K, Cortes AR, Ramırez MC et al (2008) Pharmacological study of antispasmodic activity of Mirabilis Jalapa Linn flowers. J Ethnopharmacol 116:96–101

Asima G, Anupam N, Julie B (2014) Chemical characterization of seed proteins of *Mirabilis jalapa* L (Nyctaginaceae). Int J Food Prop 17(3):559–569

Balfour IB (1879) Flowering plants and ferns. In: Philosophical Transactions of the Royal Society of London, pp 326–387

Bedi SJ (1978) Ethno-botany of the Ratan Mahal Hills, Gujarat, India. Eco Bot 32(3):278–284

Bhatia H, Manhasb RK, Rani M (2014) Traditional knowledge of poisonous plants of Udhampur district of Jammu and Kashmir, India. J Ethnopharmacol 152:207–216

BioNET-EAFRINET (2016) Invasive plants of East Africa (Kenya, Uganda and Tanzania). Lucid vol. 3.5 key and fact sheets. http://keys.lucidcentral.org/keys/v3/eafrinet/index.htm

Boulogne I, Lionel GR, Harry OL et al (2011) Ethnopharmacological survey in Les Saintes (Guadeloupe, French West Indies): a comparative studies. J Ethnopharmacol 133:1039–1050

Breeze E, Wagstaff C, Harrison E et al (2004) Gene expression patterns to define stages of post-harvest senescence in *Alstroemeria* petals. Plant Biotechnol J 2:155–168

Breyer-Brandwijk MG (1962) The Medicinal and Poisonous Plants of South Africa. E Living Stone Lond (2):801

Broome R, Sabir K, Carrington S et al (2007) Plants of the Eastern Caribbean. Barbados, West Indies. http://ecflora.cavehill.uwi.edu/index.htm

Brunt AA, Kitajima EW (1973) Intracellular location and some properties of *Mirabilis* mosaic, a new member of the cauliflower mosaic group of viruses. Phytopatho Zeithscrifht 76:265–275

Cantley WEM, Dennys NB (1886) Notes on economic plants, Straits settlements. J Malayan Branch Royal Asiatic Soc 18:293–344

Channelière S, Rivière S, Scalliet G et al (2002) Analysis of gene expression in rose petals using expressed sequence tags. FEBS Lett 515:35–38

Chen X, Hu XH, Xiao YA, Xie Q (2008) Floral syndrome and breeding system of *Mirabilis jalapa* L. Chin J Ecol 27(10):1653–1658

Chetty KM, Sivaji K, Rao KT (2008) Flowering plants of Chittoor district, Andhra Pradesh, India, 1st edn. Student Offset Printers, Tirupati

Cleall E, Wadman J, Baker J (1807) Papers in colonies and trade. In: Transactions of the Society, Instituted at London, for the Encouragement of Arts, Manufactures, and Commerce 25:143–212

Dafni A, Firmage D (2000) Pollen viability and longevity. Plant Syst Evol 222:113–132

DAISIE (2016) Delivering Alien Invasive Species Inventories for Europe. http://www.europe-aliens.org/

Della A, Latrou G (1995) New plant records from Cyprus. In: Kew Bulletin 50(2):387–396

Devi SL, Dhanamani M, Kannan S (2010) *In vitro* antibacterial activity of various extracts of *Mirabilis jalapa* stem. Pharm Lett 2(6):329–332

Douglas NA, Manos PS (2004) Molecular phylogeny of Nyctaginaceae: Taxonomy, Biogeography, and characters associated with a radiation of xerophytic genera in North America. American J Bot 94(5):856–872

Dutta PD, Mondal M, Maiti GG (2015) Pollen morphological studies on some members of the family Nyctaginaceae in India. Asian J Plant Sci Res 5(2):72–76

Encyclopedia of Life (2016) Encyclopedia of Life. In: Encyclopedia of Life. http://www.eol.org

Eneji SM, Inuwa HM, Ibrahim S et al (2011) *In vitro* assessment of bioactive components of *Mirabilis jalapa* ethanolic extract on clinical isolates of *Salmonella typhi* and *Bacillus cereus*. Afr J Biotechnol 7:16006–16011

Eneji SM, Mairo II, Salisu I et al (2011) In vitro assessment of bioactive components of Mirabilis jalapa ethanolic extract on clinical isolates of Salmonella typhi and Bacillus cereus. African J Biotech 10:16006–1601

Felter HW, Lloyd JU (2010) King's Amer dispensatory http://www.henriettesherbal.com/eclectic/kings/ipomoea-jala.html. Accessed 20 Oct 2010.

Flora do Brasil (2016) Brazilian Flora 2020 in construction. http://reflora.jbrj.gov.br/reflora/listaBrasil/ConsultaPublicaUC/ConsultaPublicaUC.do#CondicaoTaxonCP

Flora of India (2015) Four o'clock Plant. In: Four o'clock Plant. http://indiaflora.blogspot.co.uk/search?q=mirabilis

Gagoi J, Nakhuru KS, Policegoudra RS et al (2016) Isolation and characterization of bioactive components from *Mirabilis jalapa* L. radix. J Tradtnl Comple Med 6(1):41–47

Ghosh A, Nayak A, Banerji J (2014) Chemical characterization of seed proteins of *Mirabilis jalapa* L. (Nyctaginaceae). Int J Food Prop 17(3):559–569. https://doi.org/10.1080/10942912.2011.642632

Goodwin SB, Fry WE (1994) Genetic analyses of interspecific hybrids between *Phytophthora infestans* and *Phytophthora mirabilis*. Exp Mycol 18:20–32

Greenwood W (1944) Supplementary notes on the adventives and weed flora of the leeward coasts of Fiji. J Arnold Arboretum 397–405

Hamilton AW, Holttum RE (1933) A Malay garden. J Malayan Branch Royal Asiatic Soc 11(2):139–143

Hassler MEM (2012) UCLA botanical Garden https://web.archive.org/web/20120815175007/; http://www.botgard.ucla.edu/html/MEMBGNewsletter/Volume4number2/Theplantsthatlovehawkmoths.html. Accessed 31 July 2012

Heldreich T (1863) Phänologische Notizen aus Griechenland. Osterr Bot Z 13(2):48–49

Hunter DA, Steele BC, Reid MS (2002) Identification of genes associated with perianth senescence in daffodil (*Narcissus pseudonarcissus* L. 'Dutch Master'). Plant Sci 163:13–21

Hutchinson J (1969) Evolution and phylogeny of flowering plants, London, Academic Press, New York. https://www.worldcat.org/title/evolution-and-phylogeny-of-flowering-plants-dicotyledons-facts-and-theory-with-over-550-illustrations-and-maps-by-the-author/oclc/34518

Ikawati Z, Sudjadi WE, Dyah P (2003) Induction of apoptosis by protein fraction isolated from the leaves of on Hela and Raji cell-line. Orient Pharm Exp Med 3(3):151–156

Invasive species compendium (2019) Centre for Agriculture and Biological International (CABI), India. https://www.cabi.org/isc/datasheet/34254

Irena S, Phyllis GW, Abdullah G (2007) Phytoplasma infection in the four o'clock flower (*Mirabilis jalapa*). Bull Insectol 60(2):281–282

Jiang S, Liang JG, Feng H et al (2017) Identification and characterization of REC66, a Ty1-copia-like retrotransposon in the genome of red flower of *Mirabilis jalapa* L. Arch Biol Sci 69(2):315–322

Jyotchna G, Khonamai SN, Rudragoud SP et al (2016) Isolation and characterization of bioactive components from *Mirabilis jalapa* L. J Tradit Complement Med 6(1):41–47

Jyothi B, Akshmi SM, Anitha K (2013) Protective effect of *Mirabilis jalapa* leaves on anti-tubercular drugs induced hepatotoxicity. Asian J Plant Sci Res 6(3):221–224

Kakad SL, Dhembare AJ, Chakane R (2015) Evaluation of antifungal activities of some selected plant species against fungal pathogen. J Microbiol Biotechnol Res 5(1):24–27

Kaladhar D, Nandikolla SK (2010) Antimicrobial studies, biochemical and image analysis in *Mirabilis jalapa* L. Int J Farm Technol 3:683–693

Kamagaju L, Bizuru E, Minani V et al (2013) An ethnobotanical survey of medicinal plants used in Rwanda for voluntary depigmentation. J Ethnopharmacol 150:708–717

Kew Royal Botanic Gardens (2016) Millennium Seed Bank – seed list. Richmond, UK. http://apps.kew.org/seedlist/SeedlistServlet

Khurian JC (2003) Plants that heals. Oriental watchman publishing house, Pune (5th edition, Vol 1), pp 214–215

Kirtikar KR, Basu BD (1918) Indian Medicinal Plants. Archaeological Survey of India, New Delhi. https://www.indianculture.gov.in/ebooks/indian-medicinal-plants

Kress WJ, DeFilipps RA, Farr E et al (2003) A checklist of the trees, shrubs, herbs, and climbers of Myanmar. Contrib U S Natl Herbarium 45:1–590. http://botany.si.edu/pubs/CUSNH/vol_45.pdf

Kumar DCK, Usha M (2015) Cytotoxicity against tumor cell lines of a purified *Mirabilis* antiviral protein isolated from root of *Mirabilis jalapa*. World J Pharm Res 4(2):1696–1710

Kumar V, Singh I, Gauttam V et al (2010) Anti-inflammatory activity of aqueous extract of *Mirabilis jalapa* L. leaves. Pharm Res 2(6):364–367. https://doi.org/10.4103/0974-8490.75456

Ladwig-Cooper S (2012) Due diligence: four o'clocks plants. In: Due diligence: four o'clocks plants. http://permaculturenews.org/2012/08/08/duediligencefouroclockplants/

Le Bourgeois T, Carrara A, Dodet M et al (2008) Advent- OI: Principales adventices des îles du sud-ouest de l'Océan Indien. 1, Cirad, Montpellier, France. http://publish.plantnet-project.org/project/plantinvasivekruger/collection/collection/synthese/details/MIBJA

Lee YH, Chun JY (1998) A new homeodomain-leucine zipper gene from *Arabidopsis thaliana* induced by water stress and abscisic acid treatment. Plant Mol Biol 37:377–384

Lorick KL, Jensen JP, Fang S (1999) RING fingers mediate ubiquitin-conjugating enzyme (E2)-dependent ubiquitination. Proc Natl Acad Sci U S A 96:11364–11369

Mahalingam R, Bhirathidasan R, Ambikapathy V et al (2012) GC-MS determination of bioactive compounds of *Mirabilis jalapa*. Asian J Plant Sci Res 3:224–227

Mahapatra S, Bhaskar P (2013) Flower pigment analysis of *Mirabilis jalapa*. Int J Appl Eng Res 4: 1257–1271

Manandhar NP (1991) Medicinal plant-lore of Tamang tribe of Kabhrepalanchok District, Nepal. Eco Bot 45(1):58–71

Martínez del Río C, Búrquez A (1986) Nectar production and temperature dependent pollination in *Mirabilis jalapa* L. Biotropica 18(1):28–31

Mathur A (2018) Extraction and characterization of *Mirabilis jalapa* seed oil. IRE J 2(2):1–4

Miko I (2008) Gregor Mendel and the principles of inheritance. In: Experiments in plant hybridization. Nature Edu 1(1). https://www.sciencelearn.org.nz/resources/2000-mendel-s-principles-of-inheritance. Accessed 5 July 2011

Mirabilis jalapa L. (2011) Pacific Island Ecosystems at Risk (PIER) Hawaii, USA. www.hear.org/pier/species/mirabilis_jalapa.htm

Missouri Botanical Garden (2016) Tropicos database, St. Louis, Missouri, USA. http://www.tropicos.org

Missouri Botanical Garden (2017) Missouri Botanical Garden Plant Finder. St. Louis, MO, USA: Missouri Botanical Garden. http://www.missouribotanicalgarden.org/plantfinder/plantfindersearch.aspx

Mohammed M (2012) Study of some *Mirabilis jalapa* L. leaves components and effect of their extracts on growth of pathogenic bacteria. Al- Mustansiriyah J Sci 23:117–124

Muhammad A, Adeel M, Rifat NM et al (2013) Indigenous knowledge of medicinal plants from Guranwala district Pakistan. J Ethnopharmacol 148:714–723

Nair R, Kalariya T, Sumitra C (2005) Antibacterial activity of some selected Indian medicinal flora. Turk J Biol 29:41–47

Nath LR, Manjunath KP, Savadi RV (2010) Anti-inflammatory activity of *Mirabilis jalapa* L. leaves. J Basic Clin Pharm 1:93–96

Nidavani RB, Mahalakshmi AM (2014) An ethano-pharmacological review of four o' clock flower plant (*Mirabilis jalapa* linn.). J Biol Sci Opin 2(6):344–348

Noronha AB, Amelia M, Alexandre V (1993) Protection against Tobacco mosaic virus induced by some caryophyllales plant extracts. Microbios 73:75–80

Pampanini R (1911) La Tigridia pavonia Ker-Gawl. La sua storia e le sue varieta. Bullettino della R. Societa Toscana di Orticultura 16(11):306–317

Paria ND, Chattopadhyay SP (2000) Flora of Hazaribagh District Bihar I: 97–100 (Nyctaginaceae). Botanical Survey of India, Calcutta

Parrella G (2002) First report of *Parietaria* mottle virus in *Mirabilis jalapa*. Plant Pathol 51:40

Peng S, Qixing Z, Zhang C et al (2009) Phytoremediation of petroleum contaminated soils by *Mirabilis jalapa* in a greenhouse plot experiment. J Hazard Mater 168:1490–1496

PFAF (2016) Plants For A Future Database. http://www.pfaf.org/USER/Default.aspx

PIER (2016) Pacific Islands Ecosystems at Risk. Honolulu, USA: HEAR, University of Hawaii. http://www.hear.org/pier/index.htm

Prakeen R (1944) Contributions to the genetics and cytology of *Mirabilis*. Hereditans 30:201–212. https://onlinelibrary.wiley.com/doi/pdf/10.1111/j.1601-5223.1944.tb03310.x

PROTA (2016) PROTA4U web database. Wageningen, Netherlands: Plant Resources of Tropical Africa. http://www.prota4u.org/search.asp

Pumacahua RA, Demiate IM, Schnitzler E (2015) Morphological thermal and physicochemical characteristics of small granules starch from *Mirabilis jalapa* L. Thermochim Acta 602:1–7

Rakshith D, Sreedharamurthy S (2011) Endophytic mycoflora of L. and studies on antimicrobial activity of its endophytic sp. Asian J Exp Biol Sci 2(1):75–79

Reverchon J (1886) Botanizing in Texas. II. In: Bot Gaz 11(8):211–216

Rhoades MM (1941) The genetic control of mutability in maize. Cold Spring Harb Symp Quant Biol 9:138–144

Roxburgh W (1832) Descriptions of Indian plant. Serampore. https://www.biodiversitylibrary.org/creator/323/title#/titles

Royal Horticultural Society (2016) *Mirabilis jalapa* four o'clock flower. https://www.rhs.org.uk/Plants/11156/Mirabilis-jalapa/Details?returnurl

Rozina R (2016) Pharmacological and biological activities of *Mirabilis jalapa* L. Int J Pharm Res 6(5):160–168. https://doi.org/10.7439/ijpr

Rudolph D (2017) Uniformitätsregel (Mendelsche Regel). https://www.frustfrei-lernen.de/biologie/uniformitaetsregel.html

Rusby HH (1895) On the collections of Mr. Miguel Bang in Bolivia. Part II. In: Memoirs of the Torrey Botanical Club 4(3):203–274

Sand SA (1957) Phenotypic variability and the influence of temperature on somatic instability in cultures derived from hybrids between *Nicotiana langsdorfii* and *N. sanderae*. Genetics 42:685–703

Sasaki N, Adachi T, Koda T et al (2004) Detection of UDP–glucose: cyclo- DOPA 5-O-glucosyltransferase activity in four o'clocks (*Mirabilis jalapa* L.). FEBS Lett 568(1–3):1–11

Satish S, Mohana DC, Ranhavendra MP et al (2007) Antifungal activity of some plant extracts against important seed borne pathogens of *Aspergillus* sp. J Agric Technol 3(1):109–119

Schnell JD, Hicke L (2003) Non-traditional functions of ubiquitin and ubiquitin-binding proteins. J Biol Chem 278:35857–35860

SEINet (2016) SEINet, Arizona New Mexico chapter data portal. In: SEINet, Arizona-New Mexico chapter data portal. http://swbiodiversity.org/seinet/index.php

Sessa G, Carabelli M, Ruberti I et al (1994) Identification of distinct families of HD-Zip proteins in *Arabidopsis thaliana*. In: Puigdomenèch P, Coruzzi G (eds) Molecular-genetic analysis of plant development and metabolism. Springer, Berlin

Shahri W, Tahir I (2014) Flower senescence: some molecular aspects. Planta 239:277–297

Shaik S, Dhanalakshmi M, Jayasree T (2012) Nephroprotective activity of leaves of *Mirabilis jalapa* L. by acetaminophen induced nephrotoxicity. Int J Farm Technol 4(2):2075–2084

Sharma HK, Chhangte L, Dolui AK (2001) Traditional medicinal plants in Mizoram, India. Fitoterapia 72(2):146–161

Sharma J, Sumeet G, Gour RG et al (2012) The treatment of jaundice with medicinal plants in indigenous communities of Sub Himalayan of Uttarakhand India. J Ethnopharmacol 143:262–291

Shi L, Gao R, Li H et al (2010) The chromosome number and karyotype analysis of *Mirabilis jalapa*. Pratacultural Sci 1:45–53

Showalter HM (1933) Self flower-color inheritance and mutation in *Mirabilis jalapa* L. Genetics 19:568–580
Showalter AM (1935) A study of the chromosomes and degenerating microspore tissue in *Mirabilis*. Am J Bot 22:594–608
Sinha SN (2015) Bactericidal activities of flower of *Mirabilis jalapa* L. Int J Emerg Trends Pharm Sci 3(1):1–3
Smyth BB (1889) Periodicity in plants. Transactions of the Annual Meeting of the Kansas Academy of Science 12:75–81
Spitters CJT, Vosselman L, Engels JMM et al (1975) Investigations of the inheritance of flower variegation in *Mirabilis jalapa* L. and genetic system of flower variegation and speculation about its existence. Euphytica 24:323–332
Springob K, Nakajima J, Yamazaki M et al (2003) Recent advances in the biosynthesis and accumulation of anthocyanins. Nat Prod Rep 20:288–303
Srithi K, Chosie T, Prasit W et al (2012) Medicinal plants used in Hmong women's health care in Northern Thailand. J Ethnopharmacol 139:119–135
Stone S, Hauksdóttir H, Troy A et al (2005) Functional analysis of the RING-type ubiquitin ligase family of *Arabidopsis*. Plant Physiol 137:13–30
Strack D, Vogt T, Schliemann W (2003) Recent advances in betalain research. Phytochemistry 62(3):247–269
The Wealth of India (1949) A dictionary of Indian raw materials and industrial products (Industrial products—part I). Ind Med Gaz 84(10):476–477
Thilagavathy P, Saratha R (2015) *Mirabilis jalapa* flowers extract as corrosion inhibitor for the mild steel corrosion in 1M HCl. IOSR J Appl Chem 8(1):30–35
Tischler G (1928) Uber die Verwendung der Chromosomenzahl fur phylogenetische Probleme bei den Angiospermen. Biol Zentralbl 48:321–345
Tischler G (1929) Revisionen friiherer Chromosomenzahlungen und anschliessende Untersuchungen. Planta 8:663–697
Trelease W (1897) Botanical observations on the Azores. Mo Bot Gard Annu Rep 77–220
Uotila P (2011) Nyctaginaceae. In: Euro + Med Plant base – the information resource for Euro – Mediterranean plant diversity. http://ww2.bgbm.org/EuroPlusMed/PTaxonDetail.asp?NameCache=Mirabilis%20jalapa&PTRefFk=7300000
Valla JJ, Ancibor W (1978) Biología floral de *Mirabilis jalapa* L. (Nyctaginaceae). Darwin 21(2–4):407–415. ISSN: 0011-6793
van Doorn WG, Balk PA, van Houwelingen AM et al (2003) Gene expression during anthesis and senescence in *Iris* flowers. Plant Mol Biol 53:845–863
Vanker PS, Dhara B (2010) Preparation of gold nano particles from *Mirabilis jalapa* flowers. Indian J Biochem Biophys 47:157–160
Ved DK, Sureshchandra ST, Barve S et al (2016) FRLHT's ENVIS Centre on Medicinal Plants, Bengaluru. http://envis.frlht.org/plant_details.php?disp_id=1441
Vibrans H (ed) (2009) Malezas de Mexico. http://www.conabio.gob.mx/malezasdemexico/nyctaginaceae/mirabilis-jalapa/fichas/ficha.htm. Accessed 8 May 2020
Vivanco JM (1999) Antiviral and antiviroid activity of MAP-containing extracts from *Mirabilis jalapa* roots. Plant Dis 83:1116–1121
Vivanco JM, Savary BJ, Flores HE (1999) Characterization of two novel type i ribosome-inactivating proteins from the storage roots of the andean crop *Mirabilis expansa*. Plant Physiol 119(4):1447–1456. https://doi.org/10.1104/pp.119.4.1447
Walker CIB, Gabriela T, Mateus FR et al (2008) Antinociceptive activity of *Mirabilis jalapa* in mice. J Ethnopharmacol 149:685–693
Walker C, Trevisan G, Fortes R (2013) Antinociceptive effect of *Mirabilis jalapa* on acute and chronic pain models in mice. J ethnopharmacology 149(3):685–693
Wang YS, Pi LY, Chen X et al (2006) Rice XA21 binding protein 3 is an ubiquitin ligase required for full *Xa21*-mediated disease resistance. Plant Cell 18:3635–3646

Weckerle CS, Ineichena R, Huber FK et al (2009) Mao's heritage: medicinal plant knowledge among the Bai in Shaxi, China, at a crossroads between distinct local and common widespread practice. J Ethnopharmacol 123:213–228
Winge O (1938) Inheritance of species characters in *Tragopogon*: a cytogenetic investigation. C R Trav Lab Carlsberg 22:155–194
Witt A, Luke Q (2017) Guide to the naturalized and invasive plants of Eastern Africa. (ed. by Witt A, Luke Q) Wallingford, UK. http://www.cabi.org/cabebooks/ebook/20173158959
Wong RN, Nang TB, Chan SH (1992) Characterization of Mirabilis antiviral protein–a ribosome inactivating protein from *Mirabilis jalapa* L. Biochem Int 28(4):585–593
Xu X, Gookin T, Jiang C et al (2007) Genes associated with opening and senescence of the ephemeral flowers of *Mirabilis jalapa*. J Exp Bot 58:2193–2201
Yi-Fen W, Ji-Jun C, Yan Y et al (2002) Antifungal activity: A bioassay-guided fraction of an organic extract of the cell mass form manipulated plant cell culture of *Mirabilis jalapa*. L. Chim Acta 85(8):2342–2348
Zachariah SM, Vidya VA, Aleykutty NA (2012) Free radical scavenging and antibacterial activity of *Mirabilis jalapa* L. using *in vitro* models. Asian J Plant Sci Res 5(3):115–120
Zant MKV (2016) Review of the economic and ethnobotany of the family Nyctaginaceae. Atlas J Biol:249–266. https://doi.org/10.5147/ajb.2016.0139. ISSN 2158-9151
Zhou J, Zhou S, Zeng S (2012) Hypoglycemic and hypolipidemic effect of ethanolic extract of *Mirabilis jalapa* L. root on normal and diabetic mice. Evid Based Complement Alternat Med 2012:1–9
Zhou X, Ma D, Zhou Q et al (2008) The genetic toxicity of an invasive plant, *Mirabilis jalapa* L. South-West China J Agri Sci 21(1):152–155
Zipf A (1995) Mechanisms of antiviral protein activity in higher plants: antiviral proteins in higher plants. CRC Press, Boca Raton

Index

S. K. Datta, Y. C. Gupta (eds.), *Floriculture and Ornamental Plants*, Handbooks of Crop Diversity: Conservation and Use of Plant Genetic Resources,
https://doi.org/10.1007/978-981-15-3518-5

GPSR Compliance

The European Union's (EU) General Product Safety Regulation (GPSR) is a set of rules that requires consumer products to be safe and our obligations to ensure this.

If you have any concerns about our products, you can contact us on ProductSafety@springernature.com

In case Publisher is established outside the EU, the EU authorized representative is:

Springer Nature Customer Service Center GmbH
Europaplatz 3
69115 Heidelberg, Germany

Batch number: 10732021

Printed by Printforce, the Netherlands